Physical Constants and Data

Speed of light	$c = 2.997925 \times 10^8$ m/s
Gravitational constant	$G = 6.67 \times 10^{-11}$ N·m²/kg²
Avogadro's number	$N_A = 6.022 \times 10^{26}$ particles/kmol
Boltzmann's constant	$k = 1.38066 \times 10^{-23}$ J/K
Gas constant	$R = 8314$ J/kmol·K $= 1.9872$ kcal/kmol·K
Planck's constant	$h = 6.6262 \times 10^{-34}$ J·s
Electron charge	$e = 1.60219 \times 10^{-19}$ C
Electron rest mass	$m_e = 9.1095 \times 10^{-31}$ kg $= 5.486 \times 10^{-4}$ u
Proton rest mass	$m_p = 1.6726 \times 10^{-27}$ kg $= 1.007276$ u
Neutron rest mass	$m_n = 1.6749 \times 10^{-27}$ kg $= 1.008665$ u
Permittivity constant	$\epsilon_0 = 8.85419 \times 10^{-12}$ C²/N·m²
Permeability constant	$\mu_0 = 4\pi \times 10^{-7}$ N/A²
Standard gravitational acceleration	$g = 9.80665$ m/s² $= 32.17$ ft/s²
Mass of earth	5.98×10^{24} kg
Average radius of earth	6.37×10^6 m
Average density of earth	5.57 g/cm³
Average earth-moon distance	3.84×10^8 m
Average earth-sun distance	1.496×10^{11} m
Mass of sun	1.99×10^{30} kg
Radius of sun	7×10^8 m
Sun's radiation intensity at the earth	0.032 cal/cm²·s $= 0.134$ J/cm²·s

Introduction to Physics for Scientists and Engineers

Introduction to Physics for Scientists and Engineers

Third Edition

Frederick J. Bueche
Professor of Physics
University of Dayton

McGraw-Hill Book Company
New York St. Louis San Francisco Auckland Bogotá
Hamburg Johannesburg London Madrid Mexico
Montreal New Delhi Panama Paris São Paulo
Singapore Sydney Tokyo Toronto

INTRODUCTION TO PHYSICS FOR SCIENTISTS AND ENGINEERS

1 2 3 4 5 6 7 8 9 0   VHVH   8 9 8 7 6 5 4 3 2 1 0

This book was set in Century Schoolbook by York Graphic Services, Inc. The editors were John J. Corrigan and James W. Bradley; the design was done by Caliber Design Planning; the production supervisor was Dennis J. Conroy. New drawings were done by J & R Services, Inc.
Von Hoffmann Press, Inc., was printer and binder.

Cover photograph: © 1979 Rodriguez-Wasserman.

Library of Congress Cataloging in Publication Data

Bueche, Frederick, date
  Introduction to physics for scientists and engineers.

  Includes index.
  1. Physics. I. Title.
QC21.2.B83 1980   530   79-20613
ISBN 0-07-008875-6

Contents

Preface to the Third Edition

The purpose and general scope of this text have not changed from the first edition. They are outlined in the Preface to the First Edition that follows. Past editions of this text have been used widely, and many of its users have shared their wealth of experience with me. To them I am most grateful. Their suggestions for improvements in the text have been invaluable to me in preparation of the third edition.

Although the changes between this and the second edition are too numerous to list here, a few of them will be mentioned. Several chapters and topics have been rewritten and reorganized. Among these are the material on Newton's laws of motion as well as the chapters on momentum, work, and energy. Entropy and the second law are now discussed so as to show their basis in statistical mechanics. The chapters on wave motion have been rewritten so as to better introduce the subject to students with no prior knowledge of wave physics. This has made it possible to further unify the discussion of waves of all types. Much of the development of Maxwell's equations and the wave equation has been rewritten. The basic material on relativity has been placed in a single chapter (Chapter 8) so that those who wish to postpone the material until later may do so easily. Sections have been added to discuss the becquerel, gray, rad, and rem units. A discussion of radiation dangers and precautions is also given.

The number and variety of problems have been expanded greatly. There are about 30 percent more problems, and many of the previous problems have been combined and modified. In keeping with the increasing emphasis on SI units, they are now used almost exclusively. However, the British system is still presented in mechanics for those who wish to use it. Those who wish to avoid all but the SI can easily do so.

I am sincerely grateful to all those who offered comments and suggestions concerning previous editions. They have been of great value to me, and comments for improvement of this new edition will be welcomed as well. A special thanks is due Professor J. Kepes for his aid in the preparation of this edition.

FREDERICK J. BUECHE

Preface to the First Edition

Among all those enrolled in a basic university course in physics, only a few will end up as professional physicists. But in going on to become engineers, chemists, or biologists, a far greater number will need knowledge of, and facility with, physics, which is an indispensable tool for their work. There is more than one road to physics, and a realistic first course must take that fact into account.

It must recognize, as well, the importance of a feeling of accomplishment and satisfaction in the student. Too often teachers in other disciplines have reason to look upon physics as a course which they could teach better; too often do we find that our students do not grasp principles which we know are not that hard. Some of the causes of this unfortunate situation are within our control and can be removed. A main purpose of this text is to make the general physics course a more satisfactory learning experience for the student, and to accomplish this, several guidelines have been followed.

1. It is no sin to make use of the student's intuition and previous practical experience. So far as possible, each physical principle is developed according to the following steps: intuitive exposition of the physics involved so that the student obtains a "feel" for the principle; presentation of a mathematical framework by means of which the principle can be treated quantitatively; practice in the use of the principle.

2. There is a limit to the number of details one should introduce in this course. Nearly any topic in physics can be made the subject of a whole book. We omit nonessentials and subtle ramifications; it is enough if the student understands the principles and knows how to use them.

3. Too much rigor leads to rigor mortis. While we must not sacrifice accuracy, we take care not to fill a beginning text with so much strict codification that the principles are buried.

4. Physicists, and scientists in general, think and act like other people. Because journal space is precious, a published paper these days retains very little of the thought processes which gave rise to the final result. A textbook, however, should not be distilled to the point where the humanness of science is lost.

5. The general physics course is a service course for engineering and the other sciences as well as a course for future physicists. We must be sure to furnish the basic tools these other departments require.

Although I cannot claim to have been successful in realizing these objectives, it is my hope that, because of them, this book overcomes some of the difficulties mentioned above.

Concerning the text itself, it is presupposed that a calculus course is being taken concurrently. Nevertheless, the concepts of calculus are explained as they are needed. A separate appendix is devoted to the notion of the integral, since many calculus courses do not present adequately the relation between integral, area, and sum. Mathematicians apparently share in our own failing, designing courses so heavy with rigor that "feeling" for the ideas is lost. Calculus is a tool which physicists and other scientists use in their calculations, and very early in the text we present it in this light to the student, so that he will not confuse it with the physics itself.

The mathematical level and sophistication of the book increase steadily from beginning to end. It has been my experience that students find such a rise to be compatible with their development and their interests. Throughout, mathematical concepts are introduced as needed for the physics being discussed. This means that the scalar product is introduced in terms of work, while the cross product is presented somewhat later when rotation is treated. These are but two examples of how topics and concepts are taught in such a way that the student can see their relevance.

A few comments about the structure of the book. The contents are ordered in the following sequence: mechanics, heat and thermodynamics, electricity, waves and light, atomic and nuclear physics. Relativity is introduced early in the study of mechanics and is used throughout the remainder of the text. The chapters on waves and light encompass sound, mechanical, and electromagnetic waves as a unified whole. It is a rather simple matter to extend wave concepts to elementary wave mechanics, and this is done immediately. Atomic and nuclear physics are then discussed in some detail. Set off from the body of the text are a number of short sections which, it is hoped, provide interesting sidelights on the history of physics.

The discussion questions at the end of each chapter are designed to stimulate thought and should be utilized if at all possible. A special effort has been made to furnish a suitable spectrum of problems at the end of each chapter. In particular, I believe strongly that the student should be assigned a few easy problems to promote self-confidence and that "trick" problems should be used sparingly.

Many students would benefit from personalized directed drill on the meaning and use of the principles of physics, but unfortunately this is impossible in most classes. To compensate in part for this lack, a physics workbook is available for this text which contains a multitude of problems and questions with their solutions; it should give the average student considerable additional practice and aid.

FREDERICK J. BUECHE

Introduction
to Physics
for Scientists
and Engineers

1 Physics and Measurement

Scientists measure the phenomena of nature in an effort to determine the laws which govern the behavior of the universe. Their success in this endeavor is dependent largely upon four factors: the experiments they choose to do must yield data which are helpful in constructing a pattern of significant natural events; the data must be reliable so that an accurate pattern is obtained; the details of the pattern must be synthesized into a statement of natural behavior which we call a law of nature; if possible, this law must be placed in proper relation to other laws already known. From this we see that measurement is basic to all science. In this chapter we discuss some aspects important to these measurements.

1.1 SCIENCE AND TECHNOLOGY

Our modern civilization is firmly based on a technological foundation which far surpasses the visions of our ancestors a century ago. To them today's technology would appear to be a hopelessly complex assemblage of mysterious contrivances by means of which the human race can perform awesome acts of good and evil. We, too, stand in awe of our power. In times past there was some reason to think that we might destroy the human race by misuse of that power. But now, hopefully, that danger has passed as we become more responsible recipients of the knowledge we now possess.

Civilization as we know it would be impossible without the past and present close interlacing of science and engineering. In theory, the scientist discovers the laws of nature. These laws are then used by the engineer to make technological advances. In practice however, a clear-cut distinction between scientist and engineer is often impossible. Scientists frequently seek out practical applications of their newfound knowledge of nature's behavior. Engineers frequently find that they must search for still undiscovered patterns in nature before they can find a satisfactory solution to a technological problem. Because of this diffuse boundary between science and technological application, a flow of both men and ideas between the two areas is expected and encouraged.

Since technology builds upon the laws of nature discovered by the scientist, the capable engineer must be knowledgeable of nature's laws. In particular, because technological advance is usually the result of ingenious and insightful application of a known law, the successful engineer is usually one who possesses a wide and thorough knowledge of science. And, since the interchange must be two ways, the discoveries of a scientist are best utilized if the scientist has an understanding of the technology to which the discovery can be applied. Even though there is merit in "science for science's sake," the practical difficulties of funding the expensive laboratories of today necessitate a balance between "pure" and "applied" science.

What we are trying to point out is that science and engineering are inextricably intertwined. The basic course in physics, for which this text is designed, enables both scientists and engineers to gain an understanding of many of the most basic laws of nature. Without an understanding of these laws, one is excluded from creative work in vast areas of engineering and many branches of science. This is particularly important now that we begin to realize that the same physical laws apply to all of nature, whether animate or inanimate. How these laws can be used to explain the behavior of living things is still a question largely unanswered. But whether it is the vast computer of our brain or the tiny workings of an individual cell, we believe that explanation will be made eventually in terms of the basic laws of nature you will study in physics.

Since all science is based upon measurement, it is important for scientists and engineers to be able to describe their experimental results in a meaningful way to their associates. Let us now begin our study of physics by learning how scientists describe their discoveries, the results of measurements in most cases, to fellow scientists and engineers.

1.2 SCIENCE AND MEASUREMENT

Physics is an experimental science

Measurement is basic to all science. In order to discover the laws of nature, scientists must observe and measure the phenomena in the world about them. Once the measurements reveal a regularity in a process, it becomes possible to describe the process by a mathematical representation or formula. Of course, the formula is only as valid as the measurements upon which it is based. And since no measurement is completely free from error, the laws of nature, the formulas which describe nature's behavior, are known only approximately. It is for this reason, among others, that scientists consider their knowledge to be incomplete. They do not hesitate to acknowledge that future, more precise measurements or new interpretations of known facts may indicate that a currently accepted physical law is only approximately correct. All science is built upon a few basic laws. And these laws, in turn, are arrived at from an analysis of the data obtained from the measurements made during scientific experiments.

Measurements can be made only if we have tools and units. For example, suppose you wished to measure the height of the ceiling in the room in which you are reading this. A problem presents itself at once if no meter stick is available to you. You could guess the height and express your answer in terms of some agreed length unit, the meter or inch perhaps. But suppose you were from Germany and had used the metric system exclusively, while the person to whom you wished to communicate the height knew only the British system. Clearly, not only is a tool necessary for carrying out the measurement, but a unit of measurement is also needed.

Units are necessary to describe the results of measurements

It would be convenient if we all could agree on a few basic units with which to describe our measurements. Most scientists (and, indeed, most people in the world) make use of the metric system of units. Only the

TABLE 1.1 Typical Lengths and Distances (Meters)

Most distant star thus far measured	$\approx 10^{26}$
Nearest star (Alpha Centauri)	4.3×10^{16}
Average distance to sun	1.49×10^{11}
Average distance to moon	3.8×10^{8}
Earth radius	6.4×10^{6}
Typical cruising height of jet plane	1×10^{4}
Height of tall man	2×10^{0}
Thickness of sheet of paper	1×10^{-4}
Polymer spheres in latex paint	1×10^{-8}
Diameter of hydrogen atom	1×10^{-10}
Approximate "radius" of a proton	1×10^{-15}

English-speaking nations still use the British system of units widely. However, it is likely that even in these countries the metric system will eventually prevail. As we shall see, the metric system also has other advantages. For these reasons, we strongly emphasize the metric system in this text. This system is based upon several primary quantities, prominent among which are a measure of length, a measure of time, and a measure of mass.

1.3 THE LENGTH STANDARD

When we measure the length of an object, we are really comparing it with a known length. For example, when we say that a boy is 1.50 meters tall, we mean that his height is $1\frac{1}{2}$ times that of a meter stick. Similarly, when we state that the diameter of an atom is 4×10^{-10} meter (see Appendix 3 if you are not familiar with this notation), we mean that the diameter of the atom is 4/10,000,000,000 of a meter. Clearly, then, if we know how long a meter stick is, we can specify the length of other objects by comparison. The great range of lengths we might wish to measure is shown in table 1.1.

For many years a platinum-iridium-alloy metal bar was kept near Paris, and the distance between two marks upon it was defined to be exactly 1 meter. This standard of length has several objectionable properties, however. Since a metal bar changes size when its temperature is changed, the bar had to be maintained accurately at an agreed-upon temperature when it was being used. It is very difficult to maintain precisely controlled temperatures, and so some inaccuracy always resulted from this source when objects were compared with it.

In addition, the bar had to be copied so that other places in the world (and in space) could compare lengths with it. No copy of an object can be a perfect duplicate of the original, and so inaccuracy was caused by the use of secondary standards, duplicates of the bar. These secondary standards were then copied to make precise measuring sticks, and these in turn were used to make the meter sticks familiar to all of us. Of course, each time a copy is made, the copy differs slightly from the original, and so the bar near Paris remained the only valid meter length. If it were ever to be destroyed by war or a disaster of nature, the world would have to redefine the meter length.

For these reasons, among others, it became clear that a new standard of

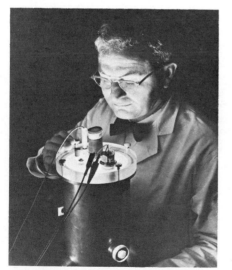

FIGURE 1.1 An NBS scientist adjusts a krypton 86 lamp in its liquid-nitrogen bath. The wavelength of the orange-red light emitted by the lamp was adopted as the international standard of length in 1960. The lamp is operated at the triple point of liquid nitrogen, 63 K, to increase the stability of the standard wavelength. (*Courtesy of U.S. National Bureau of Standards.*)

TABLE 1.2 Units of Length

1	kilometer (km)	$= 10^3$ m
1	decimeter (dm)	$= 10^{-1}$ m
1	centimeter (cm)	$= 10^{-2}$ m
1	millimeter (mm)	$= 10^{-3}$ m
1	micrometer (μm)	$= 10^{-6}$ m
1	nanometer (nm)	$= 10^{-9}$ m
1	angstrom (Å)	$= 10^{-10}$ m
1	picometer (pm)	$= 10^{-12}$ m
1	femtometer (fm)	$= 10^{-15}$ m

The standard length unit, the meter, is defined in terms of a wavelength of light

TABLE 1.3 Powers-of-Ten Prefixes

number	prefix	abbreviation
10^{18}	exa-	E
10^{15}	peta-	P
10^{12}	tera-	T
10^{9}	giga-	G
10^{6}	mega-	M
10^{3}	kilo-	k
10^{-2}	centi-	c
10^{-3}	milli-	m
10^{-6}	micro-	μ
10^{-9}	nano-	n
10^{-12}	pico-	p
10^{-15}	femto-	f
10^{-18}	atto-	a

length would be preferable. Many possibilities come to mind. The size of an atom might be used as a length standard. Unfortunately, the "size" of an atom is difficult to define in terms of measurable quantities. It is not really a solid ball of matter, and we cannot see it, in any case. Or perhaps the diameter of the earth or the sun could be used as the standard. But these change with time and are therefore not acceptable.

One of the quantities which is relatively easy to measure with extreme precision is the wavelength of light given off by an atom. In addition, each atom of an element gives off light of well-defined wavelengths, and so one can obtain these lengths quite easily any place where that element is available. Moreover, the wavelengths of light emitted by an atom are essentially constant and are rather difficult to change. It therefore seemed that a particular wavelength of light emitted by an atom would make a good length standard.

In 1960 the standard of length was redefined in terms of the light given off by krypton atoms when subjected to an electric discharge. The definition makes use of only one isotope of krypton (krypton 86) and only one of the many wavelengths of light which these atoms give off. After comparing the length of the standard meter bar with the wavelength of the orange light emitted by these atoms (using methods we shall discuss in our study of light), the meter bar was found to be 1,650,763.73 wavelengths long. It was then decided to *define* the meter to be exactly that many wavelengths long. Therefore, by definition, the meter is a length equal to exactly 1,650,763.73 of these wavelengths. A typical krypton source for obtaining this wavelength is shown in figure 1.1.

We shall see later (when we study interference of light) how the wavelength of light is used for precision measurements. For many purposes the common meter stick is a sufficiently accurate measuring tool. However, there are situations which require higher precision than the smallest rulings one can make on a scale. Clearly, one cannot measure the size of an atom using graduations on a ruler since the graduations cannot be made as close together as the comparable atoms composing the ruler. Invariably, the most accurate determinations of length are made by comparison with the wavelength of a beam of light or some other electromagnetic radiation.

Once we have defined the meter length, other units of length are defined in terms of it. A centimeter is $\frac{1}{100}$ m, while an inch is exactly 0.0254 m. A summary of the metric units of length is given in table 1.2. Notice the prefixes used in the metric system. They are summarized in table 1.3. We shall use them frequently, and you should become familiar with them.

1.4 THE TIME STANDARD

Another unit is needed if we are to state how fast an object is moving. If a girl walks 20 m in 10 sec, we say that her speed is 2.0 m/s, that is, she walks 2 meters each second. In order to convey this information, we need to know what is meant by a meter and by a second. We must therefore define a unit of time. Typical measured times are shown in table 1.4.

It was convenient historically to use the length of the day as a unit of time. As more precise measurements became possible, the day was split into hours by use of the hourglass. Also, short times can be measured by use of the human heartbeat as the timing instrument. (Galileo timed the period of

TABLE 1.4 Typical Measurable Times (Seconds)

Age of the earth	$\approx 1.3 \times 10^{17}$
Known time of man on earth	$\approx 1.6 \times 10^{13}$
Time taken for half of a sample of radium to decay	5.2×10^{10}
One year (one revolution of earth around sun)	3.2×10^{7}
One day (one revolution of earth on axis)	8.6×10^{4}
Period of heartbeat	$\approx 10^{0}$
Period of highest frequency of audible sound	5×10^{-5}
Period of typical radio waves	1×10^{-6}
Period of vibration of an O_2 molecule	2×10^{-14}
Half-life of a neutral pion	2×10^{-16}

oscillation of a pendulum by this method.) Other more precise timing devices have been developed over the years. Most of these have been based upon the fact that the vibrations of a pendulum, a spring, a solid bar, a crystal, or even a molecule are of essentially constant period.

Until 1967 the standard of time was defined in terms of the time it took the earth to circle the sun in the year 1900. It was necessary to specify the year, since the motion of the earth is not exactly constant. In fact, the motion of the earth is known to vary slightly from day to day. Clearly, a more stable and more easily used standard of time is desirable. A new standard of time was adopted in 1967. It is based upon a vibration characteristic of the atoms of cesium which have an atomic mass of 133. One second is now defined to be exactly 9,192,631,770 of these vibrations. A commercially available cesium clock is shown in figure 1.2. Much more complex devices to monitor the vibration of cesium atoms exist throughout the world. The U.S. Bureau of Standards maintains a cesium clock accurate to 10^{-12} s. By comparison with it, the Bureau furnishes radio signals to much of the world that can be used for precise time measurements. (See NBS special publication 432, "NBS Time and Frequency Dissemination Services," available from the U.S. Government Printing Office.)

The unit of time is defined in terms of a frequency of vibration associated with the Cs 133 atom

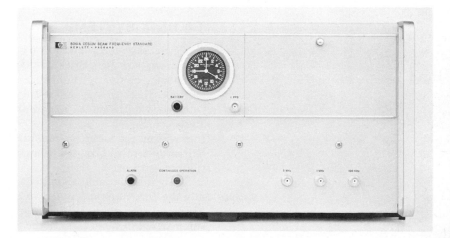

FIGURE 1.2 A commercially produced cesium standard clock. (*Courtesy of Hewlett-Packard.*)

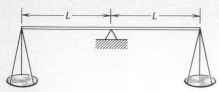

FIGURE 1.3 A schematic diagram of an equal arm balance for comparing masses. What refinements would be necessary for a workable device?

1.5 THE MASS STANDARD

There are several measurable quantities which cannot be expressed in terms of the two previously discussed measurement units, length and time. A simple example is the quantity of substance. In fact, this quantity, which defines the amount of material in a solid object, liquid, or gas, plays a prominent role in physical processes. We must therefore set up a procedure by which it can be measured.

Several possibilities come to mind at once since we are all familiar with the ways in which this quantity is measured in commerce. For example, one buys sugar at the store by weight. Liquids are often measured in terms of their volumes. Neither of these methods is of basic scientific usefulness, however, since each contains at least one factor which is clumsy or unnecessarily arbitrary. For example, volume is not a good measure of quantity of substance since it would give the same value of the quantity for equal volumes of wood and iron, as examples. And yet, for some physical experiments, two equal volumes of these substances behave in much different ways. Similarly, weight is a poor measure of quantity of substance since the weight of an object is found to vary slightly from place to place on the earth. On the moon an object weighs only about one-sixth as much as on earth.

It has been decided to define the quantity of substance (the *mass*) of a body in the following way. (This is discussed in detail in Chap. 5. We here present only an outline of the operational method.) A standard cylinder of metal, the standard *kilogram,* is kept near Paris. When any object exactly balances the standard kilogram on an equal arm balance, as shown in figure 1.3, that object is also said to possess a mass of 1 kilogram. If two such objects exactly balance a third, the third object has a mass of 2 kilograms, and so on. Similarly, by splitting a kilogram mass into two parts which exactly balance, one is able to obtain $\frac{1}{2}$-kilogram masses. By following these procedures, one is able to define any mass in terms of the standard kilogram.

The unit of mass, the kilogram, is defined as the mass of a standard metal cylinder

Other units of mass are defined in terms of the kilogram. For example, in the metric system, which we shall most frequently use, the most common units of mass are the kilogram (kg), gram (g), and milligram (mg). As their names imply, the kilogram is 10^3 g, and the milligram is 10^{-3} g.

1.6 UNIT SYSTEMS

Two general systems of units are widely used in the world, the metric and the British systems (the British is also referred to in this country as the U.S. Customary System). The metric system measures lengths in meters (and multiples thereof) whereas the British system makes use of the foot, inch, etc. Of these two, the metric system is by far the most widely used and is most always employed by scientists; the British system is falling into disuse. For that reason, the metric system will be emphasized in this text, although the British system will be used enough so that you will be able to work with it if the necessity arises.

SI units

By international agreement the metric system was formalized in 1971 into the *International System of Units,* abbreviated SI from the French *Le Système International*. There are seven basic units in the SI and they are summarized in table 1.5. For the first portion of this text we shall be interested in only three of these units, the meter, kilogram, and second. Before

TABLE 1.5 Basic SI Units

Quantity	Name	Symbol
Length	meter	m
Mass	kilogram	kg
Time	second	s
Temperature	kelvin	K
Electric current	ampere	A
Number of particles	mole	mol
Luminous intensity	candela	cd

1971 the system was often referred to as the *mks system,* a designation that made use of the first letters of these three basic units.

1.7 DERIVED QUANTITIES

All physical quantities measured by physicists can be expressed in terms of the three basic units of length, time, and mass. For example, speed is simply length divided by time. As we shall see in Chap. 5, force is actually mass multiplied by length divided by time squared. In symbols we write these quantities

$$[\text{Speed}] = \frac{L}{T} = LT^{-1}$$

$$[\text{Force}] = \frac{ML}{T^2} = MLT^{-2}$$

where [speed] is meant to indicate the units of speed, and *M, L,* and *T* represent mass, length, and time units.

Even the most complex physical quantities can be expressed in terms of these three units, or dimensions. For example, the dimensions of any kind of energy are expressible as ML^2/T^2, while the dimensions of electric charge could* be expressed as $(ML^3/T^2)^{1/2}$. However, we often define other units in order to simplify notation and facilitate communication. Clearly, it is much simpler to say that a force is 10 newtons† rather than 10 kilogram-meters per second squared, even though both statements have the same meaning. Quantities other than length, mass, and time are called *derived* quantities, and their units are *derived* units. It is sometimes necessary to express a quantity in a different unit than the one given, for example ft/s rather than m/h (meters per hour). The procedure for doing this is explained in Appendix 3.2.

Oftentimes an examination of units in an equation can lead to useful information. Because only like quantities can be equated, the units on one side of an equation must equal the units on the other side. For example, in the equation $s = vt$, which tells us how far (*s*) an object will move in time *t* if its speed is *v*, the units must be such that

$$[s] = [v][t]$$

Because $[s] = L$, $[v] = L/T$, and $[t] = T$, we have

---

*In the SI, the unit of charge is defined in terms of the ampere.
†A force of 1 newton is equivalent to a force of 0.225 pound.

$$L = \frac{L}{T}T$$

from which $L = L$.

Or suppose we are unsure about the correct form of an equation such as that between wavelength λ, frequency f, and speed v, where $[\lambda] = L$, $[f] = 1/T$, and $[v] = L/T$. We can conclude that $\lambda = v/f$ and not $\lambda = vf$ might be correct because

$$L = \frac{L}{T} \div \frac{1}{T} = L$$

but

$$L \neq (L/T)(1/T)$$

Dimensional analysis

This procedure for analyzing the units of equations is called *dimensional analysis*.

1.8 COMPLICATING FACTORS

The definitions of units given in the preceding sections are quite straightforward, and were designed for simplicity, exactness, and convenience. It would be a mistake, however, to assume that even such basic quantities as length, mass, and time are always easily visualized. With the discovery of the principles and consequences of *relativity* by Albert Einstein in 1905, it became clear that length, mass, and time are more complex than had previously been thought. We shall examine Einstein's theory of relativity in subsequent chapters. We shall see that a given object may be measured to have different lengths by two people who are moving relative to each other. In addition, the mass of the object may be found different by the two observers. Moreover, clocks carried by the two observers moving relative to each other will not measure time in the same way.

According to relativity theory, mass, length, and time measurements depend upon the state of motion of the observer

Uniting these rather disconcerting predictions of Einstein's theory of relativity with the classical physics existent prior to 1905 has been a fascinating and fruitful adventure for physicists. Many experiments have been carried out to observe these predictions, and within the limits of experimental error, the predictions have been found true. As we proceed with our study of physics we shall find that the principles which Einstein discovered are fundamental to such diverse fields as nuclear energy and interplanetary voyages of the future.

Chapter 1
**Questions
and Guesstimates**

1. If you were given a precise ruler divided into millimeters, how closely could you determine the length of this page with it? The thickness of the page?
2. Many common units of measure are based upon comparison with commonplace objects. For example, horse enthusiasts very often speak of a horse's height in "hands." What other similar measuring units can you name?
3. If you were suddenly marooned on a desert island without a watch or any other human-made measuring tool, how could you set up a measuring system for distance, time, and mass?
4. In many places in the world, accurate weighing instruments are not

available. How might grain merchants improvise an instrument suitable for their purposes?

5. What disadvantages rule out using the period of a pendulum as the time standard?

6. Rutherford first estimated the size of the nuclei of atoms by shooting small particles (α particles) at a thin sheet of atoms. The method was somewhat similar to measuring the size of a small invisible ball suspended in an open window by throwing many small pebbles through the window at random places. Explain how either method works.

7. How could you measure the height of a tree without climbing it?

8. Estimate the accuracy to which one can determine the time taken for a boy to run 100 m.

9. Suppose a person has never seen a liquid of any kind. (Do not bother with the details of how this might be!) How would you explain what a liquid was like? In your explanation, take note of the analogies you use. Physicists have a similar problem in explaining the electron, a particle which even they will never be able to see.

10. One common way of estimating distances is to pace off the distance assigning 1 meter to each pace. Estimate how accurate this method would be for you in measuring the length of a room. Test your estimate by an experiment. How could you "calibrate yourself" so that future measurements of this type would be more accurate?

11. If you throw 100 pennies onto the floor, about how many should come to rest with "heads" up? Twenty goldfish are swimming in an unobstructed rectangular pool. About how many would you expect to be in the left half of the pool at a given instant? In the left one-third? For your information, if your answer in each case is designated as N, then experiment (and theory) shows that individual experiments give a result $N \pm \sqrt{N}$ if you choose N properly. This result applies to all random statistical events. If the average of a large number of trials is N, then the deviation from the average one might find on a single trial is about $\sqrt{N}$.

12. Suppose a number of people measure the time for 100 swings of a 1-m-long pendulum using their pulse beat as a time-measuring device. Write the result you would expect to obtain in the form

$$\text{Pulse beats/100 swings} \pm \text{pulse beats}$$

You will need to make some estimates concerning pulse rates.

Chapter 1
Problems

1. Write the following numbers in powers-of-ten notation: 3407; 0.0291; 637; 10,000; 0.0001; 0.000100; 0.000137.

2. Write the following numbers without powers-of-ten notation: 1.51×10^4; 10^4; 3.01×10^{-2}; 10^{-3}; 0.51×10^{-3}; 63×10^5.

3. A certain 900-page book is 3.8 cm thick between the inside surfaces of the covers. What is the thickness of a page (in meters) expressed in powers-of-ten notation?

4. One kilomole (kmol) of atoms contains 6×10^{26} atoms. If each carbon atom has a mass of 20×10^{-27} kg, what is the mass of a kilomole of carbon atoms? (It is no coincidence that your answer is a number equal to the chemist's atomic weight for carbon.)

5. The wavelength of light used as a length standard can be measured to an

accuracy of better than 1 part in 10^9. If, instead, we used the length of a bar of iron as the standard, how large a temperature change would cause a change in length of 1 part in 10^9? Each 1°C change in temperature of an iron bar causes its length to change by 1 part in 10^5.

6. Using the fact that 1 inch (in) = 2.54 cm, find the height in feet of a 1-m-high girl. (1 ft = 12 in)

7. Give the height in meters of a 6-ft-tall man. (1 in = 2.54 cm and 12 in = 1 ft)

8. Using the facts that 1 mile = 1.609 km and 1 hour = 3600 s, show that 60 mi/h is equivalent to 26.8 m/s.

9. The cubit is an ancient length unit based upon the length of the arm from a man's elbow to the tip of his hand. It is equivalent to about 50 cm. How many cubits are there in a mile? (There are 1.609 km in a mile.)

10. An equation cannot be correct if the dimensions are not consistent. That is to say, if one side has units of mass divided by length, then the other side must have these same units. With this in mind, which of the following equations are incorrect?

$$s = at + \tfrac{1}{2}vt^2$$
$$\lambda = vt$$
$$v = st$$
$$F = m/a$$

In these expressions, s and λ are distances, m is mass, and $[a] = LT^{-2}$, $[v] = LT^{-1}$, $[F] = MLT^{-2}$.

11. Assuming s to be distance and t to be time, what must be the dimensions of the constants C_1, C_2, C_3, and C_4 in each of the following equations?

$$s = C_1 t \qquad s = \tfrac{1}{2}C_2 t^2 \qquad s = C_3 \sin C_4 t$$

(*Hint:* See the previous problem. Also, the argument of any trigonometric function must be dimensionless.)

12. The frequency of vibration f of a simple pendulum is related to the pendulum length l and the acceleration due to gravity g by one of the following relations: $2\pi f = \sqrt{l/g}$, $2\pi f = g/l$, $2\pi f = \sqrt{gl}$, $2\pi f = \sqrt{g/l}$. Using the facts that $[f] = T^{-1}$ and $[g] = LT^{-2}$, select the correct relation.

13. It is known that the time taken for a pendulum to swing back and forth increases with the pendulum length in the following way: (const) × (time)^2 = (length). Find the dimensions of the constant (i.e., mass, length, and time). What physical quantity has these dimensions? What constant of nature probably is involved? What experiments could you perform to test your guess?

14. According to Einstein's theory of relativity, an object moving with speed v appears shortened along the line of motion by a factor of $\sqrt{1 - (v/c)^2}$, where c is the speed of light, 3×10^8 m/s. Show that this factor is 0.999 999 999 9995 for a plane moving at 300 m/s (about 670 mi/h). To do this, make use of one of the expansions given in Appendix 3.4.

15. The radius of most atoms is about 1 angstrom (1 Å = 10^{-10} m). How many atoms laid side by side along a line are needed to form a line 1 cm long?

16. Using the result of problem 15 estimate the number of atoms needed to cover a surface 1 cm^2 by a layer which is one atom thick. Assume the atoms to be laid out in a rectangular pattern.

17. Using the results of problems 15 and 16, about how many atoms are

needed to fill a cube 1 cm on each side? Assume the atoms to be placed in a cubical pattern.

18. The density of a substance in g/cm^3 is defined to be the mass in grams of a 1-cm^3 volume of the substance. How large a volume of mercury (density $= 13.6\ g/cm^3$) has a mass of 1 kg?

19. The mass of an iron atom is 9.3×10^{-23} g. How many iron atoms are there in 1 cm^3 of iron? The density of iron is $7.86\ g/cm^3$ (see problem 18).

20. One can picture roughly the proton and neutron to be uniform, solid spherical balls each having a radius of about 1×10^{-15} m. The mass of each is 1.67×10^{-27} kg. What is the density of substance in these particles (see problem 18)? Compare with the density of mercury.

2 Directed Quantities; Vectors

For many physical purposes, the direction of a quantity is equally as important as its magnitude. In the preceding chapter we saw how we defined the units for expressing the magnitude of a quantity. This chapter is devoted to the use of quantities which have direction as well as magnitude. These quantities are called *vectors*.

2.1 VECTOR AND SCALAR QUANTITIES

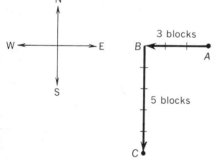

FIGURE 2.1 How far is it from point *A* to point *C*?

Suppose that you wished to draw a diagram to illustrate the statement "I went 3 blocks west and then 5 blocks south." The diagram you would draw might look somewhat like the one shown in figure 2.1. Your starting point was *A*, you first went to *B*, and your endpoint was *C*. We call a diagram such as this a *vector diagram*. It makes use of arrows to represent displacements.

In physics we use the word *displacement* in a precise way. When you move from point *A* to *B*, as indicated in the figure, you have undergone a displacement of 3 blocks west. Notice that the displacement consists of two pieces of data: (1) the straight-line distance between the two points (3 blocks in this case) and (2) the direction of the endpoint from the starting point (west in this case). Observe the importance of the words *straight-line distance*. If you go from *A* to *B* in figure 2.1 by any path, whether or not you take a side trip into a shop along the way, the "straight-line distance" from *A* to *B* is still 3 blocks. Hence *the displacement from one point to another is determined only by the positions of the start and endpoints.*

The vector diagram in figure 2.1 also shows that the displacement from *B* to *C* has a magnitude of 5 blocks and is directed southward. All displacements must have both a magnitude and a direction. There are many other quantities that we encounter besides displacements that have both direction and magnitude. A tow rope pulls on a car with a certain force, for example. This force has not only a magnitude but also a direction in which it pulls. We shall see in later chapters that, in addition to displacements and forces, such quantities as velocity, acceleration, gravitational and electric field strengths, and many others have both magnitude and direction. We call such quantities *vector quantities* or, sometimes, just *vectors* for short. A VECTOR QUANTITY *has both magnitude and direction.**

Vector quantities have direction as well as magnitude

*For reasons we shall explain in Chap. 10 when we discuss rotation, we also require that to qualify as a vector a quantity must obey the same mathematical rules that displacements and other common vectors obey.

Many quantities do not possess direction, and so they are not vectors. In this category are the number of pages in this book, the number of people in a university, the quantity of water in a pail, and so on. These quantities, which do not have direction associated with them, are called *scalar* quantities. They possess magnitude but not direction.

Scalar quantities do not have direction

It is often necessary to distinguish between symbols which represent vector and scalar quantities. When a quantity has magnitude only (a scalar quantity) we represent it by an ordinary letter (such as a) or number (such as 5). However, if we also wish to convey the fact that the quantity has direction as well as magnitude (i.e., that it is a vector), we print its symbol in boldface type, such as **a**. When writing vectors by hand, the vector **a** would be represented as $\vec{a}$. Often, when we are interested only in the magnitude of a vector quantity, such as **a**, we indicate that fact by use of the absolute magnitude notation $|\mathbf{a}|$ or, more often, by writing it as a scalar, a.

2.2 RESULTANT OF SEVERAL VECTORS

In reference again to the situation shown in figure 2.1, one might wish to know what the straight-line distance is from A to C. To put the problem in a slightly different way, if you were to walk 3 blocks west and then 5 blocks south, how far and in what direction would you have moved from your starting point? The vector representing this required quantity is shown as **R** in figure 2.2. It is called the *resultant displacement vector* (or just the *resultant*) arising from the two displacement vectors indicated.

To find **R** in figure 2.2 we make use of the pythagorean theorem, and find at once

$$R = \sqrt{(5)^2 + (3)^2} \text{ blocks} = 5.83 \text{ blocks}$$

where R is the magnitude of **R**. The direction of **R** may be stated in many ways. Perhaps the most convenient in this case is to give the value of ϕ shown in the figure.* We have

$$\tan \phi = \tfrac{5}{3} = 1.67$$

from which $\phi \approx 59°$. (If your trigonometry is rusty, you might well refer to Appendix 3, Sec. A3.3, at this time, where a short review is given. Tables of trigonometric functions are to be found in Appendix 4.)

When a displacement is at an angle other than 90° to another displacement, the addition of the displacement vectors is slightly more complicated since the pythagorean theorem only applies to right triangles. Suppose, for example, one wishes to add the two displacements shown in figure 2.3(*a*). In other words, we want the vector displacement from A to B. This is easily found if we first rephrase the question to ask, In going from A to B, how far

*ϕ is the Greek letter phi.

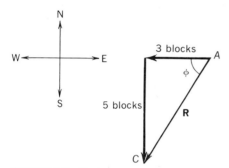

FIGURE 2.2 Find the magnitude and direction of **R**.

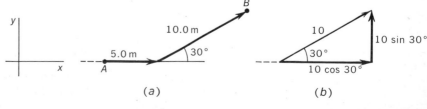

FIGURE 2.3 Notice how the components of the 10-m vector are found in (*b*).

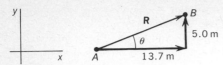

FIGURE 2.4 **R** is the resultant of the two vectors shown in (a) of figure 2.3. What are the **x** and **y** vectors called which add to give **R**?

did we go in the x direction and how far in the y direction? To answer this question we must find the x and y displacements resulting from each of the two vectors.

Since the 5-m vector is completely in the x direction, it resulted in a 5.0-m x displacement and a zero y displacement. To find the x and y displacements resulting from the 10-m vector, we refer to part (b) of figure 2.3. As seen there, the x displacement is 10 cos 30°, or 8.7 m, while the y displacement is 10 sin 30°, or 5.0 m. Until you become accustomed to the process, it is wise to add these displacements in tabular form, thus:

Vector	x	y
5 m	5.0	0
10 m	8.7	5.0
Total	13.7	5.0

We see at once that the combined effect of the two displacements has been to cause an x displacement of 13.7 m and a y displacement of 5.0 m. In other words, the two vectors shown in figure 2.3(a) are equivalent to two other vectors, the x and y vectors shown in figure 2.4. But we know how to add these two vectors at right angles since we did that in figure 2.2. Repeating the process for the vectors shown in figure 2.4, we find

$$R = \sqrt{(5.0)^2 + (13.7)^2} \text{ m}$$

or

$$R = 14.6 \text{ m}$$

In addition, $\tan \theta = 5/13.7 = 0.365$, from which $\theta = 20°$.

As we see, the x and y parts of a vector (or more nearly correctly, the x

Accuracy of Numbers

In our discussion of vectors, we have often placed definite numbers on them. For example, we have represented a displacement of 4.0 m by a vector which is 4.0 units long. However, since physics is concerned with things which can be measured, it is necessary in physics to use numbers which are not known very exactly.

Numbers such as pi (π) and the square root of 2 may be written down to an accuracy of as many digits as time and the occasion allow. They are purely numbers and there is no doubt whatsoever about the value of the ninth digit beyond the decimal point for pi; it is 3. This is not the case with numbers frequently encountered in physics, which are usually the result of measurement. Since measurements are never absolutely precise, the resulting numbers are imprecise as well.

As an example, suppose you were requested to measure the area of the rectangle shown in the figure by means of the ruler placed beside it. We see that side b is about 0.72 cm long. If we estimate the possible error in reading the ruler to be about 0.01 cm on either side of the value read, we conclude that the correct reading might have been between 0.71 and 0.73 cm. This prompts us to write our measured result as $b = 0.72 \pm 0.01$ cm. Similarly, the other dimension would be 0.50 $\pm$ 0.01 cm. The range of uncertainty in the area can be found by multiplying the extreme values together, thus:

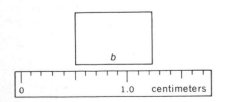

$$(0.73 \text{ cm})(0.51 \text{ cm}) = 0.3723 \text{ cm}^2$$
$$(0.71 \text{ cm})(0.49 \text{ cm}) = 0.3479 \text{ cm}^2$$

and y vectors which add to give the vector) are of great usefulness in adding vectors. We call these parts the x and y *components* of the vector. If we are given several vectors to add, the addition may most easily be accomplished by finding the x and y components of the vectors. These components, which are also vectors, are then added so as to find the x and y components of the resultant. When these are known, the resultant can be found as in figure 2.4.

A similar procedure can be followed in three dimensions. When the usual x, y (or x, y, z) right-angle coordinate system is used, the components are called *rectangular components*. However, as we shall see later, it is sometimes convenient to take the components of a vector in other than the x, y, z directions.

Before proceeding with examples to illustrate these ideas, we pause to point out that it is usual to describe the direction of a vector in two dimensions by means of a standard choice of angle. Unless otherwise stated, the angle θ of a vector will be taken as the angle the vector makes with the $+x$ axis in an xy coordinate system. Using that terminology, the vector $\mathbf{R}$ in figure 2.4 is 14.6 m at an angle of 20°.

Illustration 2.1

Find the resultant of the displacements shown in figure 2.5(*a*).

Reasoning Since the 7- and 15-m displacements are directed along the axes, their components are obvious. To resolve the other two vectors into components, we proceed as in parts (*b*) and (*c*) of the figure. In doing this we make use of the fact that sin 37° = 0.60 and cos 37° = 0.80. Now that the components of the vectors are known, we can tabulate their magnitudes (and directions) as shown. Notice that a component which points in the negative x

---

We are uncertain in our computation of the area, since its true value can only be estimated to be somewhere between these two extremes. If we had taken the average measured values, we would have obtained (0.72 cm)(0.50 cm) = 0.3600 cm². From these considerations we see that the measured area is given approximately by the result 0.36 ± 0.01 cm², or simply, 0.36 cm². It would be wrong to write the result as 0.3600 cm², since we have no idea at all that the last two digits would really be zeros if we were able to measure the area more accurately. When a scientist writes down a number 0.3600 cm², he or she is telling those who read the number that there is reason to believe that the true area is 0.3600 to an accuracy of at least 0.0009 cm². Since in our case we have 0.36 ± 0.01 cm², numbers in the third and fourth decimal places have no significance, and are therefore omitted.

As a rough rule of thumb, if two numbers are to be multiplied or divided by each other, the answer will be accurate only to the same number of digits as was the least accurate number used in the computation. For example, if we wished to multiply the area measured (0.36 cm²) by the square root of 2, we should have (0.36 cm²)(1.41421) = 0.51 cm², and not 0.5091156 cm². The third digit and those beyond it are completely uncertain because of the uncertainty in the measured area. One should never carry digits which are not significant when doing computations in physics.

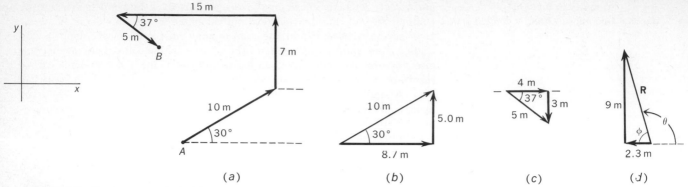

(a) (b) (c) (d)

FIGURE 2.5 Suppose one started from point A with the 5-m displacement, then underwent the 15-, 7-, and 10-m displacements in succession thereafter. Would one still end up at point B?

or y direction must be considered negative since it represents a displacement in the $-x$ or $-y$ direction.

Vector	x	y
10 m	8.7	5.0
7 m	0	7.0
15 m	-15.0	0
5 m	4.0	-3.0
Total	-2.3	9.0

Now that we know the components of the resultant, we can construct the resultant as in part (d) of figure 2.5. We therefore find

$$R = \sqrt{(2.3)^2 + (9.0)^2} = 9.3\ \text{m}$$

and
$$\tan \phi = 9/2.3 \quad \text{or} \quad \phi \approx 76°$$

From figure 2.5(d) we see that $\theta + \phi = 180°$ which gives $\theta = 180° - \phi = 180° - 76° = 104°$. The resultant displacement is therefore 9.3 m at 104°.

2.3 FORCES AS VECTORS

If you attach a rope to an object and pull on the rope, a force is exerted on the object. This force will have a certain magnitude, and it also has direction. It is a vector quantity, and is found to obey the laws we ascribe to vectors. Force vectors can be handled in the same way that we have used displacement vectors. This results from the fact that any vector can be represented by its components. Since our method of adding vector displacements is based entirely on the replacement of the vector by its components, the method will work equally well for any vector, including force vectors.

The procedure involved is easily seen from the problem posed in figure 2.6. Shown there are three ropes pulling on a fixed object with the forces indicated in part (b) of the figure. (The SI force unit used in the figure is the newton, N, whose definition will be given in Chap. 5; 1 newton = 0.225 pounds.)

We wish to find the resultant of these forces; that is to say, we wish to find what single force pulling on the object would be equivalent to the three forces shown.

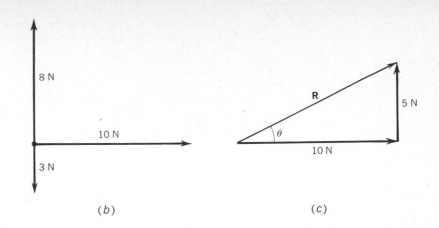

(a)   (b)   (c)

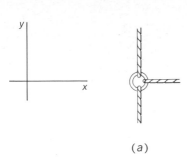

FIGURE 2.6 If we wish the ring in (a) not to move, describe a fourth force which must act upon it.

Since the resultant force must be equivalent to the combined three forces, it must have an x component equal to the sum of the three x components. A similar situation exists for the y components. Making the usual component table gives

Vector	x	y
8 N	0	8
10 N	10	0
3 N	0	-3
Total	10	5

In doing this, notice that the vectors can be taken in any order, since the order in which we add numbers, the magnitudes of the components, will not alter the result. The resultant has the components indicated, and is shown in figure 2.6(c). Therefore

$$R = \sqrt{125} \approx 11.2 \text{ N}$$

while $\tan \theta = \frac{5}{10}$ so $\theta \approx 26.5°$

The resultant is 11.2 N at 26.5°.

Illustration 2.2
Find the resultant of the forces shown in figure 2.7(a).

FIGURE 2.7 Add the forces shown in (a).

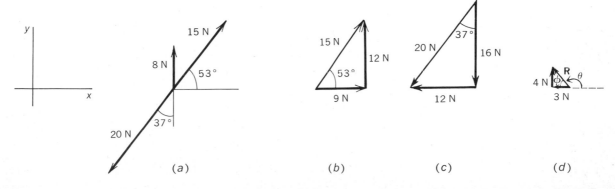

(a)   (b)   (c)   (d)

Reasoning We use the rectangular component method as before. If the forces are split into their components, the component table appears as follows:

Vector	x	y
15 N	9	12
8 N	0	8
20 N	−12	−16
Total	−3	4

Notice the use of minus signs in the component columns of the table above. The result is

$$R = \sqrt{(3)^2 + (4)^2} = 5\,\text{N}$$

and

$$\tan \phi = \tfrac{4}{3} \quad \text{so} \quad \phi = 53°$$

The resultant is 5 N at $180 - 53 = 127°$.

2.4 UNIT VECTORS; BASIS VECTORS

The representation we have used for vectors heretofore has been rather cumbersome. In particular, we have had to draw pictures to represent the vectors. Of course, we could have described the direction and magnitude of each of the vectors in words, and then a picture would not be necessary. This, too, would have been unwieldy. Clearly, a more compact method for representing vectors would be of value. We can accomplish this by use of unit vectors. Although, for ease in drawing pictures, our preceding discussion was confined to two dimensions, the method of unit vectors is easily treated even in three dimensions, as we shall see.

We define three vectors of unit length **i**, **j**, and **k.** The **i** vector is chosen to point in the $+x$ direction, and the **j** vector points in the $+y$ direction, and the **k** vector points in the $+z$ direction. They are shown in figure 2.8(a).

Since the vectors are of unit length, they may be used to give direction to numbers. For example, if we multiply the number 6 by **i** to give 6**i,** the result is a vector in the $+x$ direction, which has a magnitude of 6 units. To see the utility of this, refer to figure 2.8(b). The vector **A** has components of magnitudes 3 and 4. These component vectors can be written 3**i** and 4**j.** As we have seen previously, the component vectors add as vectors to give the resultant. We can therefore write

$$\mathbf{A} = 3\mathbf{i} + 4\mathbf{j} + 0\mathbf{k} = 3\mathbf{i} + 4\mathbf{j}$$

Notice that this is a vector addition and follows the rules we have previously used for adding vectors. Accordingly, $A = \sqrt{9 + 16} = 5$.

Now that we know what is meant by the unit vectors, we can describe any vector completely by writing it in component form as we have just done for **A.** When we equate **A** to 3**i** + 4**j**, we know at once that **A** is a vector having an x component with magnitude $+3$ and a y component of $+4$. You should examine the vectors in figure 2.9 and verify that the representation given in terms of the unit vectors is correct. Moreover, it is well to recall that in three dimensions the pythagorean theorem can be extended to give the diagonal of a rectangular box in terms of its sides. In the present context it tells us that if $\mathbf{R} = A\mathbf{i} + B\mathbf{j} + C\mathbf{k}$, then $R^2 = A^2 + B^2 + C^2$. Using this fact, the vector in

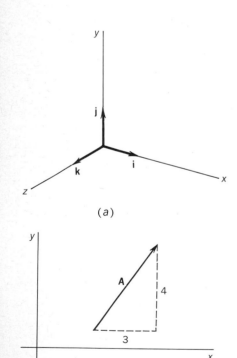

(a)

(b)

FIGURE 2.8 In terms of the unit vectors, the vector **A,** which is entirely in the xy plane, is given by $\mathbf{A} = 3\mathbf{i} + 4\mathbf{j} + 0\mathbf{k}$.

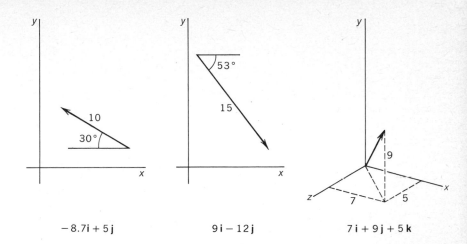

FIGURE 2.9 Each vector can be compactly represented in terms of the unit vectors.

$-8.7\mathbf{i} + 5\mathbf{j}$

$9\mathbf{i} - 12\mathbf{j}$

$7\mathbf{i} + 9\mathbf{j} + 5\mathbf{k}$

Multiplication of a scalar by a unit vector gives direction to the scalar without changing its magnitude

the last part of figure 2.9 has a magnitude $\sqrt{(7)^2 + (9)^2 + (5)^2} = 12.45$.

Unit vectors also prove useful when one is working in other than rectangular coordinate systems. Just as the unit vectors **i, j, k** describe the rectangular coordinates, other unit vectors are used to describe the coordinates in other systems. These unit vectors which are assigned to the basic coordinates are called *basis vectors*. In still other situations, some of which we shall encounter later, a unit vector is chosen to denote a direction of interest which is not one of the basic coordinate directions. Unit vectors prove to be very useful in science and engineering; we shall often make use of them.

Illustration 2.3

Find the sum of the two vectors

$$5\mathbf{i} - 2\mathbf{j} \quad \text{and} \quad -8\mathbf{i} - 4\mathbf{j}$$

Reasoning The two vectors are shown in figure 2.10. We do not really need these pictures of them, but they are shown to help us visualize the vectors. Suppose the vectors represent forces. The x component of the resultant force will be the algebraic sum of the x components $5 + (-8) = -3$. Similarly, the y component of the resultant will be $(-2) + (-4) = -6$. The resultant vector is therefore $-3\mathbf{i} - 6\mathbf{j}$. To find the magnitude of the resultant, we employ the pythagorean theorem as usual.

$$R = \sqrt{(3)^2 + (6)^2}$$

Since the resultant vector is in the third quadrant (draw it and see), it makes an angle ϕ with the negative x axis given by $\tan \phi = \frac{6}{3}$.

Illustration 2.4

Find the sum of the following vectors:

$$a_x\mathbf{i} + a_y\mathbf{j}; \ b_x\mathbf{i} + b_y\mathbf{j}; \ c_x\mathbf{i} + c_y\mathbf{j} + c_z\mathbf{k}$$

Reasons Clearly, the x, y, and z components of the three vectors are indicated by the subscripts.

The magnitude of the x component of the resultant, R_x, will be

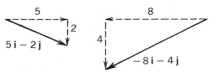

FIGURE 2.10 What is the resultant of these two vectors?

$$R_x = a_x + b_x + c_x$$

while
$$R_y = a_y + b_y + c_y$$

and
$$R_z = 0 + 0 + c_z$$

From this the resultant vector is just

$$\mathbf{R} = R_x\mathbf{i} + R_y\mathbf{j} + R_z\mathbf{k}$$

or
$$\mathbf{R} = (a_x + b_x + c_x)\mathbf{i} + (a_y + b_y + c_y)\mathbf{j} + c_z\mathbf{k}$$

2.5 MATHEMATICAL NICETIES

In the preceding sections we have been discussing vectors in a rather intuitive way. Although such procedures are commonly used by physicists, it is necessary to consider explicitly a few mathematical properties of vectors in more detail if we are to be able to use them in more complicated situations. Before doing this, it should be pointed out that much of what we have done with vectors is purely a matter of definition. To begin with, we stated the definition of a *vector quantity* to be any quantity which possesses both magnitude and direction and obeys the mathematical rules we assign to vectors. The directed line segment or arrow used to represent such a quantity is called a *vector arrow*, or simply vector for short.

The vector arrow is free to move about in space. The sole requirements are that its direction be that of the quantity which it is to represent, and that its length is to be proportional to the magnitude of the quantity. Although the vector may represent the force applied at some particular point to a body, the vector itself need not be drawn at that point. We have frequently made use of this property when we moved the vectors from place to place within a diagram.

The statement that two vectors are equal to each other is defined to mean that they are equal in both magnitude and in direction. As with algebraic quantities, this equality is written

$$\mathbf{A} = \mathbf{B}$$

If two vectors are equal, their components are equal. From the definition of equality we deduce the fact that two vectors which are equal to a third vector are equal to each other.

Addition of two vectors is defined by the following operation. To the tip of vector $\mathbf{A}$ is joined the tail of vector $\mathbf{B}$. The sum of these two vectors, $\mathbf{A} + \mathbf{B}$, is defined to be the vector drawn from the tail of $\mathbf{A}$ to the tip of $\mathbf{B}$. We have called this vector the *resultant* of the two vectors. By adding a third vector to the resultant, this definition leads directly to a rule for the addition of several vectors. It also follows from the definitions given that

$$\mathbf{P} + \mathbf{Q} = \mathbf{Q} + \mathbf{P}$$

Vectors obey the commutative and associative laws

which is called the *commutative* law, and that

$$(\mathbf{P} + \mathbf{Q}) + \mathbf{S} = \mathbf{P} + (\mathbf{Q} + \mathbf{S})$$

which is the *associative* law.

The operation of subtraction of two vectors is defined so as to make subtraction equivalent to the addition of a negative vector. That is to say, to subtract $\mathbf{B}$ from $\mathbf{A}$ we define the operation as follows:

$$\mathbf{A} - \mathbf{B} = \mathbf{A} + (-\mathbf{B})$$

In this relation, $-\mathbf{B}$ is used to denote vector $\mathbf{B}$ with its direction reversed. If $\mathbf{B}$ is expressed in component form, this means

$$\mathbf{A} - (B_x\mathbf{i} + B_y\mathbf{j}) = \mathbf{A} + (-B_x\mathbf{i} - B_y\mathbf{j})$$

Simply stated, to subtract a vector, we reverse its direction and add it.

The properties of vectors which we have just given are the result of the way in which we defined vectors and vector addition. Later on we shall have occasion to use the product of vectors, as well as to perform other mathematical operations with them. We postpone until that time the definitions of those operations since the utility of the definitions cannot be fully appreciated until their physical basis is made clear.

Chapter 2
Questions and Guesstimates

1. Which of the following items can be represented as vectors: the motion of an automobile, the progress of a nation, the number of eggs in a carton, the flow of water in a pipe, the movement of the wind, the books on a shelf, the superhighway between two cities?
2. A 100-m race is run on a circular track that is 50 m in circumference. As a runner proceeds from start to endpoint of the race, what is her total displacement during the race?
3. When you finally lie down on your bed tonight, what has been your total displacement since the time you rose from the same bed this morning?
4. Two helicopter landing sites A and B are a few kilometers apart on a perfectly flat plane. A woman takes off from A in a helicopter, circles around, and lands at B, while a man walks from B to A. Compare their displacements.
5. Can a northward-directed force balance a force which is directed eastward?
6. Is it possible to obtain a force directed straight north by combining a northeast force and an equal northwest force? What if the forces are unequal?
7. What must be true about a third force if it is to balance simultaneously a force $\mathbf{A}$ in the x direction and a force $\mathbf{B}$ in the y direction?
8. How do we know that forces obey the rules we have stated concerning manipulation of vectors?
9. Can you think of a quantity which has direction and magnitude but which does not obey the commutative law? The associative law?
10. Show how three equal-magnitude vectors would have to be oriented if they were to add to give zero. Can this be done with three unequal vectors? Two unequal vectors?
11. Give examples of physical situations in which vector quantities subtract from each other. Can they be thought of as adding rather than subtracting?
12. Represent each person in a city of 200,000 by a vector extending from toe to nose. Estimate the resultant of these vectors at (a) 12 noon and (b) 12 midnight.

Chapter 2
Problems

1. A 50-m race is run on a 100-m circumference circular track. What is the displacement of the endpoint of the race from the starting point?
2. A Ping-Pong ball is dropped into a vertical glass tube from a height of

75 cm above the tube's closed bottom. The ball bounces 8 times before coming to rest. If the average bounce height is 11 cm, what is the total displacement of the ball from the time it is dropped to the time it comes to rest?

3. A displacement of 20 m is made in the x, y plane at an angle of 70° (i.e., 70° counterclockwise from the $+x$ axis). Find its x and y components. Repeat if the angle is 120°; if the angle is 250°.

4. A force of 50 newtons (N) pulls in the x, y plane at an angle of 40°. Find its x and y components. Repeat if the force angle is 200°; if the angle is 310°.

5. A car travels 5 blocks east and 2 blocks north. Find the magnitude and direction of its resultant displacement vector.

6. It is found that an object will hang properly if an x force of 20 N and a y force of -30 N are applied to it. Find the single force (magnitude and direction) which would do the same job.

7. Find the magnitude and direction of the force which has an x component of -40 N and a y component of -60 N.

8. Find the magnitude and direction of the sum of the following two coplanar displacement vectors: 20 m at 0° and 10 m at 120°.

9. Add the following two coplanar forces: 30 N at 37° and 50 N at 180°.

10. A force **A** is added to a second force which has x and y components 3 N and -5 N. The resultant of the two forces is in the $-x$ direction and has a magnitude of 4 N. Find the x and y components of **A**.

11. Find the components of a displacement which when added to a displacement of $7\mathbf{i} - 4\mathbf{j}$ m will give a resultant displacement of $5\mathbf{i} - 3\mathbf{j}$ m.

12. Find the magnitude and direction of the vector sum of the following three vectors: $2\mathbf{i} - 3\mathbf{j}, -9\mathbf{i} - 5\mathbf{j}, 4\mathbf{i} + 8\mathbf{j}$.

13. What must be the components of a vector which, when added to the following two vectors, gives rise to a vector $6\mathbf{j}$: $10\mathbf{i} - 7\mathbf{j}$ and $4\mathbf{i} + 2\mathbf{j}$?

14. A certain room has a floor which is 5 m × 6 m and the ceiling height is 3 m. Write an expression for the vector distance from one corner of the room to the corner diagonally opposite it. What is the magnitude of the distance?

15. Find the displacement vector from the point $(0, 3, -1)$ to the point $(-2, 6, 4)$. Give your answer in $\mathbf{i}, \mathbf{j}, \mathbf{k}$ notation. Also give the magnitude of the displacement.

16. An object, originally at the point $(2, 5, 1)$, is given a displacement $8\mathbf{i} - 2\mathbf{j} + \mathbf{k}$. Find the coordinates of its new position.

17. Find the resultant displacement caused by the following three displacements: $2\mathbf{i} - 3\mathbf{k}, 5\mathbf{j} - 2\mathbf{k}$, and $-6\mathbf{i} + \mathbf{j} + 8\mathbf{k}$. Give its magnitude as well as its $\mathbf{i}, \mathbf{j}, \mathbf{k}$ representation.

18. Give the $\mathbf{i}, \mathbf{j}, \mathbf{k}$ representation and magnitude of the force which must be added to the following two forces to give a force $7\mathbf{i} - 6\mathbf{j} - \mathbf{k}$: $2\mathbf{i} - 7\mathbf{k}$ and $3\mathbf{j} + 2\mathbf{k}$.

19. Vectors **A** and **B** are in the x, y plane. If **A** is 70 N at 90° and **B** is 120 N at 210°, find (*a*) **A** − **B**, and (*b*) vector **C** such that **A** − **B** + **C** = 0.

20. If $\mathbf{A} = 2\mathbf{i} - 3\mathbf{j} + 5\mathbf{k}$ and $\mathbf{B} = -\mathbf{i} - 2\mathbf{j} + 7\mathbf{k}$, find (in component form) (*a*) **A** − **B**, (*b*) **B** − **A**, and (*c*) vector **C** such that **A** + **B** + **C** = 0.

21. A certain vector is given by $3\mathbf{i} + 4\mathbf{j} + 7\mathbf{k}$. Find the angle it makes with the $+z$ axis.

22. What must be the relation between vectors **A** and **B** if the following condition is to be true:

$$A - 2B = -3(A + B)$$

If vector $A = 6i - 2k$, what is B?

23. The vector $A = 2i - 5j + 7k$. If $5B - 2A = 3(A + 5B)$, find the vector B.

24. The vector displacements of two points A and B from the origin are

$$s_A = 3i - 2j + 5k \qquad \text{and} \qquad s_B = -i - 5j + 2k$$

Find the magnitude and i, j, k representation of the vector from point A to point B.

25. Point A is at a vector position $s_A = -2i + 6j + k$ from the origin. A displacement $4i - 2k$ is made from A to point B. Find the vector position of B from the origin.

26. What must be the relation between two vectors A and B if the magnitude of $A + B$ equals the magnitude of $A - B$, that is,

$$|A + B| = |A - B|$$

3 Bodies at Rest

Much of physics has to do with objects and systems which are at rest and which remain at rest. This portion of physics is called *statics*. It is of prime importance since the concepts which it involves permeate most fields of physical science and engineering. In this chapter we shall find that two basic conditions must be satisfied if an object is to remain at rest. When we study Newton's second law of motion in Chap. 5, it will be found that these conditions arise from application of the law to a special case.

3.1 THE FIRST CONDITION FOR EQUILIBRIUM

When an object is at rest and remains at rest, we say that the object is in static equilibrium.* Our study of objects at rest begins by determining what conditions must be true if the object is to be in static equilibrium. We shall see that only two conditions need be satisfied. As a result, the subject area of statics is based entirely on the use of the two relations which express these two conditions in a mathematical way. Once you understand these relations, the solution of statics problems should be both systematic and straightforward for you. Let us arrive at these relations by reference to situations with which we are all familiar.

Consider, for example, an object hanging from a rope as in figure 3.1. Let us say that the object weighs 100 newtons (22.5 pounds). If we examine the body in detail, we see that two distinct forces act on it. The rope pulls upward with a 100-N force, while the gravitational attraction of the earth for the object causes a 100-N downward force to pull on it. We shall call the gravitational attractive force the *pull of gravity*. This gravitational force on the object will also be designated the *weight* of the body. These forces are shown in part (*b*) of figure 3.1.

Experiment shows that the body will remain motionless as long as the vertical forces acting directly on it are equal and opposite. Notice that we are concerned only with the forces acting *on the body*. Other forces exist; for example, the rope pulls on the ceiling. However, that force does not act on the body we are talking about, and so it cannot influence the body directly. These other forces will affect the body under consideration only inasmuch as they alter the forces which act directly on the body.

As a more complicated case, consider the object shown in figure 3.2. Let

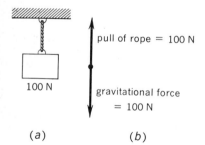

FIGURE 3.1 The body will remain at rest if the upward pull of the rope on it equals the downward pull of gravity.

For a body to remain at rest, the resultant force on it must be zero

* An object which is moving with constant velocity is also in equilibrium, but a discussion of this case is postponed until Chap. 5.

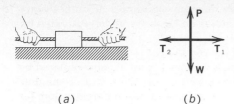

(a)   (b)

FIGURE 3.2 What must be true of the vectors shown in (b) if the object is to remain at rest?

A necessary condition for equilibrium

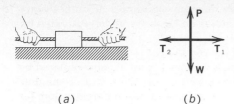

FIGURE 3.3 reference image

rope 2   rope 1

θ_2 θ_1

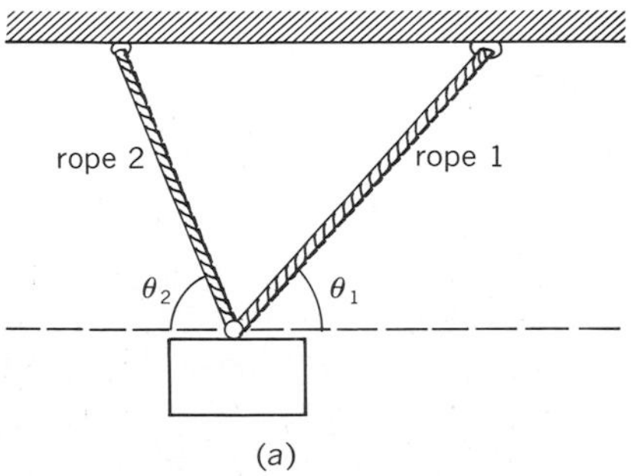

(a)

FIGURE 3.3 Under what condition will $T_1 = T_2$ when the body is in equilibrium? (See next page for an analysis.)

us say that the weight of the object is W. The forces acting *on the body* are shown in the vector diagram. If the body is not to move up or down, the upward push of the floor, **P**, must balance the downward pull of gravity, **W**. In addition, experiment shows that the tensions in the two ropes, T_1 and T_2, must balance each other if the object is not to move sideways.*

Let us consider still another case, the object shown in figure 3.3(a). The object has a weight W, and is supported by the two ropes which have tensions T_1 and T_2. In this case, also, experiment shows that equilibrium is achieved provided the y-directed forces cancel each other and the x-directed forces also cancel. Of course, in stating this, we think in terms of the components of $\mathbf{T}_1$ and $\mathbf{T}_2$, knowing from the preceding chapter that the components are fully equivalent to the original force vectors.

From these, as well as many other experiments, we can conclude that an object will remain in equilibrium only if the vector sum of the x-, y-, and z-directed forces acting on it is zero. To state this conclusion in compact form, we designate the various forces acting on the body as $F_1, F_2, \ldots, F_N$. They will have x components $F_{x1}, F_{x2}, \ldots, F_{xN}$ and similarly for the y and z components. In terms of these symbols, the condition we have found to be necessary for equilibrium is

$$\sum_N F_{xn} = 0 \qquad \sum_N F_{yn} = 0 \qquad \sum_N F_{zn} = 0 \qquad 3.1$$

We are using the usual mathematical shorthand notation in which the summation sign Σ lets us represent the equation

$$F_{x1} + F_{x2} + F_{x3} + \cdots + F_{xN} = 0$$

by

$$\sum_N F_{xn} = 0$$

Since the equations *3.1* are simply a statement that the vector sum of the forces acting on the object must be zero for equilibrium to be possible, they could be written more compactly as

$$\sum_N \mathbf{F}_n = 0$$

However, in practice we shall find that the component form given in equation *3.1* is most useful for computations. Let us now see how we make use of it.

Illustration 3.1

If the body in figure 3.3(a) is to be in equilibrium, what must be the ratio T_1/T_2?

Reasoning The physical situation in (a) will be analyzed by use of equations *3.1*. To use them, we must isolate an object for discussion. In this case, we choose the block as our isolated object. Three forces act on it, the pull of gravity W and the tensions in the two ropes, T_1 and T_2.

Having isolated an object for discussion, we draw a *free-body diagram* for it, a diagram that shows the forces on the object. This is done in part (b)

*The phrase "tension in a rope" means the force exerted by the rope on the object to which it is attached.

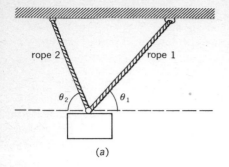

rope 2   rope 1

θ_2 θ_1

(a)

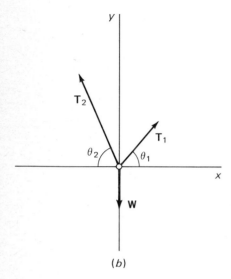

(b)

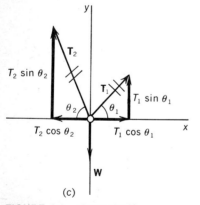

(c)

FIGURE 3.3 (Continued.)

of the figure. Notice that a free-body diagram eliminates all unnecessary detail and concentrates on the forces acting on the body we have isolated for discussion.

Because we wish to make use of the components of the forces in (b), we find them in (c). Notice that the original vectors $\mathbf{T}_1$ and $\mathbf{T}_2$, now replaced by their components, have crosshatch (or cancel) marks on them to remind us that they have been replaced. Some people, instead, use different colors for the original and component vectors. In this text we shall often indicate this replacement by making the original vector of different width than its components. In any event, we can now write down equations *3.1* in this situation by simply referring to part (c). We have that

$$\Sigma F_{xn} = 0$$

becomes

$$-T_2 \cos \theta_2 + T_1 \cos \theta_1 = 0$$

from which

$$\frac{T_1}{T_2} = \frac{\cos \theta_2}{\cos \theta_1}$$

Can you show that this implies that the more nearly vertical rope will carry the larger load? When θ_1 is 90°, this predicts that the ratio T_1/T_2 becomes infinite. What is the physical meaning of that result?

Illustration 3.2

If in figure 3.3 we have $W = 60$ N, $\theta_1 = 53°$, and $\theta_2 = 37°$, what will be the values of T_1 and T_2?

Reasoning   Again referring to part (c) of the figure, we write

$$\Sigma F_{xn} = 0 \qquad \text{which gives} \qquad T_1 \cos \theta_1 - T_2 \cos \theta_2 = 0$$
$$\Sigma F_{yn} = 0 \qquad \text{which gives} \qquad T_1 \sin \theta_1 + T_2 \sin \theta_2 - W = 0$$

Placing in the given values, we find

$$0.6T_1 - 0.8T_2 = 0$$
$$0.8T_1 + 0.6T_2 - 60 \text{ N} = 0$$

Solving these two equations simultaneously for T_1 and T_2 yields $T_1 = 48$ N and $T_2 = 36$ N.

Illustration 3.3

If the tension in rope 1 in figure 3.4(a) is 120 N (27 lb), how much does the object weigh?

Reasoning   We see at once that rope 5 supports the weight of the object. Therefore the tension in that rope, T_5, will equal the object's weight. The problem therefore reduces to finding T_5. In a complicated problem such as this we must be very careful to *isolate an object clearly*; only the forces acting on that object will be considered. If we decide to discuss the block itself, only two forces act upon it: the pull of gravity down, W, and the tension in the rope, T_5. Since the vector sum of these forces must be zero, we have $-W + T_5 = 0$, which tells us $T_5 = W$, a fact we already knew.

It proves more fruitful to talk about the junction of the ropes at A. Using that junction as our isolated body, we draw the free-body diagram for it in (b) of figure 3.4. After taking z and y components, we can write *3.1*, since the junction is at equilibrium. We find

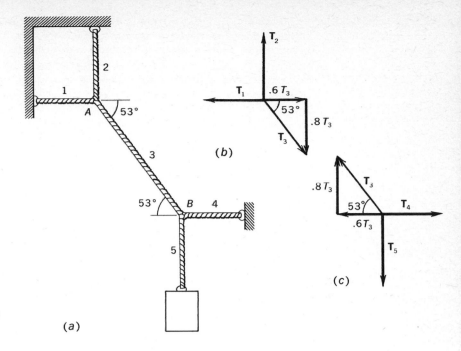

FIGURE 3.4 Equation 3.1 is written separately for the junction at A and the junction at B.

(a)

(b)

(c)

$$\Sigma F_{xn} = 0 \quad \text{gives} \quad 0.6T_3 - T_1 = 0$$
$$\Sigma F_{yn} = 0 \quad \text{gives} \quad T_2 - 0.8T_3 = 0$$

We were told that $T_1 = 120$ N; so the first equation yields $T_3 = 200$ N. Placing this value in the second equation yields $T_2 = 160$ N.

Let us now consider a new body, the junction at point B. It also is at equilibrium and must obey 3.1. The free-body diagram for this junction is shown in (c) of figure 3.4. Notice that the tension in rope 3 is still T_3, even at the lower junction. (If the rope were quite heavy, this would not be true. For now we assume the rope to have negligible weight.) Writing 3.1 for this body, we find

$$\Sigma F_{xn} = 0 \quad \text{gives} \quad T_4 - 0.6T_3 = 0$$
$$\Sigma F_{yn} = 0 \quad \text{gives} \quad 0.8T_3 - T_5 = 0$$

Since T_3 has already been found to be 200 N, we can substitute this value in the last equation, and we then find $T_5 = 160$ N. Therefore the object must weight 160 N.

3.2 THE SECOND CONDITION FOR EQUILIBRIUM

In many instances an object will not remain at rest even if 3.1 is satisfied. For example, when the rod in figure 3.5 is subjected to two equal and opposite forces as shown, the rod will not remain stationary. Instead, it will begin to rotate counterclockwise. This happens in spite of the fact that $\Sigma F_n = 0$. Clearly, some additional restriction must be made if a body is to remain at rest.

Even rather crude experiments such as those shown in figure 3.6 are sufficient to show what additional condition must be satisfied in order that an

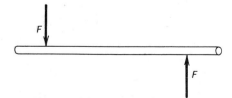

FIGURE 3.5 Even though the resultant force acting on the rod is zero, it will not remain at rest.

object remain in equilibrium. The complication apparent in figure 3.5 concerns rotation of the object. We therefore investigate what conditions must be satisfied if an object is not to rotate. Let us now examine the situations shown in figure 3.6. A very light rod is pivoted at the point indicated. The rod will not rotate when the conditions are as shown.

Investigation of such situations leads to the following conclusions: (1) A force whose line of action (the *line of the force*) passes through the pivot point will not cause rotation about the point [see part (*a*)]. (2) The rotation effect of a force such as those in (*b*) and (*c*) is proportional to the distance from the pivot. (3) As we see in (*d*), when a force is replaced by its rectangular components with one component passing through the pivot, the rotation is due entirely to the other component(s).

Extensive experimentation has confirmed the results of these simple experiments. Insofar as rotation is concerned, the important factor is the *torque* τ, caused by a force. The term torque is defined in relation to figure 3.7 for the case of rotation in a plane. We first choose a point in the plane about which we shall consider the forces in question to cause rotation. This point will be designated the *pivot point*. It is simply an arbitrary point in the plane and, as we shall see in a later section, need not be the point about which the given object is constrained to rotate.

The force **F** is applied at a point in the plane as indicated in the figure, and we shall be concerned with the radial line of length r from the pivot point to the point of application of the force. As we saw previously, the force component F_r along the radius causes no turning effect because it is in line with the pivot point. The entire turning effect is due to the component of **F** perpendicular to the radius, and we see that its magnitude is $F \sin \theta$, where θ is the angle between the force and the radial line. Moreover the turning effect is proportional to both $F \sin \theta$ and r. We therefore define the turning effect, which we call *torque*, in the following way:

Definition of torque

$$\tau \equiv \text{torque} = rF \sin \theta \qquad\qquad 3.2$$

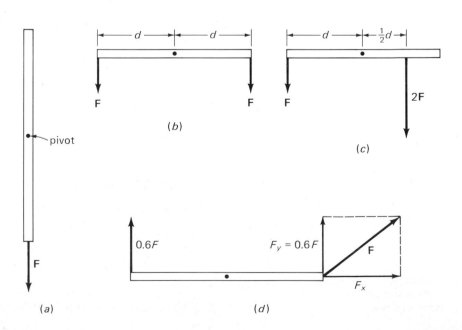

FIGURE 3.6 With the forces as shown, the rod will not rotate about the pivot through its center.

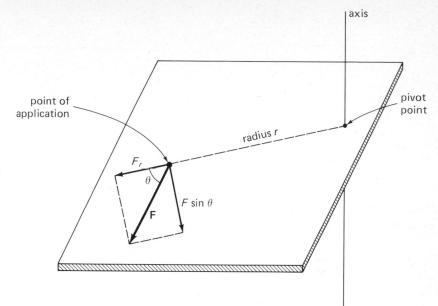

FIGURE 3.7 Only the component $F \sin \theta$ causes rotation about the axis.

where θ is the angle the force F makes with the radius r from the pivot point to the force application point.

Referring back to figure 3.6, we see in part (b) that the torque trying to turn the rod clockwise is $(d)(F)$, while the counterclockwise torque is also $(d)(F)$. In part (c), both of these torques are equal to $(d)(F)$. In part (d) both torques are 0.6 $Fd/2$. In cases such as these (where the forces all lie in the plane of rotation), the system will not rotate if the clockwise torques are equal in magnitude to the counterclockwise torques. To make the notation more compact, we shall call clockwise torques negative, and counterclockwise torques positive. Therefore, when the torques balance, the algebraic sum of the torques is zero. That is,

The second necessary condition for equilibrium

$$\Sigma \tau_n = 0$$

at equilibrium. This is the second condition for equilibrium.

We may therefore summarize our conditions for equilibrium in the case of forces which lie in a plane by*

$$\Sigma F_{xn} = 0$$
$$\Sigma F_{yn} = 0 \qquad\qquad 3.3$$
$$\Sigma \tau_n = 0$$

Before making use of these relations, it is convenient to dispose of two other matters which appear in such uses. The first has to do with the choice of the pivot point. The other allows us to deal with objects which have appreciable weight.

---

* You might well wonder what happens when torques are active in three dimensions rather than the two dimensions considered here. For example, suppose torques acting on an object are trying to rotate it about each of the three coordinate axes. As you might expect, for equilibrium to be maintained, the torques about each axis must cancel. Situations such as this are best treated by defining torque to be a vector. We shall see how this is done in Chap. 10.

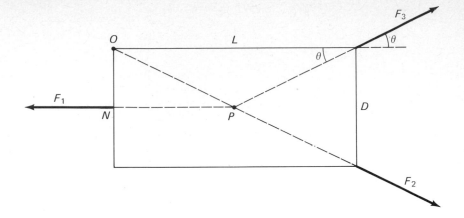

FIGURE 3.8 Find the torques about axes at O and at P.

Illustration 3.4

Find the torques for the forces shown in figure 3.8 about (*a*) P as axis and (*b*) O as axis.

Reasoning (*a*) Because the lines of all three forces pass through point P, torques due to the forces about this point as pivot point are zero.
(*b*) For an axis through O, we have as follows:

Force	Sense	Torque
F_1	Clockwise	$-F_1 \, \overline{ON} \sin 90° = -\tfrac{1}{2}DF_1$
F_2		0 [as in (*a*)]
F_3	Counterclockwise	$F_3 \, L \sin \theta = F_3 LD / \sqrt{L^2 + D^2}$

3.3 THE POSITION OF THE PIVOT POINT IS ARBITRARY

As the heading of this section implies, we can prove the following theorem for *rotation in a plane: If the sum of the torques about one pivot point is zero, the sum will be zero about any pivot point, provided the vector sum of the forces is zero.* To prove this theorem we refer to figure 3.9.

The sum of the torques due to the various forces shown is known to be

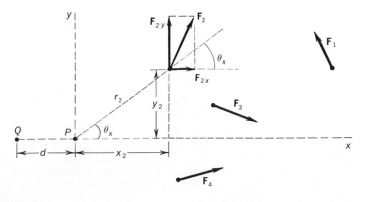

FIGURE 3.9 If the forces are in equilibrium about P, they will also be in equilibrium about Q.

zero about pivot point P. Since each force may be replaced by its x and y components, we may deal with them rather than with the force itself. If Q is any other pivot point, take the x, y coordinates as shown. The torque due to F_{2x} (about pivot point P) is

$$\tau_{2x} = -F_{2x} r_2 \sin \theta_x = -F_{2x} y_2$$

and due to F_{2y} is

$$\tau_{2y} = F_{2y} r_2 \sin (90 - \theta_x) = F_{2y} x_2$$

From which

$$\tau_2 = x_2 F_{2y} - y_2 F_{2x}$$

Notice that if the usual sign conventions are used for the F's, x's, and y's, clockwise torques will be negative, while counterclockwise torques will be positive. Our original assumption concerning the sum of the torques about P gives

$$\tau_1 + \tau_2 + \cdots + \tau_N = 0$$

where we assume N forces exist. Or, in terms of components,

$$(F_{1y} x_1 + F_{2y} x_2 + \cdots + F_{Ny} x_N) - (F_{1x} y_1 + F_{2x} y_2 + \cdots + F_{Nx} y_N) = 0$$

Now let us write the sum S of the torques about the different axis at Q:

$$S = [F_{1y}(d + x_1) + F_{2y}(d + x_2) + \cdots + F_{Ny}(d + x_N)]$$
$$- (F_{1x} y_1 + F_{2x} y_2 + \cdots + F_{Nx} y_N)$$

But the terms involving x and y are identical with those in the preceding equation, and their sum is known to be zero. We therefore find S to be

$$S = d(F_{1y} + F_{2y} + \cdots + F_{Ny})$$

At this point we recall that the theorem concerns the case when the vector sum of the forces is zero. As a result, the term in parentheses, being just the sum of the y components, must be zero. Therefore we have proved that the sum of the torques about point Q is also zero, and the theorem is proved.

3.4 CENTER OF GRAVITY

If an object has weight, this fact alone can cause the body to rotate about a pivot point. For example, the uniform rod shown in figure 3.10(a) will rotate clockwise about the pivot at its top. Only if the rod is oriented as in (b) will it remain at rest. We conclude that in (b) the pull of gravity on the rod, **W,** must in effect lie along the dotted line which passes through the pivot. In that case the torque due to the weight of the rod would be zero.

We now extend this reasoning to a more complex body, such as the one shown in part (c). When the body hangs at rest from point A as shown, the pull of gravity must act along the vertical line AB. Similarly, if the body hangs at rest as shown in (d), the pull of gravity must act along the vertical line CD. It therefore appears that the gravitational force acting on the body can be considered to act at the intersection of these two lines, namely, at M. This is confirmed by the fact that the body may be supported in any arbitrary orientation by a force equal and opposite to **W** applied at point M.

The point M in figure 3.10(d) is called the *center of gravity* of the body.

The weight of a body may be considered to be applied at the center of gravity

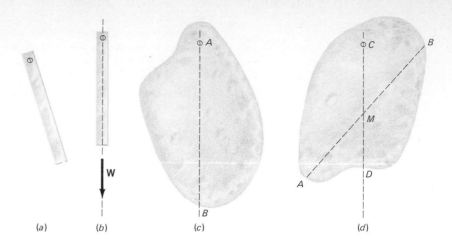

FIGURE 3.10 How to locate the center of gravity of an object.

(a)   (b)   (c)   (d)

It is the point at which the force due to gravity may be assumed to be applied to the body. For a uniform sphere, the center of gravity is at the center of the sphere. In cases of less symmetric forms, the center of gravity can always be found by the method illustrated in the figure.

The center of gravity of an object is sometimes confused with a second quantity, the center of mass. The center of mass will be defined and used in Chap. 7, and we shall see then that it is more fundamental than the center of gravity. When we relate the two quantities, we shall find that the two points coincide in the special case of an object in a uniform gravitational field.

Illustration 3.5

The horizontal board shown in figure 3.11 is uniform and weighs 200 N. Find the tension in the two ropes supporting it when a weight W is suspended as shown.

Reasoning We isolate the board as the object for discussion. The forces acting on it are drawn in part (b) of the figure. Notice that the weight of the board is taken to act at the center of gravity of the board. Since the board is at equilibrium, we know that

$$\Sigma F_{xn} = 0$$
$$\Sigma F_{yn} = 0$$
$$\Sigma \tau_n = 0$$

The first equation is automatically true because no x-directed forces act on the board. The y equation yields

$$T_1 + T_2 - W - 200\,\text{N} = 0$$

In order to write the torque equation, we must select a pivot point. As pointed out in the preceding section, all pivot points are valid. If we take the pivot at point P, the unknown force T_1 will not appear in the torque equation because the line of the force goes through the axis. This absence of one of the unknowns from the equation will simplify the algebra, and so we shall take the pivot at point P. (A pivot at Q would be equally suitable.)

The 200-N and W forces cause clockwise rotation (negative) about P, while T_2 will cause a positive torque. Therefore the torque equation is

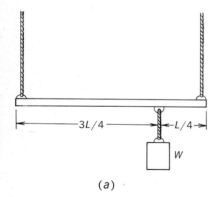

(a)

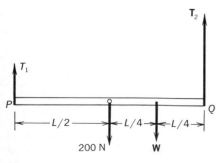

(b)

FIGURE 3.11 Why should the pivot point be taken at P or Q?

$$-(200\,\text{N})\frac{L}{2} - W\frac{3L}{4} + T_2L = 0$$

Notice that the length of the board cancels out. As a result,

$$T_2 = 100\,\text{N} + \frac{3W}{4}$$

This result can be substituted in the preceding equation involving T_1 to obtain

$$T_1 = \frac{W}{4} + 100\,\text{N}$$

Illustration 3.6

The uniform board shown in figure 3.12 weighs 200 N. Find the tension in each of the three ropes. The man weighs 800 N.

Reasoning We isolate the board for consideration and draw the forces acting on it. They are shown in the figure. Notice that it is convenient to replace force $\mathbf{T}_3$ by its two components. Since the board is in equilibrium.

$$\Sigma F_{xn} = 0 \qquad \text{giving} \qquad 0.6T_3 - T_2 = 0$$
$$\Sigma F_{yn} = 0 \qquad \text{giving} \qquad T_1 + 0.8T_3 - 200\,\text{N} - 800\,\text{N} = 0$$

To write the torque equation we must choose a pivot point. We take it at point Q since this will eliminate T_1 and T_2 from the equation. Therefore

$$\Sigma \tau_n = 0 \qquad \text{gives} \qquad -(200\,\text{N})\frac{L}{2} - (800\,\text{N})\frac{3L}{4} + 0.8T_3L = 0$$

Notice that the torque due to $0.6T_3$ is zero since the line of the force goes through the pivot.

Solving the equation for T_3 gives

$$T_3 = 875\,\text{N}$$

Substituting this in the preceding equations tells us that $T_1 = 300\,\text{N}$ and $T_2 = 525\,\text{N}$.

FIGURE 3.12 Find $\mathbf{T}_1$, $\mathbf{T}_2$, and $\mathbf{T}_3$.

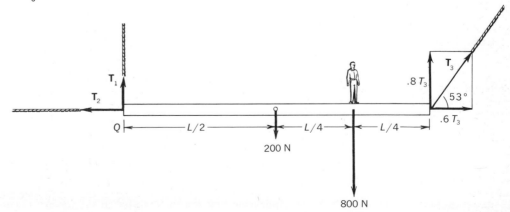

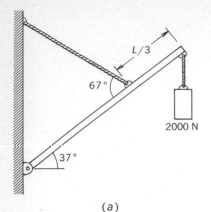

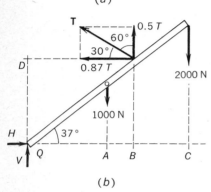

(a)

(b)

FIGURE 3.13 What are the torques of the various forces in (b)?

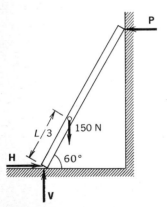

FIGURE 3.14 How large must the friction force H be if the ladder is not to slip?

Illustration 3.7

For the situation shown in figure 3.13, find the tension in the rope and the components of the force acting at the wall. The beam is uniform and weighs 1000 N.

Reasoning The forces acting on the beam are shown in part (b) of the figure. Notice that we have represented the force at the wall by its two components, H and V. Using the components of $\mathbf{T}$ as well, we have

$$\Sigma F_{nx} = 0 \qquad \text{whence} \qquad H - 0.87T = 0$$
$$\Sigma F_{ny} = 0 \qquad \text{whence} \qquad V + 0.5T - 1000\text{ N} - 2000\text{ N} = 0$$

In order to write the torque equation, we take the pivot at point Q. Why? The torques are as follows:

Force	Torque
H and V	0
1000 N	$-(\frac{1}{2}L)(1000\text{ N})\sin 53°$
2000 N	$-(L)(2000\text{ N})\sin 53°$
T	$(\frac{2}{3}L)(T)\sin 67°$

The torque equation is then

$$-(\tfrac{1}{2}L)(1000\text{ N})(0.8) - (L)(2000\text{ N})(0.8) + (\tfrac{2}{3}L)(T)(0.92) = 0$$

Solving for T gives

$$T = 3260\text{ N}$$

This value may be used in the preceding equations, and we find

$$H = 2840\text{ N} \qquad \text{and} \qquad V = 1370\text{ N}$$

Illustration 3.8

The 150-N ladder shown in figure 3.14 leans against a smooth wall. (When we say the wall is smooth, we mean that the force of the wall on the ladder is perpendicular to the wall.) How large a horizontal force must the floor provide if the ladder is not to slip? The center of gravity of the ladder is one-third of the way up from the bottom.

Reasoning The forces on the ladder are shown in the figure. Taking the axis at the intersection of H and V we obtain the torque equation

$$-(\tfrac{1}{3}L)(150\text{ N})(0.5) + (L)(P)(0.87) = 0$$

from which

$$P = 28.7\text{ N}$$

The force H is given from

$$\Sigma F_{nx} = 0 \qquad \text{to be} \qquad H = P = 28.7\text{ N}$$

It is customary to designate as friction force those forces which are parallel to a surface and are caused by rubbing against the surface. For this reason, H in the present problem is called a *friction force*. Since the wall was said to be smooth, the force parallel to it, the friction force, was zero.

3.5 LABORATORY WORK IN STATICS

Before leaving this chapter on statics, we should point out a possible complication you may encounter in laboratory experiments involving statics.

Oftentimes you may be required to make use of weights whose values are stamped in kilograms and grams rather than in newtons. These stampings give the masses of the weights. We shall learn in Chap. 5 that an object that is stamped 1 kg weighs 9.8 N on earth. Therefore, if you have a weight that has a total stamped value of 362 g (0.362 kg), then its weight is $0.362 \times 9.8 = 3.55$ N.

Chapter 3
Questions and Guesstimates

1. Observation indicates that traffic lights hung from cables stretched across a street cause the cable to sag. Why don't the workmen remove the sag when they adjust the cable?

2. Draw the free-body diagrams for a 200-N girl in the following equilibrium situations: (*a*) she stands on one foot; (*b*) she hangs from a bar with one hand; (*c*) she stands on her head; (*d*) she does a handstand with a single hand resting on a stool.

3. Frequently, people prop open a door by placing a wedge in the crack between the door and its casing near the hinge. Why does this usually result in ruining the hinge?

4. Suppose that you place your foot in a loop at one end of a rope, the other end passing over a pulley hanging from the roof and back down to you. Discuss how the tension in the rope changes as you pull harder and harder on it. Can one lift one's self in this way? Does this disprove the old concept that you cannot lift yourself by your own bootstraps?

5. A pail is hung from a tight clothesline by means of a rope similar to the clothesline. Water is slowly added to the pail. Which, if either, should break first, the rope holding the pail or the clothesline?

6. As part of his act, a performer is lifted slowly by pulling upward on the hair of his head, but he is rapidly going bald. Approximately how many hairs must remain on top of his head if the act can be performed safely?

7. You are given two measuring "instruments," a 2-m long rigid plank that is 25 cm wide, and two blocks each 1 cm × 1 cm × 20 cm. How could you use these to determine the height from the floor of the center of gravity of a person who is standing straight up?

8. A uniform block of wood sits on a board as shown in figure P3.1. The small ridge shown keeps the block from slipping. As the board is tilted further so as to increase θ, the block eventually topples over even though it does not slip. How could you predict at what angle it would topple?

9. A bricklayer notes that he can stack bricks as shown in figure P3.2 and they will not topple over. Assuming the bottom brick to be sitting freely on the surface and the bricks to be cemented together, can the bricklayer pile an indefinite number of bricks in this way without them toppling?

10. Hold your body rigid and straight and try to incline your body at an angle θ to the vertical. Estimate how large θ can become before you become unstable and fall over. What features of your body influence θ most?

11. Your arm bones above and below the elbow are much like two pivoted boards. The tendons of your arm act like a cable connecting the two bones. Estimate the tension in these tendons when you hold a 20-N weight in your hand at several different orientations of the arm.

12. Examine the situation of a person bending over insofar as the muscles in his back and the compressive forces in the backbone are concerned.

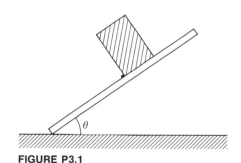

FIGURE P3.1

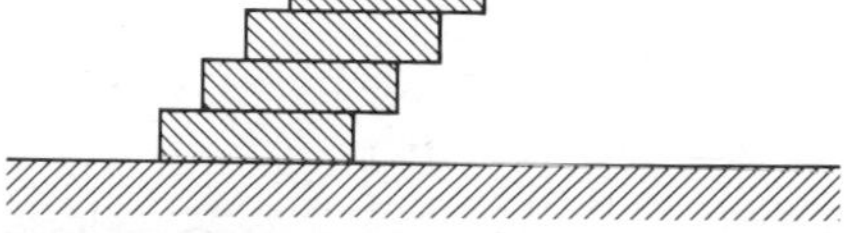

FIGURE P3.2

Explain why back injury can easily occur if heavy weights are lifted by a person who bends over rather than squats down.

13. If a door is properly hung, it will remain standing open at any angle at which it is placed. Many doors will swing farther open or closed when released. Why?

Chapter 3
Problems

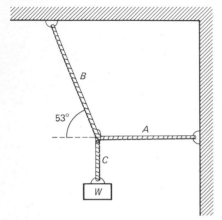

FIGURE P3.3

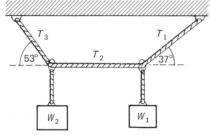

FIGURE P3.4

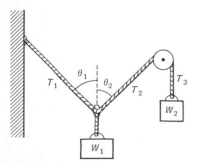

FIGURE P3.5

1. A 160-N child is sitting on a swing of negligible weight. How large a horizontal force must be applied to the child if the swing rope is to be held at an angle of 50° to the vertical?

2. In designing a huge display pendulum, an engineer decided to support the 200-N pendulum ball from a cable which has a strength of 400 N. If the pendulum ball is pulled sideways by a horizontal force, what will be the angle the pendulum makes with the vertical when the cable breaks? To see if the engineer is in trouble, calculate how large a horizontal force is needed to do this.

3. Three ropes are attached to a 120-N ball. Rope A suspends it from the ceiling. Rope B pulls on it horizontally with a force of 50 N while rope C pulls straight down with a force of 30 N. Find the tension in rope A and the angle it makes to the vertical. (*Hint:* Note that $\tan\theta = \sin\theta/\cos\theta$.)

4. A 200-N ball is supported from the ceiling by rope A. Rope B pulls downward and to the side on the ball. If the angle of A to the vertical is 20° and if B makes an angle of 50° to the vertical, find the tensions in ropes A and B.

5. Repeat problem 4 if rope B pulls upward at an angle of 37° to the vertical on the ball.

6. A boy of weight W hangs from the center of a clothesline and distorts the line so that it makes 20° angles with the horizontal at each end. Find the tension in the clothesline in terms of W.

7. Refer to figure P3.3. At equilibrium, what will be the tension in rope A expressed in terms of the weight W of the suspended object?

8. In figure P3.3, suppose the suspended object is a bucket partly filled with sand. As more sand is added, one of the identical ropes (A, B, or C) breaks. Which rope will break first? What will be the tension in it (in terms of W) when it breaks?

9. The weight W_1 shown in figure P3.4 is 300 N. Find T_1, T_2, T_3, amd W_2.

10. Referring to figure P3.4, obtain a numerical relation between W_1 and W_2 if the system is to hang in equilibrium as shown.

11. Given that $\theta_1 = \theta_2$ in figure P3.5, what can you say about T_1, T_2, T_3, W_1, and W_2 provided the pulley is frictionless?

12. Referring to figure P3.5, if $\theta_1 = 53°$ and $\theta_2 = 37°$, how large is W_1 in comparison to W_2?

13. Suppose that $W_1 = W_2$ in figure P3.5 and that $\theta_1 = 53°$. Show that $0.75\sin\theta_2 + \cos\theta_2 = 1$. How could you find θ_2 itself?

14. A force $\mathbf{F} = 6\mathbf{i} + 2\mathbf{j}$ N acts at the point (0, 5 m, 0). Find its torque about the z axis. About the x axis.

15. Repeat problem 14 if the force is applied at the point (4 m, 5 m, 0).

16. Referring to figure P3.6, the plank is uniform and weighs 500 N. How large must W be if T_1 and T_2 are to be equal? Evaluate T_1.

17. Assuming the 500-N plank shown in figure P3.6 to be uniform, find T_1 and T_2 if $W = 800$ N.

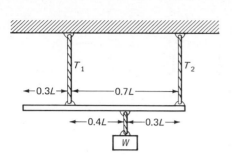

FIGURE P3.6

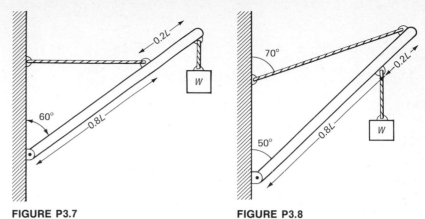

FIGURE P3.7

FIGURE P3.8

18. In figure P3.7, the beam is uniform and weighs 60 N. If $W = 200$ N, find the tension in the tie rope and the x- and y-component force that the hinge exerts on the door.

19. Neglecting the weight of the beam in figure P3.8 find the tension in the tie rope and the force components at the hinge in terms of W.

20. In figure P3.8, if the tie rope can hold a maximum tension of 1000 N and if the beam is uniform and weighs 200 N, what is the maximum weight W which can be supported as shown?

21. A person holds a 20-N weight as shown in figure P3.9. Find the tension in the supporting muscle and the component forces at the elbow. Assume the system can be approximated as shown in (*b*) where T_m is the tension in the muscle, the beam is the lower arm, and the lower arm weighs 65 N with center of gravity as shown.

22. Consider a woman lifting a 60-N bowling ball as shown in figure P3.10. Find the tension in her back muscle and the compressional force in her spine when her back is horizontal. Approximate the situation as shown in (*b*) and assume the upper part of her body to weigh 250 N with center of gravity as indicated.

23. When one stands on tiptoe and at equilibrium, the situation is much like that shown in figure P3.11. We can replace the actual situation by the

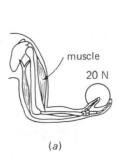

(*a*)

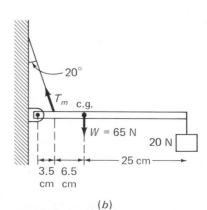

(*b*)

FIGURE P3.9

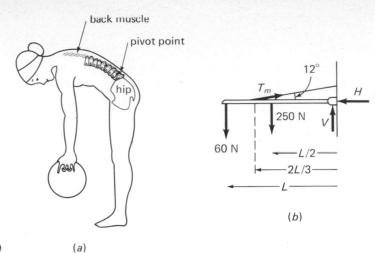

model in (b). In terms of the push F of the floor on the toe, find (a) the tension in the Achilles tendon and (b) the forces H and V at the ankle.

24. (a) Consider the object of negligible weight shown in figure P3.12. If $T_1 = 40\,$N, how large must T_2, V, and H be if the object is to remain in equilibrium? (b) Repeat if the force T_2 is pulling at an angle of 37° below the horizontal.

25. In figure P3.13, the weight of the beam is negligible. How large must W be if the system shown there is to be in equilibrium? Also, find T_1 and T_2.

26. Refer back to figure P3.1. Assume the block is uniform and twice as tall as it is wide in the plane of the page. Neglecting the height of the slight ridge that keeps it from slipping, at what angle θ will the block topple over as the board is increasingly tilted?

27. Nearly everyone knows that it is advantageous for a car to have a low center of gravity and a "wide track." Suppose the center of gravity of a car is a distance h above the roadway and the width of the car between wheel contact points with the road is d. If you try to tip the car over sideways, through how large an angle will you have to tilt it? More will be said about this in Chap. 10.

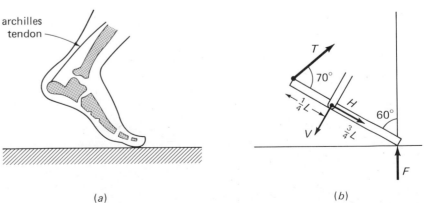

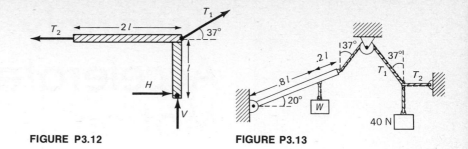

FIGURE P3.12   **FIGURE P3.13**

28. Two equal but opposite forces whose lines of action are parallel is called a *couple*. Suppose each force has a magnitude F and their lines are a distance d apart. (*a*) Show that the torque due to the couple about *any* axis perpendicular to the plane defined by the two force lines is Fd. (*b*) What is the torque about an axis that lies in the same plane as the force lines? (*c*) What is the torque about an axis that is in a plane parallel to the plane of the two forces with the distance between the two planes being y?

4 Uniformly Accelerated Motion

In the preceding chapter we discussed the behavior of objects at rest. Now we examine the motion of bodies in order to define velocity, speed, and acceleration. Once the meaning of these quantities is understood, we investigate the behavior of objects moving on a straight line. The relations found for straight-line motion are then extended so as to apply to objects undergoing projectile-type motion. The causes of motions such as those described in this chapter are the subject of the following chapter.

4.1 VELOCITY

Consider the situation shown in figure 4.1. A boy participating in a race starts at one goal line and sprints to the other. He accomplishes this in a time Δt. His displacement is shown as the vector $\Delta \mathbf{r}$ in the figure. We define a vector $\bar{\mathbf{v}}$, *Definition of average velocity* the *average velocity* of the boy, by the relation

$$\bar{\mathbf{v}} = \frac{\Delta \mathbf{r}}{\Delta t} \qquad 4.1$$

Notice that we have used a bar above $\bar{\mathbf{v}}$ as an indication that the velocity is an average. The direction of $\bar{\mathbf{v}}$ is the same as that of $\Delta \mathbf{r}.$

This definition of velocity is known to all of us since we are accustomed to reading a car's speed on a speedometer. When the speedometer reads 50 km/h (31 mi/h), we know that the car will be displaced 50 km (Δr) in a time $\Delta t = 1$ h if the car continues in a straight line with this velocity. Or, in a more reasonable situation, the car might go east a distance of 5 km in a time of 0.10 h. Then the magnitude of its average velocity would be $\Delta r \div \Delta t$, which is 5 km $\div$ 0.10 h or 50 km/h. The units of velocity are clearly those of length

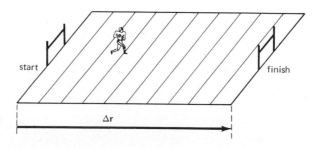

start   finish

Δr

FIGURE 4.1 How is the average velocity of the sprinter expressed in terms of $\Delta \mathbf{r}$?

divided by time. The vector portion of the definition states that the direction of the velocity is the same as the direction of the displacement.

Unfortunately, we usually cannot tell how fast a moving object was going at a particular point (or instant) from a knowledge of its average velocity. For example, if a car goes 20 km east in $\frac{1}{2}$ h, we know its average velocity was 20 km ÷ 0.5 h = 40 km/h east. But the car probably did not hold its speed exactly at 40 km/h for the whole half hour. From the data given, it is impossible to state the actual velocity of the car at any particular point in the 20-km trip.

In order to specify the velocity of the car at a particular point we must measure the distance moved by the car in a very small time interval *as it passed the point*. If we make the time interval Δt approach zero in magnitude, the car will not have changed its velocity much during that time. Our computed velocity would therefore closely approximate the true velocity of the car at that point. We call the velocity at a point the *instantaneous velocity,* and it is given by*

Definition of instantaneous velocity

$$\mathbf{v} = \lim_{\Delta t \to 0} \frac{\Delta \mathbf{r}}{\Delta t} \equiv \frac{d\mathbf{r}}{dt} \qquad\qquad 4.2$$

As indicated in the equation, this limit is, by the definition of the derivative, the derivative of $\mathbf{r}$ with respect to t.

Velocity is a vector; speed is a scalar

Notice that *velocity is a vector*. It tells us both the direction and magnitude of the rate of displacement of an object. Although in colloquial speech it is common to use the terms velocity and speed interchangeably, they have quite different meanings in physics. The *average speed* of an object is defined to be the length of the path traveled by an object Δs, divided by the time taken; i.e.,

$$\text{Speed} = \frac{\Delta s}{\Delta t}$$

It is a scalar. Let us show by example why *even the magnitude* of the average velocity usually differs from the speed.

Suppose a boy travels the path shown in figure 4.2(a). We see that he goes in the $+x$ direction from A to B and then turns around to come back as far as point C all in time Δt.

The vector displacement of the boy from his starting point is shown as the vector $\Delta\mathbf{x}$. From the definition given in *4.1* we find

$$\bar{\mathbf{v}} = \frac{\Delta\mathbf{x}}{\Delta t}$$

However, the speed of the boy is defined in terms of the total path length traveled, namely, $\overline{AB} + \overline{BC}$. Therefore

$$\text{Speed} = \frac{\overline{AB} + \overline{BC}}{\Delta t}$$

The magnitudes of average speed and velocity are often unequal

Since the magnitude of $\Delta\mathbf{x}$ (usually written $|\Delta\mathbf{x}|$, or just Δx) is much smaller than $\overline{AB} + \overline{BC}$, the value of $|\bar{\mathbf{v}}|$ and the speed will be far different. As an exercise, the student might consider the values of these two quantities if the

---

*For those who are unfamiliar with the symbolism $\lim_{\Delta t \to 0}$, it is read as "in the limiting case where Δt approaches zero." Roughly it means to use the value of $\Delta\mathbf{r}/\Delta t$ which one obtains by taking Δt small enough so that the ratio $\Delta\mathbf{r}/\Delta t$ does not change appreciably even if the measurement interval is made still smaller. By definition, this quantity is called the derivative of $\mathbf{r}$ with respect to t and is represented by $d\mathbf{r}/dt$.

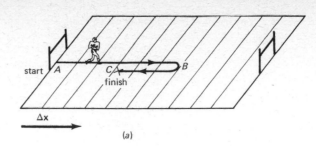

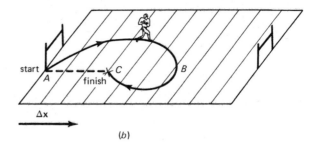

boy returned to his original starting point. In part (b) of figure 4.2 we show another situation, in which the average speed differs from the average velocity.

There is one case where the magnitudes of the average speed and velocity are the same. If an object travels in the same direction for the whole time Δt, then $|\Delta x|$ will be equal to the path length covered, and therefore the speed and velocity will be equal numerically. This is a quite common situation provided one is considering a rather small time interval, Δt. In fact, the magnitudes of the instantaneous speed and velocity are always equal. Why?

4.2 ACCELERATION

If a car speeds up from 20 to 30 km/h we say that it has accelerated. Acceleration is a measure of how rapidly the velocity of an object is changing. In particular, consider a car which starts from rest and is going down a straight road with the following speeds each second thereafter:

Time, s	0	1	2	3	4	5
Speed, m/s	0	4	8	12	16	20

We see that it speeds up by 4 m/s each second. The magnitude of its acceleration is simply how much it speeds up in unit time; in this case the acceleration is 4 m/s per second (usually written 4 m/s²).

Acceleration tells the increase in velocity per unit time

As another example, when an object is dropped from a tall building, it accelerates 32 ft/s each second. That is to say, after falling 1 s its speed is 32 ft/s, while at the end of 3 s, say, its speed is $32 + 32 + 32 = 96$ ft/s. For an object moving along a straight line, acceleration is a very simple concept. It tells us how much the speed of the object changes in unit time.

But objects do not always travel in straight-line paths. We make the concept of acceleration more useful by attributing direction to it and this

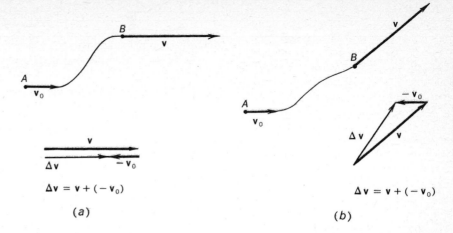

FIGURE 4.3 $\Delta \mathbf{v}$ need not be zero if $|\mathbf{v}_0| = |\mathbf{v}|$.

$$\Delta \mathbf{v} = \mathbf{v} + (-\mathbf{v}_0)$$

(a)

$$\Delta \mathbf{v} = \mathbf{v} + (-\mathbf{v}_0)$$

(b)

becomes important in non-straight-line motion. In particular *we define acceleration to be the change in velocity* (not speed, but the vector velocity) *per unit time.* Since velocity can be changed by changing the speed and/or the *direction* of motion of an object, this distinction will be of importance to us later. Let us now write the defining equation for acceleration.

Referring to figure 4.3, the object has a velocity $\mathbf{v}_0$ at point A and a velocity $\mathbf{v}$ at point B in both of the two examples given in the upper part of the figure. If the time taken to move from A to B is Δt, the average acceleration during this time is

Definition of average acceleration; acceleration is a vector quantity

$$\bar{\mathbf{a}} \equiv \frac{\mathbf{v} - \mathbf{v}_0}{\Delta t} = \frac{\Delta \mathbf{v}}{\Delta t} \qquad 4.3$$

The units of acceleration are clearly a velocity unit divided by a time unit. We see from the definition that $\bar{\mathbf{a}}$ has direction, and the direction is the same as that of $\Delta \mathbf{v}$.

In finding $\Delta \mathbf{v}$ we recall that the difference between two vectors may be written as a sum by reversing the direction of the vector being subtracted. Writing this statement as an equation gives

$$\Delta \mathbf{v} = \mathbf{v} - \mathbf{v}_0 = \mathbf{v} + (-\mathbf{v}_0)$$

This procedure for finding $\Delta \mathbf{v}$ is used in the lower parts of figure 4.3. You should examine these two cases carefully to be sure that you understand the direction of $\bar{\mathbf{a}}$ in each. It has the same direction as $\Delta \mathbf{v}$.

An interesting case, similar to that in part (*b*) of figure 4.3, will arise later in our study of motion in a circular path. Even though $|\mathbf{v}_0| = |\mathbf{v}|$, the value for $\Delta \mathbf{v}$ (and therefore of the acceleration) need not be zero. In particular, even if the length of the $\mathbf{v}_0$ and $\mathbf{v}$ vectors had been the same in part (*b*), it is clear that $\Delta \mathbf{v}$ would still not be zero. We do not pursue this matter further at the present time, but return to it in Chap. 9.

Recalling our discussion of average velocity and instantaneous velocity in the preceding section, it will be seen that an analogous situation exists with acceleration. Without repeating the arguments used there, it should be reasonably obvious that they also apply here. As a result, we can define an instantaneous acceleration, the acceleration at a particular point, by the relation

$$\mathbf{a} \equiv \lim_{\Delta t \to 0} \frac{\Delta \mathbf{v}}{\Delta t} \equiv \frac{d\mathbf{v}}{dt} \qquad\qquad 4.4$$

The reasoning behind this definition is similar to that used for instantaneous velocity. It is assumed that if the small time interval Δt (which contains the time at which the object passes a certain point) is made small enough, the average acceleration of the object will not differ appreciably from the acceleration at the point in question. In practice, of course, the allowable upper limit on Δt will depend upon the accuracy sought and the actual behavior of the object in the vicinity of the point.

4.3 MOTION IN A STRAIGHT LINE

A very important type of motion is movement of an object along a straight line. In order that an object move in a straight line along the x axis, for example, the direction of the object's acceleration must necessarily be in the plus or minus x direction. (If $\mathbf{a}$ had a component perpendicular to the axis, $\Delta\mathbf{v}$ would acquire a component perpendicular to the axis, and the particle would move away from the straight line.) We therefore see that for motion in a straight line, the displacement, velocity, and acceleration vectors all lie along the line. As a result, if only straight-line motion is being discussed, the vector equations *4.1* to *4.4* can be considerably simplified.

Only two possible directions exist for the displacement, velocity, and acceleration vectors in straight-line motion. They can point in either the plus or the minus x direction. It is therefore convenient to represent their vector nature merely by the use of plus and minus signs. If a vector points in the $+x$ direction, it will be written as a positive magnitude; if it points in the $-x$ direction, it will be negative.

With this convention, we can rewrite our defining equations

For straight-line motion, the directions of x, v, and a can be shown by plus and minus signs

$$\bar{v}_x = \frac{\Delta x}{\Delta t} \qquad \bar{a}_x = \frac{\Delta v_x}{\Delta t} \qquad\qquad 4.5$$

Notice that Δx is numerically equal to the magnitude of the displacement vector and is plus or minus, depending upon whether the vector points in the plus or minus x direction. Similarly, the quantities $\bar{v}_x$, $\bar{a}_x$, and Δv_x will be plus or minus, depending upon the direction of the vector quantity involved.

Let us further simplify the forms of *4.5* by taking the origin of coordinates at the tail of the displacement vector, so that

$$\Delta x = x - 0 = x$$

Similarly, if we take $t = 0$ when the object is at the origin, then

$$\Delta t = t - 0 = t$$

and we can rewrite *4.5*

$$x = \bar{v}_x t$$
$$v_x - v_{0_x} = \bar{a}_x t \qquad\qquad 4.6$$

where the velocity at time $t = 0$ is taken to be v_{0x}, so that $\Delta v_x = v_x - v_{0x}$. When no confusion can exist as to the direction involved, the subscripts x on the various quantities will be omitted.

Straight-line motion is conveniently summarized by graphs. For example, in figure 4.4 we see the displacement of an object as a function of time

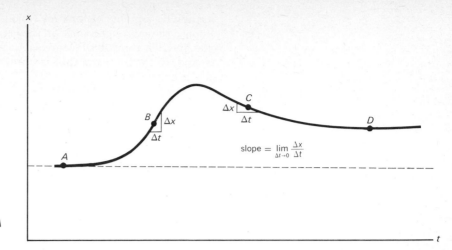

FIGURE 4.4 Where else besides in regions *A* and *D* is the speed zero? Describe what the object did between *A* and *D*.

as the object moves along the x axis. At the time of points A and D, the object was not moving because x neither increases nor decreases with time in those time ranges.

At any other point, say B, our definition of speed tells us the instantaneous speed of the object is $\Delta x / \Delta t$ in the limit of Δx and Δt much smaller than shown. But this limit of $\Delta x / \Delta t$ is the slope of the graph at point B. We therefore see that *the slope of the distance vs. time graph represents the instantaneous speed at that point.* Knowing this, it is obvious that the object's speed was zero at A and D since the slope of the graph is zero there. What does the region of negative slope near point C tell us about the object? Your answer to this question should convince you that the slope, being positive or negative, can also tell us something about the direction of the motion as well as giving its magnitude.

Instantaneous speed is the slope of the x vs. t graph

In a similar way, the graph of velocity vs. time is informative. Its slope is

$$\lim_{\Delta t \to 0} \frac{\Delta v}{\Delta t}$$

and this is the instantaneous acceleration for motion along a line. When the graph line is rising, the slope is positive and the object is accelerating. When the slope is negative, the object is slowing down, decelerating. In the next section we shall consider the very important case where the graph is a straight line, the case of constant acceleration.

4.4 UNIFORMLY ACCELERATED MOTION

Let us now consider a special type of accelerated motion, namely motion with constant acceleration. This type of motion occurs frequently and, since the acceleration is constant, is particularly simple. For constant, or *uniform,* acceleration we have from *4.6* that

For constant acceleration

$$v = v_0 + at \qquad\qquad 4.6a$$

Notice what this equation says. The velocity after a time t has passed is the original velocity plus at, the change in velocity per unit time multiplied by the number of units of time for which the velocity was changing. We have dropped the subscript x since, in the present case of motion along a line, only one direction is involved. Moreover, since the acceleration is constant, the average acceleration and the instantaneous accelerations are the same, and so $\bar{a}$ is simply a.

It will be recalled that the equation of a straight line whose intercept is b and whose slope is m is given by

$$y = b + mx$$

If we compare this with 4.6a we see that v takes the place of y, and t replaces x. The straight line representing 4.6a is shown in figure 4.5(a). Notice in particular the intercept v_0 and the slope a of the line.

One can obtain an especially simple equation for the average velocity in this special case of uniform acceleration. To do so we notice that the definition of average velocity is

$$\bar{v} = \frac{\Delta x}{\Delta t}$$

Let us consider the time interval between $t = 0$ and $t = t_N$ shown in figure 4.5(b). Then

$$\Delta t = t_N - 0 = t_N$$

As indicated in the figure, we shall divide the time interval between 0 and t_N into N very small time intervals. We make the time intervals small enough so that the velocity is not changed greatly during the time interval.

The distance moved by the object during the successive time intervals will be

$$\Delta x_1 = v_1 \, \Delta t_1$$
$$\Delta x_2 = v_2 \, \Delta t_2$$
$$\Delta x_3 = v_3 \, \Delta t_3$$
$$\cdots \cdots \cdots \cdots$$
$$\Delta x_N = v_N \, \Delta t_N$$

Since the sum of all these displacements is the total distance moved, we have

$$\Delta x = \Delta x_1 + \Delta x_2 + \cdots + \Delta x_N$$

or

$$\Delta x = v_1 \, \Delta t_1 + v_2 \, \Delta t_2 + v_3 \, \Delta t_3 + \cdots + v_N \, \Delta t_N \qquad 4.7$$

But if we refer to the figure, we see that $v_1 \Delta t_1$ is the area of the rectangle above the interval $0 \to t_1$. Similarly, $v_2 \Delta t_2$ is the area of the adjacent rectangle above the interval $t_1 \to t_2$, and so on. Therefore the sum in 4.7 is simply the total area of the rectangles from $0 \to t_N$. If the Δt's are made very small, the sum of the areas of the rectangles (which is Δx) will differ negligibly from the total area under the line from $t = 0$ to $t = t_N$, the shaded area. Therefore

$$\Delta x = \text{shaded area in figure } 4.5(b)$$

This area is found from figure 4.5(b) to be

$$\text{Area} = v_0 t_N + \tfrac{1}{2}(v_N - v_0) t_N$$

These two terms represent the area of the lower rectangle and the area of the upper triangle, respectively. Equating the area to Δx gives

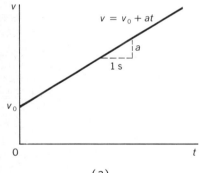

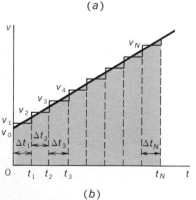

FIGURE 4.5 Since the acceleration is constant, the velocity changes by the same amount in each time interval Δt. As a result, the velocity-time graph is a straight line.

$$\Delta x = \tfrac{1}{2}(v_0 + v_N)t_N$$

To find $\bar{v}$ we use the fact that it is $\Delta x/\Delta t$ and obtain, since $\Delta t = t_N - 0$.

$$\bar{v} = \tfrac{1}{2}(v_0 + v_N)$$

We shall now drop the subscript N on the final velocity so that the final velocity is represented by v. The final result for the average velocity in uniformly accelerated motion along a straight line is

Average velocity for constant acceleration

$$\bar{v} = \tfrac{1}{2}(v_0 + v) \qquad\qquad 4.8$$

At this point we shall emphasize the fact that the v's are velocities. Since the present discussion considers only motion along a line, the vector nature of the v's is adequately taken care of by use of plus and minus signs. If the direction of motion is in the $-x$ direction, the velocity will be negative. Similarly, an acceleration in the $-x$ direction is also negative.

4.5 AVERAGES BY INTEGRATION*

The result obtained in the preceding section is more easily obtained by use of the calculus. Students who have not yet studied integration may wish to postpone this section until somewhat later. However, the process of integration is of considerable importance and its essentials can be mastered easily by studying Appendix 7.

Suppose a quantity f is a function of a variable x. Its graph may appear as shown in figure 4.6. From the definition of the average, we should add all the values of f in the interval under consideration, $c \leq x \leq d$, and divide by the total number of values added. Since f is a continuous function in this case, it takes on an infinite number of values in this interval. We simplify the situation by initially thinking of the function as discontinuous, possessing segments Δx long. There will be $(d - c)/\Delta x$ such segments in the interval $c \leq x \leq d$. The average values of f in these various segments are taken as f_1, $f_2, \ldots$, as shown in the figure.

From the definition of the average value, we write

$$\bar{f} = \frac{\text{sum of items}}{\text{number of items}} = \frac{f_1 + f_2 + \cdots}{(d - c)/\Delta x}$$

which becomes

$$\bar{f} = \frac{\Sigma f_n\, \Delta x}{d - c}$$

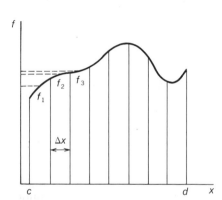

FIGURE 4.6 The average of f in the region $c \leq x \leq d$ is given by *4.9*.

where the sum extends over the values from $x = c$ to $x = d$.

It is now convenient to let $\Delta x \rightarrow 0$, so that the sum in this equation may be replaced by an integral as noted in Appendix 7. The result is

The average of a function f(x) in the interval c ≤ x ≤ d

$$\bar{f} = \frac{\displaystyle\int_c^d f\, dx}{d - c} \qquad\qquad 4.9$$

This is a useful representation for the average of a function f. Let us now apply it to find the average value of v, where v is given by the equation

$$v = v_0 + at$$

a being a constant.

*This section may be omitted without loss of continuity.

Since v is a function of t, the variable x is replaced by t in *4.9*. Thus

$$\bar{v} = \frac{\int_0^{t_f} (v_0 + at)\,dt}{t_f}$$

where we are finding the average in the interval $0 \le t \le t_f$. Carrying out the integration* gives

$$\bar{v} = \frac{1}{t_f}(v_0 t_f + \tfrac{1}{2}at_f^2)$$

so that

$$\bar{v} = v_0 + \tfrac{1}{2}at_f$$

Replacing at_f from *4.6* gives the result obtained previously.

$$\bar{v} = \tfrac{1}{2}(v_0 + v) \qquad\qquad 4.8$$

Illustration 4.1

Find the average value of x^2 in the range

$$0 \le x \le a$$

---

*To do this, write the integral as $\int_0^{t_1}(v_0 + at)\,dt = \int_0^{t_1} v_0\,dt + \int_0^{t_1} at\,dt$

Since v_0 and a are constants, we evaluate these from integrals A7.9 and A7.10 to give

$$\int_0^{t_1}(v_0 + at)\,dt = (v_0 t_1 - 0) + \left(\frac{a}{2}t_1^2 - 0\right)$$

which leads to the result given above.

An Experiment to Determine the Nature of Light

The chapters we are now studying show the importance of accurate descriptions of physical phenomena and of an understanding of the conditions affecting these phenomena. If physicists are to achieve the goal of interpreting observed natural phenomena, ambiguous or incomplete descriptions must be corrected or the interpretation will be subject to error. Often several different interpretations of (or theories about) an observation are proposed. To choose among these theories, physicists seek to predict various new observations based upon the competing theories. If the predictions from two competing theories differ, a choice between the theories can be made according to which theory predicts the correct new result. An illustration of this method for choosing the better of two competing theories is given by the important experiment described below.

Two opposing views as to the nature of light were prevalent in the early 1700s. The corpuscular, or particle, theory of light was espoused by Sir Isaac Newton. It considers a beam of light to consist of a stream of particles. To explain the bending (or refraction) of a beam of light as it enters water from air (a phenomenon discussed in detail in Chap. 29), proponents of the particle theory assumed that the light "particles" travel faster in water than in air. This view prevailed for about 125 years after the death of Newton in 1727.

A less widely accepted concept of light, the wave theory, needed to assume that the speed of light was smaller in water than in air in order to explain the observed refraction effects. Although a clear test of

Reasoning Applying *4.9,* we have

$$\overline{x^2} = \frac{\int_0^a x^2\, dx}{a}$$

The integral of x^2 is $x^3/3$. We therefore find

$$\overline{x^2} = \frac{1}{3a}\,(a^3 - 0)$$

or

$$\overline{x^2} = \frac{a^2}{3}$$

Is this value about what you would expect?

4.6 THE FIVE MOTION EQUATIONS; CONSTANT ACCELERATION

Let us now review the relations which we have found to apply to motion in a straight line with constant acceleration. They are *4.6* and *4.8,* namely,

The three basic motion equations when a is constant

$$x = \overline{v}t \qquad\qquad 4.10$$
$$v = v_0 + at \qquad\qquad 4.11$$
$$\overline{v} = \tfrac{1}{2}(v + v_0) \qquad\qquad 4.12$$

Although these equations are sufficient to solve all problems dealing with this type of motion, it is convenient to obtain two other equations from

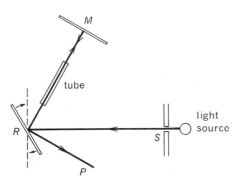

the competing theories could be made by measuring the speed of light in air and in water, this measurement was not readily carried out because of the exceedingly high speed of light. It was not until 1850 that J. L. Foucault (1819–1868) succeeded in measuring the speed of light in water so that a definite experimental test of the two opposing theories could be made.

A greatly simplified diagram of his apparatus is shown in the figure. Light from slit S is reflected by mirror R through a hollow glass tube to mirror M. Upon reflection from mirror M the light retraces its path and comes back to slit S provided that mirror R is stationary. During the actual experiment, mirror R is rotating at high speed, and when the light reflected by mirror M reaches R on its return trip, mirror R is now in the dotted position shown. As a result, the light is now reflected to point P rather than back to S. In practice, the distance between P and S is no more than a millimeter, and so our diagram is greatly exaggerated.

Foucault carried out the experiment with the tube first filled with air and then with water. He found that the point P was farther from S when water was used than when air filled the tube. We must therefore conclude that mirror R rotated farther during the time taken for the trip through water than through air. Hence the speed of light in water is less than in air, as predicted by the wave theory. This fact was one of the major reasons for the final acceptance of the wave theory of light, as opposed to Newton's concept of a corpuscular form of light. But we shall see in Chap. 34 that Newton was not completely in error in this respect.

them. For example, suppose we knew the initial velocity of a car and how far it went before coming to rest under the condition of uniform acceleration. How could we find the acceleration? None of the above equations is suitable for finding a directly when x, v, and v_0 are known. To obtain an equation involving only these variables, we should substitute *4.12* in *4.10* to give

$$x = \tfrac{1}{2}(v + v_0)t$$

We should then replace t by use of *4.11*, giving

$$x = \frac{\tfrac{1}{2}(v + v_0)(v - v_0)}{a}$$

which simplifies to

A derived relation for constant a

$$v^2 - v_0{}^2 = 2ax \qquad\qquad 4.13$$

This equation could be used directly to find a if x, v, and v_0 were known.

Another frequent situation is the need to find displacement x when v_0, a, and t are known. An equation relating these quantities is easily found by using *4.11* to replace v in the relation

$$x = \tfrac{1}{2}(v + v_0)t$$

We then find

$$x = \tfrac{1}{2}(2v_0 + at)t$$

which gives

$$x = v_0 t + \tfrac{1}{2}at^2 \qquad\qquad 4.14$$

In summary, we have five equations:

Five useful motion equations for constant a

$$x = \bar{v}t \qquad\qquad 4.10$$
$$v = v_0 + at \qquad\qquad 4.11$$
$$\bar{v} = \tfrac{1}{2}(v + v_0) \qquad\qquad 4.12$$
$$v^2 - v_0{}^2 = 2ax \qquad\qquad 4.13$$
$$x = v_0 t + \tfrac{1}{2}at^2 \qquad\qquad 4.14$$

where it is assumed that $x = 0$ when $t = 0$.

The last four of these equations were derived assuming constant acceleration and are valid only in that case.

Illustration 4.2

A certain car is said to be able to accelerate from rest to a speed of 10 m/s in a time of 8 s. What is its acceleration and how far will it go in this time? Assume the acceleration to be constant.

Reasoning It is well to be quite systematic in answering questions of this type. Let us take the motion to be in the $+x$ direction. We are given the following quantities:

$$v_0 = 0$$
$$v = 10 \text{ m/s}$$
$$t = 8 \text{ s}$$

We can find a from *4.11*. It is

$$a = \frac{v - v_0}{t}$$

which gives

$$a = \frac{10 \text{ m/s}}{8 \text{ s}} = 1.25 \text{ m/s}^2$$

Since the average velocity is 5 m/s, we can use *4.10* to find

$$x = \bar{v}t$$

or
$$x = (5\,\text{m/s})(8\,\text{s}) = 40\,\text{m}$$

Illustration 4.3

An electron moving at a speed of 5×10^6 m/s is shot through a sheet of paper which is 2.1×10^{-4} cm thick. The electron emerges from the paper with a speed of 2×10^6 m/s. Find the time taken by the electron to pass through the sheet.

Reasoning We have no alternative but to assume as an approximation that the acceleration is uniform. The known quantities are, assuming the motion to be in the $+x$ direction,

$$v_0 = 5 \times 10^6\,\text{m/s}$$
$$v = 2 \times 10^6\,\text{m/s}$$
$$x = 2.1 \times 10^{-4}\,\text{cm} = 2.1 \times 10^{-6}\,\text{m}$$

We change the centimeters to meters at once in order to avoid two sets of length units in the problem. Let us first find the acceleration from *4.13*.

$$(4 \times 10^{12} - 25 \times 10^{12})\,\text{m}^2/\text{s}^2 = 2a(2.1 \times 10^{-6}\,\text{m})$$

which gives
$$a = -5 \times 10^{18}\,\text{m/s}^2$$

The minus sign indicates that the acceleration is in the negative x direction and is therefore a deceleration; that is to say, the electron is slowing down.

To find t we can now use *4.11*. It is

$$v = v_0 + at$$
$$2 \times 10^6\,\text{m/s} = (5 \times 10^6\,\text{m/s}) - (5 \times 10^{18}\,\text{m/s}^2)t$$

which yields
$$t = 6 \times 10^{-13}\,\text{s}$$

How could t have been found without first computing a?

4.7 FREELY FALLING BODIES

It was first proved by Galileo (1564–1642) that if one can ignore the effect of air friction, all bodies fall to earth with the same acceleration. We now know that this acceleration is about 9.8 m/s², or 32.2 ft/s². It is denoted by the letter g and is called the *acceleration due to gravity*. We know from accurate experiments that g varies slightly from place to place on the earth. Some of the reasons for this are pointed out in later chapters. Typical values for g are given in table 4.1.

Like all accelerations, the acceleration due to gravity is a vector. It is directed downward toward the earth. Consequently, it causes falling objects to speed up and rising objects to slow down. In the latter case it is in the opposite direction of the motion and therefore acts to decelerate the motion. Care must be taken in dealing with motions which involve both up and down motion since the proper use of signs is of great importance. This will become clear in the examples which follow.

We are already prepared to deal with vertical motion under the effect of gravity because it is uniformly accelerated motion in a straight line. Our five motion equations therefore apply to it. Of course, in this case, $a = g$, and so the acceleration is always known to be about 9.8 m/s² directed downward.

The acceleration due to gravity is 9.8 m/s² (32 ft/s²) directed toward the earth's center

TABLE 4.1 Acceleration due to gravity, g

Place	Elevation (m)	g (m/s²)
Beaufort, N.C.	1	9.7973
New Orleans	2	9.7932
Galveston	3	9.7927
Seattle	58	9.8073
San Francisco	114	9.7997
St. Louis	154	9.8000
Cleveland	210	9.8024
Denver	1638	9.7961
Pikes Peak	4293	9.7895

Illustration 4.4

An object is thrown straight up from the ground with an initial velocity of 20 m/s. How high does it rise?

Reasoning Let us replace x by y in the five motion equations and consider vectors to be positive when pointing upward. We then know, taking the highest point as the endpoint for the motion problem,

$$v_0 = 20 \text{ m/s (upward and positive)}$$
$$v = 0$$
$$a = -9.8 \text{ m/s}^2 \text{ (downward and negative)}$$

We write that $v = 0$ for the following reason. As the object rises, it slows down, and eventually stops for an instant, at which time $v = 0$. It is then at the top of its path and is momentarily at rest. To find y we make use of

$$v^2 - v_0{}^2 = 2ay$$
to give $0 - 400 \text{ m}^2/\text{s}^2 = 2(-9.8 \text{ m/s}^2)y$
yielding $y = 20.4 \text{ m}$

Illustration 4.5

An object is thrown upward from the edge of a building with a velocity of 20 m/s, as shown in figure 4.7. Where will the object be at 3 s after it is thrown? After 5 s?

Reasoning Let us take the upward direction as positive. Then

$$v_0 = 20 \text{ m/s}$$
$$t = 3 \text{ s}$$
$$a = -9.8 \text{ m/s}^2$$

Notice that a is negative because the acceleration due to gravity is downward. To find y we use

$$y = v_0 t + \tfrac{1}{2}at^2$$
$$= 60 - 44 = 16 \text{ m}$$

Since y is positive, the object is above the throwing point. To see if it is still going upward or is on its way down, we compute v.

$$v = v_0 + at$$
$$\approx 20 - 29 \approx -9 \text{ m/s}$$

The negative sign tells us the final velocity is downward, and so the situation is as shown in figure 4.7(b).

To find the object after 5 s we again substitute

$$y = v_0 t + \tfrac{1}{2}at^2$$
$$= 100 - 122 = -22 \text{ m}$$

The negative sign indicates that the situation is as shown in figure 4.7(c).

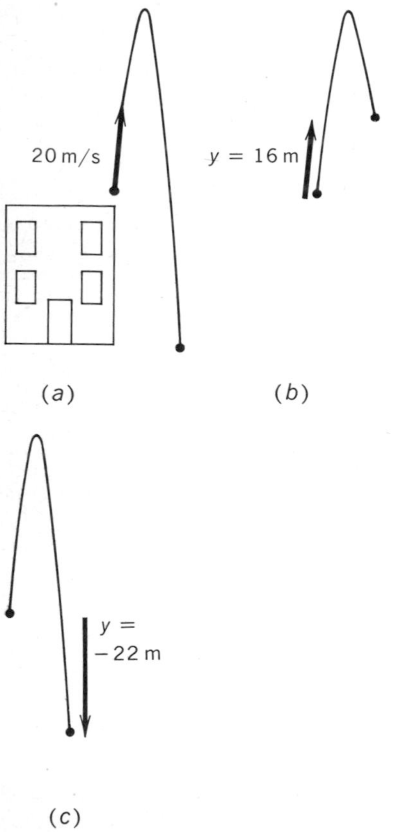

4.8 PROJECTILE MOTION

As we learned in the preceding section, an object free to move in a vertical direction experiences the downward acceleration due to gravity. We now

20 m/s

(a)

$y = 16$ m

(b)

$y = -22$ m

(c)

FIGURE 4.7 Notice that the displacement y is the vector from the starting point to the endpoint. Consider the horizontal motion to be negligible.

consider the motion of an object which is moving both vertically and horizontally on the earth at the same time. By sliding an object across a nearly frictionless horizontal tabletop, we can surmise that, in the absence of friction, the object would travel at constant horizontal velocity. This same conclusion is reached after carrying out more extensive experiments. From these experiments we conclude: In the absence of friction forces or other retarding effects in the horizontal direction, the horizontal component of the velocity of an object remains constant. Further, experiment shows that the vertical and horizontal motions of an object above the earth are independent of each other. We can therefore make the following statement:

The horizontal and vertical motions of a projectile are separable

An object (a projectile) undergoing both vertical and horizontal friction-free motion above the earth undergoes two simultaneous independent motions. It moves horizontally with constant speed. At the same time, it moves vertically in a way a similar object not undergoing horizontal motion would move.

Suppose an object is shot parallel to the earth with velocity $v_{0x}\mathbf{i}$, where $\mathbf{i}$ is the unit vector in the x direction, parallel to the earth's surface. The situation is shown in figure 4.8. If we ignore air friction, the object will move as follows. The *horizontal* motion of the object will *not* be accelerated. The object continues to move in the $+x$ direction with constant velocity v_{0x}. As a

a = 0 for the horizontal motion

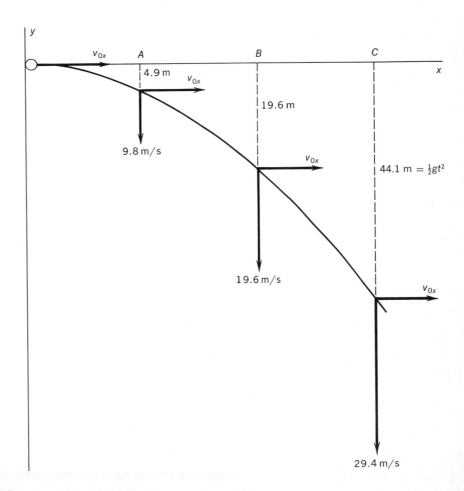

FIGURE 4.8 The body undergoes two independent motions at right angles to each other.

result, $v_{0x} = v_x = \bar{v}_x$, and the only important motion equation for the horizontal nonaccelerated motion is

$$x = \bar{v}_x t$$

The other motion equations tell us nothing new in this case of $a_x = 0$.

The vertical motion is that of a freely falling body

Since gravity is pulling on the object in the $-y$ direction, the object will accelerate downward with acceleration g. This type of motion has already been discussed in the preceding section, where we considered freely falling bodies. From what we learned there we can predict that the object shown in figure 4.8 will fall in such a way that its vertical velocity is 9.8 m/s downward after 1 s, 19.6 m/s after 2 s, etc. The path of the object will be as shown in the figure, and v_x and v_y at 1-s intervals along the path are as indicated.

The important conclusion to be drawn is this: The motion of a projectile near the earth is really two superposed motions. An exceedingly simple horizontal motion at constant horizontal velocity is combined with the usual motion of a freely falling object. Of course, we have idealized the situation by ignoring the friction effects of the air.

Illustration 4.6

An object is thrown horizontally with a velocity of 10 m/s from the top of a 20-m-high building as shown in figure 4.9. Where does the object strike the ground?

Reasoning We consider the horizontal and vertical problems separately. In the vertical problem, if the downward direction is taken to be positive.

$$v_{0y} = 0$$
$$a_y = 9.8 \, \text{m/s}^2$$
$$y = 20 \, \text{m}$$

We find the time taken to reach the ground from

$$y = v_{0y}t + \tfrac{1}{2}a_y t^2$$
giving
$$t = 2.02 \, \text{s}$$

Now let us consider the horizontal motion. We have just found that the object will be in the air for 2.02 s. The known quantities are

$$v_{0x} = v_x = \bar{v}_x = 10 \, \text{m/s}$$
$$t = 2.02 \, \text{s}$$

We may therefore use $x = \bar{v}_x t$ to find the horizontal distance it moves. The result is

$$x = 20.2 \, \text{m}$$

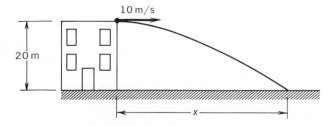

FIGURE 4.9 How large is x?

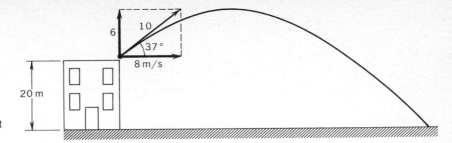

FIGURE 4.10 Where does the object hit the ground?

Illustration 4.7

Suppose that in the illustration 4.6 the object had been thrown upward at an angle of 37° to the horizontal with a velocity of 10 m/s. Where would it land?

Reasoning The situation is shown in figure 4.10. We immediately resolve the initial velocity into x and y components in an effort to separate the two motions. The known quantities in the y-directed motion (taking down positive) are

$$v_{0y} = -6 \, \text{m/s (negative since down is positive)}$$
$$a_y = 9.8 \, \text{m/s}^2$$
$$y = 20 \, \text{m}$$

To find the time of flight we can use

$$y = v_{0t}t + \tfrac{1}{2}a_y t^2$$

which gives a quadratic equation for t since $v_{0y} \neq 0$ in this case. To avoid use of the quadratic formula, let us first find v_y by use of

$$v_y{}^2 - v_{0y}{}^2 = 2a_y y$$

We then find $v_y = 20.7 \, \text{m/s}$. Then, by using

$$v_y = v_{0y} + a_y t$$

we find t to be 2.73 s. It is instructive for the student to verify this result by using the quadratic equation for t.

Turning now to the motion in the x direction, we have

$$v_{0x} = v_x = \bar{v}_x = 8 \, \text{m/s}$$
$$t = 2.73 \, \text{s}$$

Substitution in $x = \bar{v}_x t$ gives $x = 22 \, \text{m}$.

Illustration 4.8

Refer to the situation shown in figure 4.11: where will the ball hit the wall?

Reasoning Here, too, we begin by splitting the initial velocity into its two components. In the horizontal motion we know that

$$v_{0x} = v_x = \bar{v}_x = 16 \, \text{m/s}$$
$$x = 32 \, \text{m}$$

from which, by use of $x = \bar{v}_x t$, we find

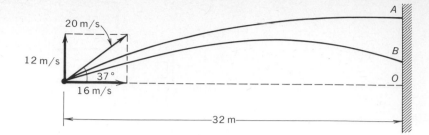

FIGURE 4.11 Where does the object strike the wall?

$$t = 2.0\,\text{s}$$

The object strikes the wall after 2 s.

To find its vertical height at this time, we take the upward direction as positive and write what is known:

$$v_{0y} = 12\,\text{m/s}$$
$$a_y = -9.8\,\text{m/s}^2$$
$$t = 2.0\,\text{s}$$

The vertical displacement y at the end of this time can be found from

$$y = v_{0y}t + \tfrac{1}{2}a_y t^2$$

which gives $y = 4.4$ m. Since the answer is positive and since up was taken as positive, we know that the object hits the wall 4.4 m above point O.

It is of interest to determine whether the object is still going up (as at point A) or going down (as at point B) when it strikes the wall. This is easily done by finding the vertical velocity just before it hits. We use

$$v_y = v_{0y} + a_y t$$

to give
$$v_y = -7.6\,\text{m/s}$$

Since the final velocity is negative and since upward is taken as positive, the object must be going down, as at point B, when it strikes the wall.

Chapter 4
Questions and
Guesstimates

1. Give an example of a case where the velocity of an object is zero but its acceleration is not zero.
2. Sketch a graph of the velocity and acceleration as a function of time for a car as it strikes a telephone pole.
3. Consider a small rocket which is shot straight upward. Assume it to start from rest and accelerate uniformly until its fuel is exhausted. Sketch a graph of the velocity and acceleration as a function of time for the rocket as it rises and then falls to the earth.
4. Are any of the following statements true? (*a*) An object can have a constant velocity even though its speed is changing. (*b*) An object can have a constant speed even though its velocity is changing. (*c*) An object can have zero velocity even if its acceleration is not zero. (*d*) An object subjected to a constant acceleration can reverse its velocity.
5. The average value of $\sin\theta$ in the range $0 \le \theta \le 2\pi$ is zero. How can one infer this from a graph of $\sin\theta$ versus θ without lengthy computation?

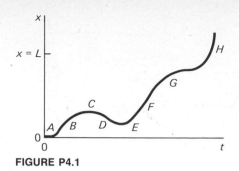

FIGURE P4.1

What must be true of the graph of a function if the average of the function is to be zero during a certain interval?

6. The acceleration due to gravity on the moon is only about one-sixth that on earth. Compare the heights to which a ball would rise on earth and the moon if it was thrown upward with the same speed in the two places.

7. A cat enters the end of a drain pipe which has ends at $x = 0$ and $x = L$. Its motion from that instant on is shown in figure P4.1. Describe the cat's motion in words.

8. If you were to tell someone how to broad jump best, what would be your recommendations?

9. Airplane enthusiasts sometimes hold meets where they try to show their skills. One event is to drop a sack of sand exactly in the center of a circle on the ground while flying at a predetermined height and speed. What is so difficult about that? Don't they just drop the sack when they are directly above the circle?

10. How is the integral of a function with respect to a variable x related to the graph of the function against x?

11. Under what condition is it wrong to say that the acceleration of an object is negative when the object is thrown upward? Does the sign of the acceleration depend at all upon the direction of motion? Can the acceleration of an object be positive even though the object is slowing down?

12. Describe the motion of a coin as seen by the man who drops it if the man is in (*a*) a car moving at constant velocity along a level road; (*b*) an elevator which is falling freely just after its cable broke; (*c*) a car going around a sharp corner at constant speed.

13. Sketch a graph showing the time variation of the velocity and acceleration of a baseball as it approaches the bat and is hit by it.

14. The driver of a car going 60 km/h (16.7 m/s) suddenly sees a child dart into the roadway. About how far will the car go before he hits the brake?

15. Estimate the maximum average acceleration capability of a standard car in the range 0 to 65 km/h.

16. Estimate the largest possible average deceleration for a man running at top speed along a level asphalt road provided he stops without hitting anything or falling over.

Chapter 4
Problems

1. A race car makes one complete trip around a 2-mi oval-shaped race track in 90 s. For the complete trip around the track, find the car's (*a*) speed and (*b*) velocity.

2. To get to school in the morning, a small boy goes 3 blocks east and 4 blocks north. It takes him about 20 min. Find his average (*a*) speed and (*b*) velocity.

3. A young woman borrowed the family car to go to the store. Her father noticed that the odometer read 13,597 mi when she left and 13,721 when she returned 2 h later. Compute the woman's average speed and velocity for the 2-h trip.

4. A boy begins to walk eastward along a street in front of his house, and the graph of his displacement from home is shown in figure P4.2. Find his average speed for the whole time interval as well as his instantaneous speed at points *A*, *B*, and *C*.

5. Repeat problem 4 but find velocities instead of speed.

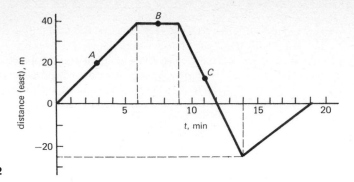

FIGURE P4.2

6. According to the performance data for a certain car, it is capable of accelerating from rest to 12 m/s in 8 s. Assuming constant acceleration, compute its acceleration and the distance it goes in 8 s.

7. A car moving at 30 m/s strikes a huge tree and comes to rest in a distance of 1.5 m. Assuming uniform deceleration, find the time taken to stop and the average deceleration.

8. Alpha particles (the nuclei of helium atoms) are shot with speed 4×10^7 m/s into air. They travel 8 cm before stopping. Assuming their deceleration to be uniform, find their deceleration and the time taken to stop.

9. A proton moving with speed 1×10^7 m/s passes through a 0.020-cm-thick sheet of paper and emerges with a speed of 2×10^6 m/s. Assuming uniform deceleration, find the deceleration and the time taken to pass through the paper.

10. The driver of a car that is going 25 m/s suddenly notices a train blocking the road. At the instant the brakes are applied, the train is 60 m away. The car decelerates uniformly and strikes the train 3 s later. (*a*) How fast was the car moving on impact, and (*b*) what was its deceleration during the 3 s?

11. Just as a car starts to accelerate from rest with acceleration 1.4 m/s², a bus moving with constant speed of 12 m/s passes it in a parallel lane. (*a*) How long before the car overtakes the bus? (*b*) How fast will the car then be going? (*c*) How far will the car then have gone?

12. Two boys start running straight toward each other from two points that are 100 m apart. One runs with a speed of 5 m/s while the other moves at 7 m/s. How close are they to the slower one's starting point when they reach each other?

13. Two trains are headed toward each other on the same track with equal speeds of 20 m/s. When they are 2 km apart, they see each other and begin to decelerate. (*a*) If their decelerations are uniform and equal, how large must the decelerations be if the trains are to barely avoid collision? (*b*) If only one train slows with this deceleration, how far will it go before collision occurs?

14. A police car is at rest alongside a road monitoring passing cars when one passes at a constant speed of 32 m/s. Five seconds later, the police car accelerates from rest with an acceleration of 1.6 m/s². (*a*) If it could maintain this acceleration, how far would it move before catching the car? (*b*) What would its speed then be?

15. A car is traveling east along a highway. Its velocity is shown in figure

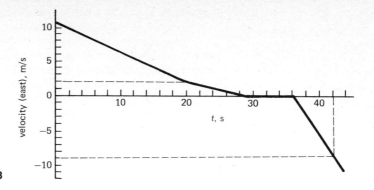

FIGURE P4.3

P4.3. (*a*) What was the car's acceleration at $t = 10$ s, $t = 25$ s, $t = 40$ s? (*b*) How far did the car travel during the first 20 s shown in the figure?

16. From the data of figure P4.3, taking the car's position at $t = 0$ to be zero, how far from the zero point will the car be at the following times: 20 s, 29 s, 36 s, 42 s?

17. A stone is thrown straight upward with a speed of 20 m/s. How high does it go? How long does it take to rise to its maximum height?

18. An object is thrown straight upward and falls back to the thrower after a round trip time of 0.80 s. How fast (in m/s) was the object thrown?

19. A man is holding a flower pot out of a window 12 m from the ground. He tosses the pot upward with a speed of 5 m/s. How long does it take the pot to reach the ground, and how fast is it moving just before it hits?

20. Answer problem 19 if the man is leaning over the edge of a basket being lifted by a balloon which is ascending with a speed of 3 m/s. The pot is still tossed at a height of 12 m above the ground.

21. A ball is dropped from the top of a building. The ball takes 0.50 s to fall past the 3-m length of a window some distance from the top of the building. (*a*) How fast was the ball going as it passed the top of the window? (*b*) How far is the top of the window from the point at which the ball was dropped?

22. A ball is thrown straight upward with a speed v from a point h m above the ground. Show that the time taken for the ball to strike the ground is

$$\frac{v}{g}\left(1 + \sqrt{1 + \frac{2hg}{v^2}}\right)$$

23. A monkey in a perch 20 m high in a tree drops a coconut directly above your head as you run with speed 1.5 m/s beneath the tree. (*a*) How far behind you does the coconut hit the ground? (*b*) If the monkey had really wanted to hit you, how much earlier should the coconut have been dropped?

24. Two balls are dropped to the ground from different heights. One ball is dropped 2 s after the other but they both strike the ground at the same time, 5 s after the first is dropped. (*a*) What is the difference in the heights at which they were dropped? (*b*) From what height was the first dropped?

25. A small ball rolling at 30 cm/s rolls off the edge of a 70-cm-high table. How far from the point directly below the edge of the table does the ball strike the floor?

26. How fast must a ball be rolled along a 70-cm-high table so that when it rolls off the edge it will strike the floor at this same distance (70 cm) from the point directly below the table edge?

27. In an ordinary TV set, the electron beam consists of electrons shot horizontally at the TV screen with a speed of about 5×10^7 m/s. How far does a typical electron fall as it moves the approximately 40 cm from the electron gun to the screen? For comparison how far would a drop of water shot horizontally at 2 m/s from a hose drop as it moves a horizontal distance of 40 cm?

28. A projectile is shot at a speed of 30 m/s at an angle of 50° above the horizontal. How far away will it be when it reaches the shooting level once again?

29. Suppose the projectile of the previous problem hits a vertical wall at a height of 20 m above the shooting level. Give the two possible locations of the wall.

30. A ball is thrown with a speed of 20 m/s at an angle of 37° above the horizontal. It lands on the roof of a building at a point displaced 24 m horizontally from the throwing point. How high above the throwing point is the roof?

31. Projectile A is shot with speed 150 m/s at an angle of 53° above the horizontal. Simultaneously a second projectile is shot straight up from point B which lies below A's line of flight. If the two projectiles are to collide, how fast must the B projectile be shot?

32. A projectile is to be shot at 50 m/s over level ground in such a way that it will land 200 m from the shooting point. At what angle should the projectile be shot? (*Hint:* $2 \sin \theta \cos \theta = \sin 2\theta$.)

33. Show that a projectile shot with speed v_0 at angle θ above the horizontal will land at a distance R away, where

$$R = \frac{v_0{}^2 \sin 2\theta}{g}$$

Notice that maximum R occurs when $\theta = 45°$. Does the answer for $\theta = 90°$ make sense? See *Hint* for problem 32.

34. A cart is moving horizontally along a straight line with constant speed 30 m/s. A projectile is to be fired from the moving cart in such a way that it will return to the cart after the cart has moved 80 m. At what speed (relative to the cart) and at what angle (to the horizontal) must the projectile be fired?

35. Using calculus, find the speed and acceleration at $t = 4$ s for a particle moving along the x axis with its x coordinate being given by

$$x = 5 + 18t - 2t^2 \qquad \text{m}$$

Is it speeding up or slowing down at that instant?

36. The speed of a certain particle moving along a straight line is given by

$$v = 45 + 3t \qquad \text{m/s}$$

Find its acceleration and its average velocity during the time interval $0 \le t \le 10$ s. You may use calculus but this can be done without its use.

37. Using calculus, find an expression for the acceleration of a particle moving along a straight line if its velocity is given by $v = 50 + 8t^2$ m/s. What is its average velocity in the time interval $0 \le t \le 10$ s?

38. We found the vertical motion of a projectile was given by

$$y = v_0 t + \tfrac{1}{2}gt^2$$

Using calculus, find the vertical velocity and acceleration of the particle as a function of time.

39. When a ball is thrown upward with initial speed 19.6 m/s², its height y as a function of time is given by

$$y = 19.6t - 4.9t^2$$

where t is in seconds and y is in meters. (a) Find the averge height of the ball in the time interval $0 \leq t \leq 4$ s. (b) How does this compare to the maximum height reached by the ball?

5 Newton's Laws of Motion

Sir Isaac Newton (1642–1727) is well known to all physics students as the discoverer of several fundamental laws of nature. Three of these, his laws of motion, are usually grouped together. As we shall see in this and following chapters, they form the foundation upon which much of physics is based.

5.1 INERTIA AND THE FIRST LAW

Nearly everyone understands the qualitative aspects of inertia. Even a child knows that an empty can is easily kicked aside while the same can, when full with liquid, is much less easily set in motion. We see many examples in everyday life that illustrate the fact that, the more massive an object, the more difficult it is to cause it to move. Objects at rest require forces to set them into motion and, the more massive the object, the larger the force that is required.

Another aspect of inertia involves the stopping of objects that are in motion. Stopping forces are needed if a moving object is to be slowed. Again, the magnitude of the force required depends on the massiveness of the moving object. It is much easier to catch an empty can that has been thrown at you than it is to stop the same can when it is full. Objects in motion tend to remain in motion, and objects that are at rest tend to remain at rest. *Inertia* is the tendency of an object at rest to remain at rest and of an object in motion to remain in motion in the same direction.

Newton's first law, his law of inertia, can be stated as follows:*

Newton's first law

In the absence of unbalanced forces acting on it, an object at rest will remain at rest and an object in motion will remain in motion with the same velocity.

Notice that the motion part of the law involves unchanging *velocity,* not just speed. An object will continue moving in the same direction as well as with unchanging speed if no unbalanced force acts on it. Only an unbalanced force acting on an object can slow it or deflect it from straight line motion.

---

*Newton's laws are true only in inertial reference frames, frames that are not accelerating. This fact, not of present importance to us, will be discussed in Chap. 8.

5.2 ACTION AND REACTION: THE THIRD LAW

Newton's third law

The third of Newton's laws involves the concept of *action and reaction* forces. It may be stated as follows: *For every force which acts on a body (the action force) there is an equal and opposite force (the reaction force) which acts on some other body.* That is to say, if a body A exerts a force F on body B, then body B exerts an equal and opposite force $-F$ on body A. In many everyday situations this law is quite easily justified. For example, a wall pushes back upon you with the same force with which you push on it. Similarly, a block sitting on a table pushes down on the table with a force identical with the force the table exerts on it. You can certainly supply many other examples of this type. Notice that two bodies are always involved; the action force acts on one while the reaction force acts on the other.

In spite of the apparent simplicity of the third law, this law is far from being self-evident. In fact, it is open to serious question when one is considering forces which change with time and which act at a distance. For example, two positively charged balls repel each other as shown in figure 5.1(a). The forces of repulsion are equal and opposite while the system is at rest, as indicated in figure 5.1(a). However, if one ball is suddenly moved, the forces must change. But it takes a certain time for the electromagnetic signal to travel from one ball to the other, and this signal must reach the second ball before the force on it can reflect the movement of the first ball. During this time interval, the validity of Newton's third law is not obvious. Although we cannot pursue this complicated matter further here, it is hoped that you will investigate this aspect at a later time when your background in electromagnetism will have been more extensive.

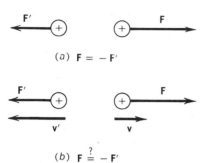

(a) $\mathbf{F} = -\mathbf{F}'$

(b) $\mathbf{F} \overset{?}{=} -\mathbf{F}'$

FIGURE 5.1 When the charged bodies are at rest, as in (a), the action and reaction forces are equal in magnitude. This is not necessarily true, however, when the bodies are moving rapidly relative to each other.

5.3 NEWTON'S SECOND LAW

In order to cause an object to accelerate, an unbalanced force must be applied to it. To cause this book to begin moving along the table top, you must push sideways on it. The larger the force with which you push, the faster the book will be caused to move. Newton recognized that unbalanced forces cause changes in motion and his second law relates the resultant (or net) external force $\mathbf{F}_{net}$ acting on an object to the change in motion of the object.

Because an unbalanced force on an object can not only change its speed but deflect it as well, Newton related $\mathbf{F}_{net}$ to the change in *velocity* of the object. Moreover, because the more massive the object, the larger the force required to change its motion, Newton knew that all three quantities— unbalanced force, mass, and change in velocity—were interrelated. The mass of an object, represented by m, is a measure of its inertia. We shall discuss it in detail soon. Newton chose to write his second law in terms of the change of a quantity called momentum, $m\mathbf{v}$, where m is the mass of the object subjected to the unbalanced force and $\mathbf{v}$ is its velocity. His second law is as follows:

$$\text{Resultant external force on object} = \frac{\text{change in momentum}}{\text{duration time of force}}$$

or

$$\mathbf{F}_{net} = \frac{\Delta(m\mathbf{v})}{\Delta t} \qquad \qquad 5.1$$

The quantity $m\mathbf{v}$, which is the momentum of the object, is often represented as the vector $\mathbf{p}$ and we shall discuss it in more detail in Chap. 7.

Equation *5.1* gives the average force required to cause a momentum change $\Delta(m\mathbf{v})$ during the time interval Δt. If the time interval is made very small so that $\Delta t \to 0$, then the derivative notation from calculus applies and one can write

Newton's second law

$$\mathbf{F}_{net} = \frac{d(m\mathbf{v})}{dt} \qquad 5.2$$

and the force so given is the instantaneous resultant force acting on the object.

Newton's second law is more familiar when written in a less general form. If the mass of the object subjected to the force $\mathbf{F}_{net}$ does not change (a condition that usually applies), then the change in momentum $\Delta(m\mathbf{v})$ is due entirely to the change in velocity. For the case of constant mass, then,

$$\mathbf{F}_{net} = \frac{\Delta(m\mathbf{v})}{\Delta t} = m\frac{\Delta \mathbf{v}}{\Delta t}$$

But in the preceding chapter we defined $\Delta\mathbf{v}/\Delta t$ to be the acceleration $\mathbf{a}$ of the object. Substitution of this value yields the most often used form of Newton's second law.

Newton's second law when m is
constant

$$\mathbf{F}_{net} = m\mathbf{a} \qquad 5.3$$

Equation *5.3* summarizes the experimental fact that if the resultant external force acting on an object of constant mass is $\mathbf{F}_{net}$, the object will accelerate in the direction of $\mathbf{F}_{net}$. The value of the acceleration is given by *5.3,* and we note that a given force will produce a larger acceleration on the less massive of two objects. It is for this reason that football teams choose massive players in their efforts to stop the motion of the opposing team. To set a massive lineman into motion requires a large force, and so he is not easily removed from the path of the ball carrier. As discussed previously, massive objects possess a large amount of inertia. We also notice that Newton's first law is contained within the second law, as stated in *5.3*. If the body has no external unbalanced force acting on it, $\mathbf{F}_{net} = 0$, and therefore $\mathbf{a} = 0$. In the event that the body is at rest, it will remain so. Or if it is in motion, it will continue to move with the same velocity.

5.4 UNITS OF FORCE

To use Newton's second law in quantitative applications we must define the units to be used in it. Two of the three quantities in $F_{net} = ma$ were assigned SI units in previous chapters. Mass m is to be measured in kilograms while acceleration is to be expressed in m/s^2. We next use Newton's law to *define* the concept of force and its units. Assuming an object of constant mass m to be accelerated by a force F, we can use $F_{net} = ma$ to define a force unit we call the *newton* (N):

>*A net force of one newton acting on a 1 kilogram mass will give it an acceleration of 1 m/s².*

Similarly, a net force of 2 N will give the standard 1 kg mass an acceleration of $2\,m/s^2$, and so on. The magnitude of a force can, in principle at least, be determined by measuring the acceleration it gives to the standard 1 kg mass. The direction of the force is the same as that of the acceleration. We summarize our considerations as follows:

The SI force unit is the newton. When using $F_{net} = ma$, F_{net} must be in newtons, m must be in kilograms, and a must be in m/s².

5.5 OTHER UNIT SYSTEMS

Those who work with the British system of units use the *pound* (lb) as the force unit. The pound is defined in terms of SI units in such a way that 1 lb = 4.448199 N. Then use is made of $F_{net} = ma$ with *a* in ft/s² to define a unit of mass called the *slug*. The slug is equivalent to a mass of 14.593881 kg. Therefore we can state that:

The British engineering force unit is the pound (lb). When using $F_{net} = ma$, if F_{net} is in pounds, then a must be in ft/s², and m must be in slugs.

There is another metric-based unit system; it makes use of the centimeter, gram, and second (the cgs system). It measures mass in grams, acceleration in cm/s², and the resultant force unit is the *dyne*. It turns out that 1 dyne is exactly equal to 10^{-5} N. We shall not use this system because data given in it are easily changed to the SI. The three units systems we have discussed are summarized in table 5.1.

5.6 MASS RELATED TO WEIGHT

We have defined mass in terms of the 1 kg standard mass kept near Paris. Other masses are defined by comparison with this standard mass. Suppose a given force is applied first to the standard 1 kg mass and then to an unknown object. If the force gives the same acceleration to the two objects, assuming them free of all other unbalanced forces, then the masses of the two objects are the same. This follows directly from $F_{net} = ma$ for if the two *F*'s and *a*'s are the same, then the masses must also be the same. Similarly, an object of mass *n* kg will acquire an acceleration of only $1/n$th that imparted to the standard kilogram by the same force. We see, then, that the unknown mass of an object can be determined by comparing its acceleration to that of the standard kilogram when subject to the same force.

In practice, however, we usually determine masses in more convenient ways based on weighing. Of course, mass is a measure of the inertia of an object while the weight of the object is the force with which gravity pulls on it. Like apples and eggs, mass and weight are entirely different quantities. But they are obviously related in some way because massive objects are heavy. Let us now investigate this apparent relation between mass and weight.

There is a simple experiment that we can do which tells us the relation between mass and weight. It involves the free fall of an object as shown in

TABLE 5.1 Consistent Force Units

System	F	m	a	g (on earth)
SI	newtons (N)	kilograms (kg)	m/s²	9.80 m/s²
cgs	dynes (dyn)	grams (g)	cm/s²	980 cm/s²
British engineering	pounds (lb)	slugs	ft/s²	32.2 ft/s²

1 N = 10^5 dyn = 0.225 lb

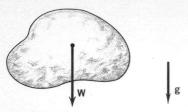

FIGURE 5.2 Write Newton's second law for a body in free fall.

figure 5.2. We know that in free fall an object accelerates downward with the free-fall acceleration g, the acceleration due to gravity. Moreover, the force that causes the object to accelerate downward is the pull of gravity on the object, the weight W of the object. And of course the object must obey Newton's second law, $F_{net} = ma$. For the free-fall situation shown in figure 5.2, we have that F_{net} is simply W, the weight of the object. Also, the acceleration a is g in this case, the free-fall acceleration. Hence, in this experiment, $F_{net} = ma$ becomes

$$W = mg \qquad\qquad 5.4$$

This is an extremely important relation because it tells us how the weight W of an object, any object, is related to its mass. As we see, even though weight is a force and mass is a measure of inertia, these two quite different quantities are proportional.

Let us now point out several important conclusions we can deduce from equation *5.4*. First, it tells us that the weight, on earth, of 1-kg mass is g *newtons* (N). This is obtained by substitution in *5.4*, thus:

On earth, 1 kg weighs 9.8 N

$$W = (1\,\text{kg})(g\,\text{m/s}^2)$$
$$= g\,\text{kg}\cdot\text{m/s}^2 = g\ \text{newtons}$$

We see that the newton (N) is really a $\text{kg}\cdot\text{m/s}^2$ and is therefore clearly defined by our basic units. Moreover, this same example shows (as experiment confirms) that the standard kilogram mass weighs different amounts at various places on the earth. Since g varies from place to place, so too must W. Following a similar procedure for the other units, we reach the following approximate conclusions for measurements conducted by a person on the surface of the earth:

1. A kilogram mass weighs 9.8 N.
2. A gram mass weighs 980 dyn.
3. An object which weighs 1 lb has a mass of 1/32.2 slugs.
4. $1\,\text{N} = 10^5$ dyn exactly, anywhere.
5. A kilogram weighs 2.21 lb.

$m = \dfrac{W}{g}$

Use is often made of *5.4*, especially by engineers who work in the British unit system. They frequently replace m by W/g and thereby eliminate m from *5.3*. As a result, they do not often use the mass unit, the slug. Instead, they replace the mass wherever it occurs by W/g. One difficulty with this procedure has to do with the fact that both g and W vary from place to place on the earth and in space. The ratio W/g must be (and is found experimentally to be) a constant, the mass. Therefore the difficulty can be overcome if the values used for W and g are those measured at the same place. One cannot use the earth value of g together with the weight measured on the moon, for example.

5.7 APPLICATION OF THE SECOND LAW

Newton's second law tells us that the acceleration of an object is in the direction of the resultant force on the object; the numerical magnitude of the force required to cause the acceleration is m times larger than the acceleration. Since the law concerns the force on an object, we must be careful to *isolate a body* for discussion when applying the law so that we know which object we are talking about. That is, we must choose a body to which the law

Isolate a body, draw the forces on it, and write, $\mathbf{F}_{net} = m\mathbf{a}$

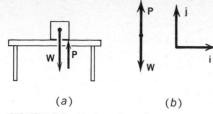

FIGURE 5.3 Under what circumstances is **P** = −**W**?

will be applied. After the choice has been made, we must acquaint ourselves with the forces acting on the body so that the resultant force may be found. The resultant (or net) force on the isolated body having been determined, we can then write 5.3 for the body.

As an illustration, suppose a block is on a table as in figure 5.3(a). We wish to find the acceleration of the block. Since our concern is with the block, we isolate it and consider only the forces acting on it. Other forces exist, the force of the table legs on the floor, for example. But since they do not act on the body we have isolated, we are not concerned with them when applying $F = ma$. Only the forces on the body which we have isolated are of importance to us, and only these are drawn in the figure. They are the pull W of gravity down on the body (its weight) and the push P of the surface of the table upward on the block.

Once the forces on the isolated body have been found, it is convenient to disregard all the extraneous physical details of the situation and show the important features in a *free-body diagram*. Such a diagram is drawn in part (b) of figure 5.3. It contains only the forces acting on the body which we have isolated. The resultant of these forces, **P** + **W**, is the force **F** which appears in 5.3. Therefore

$$\mathbf{F}_{net} = m\mathbf{a}$$

yields
$$\mathbf{P} + \mathbf{W} = m\mathbf{a}$$

We can conveniently write this in terms of the unit vectors shown in the figure. Making use of the fact that $\mathbf{P} = P\mathbf{j}$ and $\mathbf{W} = -W\mathbf{j}$, we have

$$(P - W)\mathbf{j} = m\mathbf{a}$$

As a result, it is clear that if the push of the table P is equal to the weight of the block W, the block will not accelerate, $a = 0$. It would remain at rest. However, if W exceeded P, which might be the case if the legs on the table were not too strong, then $(P - W)$ would be a negative number and the block would accelerate downward.

Of course, we have made this problem much too difficult. We all know without going into detail that when $P = W$ the block will remain at rest. The detailed analysis was carried through to provide practice in a very simple situation. Let us analyze another, similar case before assuming that we can proceed intelligently to more complicated situations.

Consider the object shown at B in figure 5.4 which is subjected to, let us say, N forces, only one of which is shown, $\mathbf{F}_n$. Applying $F = ma$ to this body, we have

$$\mathbf{F}_1 + \mathbf{F}_2 + \cdots + \mathbf{F}_n + \cdots + \mathbf{F}_N = m\mathbf{a}$$

However, each one of these forces may be represented by its components in the same way $\mathbf{F}_n$ is shown in the figure. Replacing the forces in that way, we have, after combining like terms,

$$(F_{1x} + F_{2x} + \cdots + F_{Nx})\mathbf{i} + (F_{1y} + F_{2y} + \cdots + F_{Ny})\mathbf{j} \\ + (F_{1z} + F_{2z} + \cdots + F_{Nz})\mathbf{k} = m(a_x\mathbf{i} + a_y\mathbf{j} + a_z\mathbf{k})$$

where we have also written the acceleration vector in terms of its components.

But this resulting equation simply equates two vectors, the resultant force vector and m times the acceleration vector. Since the vectors are equal, their components are equal. Therefore, equating components, we obtain

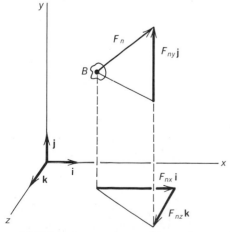

FIGURE 5.4 The force indicated may be written, $\mathbf{F}_N = F_{nx}\mathbf{i} + F_{ny}\mathbf{j} + F_{nz}\mathbf{k}$.

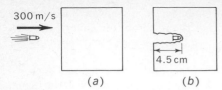

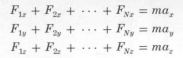

$$F_{1x} + F_{2x} + \cdots + F_{Nx} = ma_x$$
$$F_{1y} + F_{2y} + \cdots + F_{Ny} = ma_y$$
$$F_{1z} + F_{2z} + \cdots + F_{Nz} = ma_z$$

300 m/s

4.5 cm

(a)   (b)

FIGURE 5.5 How large was the average force which stopped the 10-g bullet?

These equations may be written more compactly

$$\sum_{1}^{N} F_{nx} = ma_x$$

$$\sum_{1}^{N} F_{ny} = ma_y \hspace{3cm} 5.5$$

$$\sum_{1}^{N} F_{nz} = ma_z$$

Equation 5.5 is another, fully equivalent, way of writing the second law. In practical situations, this form is often easiest to use as we shall see in the following examples.

Illustration 5.1

A 10-g bullet is shot with an initial speed of 300 m/s into a block of wood, as shown in figure 5.5. It stops after penetrating 4.5 cm into the wood. Find the force needed to stop it. Assume uniform deceleration.

Reasoning We shall assume the motion to take place in the +x direction. In order to find the force acting on the bullet, we shall need the acceleration of the bullet. This may be found by means of the motion equations. We know $v_0 = 300$ m/s, $v = 0$, $x = 0.045$ m, and we shall use all quantities in the SI.

Making use of

$$v^2 - v_0{}^2 = 2ax$$

gives $a = -1 \times 10^6 \text{ m/s}^2$

where the minus sign indicates that the bullet is slowing down.

Let us now find the applied force by use of 5.5. Then

$$\sum F_{nx} = ma_x$$

or

$$F_x = (0.010 \text{ kg})(-1 \times 10^6 \text{ m/s}^2)$$
$$= -1 \times 10^4 \text{ N}$$

where use has been made of the fact that a kg · m/s² is a newton. The minus sign tells us the force is in the $-x$ direction, as we knew it would be.

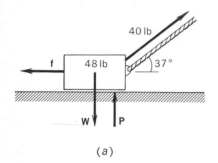

40 lb

48 lb

f

37°

W   P

(a)

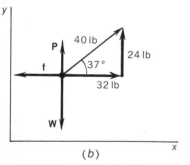

y

P

40 lb

24 lb

f

37°

32 lb

W

(b)

x

FIGURE 5.6 The friction force on an object is always parallel to the sliding surface and opposite to the motion of the object.

The friction force always opposes sliding

Illustration 5.2

Consider the situation shown in figure 5.6(a). The applied force due to the rope is 40 lb, and the block weighs 48 lb. If the block accelerates at 5.0 ft/s², how large a friction force must be retarding its motion?

Reasoning The object we should isolate is clearly the block. We have already drawn the forces acting on it in part (a) of the figure. They are the 40-lb pull of the rope, the weight (48 lb) of the object, the push of the table up on the object **P**, and the friction force **f**. *Notice that the friction force is always parallel to the surface on which the object slides and is in a direction opposite to the motion.*

All these forces are drawn on the free-body diagram of part (b) of the figure. Since we shall write $F = ma$ in its component form (5.5), the components of the forces are also shown. No motion occurs in the y direction, and so we shall not be concerned with the y forces. We have for the x direction

$$\sum F_{nx} = ma_x$$

$$32 - f = \frac{48}{g}(5.0)$$

all units being those of the British engineering system. In writing this, we have followed the engineering practice of replacing m by W/g. Taking $g = 32 \text{ ft/s}^2$ and solving for f, we find

$$f = 24.5 \text{ lb}$$

Illustration 5.3

An electron ($m = 9.1 \times 10^{-31} \text{ kg}$) experiences the following force in a particular device:

$$\mathbf{F} = (2 \times 10^{-15}\mathbf{i}) - (3 \times 10^{-15}\mathbf{j}) \qquad \text{N}$$

Find its velocity 10^{-9} s after being released from rest.

Reasoning We shall use 5.5 to find the accelerations in the x and y directions. Knowing these, we can find the velocity components after 10^{-9} s, and then the velocity.

Writing 5.5, once each for the x and y components,

$$\Sigma F_{nx} = ma_x \qquad\qquad \Sigma F_{ny} = ma_y$$
$$2 \times 10^{-15} \text{ N} = (9.1 \times 10^{-31} \text{ kg})a_x \qquad -3 \times 10^{-15} \text{ N} = (9.1 \times 10^{-31} \text{ kg})a$$
$$a_x = 2.2 \times 10^{15} \text{ m/s}^2 \qquad\qquad a_y = -3.3 \times 10^{15} \text{ m/s}^2$$

We see that the a's are constant, so the uniform motion equations are applicable.

Now making use of $v = v_0 + at$ and recalling that $v_{0x} = v_{0y} = 0$, we find

$$v_x = 2.2 \times 10^6 \text{ m/s} \qquad \text{and} \qquad v_y = -3.3 \times 10^6 \text{ m/s}$$

The resultant velocity is therefore

$$\mathbf{v} = (2.2 \times 10^6\mathbf{i}) - (3.3 \times 10^6\mathbf{j}) \qquad \text{m/s}$$

This velocity vector is shown in figure 5.7.

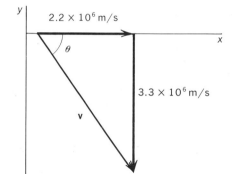

FIGURE 5.7 What are the values of θ and $\mathbf{v}$?

5.8 MOTION ON AN INCLINE

The object shown in figure 5.8 will slide down the incline if the friction force between it and the incline is small. Let us investigate the cause of this motion. To that end, we isolate the block and draw the forces on it, as shown in part (a) of the figure. Notice that the force exerted on the block by the surface of the incline is divided into two parts, a friction force f and the force Y (the *normal*, or perpendicular, force). Three forces act on the block, Y, f, and W, the weight of the block.

The three forces acting on the block are shown in the free-body diagram

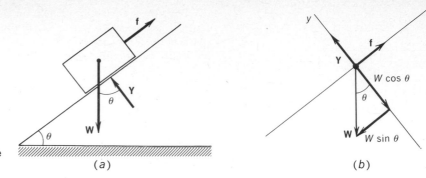

FIGURE 5.8 Notice the choice of the x and y axes, along which components are taken.

(a)   (b)

The x and y coordinates are taken parallel and perpendicular to the incline

of figure 5.8(b). Since motion will occur along the incline, it is convenient to take the x axis parallel to the incline, as indicated. To write the component equations for Newton's second law, the weight must be split into its x and y components. Since $\mathbf{W}$ is perpendicular to the base of the incline and $\mathbf{Y}$ is perpendicular to its surface, the angle between $\mathbf{Y}$ and $\mathbf{W}$ is equal to the angle of the incline θ. We can express the components of $\mathbf{W}$ in terms of θ as shown in the figure.

Now that the forces acting on the block are expressed in terms of their x and y components, we can write

$$\sum F_{nx} = ma_x$$

$$\sum F_{ny} = ma_y$$

The first of these yields

$$-W \sin \theta + f = ma_x$$

where m is the mass of the block. Solving for a_x, we find

$$a_x = \frac{1}{m}(-W \sin \theta + f)$$

Since f cannot exceed $W \sin \theta$ (for if it did, the block would move up the incline, and friction forces always oppose the sliding motion of one surface across another rather than cause it), we see that a_x is negative. This indicates that the acceleration is down the incline in the $-x$ direction, as it should be.

A special case of interest occurs if f is small enough to be negligible. Then

$$a_x = -\frac{1}{m} W \sin \theta$$

But $W = mg$, and so

$$a_x = -g \sin \theta$$

This is a *very specialized relation,* and is not true when f is not negligible. Notice that it predicts $a_x = -g$ when $\theta = 90°$ and $a_x = 0$ when $\theta = 0°$. Are these results reasonable?

The y-equation part of Newton's second law is really rather trivial in the present case. It says

$$\sum F_{ny} = ma_y$$

Discovery of X-rays

Many of the laws of physics have been discovered by interpreting the results of experiments carefully designed to determine these laws. This was the case in Galileo's discovery of the law of falling bodies. Newton's laws are also of this same general type. However, sometimes in physics a curious, unexpected laboratory circumstance leads to the discovery of an important phenomenon. This was the way in which Wilhelm Konrad Roentgen (1845–1923) discovered the existence of x-rays.

He was carrying out experiments with high-voltage discharges or sparks, and he applied a potential difference of several thousands of volts to electrodes in the two ends of a partly evacuated tube. (This is called a discharge tube.) Under such conditions, a discharge, or glow, much like that observed in neon signs, occurs. However, if the pressure of the gas in the tube is reduced to a low enough value, the glow nearly ceases. But Roentgen observed, while carrying out experiments with a highly evacuated discharge tube in his poorly lit laboratory, in 1895, that a nearby fluorescent screen (much like that on the end of present-day TV tubes) was also glowing in the darkness. By moving the screen about the room, he was able to show that the light given off by the fluorescent screen resulted from something taking place in the discharge tube. Since a light-tight cover could be placed over the tube without greatly affecting the glow on the screen, it was clear that the fluorescent glow was caused by something other than the light given off by the tube. Roentgen named this unknown radiation striking the screen and coming from the tube x-rays. He carried out many experiments with the rays from the tube and found that they were highly penetrating, even being able to pass through a book. However, heavy metals and bone, among other materials, were not nearly so transparent to the rays as were carbonaceous materials, such as wood and paper. He was even able to cast a shadow with these rays. Shadows could be produced showing the bones of his wife's hand and the ring on it. Because of this ability to produce shadows, Roentgen held open the possibility that these rays were short-wavelength light. However, he was reluctant to state this as fact because, unlike light, the rays were not deflected (i.e., refracted) appreciably when passed from air to water.

As we shall see in later chapters, the x-rays discovered by Roentgen are indeed similar to light waves but have much shorter wavelengths.

Unfortunately, as we know today, x-rays damage human tissue and cause severe burns through overexposure. Even when visible burns do not occur, other damage may be present. Many of the early workers in the field of x-rays suffered deteriorating health, and even death, because of their ignorance of the harmful properties of these little-understood rays.

The structure of a jet aircraft engine, laid bare in a single x-ray photo. (Courtesy *Eastman Kodak* Company.)

and since
$$a_y = 0$$
we have
$$Y - W \cos \theta = 0$$
which simply tells us how large the normal force is.

Illustration 5.4

Consider the situation shown in figure 5.9(a). If the friction force on the

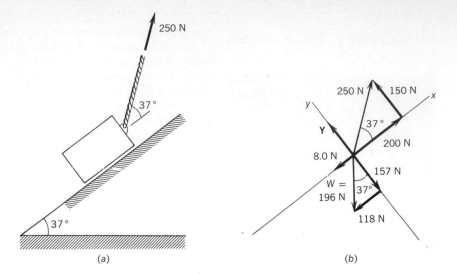

FIGURE 5.9 Once inclined axes such as these are chosen, be careful not to inadvertently use the vertical and horizontal axes.

(a)

(b)

20-kg block is 8.0 N and the tension in the rope is 250 N, what will be the acceleration of the block?

Reasoning The appropriate free-body diagram is shown in part (*b*). Since the 250-N force will cause the block to move up the incline, the friction force will be directed *down* the incline so as to oppose the motion. Notice that the weight is $mg = 20 \times 9.8$ N. Once again the *x* and *y* axes are chosen parallel and perpendicular to the incline. In order to use the component form of $F = ma$, we split the forces into components, as indicated. Since we are interested only in the motion of the system, and since this motion will occur along the *x* axis, we write

$$\sum F_{nx} = ma_x$$

to give

$$-8 - 118 + 200 = 20a_x$$

where of course the forces are in newtons. Solving gives

$$a_x = 3.7 \text{ m/s}^2$$

5.9 MOTION OF SEVERAL CONNECTED BODIES

An interesting situation results when two or more objects are constrained to move in various ways by attachments between the bodies. For example, the block on the table in figure 5.10 is pulled along by the weight hanging over the pulley. Assuming the mass of each object to be known, and that the string and pulley have negligible mass and friction effects, we should be able to find the acceleration of the weight as it falls downward. To make the problem more general, let us say that a friction force *f* impedes the motion of the block along the table.

Before beginning quantitative work on this situation let us first notice a few qualitative points. If there were no friction force, the weight of mass m_2 would cause it to fall no matter how large m_1 was. This follows from the fact that only the friction force *f* is trying to hold m_1 from moving. The weight of

m_1 merely holds the mass tightly on the tabletop; it has no component to the left, and so it cannot hold the mass from moving to the right.

A second point to notice is that if the string between the masses were to break, mass m_2 would be a freely falling body. Only then would its acceleration become as large as g. With the string unbroken, the tension in the string will tend to hold the mass up, and so its downward acceleration would be less than g. If the tension in the string were equal to the weight of m_2, namely, $m_2 g$, then m_2 would remain motionless since the resultant force on it would be zero. Clearly, if m_2 is to fall, the tension T in the string must be less than $m_2 g$. With these facts in mind, let us now approach the problem quantitatively.

Newton's second law is concerned with the forces on and the acceleration of a body. We must at the outset state what body we are isolating for consideration. Isolating m_2 first, we draw the forces on it in part (b) of figure 5.10. It is convenient to take the direction of motion to be the positive direction since then the acceleration will be positive. With that sign convention we write

Isolate each of the several bodies in turn and write $F = ma$ for each

$$\sum F_y = ma$$

for the mass m_2 and obtain

$$m_2 g - T = m_2$$

This equation contains two unknowns, T and a.

Turning now to m_1 and its free-body diagram as shown in part (c), we write

$$\sum F_x = ma$$

or

$$T - f = m_1 a$$

where the a's for the two masses must be the same since they are tied together by a string of fixed length. If we add this equation to the one involving m_2, we eliminate T to obtain

$$m_2 g - f = m_2 a + m_1 a$$

Solving for a, we find

$$a = \frac{m_2 g - f}{m_2 + m_1}$$

As a check on this equation, notice that if f and m_1 are zero, $a = g$. Why should this be true? On the other hand, if m_1 is extremely massive in comparison with m_2, its inertia will be large enough to hold the whole system nearly at rest, and $a \to 0$. Does the case $f > m_2 g$ have any physical significance?

Illustration 5.5

Find the acceleration of the 20-kg block in figure 5.11 if the friction forces are negligible. Also find T_1 and T_2.

Reasoning Let us first isolate the 5-kg block and consider the direction of motion to be positive. Its free-body diagram is shown in part (b) of the figure, and since motion occurs in the direction of T_1 only, the forces Y_1 and W_1 must cancel each other. Writing $F = ma$ for this block,

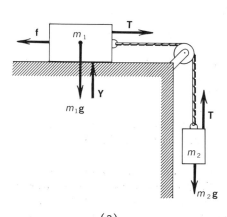

(a)

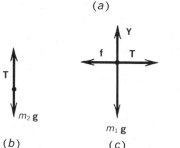

(b)   (c)

FIGURE 5.10   How large must m_2 be if the blocks are to move?

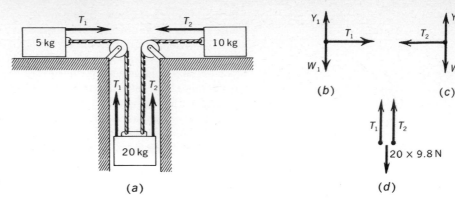

FIGURE 5.11 If we ignore friction, which will be smaller, T_1 or T_2?

(a)

(b)

(c)

(d)

$$T_1 = 5a$$

Similarly, the 10-kg block gives

$$T_2 = 10a$$

where both blocks will have the same acceleration. Making use of the free-body diagram for the 20-kg block shown in part (d), we have

$$196\,\text{N} - T_1 - T_2 = 20a$$

We therefore have three equations involving the three unknowns T_1, T_2, and a. Solving them simultaneously, we find

$$a = 5.6\,\text{m/s}^2$$
$$T_1 = 28\,\text{N}$$
$$T_2 = 56\,\text{N}$$

As a crude check on our answers, we notice that $T_1 + T_2$ is less than the weight of the 20-kg block, and so it will fall.

Illustration 5.6

In figure 5.12 the friction force between the block and plane is 3.0 lb. Find the acceleration of the system and the tension in the rope.

Reasoning Let us first decide which way the system would move if the friction force were zero. The free-body diagrams are shown in parts (b) and (c). As shown in part (b), the component of the 20-lb weight pulling down the incline is 12 lb. This force would tend to pull the 18 lb weight upward. However, this latter block is pulled downward by an even larger force, 18 lb, and so it will actually fall. We see then that the block on the incline will move up the incline. The 3-lb friction force will tend to oppose this motion, and so it is directed down the incline.

Referring to part (c), we can write $F = ma$ by using the direction of motion as positive and recalling that $m = W/g$, where $g = 32\,\text{ft/s}^2$ in this case. Therefore

$$18 - T = \tfrac{18}{32}a$$

Similarly for the situation shown in part (b)

$$T - 12 - 3 = \tfrac{20}{32}a$$

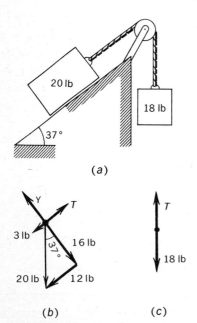

(a)

(b)   *(c)*

FIGURE 5.12 How large would the friction force have to be if the system were to remain at rest?

where the units in these equations should be supplied by the student. Adding the two equations yields

$$a = 2.52 \text{ ft/s}^2$$

Placing this value in either of the equations gives

$$T = 16.6 \text{ lb}$$

5.10 WEIGHT AND WEIGHTLESSNESS

A fascinating physical phenomenon called weightlessness is sometimes observed when objects are accelerating. Although we shall postpone a discussion of weightlessness in spaceships orbiting the earth until we have studied about motion in a circle, other examples of weightlessness also exist. A great deal of insight into the reasons for this phenomenon can be obtained by considering an object suspended from the ceiling of an elevator. The situation is shown in figure 5.13. Notice that the reading of the spring scale is the quantity which is commonly called the weight of the object. However, since we have defined the weight of an object to be the force of gravity which acts on the object, we call the spring-scale reading the *apparent weight* of the object.

Definition of apparent weight

The free-body diagram shown in part (b) of figure 5.13 shows the forces acting on the object. There are only two, the pull of gravity (the weight of the object), W, and the tension in the light rope, T. This tension is equal to the reading of the spring scale, and is therefore equal to the *apparent* weight of the object.

Case 1 Elevator at Rest
In this case the acceleration is zero. Applying $F = ma$ to the free-body diagram, we find at once that $T = W$. The scale will read W, and the apparent weight of the object equals the gravitational force on the object.

Case 2 Elevator with Constant Velocity
From the definition of acceleration as dv/dt, we see that here, too, $a = 0$, since v is a constant. Once again $F = ma$ yields the fact that $T = W$, and so the apparent weight of the object is equal to the gravitational force on it.

Case 3 Elevator Accelerating Upward
Let us call the acceleration a, and, taking up as the positive direction, Newton's second law gives

$$T - W = ma$$

from which

$$\text{Apparent weight} = T = W + ma$$

We see that the apparent weight of the object is increased over its usual rest value. The support, in this case the rope, must not only balance the gravitational force but must also provide an unbalanced force $T - W$ which causes the upward acceleration.

Case 4 Elevator Accelerating Downward
Still taking up as the positive direction, the acceleration now becomes negative. Newton's second law is now written

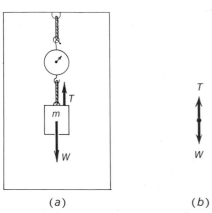

FIGURE 5.13 The spring scale reads the tension T in the rope, and this is equal to the apparent weight of the object.

(a) (b)

$$T - W = -ma$$

from which

$$\text{Apparent weight} = T = W - ma$$

Clearly, the apparent weight of the object is less than the gravitational attraction for the object.

An interesting case occurs when $a = g$, the acceleration due to gravity. In that case, since $W = mg$, we find

$$T = mg - mg = 0$$

An example of weightlessness

and the object appears weightless. This is not surprising when we consider that an object can acquire a downward acceleration of g only if it is in free fall. In that case the rope cannot be pulling upward on it, and so the support needs to exert no force on the object to hold it in place. The object therefore appears to be weightless.

Although the elevator situation outlined here is perhaps rather artificial, it does serve to show that the apparent weight of an object depends critically on its acceleration. We shall see in Chap. 9 that these ideas may be carried over to motion of an object in a circular path. It will then become clear why objects in orbit appear to be weightless. Indeed, a spaceship coasting anywhere in space is like the freely falling elevator; all objects within it appear to be weightless.

5.11 FRICTION FORCES

Until now we have made use of friction forces without discussing their nature. Although the molecular origin of friction forces between solids is a very complicated phenomenon still not completely understood, there are several practical facts concerning them which we shall discuss. Let us consider a block resting on a surface as shown in figure 5.14.

When we try to slide the block to the right by pulling with a force F, the block will not slide unless F exceeds a certain critical value, f_c. Apparently the contact between the two surfaces is capable of furnishing a friction force equal to F so long as F remains smaller than f_c. But if F exceeds f_c, the block begins to slide. Once the block is in motion, a force F smaller than f_c will keep it in motion with constant velocity. We see from this that the friction force that opposes sliding has a value $f < f_c$ after the block is moving. Experiment shows that a rough value for f can be found in the following way.

Suppose that we separate into two components the force that the supporting surface exerts on the object. As shown in figure 5.15, one component is the friction force f, and it is directed parallel to the supporting surface. The other component is perpendicular to the supporting surface and is shown as Y in figure 5.15. It is called the *normal* force (where normal means perpendicular in this context). For an object sliding across the surface,

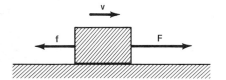

FIGURE 5.14 The friction force **f** opposes the motion of the block caused by the force **F**.

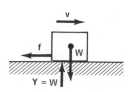
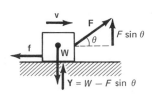
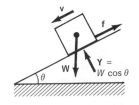

FIGURE 5.15 Notice that the normal force **Y** need not be equal to *W*.

TABLE 5.2 Kinetic Coefficients of Friction

Surface 1	Surface 2	μ_k
Wood	snow	0.06
Brass	ice	0.02–0.1
Metal	metal (lubricated)	0.07
Oak	oak	0.25
Rubber	concrete (wet)	0.5–0.9
Rubber	concrete (dry)	0.7–1.0

experiment shows that f is approximately proportional to the normal force. That is to say, $f = \mu_k Y$ where μ_k is called the kinetic coefficient of friction. Typical values for μ_k are given in table 5.2. These values are only approximate because the friction force depends strongly on the exact nature of the surfaces involved.

The values for μ_k given in table 5.2 are called *kinetic* coefficients of friction because they apply only to an object that is already sliding. Because f_c, the friction force when the object is just on the verge of sliding, is larger than the value of f after sliding has begun, another friction coefficient μ_s, the *static* coefficient of friction, is also used. It is defined by the relation $f_c = \mu_s Y$ and μ_s is invariably larger than μ_k for the same conditions. Because of the importance of the normal force Y in this definition, you may wish to check through the various parts of figure 5.15 to make sure you know how Y is calculated.

Illustration 5.7

A box sits in the rear of a truck as the truck decelerates at $3\ \text{m/s}^2$. How large must be the static friction coefficient between the box and the bed of the truck if the box is not to slide?

Reasoning To give the box of mass m a deceleration of $3\ \text{m/s}^2$ a stopping force given by $F_{\text{net}} = ma$ is required.

$$\text{Stopping force} = m\,\frac{3\ \text{m}}{\text{s}^2}$$

This force must be supplied by the friction force between box and bed. If the box is just on the verge of slipping,

$$f_c = \text{stopping force} = m\,\frac{3\ \text{m}}{\text{s}^2}$$

But

$$\mu_s = \frac{f_c}{Y} = \frac{m(3\ \text{m/s}^2)}{mg} = \frac{3}{9.8} = 0.31$$

and so the static coefficient must be at least 0.31.

Chapter 5
Questions
and Guesstimates

1. Why do you have a tendency to slide across a car seat as the car quickly turns a corner? Why will a carton of eggs fall from the seat if the car stops too quickly?
2. Distinguish clearly between mass, weight, and inertia.
3. Clearly identify the action-reaction forces in each of the following: A boy

kicks a can; the sun holds the earth in orbit; a ball breaks a window; a parent spanks a child; a Ping-Pong ball bounces on a table; a boat tows a water skier.

4. The acceleration due to gravity for objects on the moon is about 1.6 m/s². An object that has a measured mass of 2 kg on earth will weigh how much on the moon? On the earth? What will be its mass on the moon?

5. Because objects weigh only about one-sixth as much on the moon as on the earth, you would almost certainly be able to lift Muhammad Ali if the two of you were on the moon. Could you easily stop him if he was running at a fair rate across the moon's surface?

6. Describe how the force exerted on you by the seat in a roller coaster varies as you go down a steep hill, rise up the next hill, and go over its top at a high speed.

7. Is it possible for an object to accelerate downward on the earth at a rate greater than g?

8. It is generally believed that, on the average, a drunk would be less injured after a fall from a window than a sober man would be. Can you explain why the belief might be valid?

9. Suppose a brick is dropped from a height of several inches onto your open hand. Why may severe injury to your hand ensue if your hand is laying flat on a tabletop at the time even though you can easily catch a falling brick without injury under other circumstances?

10. An object is being weighed in an elevator. If the elevator suddenly begins to accelerate upward, explain what would happen if the weighing device were (a) a spring balance, (b) an analytical two-pan beam balance, (c) a triple-beam balance (i.e., an unequal arm balance).

11. A car at rest is struck from the rear by a second car. The injuries (if any) incurred by the drivers of the two cars will be of distinctly different character. Explain what will happen to each driver.

12. A man standing in an elevator with a pipe in his hand is so surprised when the elevator cable breaks that he drops his pipe. What will the pipe do? (Do not idealize the situation; be practical.)

13. Estimate the minimum distance in which a car can be accelerated from rest to 10 m/s if its motor is extremely powerful.

14. When a high jumper leaves the ground, where does the force come from which accelerates him upward? Estimate the force which must be applied to him in a 2-m-high jump.

15. Estimate the force your ankles must exert as you strike the floor after jumping from the top of a 2.0-m-high ladder. Why should you let your legs flex in such a situation?

16. Consider the large type mops used to sweep the halls in a school. It is easy to slide the mop along the floor if the mop handle makes only a small angle with the floor. But if the angle between handle and floor is too large, you can't push the mop along no matter how hard you push. Explain why. Can you find a relation between the critical angle for sliding and the friction coefficient between floor and mop?

Chapter 5
Problems

1. A horizontal force of 20 N is required to move a box along a level floor at a constant speed of 0.30 m/s. How large is the friction force which opposes the motion?

2. When a person falling through the air falls far enough, his velocity reaches a constant maximum value called his *terminal* velocity. A typical terminal velocity for a 150-lb (670 N) man would be 250 ft/s. While falling with that velocity, how large an upward force is exerted on him by the rush of the air past him?

3. A horizontal force of 5000 N is required to give a 1000-kg car an acceleration of 0.20 m/s² on a level road. How large is the friction force which opposes the motion?

4. A certain 1500-kg car can accelerate from rest to 20 m/s in 8 s. How large an unbalanced force must act on the car to cause this acceleration? Assume uniform acceleration.

5. A rough rule of thumb states that the friction force between dry concrete and a skidding car's tires is about equal to nine-tenths the car's weight. If the skid marks left by a car in coming to rest are 20 m long, about how fast was the car going just before the brakes were applied? Justify the rule of thumb.

6. If the coefficient of friction between a car's wheels and a roadway is 0.70, what is the least distance in which the car can accelerate from rest to a speed of 15 m/s?

7. An electron ($m = 9.1 \times 10^{-31}$ kg) is accelerated from rest to a speed of 5×10^7 m/s in a distance of 0.80 cm in a TV tube. Find the average accelerating force on the electron. How many times larger than mg is it?

8. Write in vector-component form the force needed to give a proton ($m = 1.67 \times 10^{-27}$ kg) an acceleration $2 \times 10^9 \mathbf{i} - 3 \times 10^9 \mathbf{j}$ m/s².

9. An object of mass m is subject to the following force: $F_x\mathbf{i} + F_y\mathbf{j}$. (*a*) Show that the magnitude of its acceleration is $\sqrt{F_x^2 + F_y^2}/m$. (*b*) In what direction is the acceleration?

10. A 200-g object is subjected to a force $0.30\mathbf{i} - 0.40\mathbf{j}$ N. Starting from rest, what will be the velocity of the object after 6 s?

11. It is observed that when the brakes of a 2500-lb car are released, the car rolls 40 ft down a uniform incline in 20 s. How large was the resultant force which caused it to roll down the incline?

12. A 1500-kg car moving at 20 m/s begins to coast up an incline. It rolls 80 m up along the steady incline before stopping. Find the resultant force which acted to slow it.

13. How large a force parallel to a 30° incline is needed to give a 5.0-kg box an acceleration of 0.20 m/s² up the incline if friction is negligible?

14. An 8.0-kg box is released on a 30° incline and accelerates down the incline at 0.30 m/s². Find the friction force impeding its motion. How large is the coefficient of friction in this situation?

15. How large a force pulling upward at an angle of 30° to the surface of a 37° incline is needed to pull an 8-kg box up the incline at constant speed? Assume the friction force to be 20 N. How large a coefficient of friction would apply in this case?

16. For the situation described in problem 15, how large must the force be if the box is to accelerate down the incline at 0.20 m/s²? Assume (unrealistically) that the friction force is still 20 N.

17. A book sits on a horizontal top of a car as the car accelerates horizontally from rest. If the static coefficient of friction between car top and book is 0.45, what is the maximum acceleration the car can have if the book is not to slip?

18. A carton of eggs rests on the seat of a car moving at 20 m/s. If the car is to be uniformly slowed to a stop, what is the least distance in which the

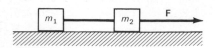

FIGURE P5.1

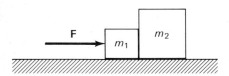

FIGURE P5.2

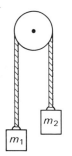

FIGURE P5.3

car can stop if the eggs are not to slide on the seat? The value of μ_s between carton and seat is 0.15.

19. A cement block sits on the floor of a station wagon as it goes down a 20° incline decelerating at 1.5 m/s². How large must the static friction coefficient be between floor and block if the block is not to slide?

20. Referring to figure P5.1, find the acceleration of the blocks and the tension in the connecting string if the applied force is F and the friction forces on the blocks are negligible.

21. In figure P5.1, if $F = 20$ N, $m_1 = m_2 = 3$ kg, and the aceleration is 0.50 m/s², what will be the tension in the connecting cord provided the friction forces on the two blocks are equal? How large is the friction force on either block?

22. Two blocks with masses $m_1 = 3$ kg and $m_2 = 4$ kg are in contact on a frictionless table as shown in figure P5.2. If the force F shown acting on m_1 is 5 N, (a) what is the acceleration of the two blocks and (b) how hard does m_1 push against m_2? (c) Repeat (a) and (b) if the applied force F is in the reverse direction and pushes on m_2 rather than on m_1.

23. A uniform ladder of mass m and length L leans against a smooth wall in such a way that it makes an angle θ with the floor. What must be the value of μ_s between floor and ladder if the ladder is not to slide?

24. The device shown in figure P5.3 is called an *Atwood's machine*. In terms of m_1 and m_2 with $m_2 > m_1$, (a) how far will m_2 fall in time t after the system is released? (b) What is the tension in the light cord that connects the two masses? Assume the pulley to be frictionless and massless.

25. A rectangular block of mass m sits on top of another similar block, which in turn sits on a flat table. The maximum possible friction force of one block on the other is 2.0m N. What is the largest possible acceleration which can be given the lower block without the upper block sliding off? What is the coefficient of friction between the two blocks?

26. A passenger in a large ship sailing in a quiet sea hangs a ball from the ceiling of his cabin by means of a long thread. Whenever the ship accelerates, he notes the pendulum ball lags behind the point of suspension and so the pendulum no longer hangs vertical. How large is the ship's acceleration when the pendulum stands at an angle of 5° to the vertical?

27. For the situation shown in figure P5.4, how large is the friction force on m_2 if the tension in the connecting cord is 0.60 m_1g and $m_1 = m_2 = 2$ kg.

28. If in figure P5.4 it takes 2.0 s for m_1 to drop 50 cm after the system is released, how large is the friction force on m_2? ($m_1 = 0.40$ kg, $m_2 = 1.80$ kg.)

29. It is observed that the system shown in figure P5.5 can be set in motion in

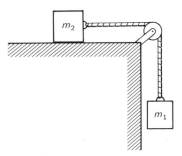

FIGURE P5.4

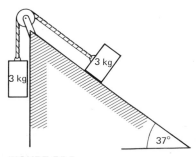

FIGURE P5.5

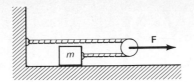

FIGURE P5.6

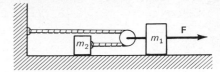

FIGURE P5.7

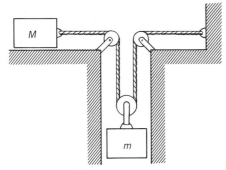

FIGURE P5.8

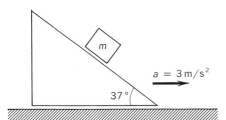

$a = 3\,\text{m/s}^2$

$37°$

FIGURE P5.9

such a way that the mass slides on the incline with constant speed with no other force than friction acting on the system. How large is the friction force and the friction coefficient between block and plane?

30. For the situation of figure P5.5, if the block slides with acceleration $0.20\,\text{m/s}^2$ when the system is released, how large is the friction force?

31. A boy who normally weighs 80 lb on a bathroom scale crouches on the scale and suddenly jumps upward. His companion notices that the scale reading momentarily jumps up to 110 lb as the boy springs upward. Estimate the boy's maximum acceleration in this process.

32. A car coasting at 20 m/s starts up a 37° incline at the same instant that another car 100 m up the incline starts to coast from rest down the incline. Ignoring friction forces, (a) how far from the bottom of the incline will the cars be when they coast past each other? (b) What will be the velocity of the originally upward moving car when they pass?

33. In figure P5.6 the pulley is assumed massless and frictionless. Find the acceleration of the mass m in terms of F if there is no friction between the surface and m. Repeat if the friction force on m is f.

34. In figure P5.7, assume there is negligible friction between the blocks and table. Compute the tension in the cord and the acceleration of m_2 if $m_1 = 300$ g, $m_2 = 200$ g, and $F = 0.40$ N. (*Hint:* Note that $a_2 = 2a_1$.)

35. In figure P5.8, when m is 3.0 kg, the acceleration of the block m is $0.6\,\text{m/s}^2$, while $a = 1.6\,\text{m/s}^2$ if $m = 4.0$ kg. Find the friction force on block M as well as its mass. Neglect the mass of the pulleys.

36. A block sits on an incline as shown in figure P5.9. (a) What must be the friction force between block and incline if the block is not to slide along the incline when the incline is accelerating to the right at $3\,\text{m/s}^2$? (b) What is the least value μ_s can have for this to happen?

37. If the inclined plane shown in figure P5.9 is accelerating to the right at $3\,\text{m/s}^2$, will the block m accelerate up or down the incline? Find its acceleration relative to the incline. Neglect friction forces.

6 Work and Energy

It often proves advantageous in science to define quantities that are conceptual in nature. One of the most useful of these quantities is the concept of energy and we shall begin an investigation of it in this chapter. Our attention here will be focused on mechanical energy and how it is related to the concept of work. In later chapters we shall investigate other types of energy such as heat, electric, and nuclear energy.

6.1 THE DEFINITION OF WORK

There are many colloquial meanings for the word *work*. We go to work in the morning; we work at our studies; we do all kinds of work. Because of this fact, it becomes necessary for a scientist to state exactly what he or she means by the word. The meaning adopted universally in science is as follows:

Definition of work

The work ΔW done by a force $\mathbf{F}$ that acts on an object as the object moves through a small displacement $\Delta \mathbf{r}$ is

$$\Delta W = F_r \, \Delta r \qquad\qquad 6.1a$$

where F_r is the component of $\mathbf{F}$ in the direction of the displacement.

From the definition we see that the units of work are as follows for our usual systems:

The SI unit of work is the joule

System	Compound unit	Name
SI	newton · meter (N · m)	joule (J)
cgs	dyne · centimeter (dyn · cm)	erg
British engineering	pound · foot (lb · ft)	foot · pound (ft · lb)

Work has no direction and is therefore a scalar quantity. The student should verify that 1 joule = 10^7 ergs. Another unit of work which is frequently used in nuclear physics and electronics is the *electron volt* (eV). It is defined by $1 \text{ eV} = 1.602 \times 10^{-19} \text{ J} = e$ joules where e is the magnitude of the charge on the electron.

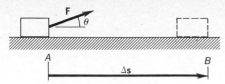

FIGURE 6.1 The work done in moving the object from A to B is $(F \cos \theta) \Delta s$.

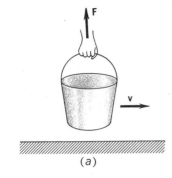

(a)

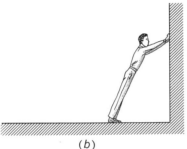

(b)

FIGURE 6.2 Why is no work being done on the pail or on the wall?

To comprehend the meaning of this definition more clearly, let us refer to figure 6.1. An object is being pulled along the table by a constant force **F** at an angle θ to the direction of motion. When the object is moved from A to B, a vector displacement $\Delta\mathbf{s}$, the work done is

$$\Delta W = (F \cos \theta) \Delta s \qquad 6.1b$$

The quantity $F \cos \theta$ is the component of the force in the direction of motion, and $\Delta\mathbf{s}$ is the displacement.

Notice that two things are necessary if work is to be done on an object: (1) a component of the force must exist in the direction of the displacement, and (2) the object must move. This means that no work is being done on the pail in figure 6.2(a) or on the wall in figure 6.2(b). In the first case, the force has no component in the direction of motion; θ is therefore 90°, and $\cos \theta = 0$. In the second case, no motion of the wall occurs.

Very often *6.1* is written in terms of the *scalar product* of two vectors. This product is defined in the following way. Suppose two vectors **A** and **B** are oriented such that the angle between them is θ. We then define

$$\text{Scalar product of } \mathbf{A} \text{ and } \mathbf{B} \equiv \mathbf{A} \cdot \mathbf{B} \equiv AB \cos \theta$$

Notice that the result is a scalar, hence the name *scalar product*. In terms of this shorthand notation, the work ΔW done by a force **F** in moving an object through a displacement $\Delta\mathbf{s}$ is given by

$$\Delta W = \mathbf{F} \cdot \Delta\mathbf{s} = (F \cos \theta)(\Delta s) \qquad 6.1c$$

By virtue of the fact that $(F \cos \theta)(\Delta s) = (\Delta s)(F \cos \theta)$ we see at once that the order in which a scalar product is written is unimportant. Therefore

$$\mathbf{F} \cdot \Delta\mathbf{s} = \Delta\mathbf{s} \cdot \mathbf{F}$$

and the scalar product is *commutative*. It should be remembered that the scalar product defined by $\mathbf{A} \cdot \mathbf{B} = AB \cos \theta$ is simply a shorthand way of writing the quantity $AB \cos \theta$.

6.2 THE DEFINITION OF POWER

Power is defined to be the rate at which work is being done. This is to say,

$$\text{Power} = \frac{\text{work done}}{\text{time taken}} \qquad 6.2a$$

If we recall that the work done on an object in time Δt as the applied force **F** moves the object a distance $\Delta\mathbf{s}$ is given by $\mathbf{F} \cdot \Delta\mathbf{s}$, then the power can be written

$$P = \frac{\mathbf{F} \cdot \Delta\mathbf{s}}{\Delta t} \qquad 6.2b$$

However, since $\Delta\mathbf{s}/\Delta t = \mathbf{v}$, the velocity of the object, the power can be written, alternatively,

$$P = \mathbf{F} \cdot \mathbf{v} \qquad 6.2c$$

All these relations are equivalent.

The units of power are, for our three usual unit systems:

System	A compound unit	Name
SI	joule/second (J/s)	watt (W)
cgs	erg/second (erg/s)	—
British engineering	foot · pound/second (ft · lb/s)	—

The SI unit of power is the watt

A common unit, called the *horsepower* (hp), is used, even though it belongs to none of these systems. One horsepower is defined to be 550 ft · lb/s, which makes it equivalent to about 746 W.

As we shall see later in our study of electricity, work is sometimes expressed in terms of a unit called the *kilowatt · hour* (kWh). This unit is based upon *6.2a.* If power is measured in kilowatts (1 kW = 1000 W) and time measured in hours, the resultant work unit is the kilowatt · hour. Of course it does not belong to our three basic unit systems, and should therefore be used with caution.

Illustration 6.1

What minimum horsepower motor is needed to lift an 80-kg man at a constant speed of 0.20 m/s?

Reasoning The man weighs (80)(9.8) N, and so a force this large is needed to lift him at constant speed. Using *6.2c,* we find

$$P = \mathbf{F} \cdot \mathbf{v} = [(80)(9.8)\,\text{N}](0.20\,\text{m/s})$$
$$= 157\,\text{W}$$
$$= 157/746 = 0.21\,\text{hp}$$

Illustration 6.2

What power motor is needed to accelerate the 200-kg cart shown in figure 6.3 up the incline at 0.80 m/s² if friction forces are negligible?

Reasoning Referring to the figure, since the cart is accelerating up the incline, we see that *T* must be larger than 0.6*W*. Applying Newton's second law, we have

$$T - 0.6W = ma$$

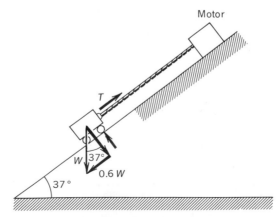

FIGURE 6.3 Why must the motor produce more power if the cart is accelerating than if it is not when the speed of the cart has a certain value?

Noticing that $W = 1960\,\text{N}$, $m = 200\,\text{kg}$, and $a = 0.80\,\text{m/s}^2$, we find $T = 1340\,\text{N}$. This is the force with which the motor must pull on the object, and 6.2c can now be used to find the power.

$$P = 1340v \text{ watts}$$

where v is the speed of the cart at the instant being considered. We see that the power required increases as the speed of the cart increases, as one would expect.

6.3 WORK DONE BY A VARIABLE FORCE

Let us consider a system such as that shown in figure 6.4, where an object is being pushed along a nonuniform surface. Suppose the x-component force needed to push it from point $x = a$ to $x = b$ varies as shown in part (b) of the figure. The variation in force could easily arise if dirt on the surface made the object harder to push at some points than at others.

When the object moves through the very small distance Δx_5 under the action of the approximately constant x-directed force F_{x5}, the small amount of work done is

$$\Delta W = F_{x5}\,\Delta x_5$$

We analyze the motion of the body from a to b by considering its motion through N successive small movements such as Δx_5. Assume these movements to be much smaller and more numerous than shown. The total work done will therefore be

$$W = F_{x1}\,\Delta x_1 + F_{x2}\,\Delta x_2 + \cdots + F_{xN}\,\Delta x_N$$

It is of interest (and importance) to notice at this point that $F_{x5}\,\Delta x_5$ represents the area of the small heavily shaded rectangle in figure 6.4(b). Each term of the expression for the work is a similar area.* *The work* itself, being the sum of all these areas, *is the area under the F_x versus x curve.* This will prove a convenient fact to remember when we compute the work needed to stretch a spring, in a later section.

If we call the typical term in the work equation $F_{xi}\,\Delta x_i$, we can rewrite it

$$W = \sum_{i=1}^{N} F_{xi}\,\Delta x_i$$

Further, if the Δx_i are allowed to approach zero in size (thereby N becoming very large), we obtain the limit as $\Delta x \to 0$ for the sum. This is by definition what we call an integral. It is represented by†

$$W \equiv \int_{x=a}^{x=b} F_x\,dx$$

where Δx_i is replaced by dx, and the limits $x = a$ and $x = b$ are used to indicate that the sum is to extend over the interval $a \le x \le b$.

In the present case, the equation for F as a function of x is not simple, and so the integral is not easily evaluated. However, we know that it repre-

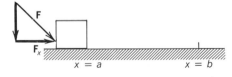

(a)

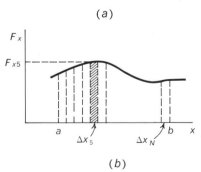

(b)

FIGURE 6.4 How is the work done by the force F_x obtained from the graph?

*This assumes, of course, that the rectangles are much narrower than shown.

†An introduction to the use of integrals is given in Appendix 7.

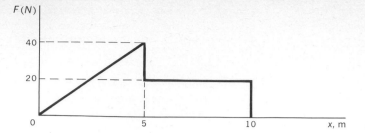

FIGURE 6.5 How much work is done by the force?

scnts the area under the curve in the range $a \leq x \leq b$, and so the sum, or integral, could, if necessary, be carried out graphically. In general, the work can be written in the following ways:

$$W = \sum F_{si} \, \Delta s_i = \sum \mathbf{F}_i \cdot \Delta \mathbf{s}_i \qquad 6.3$$

$$W = \int \mathbf{F} \cdot d\mathbf{s} \qquad 6.4$$

Work done is the area under the F_s versus s curve

for the case of variable force. To write this we have made use of the fact that $\mathbf{F} \cdot d\mathbf{s} = F_s \, ds$, where F_s is the force component in the direction of $d\mathbf{s}$.

Illustration 6.3

Find the work done by the force shown graphically in figure 6.5 as it acts through the distance $0 \leq x \leq 10 \, \text{m}$.

Reasoning The work is equal to the area under the curve. We can find the area by adding the areas of the triangle $0 \leq x \leq 5$ and the rectangle $5 \leq x \leq 10$. The triangle area is $(\frac{1}{2})(40)(5) \, \text{N} \cdot \text{m}$, or $100 \, \text{J}$. The rectangular area is also $100 \, \text{J}$. Therefore the total work done is $200 \, \text{J}$.

Illustration 6.4

A force which pushes an object in the x direction is given by

$$F_x = 3x + 2 \, \text{newtons}$$

Find the work done by this force as it pushes the object from $x = 0$ to $x = 4 \, \text{m}$.

Reasoning Since the force and displacement are in the same direction, *6.4* becomes in this case*

$$W = \int_0^4 (3x + 2) \, dx$$

where dx replaces ds. The limits on the integral are determined from the starting and end points given in the problem. We can write this integral as two separate integrals.

$$W = \int_0^4 3x \, dx + \int_0^4 2 \, dx$$

After using the facts that (see Appendix 7)

$$\int cx \, dx = \tfrac{1}{2}cx^2 \qquad \text{and} \qquad \int c \, dx = cx$$

*This problem can be solved without calculus. Use the method of the previous illustration.

86 WORK AND ENERGY

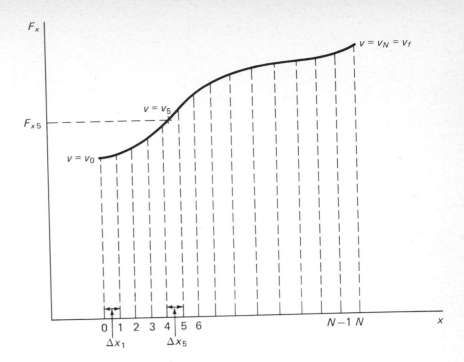

FIGURE 6.6 During the displacement Δx_5, the work done is $F_{x5}\,\Delta x_5$.

we have that

$$W = \tfrac{1}{2} \cdot 3x^2 \Big|_0^4 + 2x \Big|_0^4 = (24 - 0) + (8 - 0)$$

$$= 32\,\mathrm{J}$$

6.4 KINETIC ENERGY

Suppose at a certain instant an object has a speed v_0. If now a force **F** is applied to the object in the direction of motion, the force will accelerate the object to a still higher speed. We wish to find a relation between the work done by this force and the change in speed of the object. To do this most conveniently let us restrict ourselves to motion along a straight line with F_x being the force component along that line. Perhaps the graph of F_x versus distance along the line x is as shown in figure 6.6.

Consider the work done during the small displacement Δx_5 by the force F_{x5}. We have that

$$\Delta W_5 = F_{x5}\,\Delta x_5$$

But if Δx_5 is very small (much smaller than shown), then F_x will be essentially constant during that portion of the displacement. Because F is constant, the acceleration of the object during Δx_5 will also be constant. We can then make use of the uniform acceleration formula $v^2 - v_0^2 = 2as$. In the present case $s = \Delta x_5$ with $v_0 = v_4$ and $v = v_5$. Therefore

$$2a_5\,\Delta x_5 = v_5^2 - v_4^2$$

But according to $F = ma$, $a_5 = F_{x5}/m$ and so

$$\Delta x_5 = \tfrac{1}{2}\frac{m}{F_{x5}}(v_5{}^2 - v_4{}^2)$$

Substitution then gives

$$\Delta W_5 = \tfrac{1}{2}m(v_5{}^2 - v_4{}^2)$$

Similar expressions would exist for ΔW_1, ΔW_2, ΔW_3, etc., the work done in each of the displacements of figure 6.6.

The total work done by the force is the sum of all these incremental works; hence

$$\begin{aligned}
W &= \Delta W_1 + \Delta W_2 + \Delta W_3 + \cdots + \Delta W_N \\
&= \tfrac{1}{2}m(v_1{}^2 - v_0{}^2) + \tfrac{1}{2}m(v_2{}^2 - v_1{}^2) + \cdots + \tfrac{1}{2}m(v_N{}^2 - v_{N-1}^2) \\
&= \tfrac{1}{2}m[(v_1{}^2 - v_0{}^2) + (v_2{}^2 - v_1{}^2) + (v_3{}^2 - v_2{}^2) + \cdots + (v_N{}^2 - v_{N-1}^2)] \\
&= \tfrac{1}{2}m[v_N{}^2 - v_0{}^2] = \tfrac{1}{2}mv_f{}^2 - \tfrac{1}{2}mv_0{}^2
\end{aligned}$$

And so we conclude that

$$\text{Work done by external force} = \tfrac{1}{2}mv_f{}^2 - \tfrac{1}{2}mv_0{}^2 \qquad\qquad 6.5$$

This extremely important equation tells us how the work done by the resultant force on an object changes the motion of the object. It is often called the *work-energy theorem*. The motion is characterized by the quantity $\tfrac{1}{2}mv^2$, and this quantity is called the translational *kinetic energy K* of the object. Its units are the same as those of work, namely joules, ergs, or foot · pounds. We can summarize the work-energy theorem in the following way: *the work done by the resultant force on an object equals the change in kinetic energy of the object.*[*]

Translational kinetic energy is $\tfrac{1}{2}mv^2$

A special case worth noting is that of a stopping force acting on the object. In that case $\mathbf{F}$ and $\Delta\mathbf{s}$ are in opposite directions so that

$$\Delta W = (F\cos\theta)\,\Delta s = (F\cos 180°)\,\Delta s = -F\,\Delta s$$

We see that a stopping force does negative work on an object. In keeping with the fact that a stopping force slows an object, the change in K caused by it is negative; the kinetic energy is decreased by a stopping force. Moreover, the object being stopped exerts an equal-magnitude reaction force on the stopping agent. This reaction force acts through the same distance as the stopping force does, and so the work it does on the stopping agent is equal to its loss in kinetic energy. As we see, work can be converted to kinetic energy and kinetic energy can be converted to work. A given number of joules of kinetic energy is able to do the same number of joules of work. Or, conversely, a given amount of work can produce the same amount of kinetic energy.

Illustration 6.5

An α particle (i.e., a helium nucleus) enters a chamber filled with gas. Its initial speed is 0.80×10^7 m/s, and after traveling 5.0 cm, its speed is 0.30×10^7 m/s. How large is the average force exerted by the gas on the α particle? ($m_\alpha = 6.7 \times 10^{-27}$ kg)

Reasoning The decrease in kinetic energy of the α particle was caused by the work done on it as a result of the average stopping force F of the gas through which it passed. From the work-energy theorem, we write

---

[*] The name kinetic energy actually comes from a consideration of the work done by a moving object. *Energy* is the ability to do work, while *kinetic* comes from the Greek word meaning "to move." The use of these words in the present instance tells us that kinetic energy is the ability to do work because of the motion of the object.

Change in K = work done

$$\tfrac{1}{2}mv^2 - \tfrac{1}{2}mv_0^2 = Fs$$

Placing in the values, using SI throughout,

$$(\tfrac{1}{2})(6.7 \times 10^{-27})(9 \times 10^{12}) - (\tfrac{1}{2})(6.7 \times 10^{-27})(64 \times 10^{12}) = F(0.050)$$

from which

$$F = -3.7 \times 10^{-12} \text{ N}$$

The force is negative since it is opposing the motion.

6.5 CONSERVATIVE FORCE FIELDS

We know that each body on the earth is attracted by the earth. This gives rise to the weight of the object. It is customary to say that a *gravitational force field* exists in the neighborhood of the earth, and this terminology simply means that objects experience gravitational forces in this region. The details of the gravitational force field are explored in more detail in Chap.9; here we examine only the general features.

Let us refer to figure 6.7, in which three different paths are shown for lifting a mass m from point A to point B. Consider first path 1. The force needed to support the object is equal to its weight, just mg. This force is directed upward on the object. No work is done by the applied force in carrying the object from A to D because the motion is perpendicular to the force direction, that is, $\cos\theta = 0$. When the object is lifted with constant velocity through the distance h from D to B, however, the applied force of magnitude mg is in the direction of motion, and so the work done by it is mgh. Hence, the work done in carrying the mass from A to B by path 1 is

$$W_{ADB} = 0 + mgh$$

If we consider the path ACB, we see that no work is done by the applied force in the movement from C to B. However, the work done by the applied force in going from A to C is mgh. So the total work done by the applied force is

$$W_{ACB} = mgh + 0$$

which is identical with that found using the previous path.

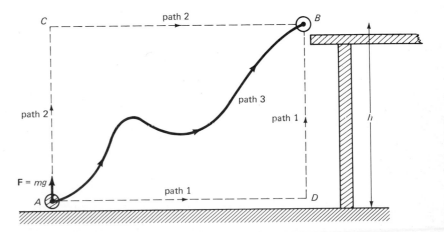

FIGURE 6.7 The work done in lifting an object from A to B in the earth's gravitational field is independent of the path followed. What kind of force field do we call this?

Before considering the work done by the applied force (against the force of gravity) for the third path, let us consider the reverse motion along path 2. How much work is done by the force supporting the object as the object is moved from B to A by path BCA? In going from B to C, no work is done since

$$\mathbf{F} \cdot \mathbf{s} = Fs \cos \theta = Fs \cos 90° = 0$$

However, in going from C to A we notice that the force supporting the object is directed upward, while the direction of motion is downward, thereby making $\theta = 180°$. As a result, the work is

$$Fs \cos \theta = mgh \cos 180° = -mgh$$

and so $$W_{BCA} = 0 - mgh = -mgh$$

Two interesting facets of this result should be noticed. First, the result is the negative of the result obtained using the reverse path. We shall see that this is always the situation in the case of a force such as this. And second, the work is negative. These two results tell us an easily understood fact. When we lift an object against the pull of gravity, we do work against that force. However, when we allow the object to be pulled with constant speed in the direction of the gravitational force, the gravitational force does work on us as we restrain the object from accelerating. Put more simply, when we move an object a distance h against the pull of gravity, we must do work; but when we restrain the object to move with constant speed the same distance h in the direction of the gravitational pull, we are given back the same amount of work. Hence, if we lift an object a distance h, we do mgh amount of work. If we now move the object back down to its original level, an amount of work mgh is given back to us. Our work *output* is therefore $(mgh) + (-mgh) = 0$. Let us now use this idea to compute the work done in the third example shown in figure 6.7.

In order to understand the method for computing the work done by the applied force when following the third path, let us in our mind split the path into extremely small x and y steps, as illustrated in figure 6.8. We shall later let the size of these steps approach zero, and so there will be no detectable difference between the smooth and jagged paths. The advantage of this approach is that we can readily compute the work done in going from A to B along the jagged path. We know at once that no work is done during each of the x (or horizontal) steps since the applied force is perpendicular to the motion.

The work done by the supporting force on a step in the y direction is simply $mg \, \Delta y$, where Δy is to be given its proper sign, $+$ if the displacement is upward and $-$ if downward. If we sum the work done on all the y steps, we find it to be

$$y \text{ step } W = mg(\Delta y_1 + \Delta y_2 + \cdots + \Delta y_N)$$

But the sum of all the Δy's must add up to h; all the wiggles in the path cancel out since the Δy's are negative during the downward part of the wiggle and positive otherwise. As a result, the total work done in going from A to B via path 3 is merely

$$W_1 = mgh$$

Since path 3 is a general path, we conclude that work done in moving an object from one point to another in the earth's gravitational force field is *independent of the path followed*. Any force field for which this is true, the work done being independent of the path, is called a *conservative force field*.

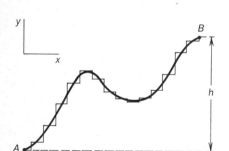

FIGURE 6.8 A smooth path may be replaced by a series of infinitesimal x and y displacements. Work is done only during the y displacements.

The gravitational field is a conservative field

The force associated with this field is called a *conservative force*. We have seen that the gravitational force is a conservative force. When you study electricity in this text, you will find that the electrostatic force between charges is another very important conservative force.

Of course, not all forces are conservative. Friction forces are typical of the so-called *nonconservative forces*. For example, the work done in sliding this book along the table (against the friction force) is obviously dependent on the path used. Move the book about and return it to the original position. You must always push in the direction of motion so that $\mathbf{F} \cdot \Delta \mathbf{s}$ is always positive and no cancellation occurs. One way to move the book from its original position and back again is to not move it at all. Clearly, no work is done. (If you object to this, move the book 0.0001 mm away and then back. Within experimental error the result will be the same.) However, if you move the book 50 cm away and back, considerable work is done. The work done against the friction force, definitely, is path-dependent. It is not a conservative force. We call the friction force and others of this type nonconservative forces.

Friction is a nonconservative force

6.6 GRAVITATIONAL POTENTIAL ENERGY U_g

An object which is capable of falling in the earth's gravitational field is capable of doing work. It therefore possesses energy, i.e., the ability to do work. For example, a kilogram mass held above a table and released is capable of lifting another mass as it falls. In so doing, it does work. We say that the object possesses *gravitational potential energy*. The energy is potentially available in the sense that the object must be released before the work can be done. Gravitational potential energy is the ability to do work because of the position of the object in the gravitational force field. Let us now assign a value to this energy.

Definition of U_g, gravitational potential energy

We saw in the preceding section that the work done in lifting an object of mass m a distance h is mgh, independent of the path used. Furthermore, we saw that the object was capable of doing a quantity of work mgh as it moved back down through this same distance h. As a result we conclude that the work which an object at height h is capable of doing as it falls through this distance is mgh. We define this to be the gravitational potential energy of the object,* U_g. Therefore

$U_g = mgh$

$$U_g = mgh \qquad\qquad 6.6$$

The units of U_g are clearly the same as the units of work.

From the way we have defined the potential energy, it is seen that work and potential energy are interconvertible. By this we mean that when a certain number of joules of work is done on an object against the gravitational force, the potential energy of the object increases by this same number of joules. Moreover, if the object is now allowed to return to its original

---

*For motions through distances so large that g cannot be considered constant, i.e., too far above the earth, this definition must be replaced by

$$U_g = mg_1 \, \Delta h_1 + mg_2 \, \Delta h_2 + \cdots + mg_N \, \Delta h_N$$

The height increments are taken very small, so that g is essentially constant in each, and the sum of the increments must add to give h. Replacing the sum by an integral in the usual way, this becomes

$$U_g = m \int g \, dh$$

where g would be a function of h.

height, it is capable of doing the same number of joules of work. We shall often make use of this important property of a conservative force and the potential energy associated with it.

It should also be pointed out that the potential energy of an object is not an absolute quantity. Different people may assign values to it which differ by an additive constant. For example, consider the object shown in figure 6.9. If one considers the height of the object above the table, its potential energy could be written mgh_1. But with respect to the floor, its potential energy is mgh_2. We shall see later why we do not bother to exclude either of these values, and simply consider both correct. The reason is associated with the fact that we are always interested in differences in heights (or potential energies), and so additive constants such as mgh_3 in

$$mgh_1 = mgh_2 - mgh_3$$

will always cancel out.

Furthermore, it is possible to obtain negative potential energies. For example, suppose one chooses to measure distances from the top of the table in figure 6.9. If the object is at a distance y above the table, its potential energy is mgy. As it is lowered to the tabletop, its U_g decreases to zero. When it is lowered still farther, its y coordinate becomes negative, and so U_g becomes negative. This simply means that the object has less potential energy below the table than it has at the tabletop, the position arbitrarily taken as the zero level. In order to restore the object to the zero level, it must be lifted. As indicated earlier, this will cause us no difficulty, since we are always interested only in differences in U_g, that is, in the work the body can do in moving from one level to another.

The zero level for U_g is arbitrary

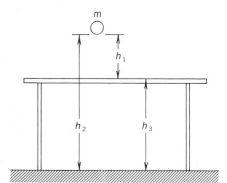

FIGURE 6.9 U_g can be considered to be either mgh_1 or mgh_2. Notice that they differ by the additive constant mgh_3.

6.7 CONSERVATION OF $U_g + K$

One of the properties of motion of an object in a conservative force field is that the sum of the kinetic and potential energies of the object is a constant. This fact follows directly from the definitions of potential energy and the work-energy theorem. The potential energy of an object at a point is equal to the work done in lifting it to that point from some reference point. When an object loses potential energy by moving from one point to another of lower potential energy, the energy loss is equal to the work done by the conservative force (gravitational force in this case) on the object as it moves from the one point to another.

In the absence of other forces, the work done on the object by the gravitational force must appear as kinetic energy of the object. As a result, the decrease in U_g is equal to the increase in K, and so the sum of U_g and K is a constant. A similar reasoning leads to the fact that $U_g + K$ is a constant when U_g is increased. Notice that this conservation law for the quantity $U_g + K$ is true only if (1) the force field is conservative, and (2) no other forces do work on the object. Other forces can restrain the motion of the object, but they must have no component in the direction of motion if condition 2 and the conservation law are to be true.

Work, U_g and K are interconvertible

Illustration 6.6

A ball drops from a height of 30 m above the ground. Find its speed just before it strikes the ground. Ignore friction forces.

Reasoning We could, of course, apply the usual motion equations to find the desired result. However, it is a much quicker way to use the energy conservation concept. At the start, the ball has only potential energy $U_g = mgh$, while at the end the ball has only kinetic energy $\frac{1}{2}mv^2$. Using the conservation of $U_g + K$, we find

$$(U_g + K)_{\text{start}} = (U_g + K)_{\text{at end}}$$
$$mgh + 0 = 0 + \tfrac{1}{2}mv^2$$

which gives
$$v = \sqrt{2gh}$$

or in terms of our values.

$$v \approx 24.5 \text{ m/s}$$

Illustration 6.7

A bead slides on the frictionless wire shown in figure 6.10. If the speed of the bead is 2.0 m/s when it is at point A, how fast will the bead be going at point B? At C?

Reasoning Since there is no friction, the force exerted on the bead by the wire is always perpendicular to the bead motion, and so it does no work. We are justified, therefore, in writing

$$(U_g + K)_{\text{at } A} = (U_g + K)_{\text{at } B}$$

If we take the lowest point, B, to have zero gravitational energy, we have, after substituting in this relation,

$$mgh_A + \tfrac{1}{2}mv_A{}^2 = 0 + \tfrac{1}{2}mv_B{}^2$$

Notice that the mass of the bead is of no importance since it cancels out. Solving for v_B gives

$$v_B = \sqrt{2gh_A + v_A{}^2}$$

Placing in the SI numbers given, we find

$$v_B = 3.71 \text{ m/s}$$

It is also of interest to square the equation for v_B and rearrange it to read

$$v_B{}^2 - v_A{}^2 = 2gh_A$$

This is merely the motion equation $v^2 - v_0{}^2 = 2ax$. Therefore the speed of the bead at B is the same as if it had been thrown straight down from a point 0.50 m above B, with the initial speed 2 m/s. We might have expected this

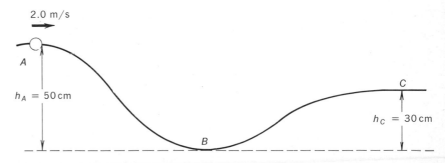

FIGURE 6.10 What can be said about the sum $U_g + K$ for the bead at A, B, and C?

since the change in speed is entirely due to the work done on the bead by the earth's gravitational force. The wire serves only to direct the motion of the bead.

To find the speed at C we proceed in the same way and write

$$(U_g + K)_A = (U_g + K)_C$$

or $$mgh_A + \tfrac{1}{2}mv_A{}^2 = mgh_C + \tfrac{1}{2}mv_C{}^2$$

Placing in the given numbers, we find $v_C \approx 2.8 \, \text{m/s}$.

6.8 EXTERNAL FORCE EFFECTS

The powerful method discussed in the preceding section can be extended to situations in which external forces are present to retard or accelerate the motion. We need only recognize that the external forces will do work on the object, and this work will appear as an increase or decrease of the total energy of the object. For example, if a force acts in such a way as to accelerate the object, the total energy of the object will increase by an amount equal to the work done by the force. Or if the force acts so as to stop the object, the object will lose an energy equal to the work done by the stopping force.

The basic equation for each situation can perhaps best be written as follows:

An energy-conservation relation

$$(U_g + K)_{\text{start}} + \text{energy input} = (U_g + K)_{\text{end}} + \text{energy output}$$

As before, $U_g + K$ represents the sum of potential and kinetic energies. When an external force other than gravity* does positive work on the object (i.e., accelerates it), the energy-input term must appear and is, of course, equal to this work. If the object does work against friction or other nonconservative forces, an energy loss occurs, and part of the original energy of the system appears as an energy output equal to this work.

Basically, the above equation represents an energy balance sheet. We have a certain quantity of energy supplied by the external forces aiding the motion. This energy must be accounted for at the end, and our equation says that part is still present, while some has been assigned as energy output. A frequent type of energy output is work done against friction forces. This friction-work energy loss eventually appears as heat energy.

Illustration 6.8

A 200-g ball falls from rest through the air. If its speed is 15 m/s after having fallen through a distance of 20 m, how much energy was lost as friction work against the air?

Reasoning At the start the ball possessed potential energy only. No non-gravitational force acted so as to accelerate the object, and so the energy input is zero. At the end, the object has kinetic energy only. In addition, friction work has been done against the air, and this appears as heat energy output. The basic equation is

$$(U_g + K)_{\text{start}} + \text{energy input} = (U_g + K)_{\text{end}} + \text{energy output}$$

Placing in the values,

*The work done by (or against) the gravitational force is already included by the U_g term.

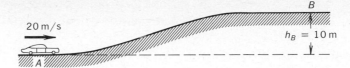

FIGURE 6.11 If the car stops when it gets to *B*, what has happened to its original energy?

$$mgh + 0 + 0 = 0 + \tfrac{1}{2}mv^2 + \text{friction work}$$

Solving for the friction work gives

$$\text{Friction work} = m(gh - \tfrac{1}{2}v^2) \approx 17\,\text{J}$$

Illustration 6.9

A 2000-kg car starts to coast up a hill illustrated in figure 6.11. Its speed initially is 20 m/s at point *A*. If its speed is 5.0 m/s at point *B*, how large an average friction force retarded its motion? The distance along the road from *A* to *B* is 40 m.

Reasoning As before, we have

$$(U_g + K)_{\text{start}} + \text{energy input} = (U_g + K)_{\text{end}} + \text{energy output}$$

or

$$0 + \tfrac{1}{2}mv_A{}^2 + 0 = mgh_B + \tfrac{1}{2}mv_B{}^2 + fs$$

where *f* is the friction force, and *s* is the distance along the road from *A* to *B*. Placing in the numerical quantities, we find $f \approx 4500\,\text{N}$.

6.9 ELASTIC ENERGY

Thus far in this chapter we have dealt primarily with two forms of energy, kinetic and gravitational potential energy. Another form of energy often encountered in mechanical systems is the energy stored in springs. For example, a compressed spring on a pop gun can provide energy to a pellet and shoot it from the gun. Even the simple slingshot uses the elastic energy stored in a rubber band to shoot its projectile. There are springlike devices of widely varying types, and each is capable of storing elastic (or spring) potential energy. We shall now examine the behavior of springlike systems.

All solids (and even liquids and gases) possess elasticity. It is for this reason that solid objects rebound in a collision. When two balls collide, it is much as though there were a spring between the balls. During the collision the spring is compressed. Eventually, the spring throws the balls apart again. In a case such as this, the spring involved is a very complex one. Its motion consists in the distortions of many varied molecular bonds within the solid. In spite of this, for small deformations the solid will obey the law of springs, *Hooke's law*. Let us now see how a hookean spring behaves.

Consider the spring shown in figure 6.12. Although the following discussion is made in terms of a common mechanical spring, it has much wider validity than this. The only restriction which is made is that the spring system must obey a linear force-extension law. All common elastic systems obey such a law provided the extension is not too large. (This fact was first stated precisely by Robert Hooke, a contemporary of Newton, and is known as Hooke's law.)

In order to pull the end of the spring in figure 6.12(*a*) a distance *x* from

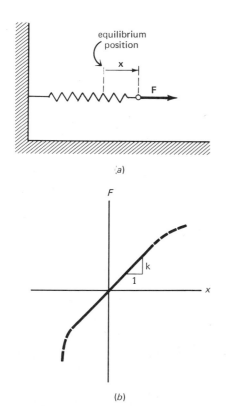

FIGURE 6.12 When Hooke's law is obeyed, the restoring force is a linear function of the strain.

its equilibrium position, a stretching force $\mathbf{F}$ is required. Experiment shows that $\mathbf{F}$ is proportional to the distance the spring is stretched, x. As an equation,

$$\mathbf{F} = k\mathbf{x} \qquad\qquad 6.7$$

For a Hooke's law spring, F = kx

This is a mathematical statement of *Hooke's law*. A plot of this relation is given as the solid line in figure 6.12(*b*). As seen from *6.7*, the slope of this line is k, where k is called the *spring constant* of the spring. Its units are newtons per meter, and as the units imply, it states the force required to produce unit elongation of the spring. Stiff springs have large values for k.

All springlike systems depart from Hooke's law if the spring is stretched too far. The bond holding two hydrogen atoms together in a hydrogen molecule stretches linearly with force up to a certain point. As higher elongations are imposed, the restoring force departs from the linear relation, and the springlike bond appears to have smaller value of k as the elongation increases. When the molecule is compressed by pushing the two atoms together, the F versus x relation is still linear even at negative x values (compressions), as shown in figure 6.12(*b*). However, at large compressions the force rises more rapidly than the linear relation would predict, and the spring appears stiffer. This behavior is typical of many (but not all) springlike systems. The departures from linearity are shown by the broken curve in figure 6.12(*b*).

When a spring is stretched or compressed, work is done on the spring. The work done in stretching a spring from its unstretched position, $x = 0$, through a distance x is obtained from the definition of work. We have that

$$W = \sum \mathbf{F}_i \cdot \Delta \mathbf{s}_i$$

According to Hooke's law, the stretching agent must pull in the direction of the displacement or elongation $\mathbf{s}$ with a force ks if the spring is to be held at that elongation. Therefore the stretching force is in the same direction as $\Delta \mathbf{s}$ and has a magnitude ks. This force is shown in figure 6.13 as a function of $\mathbf{s}$.

The work equation therefore becomes, since $\mathbf{F}_i$ and $\Delta \mathbf{s}_i$ are in the same direction,

$$W = \sum F_i \, \Delta s_i$$

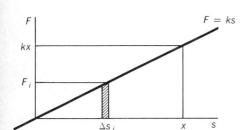

FIGURE 6.13 The force needed to stretch the spring is opposite in sign to the spring's restoring force.

where s is to range from $0 \le s \le x$. Notice that each term in the sum is the area of a rectangle such as the one shown shaded in figure 6.13. Adding all such rectangles from $s = 0$ to $s = x$ gives the area of the triangle under the curve. This area is just $\frac{1}{2}(x)(kx)$, where kx is the value of F at $s = x$. Therefore the work done in stretching the spring a distance x is*

$$W = \tfrac{1}{2}kx^2 \qquad\qquad 6.8$$

Work done in stretching a spring is $\frac{1}{2}kx^2$

Since the spring exerts a restoring force equal and opposite to the stretching force, the spring is capable of doing this same amount of work when the spring contracts. Therefore a stretched (or compressed) spring has the ability to do work, and possesses potential energy. We represent the potential energy stored in a spring by the symbol U_s, and it is equal to W in *6.8*. Therefore

Energy stored in a stretched spring, $U_s = \frac{1}{2}kx^2$

---

*We could arrive at *6.8* by simply replacing the sum by an integral. In this way

$$W = \sum F_i \, \Delta s_i = \int_0^x ks \, ds = \tfrac{1}{2}kx^2$$

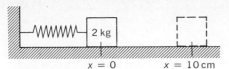

FIGURE 6.14 If the mass is released from the 10-cm point, how fast will it be moving at $x = 0$?

$$U_s = \tfrac{1}{2}kx^2 \qquad\qquad 6.9$$

where x is the distance which the spring has been stretched or compressed. This energy, like U_g, the gravitational potential energy, can be used to do an equivalent quantity of work and to provide the same quantity of kinetic energy.

Illustration 6.10

A certain spring stretches 25 cm when a 500-g mass is hung from its end. It is now used as shown in figure 6.14, with a 2-kg mass at one end placed on a frictionless surface. Suppose the mass is pulled to the right until the spring is stretched 10 cm and then released. How fast will the mass be moving as it moves through the equilibrium (unstretched) position?

Reasoning We are told that the spring stretches 0.25 m under a force of $(0.5)(9.8)$ N. Since $F = kx$, the spring constant k is 19.6 N/m. When the mass is first released at the 10-cm mark, its speed will be zero, and so it possesses no kinetic energy. However, the stretched spring has a potential energy $\tfrac{1}{2}kx^2$. All this energy will be lost by the time the mass has reached the equilibrium point, $x = 0$. It will all have been transformed to kinetic energy of the mass. Our basic equation will therefore read

$$(U_s \text{ at } 0.10\,\text{m}) = (K \text{ at } x = 0)$$

or

$$(\tfrac{1}{2})(19.6\,\text{N/m})(0.10\,\text{m})^2 = (\tfrac{1}{2})(2.0\,\text{kg})v^2$$

from which

$$v \approx 0.31\,\text{m/s}$$

Illustration 6.11

Suppose the mass of the preceding example is actually found to have a speed of 0.20 m/s when it reaches $x = 0$. How large an average friction force retarded its motion?

Reasoning The energy stored in the spring when $x = 10$ cm now is lost to two mechanisms. Part goes to kinetic energy of the mass, and the rest is lost doing friction work; i.e., it is changed to heat energy. The basic equation is therefore

$$U_s \text{ at } 0.10\,\text{m} = K + \text{friction work}$$

or

$$(\tfrac{1}{2})(19.6\,\text{N/m})(0.10\,\text{m})^2 = (\tfrac{1}{2})(2.0\,\text{kg})(0.20)^2 + f(0.10\,\text{m})$$

where f is the average friction force. Solving, we find $f = 0.58$ N.

Illustration 6.12

The 2.0-kg block shown in figure 6.15 experiences an 8-N friction force. Its speed at point A on the way down the incline is 3.0 m/s. On arriving at B, it compresses the spring 20 cm, stops, and then rebounds up the incline. Find k for the spring and the height to which the mass rebounds.

Reasoning At point A the block has both potential and kinetic energy. This is the total energy of the system, since the spring is unstretched. Let us take the lowest position reached by the block as the zero level for U_g. In that way we ensure that $U_g = 0$ when the block stops at the bottom. At that time, all the original energy of the block will have been changed to potential energy of the spring plus friction work. Our basic equation is

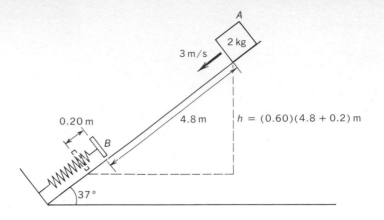

FIGURE 6.15 Find k for the spring if it compresses 20 cm and the friction force is 8 N.

In the figure: A, 2 kg, 3 m/s, 0.20 m, B, 4.8 m, $h = (0.60)(4.8 + 0.2)$ m, 37°

$$(K + U_g) \text{ at top} = (U_s + \text{friction work}) \text{ at end}$$

This becomes

$$(\tfrac{1}{2})(2 \text{ kg})(3 \text{ m/s})^2 + (2 \text{ kg})(9.8 \text{ m/s}^2)(3.0 \text{ m}) = \tfrac{1}{2}k(0.20 \text{ m})^2 + (8 \text{ N})(5 \text{ m})$$

Solving for k, we find it to be 1390 N/m.

To find how high the mass rebounds, we equate the energy stored in the spring at the bottom to the friction work done plus the potential energy of the block when it stops at its highest point. Therefore

$$U_s \text{ at bottom} = (U_g + \text{friction work}) \text{ at end}$$

$$(\tfrac{1}{2})(1390 \text{ N/m})(0.20 \text{ m})^2 = (2 \text{ kg})(9.8 \text{ m/s}^2)(0.60s) + (8 \text{ N})s$$

where s is the distance up along the incline which the block rises. Solving for s, we find it to be 1.4 m. The final value for h is therefore 0.6×1.4, or 0.84 m.

Chapter 6
Questions
and Guesstimates

1. The moon and most human-made earth satellites circle the earth in essentially circular paths with center at the earth's center. How much work does the earth's gravitational force do on them?

2. Reasoning from the standpoint of kinetic energy, why is a loaded truck likely to be much more damaging than a Volkswagen in a collision with a massive stationary object? Assume equal initial speeds.

3. A ball hangs at the end of a thread and the system swings as a pendulum. Describe what happens to the kinetic energy of the ball as the pendulum swings back and forth. Repeat for U_g. How are the two related? What happens to the energy of the system as the pendulum loses its energy? Does the pull of the thread on the ball do any work?

4. A person holds a bag of groceries while standing still talking to a friend. A car sits stationary with its motor running. From the standpoint of work and energy, how are these two situations similar?

5. A conscientious hobo in a boxcar traveling from Chicago to Peoria pushes on the front wall of the car all the way. Having at one time been a physics student, he thinks his pushing did a great deal of work since both F_s and s were large. Where was the flaw in his reasoning?

6. Since the earth is moving with respect to the sun, everything on it has kinetic energy, at least in the opinion of an observer who thinks the sun

is at rest. Why did we not need to consider this when working the examples in this chapter?

7. Is kinetic energy a vector or a scalar quantity?

8. A ball of mass m is held at a height h_1 above a table. The table top is a height h_2 above the floor. One person says that the ball has $U_g = mgh_1$, but another person says that its U_g is $mg(h_1 + h_2)$. Who is correct?

9. As a rocket reenters the atmosphere, its nose cone becomes very hot. Where did this heat energy come from?

10. If a piece of chalk falls off a table, its speed just before it hits the floor can be found by equating the original U_g to its total energy just before striking the floor. Show that the same result is obtained if the zero of U_g is taken as (a) the floor, (b) the table top, (c) the ceiling.

11. Automobiles, tractors, etc., have gear systems which can be changed by shifting. Considering these to be ideal machines, discuss why the shifting process is used.

12. It has been suggested that tides flowing in and out of harbors could be used as sources of energy. Another suggestion is to use the ocean waves for this purpose. Discuss the pros and cons of either proposal from a practical standpoint. Where does the energy come from in each case?

13. Reasoning from the interchange of kinetic and potential energy, explain why the speed of a satellite in a noncircular orbit about the earth keeps changing. Is its speed largest when it is at apogee (farthest point from the earth) or when it is at perigee (the closest point)?

14. About what horsepower is a human being capable of producing for a short period, as in climbing a long flight of stairs? How large a continuous power can a person produce?

15. Estimate the amount of useful work (as defined in physics) which an average human being might perform in 1 day. For comparison purposes, a typical diet might furnish the person with 2000 kcal ($\approx 8.4 \times 10^6$ J) of energy each day. Where does the rest of the energy go?

16. Estimate the force a driver experiences when the car being driven hits another car head on. Assume both cars to be similar and traveling at 25 m/s (90 km/h). Discuss the effect of seat belts, position of the person in the car, and similar factors.

Chapter 6
Problems*

1. How much lifting work is done by a 50-kg woman as she climbs a 4-m-high flight of stairs?

2. In order to lift a 5.0-kg child through a vertical distance of 40 cm, how much work must be done?

3. The 20-kg window counterweight shown in figure P6.1 does how much work as the window is lifted 30 cm? What is the function of the counterweight?

4. A child pushes his toy box 4.0 m along the floor by means of a force of 6 N directed downward at an angle of 37° to the horizontal. How much work does he do? Would you expect him to do more or less work for the same displacement if he pulls upward at the same angle to the horizontal?

5. Show that if $\mathbf{A} = A_x\mathbf{i}, \mathbf{B} = B_x\mathbf{i} + B_y\mathbf{j} + B_z\mathbf{k},$ and $\mathbf{C} = C_x\mathbf{i} + C_y\mathbf{j} + C_z\mathbf{k},$ then (a) $\mathbf{A} \cdot \mathbf{B} = A_x B_x$ and (b) $\mathbf{B} \cdot \mathbf{C} = B_x C_x + B_y C_y + B_z C_z.$

*In solving these problems, use energy methods wherever it is practical to do so.

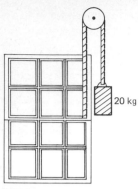

FIGURE P6.1

6. A constant resultant force $\mathbf{F} = F_x\mathbf{i} + F_y\mathbf{j} + F_z\mathbf{k}$ acts on an object to give it a displacement from the origin $\mathbf{s} = x\mathbf{i} + y\mathbf{j} + z\mathbf{k}$. Show that the work done on the object is $F_x x + F_y y + F_z z$. Is that the same as $Fs\cos\theta$ where θ is the angle between $\mathbf{F}$ and $\mathbf{s}$?

7. A 200-g ball is thrown straight upward with a speed of 12 m/s. (*a*) How high will it rise if friction forces are negligible? (*b*) If it only rises 6.0 m, how large an average friction force impeded its motion? (*c*) Under the average friction force of (*b*), how fast will it be moving when it returns to the thrower?

8. A 500-kg block is at the top of a 30° incline when it is released and slides a distance 140 cm down along the incline to the bottom. What will be the speed of the block when it reaches the bottom (*a*) if friction is negligible? (*b*) If the friction force is 0.90 N? (*c*) In (*b*), how large is the friction coefficient?

9. A certain tractor is said to be capable of pulling with a steady force of 14,000 N while moving at a speed of 3.0 m/s. How much power in watts and in horsepower is the tractor developing under these conditions?

10. A small electric motor is needed to lift a 200-g mass at a steady rate of 5.0 cm/s. Give the power in watts and horsepower that the motor must be capable of producing.

11. It takes 8.0 s for a 30-kg girl to climb a 9-m-high flight of stairs. How much horsepower is the girl developing?

12. A pump is needed to lift water through a distance of 6 m at a steady rate of 3 kg/s. What is the minimum horsepower motor that could operate the pump if (*a*) the velocity of the water is negligible at both intake and outlet? (*b*) The velocity at intake is negligible but at the outlet the water is moving with a speed of 9.0 m/s?

13. At high speeds the friction forces acting on a car increase in proportion to v^2 where v is the car's speed. Considering this to be the major factor involved, if a car is rated at 32 km (20 mi) per gallon of gas at 80 km/h (50 mph), what would its mileage rating be at 110 km/h (68 mph)?

14. At low speeds the friction force tending to stop a car is due mainly to energy loss in the rolling tires. For a typical compact car moving at 9.0 m/s, the total force resisting motion is about 200 N. (*a*) How much useful power must the motor provide to maintain this speed? Express your answer in horsepower. (*b*) At 36 m/s the combined rolling and air friction has risen to 1100 N for the same car. What horsepower is needed at this speed?

15. How large a force is required to accelerate an electron $(m = 9.1 \times 10^{-31}\,\text{kg})$ from rest to a speed of 2×10^7 m/s in a distance of 0.50 cm?

16. How large a force is required to stop a 100-kg man in a distance of 1.5 m if he is moving with a speed of 2.0 m/s?

17. How large a friction force between its two rear tires and the pavement is needed to accelerate a 2000-kg car from rest to a speed of 20 m/s in a distance of 80 m? How long will it take? Assume constant acceleration. How large must μ be for this to be possible?

18. Assuming a coefficient of friction μ between car and roadway, use energy methods to show that the stopping distance for a car traveling with speed v is given by $v^2/2\mu g$. Note that the important quantities are speed squared and friction coefficient.

19. A 3-kg mass starts at rest at the top of a 37° incline which is 5.0 m long.

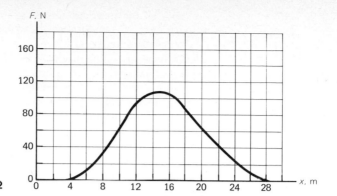

FIGURE P6.2

Its speed as it reaches the bottom is 2.0 m/s. Use energy methods to find the average friction force which retarded its motion.

20. A 3-kg mass has a speed of 5.0 m/s at the bottom of a 37° incline. How far up along the incline will the mass slide before it stops if the friction force retarding its motion is 20 N?

21. If a small child on a swing is to rise 8.0 m above the bottom position of the swing, how fast must the child be moving at (*a*) the lowest position? (*b*) When the child is 4 m above the bottom? Discuss the kinds of energy the child possesses at various points during the swinging motion.

22. Typically, a rubber ball dropped from a height of 100 cm onto a hard floor will rebound to about a height of 70 cm. What fraction of its energy is lost in the impact with the floor? Where does the energy go?

23. If the force exerted on a cart by a boy varies with position as shown in figure P6.2, how much work does the boy do on the cart? (*Hint:* Note that the area of each square is 40 J.)

24. A particle of mass *m* is subjected to an *x*-directed force given by $F = 3.0 + 0.50x$ N. Find the work done by the force as the particle moves from $x = 0$ to $x = 4.0$ m.

25. An *x*-directed force $F_x = 21 - 3x$ N displaces an object from $x = 0$ to $x = 7$ m. (*a*) Find the work done by the force. (*b*) Repeat for $x = 0$ to $x = 14$ m. Explain your (perhaps) peculiar result.

26. A 2-kg object has a velocity of 8 m/s in the +*x* direction as it passes the origin of coordinates. It is subjected to a retarding force such that $F_x = -0.50$ N. What will be its *x* coordinate when it stops?

27. An object of mass *m* has a speed *v* as it passes through the origin on its way out along the +*x* axis. It is subjected to a retarding force given by $F_x = -Ax$. Find its *x* position when it stops.

28. A bead of mass *m* starts at rest at point *A* in figure P6.3 and begins to

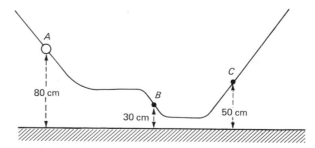

FIGURE P6.3

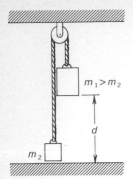

FIGURE P6.4

slide along the frictionless wire shown. Find its speed at points B and C.

29. Repeat problem 28 if the bead has an initial velocity of 2.0 m/s *up* the wire at point A.

30. If the masses in figure P6.4 are released from the position shown, (a) find an expression for the speed of either mass just before m_1 strikes the floor. Ignore the mass and friction of the pulley. (b) Repeat if m_1 has a downward velocity v_0 at the instant shown in the figure.

31. A 3-kg block starts to slide up a 20° incline with an initial speed of 200 cm/s. It stops after sliding 37 cm up the incline and then slides back down. Assuming the friction force impeding its motion to be constant, (a) how large is the friction force, and (b) what is the block's speed as it reaches the bottom?

32. The pendulum of length L shown in figure P6.5 is released from point A. As it swings down, the string strikes the peg at B and the ball swings through point C. (a) How fast is the ball moving as it passes through C? (b) Neglecting friction, the ball will approach a limiting speed as the string winds up on the pin. What is that speed?

33. A 3-kg object initially at rest is subjected to a force $\mathbf{F} = 20\mathbf{j}$ N that acts through a displacement $\mathbf{s} = 7\mathbf{i} + 4\mathbf{j}$ m. (a) What is the final velocity of the object (in component form)? (b) Repeat for the case $\mathbf{F} = 30\mathbf{i} + 20\mathbf{j}$ N.

34. A constant force $F_x\mathbf{i} + F_y\mathbf{j} + F_z\mathbf{k}$ acts on an object of mass m. If it acts through a displacement $s_x\mathbf{i} + s_y\mathbf{j} + s_z\mathbf{k}$, (a) prove that the gain in translational kinetic energy of the object is $F_x s_x + F_y s_y + F_z s_z$. (b) If the object was originally at rest, show that its final $v_x = \sqrt{2F_x s_x/m}$.

35. When a 300-g mass is hung from the end of a vertical spring, the spring's length is 40 cm. With 500 g hanging from it, its length is 50 cm. What is the spring constant of the spring?

36. A spring which stretches 10 cm under a load of 200 g requires how much work to stretch it 5 cm from its equilibrium position? How much work is required to stretch it the next 5 cm?

37. How much energy is required to change the elongation of a spring from 10 to 20 cm if a load of 80 g elongates it 4.0 cm? All elongations are measured from its unstretched position.

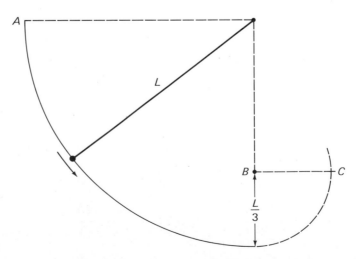

FIGURE P6.5

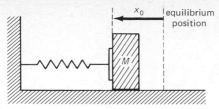

FIGURE P6.6

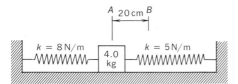

FIGURE P6.7

38. The spring shown in figure P6.6 has a constant k and has been compressed a distance x_0. The mass M is free to leave the end of the spring and experiences negligible friction with the table. How fast will the mass be moving when it leaves the spring if the system is released?

39. The mass M shown in figure P6.6 is fastened to the end of the spring and moves without friction. When released from the position shown, the system oscillates between $\pm x_0$ from the equilibrium position. Find the speed of the mass when the elongation of the spring is zero. Repeat for an elongation of $2x_0/3$. The spring constant is k.

40. A vertical spring with constant 200 N/m has a light platform on its top. When a 500-g mass is set on the platform, the spring compreses 0.0245 m. The mass is now pushed down 0.0755 m further and released. How far above this latter position will the mass fly?

41. Suppose a 300-g mass is dropped from a height of 40 cm onto the spring described in problem 40 and sticks to the platform. (a) How far will the spring compress? (b) How far will the spring be stretched as the mass and spring rebound?

42. Referring to figure P6.7, neither spring is distorted in the position shown. If now the mass is displaced 20 cm to point B and released, find (a) the speed of the block as it passes through point A and (b) how far the block goes to the left before stopping.

43. (Integration problem) Newton's second law can be written as $F_x = m(\Delta v_x/\Delta t)$ for m constant. The work done by F_x in a small displacement Δx is

$$\Delta W = F_x \, \Delta x = m \, \Delta v_x \frac{\Delta x}{\Delta t}$$

from which $\Delta W = mv_x \, \Delta v_x$. Pass to the limit of $\Delta v_x \to 0$ and then integrate over the range $v_0 \le v_x \le v_f$ to find the work done. What is the meaning of your result?

44. (Work and K for moving observers) A block of mass m sits at rest on a frictionless table in a rail car that is moving with speed v_c along a straight horizontal track. A person riding in the car pushes on the block with a horizontal force F for a time t in the direction of the car's motion. (a) What is the final speed of the block according to a person in the car (i.e., the speed relative to the tabletop)? (b) According to a person standing on the ground outside the train (i.e., relative to the ground)? (c) How much did K of the block change according to the person in the car? (d) According to the person on the ground? (e) In terms of F, m, and t, how far did the force displace the object according to the person in the car? (f) According to the person on the ground? (g) How much work does each say the force did? (h) Compare the work done to the K gain according to each person. (i) What can you conclude from this computation?

7 Momentum and the Motion of Systems

When two or more objects interact, as in a collision, the motion of the system can be simplified greatly if one makes use of the concepts of energy, momentum, and mass center. In this chapter we shall learn how to use energy and momentum to describe collisions and other complex situations. It will be found that momentum, as well as energy, obeys a conservation law.

7.1 IMPULSE AND MOMENTUM

If a particle of mass m moving with velocity $\mathbf{v}$ is subjected to a force $\mathbf{F}$, the particle will undergo acceleration; that is to say, its velocity will be changed. We defined the instantaneous force in Chap. 5 through the relation $\mathbf{F} = d(m\mathbf{v})/dt$. Or, for the average force acting in a time interval Δt, we had

$$\bar{\mathbf{F}} = \frac{\Delta(m\mathbf{v})}{\Delta t} \qquad 7.1$$

It will be recalled that $\Delta(m\mathbf{v})$ is the change which occurs in the quantity $m\mathbf{v}$ during the time interval Δt. The name given to the quantity $m\mathbf{v}$ is *linear momentum*. Since we are accustomed to thinking of a massive object moving with high speed as having large momentum, its representation as the product of mass and velocity should not be surprising. Notice that momentum is a vector and its direction is that of the motion. We frequently represent the momentum $m\mathbf{v}$ by the symbol $\mathbf{p}$.

Definition of linear momentum

Momentum is a vector

It will become apparent as we progress through this chapter that there are many situations where only the product $\bar{\mathbf{F}} \Delta t$ is of importance. We call this quantity the *impulse*. From its definition as $\bar{\mathbf{F}} \Delta t$ we can see that impulse is a measure of a force's effect on an object. For example, the impulse given to a baseball by a bat is the average force on the ball multiplied by the time of contact between ball and bat. The larger the impulse on the ball, the greater will be the change in momentum of the ball. Unfortunately, often in situations such as this where a collision occurs, the complicated nature of the collision prevents us from being able to measure or compute $\bar{\mathbf{F}}$ or Δt individually. However, their product, the impulse, is easily related to measurable quantities and is of great importance.

Definition of impulse

To see the importance of impulse, $\bar{\mathbf{F}} \Delta t$, we need only rearrange *7.1* to read

$$\bar{\mathbf{F}} \Delta t = \Delta(m\mathbf{v}) \qquad 7.2a$$

or, in terms of the initial and final momenta,

Impulse $\bar{\mathbf{F}} \Delta t$ equals $m(\mathbf{v} - \mathbf{v}_0)$

$$\bar{\mathbf{F}} \Delta t = (m\mathbf{v})_{\text{final}} - (m\mathbf{v})_{\text{initial}} \qquad 7.2b$$

$$= \mathbf{p}_{\text{final}} - \mathbf{p}_{\text{initial}}$$

Clearly, the impulse is equal to the change in momentum of the object. When the bat strikes a pitched ball, it reverses the momentum of the ball, and 7.2 tells us that this change in momentum is equal to the impulse exerted by the bat on the ball.

Illustration 7.1

An electron in a TV tube is subjected to a force of about 2×10^{-14} N for a time interval of about 3×10^{-9} s as the electron beam is accelerated from rest. Find the final speed of the electron.

Reasoning We can make use of 7.2 since $F = 2 \times 10^{-14}$ N, $\Delta t = 3 \times 10^{-9}$ s, and the mass of an electron is 9.1×10^{-31} kg. From 7.2 we have

$$(2 \times 10^{-14}\,\text{N})(3 \times 10^{-9}\,\text{s}) = (9.1 \times 10^{-31}\,\text{kg})v - 0$$

We have dropped the vector notation since the force and motion are all in the same direction. Solving we find

$$v = 6.6 \times 10^7\,\text{m/s}$$

This speed is still quite small compared to the speed of light, 30×10^7 m/s. Even so, we shall see in the next chapter that the above result is in error slightly. At such high speeds, the relativistic methods of the next chapter should be used.

7.2 IMPULSES INVOLVING NONCONSTANT FORCES

Very seldom is a constant force experienced by a particle in a collision. When a ball collides with a bat or a wall the force involved must vary in a complicated way with time. For example, the ball in collision with a bat in figure 7.1(a) might experience a force varying with time in the way shown in figure 7.1(b). Let us discuss this situation in some detail.

From 7.2 we have the following relation between impulse and the change in momentum if $\mathbf{F}$ is assumed constant:

$$\mathbf{F} \Delta t = m \Delta \mathbf{v}$$

In the case of the batted ball, F will change with time. Let us write this equation for a small enough time Δt so that the value of F is nearly constant during it. For example, in figure 7.1(b), during the small time interval Δt_n the force on the ball is approximately F_n. To obtain the total effect of the bat, we must add all the changes in momentum caused by these subdivisions of the total impulse. Let us assume the direction of $\mathbf{v}_0$ to be along the $+x$ axis. The bat, opposing the motion of the ball, will be assumed to hit the ball squarely, so that $\mathbf{F}$ will be in the $-x$ direction. We can write the above equation for this small portion of the total impulse as

$$F_n \Delta t_n(-\mathbf{i}) = m \Delta \mathbf{v}_n$$

The magnitude of the small impulse exerted by the force F_n during the time interval Δt_n is just $F_n \Delta t_n$.

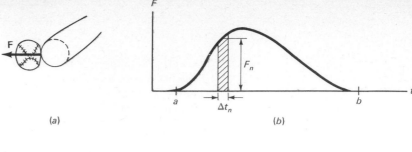

(a)

(b)

(c)

FIGURE 7.1 As the ball is hit by the bat, the force on the ball might be as shown in (b). What is the impulse exerted by the bat on the ball?

Notice that $F_n \, \Delta t_n$ is only a small portion of the total impulse, and therefore Δv_n is only a small part of the total change in velocity. Moreover, $F_n \, \Delta t_n$ is the area of a rectangle of height F_n and width Δt_n; if Δt_n is made small, then this area is essentially equal to the shaded area shown in figure 7.1(b). To obtain the total impulse on the ball, we subdivide the impulse into numerous tiny impulses similar to the impulse $F_n \, \Delta t_n$ we have been discussing. This is shown schematically in part (c) of the figure. However, the widths of the small sections would be taken much smaller than shown so that the area of a typical region of width Δt_n would be the rectangle area $F_n \, \Delta t_n$.

We now obtain the total impulse by adding the areas of all the small (nearly) rectangles shown schematically in figure 7.1(c). That is,

$$\text{Total impulse} = F_1 \, \Delta t_1 + F_2 \, \Delta t_2 + \cdots + F_n \, \Delta t_n + \cdots + F_N \, \Delta t_N$$

Or, using the summation symbol,

$$\text{Total impulse} = \sum_1^N F_n \, \Delta t_n$$

But if we let $\Delta t_n \to 0$ so that the shaded and rectangle areas become essentially identical, then this sum becomes what is defined to be an integral. It is then written

$$\text{Total impulse} = \int_a^b F \, dt$$

where the limits a and b on the integral tell us to take the sum of rectangles in the range from $t = a$ to $t = b$, the time limits of the impulse as we see from figure 7.1. Do not let the notation mystify you; the integral is simply the sum which was written out in detail above.

As we have seen, the total impulse is obtained by taking the sum of all the little impulses $F_n \, \Delta t_n$. This sum turned out to be the area under the force vs. time graph. Therefore

Impulse equals area under the F versus t curve

$$\text{Total impulse} = \sum_1^N F_n \, \Delta t_n$$

$$= \int_a^b F \, dt$$

$$= \text{area under } F \text{ versus } t \text{ graph}$$

All three of these expressions for the impulse are the same, of course, but one form may be found more useful than another in a particular situation.

The total impulse must also be equal to the sum of all the terms on the other side of the equation, a typical term being $m \, \Delta v_n$, where $\Delta \mathbf{v}_n$ is the change in velocity caused by the little portion of impulse $\mathbf{F}_n \, \Delta t_n$. Thus, assuming all the $\Delta \mathbf{v}_n$ to be in line so the vector notation can be dropped,

$$\text{Total impulse} = m \, \Delta v_1 + m \, \Delta v_2 + \cdots + m \, \Delta v_n + \cdots + m \, \Delta v_N$$

$$= m(\Delta v_1 + \Delta v_2 + \cdots + \Delta v_N + \cdots + \Delta v_N)$$

But the sum of all the changes in velocity, the quantity in parenthesis, must be the total change in velocity, namely, $v_{final} - v_{initial}$. We therefore have

$$\text{Total impulse} = m v_{final} - m v_{initial}$$

This equation should not seem new to us. It simply says that the impulse (in this case, a complicated one) results in a change in momentum and the two are equal.

As we see, several forms exist for the total impulse even in the case when the force is not constant. Depending upon the usefulness in a given situation, we can choose to equate any of these expressions for the impulse. In addition there is still another way of writing the total impulse. It is based upon defining an average force during impact to be the same as that constant force which would cause the same momentum change as the actual force does. Calling this average force $\bar{\mathbf{F}}$, we have as its definition,

$$\text{Total impulse} = \bar{\mathbf{F}}(b - a) = m(\mathbf{v}_{final} - \mathbf{v}_{initial}) \qquad 7.3$$

where $b - a$ is the time duration of the impulse.

Illustration 7.2

An airplane traveling at 300 m/s (about 650 mi/h) strikes a bird, which can be assumed to be at rest relative to the earth. Assuming the bird to have a mass of 0.50 kg and to be about 20 cm in size, estimate the average force exerted by the bird on the airplane.

Reasoning Let us view the collision using the earth as our reference frame. The original momentum of the bird was negligibly small, and so we take it to be zero. After collision, the bird is moving with the speed of the plane, or so we shall assume. Therefore, from 7.2, or 7.3,

$$\bar{F} \, \Delta t = \Delta(mv)$$
$$= (0.50 \text{ kg})(300 \text{ m/s}) - 0$$
$$= 150 \text{ kg} \cdot \text{m/s} = 150 \text{ N} \cdot \text{s}$$

This is the impulse. To find $\bar{F}$ we must estimate Δt.

The collision will take place in a time about equal to the time taken for the plane to cover a distance comparable with the size of the bird. Therefore, using $x = \bar{v}t$,

$$\Delta t \approx \frac{0.20 \text{ m}}{300 \text{ m/s}}$$

$$\approx 7 \times 10^{-4} \text{ s}$$

Using this value in the impulse equation gives $\bar{F} \approx 2 \times 10^5$ N, which is close to a force of 20 tons. Although our estimate of the force is quite crude, the order of magnitude must be correct, and it indicates that in this type of collision large forces are developed.

(What factors have been omitted from our analysis that could have a large effect on the computed force?)

7.3 MOMENTUM CONSERVATION

Suppose two particles with original momenta $m\mathbf{v}_0$ and $m'\mathbf{v}_0'$ collide at point P, as shown in figure 7.2(a). After collision they will have velocities of $\mathbf{v}$ and $\mathbf{v}'$. We wish to relate the momenta before collision to the momenta after collision. This relation will be found by considering the impulse on each particle during the collision.

The impulse experienced by m will be

$$\mathbf{F} \, \Delta t = m\mathbf{v} - m\mathbf{v}_0$$

according to 7.2. Similarly, the impulse experienced by m' will be

$$\mathbf{F}' \, \Delta t = m'\mathbf{v}' - m'\mathbf{v}_0'$$

where the Δt are taken the same in the two cases since it is the time of contact of the objects. Of course, the forces involved are the average forces during Δt.

According to Newton's law of action and reaction, $\mathbf{F} = -\mathbf{F}'$ as illustrated in part (b) of figure 7.2. As a result we can equate one of the impulses to the negative of the other to find

$$m\mathbf{v} - m\mathbf{v}_0 = -(m'\mathbf{v}' - m'\mathbf{v}_0')$$

which simplifies to give

$$m\mathbf{v}_0 + m'\mathbf{v}_0' = m\mathbf{v} + m'\mathbf{v}'$$

In words, this equation is simply stated as

Total momentum before collision = total momentum after collision

We have therefore proved that the *vector* sum of the momenta before the collision of two objects is equal to the *vector* sum of the momenta after collision. That is to say, the total momentum does not change, it is *conserved*, in a collision. This is called the *law of conservation of momentum*.

In the absence of unbalanced external forces, momentum is conserved

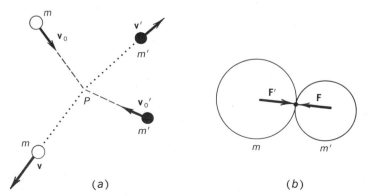

FIGURE 7.2 Two quantities do not change during the contact of the particles. What are they?

(a) (b)

It is a rather simple matter to extend this proof to collisions involving three or more particles. One merely applies the above reasoning to each point of contact, thereby obtaining two equations for each point of contact. These pairs of equations will be of the form

$$\mathbf{F} \, \Delta t = m \, \Delta \mathbf{v}$$
$$\mathbf{F}' \, \Delta t = m' \, \Delta \mathbf{v}'$$

After recognizing that $\mathbf{F} = -\mathbf{F}'$, we have

$$m \, \Delta \mathbf{v} + m' \, \Delta \mathbf{v}' = 0$$

Since each contact point will give an equation of this form, clearly the change in momentum for each pair of particles is zero. As a result, the total change in momentum because of all the collisions will be zero.

It must be pointed out that momentum is conserved in *all* collisions. However, in considering the momentum interchanges during the collision, one must take account of all the particles which participate in the collision. All these particles taken together are called a *system,* and the momentum of the whole system is the topic of the conservation law. If we consider a system of particles upon which no* outside forces act, an *isolated system,* it is clear that the momentum of the system will remain constant. As we have seen, the forces resulting from collisions within the system do not change the momentum of the system. In the absence of external forces the momentum will therefore not change. We summarize the law of conservation of momentum as follows: *The momentum of an isolated system of particles is constant.*

Law of conservation of momentum

Illustration 7.3

A radium atom at rest suddenly emits an alpha particle (α particle). If the speed of the α particle is 1.5×10^7 m/s, what will be the speed of the nucleus just after the α particle is ejected?

Reasoning The situation is shown in figure 7.3, where the masses of the various particles are given in atomic mass units ($1 \, u = 1.66 \times 10^{-27}$ kg). Since no external forces act upon the system, we know that

Momentum before disintegration = momentum after disintegration

The radium atom was at rest initially, and so the initial momentum is zero. Using the symbols shown in the figure, this equation becomes

$$0 = m_1\mathbf{v}_1 + m_2\mathbf{v}_2$$

Rearranging gives
$$\mathbf{v}_1 = -\frac{m_2}{m_1}\mathbf{v}_2$$

*If the resultant of any existing forces from outside is zero, these statements will also be true.

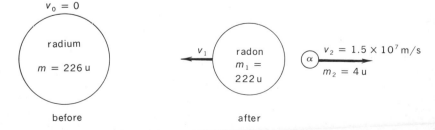

FIGURE 7.3 What is the momentum of the system after the radium nucleus has decayed to radon and an α particle?

The two vectors, $\mathbf{v}_1$ and $-(m_2/m_1)\mathbf{v}_2$, must be equal both in magnitude and direction. Since the direction of $-(m_2/m_1)\mathbf{v}_2$ is opposite to $\mathbf{v}_2$ itself, we see that $\mathbf{v}_1$ and $\mathbf{v}_2$ are in opposite directions. Placing in the numerical values and noting that only the ratio of the masses is needed, we find

$$v_1 = -(4/222)(1.5 \times 10^7 \text{ m/s})$$
$$= -2.7 \times 10^5 \text{ m/s}$$

Illustration 7.4

A truck and a car collide head on. The speed of the truck was 20 m/s, and that of the car was 30 m/s. The truck has a mass 5 times that of the car. If they stick together after collision, how fast are they moving, as a unit, just after the collision?

Reasoning During the collision the force of the car on the truck and of the truck on the car are the major forces involved in changing the velocities of the two vehicles. The vertical forces of the ground on the car and truck do not participate in the changes in momentum since they are balanced by the downward pull of gravity. Although the friction forces at the road do affect the velocities of the vehicles, this effect will be small (because these forces are relatively small) if we wish to find the velocities just an instant after collision. (It is true that, in time, the friction forces will alter the velocities considerably. But this effect will follow after the impact itself.)

We have, from the law of conservation of momentum,

Momentum before collision = momentum after collision
$$m\mathbf{v}_0 + M\mathbf{V}_0 = (m + M)\mathbf{v}'$$

where m and $\mathbf{v}_0$ refer to the car, and M and $\mathbf{V}_0$ to the truck, and $\mathbf{v}'$ is the velocity after the collision of the truck and car which are stuck together. Since $\mathbf{v}_0 = v_0\mathbf{i}$ and $\mathbf{V}_0 = -V_0\mathbf{i}$, if we assume the original direction of motion to be the x-axis direction, the relation can be written

$$(mv_0 - MV_0)\mathbf{i} = (m + M)\mathbf{v}'$$

Using the fact that $M = 5m$, we find

$$6\mathbf{v}' = -70\mathbf{i} \qquad \text{m/s}$$
or
$$\mathbf{v}' = -11.7\mathbf{i} \qquad \text{m/s}$$

In other words, the combination is moving in the direction of motion of the truck, and the speed of the truck is a little more than half its original speed.

7.4 COLLISIONS IN THREE DIMENSIONS

Suppose the momentum of one particle is $\mathbf{p}_0$ and another is $\mathbf{p}_0'$ before collision and $\mathbf{p}$ and $\mathbf{p}'$ after collision. The law of conservation of momentum tells us

$$\mathbf{p}_0 + \mathbf{p}_0' = \mathbf{p} + \mathbf{p}'$$

Expressing each of these vectors in x, y, and z coordinates gives

$$(p_{0x}\mathbf{i} + p_{0y}\mathbf{j} + p_{0z}\mathbf{k}) + (p_{0x}'\mathbf{i} + p_{0y}'\mathbf{j} + p_{0z}'\mathbf{k})$$
$$= (p_x\mathbf{i} + p_y\mathbf{j} + p_z\mathbf{k}) + (p_x'\mathbf{i} + p_y'\mathbf{j} + p_z'\mathbf{k})$$

Collecting terms,

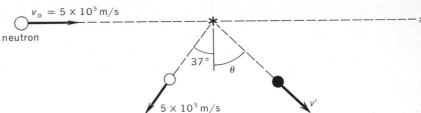

FIGURE 7.4 From the data given in the figure, find θ and v'.

In an isolated system, the rectangular component momenta are conserved

$$(p_{0x} + p'_{0x})\mathbf{i} + (p_{0y} + p'_{0y})\mathbf{j} + (p_{0z} + p'_{0z})\mathbf{k}$$
$$= (p_x + p'_x)\mathbf{i} + (p_y + p'_y)\mathbf{j} + (p_z + p'_z)\mathbf{k}$$

Now if two vectors are equal, their components must be equal. Therefore

$$p_{0x} + p'_{0x} = p_x + p'_x$$
$$p_{0y} + p'_{0y} = p_y + p'_y$$
$$p_{0z} + p'_{0z} = p_z + p'_z$$

These three equations tell us that, in a collision, the momentum in the x direction is conserved, and similarly for the y and z directions. We see, therefore, that the components of momentum are also conserved in a collision. This allows us to focus attention on momentum along any arbitrary line if it proves to be convenient.

Illustration 7.5

A neutron ($v_0 = 5 \times 10^5$ m/s, $m = 1.0$ u) collides with a helium nucleus (3.75×10^5 m/s, $m = 4.0$ u), as shown in figure 7.4. If the speed of the neutron after collision is 5×10^5 m/s, and it is moving in the direction shown, find the speed and direction of motion of the He nucleus after collision.

Reasoning We know that both the x and y components of the momentum must be conserved independently. For the x components we have

$$(1)(5 \times 10^5) = (1)(-0.6 \times 5 \times 10^5) + (4)(v' \sin \theta)$$

and for the y components

$$(4)(3.75 \times 10^5) = (1)(0.8 \times 5 \times 10^5) + (4)(v' \cos \theta)$$

Solving these equations we find

$$\tan \theta = 0.73 \qquad v' = 3.4 \times 10^5 \text{ m/s}$$
from which
$$\theta = 36°$$

Why was it allowable to use the masses in atomic mass units in these equations?

7.5 ELASTIC AND INELASTIC COLLISIONS

Unlike the situations we have thus far dealt with, the general collision problem requires more than momentum considerations for its solution. For

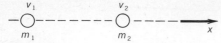

FIGURE 7.5 When the two balls move freely along the x axis and collide, we know that their combined momentum must remain unchanged.

example, let us consider the case of a so called *head-on collision,* one in which all the motion occurs along a straight line. A collision of this type is shown in figure 7.5 where two balls are moving along the x axis in such a way as to collide. Their masses are m_1 and m_2, while their initial velocities are v_{1i} and v_{2i}. The vector nature of the velocities can be most easily handled by the use of $+$ and $-$ signs, since the motion is along a line. If the motion is in the $+x$ direction, v will be positive, and it will of course be negative if the motion is in the $-x$ direction.

When the balls in figure 7.5 collide, momentum must be conserved, and so we write

$$m_1 v_{1i} + m_2 v_{2i} = m_1 v_{1f} + m_2 v_{2f} \qquad 7.4$$

where v_{1f} and v_{2f} are the final velocities after collision. Usually, the balls will bounce apart after collision, and so v_{1f} and v_{2f} are both unknown. This means that the motion of the balls after collision cannot be found from momentum considerations alone since *7.4* is a complete statement of momentum conservation in this case. It provides only one equation, and since there are two unknowns, another equation is needed. (Previously we considered only those cases where the objects stuck together after collision. As a result, $v_{1f} = v_{2f}$, and *7.4* was sufficient for the determination of the final motion.)

If we are to proceed further with the solution of the general collision between two bodies, we must obtain another equation. This can be done *provided* one is able to say something about the kinetic energy before and after the collision. In most cases, considerable kinetic energy will be lost during the collision. Although some kinetic energy is often changed to sound energy, most often kinetic energy is lost doing work against friction forces. That is to say, during the collision, molecules are displaced by the impact, and in so doing heat energy is generated through friction effects between molecular layers in the solid. The total amount of heat and sound energy generated is equal to the kinetic energy lost, and so the total energy remains constant. In the collision of molecules and atoms, kinetic energy may be changed to excitation energy within the molecule or atom. (By this we mean that the electrons and atoms within a molecule could acquire additional energy, which would not then be available for translational kinetic energy of the molecule.)

To illustrate how large this loss in kinetic energy can become, let us for a moment consider the case where the two balls stick together after collision. Then, $v_{1f} = v_{2f} = v$, and we can use *7.4* to show that

$$v = \frac{m_1 v_{1i} + m_2 v_{2i}}{m_1 + m_2}$$

Suppose, further, that $v_{2i} = 0$; that is, the second ball was standing still before the collision. The kinetic energies before and after the collision are therefore given by

$$K_i = \tfrac{1}{2} m_1 v_{1i}{}^2 \qquad \text{and} \qquad K_f = \tfrac{1}{2}(m_1 + m_2)v^2$$

Placing in the above value for v and forming the ratio K_f/K_i, we find that the fraction of the kinetic energy left after the collision is

$$\frac{K_f}{K_i} = \frac{m_1}{m_1 + m_2}$$

In other words, if $m_1 \ll m_2$, nearly all the kinetic energy would be lost.

For the general case, a fraction β of the original kinetic energy will exist after the collision. We can write, therefore, that

$$\beta(m_1 v_{1i}^2 + m_2 v_{2i}^2) = m_1 v_{1f}^2 + m_2 v_{2f}^2 \qquad 7.5$$

where the factor $\frac{1}{2}$ common to all terms has been canceled. If $\beta = 1$, all the kinetic energy is conserved in the collision, and the collision is said to be *perfectly elastic.* In all other instances, for $\beta < 1$, the collisions are *inelastic,* and kinetic energy is lost. As we shall see in problem 34 at the end of this chapter, the value of β depends upon the reference frame we use in describing the motion.

In a perfectly elastic collision, no kinetic energy is lost

Illustration 7.6

A *ballistic pendulum* is a device for finding the speed of bullets. The bullet of mass m is shot into a block of mass M, as shown in figure 7.6. After collision, the center of mass of the block swings to a height h. Find the speed of the bullet, v, in terms of m, M, and h.

Reasoning The law of conservation of momentum applies to the collision. We can therefore write

$$mv = (m + M)V$$

where V is the speed of the bullet-block combination *just* after the collision. We wish to find v but V is unknown. Since kinetic energy was lost to friction work as the bullet tore into the block, we cannot write an energy equation for the collision. However, we know what happened to the energy after the collision.

After the collision, the block and bullet have a kinetic energy $\frac{1}{2}(m + M)V^2$. This will be lost to potential energy as the block swings upward and stops. Therefore, after the collision,

$$\tfrac{1}{2}(m + M)V^2 = (m + M)gh$$

which gives
$$V = \sqrt{2gh}$$

Substitution of this value in the momentum equation gives

$$mv = (m + M)\sqrt{2gh}$$

or
$$v = \frac{m + M}{m}\sqrt{2gh}$$

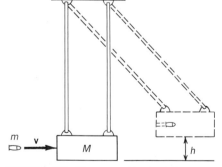

m v

M h

FIGURE 7.6 How can this device be used to find the speed of the bullet?

We therefore are able to find the bullet's speed in terms of the height to which the pendulum swings.

Notice that the kinetic energy of the bullet before collision is *not* equal to the kinetic energy of the block and bullet after collision. A large amount of energy was lost doing friction work as the bullet plowed into the block. This appeared as heat energy. Often in such an experiment the bullet becomes warm enough to melt partly.

7.6 PERFECTLY ELASTIC COLLISIONS IN ONE DIMENSION

To find the final speeds of two colliding particles in the special case of a *perfectly elastic collision,* we solve 7.4 and 7.5 simultaneously, after setting $\beta = 1$. This is conveniently done by rearranging 7.5 to give

$$m_1(v_{1i}^2 - v_{1f}^2) = m_2(v_{2f}^2 - v_{2i}^2)$$

and factoring so that

$$m_1(v_{1i} - v_{1f})(v_{1i} + v_{1f}) = m_2(v_{2f} - v_{2i})(v_{2f} + v_{2i})$$

We further notice that *7.4* can be arranged to give

$$m_1(v_{1i} - v_{1f}) = (v_{2f} - v_{2i})m_2$$

which can then be divided into the previous equation. The result is

$$v_{1i} + v_{1f} = v_{2i} + v_{2f} \qquad\qquad 7.6$$

This equation can be arranged to give

$$v_{1i} - v_{2i} = v_{2f} - v_{1f} \qquad\qquad 7.6a$$

The quantity on the left is the speed of approach of the two objects, while the right-hand side is the speed with which the objects recede from each other. These two speeds are equal in a *linear, perfectly elastic* collision.

Equation *7.6* (or *7.6a*) is a convenient alternative equation to replace *7.5* in the treatment of perfectly elastic linear collision problems. Its main advantage is obviously its simplicity. We shall make use of it. However, it should be remembered that it is a consequence of the conservation of *K*, the kinetic energy, in a perfectly elastic collision.

If we now solve *7.4* and *7.6* simultaneously, we find

$$v_{1f} = \frac{(m_1 - m_2)v_{1i} + 2m_2 v_{2i}}{m_1 + m_2}$$
$$v_{2f} = \frac{(m_2 - m_1)v_{2i} + 2m_1 v_{1i}}{m_2 + m_1} \qquad\qquad 7.7$$

It is of interest to examine this result in several limiting cases.

Case 1 $m_1 = m_2$; $v_{2i} = 0$

It is readily seen in this case that $v_{1f} = 0$ and $v_{2f} = v_{1i}$. In other words, in a head-on collision of two identical bodies, one originally at rest, the moving body stops and the other takes on its initial speed. This is of importance in various nuclear applications (reactors, for example), where it is desired to slow the motion of neutrons. Clearly, this is best done by using a material such as water or a hydrocarbon which contains many hydrogen nuclei since the neutron and proton (the hydrogen nucleus) have essentially the same mass.

Case 2 $m_1 = m_2$

For this case *7.7* yields $v_{1f} = v_{2i}$ and $v_{2f} = v_{1i}$. The particles merely exchange velocities upon collision. Notice that Case 1 was a special example of this case.

Case 3 $m_2 \gg m_1$; $v_{2i} = 0$

Equation *7.7* shows, in this instance, that $v_{1f} = -v_{1i}$ and $v_{2f} = 0$. In other words, the small object merely bounces back off the very large stationary object. The large object remains essentially motionless after the collision.

Case 4 $m_2 \ll m_1$; $v_{2i} = 0$

This result can be guessed from our common experience. The large object will not be slowed appreciably during the collision. The small object will be given a speed twice that of the large object. If a person were to "stand on" the large object m_1, it would appear that the large object was at rest and that the small object m_2 was approaching with speed v_{1i}. In his view, the problem is similar to that posed in Case 3, and he sees the ball "rebound" with speed v_{1i}. This speed must be added to his own, v_{1i}, to obtain the actual speed of the ball, $2v_{1i}$.

7.7 ELASTIC COLLISION IN TWO DIMENSIONS

In a collision such as the one shown in figure 7.7, we have shown in preceding sections that the total momentum in each of the x, y, and z directions is conserved. We can therefore write three equations to state the conservation of momentum in the three coordinate directions. For a collision confined to a plane, such as the one in figure 7.7, only two of these equations are of value:

Momentum before = momentum after

x direction $$m_1 v_{1i} = m_1 v_{1fx} + m_2 v_{2fx} \qquad 7.8a$$

y direction $$0 = m_1 v_{1fy} + m_2 v_{2fy} \qquad 7.8b$$

These two equations contain four unknowns if we assume the initial conditions and masses to be known. Two other equations, or pieces of information, are needed to evaluate these four unknowns. If the collision were perfectly elastic, we could write

$$\tfrac{1}{2}m_1 v_{1i}^2 = \tfrac{1}{2}m_1 v_{1f}^2 + \tfrac{1}{2}m_2 v_{2f}^2$$

Or in terms of components, this is written

$$m_1 v_{1i}^2 = m_1(v_{1fx}^2 + v_{1fy}^2) + m_2(v_{2fx}^2 + v_{2fy}^2) \qquad 7.9$$

We therefore have three equations involving four unknowns. To proceed further, one additional piece of data is needed. This might be one of the final velocity components, or perhaps θ_1 or θ_2. Given any of these, we should be able to complete the solution of the problem.

Illustration 7.7
Referring to figure 7.7, suppose particle 1 is a neutron with a speed of 5×10^6 m/s. Particle 2 is a proton. If $\theta_1 \approx 37°$, how large are the final speeds of the two particles? Assume the two particles to have identical masses.

Reasoning Since both the x and y momenta are conserved, we can write

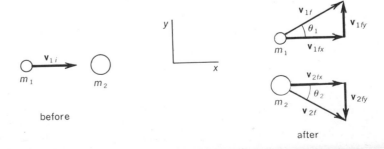

before

after

FIGURE 7.7 What can be stated concerning this collision even if it is inelastic?

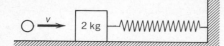

FIGURE 7.8 How can one find v from the distance the spring compresses?

$$5 \times 10^6 \, \text{m/s} = 0.8v_{1f} + v_{2fx}$$
$$0 = 0.6v_{1f} + v_{2fy}$$

where m_1 and m_2 for the proton and neutron have been canceled. A collision such as this is usually perfectly elastic and so the conservation of kinetic energy equation can be used to give

$$25 \times 10^{12} \, \text{m}^2/\text{s}^2 = v_{1f}^2 + (v_{2fx}^2 + v_{2fy}^2)$$

These equations contain three unknowns. Solving them simultaneously, we find

$$v_{1f} = 4 \times 10^6 \, \text{m/s}$$
$$v_{2fy} = -2.4 \times 10^6 \, \text{m/s}$$
$$v_{2fx} = 1.8 \times 10^6 \, \text{m/s}$$

The negative sign on v_{2fy} indicates it to be in the $-y$ direction, as shown in the figure. Knowing v_{2fx} and v_{2fy}, one could find v_{2f} and θ_2 if one wished.

Illustration 7.8

As shown in figure 7.8, a 2.0-kg block is placed at the end of an unstretched spring for which $k = 2000 \, \text{N/m}$. When a 40-g bullet is shot into and lodges in the block, the block compresses the spring a distance of 10 cm. How fast was the bullet moving? Ignore the small friction force between block and table.

Reasoning For the collision process we can write the fact that momentum is conserved. That is,

Momentum of bullet before collision
$$= \text{momentum of block} + \text{bullet after collision}$$

Placing in the appropriate values for the masses and calling v the bullet's speed and V the speed of the block and bullet after collision, we have

$$(0.040 \, \text{kg})v = (2.04 \, \text{kg})V$$
from which $$v = 51V$$

Just after collision, the block and bullet have kinetic energy. This energy will be lost to potential energy of the spring as the spring is compressed by the block. Therefore

$$K \text{ of system just after collision} = U_s \text{ at end}$$

Placing in the values, remembering that $U_s = \frac{1}{2}kx^2$,

$$(\tfrac{1}{2})(2.04 \, \text{kg})V^2 = (\tfrac{1}{2})(2000 \, \text{N/m})(0.10 \, \text{m})^2$$
so that $$V \approx \sqrt{10} \, \text{m/s}$$

Using this in the expression for v, we find v to be 161 m/s.

7.8 ROCKET PROPULSION

An intersting problem in momentum occurs when a rocket is accelerated by ejecting mass (the gases resulting from combustion of fuel) from itself. Suppose the rocket has a velocity $V\mathbf{i}$ relative to the earth and mass m. In time Δt it will eject a mass Δq of fuel with a velocity $-v\mathbf{i}$ with respect to the rocket. Relative to the ground, the ejected fuel will move with velocity $(V - v)\mathbf{i}$.

Let us assume the rocket to be moving parallel to the earth or in distant space so that gravitational forces can be neglected. Applying the law of conservation of momentum to the system (rocket + fuel), we have that the momentum at the start of the interval Δt must equal the momentum at the end of the interval. Therefore

$$mV\mathbf{i} = (m - \Delta q)(V + \Delta V)\mathbf{i} + (\Delta q)(V - v)\mathbf{i}$$

After dropping the vector notation, since each term has the same direction, and after collecting terms, we find

$$0 = m\,\Delta V - v\,\Delta q$$

where the second-order term $(\Delta q)(\Delta V)$ has been neglected. We notice that the quantity of fuel ejected, Δq, is equal to the loss in mass of the rocket, $-\Delta m$. The negative sign is appended to Δm to indicate it is a decrease in m.

After making this substitution in the above equation, one has, upon rearranging,

$$\Delta V = -\frac{\Delta m}{m}v$$

If we pass to the limit of $\Delta t \to 0$, then ΔV and Δm can be replaced by dV and dm to yield

$$dV = -\frac{dm}{m}v$$

This equation gives us the increment in the rocket's speed for a loss of fuel equal to dm. To find the final speed of the rocket itself, we need to sum all these increments from the starting of the rocket, at which time the rocket had a total mass $M + m_0$, until the time when all the fuel is gone. We designate the original amount of fuel in the rocket as m_0 and the rocket, less fuel, to have mass M. Since the sum of all the ΔV's is the change in velocity $V - V_0$, we have

$$V - V_0 = \sum \Delta V = -\sum \frac{\Delta m}{m}v$$

or in integral notation and after recalling that v is a constant,

$$V - V_0 = -v \int_{m=M+m_0}^{M} \frac{dm}{m}$$

Notice the limits on m; they are the initial and final mass of the rocket.

At this point the physics is completed, and if the integral is not known, it can be looked up in a set of integral tables. We find there that

$$\int \frac{dx}{x} = \ln x$$

where $\ln x$ is the logarithm to the base e. You may recall that base e logarithms are related to base 10 logarithms by

$$\ln x \approx 2.30 \log_{10} x$$

We then find $\qquad V - V_0 = v[\ln (M + m_0) - \ln M]$

If we recall that $\ln a - \ln b = \ln (a/b)$, this may be put in a more convenient form:

$$V - V_0 = v \ln \left(1 + \frac{m_0}{M}\right)$$

This equation points out clearly what factors are important if the rocket is to achieve high speeds. Since the final rocket speed V is proportional to the speed v with which fuel is ejected from the rocket, the speed at which the combustion gases leave the rocket should be made as high as possible. To achieve this, extremely high temperatures in the combustion chamber are desirable.

The other quantity of importance is the ratio m_0/M, the mass of fuel to the mass of the nonfuel portion of the rocket. As we should expect, the more fuel, the better. However, we should also notice that the type of fuel is critical. The best of several fuels will give the highest ejection velocity provided each fuel has the same density. If two fuels give the same value for v but differ in density, which would be preferable?

Illustration 7.9

A typical rocket shoots gas from its engine at a rate of 1300 kg/s at a speed of 50,000 m/s relative to the rocket. How large a total mass can the rocket lift slowly from the earth?

Reasoning The *thrust* (or pushing force) on a rocket is the reaction force on the rocket as the gas is shot out its rear. Assuming the rocket to be rising very slowly from the earth so as to make acceleration effects negligible, the thrust must just counterbalance the weight of the rocket.

But the force needed to eject the gas from the rocket can be obtained from the impulse equation: $F \Delta t = \Delta(mv)$. A mass of 1300 kg is accelerated from essentially rest to a speed of 50,000 m/s during each second as gas is expelled to the rear. Therfore we can take, for $\Delta t = 1$ s,

$$\Delta(mv) = (1300 \text{ kg})(50,000 \text{ m/s} - 0 \text{ m/s}) = 6.5 \times 10^7 \text{ kg} \cdot \text{m/s}$$

Placing these values in the impulse equation yields $F = 6.5 \times 10^7$ N as the force exerted on the gas to expel it. The thrust, being the reaction force to the expulsion force, also has a magnitude of 6.5×10^7 N.

Because the thrust simply supports the weight Mg of the rocket, in this case of slow lift-off, we have that $M = (\text{thrust})/g = 6.6 \times 10^6$ kg and this is the total mass of the rocket.

7.9 CENTER OF MASS

Oftentimes we wish to find the acceleration or other motion of a complex object under the action of an external force that causes the object to rotate as well as translate. In this and many other complex situations the motion is most easily described in terms of the motion of the *center of mass* of the object or of some collection of objects. Let us now see what is meant by the mass center.

The position of the center of mass is defined in the following way. Suppose an object or system of objects consists of N particles having masses $m_1, m_2, \ldots, m_N$. A few of these particles are shown in figure 7.9. If the x

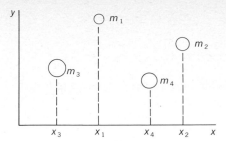

FIGURE 7.9 Find the center of mass of these four particles.

coordinates of these particles are $x_1, x_2, \ldots, x_N$, then the x coordinate of the center of mass is defined by

$$x_{cm} \equiv \frac{x_1 m_1 + x_2 m_2 + \cdots + x_N m_N}{m_1 + m_2 + \cdots + m_N}$$

Similar expressions can be written for the y and z coordinates of the mass center. Writing these in terms of sums, we have

$$x_{cm} = \frac{\sum_1^N x_i m_i}{\sum_1^N m_i} \qquad y_{cm} = \frac{\sum_1^N y_i m_i}{\sum_1^N m_i} \qquad z_{cm} = \frac{\sum_1^N z_i m_i}{\sum_1^N m_i} \qquad 7.10$$

This definition of the center of mass agrees with our general intuitive ideas concerning the mass center. The mass center of a sphere is at its geometrical center, and similarly for a cylinder, a rod, an ellipsoid, a plate, and other simple shapes. For example, consider the mass center of the object shown in figure 7.10. To find its position, we use 7.10 to give

$$x_{cm} = \frac{mb + (2m)(b + a) + (3m)b}{m + 2m + 3m} = b + \frac{a}{3}$$

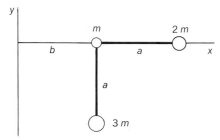

FIGURE 7.10 Can you guess the position of the center of mass of this system?

and

$$y_{cm} = \frac{0 + 0 + (3m)(-a)}{m + 2m + 3m} = -\frac{a}{2}$$

If you examine the object carefully, you will see that the mass center we have found is reasonable.

If the system under consideration is actually a solid object, it is convenient to consider the object to be separated into small masses. As a typical example, suppose it is necessary to calculate the position of the center of mass for the *nonuniform* rod shown in figure 7.11. Suppose the mass per unit length of the rod is said to be λ, where λ is the following function of x:

$$\lambda = \beta x^2$$

In other words, a small length of rod Δx will have a mass $\lambda \Delta x$. (If λ is 200 g/cm, then the mass of a length $\Delta x = 0.10$ cm long will be 20 g.)

In order to apply 7.10, we split the rod into N small masses, each of length Δx, as shown in figure 7.11(b). Since we know $y_{cm} = z_{cm} = 0$, we are interested only in x_{cm}; it is

$$x_{cm} = \frac{\sum_1^N x_i m_i}{\sum_1^N m_i}$$

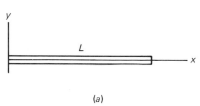

(a)

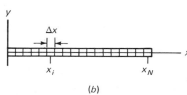

(b)

FIGURE 7.11 The rod is not uniform but has a mass per unit length $\lambda = \beta x^2$. Where is its mass center?

The mass m_i of the little length Δx_i is simply $\lambda_i \Delta x_i$, where λ_i is the value of λ at x_i, namely $\lambda_i = \beta x_i^2$. Placing this value in the equation gives

$$x_{cm} = \frac{\sum_1^N \beta x_i^3 \Delta x_i}{\sum_1^N \beta x_i^2 \Delta x_i}$$

If we pass to the limit of $\Delta x \to 0$, the sum may be replaced by an integral, and we have

$$x_{cm} = \frac{\int_{x=0}^{x=L} \beta x^3 \, dx}{\int_{x=0}^{x=L} \beta x^2 \, dx}$$

The limits on the integrals are obtained by considering the values of x at the two extreme ends of the rod.

Since β is a constant, it may be taken outside the integrals and canceled from numerator and denominator. We then make use of the facts that

$$\int x^3 \, dx = \tfrac{1}{4}x^4 \qquad \text{and} \qquad \int x^2 \, dx = \tfrac{1}{3}x^3$$

to obtain
$$x_{cm} = \frac{\tfrac{1}{4}(L^4 - 0)}{\tfrac{1}{3}(L^3 - 0)} = \tfrac{3}{4}L$$

Since the mass of the rod increases as x^2, we are not surprised to find that the center of mass of the rod is to the right of its center. To test your understanding, can you show that $x_{cm} = L/2$ for a uniform rod (that is, $\lambda = \beta$) and $x_{cm} = 2L/3$ if $\lambda = \beta x$?

You may already have noticed that the center of mass coincides with the center of gravity of the objects we have thus far considered. Except in one case, to be mentioned soon, this is always true. We shall now prove that the center of gravity as well as the center of mass is given by our equations for finding the mass center.

To do so, let us consider the center of mass of an object to coincide with the coordinate origin so that $x_{cm} = y_{cm} = z_{cm} = 0$. Then from the defining equation 7.10 we have $x_1 m_1 + x_2 m_2 + x_3 m_3 + \cdots + x_N m_N = 0$. But if we multiply this equation by the acceleration due to gravity, g, it takes on new meaning since $m_1 g$, $m_2 g$, etc., are the weights of the tiny masses making up the whole object. Then

$$x_1(m_1 g) + x_2(m_2 g) + x_3(m_3 g) + \cdots + x_N(m_N g) = 0$$

Now $x_1(m_1 g)$ is the torque exerted by the weight $m_1 g$ about the z axis and similarly for the other terms. Since the sum of all these torques is zero according to the equation, the equation states that we have chosen the coordinate axes in such a way that gravity exerts no net torque on the object about this axis. In the same way, we can show that gravity exerts zero torque on the object about the y and x axes as well.

Coincidence of centers of mass and gravity

Clearly, by choosing our coordinate origin at the mass center, we have chosen the center point at which all gravitational forces can be considered to act. That is to say, we can balance the object by use of a pivot placed at the mass center. But this point is the point we have defined to be the center of gravity of the object. Hence, *the center of mass and center of gravity coincide*. Only one exception occurs to this rule. If g cannot be considered constant over the total expanse of the object, then our original equation has no simple meaning. In that very unusual case, the two centers do not coincide.

7.10 MOTION OF THE MASS CENTER

Suppose once again that we have a system of N small masses. They need not be connected. The x coordinate of the mass center is

$$x_{cm} = \frac{\sum\limits_{1}^{N} x_i m_i}{\sum\limits_{1}^{N} m_i}$$

Since the denominator is merely the sum of all the masses, it will be the total mass of the system M. After multiplying through by it, we obtain

$$Mx_{cm} = x_1 m_1 + x_2 m_2 + \cdots + x_N m_N \qquad 7.11$$

Let us now differentiate this equation with respect to time. We consider all the masses to be constant, and so we have

$$M\frac{dx_{cm}}{dt} = m_1 \frac{dx_1}{dt} + \cdots + m_N \frac{dx_N}{dt}$$

But dx/dt is simply the x-directed speed, and so this equation can be written

The system's momentum is Mv_{cm}

$$Mv_{xcm} = m_1 v_{x1} + \cdots + m_N v_{xN} \qquad 7.12$$

This equation tells us that *the total x-directed momentum of the system is the total mass multiplied by the x-directed speed of the mass center.* We therefore conclude that the total momentum of the system can be described in terms of the motion of the mass center.

If we differentiate *7.12* with respect to time, we obtain terms in dv/dt. But this is merely the rate of change of velocity, which is the acceleration a. Therefore, after differentiation, *7.12* becomes

$$Ma_{xcm} = m_1 a_{x1} + m_2 a_{x2} + \cdots + m_x a_{xN} \qquad 7.13$$

We may now use Newton's second law, $F_x = ma_x$, to replace the terms on the right-hand side of *7.13*. This then gives

$$Ma_{xcm} = F_{x1} + F_{x2} + \cdots + F_{xN} \qquad 7.14$$

where F_{xi} is the sum of the x-directed forces acting on the ith mass.

It is now appropriate to notice that two types of forces may act on a mass. One type of force is exerted by external means, i.e., from outside the system. These are called *external forces*. The other type of force is exerted on a mass by some other mass *within the system*. These latter forces are called *internal forces*. According to Newton's law of action and reaction, each internal force on one mass is accompanied by an equal and opposite force on one of the other masses. As a result, when we add all the internal forces, the action forces will be exactly canceled by the reaction forces (even though

The internal forces cancel they act on different points), and so the *sum of the internal forces is zero.* Therefore the sum of the forces in *7.14* is equal to the total x component of the *external* forces acting on the system. Equation *7.14* therefore becomes

$$F_x = Ma_{xcm}$$

This result can be duplicated for the y and z coordinates. The complete set of results can then be summarized in the vector equation

Motion of the mass center obeys $F = Ma$

$$\mathbf{F} = M\mathbf{a} \qquad 7.15$$

FIGURE 7.12 Even if the projectile explodes at point *B*, its center of mass will still follow the dotted path.

where M = mass of system

$\quad\quad$ **a** = acceleration of mass center

$\quad\quad$ **F** = resultant *external* force on system

We see that Newton's second law applies to a system of particles, as well as to an individual particle, provided the motion of the system is described in terms of the motion of the mass center.

Illustration 7.10

Suppose a rocket is in outer space, at rest relative to the fixed stars. Assume that no appreciable external forces act on the rocket. What can one say about the rocket's motion if it suddenly explodes?

Reasoning Since $F = 0$ in *7.15*, the mass center of the rocket cannot accelerate. When the rocket explodes, the center of mass remains fixed at the same point in space. Although rocket fragments fly in every direction, their center of mass must remain at the position it had before the explosion. The internal forces during the explosion were certainly very large. However, they cancel in pairs because of action and reaction, and therefore can cause no motion of the mass center.

Illustration 7.11

Suppose a projectile is shot as shown in figure 7.12. When it reaches the point B it explodes. What can one say about its motion? (Ignore air friction.)

Reasoning If the projectile had not exploded, it would have followed the dotted path shown. Since the same external forces act on the projectile whether it explodes or not, the center of mass will move on the same path even if it explodes. Therefore the center of mass will follow the dotted path.

7.11 ENERGY DIAGRAMS

Often one can tell a great deal about the motion of a system from a graph which shows its potential energy as a function of a coordinate. For example, consider the pendulum in figure 7.13(a). We know that if it is released at the point indicated, $x = -x_m$, the pendulum will oscillate back and forth between $-x_m \leq x \leq x_m$ provided friction effects can be neglected. The pendulum has a constant amount of energy, but the proportions of kinetic (K) and potential (U_g) energy vary as the pendulum moves from point to point. At $x = x_m$ and $x = -x_m$, all the energy is potential. As the ball approaches $x = 0$, the potential energy decreases, and is all changed into kinetic energy when the pendulum reaches $x = 0$.

FIGURE 7.13 The total energy of the pendulum remains constant. At a point such as *A*, the division of the energy between kinetic (*K*) and potential (*U_g*) is as shown.

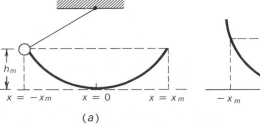

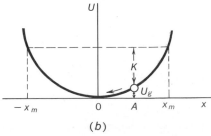

(a) $\quad\quad\quad\quad\quad\quad\quad\quad\quad\quad$ (b)

Since $U_g = mgh$ for this system, the path of the pendulum itself, being a plot of h, is essentially a plot of U_g as well. We have replotted U_g in part (b) of figure 7.13. Two things should be noted about this plot:

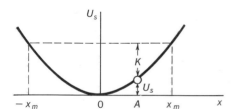

$U_g + K = const$

1. Since the sum $U_g + K$ must be a constant in the absence of friction, the values of K and U_g are immediately obvious for the system at any point such as A. The total energy of the system is equal to the potential energy the system had at the starting point, $x = -x_m$. At point A, the values of U_g and K are as indicated in the figure. Clearly, K is a maximum at $x = 0$, where U_g has its minimum value.

2. The x-coordinate motion of the system is the same as the x-coordinate motion of a bead which uses the curve in figure 7.13(b) as a frictionless wire upon which it slides. This follows from the fact that both the bead and the system represented by the graph obey the same energy relations. Both will have the same (or a proportionate) value for K and U_g at any point on the curve. As a result, the motion of the two systems will be similar.

The system–bead-on-a-wire analogy

Extending the bead-on-a-wire analogy, we see that the bead, and therefore the system, will oscillate back and forth between $\pm x_m$. In addition, if friction forces are present, the oscillations will gradually diminish in amplitude. Eventually, the bead (or the system) will come to rest at $x = 0$. Other aspects of this analogy will be pointed out as we proceed.

7.12 ENERGY DIAGRAM FOR A SPRING SYSTEM

Consider a mass oscillating in the horizontal direction at the end of a spring. We have already found the potential energy to be (6.9)

$$U_s = \tfrac{1}{2}kx^2$$

when the spring is stretched or compressed a distance $\pm x$. This energy function is plotted in figure 7.14, and it is a parabola.

Suppose the spring is compressed to $-x_m$ and released. The system may be thought of as a bead starting from rest on the curve at $x = -x_m$. It will oscillate back and forth between $\pm x_m$. When the system is at point A on the curve, it has the indicated values of K and U_s. Their sum must equal the potential energy at x_m, namely, $\tfrac{1}{2}kx_m^2$. In the presence of a small dissipative mechanism such as friction, the system will slowly decrease its amplitude of oscillation, and will finally stop at $x = 0$.

FIGURE 7.14 The potential energy function for a spring system.

$U_s + K = const$

7.13 FORCES AND THE ENERGY DIAGRAM

The curves shown in figures 7.13 and 7.14 show the potential energy of the systems as a function of one of the coordinates. If a positive increment $\Delta\mathbf{x}$ in the coordinate (an *increase* in x) results in an increase in the potential energy of the system, the distortion $\Delta\mathbf{x}$ must be in the same direction as the applied force causing the distortion. But since the increase in potential energy, ΔU, is equal to the work done by the applied force, $\mathbf{F}_{app}$, we have

$$\Delta U = \mathbf{F}_{app} \cdot \Delta\mathbf{x}$$

It is frequently more convenient to think of the motion of the system once the applied forces have been removed. In that case, any change in U is caused by the *internal* forces doing work on the system.

For example, for a mass at the end of a spring, the restoring force of the spring does work on the mass. This work results in the observed interchange between potential and kinetic energy. For this reason, it is more common to express U in terms of the internal forces, i.e., forces exerted by one part of the system on another part. They are exactly equal and opposite to the applied force needed to hold the system at some fixed distortion. (For example, to stretch a spring to an elongation x, we must exert an external force equal and opposite to the internal restoring force of the spring.) Representing the *internal* force by $\mathbf{F}$, we have

$$\Delta U = +\mathbf{F}_{app} \cdot \Delta \mathbf{x} = -\mathbf{F} \cdot \Delta \mathbf{x}$$

or

$$\Delta U = -F_x \, \Delta x$$

If we divide through by Δx, we find

$$F_x = -\frac{\Delta U}{\Delta x}$$

and if Δx is taken small enough, we can write this as

$$F_x = -\left(\frac{dU}{dx}\right)_{y,z} \qquad\qquad 7.16$$

The subscripts y, z are placed on the derivative in order to convey the fact that only the x coordinate is being varied and that any other coordinates, such as y and z, are being held constant. This is also frequently denoted by the *partial derivative* symbolism, namely,

$$F_x = -\partial U/\partial x$$

In any event, $\Delta U/\Delta x$ or $\partial U/\partial x$ is merely the slope of the U versus x curve. Therefore we see that F_x is just the negative of the slope of the potential energy curve. This will give us another tool for interpreting these curves. For example, in both figures 7.13 and 7.14, the curves have zero slope at $x = 0$. This tells us at once that the restoring force is zero when the system is at that point. Further, when the system is at positive x values, the slope is positive, and so F_x is negative. The system is being pulled back to smaller x values by its internal forces. Similarly, when x is negative, the slope is negative, and F_x will be pushing the system toward $x = 0$. All oscillating systems behave in this same general way, as we shall see in the next section.

7.14 STABLE AND UNSTABLE EQUILIBRIUM

Suppose a certain system has a potential energy curve such as the one shown in figure 7.15. Using the analogy of a bead on a wire, the system will oscillate back and forth around point C, provided the system is in the region BCD and has a total energy less than U_m. If the system is originally to the right of point B and has an energy less than U_m, it can never reach the region to the left of B since it does not have sufficient energy to rise to point U_m on the energy curve. If the system is released from rest at point G, it will move to the left, acquiring kinetic energy as it moves. Beyond E, to the left, its kinetic energy will be large and constant because the potential energy is no longer changing.

From the fact that the internal force on the system is the negative of

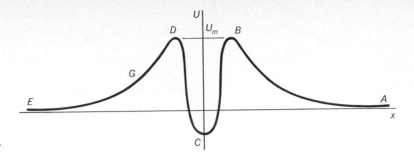

FIGURE 7.15 Where is the system in stable equilibrium? Unstable equilibrium?

the slope of the curve, we can make the following statement. To the left of *D*, the force is in the negative direction and will drive the system to the left. To the right of *B*, the force is positive and drives the system to the right. In the regions *CD* and *BC* the force drives the system toward *C*.

Consider the system when it is at point *C*. The force on it is zero since the slope of the curve is zero. If the system is displaced slightly toward *B*, or toward *D*, the force will try to return it to *C*. Therefore the system is stable around and at point *C*, tending to return to *C* if displaced slightly. We say that *C* is a point of *stable equilibrium*.

If the system is at point *B*, it also has no force acting on it, since here too the slope is zero. Therefore *B* is also an equilibrium point. However, if the system is displaced slightly from *B*, the internal forces are in such a direction as to encourage further displacement. The system will therefore continue to move away from point *B* if it is slightly displaced. Such a point is called a

Definition of stable and unstable point of *unstable equilibrium*. Can you show that *D* is similar to *B*? All these
equilibrium conclusions agree with what our bead analogy would predict.

Chapter 7
Questions
and Guesstimates

1. When a large cannon is fired, it recoils for some distance against a cushioning device. Why is it necessary to make the support so that it "gives" in this way?
2. A wad of gum is shot at a block of wood. In which case will the gum exert the larger impulse on the block: if it sticks to it or rebounds from it?
3. When a balloon filled with air is released so that the air escapes from it, the balloon shoots off into the air. Explain why this happens. Would it happen also if it were released in a vacuum?
4. Explain why a rocket is capable of accelerating even in outer space where there is no air for it to push against.
5. Suppose a nucleus explodes into several smaller particles. This happens when uranium 235 undergoes fission. What can be said about the motion of the particles?
6. An inventor constructs a sailboat with a large electric fan mounted on it. He directs the fan at the sail and blows air at the sail, expecting thereby to move in the direction of this artificial wind. To his surprise, the boat moves slowly in the opposite direction. Can you tell him why it does so?
7. A ball dropped onto a hard floor has a downward momentum, and after it rebounds, its momentum is upward. Clearly the momentum of the ball is not conserved in the collision, even though the ball may rebound to the height from which it was dropped. Does this contradict the law of momentum conservation?

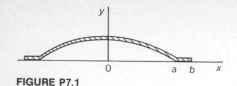

FIGURE P7.1

8. Reasoning from the impulse equation, explain why it is unwise to hold your legs rigidly straight when you jump to the ground from a wall or table. How is this related to the commonly held belief that a drunken person has less chance of being injured in a fall than one who is sober?

9. Explain, in terms of the impulse equation, the principle of operation of impact-absorbing car bumpers and similar impact-absorbing devices.

10. A baseball player has the following nightmare. He is accidentally locked in a railroad boxcar. Fortunately, he has his ball and bat along. To start the car moving, he stands at one end and bats the ball toward the other. The impulse exerted by the ball as it hits the end gives the car a forward motion. Since the ball always rebounds and rolls along the floor to him, the player repeats this process over and over again. Eventually the car attains a very high speed, and the player is killed as the boxcar collides with another car sitting at rest on the track. Analyze this dream from a standpoint of the physics involved.

11. Explain how a Mexican jumping bean can cause itself to jump.

12. Contrive a device which, momentarily at least, can have kinetic energy but no momentum. Is it possible to design a device having momentum but no kinetic energy?

13. Two blocks of unequal mass are connected by a spring, the whole system lying on an essentially frictionless table. The blocks are pushed together and tied together with a string so that the spring is compressed. If now the string is cut or burned, describe the motion of the blocks.

14. Using a energy diagram, explain why it is impossible to balance a needle on its tip.

15. A ball just fits into a metal tube, and the tube is bent into the form shown in figure P7.1. The ends of the tube are sealed tight. Draw the energy diagram for the ball in the tube, including the range of x just slightly larger than b. Discuss the motion of the ball in the tube.

16. How would you go about taking measurements to find whether the collision of a ball with the floor was perfectly elastic? The collision of two balls with each other?

17. Referring to the potential energy diagram for a particle shown in figure P7.2, (a) which points are points of stable equilibrium; (b) which points are points of unstable equilibrium; (c) in what range is the force on the particle positive?

18. Suppose you lay your hand flat on a table top and then drop a 1.0-kg laboratory mass squarely on it from a height of 0.50 m. Estimate the average force exerted on your hand by the mass. Why is injury very likely in this case even though you can catch the mass easily when dropped from this height?

19. A 100-kg man jumps from a roof 10 m above the ground. About how large a force must his legs withstand when he lands on the ground?

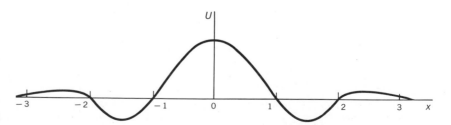

FIGURE P7.2

20. A 12 kg child falls from a window and is caught by a man 10 m below. Estimate the force experienced by the child as it is caught.

21. How large a force does a moderate size raindrop exert on the head of a bald man?

22. Estimate the force exerted by a man's head upon his neck if his stationary car is hit from the rear by a loaded truck going 30 km/h. Why does this type of accident lead to the so-called *whiplash injury?*

Chapter 7
Problems

1. A ball of mass m falls straight down to the floor and rebounds straight up. Just before hitting the floor its speed is v_0, and just after its speed is v_f. What was the change in momentum of the ball due to the collision with the floor?

2. An earth satellite of mass M circles the earth with speed v. By how much does its momentum change as it goes halfway around the earth? As it goes all the way around? Ignore the earth's rotation.

3. A baseball of mass m is thrown with speed v_0 along the third-base line (the x axis) toward the batter (at the coordinate origin) and is deflected by him along the first-base line (the y axis) with speed v_f. Using **i, j, k** notation, write the (*a*) original momentum, (*b*) final momentum, and (*c*) change in momentum of the ball.

4. A billard ball of mass m is rolling with speed v_0 parallel to one edge of a rectangular billiard table when it strikes another ball and bounces off at a 30° angle to its original direction with speed v_f. Taking its original direction to be that of the x axis, use **i, j, k** notation to represent its (*a*) original momentum, (*b*) final momentum, and (*c*) change in momentum.

5. A car of mass M and speed v strikes a tree head on and stops. (*a*) What is the magnitude of the impulse the tree exerts on the car? (*b*) If the time taken to stop is t_s, how large an average force is exerted on the car by the tree? (*c*) Evaluate the average force for a 2000-kg car with speed 5.0 m/s which stops in 0.4 s.

6. While waiting in his car at a stoplight, an 80-kg man and his car are suddenly accelerated to a speed of 5 m/s as the result of a rear-end collision. Assuming the time taken to be 0.3 s, find (*a*) the impulse on the man and (*b*) the average force exerted on him by the back of the seat of his car.

7. A 500-g ball moves along the x axis with a speed of 20 m/s. It hits a bat and reverses its direction so its speed along the x axis is now 30 m/s. Find its change in momentum and its change in kinetic energy.

8. An unlucky bystander finds himself in the center of a shootout between the good guys and bad guys. A 5.0-g bullet moving at 100 m/s strikes him and lodges in his shoulder. Assuming the bullet undergoes uniform deceleration and stops in 6.0 cm, find (*a*) the time taken to stop, (*b*) the impulse on the shoulder, and (*c*) the average force experienced by the man.

9. A stream of water from a hose is hitting a window. The window is vertical, the stream is horizontal, and the water stops when it hits. About 10 cm^3 (that is, 10 g) of water with speed 2.0 m/s strikes the window each second. Find (*a*) the impulse on the window exerted in time t and (*b*) the force exerted by the stream on the window.

10. During a switching operation, a train car of mass M_1 coasting along a straight track with speed v, strikes and couples to a car of mass M_2 sitting at rest. Find their speed after coupling.

11. While engaging in target practice, a woman shoots a 3.0-g bullet with horizontal velocity 250 m/s into a 5.0-kg watermelon sitting on top of a post. The bullet lodges in the watermelon. With what speed does the watermelon fly off the post?

12. While coasting along a street at a constant velocity of 0.50 m/s, a 20-kg boy in his 5-kg wagon sees a vicious dog in front of him. He has with him only a 3.0-kg bag of sugar which he is bringing from the grocery and he throws it at the dog with a forward velocity of 4.0 m/s relative to his original motion. How fast is he moving after he throws the bag of sugar?

13. Near the Fourth of July, a girl places a firecracker (of negligible mass) in an empty soup can (mass, 40 g) and plugs the end with a wooden block (mass 200 g). After igniting the firecracker, she throws it straight up and it explodes at the top of its path. If the block shoots out with a speed of 3.0 m/s, how fast will the can be going?

14. Two roads at right angles to each other are carrying a 20,000-kg truck moving at 10 m/s and a 1000-kg car moving at 20 m/s toward a collision at the intersection. After collision they stick together. Taking x and y coordinates along the original directions of motion of the truck and car, respectively, express the final velocity in terms of the unit vectors $\mathbf{i}$ and $\mathbf{j}$.

15. A particle of mass m traveling with speed v_0 along the x axis suddenly shoots out one-third its mass parallel to the y axis with speed $2v_0$. Express the velocity of the remainder of the particle in $\mathbf{i}, \mathbf{j}, \mathbf{k}$ notation.

16. A particle of mass M moves along the x axis with speed v_0 and collides and sticks to a particle of mass m moving with speed v_0 along the y axis. Assuming $\mathbf{i}$ and $\mathbf{j}$ to be in the directions of motion, express the velocity of the combined particle after collision in $\mathbf{i}, \mathbf{j}, \mathbf{k}$ notation.

17. An object at rest in space suddenly explodes into three equal parts. The velocities of two parts are $v_0\mathbf{i}$ and $2v_0\mathbf{j}$. Find the velocity of the third part.

18. Three equal-size particles with velocities $v_0\mathbf{i}$, $-3v_0\mathbf{j}$, and $5v_0\mathbf{k}$ collide successively with each other in such a way that they form a single particle. Find the velocity of the resultant particle in $\mathbf{i}, \mathbf{j}, \mathbf{k}$ form.

19. Two boys in never-never land each have a mass m and are at rest on a railroad flatcar of mass M which also is at rest but can move in the $\pm\mathbf{i}$ direction with no friction. The boys toss a ball with mass B back and forth with velocity $\pm v\mathbf{i}$. Find the velocity of the flatcar when the ball has a velocity (a) $v\mathbf{i}$, (b) $-v\mathbf{i}$, (c) zero, and (d) if one of the boys misses the ball and it lands on the ground beside the track.

20. A railroad flatcar of mass M is coasting along a track at speed v when a large machine of mass m topples off a platform and falls straight down onto the car. How fast is the car moving after the machine comes to rest on it?

21. A railroad car of mass M is coasting with speed v_0 past a vertical grain chute when grain begins to fall into the car at the constant rate of $\Delta m/\Delta t = k$, where Δm is the mass of grain which falls in time Δt. Find an expression for the car's speed as a function of time, t, measured from the instant the grain begins to hit the car.

22. An 80-kg astronaut finds (himself/herself) at a distance of 5 m from (his/her) space capsule without a connecting cable and at rest relative to it. Flailing of arms and legs is to no avail. Luckily (he/she) has a can of

hair spray in one pocket. Assuming there is 100 g of spray and it sprays out at 50 cm/s, how can the spray be used to get the astronaut back to the capsule? If the contents are sprayed out almost instantaneously, how long will it take the astronaut to get back to the capsule? Is this the quickest way to get back?

23. A space capsule of mass M originally at rest is subjected to a force given by $F = At$ where A is a constant. Find (a) the impulse to which the capsule is subjected in the time interval $0 \leq t \leq t_0$ and (b) the speed of the capsule at time t_0.

24. If you have some experience with calculus, repeat problem 23 in the case where $F = At^3$.

25. The separation between the centers of the hydrogen and chlorine atoms in the hydrogen chloride molecule (HCl) is about 1.30 Å (1.30×10^{-10} m). How far from the center of the hydrogen atom is the mass center of the molecule? The atomic masses of H and Cl are 1 and 35 u, respectively.

26. A 45-kg boy is sitting on one end of a teeter-totter while a 15-kg girl sits on the other. They are 4 m apart. How far from the 15-kg girl is the mass center? Where should the system be pivoted? Neglect the boards' weight.

27. Find the mass center of the system shown in figure P7.3.

28. Find the mass center of the system shown in figure P7.4.

29. Prove that the center of mass of a uniform rod is at its center.

30. A straight rod of length L has one of its ends at the origin and the other at $x = L$. If the mass per unit length of the rod is given by Ax where A is a constant, where is its mass center?

31. Repeat the previous problem if the mass per unit length is $Ax + B$.

32. A thin uniform rod of length L and mass M_1 has a uniform disk of radius a and mass M_2 fastened to the rod so that the disk and rod are coplanar. The center of the disk is at the rod's end. Find the distance of the disk's center from the mass center of the combination.

33. A 500-g pistol lies at rest on an essentially frictionless table. It accidentally discharges and shoots a 10-g bullet parallel to the table. How far has the pistol moved by the time the bullet hits a wall 5 m away?

34. As seen by a technician in the laboratory, two equal masses are shot at each other with equal speeds. After the collision, they stick together. What fraction of their kinetic energy is lost according to (a) an observer in the laboratory and (b) an observer riding on one of the masses?

35. A neutron ($m = 1$ u) moving with speed v_0 strikes a stationary particle of unknown mass and rebounds elastically straight back along its original path with speed $v_0/3$. What is the mass of the particle it struck?

36. A neutron ($m = 1$ u) moving with speed v_0 strikes a gold nucleus ($m = 197$ u) and rebounds straight back in a perfectly elastic collision. Find the speed of the gold nucleus after the collision if it is free to move and was originally at rest.

37. Two protons are moving along the x axis, one with velocity v_0 and the other with velocity $-v_0$. They undergo a perfectly elastic collision. After collision, one goes off at an angle of 37° to the $+x$ axis in the xy plane. What happens to the other? What are their speeds after collision?

38. A proton ($m = 1$ u) moving with speed v_0 along the x axis strikes a deuteron ($m = 2$ u) originally at rest. The proton glances off elastically at an angle of 37° to the $+x$ axis. Find the speed of the proton after the collision.

39. A particle of mass m has a velocity $-v_0\mathbf{i}$ while a second particle of the same mass has a velocity $v_0\mathbf{j}$. After colliding, one particle is found to

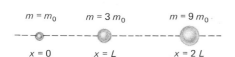

$m = m_0$　　$m = 3\,m_0$　　$m = 9\,m_0$

$x = 0$　　$x = L$　　$x = 2\,L$

FIGURE P7.3

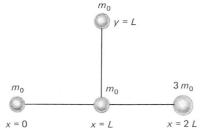

m_0

$y = L$

m_0　　　m_0　　　$3\,m_0$

$x = 0$　　$x = L$　　$x = 2\,L$

FIGURE P7.4

have a velocity $-\tfrac{1}{2}v_0\mathbf{i}$. Find the velocity of the other. Was the collision perfectly elastic?

40. A particle of mass m is moving in along the x axis and strikes an identical stationary particle at the origin. After collision, the particles move away from the origin making angles of 37° and 53° with the $-x$ axis. What fraction of the original energy remains after the collision?

41. According to a police report a car was sitting at rest waiting for a stoplight when it was hit from the rear by an identical car. Both cars had their brakes on and, from their skid marks, it is surmised that they skidded together about 8 m in the original direction of travel before coming to rest. Assuming a stopping force of about 0.7 times the combined weights of the cars (that is, $\mu = 0.7$), about what must have been the speed of the oncoming car?

42. A 500-g block sits over a hole in a tabletop. When a 5-g bullet is shot up through the hole and lodges in the block, the block flies 60 cm above the tabletop. How fast was the bullet moving?

43. Two blocks of masses 200 g and 500 g sit on a frictionless table with an essentially massless spring placed between them. They are pushed together until an energy of 3.0 J is stored in the spring. When released, the masses shoot off in opposite directions. What is the speed of (a) the mass center and (b) the 500-g block?

44. Mass m_1 in figure P7.5 is moving along a frictionless surface with speed 30 cm/s when the spring shown in the figure strikes and becomes fastened to m_2 where $m_2 = 3m_1$. (a) How fast will the center of mass be moving after the collision? (b) If the spring constant is k, what will be the maximum compression of the spring?

45. The potential energy per kilogram for a mass m in the earth's gravitational field is shown in figure P7.6, where r is the distance of the object from the earth's center and $R_e = 6.4 \times 10^6$ m is the earth's radius. Zero potential energy is taken for $r \to \infty$. (a) Ignoring air friction, how much energy is required to free a 1-kg object from the earth (i.e., to move it from $r = R_e$ to $r \to \infty$)? (b) With what speed must an object be shot away from the earth if it is to escape? (This is called the *escape velocity*.)

46. The gravitational potential energy curve for an object of mass m at a distance r from the earth's center is shown in figure P7.6. Zero energy is taken for an infinite separation. If an object is released far from the earth, find its speed at a distance $r = 2R_e$ from the earth's center, where R_e is the earth's radius. Why is it allowable to ignore air friction effects?

30 cm/s

FIGURE P7.5

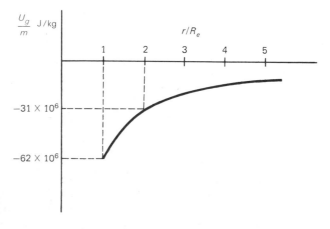

FIGURE P7.6

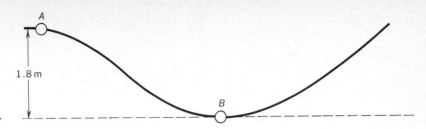

FIGURE P7.7

47. Ball A in figure P7.7 is released from the point shown. It slides along the frictionless wire and collides with ball B. If the collision is perfectly elastic, find how high ball B will rise after the collision ($m_A = m_B/2$).

48. Consider a proton (positively charged) shot head on from a large distance at a heavy nucleus (also positively charged). To a first approximation, the heavy nucleus remains motionless as the proton approaches. Repulsion between the two charges slows the proton as it approaches. Eventually the proton stops at r_0 and reverses its motion. If the potential energy of the system varies with distance between particle centers (r) as shown in figure P7.8, (a) what energy must the proton be shot with if it is to reach r_0? (b) What will be the proton speed when it is first shot? (c) When it is at $2r_0$? (m of proton is 1.67×10^{-27} kg; 1 eV $= 1.6 \times 10^{-19}$ J.)

49. The potential energy curve for a proton in the vicinity of a large nucleus is shown in figure P7.8. If the proton is released from the position $r = r_0$, (a) what will be its speed at a large distance from the nucleus? (b) at $2r_0$? Assume the heavy nucleus remains stationary (see problem 48 for numerical data).

50. The potential energy of a certain particle is given by $U = 20x^2 + 35z^3$. Find the vector force exerted on it.

51. A particle in a certain conservative force field has a potential energy given by $U = 20xy/z$. Find the vector force exerted on it.

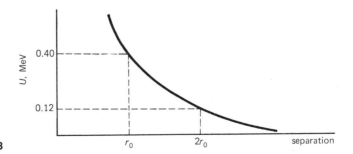

FIGURE P7.8

8 Relativistic Effects *

Our direct experience in everyday life is confined to objects moving with speeds much smaller than the speed of light. During the present century, motions at very high speeds have become of more importance, and the ability to produce and observe such speeds has greatly increased. We shall see in this chapter that the domain of high speeds does not conform to our rules for time and distance which we have inferred from observable motions at low speeds.

8.1 TWO BASIC POSTULATES

Over the centuries, multitudes of experiments have been carried out to learn the laws of nature. As early as 1905 it became apparent that the results of these experiments forced us to accept two inescapable features of the physical universe which have profound and startling significance. The existence of these two fundamental facts about nature was first faced up to by Albert Einstein in 1905. He followed them to their logical conclusions and showed that many unexpected facets of the world about us were yet to be discovered.

Einstein based his famous theory of relativity upon the following two conclusions about nature which are forced on us by the results of experiments. They are:

Two basic postulates

1. Accurate identical experiments performed in any inertial reference frame will give identical results.
2. In an inertial reference frame, accurate measurements of the speed of light in vacuum will give the same result, *c,* independent of whether or not the light source is moving relative to the reference frame.

These are the two basic postulates of relativity theory. Since they are couched in terms of what we refer to as an *inertial reference frame,* let us begin our study of relativity by discussing the meaning of this terminology and the effects of motion upon measurements.

8.2 MOVING REFERENCE FRAMES

When we take measurements, we frequently need to specify the location of an object. To do so, we make use of a set of coordinates by means of which we

---

* If your instructor postpones this chapter until later, you may wish to acquaint yourself with its major conclusions by reading Appendix 8A.

can assign a position to the object. Such a set of coordinates is called a *reference frame*. Often we generalize this meaning and take as the reference frame the room or building or planet in which the coordinates are at rest. It is of interest to consider the outcome of simple experiments in various reference frames.

Suppose a man is sealed inside a windowless boxcar traveling with constant speed, without noise, on a level, straight track. He does not know at the outset whether or not the car is in motion. In an effort to find out, he contrives several experiments. For the purposes of describing his measurements, he takes the intersection lines of the two walls and the floor at one corner of the car as his *x, y,* and *z* axes. These axes, his *reference frame,* are fixed in the boxcar and move along with it.

First the man drops a ball from his hand, and after careful observation, concludes that it fell straight down to the floor. Since this is exactly the behavior he would predict if the car were standing still on the earth, he at first concludes the car is at rest. But additional thought shows him that, even if the car were moving with constant speed, the ball would still fall straight down to the floor in his reference frame. His reasoning is easily understood by considering figure 8.1.

If the car is moving with speed v_0 relative to the earth, everything within it, the ball and man included, will have a horizontal velocity v_0 with respect to the earth. When released, the ball will begin to fall to the earth. However, as is the case for projectile motion, it will continue to move horizontally with the same speed v_0. Since the man and car are also moving horizontally with this speed, the ball will not move away from the man as it falls to the floor. It will strike the floor directly below his hand. To the man it will appear to fall straight down, although an observer outside the car standing on the ground would see both ball and man move horizontally with speed v_0. Clearly, the experimental result observed by the man in the car is independent of any constant-velocity motion the car may have. He cannot, from this experiment, determine whether or not his reference frame, the car, is in motion with constant velocity.

If we analyze this experiment, we can easily see why the experimental result does not depend upon the constant-velocity motion of the car. No force is required to maintain an object in motion with constant velocity. Originally the object appeared to the man in the car to be at rest. The only unbalanced force acting on it when it is dropped is the pull of gravity. It accelerates downward because of this force, and so he sees the object fall straight down

Only relative velocities can be measured

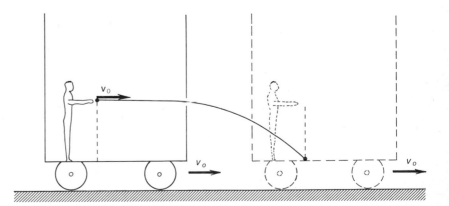

FIGURE 8.1 If the car moves at constant speed, the ball, traveling horizontally with the same speed, will travel just as far as the man as it falls to the floor.

to the floor. As long as the observer in the moving reference frame perceives that his experiments are being influenced by all the forces involved, he will expect the experimental result he actually obtains.

However, if some external force on the outside of the car was causing the car to accelerate horizontally, the unknowing observer in the car would see very strange effects. Objects would not fall straight down when released in his reference frame since no horizontal force would be available to accelerate the freely falling object. Its horizontal motion would therefore differ from that of the man and car. Another similar situation was encountered in problem 26 of Chap. 5, where it was seen that a pendulum does not hang straight down in an accelerating reference frame. Clearly, the *acceleration* of a reference frame can be detected easily.

But, as we have seen, constant-*velocity* motion requires no external force and this is why experiments restricted to a given reference frame cannot detect the constant-velocity motion of the frame. If one is allowed to look beyond the frame, motion relative to some other frame can be detected if it exists. In such a measurement one only measures relative motion—how fast one is moving relative to the other. However, since no reference frame is known to be at rest, the absolute motion of a reference frame cannot be learned. In fact, the term "to be at rest" loses much of any meaning under these circumstances.

Absolute acceleration can be measured

8.3 INERTIAL REFERENCE FRAMES

Definition of inertial reference frame

A reference frame in which measurements give results which conform to Newton's first law is called an *inertial reference frame*. It derives this name from the fact that the first law is often called the law of inertia. Any reference frame which is moving with constant velocity relative to the distant stars of the universe is an inertial reference frame. In such a reference system, Newton's first law is obeyed. In an inertial reference frame, a body at rest will remain at rest in the absence of any unbalanced force acting on it.

For most purposes the surface of the earth can be considered an inertial reference frame. However, since the earth is rotating, a coordinate system fixed on the earth's surface does not move in a straight line. The direction of its velocity changes during the day as the earth rotates. Because of this rotation of a coordinate system fixed on the earth's surface, forces are required to hold an object at rest in this rotating reference frame. Strictly speaking, an object at rest in such a rotating frame will not remain at rest if there is no unbalanced force acting on it. It is fortunate for us that the rotation effect is small enough so that Newton's laws are applicable to our earth-based experiments. To a very good approximation "objects at rest remain at rest . . ." on the earth's surface even though it is not strictly an inertial reference frame.

The earth is nearly, but not exactly, an inertial reference frame

8.4 CONCERNING THE POSTULATES

We have stated the two postulates of the theory of relativity in Sec. 8.1. The above discussion concerning experiments in inertial reference frames has pointed out the basic reason why absolute velocity is not measurable. Only relative motion can be determined, provided the motion is at constant velocity. This must mean that the basic laws of physics are the same in all inertial reference frames. If this were not true, the change in the law and the resulting change in the observed experimental results could be used to determine the absolute motion of a system.

The first postulate of relativity, that identical experiments in all inertial reference frames will give identical results, is not too difficult to accept. There are many simple experiments we can perform which support it. To say that no experiment will ever be found which disagrees with it seems to be a reasonable assertion. By 1905 most scientists agreed that the experimental evidence was sufficient to convince them of the validity of the first postulate.

But the situation was less certain in regard to the second postulate. You will recall that it asserts the constancy for the measured speed of light in vacuum for all observers. For many years prior to 1905, light was thought to be a disturbance which traveled through a material (called the *ether*) that filled all of space, including vacuum. Like sound, which is a disturbance that travels through the air, light was thought to be a disturbance that travels through the ether. Since sound is known to travel faster in the direction of the wind than against it, light too was thought to have different speeds in different directions on the earth since the earth must be flying through the ether in its journey around the sun.

When experiments finally became feasible to measure this supposed change in light speed, no such variation in speed could be detected. The first really definitive results were obtained by Michelson and Morley in 1881 to 1887. They measured the speed of light simultaneously in various directions on the earth by use of the Michelson interferometer, a device described in Chap. 30. They found the speed of light to be the same in all directions and at different places on the earth. Subsequently, many very precise measurements have been carried out to test their unexpected results. All these experiments led to the conclusion which Einstein stated as his second postulate of relativity: The speed of light in vacuum is independent of the state of motion of the source or of the observer. This speed is designated as c and is approximately 2.998×10^8 m/s.

8.5 THE SPEED c AS A LIMITING SPEED

In science as in cooking we take the attitude that the "proof is in the pudding." For a theory to be of value, it must lead to good (i.e., correct) results. Whether or not Einstein's postulates are of value is determined by their usefulness. If we can use them to predict a facet of nature still unknown to us and if subsequent experiments find the prediction to be correct, then we chalk up one point for his theory. It only takes one wrong prediction to destroy a theory. But a correct theory leads to our confidence in its correctness only after a large number of successes have been chalked up for it. The theory of relativity has not been discredited by failure. It has a large number of successful predictions to its credit.

There are several startling predictions we can make based upon simple reasoning and the two postulates of relativity theory. One prediction is that no object (or other energy-carrying entity) can be accelerated to a speed larger than c, the speed of light in vacuum. To make this prediction we shall use a favorite method of the logician. In Latin the method is called *reductio ad absurdum* (reduction to absurdity). To use it, one assumes the correctness of the idea which one wishes to prove incorrect. Reasoning from this incorrect

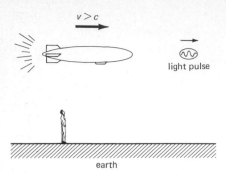

v > c

light pulse

earth

FIGURE 8.2 Both the earth and spaceship observers can measure that the light pulse is moving faster than the spaceship.

initial assumption, one arrives at a result which is known to be wrong. This then establishes the fact that the initial assumption must be wrong.

We wish to prove that no object can move faster than the speed of light, c. Using the method outlined, we will at the outset assume (incorrectly) that the spaceship shown in figure 8.2 is moving in a straight line past the earth at a constant speed *in excess* of the speed of light. Our object will be to show that this assumption leads to a result which contradicts one of the two basic postulates of relativity.

Let us suppose that, by mutual agreement with a man on earth, the woman in the spaceship sends out ahead of the ship a pulse of light from a searchlight as the ship cuts a wire above the earth's surface. As shown in figure 8.2, the light pulse will travel away out in front of the ship since, according to postulate 1, that is the way the observer in the ship (an inertial frame) must see it. Therefore both the woman in the ship and the man on earth agree that figure 8.2 is a correct representation. But the man on earth can easily measure the speed of the ship and, by our initial assumption, find its speed to be larger than c. Since he can detect that the light pulse is traveling away from the front of the ship, its speed is even larger than the speed of the ship. As a result, the man on the earth will measure the speed of light to be greater than c.

But this contradicts postulate 2 of relativity: The speed of light in vacuum must always measure to be c. Since we believe in the validity of the postulate, something must be wrong with our spaceship example. We trace the error back to the original assumption since we see that, whenever we assume a speed greater than c for the spaceship, this contradiction with the second postulate will result. This line of reasoning can be extended to include any material object. Thus we are forced to conclude:

The speed of light is a limiting speed

No material object can be accelerated to a speed larger than c, the speed of light.

This conclusion has been confirmed by experiment, as we shall see later. At that time we will find out why it is impossible to accelerate an object to a speed in excess of c.

8.6 SIMULTANEOUS EVENTS ARE NOT ALWAYS SIMULTANEOUS

We shall see in this section that the basic postulates of relativity force us to conclude that events which are simultaneous in one inertial reference frame may not be simultaneous in another. To show this simply, we will again resort to a thought experiment. The progress of a light pulse as noted by two inertial observers will form the basis for our experiment.

As shown in figure 8.3(a), suppose a boxcar is traveling to the right at a very high constant velocity. At the exact center of the car is a high-speed flashbulb which has reflectors so it will send out light pulses to the right and left when it explodes. The boxcar is fitted out with photocells at each end so a woman in the boxcar can detect when the light pulses strike the ends of the car. By some ingenious device, a man at rest on the earth is also able to measure the progress of the two pulses. Notice that both observers are in inertial reference frames (one is the moving boxcar, the other is the earth),

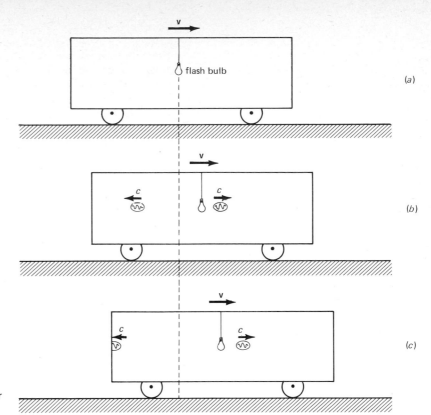

FIGURE 8.3 Unlike the inertial observer in the moving frame, the observer stationary on earth does not observe that the light pulses strike the ends of the car simultaneously.

and so they must both see the light pulses behave "normally" in their reference frames. Of course, "normal" for the man on earth is that the light pulses travel with speed c to the right and left from the flashbulb. "Normal" for the woman in the car is that the two light pulses strike the detectors at opposite ends of her car simultaneously.

Consider first the woman in the car. To her, the experiment is very simple. The flashbulb is at rest relative to her in the center of her car. When the bulb explodes, two pulses travel the equal distances to the two ends of the car in equal times. (Remember, for her the experiment must be the same whether or not the car is moving since she cannot tell.) Hence *the light pulses hit the two ends of the car simultaneously.*

Now let us consider how the man stationary on the earth sees the experiment. His measurements show the experiment to proceed "normally" (for him) and so the situation progresses as shown in (*b*) and (*c*) of figure 8.3. Notice that the pulses travel equal distances in equal time to the right and left. But since the boxcar is moving to the right, the distance to the left end is shortened. As a result, the observer stationary on the earth measures the pulse on the left to strike the end of the boxcar before the other pulse strikes the opposite end. According to him, *the light pulses do not hit the two ends of the car simultaneously.*

We must therefore conclude that time is not a simple quantity because:

Simultaneity is reference frame dependent

Events which are simultaneous in one inertial system may not be simultaneous in another.

Further considerations show that this situation exists only if the two events occur at different locations. In the present case, one event took place at one end of the car and the other was at the opposite end.

From the startling predictions we have already been able to make using Einstein's postulates, one might suspect that a quantitative approach would be of value. As we shall see, only elementary algebra and geometry will be needed to obtain the basic equations of relativity in the special case of nonaccelerating reference frames (so-called *special relativity*). Before obtaining them, though, let us first see what the basic equations were thought to be before Einstein's theory was presented.

8.7 GALILEAN RELATIVITY

To describe the motion and other behavior of objects and waves, we must possess a coordinate system or reference frame. As we have seen, observers in reference frames which are in relative motion describe events in different ways. To understand their descriptions, we must know how their coordinate systems are related. As we shall see, once this interrelation is known, the results of Einstein's theory of special relativity are easily obtained.

Before 1900, our experience with moving objects led us to expect the behavior depicted in figure 8.4. There we see two reference frames. The x, y, z frame is assumed at rest on the tabletop, while the x', y', z' frame is moving to the right with a speed v, the speed of the pan of water. Since the moving system's origin is not being displaced in the y or z direction, the y' and z'

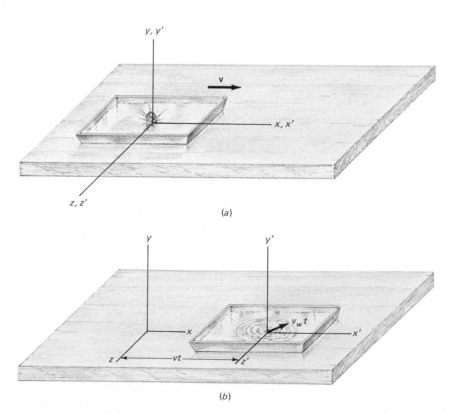

FIGURE 8.4 In the case of water waves, the waves are carried along by the moving water.

coordinates of a point will be the same as the y and z coordinates. Hence we see that for a given point, its coordinates in the two reference frames are related through

$$y = y'$$
$$z = z'$$

However, since the moving frame is being displaced to the right with speed v_x, its $x' = 0$ point is at the point $v_x t$ at a time t after the origins were coincident. Therefore the x and x' coordinates of a point are related through

$$x = x' + v_x t$$

These three equations, which were thought to relate the coordinates of a stationary reference frame to the coordinates of a moving frame, are called the *galilean transformation relations*. In terms of them it is a simple matter to describe the motion of objects as well as their positions. For example the x-direction motion of the water wave in figure 8.4 is simply

$$x' = v'_w t$$

in the moving coordinates where v'_w is the speed of the water wave measured relative to the water-pan coordinate system. Or, since $x = x' + v_x t$, the progress of the wave in the stationary system of coordinates is simply

$$x = v_x t + v'_w t$$

or

$$x = (v_x + v'_w)t$$

In other words, the velocity of an object in the stationary frame is equal to the sum of its velocity relative to the moving frame and the velocity of the frame itself.

Another example of this is shown in figure 8.5. If the ball leaves the boy's hand with speed v'_b, then its speed relative to the ground is "obviously" $v + v'_b$. We place the word *obviously* in quotation marks because it is obvious only to one who does not know about Einstein's results. Let us summarize the galilean transformations, the assumed correct relations until Einstein showed that they were not correct at very high speeds.

$$y' = y \qquad z' = z$$
$$x' = x - v_x t \qquad\qquad 8.1$$
$$u'_x = u_x - v_x$$

where u'_x and u_x are the speeds of an object expressed in the moving and stationary frames, respectively. Moreover, galilean machanics assumes that clocks in both systems will read alike and so $t = t'$.

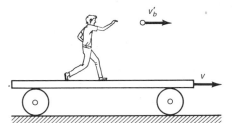

FIGURE 8.5 According to the galilean transformations, the speed of the ball relative to the ground is $v + v'_b$.

The galilean transformation relations

8.8 BREAKDOWN OF THE GALILEAN TRANSFORMATIONS

In spite of the fact that the galilean transformations agree with our everyday experience, they are wrong. Of course they must be nearly right under common situations or they would not agree with our experience. The difficulty arises under rather exceptional circumstances, chiefly when objects move with speeds close to the speed of light. Since the water wave of figure 8.4 and the object of figure 8.5 travel much less fast than light, the difficulty was not apparent in those experiments. However, if we had been using light waves instead of water waves or light pulses instead of balls, the speed would be large enough to show the effect.

One of the easiest ways to show the failure of the galilean transformations is to consider the modified experiment shown in figure 8.6. There we see a light pulse sent out along the line of motion by a man on a moving cart. As with the ball in figure 8.5, the galilean prediction is that the speed of the pulse relative to the earth should be the sum of two speeds. Since the man on the cart would measure its speed to be c, and since he is moving with speed v relative to the ground, its speed relative to the ground "should" be $c + v$.

But this result is wrong! According to the second postulate of relativity, the man on the earth must also measure the speed of the pulse to be c. Its speed relative to the earth is c, not $c + v$. Even though the galilean result $c + v$ seems more "reasonable" to us, it is wrong. We usually do not notice it to be wrong since $c + v$ and c are appreciably different only if v is large enough to be comparable to c. Only seldom, and perhaps never, do we encounter an object moving with speed v large enough to be comparable to the speed of light c. For this reason, our common experience has not confronted us with situations where the galilean velocity-addition relation is incorrect.

There can be no doubt about it, however. The galilean transformations are only an approximation. They are a good approximation for $v \ll c$, as we shall see. We wish now to find the correct transformations between the coordinates of reference frames in relative motion.

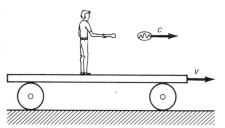

Breakdown of the galilean transformations

FIGURE 8.6 Galilean relativity predicts the speed of the light pulse relative to the ground to be $v + c$. Why cannot this result be correct?

8.9 THE LIGHT SPHERE

We shall derive the correct transformation equations between coordinate systems by considering an experiment very similar to the water-wave experiment of figure 8.4. That experiment led us to the galilean transformation equations. As we have seen, trouble arises if we consider the motion of objects at speeds close to the speed of light. For that reason, we will replace the water-wave experiment by a similar experiment which makes use of light waves.

Suppose the cart, shown in figure 8.7(a), moving on a table carries a flashbulb to the right with a speed v which is comparable to the speed of light c. We take two coordinate systems, one at rest relative to the table (x, y, z) and one moving with the cart (x', y', z'). At the instant shown in part (a) the reference frames are coincident and we will define $t = t' = 0$ at that instant. Let us assume the flashbulb explodes just at the instant depicted in part (a). We will now consider what happens thereafter as measured by an observer at rest relative to the table. Therefore, we will say that the coordinate system x, y, z is at rest while the x', y', z' system is moving to the right with speed v. (We should note, however, that since only relative motion can be determined, an equally valid choice would be to consider the cart and bulb to be at rest and the table moving to the left with speed v. In fact, later we shall examine the situation from that viewpoint.)

Since the stationary observer (the x, y, z system observer) is in an inertial reference frame, his measurements must conform with the postulates of relativity. To him the situation at a time t (that is, t after the bulb explodes) is as shown in figure 8.7(b). The cart has moved to $x = vt$, and the light from the bulb has spread out in all directions as a spherical wave centered on the position of the point where the bulb exploded. (Recall that

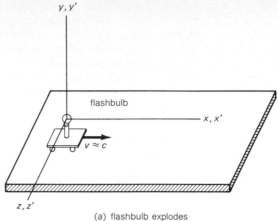

(a) flashbulb explodes

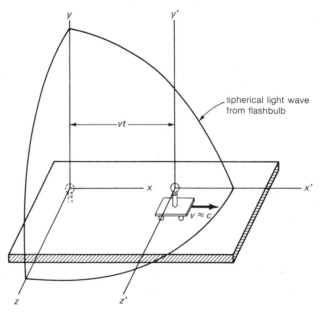

spherical light wave
from flashbulb

FIGURE 8.7 The experiment as seen by the observer who is at rest in the x, y, z frame.

(b) t seconds later. Radius of sphere is ct

postulate 2 says the light behaves the same whether or not the bulb is moving.) Light travels with speed c, and so the radius of the light-wave sphere is given by $r = ct$.

You may recall from geometry that the equation of a sphere is $r^2 = x^2 + y^2 + z^2$. In this case, $r = ct$ and so the stationary observer writes the equation for the light-wave sphere to be

Stationary observer: $\qquad x^2 + y^2 + z^2 = c^2 t^2 \qquad\qquad$ 8.2

This is his description of the light wave as he observes it in his coordinate frame.

Now comes the almost unbelievable part. As unlikely as it may seem, an observer on the cart does not see the light wave to be as shown in figure 8.7(b). She, too, is in an inertial frame and so everything must appear normal to her in the x', y', z' system. In particular, her measurements cannot tell her she is moving (which postulate says this?) and so her results will be those one would obtain if she were at rest. To her, the light wave spreads out in all directions from the bulb on her cart (and she might just as well think her cart is at rest since she cannot tell). The wave is a sphere centered on her light bulb. So she writes the equation of the light wave in terms of her coordinate system

$$r'^2 = x'^2 + y'^2 + z'^2$$

Of course light travels with speed c according to her too, and her clock on the cart measures the time since the flashbulb exploded to be t'. Therefore, she knows $r' = ct'$. Thus, to the observer on the cart, whom we will call the moving observer, the equation of the light-wave sphere is

Moving observer: $$r'^2 = x'^2 + y'^2 + z'^2 \qquad\qquad 8.3$$

In the view of the stationary observer, the obervations of the observer on the cart are definitely not galilean. Although the stationary x, y, z observer measures the situation to be that shown in figure 8.7(b), the observer on the cart (x', y', z') measures the situation to be quite different. For her, the light sphere is centered on the position of the bulb in part (b), not upon the position shown in the figure.

As contradictory as this may seem, the postulates of relativity tell us that the measurements of the two observers will yield this result. Both equations, *8.2* and *8.3*, are correct representations of the experimental results of the two observers. They express the fact that the two observers measure the same light-wave sphere to have its center at two different points. But if they do accurate measurements and obtain different results, then there must be some discrepancy in their measuring tools, their clocks, and measuring sticks.

Before discussing the discrepancy, though, we must be careful to point out that both observers are correct. Although the x, y, z observer has been called the stationary observer by us, the postulates of relativity tell us that only relative velocities can be determined. In actuality, the cart may be at rest and the table could be flying to the left. Or perhaps both are moving. All we can really say is that one is moving with speed v relative to the other.

8.10 THE LORENTZ–EINSTEIN TRANSFORMATION EQUATIONS

Since both equations *8.2* and *8.3* are correct, it should be possible to obtain a relation between the x, y, z, t measurements of the stationary observer and the x', y', z', t' measurements of the moving observer. We recall that galilean relativity gave these relations as equation *8.1*, namely, $y' = y$, $z' = z$, $t' = t$, and $x' = x - vt$ with $v \ll c$ in this case. If *8.2* and *8.3* are correct, then the galilean relations must be wrong. For if we substitute from them for x', y', z', and t' in equation *8.3*, we do not obtain *8.2* but instead

$$x^2 + y^2 + z^2 - 2xct = 0$$

We therefore seek the relations between x, y, z, t and x', y', z', t' which yield equation *8.2* from *8.3*. These then should be the correct transformation equations.

The algebra involved in obtaining the correct transformation equations is lengthy and we relegate it to Appendix 8B. They are found to be as follows:

Lorentz-Einstein transformation

$$y' = y \qquad\qquad 8.4a$$

$$z' = z \qquad\qquad 8.4b$$

$$x' = \frac{x - vt}{\sqrt{1 - \beta^2}} \qquad\qquad 8.4c$$

$$t' = \frac{t - (vx/c^2)}{\sqrt{1 - \beta^2}} \qquad\qquad 8.4d$$

where $\beta = v/c$. You may wish to confirm the correctness of these by substituting them in equation *8.3* to obtain *8.2*. They were first derived by H. A. Lorentz, but their full importance and meaning were first pointed out by Einstein. We call them the Lorentz-Einstein transformation.

Two very important points should be made concerning the transformation relations. First, since either observer could have been called "stationary," equations *8.4* could equally well be written for the moving observer. In that case, x would replace x' and so on. Further, since the observer's motion would reverse in direction, the sign of v would be changed.

When $v \ll c$ the Lorentz and galilean transformations are equivalent

Second, the Lorentz-Einstein relations reduce to the galilean relations when the relative motion of the coordinate systems is small. This is easily seen by noting that if $v/c \ll 1$ then the terms involving v/c and β can be neglected. Therefore, at low speeds such as those we usually encounter, the relations agree with our everyday experience as they must.

With the aid of the Lorentz-Einstein transformation relations, we can proceed to deduce some extremely important conclusions about the physical universe. We have already found two of these: (1) the speed of light is a limiting speed and (2) simultaneous events in one reference frame may not be simultaneous in another. The deductions we shall now make concern our basic concepts of length, time, and mass.

8.11 LORENTZ–FITZGERALD LENGTH CONTRACTION

If we look at figure 8.7, it is apparent that the stationary observer (whom we will call A) will find it difficult to reconcile his observations with those of the moving observer (whom we will call B). The moving observer on the cart measures that she is at the center of the light-wave sphere shown. In the opinion of the stationary observer A, the only way B could arrive at such a result is if there is something strange about B's measuring instruments. Basically, these instruments will be length- and time-measuring devices. Let us first determine what properties A must attribute to B's meter sticks.

Suppose B in the moving reference frame has a rod of length L' which she lays along the y axis. Its two ends will be at y'_1 and y'_2, so we have

$$L' = y'_2 - y'_1$$

But according to observer A, the rod ends are at y_2 and y_1 in his coordinate system so he measures the length of the rod to be

$$L = y_2 - y_1$$

Since the transformation equations tell us $y = y'$, we see that $y_2 - y_1 = y'_2 - y'_1$ and so $L = L'$. Hence both the moving and stationary observer agree that the rod has the same length L when oriented in the y direction. A similar result is obtained if the rod is oriented in the z direction.

When the rod is placed along the x direction so that it is lined up with the direction of motion, a different result is obtained. In B's system we have

$$L' = x_2' - x_1'$$

while for A

$$L = x_2 - x_1$$

From the transformation relations we see that $x' = (x - vt)/\sqrt{1 - \beta^2}$. Substituting this in the relation for L' yields

$$L' = \frac{x_2 - x_1}{\sqrt{1 - \beta^2}}$$

Or, since $x_2 - x_1 = L$, this becomes

$$L = L'\sqrt{1 - \beta^2} \qquad\qquad 8.5$$

where $\beta = v/c$.

This tells us that if a rod is moving with observer B and is oriented along the line of motion, then if B measures its length to be L_0, a stationary observer (A) will measure its length to be $L_0\sqrt{1 - \beta^2}$. Since the rod is at rest relative to B, we say that B has measured its *proper length* (i.e., the length measured in a reference frame in which the object is at rest). If the rod is moving lengthwise with speed v relative to some other observer, the moving rod's length will be measured as $L_0\sqrt{1 - \beta^2}$. In other words,

An observer past whom a system is moving with speed v measures objects in the moving system to be shortened along the direction of motion by a factor $\sqrt{1 - \beta^2}$.

This statement summarizes what is called the *Lorentz-Fitzgerald* contraction after the men who first postulated it. Notice that only the dimension along the line of motion is changed. The other two perpendicular dimensions are unchanged. It should also be remembered that the effect occurs only if the object is moving.

As a simple example of this effect, consider the hypothetical case of a woman and her meter stick traveling in a spaceship close to the speed of light past the earth. Since she is in the spaceship with the meter stick, the stick is essentially at rest with respect to her. The woman therefore measures its proper length no matter what the stick's orientation. Carefully notice that, to her, no contraction effect exists. (If it did, she could tell she was in motion.) However, someone on earth taking measurements of the meter stick as it flies past would measure it to contract by the factor $\sqrt{1 - \beta^2}$ as it is rotated from perpendicular to parallel to the line of motion.

Illustration 8.1

How fast would a spaceship have to be moving relative to the earth if an observer on earth were to measure it to be only half as long as an observer moving with the spaceship would measure it?

Reasoning The length-contraction factor is $\sqrt{1 - \beta^2}$, where $\beta = v/c$. If this factor is to be 0.50, we have

$$0.25 = 1 - \beta^2$$

144   RELATIVISTIC EFFECTS

$$\text{and} \qquad\qquad \beta = \frac{v}{c} = \sqrt{0.75}$$

from which
$$v = 0.867c$$
$$= 2.6 \times 10^8 \, \text{m/s}$$

Notice that the length contraction becomes appreciable only for objects moving near the speed of light. It is not observable for ordinary motions on the earth, except if the objects being observed are of atomic size. Only for these can speeds approaching c be realized.

Illustration 8.2

As measured from the earth, the distance from the earth to Alpha Centauri, the nearest star, is 4.3×10^{16} m. A spaceship observer moving at a speed $v = 0.990c$ from earth to the star will measure the distance to be how large?

Reasoning The man in the spaceship and the line from earth to star (a fictitious rod if you like) have a relative speed of $0.990c$. As a result, the man will not measure the proper length of this distance but will, instead, find its length shortened by the factor $\sqrt{1 - \beta^2}$. But

$$\sqrt{1 - \beta^2} = \sqrt{1 - (0.990)^2} = \sqrt{0.02} = 0.14$$

He will therefore measure the distance to be

$$(0.14)(4.3 \times 10^{16} \, \text{m}) = 0.61 \times 10^{16} \, \text{m}$$

In practice, speeds for spaceships are far lower than this. However, if ships could travel at such high speeds, the trip would be shortened for its occupants by this effect.

8.12 TIME DILATION

We saw in the previous section that an observer measures objects which are flying past him to have shrunk along the line of motion. Therefore we see that the basic concept of length is complicated by relativistic effects. The next basic concept we wish to investigate is time. In particular, we shall discuss the behavior of clocks as they tick out time intervals in two different reference frames.

Let us begin with two identical clocks which tick out time identically when at rest side by side. Returning to the hypothetical example shown in figure 8.7, we place one clock at the origin of A's coordinate system and the other at B's. At the instant when the two origins are coincident, both clocks are started. Thereafter A's clock will tick out time t in the x, y, z coordinate system while B's clock ticks out time t' in the x', y', z' system with which it is moving. The transformation equations 8.4 tell us how t and t' are related.

In order to compare the behavior of the two clocks, let us view a particular experiment from A's coordinate system. By prior agreement, B is to stop her moving clock as she passes a particular point in A's reference frame. A will stop his clock at the same instant. Both clocks will read the time taken for B to move from the point where the origins were coincident to the previously agreed upon point. Since in our view A's clock is at rest, its reading for this trip will be t. The time for the trip read by B's clock is t'. How do t and t' compare?

To find the relation between t and t' we need only refer to the transformation equation given in 8.4. It is

$$t' = \frac{t - (vx/c^2)}{\sqrt{1 - \beta^2}}$$

As shown in figure 8.7, according to A the distance x is simply vt. Upon substitution of this value and factoring out t, we find

$$t' = t\frac{1 - (v/c)^2}{\sqrt{1 - \beta^2}}$$

But $v/c = \beta$ and so the time read by the moving clock t' is related to the time t by

$$t' = t\sqrt{1 - \beta^2} \qquad\qquad 8.6$$

The moving clock says less time has passed than is recorded by the stationary clock.

We are therefore forced to conclude that when a moving clock is compared with a clock at rest in our reference frame, the moving clock will be found to tick out time too slowly. When our stationary clock ticks out a time of t, the moving clock will have ticked out a fewer number of seconds, namely, $t\sqrt{1 - \beta^2}$. Moving clocks lengthen the time between their ticks. It is for this reason that the effect is referred to as *time dilation*. To summarize this effect:

Time dilation

A clock moving with speed v relative to an observer will appear to run too slowly by a factor $\sqrt{1 - \beta^2}$ when compared with a clock stationary with respect to the observer.

Note that the effect is stated entirely in terms of relative motion. Any observer at rest relative to a clock measures time by means of the stationary clock. A clock moving with speed v relative to the observer ticks out time more slowly. Neither clock can be said to be right or wrong. The same clock can read a certain time for a man carrying it while reading a shorter time for a man rushing past it.

Illustration 8.3

A pi-meson (π-meson, or pion) is an unstable particle formed in large numbers in reactions carried out in the larger nuclear accelerators. It has a mass about 270 times larger than an electron's mass and carries a charge equal to that of an electron. When a number of these particles are observed at rest in the laboratory, it is found that they disintegrate to form other particles. After a time of 2×10^{-8} s, half the original mesons will be found to have disintegrated. Suppose these mesons, instead of being at rest, were moving through the laboratory with a speed $v = 0.98c$. Speeds even larger than this (but never larger than c!) are often observed for the mesons rushing from the target which is being bombarded by a nuclear accelerator. How long will it take, according to the laboratory timers, for half of these mesons to disintegrate?

Reasoning We know very little about how a meson disintegrates. However, the physical situation involved must be dependent on the rate at which processes are occurring within the meson. These processes act as a clock, and this "clock" moves with the meson. As seen from the laboratory reference frame, the meson "clock" will be running too slowly by a factor of

$$\sqrt{1 - \left(\frac{0.98c}{c}\right)^2} = 0.20$$

Since the mesons disintegrate as a result of their own internal processes, their "clock" must read 2×10^{-8} s when half have been lost. The laboratory clock will read this time to be $1/0.20 = 5$ times larger. As a result, half of the moving mesons will disintegrate after a time of 10×10^{-8} s, which is a time 5 times larger than observed when the mesons were at rest in the laboratory. Measurements such as those indicated by this illustration have been carried out, and the results predicted by our computations have been well confirmed.

Illustration 8.4

The star closest to our solar system is Alpha Centauri, which is 4.3×10^{16} m away. Since light moves with a speed of 3×10^8 m/s, it would take a pulse of light 1.43×10^8 s, or 4.5 years, to reach there from the earth. (We say that the distance to the star is 4.5 *light-years*.) How long would it take according to earth clocks for a spaceship to make the *round trip* if its speed is $0.9990c$? According to clocks on the spaceship, how long would it take?

Reasoning To a good approximation, we can take the spaceship speed to be c for this computation, and so the *round trip* would require 9.0 years according to earth clocks.

The spaceship clocks will appear to run too slow by the relativistic factor

$$\sqrt{1 - (0.999)^2} \approx 0.045$$

Therefore the spaceship clocks will read the 9.0 years as $(0.045)(9.0)$, or about 0.4 year. As a result, the journey would seem to take only about 5 months according to the crew of the spaceship—far more tolerable than the 9.0 years which people on earth would record.

Twin paradox Incidentally, the twin of one of the crew who was left behind on the earth would age 9.0 years during the time of the voyage. However, his twin brother, a member of the spaceship crew, would only age 5 months. This phenomenon, the so-called *twin paradox,* has been discussed at length by scientists. They generally agree that this result is valid and that the two twins actually will age differently.

8.13 THE RELATIVISTIC FACTOR

We have seen in the previous sections that the factor $\sqrt{1 - \beta^2}$ with $\beta = v/c$ is of importance in relativity. It is called the *relativistic factor,* and it is a measure of departures from galilean or classical physics. If v/c is much smaller than unity as it is in all everyday situations, then β^2 is so small that the relativistic factor is essentially unity. Under those conditions, relativity and classical physics predict nearly identical results.

However, when v becomes comparable to c (as a rough estimate, when v is larger than about 3×10^7 m/s), then the galilean transformations are noticeably wrong. Classical physics no longer applies and one must use the correct relativistic relations. Since the relativistic equations reduce to the classical ones when $v \ll c$, we still make use of the simpler classical equations in everyday situations. This will become more apparent to us in later sections.

In computations using the relativistic factor there is a mathematical approximation which often shortens the arithmetic greatly. It is based upon the series expansion for the quantity $\sqrt{1-x}$. From Sec. A3.4 of Appendix 3 we find for $x \ll 1$

$$\sqrt{1-x} = 1 - \tfrac{1}{2}x - \tfrac{1}{8}x^2 - \cdots \cong 1 - \tfrac{1}{2}x$$

and

$$\frac{1}{\sqrt{1-x}} = 1 + \tfrac{1}{2}x + \tfrac{3}{8}x^2 + \cdots \cong 1 + \tfrac{1}{2}x$$

By means of these we easily see that the relativistic factor is $1 - \tfrac{1}{2}\beta^2$ if β^2 is much smaller than unity. Even for a speed so high that $v/c = 0.10$, the factor is very close to unity, namely, 0.995. As we see, relativistic effects only become appreciable at the very highest speeds.

8.14 VELOCITY ADDITION

As another example of the use of the transformation equations, let us determine the relativistic velocity-addition equations. Suppose in the situation depicted in figure 8.8 that man B is moving with speed v in the $+x$ direction relative to man A. If B throws an object in the $+x$ direction with speed u' relative to himself, how fast will the object be moving in A's view? We saw previously that the "reasonable" answer $v + u'$ given by the galilean transformations is incorrect. It predicts speeds in excess of c. The correct result can only be obtained by use of the Lorentz-Einstein transformation.

If we have both A and B measure the speed of the object by noting the distance it travels in time t or t' after passing their instantaneously coincident origin of coordinates, we have

$$u' = \frac{x'}{t'} \qquad \text{and} \qquad u = \frac{x}{t}$$

where u is the speed A notes for the object.

Taking the ratio of x'/t' from 8.4, we find

$$u' = \frac{x - vt}{t - x(\beta/c)}$$

After dividing numerator and denominator by t and recognizing that $x/t = u$, we have

$$u' = \frac{u - v}{1 - u(\beta/c)}$$

FIGURE 8.8 Although B sees the ball's speed to be u', A observes it to be

$$u = \frac{v + u'}{1 + vu'/c^2}$$

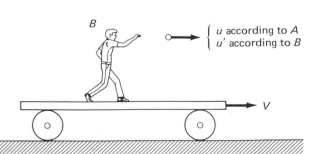

u according to A
u' according to B

This equation can now be solved to give u, and after using $\beta = v/c$, the result is

Relativistic velocity addition in parallel motions

$$u = \frac{v + u'}{1 + vu'/c^2} \qquad (u' \text{ and } v \text{ parallel}) \qquad 8.7$$

We have arrived at a result telling us how fast A considers the object to move if B claims he threw it with speed u' along the direction in which A sees him moving at speed v. What does it predict if B's speed v is $0.8c$ and the thrown object appears to B to have a speed u' of $0.6c$? Substitution of these values yields

$$u = \frac{0.8 + 0.6}{1 + 0.48}c = 0.947c$$

and so A still sees it as moving with a speed less than c. In fact, only when $v = u' = c$ does $u = c$.

We may proceed in a similar way to find the speed seen by observer A when observer B throws an object with speed u'_y perpendicular to the x direction, the direction of relative motion. As before, we write

$$u'_y = \frac{y'}{t'} \qquad \text{and} \qquad u_y = \frac{y}{t}$$

But we have previously found (8.4) that

$$y' = y \qquad \text{and} \qquad t' = \frac{t - \beta x/c}{\sqrt{1 - \beta^2}}$$

from which

$$u'_y = \frac{y\sqrt{1 - \beta^2}}{t - \beta x/c}$$

After dividing numerator and denominator by t and recognizing that $y/t = u_y$ and $x/t = u_x$, we find

Relativistic velocity addition for perpendicular motion

$$u'_y = \frac{u_y \sqrt{1 - \beta^2}}{1 - \beta u_x/c} \qquad 8.8$$

By reversing the process and solving for u_y, one finds (as one would expect from the equivalence of the two observers)

$$u_y = \frac{u'_y \sqrt{1 - \beta^2}}{1 + \beta u'_x/c} \qquad 8.9$$

When v is much smaller than c, 8.7 says

$$u \approx v + u'$$

and so the classical answer is correct for objects not approaching the speed of light. We therefore see once again that the results of special relativity agree with the classical results for motion at speeds much less than the speed of light. Einstein's theory has therefore shown us that the classical results are correct in the situations for which they were postulated, namely, motion at ordinary speeds. However, his theory shows clearly how we must treat time and distance in the less familiar domain of very high speeds.

8.15 MOMENTUM AND RELATIVISTIC MASS

Let us now consider the effect of very high speeds on mass, momentum, and force. It is easy to see that these quantities are made more complicated by relativistic effects. For example, if we were to apply a constant force F to an object for a long time, the simple form of the second law, $F = ma$, would indicate that the object should continue to accelerate with $a = F/m$. But

$$v = v_0 + at$$

and so the speed v should increase indefinitely. However, we have seen that v cannot exceed c. As a result, the acceleration cannot remain constant, or this would certainly happen. Clearly, something is wrong with $F = ma$ at high speeds.

Basically, Newton's second law makes a statement relating forces and the time rate of change of momentum. To discuss its behavior directly at high relative speeds is not a judicious way to approach the problem since it involves two quantities, force and momentum, which are both uncertain. It would be far better to discuss, first, the behavior of momentum alone at high speeds. After that, the definition of force could be more intelligently decided upon. For that reason, we shall first investigate the momentum of objects traveling at high relative speeds.

One of the basic laws of newtonian mechanics which we should like to preserve if possible is the law of conservation of momentum. Let us now examine a collision in the relativistic range and see if momentum is still conserved. For this purpose, consider the following thought experiment.

Two railroad flatcars, A and B, approach each other on parallel straight tracks. The speed of each car relative to the ground we denote by v_g. A top view of the situation is shown in figure 8.9. Available to man A and man B on the two cars are identical balls, each of which had mass m_0 when they were compared at rest on the earth. Both men are required to throw their balls *perpendicular to the direction of motion* with identical speeds v_0 (as measured by the man doing the throwing). In addition, they are to be thrown in such a way that they collide at D and rebound, as indicated in figure 8.9. The paths shown there are the paths seen by an observer on the ground. Each man sees his own ball to travel exactly perpendicular to the tracks, i.e., in the $\pm y$ direction, with speed v_0. (Notice that two different speeds are being discussed here. For simplicity, the speed v_0 of the ball relative to the thrower is assumed rather small, while the speed of the car v_g may be very large.)

This collision as seen by the observer on the ground is quite normal. After collision the x-component momentum of each ball is the same as before collision. The x-directed momentum is therefore conserved. In addition, the

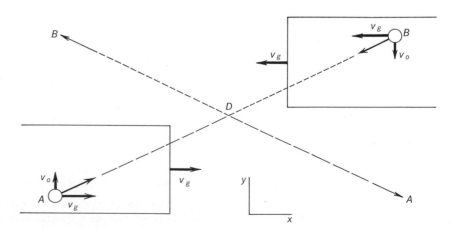

FIGURE 8.9 The collision of two identical balls as observed by a man stationary near point D.

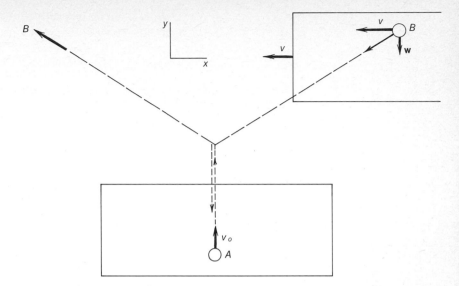

FIGURE 8.10 The collision as seen by man A.

y-directed momentum of the system is zero both before and after the collision since the A and B balls have equal but opposite y momentum. As a result, the observer on the earth concludes that momentum is conserved in the collision.

Let us now view the collision from the standpoint of man A. Assuming himself to be at rest, he sees B's car approaching with speed v. The collision seen by him is shown in figure 8.10. Notice that he sees his ball to move in the y direction, with its velocity v_0 being reversed by the collision. He also sees B's ball to move with a velocity whose y component is w; this component is reversed by the collision.

For man A, the law of conservation of y-directed momentum in the collision may be written

Momentum before collision = momentum after collision
$$m_0 v_0 - mw = -m_0 v_0 + mw$$

where m_0 and v_0 are the mass and speed of A's ball, and m and w are the similar values for B's ball. Although both A's and B's balls had equal masses to begin with, A decides (wisely) that B's ball may have changed, and so he calls its mass m. We use m_0 for the mass of A's ball since it is moving slowly relative to him. Under these conditions we know its mass is essentially the same as when it was at rest.

Now we know that B, following instructions carefully, threw his ball with a y velocity of v_0 *as measured by himself, B*. However, w is the speed which A sees B's ball to have. From Sec. 8.14 we know how these two observations compare. From *8.9* we find

$$w = \frac{v_0 \sqrt{1 - (v/c)^2}}{(1 + \beta v_{0x}/c)}$$

where v_{0x} is the x component of the ball's velocity as seen by B. Since $v_{0x} = 0$,

$$w = v_0 \sqrt{1 - \left(\frac{v}{c}\right)^2}$$

Notice that no assumption is made concerning how fast the car is moving with respect to A.

If the relation for w is substituted in A's equation for the conservation of momentum, we obtain

$$2m_0 v_0 = 2mv_0 \sqrt{1 - \left(\frac{v}{c}\right)^2}$$

Upon solving for m, the mass A actually concludes B's ball to have, we find

$$m = \frac{m_0}{\sqrt{1 - (v/c)^2}} \qquad\qquad 8.10$$

Of course, B sees a similar situation and concludes that A's ball has the mass given by 8.10.

This tells us that, if the law of conservation of momentum is to be preserved for observers in relative motion, moving objects must be considered to possess a variable mass. An object moving with speed v past an observer appears to that observer to have a larger mass (i.e., more inertia) because of its motion. In particular, if the mass of the object is m_0 (the so-called *rest mass*), when the object is at rest relative to the observer, its mass m will be given by 8.10 when it is moving with speed v past the observer. Notice that this effect is very small except when v is comparable with the speed of light, c. However, as the speed $v \rightarrow c$, the value of $m \rightarrow \infty$, as the equation shows. A graph showing m as a function of v/c is given in figure 8.11.

To preserve the momentum conservation law, the mass m of a moving object must be related to its rest mass m_0 by 8.10

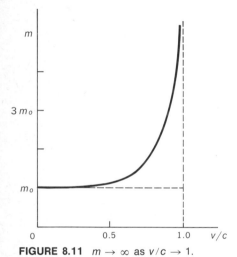

FIGURE 8.11 $m \rightarrow \infty$ as $v/c \rightarrow 1$.

Illustration 8.5

How fast must a proton accelerated by a nuclear accelerator be moving if its mass is to be twice its rest mass?

Reasoning Using 8.10 directly, we find

$$2 = \frac{1}{\sqrt{1 - (v/c)^2}}$$

Squaring, solving for $(v/c)^2$, and then taking the square root gives

$$\frac{v}{c} = \frac{\sqrt{3}}{2} \approx 0.867$$

The student may find it instructive to find this ratio for $m/m_0 \approx 100$.

8.16 FORCE IN RELATIVISTIC MECHANICS

It is quite natural to ask if the mass increase indicated by 8.10 is "real." In order to answer this question, we must decide what we mean by "real." If we mean that the increased mass of the particle is directly observable, then the answer is, yes, the mass is real. For example, in the experiment performed by A described in the preceding section, the moving particle (or ball) was seen to have more momentum than one would expect from the relation $m_0 v$. There are more direct observations which emphasize this fact more clearly.

For example, as you will learn later, if you do not know already, when a charged particle such as an electron is shot into a region where a magnetic field exists, its motion is quite simple. If the magnetic field is perpendicular to the page in figure 8.12, the electron will travel in a circular path as indicated. The radius of this circle is proportional to the momentum of the particle, mv.

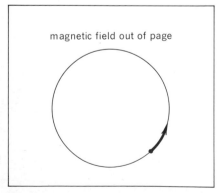

magnetic field out of page

FIGURE 8.12 The radius of the circular path of the electron is a measure of the electron's momentum.

Carrying out such an experiment with electrons of different velocities near c, one finds that m must indeed increase as indicated by *8.10*.

Many other experiments can be carried out to detect the predicted change in mass. The predicted change is always found. We can therefore state that the mass change is "real" in the sense that all known experiments which are capable of detecting effects of the mass change have led to the expected result. This fact is a direct consequence of the way in which we have *defined mass* at high speeds in order to preserve the law of conservation of momentum, another defined quantity. The fact that this agreement is found simply indicates that our definitions are not mutually contradictory.

A similar situation arises in the definition of force. In newtonian mechanics it was defined to be

The usual definition of force, 8.11, is applicable to the relativistic region

$$\mathbf{F} = \frac{d(m\mathbf{v})}{dt} \tag{8.11}$$

This definition agrees with our usual ideas about what a force is, based on easily observed experiments. It is convenient that we have defined force in such a way that it agrees with our qualitative, colloquial use of the word.

It proves convenient to retain *8.11* as the definition of force even at relativistic speeds. It is precisely for this reason that we prefer *8.11* to $F = ma$ since the latter form can be obtained from *8.11* only by assuming m to be constant. Although this definition of force leads to force being a rather complex quantity in the relativistic range, it has the advantage of being identical with the newtonian definition. We must recognize, however, that m in *8.11* is not necessarily constant.

8.17 RELATIVISTIC ENERGY

In this section we shall see that the theory of relativity altered profoundly our ideas of mass and energy. The new insights led us to discover a previously unknown energy source, nuclear energy. Let us now see how kinetic energy should be defined at high speeds.

As in the nonrelativistic case, we begin by defining work done by a force $\mathbf{F}$ in a small displacement $\Delta\mathbf{s}$ by the relation

$$\Delta W = \mathbf{F} \cdot \Delta\mathbf{s}$$

If the force is in the direction of motion, this becomes

$$\Delta W = F\,\Delta s \tag{8.12}$$

But we have defined force through the relation

$$F = \frac{d(mv)}{dt}$$

which is, for the present case,

$$F = \frac{\Delta(mv)}{\Delta t} \tag{8.13}$$

If we now substitute *8.13* in *8.12*, we have

$$\Delta W = \Delta(mv)\frac{\Delta s}{\Delta t}$$

After recognizing that $\Delta s / \Delta t$ is merely v, this becomes

$$\Delta W = v\, \Delta(mv) \qquad\qquad 8.14$$

To find $\Delta(mv)$ we notice that m is given in terms of the rest mass to be

$$m = \frac{m_0}{\sqrt{1 - (v^2/c^2)}} \qquad\qquad 8.15$$

Upon squaring both sides, clearing fractions, and rearranging, this becomes

$$m^2 v^2 = m^2 c^2 - m_0^2 c^2$$

From this we obtain

$$mv = c\sqrt{m^2 - m_0^2} \qquad\qquad 8.16$$

It is possible by the use of algebra or calculus to obtain $\Delta(mv)$ from this equation for mv. We shall not carry through the rather lengthy algebra. The end result is

$$\Delta(mv) = \frac{c^2\,\Delta m}{v}$$

This value is now suitable for substitution into the equation for ΔW, *8.14.* We then find

$$\Delta W = c^2\,\Delta m \qquad\qquad 8.17$$

To find the total work done, we sum over the increments in Δm as the mass changes from its rest mass m_0 to its final value m. Therefore

$$W = c^2 \sum \Delta m_i$$

But the sum of all the Δm_i is simply the change in mass as the object is accelerated from rest (where its mass is m_0) to high speed (where its mass is m). Therefore we can replace $\Sigma\, \Delta m_i$ by $m - m_0$, from which

$$W = c^2(m - m_0)$$

or

$$W = mc^2 - m_0 c^2 \qquad\qquad 8.18$$

This may also be written in the following way by replacing m, using *8.15*:

$$W = m_0 c^2 \left(\frac{1}{\sqrt{1 - (v/c)^2}} - 1 \right) \qquad\qquad 8.19$$

Equations *8.18* and *8.19* give us expressions for the change in motion of an object in terms of the work done on the object. Since we have ignored dissipative (i.e., friction) forces and have assumed the potential energy to be constant, this work must appear as kinetic energy of the object. We therefore tentatively conclude that the kinetic energy of an object is given by *8.18* and *8.19* to be

$$K = m_0 c^2 \left(\frac{1}{\sqrt{1 - (v/c)^2}} - 1 \right) \qquad\qquad 8.20$$

or its equivalent,

$$K = mc^2 - m_0 c^2 \qquad\qquad 8.21$$

To see if this interpretation is consistent with our previous definition of K for objects moving at nonrelativistic speeds, let us expand *8.20* in the limit of $v/c \ll 1$. This may be done if we recall the series expansion formula

$$\frac{1}{\sqrt{1-x}} = 1 + \frac{1}{2}x + \frac{3}{8}x^2 + \frac{5}{16}x^3 + \cdots$$

In our case $(v/c)^2 = x \ll 1$, and so terms beyond x can be neglected. Equation 8.20 then becomes

At $v \ll c$, 8.21 reduces to $\frac{1}{2}m_0v^2$

$$K \approx m_0c^2\left[1 + \frac{1}{2}\left(\frac{v}{c}\right)^2 - 1\right] \qquad \frac{v}{c} \ll 1$$

$$= \tfrac{1}{2}m_0v^2 \qquad \frac{v}{c} \ll 1$$

This is the familiar equation for kinetic energy, and so it is plausible to conclude that 8.20 and 8.21 do indeed give the kinetic energy of a body.

Equation 8.21 is of particular interest. It tells us that the kinetic energy is merely the change in mass of a body multiplied by c^2. In other words, whenever an object acquires kinetic energy, its mass increases according to the formula

$$m = \frac{m_0}{\sqrt{1 - (v/c)^2}}$$

The quantity $(m - m_0)c^2$ is the kinetic energy of the object. This representation for K is valid at all speeds and reduces to the nonrelativistic value $\frac{1}{2}m_0v^2$, if $v \ll c$.

Although we cannot prove it here, 8.21 is a special case of a more general equation. That equation is

$$\text{Energy} = mc^2 \qquad\qquad 8.22$$

Whenever the energy of an object is increased, the mass of the object, according to 8.22, will increase. If a body is lifted in the earth's gravitational field so that its potential energy is increased, the increase in energy is related to the increase in mass of the body by

Mass and energy are interconvertible

$$\Delta(\text{energy}) = (\Delta m)c^2$$

From this it is seen that mass and energy are directly related through the factor c^2. As a result, the term m_0c^2 in 8.21 may be regarded as the energy of the object resulting from the rest mass of the object. It is called the *rest mass energy,* or rest energy. One suspects that this rest mass energy might be made available to do work under certain circumstances. This supposition turns out to be true experimentally. The most spectacular evidence is the explosion energy of a nuclear bomb, which results from the loss of part of the rest mass of the material composing the bomb.

The reverse reaction, the creation of rest mass from energy, is also observed. For example, material particles, an electron and a positron (a positive electron), are produced from pure energy in a high-energy reaction called *pair production*. It is not true to say that rest mass is conserved since it can be interchanged with energy. We arbitrarily force energy to be conserved by defining a mass m to have an energy mc^2.

Mass is not conserved

Illustration 8.6

An electron (or any other particle with the same charge) accelerated through a potential difference of V volts (V) is said to have a kinetic energy of *V electron volts* (eV). (You will learn in your study of electricity that $1\text{ eV} = 1.6 \times 10^{-19}\text{ J}$ and that this conversion factor is simply the magnitude of the charge on the electron.) What is the mass of an electron which has been accelerated through 10^6 V? What is its speed? ($m_0 = 9.1 \times 10^{-31}$ kg.)

Reasoning We saw in the preceding section that

$$K = (m - m_0)c^2$$

Stellar Energy Sources

Tremendous amounts of energy are being generated constantly in our sun and in the stars. The source of this energy remained a puzzle until the existence of nuclear fusion reactions became known. These reactions are striking evidence for Einstein's concept of the interrelation between mass and energy. Accurate measurements of the masses of atomic nuclei show that the mass of the helium nucleus is less than the combined masses of the two neutrons and two protons of which it is composed. According to Einstein, this loss in mass, Δm, which occurs when helium nuclei form, must appear as energy, the relation being

$$\text{Energy} = (\Delta m)c^2$$

where c is the speed of light.

It is now believed that the energy generated in the sun and stars is obtained chiefly through the following set of reactions. A proton (i.e., a hydrogen atom nucleus) p, upon collision at high energy with another proton, forms a nucleus ^{2}H composed of a proton and neutron in the following way:

$$p + p \rightarrow {}^2\text{H} + e^+ + \nu$$

The other particles formed in the process are the positron e^+ (a positive electron) and a neutrino ν (a neutral particle which has no rest mass). We call the heavy hydrogen atom ^{2}H whose nucleus is produced in this reaction deuterium.

In turn, the deuterium nucleus collides with another proton to form a light form of helium nucleus ^{3}He, according to the reaction

$$^2\text{H} + p \rightarrow {}^3\text{He}$$

The nucleus ^{3}He consists of two protons and one neutron. Subsequently, two ^{3}He nuclei collide to cause the following reaction:

$$^3\text{He} + {}^3\text{He} \rightarrow {}^4\text{He} + 2p$$

The product ^{4}He is the common helium nucleus, which is stable and consists of two protons and two neutrons.

We see that the net effect of these reactions is to "burn" protons to form helium nuclei and more protons. In the process, 4.5×10^{-29} kg of mass is lost. According to Einstein, this mass should appear as

$$(\Delta m)c^2 = (4.5 \times 10^{-29})(3 \times 10^8)^2 \approx 4 \times 10^{-12} \text{ joule of energy}$$

Since there are 6×10^{23} hydrogen atoms (or protons) in 1 gram of hydrogen, this means that 1 gram of hydrogen will give rise to an energy of

$$(\tfrac{1}{4})(6 \times 10^{23})(4 \times 10^{-12}) = 0.6 \times 10^{12} \text{ joules of energy}$$

for relativistic particles (as well as for slower-moving particles, of course). Since the electron will be moving at speeds near that of light, $\frac{1}{2}m_0 v^2$ will not be applicable. From the data given, we know $K = (1.6 \times 10^{-19})10^6$ J and so

$$1.6 \times 10^{-19} \times 10^6 \, \text{J} = (m - m_0)(9 \times 10^{16} \, \text{m}^2/\text{s}^2)$$

which gives $\qquad m - m_0 = 1.78 \times 10^{-30} \, \text{kg}$

(Photograph from Mount Wilson and Palomar Observatories.)

(The factor $\frac{1}{4}$ enters because four protons are consumed in the reaction.) This energy appears chiefly as kinetic energy of the products and as x radiation. Most of the kinetic energy appears as heat. It should be noted that this is a tremendous amount of energy. For comparison purposes, when 1 gram of hydrogen is burned in oxygen to form water, the energy liberated is of the order of 10^{-7} as large as the energy given off when 1 gram of hydrogen is fused to form helium.

If it were possible to control this stellar fusion reaction on the earth, it would furnish us with an almost inexhaustible energy source since the seas are filled with protons (hydrogen nuclei). Unfortunately, no practical means have yet been found to control the reaction. The difficulty occurs from the fact that like charges repel each other. Since protons are positively charged, one must shoot them together at high energies in order for them to get close enough so that they will react. In the extremely hot interior of the sun and stars, the thermal kinetic energy of the protons is high enough to cause the required high-energy collisions. However, no practical means have been found for containing materials at such high temperature on earth. Only in the nuclear bomb, where a controlled reaction is not required, has this fusion reaction been successfully utilized on the earth. These topics are discussed more fully in Chaps. 37 and 38.

Or since $m_0 = 0.91 \times 10^{-30}$ kg, we find

$$m = 2.69 \times 10^{-30} \text{ kg}$$

In other words, at this energy the electron has a mass approximately three times as large as its rest mass.

To find the speed of the electron we recall that

$$m = \frac{m_0}{\sqrt{1 - (v/c)^2}}$$

Squaring both sides of the equation and placing in the values for m and m_0, we find v to be given as $0.94c$, or 2.8×10^8 m/s.

Chapter 8
Questions and Guesstimates

1. Suppose an astronaut has perfect pitch so that he can recognize at once that a particular tuning fork gives off a sound of middle C when struck. What would he hear if he listened to the tuning fork inside his spaceship while traveling through space at a speed of $0.9c$?

2. The following argument is sometimes used to show that it is possible to tell if a body is in absolute motion with constant velocity: Since the frequency of a pendulum of length L is given by the formula $(1/2\pi) \cdot \sqrt{g/L}$, simply measure the frequency. If the frequency is right, the time dilation must be zero, and so the pendulum is at rest. What is wrong with this argument?

3. Since the earth is not an inertial reference frame, a rocket shot straight up from it will be observed by a person on earth to follow a curved path instead. Explain why.

4. Before leaving the earth, a vibrating tuning fork in a spaceship is used to mark time intervals by printing out a number on a chart each time it passes its center point. A similar tuning-fork system with an identical period is kept on the earth as the spaceship goes out into space, travels around for a few years at nearly the speed of light, and returns to earth. How will the two charts for the two tuning forks compare when the ship returns to earth?

5. Observer A at rest on the earth has proved that the length of objects in B's moving spaceship shrinks only in the direction of motion. For that reason, B's spaceship must be deformed. Should not B be able to tell this by measuring the length and width of his ship?

6. Suppose the speed of light were 20 m/s. Discuss how our lives would be changed.

7. What is wrong with the following scheme for contradicting the principle of the limiting speed of an object? A rocket in space ejects fuel and speeds up by an amount v_1. Since it is now coasting at constant speed, its occupants cannot tell they are moving. They repeat the process, and their speed will then be $2v_1$. This process can be repeated indefinitely, and so a speed in excess of c, the velocity of light, can be attained.

8. Most human beings live less than 100 years. Since the maximum velocity one can acquire relative to the earth is c, the speed of light, it is impossible for a person on earth to travel farther than 100 light-years* into space before he becomes 100 years old. Does this necessarily mean that no person from earth will ever be able to travel farther from earth than 100 light-years?

*A light-year is the distance light travels in a time of one year, 9.46×10^{15} m.

9. When a book is rotated in different directions in a spaceship traveling close to the speed of light, does a man in the spaceship see the book change shape?

10. If a man in a spaceship which is moving very close to the speed of light shoots a bullet in the direction of his motion, what will a stationary observer conclude? Will the bullet possibly appear to just slowly leave the gun?

11. The neutrino, a particle with zero charge and mass much smaller than the neutron, is given off in many nuclear reactions. Because of its zero charge and extremely small mass, it does not interact appreciably with matter, and was not observed conclusively until the mid-1950s. If its rest mass is truly zero (as we now know it to be), what is the smallest speed with which this particle can move?

12. Newton pictured a beam of light to consist of a stream of particles. In view of the fact that light, and therefore these particles, travel at the speed of light c, what can we conclude about the rest mass of these particles if they exist?

13. As a train travels with constant velocity along a track, a man inside one of the cars pushes on the wall at the front of the car. Show that the amount of work he does is the same in both reference systems, the boxcar and the earth. (*Hint:* His hands push; what about his feet?)

14. A student presents the following argument. It is possible to apply a force to a particle and do work on it indefinitely. This work must appear as kinetic energy if the work-energy theorem is correct. Therefore the kinetic energy can increase indefinitely. Since kinetic energy is $\frac{1}{2}mv^2$, the velocity must increase indefinitely. This contradicts the observed fact that a particle cannot move faster than the speed of light.

15. In a certain phenomenon called *pair production* a photon (which is a packet of electromagnetic energy having zero rest mass) changes into a positive and negative electron, called an electron-positron pair. What restrictions do the laws of conservation of momentum and energy impose upon this reaction?

16. When a radioactive radium nucleus emits an α particle, the combined mass of the remaining nucleus and α particle is less than the original mass. How would you compute this change in mass from the speed of the emitted particle?

17. It is possible for a particle to have zero rest mass and yet have energy and momentum. How fast must it be going? A neutrino, a particle given off in nuclear reactions, is of this type. (Do not confuse it with the neutron, which has mass of 1 u.)

18. A proton is shot at another proton so as to collide head on. Both particles are free in space and repel each other since they have the same charge. Discuss the collision from the following reference frames: (*a*) the laboratory frame in which one particle is initially at rest; (*b*) the frame of the originally stationary particle; (*c*) the frame of the particle which was shot.

Chapter 8
Problems

1. An automobile moves with speed 30 m/s past a corner. By what percent will an observer on the corner measure the length of the car to be contracted?

2. Suppose a spaceship could travel at a speed of 0.50c relative to an

observer on the earth. If the ship was traveling from New York to London, a straight-line distance D according to the usual maps, how large would a spaceship occupant measure this distance to be? Neglect the curvature of the earth. Repeat for a speed of $0.990c$.

3. According to a man on the ground, how many seconds are ticked out by a clock in an airplane moving overhead at 300 m/s while his clock ticks out exactly 1000 s?

4. In the streams of atoms and other particles which shoot out from the sun in solar flares, very high speeds are often achieved. By what factor, according to us on earth, are the internal workings of such an atom slowed as it shoots towards the earth with a speed of $0.90c$?

5. Suppose superior beings on a planet near Alpha Centauri, which is 4.3×10^{16} m away, send a spaceship to us at a speed of $0.9990c$. It is contaminated by a pair of microbes which reproduce on earth in such a way that the population doubles every 6.4×10^5 s. How many microbes will be on board when it hits the earth?

6. A queer particle is shaped like a cube, with edge length b. If it is set into motion in a direction parallel to one of its edges with a high speed v, what will be its volume measured while it hurtles through the laboratory?

7. The insignia on a particular spaceship is a square with a dot in its center. How fast must the ship be moving past the earth so that the people on earth would measure the square to be the insignia of space invaders who use a rectangle with sides in the ratio $1:2$?

8. The straight-line distance between the earth and the star Alpha Centauri is about 4.3×10^{16} m. Suppose a spaceship could be sent to the star with a speed of 2×10^8 m/s. (*a*) How long will the trip take according to earth clocks? (*b*) How large a time will the spaceship clocks record this journey to take? (*c*) How large will the spaceship occupants measure the earth to star distance to be? (*d*) How fast will the spaceship occupants compute their speed to be from the results of (*b*) and (*c*)?

9. On the average, a pi-meson (see illustration 8.3) lives about 2×10^{-8} s when traveling at low speed through the laboratory. How fast must it be going if it is to have a fifty-fifty chance of living as it travels the full length of a 20-m-long laboratory?

10. A spaceship traveling at a speed of $0.90c$ relative to the earth shoots a projectile out from its nose in line with its motion. If the projectile is designed to shoot from the ship with a speed of 0.60c, what will be the projectile's speed relative to the earth?

11. It is likely that ion propulsion engines will power space flights in the distant future. Suppose one such engine ejects ions out the ship's rear at a speed $0.90c$. If the ship is traveling away from the earth at speed $0.90c$, what will be the speed of the ions relative to the earth? What if the ejection speed is $0.70c$?

12. A rocket ship moving with speed $0.80c$ relative to the earth shoots a pulse of light out along its line of motion. Find the speed of the pulse relative to the earth.

13. Repeat problem 12 if the pulse of light is shot out perpendicular to the line of the ship's motion. (*Hint:* If you use the velocity transformations, note that both u_x and u_y exist.) At what angle to the ship's flight line will the earth observer see the pulse to travel?

14. A 30-year-old astronaut marries a 10-year-old girl just before setting out on a space voyage. When he returns to earth, she is 25 and he is 32. How

long was he gone according to earth clocks and what was his average speed during the trip?

15. According to an observer on the earth, a spaceship is going east with a speed 0.60c and is going to collide in 5 s head on with a comet going west at 0.80c. (a) How fast does the spacecraft see the comet to be approaching? (b) How much time do they have according to their clocks to get out of the way?

16. By direct substitution show that the Lorentz-Einstein equations are compatible with the two equations for a spherical wave as given in equations 8.2 and 8.3.

17. The most distant stars we can observe are about 10×10^9 light-years away from us and are receding from us at a speed in the vicinity of 0.99c. (One light-year is the distance light travels in one year, 9.46×10^{15} m.) (a) How long will we have to wait for the light they are emitting at present to reach us? (b) If their speed relative to us is 0.990c, how far do they think we are from our nearest star whose distance we measure to be 4.3×10^{16} m? Assume all distances along the same straight line. (c) Assuming their recession speed is 0.990c, how fast relative to them must they eject a rocket towards us if it is to reach us before 100×10^9 years have elapsed on earth?

18. The rest mass of an electron is $m_0 = 9.1 \times 10^{-31}$ kg. Find m/m_0 for it when its speed is (a) 3×10^5 m/s, (b) 3×10^7 m/s, (c) 2.0×10^8 m/s, (d) 2.9×10^8 m/s.

19. How fast must a particle be going (in relation to c) if its $m = 1000m_0$?

20. In the radioactive decay of a radium nucleus, the nucleus emits an alpha particle (i.e., a helium nucleus). The kinetic energy of the resultant particles is about 4.9 MeV (million electron volts). Recalling that 1 eV of energy is equivalent to 1.6×10^{-19} J, how much rest mass is converted to kinetic energy in the decay process? For comparison purposes, the mass of the original radium nucleus is about 3.8×10^{-25} kg.

21. Find the kinetic energy of an electron which is moving with a speed 0.95c ($m_e = 9.1 \times 10^{-31}$ kg). Express your answer in both joules and eV ($1 \text{ eV} = 1.6 \times 10^{-19}$ J).

22. In a TV tube, the electrons in the electron beam are accelerated through a potential difference of about 20,000 V and thereby acquire an energy of 20,000 eV. Find the ratio of m/m_0 for these fast-moving electrons ($1 \text{ eV} = 1.6 \times 10^{-19}$ J).

23. To melt 1 g of ice at 0°C requires an energy of 80 cal ($= 4.184 \times 80$ J). By how much does the mass of the ice increase because of the energy added to melt it?

24. Chemists sometimes say "the mass of the reactants equals the mass of the products" in a chemical reaction. When 2 g of hydrogen is burned with 16 g of oxygen to form 18 g of water, the reaction gives off about 60,000 cal of heat energy. How much mass is lost in the process ($1 \text{ cal} = 4.184$ J)?

25. The hydrogen bomb makes use of the fusion reaction in which the nuclei of hydrogen atoms are fused together to form more massive nuclei. In the process, for every gram of hydrogen used, about 0.006 g of mass is lost. Find the ratio of the energy released in this reaction to the energy released when an equal quantity of hydrogen is burned to form water. When burned, 1 g of hydrogen releases an energy of 1.3×10^5 J.

26. The sun generates its heat by means of a fusion reaction similar to that

in the hydrogen bomb. It radiates energy out into space at a rate of about 5×10^{26} J/s. At what rate is the mass of the sun decreasing because of this? Compare this with the mass of the sun, 2×10^{30} kg.

27. In a large nuclear accelerator, protons are given energies of 20 GeV (that is, 20,000 MeV). Find both the ratio m/m_0 for the protons and their velocity (proton mass $= 1.67 \times 10^{-27}$ kg).

28. Show that the momentum p of a particle is related to the total energy (i.e., rest mass energy included) of the particle through the relation

$$(\text{Total energy})^2 = p^2 c^2 + m_0{}^2 c^4$$

29. Show that the relativistic relation between kinetic energy K and momentum of a particle is given by

$$(K + m_0 c^2)^2 = p^2 c^2 + m_0{}^2 c^4$$

9 Rotational and Orbital Motion

In this chapter we begin our study of particles which move on circular paths. We shall find that five motion equations, similar to our linear motion equations, prove useful. In addition, the concept of radial acceleration and the force needed to produce it are discussed. After a presentation of Newton's law of universal gravitation, the orbital motion of satellites and planets is considered. The concept of weightlessness is also investigated.

9.1 ANGULAR MEASURE

Angles are measured in radians, degrees, and revolutions:

angle in radians = arc length/radius

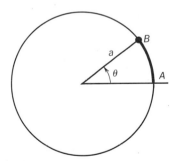

FIGURE 9.1 The value of θ in radian measure is $\overline{AB}/a$.

The radian is the preferred unit

When a particle travels along the arc of a circle, it is convenient to describe its position in terms of the angle between a fixed radius of the circle and a radius from the center of the circle to the particle. For example, the angle θ in figure 9.1 may be used as the coordinate in the description of the location of the particle shown at B in the figure.

Three units are customarily used to express θ, and we shall use all of them: degrees, revolutions, and radians. Only the last of these units is unfamiliar enough to require explanation. An angle in radians is the arc length subtended by the angle divided by the radius of the arc. In figure 9.1 the angle θ in radians is $\overline{AB}/a$. Notice that the radian unit is really no proper unit at all since it is merely the ratio of two lengths. In spite of this we designate angular measurements by the terms θ rad, or θ deg, in order to express the way in which the angle is being measured. But we must not expect these designations to behave as units in equations.

When θ becomes large enough to fill the whole circle (that is, 360°) it subtends an arc equal to the circumference $2\pi a$. Therefore 360° is equivalent (from definition of radian measure) to $2\pi a/a$, or 2π rad. This conversion factor, 2π rad = 360°, can now be used to convert any angle between these two unit systems. In addition, since one revolution is the same as 360°, we find at once that 1 rev = 2π rad. We shall see later that the radian is the preferred unit of angular measure. However, the other two will often be used.

9.2 ANGULAR QUANTITIES AS VECTORS

At first sight one might insist that angles and rotations cannot be vectors. However, since a vector is any quantity having magnitude and direction and which obeys the laws of vector addition, we can, *if we choose,* give direction to

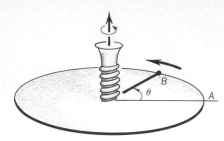

FIGURE 9.2 The vector representing a rotational quantity points in the direction of motion of a right-hand screw which is rotated in the same sense as the rotational quantity.

The rotation vector follows the right-hand-screw rule

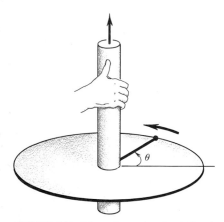

FIGURE 9.3 When the axis of rotation is grasped with the right hand, fingers circling in the direction of rotation, the thumb points in the direction of the rotation vector.

angular quantities and thereby make them vectors. Whether or not we do this will depend upon the usefulness of the concept. As we shall see later, it does prove useful to be able to treat angular quantities as vectors, and therefore we often do so.

The question now arises, What direction should we assign to the rotation from *A* to *B,* as shown in figure 9.1? Since the direction of the tangent to the circle will change continuously as point *B* moves to larger θ values, its direction is far too complicated for the description of θ. Only two simple directions exist for the description of θ, a vector perpendicular to the page, either into or out of the page. The usual choice for the direction of a rotational quantity such as θ is made in the following way.

If we refer to figure 9.2, it is seen that we make an analogy between the rotational quantity and a common right-hand screw. The radius from the center of the circle to the particle sweeps along a plane as the rotation occurs. We now consider a right-hand screw sticking into this plane. If the screw is turned as shown in figure 9.2 (in a direction such that the notch in the top of the screw rotates in the same way as the radius to the particle), the screw will come up and out of the plane. This direction, the direction of motion of a right-hand screw performing the same rotation, is taken as the defined direction of a rotation. It is this direction which we assign to the vector representing the rotation quantity. We call this the *right-hand-screw rule*.

For those who are not very familiar with the motion of screws, another rule, the right-hand rule, often proves useful for this purpose. As shown in figure 9.3, if one grasps the axis of rotation in the right hand in such a way that the fingers circle the axis in the direction of rotation, the thumb points in the direction of the vector assigned to the rotation. Although the rule assigning direction to rotation is purely arbitrary, we shall see many other instances in our study of physics where this same rule has been applied to other situations. It is therefore important that you become familiar with it at this time.

Now that we have agreed on a direction for the rotational quantity, we must show that the vector so defined obeys the laws of vector addition. In particular, the sum of the two vectors representing two successive rotations, $\boldsymbol{\theta}_1$ and $\boldsymbol{\theta}_2$, must give the same result, independent of the order in which they are taken. That is to say, they must commute. In symbols,

$$\boldsymbol{\theta}_1 + \boldsymbol{\theta}_2 = \boldsymbol{\theta}_2 + \boldsymbol{\theta}_1$$

Unfortunately, this law is not followed in the case of finite rotations except in special cases. This fact is easily demonstrated.

Consider the two successive 90° rotations of the block shown in figure 9.4. Clearly, the order in which the rotations are performed is important since the results are different in the two cases. Therefore rotational vectors so defined do not commute, and so such a definition of the vectors is impractical. However, if we restrict ourselves to small rotations, the order in which they are taken is unimportant, as we show in figure 9.5. In that case, as well as for finite rotations whose vectors are all perpendicular to a given plane, our vector definition will be valid and useful. We shall therefore restrict our vector rotations to infinitesimal angles unless the rotation vectors are all perpendicular to the same plane. As we shall soon see, angular velocities and accelerations are defined in terms of small rotations, and so they too can be considered vectors.

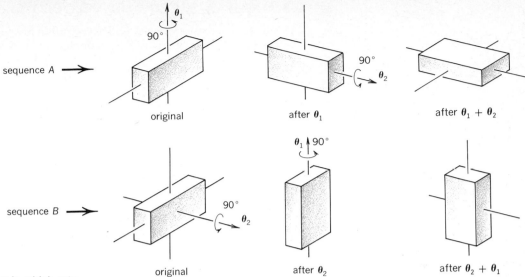

FIGURE 9.4 The order in which rotations are performed is important since, as shown, $\theta_1 + \theta_2 \neq \theta_2 + \theta_1$.

FIGURE 9.5 As the sphere undergoes two successive rotations, the black dot is displaced. If, as indicated, the rotation angles are very small, the spot ends up at the same point, independent of the order in which the rotations are taken.

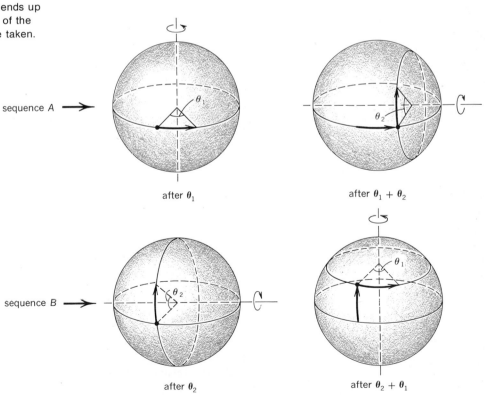

9.3 ANGULAR SPEED AND VELOCITY

The angular quantities velocity and acceleration are defined in a way much like their linear counterparts. For example, linear velocity was defined as

$$\mathbf{v} = \frac{\Delta \mathbf{s}}{\Delta t}$$

where $\Delta \mathbf{s}$ is the vector displacement in time Δt. Similarly, if an angle $\Delta \boldsymbol{\theta}$ is swept out in time Δt, the angular velocity ω (omega) is defined to be

Definition of angular velocity

$$\omega = \frac{\Delta \boldsymbol{\theta}}{\Delta t} \qquad \text{or} \qquad \omega = \frac{d \boldsymbol{\theta}}{dt} \qquad\qquad 9.1a$$

Notice that the units of ω are those of (time)$^{-1}$. However, to indicate the angular measure being used, we shall state ω to be in radians per second, revolutions per minute, or some other similar form. Notice that ω is defined in terms of an *incremental* rotation, and so it obeys the usual vector addition laws.

If the value of $\Delta t \rightarrow 0$, then *9.1a* is the definition of instantaneous angular velocity. For larger time intervals, ω becomes an average over that time. Because the rotation vectors we consider in this chapter are all parallel or antiparallel to the same direction, we are not concerned with the vector nature of the quantity and merely use the magnitudes of the quantities to give

$$\Delta \theta = \omega \, \Delta t \qquad\qquad 9.1b$$

This equation will then define the angular speed ω. As we shall see, the distinction between speed and velocity is less often made in angular than in linear motion.

9.4 ANGULAR ACCELERATION

Like angular velocity, angular acceleration is defined in analogy to the similar linear motion quantity. In the case of linear motion we had

$$\mathbf{a} = \frac{\mathbf{v} - \mathbf{v}_0}{t}$$

where t is the time taken for the velocity to change from $\mathbf{v}_0$ to $\mathbf{v}$. The angular acceleration $\boldsymbol{\alpha}$ is defined by

Definition of angular acceleration

$$\boldsymbol{\alpha} = \frac{\omega - \omega_0}{t} \qquad \text{or} \qquad \boldsymbol{\alpha} = \frac{d\omega}{dt} \qquad\qquad 9.2$$

where t is the time taken for the angular velocity to change from ω_0 to ω. Since the units of ω are 1/time, the units of α must be 1/(time)^2. But we shall state α as rad/s^2, etc., in order to indicate the angular measure as well.

You will recall from the discussion of linear acceleration that a distinction is made between instantaneous and average acceleration. Here, too, the same distinction can be made, depending upon whether or not $t \rightarrow 0$. However, since we shall usually be concerned with constant acceleration, this distinction will not often be of importance for us. One must remember, though, that the equations we shall next obtain are valid only for the case of α constant.

When the angular acceleration is constant, the average angular speed can be related to the initial and final speeds by the relation

$$\bar{\omega} = \tfrac{1}{2}(\omega + \omega_0) \qquad\qquad 9.3$$

This relation is obtained in exactly the same way as we obtained the average velocity equation in Chap. 4. We shall not give the details again here.

If we now rewrite *9.1b* as

$$\theta = \bar{\omega}t \qquad\qquad 9.1$$

we can combine it with *9.2* and *9.3*, namely,

Equations 9.1 to 9.5 are the five rotational motion equations if $\alpha = const$

$$\omega = \omega_0 + \alpha t \qquad\qquad 9.2$$

$$\bar{\omega} = \tfrac{1}{2}(\omega + \omega_0) \qquad\qquad 9.3$$

to obtain the following analogs to the linear motion equations:

$$\omega^2 - \omega_0{}^2 = 2\alpha\theta \qquad\qquad 9.4$$

$$\theta = \omega_0 t + \tfrac{1}{2}\alpha t^2 \qquad\qquad 9.5$$

They are, of course, valid only for uniformly accelerated motion since *9.3* is used in their derivation. With these five equations we can easily describe uniformly accelerated angular motion, as we shall soon see. Notice that the equations are simply the linear motion equations, with θ, ω, and α replacing the quantities x, v, and α, respectively.

9.5 TANGENTIAL QUANTITIES

As a particle moves on a circle, it not only sweeps out an angle θ but it also covers an arc of length s. Often one wishes to convert from angular quantities to linear quantities in order to describe a motion fully. This is rather easily done, provided angles are measured in radians. For, from the definition of radian measure, we know that an angle in radians is related to the arc length s and radius r through

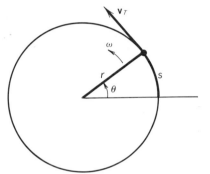

FIGURE 9.6 What is the factor of proportionality between the angular and tangential quantities?

$$\theta = \frac{s}{r}$$

Arc length is $r\theta$, with θ in radians

from which $\qquad\qquad s = r\theta \qquad (\theta \text{ in radians}) \qquad\qquad 9.6$

These quantities are shown in figure 9.6.

If the particle moves on the circumference of the circle, it has a linear velocity $\mathbf{v}_T$, which is always tangential to the circle. This velocity, shown in figure 9.6, is called the *tangential* velocity of the particle. To find it we need only take the time derivative of *9.6*, recalling that r is constant, to give

$$\frac{ds}{dt} = r\frac{d\theta}{dt}$$

or, since $v_T = ds/dt$ and $\omega = d\theta/dt$,

Tangential speed is $r\omega$

$$v_T = r\omega \qquad (\omega \text{ in radian measure}) \qquad\qquad 9.7$$

Here, too, radian measure must be used for ω because of the restriction on *9.6*.

To find the tangential acceleration of the particle, a_T, we take the time derivative of *9.7*, again holding r constant. It gives, since $a = dv/dt$ and $\alpha = d\omega/dt$,

$$\frac{dv_T}{dt} = r\frac{d\omega}{dt}$$

Tangential acceleration is $r\alpha$ or

$$a_T = r\alpha \qquad (\alpha \text{ in radian measure}) \tag{9.8}$$

Once again radian measure must be used.

It should be noticed that *9.6* to *9.8* are useful in many situations. For example, if a rope is being wound on a wheel of radius *r*, these equations give the length, speed, and acceleration of the rope being wound. Or if a wheel is rolling along the ground, these equations give the distance the center of the wheel has moved, as well as its velocity and acceleration. It is left as an exercise for the student to justify these assertions.

Illustration 9.1

A wheel is turning at a speed of 2.0 rev/s and is then allowed to coast to rest. If it does so in 30 s, how far did it turn in the process? Assume uniform deceleration to be a good approximation.

Reasoning This is an angular-motion problem in which $\omega_0 = 2.0$ rev/s, $\omega = 0$, and $t = 30$ s. To find θ, the total angular distance turned, we use

$$\bar{\omega} = \tfrac{1}{2}(\omega_0 + \omega)$$

to find $\bar{\omega} = 1.0$ rev/s, and then use

$$\theta = \bar{\omega}t$$

to find $\theta = 30$ rev. Since 1 rev $= 2\pi$ rad $= 360°$, this answer may be written, alternatively,

$$\begin{aligned}\theta &= 60\pi \text{ rad} \\ &= 10{,}800 \text{ deg}\end{aligned}$$

Illustration 9.2

A vehicle with 50-cm-diameter wheels is moving at a speed of 20 m/s. (*a*) How fast are its wheels turning? (*b*) How many revolutions will the wheels turn if the car coasts uniformly to rest in 30 s?

Reasoning (*a*) The speed of the wheel along the ground is equal to its tangential speed, as pointed out before; so the tangential speed is 20 m/s. To find the angular speed we use

$$v_T = \omega r$$

where ω will be in radian measure. This yields $\omega = 80$ rad/s. To find the speed in rev/s, we recall that 1 rad $= 1/2\pi$ rev, and therefore find

$$\omega = (80 \text{ rad/s})\left(\frac{1}{2\pi} \text{rev/rad}\right) = \frac{40}{\pi} \text{rev/s}$$

(*b*) In the motion problem we know $\omega_0 = 40/\pi$ rev/s, $\omega = 0$, and $t = 30$ s. Using

$$\bar{\omega} = \tfrac{1}{2}(\omega + \omega_0)$$

we find $\bar{\omega} = 20/\pi$ rev/s, from which

$$\theta = \bar{\omega}t$$

gives $\theta = 600/\pi$ rev.

9.6 RADIAL ACCELERATION

It is possible, and often convenient, to write down the x and y coordinates of a particle moving on a circle. This is easily done if we refer to figure 9.7(a). Taking the origin of coordinates at the center of the circle, we see at once that the particle's coordinates are

$$x = r \cos \theta$$
$$y = r \sin \theta$$

If the particle stays on the circle, and we shall assume that it does, r will be a constant. However, θ will be a function of time.

Suppose the particle travels around the circle at constant angular speed ω. In that case $\bar{\omega} = \omega$ and 9.1 gives

$$\theta = \omega t$$

Upon substitution of this in the equations for x and y, we find

$$x = r \cos \omega t$$
$$y = r \sin \omega t \qquad \qquad 9.9$$

As we should expect from figure 9.7, the x and y coordinates of the particle oscillate back and forth sinusoidally on the x and y axes, with $-r \le x \le r$ and $-r \le y \le r$. Furthermore, the two oscillations are not in phase (i.e., do not arrive at maximum at the same time).

Since the x component of the velocity of the particle is $v_x = dx/dt$, and since $v_y = dy/dt$, we can easily find these components from 9.9. In general, of course, ω would not be constant. Let us, however, restrict ourselves to the case where the particle travels around the circle at constant speed, so that $\alpha = d\omega/dt = 0$. For that case the derivative of 9.9 gives

$$v_x = -r\omega \sin \omega t$$
$$v_y = r\omega \cos \omega t \qquad \qquad 9.10$$

where use has been made of the calculus formulas

$$\frac{d(\sin ax)}{dx} = a \cos ax$$

$$\frac{d(\cos ax)}{dx} = -a \sin ax$$

These velocity components are shown in figure 9.7(b).

In 9.10 are given the x and y components of the velocity of the particle as it travels around the circle at constant angular speed. Notice that both v_x and v_y are functions of the time. Since the velocity components change with time, *the particle must be accelerating even though its speed in the circle is constant.* The reason for this is quite simple. Velocity is a vector, and clearly the velocity of the particle changes direction as the particle traverses the circle. It is this directional change in velocity which gives rise to an acceleration for the particle. This is discussed in more detail in the next section.

To find the x and y components of the acceleration, we make use of the fact that

$$a_x = \frac{dv_x}{dt} \qquad \text{and} \qquad a_y = \frac{dv_y}{dt}$$

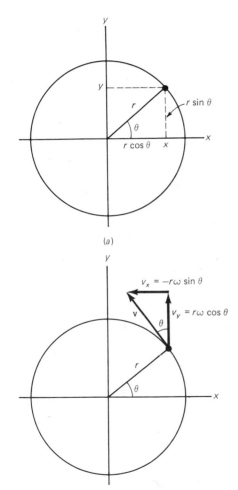

(a)

(b)

FIGURE 9.7 As the particle moves around the circle, the direction of **v** changes. The particle is therefore accelerating even though its speed is constant.

These quantities can be obtained by taking the derivative of 9.10 to give

$$a_x = -r\omega^2 \cos \omega t$$
$$a_y = -r\omega^2 \sin \omega t \qquad \text{9.11}$$

It is of interest to notice that the components of the acceleration are proportional to the negative of the displacements, as given in 9.9. This means that the displacement vector, which is the radius vector from the center of the circle to the particle, must be proportional to the acceleration vector but opposite in direction. We therefore conclude that the acceleration of the particle in the circle is a vector which points radially *inward* from the particle to the center of the circle. For this reason the acceleration is called the *radial acceleration,* and we represent it by the symbol a_R.

The acceleration is toward the center of the circle

Now that the components of the acceleration are known, the resultant total acceleration is easily found. We know that the resultant R of two component vectors R_x and R_y is given by

$$R^2 = R_x{}^2 + R_y{}^2$$

and so the radial acceleration a_R is given by

$$a_R{}^2 = a_x{}^2 + a_y{}^2$$
$$= r^2\omega^4 \cos^2 (\omega t) + r^2\omega^4 \sin^2 (\omega t)$$

Making use of the fact that $\cos^2 \theta + \sin^2 \theta = 1$, we find

The radial acceleration is $r\omega^2$

$$a_R = r\omega^2 \qquad \text{9.12}$$

Let us emphasize once again that even though the particle is traveling around the circle at constant speed, its velocity vector is changing direction. The rate of change of this velocity vector is the radial acceleration of the particle. As its name implies, the particle is accelerated along a radius toward the center of the circle. The magnitude of the acceleration is given by 9.12.

9.7 CENTRIPETAL FORCE

We found in the preceding section that a particle which is moving around a circle at constant speed is also accelerating inward toward the center of the circle. This fact can be made more obvious from a consideration of figure 9.8. As the particle moves with constant speed from point A to B, its velocity vector changes direction as shown. We are interested in the acceleration

$$\mathbf{a}_R = \frac{\mathbf{v}_B - \mathbf{v}_A}{\Delta t}$$

where Δt should be made small enough so that points B and A are close together.

We recall that to subtract $\mathbf{v}_A$ from $\mathbf{v}_B$ we add $-\mathbf{v}_A$ to $\mathbf{v}_B$, and this is done in the figure. The resultant $\Delta \mathbf{v}$ is just $\mathbf{v}_B - \mathbf{v}_A$, and will have the direction of $\mathbf{a}_R$. Clearly, as points B and A are moved close together, $\Delta \mathbf{v}$, and therefore $\mathbf{a}_R$, will be directed radially inward. As a matter of fact, a consideration of the geometry of figure 9.8 can be used to develop the expression we have found previously for $\mathbf{a}_R$. Since such a derivation adds nothing new, we shall not pursue it further here, although the interested student should be able to supply it.

In order for a particle of mass m to accelerate, an unbalanced force must act upon it. This force is given by Newton's second law, $\mathbf{F} = m\mathbf{a}$. In our

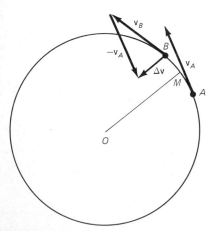

FIGURE 9.8 Notice that $\Delta \mathbf{v} = \mathbf{v}_B - \mathbf{v}_A$ is nearly parallel to the line *OM*. What does this tell us about the acceleration of the particle?

case **a** is the radial acceleration $\mathbf{a}_R$, which was found in the last section to be $r\omega^2$. Therefore,

$$\mathbf{F}_R = m\mathbf{a}_R$$

or

$$F_R = m\omega^2 r \qquad 9.13$$

where F_R is the force which causes the radial acceleration. We call this force, the force needed to hold a particle in a circular path, the *centripetal force*. It is not at all surprising that such a force is required to constrain a particle to move in a circular path since Newton's first law tells us that, without such a force, the particle will move in a straight line. Since $\mathbf{F}_R$ and $\mathbf{a}_R$ have the same direction, the centripetal force is as shown in figure 9.9. Often it is convenient to replace ω in *9.13* by v_T/r to obtain

$$F_R = \frac{mv_T^2}{r} \qquad 9.14$$

Whenever an object travels in a circular path, a centripetal force must act upon it to cause it to bend from straight-line motion. In the case of the moon circling the earth, the centripetal force is the gravitational force of attraction of the earth for the moon. Although this force is less easily visualized than the tension in a string which holds a ball in a circular path, it is no less real. Similarly, the gravitational force needed to hold an earth satellite in orbit is also given by *9.14*. We examine the gravitational force law in more detail in the next section.

Illustration 9.3

A ball of mass m is rotated at constant speed v in a vertical circle (radius r), as shown in figure 9.10. Find the tension in the cord when the ball is at position A and at position B.

Reasoning We know that a force equal to the centripetal force is needed to hold the particle in its circular path. When the particle is at point A, this force is the sum of the forces T_A, the tension in the cord, and mg, the weight of the particle. Both are directed radially inward and constitute the centripetal force. Therefore

$$T_A + mg = \frac{mv^2}{r}$$

or

$$T_A = m\left(\frac{v^2}{r} - g\right)$$

It is interesting to notice that the tension in the cord becomes zero when $v^2/r = g$. Under that condition the gravitational attraction of the earth produces the required radial or centripetal acceleration at A since the centripetal acceleration v^2/r is then equal to the gravitational acceleration g. If v is so small that $v^2/r < g$, the pull of gravity exceeds the required centripetal force and the object will fall out of the circular path before reaching A.

When the ball is at point B, the total radial force on the ball must still be mv^2/r if it is to move on the circle. As seen from figure 9.10, the total radial force is $T_B - mg$, and so

$$T_B - mg = \frac{mv^2}{r}$$

or

$$T_B = m\left(\frac{v^2}{r} + g\right)$$

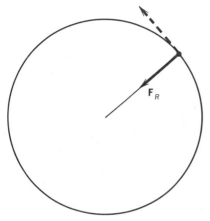

A centripetal force $m\omega^2 r$ is needed to cause the radial acceleration

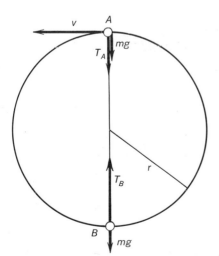

FIGURE 9.9 If the centripetal force $\mathbf{F}_R$ were not present, the particle would follow the dashed path.

FIGURE 9.10 What are the values of T_A and T_B?

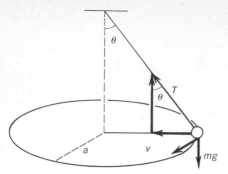

FIGURE 9.11 The tension in the cords accomplishes two things.

Notice in this case that the tension in the cord must do two things: support the weight of the ball mg and furnish the centripetal force mv^2/r.

Illustration 9.4

The ball of mass m shown in figure 9.11 moves in a circle of radius a with speed v. (This is called a *conical pendulum*.) Find the tension in the cord.

Reasoning To hold the ball in a circular path, the centripetal force needed is mv^2/a. This radial force is furnished by the component of **T** in the plane of the circle, namely, $T \sin \theta$. Therefore

$$T \sin \theta = \frac{mv^2}{a}$$

Moreover, the vertical component of **T** must balance the gravitational force mg, so that

$$T \cos \theta = mg$$

Either of these two equations can be used to find T. They also indicate, upon division of one by the other, that

$$\tan \theta = \frac{v^2}{ag}$$

so that the mass of the ball is unimportant in determining the angle at which the ball will hang.

9.8 LAW OF GRAVITATION

One of the most interesting examples of nearly circular motion, planetary motion, was the subject of intensive study by astronomers many centuries ago. Astronomy was a highly developed but largely empirical science long before the time of Newton. When Newton began his study of the action of forces, he had available highly precise data concerning the orbits of the planets in the solar system. Astronomers of the time could accurately predict the future motions of the planets, but they were unable to justify the rules they had found applicable over the centuries. Two keys were needed before the planetary motions could be explained.

The first of these necessary keys was the law relating force and acceleration. This key was found when Newton discovered his three laws of motion. However, no progress could be made until the forces acting on the planets had been discovered. This discovery was also made by Newton, and is called the *law of universal gravitation*.

Considering the planets and the sun to act as point masses, Newton was able to show that the planetary motions corresponded to the following force law. If two point masses, m_1 and m_2, are separated by a distance r, as shown in figure 9.12, the force exerted by one mass on the other is

FIGURE 9.12 What law of physics tells us that the two gravitational forces must be equal and opposite?

Law of gravitation

$$F = G \frac{m_1 m_2}{r^2} \qquad 9.15$$

where G is a constant of nature. The force is one of attraction along the line joining m_1 and m_2. From Newton's third law we know that for every force on one body there must be an equal and opposite force on some other body. In the present case we have only two masses, and so the action and reaction forces are obviously those shown in figure 9.12.

We have already defined all the quantities in *9.15*, except G. As a result, we cannot choose G at will, but must use the value given by direct experimental test of *9.15*. Such an experiment is not easily performed with high accuracy since, for laboratory-size masses, the gravitational force is extremely small (see problem 20 at the end of this chapter). The first accurate measurement of G was made by Cavendish in 1798, using what is now called, naturally enough, the Cavendish balance. It is illustrated schematically in figure 9.13.

The Cavendish experiment to measure G

Two small balls at the ends of a light rod are suspended by a very delicate fiber. When the fiber is twisted, the restoring torque is extremely small. The twist of the fiber can be observed by means of a light beam reflected from a mirror attached to the rod. If two large masses M are now brought close to the small masses as shown, the attraction force due to the masses will cause the fiber to twist. Equilibrium is reached when the restoring torque due to the twisted fiber equals the torque due to the gravitational attraction. If the fiber has been calibrated, the gravitational attraction torque, and therefore the gravitational force, can be found.

The presently accepted value for G is

$$G = 6.6720 \times 10^{-11}\,\text{N} \cdot \text{m}^2/\text{kg}^2$$

To within this accuracy, the gravitational force is independent of the kind of mass involved. No dependence of G on the type of matter has ever been found.

The superposition principle is valid for gravitational forces

Another important property of the gravitational force is embodied in the *superposition principle*. This principle states that the gravitational force on a body due to several other bodies is the vector sum of the individual forces due to the several masses. Although this may seem nearly self-evident, it need not have been true. For example, it leads us to conclude that one mass will experience the same attraction force from a second body, independently of whether or not a third mass is placed between them.

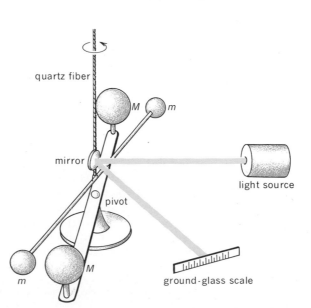

FIGURE 9.13 A schematic diagram of the Cavendish balance. Observe how the beam of light is used to detect the twist of the fiber.

Illustration 9.5

Assuming the orbit of the earth about the sun to be circular (it is actually slightly elliptical) with radius 1.5×10^{11} m, find the mass of the sun.

Reasoning The centripetal force needed to hold the earth in an orbit of radius R is furnished by the gravitational attraction of the sun. We therefore have (using m_e and m_s for the mass of the earth and sun, respectively)

$$\text{Centripetal force} = \text{gravitational force}$$

or, in symbols,

$$\frac{m_e v^2}{R} = G \frac{m_e m_s}{R^2}$$

which gives

$$m_s = \frac{v^2 R}{G}$$

where v is the speed of the earth in its orbit around the sun. Since the earth travels around its orbit once each year or in a time of 3.15×10^7 s, we have

$$v = \frac{2\pi(1.5 \times 10^{11}\,\text{m})}{3.15 \times 10^7\,\text{s}} = 3.0 \times 10^4\,\text{m/s}$$

from which

$$m_s = 2.0 \times 10^{30}\,\text{kg}$$

9.9 THE GRAVITATIONAL FORCE AND WEIGHT

Every object on the earth is attracted by the mass of the earth. Since the earth's mass is so large, this attraction force is many orders of magnitude larger than the attractive force between objects on the earth's surface. Let us see what we can conclude from a knowledge of the gravitational force.

It can be shown* that the attraction force which a uniform spherical mass exerts on an object outside the sphere can be computed by assuming all the mass of the sphere to be concentrated at its center. If we call the earth's mass m_e, an object of mass m on the earth's surface will experience a force

$$F = G \frac{m m_e}{R^2} \qquad\qquad 9.16$$

where R is the radius of the earth. This attractive force which the earth exerts on an object is simply the weight of the object, mg.

Replacing F in *9.16* by *mg,* we find the following value for *g,* the acceleration due to gravity:

Relation between g and G

$$g = \frac{G m_e}{R^2} \qquad\qquad 9.17$$

Notice that we could use this equation to find the mass of the earth if its radius were known.

As was pointed out in Chap. 5, the weight of an object of mass m depends upon its position. We notice from *9.17* that g., and therefore the weight, will vary with distance from the center of the earth. Since the earth bulges somewhat at the equator, slight variations of g and of weights from

---

*One method for doing this is much like the procedure utilized in Sec. 9.10. However, it involves fairly complex calculus and so we shall not reproduce it here. An analogous problem in electrostatics is solved completely in Chap. 18.

place to place on the earth are expected. (In addition, the rotation of the earth causes the apparent weight of an object to be less at the equator than at the poles.)

Illustration 9.6

Compare the earth's attraction for an object at the following two locations: (a) at the earth's surface and (b) 160 km (100 mi) above the earth. ($R_e = 6.37 \times 10^6$ m.)

Reasoning We make use of equation 9.16 to find the force of attraction the earth exerts on the object. Writing the equation twice, once for each of the two radii, we can divide one equation by the other to obtain

$$\frac{F_{160}}{F_e} = \frac{R_e^{\,2}}{R_{160}^{\,2}} = \frac{40.6 \times 10^{12}}{42.7 \times 10^{12}}$$

or

$$\frac{F_{160}}{F_e} = 0.952$$

We see that the force of the earth's gravity is reduced only about 5 percent at a height of 160 km.

9.10 ATTRACTIVE FORCE BY INTEGRATION

Newton's law of gravitation applies to point masses. For more complicated situations involving extended masses, the distance r to be used in 9.15 is usually not obvious. One exception exists, however. As we mentioned previously, the mass of a uniform spherical shell and of spheres made up of such shells gives rise to gravitational forces outside the sphere identical with the force exerted by a point mass at the center of the sphere. In all other cases, the situation is less simple. To illustrate the procedure one must follow in these more complicated cases, let us consider the attraction between a point mass m and a rod of length L whose mass per unit length is μ (Greek mu).

The situation is shown in figure 9.14. We can compute either the force exerted by the point mass on the bar or the force exerted by the bar on the point mass. Newton's law of action and reaction tells us they will be equal and opposite. Let us compute the force due to the bar on the point mass. To do this we need to employ 9.15, which applies only to point masses. This necessitates that we split the rod into essentially point masses. We shall find the force on m due to each of these point masses and shall add the forces to obtain the total force.

For discussion purposes, let us split the bar into N pieces of lengths Δx_1, $\Delta x_2, \ldots, \Delta x_i, \ldots, \Delta x_N$. Since the mass per unit length is μ, the mass of the length Δx_i will be* $\mu \Delta x_i$. The small force ΔF_i on the mass m due to the essentially point mass $\mu \Delta x_i$ is given by the gravitation law as

$$\Delta F_i = G \frac{(\mu \Delta x_i) m}{x_i^{\,2}}$$

where x_i and ΔF_i are as shown in the figure.

We notice that all the little pieces of the rod will attract m toward the

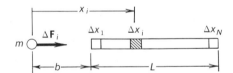

FIGURE 9.14 The essentially point mass at x_i attracts m with a force ΔF_i.

---

*For example, if $\mu = 8$ kg/m and $\Delta x_i = 0.010$ m, then the mass of the segment is $(8$ kg/m$)(0.010$ m$) = 0.08$ kg.

right. Hence all the ΔF's will be in the same direction, and so we can add them to find the total force.

$$F = \sum_1^N \Delta F_i = \sum_1^N Gm\mu \frac{\Delta x_i}{x_i{}^2}$$

If we now pass to the limit of N being a very large number, so that $\Delta x \to 0$, the sum may be replaced by an integral. We then have

$$F = \int_b^{L+b} Gm\mu \frac{dx}{x^2}$$

Be particularly careful to notice the limits on the integral. They must correspond to the value of x at the two end masses, at x_1 and x_N. Since Δx_1 is at a distance $x_1 = b$ from m, the lower limit is b. Similarly, the Nth mass is at a distance $x_N = b + L$ from m, and so the upper limit is $b + L$. In addition, we notice that $Gm\mu$ are constants, and so they may be removed from the integral. We then find

$$F = Gm\mu \int_b^{L+b} \frac{dx}{x^2}$$

This integral is found in all integral tables, and is

$$\int \frac{dx}{x^2} = \frac{-1}{x}$$

Therefore, after substituting the limits, we find

$$F = \mu Gm \left(\frac{1}{b} - \frac{1}{L+b} \right)$$
$$= \frac{Gm\mu L}{b(L+b)}$$

Since μL is just the mass of the bar, call it M, we have, finally,

$$F = \frac{GmM}{b(L+b)}$$

This is to be compared with Newton's law of gravitation, namely, $F = Gm_1m_2/R^2$. Notice that the length analogous to R is not the distance from m to the center of mass of the bar, since that would be $b + \frac{1}{2}L$. Instead, it is some other type of average length, namely, $\sqrt{b(L+b)}$. It would be difficult to guess this average length without carrying out the computation we have made.

9.11 WEIGHTLESSNESS

We have previously defined the *apparent* weight of an object to be equal to the force needed to support it. For example, a book lying on a table requires a force mg exerted by the table to support it, and this is the book's apparent weight. The apparent weight of an object need not be mg, however, even on the earth, as pointed out in Chap. 5. To see this, consider the situation shown in figure 9.15.

In figure 9.15 we see a mass m supported from a spring scale in an elevator. When the elevator is at rest with respect to the earth, the tension in

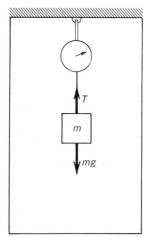

FIGURE 9.15 What is the elevator doing if the scale reads *mg*? Less than *mg*? Greater than *mg*?

the rope supporting it is *mg,* and the scale will read this value. Even if the elevator is moving with constant speed relative to the earth, the tension and scale reading will still be the same, namely, *mg.* This follows from the fact that, since the system is not accelerating, the vertical forces acting on the mass must add to give zero. Therefore $T - mg = 0$ and $T = mg.$

Suppose, however, that the elevator is accelerating downward with acceleration a. We know that an observer in the accelerating system, the elevator, will not generally be able to understand his experimental results, using the ordinary laws of physics, unless he makes some allowance for the fact that he is accelerating. However, an observer watching the elevator from a stationary point on the earth will be able to describe phenomena occurring in the elevator in terms of the usual laws of physics, since he is at rest in an inertial frame, the earth. (As discussed previously, the earth is not strictly an inertial frame, but the deviation will be negligible in the present case.) If the observer on the earth applies $F = ma$ to the object in the elevator, he will write, for the situation of figure 9.15, taking down as positive,

$$mg - T = ma$$

from which
$$T = m(g - a)$$

When its downward acceleration is a, an object of mass m appears to weigh $m(g - a)$

We see at once that the tension in the rope is less than *mg;* so the scale will read $m(g - a)$, and not *mg.* The object appears to weigh less in the downward-accelerating elevator. Moreover, if the elevator is freely falling, $a = g$, and T will be zero. The object would appear to have zero weight. This is almost obvious; for if an object is to be freely falling so as to have a downward acceleration g, then no force can be supporting it. If the object hangs from a string or rests on your hand, this source of support must not exert a force on the object. Therefore the object appears to be weightless. This condition of weightlessness in a freely falling elevator is quite analogous to the weightlessness astronauts observe while circling the earth in a satellite ship. Let us now extend our discussion to this new situation.

From a qualitative standpoint, the situation involving an earth satellite is quite simple. To understand what is happening, let us consider a projectile which is shot parallel to the earth's surface at a height sufficient to allow us to neglect the effect of the very sparse atmosphere. (One hundred kilometers above the earth the atmosphere is only one-millionth as dense as it is at sea level, and so air friction at that height should be negligible.) The situation for various initial speeds is shown in figure 9.16. As with ordinary projectile motion, the projectile is caused to fall to the earth by the pull of gravity. However, as shown in the figure, if the speed of the projectile is just large enough, the freely falling projectile will simply circle the earth! Although the projectile is falling continuously toward the earth's center, the curvature of the earth is the same as the curvature of the projectile's path. The speed of the projectile in its circular orbit is just proper so that the required centripetal acceleration is equal to the gravitational acceleration provided by the earth. We shall soon see how fast the projectile must be shot to achieve such an orbit.

Before doing that, though, we should notice that the projectile might in fact be a spaceship. As it circles the earth, it is a freely falling object. Everything within it is accelerating toward the center of the earth with the acceleration due to gravity. (Notice that this is a centripetal acceleration since the object is circling the earth.) As in the case of the freely falling elevator discussed previously, no supporting force is needed to hold an object

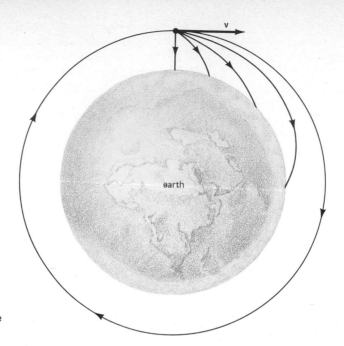

FIGURE 9.16 If a projectile is shot fast enough and parallel to the surface of the earth, it will circle the earth.

within the freely falling spaceship. Hence all objects in the spaceship appear to be weightless.

A spaceship need not be circling the earth for this weightless condition to exist. Whenever its rocket motors are not operating, the ship is falling freely through space. To remain in place in the ship, all objects within the ship must be freely falling as well. Hence there can be no supporting force acting on them. They thus appear to be weightless.

Illustration 9.7

How fast must a spaceship of mass m be moving if it is to circle the earth at a height of 160 km (100 mi)?

Reasoning For the spaceship to circle the earth, the centripetal force mv^2/R must be furnished by the gravitational force GmM_e/R^2. Since the radius of the earth is 6.37×10^6 m, we have $R = 6.53 \times 10^6$ m. Further, the mass of the earth is 6.0×10^{24} kg. Equating centripetal and gravitational forces yields

$$\frac{mv^2}{R} = \frac{GmM_e}{R^2}$$

or

$$v = \sqrt{\frac{GM_e}{R}}$$

from which

$$v = 7.8 \times 10^3 \text{ m/s}$$

Notice that the mass of the satellite spaceship canceled out. Hence our calculation applies to any object orbiting at this height.

It is of interest to find the time it would take for a satellite to orbit the earth. Since $t = 2\pi R/v$, one finds $t = 5260\,\text{s} = 87.7\,\text{min}$. Our calculation is not quite correct since it ignores the rotation of the earth beneath the satellite.

9.12 ACCELERATING REFERENCE FRAMES

We saw in the previous section that peculiar effects such as weightlessness can occur in accelerating systems. For this reason, among others, it is sometimes convenient to discuss a situation with reference to an accelerating reference frame. We know at once, since such a reference frame is not inertial, that the usual laws of physics may not be obeyed when described in this frame. Let us now see how Newton's second law, $F = ma$, must be modified if it is to be used in an accelerating system.

Suppose a railroad car is accelerating in the $+x$ direction with acceleration a_x relative to the earth. Each object at rest within the car must also be accelerating in the same way. As a result, an observer on the earth knows that a force

$$F_x = ma_x$$

must be exerted on the object in order to cause this acceleration. However, an observer within the car sees the object to be at rest with respect to the car. She therefore concludes, in her reference frame, that the acceleration of the object is zero. In spite of this, she notices that a force F_x is needed to keep the object at rest within the car. If it is not provided, the object accelerates in the negative x direction, and its acceleration is $-a_x$; that is to say, the object would then have zero acceleration relative to the earth.

From this we see that a woman in the accelerating reference system will have to apply forces to objects to hold them at rest in her reference system. It is as though a force were trying to accelerate the object in the reverse direction in her accelerating reference frame. When one attempts to describe the motion of objects in accelerating reference frames, it is always found that fictitious forces such as this must be invoked in order to explain nonconformity to $F = ma$ in such a noninertial system.

A very interesting case arises when one attempts to describe motion in terms of a rotating reference frame. For example, suppose an object of mass m is constrained by a rope to rotate with a disk as shown in figure 9.17. The length of the rope is r, and the angular speed is ω, a constant. An observer at rest on the earth knows full well that the rope must supply the centripetal force $m\omega^2 r$ to hold the mass in its circular path. This is the cause of the tension T in the rope, and $T = m\omega^2 r$.

However, an observer at rest relative to the rotating disk may not recognize that the disk is rotating. In his noninertial reference frame both the disk and mass are at rest. Yet he notices that a force T is needed to hold the mass in place. To him it appears that some fictitious force is trying to throw the mass radially out from the center of the disk. This force is equal and opposite to T, and is therefore equal and opposite to the centripetal force. This fictitious force is sometimes called the *centrifugal force*. Notice that it arises only because the man on the disk insists on describing the situation in reference to his noninertial system. A man at rest on the earth has no use for such a fictitious force since he is describing the event in terms of an inertial reference frame.

To avoid confusion, let us emphasize that the fictitious forces we have mentioned do not appear when inertial reference frames are used. They arise only when one insists on describing events by reference to a noninertial frame. It is preferable to avoid the use of noninertial reference frames and fictitious forces if possible. In advanced work, however, rotating reference

In accelerating reference frames, one must invoke fictitious forces

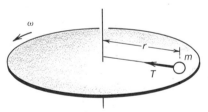

FIGURE 9.17 What is the meaning of centrifugal force in a situation such as this?

frames are sometimes used since their use simplifies other aspects of the situation. We do not make further use of fictitious forces and noninertial reference frames in this text.

9.13 THE EQUIVALENCE PRINCIPLE

When we wrote down Newton's law of gravitation, it was stated that the gravitational force is proportional to the product of the masses of the two bodies. Upon reflection, this is a most curious result. It will be recalled that Newton had previously discovered another basic law of nature, which involves mass, namely, $F = ma$. If we chose to use force as a standard rather than mass, either of these two laws could be used to define what we mean by mass. The question then arises as to whether the two definitions would be completely equivalent.

Newton spent some time investigating this point. Suppose we apply a force F to an object of mass m_0. Its acceleration will be given by

$$F = m_0 a_0$$

We can apply this force by using the gravitational force of the earth on the mass m_0. Then we have

$$G\frac{m_0' M_e}{r^2} = m_0 a_0$$

where M_e is the mass of the earth. In this expression we use m_0' in place of m_0 since the gravitational mass m_0' may not be the same as the inertial mass m_0.

If we repeat this experiment for a second mass, we shall have

$$G\frac{m_1' M_e}{r^2} = m_1 a_1$$

Taking the ratio of these two equations, we find

$$\frac{m_0'}{m_1'} = \frac{m_0}{m_1}\frac{a_0}{a_1}$$

If we find $a_0 = a_1$ experimentally, we shall know that the ratios of the inertial and gravitational masses are the same. In our present experiment, the a's will be the acceleration due to gravity, and we know that, to the experimental accuracy available, g is independent of m. Therefore we can conclude that, to high precision, the inertial and gravitational masses can be taken as identical, provided we choose our units properly. Other experiments more accurate than the one suggested here have confirmed this result.

This apparent identity of masses suggested to Einstein that gravitational forces could be considered to be the result of acceleration or some other type of inertial effect. In conformity with this, we have already seen in this chapter as well as earlier ones that weight and acceleration are related. Consider, for example, a man of mass m sitting at rest in a rocket ship. Before takeoff he exerts a force mg on his seat. We say he weighs mg. If he were to travel into outer space where gravitational forces were negligible, his apparent weight would be zero. However, if the rocket ship were now given a forward acceleration g by its rockets, the seat would have to push with a force mg on the man to give him this acceleration. The *principle of equivalence* states that this force and the weight of the man are indistinguishable as far as their variation with m and acceleration is concerned. This was one of the basic assumptions made by Einstein in his general theory of relativity. To the best of our present knowledge, gravitational and inertial forces and masses are identical. We are therefore led to believe that gravitational effects may be intimately related to accelerations.

Gravitational and inertial mass are identical

Chapter 9
Questions
and Guesstimates

1. One point is on the rim of a wheel while a second point is two-thirds of the way from the center to the rim. As the wheel spins on an axle through its center, compare the following for the two points: (*a*) tangential velocity; (*b*) angular velocity; (*c*) angular acceleration; (*d*) centripetal acceleration; (*e*) tangential acceleration.

2. Describe the angular velocity vector for each of the following: (*a*) earth rotating on its axis; (*b*) earth circling the sun; (*c*) total orbital motion of the planets around the sun.

3. How would you find the sum of two vector angular velocities if the velocities were equal and (*a*) parallel, (*b*) antiparallel, (*c*) perpendicular?

4. The earth attracts a bowling ball with a much larger force than it does a beach ball of the same size. Why, then, do they both fall at the same rate?

5. Can we obtain the masses of the planets from a knowledge of the radius of their orbits, and the mass of the earth?

6. According to the postulates of relativity, one's body should experience no sensation resulting from constant-speed motion along a straight line. Why is this not true for constant-speed motion in a circle? Analyze a few of the effects one can feel because of such motion.

7. When an airplane pilot pulls out of a steep dive, he experiences a "force of several *g*'s"; that is, his support in the plane must push on him with a force several times that of his weight. Why must this force exist? If the pullout is too quick, the pilot blacks out. Why?

8. The gravitational attraction (i.e., weight) caused by the moon on a person on the moon is 1.7*m* N, where *m* is the person's mass. In what ways will this make a person's life on the moon different from life on earth?

9. A woman is sealed in a room with no measuring tools of any kind. Could she tell whether or not the room was rotating about a vertical axis?

10. When an automobile decelerates rapidly, its occupants sometimes assert that they were "thrown forward" in the car. Is there any justification for such an assertion? In your answer, explain clearly why they believe that some force pushed them ahead.

11. A woman weighs herself daily on a spring-type bathroom scale. Suppose that the earth stopped rotating about its axis. Would she weigh more, less, or the same? Would it matter where she was on the earth?

12. An insect is sitting on a smooth, flat wheel which can be rotated about a vertical axis perpendicular to the plane of the wheel and through its center. Describe qualitatively the motion of the insect as the wheel begins to rotate. Assume that the insect is quite close to the axis and that there is some, but not much, friction between it and the wheel. (The wheel might be a phonograph turntable, for example.)

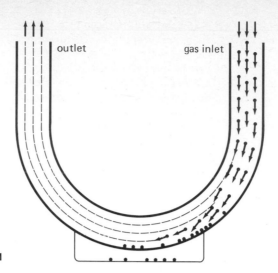

outlet   gas inlet

FIGURE P9.1

13. Figure P9.1 shows a simplified version of a cyclone-type dust remover. It is widely used to purify industrial waste gases before venting them to the atmosphere. The gas is whirled at high speed around a curved path, and the dust particles collect at the outer edge and are removed by a water spray or by other means. Explain the principle behind this method for removing particulate matter from dirty air.

14. A bug sits on the very top of a freshly waxed bowling ball. It loses its footing and slides freely down the surface of the ball. Explain why the bug will leave the surface *before* it falls halfway down the ball.

15. From the fact that the moon circles the earth at a radius of about 3.8×10^8 m, estimate the mass of the earth.

16. About how fast can a car be going if it must negotiate a turn from one street into a perpendicular street? Assume both streets to be made of concrete and to be of average size, carrying one lane of traffic each way.

17. During the 1970 flight of Apollo 13 to the moon, serious trouble developed when the ship was about halfway, and it returned to earth without executing its moon mission. However, the ship continued toward the moon, passed on the other side of it, and only then returned to earth. Why didn't they simply turn around at the time the decision was taken to return instead of continuing toward the moon?

Chapter 9
Problems

1. Express the following angles in degrees, radians, and revolutions: 20°; 0.40 rad; $\frac{1}{3}$ rev.

2. Express the following angular speeds in deg/s, rad/s, and rev/s: 0.020 rev/s; 30°/s; 1.40 rad/s.

3. (a) What is the angular velocity of the minute hand on a watch? (b) The second hand? (c) What is the vector sum of these two velocities? (d) Two grandfather clocks sit in a room; one is against the north wall and the other is against the east wall. What is the vector sum of the angular velocities of the minute hands on the two clocks?

4. A sphere is rotating about its axis in such a way that its angular velocity is given by $\omega = 3\mathbf{i} + 5\mathbf{j}$ rad/s. Find the angle between its axis and the x axis.

5. A roulette wheel originally turning at 0.80 rev/s coasts to rest in 20 s. What was the deceleration of the wheel? Through how many revolutions did it turn in the process? (Assume uniform deceleration.)

6. In 7 s a car accelerates uniformly from rest to such a speed that its wheels are turning at a rate of 6.0 rev/s. What was the angular acceleration of a wheel? Through how many revolutions did the wheel turn?

7. A car accelerates uniformly from rest to a speed of 15 m/s in a time of 20 s. Find the angular acceleration of one of its wheels and the number of revolutions turned by a wheel in the process. The radius of the car wheel is $\frac{1}{3}$ m.

8. A belt runs on a wheel of 30 cm radius. During the time that the wheel coasts uniformly to rest from an initial speed of 2.0 rev/s, 25 m of belt length passes over the wheel. Find the deceleration of the wheel and the number of revolutions it turns while stopping.

9. A 90-cm-radius roulette wheel initially turning at 3.0 rev/s slows uniformly and stops after 26 rev. (a) How long did it take to stop? (b) What was its angular deceleration? (c) What was the initial tangential speed of a point on its rim? (d) The initial radial acceleration of a point on the rim? (e) The magnitude of the resultant initial acceleration of a point on its rim?

10. A cylinder of 5 cm radius is rolling along the floor with a constant speed of 80 cm/s. (a) What is the rotation speed of the cylinder about its axis? (b) What is the magnitude and direction of the acceleration of a point on its surface? (c) At the instant a certain point on its surface is at the top of the cylinder, what is the velocity of the point? (d) Repeat if the point is at the contact with the floor. (e) Repeat if the point is midway between top and floor and at the foreward surface of the cylinder.

11. In an ultracentrifuge a solution is rotated with an angular speed of 3000 rev/s at a radius of 10 cm? How large is the radial acceleration of each particle in the solution? Compare the centripetal force needed to hold a particle of mass m in the circular path with the weight of the particle mg.

12. The red blood cells and other particles suspended in blood are too light in weight to settle out easily when the blood is left standing. How fast (in rev/s) must a sample of blood be rotating at a radius of 10 cm in a centrifuge if the centripetal force needed to hold one of the particles in a circular path is 10,000 times the weight of the particle, mg? Why do the particles separate from the solution in a centrifuge?

13. A certain car of mass m has a maximum friction force of 0.7 mg between it and pavement as it rounds a curve on a flat road ($\mu = 0.7$). How fast can the car be moving if it is to successfully negotiate a curve of 15-m radius?

14. The rotation speed of the earth is 1 rev/day or 1.16×10^{-5} rev/s, and the earth's radius is 6.37×10^6 m. If a man at the equator is standing on a spring scale, by what percent will his apparent weight increase if the earth were to stop rotating? A man at the north pole?

15. When constructing a roller coaster, the designer wishes the riders to experience weightlessness as they round the top of one hill. How fast must the car be going if the radius of curvature at the hill top is 20 m?

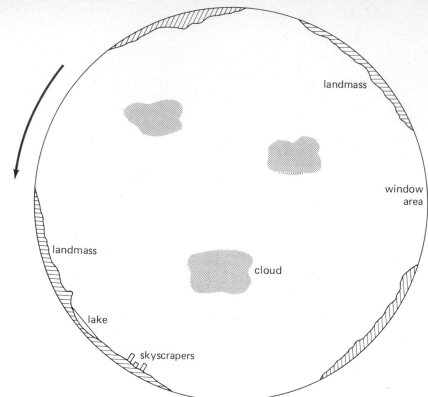

FIGURE P9.2

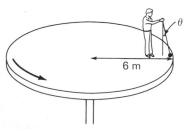

FIGURE P9.3

16. Figure P9.2 shows a possible design for a space colony of the future. It consists of a 6-km-diameter cylinder of length 30 km floating in space. Its interior is provided with an earthlike environment. To simulate gravity, the cylinder spins on its axis as shown. What should be the rate of rotation of the cylinder (in rev/h) so that a person standing on the land mass will press down on the ground with a force equal to his or her weight on earth? (For details, see G. K. O'Neill, *Physics Today,* September 1974, p. 32.)

17. As shown in figure P9.3, a boy on a rotating platform holds a pendulum in his hand. The pendulum is at a radius of 6.0 m from the center of the platform. The rotation speed of the platform is 0.020 rev/s. It is found that the pendulum hangs at an angle θ to the vertical as shown. Find θ.

18. A particle is to slide along the horizontal circular path on the inside of the funnel shown in figure P9.4. The surface of the funnel is frictionless. How fast must the particle be moving (in terms of r and θ) if it is to execute this motion?

19. In figure P9.5 the mass m is held by two strings and the system is rotating with angular velocity ω. Find the tensions in the two strings in terms of m, ω, r, and θ.

20. Two identical coins of mass 8 g are 50 cm apart on a tabletop. How many times larger is the weight of one coin than the gravitational attraction of the other coin for it?

21. Three identical point masses lie along the x axis at $x = 0$, 0.20 m, and

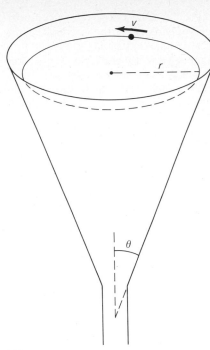

FIGURE P9.4

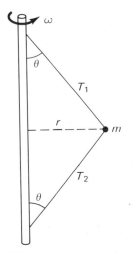

FIGURE P9.5

0.70 m. Calling each mass M, find the gravitational force F_x on the center mass.

22. Three identical point masses M lie in the xy plane at points $(0, 0)$, $(0, 0.20\text{ m})$, and $(0.20\text{ m}, 0)$. Find the components of the gravitational force on the mass at the origin.

23. A straight rod of length L extends from $x = a$ to $x = L + a$. Find the gravitational force it exerts on a point mass m at $x = 0$ if the mass per unit length of the rod is $\mu = A + Bx^2$.

24. Repeat problem 23 if $\mu = Ax + Bx^2$.

25. Knowing that the force with which a mass m is attracted to the earth is given by GM_em/r^2 where M_e is the earth's mass, (*a*) show by integration that the work needed to carry the mass from the earth's surface to $r \to \infty$ is given by GM_em/R_e. (*b*) How much kinetic energy must the mass have if it is to escape completely from the earth when shot straight upward? (*c*) Show that the speed of the particle must be $\sqrt{2GM_e/R_e}$. This is called the *escape velocity*. Notice that it is the same for all masses.

26. An engineer who usually uses his calculator to evaluate $10 \div 2$ decides to check his car's speedometer in the following way. While rounding a curve of 200 m radius he measures that a pendulum in the car hangs at an angle of $15°$ to the vertical. What should the speedometer read (in km/h)?

27. A huge pendulum consists of a 200-kg ball at the end of a cable 15 m long. If the pendulum is drawn back to an angle of $37°$ and released, what maximum force must the cable withstand as the pendulum swings back and forth?

28. A mass m is fastened to the axis of rotation of a horizontal disk by a spring of equilibrium length a and spring constant k. Show that if the disk is rotating with frequency ω, the mass will come to equilibrium at a radius r given by $r = a + m\omega^2a/k$ provided $m\omega^2/k \ll 1$.

29. Communication satellites are placed in orbit above the equator in such a way that they remain stationary above a given point on earth below. The satellite orbits the earth at the same rate that the earth turns and so, relative to a point on earth below it, the satellite is at rest. How high above the surface of the earth is the orbit of such a satellite? The satellite is said to be in *synchronous orbit*. ($R_e = 6.4 \times 10^6$ m, $M_e = 5.98 \times 10^{24}$ kg.)

30. The bug shown in figure P9.6 has just lost its footing while near the top of the bowling ball. It slides down the ball without appreciable friction. Show that it will leave the surface of the ball at the angle θ shown, where θ is given by $\cos \theta = \frac{2}{3}$.

31. During the initial part of its acceleration, the angle through which a car's wheel turns as a function of time is given by

$$\theta = Bt + Ct^2$$

where B and C are constants. Find the linear displacement of the car and its speed as a function of time. The radius of the car wheel is R.

32. The angle which a pendulum of length L makes with the vertical varies with time according to the following equation

$$\theta = \theta_0 \sin (2\pi ft)$$

where θ_0 is the maximum angle of swing and f is the frequency of the pendulum, both constants. By differentiation, find the speed and acceleration of the pendulum ball as a function of time.

33. The acceleration due to gravity on the moon is only one-sixth that on

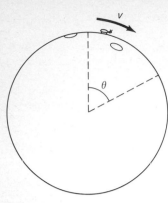

FIGURE P9.6

earth. Assuming the earth and moon to have the same average composition, what would you predict the moon's radius to be in terms of the earth's radius R_e? In fact, the moon's radius is $0.27R_e$.

34. Knowing that the earth-moon distance is 3.8×10^5 km, compute the time (in days) it takes the moon to circle the earth. ($M_e = 5.98 \times 10^{24}$ kg.)

35. Sometimes a very complicated problem can be solved quite simply if one uses some insight (others say, by use of a trick!). One such problem, frequently given to graduate students, is as follows. In figure P9.7 is shown a uniform sphere of original total mass M in which a spherical hole of diameter R has been formed. Show that it attracts the mass m with a force given by

$$\frac{GMm}{D^2}\left[1 - \frac{1}{8}\left(1 - \frac{R}{2D}\right)^{-2}\right]$$

FIGURE P9.7

10 Rotational Dynamics

In the preceding chapter we discussed the motion of a simple particle in a perfect circle. We dealt only with the motion itself, and not with the forces which gave it its motion. The forces which cause rotation of both particles and extended rigid bodies are the subject of this chapter. It will be seen that the inertia of a body for rotational motion depends not only upon the mass of the body, but upon the distribution of the mass within the body. A rotational analog to $F = ma$ will be found and used to discuss rigid-body rotation.

10.1 TORQUE AND ROTATION

Suppose a particle of mass m is subject to a force $\mathbf{F}$, as shown in figure 10.1. We wish to describe the motion of this particle under the action of the force. The general motion of such a particle in three dimensions can become quite complicated. Fortunately, many of the situations one encounters in practice can be reduced to motion of the particle in a circle. In such cases, r is constant.

As you doubtless know, mathematical situations involving circles are best handled using polar coordinates. To refresh your memory, the two polar coordinates r and ϕ are taken as shown in figure 10.1. In our case, ϕ will change as the particle traces out a circular path with r constant. It is convenient in such situations to use r and ϕ components for vectors rather than the x and y components we have used previously. To that end, we define the unit vectors $\hat{\mathbf{r}}$ and $\hat{\boldsymbol{\phi}}$ shown in figure 10.2.* Notice that, unlike the unit vectors $\mathbf{i}$, $\mathbf{j}$, and $\mathbf{k}$, these new unit vectors change direction as the particle moves from place to place. Notice also that $\hat{\mathbf{r}}$ and $\hat{\boldsymbol{\phi}}$ are perpendicular to each other.

Before proceeding to discuss the motion of the particle in terms of these coordinates, let us practice using them by writing in component form the vectors $\mathbf{r}$ and $\mathbf{F}$ shown in figure 10.3. You should examine figures 10.2 and 10.3 to make sure that you understand why we can write $\mathbf{r}$ and $\mathbf{F}$ as follows:

$$\mathbf{r} = r\hat{\mathbf{r}} \qquad \text{and} \qquad F = F_r\hat{\mathbf{r}} + F_\phi\hat{\boldsymbol{\phi}}$$

where $F_r = F\cos\theta$ and $F_\phi = F\sin\theta$. Now that we understand how to represent vectors in polar coordinates, let us return to a discussion of the motion of a particle in a plane with r constant, i.e., circular motion.

* We read $\hat{\mathbf{r}}$ as "r hat" or "r caret." Other designations for the unit vectors are $\mathbf{l}_r$, $\mathbf{l}_\phi$ and $\mathbf{u}_r$, $\mathbf{u}_\phi$.

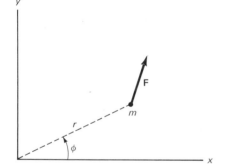

FIGURE 10.1 The motion of the mass m caused by the force shown is described in terms of the polar coordinates r and ϕ.

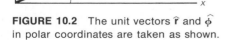

FIGURE 10.2 The unit vectors $\hat{\mathbf{r}}$ and $\hat{\boldsymbol{\phi}}$ in polar coordinates are taken as shown.

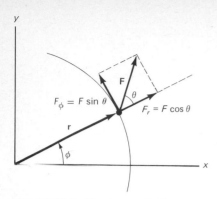

FIGURE 10.3 The tangential force component F_ϕ causes a tangential acceleration $a_\phi = a_T$. The torque exerted by it about the origin is rF_ϕ. What is the torque due to F_r?

Consider the effects of the force **F** shown in figure 10.3 on the particle of mass m. Newton's law tells us that $\mathbf{F} = m\mathbf{a}$. Or, in component form,

$$F_r\hat{\mathbf{r}} + F_\phi\hat{\boldsymbol{\phi}} = ma_r\hat{\mathbf{r}} + ma_\phi\hat{\boldsymbol{\phi}} \qquad 10.1$$

where a_r and a_ϕ are the components of **a** in polar coordinates. We see at once that the component equations are simply

$$F_r = ma_r \qquad \text{and} \qquad F_\phi = ma_\phi \qquad 10.2$$

What do these equations mean for the motion of a particle on a circle?

The ϕ equation is perhaps easiest to understand. It tells us that the component of the force tangential to the circular path causes the particle to speed up or slow down in its circular motion. But this speeding up or slowing down is a change in the tangential velocity of the particle, and so we see that the ϕ component of the force gives rise to a tangential acceleration. We denote it by a_T in a way analogous to our designation of tangential velocity as v_T. Therefore a_ϕ is the same as a_T.

The radial component of the force F_r causes an acceleration along the radius vector. This is, of course, the radial acceleration we discussed in the previous chapter. You will recall that when a particle is moving in a circle, it is accelerating radially and a_r is this radial or centripetal acceleration. We shall not be concerned further with F_r and a_r in our discussion.

Let us now write F_ϕ from equation *10.2* in terms of the angular quantity α instead of a_T. Since we had from equation *9.8* that $a_T = \alpha r$, we can write

$$F_\phi = mr\alpha \qquad 10.3$$

But, as we see from figure 10.3, we can write F_ϕ as $F\sin\theta$ where θ is the angle between **F** and **r**. Therefore we have

$$F\sin\theta = mr\alpha \qquad 10.4$$

For reasons which will become apparent, it is usual to multiply *10.3* by r and write it as

$$rF\sin\theta = mr^2\alpha \qquad 10.5$$

The quantity on the left-hand side of this equation was encountered in Chap. 3. In figure 10.3 we see that θ is the angle between **r** and **F** and so $rF\sin\theta$ is the torque about the origin due to F. Calling the torque τ, equation *10.5* becomes

For a single particle moving in a circle,
$$\tau = mr^2\alpha$$

$$\tau = mr^2\alpha \qquad 10.6$$

where the units of torque are newton $\cdot$ meters.

We notice at once a similarity between the equations $F = ma$ and $\tau = (mr^2)\alpha$. The linear acceleration a has been replaced by the angular acceleration α. In linear motion, the unbalanced force F is the cause of the linear acceleration. As we might expect, in rotational motion the angular acceleration is caused by an unbalanced torque. Therefore τ replaces F in the analog to $F = ma$. Inertia in the case of linear motion is represented by the mass of the object m. Likewise, in rotational motion we would expect an object to show inertia to being set into rotation about an axis. Apparently the rotational inertia for a point mass m rotating at a radius r from an axis is mr^2 since this quantity replaces m in the rotational analog to $F = ma$.

This means that the farther the particle is from the axis of rotation the more difficult it is to give it a specified angular acceleration. This is not too surprising, perhaps, since the equivalent linear acceleration a_ϕ is given by αr, and therefore increases with r, even though α is constant. We shall investigate this matter further in the sections which follow.

Illustration 10.1

A particle of mass 50 g is attached to an axis by a very light rod of length 80 cm, similar to the situation shown in figure 10.1. The particle is free to rotate about the axis in the ϕ direction. If a force of 0.20 N is applied to the mass in a direction perpendicular to the rod, how long will it take the mass to reach a rotation speed of 4.0 rev/s?

Reasoning The force is in the ϕ direction, i.e., perpendicular to the rod, and so *10.6* gives

$$(0.80 \, \text{m})(0.20 \, \text{N}) = (0.050 \, \text{kg})(0.80 \, \text{m})^2 \alpha$$

Solving, $\qquad\qquad\qquad \alpha = 5.0 \, \text{rad/s}^2$

Notice the units on α; why radians?

To find the time taken to reach a speed of 4.0 rev/s $= 8.0\pi$ rad/s, we use the motion equation

$$\omega = \omega_0 + \alpha t$$

to give $\qquad\qquad\qquad t = \dfrac{8.0\pi - 0}{5.0} \approx 5 \, s$

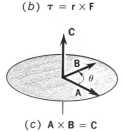

(a) $\tau = mr^2\boldsymbol{\alpha}$

(b) $\boldsymbol{\tau} = \mathbf{r} \times \mathbf{F}$

(c) $\mathbf{A} \times \mathbf{B} = \mathbf{C}$

FIGURE 10.4 The vector (or cross) product carries a direction as well as a magnitude. What is the magnitude of τ in terms of $\mathbf{r}$ and $\mathbf{F}$? What is the magnitude of $\mathbf{C}$?

The torque direction is given by the right-hand-screw rule

10.2 THE VECTOR (OR CROSS) PRODUCT

It was pointed out in the preceding chapter that sometimes it is convenient to represent angular quantities as vectors. The angular acceleration α in *10.6* is such a quantity, and we should therefore like to write *10.6* in vector form. Since we have not heretofore assigned direction to the torque τ, we can use *10.6* to define a direction for it. In that case we should write

$$\boldsymbol{\tau} = mr^2\boldsymbol{\alpha} \qquad\qquad\qquad 10.7$$

The relationships between these vectors and the axis of rotation are shown in figure 10.4(a). Notice that the direction of $\boldsymbol{\alpha}$ is given by the right-hand-screw rule. (The rotational acceleration is caused by F_ϕ, and so one should grasp the axis with fingers circling in the direction of F_ϕ.)

However, if we assign a vector to the torque, it must be consistent with our previous definition of torque, namely,

$$\tau = rF \sin\theta \qquad\qquad\qquad 10.8a$$

This means we must assign a direction to $rF \sin\theta$. The various vectors involved are shown in figure 10.4(b). We wish the product of the two coplanar vectors $\mathbf{r}$ and $\mathbf{F}$ with the sine of the angle between them to give a vector perpendicular to their plane. This can be done by use of the *vector product* notation illustrated in figure 10.4(c).

The *vector, or cross, product* of two vectors $\mathbf{A}$ and $\mathbf{B}$ is defined in the following way:

Definiton of the cross product $\mathbf{A} \times \mathbf{B}$

$$\mathbf{A} \times \mathbf{B} = \mathbf{C} \qquad \text{with} \qquad C = AB \sin\theta \qquad\qquad 10.9$$

where θ is the angle between $\mathbf{A}$ and $\mathbf{B}$. These three vectors follow the right-hand-screw rule if we interpret *10.9* to say: *Rotate the vector* $\mathbf{A}$ *through the smallest possible angle to make it point in the direction of* $\mathbf{B}$; *the*

$$|\mathbf{A} \times \mathbf{B}| = AB \sin \theta \text{ and } \mathbf{A} \cdot \mathbf{B} = AB \cos \theta$$

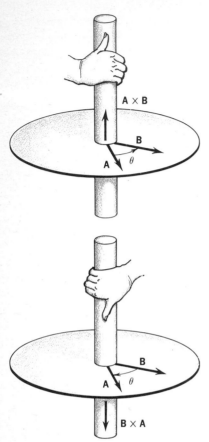

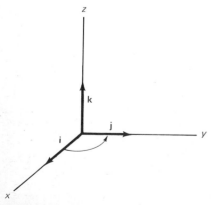

FIGURE 10.5 $\mathbf{A} \times \mathbf{B} = -\mathbf{B} \times \mathbf{A}$

FIGURE 10.6 Since i, j, and k are unit vectors, $\mathbf{i} \times \mathbf{j} = \mathbf{k}$.

resultant vector **C** *points in the direction given by the right-hand-screw rule for such a rotation.*

If we are interested in the magnitude of the result only, the symbolism **A** × **B** merely means $AB \sin \theta$, where θ is the angle between A and B. In addition, however, **A** × **B** conveys direction, and therefore results in a vector. It is for this reason that it is called a *vector* product. Notice that it differs from the scalar, or dot, product **A** · **B**, which resulted in a scalar, $AB \cos \theta$. Moreover, unlike the scalar product, the order of the vectors in a vector product is important. As we see from figure 10.5, **A** × **B** is not equal to **B** × **A**. The two results are numerically equal but point in opposite directions. It is therefore important to keep the proper order of the quantities in a vector product.

To summarize our results, we can express the torque due to a force **F** at a radial distance **r** from the axis by

$$\boldsymbol{\tau} = \mathbf{r} \times \mathbf{F} \qquad\qquad 10.8b$$

This simply means

$$\tau = rF \sin \theta$$

where θ is the angle between **r** and **F**. In addition, however, it specifies the direction of τ as being perpendicular to the plane of **r** and **F**, the proper direction being given by the right-hand-screw rule.

Illustration 10.2

Evaluate $\mathbf{i} \times \mathbf{i}$, $\mathbf{i} \times \mathbf{j}$, $\mathbf{j} \times \mathbf{i}$, and $\mathbf{j} \times \mathbf{j}$, where **i** and **j** are the usual unit vectors.

Reasoning By definition,

$$|\mathbf{i} \times \mathbf{i}| = i^2 \sin \theta$$

where θ is the angle between **i** and **i**, namely, 0°. But $\sin 0 = 0$ and so $\mathbf{i} \times \mathbf{i} = 0$. Similarly $\mathbf{j} \times \mathbf{j} = 0$. (Obviously $\mathbf{k} \times \mathbf{k} = 0$, too.)

Also from the definition of the cross product,

$$|\mathbf{i} \times \mathbf{j}| = ij \sin (90°) = 1$$

since **i**, **j**, and sin 90° are all unity. (Remember, **i** and **j** are unit vectors.) The direction of $\mathbf{i} \times \mathbf{j}$ is seen from application of the right-hand rule to figure 10.6 to be in the **k** direction. Since **k** is a unit vector we can write

$$\mathbf{i} \times \mathbf{j} = \mathbf{k}$$

Further, as with the situation shown for A and B in figure 10.5, we have

$$\mathbf{j} \times \mathbf{i} = -\mathbf{k}$$

Reasoning similar to that used here will show you that

$$\mathbf{i} \times \mathbf{i} = \mathbf{j} \times \mathbf{j} = \mathbf{k} \times \mathbf{k} = 0$$

and	$\mathbf{i} \times \mathbf{j} = \mathbf{k}$	$\mathbf{j} \times \mathbf{k} = \mathbf{i}$	and	$\mathbf{k} \times \mathbf{i} = \mathbf{j}$
while	$\mathbf{j} \times \mathbf{i} = -\mathbf{k}$	$\mathbf{k} \times \mathbf{j} = -\mathbf{i}$	and	$\mathbf{i} \times \mathbf{k} = -\mathbf{j}$

10.3 RIGID–BODY ROTATION

Let us now discuss the motion of a solid object which is constrained to rotate about a stationary axis. That is to say, the various pieces of matter making

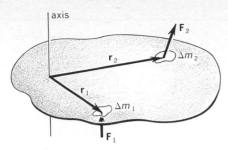

FIGURE 10.7 The object is constrained to rotate rigidly about the axis.

up the object always remain at the same radial distances from the axis, and therefore move in circles. A typical situation is shown in figure 10.7. In addition, we assume the object to be rigid, so that the size, shape, and mass distribution within the object remain constant. This type of situation is one usually encountered in practical applications. Although the object is shown as flat, this is not essential. For a three-dimensional object the various $\mathbf{r}$ vectors would not be coplanar. They would all be perpendicular to the axis, however.

It is convenient to subdivide the rigid object, mentally, into N small masses, $\Delta m_1, \Delta m_2, \ldots, \Delta m_N$. Two of these are shown in figure 10.7. Calling $\mathbf{F}_1$ the resultant force on Δm_1, and so on for the rest, we can write 10.7 for each mass in turn to give

$$\mathbf{r}_1 \times \mathbf{F}_1 = \boldsymbol{\tau}_1 = (\Delta m_1) r_1{}^2 \boldsymbol{\alpha}_1$$
$$\mathbf{r}_2 \times \mathbf{F}_2 = \boldsymbol{\tau}_2 = (\Delta m_2) r_2{}^2 \boldsymbol{\alpha}_2$$
$$\cdots\cdots\cdots\cdots\cdots\cdots\cdots$$
$$\mathbf{r}_N \times \mathbf{F}_N = \boldsymbol{\tau}_N = (\Delta m_N) r_N{}^2 \boldsymbol{\alpha}_N$$

Since the body is rigid, all the angular acceleration vectors must be identical. Therefore, summing all these equations yields

$$\sum_1^N \boldsymbol{\tau}_i = \boldsymbol{\alpha} \sum_1^N r_1{}^2 \, \Delta m_i \qquad\qquad 10.10$$

The sum in 10.10 involving the torques includes torques due to two causes. First, external forces acting on the body cause torques, and experience tells us these torques are the cause of any rotational acceleration given the object. In addition, a second type of torque exists. This latter type is due to the action of internal forces within the object. That is to say, mass Δm_1, for example, experiences forces resulting from the neighboring masses. These internal forces and their resultant torques cancel each other because they are the result of the in-line equal and opposite action and reaction forces within the body.

We therefore conclude that the sum of the torques in 10.10 is merely the total resultant torque due to the external forces. As a result, 10.10 becomes

$$\boldsymbol{\tau}_{\text{ext}} = \boldsymbol{\alpha} \sum_1^N r_i{}^2 \, \Delta m_i \qquad\qquad 10.11$$

The remaining sum in 10.11 is of considerable interest, and we shall now discuss it. Before doing that, however, we must point out that the fixed axis of rotation to which we assume our object attached has a special importance for the form of the external torque in 10.11.

For an object constrained to rotate about a fixed axis, α and τ are along the axis

If the object is rigidly attached to the axis, then $\boldsymbol{\alpha}$ can take on only two possible directions. The object can only rotate clockwise or counterclockwise about the axis. As a result, $\boldsymbol{\alpha}$ will always be directed along the axis. Since $\boldsymbol{\alpha}$ and $\boldsymbol{\tau}_{\text{ext}}$ are parallel, the resultant torque on the object must be along the axis. External forces imposed by the axis upon the object constrain the object to rotate in the way indicated. The forces exerted upon the object by the axis cause a torque which exactly cancels any nonaxial torque resulting from other external forces. For this reason, when one uses 10.11, which is applicable to a body constrained to rotate about an axis, only the axial components of the external torques need be considered.

10.4 MOMENT OF INERTIA

Let us return to the sum in *10.11*. Insight into its function can be found by rewriting the equation as

$$\sum_{1}^{N} r_i{}^2 \, \Delta m_i = \frac{\tau}{\alpha}$$

where the vector properties of τ and α have been omitted, since they are both axial. The sum is equal to the torque needed to cause unit angular acceleration. As such, it is large if the object is difficult to accelerate, and small if the object is easily accelerated. In other words, the above sum *is a measure of the rotational inertia* of the object about the chosen axis. We call it the *moment of inertia* and represent it by the symbol I. Therefore the moment of inertia I is defined to be

Definition of moment of inertia I

$$I \equiv \sum_{1}^{N} r_i{}^2 \, \Delta m_i$$

Notice that r_i is defined to be the *length of the perpendicular* dropped from Δm_i to the axis.

The basic equation, analogous to $F = ma$, for a rotating object is obtained from *10.11* by use of the definition of I, and is

$\tau = I\alpha$ is analogous to $F = ma$

$$\tau = I\alpha \qquad\qquad 10.12$$

This is the rotational analog to $\mathbf{F} = m\mathbf{a}$. Once the values of I and τ are known for a body rotating about a specified axis, the rotational acceleration of the body may be found from *10.12*.

Two basic facts concerning the moment of inertia should be noted if one is to obtain an understanding of it. First, I is proportional to the total mass of the object. The more massive the object, the more difficult it will be to start it rotating or stop it from rotating. Second, for a given mass, its moment of inertia depends upon the distribution of the mass from the axis. The larger the distance of the mass from the axis, the larger will be its rotational inertia. We should therefore expect a ring of mass M to have a larger I than a disk with the same mass and radius. We shall find quantitative relations for I in the following illustrations.

Illustration 10.3

Find the moment of inertia of the system shown in figure 10.8 about the axis OO'. Assume the mass of the rigid rods connecting the masses to be zero, and consider the masses to be point masses.

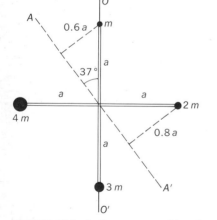

FIGURE 10.8 Is the moment of inertia about OO' as axis the same as about AA'?

Reasoning From the definition of I we have (noting that r is the length of a perpendicular from the mass to the axis)

$$I = \sum r_i{}^2 \, \Delta m_i$$

$$= (0)(m) + (a^2)(2m) + (0)(3m) + (a^2)(4m)$$

$$= 6ma^2$$

If we write the total mass of the object as M and recall that it is $10m$, we can write I as

$$I = M(0.6a^2) \equiv Mk^2$$

The quantity k^2 multiplying the total mass, $0.6a^2$ in this case, is often called the square of the *radius of gyration*. It is a measure of the average square distance from the axis to the mass of the object.

Illustration 10.4

For the object of figure 10.8, find the moment of inertia about AA' as axis.

Reasoning From the definition of I, we find (remembering that r_i is the *perpendicular* distance from Δm_i to the axis)

$$I = \sum_{1}^{N} r_i^2 \, \Delta m_i$$

$$= (0.6a)^2(m) + (0.8a)^2(2m) + (0.6a)^2(3m) + (0.8a)^2(4m)$$

$$= 5.3a^2 m$$

$$= M(0.53a^2)$$

What is the radius of gyration about this axis?

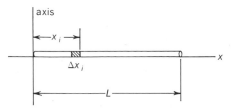

FIGURE 10.9 In the text the moment of inertia of this rod is found assuming λ, the mass per unit length, to be constant. How would you modify the computation if λ were a known function of x?

10.5 MOMENT OF INERTIA OF A THIN ROD

Let us compute the moment of inertia of the thin uniform rod shown in figure 10.9. As axis we shall use a line perpendicular to the rod and at its end. Since the general expression for I is

$$I = \sum r_i^2 \, \Delta m_i$$

we split the continuous rod into a series of segment masses of length Δx_1, $\Delta x_2, \ldots, \Delta x_N$ at radii $x_1, x_2, \ldots, x_N$ from the axis. If each meter length of the rod has a mass λ kg/m, then the mass of the length Δx will be $\lambda \, \Delta x$. (For example, if $x = 0.1$ m, the mass of that length would be 0.1λ.) Therefore we have sectioned the rod into N tiny masses, $\lambda \, \Delta x_1, \lambda \, \Delta x_2, \ldots, \lambda \, \Delta x_N$.

If we place these quantities in the expression for I, we find

$$I = \sum_{1}^{N} x_i^2 (\lambda \, \Delta x_i)$$

Assuming λ to be constant, we factor it to obtain

$$I = \lambda \sum_{1}^{N} x_i^2 \, \Delta x_i$$

We now pass to the limit of $\Delta x \to 0$ and $N \to \infty$, so that the above sum can be replaced by an integral. This gives

$$I = \lambda \int_{x=0}^{x=L} x^2 \, dx$$

The limits on the integral are obtained by noticing that the mass closest to the axis has an $x = 0$, while the farthest has $x = L$.

The integral in the expression for I is easily evaluated (or found in tables) and gives

$$I = \lambda \left. \frac{x^3}{3} \right|_0^L$$

from which

$$I = \tfrac{1}{3}\lambda L^3$$

If we notice that the total mass of the rod is λL, we can write this

$$I = Mk^2$$

where

$$k^2 = \frac{L^2}{3}$$

10.13

As indicated previously, the quantity k is called the radius of gyration of the rod about the axis. It is a rough measure of the distance (on the average) of the mass of the object from the axis. Can you show that the radius of gyration for the rod about an axis coincident with the axis of the rod (i.e., the x axis) is essentially zero? Why is not the moment of inertia the same for an axis parallel to the axis through the end but passing instead through the center of the rod?

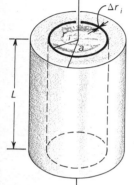

FIGURE 10.10 The thin cylindrical shell has mass Δm_i, all of it at a distance r_i from the axis.

10.6 MOMENT OF INERTIA OF A CYLINDER

Consider the solid cylinder of figure 10.10, which has a constant mass per unit volume ρ. To find the moment of inertia of it about the axis natural to the cylinder, we use the definition of I, namely,

$$I = \sum_1^N r_i^2 \, \Delta m_i$$

We look for a small piece of mass Δm_i, which is all at the same distance r_i from the axis. A cylindrical shell of radius r_i, thickness Δr_i, and length L satisfies this condition, and is shown in the figure.

The volume of the shell is the area of its cylindrical surface, $2\pi r_i L$, multiplied by the thickness of the shell, Δr_i. Since the mass per unit volume is ρ, the mass of this shell is just

$$\Delta m_i = \rho(2\pi r_i L \, \Delta r_i)$$

Substitution of this value in the expression for I yields, if ρ is constant,

$$I = 2\pi \rho L \sum_1^N r_i^3 \, \Delta r_i$$

If we now pass to the limit, this becomes an integral, with limits determined by the fact that $0 \leq r \leq a$, and we have

$$I = 2\pi \rho L \int_0^a r^3 \, dr$$

Evaluating this integral and placing in the limits gives

$$I = \tfrac{1}{2}\pi \rho L a^4$$

Since the total mass M of the cylinder is $\rho(\pi a^2 L)$, we can write this

$$I = \tfrac{1}{2}a^2 M$$

10.14

We see that the radius of gyration, k, of the cylinder is merely $a/\sqrt{2}$.

In principle, the moment of inertia of any body about any axis can be found in the way we have done this example. The answers for several common shapes are given in figure 10.11. The student should examine these results to see if the radius of gyration k seems reasonable in each case.

10.7 THE PARALLEL AXIS THEOREM

Most all the moments of inertia in figure 10.11 are for an axis passing through the center of mass of the object. We could greatly increase our knowledge if we could find a simple relation between these values of I and the

FIGURE 10.11 Some frequently used moments of inertia.

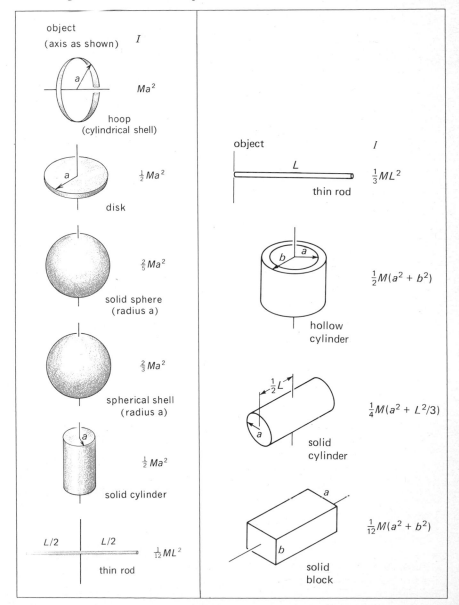

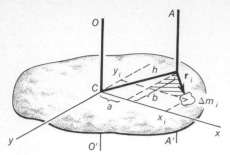

FIGURE 10.12 What does the parallel axis theorem tell us about the moments of inertia for the parallel axes *OO'* and *AA'* (the point *C* is the center of mass)?

value of *I* for some other axis but one which is still parallel to the axis shown in the figure. Such a relation does indeed exist, and we shall now find it.

Consider the object shown in figure 10.12. The axis *OO'* is assumed to pass through the center of mass of the object, and the axis *AA'* is parallel to *OO'*. Call the distance between the axes *h*. Let us now compute the moment of inertia *I* for the body about axis *AA'*. It is, using the Δm_i shown as typical,

$$I = \sum r_i^2 \, \Delta m_i$$

We see from the figure that the two sides of the shaded right triangle containing $\mathbf{r}_i$ as hypotenuse are $b - y_i$ and $x_i - a$. Therefore

$$r_i^2 = (b - y_i)^2 + (x_i \quad a)^2$$

and so

$$I = \sum [(b - y_i)^2 + (x_i - a)^2] \, \Delta m_i$$

Upon expanding and regrouping terms, this becomes

$$I = \sum [(y_i^2 + x_i^2) \, \Delta m_i] + (a^2 + b^2) \sum \Delta m_i - 2b \sum y_i \, \Delta m_i - 2a \sum x_i \, \Delta m_i$$

But the last two sums of this expression must be zero, since the definitions of the coordinates of center of mass are (Chap. 7)

$$x_{cm} = \frac{\sum x_i \, \Delta m_i}{M} \quad \text{and} \quad y_{cm} = \frac{\sum y_i \, \Delta m_i}{M}$$

and we have taken the center of mass at the origin, so that $x_{cm} = y_{cm} = 0$.

In addition, we notice that $y_i^2 + x_i^2$ is merely the square of the radius from the *OO'* axis to the mass Δm_i. Therefore the term

$$\sum (y_i^2 + x_i^2) \, \Delta m_i = I_{cm}$$

where I_{cm} is the moment of inertia of the object about the axis *OO'* through the center of mass.

Finally, we notice that $a^2 + b^2$ is merely h^2, the square of the distance between the axes. The expression for *I* therefore becomes

The parallel axis theorem

$$I = I_{cm} + Mh^2 \qquad\qquad 10.15$$

This equation is known as the *parallel axis theorem*. It tells us that the moment of inertia of an object about an axis parallel to the axis through the center of mass and a distance *h* away is $I_{cm} + Mh^2$. Notice that Mh^2 is just the moment of inertia about the new axis of a point mass *M* placed at the center of mass of the object. By use of *10.15*, the relations given in figure 10.11 can be extended to cover many other situations.

10.8 EXPERIMENTAL DETERMINATION OF *I*

Equation *10.12*, $\tau = I\alpha$, can be used to determine by experiment the moment of inertia of any device which can be rotated about an axis. For example, the wheel of unknown moment of inertia in figure 10.13 can be caused to rotate about its axis by applying a force *F* to a rope wound on the wheel. Calling the radius of the wheel *b*, we see at once that

$$\text{Torque} = |\mathbf{r} \times \mathbf{F}| = rF \sin \theta = bF$$

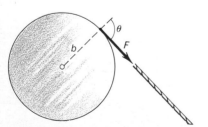

FIGURE 10.13 What must be measured in this experiment in order to determine *I*?

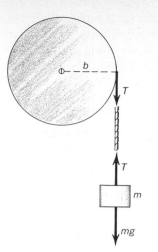

FIGURE 10.14 Why doesn't $T = mg$?

This torque will cause the wheel to accelerate with angular acceleration α. We know from *10.12* that

$$\tau = I\alpha$$

Placing the values found from experiment, namely, b, F, and α, in this relation and solving for I, one finds

$$I = \frac{bF}{\alpha}$$

What actual data would one need to take in order to find α?

Although the method indicated in figure 10.13 for determining I is a possible one, in practice the constant force F is most easily obtained by use of a falling mass, as indicated in figure 10.14. The tension in the cord shown there, T, is *not* equal to mg. (We can easily see this by noting that when the system is released the mass will accelerate downward. Therefore it is clear that T does not balance mg but instead $T < mg$.) If the wheel is free to rotate, it will turn under the action of the torque bT according to the relation

$$\tau = I\alpha$$

or

$$bT = I\alpha$$

Since both T and I are unknown, we need yet another equation. If we isolate the mass m and write $F = ma$ for it, we find

$$mg - T = ma$$

Since a and α are related through $a = r\alpha$, this equation and the preceding one can be used to eliminate T. Solving for I, one finds

$$I = b^2 m \left(\frac{g}{a} - 1 \right)$$

Therefore, if one measures the acceleration of the falling mass, I can be computed.

10.9 EQUILIBRIUM OF A RIGID BODY

In Chap. 3 we discussed the conditions necessary for a body to remain at rest. At that time we stated that the sum of the forces on the body must be zero and that the clockwise-turning torques must be balanced by counterclockwise torques if the body is to remain at rest. We are now in a position to generalize these conditions to three dimensions. As shown in Chap. 7, for an object of mass M, the mass center motion is given by

$$\mathbf{F} = M\mathbf{a}$$

where $\mathbf{F}$ is the vector sum of the forces on the object. When $\mathbf{F} = 0$, then also $\mathbf{a} = 0$. Notice that $\mathbf{v}$, the velocity of the object, need not be zero; it may have any constant value, including zero.

Similarly, since

$$\tau = I\alpha$$

At equilibrium, both α and a are zero

where τ is the sum of all the torques acting on a body, when $\tau = 0$, then also $\alpha = 0$. This does not imply that $\omega = 0$, but merely that ω is a constant. We define a body to be at equilibrium when $\tau = 0$ and $\mathbf{F} = 0$. This includes the case where the body is moving with constant velocity. Even

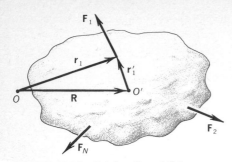

FIGURE 10.15 Origins O and O' are equally suitable for computing the torque if the body is in equilibrium. Then the torques about both origins are the same.

though a body can be rotating at constant ω and still satisfy these conditions for equilibrium, we usually restrict the term equilibrium to apply to nonrotating bodies. However, this is merely a matter of semantics.

Referring to figure 10.15, the body shown there will be in equilibrium if (and only if)

$$\mathbf{F}_1 + \mathbf{F}_2 + \cdots + \mathbf{F}_N = 0$$

and

$$\boldsymbol{\tau}_1 + \boldsymbol{\tau}_2 + \cdots + \boldsymbol{\tau}_N = 0$$

In order to compute the torques due to the various forces, an origin from which we measure the $\mathbf{r}$'s must be chosen. The position of this origin is of no consequence, *provided the external forces have zero resultant*. This we shall now show.

Consider the two possible origins at O and O' in figure 10.15. The torque about origin O is

$$\boldsymbol{\tau}_0 = (\mathbf{r}_1 \times \mathbf{F}_1) + (\mathbf{r}_2 \times \mathbf{F}_2) + \cdots + (\mathbf{r}_N \times \mathbf{F}_N)$$

and about O' is

$$\boldsymbol{\tau}_0' = (\mathbf{r}_1' \times \mathbf{F}_1) + (\mathbf{r}_2' \times \mathbf{F}_2) + \cdots + (\mathbf{r}_N' \times \mathbf{F}_N)$$

We notice from the vector diagram in figure 10.15 that

$$\mathbf{R} + \mathbf{r}_1' = \mathbf{r}_1 \qquad \text{or} \qquad \mathbf{r}_1' = \mathbf{r}_1 - \mathbf{R}$$

and similar forms exist for the other $\mathbf{r}$'s. Making use of these in $\boldsymbol{\tau}_0'$ gives

$$\boldsymbol{\tau}_0' = (\mathbf{r}_1 \times \mathbf{F}_1 + \mathbf{r}_2 \times \mathbf{F}_2 + \cdots + \mathbf{r}_N \times \mathbf{F}_N) - \mathbf{R} \times (\mathbf{F}_1 + \mathbf{F}_2 + \cdots + \mathbf{F}_N)$$

Notice that the term in parentheses multiplying $\mathbf{R}$ is merely the sum of the forces acting on the body. If the body is at equilibrium, this sum is zero, and so

$$\boldsymbol{\tau}_0' = \mathbf{r}_1 \times \mathbf{F}_1 + \mathbf{r}_2 \times \mathbf{F}_2 + \cdots + \mathbf{r}_N \times \mathbf{F}_N$$

For the case $\Sigma\mathbf{F} = 0$, the torque about any origin is the same

which is exactly the same as the expression for $\boldsymbol{\tau}_0$. Therefore $\boldsymbol{\tau}_0 = \boldsymbol{\tau}_0'$, and our point is proved; namely, the origin taken for computation of torques is of no importance provided $\Sigma\mathbf{F} = 0$.

In practical numerical computation of bodies at equilibrium, the vector formalism employed in this section is usually abandoned. One merely replaces the vector torque equation

$$\boldsymbol{\tau} = 0$$

by the three scalar equations

$$\tau_x = 0 \qquad \tau_y = 0 \qquad \tau_z = 0$$

where τ_x, for example, is the resultant torque in the x direction. If the forces are all coplanar, one of the coordinate axes, say, the x axis, may be taken perpendicular to this plane, so that the torques all lie along that axis. As a result, there is only one torque equation of importance:

$$\tau_x = 0$$

This equation, combined with the force equations

$$F_x = F_y = F_z = 0$$

then constitute a statement of the conditions for equilibrium. These conditions are the same as those employed in Chap. 3. As a result, we see that the methods of Chap. 3 are still the practical methods for mundane problems involving bodies in equilibrium. The advantage of the more elegant vector methods used in this section becomes apparent when one compares the proofs given here and in Chap. 3 concerning the arbitrary nature of the origin. Vector methods are distinctly preferable where general considerations are involved.

Chapter 10
Questions
and Guesstimates

1. Which would be easier to stop, a rotating bicycle tire filled with air or one filled with water? (Methods involving rotation are preferred.)

2. The shapes shown in figure P10.1 are cut from a uniform piece of wood. (*a*) Which would have the largest radius of gyration about an axis through their geometrical centers and perpendicular to the page? The smallest? Rank them in proper order. (*b*) Compare their moments of inertia about an axis through their geometrical centers and perpendicular to the page.

3. How large a torque does the sun exert on the earth? The earth on the moon? In view of this, why does the moon's speed change as it orbits the earth? Its speed is greatest when it is closest to the earth in its elliptical orbit.

4. Suppose that you wish to design a rotating device that is as close to freely rotating as possible. What considerations will influence your choice of axle and support of the axle? Why is an axle tapered to a pin point at the support sometimes used? If a cylindrical bearing is to be used, what effect will shaft diameter have?

5. Various devices such as windmills and children's pinwheel toys are caused to rotate by the wind. Explain how a wind blowing against them causes the rotation.

6. A large irregular rock is to be lifted by a crane. How must the crane be attached to the rock so that the rock will not rotate as it is lifted from the level ground upon which it sits?

7. List as many objects as you can for which the moment of inertia can be stated easily without lengthy computation or reference to tables of moments of inertia.

8. An arbitrary object rests on a table. For an axis perpendicular to the tabletop, where should the axis be taken if the moment of inertia of the object is to be smallest? Repeat for an axis parallel to the surface of the table.

9. An axis is taken through the center of mass of an object and $\mathbf{r}_i$ is the radius from the axis to the incremental mass Δm_i. Summing over all the masses of the body, (*a*) what is the value of $\Sigma r_i \, \Delta m_i$? (*b*) What does the quantity $\Sigma \mathbf{r}_i{}^2 \, \Delta m_i$ give? (*c*) Can the answers to (*a*) and (*b*) ever be the same?

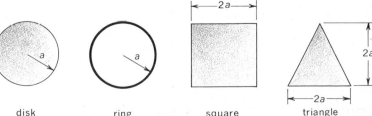

FIGURE P10.1   disk   ring   square   triangle

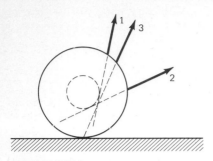

FIGURE P10.2

10. Can we say, The moment of inertia of a rigid object about an axis is equal to the sum of the moments of inertia of the parts of the object about the same axis?

11. In computing the moment of inertia of a sphere about an axis which does not pass through the center of the sphere, can one replace the sphere by an equal mass located at its center of mass?

12. Referring to Sec. 10.9, if the sum of the external torques on a body is zero about one axis, will the sum be zero with respect to any axis?

13. The spool shown in figure P10.2 moves to the left when the cord is in position 1 and to the right when the cord is in position 2. Why? What happens when the cord is in position 3?

14. Estimate the moment of inertia of an average 80-kg man about an axis passing through his head and center of mass (i.e., a vertical axis when the man is standing). Repeat for an axis through his center of mass and perpendicular to his stomach.

15. A meter stick rests on a frictionless table. To it is applied a force parallel to the tabletop. (*a*) Compare the mass center motions imparted to it if the force is applied (1) perpendicular to the rod at its mass center, (2) at the end of and in line with the rod, (3) perpendicular to the rod at its end. (*b*) Compare the rotational accelerations given to the rod in these same situations.

Chapter 10
Problems

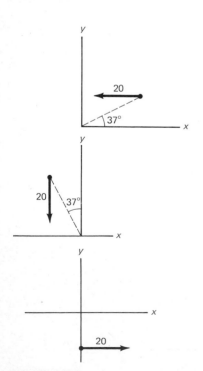

FIGURE P10.3

1. Find the **r** and ϕ components of each of the vectors shown in figure P10.3.

2. A pendulum consists of a small mass m at the end of a string of length L. The pendulum is pulled aside to an angle θ with the vertical and released. At the instant of release, using the suspension point as axis, find (*a*) the torque on the pendulum and (*b*) its angular acceleration.

3. Find $\mathbf{A} \times \mathbf{B}$ if $\mathbf{A} = 3\mathbf{i} - 4\mathbf{j}$ and $\mathbf{B} = 2\mathbf{j} + 6\mathbf{k}$. Repeat for $\mathbf{B} \times \mathbf{A}$.

4. Find the torque $\mathbf{r} \times \mathbf{F}$ if $\mathbf{r} = 2\mathbf{j} - 6\mathbf{k}$ and $\mathbf{F} = 3\mathbf{i} + 4\mathbf{k}$. Repeat if $\mathbf{r} = 5\mathbf{i} + 2\mathbf{j} - 6\mathbf{k}$.

5. The four tiny masses shown in figure P10.4 are connected by a rod of negligible mass. Find the moment of inertia and gyration radius for the system about AA' as axis. Repeat for an axis BB'.

6. The four masses in figure P10.5 are held rigid by the very light circular frame shown. Find the moment of inertia and gyration radius of the

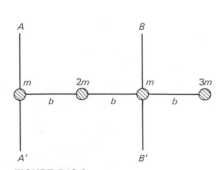

FIGURE P10.4

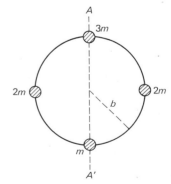

FIGURE P10.5

system for an axis through the center of the circle and perpendicular to the page. How large a torque must be applied to the system to give it an angular acceleration α about this axis provided it is free to turn? Repeat for the axis AA'.

7. Using integration, compute the moment of inertia of a uniform, thin rod of length L about an axis perpendicular to the rod at a point $L/4$ from one of its ends. Show how the same result could be obtained by use of the parallel axis theorem and the results given in figure 10.11.

8. Repeat problem 7 for a similar axis through a point $L/3$ in from one of the ends of the rod.

9. Knowing that $I = \frac{2}{5}Mr^2$ for a sphere with axis through its center, find I for an axis tangent to the sphere.

10. Two thin hoops of masses m_1 and m_2 have radii a_1 and a_2. They are mounted rigidly on a frame of negligible mass as shown in figure P10.6. Find the system's moment of inertia about an axis through the center and perpendicular to the page.

11. Four coplanar large irregular masses are held by a rigid frame of negligible mass as shown in figure P10.7. Taking an axis through P and perpendicular to the page, show that the system's moment of inertia is $I = I_1 + I_2 + I_3 + I_4$ where I_1 is the moment of inertia of object 1 alone about the axis and similarly for the others. What general rule could you prove in this way?

12. A uniform hollow cylinder has a density ρ, length L, an inner radius a, and an outer radius b. Show that its moment of inertia about the axis of the cylinder is $I = \frac{1}{2}\pi\rho L(b^4 - a^4) = \frac{1}{2}M(b^2 + a^2)$ where M is the mass of the cylinder.

13. Three children are sitting on a teeter-totter in such a way that it balances. A 20- and a 30-kg boy are on opposite sides at a distance of 2.0 m from the pivot. If the third boy jumps off, thereby destroying the balance, find the initial angular acceleration of the board. (Neglect the weight of the board.)

14. A nearly massless rod is pivoted at one end so it can swing freely as a pendulum. Two masses are attached to it, $2m$ and m at distances b and $3b$, respectively, from the pivot. The rod is held horizontal and then released. Find its angular acceleration at the instant it is released.

15. The uniform wheel of moment of inertia I shown in figure P10.8 is pivoted on a horizontal axis through its center so its plane is vertical. As shown, a small mass m is stuck on the rim of the wheel. Find the angular acceleration of the wheel when the mass is at point A. Repeat for points B and C. (Assume I not to be changed much by the presence of m.)

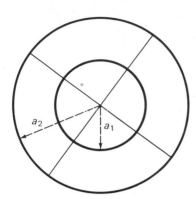

FIGURE P10.6

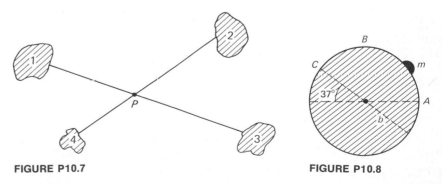

FIGURE P10.7   **FIGURE P10.8**

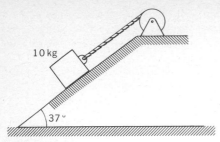

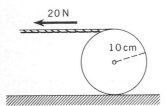

FIGURE P10.9

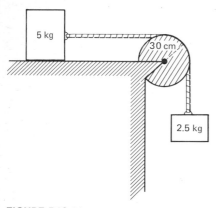

FIGURE P10.10

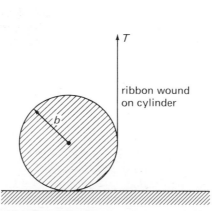

FIGURE P10.11

16. A certain auto wheel has a moment of inertia of 4.0 kg · m². When given a rotation speed of 0.50 rev/s, it coasts to rest in 8.0 s. How large is the friction torque which slows it?

17. A torque of 0.70 N · m is required to accelerate a wheel from rest to 5.0 rev/s in a time of 8.0 s. Find the moment of inertia of the wheel.

18. To determine *I* for a huge motor, the motor is brought up to its operating speed of 20 rev/s and then allowed to coast to rest twice. First, when only the friction torque stops it, the time taken is 240 s. When an extra retarding torque of 500 N · m is added, it takes 40 s to stop. What is the moment of inertia of the motor?

19. A wheel with 20 cm radius of gyration and mass 40 kg has a rim radius of 30 cm and is mounted vertically on a horizontal axis. A 2-kg mass is suspended from the wheel by a rope wound around the rim. Find the angular acceleration of the wheel when the system is released.

20. A wheel of rim radius 40 cm and mass 30 kg is mounted on a frictionless horizontal axle. When a 0.100-kg mass is suspended from a cord wound on the rim, the mass drops 2.0 m in the first 4.0 s after the mass is released. What is the radius of gyration of the wheel?

21. An Atwood's machine consists of two masses $m_1 > m_2$ hung over a frictionless pulley wheel of radius *b*. If m_1 accelerates downward with acceleration *a* after the system is released, what is the moment of inertia of the pulley wheel?

22. The moment of inertia of the wheel in figure P10.9 is 8.0 kg · m². Its radius is 40 cm. Find the angular accceleration of the wheel caused by the 10.0-kg mass if the friction force between the mass and the incline is 30 N.

23. Refer to the wheel of problem 20. Its moment of inertia is 0.61 kg · m². A 2.5-kg mass is suspended from the cord and the wheel is rotating at 2.0 rev/s in such a way as to lift the mass. If the wheel is coasting, how high will it lift the mass before stopping?

24. The rope shown in figure P10.10 is wound around a cylinder of mass 4.0 kg and $I = 0.020$ kg · m², where *I* is the moment of inertia about an axis along the cylinder axis. If the cylinder rolls without slipping, what is the linear acceleration of its mass center? Repeat for the case where no friction force exists between the table and the cylinder.

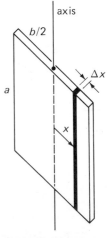

FIGURE P10.12

FIGURE P10.13

25. The friction force between the block and table in figure P10.11 is 20 N. If the moment of inertia of the wheel is 4.0 kg · m², find how long it will take the block to drop 60 cm after the system is released.

26. For the cylinder shown in figure P10.12, $I = \frac{1}{2}Mb^2$ and $T < Mg$. Describe the translational and rotational motion of the cylinder if (a) the cylinder doesn't slip on the floor and (b) if there is no friction between floor and cylinder.

27. A uniform thin rod (such as a meter stick) of length L stands vertically on one end on the floor. Its top is now given a tiny push so the rod begins to topple over. Assuming the base of the rod not to slip, find (a) the angular acceleration of the rod when it makes an angle θ with the floor, (b) the tangential acceleration of the upper tip of the rod, (c) the vertical acceleration of the tip. (d) Explain why the answer to (c) can exceed the free fall acceleration g.

28. The *very thin* uniform sheet shown in figure P10.13 has sides a and b. Its total mass is M. Show that its moment of inertia about the axis indicated is $Mb^2/12$. (*Hint:* You may first wish to find the moment of inertia about the axis of the element shown.)

11 Rotational Momentum and Energy

Momentum and kinetic energy were found to play a predominant role in linear motion. One might suspect that rotational motion might also be concerned with analogous quantities. This in fact proves to be true. In this chapter we consider kinetic energy of rotation and rotational or angular momentum. Conservation theorems for these quantities are stated, and their importance is illustrated.

11.1 ROTATIONAL ENERGY OF A PARTICLE

Let us first investigate the rotational portion of the motion of a free particle. To that end, consider the particle of mass m shown in figure 11.1, subjected to the force $\mathbf{F}$. In order to display the rotational character of the situation, split the force into two components—one along the radius vector F_r, and the other perpendicular to the radius vector F_ϕ. Notice that

$$F_r = F \cos \theta$$
$$F_\phi = F \sin \theta$$

Clearly, the rotation about the origin will be a consequence of F_ϕ while F_r will merely cause radial motion out along r. *Let us restrict our discussion to the case where r is held constant.*

Writing Newton's second law for each of these forces gives

$$F_r = ma_r$$
$$F_\phi = ma_\phi = ma_T \qquad \textit{11.1}$$

where a_ϕ has been rewritten a_T, the tangential acceleration used in Chap. 9. Since the particle is held at a fixed radius r, F_r does no work.

The ϕ component of the force will do work as the particle is displaced through an arc $r\,\Delta\phi$ caused by a small angular displacement $\Delta\phi$. In particular, the work done by the force F_ϕ in a displacement $r\,\Delta\phi$ as shown in figure 11.2 is, from *11.1,*

$$\Delta W_\phi = F_\phi r\,\Delta\phi = ma_T r\,\Delta\phi \qquad \textit{11.2}$$

But the tangential acceleration $a_T = r\alpha$, and so

$$\Delta W_\phi = mr^2\alpha\,\Delta\phi$$

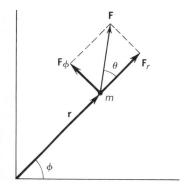

FIGURE 11.1 The force on a particle resolved into its radial and ϕ component.

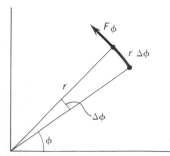

FIGURE 11.2 When ϕ changes by $\Delta\phi$, the force F_ϕ moves through a distance r $\Delta\phi$.

Using the fact that $\alpha = \Delta\omega/\Delta t$, this can be written

$$\Delta W_\phi = mr^2 \Delta\omega \frac{\Delta\phi}{\Delta t}$$

where the Δt has been placed beneath $\Delta\phi$, rather than beneath $\Delta\omega$. Since $\Delta\phi/\Delta t$ is just ω, we find

$$\Delta W_\phi = mr^2\omega \, \Delta\omega \qquad\qquad 11.3$$

This is the work done in a single displacement. To find the total work done during a series of such displacements, we need only sum the $\Delta\omega$ to give, for the total work,

$$W = \sum (mr^2\omega_i \, \Delta\omega_i)$$

If we now pass to the limit of small increments, the sum can be replaced by an integral, and we find

$$W = \int mr^2\omega \, d\omega \qquad\qquad 11.4$$

where the limits are to be from the initial situation to the final situation of the particle.

Rewriting 11.4 with appropriate limits, ω_0 and ω_f, the initial and final angular speed, and noting that m and r are constants,

$$W = mr^2 \int_{\omega_0}^{\omega_f} \omega \, d\omega$$

which integrates directly to give

$$W = \tfrac{1}{2}mr^2\omega_f^2 - \tfrac{1}{2}mr^2\omega_0^2 \qquad\qquad 11.5$$

Kinetic energy of a particle moving in a circle is $\tfrac{1}{2}mr^2\omega^2$

We call the quantity $\tfrac{1}{2}mr^2\omega^2$ the *rotational kinetic energy* of the particle. Expression 11.5 is equal to the work done by the force F_ϕ on the particle, and represents the change in kinetic energy of the particle due to this force.

We can show even more clearly that 11.5 represents the angular kinetic energy change by recalling that the tangential velocity of the particle, v_T, is equal to $r\omega$. Making this substitution in 11.5, we find at once

$$W = \tfrac{1}{2}mv_{Tf}^2 - \tfrac{1}{2}mv_{T0}^2$$

This is obviously the change in kinetic energy due to the tangential motion of the particle. We shall next make use of 11.5 to find the kinetic energy of rotation for a rigid body composed of many particles.

11.2 ROTATIONAL KINETIC ENERGY OF A RIGID BODY

Consider now a rigid body which is constrained to rotate about some fixed axis, as shown in figure 11.3. The body could be attached to the axis by a rigid rod, as shown, or constrained in some other way, so that the distance of each part of the body from the axis remains constant. We found in the preceding section that the rotational kinetic energy K_{rot} of the ith particle in the body (mass $= \Delta m_i$) is merely $\tfrac{1}{2}r_i^2(\Delta m_i)\omega_i^2$. If we consider the body to be made up of

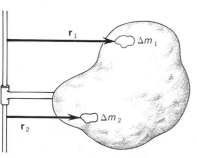

FIGURE 11.3 The rigid body is constrained to rotate about the axis shown.

N such small particles, its rotational kinetic energy is

$$K_{\text{rot}} = \tfrac{1}{2}r_1^2(\Delta m_1)\omega_1^2 + \tfrac{1}{2}r_2^2(\Delta m_2)\omega_2^2 + \cdots + \tfrac{1}{2}r_N^2(\Delta m_N)\omega_N^2$$

Because the body is constrained to rotate as a whole about the axis, the values of ω will be the same for all the particles. Therefore

$$K_{\text{rot}} = \tfrac{1}{2}\omega^2 \sum_1^N (\Delta m_i)r_i^2 \qquad\qquad 11.6$$

But the sum in *11.6* is the quantity defined in the preceding chapter as the moment of inertia I of the body about the axis of rotation. As a result, *11.6* becomes

Two analogous relations:
$$K_{\text{rot}} = \tfrac{1}{2}I\omega^2$$
$$K_{\text{trans}} = \tfrac{1}{2}mv^2$$

$$K_{\text{rot}} = \tfrac{1}{2}I\omega^2 \qquad\qquad 11.7$$

Notice that since I depends upon the choice of axis, so too will K_{rot}.

Equation *11.7* might well have been guessed without any calculation. It will be recalled that the translational kinetic energy is given by

$$K_{\text{trans}} = \tfrac{1}{2}mv^2$$

As we might have suspected, in the rotational case the mass m is replaced by I, the moment of inertia. In addition, the linear velocity v is replaced by the angular velocity ω. We shall find that this correspondence also applies in the case of the formulas for linear and angular momentum.

K_{rot} and work are interconvertible

In deriving *11.5* and *11.7* we have shown that the work-energy theorem of Chap. 6 also applies to rotational kinetic energy. As *11.5* shows, the translational and rotational kinetic energies arise in exactly the same way. Therefore all the considerations we have made in regard to production and loss of translational kinetic energy apply to rotational kinetic energy as well. In particular, interconversion among work, K_{trans}, K_{rot}, friction work, and potential energy can occur. We have simply added another category, namely, rotational kinetic energy, into which energy can go or from which energy can be obtained.

Illustration 11.1

The mass shown in figure 11.4 starts from rest and falls a height h. Find the final angular speed of the wheel in terms of the radius of the wheel and I, m, h, and t.

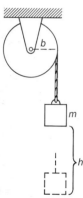

FIGURE 11.4 The rope unwinds from the wheel as the mass falls a distance h in the time t.

Reasoning We have solved an example much like this in the preceding chapter by use of $F = ma$ and $\tau = I\alpha$. Now, however, let us apply energy considerations to obtain a solution. The mass loses potential energy as it falls. This energy will be converted into K_{trans} of the mass and K_{rot} of the wheel. Our basic relation will be

$$\text{Loss in } U_g \text{ of mass} = K_{\text{trans}} + K_{\text{rot}} \text{ at end}$$

Placing in the values,

$$mgh = \tfrac{1}{2}mv^2 + \tfrac{1}{2}I\omega^2$$

The linear velocity v is really the tangential velocity of the wheel, and so $v = b\omega$. Since we wish to find ω, we shall replace v. Solving for ω then gives

$$\omega = \left(\frac{2mgh}{mb^2 + I}\right)^{1/2}$$

What units will ω be in?

11.3 WORK AND POWER IN TERMS OF TORQUE

It is sometimes convenient to have an expression for the work done on a body by a torque. For that reason we digress for a moment to obtain the requisite expression. We restrict the discussion to a rigid body constrained to rotate about a fixed axis. Let us consider any one of the several external forces acting upon the body.

The force $\mathbf{F}_i$ can be split at once into a radial and an angular component, as was done in figure 11.1. Since the radii of the various points in the body are fixed relative to the axis, no radial motion can occur, and so no work is done by that component. However, the ϕ component of the force does the following amount of work in a small rotation $d\phi$ of the body:

$$dW_i = F_{i\phi} r_i \, d\phi$$

In this expression, $r_i \, d\phi$ is the small distance through which $F_{i\phi}$ acts during the rotation $d\phi$.

The total work on the body will be the sum of all such contributions due to the various forces. Since $d\phi$ will be the same for all, the right side of the equation will be the total external torque τ multiplied by $d\phi$. We then have for the work done on the body in a small angular displacement $d\phi$

$dW = \tau \, d\phi$
$$dW = \tau \, d\phi \qquad\qquad 11.8$$

To find the power expended by the applied forces, we recall that

$$\text{Power} = \frac{dW}{dt}$$

In the present case this becomes

$$\text{Power} = \frac{\tau \, d\phi}{dt}$$

$Power = \tau\omega$ or
$$\text{Power} = \tau\omega \qquad\qquad 11.9$$

Illustration 11.2

A constant torque of $20 \, \text{N} \cdot \text{m}$ is applied to a wheel pivoted on a fixed axis. At what rate is power being furnished to the wheel when it is rotating at $\frac{1}{2}$ rev/s?

Reasoning By straightforward application of *11.9*, we find

$$\text{Power} = (20 \, \text{N} \cdot \text{m})(0.5 \times 2\pi \, \text{rad/s})$$
$$= 62.8 \, \text{W}$$

11.4 ROTATION AND TRANSLATION COMBINED

In the preceding section we discussed the motion of a rigid body constrained to rotate about a fixed axis. Although this is frequently the case, often objects are translating while rotating. A wheel rolling along the ground is a case in point. In analyzing situations such as these, it is convenient to discuss the rotation of the body about its center of mass. We shall see that this choice gives a particularly simple result.

Consider the object shown in figure 11.5. Its center of mass is at O, and we shall call the velocity of the center of mass relative to the fixed, stationary reference frame $\mathbf{v}_0$. In addition, a small particle in the object, Δm_i, for

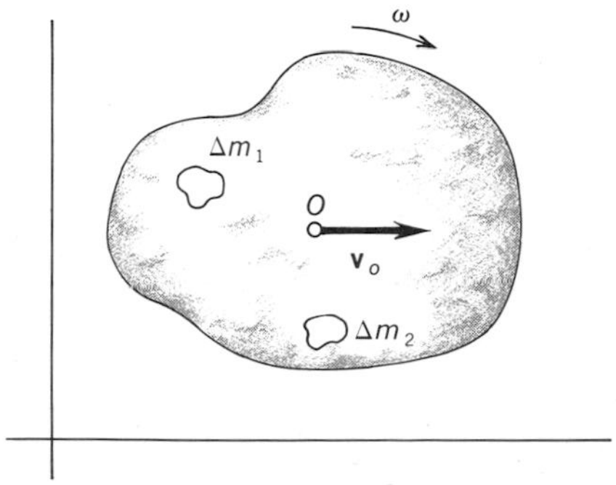

FIGURE 11.5 How can one express the kinetic energy of this rigid body?

example, will have velocity $\mathbf{v}_i'$ *relative to the center of mass*. The velocity of the particle *relative to the fixed reference frame* will therefore be $\mathbf{v}_i$, given by

$$\mathbf{v}_i = \mathbf{v}_0 + \mathbf{v}_i'$$

The kinetic energy of this particle will be given by

$$\tfrac{1}{2}(\Delta m_i)v_i{}^2$$

It is worthwhile at this point to emphasize that

$$\mathbf{v}_i{}^2 = \mathbf{v}_i \cdot \mathbf{v}_i = v_i{}^2 \cos \theta = v_i{}^2$$

Therefore, when we replace $\mathbf{v}_i$ by $\mathbf{v}_0 + \mathbf{v}_i'$, we shall have

$$\tfrac{1}{2}(\Delta m_i)v_i{}^2 = \tfrac{1}{2}(\Delta m_i)(\mathbf{v}_0 + \mathbf{v}_i') \cdot (\mathbf{v}_0 + \mathbf{v}_i')$$

which, upon expansion, gives

$$\tfrac{1}{2}(\Delta m_i)v_i{}^2 = \tfrac{1}{2}(\Delta m_i)(v_0{}^2 + v_i'{}^2 + 2\mathbf{v}_i' \cdot \mathbf{v}_0)$$

The total kinetic energy of the object is the sum of all the expressions such as this for all the N particles comprising the object. We then have

$$K = \sum_1^N \tfrac{1}{2}(\Delta m_i)(v_0{}^2 + v_i'{}^2 + 2\mathbf{v}_i' \cdot \mathbf{v}_0)$$

This may be rewritten

$$K = \tfrac{1}{2}\left(\sum_1^N \Delta m_i\right)v_0{}^2 + \sum_1^N \tfrac{1}{2}(\Delta m_i)v_i'{}^2 + \sum_1^N \Delta m_i \mathbf{v}_i' \cdot \mathbf{v}_0 \qquad 11.10$$

We interpret these terms in the following way. Since the value of $\sum_1^N \Delta m_i$ is simply the total mass of the object, the first term is just the kinetic energy of translation for the total mass of the object considered localized at the center of mass and traveling with the velocity of the mass center. The second term is the kinetic energy of the particle's motion *relative to the mass center*. It will now be shown that the third term is zero.

The third term can be rewritten

$$\mathbf{v}_0 \cdot \sum_1^N (\Delta m_i)\mathbf{v}_i'$$

However, if $\mathbf{r}_i$ is the radius vector from the center of mass O to mass Δm_i, then

$$\mathbf{v}_i' = \frac{d\mathbf{r}_i}{dt}$$

As a result, we can write the third term

$$\mathbf{v}_0 \cdot \sum_1^N (\Delta m_i)\frac{d\mathbf{r}_i}{dt}$$

But since the Δm_i do not depend on time, we can place them inside the derivative without altering the result, and so this expression becomes

$$\mathbf{v}_0 \cdot \frac{d}{dt}\left(\sum_1^N \Delta m_i \mathbf{r}_i\right)$$

Let us now recall that the center of mass is defined in such a way that

$$\sum_{1}^{N} \Delta m_i \mathbf{r}_i = 0$$

and therefore we conclude this term must be zero.

Making use of these facts, we arrive at the following expression for the kinetic energy of the object:

$$K = \tfrac{1}{2}Mv_0^2 + \tfrac{1}{2}\sum_{1}^{N}(\Delta m_i)v_i'^2$$

In this equation M is the total mass of the object, and v_0 is the velocity of the mass center. It will be recalled that v_i' is the velocity of the ith particle relative to the center of mass. For a rigid body, this will be the velocity due to the rotation of the body, i.e., the tangential velocity.

If we look once again at figure 11.5 we see that $\boldsymbol{\omega}$, the vector velocity of rotation, is directed perpendicular to the page. Let us choose an axis through the center of mass and in line with $\boldsymbol{\omega}$. (We call this the *spin axis* of the object.) With this choice of axis, v_i' is simply $r_i\omega_i$, and so we may write

$$\tfrac{1}{2}\sum_{1}^{N}(\Delta m_i)v_i'^2 = \tfrac{1}{2}\sum_{1}^{N}(\Delta m_i)r_i^2\omega_i^2$$

Since the body is rigid, all the ω_i will be the same. Therefore this becomes

$$\tfrac{1}{2}\omega^2\sum_{1}^{N}(\Delta m_i)r_i^2$$

We recognize this sum as being the moment of inertia of the object, and so K may finally be written

$$K = \tfrac{1}{2}Mv_0^2 + \tfrac{1}{2}I\omega^2 \qquad\qquad 11.11$$

The translational and rotational energies are additive for ω about the mass center

where I is the moment of inertia about an axis that passes through the mass center and that is in line with $\boldsymbol{\omega}$. This is a very important relation because it tells us that *the kinetic energy of an object is expressible as the sum of (1) the translational kinetic energy of a particle having the total mass of the object moving with the speed of the center of mass and (2) the rotational kinetic energy due to the rotation of the object about its mass center.* Use will be made of this fact in the following examples.

Illustration 11.3

A solid sphere of radius b and mass m is released from the top of an incline as shown in figure 11.6. If it rolls without slipping, what will be its linear speed when it reaches the bottom of the incline?

Reasoning At the start, the sphere has gravitational potential energy U_g. When it reaches the bottom, all of U_g will be changed to rotational and translational kinetic energy. We therefore have

$$U_g \text{ at start} = K_{\text{rot}} + K_{\text{trans}} \text{ at end}$$
$$mgh = \tfrac{1}{2}I\omega^2 + \tfrac{1}{2}mv^2$$

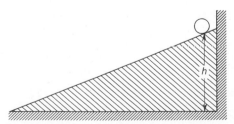

FIGURE 11.6 How fast will the sphere be rotating when it reaches the bottom?

In writing this latter relation we have used the relation proved in the preceding section concerning rotational and translational kinetic energies. Of course, I must be the moment of inertia of a sphere about an axis through its mass center. From the preceding chapter (figure 10.11) this is $I = \frac{2}{5}mb^2$. Moreover, $\omega = v/b$, and so we find

$$mgh = (\tfrac{1}{2})(\tfrac{2}{5})mv^2 + \tfrac{1}{2}mv^2$$

Notice that the rotational kinetic energy is 40 percent as large as the translational in this case. Moreover, neither the mass nor the radius of the sphere is of importance since they cancel. Solving for v, we find

$$v = \sqrt{\frac{10gh}{7}}$$

For comparison purposes, you may recall that a block sliding on the plane would have $v = \sqrt{2gh}$. The fact that some of the kinetic energy is now rotational causes the translational velocity to be decreased.

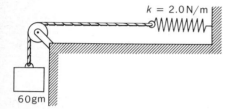

Illustration 11.4

For the situation shown in figure 11.7, the spring has a constant $k = 2.0 \, \text{N/m}$, the moment of inertia of the wheel is $0.50 \, \text{kg} \cdot \text{m}^2$, and the radius of the wheel is 30 cm. Find the speed of the 60-g mass after it falls 40 cm. Assume the mass to start at rest, with the spring in the unstretched position.

Reasoning The mass will lose a potential energy mgh with $h = 0.40 \, \text{m}$ as it falls. This energy will go into kinetic energy of the mass and wheel, plus potential energy in the spring. We have

$$U_g \text{ of block} = K_{\text{rot}} + K_{\text{trans}} + U_s$$
$$mgh = \tfrac{1}{2}I\omega^2 + \tfrac{1}{2}mv^2 + \tfrac{1}{2}kx^2$$

with $x = h$. Placing in the values and setting $\omega = v/0.3 \, \text{m}$, we find

$$(0.060 \, \text{kg})(9.8 \, \text{m/s}^2)(0.40 \, \text{m})$$
$$= \tfrac{1}{2}(0.50 \, \text{kg} \cdot \text{m}^2)(v^2/0.090 \, \text{m}^2) + \tfrac{1}{2}(0.060 \, \text{kg})v^2 + \tfrac{1}{2}(2.0 \, \text{N/m})(0.40 \, \text{m})^2$$

Solving for v, we find it to be 16 cm/s.

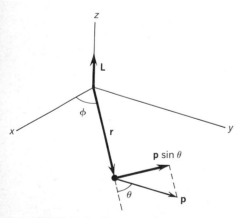

FIGURE 11.7 How fast will the mass be falling after it has dropped 40 cm?

11.5 ANGULAR MOMENTUM OF A PARTICLE

We found in our discussion of linear motion that $\mathbf{p}$, the linear momentum, proved to be a basic and useful quantity. Since useful rotational analogs to mass, force, and energy have been found, it is natural to suspect that an important quantity will be the rotational analog to linear momentum. This is indeed the case, as we shall soon see.

Consider the particle with linear momentum $\mathbf{p}$ shown in figure 11.8. It is at a vector distance $\mathbf{r}$ from the origin of an inertial reference frame. We define the angular momentum $\mathbf{L}$ of the particle by the relation

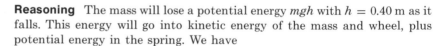

FIGURE 11.8 Show that $\mathbf{L} = \mathbf{r} \times \mathbf{p}$ actually has the direction shown.

Definition of angular momentum $\mathbf{L}$

$$\mathbf{L} = \mathbf{r} \times \mathbf{p} \qquad\qquad 11.12$$

Notice that $\mathbf{L}$ has direction, and the direction is determined by the usual right-hand rule. In figure 11.8, the rotation is about the z axis. If we grasp the

z axis with the right hand in such a way that our fingers circle the axis in the direction of rotation, the thumb will point in the $+z$ direction, as indicated. The angular momentum **L** is therefore in that direction. We often call a line in the direction of **L** through the origin the *axis of rotation*. In figure 11.8 it is the z axis.

Our defining equation for **L**, *11.12,* has been chosen in such a way that the analogy between rotational and linear laws will persist. You will recall that we had the two analogous equations,

$$\mathbf{F} = m\mathbf{a} \quad \text{and} \quad \boldsymbol{\tau} = I\boldsymbol{\alpha}$$

In addition, however, Newton's second law could be written more precisely as

$$\mathbf{F} = \frac{d(m\mathbf{v})}{dt}$$

or

$$\mathbf{F} = \frac{d\mathbf{p}}{dt}$$

This form is true even when m is not constant.

If we take the cross product of **r** with this latter equation, we have

$$\mathbf{r} \times \mathbf{F} = \mathbf{r} \times \frac{d\mathbf{p}}{dt}$$

We should like at this point to place **r** inside the derivative so as to obtain **r** × **p**, which is just **L**. Since **r** is not constant, this must be justified in some other way. We can do it by noticing that, from the rule for the derivative of a product,

$$\frac{d}{dt}(\mathbf{r} \times \mathbf{p}) = \frac{d\mathbf{r}}{dt} \times \mathbf{p} + \mathbf{r} \times \frac{d\mathbf{p}}{dt}$$

But

$$\frac{d\mathbf{r}}{dt} = \mathbf{v} \quad \text{and} \quad \mathbf{p} = m\mathbf{v}$$

and so

$$\frac{d}{dt}(\mathbf{r} \times \mathbf{p}) = \mathbf{v} \times \mathbf{v}m + \mathbf{r} \times \frac{d\mathbf{p}}{dt}$$

However, $\mathbf{v} \times \mathbf{v} = 0$ since the angle θ between **v** and itself is zero. Therefore

$$\mathbf{r} \times \frac{d\mathbf{p}}{dt} = \frac{d}{dt}(\mathbf{r} \times \mathbf{p})$$

and we are justified in placing **r** inside the derivative.

We therefore find

$$\mathbf{r} \times \mathbf{F} = \frac{d}{dt}(\mathbf{r} \times \mathbf{p})$$

But **r** × **F** is merely $\boldsymbol{\tau}$, and **r** × **p** is **L**. This relation therefore becomes

$$\boldsymbol{\tau} = \frac{d\mathbf{L}}{dt} \qquad\qquad 11.13$$

Two analogous relations: which is entirely analogous to the linear relation

$$\boldsymbol{F} = \frac{d\boldsymbol{p}}{dt}$$

$$\mathbf{F} = \frac{d\mathbf{p}}{dt}$$

$$\boldsymbol{\tau} = \frac{d\mathbf{L}}{dt}$$

Obviously, our definition of angular momentum **L** has preserved the analogy between rotational and linear forms of Newton's second law.

11.6 CONSERVATION OF ANGULAR MOMENTUM: SINGLE PARTICLE

Equation *11.13* leads at once to the fact that, in the absence of a torque about a given axis, the angular momentum about that axis is a constant. For if $\tau = 0$, then *11.13* gives

If $\tau = 0$, angular momentum is conserved

$$\frac{d\mathbf{L}}{dt} = 0$$

or

$$\mathbf{L} = \text{const}$$

This is a statement of the *law of conservation of angular momentum*. Notice, however, that even if the torque about a particular axis is zero, it need not be zero about all axes.

This is easily seen to be the case in figure 11.9. The torque about the x and y axes is zero (that is, τ has no component in these directions), but a torque does exist in the z direction about the z axis. Moreover, since the torque depends upon r, the choice of the coordinate origin will also be of importance.

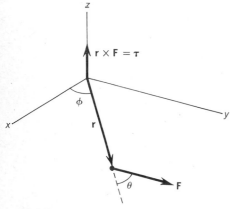

The conservation of angular momentum of a particle has particular importance for the motion of the planets. Consider, for example, the earth as a particle moving about the sun, as shown in figure 11.10. The major force on the earth is due to the gravitational force exerted on it by the sun. This force is always directed toward the sun as center. It is called a *central force*. If we take an axis through the sun, this force exerts no torque about the axis. We therefore conclude that the angular momentum of the earth about an axis through the sun is constant.

FIGURE 11.9 Since **r** and **F** are both in the xy plane, $\tau_x = \tau_y = 0$ and L_x and L_y are constant. What about L_z?

The reasoning is not restricted to the earth and sun, of course. It applies equally well to any of the planets (ignoring minor effects due to moons of planets, etc.). The orbital motion of satellites also must conform to this conclusion. In fact, the orbital motion of any body under the action of a central force must have a constant angular momentum about an axis passing through the force center. When an object moves in an elliptic orbit, as do the planets and most satellites, this means that the tangential speed of the object in orbit must vary. Since at two points in the orbit,

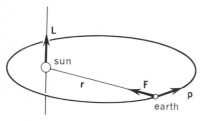

FIGURE 11.10 The central force **F** acting on the earth causes no torque about the axis shown. Therefore **L** must be constant. The orbit of the earth is much more nearly circular than indicated.

$$\mathbf{L}_1 = \mathbf{L}_2$$

and therefore

$$\mathbf{r}_1 \times \mathbf{p}_1 = \mathbf{r}_2 \times \mathbf{p}_2 \qquad 11.14$$

the linear momentum **p** must vary as the radius from the force center changes. When the object is farthest from the force center, its speed is least; when it is closest, its speed is greatest.

Two basic facts involving motion under a central force can be seen by reference to *11.14* and figure 11.11. First, since $\mathbf{L}_1 = \mathbf{L}_2$, the angular momentum is always in the same direction, and so the orbit lies in a plane and **L** will be perpendicular to it. Second, we shall now show that the area swept out in unit time by the radius vector to the particle will be a constant no matter where in the orbit the particle is. (This is one of Kepler's three laws of planetary motion.)

Kepler's law of equal areas

Consider the shaded area Δs in figure 11.11 swept out by the radius vector to the planet as the planet moves from A to B in time Δt. If $\Delta r/r \ll 1$, the area will be, to good approximation, the area of a triangle of altitude r and base $\Delta l \sin \theta$ so

$$\Delta(\text{area}) = \Delta s = \tfrac{1}{2} r \, \Delta l \sin \theta$$

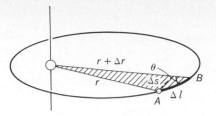

FIGURE 11.11 What can one say about $\Delta s/\Delta t$ as the planet moves along its orbit?

where θ is the angle between $\mathbf{r}$ and the tangential velocity. We shall ensure the validity of this approximation by considering the motion from A to B only if the time taken, Δt, is very small. Dividing by Δt gives

$$\frac{\Delta s}{\Delta t} = \tfrac{1}{2} r \frac{\Delta l}{\Delta t} \sin\theta$$

But $\Delta l/\Delta t$ is merely v, the tangential velocity of the planet. After multiplying through by m, our equation therefore becomes

$$m \frac{\Delta s}{\Delta t} = \tfrac{1}{2} rmv \sin\theta$$

Since mv is simply p, we can write this

$$m \frac{\Delta s}{\Delta t} = \tfrac{1}{2} |\mathbf{r} \times \mathbf{p}|$$

where $\|$ means to take the magnitude only. Therefore

$$m \frac{\Delta s}{\Delta t} = \frac{L}{2}$$

and since L is constant in a situation such as this, the rate at which area is swept out, $\Delta s/\Delta t$, must also be constant.

Illustration 11.5

A particular earth satellite orbits the earth in an elliptic orbit whose distance from the earth surface at apogee (farthest point from earth) is 1.000×10^6 m and whose distance at perigee (closest point to earth) is 1.00×10^5 m. Taking the radius of the earth to be 6.37×10^6 m, what is the ratio of the speeds of the satellite at these two points?

Reasoning From the law of conservation of angular momentum (which is applicable in this central-force situation), we have

$$\mathbf{L}_{\text{apogee}} = \mathbf{L}_{\text{perigee}}$$

or

$$(m\mathbf{r} \times \mathbf{v})_a = (m\mathbf{r} \times \mathbf{v})_p$$

At the two extremes of the ellipse, $\mathbf{v}$ and $\mathbf{r}$ are perpendicular; so we have

$$r_a v_a = r_p v_p$$

Taking ratios and placing in the numbers, we find

$$\frac{v_a}{v_p} = \frac{r_p}{r_a} \approx 0.88$$

What additional data would you need to find an approximate average speed for the satellite?

11.7 ANGULAR MOMENTUM OF A SYSTEM OF PARTICLES

We define the angular momentum $\mathbf{L}$ of a system of particles to be the sum of their individual angular momenta. (The system might, as a special case, be a rigid body.) Therefore

$$\mathbf{L} = \mathbf{r}_1 \times \mathbf{p}_1 + \mathbf{r}_2 \times \mathbf{p}_2 + \cdots + \mathbf{r}_N \times \mathbf{p}_N \qquad 11.15$$

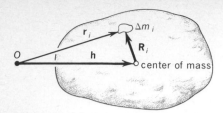

FIGURE 11.12 To transform from the origin at O to an origin at the center of mass, replace $\mathbf{r}_i$ by $\mathbf{h} + \mathbf{R}_i$.

The angular momentum of a system can be written as the sum of two parts

Definition of spin angular momentum

In writing this we are referring to an arbitrary origin in an inertial reference frame. If this arbitrary origin is at a vector distance $\mathbf{h}$ from the center of mass, we can replace $\mathbf{r}_i$ by $\mathbf{R}_i + \mathbf{h}$, as shown in figure 11.12, to give

$$\mathbf{L} = \mathbf{R}_1 \times \mathbf{p}_1 + \mathbf{R}_2 \times \mathbf{p}_2 + \cdots + \mathbf{R}_N \times \mathbf{p}_N + \mathbf{h} \times (\mathbf{p}_1 + \mathbf{p}_2 + \cdots + \mathbf{p}_N)$$

The sum involving the $\mathbf{R}_i$'s is merely the angular momentum of the system about the center of mass $\mathbf{L}_{cm}$. In addition, the sum of the individual momenta is just the total linear momentum of the system of particles, $\mathbf{p}$. Therefore

$$\mathbf{L} = \mathbf{L}_{cm} + \mathbf{h} \times \mathbf{p} \qquad\qquad 11.16$$

We conclude from this that the angular momentum of a system of particles about any origin fixed in an inertial system can be expressed as the sum of two terms: (1) the angular momentum of the system about its mass center and (2) the angular momentum of a particle at the mass center which has the same linear momentum as the system as a whole. Frequently, we choose the origin at the mass center in order to eliminate this latter term. The angular momentum of an object about an axis through the mass center and in line with $\mathbf{L}_{cm}$ is often called the *spin* angular momentum of the object.

If we take the time derivative of *11.15*, we find

$$\frac{d\mathbf{L}}{dt} = \frac{d}{dt} \sum_1^N (\mathbf{r}_i \times \mathbf{p}_i)$$

$$= \sum_1^N \left(\frac{d\mathbf{r}_i}{dt} \times \mathbf{p}_i + \mathbf{r}_i \times \frac{d\mathbf{p}_i}{dt} \right)$$

Now
$$\frac{d\mathbf{r}_i}{dt} = \mathbf{v}_i \qquad \text{and} \qquad \frac{d\mathbf{p}_i}{dt} = \mathbf{F}_i$$

The latter time derivative is merely Newton's second law. Further, we note that $\mathbf{p}_i$ is $\Delta m_i \mathbf{v}_i$, and so the first term in parentheses is zero because

$$\mathbf{v}_i \times \mathbf{p}_i = 0$$

since they are in the same direction. Therefore

$$\frac{d\mathbf{L}}{dt} = \sum \mathbf{r}_i \times \mathbf{F}_i$$

The quantity $\mathbf{r}_i \times \mathbf{F}_i$ is $\boldsymbol{\tau}_i$, the torque on the ith particle. As discussed previously, in a situation such as this, the internal torques will cancel. We see, then, that the sum is simply the total external torque acting on the system, and so we can write as our final relation

True for a system as well as a particle

$$\boldsymbol{\tau} = \frac{d\mathbf{L}}{dt} \qquad\qquad 11.17$$

This equation is similar to *11.13* obtained for a single particle.

We see at once that *11.17* states the law of conservation of angular momentum for a system. If the torque about a given axis is zero, the angular momentum about that axis is constant. Nearly everything we said concerning *11.13* will apply to *11.17* as well. As pointed out in *11.16*,

$$\mathbf{L} = \mathbf{L}_{cm} + \mathbf{h} \times \mathbf{p}$$

The motion of the system corresponds to the motion of a particle of equal mass placed at the center of mass plus the motion about the center of mass.

TABLE 11.1 Analogous Relations

Linear	Rotational
$\mathbf{F} = m\mathbf{a}$	$\boldsymbol{\tau} = I\boldsymbol{\alpha}$
$\mathbf{F} = \dfrac{d\mathbf{p}}{dt}$	$\boldsymbol{\tau} = \dfrac{d\mathbf{L}}{dt}$
$dW = \mathbf{F} \cdot d\mathbf{s}$	$dW = \boldsymbol{\tau} \cdot d\boldsymbol{\phi}$
$K_{\text{trans}} = \frac{1}{2}mv^2$	$K_{\text{rot}} = \frac{1}{2}I\omega^2$
Power $= \mathbf{F} \cdot \mathbf{v}$	Power $= \boldsymbol{\tau} \cdot \boldsymbol{\omega}$
$\mathbf{p} = m\mathbf{v}$	$\mathbf{L}_{cm} = I_{cm}\boldsymbol{\omega}$

In effect, the system differs from a single particle only because the mass is not concentrated at a point. As a result, the moment of inertia of the system about the center of mass is not zero, and so rotation of the system about its own center of mass gives rise to appreciable angular momentum and rotational kinetic energy.

11.8 SUMMARY OF ROTATIONAL AND LINEAR RELATIONS

As we have shown in the preceding two chapters, nearly all our equations which apply to straight-line motion correspond to analogous equations in rotational motion. The most obvious correspondence occurred with the five linear motion equations. By replacing x by θ, v by ω, and a by α, the motion equations could be obtained. We have now found several other analogous relations and these are listed in table 11.1.

The last equation in table 11.1 has not previously been stated as such. However, it follows at once from *11.16*. We there defined

$$\mathbf{L}_{cm} = \mathbf{R}_1 \times \mathbf{p}_1 + \mathbf{R}_2 \times \mathbf{p}_2 + \cdots + \mathbf{R}_N \times \mathbf{p}_N$$

Since

$$p_i = m_i v_i = m_i R_i \omega_i$$

this becomes

$$\mathbf{L}_{cm} = \sum_1^N m_i R_i{}^2 \omega_i$$

(By drawing a picture showing $\mathbf{R}_i \times \mathbf{p}_i$ or $\mathbf{R}_i \times \mathbf{v}_i$, you can see at once that $\boldsymbol{\omega}_i$ is in the direction given by the cross product.)

If the body is rigid, and since the origin is taken at the center of mass, all the $\boldsymbol{\omega}_i$ are equal. This gives

$$\mathbf{L}_{cm} = \omega \sum_1^N m_i R_i{}^2$$

or from the definition of I,

$$\mathbf{L}_{cm} = I_{cm}\boldsymbol{\omega} \qquad\qquad 11.18$$

Notice that this relation is restricted to the case of a rigid body, and does not apply to a general system.

In addition to these relations we have the following important definitions and simplifications:

$$\boxed{\begin{aligned} I &= \sum (\Delta m_i) r_i{}^2 \\ I &= I_{cm} + Mh^2 \quad \text{(parallel axes)} \\ \boldsymbol{\tau} &= \mathbf{r} \times \mathbf{F} \\ \mathbf{L} &= \mathbf{r} \times \mathbf{p} \\ \mathbf{L} &= \mathbf{L}_{cm} + \mathbf{h} \times \mathbf{p} \end{aligned}}$$

We shall now proceed to make use of these equations in various examples.

Illustration 11.6

The device shown in figure 11.13 rotates on the vertical axle as shown. Consider the mass of the frame to be negligible in comparison with the four masses, each of mass m. When the object is as shown, its angular velocity is

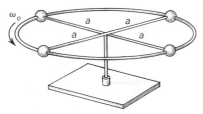

FIGURE 11.13 What will happen to the angular velocity if the radii are made larger through some internal mechanism?

ω_0. If an internal mechanism now causes the spokes in the frame to lengthen so that the radii of the masses become $2a$, what will be the new angular velocity of the system?

Reasoning If we ignore friction torques at the axle, there will be no torque on the system. The law of conservation of angular momentum tells us that its angular momentum before the spokes are lengthened must be the same as after. Therefore

$$\mathbf{L}_{\text{before}} = \mathbf{L}_{\text{after}}$$
$$I_{\text{before}}\omega_0 = I_{\text{after}}\omega$$

From the definition of I we have

$$I = \sum \Delta m_i r_i^2 = 4mr^2$$

Therefore

$$4ma^2\omega_0 = 4m(2a)^2\omega$$

from which

$$\omega = \frac{\omega_0}{4}$$

The reverse of this effect is well known to astronomers. Under the action of gravitational forces, clouds of dust may collapse to form more or less compact bodies. Any angular momentum of the cloud before collapse must be preserved. As a result, the very slow rotation of the cloud as a whole will be replaced by a much higher angular velocity for the matter in its collapsed form.

Illustration 11.7

A drop of water of mass m and speed v falls onto a paddle of a water-wheel (moment of inertia I), as shown in figure 11.14. Find the angular velocity of the paddle wheel after it is hit by the drop. (Assume the drop to remain attached to the wheel after it hits.)

Reasoning Considering the droplet and wheel as a system and assuming the gravitational force on the droplet is balanced by air friction, the torque on the system is zero. Therefore the angular momentum before the drop strikes must be equal to the angular momentum after the drop hits. Taking the center of the wheel as axis, we have, after noting that for the droplet $|\mathbf{r} \times \mathbf{p}| = bmv$,

$$I\omega_0 + mvb = I\omega + mv'b$$

Since the droplet moves with the wheel after collision, we have $v' = \omega b$, and so

$$I\omega_0 + mvb = I\omega + mb^2\omega$$

From this

$$\omega = \frac{I\omega_0 + mvb}{I + mb^2}$$

Notice that if the tangential speed of the wheel, $\omega_0 b$, is equal to the speed of the droplet, v, the speed of the wheel will remain unchanged. Is this reasonable?

Illustration 11.8

The object shown in figure 11.15 has a mass M, and I_{cm} is its moment of inertia about its center of mass. If it is released from the position shown, find how fast its center of mass is moving when it reaches the dashed position.

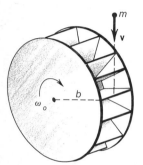

FIGURE 11.14 How much is the wheel speeded up by the droplet?

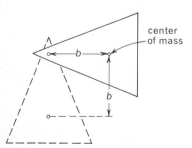

FIGURE 11.15 How fast is the center of mass moving when the object reaches the dashed position?

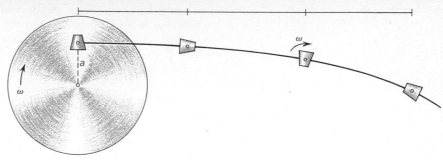

FIGURE 11.16 After the attached piece breaks loose from the wheel, its center of mass acts like a rotating projectile.

Reasoning When the object falls it loses potential energy Mgh. This potential energy appears as kinetic energy. Therefore, using 11.11,

$$Mgb = \tfrac{1}{2}I_{cm}\omega^2 + \tfrac{1}{2}Mv_{cm}^2$$

Since $v_{cm} = \omega b$, we find

$$Mgb = \frac{\tfrac{1}{2}(I_{cm} + Mb^2){v_{cm}}^2}{b^2}$$

Solving for v_{cm} gives

$$v_{cm} = \left(\frac{2Mgb^3}{I_{cm} + Mb^2}\right)^{1/2}$$

Why is the angular momentum of the object not constant?

Illustration 11.9

An irregular object is cemented to the surface of a wheel as shown in figure 11.16. When the wheel is rotating with frequency ω, the object breaks loose and flies off. Describe its motion, assuming the object breaks off in the position shown. The plane of the wheel is vertical.

Reasoning The object's center of mass has a speed $v = \omega a$ at the instant of breaking loose. Now the particle is free, and its center of mass will move like any other projectile, with an initial horizontal speed v. In addition, the object is rotating. When attached to the wheel, it rotated once every time the wheel turned around. Therefore its angular speed about its own center of mass is also ω. Only the gravitational force acts on the object after it breaks loose, and this causes no resultant torque about the center of mass of the object. (Can you prove this statement?) Therefore the angular momentum of the object about its center of mass must remain constant. The object will therefore fly through space, rotating at constant angular velocity ω about the center of mass, and its **L** vector will remain perpendicular to the page in figure 11.16.

11.9 SPINNING TOPS

As indicated in illustration 11.8 and figure 11.15, a rigid object will ordinarily fall in such a way that the center of mass lies directly below the pivot point. There is one very important instance when this is not true, namely, when the object is spinning. A case in point is the spinning top. We shall now examine its behavior in simple situations.

First let us try to see why the top does not fall as long as it is spinning swiftly. If we refer to figure 11.17, we see that the top is pivoted at the origin of coordinates. Two forces act upon it, the weight **W**, which acts at the mass center, and **F**, the reaction at the pivot. Our intuition tells us that **W** should pull the center of mass down, and the top would fall. However, we must be careful, since we know that, if the angular momentum $\mathbf{L} = I\omega$ is large, it will require a very large torque to change its orientation in space.

In fact, if we take the origin of coordinates as the center for torques, we see that **F** exerts no torque. The torque due to **W** is τ, and it will cause a change in angular momentum (see *11.13*).

$$\tau = \frac{\Delta \mathbf{L}}{\Delta t} \qquad\qquad 11.17$$

Notice that τ, and therefore $\Delta\mathbf{L}$, is perpendicular to the xz plane. If the top were not spinning, the increment $\Delta\mathbf{L}$ would cause the top to fall toward the x axis, as we normally expect.

*Since $\Delta\mathbf{L}$ is perpendicular to **L**, the **L** vector precesses*

However, if the top is spinning, $\Delta\mathbf{L}$ must be added to the already existing **L** if we are to find out what the top will do. To see the result more clearly, the angular momentum vectors are drawn separately in figure 11.18. As shown there, the resultant angular momentum is merely $\mathbf{L} + \Delta\mathbf{L}$, and if $\Delta\mathbf{L} \ll \mathbf{L}$, the magnitude of **L** will remain essentially constant. Only its direction will change by rotation through an angle $\Delta\phi$ about the z axis. The torque τ will always be perpendicular to the plane defined by the axis of the top and z axis. It will therefore rotate around the z axis. Successive repetition of this picture causes the **L** vector to rotate around the z axis on the surface

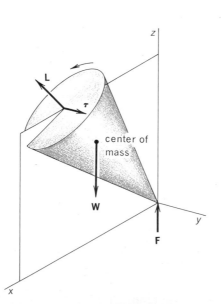

FIGURE 11.17 At the instant shown, the axis of the top, together with vectors **F**, **W**, and **L**, lies in the xz plane, whereas τ is parallel to the y axis.

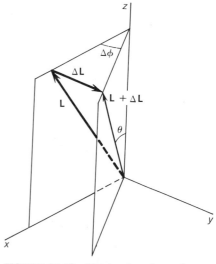

FIGURE 11.18 The torque shown in figure 11.17 causes an increment $\Delta\mathbf{L}$ in **L** as shown. This rotates the angular momentum vector through an angle $\Delta\phi$. If $L \gg \Delta L$, only the direction, and not the magnitude of **L**, will be changed appreciably.

218   ROTATIONAL MOMENTUM AND ENERGY

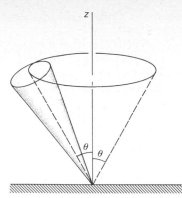

FIGURE 11.19 The axis of the top sweeps out the surface of a cone as it precesses around the z axis. When properly disturbed, the top also nutates, and the center of its upper surface wobbles from the perfect circle shown.

of a cone whose axis is the z axis and whose vertex angle is 2θ. Since the angular momentum vector remains coincident with the axis of the top when **L** ≪ Δ**L**, this means that the axis of the top itself will rotate around the z axis, as shown in figure 11.19. The top is said to *precess* about the z axis.

In order to find the precession angular velocity, or frequency ω_p, we need only evaluate $\Delta\phi/\Delta t$, to which it is equal. From figure 11.18 we see that

$$\sin\frac{\Delta\phi}{2} = \frac{(\Delta L)/2}{L\sin\theta}$$

Or since Δφ is very small and $\sin(\Delta\phi/2) \approx \Delta\phi/2$, we have

$$\Delta\phi = \frac{\Delta L}{L\sin\theta}$$

Dividing through by Δt, we find

$$\frac{\Delta\phi}{\Delta t} = \frac{\Delta L}{\Delta t}\frac{1}{L\sin\theta}$$

We notice that $\Delta\phi/\Delta t$ is merely the precession frequency ω_p, and from *11.17* we have $\Delta L/\Delta t$ to be τ. Therefore the precession frequency, or precession angular velocity, is given by

$$\omega_p = \frac{\tau}{L\sin\theta} \qquad\qquad 11.19$$

The precession speed varies as τ/L

Since L is the angular momentum of the top due to its spinning about its axis, it is often called the *spin angular momentum*. Clearly, the larger the spin angular momentum, the more slowly the top will precess. When the weight of the top is large, the torque acting on it will be large, and its precession frequency will become large. However, we should remember that *11.19* applies only when L is large enough so that the angular momentum due to the precession does not cause the total angular momentum vector to depart much from the axis of the top. This assumption was made in arriving at *11.19*.

In practice, it is difficult to cause the top to move smoothly about the circle of figure 11.19 because of difficulty in adjusting the axis to the proper θ. Usually, the top wobbles slightly around this path. The top is said to *nutate* in such a case. This aspect of the motion is rather complicated, however, and is not pursued further here.

Chapter 11
Questions and Guesstimates

1. What is meant by rotational inertia? In what ways is it similar to inertia for translational motion? How is it related to rotational momentum and energy?
2. List as many factors as you can which might cause the earth to change its rate of rotation about its own axis as the years pass by.
3. A large hollow sphere is filled with viscous fluid. As the sphere rotates freely about an axis through its center, an internal mechanism starts the liquid inside it to rotate in the opposite direction. What will happen to the rotation of the sphere? If the mechanism is shut off and the liquid slowly loses its rotation relative to the sphere because of viscous energy losses, what will the final speed of rotation for the sphere be?
4. It is possible to tell a hard-boiled egg from a fresh egg by spinning it. Explain how this works.

5. A bowling ball is thrown along a slippery floor. Under what circumstances will it roll rather than both slide and roll?

6. Is there any way that we could make use of the earth's rotational energy about its axis or about the sun to provide an energy source to the earth?

7. In order to keep a football or any other projectile from wobbling, the projectile is caused to spin about an axis in line with its direction of motion. Explain why the spin is of aid in this respect.

8. A woman suddenly finds herself standing on a frictionless surface. As she falls with wild flaying of her arms, what happens to her center of mass? Does it matter if the woman was originally moving sideways instead of being at rest?

9. One proposal for storing energy is by means of a massive, fast rotating flywheel. Discuss its pros and cons as (a) an automobile power source and (b) a means for the power company to equalize its power production during various hours of the day.

10. A "do it yourselfer" builds a helicopter with a single propeller on a vertical axis. In its maiden flight, the operator becomes sick because the whole helicopter tends to spin about a vertical axis. What went wrong? How is this difficulty overcome in more sophisticated machines?

11. Describe the angle between the following two vectors for a moving free particle: its linear momentum **p** and its angular momentum **L**.

12. Can a system have rotational energy even though it has zero angular momentum? Can it have angular momentum and zero K_{rot}?

13. Can the torque on a system be zero even though a net force acts on the system? Can it have zero net force even if the torque on it is not zero?

14. In order for an object (such as a wheel with a chunk broken from it) not to wobble (or at least try not to wobble) when rotated on a fixed axle, the axle should pass through the center of mass of the wheel. Explain why. How is this related to the fact that car wheels require balancing?

15. Each atom has angular momentum, which we can picture crudely to be the result of rotations of the electrons about the nucleus and spinning of the electrons about their own axis. By placing a solid in a magnetic field, it is possible to partly align the originally disoriented angular momentum vectors. What must happen to the object in the process if it is freely suspended?

16. One suggestion for providing an "effective weight" for objects in an earth satellite ship is to cause the ship to spin rapidly about an axis through its center along its line of motion. Explain why rotation of the ship could cause objects to have weight. Could a ship be caused to spin while in orbit? If so, what effect would this have on its orbital motion?

17. How could you determine if a sphere was hollow or not without drilling a hole in it or measuring the density of the material from which the sphere was made?

18. The spool shown rolling down the incline in figure P11.1 takes off with greatly increased translational motion as soon as the large-diameter disks on the sides of the spool touch the floor. Explain why, paying particular attention to the way the kinetic energy is apportioned.

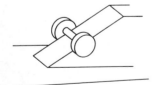

19. Explain how a skater controls her rotation speed as she stands on the tip of one skate and spins.

20. Suppose that the sun's attraction for the earth suddenly doubled. What can you say about the rate of rotation and orbit of the earth about the sun?

21. Suppose that an internal explosion suddenly opened a huge cavity in the earth by pushing its surface outward. How would this affect the rotation of the earth about its axis and about the sun?

22. A child sitting on a swing can cause the swing to rise higher by "pumping." Doesn't this contradict the law of conservation of angular momentum because the angular momentum increases without an external agent pushing on the child-swing system? Explain how the "pumping" action achieves its goal. [See S. M. Curry, "How Children Swing," *Am. J. Phys,* **44,** 924 (1976).]

23. When a car rolls down a hill, about what fraction of its total kinetic energy is the rotational energy of the wheels?

24. An automobile is facing east on an east-west highway. It accelerates from rest to its top speed. Is the rotation of the earth slowed or speeded up by this? Estimate the fractional change in the earth's rotation speed.

25. A 180-cm, 70-kg woman stands on the axis of a freely rotating stand and is turning at a rate of 0.40 rev/s. In each of her hands, held close to her body, is a 5-kg mass. (*a*) Estimate her moment of inertia. (*b*) If she now extends her arms straight out from her body, approximately what is her new moment of inertia? (*c*) What is her rotation speed after extending her arms?

Chapter 11
Problems

Wherever practicable use energy and momentum methods to solve these problems.

1. A nitrogen molecule can be thought of as two point masses (*m* of each $= 14 \times 1.67 \times 10^{-27}$ kg) separated by a distance of 1.3×10^{-10} m. In air at room temperature the average rotational kinetic energy of such a molecule is about 4×10^{-21} J as we shall see later. Find the moment of inertia of such a molecule about its mass center and its speed of rotation in rev/s.

2. Find the rotational energy of the earth about the sun due to its orbit about the sun. Data: $M_e = 6 \times 10^{24}$ kg, orbit radius $= 1.5 \times 10^{11}$ m, time for rotation $= 365$ days $= 3.2 \times 10^7$ s.

3. Each of the wheels on a certain four-wheel vehicle has a mass of 30 kg and a radius of gyration of 30 cm. When the car is going forward and the wheels are turning at 5.0 rev/s, what is the rotational kinetic energy stored in the four wheels? What is the angular momentum of the vehicle about an axis parallel to the wheel axis and through the mass center? Is the angular momentum vector directed towards the driver's right or left?

4. Find the rotational energy and angular momentum about its axis of the earth due to its daily rotation. Data: $M_e = 6 \times 10^{24}$ kg, $R_e = 6.4 \times 10^6$ m, $\omega = 1/86,400$ rev/s. Assume the earth to be a uniform sphere.

5. A wheel with $I = 20$ kg $\cdot$ m² is spinning at 3.0 rev/s on its axis. How large is the friction torque if it coasts 40 rev before stopping?

6. A uniform sphere of radius 5.0 cm and 2.0 kg mass is rolling along level ground at a speed of 30 cm/s. It rolls to rest in a distance of 15 m. How large a stopping force acted on it?

7. A cylinder of radius 20 cm is mounted on an axle coincident with its axis so as to be free to rotate. A cord is wound on it and a 50-g mass is hung

from it. If, after being released, the mass drops 100 cm in 12 s, find the moment of inertia of the cylinder.

8. A steel ball bearing of 0.50 cm radius is rolling along a table at 20 cm/s when it starts to roll up an incline. How high above the table level will it rise before stopping? Ignore friction losses.

9. A uniform sphere and a uniform disk are rolled down an incline from the same point. Find the ratio of the disk's speed to that of the sphere at the bottom of the incline. Ignore friction losses.

10. An Atwood's machine consists of a frictionless pulley (radius = b) mounted on a horizontal axis, and two masses $m_1 > m_2$ hanging at ends of a cord that passes over the pulley. Find the moment of inertia I of the pulley in terms of m_1, m_2, b, and the time t taken for m_1 to fall a distance h after the system is released.

11. Refer to figure P11.2. If the wheel shown there is rotating at a speed of 0.30 rev/s and the mass comes to rest at a vertical distance 80 cm above the position shown, how large is the moment of inertia of the wheel? Ignore friction loss.

12. Suppose the apparatus of figure P11.2 is at rest. When the system is released, the 200-g mass slides down the incline against a friction force of 0.50 N. If the moment of inertia of the wheel is 0.80 kg · m², how fast will the block be moving after it has slid 100 cm along the incline?

13. A pendulum consists of a solid sphere (radius = b) at the end of a cord of length $L - b$. Show that the speed of the ball when at the bottom of its swing is given by $\sqrt{2gh/(1 + 0.4b^2/L^2)}$ if it is released at a height h. In most cases, b is small compared to L and so the term containing b can be neglected.

14. The rigid rod joining the three masses in figure P11.3 has negligible mass. It is pivoted at one end as indicated. If it is released from the position shown, (a) how fast will the bottom mass be moving when the rod is vertical? (b) What will then be ω for the rod rotating about the pivot?

15. A meter stick is pivoted at one end on a horizontal axis. The stick is held horizontal and released. How fast is the stick rotating about the pivot (in rev/s) when the stick is vertical?

16. It is proposed to use a uniform disk 50 cm in radius turning at 300 rev/s as an energy-storage device in a bus. How much mass must the disk have if it is to be capable, while coasting to rest, of furnishing the energy equivalent of a 100-hp motor operating for 10 min?

17. A small motor delivers 0.20 hp when its shaft is turning at 1400 rev/min. How large a torque is it capable of providing?

18. In figure P11.4 the large hoop is fastened rigidly to the table while the small hoop (mass = m, radius = b) rolls without slipping inside the large hoop. It is released from the position shown. What is the speed of the center of the hoop as it passes through its lowest point? Repeat for a spherical shell of the same radius rolling inside the hoop.

19. A drum filled with sand sits upright on a dolly that can be moved across the floor with negligible friction. The mass of the drum is M and its radius is R. Around the drum is wound a rope by means of which a constant tangential force F is applied to the drum. Neglecting the mass of the dolly, find the linear and angular accelerations of the drum due to this force. The drum can be assumed to be a uniform cylinder.

20. The uniform cylinder of mass M shown in figure P11.5 is accelerated from rest by the force F applied to a ribbon wound around the cylinder. How

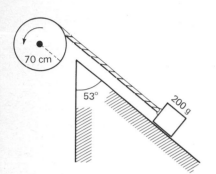

FIGURE P11.2

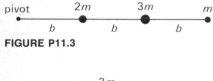

FIGURE P11.3

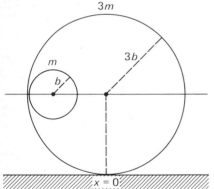

FIGURE P11.4

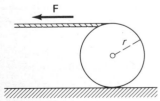

FIGURE P11.5

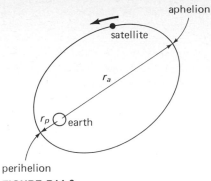

aphelion

satellite

r_a

r_p earth

perihelion

FIGURE P11.6

fast is the cylinder rotating (in rad/s) after the cylinder has rolled without slipping through a distance D?

21. Consider the earth satellite circling the earth as shown in figure P11.6. Find the ratio of its speed at perihelion to that at aphelion.

22. In figure P11.7 the disks can rotate without friction on the axes shown. The wheel on the left is rotating with angular speed ω_0 and the one on the right is at rest. Now the two disks are moved together so that their two rims touch. After a short time the disks, now in contact, are rotating without slipping. What is the final angular speed of the disk on the right?

23. A woman stands over the center of a horizontal platform that is rotating freely with speed 2.0 rev/s about a vertical axis through the center of the table and straight up through the woman. She holds two 5-kg masses in her hands close to her body. The combined moment of inertia of table, woman, and masses is $1.2 \text{ kg} \cdot \text{m}^2$. The woman now extends her arms so as to hold the masses far from her body. In so doing, she increases the moment of inertia of the system by $2.0 \text{ kg} \cdot \text{m}^2$. (a) What is the final rotation speed of the table? (b) Was the kinetic energy of the system changed during the process? Explain.

24. An ice skater moving past a post with speed v_0 grabs onto the end of a rope tied to the post. The original length of the rope is L_0 but the rope shortens as the skater circles around the post thereby winding up the rope. Assuming that the skater coasts and does not try to stop, how fast will he be moving when the length of rope (the circle radius) is L?

25. It is surmised that the sun was formed in the gravitational collapse of a dust cloud which filled the space now occupied by the solar system and beyond. Assuming the original cloud to be a uniform sphere of radius R_0 with an average angular velocity of ω_0, how fast should the sun be rotating now? For the present purposes, ignore the small mass resident in the planets and assume the sun to be a uniform sphere of radius R_s.

26. A children's merry-go-round in a park consists of an essentially uniform 200-kg solid disk rotating about a vertical axis. The radius of the disk is 6.0 m, and a 100-kg man is standing on its outer edge when it is rotating at a speed of 0.20 rev/s. How fast will the disk be rotating if the man walks 3.0 m in toward the center along a radius? What will happen if the man drops off the edge?

FIGURE P11.7

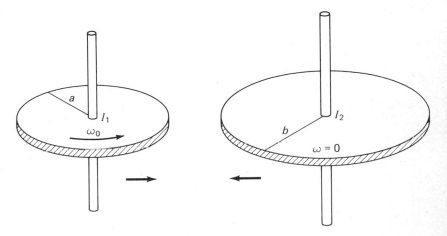

27. Suppose the merry-go-round of problem 26 has no one on it but is rotating at 0.20 rev/s. If now a 100-kg man quickly sits down on the edge of it, what will be its new speed?

28. Bohr's model of the hydrogen atom pictures the electron to circle the nucleus in an orbit of radius r. To find the radius as a function of electron speed, he equated the attractive force between positive nucleus and negative electron, namely,

$$F = (9 \times 10^9) \frac{e^2}{r^2} \quad \text{N}$$

where e is the charge on the electron, to the centripetal force. He also postulated that the electron's orbital angular momentum had a value $h/2\pi$ where h is a constant of nature called Planck's constant. Prove that

$$r = \frac{h^2}{36\pi^2 \times 10^9 e^2 m}$$

where m is the electron mass. Using the table of physical constants given in Appendix 2, show that the radius so obtained is reasonable.

29. A 20-kg child stands on and near the edge of a small, merry-go-round with the system at rest. The system's total moment of inertia about the center is $120 \text{ kg} \cdot \text{m}^2$. The child, at a radius of 2.0 m, jumps off the merry-go-round in a tangential direction with a speed of 1.5 m/s. How fast will the merry-go-round be rotating after she leaves it?

30. A simple Yo-Yo is shown in figure P11.8. The light string is wound on an inner sprocket and, as the Yo-Yo falls, the string unwinds. Call the sprocket radius b and the moment of inertia $0.7mb^2$ where m is the Yo-Yo's mass. After unwinding from rest so as to have fallen a distance h, how fast (in rev/s) will the Yo-Yo be rotating?

31. Referring back to figure P11.4, we see two hoops. The inner one rolls without slipping inside the other and that one, in turn, is free to move without friction on the table. The mass of the smaller one is m and it is $3m$ for the larger one. They start from rest in the position shown. After release, the inner hoop rolls around with decreasing amplitude and finally comes to rest at the bottom. What is the x coordinate of the center of the hoop when the system comes to rest?

32. Suppose in figure P11.4 that the two hoops are rigidly fastened together. The system is released from the position shown and there is negligible friction between the large hoop and the table. How fast relative to the table is the center of the large hoop moving when the centers of the two hoops lie on a vertical line?

33. A uniform cylinder with radius b rolls along the path shown in figure P11.9. From how high a distance h must it be released if it is to be able to hoops lie on a vertical line?

34. A large wooden wheel of radius R and moment of inertia I is mounted on an axle so as to rotate freely. A bullet of mass m and speed v is shot tangential to the wheel and strikes its edge, lodging in the rim. If the wheel was originally at rest, find its rotation rate just after collision.

35. The system shown in figure P11.10 is released from rest with the spring in the unstretched position. If friction is negligible, how far will the mass slide down the incline?

36. In the situation outlined in the previous problem, what will be the speed of the mass when the mass has slid 1.0 m down the incline?

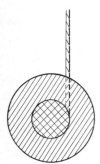

FIGURE P11.8

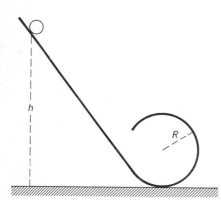

FIGURE P11.9

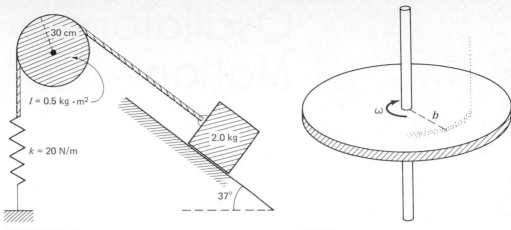

FIGURE P11.10

FIGURE P11.11

37. For the siutation outlined in problem 35, how far will the mass have slid when its speed is maximum? What will be its speed then?

38. As shown in figure P11.11, sand drops onto a disk rotating freely about an axis. The moment of inertia of the disk about this axis is I and its original rotation rate was ω_0. What is its rotation rate after a mass M of sand has accumulated on the disk?

39. Referring back to figure P11.4, the inner hoop rolls without slipping inside the larger one. The large one moves freely on the frictionless table. The mass of the smaller one is m while the larger one has a mass $3m$. They start from rest in the position shown and the inner hoop rolls down to the bottom of the large hoop with negligible energy loss for the system as a whole. How fast, relative to the table, is the center of the large hoop moving when the small hoop has its center directly below the center of the large hoop? Is the large hoop rolling clockwise or counterclockwise?

12 Oscillatory Motion

Our study of physics started with the discussion of bodies at rest. We then considered the simple case of motion along a straight line. In the preceding three chapters we investigated the rotational behavior of particles and systems. This chapter is devoted to the study of still another frequently encountered type of motion, vibratory, or oscillatory, motion. The techniques acquired in this study will be of value again later, when we investigate the behavior of waves and oscillating electric circuits.

12.1 HOOKE'S LAW AND VIBRATION

Any system which obeys Hooke's law will vibrate in a unique and simple manner, called *simple harmonic motion*. In Chap. 6 we discussed Hooke's law in some detail when describing the energy stored in springlike systems. It will be recalled that any system which stretches, bends, twists, or in some way distorts elastically can be said to obey Hooke's law if the following statement is true for it: *Hooke's law: The magnitude of the distortion is directly proportional to the magnitude of the distorting force, and their directions are the same.*

Hooke's law

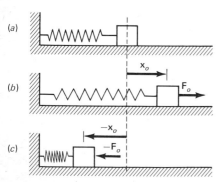

FIGURE 12.1 Since $\mathbf{F}_{ext} = k\mathbf{x}$ for the spring, it obeys Hooke's law.

As a concrete example, consider the behavior of a spring such as the one shown in figure 12.1. When left to itself, the spring takes up the position shown in (*a*). If a force $\mathbf{F}_0$ is applied to it, the spring stretches a distance $\mathbf{x}_0$, as shown in (*b*). Under the action of a force $-\mathbf{F}_0$ it is compressed a distance $-\mathbf{x}_0$. We can therefore write, if Hooke's law applies,

$$\mathbf{F}_{ext} = k\mathbf{x} \qquad\qquad 12.1$$

where k is a proportionally constant called the *spring constant*. Its units are newtons per meter in the SI, and as its units imply, it is a measure of the force needed to produce unit elongation if the spring were completely extensible. We emphasize the fact that the force in *12.1* is an externally furnished force by writing it $\mathbf{F}_{ext}$.

In a more general sense, *12.1* applies to many other systems which distort elastically. The external deforming force is proportional to the deformation which is there represented as **x.** Nearly all materials and systems which do not exhibit irreversible distortion under small applied forces will show some elasticity and will therefore obey *12.1* for small deformations. Whether the system is a molecule, a rubber band, or a steel rod, it will conform to *12.1* provided a certain critical deformation is not exceeded. As a

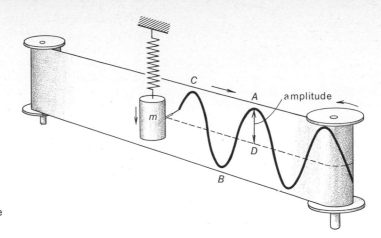

FIGURE 12.2 As the mass vibrates up and down, it leaves a record of its path on the paper moving at constant speed past it. The rest position is shown by the dashed line.

result, even though we shall employ *12.1* in a discussion of springs, our conclusions will apply to the multitude of systems which obey Hooke's law.

12.2 TERMINOLOGY

All of us are familiar with the vibration of a mass at the end of a spring. The general behavior is seen by reference to figure 12.2. In the case shown, the mass at the end of the spring leaves a record on the paper which shows how the mass oscillates up and down. *One complete vibration, or cycle,* of the mass occurs when it vibrates from the position indicated by point *A* to the point indicated by *C* or by any two other similar points. We denote this type of motion *periodic,* or *vibratory,* motion.

We call the time taken for the system to undergo one complete vibration the period *T* of the system. Since the system will undergo $1/T$ complete vibrations in unit time, this quantity is called the *frequency of the vibration* (*f* or *ν*), and we have

$$f = \frac{1}{T} \qquad\qquad 12.2$$

Strictly speaking, the units of frequency are (time)$^{-1}$. We usually express it as cycles (or vibrations) per second. One cycle per second (cps) is denoted as one hertz (Hz).

We denote the distance *AD* shown in figure 12.2 as the *amplitude* of the vibration. Notice that it is the distance from the equilibrium position (shown by the dashed line) to the position of maximum displacement. It is only half as large as the total vertical distance traversed by the mass.

Vibratory motion of a system which obeys Hooke's law, such as the spring-mass system shown in figure 12.2, is called *simple harmonic motion.* We shall see that the curve drawn by the mass as it vibrates is sinusoidal in form, and so this type of motion is often called *sinusoidal* motion. Let us now investigate this type of motion quantitatively.

A system which obeys Hooke's law undergoes simple harmonic motion

12.3 SIMPLE HARMONIC MOTION

Consider the mass *m* at the end of the spring (constant *k*) shown in figure 12.3. We assume it to move without friction on the horizontal surface. When

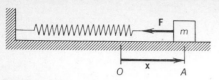

FIGURE 12.3 Since the spring obeys Hooke's law, we have $\mathbf{F} = -k\mathbf{x}$. Explain the presence of the minus sign.

it is at position O, the spring is unstretched, and the unbalanced force on it is zero. However, when the mass is at position A, the spring has been stretched by a displacement $\mathbf{x}$. In that position, the tension in the spring is kx, and therefore the spring exerts a force $-k\mathbf{x}$ on the mass. Notice that the force vector is opposite in direction to $\mathbf{x}$, and it is for this reason that the negative sign appears. We have

$$\mathbf{F} = -k\mathbf{x}$$

where $\mathbf{F}$ is the force exerted by the spring on the mass, an internal force of the system.

Since the motion will occur along the x axis, we shall drop the vector notation, and show direction by use of plus and minus signs. This unbalanced force, $-kx$, on the mass m will cause the mass to accelerate according to Newton's law. Therefore

$$F = ma$$

becomes $$-kx = m\frac{d^2x}{dt^2} \tag{12.3}$$

where use has been made of the fact that

$$a = \frac{dv}{dt}$$

$$= \frac{d}{dt}\left(\frac{dx}{dt}\right) = \frac{d^2x}{dt^2}$$

In *12.3* we have one example of what is called a *differential equation*. We now wish to find its solution. That is to say, we wish to find a functional form for x which will represent our physical situation and which will obey *12.3*. Clearly, x will be a function of time, which, when differentiated twice, must result in $-(k/m)$ multiplied by the same function before differentiation. In addition, the function when graphed against time must have the same general form as shown in figure 12.2. Inspection of the graph suggests it to be a sine or cosine function. We therefore try the equation

$$x \overset{?}{=} A \sin(\alpha t) + B \cos(\beta t) \tag{12.4}$$

where the question mark indicates we are still uncertain as to the correctness of this form for x. In this relation, A, B, α, and β are constants whose values are as yet unknown.

To find whether this functional form for x can satisfy the differential equation, we must first compute d^2x/dt^2 and substitute it and x into *12.3*. If the vibration of the mass can be described by *12.4*, then this substitution will result in an identity.

In order to compute d^2x/dt^2, we need to make use of the differentiation formulas

$$\frac{d}{dy}(\sin ay) = a \cos ay$$

and $$\frac{d}{dy}(\cos ay) = -a \sin ay$$

Using these relations, we can differentiate *12.4* twice with respect to time. We then find

$$\frac{d^2x}{dt^2} = -\alpha^2 A \sin(\alpha t) - \beta^2 B \cos(\beta t)$$

We now substitute these values in *12.3* and find

$$-k[A \sin(\alpha t) + B \cos(\beta t)] \overset{?}{=} m[-\alpha^2 A \sin(\alpha t) - \beta^2 B \cos(\beta t)]$$

After regrouping terms, this becomes

$$A(m\alpha^2 - k) \sin(\alpha t) + B(m\beta^2 - k) \cos(\beta t) \overset{?}{=} 0$$

This equation would be true if the quantities in parentheses were zero, i.e., if

$$m\alpha^2 - k = 0 \quad \text{and} \quad m\beta^2 - k = 0$$

which implies

$$\alpha = \beta = \sqrt{\frac{k}{m}}$$

Since α and β were just arbitrary constants in the first place, we can certainly assign this value to them. Therefore we find that the following functional form for x satisfies our basic differential equation and is a possible description of the vibrating mass:

The equation of motion for a mass on a spring

$$x = A \sin\left(\sqrt{\frac{k}{m}}t\right) + B \cos\left(\sqrt{\frac{k}{m}}t\right) \qquad 12.5$$

We still have two arbitrary unknown constants, A and B, in our solution. It is shown in courses in differential equations that an equation such as *12.3* can have no other periodic solution than *12.5*. Therefore, x as given there must be capable of representing the motion in our particular situation. It merely remains to determine what values of A and B are necessary to describe the motion. First, a word of warning! The angle $t\sqrt{k/m}$ is in radians, *not* degrees.

12.4 FITTING THE BOUNDARY CONDITIONS

In order to assign values to the two arbitrary constants A and B in *12.5*, we must have two independent pieces of information (the so-called *boundary conditions*) concerning our particular situation. Most frequently these two bits of information answer the following questions:

1. When was x zero?
2. What is the maximum value of x?

We shall now consider three separate cases which provide different answers to these questions.

Case 1 Shown in figure 12.4(a)
The mass was started from point P at time $t = 0$. Its subsequent motion is illustrated in the graph. Notice that the displacement x is of the form of a cosine curve since it has its maximum value at $t = 0$. Moreover, the maximum value taken on by x is x_0. We can obtain such a representation for x by use of *12.5* provided we set $A = 0$ and $B = x_0$. Then

$$x = x_0 \cos\left(\sqrt{\frac{k}{m}}t\right) \qquad 12.6$$

Since $\cos(0) = 1$, we see that x has its maximum value of x_0 when $t = 0$. Therefore *12.6* is the correct form for x in this particular case.

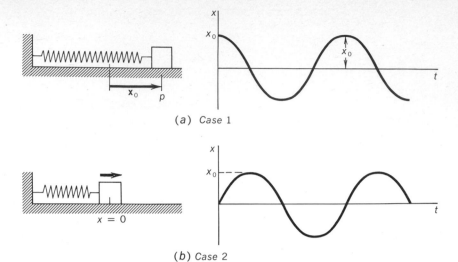

FIGURE 12.4 Both these functions are solutions of the basic differential equation. What determines which, if either, is the solution to our particular problem?

(a) Case 1

(b) Case 2

Case 2   Shown in figure 12.4(b)

At time $t = 0$ the mass was at $x = 0$ and was moving toward $+x$ values. Its subsequent motion is indicated in the accompanying graph. As we see, this is just a sine curve, and so the first term in *12.5* should be retained and B should be set equal to zero. Moreover, the amplitude of the motion is x_0, and so $A = x_0$. Therefore, in this case,

$$x = x_0 \sin\left(\sqrt{\frac{k}{m}}\, t\right)$$

12.7

Before considering the third, more general, case it will profit us to convert *12.5* to another form and to discuss that form. We make use of the trigonometric identity

$$A \sin\theta + B \cos\theta = C \sin(\theta + \phi)$$

12.8

where $\qquad C = \sqrt{A^2 + B^2} \qquad$ and $\qquad \sin\phi = \dfrac{B}{C}$

This shows at once that *12.5* for x will always give a sinusoidal form for x. Using *12.8* to transform *12.5*, we have the equivalent form

$$x = C \sin\left(\sqrt{\frac{k}{m}}\, t + \phi\right)$$

12.9

An alternative representation for simple harmonic motion

Notice that *12.9* reduces down to the result of Case 2 if we set $C = x_0$ and $\phi = 0$. To obtain Case 1, we recall that $\sin(\theta + \pi/2)$ is equal to $\cos\theta$, and so $C = x_0$ and $\phi = +\pi/2$ in that case. Clearly, the two new arbitrary constants C and ϕ replace A and B. They must be determined from the two boundry conditions involving x_0 and t. Of course, ϕ is just an angle, and it is called the *phase angle*. Its value is determined by when the timing clock is started. In order to see its significance more clearly, we have plotted *12.9* in figure 12.5. Notice that we have plotted $t\sqrt{k/m}$ rather than t on the axis, and we have given ϕ two different values in parts (a) and (b).

Referring first to part (a) of figure 12.5, since $\phi = 0$, we have that this is a plot of

$$x = x_0 \sin\left(\sqrt{\frac{k}{m}}\, t\right)$$

We know, of course, that $\sin \theta = 0$, when $\theta = 0, \pi, 2\pi$, and so on. Moreover, as shown on the graph, when θ, or in this case ($\sqrt{k/m}\, t$), goes from 0 to 2π, the motion completes one cycle. The time taken for this is the period of the motion. Therefore, when $t = T$, the quantity $\sqrt{k/m}\, T$ must be 2π. We have

$$T\sqrt{\frac{k}{m}} = 2\pi$$

and we conclude that the period of the motion is

Period of the motion is $2\pi\sqrt{m/k}$

$$T = 2\pi\sqrt{\frac{m}{k}} \qquad\qquad 12.10$$

This equation for the period, the time taken for one complete oscillation of the mass, is most important. It applies to any system moving under a Hooke's law force provided the mass moved and restoring force can be simply related through $F = ma$. Notice that *12.10* agrees with our common experience. If the mass is very large, the vibration will be slow, and T will be large. If the spring is very stiff, k will be large, the vibration will be fast, and T will be small. Since f, the frequency of oscillation, is merely $1/T$, *12.9* can be, and often is, written in the alternative forms

Alternative forms for the simple harmonic motion equation

$$x = C \sin\left(\sqrt{\frac{k}{m}}\, t + \phi\right)$$

$$x = C \sin\left(2\pi \frac{t}{T} + \phi\right) \qquad\qquad 12.11$$

$$x = C \sin\left(2\pi f t + \phi\right)$$

$$x = C \sin\left(\omega t + \phi\right)$$

where $T = 1/f = 2\pi\sqrt{m/k}$, and ω, the *angular frequency,** is defined to be $2\pi f$. Remember also, the angles are to be measured in radians.

---

* It is easily shown that the x coordinate of a particle moving in a circle about the origin in the xy plane is given by *12.11*. The angular frequency of the particle on the circle is ω. Hence this terminology.

FIGURE 12.5 Plots of $x = x_0 \cdot$ $\sin[(\sqrt{k/m})t + \phi]$. Notice the effect of the phase angle ϕ. You should be able to show from these graphs that $T\sqrt{k/m} = 2\pi$, where T is the period of the vibration.

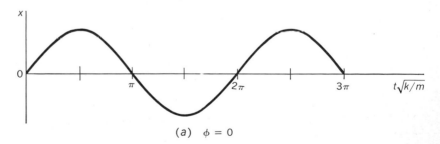

(a) $\phi = 0$

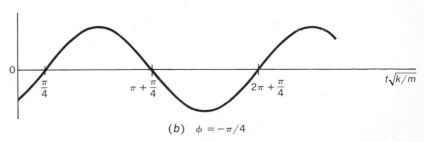

(b) $\phi = -\pi/4$

As shown in part (*b*) of figure 12.5, which represents Case 3, the phase angle ϕ merely shifts the vibration curve back and forth along the time axis. In effect, it tells us when the vibration was started. Usually, ϕ will be of no real importance unless we are comparing oscillations. When comparing the oscillations of two quantities or systems, we shall often be interested in whether or not they vibrate *in phase,* that is to say, whether or not they reach their maximum values at the same time. For two motions to be in phase, the phase angle ϕ must be the same for them. Why?

To summarize our results, we have found that a mass which moves under a Hooke's law force undergoes simple harmonic motion, the motion described by *12.5* and *12.11.* These equations are really identical except for the choice of the two arbitrary constants. We use the boundary conditions which tell us how the vibration was started in order to evaluate the constants. The period (and frequency) of the oscillation is related to the mass and spring constant by *12.10.*

12.5 USES OF THE EQUATION OF MOTION

In the preceding section we found the displacement equation, the *equation of motion,* for a body undergoing simple harmonic motion. Any such body, or system, is referred to as a *harmonic oscillator.* Since we know its equation of motion, we are in a position to describe its motion completely. This we now do. The student should be warned beforehand, though, that the equations derived in this section, like most derived equations, are not usually memorized by physicists since they can be derived quickly when needed.

We found that the displacement of the mass from equilibrium is

$$x = x_0 \sin 2\pi ft$$

where ϕ has been set equal to zero since it will be of no importance for the current discussion. To find the velocity of the mass we recall that

$$v = \frac{dx}{dt}$$

or, upon taking the derivative of x,

$$v = 2\pi f x_0 \cos 2\pi ft \qquad\qquad 12.12$$

or

$$v = v_0 \cos 2\pi ft$$

The mass has maximum speed when going through the equilibrium position

Notice that the velocity is a maximum when the displacement is a minimum. For example, at $t = 0$, the value of $x = 0$; so the mass was passing through its equilibrium position. At that time $\cos 2\pi ft = 1$, and so v had its maximum value.

This same result could have been obtained from energy considerations. When the mass is at its largest x value, x_0, the spring is fully stretched, and the mass is momentarily at rest. The energy stored in the spring is $\frac{1}{2}kx_0{}^2$, and this is the total energy of the spring-mass system. At any other elongation the sum of the potential energy in the spring, $\frac{1}{2}kx^2$, and the kinetic energy of the mass, $\frac{1}{2}mv^2$, must equal $\frac{1}{2}kx_0{}^2$. Therefore

$$\tfrac{1}{2}kx_0{}^2 = \tfrac{1}{2}kx^2 + \tfrac{1}{2}mv^2$$

After transposing and recalling that $x = x_0 \sin 2\pi ft$, we find

$$v^2 = \frac{k}{m} x_0^2 [1 - \sin^2 2\pi ft]$$

$$= \frac{k}{m} x_0^2 \cos^2 2\pi ft$$

Or, after taking the square root and using *12.10* and the fact that $T = 1/f$, we find

$$v = 2\pi f x_0 \cos 2\pi ft$$

which is the result found by direct differentiation. The advantage of this latter method is that it emphasizes the fact that potential and kinetic energy interchange as the oscillation takes place. This fact was discussed at length for the spring system in Chap. 7. We now recognize that the considerations discussed there apply to all harmonic oscillators.

Finally, we should remark that the equation for v can be written somewhat differently by noting that

$$\cos \theta = \sin \left(\theta + \frac{\pi}{2} \right)$$

Then v can be written

$$v = 2\pi f x_0 \sin \left(2\pi ft + \frac{\pi}{2} \right)$$

In other words, the time dependence of v differs from that for x by a phase angle of $\pi/2$. We say that v and x are $\pi/2$, or 90°, *out of phase*. The phase relationship between x and v is shown in figure 12.6.

To find the acceleration of the mass, we can make use of

$$a = \frac{dv}{dt}$$

Differentiation of *12.12* gives

$$a = -4\pi^2 f^2 x_0 \sin 2\pi ft \qquad\qquad 12.13$$

Or, after noticing that $x_0 \sin 2\pi ft$ is simply x, we have

$$a = -4\pi^2 f^2 x \qquad\qquad 12.14$$

FIGURE 12.6 Phase relations between x, v, and a for a harmonic oscillator.

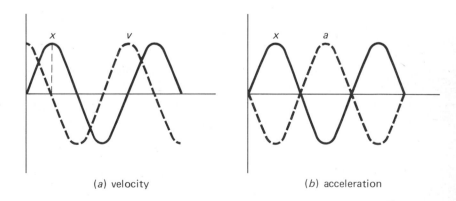

(a) velocity

(b) acceleration

Another way to obtain this result is to use $F = ma$ and $F = -kx$. Upon equating these two expressions for F, we find

$$a = -\left(\frac{k}{m}\right)x$$

which yields *12.14*,

$$a = -4\pi^2 f^2 x$$

after making use of *12.10* for $T = 1/f$.

Since $a \sim -x$, the magnitude of a is largest when the displacement x is largest. This is reasonable since $a \sim F$ and Hooke's law tells us $F \sim -x$.

The acceleration is maximum at the ends of the path

Further, we see from figure 12.6 that a is *minimum* when x is *maximum* and vice versa; the curve for a is shifted one-half cycle (or π rad or 180°) from the x curve. The acceleration is said to be π rad, or 180°, *out of phase* with the displacement. It is, of course, $\pi/2$, or 90°, out of phase with the velocity.

Illustration 12.1

A certain hookean spring stretches 20 cm when 40 g is hung from it. If a mass of 60 g is hung at its end, pulled out 20 cm from equilibrium, and released, find (*a*) the frequency of oscillation; (*b*) the equation of motion of the mass; (*c*) the speed and acceleration of the mass when it is 10 cm from the equilibrium position.

Reasoning Prior to this we have been considering horizontal springs so that the weight at the spring's end does not cause the spring to stretch. In the case of a vertical spring, the mass gives the spring an initial stretch. However, Hooke's law is a linear relation between F and x, and so an increment in F causes the same increment in stretch independently of any previous stretching of the spring. Hence the stretching caused by the mass at the end of a vertical spring does not affect the oscillatory behavior of the spring under an additional force.

First let us find the spring constant k. Since 0.040×9.8 N stretches it 0.20 m, we have (from $k = F/x$)

$$k = 1.96 \text{ N/m}$$

We now use *12.10* to find T, the period, and then take $1/T$ to find f. Thus

$$T = 2\pi \sqrt{0.060/1.96} \text{ s}$$

and

$$f = 0.91 \text{ Hz (or cps)}$$

The amplitude of the motion is 0.20 m, and it has maximum displacement at $t = 0$. Therefore x is a cosine function and is

$$x = x_0 \cos 2\pi f t$$

which gives

$$x = 0.20 \cos 5.7t \quad \text{m}$$

To find the equations of the speed and acceleration we take derivatives and find

$$v = -1.14 \sin 5.7t \quad \text{m/s}$$

and

$$a = -6.5 \cos 5.7t \quad \text{m/s}^2$$

We must now evaluate these when $x = 0.10$ m.

Placing this value for x in the equation for x, we have at once

$$\cos 5.7t = 0.50$$

from which we see that the angle $5.7t$ is $60°$. Using this angle for $5.7t$ in the equations for v and a will yield these quantities when $x = 0.10$ m. We find

$$v = -1.14 \sin 60° = -1.00 \, \text{m/s}$$

and

$$a = -6.5 \cos 60° = -3.25 \, \text{m/s}^2$$

12.6 AVERAGE QUANTITIES

In Chap. 4 we discussed how one could take the average of a function $\mathcal{F}(y)$ over a range of variable $c \leq y \leq d$. The result was stated as 4.9, namely,

$$\langle \mathcal{F} \rangle = \frac{\int_c^d \mathcal{F}(y) \, dy}{d - c} \qquad \textit{4.9}$$

where we use brackets $\langle \ \rangle$ to represent an average value of a quantity. We shall now make use of this relation to find several average values of interest for a harmonic oscillator.

First, to check our method, let us find the average of x itself over one cycle. We know that the result should be zero since we see from figure 12.4 that x has equal and opposite values during the two halves of the cycle. In this case $\mathcal{F}(t) = x = x_0 \sin 2\pi ft$, where y is replaced by t covering the range of one cycle, $0 \leq t \leq T$. Using 4.9, we have

$$\langle x \rangle = \frac{1}{T} \int_0^T x_0 \sin (2\pi ft) \, dt$$

Since

$$\int \sin (a\theta) \, d\theta = -\frac{1}{a} \cos a\theta$$

this yields

$$\langle x \rangle = \frac{x_0}{2\pi fT} \left| - \cos \left(\frac{2\pi t}{T} \right) \right|_0^T$$

where the frequency f has been replaced by $1/T$.

Placing in the limits, we find

$$\langle x \rangle = 0$$

as it should.

Case 1 $\langle x^2 \rangle$

Let us next find the average of the square of the distance of the mass from its equilibrium position. This quantity, being a squared number, will always be positive, and so its average cannot be zero. In this case

$$\mathcal{F} = x^2 = x_0^2 \sin^2 2\pi ft$$

Placing it in 4.9 and integrating over one cycle gives

$$\langle x^2 \rangle = \frac{1}{T} \int_0^T x_0^2 \sin^2 (2\pi ft) \, dt$$

To evaluate this integral we first change variables by the following substitution. Let $\theta = 2\pi ft$, in which case

$$d\theta = 2\pi f \, dt$$

The integral then becomes

$$\langle x^2 \rangle = \frac{1}{T} \int_0^{2\pi fT} x_0^2 \sin^2 \theta \, \frac{d\theta}{2\pi f}$$

where the new limits are found by substituting the limit values in the relation $\theta = 2\pi ft$. Notice that since $f = 1/T$, the upper limit is just 2π. We then have

$$\langle x^2 \rangle = \frac{x_0^2}{2\pi} \int_0^{2\pi} \sin^2 \theta \, d\theta$$

This integral may be found in the usual integral tables, and upon evaluating, its value is simply π. We therefore find

$\langle x^2 \rangle = \frac{1}{2}x_0^2$

$$\langle x^2 \rangle = \frac{1}{2}x_0^2$$

It is usual to call the square root of $\langle x^2 \rangle$ the rms (or root mean square) value of x. Thus

$x_{\text{rms}} = \dfrac{x_0}{\sqrt{2}}$

$$x_{\text{rms}} = \sqrt{\langle x^2 \rangle} = \frac{x_0}{\sqrt{2}}$$

One often encounters situations in science and engineering where the rms value of a sinusoidal function is required. We see that the result is always the amplitude divided by $\sqrt{2}$.

Case 2 $\langle U_s \rangle$

To find the average energy stored in the spring we recall from Chap. 6 that

$$U_s = \tfrac{1}{2}kx^2$$

Therefore

$$\langle U_s \rangle = \tfrac{1}{2}k\langle x^2 \rangle$$

But we just found $\langle x^2 \rangle$ to be $x_0^2/2$, and so

$$\langle U_s \rangle = \tfrac{1}{4}kx_0^2$$

Notice that the total energy of the spring system is $\tfrac{1}{2}kx_0^2$, and so

$$\langle U_s \rangle = \tfrac{1}{2}(\text{total energy of system})$$

Case 3 $\langle E_k \rangle$

To find the average kinetic energy we set

$$E_k = \tfrac{1}{2}mv^2$$

with

$$v^2 = 4\pi^2 f^2 x_0^2 \cos^2 2\pi ft$$

Therefore

$$\langle E_k \rangle = 2\pi^2 f^2 m x_0^2 \langle \cos^2 2\pi ft \rangle$$

You should be able to show that, like $\sin^2 \theta$, the average value of $\cos^2 \theta$ over one cycle is $\tfrac{1}{2}$. Making use of this value gives

$$\langle E_k \rangle = 2\pi^2 f^2 m x_0^2 \left(\frac{1}{\sqrt{2}} \right)^2$$

Recalling that $f = 1/T$ and using *12.10* for T, we find

$\langle U_s \rangle = \langle E_k \rangle = \tfrac{1}{4}kx_0^2$

$$\langle E_k \rangle = \tfrac{1}{4}kx_0^2$$

which is identical with the value found for $\langle U_s \rangle$. Therefore, on the average, half the energy of the system is kinetic, while half is potential. This is a property of all harmonic oscillators.

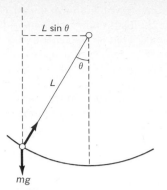

FIGURE 12.7 The torque is $mgL \sin \theta$. Under what conditions will this system undergo simple harmonic motion?

12.7 THE PENDULUM

We are all aware that the common simple pendulum undergoes oscillatory motion. Whether or not the motion is simple harmonic, i.e., is sinusoidal in form, depends upon whether or not the system obeys a differential equation similar to *12.3*. Let us examine the simple pendulum of mass m shown in figure 12.7.

If we take an axis at the point of suspension of the pendulum, the only force which exerts a nonzero torque on the pendulum is that due to the weight of the ball, mg. Because torque $= rF \sin \theta$, we have in this case that

$$\tau = -mgL \sin \theta$$

about the axis. The minus sign is used since this is a restoring torque, i.e., a torque which forces the system toward $\theta = 0$. Substituting this in the equation

$$\tau = I\alpha$$

we have

$$-mgL \sin \theta = +I\alpha$$

In these equations I is the moment of inertia of the pendulum.

But α is $d\omega/dt$, and since $\omega = d\theta/dt$, we have

$$-mgL \sin \theta = I\frac{d^2\theta}{dt^2} \qquad 12.15$$

This is to be compared with the differential equation for simple harmonic motion, *12.3*,

$$-kx = m\frac{d^2x}{dt^2} \qquad 12.3$$

Clearly, the linear distance x is to be replaced by the angular distance θ. Even so, the equations are dissimilar because of the presence of $\sin \theta$ rather than θ in *12.15*. Therefore we must conclude that, in general, a pendulum does not undergo simple harmonic motion.

However, if θ *is small,* we know that $\sin \theta \approx \theta$, and so under that condition *12.15* becomes

$$-mgL\theta = I\frac{d^2\theta}{dt^2}$$

If we notice that I for the pendulum is mL^2, we can write this

$$-g\theta = L\frac{d^2\theta}{dt^2}$$

Since this is identical in form with *12.3*, we conclude that the vibration of a pendulum is simple harmonic provided θ is small.

Because of their similar forms, all the equations derived from *12.3* will apply to the pendulum provided we replace x by θ, k by g, and m by L. We can therefore write the following results at once:

A pendulum undergoes simple harmonic motion if θ is small

$$\theta = \theta_0 \sin (2\pi ft + \phi) \qquad 12.16$$

with

$$\frac{1}{f} = T = 2\pi \sqrt{\frac{L}{g}}$$

Notice that the period of the pendulum depends only upon L and g. The acceleration due to gravity, g, can be quite accurately measured by use of this relation.

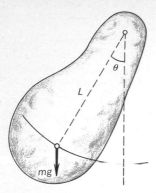

FIGURE 12.8 What data need one take to determine the moment of inertia of this compound pendulum about the pivot? About an axis through its mass center?

A rigid object suspended as a pendulum from a pivot as shown in figure 12.8 is also easily treated to this approximation. (This is called a *compound pendulum.*) Since the force of gravity can be considered as concentrated at the center of gravity, the torque on the object is

$$\tau = -mgL \sin \theta$$

where L is now the distance from the pivot to the center of gravity. Calling the moment of inertia about the pivot point I, we can write 12.15 down at once. By restricting the discussion to small angles we find

$$-mgL\theta = I\frac{d^2\theta}{dt^2}$$

This corresponds to 12.3 provided we place $k = mgL$ and $m = I$. The solution is then

$$\theta = \theta_0 \sin(2\pi ft + \phi) \qquad 12.17$$

where

$$\frac{1}{f} = T = 2\pi\sqrt{\frac{I}{mgL}}$$

One can therefore measure the moment of inertia of a rigid body by measuring its period when suspended as a pendulum.

12.8 THE TORSION PENDULUM

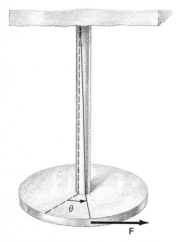

FIGURE 12.9 Is the simple harmonic vibration of a torsion pendulum restricted to small θ?

A typical torsion pendulum is shown in figure 12.9. It obeys Hooke's law in the sense that an applied torque causes the disk to twist through an angle θ, the torque being proportional to θ. We therefore write for the restoring torque

$$\tau = -k\theta$$

where k is the so-called *torsion constant* of the system. (How could you measure k?) Calling I the moment of inertia of the system about the axis of the suspension rod, we have in place of $\tau = I\alpha$

$$-k\theta = I\frac{d^2\theta}{dt^2} \qquad 12.18$$

The negative sign appears since the torque is always directed toward smaller angles.

Since 12.18 is similar in form to 12.3, we can write down the solution for the torsional vibration at once. It is

$$\theta = \theta_0 \sin(2\pi ft + \phi) \qquad 12.19$$

Equations for the torsion pendulum   with

$$\frac{1}{f} = T = 2\pi\sqrt{\frac{I}{k}}$$

The period of a torsion pendulum can obviously be used to measure moments of inertia.

Illustration 12.2

A disk having moment of inertia I_1 is suspended from a light torsion wire as shown in figure 12.9. The period of the torsion pendulum is now T_1. A second object is set on the disk, and the period becomes T_2. Find I_2, the moment of inertia of the object about an axis coincident with the torsion wire. (In practice, the center of mass of the object should lie on the axis, or the system may wobble.)

Reasoning We have from *12.19* that

$$I_1 = \frac{kT_1^2}{4\pi^2}$$

Since, from the definition of I, the moment of inertia of the combined system is

$$I = \sum_{\text{disc}} (\Delta m_i) r_i^2 + \sum_{\text{object}} (\Delta m_i) r_i^2$$

we have

$$I = I_1 + I_2$$

For the combined system

$$I_1 + I_2 = \frac{kT_2^2}{4\pi^2}$$

Dividing this by the expression for I_1 gives

$$I_2 = I_1 \left(\frac{T_2^2}{T_1^2} - 1 \right)$$

and so I_2 can be determined. To find the object's moment of inertia about an axis through axes other than its center of mass, use could be made of the parallel axis theorem.

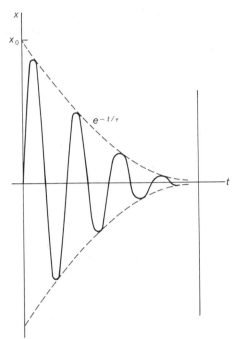

FIGURE 12.10 The behavior of a damped harmonic oscillator. We say that the oscillator is underdamped in this case.

Friction losses cause damping in a freely oscillating system

12.9 THE DAMPED HARMONIC OSCILLATOR

In the preceding sections we discussed the oscillation of various systems, implicitly assuming that friction forces were absent. This is not true in practice. We know that the oscillating system, if left to itself, will oscillate with decreasing amplitude and eventually stop. We say that the amplitude *damps down*. The reason for this is quite obvious of course. Some of the vibration energy is lost during each oscillation because of various friction-type loss mechanisms. The oscillation therefore damps down as shown in figure 12.10.

If we are to be able to describe this damped motion of the oscillator mathematically, we must know how the friction force behaves in such systems. In the case of a pendulum swinging in the air, much of the loss in vibration energy occurs because of friction with the air. Or alternatively, for a mass vibrating in a fluid as shown in figure 12.11, the viscous friction forces resulting from the motion through the liquid cause the oscillation to damp down. Experiment shows that the magnitudes of friction forces such as these are proportional to the speed of the object through the liquid or gas provided the speed is not too high. We therefore have

$$\text{Friction force} = -\eta \frac{dx}{dt} \qquad\qquad 12.20$$

where dx/dt is the speed of the vibrating object and η is a proportionality constant which depends upon the actual system being considered. Notice that a minus sign is used to show that the friction force retards the object's motion.

A damped vibrating object therefore has two forces acting on it, the friction force, *12.20*, and the restoring force $-kx$. As a result, $F = ma$ for such

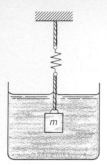

FIGURE 12.11 If the mass m is displaced vertically a distance x_0 from its equilibrium position and released, what type of motion will it undergo?

Meaning of the decay time τ

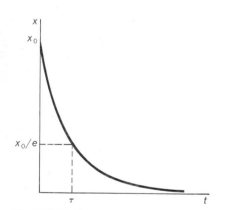

FIGURE 12.12 Exponential decay in a highly damped mass-spring system. In this overdamped case $\tau' = \eta/k$.

an object becomes

$$-\eta \frac{dx}{dt} - kx = m \frac{d^2x}{dt^2} \qquad 12.21$$

where d^2x/dt^2 has been written in place of a. This differential equation of motion is easily solved after one has had some experience with differential equations. However, since most of you have not yet progressed that far in your study of mathematics, we shall not go through the details of its solution. The qualitative features of the mathematical result are easily understood, though, and so we shall discuss them.

If the damping (or friction) effects are small, we know that the oscillation will damp down, as shown in figure 12.10. One sees that the envelope of the oscillation curve (the dashed lines in figure 12.10) appears to be a simple type of function. Solution of *12.21* shows that this is indeed true. The equations of the two dashed curves are given by the relations

$$x = x_0 e^{-t/\tau} \qquad \text{and} \qquad x = -x_0 e^{-t/\tau} \qquad 12.22$$

where

$$\tau = \frac{2m}{\eta}$$

Notice that since t/τ must be dimensionless (exponents cannot have units), the quantity τ has the units of time. It is called the *decay time* for the oscillation, and measures the time taken for the oscillation to damp down to $e^{-1} = 1/2.718$ of its original vibration amplitude. As we should expect, if η (and therefore the friction force) is large, the decay time τ is short. Why should τ be large if m is large?

Figure 12.10 does not represent the only type of motion possible for the mass shown in figure 12.11. If the liquid is extremely viscous, tar, for example, the mass will not oscillate at all. Instead, if the mass is displaced a distance x_0 from its equilibrium position, the variation of the displacement x with time will be as shown in figure 12.12. Solution of *12.21* in the case of η very large (i.e., a very large friction force) gives the result

$$x = x_0 e^{-t/\tau'} \qquad 12.23$$

where

$$\tau' = \frac{\eta}{k}$$

in this, the *overdamped* case. This equation is graphed in figure 12.12. As in the *underdamped* case, where the oscillations slowly decrease in amplitude, we have a characteristic decay time, τ' in this case. Is it reasonable that the time taken for the mass to reach the position $x = x_0/2.718$ should increase as η increases and decrease as k increases?

Damped oscillatory motion is extremely common. Only when external driving forces are applied to an oscillating system will the system vibrate with constant amplitude. The behavior of driven oscillatory systems is discussed in the following section.

12.10 THE DRIVEN HARMONIC OSCILLATOR

If one wishes a pendulum or a child in a swing or a mass at the end of a spring to vibrate with constant amplitude, one must push on the oscillating system so as to supply it with energy to compensate for the energy lost to friction forces. We all are aware of the fact that there is a right and a wrong way to push on a child in a swing if his oscillation is to be sustained. If one pushes so as to oppose the motion of the swing, the swing does work on the pusher and loses some of its energy of oscillation. To give maximum energy to the swing, one must push in step with the oscillation; i.e., one must push with maximum force when the swing is moving in the direction of the applied force.

The simplest type of applied force to treat mathematically is a force which varies sinusoidally, such as

$$F_x = F_0 \sin \omega t \qquad\qquad 12.24$$

An applied force is most effective if

$\omega = \omega_0$

where the angular frequency ω is written in place of $2\pi f$. In order for the driving force *12.24* to remain in step (or in phase) with the oscillating system to which it is applied, the force must reverse its direction each time the vibrating object reverses its motion. This means that the applied force will be most effective when ω, its frequency, is equal to the free vibration frequency ω_0 of the oscillating system. We then term the force to be in resonance with the system.

When the force given in *12.24* is applied to the vibrating system properly, it will do positive work on the system and thereby give it energy. If the system was originally at rest, the force will cause it to begin to vibrate with increasing amplitude. Since the driving force furnishes energy to the oscillating system on each cycle, the system's vibration energy should increase, and the system should vibrate with increasing amplitude if there are no friction forces. However, since the friction forces acting on the system are never zero, and increase with increasing oscillation amplitude, the oscillator will eventually reach an amplitude large enough so that the friction-type energy loss per oscillation will exactly equal the energy furnished during each oscillation by the driving force. At that time, the oscillation will continue at constant amplitude. We say that a *steady-state oscillation* has been achieved.

Even when the applied force is not of the same frequency as the free vibration frequency of the system, similar considerations apply. However, since such forces do not act in compliance with the natural vibration of the oscillatory system, they do not supply as much energy to it. Consequently, the steady-state amplitude of vibration decreases as the frequency of the driving force departs from the natural, or *resonance,* frequency of the oscillating system. This behavior is shown in figure 12.13. The steady-state

FIGURE 12.13 Response curve for a driven harmonic oscillator. Notice that friction forces reduce the response and shift its maximum in frequency.

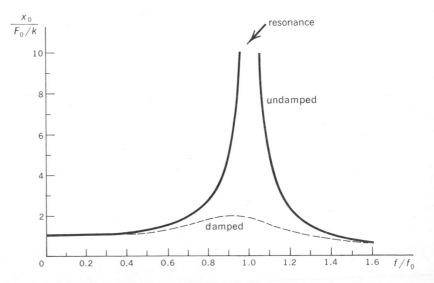

amplitude of oscillation x_0 is shown as a function of the ratio of the driving frequency f to the natural resonance frequency of the undamped system f_0. Notice that the undamped motion (motion without friction loss) becomes very large at the resonance frequency. As shown by the dashed curve, the damped oscillator behaves as one would expect from the fact that the energy loss increases with amplitude of oscillation. Let us now treat the problem analytically.

Suppose an undamped oscillator is caused to vibrate by a driving force

$$F_x = F_0 \sin \omega t$$

Newton's second law, $F = ma$, applied to this case becomes

$$-kx + F_0 \sin \omega t = m \frac{d^2x}{dt^2} \qquad 12.25$$

The solution of this differential equation will represent the motion of the mass under the action of the applied force. Although the general solution to this equation is quite complicated if one considers effects which occur just after the force is applied, the steady-state motion which occurs at longer times is usually of most importance, and we shall be concerned with it alone.

The steady-state motion must be of the same frequency as the applied force, or it could not be steady. We therefore try a sinusoidal solution of form

$$x \overset{?}{=} A \sin (\omega t) + B \cos (\omega t)$$

By differentiating twice we find

$$\frac{d^2x}{dt^2} = -\omega^2 A \sin (\omega t) - \omega^2 B \cos (\omega t)$$

Substitution of these values in the differential equation *12.25* yields, after grouping like terms,

$$(F_0 - kA + \omega^2 Am) \sin (\omega t) + (-kB + \omega^2 mB) \cos (\omega t) \overset{?}{=} 0$$

Unlike the case of the freely vibrating harmonic oscillator, the frequency ω is not unknown. It is the frequency of the driving force. The $\cos \omega t$ term, therefore, can only be made zero for all t values by placing $B = 0$. To make the sin ωt term zero, we must have

$$F_0 - kA + \omega^2 mA = 0$$

which gives

$$A = \frac{F_0}{k - \omega^2 m}$$

This can be simplified if we note that the frequency of the freely vibrating system is (from *12.10*)

$$f_0 = \frac{1}{T} = \frac{\sqrt{k/m}}{2\pi} = \frac{\omega_0}{2\pi}$$

Using this, we can write A as

$$A = \frac{F_0/m}{\omega_0{}^2 - \omega^2}$$

The response of the system is therefore given by

$$x = \frac{F_0/m}{\omega_0{}^2 - \omega^2} \sin \omega t \qquad 12.26$$

Notice, as one would expect, that the amplitude of the response, namely,

$$x_0 = \left| \frac{F_0/m}{\omega_0{}^2 - \omega^2} \right| = \left| \frac{F_0/k}{1 - (\omega/\omega_0)^2} \right| \qquad 12.27$$

depends upon the frequency of the applied force. The amplitude plotted as a function of the ratio of the driving frequency to the natural frequency is shown in figure 12.13.

We see at once that the amplitude of the undamped system approaches infinity as $\omega \to \omega_0$. The natural frequency of vibration of the undamped system,

Resonance frequency

$$\frac{\omega_0}{2\pi} = f_0 = \frac{1}{2\pi} \sqrt{\frac{k}{m}} \qquad 12.28$$

is called the *resonance frequency* of the system. When the driving force approaches that frequency, large oscillations occur. As pointed out previously, this occurs because the driving force is always in phase with the natural vibration of the system in this case. As a result, the force is always adding energy to the system. In the absence of friction forces, the energy must appear as vibrational energy. The actual behavior of a fairly heavily damped system is shown as the dashed curve in figure 12.13.

At driving frequencies other than the resonance frequency, the applied force does not supply energy to the system as efficiently. Hence the system does not build up a large oscillation. In fact, when steady state is reached and friction forces are absent, the driving force furnishes zero total energy to the oscillator during a complete cycle. This is easily seen from the conservation of energy law. (How?) Or it may be shown mathematically as follows.

The work done by the applied force in one cycle is

$$\frac{\text{Work}}{\text{Cycle}} = \oint F_x \, dx$$

where the circle on the integral sign means to take the integral over a full cycle. From *12.26* and *12.27* we have, after taking the derivative of *12.26*,

$$dx = \frac{\omega F_0/k}{1 - (\omega/\omega_0)^2} \cos \omega t \, dt$$

Therefore

$$\frac{\text{Work}}{\text{Cycle}} = \frac{\omega F_0{}^2}{k(1 - \omega^2/\omega_0{}^2)} \oint \sin \omega t \cos \omega t \, dt \qquad 12.29$$

However, during a whole cycle, the quantity under the integral is positive exactly as much as it is negative. Its integral is therefore zero, as one could easily verify by carrying out the integration. The special case $\omega = \omega_0$ is unique in that we have zero divided by zero. As we have seen, the system is said to resonate at this frequency. This property, resonance, is widely used, and we shall encounter it frequently.

Chapter 12
Questions
and Guesstimates

1. How can one determine whether or not a system will vibrate with simple harmonic motion?
2. A pendulum undergoes simple harmonic motion only if the angle of its swing is small. Is this also true for a torsion pendulum?
3. A pendulum swings back and forth with constant amplitude. Where is it

in each of the following cases? (*a*) Speed is maximum. (*b*) Speed is zero. (*c*) Acceleration is zero. (*d*) Acceleration is maximum. (*e*) Energy is all potential. (*f*) Energy is all kinetic. (*g*) Energy is half kinetic, half potential.

4. For a linear simple harmonic oscillator, can the acceleration and displacement ever have the same direction at the same time? The acceleration and the velocity? The displacement and velocity?

5. A precocious student says she can predict the frequency of a spring-mass system even though she knows neither the spring constant nor the mass. All she needs to know is how far the spring stretches when the mass is hung from it. Should you bet money that she cannot do this?

6. An unknown mass hangs at the end of a vertical spring. Can the spring constant of the spring be found by simply measuring how much an additional known mass stretches it?

7. We usually assume the mass of a spring is negligible in comparison to the mass at its end. But if not negligible, does the mass of the spring increase or decrease the period of motion?

8. When driven by a sinusoidal type force, a harmonic oscillator will only resonate if the force has the resonance frequency f_0 of the oscillator. And yet a child on a swing will resonate if the person pushing pushes at a frequency of $\frac{1}{2}f_0$, or $\frac{1}{3}f_0$, or $\frac{1}{4}f_0$, and so on, as well as with a frequency f_0. Why this difference?

9. Consider a pendulum consisting of a hollow sphere attached to a long string. In which case should the pendulum slow down and stop soonest: when the ball is full of water, half full, or empty?

10. Describe what would happen to the period of oscillation of a pendulum if the ball were initially full of water but the water slowly leaked out.

11. Two equal weights hang at the end of a spring, and the system is set vibrating. What happens to the amplitude, frequency, and maximum speed of the end of the spring if one of the weights falls off when (*a*) the spring is at its largest extension and (*b*) the mass is passing through the center position?

12. Consider a disk supported as a torsion pendulum by a thin vertical wire fastened at the center of the disk. As the horizontal disk oscillates back and forth, a drop of water falls onto the disk each time it is at the end of its swing. What effect, if any, will this have on the oscillation? What if the water drops on the disk as it is going through its equilibrium position?

13. How will the period of a pendulum change if the pendulum is in an accelerating elevator? Consider the cases of accelerations both upward and downward. What would happen to the period of a torsion pendulum under these same circumstances?

14. A common method used to free a stuck car is to rock it back and forth, with proper shifting of the gears alternately from forward to reverse and back to forward many times over. Discuss the physical basis for this method, paying particular attention to energy transfer.

15. Various portions of one's body have characteristic vibration frequencies: a freely swinging arm or leg could be cited. Discuss how these natural frequencies influence how one walks or runs. Estimate the frequencies.

16. How could you compute the up and down resonance frequency of a car using data obtained from the lowering of the car with increased load? Estimate this frequency for an automobile. When might it be important?

17. Sometimes during the spin-dry portion of a washer's cycle, the washer vibrates very strongly. Why? Is the unbalance of the load the whole

story? What should a washer designer do to minimize this problem?

18. Discuss the influence of design characteristics on the suitability and performance of a diving board or trampoline.

19. A glass tube (inner diameter = 1.0 cm) is bent to form a U-tube with each side tube 40 cm long and the bottom of the U about 10 cm long. Estimate the natural up-and-down frequency of vibration for the liquid column when the U is filled to a height of 30 cm with (a) water and (b) mercury.

Chapter 12
Problems

1. A certain spring of negligible mass stretches 10 cm when a 50-g mass is hung from it. What will be the frequency of vibration of the spring when 80 g hangs at its end?

2. Two children notice that they can make a car vibrate up and down by periodically pushing downward on it. The car vibrates through eight complete cycles in 13.0 s. (a) Assuming the car to have a mass of 1800 kg, find the spring constant for its suspension system. (b) If this value is correct, about how much should the car lower as a 70-kg person enters the car and sits in its front seat?

3. The equation of motion for a particular mass at the end of a spring is

$$x = 0.40 \cos (0.70t - 0.30) \qquad \text{m}$$

For this vibration find (a) the amplitude, (b) the frequency, (c) the period, (d) the phase angle in degrees, (e) the ratio k/m.

4. The equation of motion for a certain simple pendulum is

$$\theta = 20 \sin (3\pi t + 0.60) \qquad \text{degrees}$$

For the vibration find (a) amplitude, (b) frequency, (c) period, (d) phase angle in degrees, (e) L of pendulum.

5. The ball at the end of a simple pendulum moves according to the following equation:

$$\theta = 20 \sin (0.10t) \qquad \text{rad}$$

Find the angular speed and angular acceleration of the ball.

6. The equation of motion for a mass at the end of a particular spring is

$$y = 0.30 \cos (0.50t) \qquad \text{m}$$

Find the velocity and acceleration of the mass.

7. Given that the equation of motion of a mass is $x = 0.20 \sin (3.0t)$ m, find the speed and acceleration of the mass when the object is 5 cm from its equilibrium position. Repeat for $x = 0$.

8. A disk rotates as a torsion pendulum. The relation $\theta = 120 \sin (2t)$ degrees applies to it. Find the angular speed and acceleration of the disk when $\theta = 0°$. Repeat for $\theta = 30°$.

9. A 200-g mass is suspended at the end of a spring which stretches 30 cm under a load of 0.50 N. Write the equation of motion of the mass when it alone is attached to the end of the spring. Assume the mass was released at time $t = 0$ when the spring was compressed 5.0 cm from the equilibrium position for the system.

10. A simple pendulum 60 cm long has a 200-g ball. The pendulum is pulled aside 15° and released. If the timing clock is started just as the ball moves through its lowest position, write the equation of motion for the pendulum in terms of the pendulum angle in degrees.

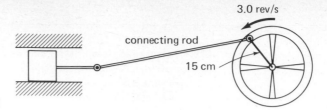

FIGURE P12.1

11. For the situation shown in figure P12.1, (*a*) write the equation of motion for the piston in the form $x = x_0 \cos \omega t$ with x_0 and ω given numerical values. Find the magnitude of the acceleration of the piston when (*b*) the connecting rod is horizontal, (*c*) when the radius to the pivot point is vertical, and (*d*) when the radius makes an angle θ to the horizontal.

12. A circle in the xy plane with center coincident with the origin of coordinates has a radius A. A particle moves around the circle with constant speed ω. Show that the particle's x coordinate is given by $x = A \cos (\omega t + \phi)$ where ϕ is the angle the particle was at when $t = 0$. How is the angular frequency ω of the particle in the circle related to f, the frequency with which the x coordinate oscillates?

13. Show that the maximum speed of a pendulum bob is given by

$$v = \sqrt{2gL(1 - \cos \theta)}$$

if the pendulum starts to swing from an angle of θ to the vertical.

14. A pendulum is drawn aside to a certain angle and released. When the bob passes the center point, the tension in the string is twice the weight of the bob. Show that the original displacement angle was 60°.

15. Find the period of motion for a torsion pendulum consisting of the following. A uniform 400-g sphere of radius 8.0 cm is suspended from a torsion wire 20 ft long. The wire twists 90° when a torque of 0.40 N · m is applied to it.

16. Find the frequency of oscillation when a meter stick is hung as a compound pendulum with the pivot at its 90-cm mark. Repeat for the pivot at the 50.10-cm mark.

17. In figure P12.2 are shown three different spring combinations. Assuming the springs to have negligible mass and friction to be absent, what is the frequency of vibration of the mass in each case? Give your answer in terms of the k's and m.

18. A uniform spring of length L_0 has a spring constant K_0. It is now cut in half and the resultant two springs are used to support a mass M. (The mass now has two vertical supporting springs.) Find the ratio of the

FIGURE P12.2

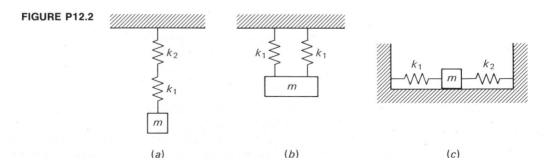

(*a*)　　　　　(*b*)　　　　　(*c*)

frequency of vibration of the mass hanging from the original spring to its frequency as now supported.

19. A motor vehicle to carry astronauts on the surface of the moon has a spring suspension and has a natural up-and-down frequency of 0.40 Hz when fully loaded on the earth. Find its natural frequency on the moon where it and everything in it will weigh only about one-sixth as much. (*Hint:* Since $F = kx$ for a spring, a mass vibrates in the same way at the end of a spring even in the presence of a constant additional stretching force.)

20. A certain pendulum clock keeps good time on the earth. If the same clock were placed on the moon where objects weigh only one-sixth as much as on earth, how many seconds will the clock tick out in an actual time of 1 minute?

21. Consider a spring with force constant 5.0 N/m. (*a*) How large a mass must be hung from the spring to stretch it 20 cm? (*b*) How much more is needed to stretch it 20 cm more? (*c*) What is the frequency of vibration of the combined masses when the spring-mass system is horizontal? (*d*) Vertical?

22. In a certain mechanical device a piston moves up and down in sinusoidal fashion with amplitude 10 cm. By accident, a washer was left sitting on the piston. As the speed of the device is increased, at what frequency of motion of the piston will the washer no longer sit passively on the piston? (*Hint:* What is the maximum possible downward acceleration of the washer?)

23. A block oscillates on a horizontal surface in simple harmonic motion with amplitude 5.0 cm and frequency f. A smaller block sits on top of this one, and the friction force between the two blocks is $4.0m$ where m is the mass of the upper block. At what minimum frequency will the upper block slip loose from the lower one?

24. Suppose a cylindrical hole is bored through the earth along a diameter. The gravitational force upon a mass m within the hole is *not* GmM_e/r^2 since the mass m is only attracted toward the earth's center by a portion of M_e. The force turns out to be approximately $1.5 \times 10^{-6}mr$ N. It is directed toward the earth's center. Suppose the mass m is now dropped down the hole from the earth's surface. Show that the mass will undergo simple harmonic motion in the hole and find the period of the motion.

25. To determine my moment of inertia about an axis passing through my center of mass from side to side (not front to back), I took the following measurements. When hanging from a fixed bar by my hands and letting my 68-kg body swing as a compound pendulum, the time for one complete swing is about 2.5 s. My horizontal body balances when I lie on a bar 105 cm from the position of my outstretched hands. Find the moment of inertia of my body about the bar from which I was swinging and then use the parallel axis theorem to find the value of I about the axis through my mass center.

26. A 10-g bullet is shot with speed 80 m/s at a 500-g wooden block. The block lies on a frictionless table and is at equilibrium at the end of a spring ($k = 5.0$ N/m) lying on the table. When the bullet moving along the line of the spring hits the block, it lodges in it and compresses the spring. With what amplitude does the block vibrate after the collision?

27. The masses in figure P12.3 slide on an essentially frictionless table. The spring constant of the spring is 200 N/m, and m_1, but not m_2, is fastened to the spring. If now m_1 and m_2 are pushed to the left so that the spring is

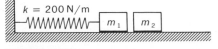

FIGURE P12.3

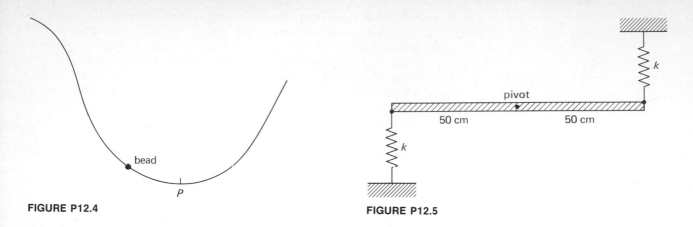

FIGURE P12.4

FIGURE P12.5

compressed 40 cm, what will be the amplitude of the oscillation of m_1 after the spring system is released?

28. A bead of mass m slides on a frictionless wire as shown in figure P12.4. Because the shape of the wire near P can be approximated as a parabola, the potential energy of the bead is given by $U = cx^2$, where x is measured from P and c is a constant. The bead will oscillate about point P if displaced slightly from P and released. Recalling that $F_x = -\partial U/\partial x$, show that the bead oscillates with simple harmonic motion and its period is $2\pi \sqrt{m/2c}$. This same line of reasoning leads to the fact that any conservative system will oscillate with simple harmonic motion about a minimum in its potential energy curve provided the oscillation amplitude is small enough.

29. The uniform stick in figure P12.5 has mass m and length L and is pivoted at its center. It is held in place by identical stretched springs each of constant k. In the equilibrium position shown, each spring is stretched a distance b. (a) Show that the stick will undergo simple harmonic motion when turned through a small angle θ_0 from the position shown and released. (Assume the springs to be longer than shown.) (b) What is the frequency of the motion? (c) How fast will the tip of the stick be moving when the stick passes through the horizontal?

30. A mass m at the end of a spring of constant k is vibrated about its equilibrium position by an external unbalanced force $F = F_0 \sin \omega t$ where ω is half as large as the resonance frequency ω_0 of the spring-mass system. Assuming damping to be negligible, (a) what is the amplitude of the motion at this frequency? (b) At what higher frequency will the amplitude be reduced to 0.10 this value?

31. If $x = x_0 \sin \omega t$, find $\langle x^3 \rangle$. Repeat for $\langle x^4 \rangle$.

13 Continuum Mechanics

In previous chapters we discussed the statics and dynamics of particles and objects. Ideally, we should continue the discussion to include the behavior of the large aggregates of particles which we refer to as fluids and solids. But the complexity involved in the motion of the individual atoms of such large aggregates leads us to seek a new mode of approach. In continuum mechanics, fluids and solids are discussed from a different aspect. We shall obtain an introduction to this way of describing nature in this chapter.

13.1 CHARACTERIZATION OF MATERIALS

To properly describe a material, we should provide a description of the motion of each atom of which it is composed. Such a description would be worthless for most purposes. It would be far too complicated and detailed for the everyday uses to which materials are put. The engineer who wishes to use a certain type of steel in construction or the surgeon who needs a plastic filament for a suture neither wants nor requires an atomic description of the material. Only rarely are situations encountered where more than the gross, overall properties of a material are needed. Usually, it is sufficient to give a nonatomic description of the material.

Two types of materials: fluids and solids

It is customary to divide materials into two broad categories, fluids and solids. As the names imply, fluids are substances which can flow, such as water, air, and tar; solids retain their geometric form even under the action of rather large forces, and we could classify rock, wood, and solid iron in this category. However, the two categories are not completely distinct; for example, glass, a solid in the opinion of most people, does flow slowly over the centuries, and so it could be termed a liquid. There are many other examples. Scientists and engineers therefore waste no time arguing about it. If the description of a material must be precise, we give it in terms of numbers, not words, as we shall see later.

There are really two types of fluids. One of these, which encompasses gases, is characterized by the fact that the fluid expands to fill the whole container in which it is kept. Members of the other type, the liquids, settle to the bottom of any container in which they are placed. Here, too, the dividing line is not precise. It will be sufficient if we associate materials such as air with gases and call waterlike materials liquids.

Gases and liquids are fluids

As we have stated, word descriptions of materials are usually too unprecise to be useful in quantitative applications. For this reason we seek ways of describing substances in terms of numerical quantities. These quantities can then be tabulated so that potential users can ascertain the suitability of the material for their application. Moreover, when scientists attempt a molecular description of a material, their theory can be checked quantitatively by comparing its prediction with the actual value for the material property being evaluated. Let us now see how scientists and engineers describe the continuum properties of materials which they use.

13.2 DENSITY

We define density in such a way that it agrees with our common experience; it is a measure of the mass contained in a unit volume of the substance. Precisely, given a volume V of substance with mass m, we define its density ρ (Greek rho) by the equation

Definition of density

$$\rho \equiv \frac{\text{mass}}{\text{volume}} = \frac{m}{V} \qquad\qquad 13.1$$

The units of density are kg/m³ in the SI, although g/cm³ are frequently used. Typical values are 1000 kg/m³ for water and 13,600 kg/m³ for mercury. These values change somewhat with temperature since the volume of an object changes slightly as the temperature is changed. (More will be said about this in Chap. 15.) The values given in table 13.1 are for the temperatures specified. Sometimes a quantity called the *weight density* of the substance is used. It is defined as the weight per unit volume rather than the mass per unit volume. Since the weight of an object changes as the object is moved from place to place (from earth to moon, for example), weight density is not as fundamental a means of characterization as mass density. Our equations will always make use of mass density ρ.

Definition of specific gravity

Another similar quantity frequently used is the relative density or *specific gravity* of a substance. It is defined as the ratio of the density of the substance to the density of water. Being a simple ratio, it has no units. However, it varies with temperature since the densities are temperature-dependent. Hence, to be meaningful, a specific gravity value must be accompanied by a statement of the temperature at which it was evaluated.

TABLE 13.1   Densities of Substances

Substance	Temperature (°C)	ρ^* (kg/m³)
Air (normal pressure)	0.0	1.29
Benzene	20.0	879
Water	20.0	998
Water	3.98	1,000
Blood	20.0	$\approx 1,060$
Bone	20.0	$1.7\text{--}2.0 \times 10^3$
Aluminum	20.0	2,700
Iron	20.0	7,860
Copper	20.0	8,920
Lead	20.0	11,340
Mercury	0.0	13,600

* 1000 kg/m³ = 1 g/cm³.

Illustration 13.1

As an approximation, consider the air to be completely nitrogen. The mass of a nitrogen molecule is $28 \times 1.66 \times 10^{-27}$ kg. How many nitrogen molecules are there in $1\,\text{cm}^3$ of air?

Reasoning The mass of a volume V of a substance is given by equation *13.1* to be

$$m = \rho V$$

Therefore, $1\,\text{cm}^3$ of air has a mass of

$$m = (1.29\,\text{kg/m}^3)(1 \times 10^{-6}\,\text{m}^3) = 1.29 \times 10^{-6}\,\text{kg}$$

Each molecule has a mass $28 \times 1.66 \times 10^{-27}$ kg, and so the number of molecules in $1\,\text{cm}^3$ is

There are of order 10^{19} molecules in $1\,\text{cm}^3$ of air

$$\text{Number/cm}^3 = \frac{1.29 \times 10^{-6}\,\text{kg/cm}^3}{28 \times 1.67 \times 10^{-27}\,\text{kg}} = 2.8 \times 10^{19}$$

Notice what a tremendously large number of molecules there are in a cubic centimeter of air.

13.3 MODULUS OF ELASTICITY

Another property of a material which is of great practical importance is its stretchability. A rubber band and a steel strand differ markedly in this respect. We characterize this property of a substance by a number of quantities, each of which is called a *modulus*. All types of moduli are defined in terms of a stress and a strain. If we refer to figure 13.1 we can see what is meant by this terminology.

As shown there, a bar of the material is subjected to a stretching (or *tensile*) force F. We are interested in how much the bar stretches ΔL under a given force F. However, since the force must be doubled if we are to produce the same elongation for a bar of twice this cross-sectional area, we eliminate this geometrical complication by discussing the force per unit area, F/A. This *Definition of stress* quantity is called the *stress*. Similarly, the amount of stretch ΔL depends upon the original length of the bar, L_0. We eliminate this geometrical com- *Definition of strain* plication by use of the *strain* defined as elongation per unit length, $\Delta L/L_0$.

FIGURE 13.1 The modulus defined in a tensile experiment such as this is Young's modulus.

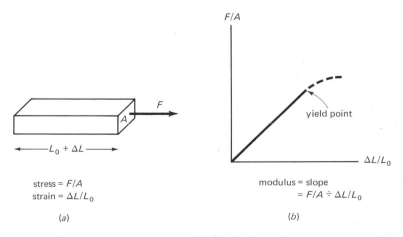

stress = F/A
strain = $\Delta L/L_0$

(a)

modulus = slope
$= F/A \div \Delta L/L_0$

(b)

TABLE 13.2 Young's Modulus Y and Bulk Compressibility k

Material	Young's modulus* $Y \times 10^{-10}$ (N/m²)	Compressibility $k \times 10^{11}$ (m²/N)
Tungsten	35	0.5
Steel	19–20	0.6
Iron (wrought)	18–20	0.7
Femur bone	≈ 14	. . .
Copper	10–13	0.8
Brass (cold rolled)	9	1.6
Iron (cast)	8–10	1.0
Aluminum	5.6–7.7	1.5
Polystyrene	≈ 0.14	20
Rubber	0.0004	40
Water	. . .	50
Benzene	. . .	100
Normal air	. . .	$\approx 1 \times 10^6$

*The label $Y \times 10^{-10}$ at the head of a column means that all the values given for Y in that column have been multiplied by 10^{-10}. Hence, Y for tungsten is 35×10^{10} N/m².

The modulus of the material is defined to be

Definition of modulus

$$\text{Modulus} \equiv \frac{\text{stress}}{\text{strain}} \qquad 13.2$$

In the present type of experiment we obtain the tensile modulus (also called *Young's modulus*). Using the symbols of figure 13.1 we have

Young's modulus

$$Y = \text{Young's modulus} \equiv \frac{F/A}{\Delta L/L_0}$$

Its units are those of force per unit area. Typical values are given in table 13.2.

Thus far we have been assuming implicitly that the material is elastic. In other words, we assume that the material retracts to its original length once the stress has been removed. Most solids conform to this assumption within certain limits. When the experimental stress-strain curve for a solid is plotted, the result usually appears like the graph shown in figure 13.1(*b*). If the applied stress is at any value in the linear portion of the curve, the material retracts completely when the stress is removed. However, if the stress is too large, the material "gives" (or yields) too much and the stress-strain curve departs from linearity as shown by the dashed portion. The point of departure is called the *yield point*. Materials strained beyond this point usually will not retract to their original length when the stress is removed. Our discussion will assume that the yield point is not exceeded.

Another type of modulus, the *shear modulus*, describes the rigidity of a material in the shearing-type experiment shown in figure 13.2. Notice that the force is tangential to the area A, unlike the situation of the tensile modulus. Using the symbols in the figure, the definition is as follows:

Shear modulus

$$\text{Shear modulus} \equiv \frac{\text{stress}}{\text{strain}} = \frac{F/A}{\Delta L/L_0}$$

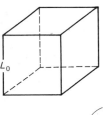

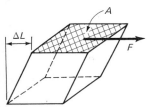

FIGURE 13.2 When ΔL is not nearly as big as indicated, the shear modulus is given as $\dfrac{F/A}{\Delta L/L_0}$.

A third type of modulus, the *bulk modulus,* measures the resistance a material presents to having its volume compressed. Suppose we have a cube of the material (volume V_0) with A being the area of each face of the cube. If a force F is now applied inward on each face of the cube, the stress is taken to be F/A, causing a volume change ΔV. We define the bulk modulus to be

Bulk modulus

$$\text{Bulk modulus} \equiv \frac{\text{stress}}{\text{strain}} = \frac{F/A}{|\Delta V|/V_0} \qquad 13.3$$

The quantity F/A is called the *pressure* when the material is a fluid. More will be said about it later. Notice that the experiment used to determine the bulk modulus can be carried out with fluids as well as solids. Hence the bulk modulus of a fluid is a useful quantity.

The reciprocal of a modulus measures the opposite of rigidity. For this reason, we call the inverse of the tensile and shear moduli the tensile and shear *compliance* of these materials. The inverse of the bulk modulus is called the *compressibility* (k) of the material. Typical compressibilities are given in table 13.2.

Compliance and compressibility

Although there are other types of moduli we could define and which are used, the three we have given are the most important. However, we have assumed that the material being measured is isotropic, i.e., the same in all directions. For many materials this is not true. Wood, for example, has a different Young's modulus along the grain than perpendicular to the grain. To describe such materials, use is made of modulus *tensors.* In effect, each modulus is treated as a vectorlike quantity having many components. This subject is vital to engineers and scientists who work in materials science and its applications. Indeed, the subject of tensors constitutes a rather large area of applied mathematics.

13.4 VISCOSITY

Viscosity measures resistance to flow

When dealing with fluids, we often need to specify how much resistance the liquid presents to flow. Liquids such as syrup and heavy oil flow much less easily than water. They are said to be very viscous liquids. In conformity with this terminology, we define a quantity called the viscosity of a liquid which is large for very-slow-flowing, syrupy liquids and which is small for very fluid, freely flowing substances.

To give quantitative meaning to viscosity, we refer to the hypothetical shear-type experiment shown in figure 13.3. There we see two parallel plates, each of area A, separated by a distance L. The region between the plates is

FIGURE 13.3 As the upper plate moves, layers of the fluid slide over each other.

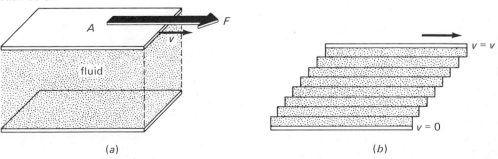

(a)

(b)

TABLE 13.3 Viscosities of Liquids and Gases at 30°C

Material	Viscosity (mPl)*
Air	0.019
Acetone	0.295
Methanol	0.510
Benzene	0.564
Water	0.801
Ethanol	1.00
Blood	≈ 1.6
SAE no. 10 oil	200
Glycerin	629
Glucose	6.6×10^{13}

*1 mPl = 10^{-3} Pl = 10^{-3} kg/m·s = 1 cP.

filled with a fluid whose viscosity we shall denote by η (Greek eta). In order to move the top plate with speed v relative to the bottom one, a force F is required. The force will be large if the fluid has a large viscosity. We define η by

$$\eta = \left(\frac{F}{A}\right)\left(\frac{L}{v}\right) \qquad 13.4$$

In the SI, the units of viscosity are, from equation 13.4, newton-seconds per square meter and this unit is given the name *poiseuille* (Pl). Other common units are the poise (P), where 1 P = 0.1 Pl, and the centipoise (cP). This latter unit is easily remembered because it is equal to a millipoiseuille, 1 cP = 1 mPl. The viscosities of several fluids are given in table 13.3.

Notice the tremendous range of viscosities in this table. From the almost glasslike glucose to the essentially frictionless air, the range is over a factor of 10^{15}. It should also be mentioned that the viscosity of liquids is very temperature-sensitive. For most liquids, η decreases with increasing temperature. This is a reflection of the fact that the molecules are less tightly bound together at the higher temperatures, and hence the friction between them is less. For water at 0, 50, and 100°C, the viscosity is 1.79, 0.55, and 0.28 mPl, respectively.

Viscosities are often measured by devices considerably simpler than that shown in figure 13.3. For example, in a capillary viscometer, the time taken for a fixed quantity of liquid to flow through a capillary tube is measured. Since a liquid having twice the viscosity of a standard fluid will require twice as long to flow through the tube, the viscosity of an unknown liquid can be found by comparing its flow time with the flow time of a known liquid. This is the common practice.†

We can gain further insight into the meaning of viscosity by examining figure 13.3(*b*). Notice that the fluid layers next to the two plates remain attached to the plates. We can think of the fluid between the plates as consisting of many thin layers, many more than shown. As the upper plate moves, these layers must slide over each other. In a high-viscosity fluid, the layers do not slide easily. A large amount of friction work is done as the layers are made to slide past each other. It is for this reason that *work done against viscous forces is equivalent to friction work.*

13.5 PRESSURE IN FLUIDS

In applications involving fluids, the forces within the fluid are often discussed most effectively in terms of the pressure within the fluid. To define what we mean by pressure, let us refer to an experiment. One possible type of pressure-measuring device is shown in figure 13.4. As shown there, application of a force to a piston causes the piston to compress a spring. The displacement of the piston can be used as a measure of the force on the piston.

If this device is now placed in a fluid, the force exerted by the fluid perpendicular to the face of the piston can be measured. This force F can then be divided by the area of the piston A to give what we shall call the

---

†If the two liquids being compared differ in density, a correction must be made to the flow times to take into account the larger gravitational force the denser liquid experiences. As a result, the force driving the liquid through the capillary may be different in the two cases.

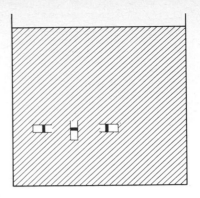

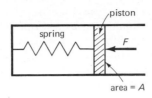

spring

piston

F

area = A

pressure – measuring device

FIGURE 13.4 The pressure-measuring device measures the same pressure, F/A, at all orientations near the same point in a liquid.

average pressure at the position of the measuring device. In symbols,

Definition of pressure

$$\text{Fluid pressure} = P \equiv \frac{F}{A} \qquad\qquad 13.5$$

By making the area of the piston sufficiently small, we can obtain the pressure very close to any point within the fluid. It is this quantity which we shall mean when we speak of the pressure at a point in a fluid. Notice that the force involved is perpendicular to the surface of the piston. The SI unit of pressure is the *pascal* (Pa). It is $1\,\text{N/m}^2$, a force per unit area.

As indicated in figure 13.4, the pressure at a point in a fluid at rest is found to be independent of the orientation of the piston. Hence the pressure as we have defined it is not direction-dependent. It is therefore convenient to define it to be a scalar quantity. Even though it has the magnitude of a force divided by an area, we do not give it the direction of the force, or any direction at all. We simply make use of it to tell us how large a force will be exerted on a surface of area A placed in the fluid. The direction of the force due to the stated pressure is always perpendicular to the surface, and its magnitude is given by *13.5*.

Let us investigate several important facts concerning the pressure within a fluid at rest. To understand two of them, refer to the container partly filled with liquid shown in figure 13.5. The liquid is assumed to be at rest and to remain that way. Let us now consider a little cubical volume element of the liquid similar to, but much smaller than, the volume V_2 shown. Since the liquid in V_2 does not move, we know that the sum of the x forces on it must be zero. The force toward the right on it is exerted by V_1 and is (from *13.5*) equal to the pressure at the left side of V_2 (call it P_1) multiplied by the side area of the cube (call it A). Thus

$$F \text{ to right on } V_2 = P_1 A$$

Similarly,

$$F \text{ to left on } V_2 = P_2 A$$

where P_2 is the pressure at the right side of V_2. Since the forces must balance, we see that $P_1 = P_2$.

Using this same line of reasoning, we can repeat this process over and over again to show that $P_1 = P_2 = P_3 = P_4$ and so on. This must mean that the pressure on a horizontal plane through V_2 must everywhere be the same. Notice that the pressure is independent of the shape of the container. Our only restriction was that the volume elements of the liquid be in equilibrium.

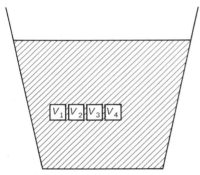

FIGURE 13.5 For a liquid at rest as shown, the pressure is the same at all points on any given horizontal plane.

Let us reiterate our conclusion: *The pressure is the same at every point on a horizontal plane through a liquid at equilibrium.* This result also leads us to conclude that a liquid at equilibrium in connected containers has its surfaces at the same level. Can you devise a line of reasoning to show that this is true?

Another fact of importance concerns the behavior of a liquid which completely fills a closed container. To see what is involved, suppose now that we have a liquid in a closed container like that shown in figure 13.6. This device is actually one modification of a hydraulic press. The two pistons shown have cross-sectional areas A_1 and A_2. If there are no forces on the pistons, and if the weight of the pistons is negligible, the liquid will stand at the same height in each of the tubes.

When an external force F_1 is applied to piston 1, the other piston would be pushed up unless a force F_2 were applied to it. If F_2 were made just large enough for no movement to occur, the liquid in the container would remain at rest. Clearly, then, the extra pressure at piston 1 must be balanced by equal pressures everywhere within the fluid, for if this were not true, the unbalanced pressure would cause the liquid to flow. This is an example of Pascal's principle. The principle may be stated as follows:

> *If a pressure is applied to a confined liquid, the pressure is transmitted to every point within the liquid.*

Of course, the liquid must remain at rest for this to be true.

It is instructive to compute the force F_2 needed to balance the force F_1. The pressure in the liquid resulting from F_1 is

$$P = \frac{F_1}{A_1}$$

The total force exerted by the liquid upon piston 2 is PA_2, or

$$F_2 = PA_2$$

$$= \frac{A_2}{A_1} F_1$$

If the area of the second piston is much larger than the area of the first, F_2 will be much greater than F_1. Hence, a device such as this is capable of lifting a large weight by the exertion of a small force. You should prove to yourself, though, that the small force F_1 must still do an amount of work equal to the work F_2 does as it lifts the second piston a distance s. If this were not the case, the device could be made into a perpetual-motion machine.

Before leaving this section we shall obtain a quantitative expression for the pressure within a liquid due to the liquid above it. Since we have already seen that the pressure will be the same independent of the container shape, let us choose the simple cylinder shown in figure 13.7 for our calculation. The pressure at the depth h below the surface of the liquid is responsible for the force which supports the liquid above it. Consider the liquid in the shaded area. Its volume is Ah where A is the cross-sectional area of the cylinder. But from 13.1 we have $m = \rho V$, and so the mass of this volume of liquid is $\rho(Ah)$ where ρ is the density of the liquid. To find the weight of this mass of liquid we recall $W = mg$ and write

$$\text{Weight of shaded volume} = \rho ghA$$

However, the liquid in the shaded region is supported by the upward force **F**. This force is due to the pressure of the lower liquid on the liquid above it and is simply PA. If we now equate this supporting force to the

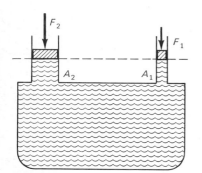

FIGURE 13.6 A small force on the small piston can balance a large force on the large piston.

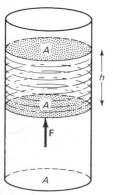

FIGURE 13.7 The pressure due to the liquid at a depth h in the liquid is ρgh.

weight of the liquid, we find

$$P = \rho g h \qquad\qquad 13.6$$

where A has been canceled from the equation. This is a very important equation since it allows us to find the pressure at any depth due to the liquid above.

Illustration 13.2

Water and oil are placed in the two arms of a glass U-tube as shown in figure 13.8. If they come to rest as shown, what is the density of the oil? Assume temperature = 20°C.

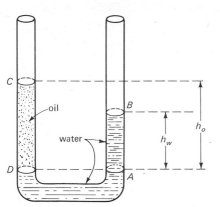

FIGURE 13.8 Since the water column *BA* balances the oil column *CD,* the density of the oil can be determined.

Reasoning Consider first the water below D and A in the tube. If the pressures at D and A were not equal, the water would flow. Since it does not, we conclude that P at point D equals P at point A due to the fluids. From *13.6* we have $P_D = \rho_{oil}gh_o$ and $P_A = \rho_w gh_w$. Equating, we find

$$\rho_{oil} = \rho_w \frac{h_w}{h_o}$$

Since $\rho_w = 1000 \text{ kg/m}^3$, we can evaluate ρ_o once h_w and h_o have been measured. As a point of interest, notice that the levels A and B are not the same. Hence, when two liquids are involved, a connected fluid does not always rise to the same level. Moreover, the pressure on a horizontal plane is not always the same. The reasoning used to prove the facts of Sec. 13.5 breaks down in the case of two or more liquids. Can you point out where the reasoning fails in this latter case?

13.6 ARCHIMEDES' PRINCIPLE; BUOYANCY

We are all familiar with the fact that objects can float in a liquid. The law of fluids which is of primary importance in such situations is called *Archimedes' principle*. It concerns the force which the surrounding fluid exerts upon an object wholly or partially immersed in it. The principle can be arrived at by referring to figure 13.9.

Consider an object A immersed in the fluid. The pressure within the fluid causes forces all over its surface as indicated. We wish to find the resultant of all these forces, a force which we shall refer to as the *buoyant force*. It can be found easily if we resort to the following reasoning process.

If everything is at rest and remains so, then the system is at equilibrium. The forces on the object due to the fluid will depend only upon the pressure within the fluid. Since this is simply $\rho g h$ at a point, ρ being the fluid density, the pressure does not depend upon the material of which the object is composed. Therefore the buoyant force will be the same on all objects of the same size and shape. Let us therefore consider the case where the object is made of the same material as the fluid. (If you wish, think of an extremely thin container filled with the same fluid.) Since the fluid remains at rest, the resultant force on the object (a portion of the fluid now) must be zero. Two forces act on it, its weight and the buoyant force. Hence the buoyant force must be equal to the weight of the fluid which occupies the volume of the object. Moreover, the buoyant force must be directed upward.

To summarize, we have found that the buoyant force exerted by a fluid on an object immersed in it can be found in the following way. If the object displaces a volume V of the fluid in which it is immersed, then the buoyant

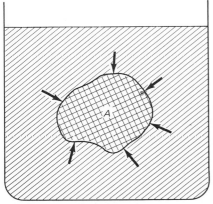

FIGURE 13.9 What does Archimedes' principle tell us about the buoyant force on the object?

force will be equal to the weight of this same volume V of displaced liquid. This is Archimedes' principle as usually stated:

Archimedes' principle

A body partially or wholly immersed in a fluid is buoyed up by a force equal to the weight of the fluid which it displaces.

You should go back through our reasoning to assure yourself that, indeed, we made no use of the fact that the object in figure 13.9 was *wholly* immersed.

One further point should be noted. Since the buoyant force supports our hypothetical "fluid object" without causing it to rotate, the buoyant force must act upward directly through the center of gravity of the object. The center of gravity of the "fluid object" is called the *center of buoyancy.* The actual object will float in equilibrium only if its own center of gravity is on the vertical line through the center of buoyancy. Can you explain why? Must the center of gravity be below the center of buoyancy for stable equilibrium?

Illustration 13.3

A queen's gold crown weighs 1.30 kg in air and 1.14 kg when completely immersed in water. Is the crown solid gold?

Reasoning Enough data are given so that we can compute the density of the material of the crown. The mass of the crown is 1.30 kg. Next we need its volume. To find it, we notice that the buoyant force on it was $(1.30 - 1.14)(9.8)$ N or $(0.16)(9.8)$ N. The weight of the displaced water $m_w g$ is therefore $(0.16)(9.8)$ N, and so $m_w = 0.16$ kg. But the density of water is 1000 kg/m^3, and so $m_w = \rho V_w$ gives the volume of displaced water to be 0.16×10^{-3} m^3. This is also the volume of the crown. Therefore

$$\rho_{\text{crown}} = 1.30 \text{ kg}/0.16 \times 10^{-3} \text{ m}^3 = 8100 \text{ kg/m}^3$$

Looking in a suitable handbook (the "Handbook of Chemistry and Physics" is excellent) we note that the density of gold is 19.3 g/cm^3 or 19,300 kg/m^3. The crown's density is less than half this, so it must be hollow or made of a gold alloy.

13.7 FLUID FLOW

Until this point we have been concerned mostly with fluids at rest. This area of study is often called *hydrostatics* or, more properly, fluid statics. Now we wish to discuss the motion of fluids; this area of study is called *fluid dynamics* (or *hydrodynamics*). We shall restrict our discussion to the simple situation of *steady, streamline* flow. This is the type of flow which exists in smoothly flowing fluids. To be more precise, it exists if the velocity of the fluid at each point in the fluid stream does not change with time. Our meaning can be seen more clearly by reference to figure 13.10.

In figure 13.10 we see a fluid undergoing steady, streamline flow in a tube of nonuniform cross section. For such flow to exist, any fluid particle which happens to be at A must always have the same velocity as any other particle at A no matter when it was at A. As it flows to B, its velocity may change. But at any point along its path it will have a velocity identical to that which all other fluid particles had as they flowed through that point. Moreover, every particle which flows through point A will traverse the same path in reaching point B. We call this path a flow *streamline*. Another flow

Definition of streamline

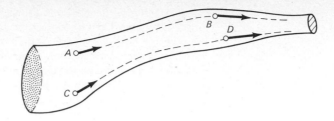

FIGURE 13.10 For steady streamline flow the particles follow definite paths (the streamlines) as they move; any particle at *A* will reach *B* by following the streamline shown.

streamline is shown from *C* to *D*. It is the path a particle traverses as it moves from *C* to *D*.

Flow is not always this simple. Suppose a bubble of air were to flow down the pipe with a liquid. The region about the bubble would not follow the streamline as the bubble compresses and expands in the nonuniform pipe. Or, perhaps an obstruction in the pipe causes the fluid to undergo large accelerations as it zips around its sharp edges. The forces needed to cause such large accelerations sometimes literally tear the liquid apart, causing cavities within it, a process called *cavitation*. Such a process is erratic and results in nonstreamline flow. Frequently, high-speed fluid flow spontaneously initiates turbulences which prevent streamline flow.

Returning now to the case of streamline flow, it is often convenient to notice that the streamline can be used to define tubes of flow. For example, in figure 13.10 we could envisage a set of streamlines which enter only a portion of the area of the left end of the pipe. This is done in figure 13.11. Those streamlines entering A_1 leave at A_2 and define a tube of flowing fluid. All the fluid which enters its left end flows out through its right end. Tubes of flow may therefore be considered to act as conduits or pipes through which the fluid is transported. As we shall see in the next sections, important deductions can be made concerning steady flow through pipes. These results will also apply to tubes of flow.

13.8 EQUATIONS OF HYDRODYNAMICS

To undertake a mathematical description of hydrodynamics, two fundamental equations of flow are required. One of these is the analog to $F = ma$ and expresses how a difference in pressure in a fluid causes the speed of a fluid particle to change. In its most general form, it is a very complex differential equation. Even in the case of stationary (or steady) streamline flow it is not simple. Fortunately we need not use it to obtain the important hydrodynamic results we shall discuss in this text.

The second hydrodynamic equation expresses the conservation of matter. It is called the *equation of continuity*. We can state it quite simply in the case of streamline flow shown in figure 13.11. Suppose the fluid involved has been flowing steadily for some time so that, if the fluid is a compressible

FIGURE 13.11 The bundle of streamlines indicated defines a tube of flow.

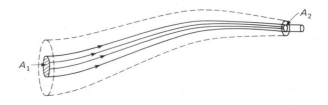

gas, for example, the mass of fluid within the section of the pipe shown is no longer changing. Then we know that the mass of fluid entering the left end of the pipe (or flow tube) equals the mass of fluid leaving the right end.

Let us consider the flow tube so that our result will be more general. (The flow tube could be the pipe if we so choose.) In an instant Δt, the liquid near the left end will move a distance $v_1 \Delta t$ where v_1 is the flow speed at the left end. The volume of fluid flowing through A_1 into the portion of the tube shown will be equal to the volume of a cylinder with base A_1 and tiny height $v_1 \Delta t$. Therefore the volume flowing into the tube is $(v_1 \Delta t)A_1$. According to the definition of density (13.1), the mass of fluid in this volume will be $\rho_1 v_1 A_1 \Delta t$ where ρ_1 is the fluid density at A_1. Similarly, the mass of fluid flowing out of the tube at A_2 is $\rho_2 v_2 A_2 \Delta t$.

According to our assumption about the situation, the mass of fluid in the tube is no longer changing, and so the inflow must equal the outflow. Therefore we can write

$$\rho_1 v_1 A_1 \Delta t = \rho_2 v_2 A_2 \Delta t$$

Equation of continuity   or $$\rho_1 v_1 A_1 = \rho_2 v_2 A_2 \qquad\qquad 13.7$$

This is a simple form of the *equation of continuity*. In a more general situation it is written as a differential or integral equation. The present form, *13.7*, is restricted to steady, streamline flow through a flow tube within which there are no sources (or sinks) for the fluid. Notice that if the fluid is incompressible so that $\rho_1 = \rho_2$, then *13.7* says $v_2 = v_1(A_1/A_2)$. As we might guess, the fluid must flow fastest where the tube is narrowest. In the next section we shall use *13.7* to obtain a useful equation for dealing with fluid-flow problems.

13.9 BERNOULLI'S EQUATION

Let us consider the steady, streamline motion of a fluid through a flow tube or pipe. We shall assume that the viscous-energy losses within the fluid are negligibly small. (You will recall that viscosity measures the difficulty successive layers of fluid have in flowing over each other.) If the flow is not too fast and if the viscosity is low, this assumption will be reasonable. Given such a flowing fluid, what can we say about the energy resident in the fluid? This is the question we shall now investigate.

Qualitatively, the situation is quite simple. The energy of the fluid presently in the section of flow tube from C to D in figure 13.12 will change as this portion of the fluid moves down the tube. Consider what will happen to it in time Δt. The leading portion of the fluid will need to push the fluid ahead of it out of the way. In so doing, it will do work. Compensating somewhat for this, the fluid to the left of point C will push the fluid originally between C and D along, thereby doing work on it. Thus the net energy gain of the fluid as a result of work done at the two ends will be the work done at C minus the work done at D. What did this work accomplish?

In the absence of friction losses, which we have agreed to ignore, the work resulted in two changes in energy. It changed the potential energy of a portion of the fluid since, in effect, the fluid originally at C has been lifted to the head of the flowing segment. (Notice that the new position taken up by the fluid originally between C and D is shown as $C'D'$ in figure 13.12. The fluid between points C' and D is fully equivalent from an energy standpoint to the

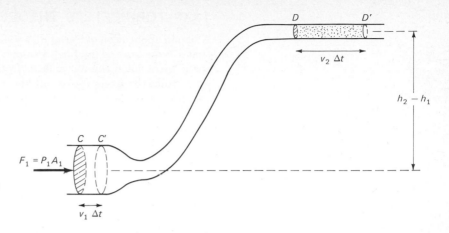

FIGURE 13.12 In effect, the fluid between C and C' was transferred to the region between D and D' during the time Δt.

fluid originally there. As a result, the effect of the displacement along the tube has been equivalent to transferring the fluid originally between C and C' to the region between D and D'.) In addition, the fluid at D is moving more swiftly than that at C. Therefore the effective transfer of fluid from C to D has increased both the kinetic and potential energy of the fluid transferred.

To make these ideas quantitative, consider the motion which occurred in Δt as the fluid moved from C to C'. Notice that the work done *on* the fluid near C was (force) $\times$ (distance) or $(P_1 A_1)(v_1 \Delta t)$. The work done *by* the fluid at D was $P_2 A_2 (v_2 \Delta t)$. Hence the energy gain of the fluid due to work done at the ends was

$$(P_1 A_1 v_1 - P_2 A_2 v_2)\, \Delta t$$

If the mass of fluid originally in the region CC' was M, then the gain in potential energy of the fluid in time Δt was $Mg(h_2 - h_1)$. The gain in kinetic energy of the fluid transferred was $\frac{1}{2}M(v_2{}^2 - v_1{}^2)$.

The work-energy theorem now allows us to equate the work done on the fluid to its energy gain. We have

$$(P_1 A_1 v_1 - P_2 A_2 v_2)\, \Delta t = Mg(h_2 - h_1) + \tfrac{1}{2}M(v_2{}^2 - v_1{}^2)$$

But $M, A,$ and v are related by the equation of continuity and the fact shown in the previous section that $M = \rho_1 v_1 A_1 \Delta t$. Substitution in the above equation gives, after some rearrangement,

$$(P_1 + \rho_1 g h_1 + \tfrac{1}{2}\rho_1 v_1{}^2)v_1 A_1 = (P_2 + \rho_2 g h_2 + \tfrac{1}{2}\rho_2 v_2{}^2)v_2 A_2$$

In the special case where $\rho_1 = \rho_2 = \rho$, the case of an incompressible fluid, the equation of continuity tells us further that $v_1 A_1 = v_2 A_2$. For that case, the above equation reduces to what is called *Bernoulli's equation,*[*] which is

Bernoulli's equation

$$P_1 + \rho g h_1 + \tfrac{1}{2}\rho v_1{}^2 = P_2 + \rho g h_2 + \tfrac{1}{2}\rho v_2{}^2 \qquad 13.8$$

We can state this another way if we note that points 1 and 2 were general points. Equation *13.8* then implies that the quantity $P + \rho g h + \tfrac{1}{2}\rho v^2$ is a constant for all points within the fluid. Let us now apply Bernoulli's equation to several practical problems.

[*] After Daniel Bernoulli, who published it in 1738.

13.10 TORRICELLI'S THEOREM

A simple application of Bernoulli's equation is shown in figure 13.13. Suppose that a large tank of fluid has a small spigot on it, as shown. Let us find the speed with which the water flows from the spigot at the right.

Since the spigot is so small, the efflux speed v_2 will be much larger than the speed v_1 of the top surface of the water. We can therefore approximate v_1 as zero. Bernoulli's equation can then be written as, since $P_1 \cong P_2 \cong$ atmospheric pressure P_0,

$$P_0 + \rho g h_1 = P_0 + \tfrac{1}{2}v_2^2\rho + \rho g h_2$$

Rearrangement of this equation gives

$$v_2 = \sqrt{2g(h_1 - h_2)}$$

Torricelli's theorem

This is *Torricelli's theorem*. Notice that the speed of the efflux liquid is the same as the speed of a ball which falls through a height $h_1 - h_2$. This points out the fact that when a little liquid flows from the spigot, it is as though the same amount of liquid had been taken from the top of the tank and dropped to the spigot level. The top level of the tank has decreased somewhat, and the potential energy lost has gone into kinetic energy of the efflux liquid. If the spigot had been pointed upward, as at the left in figure 13.13, this kinetic energy would allow the liquid to rise to the level shown before stopping. In practice, viscous-energy losses would alter the result somewhat.

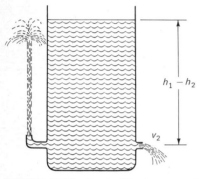

FIGURE 13.13 Torricelli's theorem tells us how fast the liquid is moving as it flows out of the spigot.

13.11 PRESSURE–VELOCITY VARIATION

The Bernoulli equation can be used to show that the pressure within a fluid flowing through a pipe is reduced at a constriction in the pipe. This perhaps surprising result can be obtained by referring to figure 13.14. Let us apply *13.8* to the two positions of the pipe shown. Since $h_1 = h_2$ for the average streamline, we have

$$P_1 + \tfrac{1}{2}\rho v_1^2 = P_2 + \tfrac{1}{2}\rho v_2^2$$

But the equation of continuity tells us that

$$v_2 = v_1 \frac{A_1}{A_2}$$

and so we find

$$P_2 = P_1 - \tfrac{1}{2}\rho v_1^2 \left[\left(\frac{A_1}{A_2} \right)^2 - 1 \right]$$

Pressure is least where flow is fastest

Notice that P_2, the pressure in the constriction, is less than P_1. This effect is used in such diverse ways as measuring the speed of a fluid from a knowledge of the two pressures (the Venturi meter) and in producing a partial vacuum by use of a water aspirator. If you are observant you may have

FIGURE 13.14 Where is the velocity of the fluid largest? Where is the pressure smallest?

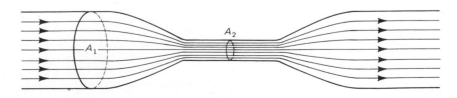

noticed that this expression for P_2 can become negative. What meaning does this have in the case of an incompressible fluid? Why is the result inaccurate if the fluid is a gas?

Although an inexperienced person might easily guess that the pressure would be largest at the constriction in the pipe shown in figure 13.14, simple qualitative reasoning justifies the opposite result we have found. If we consider the motion of a fluid particle along the center streamline, we know that the particle must speed up as it approaches the constriction. Hence an unbalanced forward force must exist upon it. Since the pressure is the only horizontal force present, the pressure ahead of it must be smaller than the pressure behind. This leads us to conclude that the pressure keeps on decreasing as we approach the constriction. Similar reasoning can be used to show that the pressure increases once again as we move beyond the constriction. It is clear, therefore, that the pressure is smallest at the constriction. The equation given above for P_2 expresses this fact in quantitative form.

Chapter 13
Questions and Guesstimates

1. There are almost exactly as many atoms in a cubic centimeter of aluminum as there are in a cubic centimeter of gold, yet the density of gold is 19,300 kg/m³ while that of aluminum is 2700 kg/m³. What can you conclude from this large difference in density? Justify your conclusion by independent data.
2. Estimate the average density of the human body. How could you measure your density to within 1 percent using simple equipment at a swimming pool?
3. Some people are able to float easier than others. Explain what factors are involved. Does it matter whether or not the person involved takes in a deep breath?
4. How can one determine the density of an irregular solid object which sinks in water? Which floats in water? The density of a liquid? Of a gas?
5. How could you measure the shear modulus of a gelatin dessert? Why would the modulus depend upon whether or not the gelatin has pieces of fruit in it? Reinforced plastics such as fiber glass have glass fiber embedded in the plastic. What effect does the glass fiber have on the mechanical properties of the plastic? (Bone is reinforced by collagen fibers in this same way. As a result, the tensile modulus and strength of bone is higher than it otherwise would be.)
6. Multigrade motor oil contains loosely coiled, long, stringlike polymer molecules dissolved in the oil. As the oil heats up, the polymer chains become more loosely coiled so that each chain spreads out over a larger volume. Although the viscosity of the pure, hot oil is less than that for cool oil, the oil-polymer solution has a viscosity that is nearly temperature-independent. Explain why the solution differs from the pure oil.
7. In a falling-ball viscometer, one times how long it takes for a small sphere to drop between two marks through a large container of the liquid in question. How could you calibrate and use such a viscometer?
8. A block of wood floats in a jar of water with half its volume submerged. The jar is sitting on the earth. How would the wood float if the jar were sitting on the moon where things weigh only one-sixth as much as on earth? In an elevator moving up at constant speed? In an elevator accelerating upward?

9. The density of ice is 0.917 g/cm³, and so 1 cm³ of ice melts to form only 0.917 cm³ of water. A piece of ice floats in water in a graduated cylinder. When the ice melts, what will happen to the water level in the cylinder? Does this have any implication for what will happen to the sea level if the polar ice caps were to melt?

10. A cylindrical measuring graduate is partly filled with water. In it is floated a narrow test tube that has a little mercury in its bottom. The mercury slowly leaks out of the tube through a small hole in its bottom. Does the water level in the graduate rise, fall, or remain the same?

11. From your own experience in floating in a lake or pool, estimate the density of your body. Why is it easier to float in seawater than in a pool?

12. Does the water pressure at the base of a dam depend on the size of the lake behind the dam?

13. A large tank of water is placed on a scale. Can a 60-kg woman already floating in the water cause the weight registered by the scale to change? If so, estimate the magnitude of the effect.

14. A glass filled to the brim with water sits on a scale. A block of wood is gently placed in the water so it floats in the glass. Some of the water overflows and is wiped away but, at the end, the glass is still filled to the brim. Compare the initial and final readings of the scale.

15. To purge a patient's rectum, nurses sometimes administer an enema. The process consists of inserting a tube into the rectum and using gravitational force to cause liquid to flow into the rectum from a container held at a higher level. Nurses are sometimes warned not to lift the can too high. Is there any basis for the warning or is this just a superstition?

16. Using the facts that the density of air is 1.29 kg/m³ and that atmospheric pressure at the surface of the ocean is 1×10^5 Pa, estimate an average height of the atmosphere above the earth. You will see in the next chapter how such an estimate differs from reality.

17. A siphon will not work unless the siphon tube is full and the exit end is lower than the entrance end. Explain these facts by describing the physics of the operation of the siphon.

18. Discuss the physical meaning of Bernoulli's equation when the fluid is not moving.

19. When you hold the end of a long strip of paper in your fingers, the strip will droop. But if you blow horizontally over your fingers and above the strip, the strip will rise. Explain what lifts it against the force of gravity.

Chapter 13
Problems

1. From the density of benzene, how many molecules of benzene exist in 1 cm³? The molecular weight of benzene is 78; each molecule has a mass $78 \times 1.66 \times 10^{-27}$ kg, where the latter number is the conversion from atomic mass units to kilograms.

2. Most of the mass of a hydrogen atom is in the nucleus of the atom, which is called a proton. By shooting particles at the proton, we infer it to be crudely representable as a sphere of radius 1.2×10^{-15} m. What is its approximate specific gravity? (m of proton $= 1.67 \times 10^{-27}$ kg.)

3. According to the widely accepted "big bang" theory of the universe, about 15×10^9 years ago the universe was compressed into an intensely hot fireball with diameter about 3×10^{10} m. It is thought that the mass of the universe is in the neighborhood of 1×10^{52} kg. Using these figures, what was the density within the fireball?

4. Given a 2.0-m length of steel wire with 1.0 mm diameter. About how much will the wire stretch under a 5.0-kg load?

5. About how large a force is required to stretch a 2.0-cm-diameter steel rod by 0.01 percent?

6. How large a pressure (in Pa) must be applied to water if it is to be compressed by 0.1 percent? What is the ratio of this pressure to atmospheric pressure, $1.01 \times 10^5 \, \text{N/m}^2$?

7. By what fraction will the volume of a steel bar increase as the air is evacuated from a chamber in which it rests? Standard atmospheric pressure $= 1.01 \times 10^5 \, \text{Pa}$.

8. What is the pressure due to the water 1 mi beneath the ocean's surface assuming the density of seawater to be $1025 \, \text{kg/m}^3$? If its compressibility is the same as pure water, by what percent has the density changed in going from the surface to this depth ($1 \, \text{mi} = 1609 \, \text{m}$)?

9. If the blood vessels in a human acted as simple pipes (which they do not), what would be the difference in blood pressure between the blood in a 1.80-m-tall man's feet and in his head when he is standing?

10. When a doctor quotes your blood pressure as 120/80, the two values pertain to the maximum (systolic) and minimum (diastolic) pressure in the major blood vessels during the heart beat cycle. The numbers represent the difference in pressure between the blood and the atmospheric pressure in the room. It is measured in millimeters of mercury, i.e., the pressure needed to balance a column of mercury of the stated height. Find the ratio of the systolic blood pressure given above to atmospheric pressure. Standard atmospheric pressure is $1.01 \times 10^5 \, \text{Pa}$.

11. A uniform glass tube is bent into a U shape such as shown in figure 13.8. Water is poured into the tube until it stands 10 cm high in each tube. Benzene is then added slowly to the tube on the left until the water rises 4 cm higher on the right. What length is the column of benzene when that situation is reached? (Water and benzene do not mix.)

12. Atmospheric pressure is $1.01 \times 10^5 \, \text{Pa}$. How high a column of water could be supported by this pressure? Of mercury?

13. The dam shown in figure P13.1 holds water at a depth H behind its rectangular face. Its dimension perpendicular to the page is L. (a) What is the total force on the dam due to the water? (b) How large is the torque about an axis perpendicular to the page at B caused by the water pressing against the dam?

14. Assuming the principle of flow to be the same in the two cases, find the time taken for blood to flow through a vertical capillary tube if an equal volume of water requires 90.0 s. The specific gravity of blood can be taken to be 1.06.

15. *Stokes' law* is often derived in advanced hydrodynamics texts. It states that the net force required to pull a sphere of radius b through a fluid with viscosity η at a speed v is given by $6\pi\eta bv$. How fast will an aluminum

FIGURE P13.1

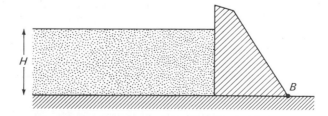

sphere of radius b fall through water once its terminal speed has been achieved?

16. An irregular piece of metal weighs 10.00 g in air and 8.00 g when submerged in water. (*a*) Find the volume of the metal and its density. (*b*) If the same piece of metal weighs 8.50 g when immersed in a particular oil. Find the density of the oil.

17. A certain pycnometer (a small flask used for density measurements) weighs 20.00 g when empty, 22.00 g when filled with water, and 21.76 g when filled with benzene. (*a*) Find the density of benzene. (*b*) For very accurate measurements of density, the weight of air in the empty flask must be taken into account. What mass of air fills the pycnometer?

18. A piece of wood weighs 10.0 g in air. When a heavy piece of metal is suspended below it, the metal being submerged in water, the weight of wood in air plus metal in water is 14.00 g. The weight when both wood and metal are submerged in water is 2.00 g. Find the volume and the density of the wood.

19. A beaker partly filled with water weighs 20.00 g. If a piece of wood having a density of 0.800 g/cm³ and volume 2.0 cm³ is floated on the water in the beaker, how much will the beaker weigh (in grams)?

20. A beaker, partly filled with water, weighs 20.00 g. If a piece of metal with density 3.00 g/cm³ and volume 1.00 cm³ is suspended by a thin string so that it is submerged in the water but does not rest on the bottom of the beaker, how much does the beaker weigh?

21. The density of ice is 917 kg/m³, and the approximate density of the seawater in which it floats is 1025 kg/m³. What fraction of an iceberg is beneath the water surface?

22. What is the minimum volume of a block of wood (density = 850 kg/m³) if it is to hold a 50-kg woman entirely above the water when she stands on it?

23. A block of material has a density ρ_1 and floats three-fourths submerged in a liquid of unknown density. Show that the density ρ_2 of the unknown liquid is given by $\rho_2 = 1.33\,\rho_1$.

24. A tiny glass sphere has a radius 0.50 mm and a density 2600 kg/m³. It is let fall through a vat of oil ($\rho = 950$ kg/m³, $\eta = 0.21$ Pl). Find (*a*) the buoyant force on the sphere, (*b*) the gravitational force on it, and then (*c*) use Stokes' law (see problem 15) to find the terminal speed of the sphere as it falls through the oil.

25. To determine the size of tiny particles, their sedimentation speed is often measured. Show that a particle of density ρ which reaches a terminal speed v while falling through a fluid of density ρ_f and viscosity η has a radius b given by

$$b = \sqrt{\frac{9\eta v}{2g(\rho - \rho_f)}}$$

Make use of Stokes' law (see problem 15).

26. A rectangular dish of length L contains water to a depth H when the dish sits at rest on a table. The dish is now given a horizontal acceleration a along the lengthwise dimension of the dish. Show that when a steady state is reached the pressure at the bottom of the dish now increases by an amount $0.01\rho_w a$ for each cm along the bottom of the dish toward its rear. How much higher does the water stand at the rear of the dish than at its front?

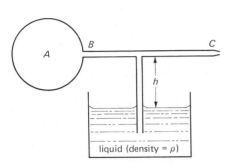

FIGURE P13.2

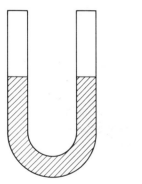

FIGURE P13.3

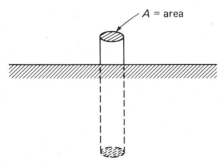

FIGURE P13.4

27. A glass of water is partly filled with water. It is now spun with angular speed ω about its vertical axis through its center. Measuring radial distances r from this central axis, show that the height y of the water in the glass varies in the following way: $dy/dr = \omega^2 r/g$.

28. In a certain centrifuge the liquid is rotated at 20 rev/s at a radius of 10 cm from the axis of rotation. Tiny spherical particles of radius b and density $1020 \, \text{kg/m}^3$ in a dilute water solution (density $1000 \, \text{kg/m}^3$) are placed in the centrifuge. Find the terminal speed with which they settle out of the solution. (*Hint:* See problem 15 for Stokes' law. Ignore the effect of gravity.)

29. A horizontal pipe system connects a pipe of cross-sectional area A_1 to a pipe of area A_2, and the other end of this latter pipe opens to the air. If atmospheric pressure is P_0 and if viscous effects can be ignored, how large a pressure is required in the first pipe to cause the water to flow at a speed v_2 out of the open end? What will be the speed of the water in the first pipe? How much water will flow out of the pipe in a time Δt? Express your answers in terms of P_0, v_2, A_1, A_2, and Δt.

30. The pipe near the lower end of a large water-storage tank springs a small leak, and a stream of water shoots from it. If the top of the water in the tank is 20 m above the point of the leak, with what speed does the water gush from the hole? If the hole has an area of $1 \times 10^{-6} \, \text{m}^2$, how much water flows out in 1 s?

31. A valve in a storage tank is 3 m below the top of the water in the tank. (*a*) Neglecting friction, with what speed will the water flow out of the open valve? (*b*) If the opening at the valve is directed straight upward, to what height will the spurting water rise?

32. Water is flowing smoothly through a closed-pipe system. At one point the speed of the water is 3.0 m/s, while at another point 1.0 m higher the speed is 4.0 m/s. If the pressure is $2.0 \times 10^4 \, \text{Pa}$ at the lower point, what is the pressure at the upper point? What would the pressure at the upper point be if the water were to stop flowing and the pressure at the lower point were $1.8 \times 10^4 \, \text{Pa}$?

33. In figure P13.2 we see a crude type of perfume atomizer. When the bulb at A is compressed, air flows swiftly through the tiny tube BC, thereby causing a reduced pressure at the position of the vertical tube. Liquid then rises in the tube, enters BC, and is sprayed out. Calling the pressure in the bulb $P_a + P$ where P_a is atmospheric pressure, and calling v the speed of the air in BC, find the approximate pressure in BC. How large would v need to be to cause the liquid to rise to BC?

34. The U-tube device shown in figure P13.3 contains M kg of a zero viscosity fluid of density ρ. The diameter of the tube is b. When the liquid is subjected to a disturbance so as to raise the level in one tube and lower it in the other, the liquid oscillates back and forth when allowed to do so freely. Show that the height of the liquid oscillates sinusoidally and find the frequency of the oscillation.

35. The cylindrical block of wood shown in figure P13.4 floats in a zero viscosity fluid of density ρ_f. If the block is pushed down slightly and released, it will vibrate up and down. Show that the motion is sinusoidal and find the frequency of the motion. The wood has a density ρ_w. Its lower end is loaded with metal so the block floats upright as shown. The total mass of the loaded block is M.

14 Ideal Gases

We begin in this chapter the transition from that branch of physics called *mechanics* to the branch called *heat*. As we shall see, the subject of heat involves mechanics on a molecular scale. Since we cannot easily describe the mechanics of the molecules individually, we examine their collective effects. In the case of a gas, these collective effects, namely, volume, pressure, and temperature, are rather easily related to molecular properties, and they are examined in this chapter.

14.1 GAS PRESSURE: GENERAL CONSIDERATIONS

Many of the properties of gases can be well represented by a theory, the so-called **kinetic theory**, which treats the gas molecules as tiny balls. We need not specify their exact size, shape, or internal construction. In our first consideration of the behavior of gases, we shall therefore merely state that they are small pieces of matter having mass m_0. Since experience shows us that gases quickly fill vacuums and permeate through space when allowed to do so, we surmise that the gas molecules are in motion. Moreover, since the usual gases tend to fill a room rather than settle to the floor, we conclude that the kinetic energy of the molecules is much larger than the potential energy required to lift them a few meters.

Let us then consider a gas composed of swiftly moving molecules having nearly negligible weight. Furthermore, consider only a gas which is so dilute that collisions between the particles is a rare event. This will not be the case for air under normal conditions; in fact it is not necessary to ignore collisions, but we shall return to the more practical cases later. Our gas will be assumed confined to a cubical container of side L and volume $V = L^3$. In this container we assume νL^3 gas molecules to be placed. ν, Greek nu, is the number of molecules per unit volume.

When one of these molecules strikes the wall of the container, it will exert a force on the wall. The cumulative effect of many such molecules striking against the wall leads to a more or less steady force on the wall. Just how steady this force is depends upon the number of molecules hitting the wall per unit time. Probability theory tells us that if the average number hitting a wall per second is N, then the deviation one might expect from this

Gas pressure is the cumulative effect of the collisions of the gas molecules with the container wall

value will be of the order of $\sqrt{N}$. The fractional deviation is therefore

$$\frac{\sqrt{N}}{N} = \frac{1}{\sqrt{N}}$$

As a result, if N is very large, we should expect the deviations from the average force to be negligible.

We can go further in our consideration of a molecule hitting the wall. If the gas is at equilibrium in the container (i.e., its pressure, volume, and temperature are not changing), its total kinetic energy should neither increase nor decrease. For this to be true, a molecule must, on the average, rebound from the wall with a speed equal to its speed before the collision. Using this fact as a rationale, we shall assume the molecule to rebound with the same speed as it had before it hit the wall. This is not a necessary assumption, but it will be made, since its use simplifies the computation considerably. Let us now compute the pressure due to a dilute gas in a container.

14.2 PRESSURE OF A DILUTE GAS

First we shall focus our attention on a single molecule of the gas. As shown in figure 14.1, it has a velocity component v_{ix} in the x direction. Assuming that collision with the wall merely reverses the velocity component perpendicular to the wall, the particle will follow the path indicated. The time Δt_i between two successive collisions of the same particle with the right-hand wall is, from $x = v_x t$,

$$\Delta t_i = \frac{2L}{v_{ix}} \qquad 14.1$$

You will recall from Chap. 7 that the impulse exerted on a particle such as this is

$$\text{Impulse} = \Delta(mv)$$

during a collision. In this case, the x momentum before collision is $m_0 v_{ix}$, and after collision it is $m_0(-v_{ix})$. Therefore

$$\text{Impulse}_x = m_0(-v_{ix}) - m_0 v_{ix}$$
$$= -2m_0 v_{ix}$$

The minus sign indicates that the direction of the impulse exerted on the molecule by the wall is toward the left. An equal and opposite impulse is exerted by the molecule on the wall.

We can, if we wish, define an average force F_{ix} exerted by the molecule on the wall during the time Δt_i. (Usually, it is convenient to take Δt equal to the time of contact with the wall. However, in the present instance, we choose the actual time between collisions, since this will give us an average force applicable to the molecule for the whole time interval from one collision to the next.) We then have, from the definition of impulse

$$F_{ix}\,\Delta t_i = 2m_0 v_{ix}$$

Or after solving for F_{ix} and putting in the value found in 14.1 for Δt_i, we have

$$F_{ix} = \frac{m_0 v_{ix}^2}{L} \qquad 14.2$$

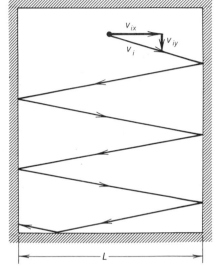

FIGURE 14.1 We assume the molecule to follow the path shown.

The impulse exerted on the wall is $2m_0 v_{ix}$

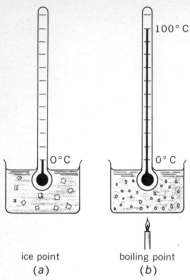

ice point
(a)

boiling point
(b)

FIGURE 14.2 Calibration of a mercury centigrade thermometer.

Pressure is the product of two-thirds the number of molecules in unit volume and the kinetic energy of each

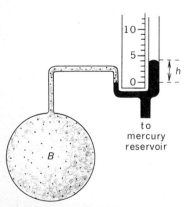

FIGURE 14.3 The constant-volume gas thermometer measures the pressure of a gas confined to the volume *B*. What experimental precautions must be taken in the design of such a device?

The total force on the wall due to the νL^3 molecules in the box will be the sum of the average force due to each. Therefore the x component of the force on the wall being considered due to the νL^3 molecules is

$$F_x = \frac{m_0}{L}[v_{1x}^2 + v_{2x}^2 + \cdots + v_{(\nu L^3)x}^2]$$

But the term in brackets is just νL^3 times the average of the square of the x-component velocities of the molecules. Calling this $\langle v_x{}^2 \rangle$, we have

$$F_x = \frac{m_0}{L} \nu L^3 \langle v_x{}^2 \rangle$$

We define the pressure of the gas on the wall to be the perpendicular force on the unit area. Since F_x is perpendicular to the area, which is L^2, we find the pressure on the wall to be

$$P = \frac{F_x}{L^2} = m_0 \nu \langle v_x{}^2 \rangle$$

This may be put in more convenient form by noting that, since the number of molecules is very large, and since there is no preferred direction in the box,

$$\langle v_x{}^2 \rangle = \langle v_y{}^2 \rangle = \langle v_z{}^2 \rangle = \frac{\langle v^2 \rangle}{3}$$

where use has been made of the fact that

$$v^2 = v_x{}^2 + v_y{}^2 + v_x{}^2$$

and where $\langle v^2 \rangle$ is the average of the squared velocities of the molecules. Its square root is called the rms (or root mean square) velocity. The pressure then becomes

$$P = \tfrac{1}{3} m_0 \nu \langle v^2 \rangle$$

which may be rewritten

$$P = \tfrac{2}{3}\nu(\tfrac{1}{2}m_0\langle v^2 \rangle) = \tfrac{2}{3}\nu\langle \tfrac{1}{2}m_0 v^2 \rangle \qquad 14.3$$

In other words, the pressure is two-thirds the number of molecules per unit volume times the average translational kinetic energy per molecule. Gas pressure therefore depends on two things: number of molecules per unit volume and average translational kinetic energy per molecule.

14.3 DEFINITION OF TEMPERATURE

Long before the physical basis for heat and temperature was known, practical methods for measuring these quantities had been found. As our understanding of these concepts increased, it became necessary to relate these empirical measurement methods to more fundamental measurements. For example, various types of thermometers were invented to enable comparison of temperatures. A common one, the centigrade (or Celsius), used to be based upon the freezing and boiling points of water under certain standardized conditions, i.e., standard atmospheric pressure. This was done in the following way.[*]

A uniform capillary tube is attached to a bulb filled with mercury. When the bulb is immersed in ice water, as shown in figure 14.2(*a*), the

[*]Later we shall describe a more acceptable method.

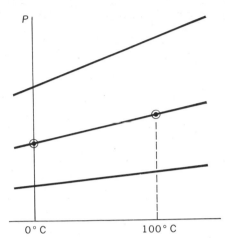

FIGURE 14.4 The pressure in a constant-volume gas thermometer can be used to set up a temperature scale.

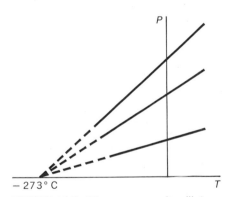

FIGURE 14.5 The pressure of a dilute gas decreases as the temperature is lowered. Extrapolating the data in the linear range, one finds an apparently unique temperature at which the pressure is zero.

When P for an ideal gas is extrapolated, P becomes zero at T = −273.15°C

mercury is at a certain height in the tube. This level is designated 0°C. The bulb is then transferred to a bath of boiling water. Since mercury expands with increasing temperature, the liquid rises to a higher level in the capillary. This level is designated 100°C. Subsequently, the tube is subdivided so as to give 100° between these two values. In addition, one can extrapolate the scale to temperatures below 0°C and above 100°C. Such a scale (with slight calibration modification) is used widely and is commonly called the centigrade (or Celsius) scale.

Difficulty arises in the calibration procedure outlined above, for two reasons. First, mercury freezes solid at −39°C. Second, when liquids other than mercury are used in this procedure, the resultant thermometer agrees with the mercury thermometer at 0 and 100° but not at other temperatures. This disagreement is the result of differences in the way the two liquids expand. As a result, the temperature scale defined in this way is not as general as one would like. We therefore search for a more universal calibration procedure. This may be done by the use of a dilute gas confined to a fixed volume, as we shall now see.

A so-called constant-volume gas thermometer is shown diagrammatically in figure 14.3. The pressure of the gas in the bulb B is measured by means of the manometer shown.* By always adjusting the mercury column on the left to read zero, the volume in the bulb is maintained constant. One can then measure the pressure of the gas by use of the manometer, and this pressure can then be used as a measure of the temperature in the following way.

Suppose one measures the pressure of the gas in the gas thermometer of figure 14.3 when the bulb is in ice water and then in boiling water. Calling the ice temperature 0°C and the boiling-water temperature 100°C, one can plot the two-point graph shown in figure 14.4. The line drawn through these points can be used to define other temperature values. We merely subdivide the x axis into equally spaced degrees, running from 0 to 100, and also to larger and smaller values. When the pressure of the thermometer is known, the corresponding temperature can be read directly from the graph. The other lines on the graph correspond to different quantities of gas in the volume V.

This type of thermometer agrees quite well, but not perfectly, with the mercury thermometer in the range 0 to 100°C. It is preferable to the mercury thermometer in that, when helium is used as the gas, the thermometer is operable to temperatures of approximately −270°C. In addition, provided the gas in the thermometer is kept very dilute and is not near its condensation temperature, all gas thermometers agree very well with each other. Moreover, changing the starting pressure in the bulb slightly merely changes the slope of the line in figure 14.4, as indicated, and the thermometer still reads the same temperatures. Why?

Another interesting feature of the gas thermometer has to do with its temperature intercept. This is shown in figure 14.5. Different gas thermometers using various gases give the results shown there. Notice that if one extrapolates the data to very low temperatures, the pressures of the gases all reach zero at the same temperature, approximately −273.15°C. Even though the gases cannot be cooled near to this temperature without condensing, it is a very distinctive temperature, and we shall see its significance in the next

---

*The manometer is the U-tube device in figure 14.3 which contains mercury. The pressure inside the bulb B is equal to atmospheric pressure plus the pressure due to a height h of mercury. Why?

Atomic Theory of Matter

The theory that all matter is composed of tiny pieces, or atoms, was postulated several times in the early history of physics. However, not until about 1800 did the atomic theory of matter take on a quantitative form. At that time John Dalton showed that much of chemistry then known could be explained if one assumed the existence of atoms which varied in character from element to element. He also inferred that the elements had atomic masses which were integer multiples of the hydrogen mass. At the same time, Avogadro introduced the idea that an atomic weight of any element contains the same number of atoms. These concepts were used successfully and continuously by the chemists.

In the mid-1800s, the kinetic theory of gases was developed by Joule (1848), Kronig (1856), Clausius (1857), Maxwell (1860), and especially by Boltzmann (1872). The atomistic theories proposed by these men fitted the experimental data then available and gave a highly detailed picture of gas behavior. However, almost all experiments at that time were determinations of macroscopic properties of gases and solids. No direct evidence existed for the atoms and molecules the theories assumed to exist. Because of this fact, many physicists prior to 1900 took the attitude that atoms did not exist. Or if they did exist, they were too small to be observed and should therefore be of no concern to physicists.

Leaders in the opposition to the atomistic approach were the noted scientist-philospher of physics E. Mach and the highly regarded physicist W. Ostwald. To sum up their reasoning in the words of Ostwald, one should try to free science "from hypothetical conceptions which lead to no immediately experimentally verifiable conclusions." He termed the atomistic approach "those pernicious hypotheses" which place "hooks and points upon the atoms." Of course, chemists continued to use the atomic concepts with great success.

In the years near 1900, Ludwig Boltzmann (1844–1906) struggled against these attacks on the kinetic theory which he had so greatly advanced. It was discouraging work, and in 1898 he wrote, "I am conscious of being only an individual struggling weakly against the stream of time." Partly, at least, as a result of this opposition to his work, he became severely depressed and, in 1906, committed suicide.

Shortly thereafter, the atomic theory was verified by direct experimental evidence. In 1908 Perrin showed that Brownian motion could be explained in terms of atomistic concepts. A year later, in 1909, Millikan proved the atomicity of charge by direct measurement. In the following years a flood of experiments showed the validity of the atomistic approach. By 1926, when Otto Stern first measured directly the distribution of molecular speeds in a gas and found perfect agreement with the kinetic-theory predictions, the atomistic approach had been so widely accepted as to make Stern's result an anticlimax. We should not forget that the physics we know today has had a very human past, and has involved the emotions as well as the cold scientific reasoning of those who have given us the presently accepted laws of physics.

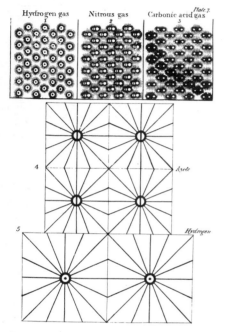

Particles of gases surrounded by a repelling atmosphere of caloric, from Dalton's "New System of Chemical Philosophy," 1810. "Azote" denotes nitrogen. (*Photo Science Museum, London.*)

section. It appears to be a natural fixed point for calibration of a thermometer, and the currently accepted temperature scale uses it as such.

In order to set up the presently accepted *absolute,* or *Kelvin,* temperature scale, the apparent zero pressure point for a thermometer is designated as 0 K, and is called *absolute zero.* The other fixed point is taken to be that unique temperature at which ice, water, and water vapor can all coexist at equilibrium. This temperature is called the *triple point* of water and is assigned the value 273.16 K (read "273.16 Kelvins"). The reason for this latter choice is to make the Kelvin degrees very closely equivalent in size to the degrees used on the centigrade scale. To ensure no ambiguity, the preferred name for the centigrade scale is now the *Celsius* scale, and the Celsius temperature T_c is defined to be

The apparent P = 0 point is taken as T = 0 on the Kelvin scale

The triple point of water is set equal to 273.16 K

$$T_c = T_K - 273.15 \text{ K} \qquad\qquad 14.4$$

where T_K is the Kelvin temperature. Notice that absolute zero, 0 K, is $-273.15°$C. The freezing point (or ice point) of water at standard pressure is found to be 0.00°C. From this fact we note that the triple point of water, 273.16 K, is very close to the ice point, 273.15 K. The student should be careful not to confuse these two temperatures.

14.4 THE MOLE, AVOGADRO'S NUMBER, AND MOLECULAR WEIGHT

We have now seen how both temperature and molecular kinetic energy influence the pressure of a gas. There turns out to be a very important relation between kinetic energy and temperature that leads us to an understanding of the meaning of temperature. We shall obtain that relation in the next section. But first we must define a few quantities in terms of which this relation is usually stated. You may already be familiar with them but this will serve as a concise review.

The unit called the *mole* (abbreviated *mol*) is a measure of number of particles. It is defined in terms of the number of atoms in 12 g of a particular type of carbon, the isotope designated carbon-12. (We shall learn about isotopes in Chap. 37.) The definition is as follows:

> *A mole of substance contains as many particles as there are atoms in 12 g of carbon-12.*

We shall most often use the kilomole (kmol), the quantity of substance that contains as many particles as there are atoms in 12 kg of carbon-12.

Using techniques that you will encounter later, it is possible to measure the number of particles in 12 g of carbon-12. This number is called Avogadro's number, N_A, and its measured value is 6.022×10^{23} particles per mole or, as we shall more frequently use it, 6.022×10^{26} particles per kilomole.

$$N_A = 6.022 \times 10^{23} \text{ mol}^{-1} = 6.022 \times 10^{26} \text{ kmol}^{-1}$$

For example, 1 mole of sugar contains N_A sugar molecules while 1 mole of baseballs contains N_A baseballs. As you see, the mole is not a measure of mass but, instead, is a measure of number of entities.

The *molecular weight* (or molecular mass) of a substance is defined in relation to the mass of a carbon-12 atom. By definition, the unit of mass we call the *atomic mass unit* (u) is one-twelfth the mass of a carbon-12 atom. We employ this unit to define the quantity M, the atomic or molecular weight of a substance.

> *The mass of one atom (or molecule) of a substance in atomic mass units is numerically equal to the atomic (or molecular) weight M of the substance.*

For example, the atomic weight of carbon-12 is exactly 12 u/atom. For the isotope hydrogen-1 it is 1.007825 u/atom, while the molecular weight of benzene (C_6H_6) is 78.11 u/molecule.

But 12 kg of carbon-12 atoms contains 1 kmol of atoms and 1.007825 kg of hydrogen-1 atoms contains 1 kmol. Hence we can restate the meaning of M as follows:

> *The atomic (or molecular) weight M of a substance is numerically equal to the number of kilograms in 1 kilomole of the substance.*

In keeping with our use of the SI, we shall always use N_A in particles/kmol and M in kg/kmol (or u/particle). Using these units, the familiar values of M (1 for H, 12 for C) are valid.

As we shall see later, the atomic weights used by chemists are the average values for the combinations of isotopes found in nature. For example, several isotopes of carbon exist in nature and ordinary carbon has an average mass per atom of 12.01115 u/atom rather than being exactly 12. When dealing with large quantities of atoms, these average values of M are appropriate and these are the values given in the atomic chart.

Now that we understand the meaning of N_A and M, it is possible to compute several important atomic quantities. We first note that atomic weight M and Avogadro's number N_A tell us the following facts at once. Because the molecular weight of nitrogen gas (N_2) is 28 kg/kmol, we know that 28 kg of nitrogen gas (a kmol) contains 6.022×10^{26} nitrogen molecules. Similarly, because methane (CH_4) has $M = 16$ kg/kmol, we know that 16 kg of methane contains 6.022×10^{26} methane molecules. The mass of water (H_2O) that contains 6×10^{26} molecules is 18 kg, and so on.

It is now a trivial matter to find the mass of an atom or molecule. For example, the molecular weight of benzene is 78. Therefore the mass of one benzene molecule is simply M/N_A or

$$\text{Benzene molecule mass} = \frac{78 \text{ kg/kmol}}{6 \times 10^{26} \text{ molecules/kmol}}$$

$$= 1.3 \times 10^{-25} \text{ kg/molecule}$$

The volume associated with a molecule can also be found from a knowledge of N_A. For example, we know that 1 m^3 of mercury (Hg) has a mass of 13,600 kg from its density, 13,600 kg/m^3. Furthermore, since Hg has an atomic weight of 201, we know that 1 m^3 of Hg contains a fraction 13,600/201 of a kilomole. Hence,

$$\text{No. of atoms/m}^3 = \frac{13,600 \text{ kg/m}^3}{201 \text{ kg/kmol}} 6 \times 10^{26} \text{ atoms/kmol}$$

$$= 4.08 \times 10^{28} \text{ atoms/m}^3$$

Taking the reciprocal of this we find that

$$\text{Volume/atom} = 24.5 \times 10^{-30} \text{ m}^3/\text{atom}$$

If we consider the atom to be a cube, each side of the cube would be

$$\sqrt[3]{24.5} \times 10^{-10}\,\text{m} = 2.9 \times 10^{-10}\,\text{m}$$

so we conclude that even one of the largest atoms in the periodic table has a diameter of order 30 nm.

As we continue with our studies, we shall see many ways in which the quantities N_A, M, and the mole unit are important. They allow us to relate atomic-size quantities to gross bulk properties of materials. In the next section we shall see how they are used to relate temperature to the average energies of the molecules in a gas.

14.5 PHYSICAL BASIS OF TEMPERATURE

One of the beauties of our definition of temperature in terms of the behavior of dilute gases is that a physical meaning is thereby given to the concept. This is most easily appreciated by reference to figure 14.5. It is seen there that the gas pressure is linearly related to the temperature as we have defined it. Experiment shows that if a mass m of a gas of molecular weight M is confined to a volume V, the following relation applies for the pressure P:

$$PV = \frac{m}{M}RT \qquad\qquad 14.5$$

This is very often referred to as the *ideal gas law*. In this relation T is the absolute, or Kelvin, temperature, and R is an experimentally determined constant of proportionality equal to 8314 J/(kmol)(K). The quantity R is often called the *universal gas constant*. [Notice that, frequently, m is measured in grams in *14.5*, in which case the units become mixed. In that case $R = 8.314$ J/(mol)(K).]

We can compare *14.5* with the expression we found in *14.3* for the pressure of a gas,

$$P = \tfrac{2}{3}\nu\langle\tfrac{1}{2}m_0 v^2\rangle$$

Multiplying each side of the equation by V, the container volume, equating to *14.5* and rearranging gives

$$T = \tfrac{2}{3}V\nu\langle\tfrac{1}{2}m_0 v^2\rangle\frac{M}{mR}$$

But νV is just the number of molecules in the volume V. This number, multiplied by m_0, is simply m, the total mass of molecules. Therefore the equation becomes

$$T = \frac{1}{3}\frac{M}{R}\langle v^2\rangle \qquad\qquad 14.6$$

From *14.6* we see at once that the absolute temperature of a gas is a measure of the average square speed of the molecules within it. We may show this more clearly by making use of the fact that one molecular weight of a substance contains Avogadro's number of molecules, $N_A = 6.02 \times 10^{26}$ molecules/kmol. Therefore the mass of a molecule is

$$m_0 = \frac{M}{N_A}$$

and we have

$$T = \frac{1}{3}\frac{N_A}{R}m_0\langle v^2\rangle$$

It is customary to designate the quantity R/N_A as *Boltzmann's constant*, $k = 1.38 \times 10^{-23}$ J/K. Then we have, after a little algebra,

$$\langle \tfrac{1}{2}m_0 v^2 \rangle = \tfrac{3}{2}kT \qquad\qquad 14.7$$

T measures the translational kinetic energy of a gas molecule

We therefore have the following physical meaning for T, the Kelvin temperature: the average translational kinetic energy of a molecule in a gas at absolute temperature T is simply $\tfrac{3}{2}kT$. Temperature is a direct measure of the molecular translational kinetic energy of a gas. *If two ideal gases have the same temperature, the average translational kinetic energies of their molecules will be the same.*

Illustration 14.1

Fourteen milligrams of nitrogen gas is placed in a container which has a volume of 5000 cm³. Find the gas pressure inside the container when the temperature is 27°C.

Reasoning The pressure will be found by use of *14.5*. From the atomic table in Appendix 6A, we see that the atomic weight of nitrogen is 14. Since nitrogen gas is diatomic, i.e., each molecule of nitrogen gas N_2 contains two nitrogen atoms, its molecular weight is 28. Substitution in *14.5* gives

$$P(5 \times 10^{-3} \text{ m}^3) = \left(\frac{14 \times 10^{-6} \text{ kg}}{28 \text{ kg/kmol}} \right) (8314 \text{ J/kmol} \cdot \text{K})(300 \text{ K})$$

or

$$P = 250 \text{ Pa}$$

Notice that 27°C was changed to 300 K before use in *14.5* since T is the absolute temperature. For comparison purposes, standard atmospheric pressure is 1.01×10^5 Pa, or 76 cm of mercury.*

Illustration 14.2

The pressure of a gas in a closed container is 5.0 cm Hg when the temperature is 0°C. What will be the pressure when the temperature is 400°C?

Reasoning Let us write down *14.5* twice, first for the original conditions and then for the final conditions:

$$P_0 V_0 = \frac{m}{M} R T_0$$

$$P_1 V_1 = \frac{m}{M} R T_1$$

Dividing one equation by the other and assuming the volume to remain constant so $V_0 = V_1$, we find

$$\frac{P_1}{P_0} = \frac{T_1}{T_0}$$

Placing in the given values, we find†

$$P_1 = (5.0 \text{ cm Hg}) \left(\frac{673 \text{ K}}{273 \text{ K}} \right) = 12.3 \text{ cm Hg}$$

---

*In the previous chapter we learned that the pressure due to a height h of fluid is $\rho g h$. Standard atmospheric pressure is equal to the pressure exerted by a column of mercury with $h = 0.76$ m.

†Notice that in using the gas law in ratio form the units cancel out. Hence any consistent set of units can be used for such purposes.

Notice here, too, the use of absolute temperatures.

Illustration 14.3

How fast is a nitrogen molecule moving on the average in air at 27°C?

Reasoning We shall find $\langle v^2 \rangle$ from *14.6*. It is $\langle v^2 \rangle = 3RT/M$.

$$\langle v^2 \rangle = 3[8314\,\mathrm{J/(kmol)(K)}]\left(\frac{300\,\mathrm{K}}{28\,\mathrm{kg/kmol}}\right)$$

$$= 2.67 \times 10^5\,\mathrm{m^2/s^2}$$

which gives the rms speed as

$$\sqrt{\langle v^2 \rangle} = 5.2 \times 10^2\,\mathrm{m/s}$$

Air molecules have speeds of about 500 m/s

At this speed, how long would it take a molecule to travel across the room in which you are working? Can you show that this speed does *not* depend upon the gas pressure?

14.6 DISTRIBUTION OF MOLECULAR SPEEDS

In the previous sections we have ignored the obvious fact that not all molecules in a gas can be moving with the same speed. Let us now examine the actual speeds of the molecules. It is interesting to notice that this topic was treated theoretically long before vacuum techniques had been developed well enough to measure the speeds of the molecules in a gas. James Clerk Maxwell, who might well be dubbed the "father of theoretical electricity," was the theoretician who first successfully examined the topic of molecular speeds. In 1860 he drived an equation which describes the way in which the speeds of gas molecules vary. This equation, together with the work of Boltzmann and other theoreticians active in this area of research, was the subject of a serious scientific controversy, as we pointed out in Sec. 14.3. Since experiments of the time were incapable of directly observing molecules, theories involving these unproven entities were suspect. Not until 1926, when Otto Stern managed to devise and assemble the necessary equipment, was the prediction made by Maxwell some 60 years earlier confirmed.

A simplified schematic diagram of Stern's experiment to determine the speeds of gas molecules is shown in figure 14.6. The entire apparatus was confined in a vacuum chamber. Mercury was vaporized in the oven O and a beam of mercury gas atoms shot out of a hole in the oven as indicated. After passing through a series of slits (not shown), the beam passed through a tiny slot in a disk D. This disk was coupled to a second similar one, D', but with

FIGURE 14.6 A schematic diagram of Stern's apparatus for measuring the speed of gas atoms. The whole apparatus is maintained in vacuum.

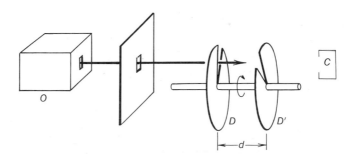

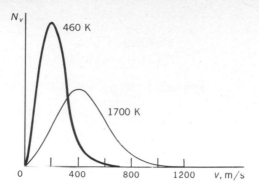

FIGURE 14.7 The distribution of molecular speeds in mercury gas at two indicated temperatures. What does the area under each curve represent?

the slit in it offset. Unless the disks were rotating, the beam could not pass beyond D' to the molecular detector at C.

However, if the disk was rotating, then molecules traveling just fast enough to move from D to D' in the time taken for the slit in D' to line up with D's original position would pass through D' and reach the detector. In this way, the number of molecules having a given speed could be measured. The results obtained by Stern accurately confirmed the prediction made by Maxwell over a half century earlier. These results are typified by the curves shown in figure 14.7.

We see there a plot of the number of molecules having various speeds. Data for two temperatures are shown. Since we already have learned that the average kinetic energy of a molecule is proportional to temperature, it is no surprise that the hotter gas has more high-speed molecules than does the cooler gas. Although these curves are meaningful in a qualitative way to all of us, they have important quantitative features as well. Since these features are characteristic of many similar curves in science and engineering, we shall digress for a moment to point them out.

The molecules in a gas have a wide distribution of speeds

The quantity plotted as N_v in the graph can be better understood by reference to figure 14.8. Consider the little section of the v axis indicated as Δv_i. If $\Delta v_i = 1$ m/s, extending from 421 to 422 m/s, for example, then N_{vi} at that position gives the number of molecules having speeds in this range. Or, more generally, if Δv_i is not exactly 1 m/s, still $N_{vi} \Delta v_i$ gives the number of molecules which have speeds in the range Δv_i.

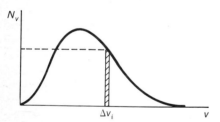

FIGURE 14.8 The number of atoms with speeds in the range Δv_i is $N_v \Delta v_i$. Although it cannot be shown on this graph without greatly expanding the vertical scale, N_v approaches 0 only when $v \to \infty$.

To find the total number of molecules used in plotting the graph, we note that $N_{vi} \Delta v_i$, the number having speeds in the range Δv_i, is the shaded portion of figure 14.8. The total number will simply be the sum of all the $N_{vi} \Delta v_i$ extending from $v_i = 0$ to $v_i \to \infty$; thus

$$N_t = \text{total number} = \sum_{v=0}^{v \to \infty} N_{vi} \Delta v_i$$

Since each $N_{vi} \Delta v_i$ is an area like the shaded area, this sum is simply the area under the N_v versus v curve. Or, if we let $\Delta v_i \to 0$, we can replace the sum by an integral to write

$$N_t = \int_{v=0}^{\infty} N_v \, dv$$

Suppose now that we ask what is the chance that a molecule has a speed in the range Δv_i centered on v_i? It will be equal to the fraction of molecules in this range. For example, if one-tenth of the molecules have speeds in this range, then our chances are only 1 in 10 that a given molecule has a speed in that range. The chance is 0.10. But, in general, the number of molecules with speed in the range Δv_i is, as we saw above, $N_{vi} \Delta v_i$. Therefore the fraction of molecules in this range will be $N_{vi} \Delta v_i \div N_t$, and this is equal to the chance that a given molecule has a speed in this range. We represent this chance, or probability, by the symbolism $P(v_i) \Delta v_i$, and so

$$P(v_i) \Delta v_i = \frac{N_{vi} \Delta v_i}{N_t}$$

Frequently we let $\Delta v = 1 \, \text{m/s}$, in which case $P(v)$ is simply the chance, or probability, that a given molecule will have a speed within the range $v \pm \frac{1}{2}\text{m/s}$. As we see, the symbolism $P(v)$ is simply $N_v \div N_t$. Hence figure 14.7 can be changed to a graph of $P(v)$ simply by dividing the N_v axis by N_t. The result is shown in figure 14.9. It is called a *probability distribution*.

We should now notice that figure 14.9 is far more useful than figure 14.7. The values of $P(v)$ have direct meaning for us. For example, referring to the 460-K curve, we note that its maximum occurs at about 200 m/s. The value for $P(v)$ there is about 4.4×10^{-3}. This tells us that the probability that a given molecule has a speed in the range 199.5 to 200.5 m/s is 4.4×10^{-3}. Or, put another way, if we are given 10^6 molecules at that temperature, the number we should expect to have speeds between 199.5 and 200.5 m/s would be

$$10^6 \times 4.4 \times 10^{-3} \qquad \text{or} \qquad 4400 \text{ molecules}$$

In symbols,

$$N(v) = P(v)N_t$$

P(v) dv gives the fraction of molecules with speeds in the range dv

Or, more generally, the probability that a given molecule has a speed within the range dv is simply $P(v) \, dv$. This concept of a probability distribution is extremely important in science and engineering and should be well understood by the student.

14.7 MAXWELL SPEED DISTRIBUTION FUNCTION

As was pointed out in the preceding section, Maxwell predicted that the speeds of the particles in a gas should be distributed among all values. The experimental speed distributions shown in figures 14.7 to 14.9 conform to

FIGURE 14.9 The normalized speed distribution function for mercury gas. Can you show from the graph that a fraction $\simeq 1.2 \times 10^{-3}$ of the atoms have speeds between 599.5 and 600.5 m/s when the temperature is 1700 K?

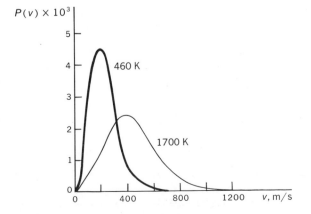

the relation derived by Maxwell. His derivation is reasonably simple, and the student may be interested in reading it.* The result is, couched in terms of $P(v)$, the fraction of particles with speed in the unit speed interval centered on v,

$$P(v) = \frac{4}{\sqrt{\pi}}\left(\frac{m_0}{2kT}\right)^{3/2} v^2 e^{-m_0 v^2/2kT} \qquad 14.8$$

where k = Boltzmann's constant
$\quad\ T$ = Kelvin temperature
$\quad m_0$ = mass of gas particle

Notice that $P(v) \rightarrow 0$ only as $v \rightarrow 0$ and as $v \rightarrow \infty$. Therefore all speeds are possible for the particles.

Certain mathematical properties of a distribution function, such as Maxwell's speed distribution given in *14.8*, are important. For example, since $P(v)\,dv$ is the fraction of particles with speed $v \pm dv/2$, the sum of $P(v)\,dv$ over all possible values of v should be unity. This follows from the fact that the fraction $P(v)\,dv$ is really the chance, or probability, that a particular particle has a speed in the range $v \pm dv/2$. Since its probability, or chance, of having a speed v somewhere between $0 \rightarrow \infty$ is unity, the sum of $P(v)\,dv$ over all v's must be unity. Expressed as an equation

$$\int_0^{\infty} P(v)\,dv = 1 \qquad 14.9$$

A probability distribution function such as $P(v)$ which satisfies the condition of *14.9* is said to be *normalized*. If *14.8* is substituted in *14.9*, the relation will be seen to be true.

We can use *14.8* to find the average value of the speed of the gas molecules $\langle v \rangle$, the average speed squared $\langle v^2 \rangle$, etc. To do this we make use of a modification of the technique developed in *4.9*. By definition of the average, we add together all the values of the quantity q_i being averaged and divide by the total number N_t of such values. Thus

$$\langle q \rangle = \frac{q_1 N_t P(v_1)\,dv_1 + q_2 N_t P(v_2)\,dv_2 + \cdots + q_n N_t P(v_n)\,dv_n}{N_t}$$

where it will be recalled that $N_t P(v_i)\,dv_i$ is the number of particles with speed $v_i \pm dv_i/2$. Or since the v's are really continuous and their values run from zero to infinity, the sum may be replaced by an integral to give (after canceling N_t)

$$\langle q \rangle = \int_0^{\infty} q P(v)\,dv \qquad 14.10$$

As an example of the use of *14.10*, we compute the value of the average square speed of the gas particles:

$$\langle v^2 \rangle = \int_0^{\infty} v^2 P(v)\,dv$$

After replacing $P(v)$ by its value from *14.8*, we have

$$\langle v^2 \rangle = \frac{4}{\sqrt{\pi}}\left(\frac{m_0}{2kT}\right)^{3/2} \int_0^{\infty} v^4 e^{-bv^2}\,dv$$

*See W. F. Magie, "A Source Book in Physics," Harvard University Press, Cambridge, Mass., 1963.

with
$$b \equiv \frac{m_0}{2kT}$$

This integral is given in most integral tables and gives

$$\langle v^2 \rangle = \frac{3kT}{m_0}$$

which is the same result found in Sec. 14.5. The rms speed is the square root of this value.

14.8 MAXWELL–BOLTZMANN DISTRIBUTION FUNCTION

Very often we are interested in how the particles of a gas are distributed in regard to their energies. It is a simple matter to transform *14.8* to yield this information in terms of translational kinetic energy K. Since $P(v)\,dv$ is the fraction of the particles having speeds in the range dv centered on v, it is convenient to represent the equivalent kinetic-energy distribution as $P(K)\,dK$. To obtain it we merely replace $\frac{1}{2}m_0 v^2$ by K in *14.8* and note that, by differentiation,

$$dK = \tfrac{1}{2}m_0\,d(v^2) = m_0 v\,dv$$

We then find that the fraction of particles with translational kinetic energies in the range dK centered on K is given by

$$P(K)\,dK = \frac{2}{(\pi^3 kT)^{3/2}}\sqrt{K}\,e^{-K/kT}\,dK \qquad 14.11$$

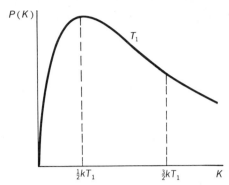

FIGURE 14.10 The kinetic-energy distribution for a gas at temperature T_1. What are the significances of the two energies labeled? How should the curve for a lower temperature look?

The energy distribution function for an ideal gas

This distribution is shown in figure 14.10. As shown in the diagram, the most probable energy [i.e., where $P(K)$ is maximum] occurs for $K = \frac{1}{2}kT$. Notice that the most probable energy is a linear function of absolute temperature, as we might have suspected.

To find the average value of K, designated $\langle K \rangle$, we use an equation identical in form with *14.10*. It is

$$\langle K \rangle = \int_0^\infty K P(K)\,dK$$

which gives, after carrying out the mathematics,

$\langle K \rangle = \frac{3}{2}kT$

$$\langle K \rangle = \tfrac{3}{2}kT \qquad 14.12$$

Since $\langle K \rangle$ is merely $\frac{1}{2}m_0\langle v^2 \rangle$, we see at once that this agrees with our previous results in *14.7*.

Until now we have only considered how the translational kinetic energy of a particle varies. A similar important situation exists for particles which can have different potential energies. For example, the molecules near the top of the atmosphere have much more potential energy than those at ground level. Since more air molecules exist at lower levels than high in the atmosphere, the probability of finding a molecule varies in some way with its potential energy. Although we cannot here derive the relation which applies to such situations, the result is simple enough so that we can state and use it. It is called the *Maxwell-Boltzmann distribution law* after the two men who contributed most to its derivation. Let us now see what the law is.

Consider a large number of particles or systems of particles which can exist in at least two ways or places—we say they can exist in at least two *states.* When the particles are in state 1, their total energy is E_1, and when in state 2, their total energy is E_2. Further, we suppose that, if $E_1 = E_2$, the particles would be equally likely to be in state 1 or state 2. In other words, the particles have no preference for either state 1 or state 2 if $E_1 = E_2$. Under these conditions, and when the number of particles in each state has become constant, the number of particles which will be found in state 1, n_1, is related to the number in state 2, n_2, by the relation

$$\frac{n_1}{n_2} = e^{-[(E_1-E_2)]/kT} \qquad 14.13$$

and this is known as the Maxwell-Boltzmann distribution. We see from this relation that if state 1 has a larger energy than state 2, $E_1 > E_2$, so that $n_1/n_2 < 1$. In other words, the particles prefer to occupy the states of lowest energy. For example, in the ridiculous situation where the particle is a rock, the rock prefers to roll downhill.

In a more practical situation, we can use *14.13* to find out how the number of molecules per unit volume of air varies with height above the earth. For convenience, consider nitrogen molecules in a very tall tank. Consider a 1-cm^3 volume at the top (this is state 1) as compared with a 1-cm^3 volume at the bottom (this is state 2) if the vertical distance between them is h. We shall consider the gas to have the same temperature throughout the tank, so that the average kinetic energies of the molecules are everywhere the same. Clearly, since the average kinetic energy K is the same, and if there were no potential energy difference between these two states, equal numbers of molecules would be found in them, since they would be fully equivalent. Therefore *14.13* should apply.

Taking the potential energy to be zero at the level of the bottom of the tank, we have

$$E_1 = K_1 + m_0gh \qquad \text{and} \qquad E_2 = K_2$$

from which
$$E_1 - E_2 = m_0gh$$

where m_0 is the mass of a nitrogen molecule and use has been made of the fact that $K_1 = K_2$.

Substituting in *14.13*, we find

$$\frac{n_1}{n_2} = e^{-m_0gh/kT}$$

Since, at equal temperatures, the pressure of a gas is proportional to the number of molecules per unit volume (n_1 and n_2 were numbers per cubic centimeter), we can also rewrite this

$$\frac{P_1}{P_2} = e^{-m_0gh/kT} \qquad 14.14$$

This latter relation is often referred to as the *law of atmospheres* since it approximates the atmospheric pressure variation with height. In practice, of course, T is not constant as one leaves the surface of the earth.

Illustration 14.4
Using nitrogen gas in a tank 10 m high and a temperature of 300 K, find the ratio of the pressures at the top and bottom of the tank. The mass of a nitrogen molecule is 4.7×10^{-26} kg.

Reasoning We substitute in *14.14*, using the fact that

$$\frac{m_0gh}{kT} = \frac{(4.7 \times 10^{-26}\,\text{kg})(9.8\,\text{m/s}^2)(10\,\text{m})}{(1.38 \times 10^{-23}\,\text{J/K})(300\,\text{K})}$$

$$= 1.11 \times 10^{-3}$$

Therefore the pressure ratio is

$$\frac{P_1}{P_2} = e^{-1.11 \times 10^{-3}}$$

This is easily evaluated without tables if we recall the expansion of e^x, namely

$$e^x = 1 + x + \frac{x^2}{2!} + \frac{x^3}{3!} + \cdots$$

from which $\quad \dfrac{P_1}{P_2} = 1 - (0.00111) + (0.0000006) - \cdots$

$$\approx 0.9990$$

Clearly, this tank is far too short to show an appreciable effect. If the gas were actually a dust composed of particles of rather large mass (compared with molecules), the effect would be more pronounced, of course. How could you use this effect to estimate the mass of dust particles?

14.9 MEAN FREE PATH

An interesting experiment can be performed in a vacuum chamber using a gas of metal atoms. For example, suppose a small oven, designated O in figure 14.11, is placed inside a glass vacuum chamber V. If copper, for example, is heated white hot inside the oven, it melts and boils off a gas of copper atoms. This gas fills the oven. Some of the gas atoms can escape through a small hole in the oven, as shown, and they fly out in straight-line paths to the glass walls of the vacuum chamber and form a copper mirror on the glass. Since the atoms travel in straight lines, they cast shadows of objects in their path, as indicated. As a result, one sees on the glass walls sharp-edged shadows cast by objects in the path of the beam.

Suppose now that air is slowly admitted into the vacuum chamber. At a certain pressure, the shadow cast by the copper atoms will cease to be sharp, and at a slightly higher pressure, the shadow will not appear at all. It is as though the copper atoms, in their flight from the oven to the wall, have become able to bend around corners and thereby appear behind obstacles. This behavior is easily explained.

In a nearly perfect vacuum, the copper atoms struck nothing in their flight from the oven to the wall. However, as more and more air molecules are admitted to the chamber, the copper atoms are more likely to strike one of them before hitting the wall. If a copper atom hits several air molecules before hitting the wall, it will have traveled a path such as the one shown in figure 14.12. Notice that the final direction of travel has little relation to the original direction of the beam. As a result, when most of the particles in a beam undergo collision before hitting the wall, sharp shadows cannot be expected. This fact is sometimes used as a rough indication of the quality of a vacuum.

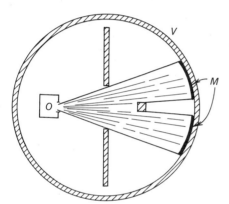

FIGURE 14.11 The beam of metal atoms from the oven forms a sharp-edged shadow of objects in its path.

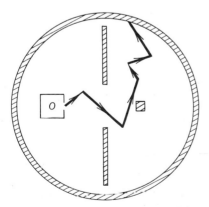

FIGURE 14.12 If the diameter of the chamber is 30 cm, about how large is the mean free path for the particle shown?

We wish now to make a quantitative estimate of this effect. To do so, we shall compute the *mean free path length* for a gas particle. The mean free path is defined to be the mean (or average) distance which a gas particle moves between collisions. For the purposes of this computation we shall consider a gas containing ν spherical particles per unit volume, each with a radius b and speed v.

Let us focus our attention on a single particle of the gas and assume all the other particles to be at rest. This particle will travel a distance vt in time t. As it moves along, it will hit any particle whose center is within a distance $2b$ of its own center, as one can easily see by looking at figure 14.13. Since the particle moves a distance vt, it will sweep out a volume nearly equal to the volume of a cylinder of radius $2b$ and length vt, as shown. In this volume, $\pi(2b)^2vt$, there will be a number of particles equal to ν times the volume. Therefore, since the particle we are considering will collide with each of them, there will be in time t the following number of collisions:

$$\text{Collisions in time } t = \nu\pi(2b)^2vt$$

from which the collision frequency (i.e., collisions per sec) is

$$f = \text{collision frequency} = 4\pi\nu b^2 v$$

To find the distance between collisions we note that the particle moves a distance v in 1 s and undergoes $4\pi\nu b^2 v$ collisions. Dividing the distance by the number of collisions gives the mean free path as

$$\ell = \text{mean free path} = \frac{1}{4\pi\nu b^2}$$

When the motion of the other particles is included in a more detailed computation, these results are changed slightly. The correct results are

and

$$f = 4\pi\sqrt{2}\,\nu b^2 v$$

$$\ell = \frac{1}{4\pi\sqrt{2}\,\nu b^2}$$

14.15

We could use *14.15* for the mean free path, along with the experiment shown in figure 14.11, to determine an approximate value for the radii of gas molecules. However, when vacuum experiments such as this became possible, good estimates of b had already been made in several other ways. Equation *14.15* was therefore used to confirm these earlier estimates.

Illustration 14.5

In a certain experiment it is necessary for the mean free path to be at least 5 cm. Assuming the gas to be nitrogen molecules with 1.7-Å radius

FIGURE 14.13 Even though the gas molecule has a radius of only b, it will strike other molecules whose centers are closer than $2b$ to the axis of the cylinder which it sweeps out.

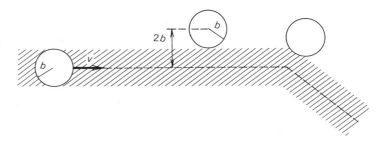

$(1.7 \times 10^{-10}\,\text{m})$, find the maximum pressure in the chamber at 300 K. The mass of a nitrogen molecule is 4.7×10^{-26} kg.

Reasoning We make use of *14.15*; placing

$$\ell = 0.05\,\text{m}$$
$$b = 1.7 \times 10^{-10}\,\text{m}$$

we find $\nu = 3.9 \times 10^{19}$ molecules/m^3. We can now proceed to find the pressure by use of *14.5*, the ideal gas law, namely,

$$PV = \frac{m}{M} RT$$

We have for $V = 1\,\text{m}^3$ that

$$m = (3.9 \times 10^{19})(4.7 \times 10^{-26})\,\text{kg} = 1.83 \times 10^{-6}\,\text{kg}$$
$$M = 28\,\text{kg/kmol}$$
$$R = 8314\,\text{J/(kmol)(K)}$$

and have, as our final result, that

$$P = 0.16\,\text{N/m}^2$$

ℓ is of order 10^{-6} cm in air at normal pressures

For comparison purposes, standard atmospheric pressure is 1.01×10^5 N/m^2. Therefore, to obtain a mean free path of the order of centimeters, the pressure must be of the order of one-millionth of atmospheric pressure. The mean free path at other pressures can be found by noting $\ell \sim 1/\nu \sim 1/P$.

Chapter 14
Questions
and Guesstimates

1. Boyle's law for gases states: "The volume of a gas varies inversely to the pressure provided the mass and temperature of the gas are maintained constant." Show that Boyle's law is a special case of the ideal gas law.
2. Charles' law for gases states: "The volume of a gas increases in direct proportion to the absolute temperature provided the pressure and mass of the gas are maintained constant." Show that this is a special case of the ideal gas law.
3. Dalton's law of partial pressure states: "The total pressure of a mixture of gases is equal to the sum of the partial pressures of the gases in the mixture." Using the ideal gas law and our kinetic theory ideas, justify Dalton's law.
4. Suppose you were the only survivor of a plane crash on a tiny island in the South Pacific. While waiting for rescue, you decide to construct a thermometer. How might you accomplish this?
5. Hydrogen and oxygen gas are sealed off in a very strong container which contains two electrodes. A spark from the electrodes ignites the gases so that the reaction

$$2\text{H}_2 + \text{O}_2 \rightarrow 2\text{H}_2\text{O}$$

results. Will the pressure in the tube be changed after the temperature has come back to its original value? (Assume the H_2O to be vapor.)
6. In outer space there is about one molecule per cubic centimeter of volume. Try to devise a method for measuring the pressure of such a thin gas.

7. The composition of the atmosphere changes with altitude. The percentage of hydrogen molecules in the air increases, and the percentage of nitrogen molecules decreases, as one goes farther above the earth. Why?

8. In practice, the collision of gas molecules with the container wall are not usually elastic; moreover, collisions occur between the gas molecules themselves. Try to justify the fact that neither of these facts changes the ideal gas equation provided the gas is far from liquefaction and is at equilibrium with the container walls.

9. What would be the ratio of the average speeds of N_2 and H_2 molecules in air? Does this mean that all hydrogen molecules in air travel faster than the nitrogen molecules?

10. We can picture the molecules of an ideal gas to act like tiny balls in continual motion. An ideal gas of colloidal-size particles can also exist. But glass beads and pool balls do not act like an ideal gas. Where (in size) does the dividing line come, and to what is it due?

11. In order to escape from the earth, a rocket must be shot out from it with at least a speed of 11,200 m/s. Explain why only a tiny amount of hydrogen exists in the atmosphere even though billions of years ago there may have been more hydrogen than nitrogen in it.

12. Sketch a probability distribution for the heights of human beings in your class.

13. Sketch a probability distribution for yearly salaries of men and women over 21 years old in the United States. Estimate the average and most probable values of the distribution. From your sketch, estimate what fraction of the people have salaries in the range from $5000 to $6000.

14. The average velocity of the molecules in a gas is zero, while the average speed is greater than zero. Explain why these are not the same.

15. A wind blowing against a window exerts more pressure against the window than still air exerts. How would one go about computing this extra force?

16. Estimate how large (in volume) a helium-filled balloon must be if it is to be able to lift 500 kg. To start at least, assume the mass of the balloon to be included in the 500 kg.

Chapter 14
Problems

1. Find the mass (in kg) of an ammonia molecule NH_3.

2. The chemical formula for a typical polystyrene molecule is $C_n H_{n+2}$ where n is of the order of tens of thousands. Find the mass (in kg) of a polystyrene molecule for which $n = 10,000$.

3. Standard conditions are 1.01×10^5 Pa and 273 K. When these conditions prevail, 22.4 m^3 of an ideal gas contains 6.02×10^{26} molecules. Under these conditions, find the following for butane gas: (a) mass of a molecule, (b) number of molecules per m^3, (c) speed (rms) of the molecules. (The formula for butane is C_4H_6.)

4. Standard conditions are 1.01×10^5 Pa and 273 K. Under these conditions 22.4 m^3 of an ideal gas contains 6.02×10^{26} molecules. Find the root mean square speed of a molecule in chlorine gas (Cl_2) under these conditions. Assume ideality.

5. Call the rms speed of the molecules in an ideal gas v_0 at temperature T_0 and pressure P_0. Find the speed if (a) the temperature is raised from 20 to 300°C; (b) the pressure is doubled and $T = T_0$; (c) the molecular weight of each of the gas molecules was three times as large.

6. While preparing a sealed-off 20-cm³ tube at low temperatures, one drop (mass 0.050 g) of nitrogen is accidentally sealed off in the tube. Find the nitrogen (N_2) pressure within the tube when the tube warms to 27°C. Assume ideality. Express your answer in atmospheres (1 atm = 1.01 × 10⁵ Pa).

7. A 20-cm³ test tube is sealed off with nitrogen gas in it under standard conditions (1.01 × 10⁵ Pa and 273 K). The tube is cooled down to about −200°C so that the nitrogen liquifies. What mass of liquid nitrogen will be obtained? What will be the volume of this mass? (Density of liquid nitrogen = 810 kg/m³.)

8. The temperature of outer space has an average value of about 3 K. Find the rms speed of a proton (a hydrogen nucleus) in space. (Mass of proton = 1.67 × 10⁻²⁷ kg.)

9. The temperature of the corona of the sun, the luminous gaseous layer we see, is about 6000 K. Find the rms speed of a proton, one of the main constituents of the corona, at this temperature. (m_p = 1.67 × 10⁻²⁷ kg.)

10. The temperature at the sun's center is estimated to be about 14 × 10⁶ K while the density there is about 100 g/cm³. Assuming the protons which constitute most of the sun's core to act like a perfect gas even at these very high densities, find the pressure of the proton gas (in atmospheres) at the sun's core. (m_p = 1.67 × 10⁻²⁷ kg.)

11. The gauge pressure* for an auto tire is 24 lb/in² or 1.65 × 10⁵ Pa and the volume of the tire is V_0. What volume of air under standard atmospheric pressure and the same temperature was used in filling the tire?

12. A car tire is filled to a gauge pressure* of 24 lb/in² (1.65 × 10⁵ Pa) when the temperature is 20°C. After running at high speed, the tire temperature rises to 60°C. Find the new gauge pressure within the tire assuming the tire's volume did not change.

13. An air bubble of volume v_0 is released by a fish at a depth h in a lake. The bubble rises to the surface. Assuming constant temperature and standard atmospheric pressure above the lake, what is the volume of the bubble just before touching the surface? The density of the water is ρ.

14. In a diesel engine, the cylinder compresses air from approximately standard pressure and temperature to about one-sixteenth the original volume and a pressure of about 50 atm. What is the temperature of the compressed air?

15. One way to cool a gas is to let it expand. Typically, a gas at 27°C and a pressure of 40 atm might be expanded to atmospheric pressure and a volume 13 times larger. Find the new temperature of the gas.

16. In the reaction $2H_2 + O_2 \rightarrow 2H_2O$, hydrogen and oxygen burn to form water vapor. Assume the temperature to be high enough so that the water is vaporized to a perfect gas. Suppose 2 mol of hydrogen gas in one tank are connected to a tank containing 1 mol of oxygen gas, both at P_0. The gases are now allowed to flow together and react. The final temperature returns to the original value. (a) How many moles of H_2O are formed? (b) What is the final pressure?

17. Butane gas burns in air according to the following reaction: $2C_4H_{10} + 13O_2 \rightarrow 10H_2O + 8CO_2$. Suppose the initial and final temperatures are equal and high enough so that all reactants and products act as perfect gases. Two moles of butane are mixed with 13 mol of oxygen and

*Tire gauges read the difference between the pressure of the atmosphere and the pressure in the tire.

then completely reacted. What will be the final pressure assuming the volume to be unchanged and the pressure before reaction to be P_0?

18. Show that the mean free path of molecules in an ideal gas can be written in terms of the gas pressure as

$$\ell = \frac{kT}{4\pi\sqrt{2}\,b^2 P}$$

19. Assuming the air to be composed of spherical nitrogen molecules with radius 1.7×10^{-10} m, find the mean free path of air molecules under standard conditions.

20. Ten small planes are flying with speed 150 km/h in an air space that is $20 \times 20 \times 1.5$ km^3 in volume. It is totally dark and you, being in one of the planes, fly at random within this space with no way of knowing where the other planes are. On the average, about how long a time will elapse between near collisions with your plane? Assume for this rough computation that a plane can be approximated by a sphere with 6-m radius.

21. Using the graph of figure 14.9, what will be the most probable speed of N_2 molecules at 460 K? [*Hint*: Note that $P(v)$ in *14.8* is essentially a function of $m_0 v^2/T$.] Mercury is monatomic.

22. Using the graph of figure 14.9, what will be the most probable speed of Hg molecules at 920 K (see *Hint* in previous problem).

23. The escape speed for an object on a planet of radius R is $v = \sqrt{2gR}$, where g is the acceleration due to gravity on the planet. What is the temperature of each of the following gases so that the rms speed of its molecules equals the escape speed: (*a*) N_2 on earth; (*b*) H_2 on earth; (*c*) N_2 on the moon where $g = 0.16g_{earth}$? (*d*) In view of this, why is it plausible that the moon has no atmosphere? $R_{earth} = 6.4 \times 10^6$ m, $R_{moon} = 1.74 \times 10^6$ m.

24. Find the temperature at which the ratio of the densities of mercury vapor at the top and bottom of a 2.0-m-high tank would be 1/2.718. (Assume an ideal gas could be obtained.)

25. A gas of dust particles fills a 2.0-m-high tank. At equilibrium (27°C), the density of particles at the top of the tank is 1/2.718 the density at the bottom. Find the mass of a typical particle and find how many times more massive it is than a nitrogen molecule.

26. One of the earliest estimates of Avogadro's number was made by Jean Perrin in 1908. He dispersed tiny, uniform particles in a clear liquid and noted that, when equilibrium had been reached, the number n of particles per unit volume decreased with height in the liquid. Show that the relation

$$n = n_0 e^{-(N_A v/RT)(\rho - \rho')gy}$$

applies where n_0 is n at $y = 0$, ρ and ρ' are the densities of the particle and liquid, respectively, and v is the volume of a particle. How would you graph your data of n versus y to obtain N_A?

27. Show that the maximum value of $P(v)$ occurs when $v = \sqrt{2kT/m}$. Use the calculus criterion for a maximum.

28. Show that the maximum value of $P(K)$ occurs when $K = kT/2$. Use the calculus criterion for a maximum.

29. Show that the Maxwell velocity distribution given in *14.8* is normalized.

$$\left(Note: \int_0^\infty x^2 e^{-ax^2}\, dx = (1/4a)\sqrt{\pi/a}. \right)$$

15 Thermal Properties of Matter

Heat energy is intimately connected with the motion of atoms and molecules. In this chapter we establish this connection and see how it gives rise to the observed thermal properties of gases, liquids, and solids. We not only investigate the mechanism of heat transfer through materials, but also find what factors influence the way in which these materials behave as heat is added to them. In particular, the concepts of phase changes and latent heats are considered.

15.1 CONCEPT OF HEAT

Man has always been able to distinguish between hot and cold. It has long been known that things which are hot can be used to heat cooler objects. However, it is only since about a century ago that any real physical understanding of the processes involved could be said to exist. Not surprisingly, the understanding of the nature of heat developed rapidly as the kinetic theory of gases was evolved. As we saw in the preceding chapter, the kinetic theory leads at once to a physical meaning for temperature. The absolute temperature T of a gas is proportional to the average translational kinetic energy of a molecule in the gas. We concluded that the average translational kinetic energy of a molecule in a gas could be found from the relation

$$\langle \tfrac{1}{2}m_0v^2 \rangle = \tfrac{3}{2}kT \qquad\qquad 14.7$$

Temperature measures translational kinetic energy of gas molecules

where $k = 1.38 \times 10^{-23}$ J/K and is called Boltzmann's constant.

Suppose now that two containers of gas originally at temperatures T_1 and T_2 ($T_1 > T_2$) are brought together and the gases are allowed to mix, as shown in figure 15.1. We know from our common experience (i.e., from experiment) that the final temperature T of the gas is intermediate between T_1 and T_2. The interpretation of this on the basis of the kinetic theory is obvious: the hot, high-energy molecules in container 1 collide with the cold, low-energy molecules in container 2; eventually, when both gases reach the same temperature, these collisions have caused the molecules in container 1 to lose enough kinetic energy to those in container 2 so that the final average molecular translational kinetic energies in containers 1 and 2 are the same.

From this and many similar considerations we conclude that, when two bodies at different temperatures are brought into contact, energy is transferred, or *flows,* from the hotter to the cooler body. *The energy which is transferred in a situation such as this is what we ordinarily refer to as heat*

$T_1 > T > T_2$

$T = T$

FIGURE 15.1 When the two gases are brought into contact, collisions cause the average molecular kinetic energy of each to change, thereby equalizing the temperatures.

energy. Heat energy is the energy which is transferred from a warm body to a cooler one as a result of the temperature difference between the two bodies. Heat energy is therefore related to molecular energy. When heat is added to a body, energy is added. When a body is cooled, it loses energy. The study of heat flow in matter is therefore the study of energy transfer.

15.2 HEAT UNITS

As we have just seen above, the fact that heat is related to molecular energy was discovered long after the use of heat had become technically important. The wide use of heat in various technical processes necessitated the existence of units for its measure. Lacking a basic understanding of the nature of heat, these units were devised purely from a practical standpoint. Since water is widely used in practical processes involving heat, it is not surprising that the heat unit is based upon the behavior of water.

The basic unit of heat energy, called the *calorie,* used to be defined in the following way: *One calorie of heat energy is the quantity of heat energy which must be added to one gram of water (under one atmosphere) to change its temperature from 14.5 to 15.5°C.* Specification of the temperature range is necessary since the amount of energy required to increase the temperature of water by 1°C varies slightly with temperature. (At 0°C the true value is 1.008 cal/g $\cdot$ °C, while at 40°C it is 0.997 cal/g $\cdot$ °C, and these are the two extremes in the range 0 to 100°C.) For most purposes this slight variation is negligible, and we shall usually assume that one calorie will change the temperature of one gram of liquid water by one degree Celsius.

Of course, since heat energy is no different from the mechanical energy we have considered heretofore, the calorie must be just another energy unit similar to the erg, the joule, or the foot $\cdot$ pound. One way of finding the relation between the calorie and the joule would be to measure the quantity of heat in calories given off when a certain number of joules of friction work is done. James Prescott Joule actually carried out such an experiment in the 1840s. We shall defer application of his experiment to one of the problems at the end of the chapter. Since that time, electrical methods for determining the relation between the calorie and joule have been perfected. They will be

Definition of the calorie

described in our study of electricity. The result found for the so-called *mechanical equivalent of heat* is

$$1 \text{ cal} = 4.184 \text{ J}$$

We can use this experimental value to translate heat-energy measurements into the more basic mechanical-energy units. In recent years the calorie has been defined to be *exactly* 4.184 J. Nutritionists use a unit called the Calorie, spelled with a capital C. By definition, 1 Calorie = 1 kcal.

15.3 SPECIFIC HEAT CAPACITY: DEFINITION

As we shall see, it is of great importance both theoretically and experimentally to know how much heat is needed to change the temperature of a material by a given amount. For this purpose we define the *specific heat capacity c* of a substance. If a mass *m* of material requires the addition to it of a quantity of heat ΔQ in order to change its temperature by an amount ΔT, then

Definition of specific heat capacity

$$c \equiv \frac{\Delta Q}{m\,\Delta T} \qquad\qquad 15.1$$

From its defining equation, c may be thought of as the quantity of heat needed to raise the temperature of unit mass by one degree. Commonly, its units will be cal/g · °C. Typical values of c for various liquids and solids are given in table 15.1. Notice that water requires more heat to increase its temperature than does any of the other substances listed. The reasons for this are pointed out later in this chapter.

In principle, c, the specific heat capacity, for a liquid or solid is rather easily determined. One common method, called the *method of mixtures,* makes use of the law of conservation of energy. Since heat energy is basically no different from mechanical energy, the conservation law should apply to it. In particular, if one or more objects of an isolated system lose heat energy, an equal amount of energy must be acquired by other objects composing the system.

To illustrate this, consider the experiment shown in figure 15.2. A mass m_x of a metal of unknown specific heat capacity c_x is initially at a temperature t_x. It is dropped into a copper calorimeter can or container (c_{Cu}, m_{Cu}) which contains a mass m_w of water, all at temperature $t_0 < t_x$. Upon mixing, everything ends with a final temperature t. Using the definition of c given in *15.1,* we have

$$\text{Heat lost by metal} = c_x m_x (t_x - t)$$

while

$$\text{Heat gained by H}_2\text{O and Cu} = (1)m_w(t - t_0) + c_{Cu}\,m_{Cu}(t - t_0)$$

Since the heat lost must equal the heat gained, we can equate these two expressions. Solving for c_x, we find

$$c_x = \frac{(m_w + c_{Cu}m_{Cu})(t - t_0)}{m_x(t_x - t)}$$

The values of t, t_x, t_0, m_w, m_{Cu}, and m_x are easily measured, while c_{Cu} can be found from the tables, for example, table 15.1. Therefore c_x is determined. In practice, however, precautions must be taken to minimize the effects of loss or gain of heat by interaction of the system with the surroundings.

TABLE 15.1 Specific Heat Capacities (cal/g · °C)
Water = 1.000

Ethanol	0.55	Glass	0.1–0.2
Paraffin	0.51	Iron	0.11
Ice	0.50	Copper	0.093
Steam	0.46	Mercury	0.033
Aluminum	0.21	Lead	0.031

Illustration 15.1

What is the minimum amount of water at 15°C which is needed to cool 200 g Hg at 80°C to a temperature of 25°C?

Reasoning Here too we can write, from the law of conservation of energy,

$$\text{Heat lost by Hg} = \text{heat gained by water}$$
$$(0.033 \text{ cal/g} \cdot °\text{C})(200 \text{ g})(55°\text{C}) = (1 \text{ cal/g} \cdot °\text{C})(m)(10°\text{C})$$

from which

$$m = 36 \text{ g}$$

Notice that, since water has a much larger specific heat than mercury, the amount of water needed is small in comparison. In practice, of course, one would have to do this experiment in a container, and provision for the effect of the container would have to be made.

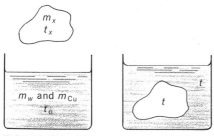

FIGURE 15.2 The method of mixtures. The calorimeter can containing the water is well insulated from the surroundings in practice.

15.4 SPECIFIC HEAT: MONATOMIC GAS; CONSTANT VOLUME

To compute from theoretical grounds what the specific heat capacity of a liquid or solid should be is an extremely difficult problem. However, for the dilute gas we have considered previously, the problem is simple and instructive. We assume as before that the molecules are like hard balls and that they can have only translational kinetic energy. This means that we consider the rotational and vibrational energy of the molecules to be negligible. Except at extremely high temperatures, a monatomic gas such as He, Ne, or Ar will satisfy these conditions.

For such a gas as this, contained in a fixed volume, the only significant way for the energy of the gas to increase is for the translational kinetic energy of the molecules to increase.* Therefore, if heat energy ΔQ is added to a mass m of such a gas, the translational kinetic energy of the molecules must increase by this same amount.

$$\Delta Q = \text{increase in } K_{trans}$$

If each molecule has a mass m_0, the total number of molecules is m/m_0, and therefore

$$\Delta Q = \frac{m}{m_0}[\tfrac{1}{2}m_0\langle v^2\rangle - \tfrac{1}{2}m_0\langle v_0^2\rangle] \qquad 15.2$$

where v_0 and v are the initial and final speeds, respectively.

But from the definition of absolute temperature T, we had

$$\tfrac{1}{2}m_0\langle v^2\rangle = \tfrac{3}{2}kT \qquad 14.12$$

where k is Boltzmann's constant. Substitution of this in 15.2 gives

$$\Delta Q = \frac{m}{m_0}\frac{3}{2}k(T - T_0)$$

or

$$\Delta Q = \frac{3k}{2m_0}m\,\Delta T \qquad 15.3$$

We defined the specific heat capacity to be

$$\Delta Q = cm\,\Delta T \qquad 15.1$$

Comparison of these two relations leads to the fact that

$$c_V = \frac{3k}{2m_0} \qquad 15.4$$

where the subscript V is placed on c in order to indicate that the volume of the gas is kept constant when the heat is added. Typical values for c_V are given in table 15.2. This relation can be put in another form by recalling from Sec. 14.5 that $k = R/N_A$, where R is the gas constant and N_A is Avogadro's number. Moreover, $N_A m_0$ is merely the molecular weight M of the gas. Therefore the specific heat capacity for a monatomic gas at constant volume is given by

$$c_V = \frac{3R}{2M} \qquad \text{or} \qquad Mc_V = \tfrac{3}{2}R \equiv C_V \qquad 15.5$$

TABLE 15.2 Specific Heat Capacities of Gases (cal/g · °C)

Gas (at 15°C)	c_V
He	0.75
Ar	0.075
O_2	0.155
N_2	0.177
H_2	2.40
CO_2	0.153
H_2O (200°C)	0.359
CH_4	0.405

*We must specify that the volume does not change as heat is added in order to be sure that no energy is lost in doing work during the process. The effect of changing volume will be considered in the next section.

Molar specific heat of a monatomic gas

The quantity Mc_V is the quantity of heat needed to raise the temperature of one mole of the gas one degree. For that reason it is sometimes called the *molar specific heat* and is represented by a capital letter, C_V. Note that in all our fundamental equations such as *15.5* we use the SI: c in J/kg · K, M in kg/kmol, R in J/kmol · K, and C_V in J/kmol · K.

Illustration 15.2

Helium gas is maintained in a 2.0-m³ tank originally at a pressure of 1.0×10^4 Pa and a temperature of 300 K. How much heat is required to raise the temperature to 400 K?

Reasoning We know the specific heat capacity of the gas under these conditions (monatomic, constant volume) to be given by *15.5*. In this case $M = 4.0$ kg/kmol and $R = 8314$ J/kmol · K. The mass of gas in the cylinder will also be needed. To find it we make use of the fact that, originally,

$$P_0 V_0 = \frac{m}{M} R T_0$$

and therefore

$$m = P_0 V_0 \frac{M}{R T_0}$$

In order to heat a substance through a range, ΔT, with the volume remaining constant, we know

$$\Delta Q = c_V m \, \Delta T = \frac{3}{2} \frac{R}{M} m \, \Delta T$$

Substituting the above value for m, we find

$$\Delta Q = \left(\frac{3}{2} \frac{R}{M} \right) \left(P_0 V_0 \frac{M}{R T_0} \right) \Delta T$$

Upon simplification, this becomes

$$\Delta Q = \frac{3}{2} \frac{P_0 V_0}{T_0} \Delta T$$

Placing in the values given, we find

$$\Delta Q = (\tfrac{3}{2})(1 \times 10^4 \text{ N/m}^2)(2.0 \text{ m}^3)(\tfrac{100}{300})$$
$$= 10^4 \text{ N} \cdot \text{m}$$

or

$$\Delta Q = 10^4 \text{ J} = 2.4 \times 10^3 \text{ cal}$$

15.5 WORK DONE BY A GAS

In the next section we compute the specific heat of a gas at constant pressure instead of at constant volume as was done in the preceding section. We shall see that the specific heats differ because of the work done as the volume increases. To show this we shall need to know how much work a gas does as it expands. We can find this work by considering the piston shown in figure 15.3.

Let us say that the pressure of the gas in the closed cylinder is P. If the piston has an area A, then the force exerted by the gas on the piston is $F_x = PA$. As the piston moves out a distance dx so that the gas expands, the

(a)   (b)

FIGURE 15.3 How much work is done by the gas as the piston moves out a distance Δx?

work done *by the gas* on the piston is

$$dW = F_x \, dx$$
$$= PA \, dx$$

However, $A \, dx$ is just the increase in volume of the gas, dV, therefore

In an expansion dV, a gas does P dV external work

$$dW = P \, dV \qquad\qquad 15.6$$

Notice that this is the work done *by* the gas. Since the gas does work, it must lose an equal amount of energy in order to do this work. If work is done *on* the gas by some external agent, then dW would be negative. Although it may seem trivial to emphasize this point, we shall find later that it is extremely important. If a system does work, then dW is positive. If work is done on the system, then dW is negative. Although the result found in *15.6* was for a special geometrical situation, it turns out to have general validity. To show this one needs to consider the arbitrary expansion of a closed surface. The mathematical detail involved is too complicated to allow its presentation here.

15.6 SPECIFIC HEAT: MONATOMIC GAS; CONSTANT PRESSURE

Let us consider what happens when the system shown in figure 15.4 is heated. Because the piston will always exert the same force (its weight) on the gas in the cylinder, the gas pressure will remain constant. As heat is slowly added to the gas, the pressure $P = (m/M)R(T/V)$ must remain constant. For this to be true, the volume must increase as T increases. Let us write the gas law twice, for the original situation and for after V has been increased to $V + \Delta V$, thus

$$P(V + \Delta V) = \frac{m}{M} R(T + \Delta T)$$

and

$$PV = \frac{m}{M} RT$$

Subtraction yields

$$P \, \Delta V = \frac{m}{M} R \, \Delta T \qquad\qquad 15.7$$

When heat is added to this constant-pressure system, the heat energy does two things. First, it heats up the gas molecules in the manner discussed in Sec. 15.4. We found there that the heat required for this purpose was

$$\Delta Q_1 = c_v m \, \Delta T$$

Second, the gas does work on the piston in order to lift it as the temperature is changed. This work is, from *15.6*,

$$\Delta W = P \, \Delta V$$

The heat energy needed for this is

$$\Delta Q_2 = P \, \Delta V$$

The total heat furnished to the gas for these purposes is therefore

$$\Delta Q = \Delta Q_1 + \Delta Q_2$$

or

$$\Delta Q = c_v m \, \Delta T + P \, \Delta V$$

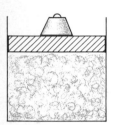

FIGURE 15.4 As the temperature of the gas increases, the movable piston maintains constant pressure. Physically, what are c_p and c_v and why is c_p larger than c_v?

Substituting for $P\,\Delta V$ from 15.7 gives, after rearrangement,

$$\Delta Q = \left(c_V + \frac{R}{M}\right) m\,\Delta T$$

If we compare this with the defining equation for specific heat capacity, the heat capacity at constant pressure c_P in this case,

$$\Delta Q = c_P m\,\Delta T$$

we have
$$c_P = c_V + \frac{R}{M} \qquad\qquad 15.8$$

c_P is larger than c_V by an amount equal to the work of expansion

Notice that c_P is larger than c_V, as we should expect, since it includes work done by the gas. Often this relation is written in terms of the molar specific heats, Mc_V and Mc_P, to give

$$C_P = C_V + R \qquad\qquad 15.9$$

We therefore have arrived at two rather fundamental relations for monatomic ideal gases. They are, in terms of molar quantities,

Important relations for an ideal monatomic gas

$$C_P - C_V = R = 8314 \text{ J/kmol} \cdot \text{K}$$

and
$$C_V = \frac{3R}{2} = 12{,}470 \text{ J/kmol} \cdot \text{K}$$

Since $R = 8314 \text{ J/kmol} \cdot \text{K} = 8.314 \text{ J/mol} \cdot \text{K}$, and since 1 cal equals 4.184 J, R can also be given the value $\approx 2.00 \text{ cal/mol} \cdot \text{K}$. In terms of this unit

$$C_P - C_V \approx 2.00 \text{ cal/mol} \cdot \text{K}$$

and
$$C_V \approx 3.00 \text{ cal/mol} \cdot \text{K}$$

The experimental values for several gases are shown in table 15.3.

We see at once that our theoretical predictions for C_P and C_V are well confirmed by experiment for the monatomic gases listed, but not for the others. This is not surprising since our derivation was restricted to monatomic gases, and many of the gases listed in table 15.3 are not of that type. In

TABLE 15.3 Experimental C_P and C_V Values
$(\text{cal/mol} \cdot \text{K})$

Gas (15°C)	C_V	C_P	$C_P - C_V$	$\gamma = \dfrac{C_P}{C_V}$
He	3.00	4.98	1.98	1.66
Ne				1.64
Ar	3.00	5.00	2.00	1.67
Kr				1.68
Xe				1.66
Hg (360°C)				1.67
O_2	4.96	6.95	1.99	1.40
N_2	4.96	6.95	1.99	1.40
H_2	4.80	6.78	1.98	1.41
CO	4.93	6.95	2.02	1.41
HCl	5.02	7.08	2.06	1.41
CO_2	6.74	8.75	2.01	1.30
H_2O (200°C)	6.46	8.46	2.00	1.31
CH_4	6.48	8.49	2.01	1.31

spite of this restriction, we note that the *difference* $C_P - C_V$ is predicted correctly for *all* the gases. The reason for this is that our discussion made no assumption concerning the internal structure of the molecules. The effect of internal structure on the specific heat is discussed in the next section.

15.7 EQUIPARTITION

When one examines the data in table 15.3, two rather striking correlations appear. First, as mentioned before, $C_P - C_V$ is 2.0 cal/mol · K for all the gases listed. This difference in specific heats is the result of the work done by the gas in expanding the volume when C_P is determined. We conclude, at least for these gases, that this work is independent of the exact molecular structure of the gas. Since the work is the result of the pressure of the gas and since the shape of the molecule was found to be of no importance when computing the pressure, this result seems reasonable.

The second feature of interest is the way in which C_V varies as we go from monatomic to diatomic gases. It was noticed by proponents of the kinetic theory that the value of C_V in cal/mol · K was related to the *number of degress of freedom* of the gas molecule involved. Let us digress for a moment to explain what we mean by a *degree of freedom*.

A point in space has three degrees of freedom. We can completely describe its behavior by giving its x, y, and z coordinates at all times. The number of independent coordinates needed to specify the motion of a system is the number of degrees of freedom of the system.

As another example, consider the two particles connected by a spring as shown in figure 15.5. Assume the spring can stretch but not bend. We can specify the linear motion of the molecule by giving the x, y, and z coordinates of its center of mass. However, the molecule could also vibrate lengthwise, and so the length between the two particles is a fourth necessary coordinate to describe its motion. In addition, the system could rotate about the three axes indicated, and therefore three angles need to be specified to tell what the system is doing. Therefore, to describe this system, we need three linear coordinates, three rotational coordinates, and one vibrational coordinate. This totals seven and so the system has seven possible degrees of freedom.

However, since, in the present case, we are considering the two particles to be point masses, their moment of inertia about the axis AA' is zero. Therefore the rotation about this axis is meaningless since it involves no energy and no observable motion of the point masses. As a result, only six coordinates are needed to describe the motion of the molecule in question, and so we assign only six, rather than seven, degrees of freedom to it.*

In general, in listing the degrees of freedom of a molecule, one ignores those coordinates which describe rotations for which the moment of inertia is negligibly small. For example, let us consider the system shown in figure 15.5 to be a diatomic molecule. Then the two particles will actually be atoms. Since most of the mass of an atom is in its nucleus, I for the molecule about the axis AA' will be negligibly small. For this reason, only two of the rotational degrees of freedom are counted, θ_1 being ignored.

From these considerations we see that a monatomic gas molecule should have three degrees of freedom. A diatomic molecule has three translational, two rotational, and one vibrational degrees of freedom. The experi-

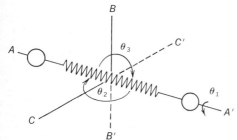

FIGURE 15.5 The system can rotate about the three mutually perpendicular axes AA' (angle θ_1), BB' (angle θ_2), and CC' (angle θ_3). It therefore has three rotational degrees of freedom. Why is θ_1 not counted for diatomic molecules?

*We shall see in Chap. 36 that a more fundamental reason exists for this neglect.

TABLE 15.4 Variation of C_V

Molecule	Degrees of freedom	C_V (cal/mol·K)
Monatomic	3	3.0
Diatomic	6	5.0

Equipartition theorem: translational and rotational degrees of freedom give $\frac{1}{2}kT$; vibrational degrees give kT

mental findings for C_V are given in table 15.3 and are summarized in table 15.4.

One of the major triumphs and, simultaneously, the major failure of the kinetic theory are found in the interpretation of the two sets of numbers shown in table 15.4.

Using the concepts of classical mechanics together with the Maxwell-Boltzmann distribution, one can calculate the average kinetic and potential energy associated with each degree of freedom of a molecule. This is done in a way analogous to that used to find the average kinetic energy of a molecule in the preceding chapter. The following important result, called the *equipartition theorem,* is then found. *The average energy associated with each translational and rotational degree of freedom is $\frac{1}{2}kT$, while an average energy kT is associated with each vibrational degree of freedom.* Let us call the number of translational, rotational, and vibrational degrees of freedom f_t, f_r, f_v, respectively. Then the average energy of a molecule should be

$$\tfrac{1}{2}kT(f_t + f_r + 2f_v)$$

Or since there are N_A molecules in a kilomole and since $N_A k$ is the gas constant R, the average energy of a kilomole should be

$$\text{Energy per kilomole} = \tfrac{1}{2}RT(f_t + f_r + 2f_v) \qquad 15.10$$

We can use this fact to obtain the molar specific heat capacity at constant volume, C_V. Since C_V is simply the increase in energy of one kilomole when the temperature is increased by one degree, we have

$$C_V = \tfrac{1}{2}R(f_t + f_r + 2f_v)(T + 1) - \tfrac{1}{2}R(f_t + f_r + 2f_v)T$$
$$= \tfrac{1}{2}R(f_t + f_r + 2f_v)$$

Or since $R = 2.0\ \text{cal/mol}\cdot\text{K}$,

C_V as given by the equipartition theorem

$$C_V = (f_t + f_r + 2f_v) \qquad \text{cal/mol}\cdot\text{K} \qquad 15.11$$

In other words, the equipartition theorem tells us that C_V should be $3 + 0 + 0 = 3\ \text{cal/mol}\cdot\text{K}$ for a monatomic gas. This agrees with the data quoted above.

However, for a diatomic molecule, one has

$$C_V = 3 + 2 + 2 = 7\ \text{cal/mol}\cdot\text{K}$$

This disagrees with the experimental result given previously, namely, $5\ \text{cal/mol}\cdot\text{K}$. The failure of the equipartition theorem in predicting the correct value for C_V when vibrational degrees of freedom are involved has fundamental importance. It cannot be explained using the concepts of classical mechanics. Only after the discovery of the laws of quantum mechanics was a satisfactory explanation for this discrepancy found.

When one discusses the mechanics of atomic-size particles, the classical mechanics we have used thus far must be extended. This more complete theory of mechanics, quantum mechanics, is discussed somewhat in later chapters. For now we point out that it shows that the molecular spring of figure 15.5 can vibrate only with certain special energies. At ordinary temperatures, the molecule is unable to acquire the necessary energy to vibrate more strongly than it does at the lowest temperatures. As a result, it does not increase its vibrational energy as the temperature increases and so the vibrational contribution to the specific heat is nonexistent.

At low temperatures, the vibrational degrees of freedom are inactive

15.8 IMPERFECT GASES AND VAN DER WAALS' EQUATION

All the gases we have discussed prior to this point have been assumed ideal, or perfect, gases. An ideal gas is one which obeys the perfect, or ideal, gas law,

$$PV = \frac{m}{M}RT$$

Most common gases obey this relation to high precision provided they are maintained at a temperature far above their liquefaction temperature. You will recall that the ideal gas law was *derived* assuming that the gas was so dilute that collisions between the molecules could be neglected. We now investigate when and why deviations from the perfect gas law occur.

A good experimental test of the gas law can be obtained by use of a gas confined in a cylinder at constant temperature T. Keeping T constant, the volume V of the gas is decreased, and the resultant pressure P is measured. Typical data are shown in figure 15.6 for a gas such as nitrogen at room temperature. The points are experimental, and the line is the relation predicted by the ideal gas law, i.e.,

$$P = \left(\frac{m}{M}RT\right)\frac{1}{V}$$

or

$$P = (\text{const})\frac{1}{V}$$

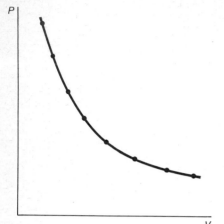

FIGURE 15.6 Under what conditions will a gas obey the ideal gas law graphed here?

It should be noted that even air (which is principally nitrogen) near normal atmosphere pressures and temperatures obeys the ideal gas law reasonably well. This is true even though the distance between collisions (the mean free path) is of the order of 10^{-5} cm under these conditions. Clearly, the restriction of *no* collisions between molecules in our derivation of the ideal gas law was too restrictive. Although we cannot show it here, much more detailed computation shows that the ideal gas law applies even when the collisions are numerous, provided certain other conditions (to be mentioned) are satisfied.

Suppose now that similar measurements are taken on the gas of figure 15.6 at lower temperatures. At a given volume, the pressure will of course decrease as the temperature decreases. Typical experimental curves are shown in figure 15.7. In addition, experiment shows that, at lower temperatures, the curve loses its ideal shape, as may be seen at T_2, and more pronouncedly at the lower temperatures. These deviations are due to two major effects.

First, as the volume of the gas decreases, the volume actually occupied by the molecules (i.e., roughly the number of molecules multiplied by the volume of each) becomes an appreciable fraction of the volume of the container. Let us consider only 1 kmol of molecules and denote by b the *covolume,* the volume we must subtract from V because of the presence of these molecules. We should then replace V in the gas law by $V - b$ to correct for the fact that the molecules are free to travel only in a volume that large.

Second, as the molecules are brought closer together, the forces between them become important. Since gases can be condensed to form liquids, we surmise that the molecules attract each other. Therefore the external pressure is actually aided by this attractive force when one tries to compress the gas. We should therefore add a term to the pressure P, which becomes larger as V becomes smaller and decreases to zero as V becomes very large.

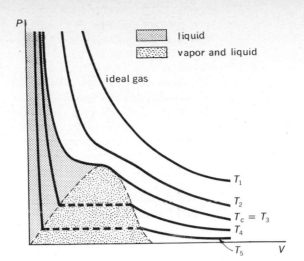

FIGURE 15.7 As the temperature is lowered from T_1 to T_5, the behavior of the gas becomes less ideal. What happens as the gas at T_4 is slowly compressed?

(This latter restriction is evident if one notices that at large volumes the molecules will be far apart most of the time, and so forces between them will, on the average, be very small.) It is found that when P is replaced by $P + a/V^2$, one is able to duplicate the data of figure 15.7 reasonably well.

The modified gas law incorporating these refinements was first stated by J. D. van der Waals and is known as his equation. It is

The van der Waals equation

$$\left(P + \frac{a}{V^2}\right)(V - b) = RT \qquad 15.12$$

where it is written for 1 mole of gas in the volume V. Although the constants a and b must be considered empirical, and are obtained by curve fitting to the experimental data, they do have sensible values. For example, in the case of nitrogen, $b = 40{,}000$ cm^3/kmol. Since each kilomole contains 6×10^{26} molecules, the covolume of a molecule should be $\frac{40}{6} \times 10^{-23}$, or about 70×10^{-24} cm^3. This is equivalent to saying that the covolume per molecule equals that of a sphere of radius 2.6 Å, which is a reasonable size since it is close to the size found for atoms in the preceding chapter by other methods.

Van der Waals' equation, *15.12*, predicts accurately the curves of figure 15.7 for T_1, T_2, and T_c. However, for temperatures and pressures where the curve enters the shaded region of the graph, *15.12* gives a complex relation between P and V. Outside the shaded region, even for these temperatures, the equation represents the data fairly well. We discuss the physical significance of this behavior in the next section.

15.9 THE LIQUID–VAPOR TRANSITION

In order to understand the deviations of a real gas from the ideal gas law, we must consider how the potential energy U_P of two molecules changes as they are brought closer together. Since the zero of potential energy is arbitrary, we follow custom and say that it is zero when the molecules are infinitely far apart. As two similar molecules, say two nitrogen molecules, are brought closer together, they attract each other. This is equivalent to saying that the system is "going downhill," i.e., to lower potential energy, as the molecules are brought together. We illustrate this in figure 15.8, where the distance between the molecules is designated r.

We see that the molecules attract each other at all separations for which $r > a_0$. (Recall that a bead placed on this curve will move in the direction in which the system, left to itself, would move. In addition, $F_r = -dU/dr$.) The two molecules repel each other strongly for separations smaller than a_0. This is merely a reflection of the fact that molecules, when very close together, act like hard balls. Clearly, if left to themselves, the two molecules would remain separated by a distance a_0. (The system, represented by a bead on the curve, is at stable equilibrium at that point.)

In practice, however, the molecules will have kinetic energy (thermal energy). As a result, they will oscillate about point a_0 much like a bead on the curve would do. For example, if the system has a maximum vibrational kinetic energy K_m as indicated on the graph, the hypothetical bead representing the system will oscillate back and forth between points P and Q. The molecules would still be bound together since they do not have enough thermal kinetic energy to go farther than the r value represented by point Q on the curve.

Suppose, however, that the thermal energy of the particles is larger than U_{PO}. In that case the bead on the curve, the system, will have enough energy to reach infinitely large values of r. The two molecules will not remain stuck together. They will be able to escape from the attraction force of one upon the other. Notice that this situation will prevail if the thermal energy (and therefore the temperature) is large.

Although the case of a large number of gas molecules subjected to pressure is much more complicated than the two-molecule situation we have just discussed, it is similar in certain very important respects. If the thermal energy of the molecules of the gas is low, the molecules are bound together, and they condense into a liquid. As one would expect, external pressure applied to the gas partly counterbalances the effect of thermal energy and favors condensation of the gas to a liquid. However, if the thermal energy of the gas molecules is too large, the molecules will tear loose from each other to

FIGURE 15.8 From this potential energy plot for two molecules, infer how the force on either molecule varies with the separation r.

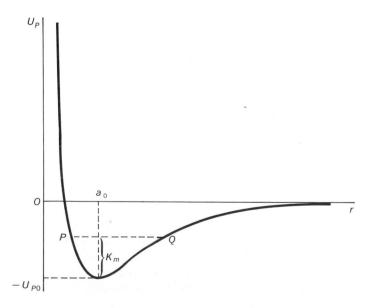

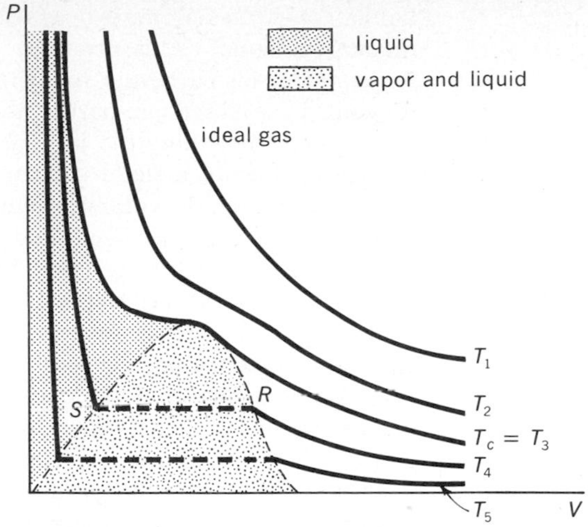

FIGURE 15.9 The gas shown cannot be liquefied if its temperature is above T_c, the critical temperature. You should be able to discuss the behavior of the substance as it is heated from very low temperature at (a) constant volume and (b) constant pressure.

form a gas no matter how high the pressure is made. These points can be illustrated by use of the experimental P versus V curves shown originally in figure 15.7 and redrawn in figure 15.9.

At a very high temperature such as T_1 the molecules have far too much thermal energy to stick together. The attraction energy is negligible, and the gas behaves ideally. At T_2 the thermal energy is small enough so that the interaction energy between molecules is not completely negligible. However, the gas still has too much energy to liquefy even at very high pressures. The particular temperature T_c, the *critical temperature,* is low enough so that the P versus V curve shows a point of zero slope. A gas at its critical temperature is borderline in that it is the highest possible temperature at which the gas may be liquefied. Critical temperatures and the pressure and density at the point of inflection are given in table 15.5 for several gases.

T_c, the critical temperature, is the highest temperature at which a gas can be liquefied

For temperatures below T_c (for example, T_4) one notices interesting behavior in the region labeled "vapor and liquid" (figure 15.9). As the gas is compressed at this temperature (going from extreme right to left along the curve), it behaves normally. However, when the pressure has been increased until the volume has decreased to point R, further attempts to increase the pressure are unsuccessful. When the volume is compressed, the pressure remains constant. In this region, some of the vapor or gas has condensed to

TABLE 15.5 Critical Point Data

Gas	T_c (°C)	P_c (atm)	Density (kg/m³)
Helium	−267.9	2.26	69.3
Hydrogen	−239.9	12.8	31.0
Neon	−228.7	25.9	484
Nitrogen	−147.1	33.5	311
Xenon	16.6	58.2	1,155
Chlorine	144.0	76.1	573
Benzene	288.5	47.7	304
Water	374.0	217.7	400
Mercury	1550	200	~400

liquid. When the volume is decreased, more gas condenses, and the pressure remains constant. This constant pressure is called the *saturated vapor pressure* for this particular temperature. It is, for example, the water vapor pressure in a closed jar partly filled with water. If some of the vapor is compressed by making the jar smaller, the pressure of the vapor does not change. Instead, a little of the vapor condenses and joins the liquid.

Eventually, the volume is reduced to the value prevailing at point S. At this point all the gas has been condensed, and the volume at point S is merely the volume of the resulting liquid. Of course, further attempts to decrease the volume will require very large changes in pressure since liquids are not easily compressed. It is for this reason that the pressure rises very steeply as one goes to smaller volumes beyond point S.

We see, then, that the curves of figure 15.9 delineate three quite well defined regions. All the region in white represents the material in the gas phase. When P, V, and T lie in this region, the material is a gas. In the center shaded region, part of the gas has condensed to liquid. For these values of P, V, and T, the two phases, liquid and gas, can exist in equilibrium with each other. However, for values of P, V, and T in the other shaded region, the material will be completely in the liquid state. You should make sure that you understand this diagram completely.

15.10 LATENT HEAT OF VAPORIZATION

Latent heat of vaporization; definition

We saw in the preceding section that attractive forces exist between molecules, and these forces cause the molecules to condense to a liquid. Conversely, if the liquid is to be separated into its individual molecules to form a gas (the process of vaporization), work must be done against these attractive forces. The total work done in separating unit mass of liquid into free molecules in the gaseous state is called the *latent heat of vaporization*. Since this is frequently accomplished by heating the liquid, the units of the latent heat are most often given in calories per unit mass. Typical values are given in table 15.6.

As one would expect, the latent heat decreases somewhat with increasing temperature. This results from the fact that at higher temperatures the molecules of the liquid contain more energy, and therefore less energy is required from outside to separate the molecules. Usually, this variation is rather small and is not important. The values given in table 15.6 are for vaporization at the normal boiling point indicated.

TABLE 15.6 Latent Heats

Substance	Boiling temp. (°C)	Vaporization (cal/g)	Melting temp. (°C)	Fusion (cal/g)
Nitrogen	−196	48	−210	6.1
Oxygen	−183	51	−219	3.3
Ethanol	+78	204	−114	25
Water	100	539	0	80
Mercury	357	65	−39	2.8
Lead			327	5.9

Illustration 15.3

Suppose 0.5 g of water evaporates from the surface of a beaker of water which originally contained 200 cm^3 of water. How much would the water be cooled if no heat exchange occurred with the surroundings?

Reasoning When a liquid evaporates, only the most energetic molecules have sufficient energy to break loose from the liquid. Those left behind will have less energy per molecule on the average than those which escape. As a result, the average molecular energy will decrease as evaporation proceeds, and the temperature will therefore decrease.

Each gram of water which evaporates requires 539 cal of heat, according to table 15.6. In our process, 0.5 g evaporated, and so the water as a whole must have furnished $\frac{1}{2} \times 539$, or about 270 cal. (Notice that we assume no heat added to the system, so that the system itself must furnish the heat of vaporization.) This heat came from the water itself, about 200 g. Using the relation between heat loss and temperature change,

$$\Delta Q = cm\,\Delta T$$

we have
$$\Delta T = \frac{270\text{ cal}}{(1\text{ cal/g}\cdot{}^\circ\text{C})(200\text{ g})}$$
$$\approx 1.35^\circ\text{C}$$

Of course, usually, the system is not well enough isolated from its surroundings to make such a computation valid.

Illustration 15.4

How much steam at 120°C is needed to heat 200 g of copper in a 60-g aluminum calorimeter can from 30 to 70°C?

Reasoning This is clearly a conservation-of-energy situation in which the energy lost by the steam equals the energy gained by the copper and aluminum. Notice that the steam first cools to 100°C, then condenses, giving off its latent heat, and the resultant water cools to 70°. Calling the mass of steam x,

$$\text{Heat lost by steam} = x(0.46\text{ cal/g}\cdot{}^\circ\text{C})(20^\circ\text{C}) + x(539\text{ cal/g})$$
$$+ x(1\text{ cal/g}\cdot{}^\circ\text{C})(30^\circ\text{C})$$
$$= 578x \quad \text{cal/g}$$

We have used the value of the specific heat of steam given in table 15.1. Similarly,

$$\text{Heat gained by Cu and Al} = (0.093\text{ cal/g}\cdot{}^\circ\text{C})(40^\circ\text{C})(200\text{ g})$$
$$+ (0.21\text{ cal/g}\cdot{}^\circ\text{C})(40^\circ\text{C})(60\text{ g})$$
$$= 1244\text{ cal}$$

Equating these two quantities gives

$$x = \tfrac{1244}{578}\text{ g} = 2.16\text{ g}$$

15.11 STATES OF MATTER

So far we have discussed two separate states of matter, gases and liquids. It is customary to group all other materials into a state called the solid state. Actually, though, this grouping is not as simple as one might think, because

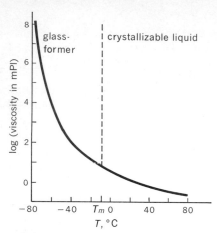

FIGURE 15.10 The variation of the viscosity of a glass-forming liquid (diethylphthalate) with temperature. Notice that as the temperature is lowered from −60 to −75°C, the substance becomes about a million times more viscous. A substance which melts at T_m would follow the dashed curve when cooled.

Glasses are low-kinetic-energy liquids

Crystalline solids possess an ordered array of atoms

Latent heat of fusion

of the existence of a material intermediate between liquids and crystalline solids, namely, glasses. As we shall see, a material can exist in two entirely different solidlike forms.

When a material is cooled from the gaseous state, it eventually reaches its boiling point and condenses. In order for it to change completely from gas to liquid, the energy of the molecules must be lowered so as to conform with this change in state. This is largely a change (a lowering) in potential energy of the molecules since work is done by the attractive forces as the molecules condense to liquid.

After the material has fully condensed, extraction of further heat energy from the liquid comes largely at the expense of the kinetic energy of the molecules. This is reflected in a decrease in temperature of the liquid. In addition, the liquid exhibits a higher viscosity as the temperature decreases. This is the result of the less rapid motion of the molecules at these lower temperatures. At first the viscosity increases rather slowly with decreasing temperature, but as the temperature is lowered further, the viscosity increases ever more swiftly, as shown in figure 15.10. Notice that the logarithm of the viscosity is plotted there. If the viscosity itself were plotted, the curve would rise much more sharply.

When the viscosity of a substance reaches a value of about 10^8 mPl (the viscosity of water is $\simeq$1 mPl), it is so viscous that it appears to be solid. For the material of figure 15.10 this occurs at a temperature near −75°C. These very viscous liquids, so viscous that they do not flow appreciably in periods of hundreds of years, are called glasses. Most transparent plastics such as polystyrene, Lucite, Plexiglas, and others are materials of this type. Ordinary window glass is also merely a very viscous liquid. It should be noted that the arrangement of the molecules in a glass is the same as that in a liquid since a glass is a liquid. This arrangement is characterized by the fact that the molecules are packed together in a more or less random way, much like grains of sand in a bucket of sand.

In many solids, however, this is not the case. When these substances are slowly cooled they suddenly transform directly to the solid state at a very definite temperature. This behavior is shown by the dashed line in figure 15.10. At the particular temperature shown as T_m the liquid suddenly becomes a solid. Conversely, when the solid is heated, it melts to a liquid at T_m. Of course, T_m is the common melting point.

Certain other things happen at T_m. The arrangement of the molecules in the material changes from the random pattern of a liquid to the highly ordered pattern characteristic of a crystal. These are the substances we refer to as crystalline solids. Typical arrangement of atoms in a few crystals are shown in figure 15.11. One would expect that the energy of the substance would change considerably in such a transformation of the atom arrangement. This is indeed true, as we shall see in the next section, and this fact aids us in distinguishing between glassy and crystalline solids.

15.12 LATENT HEAT OF FUSION

A solid crystalline material requires the addition of energy to it if its crystalline lattice structure is to be broken up so as to become liquidlike. The energy required to melt a crystal (i.e., to randomize the molecules) is called the *latent heat of fusion*. Typical values are given in table 15.6, and since this quantity represents the heat needed to melt a unit mass of crystal, its units are

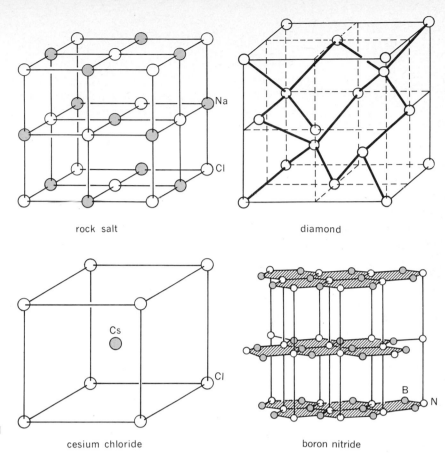

rock salt

diamond

Cs

Cl

cesium chloride

B

N

boron nitride

FIGURE 15.11 Crystal lattices of several solids.

commonly given as calories per unit mass. When a crystal melts, the heat of fusion must be added to it. When a liquid crystallizes, the substance loses the heat of fusion.

Notice the very high values of the heat of fusion for water and alcohol. These substances have H and OH groups in their molecules. Adjacent molecules are highly attracted to each other by the attraction between these two groups in neighboring molecules. This attractive force is often referred to as a *hydrogen bond*. Substances which possess hydrogen bonding between adjacent molecules need more than the usual energy to break the molecules out of an ordered pattern. Even close to the melting point the hydrogen bonds affect the behavior of water and alcohol. It is for this reason that these substances show various anomalies in this region. The anomalous variation of the density of water near the melting point is one of the better-known examples of this effect.

To summarize the behavior of a substance which crystallizes, let us consider a simple ideal experiment. Suppose 100 g of ice in a sealed container is heated in such a way that 1 cal of heat is added each second. Let us then follow the temperature variation of the ice as it is assumed to warm uniformly.*

*The procedure outlined here is very closely related to the technique followed in *differential thermal analysis,* DTA, a common tool in materials science.

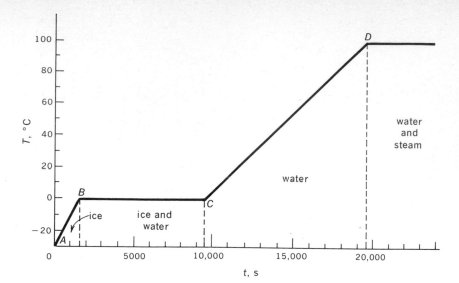

FIGURE 15.12 You should be able to explain this graph quantitatively. The 100 g of water is heated at a rate of 1 cal/s.

We shall assume the starting temperature of the ice to be $-30°C$ as shown in figure 15.12. From table 15.1 we note that the specific heat capacity of ice is 0.50 cal/g · °C. At our rate of heating, 0.01 cal/g · s, it will take 50 s for the temperature of the ice to rise 1°C. For this reason we draw the slope of the heating curve in figure 15.12 to be 2°C/100 s as long as the material is still ice. This is section AB of the curve shown.

When the ice reaches 0°C it begins to melt. All the heat energy added at this time is used to tear the ice crystals apart and to transform the ice to liquid. Not until all the ice is melted does the temperature of the material rise further. Since each gram of ice requires 80 cal to melt it, the 100 g of ice will melt in a time of 8000 s. During this melting time the temperature remains at 0°C. This is region BC in figure 15.12.

After the material has become liquid water at 0°C, it will once again begin to warm. You should be able to show that the slope in the region CD is correctly drawn. When the water reaches 100°C, it begins to boil and to change to steam. During this interval the temperature remains constant. How much heat must be added before the graph in figure 15.12 will once again begin to rise?

Illustration 15.5

How much steam at 120°C is needed to just melt 50 g of ice in a 20-g copper calorimeter can if the starting temperature is $-30°C$?

Reasoning The heat lost by the steam must equal the heat gained by the can and the ice. Using data from tables 15.1 and 15.5, we have, denoting the mass of steam by x,

$$\text{Heat loss} = \text{heat gained}$$
$$(0.46)(x)(20) + 539x + (1)(x)(100) = (0.093)(20)(30)$$
$$+ (0.50)(50)(30) + (50)(80)$$

The student should identify each term in this equation and supply its proper units. Solving for x, the quantity of steam needed, we find it to be 7.4 g.

15.13 THERMAL EXPANSION

The potential energy curve for two molecules a distance r apart was discussed in Sec. 15.9. Taking the potential energy U_P of two molecules to be zero when the molecules are far separated, a plot of U_P versus r appears as shown in figure 15.13. Recalling that the system behaves much like a bead sliding on this curve, we see that, in the absence of any kinetic energy, the system would exist at the bottom of the U_P curve. This separation between the molecules would be a_0.

In general, however, the molecules will have kinetic energy. If their kinetic energy is such that the system can oscillate back and forth from A to B on the curve, the average separation of the molecules will be a_1. That $a_1 \neq a_0$ is a reflection of the fact that the potential energy curve is not quite symmetric. Since it rises more sharply at small r values, the system will move more to large r values, and therefore the average value of r increases as the system acquires more energy, and is able to move higher up onto this curve.

Molecular separation increases as T increases

In any case, we see that $a_0 < a_1 < a_2$, indicating that the average molecular separation increases with increasing kinetic energy. Or since temperature is a measure of kinetic energy, the molecular separation increases with increasing temperature.

A similar situation exists in the more complex systems of solids and liquids. As the temperature increases, the average separation of the molecules increases. This is reflected in an increase in volume, area, and length of objects as their temperatures are increased. Experimentally, as well as theoretically, it is found that an increase in temperature ΔT causes fractional increases in volume V, area A, and length L, given by the relations

Thermal expansion relations

$$\frac{\Delta V}{V} = \gamma \, \Delta T \qquad\qquad 15.13a$$

$$\frac{\Delta A}{A} = \beta \, \Delta T \qquad\qquad 15.13b$$

$$\frac{\Delta L}{L} = \alpha \, \Delta T \qquad\qquad 15.13c$$

The quantities γ, β, and α are called the volume, area, and linear thermal expansion coefficients, respectively. They have been tabulated from experimental data, and typical values for γ and α are given in table 15.7.

FIGURE 15.13 Notice how the asymmetry of the U_P curve causes the average distance between molecules to increase as the molecular kinetic energy increases.

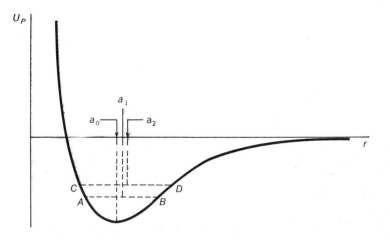

TABLE 15.7 Thermal Expansion
Coefficients (per °C at 20°C)

Substance	$\alpha \times 10^{6*}$	$\gamma \times 10^6$
Diamond	1.2	3.5
Glass (heat-resistant)	≈ 3	≈ 9
Glass (soft)	≈ 9	≈ 27
Concrete	≈ 10	≈ 30
Iron	12	36
Brass	19	57
Aluminum	25	75
Mercury		182
Rubber	≈ 80	≈ 240
Glycerin		500
Gasoline		≈ 950
Benzene		1240
Acetone		1490

$^*\alpha \times 10^6$ means that the values found in this table are α values multiplied by a factor of 10^6.

Several things should be noticed in regard to table 15.7. In the first place, the values of α and γ are only approximate, and apply only in the region near 20°C. For most materials the error involved in using these constants even at temperatures as high as 300°C will be only a few percent. A more serious difficulty arises when one notes that these equations can be written

$$\Delta V = \gamma V \Delta T$$

which are of the form

$$y = mx + 0$$

the equation of a straight line. The slope in this particular case is γV, and if we wish it to be constant, a particular V must be specified when determining γ. It is for this reason that the restriction "at 20°C" is placed on the values in table 15.7. We mean by this that V and L are to be the values measured at 20°C when these values of γ and α are used.

Notice further that the volume expansion coefficients are approximately three times as large as the linear expansion coefficients. This follows at once from the following considerations. If a cube of original edge length L_0 is heated, its edge will increase in length by ΔL. The new volume of the cube will be, assuming similar expansions in all directions,

$$V = (L_0 + \Delta L)^3 = L_0{}^3 + 3L_0{}^2\,\Delta L + 3L_0(\Delta L)^2 + (\Delta L)^3$$

This can be rewritten

$$V = L_0{}^3\left[1 + 3\frac{\Delta L}{L_0} + 3\left(\frac{\Delta L}{L_0}\right)^2 + \left(\frac{\Delta L}{L_0}\right)^3\right]$$

Since the linear expansion will be small, the factor $\Delta L/L_0$ will be much smaller than unity; so, to good approximation, we have

$$V \approx L_0{}^3\left(1 + 3\frac{\Delta L}{L_0}\right)$$

If we now recognize that $L_0{}^3 = V_0$ and that $\Delta L/L_0$ is simply $\alpha\,\Delta T$, we have, after rearranging, that

$$\frac{\Delta V}{V_0} = 3\alpha\,\Delta T$$

where ΔV is written in place of $V - V_0$. Comparing this with *15.13*, we see that

$$\gamma = 3\alpha$$

which confirms the values given in table 15.7.

Illustration 15.6

A brass sheet has a circular hole in it. The diameter of the hole is 2.000 cm at 20°C. How large will the hole diameter be when the sheet is heated to 220°C?

Reasoning The metal sheet will expand in the same way whether or not the hole is in it. Therefore the hole will expand in the same way as the circular piece of metal which originally filled the hole would expand. Its diameter would obey the relation

$$\frac{\Delta L}{L} = \alpha\,\Delta T$$

Since L is given for $20°C$, we may use the data of table 15.7 directly. Therefore

$$\Delta L = (2.000 \text{ cm})(19 \times 10^{-6}/°C)(200°C)$$

or

$$\Delta L = 0.0076 \text{ cm}$$

The new diameter of the hole would be 2.008 cm. Notice in this case that whether the value of L at 20 or $220°C$ is used to find ΔL is of no significance.

Illustration 15.7

How much will the volume of 100.0 cm^3 of benzene (measured at $10°C$) change as the benzene is heated to $30°C$?

Reasoning Since the constants in table 15.7 are given for $20°C$, we must first find the volume at that temperature, V_{20}. We have

$$\frac{\Delta V}{V_{20}} = \gamma \, \Delta T$$

Since $\Delta V = V_{20} - V_{10}$, or $V_{20} - 100 \text{ cm}^3$, this becomes

$$V_{20} - 100 \text{ cm}^3 = V_{20}(1.24 \times 10^{-3})(10)$$

from which

$$V_{20} = 101.26 \text{ cm}^3$$

Now using this value for V in *15.13*, we have

$$\Delta V = (101.3 \text{ cm}^3)(1.24 \times 10^{-3})(10)$$

$$= 1.25 \text{ cm}^3$$

Therefore the volume at $30°C$ is

$$V_{30} = 101.3 + 1.3 = 102.6 \text{ cm}^3$$

If we had not corrected V to $20°C$ before using *15.13* to compute V_{30}, we would have found $V_{30} = 102.5 \text{ cm}^3$. For most purposes this difference in the two values is unimportant, and so it is usually not necessary to correct for the temperature at which α and γ are measured, provided the temperature range involved is not far removed from $20°$.

15.14 HEAT CONDUCTION

When one portion of an object is at a higher temperature than another, heat energy flows from the higher-temperature portion to the region of low temperature. This process of heat transfer is achieved by means of molecular collisions and is called *heat conduction*. As we have seen previously, the temperature of an object is a measure of the kinetic energy of the molecules. When two groups of molecules are allowed to contact each other, they exchange and share kinetic energies as a result of collisions between the molecules. The hot, more energetic molecules lose energy to the cold, less energetic molecules. As a result, heat energy flows from the hot molecules to the cold molecules.

We can describe the flow of heat quantitatively by the following relation. Suppose a slab of material of thickness Δx (figure 15.14) connects two regions at temperature T_1 and T_2, with $T_1 > T_2$. If the area of the slab is A, one finds experimentally that the rate at which heat flows through the slab, $\Delta Q/\Delta t$, is proportional to the product of three factors: the cross-

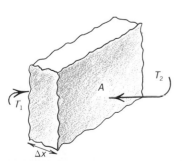

FIGURE 15.14 What is the rate of heat flow through this slab?

sectional area, the temperature difference, the reciprocal of the thickness. Each of these three factors seems reasonable. (Why?) As an equation,

$$\frac{\Delta Q}{\Delta t} = (\text{const})\frac{A(T_1 - T_2)}{\Delta x}$$

The constant of proportionality is usually represented by the Greek letter λ and is called the *thermal,* or *heat, conductivity* of the material of the slab. For most materials it is reasonably temperature-independent. Writing the temperature difference as ΔT, we then have

The heat conduction equation

$$\frac{\Delta Q}{\Delta t} = \lambda A \frac{\Delta T}{\Delta x} \qquad\qquad 15.14$$

The quantity $\Delta T/\Delta x$ is the change in temperature per unit length and is often called the *temperature gradient.* Typical values for λ are given in table 15.8.

Metals are good heat conductors

It will be noticed in table 15.8 that metals are very good heat conductors. This is a result of the fact that the valence electrons, which are rather free to move in the metal, are able to transport energy from place to place. As a result, heat is conducted through the material by both the valence electrons and the collisions of the vibrating atoms. In nonconductors there are no free electrons, and so the thermal conductivity is reduced.

TABLE 15.8 Heat Conductivities
[(cal/sec)/$^\circ$C·cm]

Material	λ
Silver	1.00
Copper	0.92
Aluminum	0.50
Brass	0.25
Glass	$\approx 20 \times 10^{-4}$
Asbestos paper	$\approx 5 \times 10^{-4}$
Rubber	$\approx 5 \times 10^{-4}$

Illustration 15.8

A rod of length L has a cross-sectional area A. It is insulated in such a way that no heat can escape from its surface except at the two ends. If one end is maintained at T_1 and the other at T_2 ($T_2 < T_1$), what is the rate of heat conduction down the rod when a steady situation prevails?

Reasoning This situation is shown in figure 15.15. Since no heat is lost out the sides, the heat-transfer rate across all cross sections must be the same when a steady state is reached. If this were not true, heat energy would accumulate at some point on the rod, and its temperature would change. We assume such things are no longer happening. Writing *15.14*, we have

$$\frac{\Delta T}{\Delta x} = \frac{1}{\lambda A}\frac{\Delta Q}{\Delta T}$$

Since we have agreed that $\Delta Q/\Delta t$ is constant at all values of x, the temperature gradient $\Delta T/\Delta x$ must be a constant. Passing to the limit of small x, we have

$$\frac{dT}{dx} = (\text{const}) = \frac{1}{\lambda A}\frac{\Delta Q}{\Delta t}$$

Let us write this

$$dT = (\text{const})\,dx$$

and integrate from one end of the rod to the other. Thus

$$\int_{T_1}^{T_2} dT = (\text{const})\int_0^L dx$$

After integrating and putting in the limits and replacing the constant by its value, we have

$$T_2 - T_1 = \frac{L}{\lambda A}\frac{\Delta Q}{\Delta t}$$

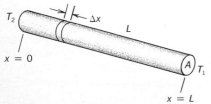

FIGURE 15.15 Under what conditions will the temperature gradient along this rod be constant?

or

$$\frac{\Delta Q}{\Delta t} = \frac{\lambda A}{L}(T_2 - T_1) \qquad\qquad 15.15$$

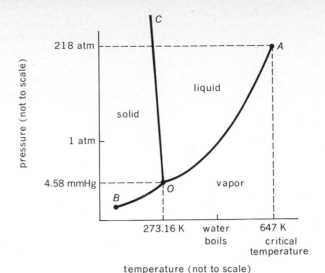

FIGURE 15.16 Point O is the triple point of water, and point A is its critical point. Be sure to understand the difference between them.

Equation *15.15* is commonly used for heat conduction through a rod. However, it must be realized that it is restricted to the case of no heat loss from the surface.

15.15 THE TRIPLE POINT

We know that if the temperature of a gas is below the critical temperature, the gas may be liquefied by application of pressure. Or it may be liquefied and subsequently solidified by lowering the temperature. The variation of the phase state (gas, liquid, or solid) of a substance is conveniently summarized by a temperature-pressure *phase diagram*. This type of phase diagram in the case of water is plotted in figure 15.16. (Notice that the temperature- and pressure-axes scales are not linear.)

When the temperature and pressure of the water are represented by a point in the region below curve BOA, the water is in the vapor form. When P and T are such as to give a point in the region bounded by OA and OC, the water is in liquid form. In the region to the left of BOC the water is in the form of ice.

We notice that there is one particular point at which the ice, liquid, and gas regions touch, point O. At this particular point $T = 273.16$ K and $P = 4.58$ mmHg. It is only at these particular values of T and P that solid, liquid, and vapor can exist in equilibrium together for water. This point is called the *triple point* of a substance. The triple point of water is of great importance since it is used as a fixed point for defining the Kelvin temperature, as we saw in the preceding chapter.

Solid, liquid, and vapor can coexist at the triple point

Chapter 15
Questions
and Guesstimates

1. How would one go about assigning a temperature to a region in space? What does it mean to say that the temperature of outer space is about 3 K? The temperature at the outer edge of the earth's atmosphere is said to be of the order of 1000 K. What does this mean?
2. Explain clearly why C_P is larger than C_V for a gas. Would this also be true for a liquid? A solid?

3. Under which condition would an ideal gas suffer the larger temperature change: when it expands from V_0 to V by pushing out a piston or when it is allowed to escape from volume V_0 into a vacuum chamber of volume V?

4. A gas is contained in a cylinder of volume V_0. If the piston closing the cylinder is quickly pushed in to decrease the volume to $V_0/10$, what will happen to the temperature of the gas? Will the change in temperature depend upon whether the gas is He or Ne? Whether it is Ne or oxygen (O_2)?

5. Although argon gas molecules (Ar) and oxygen molecules (O_2) have roughly the same molecular mass, it takes about twice as much energy to heat oxygen as argon. By reference to their internal energies, explain why this result is not unexpected.

6. In an experiment, a student is given a Thermos jug containing an unknown substance at temperature T_1. A quantity ΔQ of heat is added by adding hot water. After equilibrium is again established, the temperature is still T_1. The student concludes that the specific heat of the material in the Thermos jug is infinite. Explain why the experiment implies $c = \infty$. What is the probable explanation of these experimental results?

7. Can heat be added to something without its temperature changing? What if the "something" is a gas? A liquid? A solid? Explain.

8. A particular type of wax melts at 60°C. Describe an experiment by which one could determine the heat of fusion of the wax.

9. How would one go about computing the heat needed to raise the temperature of a mixture of known weights of two gases by 10°C?

10. It is possible to make water boil furiously by cooling a flask of water which has been sealed off when boiling at 100°C. Explain why.

11. Under what conditions is the work done by a gas equal to the decrease in the translational kinetic energy of the gas?

12. Suppose you have a container half filled with water in equilibrium with its vapor. If nitrogen gas is now added to the water vapor, will this affect the equilibrium between water vapor and water? Use molecular considerations to justify your answer, and consider the case of very high nitrogen pressure as well as the low-pressure case.

13. Temperature fluctuations are much less pronounced on land close to large bodies of water than they are in the central regions of large land masses. Explain why.

14. If you were given a closed metal container filled with an unknown substance, could you determine whether the substance was liquid, gas, crystalline solid, or glass? The mass of the empty container is unknown, but the number of calories needed to raise it 1°C is known. You are allowed to carry out any thermal experiment you wish on it. It is known that the metal has a reasonably constant specific heat.

15. In the winter, frost often forms on the inside of house windows even though the house is comfortably warm. Explain how this occurs.

16. Why does a piece of steel feel colder than a piece of wood at the same temperature?

17. What would happen if the earth were to be covered by a dense layer of smog so that the rays from the sun could not reach its surface?

18. Estimate how much the temperature of a human body would rise in 1 day if it retained the approximately 2000 large calories (kilocalories) acquired in food in one day. The value for c for a person is about 0.83 cal/g · °C.

19. It is well known that a room filled with people becomes very warm unless

it is properly ventilated. Assuming that a person gives off heat equivalent to the person's food energy in a steady way throughout the day, estimate how much the temperature of your classroom would rise in 1 h if there were no heat loss out of the room.

20. About how much water would have to evaporate from the skin of an average-size man to cool his body by 1°C? How does this fit in with what you have heard about the effect of perspiration on the body? ($c_{body} \approx 0.83 \text{ cal/g} \cdot °C$.)

21. If ice is subjected to high pressure, its melting point is decreased below 0°C. To a rough approximation, the melting temperature decreases by about 5°C for each additional 6000 N/cm² of applied pressure. Estimate the melting temperature of ice beneath an ice skater's skate.

Chapter 15
Problems

1. Forty grams of water is to be heated from 20 to 80°C. How many calories of heat is required? Joules? Repeat for 40 g of iron.

2. How much heat is lost by 300 g of mercury as it cools from 200 to 50°C? Express your answer in calories and joules.

3. How much water at 0°C is needed to cool 500 g of water at 80°C down to 20°C?

4. How much oil at 200°C must be added to 50 g of the same oil at 20°C to heat it to 70°C?

5. A 500-g piece of iron at 400°C is dropped into 800 g of oil at 20°C. If $c = 0.40 \text{ cal/g} \cdot °C$ for the oil, find the final temperature of the system. Assume no loss to the surroundings.

6. 200 g of ethanol at 20°C is poured into a 400-g aluminum container at $-60°C$. What will be the final temperature of the system if interaction with the surroundings can be neglected?

7. A 60-kg boy running at 5.0 m/s while playing basketball falls to the floor and skids along on his leg until he stops. How many calories of heat are generated between his leg and the floor? Assuming all this heat energy is confined to a volume of 2.0 cm³ of his flesh, what will be the temperature change of the flesh? Assume $c = 1.0 \text{ cal/g} \cdot °C$ and $\rho = 950 \text{ kg/m}^3$ for flesh.

8. In an extruder used to make plastic filaments, a piston applies a pressure of 10,000 lb/in² ($70 \times 10^6 \text{ N/m}^2$) to the molten plastic and forces it through a tiny nozzle at a rate of 0.0010 cm³/s. Assuming all the energy loss occurs as friction work in the plastic as it goes through the nozzle, how much friction work is done each second? How much is the temperature of the plastic raised as it is forced through the nozzle? ($c_{plastic} = 0.20 \text{ cal/g} \cdot °C$; $\rho = 1.0 \text{ g/cm}^3$.)

9. A weight attached to a string which passes over a pulley causes a paddle wheel to turn in a water container as the weight falls. The paddle is so constructed that the weight falls slowly, thereby acquiring negligible kinetic energy. Its decrease in potential energy is therefore converted entirely into friction work or heat energy in the water. Suppose the 200-g mass fell 2.0 m. How much would it heat the 100 g of water in the container if no heat were lost to the paddle or container? (This is the basic idea behind Joule's experiment for determining the mechanical equivalent of heat.)

10. A mass m of lead shot is placed at the bottom of a vertical cardboard cylinder that is 1.5 m long and closed at both ends. The cylinder is

suddenly inverted so that the shot falls 1.5 m. By how much will the temperature of the shot increase if this process is repeated 100 times? Assume no heat loss.

11. A 2.2-g lead bullet is moving at 150 m/s when it strikes a bag of sand and is brought to rest. (*a*) Assuming that all the frictional work was transferred to thermal energy in the bullet, what is the rise in temperature of the bullet as it is brought to rest? (*b*) Repeat if the bullet lodges in a 50-g block of wood that is free to move.

12. A cubical room $2.5 \times 10 \times 7$ m^3 is filled with nitrogen gas at 1×10^5 Pa and 27°C. What is the mass of gas in the room? How many calories are needed to increase the room temperature 1°C assuming all the heat goes into the gas? (Assume constant volume.)

13. A refrigeration system is needed which will reduce the temperature of a 1000-m^3 volume of air (which we will assume to be pure nitrogen at 1×10^5 Pa and 27°C) by 5°C in a certain length of time. Find the mass of gas involved and the number of calories which must be removed from it. (Assume constant volume.)

14. Twenty grams of CO_2 gas is confined to a cylinder by a piston. How much heat energy (in calories) is needed to heat this gas 5°C at constant volume? At constant pressure?

15. Assuming air to be composed of 78 percent nitrogen and 22 percent oxygen by weight, what should be the value of c_V for air?

16. One-half mole of helium gas is confined to a container at standard pressure and temperature (1.01×10^5 Pa, 273 K). How much heat energy is needed to double the pressure of the gas?

17. What would you expect c_v and c_p to be for gaseous hydrogen bromide, HBr? Give your answer in cal/g · °C.

18. Solid sodium metal has a density of 970 kg/m^3. (a) How many sodium atoms are there per cubic meter? (b) Each sodium atom contributes one electron to a "free electron gas" within the metal. Assuming it to be an ideal gas, what is the contribution (in cal/g · °C) of the electron gas to the specific heat of sodium metal?

19. How many grams of ice at 0°C must one add to a 200-g cup of coffee at 90°C to cool it to 60°C? Assume heat transfer with the surroundings to be negligible.

20. How much Coke can be cooled from 30 to 10°C by a 25-g ice cube at 0°C?

21. How much steam at 120°C is needed to heat 800 g of aluminum at 20°C to 70°C?

22. Assuming that the total heat of vaporization of water, 539 cal, can be used to supply the energy needed to tear 1 g of water molecules apart from each other, how much energy is needed per molecule for this purpose? Find the ratio of this energy to kT at the boiling point.

23. Suppose the standard of length was a 1-m-long bar of iron. What would be the maximum temperature variation of the bar if its length were to be preserved to an accuracy of 1 part per million?

24. If a car's gas tank were filled to the brim with 16 gal when the temperature was 20°C, how much gas would run out when the temperature rises to 30°C?

25. A brass sleeve is to be shrink-fitted onto a rod of 2.000 cm diameter at 20°C. How large should the internal diameter of the sleeve be at 20° if it is to just slip over the rod when the sleeve is at a temperature of 400°C?

26. A grandfather clock has a pendulum made of brass. The clock is adjusted

to tick out seconds exactly at 20°C. If operated at 30°C, how much will the clock be in error 1 week after it is set? Will it be fast or slow?

27. Show that equation *15.13a* implies the following for the temperature variation of density: $\Delta\rho = -\gamma\rho\,\Delta T$. What is the meaning of the minus sign?

28. The two ends of a horizontal iron girder 8 m long and having a solid cross section of $150\,cm^2$ are anchored firmly in massive concrete posts when the temperature is 4°C. How large a horizontal force will the girder exert on the post when the temperature is 30°C?

29. A uniform solid brass sphere of radius b_0 and mass M is set spinning with angular speed ω_0 about a diameter. If its temperature is now increased from 20 to 80°C without disturbing the sphere, what will be its new (*a*) angular speed and (*b*) rotational kinetic energy?

30. Show that the volume thermal expansion coefficient for a gas at constant pressure is $1/T$.

31. When a substance is heated and expands, it does work against the atmosphere. This work contributes a part, but not all, of the difference between c_P and c_V for the substance. Compute the difference between c_P and c_V that results from this cause in the case of benzene (density of benzene = $879\,kg/m^3$).

32. Deep bore holes into the earth show that the temperature increases about 1°C for each 30 m of depth. Assuming that the earth's crust has a thermal conductivity of about 2×10^{-3} (cal/sec)/°C cm, how much heat flows out through the surface of the earth each second for each square meter of surface area?

33. Consider a glass window of area $1\,m^2$ and thickness 0.50 cm. If a temperature difference of 20°C exists between one side and the other, how many calories would flow through the window each second? Why is this result *not* applicable to a house window on a day when the temperature difference between inside and outside is 20°C?

34. A 0.50-cm-thick sheet of brass has sealed to one face a 0.50-cm-thick rubber sheet. The other side of the brass sheet is connected to a bath maintained at 20°C. The other rubber surface is attached to a circulating bath at 80°C. Find the temperature at the rubber-brass junction.

35. The specific heat of many solids at very low temperatures varies with absolute temperature T according to the relation $c = AT^3$, where A is a constant. How much heat energy is needed to raise the temperature of a mass m of such a substance from $T = 0$ to $T = 20\,K$?

16 Thermo-dynamics

Thermodynamics, as the word implies, is the study of heat flow and motion. More precisely, it is concerned with the interrelation between heat flow, work, and the energy resident in a system. Long before our understanding of atoms and molecules became wide enough to describe thermal behavior in molecular terms, the laws of thermodynamics had been discovered. They describe the thermal behavior of systems in terms of pressure, volume, and temperature. This way of looking at phenomena is of great utility and is widely used. We shall obtain an introduction to it in this chapter.

16.1 THE FIRST LAW OF THERMODYNAMICS

The first law of thermodynamics is a statement of the law of conservation of energy. In order to state it, we consider what is called a *closed system,* i.e., a definite segregated collection of particles. This might consist of a gas confined to a cylinder by a piston, in which case the system is the molecules in the gas. Or a system may be very large; for example, the earth and everything associated with it. In any given case we must know exactly what constitutes the system if we are to be able to describe it.

Consider now a closed system. It might be the molecules of a gas confined by a piston, as shown in figure 16.1. These molecules, as well as any other system, will possess energy U. For example, this may be kinetic energy of translation, rotation, and vibration, together with potential energy of stretched molecular bonds and interaction energy between the molecules. In any case, U is used to denote the total energy possessed by the system. It is often called the *internal energy* of the system since the energy is a property of the individual molecules constituting the system.

Suppose now that an amount of heat energy đQ flows *into the system.*[*] The energy of the system must be increased since the law of conservation of energy tells us that the energy cannot just disappear. In the absence of all other changes in the system, the heat energy added must equal the *increase* in internal energy of the system, ΔU.

It is not necessary, however, that all the added heat energy will result in increased internal energy. Instead, some of it may be used to do work. For

U is the internal energy of a system

FIGURE 16.1 As heat is added to the gas, the first law tells us that $\Delta U = $ đ$Q - P\,\Delta V$.

[*] đQ, rather than ΔQ or dQ, is used to represent this energy since we do not wish to give the impression that one can define the heat Q *in* the system. The added heat energy becomes kinetic energy or results in the doing of work when it enters the system. We cannot separate a "heat portion" from the total system energy U.

example, in figure 16.1 the piston will rise as heat is added to the gas. Since the system will do work as its volume increases by ΔV, an amount of heat energy equal to this work, $\Delta W = P\,\Delta V$, will not be contributed to the internal energy.

Other examples of the work done by a system can be imagined. The energy conservation law tells us that in each case, no matter how complicated, if the system does work ΔW, that work must be done at the expense of the energy of the system. Therefore, when a system does work ΔW, its energy must change by an amount $-\Delta W$. The negative sign tells us that when the system does work it loses energy.

We can now combine these three energy changes: (1) an amount of heat $đQ$ is *added to* a system; (2) the system *does* an amount of work ΔW and therefore *loses* an energy ΔW; (3) the internal energy of the system increases by an amount ΔU. The law of conservation of energy now tells us that the increase in internal energy is equal to the heat added less the work done,

The first law is a statement of energy conservation

$$\Delta U = đQ - \Delta W \qquad\qquad 16.1$$

Equation *16.1* is a statement of the *first law of thermodynamics*. As discussed above, *if external work is done only by expansion*, then *16.1* can be rewritten

$$\Delta U = đQ - P\,\Delta V \qquad\qquad 16.2$$

This relation is actually not new to us since we derived and used it in the preceding chapter. We see now, however, that it has greater validity than the restricted case for which we derived it previously. The even more general form of the first law, *16.1*, is of great usefulness. In connection with it let us point out that ΔU can be a very complex quantity indeed. For example, it includes the internal electric, chemical, atomic, and nuclear energies of the system, as well as the mechanical energy. The first law applies to changes in U no matter what their nature may be.

Illustration 16.1

When a bar of iron is heated in the atmosphere, it expands, and therefore some of the heat energy is used to do work. Find the fraction of the heat energy which is used for this purpose.

Reasoning Let us consider a 1°C temperature rise. Using the data from tables 15.7 and 15.1, we have, from 15.13 and 15.1,

$$\Delta V = V(3.6 \times 10^{-5}/°C)(1°C) = 3.6 \times 10^{-5}V$$

and $\qquad đQ = m(0.11\ \text{cal/g} \cdot °C)(1°C) = 0.11m \qquad \text{cal/g}$

where V is the volume and m is the mass of the bar.

Assuming atmospheric pressure to be 1×10^5 Pa, we have

$$\Delta W = P\,\Delta V$$

$$= 3.6V \qquad \text{N/m}^2 = 3.6V \qquad \text{J/m}^3$$

Changing $đQ$ to consistent units by recalling that 1 cal = 4.184 J, we have

$$đQ = 0.46m \qquad \text{J/g}$$

The fraction of the heat energy used to expand the metal against the pressure of the atmosphere is therefore

$$\text{Fraction} = \frac{\Delta W}{đQ} = 7.8\frac{V}{m} \qquad \text{g/m}^3$$

Now m/V is merely the mass per unit volume or density of iron, and this is 7860 kg/m³. Changing this to grams per cubic meter by multiplying by 10^3 gives $m/V = 7.86 \times 10^6$ g/m³. Substituting, we find

$$\text{Fraction} \approx 10^{-6}$$

This means that only one-millionth of the heat energy was used to do work against the atmosphere. All the rest of the heat energy was used to increase the internal energy ΔU of the bar. It should be clear from this why we can usually ignore the energy involved in the expansion of a solid or liquid.

16.2 ISOTHERMAL VOLUME CHANGE

Isothermal changes occur at constant temperature

When a system is altered in some way without allowing the temperature of the system to change appreciably, the change is said to be *isothermal*. We shall now discuss the isothermal volume change of a perfect, or ideal, gas. This change may be accomplished by applying an external force to the piston shown in figure 16.2(a) and slowly pushing the piston down. When we push the piston down, work ΔW is done *on the system,* and the work done *by the system* is $-\Delta W$. According to *16.1* we have

$$\Delta U = đQ - (-\Delta W)$$

As we would expect, the increase in internal energy equals the sum of the added heat energy and the work done on the gas.

If the gas is ideal, there will be no appreciable average energy of interaction between the molecules. In addition, we shall carry out the compression very slowly so that we can maintain the gas at constant temperature. That is to say, the compression is isothermal. Since the kinetic energy of the molecules does not change (constant temperature) and since the potential energy is assumed essentially zero, the internal energy of the molecules U will not change during the compression. Therefore $\Delta U = 0$, and we find

$$\Delta W = -đQ$$

FIGURE 16.2 In the reversible isothermal processes shown, the system is described at any instant by the values of P and V, together with the gas law.

from the above equation.

It will be recalled that $đQ$ is the heat energy *added to* the system. We find it to be negative in this case, thereby showing that heat is given off by the system as it is compressed isothermally. This fact is obvious for the following

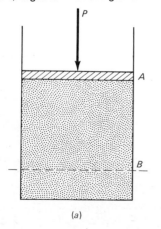

(a)

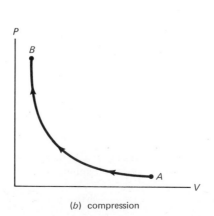

(b) compression

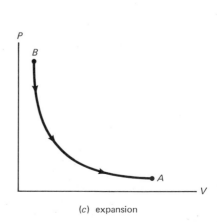

(c) expansion

reason. By compressing the piston, we do an amount of work ΔW on the gas. This gives the gas energy. Since the gas is perfect and its temperature is held constant, this added energy must flow out from the gas as soon as it is given to the gas. This outward flow of energy occurs in the form of heat flowing from the gas.

It is convenient to draw a P versus V graph for such an isothermal compression. Since the gas law tells us that

$$PV = \frac{m}{M} RT$$

and since T is held constant, we have

$$P = \frac{\text{const}}{V}$$

This type of curve, a hyperbola, is shown in figure 16.2(b). Let us now consider the reverse process.

In reverse, once the piston has been compressed to point B on the curve, the piston may be allowed to rise slowly. As the piston rises, the gas does work on it in an amount equal to $P \Delta V$ as the volume increases an amount ΔV. In order for the gas to do this work it must expend energy. The requisite energy comes from the thermal energy resident in the gas. (As the gas molecules hit the rising piston, they do work on it as it moves upward. In the process they lose translational kinetic energy, which is merely thermal energy.) Unless heat flows into the system from outside, the gas will cool because of this loss in thermal energy. We shall assume, however, that the gas is allowed to absorb heat from its surroundings and thereby make the expansion isothermal. As before

$$P = \frac{\text{const}}{V}$$

and the graph of P versus V during isothermal expansion will be as shown in figure 16.2(c). It is identical in shape with the graph in part (b) of the figure.

Reversible processes
The processes we have just described are *reversible*. By this we mean that the system changes in exactly the opposite way when going from A to B to the way it does in going from B to A. At any intermediate point the system is in exactly the same state, no matter which way the change is proceeding. In general, a reversible process requires the absence of friction forces, and further, the change must occur slowly enough so that the system is essentially at equilibrium at all stages of the change. It is only in this case that the state of the system can be readily described in terms of P, V, and T.

Definition of thermodynamic equilibrium
It is important that we now define clearly several terms that we shall be using. We define a system to be in *thermodynamic equilibrium* at a given set of measurable parameters P, V, and T if the following is true: the system is identical with a similar isolated system which has existed unchanged at these same values of P, V, and T for an infinitely long time. Strictly speaking, exact thermodynamic equilibrium can never be achieved. However, in practice, it is usually a simple matter to reduce departures of a system from thermodynamic equilibrium to negligible values. Notice that the pressure, volume, and temperature of a system are well-defined quantities if the system is in equilibrium.

Certain quantities, called *state variables,* are well-defined single-valued functions of a system in thermodynamic equilibrium. For example, temperature, pressure, and volume of a system are functions only of the particular equilibrium state of the system. No matter how the system reaches this equilibrium state, its P, V, and T will always be the same once equilibrium is attained. Since the average properties of the molecules in a system will be the same in a given equilibrium state no matter how the state is reached, the internal energy of a system is also a state variable. The work done by a system and the heat flow into the system are *not* state variables since these depend upon the past history of the system, as we shall see more clearly later.

Only when a system is in equilibrium (or very close to equilibrium) can we describe it by noting its T, P, and V values. Only then will P and T be constant (or, at least, well defined) throughout the system. The pressure and temperature of a gas undergoing turbulent flow are not sufficient to describe a gas under these highly nonequilibrium conditions. It is for this reason that we often restrict our discussion to reversible processes (often called quasi-static processes). These processes are being carried out so slowly that the system may be described by its state variables during the process.

16.3 WORK DURING ISOTHERMAL EXPANSION

As we saw in the preceding section, the work done by a gas as it expands isothermally must be compensated for by an equal amount of heat energy flowing into the gas. Otherwise the temperature of the gas would decrease as the gas did work. Let us now compute how much work a gas does as it expands isothermally from a volume V_A to V_B, as shown in figure 16.3. This work will equal the heat energy which flows into the volume in order to maintain isothermal conditions.

The work done by a gas as it expands through a volume ΔV at a pressure P has been shown to be

$$\Delta W = P\,\Delta V$$

Notice that for the volume V_i in figure 16.3, the work done in the expansion ΔV_i is just

$$\Delta W_i = P_i\,\Delta V_i$$

This increment of work is equal to the shaded area in the figure.

To find the total work done we must add all such increments of work as the volume expands from V_A to V_B. Therefore

$$W = \sum \Delta W_i$$

$$= \sum P_i\,\Delta V_i$$

Upon passing to the limit of small increments and after replacing the sum by an integral, we find

$$W = \int_{V_A}^{V_B} P\,dV \qquad\qquad 16.3$$

Notice that this work is the area under the P versus V curve between points A and B.

We cannot evaluate this integral at once since P is a function of V. However, the change is assumed to occur slowly enough so that the ideal gas

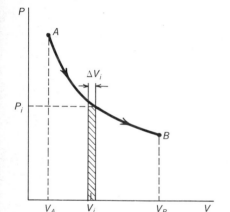

FIGURE 16.3 How much work is done as the gas expands from V_A to V_B?

The area under a PV curve represents work

law applies.* Then

$$P = \frac{mRT}{M} \frac{1}{V}$$

and since the process is isothermal, the temperature T is constant. Upon replacing P in *16.3* by this and noticing that only $1/V$ is not a constant, we find

$$W = \frac{mRT}{M} \int_{V_A}^{V_B} \frac{dV}{V}$$

This integral is easily evaluated, and it gives, for isothermal conditions,

Work done in an isothermal expansion

$$W = \frac{mRT}{M} \ln \frac{V_B}{V_A} \qquad\qquad 16.4$$

We have therefore found the work done by the gas as it expands from V_A to V_B. This is also equal to the heat energy which flowed into the volume of gas in order to replace the energy lost by the gas as it did work. It is of interest also to notice that if $V_B < V_A$, in which case the gas would have been compressed by an external agent, the work done by the gas would be negative. In that case, the energy furnished to the gas by the external agent must be lost as an outward heat flow from the volume. Notice once again that the work done is merely the area under the P versus V curve.

Illustration 16.2

How much heat is given off as 1 kg of nitrogen gas is compressed isothermally to half its original volume? The temperature is maintained constant at 0°C by immersing the apparatus in an ice-water bath.

Reasoning The molecular weight of nitrogen gas is 28, and therefore $M = 28$ kg/kmol and $m = 1.0$ kg. Placing these values in *16.4* and recalling that $V_B/V_A = \frac{1}{2}$, we have

$$W = \frac{1\,\text{kg}}{28\,\text{kg/kmol}}(8310\,\text{J/kmol} \cdot \text{K})(273\,\text{K}) \ln \tfrac{1}{2}$$

$$= 8.1 \times 10^4 \ln \tfrac{1}{2}\,\text{J}$$

But $\qquad \ln \tfrac{1}{2} = \ln 1 - \ln 2 = -\ln 2 = -2.303 \log 2 = -0.693$

and so $\qquad\qquad W = -5.6 \times 10^4\,\text{J}$

The negative sign tells us work was done *on* the gas, not *by* the gas. Of course, this amount of heat energy must flow from the gas if the gas is to remain at 0°C. We change it to calories by noting that 1 cal = 4.184 J. Therefore the heat *lost* is

$$-đQ = 1.34 \times 10^4\,\text{cal}$$

16.4 ADIABATIC VOLUME CHANGES

In the preceding two sections we have considered one extreme of the various possible ways in which temperature can be controlled during a volume change. We shall now consider a change quite different from an isothermal

---

*If the change were not quasi-static, we could not say how P and V were related during the process.

change. It is the case where no heat enters or leaves the system during the volume change. This type of change is said to be *adiabatic*. Such a situation may be realized in practice by heavily insulating the system from its surroundings or by suspending it in a mirrored vacuum chamber.

Of course, if a gas is compressed, work is done on the gas, thereby giving it energy. Ordinarily, this work will appear as internal energy in the gas. In the case of an adiabatic change, this added energy is retained by the gas, and so the temperature of the gas is raised as a result of the compression. Just the reverse occurs if the gas expands adiabatically. The gas does work, losing an equal amount of internal energy, and its temperature therefore falls.

We can apply the ideal gas law to the gas undergoing an adiabatic change. It is

$$PV = \frac{m}{M} RT$$

In the adiabatic case, unlike the isothermal, T will vary as well as P and V. Since all three quantities vary, the gas law is not particularly useful in computing behavior in adiabatic changes. We should like a relation which involves only P and V without T. Such a relation can in fact be found, as we shall now show.

Suppose an amount of work $P \Delta V$ is done by a gas. If the change is adiabatic, this work must result in a loss of internal energy from the gas, and in fact the loss $-\Delta U$ will just equal the work done $P \Delta V$. Therefore

$$-\Delta U = P \Delta V$$

However, as a result of this energy loss, the temperature of the gas will decrease by an amount $-\Delta T$. These two quantities, ΔU and ΔT, may be related in the following way. When the temperature of an ideal gas is lowered an amount $-\Delta T$, an energy

$$mc_V \Delta T$$

must be taken away from the molecules. In this case, the energy is lost, not as heat, but as work done. Nevertheless, the energy lost by the molecules is still the same, and so we can write

$$\Delta U = mc_V \Delta T$$

Equating these two expressions for ΔU, we find

$$P \Delta V = -mc_V \Delta T \qquad\qquad 16.5$$

Notice that this relation predicts our previous qualitative conclusion. If the gas does work $P \Delta V$, the temperature of the gas will decrease by the amount indicated.

We can use the ideal gas law to eliminate P from *16.5*. It then becomes

$$\frac{\Delta V}{V} \left(\frac{m}{M}\right) RT = -mc_V \Delta T$$

which simplifies to give

$$\frac{R}{Mc_V} \frac{\Delta V}{V} = -\frac{\Delta T}{T} \qquad\qquad 16.6$$

But from *15.8* we find

$$c_P - c_V = \frac{R}{M} \qquad\qquad 15.8$$

or, after dividing by c_V,
$$\frac{R}{Mc_V} = \gamma - 1$$

where
$$\gamma \equiv \frac{c_P}{c_V}$$

Using this value in 16.6, we find
$$\frac{\Delta V}{V}(\gamma - 1) = -\frac{\Delta T}{T}$$

If we pass to the limit of small increments, this equation can be rewritten
$$(\gamma - 1)\frac{dV}{V} + \frac{dT}{T} = 0$$

which integrates to (recall $\int dx/x = \ln x$)
$$(\gamma - 1)\ln V + \ln T = \text{const}$$

Or, upon combining the logarithms and taking antilogs, we find
$$TV^{\gamma-1} = \text{const} \qquad \text{or} \qquad T_1 V_1^{\gamma-1} = T_2 V_2^{\gamma-1} \qquad\qquad 16.7$$

This is a convenient relation between T and V for an adiabatic change. The values of $\gamma = c_P/c_V$ are always greater than unity (why?), and typical values are given in table 15.3.

We can use the ideal gas law to eliminate T from 16.7, and we then have

For an ideal gas undergoing adiabatic change
$$PV^{\gamma} = \text{const} \qquad \text{or} \qquad P_1 V_1^{\gamma} = P_2 V_2^{\gamma} \qquad\qquad 16.8$$

for an adiabatic change. Both 16.7 and 16.8 are of value in finding how the state variables P, V, and T change in an adiabatic expansion or compression *of an ideal gas.*

Illustration 16.3

Nitrogen gas at 300 K and atmospheric pressure in the cylinder of a diesel engine is suddenly compressed to 0.10 its original volume. Find its final temperature and pressure.

Reasoning We assume the compression to be fast enough so that no appreciable heat leaves the system. From table 15.3 we find $\gamma = 1.40$ for nitrogen. From 16.7,
$$(300\text{ K})(V_0)^{0.40} = T(0.10V_0)^{0.40}$$

from which
$$T = 750\text{ K}$$

Similarly, using 16.8 to find P, we have
$$P = 25\text{ atm}$$

The large temperature increase of a gas upon adiabatic compression is used to ignite the gas in the pistons of diesel engines, thereby eliminating the need for spark plugs. Note that we have implicitly assumed the initial and final states of the system to be equilibrium states.

16.5 CYCLIC PROCESSES

Heat engines are machines which use heat energy to do work by repeating a process over and over again. To understand the operation of such machines, we must be concerned with the thermodynamics of cyclic processes. Let us consider an ideal gas of volume V confined to a cylinder at pressure P by a piston, as shown in figure 16.4(a). Suppose that the cylinder is in contact with a constant-temperature bath maintained at a temperature T. If we carry out any change in the system slowly, the temperature of the gas will remain constant, and the change will be isothermal. Furthermore, since we assume the gas to be ideal and the piston to be frictionless, the change will be reversible.

If some of the weight is slowly removed from the piston, the pressure and volume of the gas will vary, as shown by the curve AB in figure 16.4(b). As discussed in Sec. 16.3, the work done by the gas on the piston and weights during this process is merely the area under the curve from A to B. We computed its value in Sec. 16.3, and it was given by

$$W = \frac{mRT}{M} \ln \frac{V_B}{V_A} \qquad\qquad 16.4$$

Now suppose the weights are slowly replaced on the piston so that the system retraces the curve of figure 16.4(b) from B to A. The weights and piston now do work *on* the gas by compressing it. This work is the area under the curve BA, and it will be identical in magnitude with the work done by the gas as it expanded from A to B. It will be given by *16.4*. Since this work was done *on* the system, we say the system *did* negative work.

To summarize, our system started at point A, expanded isothermally to B, and then was compressed isothermally back to the starting point A. We say that the system has gone through a *cycle*. In this particular cyclic process, the work done by the system as it expanded equaled the negative of the work done by the system as it was compressed. The total work done by the system was therefore zero. We note for future reference that the area *between* the expansion and compression curves is zero, and this is equal to the net work done by the system in the cycle.

It is possible to alter the cyclic process just considered if we do not require it to be done at constant temperature. For example, we could easily follow a cycle such as that shown as $ABCDA$ in figure 16.5(a). Starting at point A, the gas could be slowly heated. The heat comes from the external constant-temperature bath, which we shall in this case call the high-tempera-

FIGURE 16.4 As the curve in (b) is slowly traced out from A to B and back to A, the piston system in (a) is maintained at constant temperature. How much net work is done by the system in this cycle? How is it related to the area between the expansion and contraction curves?

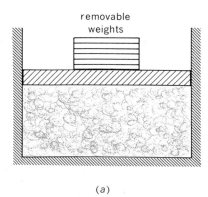

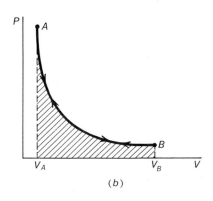

(a)

(b)

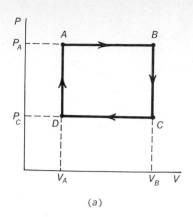
(a)

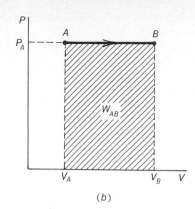
(b)

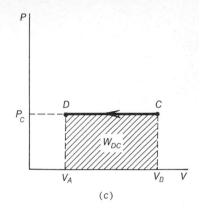
(c)

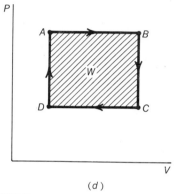
(d)

FIGURE 16.5 How can this cycle be achieved experimentally?

Work done in a cycle equals the enclosed area of the PV cycle

The Carnot cycle is the most efficient cycle

ture heat reservoir. Since the weight of the piston is held constant, the pressure of the gas remains at P_A. However, the added heat causes the gas to expand along the line AB to V_B. Once the system reaches point B, the gas is allowed to cool slowly, while weights are slowly removed from the piston so as to keep the piston stationary and the volume constant. Of course, the pressure decreases in the process, and the system follows line BC. During this process heat is given off or exhausted to the (now variable-temperature) heat reservoir.

Now the weight on the piston is kept constant as the gas is cooled further. The pressure will remain constant as the volume contracts, and so the system proceeds along the line CD. When point D is reached the gas is slowly heated as enough weight is added to the piston to keep the volume constant. The system now follows the line DA back to its original condition at A, and the cycle has been completed.

It is instructive to examine the work done by the gas during this cycle. No work is done by the gas during parts BC and DA of the cycle since V was constant. However, during part AB of the cycle the gas lifted the piston and did work W_{AB} equal to the area under the curve, as shown in part (b) of figure 16.5. During part CD of the cycle, the piston did work on the gas, and this is equal to W_{CD}, the area shown in figure 16.5(c). This positive work done by the piston is equivalent to *negative work done by* the gas. The total work done by the gas in a cycle is therefore

$$W = W_{AB} - W_{CD}$$

which is merely the difference between the two areas.

But the difference between the two areas is merely the area enclosed by the cyclic curve $ABCDA$, as shown in figure 16.5(d). We therefore have the result, which is true for any reversible process, that *the work done by the gas in any closed cycle is equal to the enclosed area of the P versus V diagram representing the cycle*. How does this check with the cycle given in figure 16.4?

16.6 THE CARNOT CYCLE

One particular type of cycle is of great importance from both a theoretical and practical standpoint. It was first discussed theoretically by Sadi Carnot in 1824 and is called the *Carnot cycle*. Carnot showed that it is the most efficient cycle possible. That is to say, the work done by the gas when following this cycle is the greatest possible for a given amount of heat energy added to the gas. The importance of this type of study arises from the fact that all engines which operate by a gas-propelled piston (such as steam and gasoline engines) can be described by their PV-cycle diagrams. We are interested in the Carnot cycle since it represents the maximum level of performance for any of these engines.

The Carnot cycle, a reversible cycle, is illustrated in figure 16.6. Originally, the gas in the cylinder is at point A on the diagram. It is allowed to expand isothermally at temperature T_1 along the curve AB. During this expansion process the ideal gas does work against the piston, and so an amount of heat Q_1 equal to this work must be furnished to the gas to keep its temperature constant. When the system reaches point B it is adiabatically expanded along curve BC. Of course, the temperature and pressure drop rapidly during this portion of the cycle.

When the system reaches point C of the cycle, it is once again held at the new constant temperature T_2 as it is compressed to point D. Since the external compression force does work on the system (less work than the system itself did along AB), heat Q_2 must flow out of the system in order for the temperature to remain constant at T_2. Finally, the gas is compressed adiabatically to point A, and the cycle is completed. Since the system is now back at its original state, $\Delta U = 0$ for the cycle.

During the full cycle the gas did an amount of work W given by the area enclosed by the curve. This work was furnished by the net heat flow $Q_1 - Q_2$ into the system since U remains unchanged. Therefore

$$W = Q_1 - Q_2 \qquad\qquad 16.9$$

It is clear from this equation that the heat exhausted from the system during the compression part of the cycle, Q_2, is detrimental since it results in no useful work. This may be seen more clearly by computing the efficiency of the Carnot engine, which gives rise to the Carnot cycle.

It is common to define efficiency in the following practical way:

Efficiency of a cycle defined
$$\text{Efficiency} = e = \frac{\text{output work}}{\text{energy input}} \qquad\qquad 16.10$$

In the present case this is

$$e = \frac{W}{Q_1} = \frac{Q_1 - Q_2}{Q_1}$$

or
$$e = 1 - \frac{Q_2}{Q_1} \qquad\qquad 16.11$$

FIGURE 16.6 The Carnot cycle. Explain what is happening in each portion of the cycle.

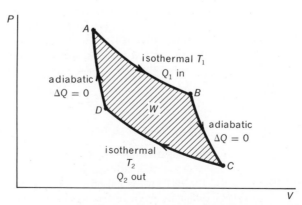

The efficiency can become unity (or 100 percent) only if no heat is exhausted to the outside, i.e., if $Q_2 = 0$. We shall see that this is an impossible condition, and so a machine which operates at 100 percent efficiency is impossible to make.

16.7 THE KELVIN TEMPERATURE SCALE

It is illuminating to express *16.11* in terms of temperatures rather than quantities of heat. To do this, we seek expressions for Q_1 and Q_2 in terms of temperatures. We have already derived the appropriate equation in *16.4*, which gives the work done W_{AB} in an isothermal expansion from V_A to V_B. It was

$$W_{AB} = \frac{mRT_1}{M}\ln\frac{V_B}{V_A} \qquad \textit{16.4}$$

where T_1 is the temperature at the time of the expansion. Since the expansion is assumed isothermal and the gas is perfect, W_{AB} is equal to the heat energy Q_1 furnished to the system during the expansion. Therefore

$$Q_1 = \frac{mRT_1}{M}\ln\frac{V_B}{V_A} \qquad \textit{16.12}$$

If we apply this result to the Carnot cycle of figure 16.6, discussed in the preceding section, we have the following results. The heat added to the system during the portion AB of the cycle is given by *16.12*. During the portion CD of the cycle an amount of heat Q_2 is expelled from the system, and this is given by

$$Q_2 = \frac{mRT_2}{M}\ln\frac{V_C}{V_D} \qquad \textit{16.13}$$

Upon dividing *16.13* by *16.12*, we find

$$\frac{Q_2}{Q_1} = \frac{T_2\ln(V_C/V_D)}{T_1\ln(V_B/V_A)}$$

If we could eliminate the volumes from this relation, we could obtain an expression for the ratio of the Q's in terms of the T's. We now show that the ratio of the logarithms is unity.

For the isothermal paths AB and CD, $PV = (m/M)RT$ tells us

$$P_A V_A = P_B V_B$$

and
$$P_C V_C = P_D V_D$$

Moreover, for the adiabatic paths BC and DA, we can write (using *16.8*)

$$P_B V_B{}^\gamma = P_C V_C{}^\gamma$$

and
$$P_D V_D{}^\gamma = P_A V_A{}^\gamma$$

If we multiply these four equations together and cancel like factors, we find

$$(V_B V_D)^{\gamma-1} = (V_C V_A)^{\gamma-1}$$

$$\frac{V_C}{V_D} = \frac{V_B}{V_A}$$

Therefore the logarithms of these two quantities are equal, and their ratio is unity, as we set out to show.

After canceling the logarithms, the ratio of the heats becomes

$$\frac{Q_2}{Q_1} = \frac{T_2}{T_1} \qquad\qquad 16.14$$

and the efficiency of the Carnot cycle (16.11) becomes*

Efficiency of a Carnot cycle

$$e = 1 - \frac{T_2}{T_1} \qquad\qquad 16.15$$

This tells us that the only way that the Carnot cycle can be made perfectly efficient is for the exhaust temperature during the compression portion of the cycle to be made zero. Notice that this is 0 K, or absolute zero.

Equation 16.15 has important meaning for the design of all heat engines. Although it applies only to the Carnot cycle, it nevertheless sets an upper limit for all other cycles, even though they are less efficient than the Carnot cycle. We see that the basic limiting factors affecting the efficiency of *For an efficient engine, T_2 should be* heat engines are the temperatures of the working gas at its hottest and *small and T_1 large* coolest. In particular, for high efficiency the gas should be as hot as possible during the portion of the cycle when it is doing work, and it should be as cold as possible when it is being compressed.

Since a steam engine cannot easily operate at a temperature much higher than 100°C, we should have $T_1 \approx 373$ K. During the cool portion of the cycle the gas cannot be at a temperature lower than that of the atmosphere, $T_2 \approx 300$ K. Therefore the best steam engine will have an efficiency of only about

$$e = 1 - \tfrac{300}{373} = 0.20, \text{ or } 20 \text{ percent}$$

Since for most practical engines $T_2 = 300$ K or higher, the engine efficiency can be made high only by increasing T_1. In gasoline engines the temperature of the gas in the piston is considerably higher than 100°C, and so their efficiencies are somewhat better than that of steam engines. Notice that this theoretical limit on engine efficiency is a very fundamental fact and cannot be altered by ingenious engine design.

Equation 16.14, relating the heat absorbed and expelled by a Carnot engine to the ratio of the absolute temperatures, affords one a method for defining the absolute temperature scale. We had

$$\frac{Q_2}{Q_1} = \frac{T_2}{T_1}$$

If we arbitrarily set the triple point of water to be $T_2 = 273.16$ on this, the Kelvin scale, then measurement of Q_2 and Q_1 allows one to measure any other temperature, T_1. The temperature scale defined in this way is called the *The Kelvin temperature scale defined* *thermodynamic,* or *Kelvin,* temperature scale. Since T_1 and T_2 have already been defined in terms of the behavior of an ideal gas, the absolute temperature scale defined previously must be the same as the thermodynamic scale. The advantage of the thermodynamic definition is that it can be applied at low temperatures, where hydrogen and helium gases do not behave ideally.

16.8 THE SECOND LAW OF THERMODYNAMICS

Someone once remarked about the universe that "left to themselves, things go from good to bad to worse." In a very crude sense, this summarizes the

---

* Although we have derived this for an ideal gas as the working fluid, the result can be shown to be true in general.

second law of thermodynamics. As we have seen, the first law is a statement of energy conservation, but it has nothing to say about the course of events in the universe. Energy is conserved when a stone falls and has its gravitational potential energy changed to kinetic energy. As the stone strikes the ground and comes to rest, its kinetic energy is changed to thermal energy. However, a stone resting on the ground never changes the thermal energy in and near it to kinetic energy and goes shooting up into the air. The first law does not rule out such a possibility since this reverse process also conserves energy. But the process does not occur.

There are many other processes in the universe which are not ruled out by the first law but which do not occur. For example, heat flows from hot to cold but not from cold to hot. Water evaporates from a saucer, but the vapor in the air does not by itself recondense into the saucer. A dead body decays and turns to dust; but the elements of the earth do not spontaneously form the body in the reverse process. *Nature has a preferred direction for the course of spontaneous events. The second law tells us what that direction is.*

During the history of thermodynamics, the *second law* has been stated in several fully equivalent ways. One of the earliest statements simply summarizes the fact that heat flows naturally from hot to cold:

Second law of thermodynamics

Heat flows spontaneously from a hotter to a colder object but not vice versa. Because of this, it is impossible for a cyclical system to transfer heat from a lower-temperature body to a higher-temperature body indefinitely unless external work is done on the system.

The fact that the second law exists is implied by our equation for the efficiency of a Carnot engine. If the heat-source temperature is the same as the exhaust temperature, the efficiency is zero. Even the most efficient heat engine can do no useful work unless there is opportunity for it to expel heat to a lower-temperature object. For example, the waters of the oceans have a huge amount of thermal energy. But we cannot use the energy unless a cooler place is found to which it can flow. As a result, for all usual purposes the thermal energy resident in the oceans is of no use to us. In the next section we shall explore the basic reason for this lack of usefulness.

16.9 ORDER VERSUS DISORDER

As any gambler knows, the odds are best for an event happening if the event can occur in many different ways. To illustrate this fact, let us consider a game in which five identical coins are tossed onto a table after being well shaken. There are only six events which can result from such a toss. They are listed as follows:

Event	Number of heads up	Number of tails up
1	0	5
2	1	4
3	2	3
4	3	2
5	4	1
6	5	0

At first guess you might think that each of these events is equally likely to occur. But that is not correct. The reason is that there is only one way that event 1 or event 6 can occur. For event 1 to occur, all coins without exception must come up tails. For event 6, all coins must come up heads.

However, for event 2 to occur, there are actually five ways in which it could happen. Calling the five coins, A, B, C, D, and E, these ways are as follows:

Way	Coin A	B	C	D	E
		Ways for one head			
1	H	T	T	T	T
2	T	H	T	T	T
3	T	T	H	T	T
4	T	T	T	H	T
5	T	T	T	T	H

Because there are five times as many ways that event 2 can happen, event 2 is five times more likely to occur than event 1. Also, event 5 can happen in five different ways. As a result, events 2 and 5 are equally likely to occur. And both these events are five times more likely than events 1 or 6.

In the same way, we can show that events 3 and 4 are equally likely and each can occur in ten different ways. Therefore, 3 and 4 are twice as likely to happen as events 2 and 5, while events 3 and 4 are ten times more likely to happen than events 1 and 6. If you were a gambler, it is obvious which events you should lay your money on if no odds are given.

We can extend this to a situation in which more coins are involved. Suppose 100 coins rather than 5 are tossed. Then, as before, there is only one way in which all the coins can come up heads (or tails). But the number of ways in which other combinations can occur becomes almost unbelievably large. The results are shown in figure 16.7. Notice that the number of ways in which 50 heads can come up is about 10×10^{28}. As you can see, the odds against all heads or all tails coming up is so small as to be negligible.

FIGURE 16.7 The number of ways in which the indicated number of heads can come up when 100 coins are tossed.

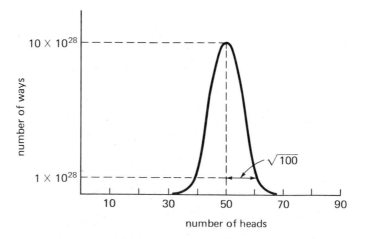

Indeed, the total number of ways for all combinations of heads and tails is about 1×10^{30}. Therefore, the chance that all coins would come up heads is 1 in 10^{30}. If you throw the coins once each 10 s for 10^{22} years, your chance of all heads coming up once is about 10 percent. For all practical purposes, there is no chance at all of all heads or all tails occurring. As we see from figure 16.7, the only really likely occurrence is for nearly equal numbers of heads and tails to occur.

If we consider 10^6 coins instead of 10^2, the situation becomes even more striking. We can summarize all such results in a very simple way. Notice in figure 16.7 that the graph line decreases to about one-tenth of its maximum value at the following two numbers of heads: 40 and 60. To give an estimate of the width of the peak, we could say it extends from 50 ± 10. In other words, if you throw 100 coins, the number of heads which should come up is about 50 ± 10. The more general result is as follows:

If one throws N coins, the expected number of heads will be about. *

$$\frac{N}{2} \pm \sqrt{N}$$

In the case of 10^6 coins, we should expect $500,000 \pm 1000$ heads to come up. Notice how very precise this estimate is. It says the expected number of heads lies between 501,000 and 499,000, a very narrow range indeed. As you can see, when the number of coins becomes very large, the percentage deviations one will find from the average are very small.

This example with the coins is typical of our universe in general. When things are left to happen by themselves, they occur by chance. As a result, the probability laws applicable to tossed coins applies to these other situations as well. For example, suppose you have a box containing gas molecules, as shown in figure 16.8. In the air, there are about 3×10^{19} molecules/cm^3. Let us say the box has 10^{20} molecules in it. We now ask: What are the chances that the molecules will all bunch up in one half of the box?

From our results with the coins, we can easily clarify this situation. To make the situations similar, call a molecule in the left side of the box a "heads" molecule. Molecules on the right will be "tails" molecules. Our general result from above tells us that the number of "heads" will be about $\frac{1}{2}N \pm \sqrt{N}$. In this case, the number of heads is about $(5,000,000,000 \pm 1) \times 10^{10}$.

Notice how small the expected deviation is. It is only about 1 part in 5 billion. For all practical purposes, the number of molecules in the two halves of the box will be the same. And, of course, there is really no chance at all that all the molecules will, spontaneously, move into one side of the box.

These considerations have fundamental importance for all spontaneous processes. Reasoning from them, we can predict that thermal motion (or other random-type disturbances) causes systems to change from order to disorder. As a crude example of this, consider the case of 100 coins again. Suppose we carefully arrange all of them with heads up. They then have a high degree of order. Now let us give them a type of motion similar to random thermal motion by shaking them. They quickly disorder and never return to their original state of order.

Similarly with the gas molecules in the box of figure 16.8. We can give the system order by placing all the molecules in one end of the box. But if we

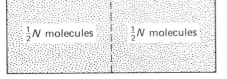

FIGURE 16.8 What is the chance that all the molecules will appear in one side of the box?

---

* Precisely, if N is large, 95.3 percent of the time the number of heads will lie in this range.

allow them to adjust with spontaneous thermal motion, they become disordered and fill the whole box. Never will they again, spontaneously, return to their original ordered state.

Basic to this discussion are the concepts of order and disorder. We can give a simple method for comparing the disorder of two states. If a state can occur in only one way, it is a highly ordered state. In such a state, each molecule (or other particle) must be placed in a single exact way. In a disordered state, however, there are many possible ways for achieving the state. With these facts in mind, we can relate disorder and probability (or number of ways) of achieving a state. That state which has the highest disorder is the most probable state; it can occur in the largest number of ways. For example, the probability of N coins all coming up heads is very small. This is a state of very low disorder. As we have seen, systems left to themselves move toward states of high disorder.

There are many examples we could give that illustrate behavior of this type. We conclude from them that *in a system composed of many molecules,*

Second law: alternative form

If a system is allowed to undergo spontaneous change, it will change in such a way that its disorder will increase or, at best, not decrease.

This law of nature, applicable to large numbers of molecules, is an alternative form of the second law of thermodynamics. In the next section we shall see yet another way in which the law can be stated.

16.10 ENTROPY

As we have just seen, a system is influenced profoundly by the number of ways in which its various possible states can occur. Each equilibrium state of the system can occur in a definite number of ways. Let us call that number Ω (capital Greek omega). A state for which Ω is very large is much more likely to occur than one for which Ω is small. Indeed, an equilibrium system composed of many molecules will always be found very close to the state which has the maximum value of Ω conforming to the constraints on the system. From this we see that Ω is an important quantity for a system. As we shall now see, it is intimately related to the state variable called entropy.

Let us define a quantity S, called *entropy,* by the following equation:

$$S = k \ln \Omega \qquad\qquad 16.16$$

where k is Boltzmann's constant and $\ln \Omega$ is the natural logarithm of Ω. It is easy to see that S, the entropy, is a state variable, a quantity that depends only on the equilibrium state of the system. Like P, V, T, and U, the entropy describes an equilibrium state unambiguously. A system of many molecules will reach equilibrium when it has achieved the state with largest Ω, the most probable state, the state of largest entropy. Hence, under a given set of constraints, a system at equilibrium has a well-defined value for S, the entropy, as well as for P, V, T, and U. This fact is of great importance and gives the entropy usefulness comparable to that of internal energy U.

Notice what equation *16.16* tells us. If a state of the system can occur in only one way, then $\Omega = 1$. But the logarithm of 1 is zero. So the entropy of such a highly unlikely state is zero. However, if a state can occur in many ways, Ω will be large. The entropy of a highly probable state is therefore large.

Let us now recall that a highly ordered state has a low number of ways it can occur while a disordered state can occur in many ways. Equation *16.16* then tells us that *entropy is a measure of disorder. The more disordered the state of a system is, the larger its entropy will be.* As we see, *entropy is a state variable which measures disorder.* Because of this, we can restate the *second law of thermodynamics* in still another way:

Second law in terms of entropy

> *If an isolated system undergoes change, it will change in such a way that its entropy will increase or, at best, remain constant.*

Because entropy depends only on the particular equilibrium state in which a system finds itself, there is an important relation that applies to it as it changes during a thermodynamic cycle. Suppose a system undergoes a series of changes that carry it around a complete cycle. Because the entropy is the same at the end of the cycle as at the beginning (the entropy depends only on the equilibrium state), the changes in entropy as the system moves around the cycle must sum to give zero. We can therefore write

$$\oint dS = 0 \qquad\qquad 16.17$$

where the circle on the integral sign means to take the sum of the entropy changes over one complete cycle. Like all other state variables, S returns to its original value at the end of a complete cycle.

Entropy has a long history and, in fact, was used many years before the molecular (or *statistical mechanics*) definition given in *16.16* was possible. Indeed, the statistical mechanics definition of entropy was chosen so as to conform to the previous definition given in thermodynamics. It is far from simple to show that the two definitions are equivalent. For that reason, we shall simply state the alternative definition of entropy.

If a system at temperature T undergoes a *reversible* change in which an amount $đQ$ of heat energy is added to the system, then the entropy change of the system is defined to be

$$dS \equiv \frac{đQ}{T} \qquad \text{(reversible process only)} \qquad 16.18$$

Moreover, if a system changes from state a to state b in a reversible way,* then

$$S_b - S_a = \int_a^b dS = \int_a^b \frac{đQ}{T} \qquad \text{(reversible process)} \qquad 16.19$$

Let us now see how entropy can be used in calculations.

Illustration 16.4

Compute the entropy change when 40 g of ice is melted at 0°C.

Reasoning Calling the ice state a and the liquid state b, we have, from *16.19* (the melting process will be reversible if carried out slowly),

$$S_{\text{liq}} - S_{\text{ice}} = \int_{\text{ice}}^{\text{liq}} \frac{đQ}{T}$$

Since T is held constant at 273 K, we have

$$S_{\text{liq}} - S_{\text{ice}} = \frac{1}{273 \text{ K}} \int_{\text{ice}}^{\text{liq}} đQ$$

---

*The restriction to reversible processes is necessitated by the fact that Q is not a state variable.

Where the integral is simply the sum of all the ΔQ's needed to melt the ice, i.e., the heat needed to melt the 40 g of ice, namely, $80 \times 40 = 320$ cal. Therefore

$$S_{liq} - S_{ice} = 3200 \text{ cal}/273 \text{ K} = 11.7 \text{ cal/K} = 49 \text{ J/K}$$

Notice that heat is added to the system, and so $\int d\!\!\!{}^{\!-}Q$ was positive. The entropy therefore increased in the process. Clearly, the disorder also increased.

Illustration 16.5

When 50 g of water is heated from 0 to 100°C, by how much does its entropy change?

Reasoning We can carry out the heating process slowly so that the process will be reversible. Therefore

$$S_{100} - S_0 = \int \frac{d\!\!\!{}^{\!-}Q}{T}$$

Now, from the definition of specific heat, we have

$$\Delta Q = mc \, \Delta T$$

which, in the present case, becomes

$$d\!\!\!{}^{\!-}Q = (50 \text{ cal/K}) \, dT$$

This gives
$$S_{100} - S_0 = (50 \text{ cal/K}) \int_{273}^{373} \frac{dT}{T}$$

from which
$$S_{100} - S_0 = 50 \ln \tfrac{373}{273} \text{ cal/K}$$

Here too the entropy, as well as the disorder, increases.

16.11 ENTROPY CHANGES IN IRREVERSIBLE PROCESSES

For an irreversible process one cannot in general represent the variation of the system by a definite path on a P versus V diagram. Usually, such processes involve nonequilibrium situations in which P, V, and T are not well defined. In spite of this, P, V, T, and S are all state variables and therefore depend only upon the state of the system. Even though a change is carried out irreversibly, these variables are completely determined by the initial and final equilibrium states of the system. We can easily specify P, V, and T in each equilibrium state. Our problem now is to find the change in S as a result of an irreversible process.

Since the initial value of S, namely, S_1, and its final value, S_2, are completely defined by the initial and final states of the system, any method for changing the system from state 1 to 2 will give identical values for S_1 and S_2. We saw in the preceding section that

$$S_2 - S_1 = \int_1^2 \frac{d\!\!\!{}^{\!-}Q}{T}$$

provided the change is carried out by means of a reversible process. No matter how a system is changed from state 1 to state 2, if we can imagine a reversible process which would accomplish the same change, that reversible process can be used to compute $S_2 - S_1$. We shall now illustrate this by considering two simple cases.

Case 1
An ampule of ideal gas, volume V_1, is broken in an evacuated bottle, volume V_2. What is the entropy change if the initial and final temperatures of the gas are the same, T?

This sudden expansion of the gas from V_1 to V_2 is definitely not carried out under near-equilibrium conditions. The motion of the gas particles is not random during the expansion, and P, V, and T are undefined during the process. A real gas would cool somewhat during an expansion such as this because work would be done against the attractive forces between the molecules. However, for an ideal gas, which we shall consider here, this effect will be missing. In addition, since the expansion took place into vacuum, no external work was done by the gas during the process.

Since $S_2 - S_1$ will depend only upon the initial and final states, we try to imagine a way of achieving the same results by a nice reversible process. This can easily be done by allowing a gas to expand isothermally against a slow-moving piston. In this reversible isothermal case we have

$$S_2 - S_1 = \frac{1}{T} \int_1^2 đQ$$

where T has been taken outside the integral since it is constant. But $\int_1^2 đQ$ is merely the work done in an isothermal expansion, and this was found in Sec. 16.3. Using the result found in *16.4,* we have

$$S_2 - S_1 = \frac{m}{M} R \ln \frac{V_2}{V_1}$$

Here once again both the entropy and disorder have increased.

Case 2
Two equal masses of the same material are at different temperatures, T_A and T_B, with $T_A > T_B$. They are connected together and allowed to come to thermal equilibrium. What is the change in entropy?

Here, too, the process is irreversible since the temperature of the system during the change is unspecified. However, we can think of slowly heating the cold material to the final temperature T_f and cooling the hot material. They can be joined after they reach the same temperature. This latter method of carrying out the change would be reversible.

We have

$$S_2 - S_1 = \int_A \frac{đQ}{T} + \int_B \frac{đQ}{T}$$

From the definition of the specific heat c for the material,

$$đQ = mc \, dT$$

This gives, for the total mass,

$$S_2 - S_1 = mc \int_{T_A}^{T_f} \frac{dT}{T} + mc \int_{T_B}^{T_f} \frac{dT}{T}$$

After carrying out the integration, we have

$$S_2 - S_1 = mc\left(\ln \frac{T_f}{T_A} + \ln \frac{T_f}{T_B}\right) = mc \ln \frac{T_f T_f}{T_A T_B}$$

or since $T_f = \frac{1}{2}(T_A + T_B)$, we shall have

$$S_2 - S_1 = mc \ln\left[\frac{(T_A + T_B)^2}{4T_A T_B}\right]$$

The quantity inside the brackets is larger than unity; therefore the logarithm is a positive number, and $S_2 > S_1$. So here also the entropy and disorder increase together.

Chapter 16
Questions
and Guesstimates

1. After heat energy has flowed into a system, where may this energy be found in the internal energy of the system? Give concrete examples of as many of these situations as you can.

2. The relation $\Delta U = \Delta Q - \Delta W$ is not always equivalent to the relation $\Delta U = \Delta Q - P\Delta V$. Give an example where the latter relation does not apply even though the former does apply.

3. In each of the following processes, point out what is meant by each quantity in the equation $\Delta U = \Delta Q - \Delta W$: an ice cube slowly melts in ice water; ice heats from -30 to $-10°C$; steam in a closed boiler cools from 120 to 110°C; solid CO_2 (dry ice) sublimates in dry air; a bottle of Coke freezes and cracks the bottle.

4. Some people term the following statements the *zeroth law of thermodynamics:* two objects in thermal equilibrium with each other must be at the same temperature; two objects, each in thermal equilibrium with a third object, are in thermal equilibrium with each other. In the case of the objects being ideal gases, justify these statements.

5. A cylinder is split into two equal cylinders by a partition. One end of the cylinder is closed by a piston. In one-half of the cylinder is gas A, and in the other is gas B, both at the same temperature and pressure. Now the partition is broken open so that the gases can mix. What happens to the pressure? If the piston is now pushed in adiabatically to the position of the partition, thereby halving the volume, what happens to P, T, and U for the assumed ideal gases?

6. What limitations, if any, apply to the statement made just below *16.3* that this work is represented by the area under the P versus V curve?

7. An ideal gas is compressed from V_0 to $\frac{1}{2}V_0$. In which case is more work done: when the compression is isothermal or adiabatic? In which case does the translational kinetic energy of the molecules become larger?

8. Can a system absorb heat energy without a change in its internal energy? Can its internal energy by increased without adding heat to it?

9. To heat a mole of gas from 0 K to a specified temperature requires a very definite quantity of heat. Why then can't we consider the gas to have a definite quantity of heat in it and thereby say that Q is a state variable?

10. Compare the internal energies of a mole of each of the following gases at the same temperature: He, H_2, N_2, and CH_4.

11. Suppose a gas cools as it expands adiabatically into a vacuum. What does this tell us about the gas?

12. Why does the existence of the first law rule out the possibility of a perpetual-motion machine?

13. Using the second law as a basis for discussion, explain why it is impossible for a refrigerator ever to cool an object to absolute zero.

14. Suppose you have a box with a porous screen separating it into halves. There are N ideal gas molecules in the box. It is easy to show (if you know a little probability theory) that the following is true. If you observe the box on a large number of occasions reasonably well separated in time, the fraction of the times that you will find all the N molecules in one of the halves of the box is $(\frac{1}{2})^{N-1}$. From this fact, what can one say about the validity of the second law?

15. Body temperature is 98.6°F. Even so, heat is carried away from the body to the environment even when the surrounding temperature is higher than this. Doesn't this contradict the fact that heat only flows from hot to cold? Similarly, a watermelon can be cooled by wrapping it in a wet cloth even on a very hot day. How can you reconcile this with the second law?

16. A baby produces a higher-ordered structure as it grows. Each molecule within it is a carefully structured entity, and the molecules are assembled in a highly ordered way. Doesn't this growth and its accompanying high degree of order contradict the second law? Ultimately, what is the energy source for the child? Repeat this question for the case of a growing plant.

17. It is said that the entropy of the universe is constantly increasing. Explain.

18. Compare the entropy change when an ideal gas is compressed to half its volume (*a*) adiabatically and (*b*) isothermally.

19. At the present time, the sun is the major source of the energy we use here on earth. Trace the energy from its source through our uses of it and show that no conflict exists with the second law. Pay particular attention to the ordering process which occurs in photosynthesis.

20. According to the "big bang" theory of the universe, at time zero the whole universe was compacted into an exploding hot ball with diameter about 10 times that of the sun. During the past 10^{10} years the universe has been expanding and cooling (to about 3 K now). In the future, gravitational forces may stop the expansion and cause the universe to fall into a duplicate of the original fireball. Would such an event be a contradiction of either the first or second law?

Chapter 16
Problems

1. To heat a certain quantity of gas from 20 to 100°C requires 400 cal when its volume is kept constant. By how much does its internal energy increase in the process? What is the maximum amount of work the gas could do in cooling back down to 20°C?

2. An ideal gas confined to a cylinder by a piston is compressed adiabatically. The compressing force does 5000 J of work against the gas. How much heat flows out of the gas? How much does the internal energy of the gas increase in the process? Repeat for an isothermal compression.

3. During a certain process, a confined ideal gas follows the *PV* curve shown in figure P16.1 and goes from *A* to *B*. How much work did the gas do? If the gas temperature at *A* was 27°C, what is its temperature at *B*?

4. An ideal gas starts at point *B* in the diagram of figure P16.1 and is compressed as shown to point *A*. If the gas temperature is the same at *B* and *A*, how much heat flowed from the gas in the process?

5. When 1 g of water at 100°C is changed to steam at 100°C the steam

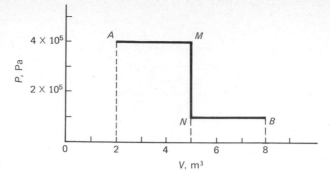

FIGURE P16.1

occupies a volume of 1670 cm^3 provided the pressure is 1 atmosphere (1.013 × 10^5 Pa). (a) How much external work does 1 g of water do as it changes to steam at 1 atm? (b) How much does its internal energy change?

6. Inside a 35-cm-long horizontal capillary tube is sealed an ideal gas at an absolute pressure of 90 cm Hg. A 13-cm-long column of mercury rests midway in the tube and blocks it. The tube is now stood on end. After temperature equilibrium is again achieved, (a) what will be the length of the gas column in the lower part of the tube? (b) Will the mercury rise or fall while temperature equilibrium is being established?

7. A quantity of nitrogen gas originally at 27°C is expanded adiabatically to a volume 20 times as large. What is the temperature of the gas after expansion? [Note: A^b = antilog (b log A).]

8. A 16,000-cm^3 cylinder is closed at one end by a piston and contains 20 g of air at 30°C. The piston is suddenly pushed in so as to change the gas volume to 1600 cm^3. The compression is adiabatic, and the final temperature of the gas is 500°C. How much work was done in compressing the gas? For air, c_V = 0.177 cal/g · °C.

9. 30 g of highly compressed air is confined to a cylinder by a piston. Its volume is 2400 cm^3, its pressure is 10 × 10^5 N/m^2, and its temperature is 35°C. The gas is now expanded adiabatically until its volume is 24,000 cm^3. During the process, 4100 J of work is done by the gas. What is the final temperature of the gas? Assume c_V = 0.177 cal/g · °C.

10. A quantity of nitrogen gas originally at 27°C is compressed adiabatically to one-twentieth of its original volume. Find its temperature after compression. (See note in problem 7.)

11. In order to compress helium gas adiabatically to one-fiftieth its original volume, by what factor must the pressure upon it be increased?

12. If a certain quantity of helium gas is expanded adiabatically to 50 times its original volume, what will be the ratio of the initial to final pressure?

13. When 5 × 10^4 cm^3 of air at STP* is isothermally compressed to 1 × 10^4 cm^3, how much heat must flow from the gas?

14. How much heat must be added to 2 × 10^5 cm^3 of air originally at STP* to cause it to expand isothermally to a volume of 7 × 10^5 cm^3?

15. Compute the work needed to compress adiabatically 20 g of nitrogen at STP* to one-fifth its original volume. (Hint: Notice that the work done will be equal to the increase in internal energy, $mc \, \Delta T$.)

16. How much work can 200 g of helium gas do as it expands adiabatically from T = 300°C to a volume 40 times as large? (See hint in problem 15.)

*STP means standard temperature and pressure, namely, 273 K and 1.01 × 10^5 Pa.

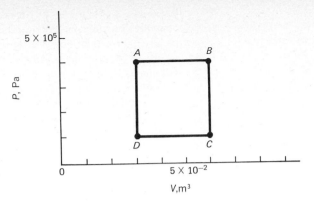

FIGURE P16.2

17. When gasoline is burned, it gives off $11,000\,\text{cal/g}$, called its heat of combustion. A certain car uses $9.5\,\text{kg}$ of gasoline per hour and has an efficiency of 25 percent. What horsepower does the car develop?

18. Starting at P_0, V_0, T_0, an ideal gas undergoes n successive adiabatic expansions with the volume being tripled on each. Show that the final temperature and pressure are given by

$$T = \frac{T_0}{3^{n(\gamma-1)}} \quad \text{and} \quad P = \frac{P_0}{3^{n\gamma}}$$

19. Referring to the reversible gas cycle shown in figure P16.2, find the work done on the gas for each of the four sections of the cycle. How much work was done by the gas during the cycle?

20. Calling the temperature of the gas T_A at A in the reversible ideal gas cycle shown in figure P16.2, find the temperatures at B, C, and D.

21. The PV diagram for a cylinder of a gasoline engine can be very roughly approximated as shown in figure P16.3 (the shaded portion is actually quite complex). About how much work is done during (a) the expansion, (b) the compression? (c) If the heat energy (in joules) added during the

FIGURE P16.3

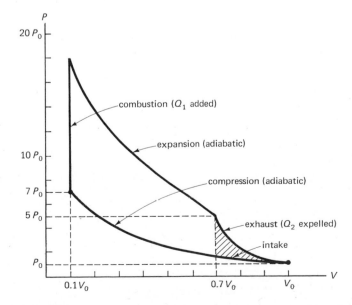

combustion process is 10 times as large as the work done during the compression, about what would be the efficiency of the engine? [*Hint:* In part (*c*), use the basic definition of efficiency = output ÷ input energy.]

22. In the reversible cycle shown in figure 16.5(*a*), take the pressure and volume at *A* to be 1 atm and 5.0 liters. At point *C* take the pressure and volume to be 0.80 atm and 20.0 liters. Find the heat absorbed by the system in going from *A* to *B*, from *D* to *A*, the work done per cycle, and the efficiency of the system. Assume the system to consist of 0.20 mole of ideal monatomic gas at 300 K at point *A*.

23. Referring back to figure 16.4(*a*), suppose the piston and cylinder are immersed in a water and ice bath maintained at 0°C. No heat can flow into or out of the bath except from the cylinder system. Weights are now removed from the piston so that the volume expands from V_0 to V_1 and the pressure decreases from P_0. After equilibrium is reached at this new volume, how much loss or gain of ice in the bath has occurred?

24. For the situation described in the preceding problem, the gas within the cylinder is carried through the following cycle: compressed isothermally from V_0 to V_1; expanded adiabatically to V_0; held at V_0 until the gas returns to 0°C. During the entire cycle, 7 g of additional ice forms in the bath. How much work was done by the external agent that manipulated the piston?

25. $\frac{1}{10}$ mol of monatomic ideal gas occupies a volume V_1 at temperature T_1. (*a*) What is its internal energy? (*b*) It expands adiabatically to V_2 at temperature T_2. What is its new internal energy? (*c*) How much work did it do? (*d*) What was its entropy change in going from V_1 to V_2 in part (*b*)?

26. Find the work done in the isothermal expansion of 1 kmol of van der Waals' gas in going from V_1 to V_2; that is, find the analog to equation *16.4*.

27. Compute the change in entropy of 30 g of ice at 0°C as it melts and is heated to water at 10°C. An answer correct to 2 percent will be sufficient.

28. Find the entropy change as 1 g of ice at 0°C melts to form water at 0°C and compare it to the entropy change as 1 g of water at 100°C vaporizes to form steam at 100°C.

29. A mass *m* of material having a constant specific heat *c* is heated from T_1 to T_2. Find its change in entropy assuming no phase change occurs in this range.

30. An ice cube is dropped into a Thermos jug of hot coffee. Assuming the jug to be perfectly insulating, find the change in entropy for the jug and contents between the instant just after the ice had dropped in and after the material within the jug has reached thermal equilibrium.

31. Calculate the entropy change that occurs in a mass *m* of ideal gas as it undergoes a reversible isothermal expansion from volume V_1 to volume V_2. (*b*) Repeat for an adiabatic expansion.

32. A Carnot engine operates between 20 and 200°C. (*a*) How much input energy is needed if the engine is to do 100 J of work? (*b*) How much waste energy is expelled by the engine in the process?

33. Two Carnot engines are connected in series such that the first operates between T_1 and T_2, while the second operates between T_2 and T_3 with $T_1 > T_2 > T_3$. If the second stage is run by the heat expelled from the first stage, show that the combined efficiency (output work/input) of the system is $1 - (T_3/T_1)$.

34. At low temperatures, the specific heat of many substances is given by $c = AT^3$. Assuming a material to have zero entropy at 0 K, find the entropy of a mass *m* at a temperature T_1.

35. When N labeled coins are tossed, there are 2^N different combinations of heads and tails possible. How many combinations are possible for (a) 3 coins, (b) for 5, (c) for 50?

36. Making use of the explanation given in problem 35, (a) what is the chance that 10 noninteracting ants will all end up in the same half of a box? (b) Repeat for the case where all but one ends up in the same half.

37. Five coins can each come up heads or tails. (a) What is the entropy of the configuration where all coins are heads? (b) For the configuration where all but one are heads?

17 The Electrostatic Field

We begin our study of electricity by considering charged particles which are at rest in an inertial reference frame. The basic experimental law which describes the interaction of such particles is called Coulomb's law. It in turn is used to describe electrical interactions in terms of electrostatic fields. The electric fields resulting from numerous charge configurations are considered, and methods for sketching electric fields are discussed. Motion of charges in an electric field are also examined.

17.1 ELECTRIC CHARGE

Great advances in our understanding of electricity were made during the nineteenth century by such men as Faraday, Ampère, and Maxwell. Their work in this field was so fruitful that, by 1900, one was able to say that the basic laws of electricity were precisely known. They are embodied in four basic equations, called Maxwell's equations.

During the years since 1900 vast changes have been seen in the field of physics. Einstein's theory of relativity and the development of quantum mechanics have greatly influenced our concepts of nature. In spite of this, Maxwell's equations remain nearly untouched by all this and are as applicable within atoms as they are in the cosmos. Although they have been augmented by quantum theory to form the subject of quantum electrodynamics, this has been more a joining of electrodynamics to particle physics than a change in Maxwell's electrodynamics.

We should not suppose, however, that Maxwell's *concepts* of electricity are still accepted today. Many of the ideas which he and his contemporaries used to visualize electrical interactions are known to be wrong. However, the mathematical relations which they set up to describe the experimental results were so well founded in experiment that they have remained valid long after the mental pictures used to obtain them have been discarded. This is not only a tribute to the excellence of the early experiments in electricity and to the clarity of analysis used to describe the results, it is also a reflection of the inherent simplicity of electrical interactions.

As an example of this simplicity we note that there appears to be a basic, smallest possible piece of charge. No one has been able to detect a charge smaller in magnitude than the charge of the electron e. In addition, the magnitudes of all other charges are found to be integer multiples of the

Charge is quantized in units of $\pm e$

magnitude of the charge on the electron.* We say that charge is quantized, or that it comes in well-defined packets. The basic packet, or quantum, of charge has magnitude e.

Only one complicating factor enters. There are two kinds of charge quanta. Although they both have the same magnitude to the experimental accuracy known today (1 part in 10^{20}), they differ in an important respect. One of the charge quanta, typified by that on the electron, attracts the other type of quantum found, for example, on the proton. However, two electrons always repel each other, as do two protons. In fact, if any charge is found to attract an electron, it always repels a proton, and vice versa. Although these two types of quanta have the same magnitude, e, we give them, arbitrarily, different signs. The electron-type quantum is called negative, while the proton quantum is designated positive. We repeat, these two electric charge quanta have equal magnitude. All known charges are integer multiples of these quanta.

Like charges repel; unlike charges attract

17.2 COULOMB'S LAW

The fundamental law governing interations between charges at rest is called *Coulomb's law*. Consider two pointlike charges at rest relative to each other as shown in figure 17.1. The force **F** exerted by charge q_1 on charge q_2 was first accurately measured by Coulomb in 1785. He found **F** to be given by

Coulomb's law

$$\mathbf{F} = (\text{const})\frac{q_1 q_2}{r^2}\hat{\mathbf{r}} \qquad 17.1$$

where q_1, q_2 = magnitudes of point charges (+ or − sign appended, depending upon sign of charge involved)
  r = distance between charges
  $\hat{\mathbf{r}}$ = unit vector along r, shown in figure

Notice that this relation is restricted to pointlike charges. The constant in the equation will depend upon the units we use for the various quantities.

For the remainder of this text only the SI units system is used. It gives rise to the practical electrical units, namely, the ampere and volt. In this set of units F must be measured in newtons, r in meters, and q is later defined in a unit called the *coulomb* (C). Although the coulomb is defined precisely in our study of forces between current-carrying wires, we state at this point that the charge quantum e is 1.6022×10^{-19} C. Making use of that set of units, *17.1* is often written as

The charge quantum is
$e = 1.602 \times 10^{-19}$ C

$$\mathbf{F} = \frac{1}{4\pi\epsilon_0}\frac{q_1 q_2}{r^2}\hat{\mathbf{r}} \qquad 17.2$$

where $\epsilon_0 = \left[\dfrac{1}{(2.998 \times 10^8)\sqrt{4\pi \times 10^{-7}}}\right]^2 \approx 8.85 \times 10^{-12}$ C^2/N · m^2

and $\dfrac{1}{4\pi\epsilon_0} \approx 9.0 \times 10^9$ N · m^2/C^2

The alert student will recognize that 2.998×10^8 is the speed of light in meters per second, and might suspect, as we later show to be true, that ϵ_0 has an important place in electrical theory. It is designated as the *permittivity of*

FIGURE 17.1 What restriction must be placed upon q_1 and q_2 if *17.1*, Coulomb's law, is to apply?

---

*According to certain currently popular theories, smaller charges, called *quarks,* exist. However, they have not been isolated experimentally as of this time.

free space, and will be found to enter in many of the basic equations of electricity. Its true significance will not become apparent, however, until we study the motion of electromagnetic waves.

Returning now to figure 17.1, we notice that when q_1 and q_2 are of like sign, the charges repel each other, as shown there. In the event that the charges are of unlike sign, the force will be reversed, since unlike charges attract. This is also borne out by *17.2*, which indicates that, if q_1 and q_2 are of unlike sign, the force will be negative, and is therefore reversed. Further, Newton's law of action and reaction tells us that an equal and opposite force will be exerted by charge q_2 on charge q_1. Of course, merely reversing the roles of q_1 and q_2 in *17.2* indicates this same fact. We emphasize once again that Coulomb's law applies only to point charges, charges so small that no question arises as to how the separation distance r is to be defined.

Illustration 17.1

The two balls shown in figure 17.2 have identical masses of 0.20 g each. When suspended from 50-cm-long strings they make an angle of 37° to the vertical. Assuming the charges on each to be the same, how large is each charge?

Reasoning Since the system is at rest, we can apply the conditions for equilibrium to the ball on the left. Notice that three forces act on the ball: its weight mg, the tension T in the string, and F, the repulsion force due to the charge on the other ball. We have the usual conditions for equilibrium:

$$\sum F_x = 0$$

from which

$$F - 0.6T = 0$$

and

$$\sum F_y = 0$$

which gives

$$0.8T - (0.2)(10^{-3}\,\text{kg})(9.8\,\text{m/s}^2) = 0$$

or

$$T = 2.45 \times 10^{-3}\,\text{N}$$

Using this to find F, we obtain

$$F = 1.47 \times 10^{-3}\,\text{N}$$

This is the Coulomb's law force.

Substituting in Coulomb's law, we have

$$1.47 \times 10^{-3} = (9 \times 10^9)\frac{q^2}{(0.60)^2}$$

FIGURE 17.2 If the mass of each ball is 0.2 g, what is the product of their two charges?

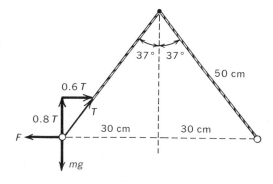

all units being SI. Solving for q, we find

$$q \approx 2.4 \times 10^{-7}\,\text{C}$$
$$\approx 0.24\,\mu\text{C}$$

where μC stands for microcoulombs, 10^{-6} C. A charge this large contains on the order of 10^{12} charge quanta.

17.3 THE SUPERPOSITION PRINCIPLE

The electric force superposition principle: Coulomb-type forces can be added vectorially

Coulomb's law applies to *two* charges. In most cases of interest, several charges exert forces on each other at the same time. We resort to experiment to see what must be done in such a case. It is found that the total force on a given charge is merely the vector sum of the forces due to the various individual charges. This fact is often referred to as the *superposition principle.*

In order to illustrate the utility of this principle, consider the situation shown in figure 17.3. We wish to find the force on the 5-μC charge due to the other two charges. These forces are drawn on the diagram. Notice that the -6-μC charge attracts the $+5$-μC charge. Using Coulomb's law, we find

$$F_1 = 0.270\,\text{N}$$
$$F_2 = 0.360\,\text{N}$$

Making use of the superposition principle, we now add the forces. This is done by splitting the forces into components, giving

$$F_x = 0.6F_2 - 0.8F_1 = 0.216\,\text{N} - 0.216\,\text{N} = 0$$

and
$$F_y = 0.6F_1 + 0.8F_2 = 0.45\,\text{N}$$

The resultant force on the 5-μC charge is 0.45 N in the upward direction.

FIGURE 17.3 Find the forces on the 5-μC charge.

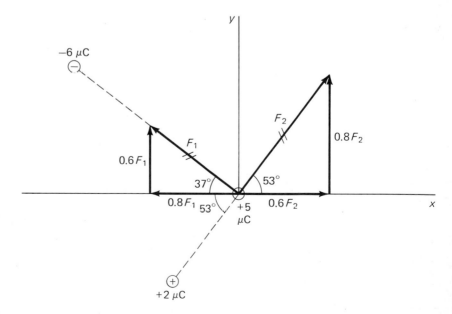

17.4 THE GRAVITATIONAL FIELD

In previous chapters we have mentioned the gravitational field of the earth. By this we meant that, where the field is strong, an object is strongly attracted toward the earth. The concept of another type of field, the electric field, is of great importance in electricity, and we shall now explore this way of looking at things in some detail. To do so, let us first refer to the gravitational field since we are more familiar with it.

We draw the gravitational field of the earth as shown in figure 17.4(a). (Only a portion of the field lines are shown. Ideally, they should come into the sphere from all directions.) The lines with arrows are called the *field lines*. Their direction is determined in the following way. If a small mass (called the *test mass*) is placed at a point such as A, it will be attracted toward the earth's center as shown. A field line drawn through point A must be tangent to the line of the force vector at A. We see, then, that the field lines show us the direction of the gravitational force on a test mass. Since we know masses are attracted toward the earth's center, the gravitational field lines must be directed radially inward as the figure shows. Unfortunately, most of us are too poor as artists to draw the field diagrams properly. Consequently we allow ourselves to draw the shorthand diagram illustrated in part (b) of figure 17.4.

Field lines show the direction of the force

A second feature of the field diagram of figure 17.4 is also important. You will notice that the field lines are closer together near the earth than they are at more distant points. Since we know that the earth's gravitational force is strongest close to the earth, we might suspect that the closeness of the lines could be used as a measure of the strength of the gravitational force. This turns out to be true. In fact we shall soon see that this is quantitatively true for inverse square law forces, i.e., forces which vary as $1/r^2$. As we know, both the gravitational and Coulomb law forces vary in this way.

The field is strongest where the field lines are closest together

Let us summarize: We say that a gravitational field exists wherever a test mass experiences a gravitational force; the field lines are drawn to show the direction of the gravitational force on a test mass. Since we know masses close together are regions of strong gravitational force.

FIGURE 17.4 The three-dimensional gravitational field shown in (a) is usually represented as in (b).

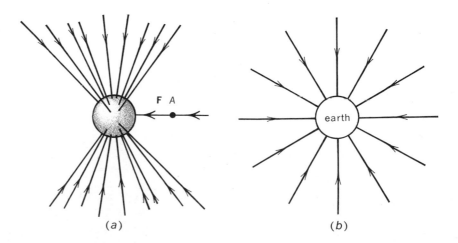

(a)   (b)

17.5 THE ELECTRIC FIELD

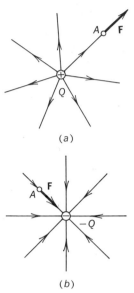

(a)

(b)

FIGURE 17.5 Electric field lines come out of positive charges and go into negative charges.

In complete analogy to the gravitational field, we determine the existence of an electric field in the following way: if, at a certain point, a tiny, stationary, *positive* charge (called the *test charge*) experiences a force of electrical origin, we say that an electric field exists at that point. The direction of the electric field is taken to be the direction of the electric force on the positive test charge. With this definition in mind, it is a simple matter to draw electric field diagrams.

For example, suppose we wish to draw a diagram for the electric field near a positive point charge Q. We do this (in our shorthand way) in the manner shown in figure 17.5(a). The positive charge Q is shown. To find the electric field at point A, we ask ourselves what sort of force a positive test charge placed at A would experience. The test charge, being positive, will be repelled by the like charge Q. Hence the force on it will be radially out from Q through A as shown. Clearly, the test charge will experience a force radially out from Q no matter where it is placed. The electric field lines must therefore run radially outward from the charge $+Q$.

Similarly, we can find the electric field near a negative charge $-Q$. This is shown in figure 17.5(b). In this case, the positive test charge placed at A is attracted by the negative charge $-Q$. Hence the force and electric field lines are directed radially inward in this case. We notice in these two examples a very important feature of electric field lines: *field lines come out of positive charges and go into negative charges.*

More complicated situations can be handled in exactly the same way. For example, suppose you wish to find the electric field in the room where you are reading this (there probably is not a large one, but let us just suppose). You would place a tiny positive test charge at any point in question and would measure the force of electric origin which acts on it. This would tell you the direction of the electric field at that point and would allow you to draw the field line through the point. By moving the test charge here and there in the room, you could obtain a set of field lines showing the electric field.

As two other examples, consider the electric field about two equal point charges. In figure 17.6(a) and (b) we show the two possible cases, unlike charges and like charges. To find the direction of the field line at a point such as A, we imagine a positive test charge to be placed at that point. The resultant force on it due to the two real charges is in the direction of the field at that point. This fact, together with our rule that lines come out of positive charges and enter negative charges, allows us to sketch the field lines quickly. You should notice that, even in a complex situation, the field lines close to the charges are directed radially outward from positive charges and radially inward toward negative charges. You should check the field lines we have drawn for these two cases to make sure that you understand why they are drawn as shown.

Until now we have been content to simply draw electric field diagrams. However, the concept of an electric field can be given much more power if we assign a quantitative significance to the term electric field. We shall define a quantity which is called the *electric field strength E*. It is defined in terms of the force F on a *positive test charge q_t*. In order to make the definition be independent of the magnitude of the test charge, we concern ourselves with the ratio F/q_t so that we will really be discussing the *force per unit positive*

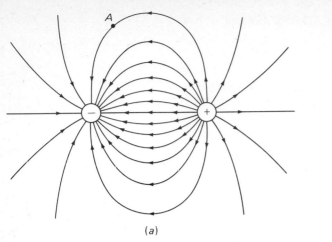

(a)   (b)

FIGURE 17.6 In each case, where is the electric field strong? Weak?

test charge. The definition is as follows: To find the electric field strength **E** at a point, suppose a positive test charge q_t to be placed at that point. The force on the test charge will be **F**. We define the electric field strength at that point to be

Defining equation for **E**

$$\mathbf{E} = \frac{\mathbf{F}}{q_t}$$

17.3

Notice that **E** has the units newtons per coulomb in keeping with our wish that it be the force per unit test charge.

Of course, the test charge is a fictitious charge. We merely ask what would be the force on it if it were placed at a point. In addition, the fictitious test charge is not allowed to influence the other charges in the region. Some texts insist that the test charge must be made infinitesimally small so that $q_t \to 0$. Then the electric field is given by

An equivalent definition of **E**

$$\mathbf{E} = \lim_{q_t \to 0} \frac{\mathbf{F}}{q_t}$$

17.4

This ensures that the real charges present will not be disturbed by the measuring process. In actual measurements of the field, this procedure must be followed. However, for conceptual experiments, we shall take the test charge to be $+1$ C and endow it with the property of not disturbing the other charges.

The electric field strength at a distance r from a point charge q is of considerable importance. We can easily find it by use of Coulomb's law. Since we want **E** at a distance r from the point charge q, we imagine a positive test charge q_t placed at that point. According to Coulomb's law, the force experienced by the test charge will be

$$\mathbf{F} = \frac{1}{4\pi\epsilon_0} \frac{qq_t}{r^2} \hat{\mathbf{r}} \qquad \text{or} \qquad \frac{\mathbf{F}}{q_t} = \frac{1}{4\pi\epsilon_0} \frac{q}{r^2} \hat{\mathbf{r}}$$

But by definition $\mathbf{E} = \mathbf{F}/q_t$, so we find at once

E due to a point charge

$$\mathbf{E} = \frac{1}{4\pi\epsilon_0} \frac{q}{r^2} \hat{\mathbf{r}}$$

17.5

(restricted to a point charge)

We emphasize that this gives the electric field due to a *point* charge. But it can be used to derive results applicable to other cases, as we shall see.

Illustration 17.2

Find the electric field strength at a distance of 50 cm from a $-2\,\mu C$ charge, that is, at point P in figure 17.7.

Reasoning The field due to a point charge, which this is, can be found by use of *17.5*. We have

$$\mathbf{E} = (9 \times 10^9)\frac{-2 \times 10^{-6}}{(0.5)^2}\hat{\mathbf{r}} = -72{,}000\hat{\mathbf{r}} \qquad \text{N/C}$$

Notice that $\mathbf{E}$ is antiparallel to $\hat{\mathbf{r}}$ and is therefore radially in toward the negative charge.

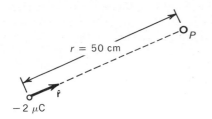

FIGURE 17.7 What is the direction of **E** at point *P*?

Illustration 17.3

Find the electric field at point P in figure 17.8(a) due to the charges shown.

Reasoning We can make use of the superposition principle here since the electric field at P due to each charge is merely the force on a unit positive test charge at P. These fields are shown in the figure and are, from *17.5*,

$$E_2 = 7.2 \times 10^4\,\text{N/C}$$
$$E_8 = 28.8 \times 10^4\,\text{N/C}$$
$$E_{12} = 43.2 \times 10^4\,\text{N/C}$$

We therefore find at P that $E_x = -36 \times 10^4\,\text{N/C}$ and $E_y = 28.8 \times 10^4\,\text{N/C}$. From this the resultant, $\mathbf{E}$, is $\sqrt{(36)^2 + (28.8)^2} = 46.1\,\text{N/C}$ at an angle ϕ for which $\tan\phi = 28.8/36$. Therefore $E = 46.1\,\text{N/C}$ at $\theta = 141°$.

FIGURE 17.8 Find **E** at the origin (P) due to the three charges.

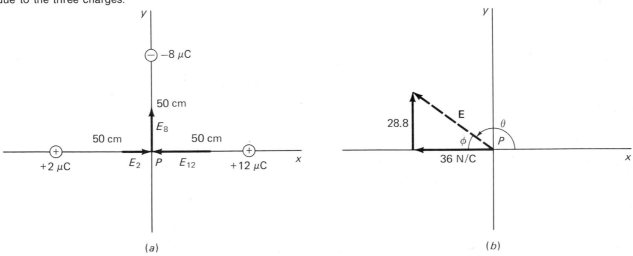

(a)

(b)

17.6 ELECTRIC FIELD DUE TO A CHARGED ROD: CASE 1

Suppose we have a uniformly charged rod such as the one shown in figure 17.9. It has a constant charge λ per unit length. In other words, each meter length of the rod has a charge λ. Of course, a length of $\frac{1}{2}$ m will have a charge $\frac{1}{2}\lambda$; similarly, $\frac{1}{4}$ m has a charge $\frac{1}{4}\lambda$; in general, a length Δx of the rod will have a charge $\Delta x \,\lambda$. We wish to find the electric field at point P due to this rod.

The only tool at our disposal for finding electric fields is Coulomb's law. But, as we have pointed out, Coulomb's law applies only to point charges, and so it cannot be applied directly in the present case. However, we can imagine the rod to be sectioned into a large number of very small segments, in which case each segment will act essentially as a point charge. For example, the little length Δx_i shown in figure 17.9 will act like a point charge of magnitude $\lambda \,\Delta x_i$. The electric field due to this at point P will be

$$\Delta \mathbf{E}_i = \frac{1}{4\pi\epsilon_0} \frac{\lambda \,\Delta x_i}{x_i{}^2} \hat{\mathbf{r}}$$

where $\hat{\mathbf{r}}$ is a unit vector in the direction shown for $\Delta \mathbf{E}_i$ in the figure.

Since the fields due to all the other segments of the rod will be in the same direction as $\Delta \mathbf{E}_i$, we can add them directly to find $\mathbf{E}$. Therefore

$$\mathbf{E} = \sum \Delta \mathbf{E}_i$$

$$= \sum \frac{\lambda}{4\pi\epsilon_0} \frac{\Delta x_i}{x_i{}^2} \hat{\mathbf{r}}$$

where the sum is to extend over all the segments of the rod.

If we make Δx sufficiently small, we can replace the sum by an integral to obtain

$$\mathbf{E} = \int_b^{L+b} \frac{\lambda}{4\pi\epsilon_0} \frac{dx}{x^2} \hat{\mathbf{r}}$$

Notice the limits on this integral. One end of the rod is at $x = b$, and the other is at $x = (L + b)$, and these values are therefore the limits on x when summing over the segments of the rod.

To evaluate this integral we notice that λ, ϵ_0, and $\hat{\mathbf{r}}$ are all constants independent of x, and so we have

$$\mathbf{E} = \hat{\mathbf{r}} \frac{\lambda}{4\pi\epsilon_0} \int_b^{L+b} \frac{dx}{x^2}$$

FIGURE 17.9 Find the electric field at point P due to the uniformly charged rod of length L. What if λ had varied as x^2?

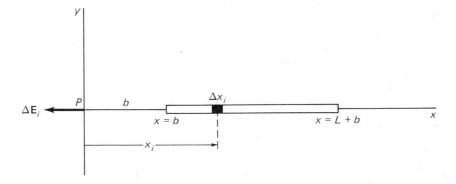

Since the integral is merely $-(1/x)$, we have as our result

$$\mathbf{E} = \hat{\mathbf{r}}\frac{\lambda}{4\pi\epsilon_0}\left(\frac{1}{b} - \frac{1}{L+b}\right) \qquad 17.6$$

17.7 ELECTRIC FIELD DUE TO A CHARGED ROD: CASE 2

Let us now consider the uniformly charged rod of the preceding example and try to find the electric field at a point P to one side of the rod, as shown in figure 17.10. As before, we section the rod into small pieces, so that it approximates a series of point charges. Using Coulomb's law for the charge $\lambda\,\Delta x_i$, we have

$$\Delta E_i = \frac{1}{4\pi\epsilon_0}\frac{\lambda\,\Delta x_i}{r_i{}^2}$$

where r_i is shown in the figure.

The important point to notice here is that, unlike the preceding case, the $\Delta\mathbf{E}$ vectors due to each little segment of the rod will point in a different direction. Since the $\Delta\mathbf{E}$'s are vectors, we must first resolve them into components before adding them. This is done on the figure. For the y component we have

$$\Delta E_{iy} = \Delta E_i \cos\theta$$

$$= \frac{\lambda}{4\pi\epsilon_0}\frac{\Delta x_i}{r_i{}^2}\cos\theta$$

Since all the y-component vectors are in the same direction, they may be added to find E_y. We have

$$E_y = \sum \Delta E_{iy} = \sum \frac{\lambda}{4\pi\epsilon_0}\frac{\Delta x_i}{r_i{}^2}\cos\theta$$

where the sum is extended over all the Δx_i composing the rod. Or, upon allowing $\Delta x_i \to 0$, we replace the sum by an integral and find

$$E_y = \int \frac{\lambda}{4\pi\epsilon_0}\frac{dx}{r^2}\cos\theta$$

FIGURE 17.10 Under what conditions will $\mathbf{E}$ at point P have no x component?

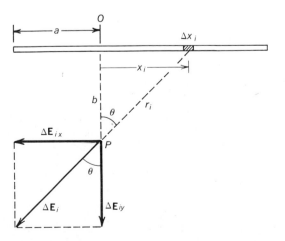

Unfortunately, there are three variables involved in this integral, x, r, and θ. However, everything can be expressed in terms of x by noting that

$$r^2 = x^2 + b^2 \qquad \text{and} \qquad \cos\theta = \frac{b}{r}$$

Substituting for r and $\cos\theta$, we find

$$E_y = \int_{x=-a}^{L-a} \frac{\lambda b}{4\pi\epsilon_0} \frac{dx}{(x^2 + b^2)^{3/2}} \qquad 17.7$$

Notice the limits on this integral. We are taking our origin for measuring x at point O. If the rod has a total length L, the largest value of x will be $L - a$.

We note that, in 17.7, the factor $\lambda b/4\pi\epsilon_0$ is constant, and so the integral involved is the remainder of this expression. This integral is given in most

---

The Electron as Observed in Faraday's Experiment

Michael Faraday (1791–1867) spent a large part of his life carrying out investigations into the nature and behavior of electricity. In 1834 he published experimental results which were fundamental to our understanding of the nature of electric charge. Using apparatus somewhat like that shown in the figure, he investigated the nature of the gases given off at the electrodes when an electric current is passed through slightly acid water.

His experiments confirmed the previous results of William Nicholson (1753–1815), showing that, for each 8 grams of oxygen given off at the positive electrode, 1 gram of hydrogen was evolved at the negative electrode. This ratio, 8:1, is the same as the ratio of chemical equivalents* for these two elements as listed by the chemists of the day. Faraday found this ratio to be independent of the acid strength, the magnitude of the current, or the time during which the current flowed.

Later experiments, using other solutions from which different elements were deposited at the electrodes, gave the following result. The same amount of electricity needed to deposit 1 mole ($= 1$ g) of hydrogen deposits one chemical equivalent of any other element as well. It was inferred from this that a definite amount of electricity is associated with a definite quantity of matter. In particular, Faraday found that 1 gram of hydrogen is liberated by what is now called 1 faraday of electric charge. In terms of the coulomb, 1 faraday of charge is approximately 96,500 coulombs.

It was considerably later, in 1857, that R. J. Clausius (1822–1888) suggested that molecules of a substance split up into their component ions in solution. In his view, the thermal vibration of the molecules broke a small number of molecules apart in solution. These broken parts, the ions, were pictured as carrying multiples of a basic electric charge and as being capable of transporting the charge through the solution. These multiples, the valences of the ions, were usually 1, 2, or 3. The minimum amount of charge carried by an ion was named the electron by Johnstone Stoney (1826–1911) in a paper presented in

---

*If m grams of a substance combines chemically with 1 gram of hydrogen, its chemical equivalent is said to be m. For example, 8 grams of oxygen combines with 1 gram of hydrogen to form water. The chemical equivalent of oxygen is therefore 8.

integral tables, and is

$$\int \frac{dx}{(x^2 + b^2)^{3/2}} = \frac{1}{b^2} \frac{x}{(x^2 + b^2)^{1/2}}$$

Placing in the limits, we find

$$E_y = \frac{\lambda}{4\pi\epsilon_0 b} \left\{ \frac{L - a}{[(L - a)^2 + b^2]^{1/2}} + \frac{a}{(a^2 + b^2)^{1/2}} \right\} \qquad 17.8$$

The x component of **E** can be found in a similar manner. We leave it as an exercise for the student to show that

$$E_x = \frac{\lambda}{4\pi\epsilon_0} \left\{ \frac{1}{[(L - a)^2 + b^2]^{1/2}} - \frac{1}{(a^2 + b^2)^{1/2}} \right\} \qquad 17.9$$

This solution for E_x has an interesting result when $a = L/2$. For that case, 17.9 gives $E_x = 0$. We could have guessed this solution at once from a

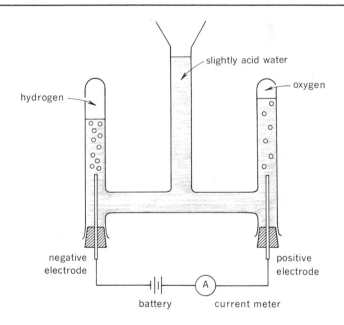

1874. Since it had already been shown by others that 1 gram of hydrogen contained about 10^{24} hydrogen atoms, Stoney concluded that the charge of the electron was about $96,500/10^{24}$, or about 1×10^{-19} coulomb. Upon using the value at present accepted for Avogadro's number of substance, the electronic charge is found to be

$$\frac{96,500}{(6.02)(10^{23})} = 1.60 \times 10^{-19} \text{ coulomb}$$

Although this concept of the basic charge unit is now recognized to be correct, it was quite speculative at the time. It was not widely accepted until other independent experiments were found to support the electron concept. Some of these later experiments are presented in following chapters.

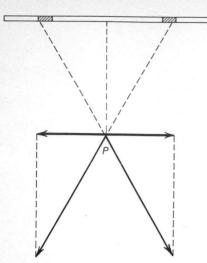

FIGURE 17.11 The x component of the electric field is zero at P.

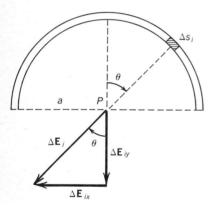

FIGURE 17.12 The semicircular arc carries a uniform charge λ per unit length. Why can one say at once that $E_x = 0$ at point P?

consideration of figure 17.11. It is seen there that the $\Delta \mathbf{E}_x$ vectors for corresponding points on the rod cancel in pairs. If the point P is on the perpendicular bisector of the rod, the left half of the rod exactly cancels the right half as far as the x component of $\mathbf{E}$ is concerned.

There is another interesting property of the result given in *17.8* and *17.9*. Suppose the rod is viewed from a large distance, so that $b \gg a$ and $b \gg L$. Under this condition you should be able to show that *17.8* and *17.9* give

$$E_y = \frac{Q}{4\pi\epsilon_0 b^2} \quad \text{and} \quad E_x \approx 0$$

where $Q = \lambda L$ is the total charge on the rod. This result is expected, since, when viewed from far away, the rod will appear to be a distant point charge. As a result, the usual Coulomb field is found at large distances.

Illustration 17.4

Find the electric field at the center of a uniformly charged semicircular arc, i.e., at point P in figure 17.12.

Reasoning We assume the charge per unit length of arc to be λ, and the radius of the arc to be a. Splitting the arc into small segments Δs_i, we have that the essentially point charge on each is $\lambda \Delta s_i$. The electric field due to the Δs_i shown is, from Coulomb's law,

$$\Delta E_i = \frac{1}{4\pi\epsilon_0} \frac{\lambda \Delta s_i}{a^2}$$

Each little portion of the arc will give a $\Delta \mathbf{E}_i$ in a different direction. We must therefore take components in order to find the total field at point P.

Before doing that, though, we notice that E_x will be zero at point P. This is the result of the fact that the ΔE_{ix} shown in figure 17.12 will be canceled by the contribution from a symmetrically placed Δs on the left half of the arc. As a result, we need only compute E_y at point P in order to find the total E. We have

$$\Delta E_{iy} = \frac{\lambda}{4\pi\epsilon_0 a^2} \Delta s_i \cos\theta$$

If we now sum over the whole arc and replace the sum by an integral in the usual way, we find

$$E_y = \frac{\lambda}{4\pi\epsilon_0 a^2} \int \cos\theta \, ds$$

The integrand involves two variables, θ and s. We can express θ in terms of s by recalling that an angle $d\theta$ subtends an arc length

$$ds = r \, d\theta \qquad\qquad 17.10$$

In the present case, $r = a$, and so we find upon substitution that

$$E_y = \frac{\lambda}{4\pi\epsilon_0 a} \int_{-\pi/2}^{\pi-2} \cos\theta \, d\theta$$

The limits on θ are obtained by noticing that we wish to sum over the whole arc. When we are considering the portion on the extreme left, $\theta = -\pi/2$, as we have defined θ. Similarly, at the extreme right, $\theta = \pi/2$.

These are therefore the limits on θ. Of course, other choices for defining θ are possible, and the limits will be different for each. Upon evaluating the integral, we find

$$E_y = \frac{\lambda}{4\pi\epsilon_0 a}[1 - (-1)] = \frac{\lambda}{2\pi\epsilon_0 a} \qquad 17.11$$

17.8 E ON THE AXIS OF A CHARGED LOOP

Consider a circular loop of radius a as shown in figure 17.13. Suppose the loop carries a uniform charge λ per unit length; the electric field near it is of interest. Unfortunately, the computation of E is very difficult except for points on the axis of the loop. We can compute the field at a point such as P without too much difficulty, and we shall now do it.

As in the preceding cases we split the charged body into a large number of essentially point charges. According to Coulomb's law, the charge $\lambda\,ds_i$ on the length ds_i gives rise to a field at P which is

$$dE_i = \frac{1}{4\pi\epsilon_0}\frac{\lambda\,ds_i}{r_i^{\,2}}$$

Here, too, the contributions to E of the various parts of the ring will be in different directions. To add these various vectors, we take components as shown in the figure. We can see from part (b) of the figure that the components of $\mathbf{E}$ perpendicular to the axis will cancel. The resultant E will therefore be due entirely to the y components. We have

$$E = \sum dE_{iy} = \sum dE_i \cos\theta$$

$$= \int \frac{\lambda}{4\pi\epsilon_0}\frac{ds}{r^2}\cos\theta$$

where we must add up the contributions of the whole ring.

FIGURE 17.13 What is the direction of the electric field at point P because of the uniform charge on the circular arc?

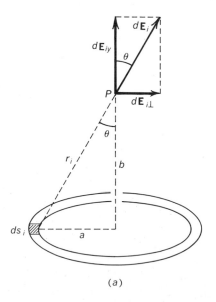

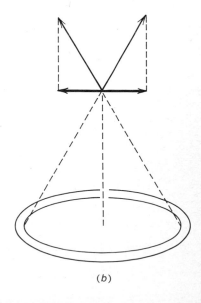

(a) $\qquad\qquad$ (b)

To simplify the integral we notice that

$$r^2 = a^2 + b^2 = \text{a constant}$$

$$\cos \theta = \frac{b}{r} = \frac{b}{\sqrt{a^2 + b^2}} = \text{a constant}$$

Therefore

$$E = \frac{\lambda b}{4\pi\epsilon_0(a^2 + b^2)^{3/2}} \int ds$$

where we are to add up ds around the whole ring. But the sum of all the ds elements is just the perimeter of the ring, $2\pi a$, and so we have

$$E = \frac{\lambda ab}{2\epsilon_0(a^2 + b^2)^{3/2}} \qquad\qquad 17.12$$

Can you show that, as $b \to \infty$, this gives the field due to a point charge? What does the result for $b = 0$ mean?

17.9 CONDUCTORS AND INSULATORS

Charges are not free to move in insulators

Electric charges and currents usually occur on or in material objects. There are two rather broad categories into which materials may be classed. *Insulators,* or *nonconductors,* are materials containing essentially no charges which are free to move. If such a material is subjected to an electric field, each electron and nucleus within the material experiences a force due to the electric field. However, these charges are bound tightly within the material, and are not free to move more than a fraction of an atomic diameter under this force. Therefore insulators are characterized by the fact that external electric fields do not cause charges to flow in the material, and therefore these materials do not conduct electric charges.

Conductors have many charges which move freely

The opposite of an insulator is a *conductor*. Metals are typical conductors. In these materials there exist electrons which are essentially free to move through the material. When an electric field is impressed upon such a material, the charges move to produce a current under the action of the field.

Of course, these are not sharply defined classifications. There exist materials which are only slightly conducting, in which only a very small proportion of the charges are relatively free to move. In fact, *all* materials have a few charges within them which can move under the action of a field. As a result, these classifications might better be designated as poor and good conductors. For most purposes, this impreciseness of classification will be of no importance.

17.10 QUALITATIVE CONSIDERATIONS

More often than not, the electric field near a material object cannot be computed without an excessive amount of time and effort. It therefore becomes important to be able to draw electric fields, approximately, merely by inspection. This may be done rather easily if we remember a few facts about the lines we draw to represent the field.

Rules for sketching electric fields

1. The lines show the direction of the force on a positive test charge.
2. The lines originate, or come out of, positive charges, and terminate, or go into, negative charges.

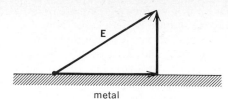

FIGURE 17.14 Why cannot the field be as shown under electrostatic conditions? Is it ever possible to have such an **E**?

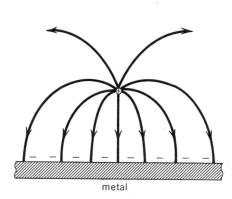

FIGURE 17.15 Where is the electric field strongest?

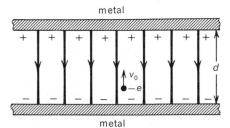

FIGURE 17.16 Are the field lines directed straight downward? Why?

3. The lines are closest together where the field is strongest. (See, for example, figure 17.5.)

4. The lines always strike the surface of a *conductor* perpendicularly under electrosatic conditions.

Statement 4 can be justified as follows. Suppose the electric field is not perpendicular to the metal or conductor surface, as shown in figure 17.14. We are assuming that we have an electrostatic situation, and so no currents can be flowing. However, the component of the field parallel to the metal surface would cause the electrons to move and would therefore give rise to a current. Since, in fact, there is *no* current, there can be no component of **E** parallel to the metal surface. Therefore **E** must be perpendicular to the metal surface.

To illustrate these points, let us sketch the electric field due to a positively charged ball above a large metal plate. This situation is shown in figure 17.15. The positive ball will attract the electrons of the plate, and so there will be a concentration of negative charge on the plate just below the ball, as shown. We know that the field lines must leave the positively charged ball and terminate on the negative charges. Further, the lines must hit perpendicularly to the metal plate. A sketch embodying these considerations is shown in figure 17.15.

As a second example, consider the electric field between two parallel infinite metal plates which carry equal and opposite charges on their adjacent faces. A cross section of this situation is shown in figure 17.16. If a positive test charge is placed between the plates, it will be repelled downward by the positive charge on the upper plate and attracted downward by the negatively charged lower plate. Since the plates are infinite, neither the left nor right direction is exceptional in any way. Therefore the electric field cannot point in one of these directions and not the other. This, of course, means that the field can have no component in either the left or right direction. The field must be vertical, as shown. Moreover, since the field lines are parallel and equally spaced, the electric field must have the same magnitude at all points between the plates.

17.11 SOME SIMPLE EXPERIMENTS

The designation plus and minus for charges was made long before the concept of atoms had been well substantiated. At that time the charge which exists on an ebonite (hard rubber) rod after rubbing with cat's fur was labeled negative. An opposite charge, positive, was found to occur on a glass rod which had been rubbed with silk. Today we know that electrons migrate from the fur to the ebonite rod, making the rod have an excess of electrons. The opposite effect occurs when glass is rubbed with silk; electrons migrate from the glass to the silk, thereby leaving an excess positive charge on the glass rod. Although these effects are very easily demonstrated, their molecular interpretation and prediction is very difficult. The theory of such electronic behavior is a subject of research in the area of solid state physics and materials science. In spite of this complexity on the atomic scale, there are many experimental situations of interest which are easily explained in simple terms. Let us examine a few of them.

Suppose a glass rod is charged positive by rubbing it with silk. If it is brought close to a metal object (or any conductor) as shown in figure 17.17(*a*), an interesting effect occurs. Most objects are neutral, i.e., un-

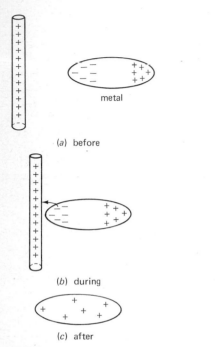

metal

(a) before

(b) during

(c) after

FIGURE 17.17 Charging by conduction. Note that the rod and charged object both have positive charges.

Charging by conduction

Charging by induction

charged. Each atom within them has as many electrons as there are protons in the nucleus, and so the object has no unbalanced or net charge. However, in metals and other conductors, a small fraction of the electrons are free from the atoms and can wander throughout the material. As a result, when the positively charged rod shown in figure 17.17(*a*) is brought close to the metal object, some of these free electrons are attracted to the side closest to the oppositely charged rod. Since the object was neutral to begin with, and must still be so, an equal positive charge (the charge on the atoms, which lost the electrons) exists on the opposite end of the object as shown. We call the charges on the two ends of the object *induced charges*.

Frequently in situations such as this people say the rod has repelled positive charge to the far end of the metal object. In most cases, certainly in metals, only the negative charges move. However, it is surprisingly difficult to prove this point experimentally and so most experiments can be "explained" either way. It is for this reason that people sometimes speak this way when the explanation is most simply stated in terms of moving positive charges.

It is easy for you to carry out an experiment similar to that shown in figure 17.17(*a*). Rub a plastic comb, pen, or other object briskly with cloth, preferably cotton or wool. It will become charged if it is not too moist. If the comb is brought close to a tiny piece of paper, it will induce charge motion within the (usually damp) bit of paper. The portion closest to the comb will have induced on it a charge with sign opposite to that of the comb. Since this induced charge is highly attracted to the comb while the more distant charge on the opposite end of the object is only repelled slightly, the bit of paper will experience a net attractive force toward the comb. It therefore darts to the comb and sticks to it. You have no doubt seen many situations in everyday life where tiny objects are caused to cling to other objects by induced charges.

Returning now to figure 17.17, notice what happens in part (*b*) as the rod is touched to the metal. The highly attracted electrons jump to the rod and escape the metal object. If the rod is now removed, as indicated in (*c*), the object is left with an excess of positive charge. These positive charges are simply the charged atoms left uncompensated by the electrons which escaped. The object has been charged positive by the *conduction* of the electrons away from the object. This general process is called *charging by conduction*. Notice that the overall phenomenon is the same as though positive charge had jumped off the charged rod onto the originally neutral metal object.

There is another way to charge an object, and this is illustrated in figure 17.18. We see there a method by which a metal ball can be charged positive by use of a negatively charged rod. As we would expect, when the negative rod is brought close, some of the electrons in the metal are repelled to the right side of the ball, leaving a positive charge on the left side. Since no charge has been added to or subtracted from the ball, it is still neutral, of course. Now, suppose that the ball is touched by another object such as one's finger. (We say in that case that the ball has been grounded, and the symbol —||| is used to show this.) The negative charges can get still farther away from the negative rod if they flow to the object touching the ball. Once this has happened, the ball will no longer be neutral since it has lost some of its negative charge. After the object touching the ball has been removed, the negative rod can be removed and the ball will have a positive charge. (Why must the touching object be removed before the rod?) This process is called *charging by induction*.

There are many other interesting situations which involve induced

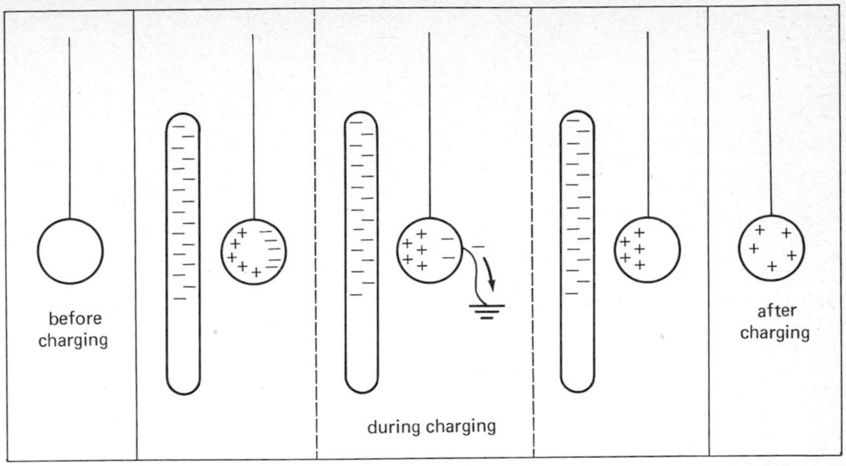

FIGURE 17.18 The method for giving a metal sphere a positive charge by induction. Note that the rod and sphere have unlike charges.

before charging

during charging

after charging

charges. A charged thundercloud moving above the earth's surface induces an opposite charge in the treetops as it passes above them. These two charges are united as a lightning flash serves as the connecting link between them. Many other natural phenomena with which you are familiar are examples of charge induction. A few of them will be pointed out in the questions at the end of this chapter. Others will be mentioned as our study of electricity continues.

17.12 MOTION OF CHARGES IN AN ELECTRIC FIELD

The motion of a charged particle in a *uniform* electric field is quite simple. It is assumed that the motion takes place in vacuum, so that collisions with other particles can be neglected. We know from the definition of field strength **E**, as the force per unit positive charge, that the force on a charge q placed in the field will be

$$\mathbf{F} = q\mathbf{E} \qquad\qquad 17.13$$

If **E** is constant, then **F** is also constant. In the nonrelativistic range of speeds, we have

$$\mathbf{F} = m\mathbf{a} = q\mathbf{E}$$

and so the acceleration of the charge also will be constant. Therefore motion in a constant uniform electric field is merely uniformly accelerated motion.

As an example, let us refer back to the charged parallel plates shown in figure 17.16. An electron ($q = -e, m = 9.1 \times 10^{-31}$ kg) is released at the lower plate as indicated. Being a negative charge, it experiences a force **F** in a direction opposite to **E**; the direction is upward. We have

$$\mathbf{F} = q\mathbf{E} = -e\mathbf{E}$$

from which

$$\mathbf{a} = \frac{\mathbf{F}}{m} = \frac{-e\mathbf{E}}{m}$$

The motion of this electron now reduces to a problem in uniformly accelerated motion. We have, taking up as positive,

$$v_0 = 0$$

$$a = \frac{eE}{m}$$

$$y = d$$

If we want the speed of the electron just before it strikes the upper plate, we have

$$v^2 - v_0^2 = 2ay$$

or

$$v^2 = \frac{2eEd}{m}$$

from which v is given by

$$v = \sqrt{\frac{2eEd}{m}}$$

This formula is not important enough to be memorized. However, the principle involved here is very important and should be well understood. Suppose $d = 1.0$ cm and $E = 10^4$ N/C. (This would correspond to a voltage of 100 V between the plates.) Putting in these numbers, we find

$$v \approx 6 \times 10^6 \text{ m/s}$$

It should be clear from this that electrons are easily accelerated to relativistic speeds.

For atomic particles the gravitational force is usually small compared with the electric force

We have ignored the gravitational force on the particle in solving this problem. To see why this is allowable, let us compare the gravitational force mg with the electric force on the particle, eE. Taking their ratio,

$$\frac{eE}{mg} \approx 1.8 \times 10^{14}$$

for the electron we have just considered.

In other words, the electric force is of the order of 10^{14} times larger than the gravitational force. This is almost always true for atomic-size particles, and therefore the gravitational force can almost always be ignored in such problems.

Illustration 17.5

An electron is shot at 10^6 m/s between two parallel charged plates as shown in figure 17.19. If E between the plates is 10^3 N/C, where will the electron strike the upper plate?

Reasoning As we saw in the preceding section, the electron will accelerate upward with a uniform acceleration

$$a = \frac{eE}{m}$$

At the same time it will travel horizontally with a constant speed of 10^6 m/s. This is clearly a projectile problem, and the path followed by the particle will be as shown.

Taking upward as positive and using $e = 1.6 \times 10^{-19}$ C, $m = 9.1 \times 10^{-31}$ kg, and $E = 10^3$ N/C, we have in the vertical problem that

$$v_{0y} = 0$$
$$a = 1.76 \times 10^{14} \text{ m/s}^2$$
$$y = 0.5 \times 10^{-2} \text{ m}$$

To find the time taken to strike the plate, we use

$$y = v_0 t + \tfrac{1}{2}at^2$$

from which

$$t = 7.5 \times 10^{-9} \text{ s}$$

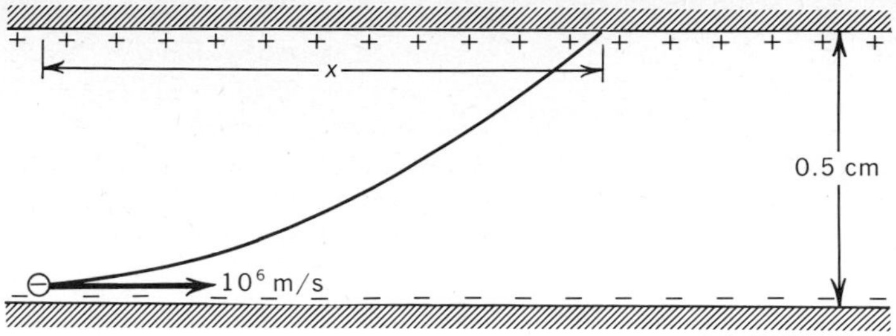

FIGURE 17.19 What would happen to the electron if the region between the plates were air-filled rather than evacuated?

Now considering the horizontal problem, we have

$$v_{0x} = v_x = \bar{v}_x = 10^6 \, \text{m/s}$$
$$t = 7.5 \times 10^{-9} \, \text{s}$$

Substitution in $x = \bar{v}_x t$ gives

$$x = 7.5 \times 10^{-3} \, \text{m} = 0.75 \, \text{cm}$$

as the horizontal distance the electron travels before hitting the plate.

Chapter 17
Questions
and Guesstimates

1. A tiny charged ball hangs from a thread. How could you tell whether or not the charge on the ball was positive? Negative?
2. You can place a static charge on nearly any dry piece of plastic by rubbing it with fabric, hair, dry toweling, or plastic wrap. How could you determine the sign of the charge placed on the plastic?
3. Static electricity causes sparks that can cause volatile gases such as ether to explode. This was a real danger in hospital operating rooms of the past. What measures can be taken to minimize this danger?
4. The dielectric strength of air is about $3 \times 10^6 \, \text{N/C}$. That is, a spark will jump through the air if the electric field strength exceeds this value. Why do sparks jump preferentially from sharp points and edges of metals? When your body becomes highly charged by walking across a deep-pile carpet in dry weather, why will a spark jump from your fingernail to a metal object such as a stove or doorknob?
5. Sketch the electric field between a positively charged thundercloud and a tree on a level expanse of earth. Where is a lightning stroke most likely to strike the earth? Benjamin Franklin understood the answer to this question and invented the lightning rod to protect property. How does a lightning rod achieve this end?
6. An infinitely long straight wire carries a uniform positive charge along its length. The wire is parallel to an infinite metal plate. Draw the electric field in the region of the wire and plate.
7. When using an electric drier to dry many manufactured fabrics, the dry clothes are found to cling together when removed from the drier. Why? What is often done to eliminate this effect?
8. Never try to wipe the dust off a phonograph record with an ordinary cotton or wool cloth. Why?
9. In dry climates one frequently notices (or hears) sparks jump when hair is combed or when clothes are removed in darkness. Frequently one experiences a shock upon touching a metal object after walking across a carpet or sliding across a car seat. Explain the causes of these effects.
10. During a thunderstorm the clouds overhead may carry a large positive

charge. Suppose a person stands knee deep in a lake as a highly charged cloud floats by. Sketch the electric field near the person and point out why the person is in danger. What should you do (and not do) during an electrical storm?

11. A metal needle is mounted like a compass needle on a nearly frictionless pivot so that the needle can point in various directions. If a charged rod is found to attract one end of the needle, can we assume that the needle is charged? What if the needle is repelled?

12. In a TV set, a beam of electrons is shot with speed about 10^7 m/s horizontally from the rear end of the TV tube toward the screen on its front end. Estimate how far the electrons drop because of gravity as they move across the tube.

13. Can two electric field lines in a proper electric field diagram ever intersect?

14. Is the electric field due to a point charge q zero at any point in space? For two equal charges $+q$ a distance d apart? For two equal and opposite charges $+q$ and $-q$? If so, show the point(s) qualitatively on a diagram.

15. If an electron escapes from the surface of the plate in figure 17.15 at the position of one of the lines, will it follow the line into the positive charge?

16. A very light metallized ball hangs from a thread. After being held in a boy's hand for a short time, the ball is released and found to be attracted by a charged plastic comb. After touching the comb, the ball is repelled by the comb. Explain these results.

17. A phonograph record is rubbed briskly with a woolen cloth and given a uniform charge over its surface. The record is left sitting on a stool outdoors. Sketch a diagram showing the electric field of the record as seen by (*a*) a tiny fruit fly sitting near its center and (*b*) a robin flying high overhead. These are truly extraordinary creatures, but let us not worry about details. Notice that the fly sees an essentially infinite plate while the bird sees a nearly point charge.

Chapter 17
Problems

1. The electron in the hydrogen atom is at a distance of about 5×10^{-11} m from the center of the proton which constitutes the atom's nucleus. Find the attractive force the nucleus exerts on the electron.

2. The uranium nucleus contains a charge 92 times that of the proton. If a proton is shot at the nucleus, how large a repulsive force does the proton experience due to the nucleus when it is 1×10^{-11} m from the nucleus center? The nuclei of atoms are of order 10^{-14} m in diameter, so the nucleus can be considered a point charge.

3. Suppose a 0.5-cm-radius solid aluminum sphere is given a positive charge of 1×10^{-8} C, a rather large charge for such a small sphere. What fraction of the electrons must be removed from the sphere to give it this charge? Each aluminum atom has a mass of 4.5×10^{-26} kg and has 13 electrons. Aluminum has a density of 2700 kg/m³.

4. Three identical point charges Q are placed along the x axis at $x = 0, 0.30$, and 0.70 m. Find the force on the center charge due to the other two.

5. Three point charges are placed at the following positions along the x axis: $x = 0, 0.30$, and 0.40 m. Their respective charges are $+Q, -Q, -Q$. Find the force on the center charge due to the other two.

6. Find the magnitude and direction of the electric field at the origin due to identical point charges Q placed at the points (0,3) and (3,0) m.

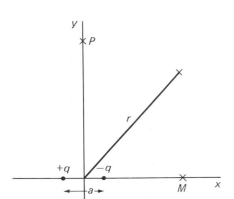

FIGURE P17.1

7. A point charge -3×10^{-6} C is placed at the point (0,2) m while a charge $+2 \times 10^{-6}$ C is placed at (2,0) m. Find the electric field at the origin. What would be the magnitude of the force experienced by a proton at the origin?

8. A point charge $+Q$ is placed at the origin of coordinates while a point charge $-2Q$ is placed on the x axis at $x = b$. Find the position on the x axis where $E = 0$.

9. A charge $-3Q$ exists at the coordinate origin. Where must a charge of $+Q$ be placed on the x axis if the combined field is to be zero at a point $x = +b$ on the axis?

10. Taken in order, the following point charges are placed at the four corners of a square: $-2Q$, $+2Q$, $-3Q$, $-2Q$. The diagonal of the square has a length $2b$. Find the magnitude of the field at the center of the square.

11. Two identical positive point charges $+Q$ are placed at the following points in the xy plane: (0,2) and (0,0) m. Find the electric field components at the point (3,0).

12. Two point charges Q_1 and Q_2 are 3 m apart and their combined charge is 20 μC. (a) If one repels the other with a force of 0.075 N, find the two charges. (b) If one attracts the other with a force of 0.525 N, find the charge magnitudes.

13. Two positive point charges a distance b apart have a combined charge Q. Calling one of the charges q, what value of q gives rise to the maximum force of one charge on the other?

14. An electric dipole consists of two equal charges q of opposite sign separated by a small distance a as shown in figure P17.1. (a) Show that the electric field at P is $(\frac{1}{4}\pi\epsilon_0)(qa/r^3)\mathbf{i}$ if $r_p \gg a$. (b) Show that the field at M is $-(\frac{1}{2}\pi\epsilon_0)(qa/r^3)\mathbf{i}$ if $r_M \gg a$.

15. The electric field $\mathbf{E}$ in a certain region of space is given by $\mathbf{E} = -0.60\mathbf{i}$ N/C. Find the force a proton will experience in this region. How much work is required to move the proton from the point (0,0) to the point (5,3) m? Express the latter answer in both joules and electron volts. (1 eV $= 1.6 \times 10^{-19}$ J.)

16. An electron experiences a force $+3.0 \times 10^{-17}\mathbf{i}$ N in a certain region of space. Find the value of $\mathbf{E}$ in this region. How much work is required to move the electron from the point (0,0) to (5,3) m in this region? Give your answer in both joules and electron volts. (1 eV $= 1.6 \times 10^{-19}$ J.)

17. Referring back to figure 17.9, suppose the charge density λ along the rod is given by $\lambda = -cx^2$. What is the value of $\mathbf{E}$ at point P?

18. A rod lies along the x axis with one end at the origin and the other at $x \to \infty$. It carries a uniform charge λ C/m. Starting from Coulomb's law, find the electric field at the point $x = -a$ on the x axis.

19. For the situation of the previous problem, find E_x and E_y for the point on the y axis where $y = b$.

20. Derive equation *17.9* of the text.

21. A thin charged rod of length L lies along the $+x$ axis with one end at the origin. Its charge per unit length $\lambda = Ax$, where A is a constant. Find the electric field at the point $x = b + L$ on the x axis. (*Hint:* Note that only A can be taken from the integral since λ is not constant. Also, $r = L + b - x$.)

22. Repeat problem 21 if the charge per unit length $\lambda = A(L + b - x)^2$.

23. A long thin rod is bent into the form of the circumference of a circle of radius b. It is uniformly charged along its length. Find the electric field caused by it at the center of the circle.

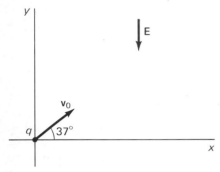

FIGURE P17.2

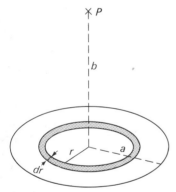

FIGURE P17.3

24. Suppose the rod of problem 23 had $\lambda = A \cos \theta$, where A is a constant and $\theta = 0$ along the x axis with the center of the circle at the coordinate origin. Find the x and y components of the field at the origin.

25. A thin rod forms a portion of the circumference of a circle of radius b. The center of the circle and the xy-coordinate origin are coincident. Measuring θ counterclockwise from the $+x$ axis, the rod extends from $\theta = 0$ to $\theta = \pi$. Its charge density is $\lambda = A \cos \theta$, where A is a constant. Find the components of $\mathbf{E}$ at the center of the circle.

26. A proton is shot out along the x axis from the origin with a speed of 10^6 m/s. In this region, a constant electric field of 3000 N/C exists in the $-x$ direction. How far from the origin does the proton get before it stops? ($m_p = 1.67 \times 10^{-27}$ kg.)

27. There is an electric field 500 N/C in the $+x$ direction in a certain region of space. A proton moving through this region in the $-x$ direction has a speed of 4×10^5 m/s as it passes point B. What will its speed be when it reaches point A which is a distance of 40 cm farther along its path? ($m_p = 1.67 \times 10^{-27}$ kg.)

28. A proton's velocity at point A is $5 \times 10^5 \mathbf{i}$ m/s. An electric field given by $\mathbf{E} = -200\mathbf{j}$ N/C exists in this region. What will be the x and y components of its velocity 1 μs later?

29. As shown in figure P17.2, a positive charge q with mass m is projected into a uniform electric field $\mathbf{E}$. Find (a) the maximum y coordinate reached by the particle and (b) the x coordinate at which it returns to the x axis.

30. A small sphere of mass m and charge q hangs as a pendulum of length L in a region where the electric field is directed downward and has magnitude E. Show that the period of the pendulum is $2\pi \sqrt{L/[g + (qE/m)]}$.

31. In figure P17.3, the surface of the circular disk (radius a) is uniformly charged with the total charge being Q. (a) Show that the electric field at P due to the ring of radius r and width dr is $(Qb/4\pi\epsilon_0 a^2)(r^2 + b^2)^{-3/2} r \, dr$. (b) What is the field at P due to the entire disk?

32. In the Bohr model of the hydrogen atom, an electron circles a proton at a radius of 5.3×10^{-11} m. How fast must the electron be moving if the centripetal force is to be supplied by the Coulomb attraction?

33. Consider the electric field on the axis of a charged loop, such as the one discussed in Sec. 17.8. Find the limiting form for E close to the plane of the loop, i.e., for $a \gg b$. Show that a charge of opposite sign with mass m and charge q will undergo simple harmonic motion on the axis. Find the frequency of this motion.

18 Gauss' Law

We saw in the preceding chapter that the forces between charges, and therefore the electric field, can be described in terms of Coulomb's law. There is another way in which the electric field can be described, and this is done by means of Gauss' law, which we discuss in this chapter. The two laws must be equivalent, as we show. However, depending upon the physical situation we wish to describe, one law is more conveniently used than the other.

18.1 LINES OF ELECTRIC FLUX

It is customary and convenient to draw pictures of electric fields. We have drawn the lines of electric force for several cases in the preceding chapter. This was done in a qualitative way only, but even so, it proved useful. Now, however, we wish to increase the utility of these lines, so-called *lines of electric flux,* by placing them on a quantitative basis. We do this by defining exactly how we shall draw them. In doing this we must realize that these lines are purely constructs and have no real physical existence. Their only justification is in their utility as aids in picturing situations and in performing computations.

As previously, the lines will be drawn in such a way as to show the direction of the electric force on a stationary positive test charge. Our only additional qualification will be that the number of lines through unit area erected perpendicular to the lines shall be equal numerically to E. This is illustrated in figure 18.1. For the case shown there, E is 16 N/C since this

E flux lines pass through unit area perpendicular to **E**

FIGURE 18.1 We agree to draw E flux lines through a unit area perpendicular to the electric field of intensity E.

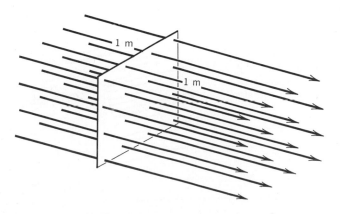

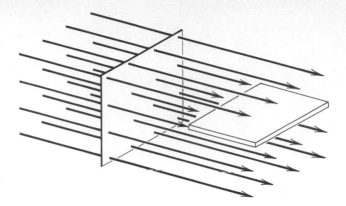

FIGURE 18.2 If the area pictured is *A*, then at the left $\phi = EA$, while at the right $\phi = 0$.

many lines go through a 1-m² area erected as shown. Although we shall not always adhere to this rule in drawing actual illustrations, it is to be understood that the illustrations are merely schematic, and the number of lines per unit area should be equal to *E*.

The number of lines which goes through an area *A* of a surface is called the *flux* through the area. We represent this number of lines, the *flux*, by the symbol Φ (Greek phi). Suppose we consider a small portion *A* of the meter-square area shown in figure 18.1. Since there were 16 lines per square meter, the number of lines through the area *A* will be 16*A* and this would be the flux Φ through *A*. Unfortunately, a complication enters in if the area is not perpendicular to the field lines. This can be understood by reference to figure 18.2.

Flux is represented by Φ

We see there an area oriented in two different ways relative to the field lines. A large number of lines go through the area when it is erect. However, the situation is quite different when the area lies horizontal. As we see, the lines simply skim the area and none goes in or out of it. Hence in this latter case $\Phi = 0$. Let us see how we can compute Φ when the area is at an arbitrary angle to the flux lines.

If we refer to figure 18.3 we see an area $L \times L = L^2$ oriented at an angle θ to the field lines as shown. We would like to know how many flux lines go through it. To that end, we notice that the same number of lines goes through the area L^2 as goes through the shaded projection area. This latter area is $L \times L \cos \theta = L^2 \cos \theta$, as you can see by reference to the figure. But since the shaded area is perpendicular to the field lines, our prescription for drawing

FIGURE 18.3 The number of lines going through the oblique area L^2 is the same as going through the shaded area. Since a number $E(L^2 \cos \theta)$ go through the shaded area, the flux through the area L^2 is also $E(L^2 \cos \theta)$. We therefore conclude that $\Phi = EA \cos \theta$.

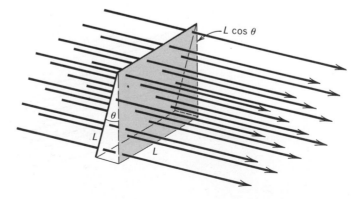

the lines (E lines per unit area) tells us that the number going through the shaded area is $(E)(L^2 \cos \theta)$. Since the same number of lines (or flux) goes through both the shaded and L^2 area, we find the desired flux to be

$$\Phi = (E)(L^2) \cos \theta$$

Notice that L^2 is simply the area through which we are trying to find the flux.

It is a simple matter to show that this result is applicable to any flat area, not just for the rectangle of area L^2. (If you wish to do it, split the area in question into tiny rectangular areas.) We then have for the flux through an area A inclined at an angle θ as indicated,

$\Phi = EA \cos \theta$ if E and θ are constant

$$\Phi = EA \cos \theta \qquad\qquad 18.1$$

This relation is reminiscent of the scalar product of two vectors. You recall that the definition of work was given as

$$\text{Work} = \mathbf{F} \cdot \mathbf{s} \equiv Fs \cos \theta$$

where θ is the angle between $\mathbf{F}$ and $\mathbf{s}$. It is convenient to define an area vector $\mathbf{A}$ to represent the area A. The only really distinctive direction associated with an area is the direction of the perpendicular to it. For this reason, we choose the area vector $\mathbf{A}$ perpendicular to the surface as shown in figure 18.4(a) and (b). Notice further that, if the area is part of an area enclosing a volume, the vector is always taken to point out from the volume.

The area vector is taken $\perp$ to the surface

If we examine figure 18.4(a) we see that the definition of the scalar product tells us that

$$\mathbf{E} \cdot \mathbf{A} = EA \cos \theta$$

where θ is the angle between $\mathbf{E}$ and $\mathbf{A}$. Because of our choice of notation, we can now write equation *18.1* as

$$\Phi = \mathbf{E} \cdot \mathbf{A}$$

where it is understood that $\mathbf{E} \cdot \mathbf{A}$ is simply shorthand for $EA \cos \theta$.*

$\mathbf{E} \cdot \mathbf{A} = EA \cos \theta$

In many cases (figure 18.4, for example) the product $\mathbf{E} \cdot \mathbf{A}$ has meaning only for a small area because $\mathbf{E}$ varies from place to place or the surface is not flat. In such cases we divide the area into small pieces as in figure 18.4(b), so

FIGURE 18.4 The area vector $\mathbf{A}$ is taken perpendicular to the area and out of the enclosed volume. In (a), the flux through an area much smaller than shown is $\Phi = \mathbf{E} \cdot \mathbf{A}$.

*Another way of looking at this expression is seen from figure 18.4(a). Notice that $E \cos \theta$ is the component of $\mathbf{E}$ perpendicular to the area in question. It contributes a flux $(E \cos \theta)A$, while the other component contributes no flux.

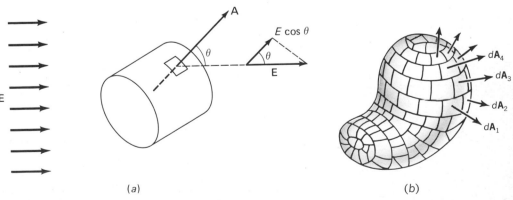

(a)

(b)

that for any one of the small pieces we can write

$$\Delta\Phi_i = \mathbf{E}_i \cdot \Delta\mathbf{A}_i \qquad\qquad 18.2$$

where $\Delta\Phi_i$ is the flux through the small area $\Delta\mathbf{A}_i$. The total flux out of the surface in figure 18.4(b), for example, would then be

$$\Phi = \Sigma \, \mathbf{E}_i \cdot \Delta\mathbf{A}_i$$

where the sum extends over all the little areas which compose the surface.

In the general case, we can allow the ΔA's to approach zero size, in which case the sum can be replaced by an integral. We then obtain the following precise definition of the electric flux passing out of an area:

Compact definition of flux

$$\Phi = \int_{area} \mathbf{E} \cdot d\mathbf{A} \qquad\qquad 18.3$$

Although this may seem to be a rather abstract relation, it is conceptually simple. It simply says that the flux coming out of a surface is equal to the number of field lines exiting through the surface. Because of our choice of the area vector, *lines coming out of a closed surface are to be taken positive while those going in through the surface are negative.*

Out is positive

18.2 FLUX FROM A POINT CHARGE

The quantitative utility of flux lines becomes clear if we examine the case of a point charge q. Let us consider an *imaginary* closed spherical surface with a point charge q at its center. The radius of the sphere will be taken as r. We picture this sphere in figure 18.5. Note that the sphere has no physical significance. It is purely an imaginary surface; i.e., we imagine it to exist in space as shown. Such surfaces in the present context are often referred to as *gaussian surfaces.*

A gaussian surface is an imaginary closed surface

We now compute the number of flux lines which leaves this spherical surface. The number of lines coming out of an area $d\mathbf{A}$ is from *18.2*,

$$d\Phi = \mathbf{E} \cdot d\mathbf{A}$$

To find the total number of lines leaving the sphere, we must sum this over the surface area of the gaussian sphere. Therefore

$$\Phi = \int_{sphere} \mathbf{E} \cdot d\mathbf{A}$$

which is merely *18.3* applied to this particular case. In this case $\mathbf{E}$ is radial, and so is the area vector $d\mathbf{A}$. Therefore, $\mathbf{E} \cdot d\mathbf{A}$ is merely $E\,dA$ because $\cos\theta = 1$. We have

$$\Phi = \int \mathbf{E} \cdot d\mathbf{A} = \int_{sphere} E\,dA$$

In addition, we notice that the charge is at the center of the gaussian sphere, and so E is constant on the surface of the sphere, and is given by Coulomb's law. It can therefore be removed from underneath the integral sign to give

$$\Phi = E \int dA = \frac{q}{4\pi\epsilon_0 r^2} \int_{sphere} dA$$

This integral just adds the increments of area which compose the surface, and is nothing more than the surface area of the sphere, $4\pi r^2$. Placing in this value gives

$$\Phi = \frac{q}{4\pi\epsilon_0 r^2} 4\pi r^2$$

or

$$\Phi = \frac{q}{\epsilon_0}$$

18.4

A charge q has q/ε₀ flux lines emerging from it

This is a very important result. It states that the flux (or number of lines) coming out of the spherical gaussian surface is q/ϵ_0 independent of the radius of the sphere. If the sphere is made very small, it encloses only the charge q. Still q/ϵ_0 lines come from it. We therefore conclude that our prescription for drawing flux lines means that each positive charge q must have q/ϵ_0 flux lines coming from it. A negative charge q will have the same number of lines going into it.

From a theoretical standpoint, the result given in 18.4 is of fundamental importance since it affords us a means for testing Coulomb's law very accurately. In fact, 18.4 is typical of all inverse square law fields, i.e., fields which vary as $1/r^2$. To see why this is true, we note in the present case that $E \sim 1/r^2$. Since the flux from the sphere is $(E)(4\pi r^2)$, we see that the r's exactly cancel and so the flux coming from the point charge is independent of r. If this exact cancellation did not occur, the results we shall derive in the following sections would be incorrect. As we shall see, experimental tests of these results end in their confirmation. We can therefore conclude that the $1/r^2$ dependence given in Coulomb's experimental law is exactly correct.

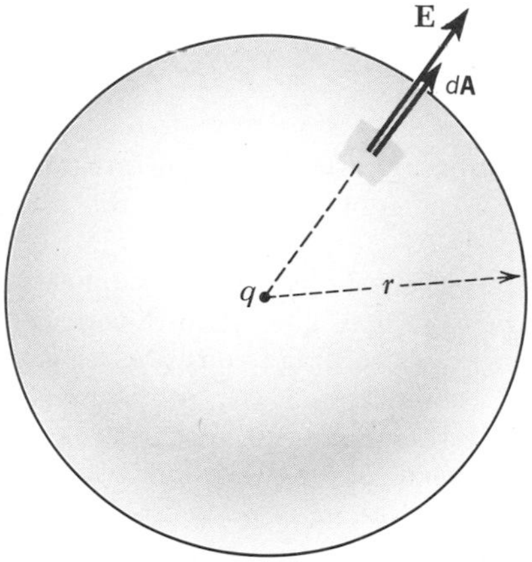

FIGURE 18.5 If **E** is 200 N/C at the radius shown, how much flux comes out of the sphere? Why is it also equal to q/ϵ_0? (After Edward M. Purcell, "Electricity and Magnetism," p. 23, vol. 2, Berkeley Physics Course, McGraw-Hill Book Company, New York, 1965. Courtesy of Education Development Center, Inc., Newton, Mass.)

18.3 GAUSS' LAW

As we saw in the preceding section, our prescription for drawing field lines means that each charge q must have q/ϵ_0 flux lines coming from it. This is the basis for an important equation referred to as Gauss' law. To obtain this equation we note the following facts:

1. If there are charges $q_1, q_2, \ldots, q_n$ inside a closed (gaussian) surface, the total number of flux lines coming from these charges will be

$$(q_1 + q_2 + \cdots + q_n)/\epsilon_0$$

2. The number of flux lines coming out of a closed surface is the integral of $\mathbf{E} \cdot d\mathbf{A}$ over that surface, $\int \mathbf{E} \cdot d\mathbf{A}$.

To obtain Gauss' law we note that the superposition principle allows us to state that the number of lines coming out of the charges inside the volume must equal the number of lines coming out through the surface enclosing the volume. For this reason, we can equate the expression given in statement 1 above to the integral in statement 2. Therefore

$$\int \mathbf{E} \cdot d\mathbf{A} = \frac{q_1 + q_2 + \cdots + q_n}{\epsilon_0}$$

or, in shorthand notation,

Gauss' law

$$\int_{\substack{\text{closed} \\ \text{surface}}} \mathbf{E} \cdot d\mathbf{A} = \frac{1}{\epsilon_0} \sum_{\substack{\text{volume} \\ \text{enclosed}}} q_i$$

18.5

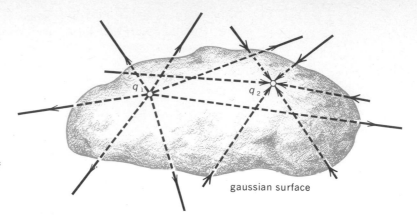

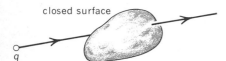

FIGURE 18.6 The total number of lines coming out of the charges is $(q_1 + q_2)/\epsilon_0$, in which q_2 is a negative number. All these lines must come out of the closed surface and their number must therefore equal $\int \mathbf{E} \cdot d\mathbf{A}$ over the surface.

gaussian surface

closed surface

q

FIGURE 18.7 Each flux line coming through a gaussian surface from a charge outside the surface both enters and leaves the surface. Its net contribution to the lines coming from the surface is zero.

Φ from a gaussian surface is due entirely to charges inside the surface

This relation is called *Gauss' law*. As the subnotes remind us, the integral is to be over the closed gaussian surface and the sum is to be over the total charge inside the surface.

Equation *18.5* is almost self-evident if we refer to figure 18.6. Suppose q_1 is a positive charge and q_2 is negative. Then the field lines should be drawn coming out of q_1 and going into q_2 as indicated. According to our prescription for drawing flux lines, the number of lines coming from q_1 is q_1/ϵ_0. Similarly, the number going into q_2 is q_2/ϵ_0. As a result, the number of lines coming *out* of the charges is $(q_1 + q_2)/\epsilon_0$ where the negative charge q_2 will contribute a negative number to this expression. Here too, the superposition principle allows us to depict the total field as the sum of its parts. Consequently, we see from the figure that the number of lines coming out of the enclosing surface is equal to the number of lines coming out of the charges within it, namely, $(q_1 + q_2)/\epsilon_0$.

However, there is another way to compute the number of flux lines coming out of the surface. In *18.3* we saw that this could be expressed as $\int \mathbf{E} \cdot d\mathbf{A}$.

If we now equate these two ways of writing the number of flux lines coming out of the surface, we find

$$\int \mathbf{E} \cdot d\mathbf{A} = \frac{1}{\epsilon_0}(q_1 + q_2)$$

Of course, *18.5* is simply this equation generalized to many charges.

Equation *18.5* is correct even if there are charges outside the gaussian surface. This is easily seen from figure 18.7. Notice that each line coming from an external charge must both enter and leave the surface if it goes through the surface at all. When it leaves the volume, it counts as $+1$, and when it enters, it counts as -1. Its total effect, as far as the number of lines coming from the surface is concerned, is zero. We therefore do not include charges outside the gaussian surface when applying Gauss' law. Let us now use Gauss' law in some examples so that we may better understand its utility.

18.4 FIELD DUE TO A UNIFORMLY CHARGED SPHERE

You will recall that Coulomb's law is capable of giving the field due to a point charge only. For larger charge distributions, one needs to use the superposition principle and sum the fields due to the little point charges which make

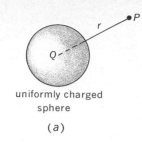

uniformly charged
sphere

(a)

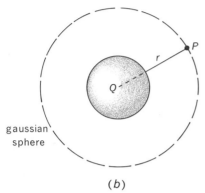

gaussian
sphere

(b)

FIGURE 18.8 Why do we take the gaussian surface to be a sphere as shown?

Coulomb's law gives the field outside a uniformly charged sphere

up the larger charge. In the case of a uniformly charged sphere carrying a total charge Q, such as the one shown in figure 18.8(a), the field at point P can be found only with considerable algebraic difficulty, using that method. However, Gauss' law is easily used in such a circumstance, as we shall now see. In fact, we shall not even require the charge distribution within the sphere to be constant and independent of r. All we require is that the charge distribution be spherically symmetric. By this we mean that, at a given value of r, the charge density shall be constant on the spherical *shell*, which has r for its radius.

Since Gauss' law is given by

$$\int_S \mathbf{E} \cdot d\mathbf{A} = \frac{1}{\epsilon_0} \sum q_i$$

we see that it is concerned with $\mathbf{E}$ on the imaginary gaussian surface S. If we wish to find E at point P in figure 18.8, we must take the surface through that point if Gauss' law is to be of value. It would be nice if E were constant everywhere on this imaginary gaussian surface through P; then E could be taken outside the integral. Fortunately, we notice that if we choose the gaussian surface to be a spherical surface through P with the charge at its center, then the perfect spherical symmetry of the system ensures that E will be constant on this gaussian surface. We therefore choose our gaussian surface as shown in figure 18.8(b).

From the spherical symmetry of the problem we know that $\mathbf{E}$ can only be radial. That is to say, since there is no reason for $\mathbf{E}$ to point to one side or the other of a radial line because of the perfect spherical symmetry, $\mathbf{E}$ must be radial. As a result,

$$\mathbf{E} \cdot d\mathbf{A} = E \, dA \cos \theta = E \, dA$$

Further, since E is a constant on the gaussian surface, it may be taken outside the integral. We therefore have, from Gauss' law,

$$E \int_S dA = \frac{1}{\epsilon_0} \sum q_i$$

But the total charge inside the closed gaussian sphere is Q. Moreover, the integral of dA over the surface of the gaussian sphere is just $4\pi r^2$. The above equation therefore becomes

$$E(4\pi r^2) = \frac{1}{\epsilon_0} Q$$

Solving for E, we find

$$E = \frac{Q}{4\pi\epsilon_0 r^2}$$

Notice that this result is identical with Coulomb's law for a point charge. This means that the field *outside* a spherically symmetric charged sphere is the same as the field due to an equal-magnitude point charge placed at the center of the sphere. We have therefore shown that *Coulomb's law applies to the region outside a spherically symmetric charged sphere as well as to a point charge.* Use of this fact will prove convenient later on in our study of electric fields.

Illustration 18.1

A spherical metal shell such as the one shown in figure 18.9(a) carries a uniform charge Q. Find the electric field at the points A and B indicated.

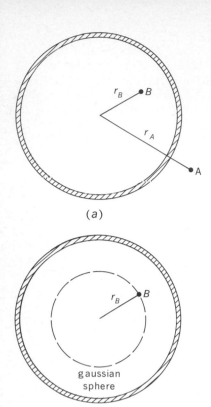

(a)

(b)

FIGURE 18.9 Qualitatively, how would one use Coulomb's law to find E at points A and B?

Under static conditions, $E = 0$ in a conductor

Reasoning We have already found the field at point A, in the preceding section. The derivation given there was independent of whether or not the sphere was hollow. It was found that the field outside the uniformly charged sphere was given by Coulomb's law, and is

$$E_A = \frac{Q}{4\pi\epsilon_0 r_A{}^2}$$

To find the field at B, we must take a gaussian surface through that point. In order to make use of the spherical symmetry, we take a gaussian sphere as shown. Since there is no charge *inside* the gaussian sphere, Gauss' law gives

$$\int_S \mathbf{E} \cdot d\mathbf{A} = \frac{1}{\epsilon_0} \sum q_i = 0$$

As reasoned before, $\mathbf{E}$ is radial and its magnitude is constant; so the integral becomes

$$E_B \int_S dA = 0$$

The integral over dA is $4\pi r_B{}^2$, and so we find

$$E_B(4\pi r_B{}^2) = 0$$

from which

$$E_B = 0$$

We therefore conclude that the field inside a uniformly charged spherical shell is zero. Later it will be shown that E is zero inside an empty hollow conductor as long as no currents are flowing.

18.5 ELECTRIC FIELD AND CHARGE IN CONDUCTORS

If no current is flowing in a conductor under steady conditions, then the electric field in the conductor is zero. This statement is easily proved in the following way. (1) Metals and other conductors have charges within them which are free to move. (2) If a resultant electric field exists in the metal, these charges will experience a force due to the field. They will move parallel (if positive) or antiparallel (if negative) to the field, and a current will flow. (3) When no current flows, the resultant force on the electrons of the metal must be zero. (4) Therefore, when no current flows, the resultant electric field in the metal must be zero. We have therefore proven that, under electrostatic conditions, the electric field in the interior of a metal or other conductor must be zero. Notice that, when we state the electric field in a metal to be zero, we are speaking about the region where the electrons are. Our proof and our conclusion do not apply to cavities in the metal.*

We may prove several interesting facts by use of Gauss' law, together with our knowledge that E is zero in a metal. For example, it is simple to prove that all the charge must reside on the outer surface of a charged conductor. (When we speak of *all* the charge, we mean the *excess* charge. Of course, the charges of the electrons and protons which make up the atoms remain in the interior. These positive and negative charges exactly cancel each other,

---

* We are speaking here of what is called the macroscopic field, the field averaged over a region which contains a large number of atoms. This will be discussed in more detail later.

FIGURE 18.10 Since no charge is inside the dashed gaussian surface, the excess charge on the metal must reside on the outer surface.

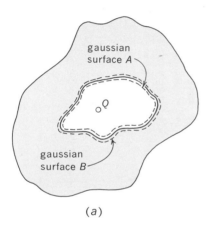

(a)

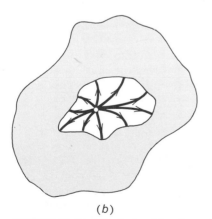

(b)

FIGURE 18.11 The lines of flux in part (b) may be used to prove that an equal and opposite charge is induced on the inner surface of the cavity.

however.) Consider the metal object in figure 18.10. It carries a charge Q, and there are no currents flowing in it. If we take a gaussian surface shown by the dashed lines, we know that $E = 0$ everywhere on it because it is in the metal, and, under static conditions, E is zero inside a metal.

Applying Gauss' law to the surface indicated, we have

$$\int_S \mathbf{E} \cdot d\mathbf{A} = \frac{1}{\epsilon_0} \sum q_i$$

But since $E = 0$, this is

$$0 = \frac{1}{\epsilon_0} \sum q_i$$

In other words, the sum of all the charges inside the gaussian surface is zero. Therefore, if a charge Q resides on the metal object, it must be outside the gaussian surface. We can, of course, place the gaussian surface just inside the metal surface. Since Q must be outside the gaussian surface, it must be on the surface of the metal object. Therefore it is shown that, under electrostatic conditions, *the excess charge on a conductor resides on the outer surface of the conductor.*

Another interesting feature peculiar to metals and other conductors becomes apparent if we consider a charge Q suspended in a cavity in a metal. This is shown in cross section in figure 18.11(a). The lines of flux coming from this charge are Q/ϵ_0 in number. They will go through gaussian surface A shown in the figure. If necessary, we can point to Gauss' law to prove this fact.

$$\int_S \mathbf{E} \cdot d\mathbf{A} = \frac{1}{\epsilon_0} \sum q_i = \frac{Q}{\epsilon_0} \qquad 18.6$$

The left-hand side of this equation is just the number of lines passing through the gaussian surface.

Suppose now that we take a second gaussian surface, surface B, just inside the metal. The field there is zero, and so the number of lines coming through surface B is zero. Gauss' law tells us that there can be no excess charge within surface B since the flux through it is zero. We therefore conclude that a charge of $-Q$ must reside on the metal surface of the cavity. The sum of this induced charge $-Q$ and the original charge Q is zero. This means that *a charge Q suspended inside a cavity in a conductor induces an equal and opposite charge $-Q$ on the surface of the cavity.*

We could easily have seen this to be true from qualitative considerations. In figure 18.11(b) the lines of flux coming from Q cannot penetrate into the metal since the field there is zero. They must therefore terminate on the surface of the cavity. But since lines terminate on negative charges only, there must be an equal negative charge on the surface of the cavity. It is often instructive to reason about the flux lines in this way to determine charge distributions.

This same line of approach can be used to show that if Q within the cavity is zero, then E must also be zero. All these proofs depend upon the fact that $E \sim 1/r^2$, according to Coulomb's law. You will recall that Gauss' law was derived by use of this assumption. In principle, then, any of these proven theorems can be used to test the inverse-square dependence. Such tests have been made, and it appears that within the range of distances r currently accessible, Coulomb's law is correct to at least 1 part in 10^{15}.

18.6 THE FIELD BETWEEN PARALLEL METAL PLATES

The electric field between two oppositely charged parallel metal plates was discussed in the preceding chapter. Now, however, we are able to obtain a quantitative estimate of E between the plates. To do this, we refer to figure 18.12. We assume the dimensions of the plate surface to be very large in comparison with the distance between them, so that, for the present purposes, the effects at the edges of the plates may be ignored. The field between the plates is then as shown in figure 18.12.

We see at once from figure 18.12 that the lines which originate on the upper plate end on the lower plate. This must mean that the charges on the two plates are equal and opposite. Let us say that the upper plate has a positive charge σ per unit area. That is to say, each square meter of surface area of the upper plate has a charge σ. Similarly, the *charge density* on the lower plate is $-\sigma$ per unit area.

If we want to find E at point P between the plates, we must choose a gaussian surface which passes through that point. We should like E to be perpendicular to the surface and constant. Clearly, a spherical gaussian surface is out of place in this problem since there is no spherical symmetry. Instead, we choose a box such as the one shown in figure 18.13 and, in cross section, by the dashed lines in figure 18.12. Notice that **E** is perpendicular to the lower surface and is constant at it. Since the upper surface is in the metal, E is zero on it. No flux goes through the sides of the gaussian box because the sides are parallel to the flux lines.

It is therefore a simple matter to evaluate the integral in Gauss' law for this gaussian box. We have

$$\int_S \mathbf{E} \cdot d\mathbf{A} = \frac{1}{\epsilon_0} \sum q_i$$

Splitting the integral into parts gives

$$\int_{\text{bottom}} \mathbf{E} \cdot d\mathbf{A} + \int_{\text{top}} \mathbf{E} \cdot d\mathbf{A} + \int_{\text{sides}} \mathbf{E} \cdot d\mathbf{A} = \frac{1}{\epsilon_0} \sum q_i$$

Putting in the values of E, you should be able to show that

$$E_P A + 0 + 0 = \frac{1}{\epsilon_0} \sum q_i$$

where A is the area of the lower surface of the gaussian box.

Since an area A of the metal plate is enclosed in the box, and since the charge per unit area is $+\sigma$, the charge within the box is σA. This is the meaning of Σq_i. We therefore have

$$E_P A = \frac{1}{\epsilon_0} \sigma A$$

Field between two large uniformly charged parallel plates

which gives

$$E_P = \frac{\sigma}{\epsilon_0} \qquad\qquad 18.7$$

This gives us the electric field between two parallel charged metal plates. Notice that the magnitude of the field is independent of position between the plates, as we had inferred from the equal spacing of the flux lines. This same value will be obtained if the gaussian box is taken below rather than above point P. Then the charge enclosed will be $-\sigma A$, and E will be negative. This merely means it is going *into* the box in this latter case.

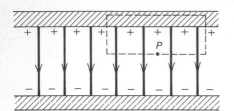

FIGURE 18.12 How do we know that the flux lines do not penetrate the metal? What can we say about the charge densities on the two plates by looking at the flux lines?

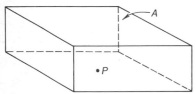

FIGURE 18.13 The complete gaussian surface seen only in cross section in figure 18.12.

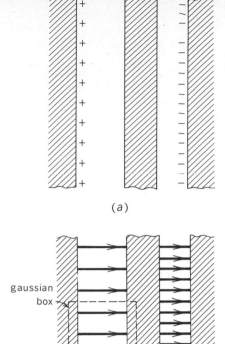

(a)

gaussian
box

(b)

FIGURE 18.14 How does the charge on the surfaces of the center metal plate compare with the charges on the side plates?

Illustration 18.2

The metal plate on the left in figure 18.14(a) carries a surface charge of $+\sigma$ per unit area. The metal plate on the right has a surface charge of -2σ per unit area. Find the charge densities on the two surfaces of the metal plate in the center. The center plate is assumed connected to the earth, so that it need not be neutral. Assume the plates to be very large.

Reasoning We can solve this problem by drawing flux lines as shown in (b). The lines coming from the plate on the left must end on the left side of the center plate. Therefore the charge density there must be $-\sigma$ per unit area. Since the lines which end on the surface of the right-hand plate originate on the right-hand surface of the center plate, the charge densities on these two surfaces must be equal and opposite. As a result, we see that the center plate carries a charge $+2\sigma$ on its right-hand side. Clearly, the center plate is not neutral; i.e., it has a net charge.

We could also carry out this proof using Gauss' law. For example, if we choose a gaussian box such as the one shown in cross section by the dashed lines in figure 18.14(b), we can easily carry out the requisite integral. It will be recalled that the integral is merely the number of lines coming out of the gaussian box. You should be able to show easily without computation that this is zero. Therefore

$$\int_S \mathbf{E} \cdot d\mathbf{A} = \frac{1}{\epsilon_0} \sum q_i$$

or

$$0 = \sum q_i$$

But this means the total charge inside the gaussian box is zero. This can be true only if the charge on the left surface of the central plate is equal and opposite to the charge on the left plate. You should be able to repeat this process to show that the answer found for the other surface is also correct. Why can you state, without computation, that the field in the center plate is zero?

18.7 A LONG UNIFORMLY CHARGED CYLINDER

Thus far we have applied Gauss' law to situations which had spherical or planar symmetry properties. Let us now consider a situation which involves cylindrical symmetry. Suppose we have a *very long* cylindrical rod which carries a charge λ per unit length. That is to say, each meter length of the rod has a charge λ on it. A length Δx will carry a charge $\lambda \Delta x$.

We insist that the rod be very long (long in comparison with the distance away from the axis, which we shall consider). The reason for this is shown in figure 18.15, where the field about a uniformly charged short rod is shown. This field is sketched by noting that the rod will appear to be a point charge if we are very far from it. Therefore, at large distances, the field will be radial. Close to the rod and far from its ends (i.e., near its center), the field should radiate perpendicular to the axis since neither direction along the axis should be especially unique. If we consider a very long rod, then for most of the rod the end effects will be negligible, and we say that the field radiates perpendicular to the rod axis.

Let us return now to the case of a point near a charged rod but not close to its ends (an infinite rod would certainly satisfy this condition). The

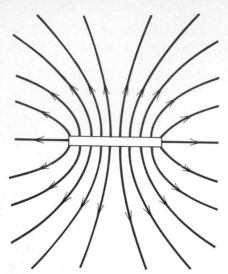

FIGURE 18.15 Close to the rod and not too close to its ends, the field lines are essentially radial from the axis.

situation is shown in figure 18.16. If we wish to find the field at point P by use of Gauss' law, we surmise that an imaginary cylindrical gaussian surface through P might be satisfactory. This type of surface is shown. Notice that it is a closed surface having flat ends as well as cylindrical sides. We call its length L.

Since the rod is assumed very long, the lines of flux are radial from it, as shown in the inset of figure 18.16. Applying Gauss' law to the gaussian cylinder in figure 18.16, the integral can be split into three parts, the integrals over the three parts of the surface. It is then

$$\int_{\substack{\text{left}\\\text{end}}} \mathbf{E} \cdot d\mathbf{A} + \int_{\substack{\text{right}\\\text{end}}} \mathbf{E} \cdot d\mathbf{A} + \int_{\substack{\text{cylindrical}\\\text{surface}}} \mathbf{E} \cdot d\mathbf{A} = \frac{1}{\epsilon_0} \sum q_i$$

The integrals over the two ends are zero since $\mathbf{E}$ is perpendicular to $d\mathbf{A}$; that is to say, none of the flux lines go through the ends, they are parallel to the ends. Since the rod is very long and uniformly charged, E should be essentially constant along the rod. In addition, since $\mathbf{E}$ is parallel to $d\mathbf{A}$ on the cylindrical part of the surface, the equation becomes

$$0 + 0 + E \int_{\substack{\text{cylindrical}\\\text{surface}}} dA = \frac{1}{\epsilon_0} \sum q_i$$

To find the charge inside the surface, we notice that it encloses a length L of the rod. The charge per unit length is λ, and so

$$\sum q_i = L\lambda$$

In addition, the integral of dA over the cylindrical surface is just the area of the surface, $2\pi rL$. We therefore find

$$E2\pi rL = \frac{1}{\epsilon_0} L\lambda$$

The field outside a long cylinder varies as 1/r

from which

$$E = \frac{\lambda}{2\pi\epsilon_0 r} \qquad\qquad 18.8$$

for the field at a distance r from a very long rod.

Notice that the length of the gaussian cylinder, a purely arbitrary quantity, does not appear in the result. This must always occur. Moreover, we notice that $E \sim 1/r$. This is to be compared with $E \sim 1/r^2$ for a point charge and $E \sim$ constant for an infinite plane.

FIGURE 18.16 The flux lines radiate out from the axis of the rod, striking the cylindrical part of the gaussian surface perpendicularly. Since the lines are parallel to the end faces of the gaussian cylinder, none go through these two surfaces.

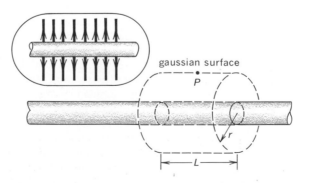

gaussian surface

P

r

L

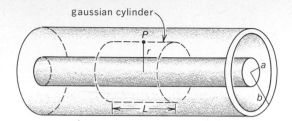

gaussian cylinder

FIGURE 18.17 If the inner metal cylinder carries a charge λ per unit length and the outer cylinder has no net charge (i.e., the sum of the charges on its inner and outer surfaces is zero), what does the electric field look like?

Illustration 18.3

A cylindrical capacitor consists of two concentric metal cylinders, as shown in figure 18.17. If the inner cylinder carries a charge λ per unit length, find E in the region between the cylinders. Assume the distance between the cylinders to be much smaller than the length of the cylinders.

Reasoning As discussed previously, *since the cylinders are very long,* the field should be radial from the cylinder axis. To find the field at point P between the cylinders, we construct the gaussian cylinder shown in figure 18.17. As before, no flux will leave the two ends, and so the integral over those parts of the gaussian surface will be zero. Further, E is constant and radial, and so we find

$$0 + 0 + E \int_{\substack{\text{cylindrical} \\ \text{surface}}} dA = \frac{1}{\epsilon_0} \sum q_i$$

The charge inside the gaussian cylinder is just λL, and so this becomes

$$E(2\pi r L) = \frac{1}{\epsilon_0}\lambda L$$

where the integral over dA has been replaced by the total area of the cylindrical gaussian surface. Solving, we find

$$E = \frac{\lambda}{2\pi r \epsilon_0} \qquad\qquad 18.9$$

This is the same result as we found for the single charged cylinder alone. Arguing from the concept of flux lines, can you show that this should be true without doing a detailed computation?

Illustration 18.4

To a very rough approximation, a heavy atom can be pictured to be a spherical nucleus with charge $+Q$ and radius a embedded in a much larger sphere of negative charge (the electrons). This negative sphere will have a radius b (the atomic radius), and its charge, distributed uniformly throughout its volume, will be $-Q$. In other words, its charge per unit volume is $-Q/\frac{4}{3}\pi b^3 = \rho$. Find E for any radius larger than a. (Notice that the volume density of charge ρ is $\Delta Q/\Delta V$, where ΔQ is the charge in volume ΔV.)

Reasoning The situation is shown in figure 18.18. For point A outside the atom, we make use of the applicability of Coulomb's law, proved earlier. Both the positive and negative charges can be considered to be at the center of the sphere, and since they are equal and opposite, their effects exactly cancel. Therefore the field should be zero at all points *external* to the atom.

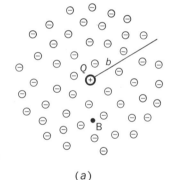

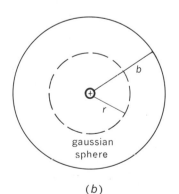

(a)

(b)

FIGURE 18.18 Approximately how does the average electric field vary with r within an atom?

At a point such as B, the field is easily found by taking a gaussian sphere as shown in part (b) of figure 18.18. The total charge within this gaussian sphere is

$$\sum q_i = +Q + \rho(\tfrac{4}{3}\pi r^3)$$

We obtain the second term by noticing that ρ is the negative charge per unit volume and that there is a volume $\tfrac{4}{3}\pi r^3$ inside the gaussian surface. Putting in $-3Q/4\pi b^3$ for the value for ρ, we have

$$\sum q_i = +Q - Q\left(\frac{r}{b}\right)^3$$

Therefore Gauss' law becomes

$$\int_S \mathbf{E} \cdot d\mathbf{A} = \frac{1}{\epsilon_0} Q\left[1 - \left(\frac{r}{b}\right)^3\right]$$

Since $\mathbf{E}$ is radial and constant, we can evaluate the integral and obtain

$$E(4\pi r^2) = \frac{Q}{\epsilon_0}\left[1 - \left(\frac{r}{b}\right)^3\right]$$

from which

$$E = \frac{Q}{4\pi\epsilon_0 r^2}\left(1 - \frac{r^3}{b^3}\right) \qquad r \le b \qquad\qquad 18.10$$

The first term corresponding to 1, in parentheses, represents the Coulomb field due to the nucleus, while the second represents the canceling field due to the electrons. As we should expect, at points near the nucleus, which are mostly inside the charged negative shell, the field is chiefly that due to the nucleus. The nucleus and innermost electrons of an atom are therefore not greatly affected by the outer electrons of the atom.

18.8 FIELD NEAR A METAL SURFACE

Now that we have obtained experience using Gauss' law, we shall use the law to find a result of great simplicity and importance. It is a general expression for the electric field just outside the surface of a metal. To find it we will need to make use of the fact that the field lines entering or leaving a metal must be perpendicular to the surface.

Suppose the pear-shaped metal object shown in figure 18.19(a) is

FIGURE 18.19 When very much enlarged, a tiny portion of the surface in (a) will appear flat as shown in (b). The object is metal.

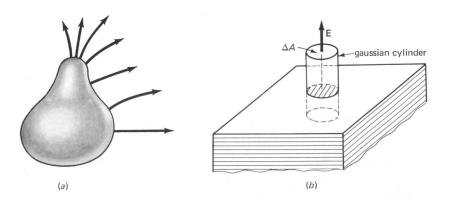

(a) (b)

charged positive. The charge will spread unevenly over its surface because of the lack of symmetry. Consider only a very small region of the surface, perhaps only a pinpoint large. This small piece of surface will appear as in part (b) of the figure when vastly enlarged. The region chosen is so small that it has negligible curvature in the area of concern. It therefore approximates a flat surface under the very high magnification of part (b).

Since the surface is metal under electrostatic conditions, the field lines must strike the surface perpendicularly as we have previously found. In the tiny region shown enlarged in (b), the charge density will not vary appreciably over such a small region. Let us call the charge density σ. The electric field lines will be parallel to the vector $\mathbf{E}$ shown. We now construct a gaussian surface such as the cylinder indicated. No flux goes out the sides since $\mathbf{E}$ simply skims the sides. No flux pass through the bottom end since $\mathbf{E}$ is zero in the metal. The flux through the tiny top-end area ΔA is $E \Delta A$ since E is constant over such a small area and is perpendicular to it. Hence the total flux out of the area is $E \Delta A$.

But according to Gauss' law, this must be equal to the charge within the gaussian cylinder divided by ϵ_0. As an equation

$$\int \mathbf{E} \cdot d\mathbf{A} = \sum \frac{q_i}{\epsilon_0} \quad \text{becomes} \quad E \Delta A = \frac{\text{charge inside}}{\epsilon_0}$$

But the charge inside is all on the metal surface and is simply $\sigma \Delta A$. Placing this in the above expression yields the following final result for the electric field just outside a metal surface:

Field just outside a metal surface is σ/ϵ_0

$$E = \frac{\sigma}{\epsilon_0} \qquad 18.11$$

This is a very important result and we shall have occasion to make reference to it.

Chapter 18
Questions and Guesstimates

1. How could one go about determining if the earth carried a large charge on its surface?
2. A charge q is placed inside an enclosure. Compare the total flux coming out of the walls of the enclosure if it has the following shapes but is closed on all sides: sphere, cube, rectangular parallelepiped, cone, hemisphere.
3. If there is a net positive charge inside a gaussian surface, does this mean that the electric field is directed out of the surface at all points? Defend your answer.
4. An unopened cereal box is placed in a uniform electric field $\mathbf{E}$. What is the value of $\int_S \mathbf{E} \cdot d\mathbf{A}$ taken over the entire surface of the box? Does it matter if the field is uniform?
5. A plastic bag is filled with air and is closed by tying its neck with a string. Inside the closed bag is a charge q_1 and outside the bag is a charge q_2. What is the value of $\int_S \mathbf{E} \cdot d\mathbf{A}$ taken over the surface of the closed bag?
6. Assuming that the test charge does not disturb the charges which exist on a conductor when it is in electrostatic equilibrium, show that no work is done against electric forces in carrying the test charge from point to point within the conductor. Show further that no work is done in moving the test charge from point to point on the surface of the metal.

7. Why is it impractical to use Gauss' law to find the electric field at a point a distance b outside a charged rod whose length is L unless $L \ggg b$?

8. Devise an experiment involving Gauss' law to determine whether or not $E \sim r^{-2}$.

9. We maintain that, in an electrostatic situation, the electric field is zero within a metal. This is not true in the strictest sense since electric fields and forces exist between the electrons and nuclei of the atoms, and perhaps even within the nuclei. How can we justify our statement about zero electric field? What field are we talking about?

10. A man is placed inside a large closed metal sphere mounted on an insulating pole. What will the man observe as charge is placed on the sphere and as the sphere becomes highly charged? What will he notice if a large charged object is brought close to the sphere? Assume he has electrical measuring equipment available.

11. Assuming the man inside the sphere described in the previous question has a metal ball that carries a charge $-q$, what will happen as he moves it about inside the sphere if there is a charge $+Q$ on the sphere? Will the charge $-q$ experience any forces? What will happen if he touches $-q$ to the inside surface of the metal sphere?

12. Using logic rather than mathematical equations, prove that when an excess charge Q is placed on an isolated metal sphere, the charge will distribute itself uniformly on the sphere's surface. Why does not the same type of proof lead to a similar result for the charge on a metal cube? What can you say about the charge on the surface of an isolated metal cube?

13. Can you tell if a metal sphere is hollow or not by the use of electrostatic experiments alone?

14. The following statement is often made: Under electrostatic conditions, all the charge resides on the surface of a metal object. The statement is wrong in at least two respects. What are they?

15. Suppose the gaussian box in figure 18.12 had been taken such that its upper surface was just below the metal plate rather than in it. Would we have been able to obtain $\mathbf{E}$ by use of it and Gauss' law?

16. Suppose a sphere is cut into halves and each half given equal but opposite uniform charges. They are then put together, but not quite touching. Explain why Gauss' law cannot easily be used to find the field outside them.

Chapter 18
Problems

1. The electric field in a certain region of space is given by $\mathbf{E} = 200\mathbf{i}$. How much flux passes through an area A if it is a portion of (a) the xy plane; (b) the xz plane; (c) the yz plane?

2. A cylinder of length L and radius b has its axis coincident with the x axis. The electric field in this region is $\mathbf{E} = 200\mathbf{i}$. Find the flux through (a) the left end of the cylinder; (b) the right end; (c) the cylindrical wall; (d) the closed surface area of the cylinder.

3. The electrical field due to a uniformly charged long wire coincident with the x axis is radial from the axis and has magnitude B/r in the region with which we are concerned. A cylinder of length L and radius b has its axis coincident with the wire. Find the flux through (a) the left end of the

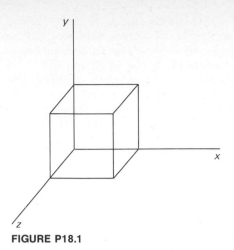

FIGURE P18.1

cylinder; (*b*) the right end; (*c*) the cylindrical wall; (*d*) the closed surface area of the cylinder.

4. A cubical box of edge *b* sits with a corner at the coordinate origin as shown in figure P18.1. An electric field $\mathbf{E} = 200\mathbf{i} + 300\mathbf{j}$ exists in this region. Find the flux through its faces which coincide with *xy*, *yz*, and *xz* planes, respectively.

5. A point charge $+Q$ is placed at the center of an uncharged spherical metal shell of inner radius *b* and outer radius *c*. Find the electric field in the regions $r < b$, $b < r < c$, $r > c$.

6. Repeat problem 5 if the metal shell has a net charge $-Q$. Where does the charge $-Q$ reside?

7. A thin, long straight wire carries a charge λ_1 per unit length. The wire lies along the axis of a long metal cylinder which carries a net charge λ_2 per unit length. The inner radius of the cylinder is *b* while its outer radius is *c*. Find the electric field in the following three regions: $r < b$; $b < r < c$; $r > c$. How much charge per unit length exists on the inner surface of the cylinder? The outer surface?

8. Two long, parallel straight wires carry charges λ_1 and λ_2 per unit length. The separation between their axes is *b*. Find the magnitude of the force exerted on unit length of one due to the charge on the other.

9. A long straight cylinder of radius *b* has a constant volume distribution of charge ρ coulombs per unit volume. Find the electric field due to this volume charge (*a*) outside and (*b*) inside the cylinder. (*c*) Where is the field strongest? (*d*) Weakest?

10. We shall see in the next chapter that when a 12-V battery is connected to two large, parallel metal plates 0.20 cm apart, the electric field between them is 6000 N/C. What will be the surface charge density σ on the plates in such a situation?

11. A large flat metal plate carries a surface charge σ per unit area on each side of the plate. Find the electric field at a distance *b* outside, but close to, one surface of the plate.

12. A sheet of charge is by definition a layer of charge that has negligible thickness. For example, the excess charge on one face of a metal plate is a sheet of charge. A very thin uniformly charged plastic sheet is another typical example. (*a*) Find the field due to a large, flat sheet of charge that has charge density σ per unit area. (*Note:* σ would be the total charge on $1\,\text{m}^2$ of a plastic sheet and includes the surface and interior charges.) (*b*) A flat metal plate has charge density $+\sigma$ on each side. Show that the total field due to the two sheets of charge is zero within the sheet and σ/ϵ_0 outside. (*Hint:* Fields obey superposition.)

13. A large, flat metal plate carries charge $+\sigma$ per unit area on one side and $-\sigma$ on the other. From a consideration of these two sheets of charge (see previous problem), show that the total field due to them is (*a*) σ/ϵ_0 inside the plate, (*b*) zero outside the plate. (*c*) How can this be reconciled with the known facts that the field inside the plate is zero and is σ/ϵ_0 outside?

14. Charge is distributed uniformly throughout a large platelike volume. (The plate is not metal.) The thickness of the plate is *b* and the charge per unit volume is ρ. Show that the electric field at a distance *x* from the center plane of the plate ($x < b/2$) is given by $(\rho/\epsilon_0)x$. If ρ is positive and a negative charge *q* with mass *m* can move freely in a small hole through the plate, find the frequency with which it will oscillate around the central plane.

15. Two concentric spherical metal shells carry net charges Q_1 and Q_2, where Q_1 is the charge on the inner sphere. If the electric field between the spheres is $3000/r^2$ N/C radially inward and the electric field outside is $2000/r^2$ N/C radially inward, what are the values of Q_1 and Q_2?

16. A rod of length L carries a uniform charge λ per unit length. Find the electric field at a distance b from the rod on its perpendicular bisector. Show that in the limit of $b/L \ll 1$ this expression reduces to the value one obtains from a very long rod. How large must the ratio b/L be if the deviation between the two results is not to exceed 2 percent? [*Note:* Recall $(1 + x)^n \approx 1 + nx$ if $x \ll 1$.]

17. The electric field at a distance b from the surface of a uniformly charged sphere of radius R is $Q/4\pi\epsilon_0 (R + b)^2$ where Q is the charge on the sphere. Show that if $b/R \ll 1$ this reduces to the result σ/ϵ_0 for the field just outside a metal plate. How large is b/R if the results are to differ by 2 percent? (See note in problem 16.)

18. A charge $+q$ and mass m is released just outside a very large metal plate which has a surface charge density σ. Find the speed of the charge when it reaches a distance b from the surface. Assume the particle to still be relatively close to the plate.

19. A nonmetallic cylinder of radius b and very long length has a variable volume charge density $\rho = k/r$. Find the electric field due to this charge (*a*) outside the cylinder and (*b*) inside the cylinder.

20. A nonmetallic sphere of radius b has a variable volume charge density $\rho = kr^2$. Find the electric field due to this charge (*a*) outside the sphere and (*b*) inside the sphere.

21. We found in this chapter that the field between two charged infinite parallel plates is σ/ϵ_0, where σ is the charge density on the plates. This is the combined field due to the charges on both plates? What is the field between the plates due to just one of the plates? Find the force on the unit area of one plate due to the charge on the other. Is it attractive or repulsive? If the above assertions are correct, why is the field outside *any* charged metal plate σ/ϵ_0?

22. A flat, very thin piece of metal is infinitely long and has a width $2a$, as shown in figure P18.2. Show that the electric field at point P in the figure is given by

$$E = \frac{2\sigma}{\pi\epsilon_0} \tan^{-1} \frac{a}{b}$$

where σ is the charge per unit area on each of its two charged surfaces. (*Hint:* Split the surface into infinite rods.)

23. Show that the result given in the previous problem reduces to the field of an infinite plate in the limit of $a \gg b$.

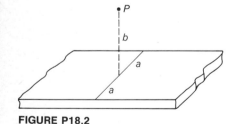

FIGURE P18.2

19 Potential

The concept of potential energy is a useful and powerful tool in mechanics. A similar situation exists in electricity. In this chapter we introduce the concept of electric potential. It will be found to be indispensable for our further work in electricity.

19.1 POTENTIAL DIFFERENCE

In our study of mechanics we introduced the concept of gravitational potential energy. Because the gravitational field is a conservative field, it makes sense to define a quantity which is equal to the work done against the field's force in carrying a mass m from one position to another in the field. You will recall that the gravitational potential energy difference of a mass m at two different points was defined to be the work done against gravity in carrying the mass from the one point to the other. In particular, if a mass m is carried from a point A to a point B, the work done against gravity in the process is mgh, where h is the height of B above A. A typical situation is shown in figure 19.1(a). We notice that mg is the force needed to pull the mass m upward through the distance h. The quantity mgh was denoted the gravitational potential energy difference for the mass m at the two locations. In other words, the work done against gravity in moving a mass from point A to point B is defined to be the change in gravitational potential energy of the mass as it is moved from A to B.

We have seen in previous chapters that the gravitational potential energy concept was extremely powerful. By use of it, together with the work-energy theorem, we have been able to analyze many rather complicated

FIGURE 19.1 The electric potential difference between points A and B is equal to the work done in carrying a unit positive test charge from A to B.

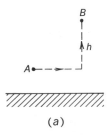

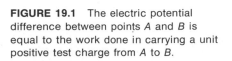

(a)

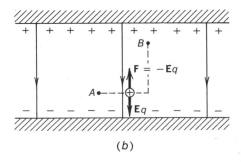

(b)

situations with a minimum of effort. Since the electrostatic field is also a conservative field (as we shall soon show), it is not surprising that the concept of potential energy proves useful in electricity as well as in mechanics. We define the electric potential energy difference between two points in a way analogous to the definition of gravitational potential energy difference. Let us state the definition as follows: *The electric potential energy acquired by a charge as it is moved from point A to point B is the work done against forces of electric origin during the process.*

Although the concept of electrical potential energy is useful, we can obtain an even more powerful tool if we go one step further. We particularize our definition to the case of a very special charge. Let us imagine carrying a *unit positive test charge* from point A to point B in figure 19.1(*b*). You will recall that the test charge is simply a mental construct. It is an imaginary positive charge which can be placed wherever we want; but it is endowed with the very special property that it does not disturb the charges to which it is brought close. Notice that we now go one step further and take the positive test charge to have a magnitude of 1 C. The practical utility of this more detailed specification will be clear once we understand the following definition.

Definition of electric potential difference

We define the *electric potential difference* between two points to be the *work done in carrying a unit positive test charge from the one point to the other.* The work is done by the person (or external agent) who carries the test charge from place to place. It will be recalled that the equation for work is

$$W = \int \mathbf{F} \cdot d\mathbf{s}$$

To find the electric potential difference between two points we must therefore find the force one must exert on a unit positive test charge to pull it from one point to the other. This force is easily found by reference to figure 19.1(*b*).

If a charge q is placed between the two plates shown (or, in fact, in any electric field), the field exerts a force $q\mathbf{E}$ on it, as shown. To hold the charge in place, or to move it from A to B, a force $-\mathbf{E}q$ must be exerted on the charge. In our particular case we are interested in a unit positive test charge, and so $q = +1$ C. The force we must exert on it to hold it and move it slowly in the electric field is $-\mathbf{E}$. Therefore the work done in carrying a unit test charge between two points, from A to B, for example, is

$$\frac{W}{\text{Unit test charge}} = -\int_A^B \mathbf{E} \cdot d\mathbf{s}$$

$V_B - V_A$ *is the work done in carrying a + 1 C test charge from A to B*

This work done to move a unit positive test charge from A to B is said to be the *electric potential difference* between A and B and is represented as

$$V_B - V_A = -\int_A^B \mathbf{E} \cdot d\mathbf{s} \qquad 19.1$$

We should be perfectly aware that this is simply the work equation and represents the work done in carrying a unit positive test charge from A to B. The force needed to carry the charge is $-\mathbf{E}$.

The units of potential difference are, of course, the units of work per charge. In the SI, the only system we shall use, this would be expressed as joules per coulomb. This unit is so frequently used that it is given a special *Potential difference is measured in volts* name and is called the *volt* (V). For example, the potential difference between the two terminals of a 12-V battery is 12 V. This in turn means that 12 J of work must be done if a 1-C charge is carried from one terminal to the other.

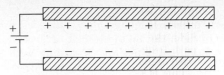

One terminal of a battery is always marked +, or positive (or sometimes it is painted red). Work must be done in carrying a positive charge outside the battery from the negative terminal of a battery to the positive terminal. When a battery is connected to two metal plates, as shown in figure 19.2, the plate attached to the positive terminal acquires a positive charge, while the other plate becomes negatively charged. Notice that a battery is designated as $\overset{+}{-}|\!\!\vdash$, where the long side is the positive terminal. Very often the + and − signs are omitted. More is said about the internal workings of batteries, later.

19.2 PARALLEL METAL PLATES

A great deal of insight into the usefulness of the concept of electric potential difference can be obtained by considering the case of two large parallel metal plates. Suppose the plates are attached to a battery, as shown in figure 19.2. The electric field between the plates will be as shown in figure 19.3(*a*). Except near the edges of the plates, the field is uniform, and this uniform region is shown in part (*b*) of figure 19.3.

If the electric field between these two plates is E, then the potential difference between the lower and the upper plate is, according to *19.1*,

$$V_B - V_A = -\int_A^B \mathbf{E} \cdot d\mathbf{s}$$

Remember, this is simply the work equation written for a unit positive test charge carried from A to B. Let us first carry out this integral (i.e., compute the work done) by going from the lower to the upper plate by way of path 1 shown in figure 19.3(*b*). Notice that, along this path, $d\mathbf{s}$ is an *upward*-directed vector, since we are going from A to B. Therefore, since $\mathbf{E}$ is directed *downward,* we have

$$-\mathbf{E} \cdot d\mathbf{s} = -E \, ds \cos 180° = E \, ds$$

Using this result in the potential difference equation, we find

$$V_B - V_A = \int_A^B E \, ds$$

But E is constant in this particular situation, and so

$$V_B - V_A = E \int_A^B ds$$

or $\qquad\qquad V_B - V_A = Ed \qquad$ (parallel plates only) $\qquad\qquad$ *19.2*

FIGURE 19.3 The potential difference between plates A and B is the work done in carrying a unit positive test charge from one plate to the other. Show that this work is independent of the path along which the charge is carried.

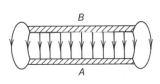

(*a*)

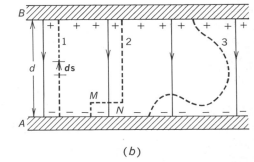

(*b*)

If E is constant, the potential difference is Ed

where d is the distance between the plates. One must be careful to realize that this is not a general relation. It gives the potential (or voltage) difference between two points a distance d apart in the direction of a *constant* electric field E. Such a field is frequently encountered when one is dealing with parallel plates. In that case, but not in others, *19.2* is valid and useful.

As in the case of the gravitational field, the same amount of work is done in carrying a test charge along *any* path from plate A to plate B. For example, when a unit positive charge is carried along path 2 in figure 19.3(*b*), no work is done in moving from M to N. This is true since the motion is perpendicular to the applied force. The total work done in carrying the unit positive charge from plate A to plate B is done in going from A to M and from N to B. This work is identical with that done by following path 1. Therefore, since *19.2* merely expresses this work, the voltage difference between plates A and B will be found to be the same by either path.

Similarly, path 3 in figure 19.3(*b*) can be split into small vertical and horizontal displacements. In carrying a unit positive charge along this path, work is done only during the vertical displacements. This work will also be *The electrostatic field is conservative* the same as by path 1. We therefore conclude that the electrostatic field is *conservative*. By this we mean that the work done in moving a charge from one point in the field to another is independent of the path followed. The interested student should be able to apply this same procedure to the field of a point charge to show that it, too, is conservative. How could that result then be used to show that *all* electrostatic fields are conservative?

Illustration 19.1

Two large parallel metal plates are separated by a distance of 0.50 cm. When a 90-V battery is connected across them, what will be the field between them and the charge per unit surface area on them?

Reasoning The potential difference between the two plates, $V_B - V_A$, is 90 V. In addition, E is constant between the two plates. Therefore the work done in carrying a unit charge from one plate to the other is (force/unit charge) $\times$ distance, and this is Ed. By definition, this is the potential difference between the plates, and so

$$V_B - V_A = Ed$$

which is merely *19.2*.

Placing in the values, the above equation yields

$$90 \, \text{V} = E(5 \times 10^{-3} \, \text{m})$$

or

$$E = 18,000 \, \text{V/m}$$

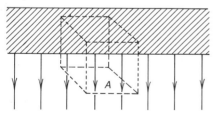

FIGURE 19.4 If E is known, how can σ, the charge per unit area of the plate, be found?

$1 \, N/C = 1 \, V/m$

You will recall that our previous unit for E was newtons per coulomb. Since one volt is one joule per coulomb, and one joule is a newton $\cdot$ meter, we find that these two units for E are the same:

$$1 \, \text{V/m} = 1 \, \text{N/C}$$

To find the charge on the positive plate, we make use of Gauss' law. We use a gaussian box, as shown in figure 19.4. No lines of flux emerge from the top of the box (since it is in the metal where the field is zero) or its sides. However, the number of lines emerging from the lower face of the box is EA, where A is the area of the bottom face of the box. You should recognize that we have just computed

$$\int_S \mathbf{E} \cdot d\mathbf{A}$$

to be EA.

This must be equated to $1/\epsilon_0$ times the charge within the box, which will be σA. Of course, σ is the charge per unit area of the plate. Equating these two quantities gives

$$EA = \frac{1}{\epsilon_0}\sigma A$$

from which

$$\sigma = \epsilon_0 E$$

This result was actually found in the preceding chapter, but we have repeated the computation here in the interest of review. Evaluating σ, we find

$$\sigma = (8.85 \times 10^{-12}\,\mathrm{C^2/N \cdot m^2})(18{,}000\,\mathrm{N/C})$$

or

$$\sigma = 1.59 \times 10^{-7}\,\mathrm{C/m^2}$$

19.3 WORK AND ENERGY IN ELECTROSTATICS

By the definition of potential difference, the work done in carrying a unit positive charge from a point A to a point B is $V_B - V_A$. For a charge twice as big, the work would of course be twice as big. In general, then, the work done in carrying a charge q from point A to point B is given by

$$\text{Work} = q(V_B - V_A) \qquad\qquad 19.3$$

The electrostatic work done in moving a charge q from A to B is $q(V_B - V_A)$

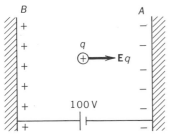

FIGURE 19.5 What will be the kinetic energy of q if it is released at B and moves to A?

Electrostatic energy can be converted to kinetic energy

To illustrate this point, let us refer to the parallel-plate system shown in figure 19.5. When a positive charge q is placed between the plates as shown, the electric field exerts a force toward the right on it. If we wish to carry the charge from plate A to plate B, the work done against this force is just that given by *19.3*. This work is stored as potential energy in the system. For, if we release the charge at plate B, it will be repelled toward plate A by the electric field. As the charge moves to plate A under the action of this force, it acquires kinetic energy. The kinetic energy acquired by the charge will be equal to the work done against the electric field force as the charge was originally carried from A to B.

The situation shown in figure 19.5 is analogous to lifting an object in a gravitational field. When we lift an object against gravity, we store gravitational potential energy in it. If the object is then released, the gravitational potential energy will be changed to kinetic energy as the object falls and accelerates. Similarly for the positive charge in figure 19.5. Work must be done to "lift" it from A to B; electrical potential energy is thereby stored in the system. Upon release at B, the charge "falls" toward A with an ensuing change of potential energy to kinetic energy. To carry the analogy further, point B is said to be at a "higher" potential than A because work is required to "lift" a positive charge from A to B. Notice, however, that this terminology is used in reference to *positive* charges. Negative charges actually fall from points of low electric potential to points of higher potential. Why? This is often summarized by saying "What is uphill for a positive charge is downhill for a negative one." As we have seen, electric potential levels are defined in terms of a *positive* test charge.

Illustration 19.2

The potential difference between the two plates in figure 19.5 is 100 V. If the system is in vacuum, what will be the speed of a proton released from plate B just before it hits plate A?

Reasoning The mass and charge of a proton are 1.67×10^{-27} kg and 1.60×10^{-19} C, respectively. When the proton is moved from plate B to plate A, it loses a potential energy $q(V_B - V_A)$, where $V_B - V_A$ is 100 V in this case. This appears as kinetic energy of the proton at plate A. The law of conservation of energy therefore tells us

Loss in potential energy = gain in kinetic energy

or
$$q(V_B - V_A) = \tfrac{1}{2}mv^2$$

Placing in the values and solving for v, we find

$$v \approx 1.4 \times 10^5 \, \text{m/s}$$

19.4 POINT CHARGES AND POTENTIAL DIFFERENCE

We have illustrated the concepts associated with potential by referring to the case of charged parallel plates. This was done because of the great simplicity of the electric field in that case. Now let us extend our computations to a somewhat different situation, namely, the field due to a point charge. For that purpose, consider the point charge q shown in figure 19.6. We wish to compute the potential difference between the points A and B shown.

The definition of potential difference between two points was given in 19.1, and is

$$V_B - V_A = -\int_A^B \mathbf{E} \cdot d\mathbf{s} = -\int_A^B \mathbf{E} \cdot d\mathbf{r}$$

In the case shown in figure 19.6, $\mathbf{E}$ is radially out from the point charge and is in the direction of the length $d\mathbf{r}$ shown. Since $\mathbf{E}$ and $d\mathbf{r}$ are both radial, we have

$$\mathbf{E} \cdot d\mathbf{r} = E \, dr$$

The expression for the potential difference is then

$$V_B - V_A = -\int_{r_A}^{r_B} E \, dr$$

To proceed further we need to know how E varies with r. This relation is well known to us *in the case of a point charge* and is given by Coulomb's law to be

$$E = \frac{q}{4\pi\epsilon_0 r^2}$$

Upon substitution of this value we find

$$V_B - V_A = \frac{-q}{4\pi\epsilon_0} \int_{r_A}^{r_B} \frac{dr}{r^2}$$

Notice in particular the limits on this integral. You should be able to justify them.

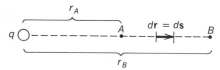

FIGURE 19.6 In computing the potential difference between A and B, we must find out how much work is done in moving a unit positive test charge from A to B. If q is positive, the work is negative; it is "downhill" from A to B.

This integral is easily evaluated using the relation

$$\int r^{-2}\, dr = -r^{-1}$$

If we use this, we obtain

Potential difference in the field of a point charge

$$V_B - V_A = \frac{q}{4\pi\epsilon_0}\left(\frac{1}{r_B} - \frac{1}{r_A}\right) \qquad 19.4$$

Let us check this result. Referring to figure 19.6, we notice that a positive charge q would repel the test charge at A. Hence, if left to itself, the test charge would fall from A to B. We say it is electrically "downhill" from A to B. Equation *19.4* also tells us this; for, if $r_A < r_B$ as we are assuming, then $(1/r_B) - (1/r_A)$ will be negative, thus making $V_B - V_A$ negative. We know from this that $V_B < V_A$ so, indeed, point B is at a lower potential than point A. In using *19.1* and *19.4* to compute potential differences, it is wise, where possible, to infer qualitatively which point is at the higher potential so that a check may be had on the algebraic sign of the answer.

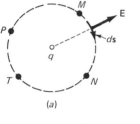

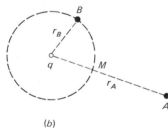

FIGURE 19.7 Describe the equipotential lines and surfaces in the field of a point charge.

19.5 EQUIPOTENTIAL LINES AND SURFACES

You should notice that the result we obtained in the previous section for the potential difference between two points outside a point charge was restricted in two ways: first, the result applied only to the region outside a *point charge;* second, the two points A and B were on the same radius vector from the point charge. As you might suspect, the first of these two restrictions can be circumvented by use of the superposition principle. This we shall examine somewhat later in this chapter. The second restriction can be removed completely. We shall now see how this may be done. At the same time we shall introduce a very useful way of looking at electrostatic situations in terms of so-called equipotential lines and surfaces.

To begin our discussion of equipotential lines and surfaces, let us consider the region outside a point charge q. Suppose we ask for the potential difference between points M and N in figure 19.7(a). By definition, the potential difference is equal to the work done in carrying a unit positive test charge from the one point to the other. Since the electrostatic field is conservative, the work done will be independent of the path we follow in moving the test charge from M to N. For simplicity, then, we choose as our path the circular one shown.

The force we must exert on the test charge as we slowly move it along the arc from M to N will be opposite in direction to $\mathbf{E}$. We must simply keep the charge from shooting away under the repulsive force from q. But the direction of motion, along $d\mathbf{s}$, for example, is perpendicular to $\mathbf{E}$. Therefore, since the force is perpendicular to the displacement, no work will be done in moving the test charge along the arc from M to N. Hence there is no difference in potential between these two points. Both M and N, as well as all points on the dashed circle, are at the same electrical level. No work is done in moving a test charge on a circle centered on the point charge q. All points on the circle shown are at the same electrical potential. The difference in potential between M, N, P, T, and all similar points is zero. We call the circle an *equipotential* (equal potential) *line.*

All points on an equipotential line have the same potential

It is easy to generalize this result to a spherical surface through M and centered on the charge q. Since $\mathbf{E}$ is radial, the restraining force applied to a test charge will be perpendicular to the spherical surface. As a result, no work will be done in moving the test charge slowly from place to place on the surface. Therefore, all points on the surface are at the same electrical level and no difference in potential exists from one to the other. We term a surface such as this an *equipotential surface*.

Let us now use this knowledge to remove the second restriction from equation *19.4*. If we refer to figure 19.7(*b*), we found in *19.4* that

$$V_M - V_A = \frac{q}{4\pi\epsilon_0}\left(\frac{1}{r_B} - \frac{1}{r_A}\right)$$

But we wish to find $V_B - V_A$. This is now done easily since we know that points B and M lie on an equipotential surface. As a result, the electrical level at B is the same as at M and so

$$V_B - V_A = V_M - V_A$$

Therefore we see that *19.4* gives us the potential difference between two points outside a point charge whether or not the points lie on the same radius vector.

There are other important aspects of equipotentials which we shall use. For example, in figure 19.8(*a*) we show the traces (or lines) of three equipotential surfaces around a point charge. Obviously there are an infinite number of equipotential surfaces near a point charge. Any spherical surface centered on the charge is an equipotential.

Another important set of equipotential surfaces is shown in (*b*) of figure 19.8. There we see two large parallel charged plates. Since the field is directed straight downward between them, no work is done in moving a test charge

Cathode Rays and the Electron

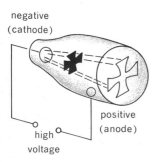

negative (cathode)

positive (anode)

high voltage

When the voltage difference between two points becomes large enough, a spark will jump from one to the other. This is what happens in thunderstorms when a large spark, a lightning bolt, jumps from the clouds to the ground. It is found that if a high voltage is applied to the two ends of a tube containing a gas, the sparking process transforms to a general glowing of the gas within the tube as the pressure of the gas is reduced. This is the glow of light given off by neon signs, for example. If the pressure is reduced still further within the tube, the glow diminishes and essentially disappears. Under these conditions the glass at one end of the glass tube begins to glow with a fluorescent greenish light, at the anode in the figure.

This phenomenon was being investigated extensively by several persons in about 1870. It was believed that the greenish fluorescence was the result of radiation of some sort traveling down the tube from the negative electrode (the cathode) to the positive electrode (the anode). These rays were called cathode rays. Whether they were waves, such as light, or particles was a question in dispute. Sir William Crookes (1832–1919), as well as others, carried out the following experiment to clarify the situation. Using a tube (a Crookes tube) such as the one shown in the figure, he found that an obstruction in the tube caused the rays to cast a shadow, as illustrated. In addition, the cathode rays were found to exert a force to the right on the obstruction. However,

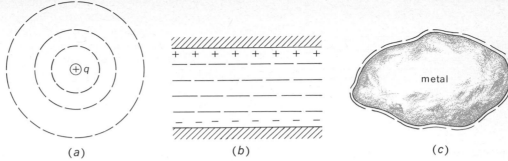

(a) (b) (c)

FIGURE 19.8 The dashed lines are equipotentials. What can one say about the entire metal body shown in part (c)?

Conductors constitute equipotential volumes in electrostatics

along one of the horizontal lines (or, in three dimensions, planes) shown. Hence each of these lines (or planes) is an equipotential. Here, too, there are an infinite number of such planes possible.

In part (c) of figure 19.8 we see a more complicated situation. The metal object may or may not be charged. We know, however, that the field lines must strike the surface perpendicularly. Therefore no work will be done in moving a test charge parallel to the surface provided we are very close to it. This follows since the motion is always perpendicular to the force with which we hold the test charge, provided we move parallel to the surface. We see, then, that the whole surface of a metal object is an equipotential surface under electrostatic conditions. Moreover, since the electric field is zero in the metal body of the object, no force is required to move the test charge from place to place within the object. Hence all the points within the body of a metal object are at the same electrical level. Metal objects therefore constitute *equipotential volumes* under electrostatic conditions.

this experiment by itself was not sufficient to decide whether the rays were the result of particles or waves.

One known characteristic of charged particles in motion is that they can be deflected by a magnet held nearby. This experiment had actually been tried by various workers, and no deflection of the rays was found. However, Crookes, in 1879, repeated the experiment at a much lower gas pressure within the tube, and showed that the cathode rays were deflected by a nearby magnet. Later, in 1895, Jean Perrin further showed that, when the deflected rays struck an electrode, they charged the electrode negatively. It was therefore concluded that the cathode rays were actually a stream of negatively charged particles.

Perrin further showed that very close to the cathode there were positive charges moving in a direction opposite to the cathode rays. This gave rise to speculation that the gas molecules in the tube were being torn apart under the action of the discharge. The oppositely charged pieces of the molecules were surmised to travel in opposite directions in the tube. It appeared that the cathode rays were the negative pieces torn from the molecules. These negative particles were investigated further by J. J. Thomson, whose work we discuss in Chap. 23.

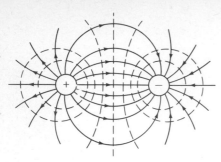

FIGURE 19.9 Notice that the electric field lines (solid) and the equipotential lines (dashed) are mutually perpendicular.

Finally, we should emphasize the relation between field lines and equipotentials. From the very definition of an equipotential, it is necessary that the field lines be perpendicular to the equipotentials. If you examine figure 19.8 and sketch in the field lines, you will see that this is true in these cases. (Recall, the field lines strike the metal perpendicularly.) Another example is shown in figure 19.9. This mutual perpendicularity of field lines and equipotentials allows us to construct the one if the other is known. Use of this will be made repeatedly throughout our study of physics.

19.6 ABSOLUTE POTENTIALS

We found it convenient in our discussion of gravitational potential energy sometimes to assign a zero level for potential energy. Since we were always concerned with differences in potential energy, the choice of the zero level was perfectly arbitrary. A similar situation exists in regard to a zero level for measurement of electric potential, or electrical level. In practical circuit work, one of the wires is usually connected to a metal pipe, and the pipe in turn is connected to the earth. The earth is taken to be at the zero potential level in that case. For other types of applications, another zero level for potential is used. We shall now discuss and use it.

Absolute potentials are assigned relative to an arbitrary zero level

It is common in work such as we have been doing to take the zero potential level at infinity, far removed from all charges. For example, in computing the potential difference between two points outside a point charge of magnitude q, we found

The zero level is often taken at infinity

$$V_B - V_A = \frac{q}{4\pi\epsilon_0}\left(\frac{1}{r_B} - \frac{1}{r_A}\right)$$

where r_B and r_A are shown in figure 19.7. In writing this, we are interested only in the potential difference between A and B, and so the zero level of potential is of no consequence. However, if point A is really at infinity, then $1/r_A \to 0$. Moreover, we *define* the potential at infinity to be zero, and so $V_A = 0$. Under that condition, i.e., zero potential at infinity, the potential at B will be obtained from the above equation, and is

$$V_B = \frac{q}{4\pi\epsilon_0 r_B}$$

where we have considered point A to be at infinity.

This is a general result for any *point charge*. It states that the *absolute potential* V at a distance r from a point charge q is

Absolute potential of a point charge with V = 0 at infinity

$$V = \frac{q}{4\pi\epsilon_0 r} \qquad\qquad 19.5$$

This is merely the potential difference between infinity and the point a distance r from the point charge. Or from the definition of potential difference, 19.5 gives the work done in moving a unit positive test charge from infinity to the point a distance r away from the charge. Since the electrostatic field is a conservative field, this work is independent of the path by which the test charge is brought in from infinity.

Illustration 19.3

Show that the absolute potential at point P in figure 19.10 is zero.

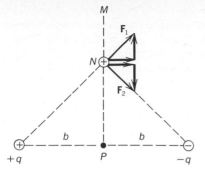

FIGURE 19.10 Can you show, directly from the definition of absolute potential, that the potential at P is zero?

The superposition principle applies to potentials

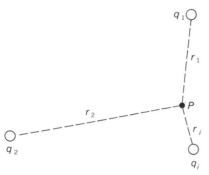

FIGURE 19.11 What is the value of the absolute potential at point P?

Reasoning Let us first compute the absolute potential at point P directly from the definition of absolute potential. We shall then show that this result can also be obtained from *19.5*. By definition, the absolute potential at P is equal to the work done in carrying a unit positive test charge from infinity to P. We may carry it along any path we choose since the field is conservative. Let us carry the test charge in from infinity along the line MP.

To find the work done, we examine the electric force on the test charge at a point such as the one indicated as N in figure 19.10. From the vector diagram, we see that since $\mathbf{F}_1$ and $\mathbf{F}_2$ are equal in magnitude, their components along the line MP cancel. Therefore the force needed to hold the test charge, as it is brought in along the line MP, is perpendicular to that line. Since there is no component of the force in the direction of motion, no work is done in bringing the test charge to point P from infinity. Therefore the potential difference between infinity and P is zero; P must have zero absolute potential.

This result can be found in another way which has more general usefulness. It is based upon the fact that the absolute potential at a point can be obtained by superposition, as we now show. In the present case, *19.5* tells us that the work done in carrying the unit positive test charge from infinity to P against the repulsion force of the $+q$ charge is

$$V_1 = \frac{+q}{4\pi\epsilon_0 b}$$

As the positive test charge is brought in from infinity, it is attracted by the negative charge $-q$, and so negative work is done in this case. Therefore the potential at P due to the $-q$ charge is

$$V_2 = \frac{-q}{4\pi\epsilon_0 b}$$

The total work done in bringing in the test charge, which is defined to be the absolute potential at P, is the sum of these two and is

$$V_P = V_1 + V_2 = 0$$

which is the result found previously. Notice that the fact that the absolute potentials are additive is a direct result of the fact that electrostatic forces and the work done against them are superposable.

19.7 POTENTIAL RESULTING FROM SEVERAL CHARGES

Suppose we are concerned with the absolute potential at a point such as P in figure 19.11. The potential at this point is, by definition, the work done in bringing a unit positive test charge from infinity up to this point. This work consists of N parts if there are N charges present, namely, the work against the repulsion of q_1, q_2, and so on up to q_N. We found the work done in bringing the test charge from infinity up to a distance r_i away from a charge q_i to be

$$V_i = \frac{q_i}{4\pi\epsilon_0 r_i}$$

The total work done in bringing the test charge in to point P from infinity is the potential at P, and will be

$$V = \sum_{i=1}^{N} \frac{q_i}{4\pi\epsilon_0 r_i} \qquad 19.6$$

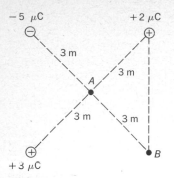

FIGURE 19.12 What is the value of $V_B - V_A$?

Notice that V is a scalar, and therefore, unlike the procedure in the computation of E, *one need not take components.*

Equation *19.6* still gives the proper potential even if some of the q's are negative. For a negative charge, the test charge is attracted in from infinity, and so negative work and potential result. This will automatically be accounted for in *19.6* if the proper signs are used for each value of q.

Illustration 19.4

For the situation shown in figure 19.12, find V_A, V_B, and $V_B - V_A$.

Reasoning This question is easily answered by use of *19.6*. We have

$$V_A = \frac{1}{4\pi\epsilon_0}\left(\frac{2 \times 10^{-6}}{3} + \frac{3 \times 10^{-6}}{3} - \frac{5 \times 10^{-6}}{3}\right) = 0$$

Therefore point A is at the same electrical level as infinity. No net work will be needed to bring a test charge from infinity up to A. Similarly,

$$V_B = \frac{1}{4\pi\epsilon_0}\left(\frac{2 \times 10^{-6}}{3\sqrt{2}} + \frac{3 \times 10^{-6}}{3\sqrt{2}} - \frac{5 \times 10^{-6}}{6}\right)$$
$$= 3140 \text{ V}$$

From this we have

$$V_B - V_A = 3140 \text{ V}$$

Notice how much more complicated this computation would have been if we had computed $V_B - V_A$ using the electric field and $\int \mathbf{E} \cdot d\mathbf{s}$.

19.8 POTENTIAL DUE TO A CHARGED ROD

Suppose the thin rod of length L shown in figure 19.13 has a uniform charge of λ per unit length. We wish to find the absolute potential at point P. To do this, we must (mentally) split the rod into a series of point charges. A typical one, for the length Δx_i, is shown. Its charge is $\lambda \Delta x_i$. The potential at P due to this small point charge is, according to *19.5*,

$$V_i = \frac{\lambda \Delta x_i}{4\pi\epsilon_0 r_i}$$

The total potential is obtained by adding the potentials due to all the little charge elements along the rod. Since V is a scalar, we need not worry

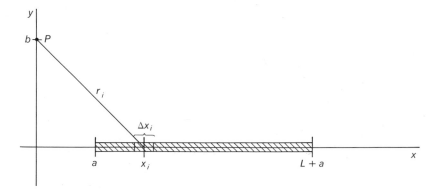

FIGURE 19.13 To find the absolute potential at P, we add the point-charge potentials due to the charge segments composing the rod.

about components while doing this. We have, for the potential at point P,

$$V = \sum \frac{\lambda \, \Delta x_i}{4\pi\epsilon_0 r_i}$$

We can eliminate r_i by noticing from figure 19.13 that

$$r_i = \sqrt{b^2 + x_i{}^2}$$

After substituting this value for r_i and letting $\Delta x \to 0$ so that the sum can be replaced by an integral, we find

$$V = \int_a^{L+a} \frac{\lambda \, dx}{4\pi\epsilon_0 \sqrt{b^2 + x^2}}$$

Notice the limits on the integral. When x is at one end of the rod, $x = a$. At the other end, $x = L + a$, and so these are the two limits on x. Since $\lambda/4\pi\epsilon_0$ is a constant, we can write

$$V = \frac{\lambda}{4\pi\epsilon_0} \int_a^{L+a} \frac{dx}{\sqrt{b^2 + x^2}}$$

This integral is listed in most integral tables and is

$$\int \frac{dx}{\sqrt{b^2 + x^2}} = \ln\left(x + \sqrt{b^2 + x^2}\right)$$

Using this to evaluate V, we obtain

$$V = \frac{\lambda}{4\pi\epsilon_0} \ln\left[\frac{(L + a) + \sqrt{b^2 + (L + a)^2}}{a + \sqrt{b^2 + a^2}}\right] \qquad 19.7$$

It is always wise to check an answer such as this in the limiting cases.

Case 1 $\quad b = 0;\ L \ll a$
This case yields, from *19.7*,

$$V_1 = \frac{\lambda}{4\pi\epsilon_0} \ln \frac{2(L + a)}{2a}$$

which is

$$V_1 = \frac{\lambda}{4\pi\epsilon_0} \ln\left(1 + \frac{L}{a}\right)$$

Now $L/a \ll 1$, and so we can make use of the expansion

$$\ln(1 + z) = z - \frac{z^2}{2} + \frac{z^3}{3} - \cdots \approx z$$

for z small. We then have

$$V_1 = \frac{\lambda L}{4\pi\epsilon_0 a}$$

But $\lambda L = Q$, the total charge on the rod, and a is the distance of point P from the far-distant rod. At such a large distance the rod appears to be a point charge. We have

$$V_1 = \frac{Q}{4\pi\epsilon_0 a}$$

which is indeed the potential of a point charge Q a distance a away. Our answer therefore checks in this limit.

Case 2 $a = 0; L \ll b$

This is essentially the same limiting case as the preceding one. We leave it as an exercise for the student to show that in this case

$$V_2 = \frac{Q}{4\pi\epsilon_0 b}$$

19.9 ELECTRIC FIELDS FROM POTENTIALS

Until now we have used *19.1*, the definition of potential difference, to find potential differences if E is known. This equation was

$$V_B - V_A = -\int_A^B \mathbf{E} \cdot d\mathbf{s} \qquad\qquad 19.1$$

It may also be used, however, to find $\mathbf{E}$ if the potential is known. To do this, one applies *19.1* to two points, A and B, separated by a very small distance $\Delta \mathbf{s}$, so that *19.1* can be written

$$\Delta V = -\mathbf{E} \cdot \Delta \mathbf{s}$$

(In doing this recall that

$$\int_A^B \mathbf{E} \cdot d\mathbf{s} \equiv \mathbf{E}_1 \cdot \Delta\mathbf{s}_1 + \mathbf{E}_2 \cdot \Delta\mathbf{s}_2 + \cdots + \mathbf{E}_n \cdot \Delta\mathbf{s}_n$$

where all the $\Delta s \to 0$. For our particular case, $\Delta\mathbf{s}_1$ is assumed to extend from A to B, so that the other terms are zero.)

Let us now write both $\mathbf{E}$ and $\Delta\mathbf{s}$ in terms of their components, so that

$$\mathbf{E} \cdot \Delta\mathbf{s} = (E_x\mathbf{i} + E_y\mathbf{j} + E_z\mathbf{k}) \cdot (\Delta x\mathbf{i} + \Delta y\mathbf{j} + \Delta z\mathbf{k})$$

After carrying out the indicated multiplication and substituting, we find

$$-\Delta V = E_x\,\Delta x + E_y\,\Delta y + E_z\,\Delta z$$

From this we conclude that

$$E_x = -\frac{\Delta V}{\Delta x} \qquad \text{if } \Delta y = \Delta z = 0$$

$$E_y = -\frac{\Delta V}{\Delta y} \qquad \text{if } \Delta x = \Delta z = 0$$

$$E_z = -\frac{\Delta V}{\Delta z} \qquad \text{if } \Delta x = \Delta y = 0$$

If we make the small increments Δx, Δy, and Δz approach zero, these equations each can be written conveniently as

E can be found from V by taking derivatives

$$E_x = -\frac{\partial V}{\partial x} \qquad E_y = -\frac{\partial V}{\partial y} \qquad E_z = -\frac{\partial V}{\partial z} \qquad 19.8$$

where the partial derivative notation, $\partial/\partial x$, etc. means to take the derivative with respect to the variable indicated while holding the other variables constant.

To illustrate the use of this, let us consider a point charge at the origin of coordinates. In this case

$$V = \frac{q}{4\pi\epsilon_0 r} = \frac{q}{4\pi\epsilon_0 \sqrt{x^2 + y^2 + z^2}}$$

In order to find **E**, we make use of *19.8* to give

$$E_x = -\frac{q}{4\pi\epsilon_0} \frac{\partial}{\partial x}\left(\frac{1}{\sqrt{x^2 + y^2 + z^2}}\right)$$

or

$$E_x = \frac{q}{4\pi\epsilon_0} \frac{x}{(x^2 + y^2 + z^2)^{3/2}}$$

Similarly,

$$E_y = \frac{q}{4\pi\epsilon_0} \frac{y}{(x^2 + y^2 + z^2)^{3/2}}$$

and

$$E_z = \frac{q}{4\pi\epsilon_0} \frac{z}{(x^2 + y^2 + z^2)^{3/2}}$$

To check this result, we recall that E for a point charge is

$$E = \frac{q}{4\pi\epsilon_0 r^2}$$

We also know that

$$E = \sqrt{E_x{}^2 + E_y{}^2 + E_z{}^2}$$

If we place in the above values for E_x, E_y, E_z and recognize that

$$r^2 = x^2 + y^2 + z^2$$

we find that the two expressions for E do agree.

Illustration 19.5
The potential at a point P a distance y above the bottom negative plate of a parallel-plate combination is $V(y) - V(0) = ky$. In this expression $V(0)$ is the potential at the lower plate, and $V(y)$ is the potential at the height y above it. Find E between the parallel plates.

Reasoning The situation can be simplified if we take $V(0) = 0$. This is allowable since the zero potential is arbitrary. We then have

$$V(y) = ky$$

Making use of *19.8*, we find

$$E_x = -\frac{\partial}{\partial x}(ky) = 0$$

$$E_y = -\frac{\partial}{\partial y}(ky) = -k$$

$$E_z = -\frac{\partial}{\partial z}(ky) = 0$$

The field therefore points downward (in the $-y$ direction) between the plates and has a constant value k.

Illustration 19.6
In the Bohr model of the hydrogen atom, the electron was pictured to rotate in a circle of radius 0.053 nm about the nucleus. (*a*) How fast should the electron be moving in this orbit? (*b*) How much energy is needed to tear the electron loose from the nucleus of the atom?

Reasoning The situation is shown in figure 19.14. In order for the electron to travel in the circular orbit, a centripetal force must be furnished to it. This

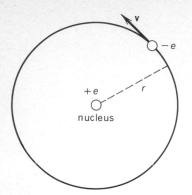

FIGURE 19.14 Bohr's picture of the hydrogen atom.

is provided by the coulomb attraction between it and the nucleus. We therefore have (after assuming that the nucleus remains stationary, a good assumption, since it is 1840 times more massive than the electron)

$$\frac{mv^2}{r} = \frac{1}{4\pi\epsilon_0}\frac{ee}{r^2}$$

In this expression m is the mass of the electron $(9.1 \times 10^{-31}\,\text{kg})$ and e is the magnitude of the charge on the electron as well as on the nucleus, $1.6 \times 10^{-19}\,\text{C}$. After placing in the values and solving for v, we find

$$v \approx 2.2 \times 10^6\,\text{m/s}$$

To find the energy needed to tear the electron loose, we need to know how much work is done in carrying the electron from its orbit out to infinity. This is easily computed because we know the absolute potential at the orbit position is $V = (1/4\pi\epsilon_0)(e/r)$. Therefore the work needed to carry a unit positive charge from infinity up to the orbit is $(1/4\pi\epsilon_0)(e/r)$. In our case, the charge in question is negative, not positive, and its magnitude is e. The work needed to pull it from the orbit and carry it to infinity is

$$\text{Work} = \left(\frac{1}{4\pi\epsilon_0}\frac{e}{r}\right)e$$

This, then, is the work needed to tear the electron loose from the atom. Placing in the numerical values, this becomes

$$\text{Work} = 4.30 \times 10^{-18}\,\text{J}$$

Actually, not all this energy need be furnished by an outside agent since the electron itself had kinetic energy when it was in the orbit. Its kinetic energy was

$$\tfrac{1}{2}mv^2 = 2.15 \times 10^{-18}\,\text{J}$$

The external energy needed to tear the electron loose is therefore

$$(4.30 - 2.15) \times 10^{-18} = 2.15 \times 10^{-18}\,\text{J}$$

This energy is referred to as the *ionization energy* of the hydrogen atom. The interested student will be able to show that

$$\frac{1}{4\pi\epsilon_0}\frac{e^2}{r} = 2(\tfrac{1}{2}mv^2)$$

and therefore the magnitude of the potential energy of the electron in orbit is twice its kinetic energy.

19.10 THE ELECTRON VOLT (eV)

When a particle of charge $e = 1.602 \times 10^{-19}\,\text{C}$ (the charge on the electron and proton) falls through a potential difference of $V_B - V_A$ volts, it acquires a kinetic energy

$$q(V_B - V_A) = e(V_B - V_A)$$

This was pointed out in Sec. 19.3, where the work-energy theorem was discussed. Placing in the values, we find

$$\text{Energy acquired} = (1.602 \times 10^{-19})(V_B - V_A) \qquad \text{J}$$

This type of situation is frequently enough encountered so that we define a new unit of energy in order to eliminate the factor 1.602×10^{-19}. We define the *electron volt* (eV) of energy to be the energy in joules divided by the magnitude of the quantum of charge, 1.602×10^{-19} C. Therefore

$$1 \text{ eV} = 1.602 \times 10^{-19} \text{ J}$$

The utility of this unit is obvious. Suppose an electron falls from a negative to a positive plate, the potential difference between the plates being 100 V. The electron's kinetic energy just before it hits the positive plate will be

$$K = q(V_B - V_A) = e(100)$$
$$= 100 \text{ eV}$$

In other words, a particle whose charge is $|q| = e$ acquires an energy of V eV when it falls through a voltage difference V. Of course, a particle whose charge is $|q| = 2e$ (for example, an α particle) will acquire an energy $2V$ eV when it falls through this same potential difference.

Illustration 19.7

A proton ($q = 1.6 \times 10^{-19}$ C) is released from a point P which is 10^{-14} m from a heavy nucleus which has a charge of $80 \times 1.6 \times 10^{-19}$ C. (This would be the nucleus of a mercury atom.) How large will the kinetic energy of the proton be when it gets far away from the nucleus? What will be its speed?

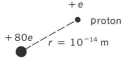

FIGURE 19.15 What will be the speed of the proton when it gets far from the nucleus?

Reasoning The situation is shown in figure 19.15. Clearly, the proton will be repelled and caused to move radially toward infinity by the positive nucleus. In effect, it falls through the potential difference between point P and infinity. This potential difference is just the absolute potential at point P and is

$$V = \frac{1}{4\pi\epsilon_0} \frac{80 \times 1.6 \times 10^{-19}}{10^{-14}} = 12 \times 10^6 \text{ V}$$

In falling through this potential difference, the proton will thus acquire a kinetic energy of

$$K = 12 \text{ MeV}$$

where the symbol MeV means million electron volts. In writing this we assume the nucleus to remain nearly at rest. Can you justify this?

If the particle's speed is not close to the speed of light, the nonrelativistic expression $\frac{1}{2}mv^2$ can be used for K. Then

$$\tfrac{1}{2}mv^2 = 12 \times 10^6 \times 1.6 \times 10^{-19} \text{ J}$$

Because the nonrelativistic mass of a proton is 1.67×10^{-27} kg, this yields $v = 4.8 \times 10^7$ m/s. This speed is small enough relative to the speed of light ($c = 3 \times 10^8$ m/s) so that relativistic effects are negligible. But for speeds larger than about 1×10^8 m/s it is imperative that relativistic methods be used. In that case the kinetic energy is $K = mc^2 - m_0c^2$, where $m_0 = 1.67 \times 10^{-27}$ kg and $m = m_0/\sqrt{1 - (v/c)^2}$. Therefore the proper equation would be

$$m_0c^2\left(\frac{1}{\sqrt{1 - (v/c)^2}} - 1\right) = 12 \times 10^6 \times 1.6 \times 10^{-19} \text{ J}$$

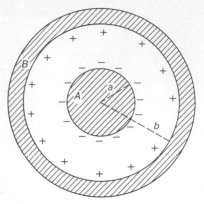

FIGURE 19.16 If the potential difference between two cylinders is V, what is the speed of an electron released at the center cylinder when it reaches the outer cylinder?

from which

$$\frac{1}{\sqrt{1 - (v/c)^2}} = 1 + \frac{1.92 \times 10^{-12}}{m_0 c^2} = 1 + 0.0128$$

Solving for v gives it to be $v = 0.158c = 4.8 \times 10^7$ m/s. As we see, the effect of relativity is negligible in this case.

19.11 TWO CONCENTRIC CYLINDERS: AN EXAMPLE

Suppose the two concentric metal cylinders shown in cross section in figure 19.16 are connected to a battery of voltage V. We assume the inner solid cylinder to be connected to the negative side of the battery, and the cylindrical shell to be connected to the positive side. As a result, the charges on the cylinders are as shown. In addition,

$$V_B - V_A = V$$

Clearly, B is at higher potential than A because work would need be done in carrying a positive test charge from A to B; the test charge is repelled by the positive charge and attracted by the negative charge.

If an electron were released at A, it would fall to B, acquiring kinetic energy in the process. Recall that even though B is at a higher potential than A (and therefore a positive charge would fall from B to A, not from A to B), the electron is a negative charge and behaves just the opposite of a positive charge. In any event, the law of conservation of energy tells us

$$K = (V_B - V_A)q = Ve$$

To find the speed of the electron, assuming it to be small enough so that relativistic effects do not enter, we write

$$\tfrac{1}{2}mv^2 = Ve$$

where m and v are the mass and speed of the electron. Solving for v, we find

$$v = \sqrt{\frac{2Ve}{m}} \qquad 19.9$$

Let us now see what voltage is needed to attain near-relativistic speeds in the case of an electron. Putting in the appropriate values of e and m, we find

$$v \approx 6 \times 10^5 \sqrt{V} \qquad \text{m/s}$$

If V is 1000 V, the electron has a speed of about 2×10^7 m/s, and is therefore approaching the relativistic range. For a proton, however, m is nearly 2000 times larger than for an electron. As a result, relativistic speeds are approached only if V is of the order of 2×10^6 V when the particle is a proton.

It is also instructive to find the charge on the cylinders in figure 19.16. This can be done by use of Gauss' law and the definition of potential difference. First let us find the electric field between the cylinders. To do this we draw a gaussian cylinder of length L, as indicated by the dashed line in figure 19.17. As indicated, no flux lines go through the two ends of the cylinder, and so the integral in Gauss' law is easily evaluated.

We have

$$\int \mathbf{E} \cdot d\mathbf{A} = \frac{1}{\epsilon_0} \sum q_i$$

which gives

$$-E(2\pi r L) = \frac{1}{\epsilon_0} \sum q_i$$

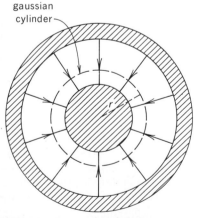

FIGURE 19.17 Since the charge on the inner cylinder is negative, the flux lines are directed inward. What can one say about the magnitude of the charge on the outer cylinder?

The negative sign arises since the flux lines go *into* the gaussian surface.

If we call $-\lambda$ the charge per unit length of the inner cylinder, the charge inside the guassian cylinder is merely $-\lambda L$. Therefore

$$-E(2\pi rL) = \frac{1}{\epsilon_0}(-\lambda)L$$

which gives the following value for E:

$$E = \frac{\lambda}{2\pi\epsilon_0 r} \qquad\qquad 19.10$$

But the potential difference between the two cylinders is, by definition,

$$V_B - V_A = -\int_A^B \mathbf{E} \cdot d\mathbf{s}$$

Placing in the values, we find, after noticing that $d\mathbf{s} = d\mathbf{r}$ and that the angle between $\mathbf{E}$ and $d\mathbf{r}$ is $180°$,

$$V = \frac{\lambda}{2\pi\epsilon_0} \int_a^b \frac{dr}{r}$$

Performing the integration, we obtain

$$V = \frac{\lambda}{2\pi\epsilon_0} \ln\frac{b}{a} \qquad\qquad 19.11$$

Solving this expression for λ, we find

$$\lambda = \frac{2\pi\epsilon_0 V}{\ln(b/a)} \qquad\qquad 19.12$$

19.12 POTENTIAL OF A UNIFORMLY CHARGED RING

As another example of the techniques we have learned in this chapter, consider the circular ring with charge λ per unit length shown in figure 19.18. We wish first to find the absolute potential at the point P indicated. To do this we split the ring into point charges of magnitude $\lambda \Delta s_i$ and make use of the fact that (remember, V is work, a scalar)

$$V = \frac{1}{4\pi\epsilon_0} \sum \frac{\Delta q_i}{r_i}$$

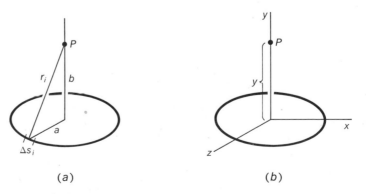

FIGURE 19.18 Point P is not a general point so far as its x and z coordinates are concerned, but its y coordinate, b, is general.

(a)　　　　　　　　(b)

In our particular case this becomes

$$V = \frac{1}{4\pi\epsilon_0} \sum \frac{\lambda \, \Delta s_i}{\sqrt{a^2 + b^2}}$$

Passing to the limit, this becomes an integral. Removing the constant factors from the integral, we have

$$V = \frac{\lambda}{4\pi\epsilon_0 \sqrt{a^2 + b^2}} \int ds$$

The integral is to extend around the ring. It is merely the sum of all the little elements of length ds around the ring and is equal, therefore, to the perimeter of the ring, $2\pi a$. We therefore find

$$V = \frac{\lambda a}{2\epsilon_0 \sqrt{a^2 + b^2}} \qquad\qquad 19.13$$

As we saw in *19.8*, it is possible to find the electric field from a knowledge of V as a function of x, y, and z. In the present instance, as shown in figure 19.18(*b*), the distance $b = y$ could have been any value of y. However, our computation is restricted to points on the axis of the ring, and so the x and z coordinates of P must always be zero if *19.13* is to be true. For this reason we are unable to state how V varies with x and z. Since E_x and E_z (according to *19.8*) can be found from V only by taking the derivative of V with respect to x and z, we cannot evaluate them from *19.13* because we do not know how V varies with x and z.

However, since *19.13* is true for all values of b (that is, y), we can write

$$V = \frac{\lambda a}{2\epsilon_0 \sqrt{a^2 + y^2}}$$

$$E_y = -\frac{\partial V}{\partial y} = \frac{\lambda a y}{2\epsilon_0 (a^2 + y^2)^{3/2}} \qquad\qquad 19.14$$

Can you show that this reduces to the proper limit (Coulomb's law) when $y \to \infty$? Can you guess E_x and E_z at point P?

19.13 POTENTIAL OF A UNIFORMLY CHARGED DISK

We can think of a uniformly charged disk as being made up of a series of uniformly charged rings. This is indicated in figure 19.19. If the charge per unit area of the disk is σ, then the charge on the indicated ring is σ multiplied by the area of the ring, $2\pi r \, dr$. Therefore the potential at point P due to the charged ring is given by *19.13* provided we recognize that the charge on the ring $2\pi a \lambda$ is equal to $\sigma(2\pi r \, dr)$. Making this replacement, the potential due to this ring will be

$$dV = \frac{\sigma(2\pi r \, dr)}{4\pi\epsilon_0 \sqrt{r^2 + b^2}}$$

Since potential is a scalar quantity, we can find the total potential at point P by summing over all the rings composing the disk. We then have

$$V = \frac{\sigma}{2\epsilon_0} \int_0^a \frac{r \, dr}{\sqrt{r^2 + b^2}}$$

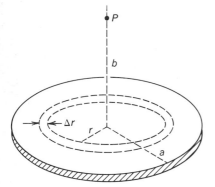

FIGURE 19.19 The charge on the ring is $(2\pi r \, dr)\sigma$.

The limits are chosen as shown because the smallest ring has zero radius and the largest has a radius of a. This integral can be evaluated, and yields

$$V = \frac{\sigma}{2\epsilon_0}(\sqrt{a^2 + b^2} - b) \qquad\qquad 19.15$$

Can you show that this gives the potential due to a point charge when $b \to \infty$? How would you find E_y from this expression?

Chapter 19
Questions
and Guesstimates

1. Two points A and B are at the same potential. Does this necessarily mean that no work is done in carrying a positive test charge from one point to the other? Does it mean that no force will have to be exerted to carry the test charge from one point to the other?
2. If V is zero at a point, must E be zero there as well?
3. What can be said about E in a region where V is constant?
4. A positive charge is released from rest. Will it move to values of higher or lower potential? Repeat for a negative charge.
5. A uniformly charged metal spherical shell carries a charge $+Q$. Is the absolute potential inside the shell zero? Is it constant? What is it? Repeat for $-Q$.
6. A metal shell with outer radius a and inner radius b carries an excess charge $+Q$. Inside and concentric to it is a solid metal sphere that has a radius c and carries an excess charge $-Q$. Sketch roughly a graph of absolute potential vs. r for $r = 0$ to $r \to \infty$.
7. Two parallel, long straight wires carry opposite charges per unit length, $+\lambda$ and $-\lambda$. Sketch the equipotentials for a cross section perpendicular to the wires. Where is the absolute potential zero?
8. Metal objects are frequently spray painted in industry by use of electrostatic methods. The sprayer is attached to one terminal of a high voltage while the metal object to be painted is attached to the other. Explain the principle of operation for this method. Why does it lead to less air pollution and to less use of paint?
9. When defining the absolute potential for a point charge, why do we not take the zero of potential to be at the center of the charge? What is the absolute potential for a point charge at $r \to 0$ if zero potential is taken at infinity?
10. A small square metal plate is suspended by a nonconducting thread. It is given a positive charge Q. Draw the approximate charge distribution, electric field, and equipotentials for the plate in the plane of the plate. Show clearly the field and equipotentials very close to and very far away from the plate.
11. A man is placed inside a large metal spherical shell. The shell is insulated from the surroundings. If charge is now added to the sphere so as to raise its potential to 10^6 V relative to the ground, will the man be able to discern this fact by carrying out measurements within the sphere? What would happen if he dropped a tiny metal sphere out through a small hole in his sphere?
12. A series of positive point charges q lie along a line, each a distance b apart. Sketch a graph showing the potential along this line, taking the midpoint between two charges as the point of zero potential. Using this graph, sketch a graph showing the force on an *electron* at various points along this line. Assume the number of charges to be near infinite.

13. Assuming an atom to consist of a uniform-density sphere of negative charge centered on the positive nucleus, sketch the absolute potential as a function of r both inside and outside the atom. Sketch also the work done in freeing an electron from the atom as a function of its original value of r in the atom.

14. Using the model of an atom described in question 13, assume that a solid consists of rows of such atoms, with their spheres somewhat squashed against each other. Sketch the potential as a function of position along one of these rows. From this sketch, deduce how the force on an electron will vary with position along such a row, and sketch the force as a function of distance.

15. Can two equipotential lines intersect?

16. The electrical strength of air is about 30,000 V/cm. By this we mean that when the electric field intensity exceeds this value, a spark will jump through the air. Using this value, estimate the potential difference between two objects where a spark is noticed to jump. A typical situation might be the spark which jumps between your body and a door handle after you have walked on a deep carpet or slid across a plastic car seat in very dry weather.

17. Referring to the data given in the previous question, about how much charge could you place on a metal sphere which has a diameter of 50 cm?

18. A simple demonstration electrostatic precipitator for removing smoke from air can be constructed in the way shown in figure P19.1. A very thin wire is placed on the axis of a much larger metal tube and a high voltage is applied to these two elements with the wire being made the negative terminal. If the wire is very thin and the voltage high, the electric field near the wire will be very high. Why? Tiny sparks (called corona) are formed near the wire due to electric breakdown (see question 16) and electrons shoot away from the wire. Why? They charge the smoke particles negatively. How? These particles then move to the outer cylinder and precipitate there. Why? As a result, the smoke is removed from the air.

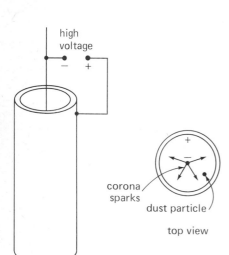

FIGURE P19.1

Chapter 19
Problems

1. It requires 5×10^{-5} J of work to carry a 2-μC charge from point R to S. What is the potential difference between the points? Which point is at the higher potential?

2. How much work is required to carry a proton from the negative to the positive terminal of a 1.5-V flashlight battery? An electron?

3. The electric field in a certain region is given by $\mathbf{E} = 5000\mathbf{i} - 3000\mathbf{j}$ V/m. Find the difference in potential $V_B - V_A$ if A is at the coordinate origin and point B is $(0, 0, 5)$ m. Repeat if point B is $(4, 0, 3)$ m. (*Hint:* Since the field is conservative, use any convenient path.)

4. For the field given in problem 3, find the potential difference $V_B - V_A$ between the two points $A = (0, 5, -1)$ and $B = (-3, 2, 2)$ m. Repeat for the points $A = (0, 5, -1)$ and $B = (3, 2, 2)$ m.

5. The electric field outside a particular charged long straight wire is given by $E = -5000/r$ V/m and is radially inward. What is the sign of the charge on the wire? Find the value of $V_B - V_A$ if $r_B = 60$ cm and $r_A = 30$ cm. Which point is at the higher potential?

6. A wire runs along the axis of a long hollow metal cylinder. The radius of the wire is r_w and the inner radius of the cylinder is r_c. Between the wire

and cylinder there exists a field $E = 200/r$ V/m radially outward. Which is at the higher potential, the wire or the cylinder? Find the voltage difference between them.

7. The electron beam in a TV tube consists of electrons accelerated from rest through a potential difference of about 20,000 V. How large an energy in electron volts do the electrons have? What is their speed? Ignore relativistic effects for this approximate calculation.

8. The potential difference between the terminals of a new flashlight battery is about 1.57 V. How fast would an electron be going if it accelerated from rest through this voltage? A proton?

9. An electron is shot from one large parallel plate toward the second plate. If its initial velocity is 5×10^6 m/s, and its velocity just before hitting the other plate is 2×10^6 m/s, what is the potential difference between the plates? Is the plate to which it is going at a higher or lower potential?

10. Two identical charges with $q = 5.0 \, \mu C$ are placed at the points $(0, 0, 0)$ and $(0, 5, 0)$ m. Find the absolute potential at point A $(0, 2, 0)$ and point B $(0, -2, 0)$ m as well as the potential difference between the points. Which is at the higher potential, A or B?

11. Two charges, $q_1 = +3 \, \mu C$ and $q_2 = +5 \, \mu C$, are 60 cm apart. Find the electric field magnitude and the absolute potential at a point midway between them.

12. Four equal point charges 3×10^{-8} C are placed at the four corners of a square which has a 60-cm diagonal. Find E and V at the center of the square.

13. Two point charges lie along the x axis; 3×10^{-6} C is at $x = 0$ while -5×10^{-6} C is at $x = 50$ cm. (a) Where on the x axis is the absolute potential zero? (b) On the y axis? (c) Where is the field zero?

14. An infinite metal plate carries a uniform surface charge 3×10^{-9} C/m². How far apart are the equipotential planes outside the plate that differ in potential by 6 V?

15. A very long straight wire carries a uniform charge 1×10^{-7} C/m. (a) Find the potential difference between the two points A and B at $r_A = 50$ cm and $r_B = 20$ cm. (b) Assuming A and B to be along the same radius, how fast must a proton be moving as it passes A if it is to reach B?

16. Two very large flat metal plates are parallel and separated by a distance D. The side of the left plate that faces the right plate has a surface charge $+\sigma/m^2$. (a) What is the field between the plates? (b) The potential difference? (c) Another metal plate, uncharged, is placed between these two without altering the charge on the original plates. Its thickness is $d < D$. What is the field in the gap between it and the left plate? (d) In the other gap? (e) What is now the potential difference between the two outer plates?

17. A solid metal sphere of radius 20 cm carries a uniform charge 0.40 μC. Find the absolute potential at the surface of the sphere assuming the sphere to be far removed from all other objects. What is the potential at the sphere's center? (Hint: Recall that a uniformly charged sphere is equivalent to a point charge under certain conditions.)

18. Sparking occurs in air when the electric field exceeds about 3×10^6 V/m. (a) How large is the surface density of charge on a metal plate when this field exists just outside it? (b) How large a charge can be placed on a smooth sphere of radius R before sparking begins? (c) What is then the absolute potential of the sphere? (See hint for problem 17.) In practice, surface roughness decreases the values achievable in (b) and (c).

19. The radius of the nucleus of the radium atom is about 6×10^{-15} m and its charge is $88 \times 1.6 \times 10^{-19}$ C. This nucleus is radioactive and ejects an alpha particle which has an energy of 4.78 MeV. The alpha particle charge is $2 \times 1.6 \times 10^{-19}$ C. Through how large a potential difference would the alpha particle have to fall in order to acquire this energy? How close to the center of the nucleus would the alpha particle have to be released if its kinetic energy was to be 4.78 MeV when far away from it? Ignore the radium atoms' electrons.

20. Assume the following approximately correct model for a radium atom. The nucleus is a uniformly charged sphere of radius 6×10^{-15} m and charge $88 \times 1.6 \times 10^{-19}$ C. Its electrons form a uniformly charged shell about the nucleus with inner radius 1×10^{-12} m and outer radius about 2×10^{-10} m. Find the potential difference between the surface of the nucleus ($r = 6 \times 10^{-15}$ m) and a point for which $r = 1 \times 10^{-13}$ m. Repeat for a point at $r = 1 \times 10^{-12}$. What does this tell us about the effect of the electrons in problems such as 19? (*Hint:* Is it allowable to ignore the electrons when finding potential differences inside the electron shell? If so, can the point charge result be used?)

21. A charged sphere obviously possesses potential energy because, if the charges are released, they would fly out to infinity under their mutual repulsion. To find its energy, consider the process of charging the sphere by bringing in tiny charge elements Δq. When the charge on the sphere is q, (a) show that the work done in adding an additional charge Δq is $q\,\Delta q/4\pi\epsilon_0 a$, where a is the radius of the sphere. (b) Show that a sphere of radius a carrying a charge Q has an energy $Q^2/8\pi\epsilon_0 a$.

22. Four equal charges q are placed at the four corners of a square with side b. Show that the potential energy of this charge configuration is $1.35q^2/\pi\epsilon_0 b$.

23. Two protons very far apart are shot straight at each other with an energy of 0.5 MeV each. Their mutual repulsions cause them to slow down, stop, and then reverse their motion. Show that each has potential energy $q^2/8\pi\epsilon_0 R$ when their centers are R apart. How close do they get to each other before stopping? (Assume R larger than the particle radius.)

24. Find the potential on the axis of a circular ring if the ring carries a charge per unit length $\lambda = A \sin \theta$, where A is a constant and $0 \le \theta \le 2\pi$. Take the radius of the ring to be a and the distance of the point from the center of the ring to be b.

25. Repeat problem 24 if $\lambda = A \sin (\theta/2)$.

26. If the potential in the region of space near the point $(-2, 4, 6\,\text{m})$ is $V = 80x^2 + 60y^2$, find the three components of the electric field at that point.

27. The potential outside a long uniformly charged wire coincident with the z axis is $A \ln(x^2 + y^2)$, where A is a constant. Find the x, y, and z components of the electric field around the wire.

28. Two point charges $+5 \times 10^{-6}$ C and -3×10^{-6} C are placed at the two points $(0, 0, 0)$ and $(2\,\text{m}, 0, 0)$, respectively. (a) Find the absolute potential due to them at a general point (x, y, z). (b) Find the x, y, z components of **E** at the point $(4, 5, -3\,\text{m})$.

29. Show that the absolute potential on the perpendicular bisector of a uniformly charged rod of length L is

$$\frac{\lambda}{2\pi\epsilon_0} \ln \left\{ \frac{L}{2b} \left[1 + \sqrt{1 + \frac{2b}{L^2}} \right] \right\}$$

where b is the distance from the rod and λ is the charge per meter.

30. Using the result quoted in the previous problem, show that the absolute potential a distance Y above the midpoint of a square, flat plate is given by

$$\frac{\sigma}{2\pi\epsilon_0} \int_{-a}^{a} \ln \left\{ \frac{a}{r} \left[1 + \sqrt{1 + \left(\frac{r}{a}\right)^2} \right] \right\} dx$$

where $r = \sqrt{x^2 + y^2}$, σ is the charge per unit area, and $2a$ is the edge length of the plate.

31. Starting from equation *19.15* for the potential outside a uniformly charged disk, compute E_y on the axis of the disk. Show that in the limit of $a \gg b$ this yields the field near the surface of the disk to be $\sigma/2\epsilon_0$. Notice that the result is only half that for the field outside a metal plate. In the case of metal objects, charges elsewhere always contribute an exactly equal field which must be added to this. Only in this way can the resultant field within the metal be zero.

32. A spherical metal shell has inner radius a and outer radius b. If the empty shell has a net charge Q, find V in the region $r > b$. Repeat for the region $r < a$. What is the potential of the metal object?

33. Suppose the spherical shell of problem 32 had a metal sphere of radius c at its center and that this sphere carries a charge Q'. Find the potential in the following three regions: $r > b$; $a < r < b$; $c < r < a$.

34. A sphere of radius b carries a uniform charge density ρ. Find the field (*a*) outside and (*b*) inside the sphere as well as the absolute potential (*c*) outside and (*d*) inside.

35. Three parallel metal plates each have thicknesses b and separations a. The field between two of the plates is $E_1\mathbf{i}$, while it is $-E_2\mathbf{i}$ between the other two. Find the potential difference between the two outer plates.

36. Charge leaks off a certain metal sphere of initial charge Q_0 at a rate $\Delta Q = -cE \, \Delta t$, where c is a constant and E is the field at the surface of the uniformly charged sphere. (The negative sign shows that the charge on the sphere is decreasing.) Show that the charge on the sphere at a time t is given by $Q = Q_0 \exp\left(-ct/4\pi\epsilon_0 a^2\right)$, where a is the radius of the sphere.

20 Circuit Elements

The preceding three chapters were concerned mainly with electrostatics, the behavior of charges at rest. In this chapter we carry over and extend the concepts of those chapters to the study of charges in motion. We begin our study of the circuits in which electric currents flow by examining the behavior of three basic elements frequently found in these circuits, namely, batteries, resistances, and capacitors.

20.1 BATTERIES AS EMF SOURCES

To set charges in motion and to maintain this motion, an energy source is needed. There are many such sources. A short list suffices to show their wide variety: car batteries, dry cells, solar batteries, thermoelectric cells, photoelectric cells, fuel cells. In addition, electric generators which convert mechanical to electric energy are perhaps the most widely used of all energy sources for electric current. The general operation of generators is discussed in later chapters. For now, we consider only those voltage sources (so-called sources of *electromotive force,* or *emf,* sources) which convert chemical to electric energy. Nearly all commonly used batteries are of this type.

Sources of emf furnish energy to electrical circuits

A simple chemical battery (or *cell*) can be obtained by immersing two dissimilar metal rods in a dilute acid solution, as shown in figure 20.1. Most metals dissolve at least slightly in the acid. When the metal dissolves, each atom of the metal leaves at least one electron behind on the rod and enters the solution as a positive ion. The electrode (or rod) from which the ion came is charged negatively by this process. Eventually, it becomes so negative that equal numbers of positive ions dissolve and are attracted back to the negative electrode; then the net number of ions leaving the electrode becomes zero. Since the electrode is negative, a potential difference will exist between it and the solution.

In figure 20.1 it is indicated that the potential difference between the electrode and solution is greater for metal B than for metal A since B dissolves more extensively than A. Since both electrodes are negative, they are both at a lower potential than the solution. However, metal B is at a lower potential than A, and so, going from electrode A to electrode B, one goes from a higher to a lower potential. This potential difference is called the electromotive force (a misnomer), or emf, of the battery. Therefore rod A will be the positive terminal of the battery, and rod B will be the negative terminal.

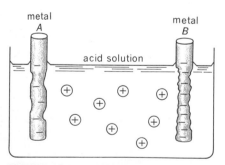

FIGURE 20.1 Positive ions go into the solution, leaving electrons behind on the electrodes. In the case shown, metal B has lost more ions per unit volume than metal A. Which electrode is at the higher potential? (The amounts dissolved are greatly exaggerated.)

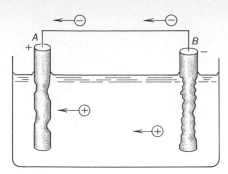

FIGURE 20.2 Will a battery such as this run forever? When and why will it stop?

Since A is at a higher potential than B, an electric field must exist between the two electrodes. Outside the solution the field will be directed from A to B, since A is the high-potential, or positive, electrode. Suppose now a metal wire is placed from electrode A to B as shown in figure 20.2. The electric field in which it is placed would cause positive charges in the wire to move toward B and negative charges to move toward A. In metals, the positive nuclei cannot move, and so only the electrons are affected by the field. Electrons therefore flow through the wire from B to A.

However, as soon as a few electrons leave electrode B, the positive metal ions, held to this electrode by the attraction of the now-missing electrons, escape into the solution. Moreover, since the electrons flowing through the wire congregate on electrode A, they attract some of the positive ions from the solution and cause them to plate out on A. Therefore, as the negative electrons go from B to A through the wire, an equal positive charge is carried from B to A through the solution by the ions. As a result, the net charge on the two electrodes remains the same. However, ions are lost from B and gained by A. In the end, electrode B completely dissolves, and its atoms plate out on electrode A. (Actually, if electrode A becomes covered with B atoms, both electrodes will, in effect, be made of the same metal. The battery voltage will then be zero. In more complicated cells, the reactions at the electrodes are not quite as simple as this.)

The commercially practical batteries are much more complicated than the one just discussed. We cannot discuss these batteries in detail here, but basically, they operate in the way outlined above. In effect, one electrode dissolves, losing chemical energy in the process. Part of this energy is regained by the other electrode as plating occurs on it. The remainder of the energy is expended by the electrons flowing through the wire which connects the electrodes. When no current is being drawn from a battery, it has a characteristic potential difference between its terminals. We call this zero current potential difference the *emf* of the battery and represent it by $\mathcal{E}$.

20.2 ELECTRIC CURRENT

A directed flow of charge is a current

As we saw in the preceding section, the electric field between two electrodes can cause charges to move in a metal wire. Such a *directed,* as contrasted with a random, motion of charge is called an *electric current*. We define the magnitude of the electric current in the following way, referring to figure 20.3: *The current moving in a wire is the charge that passes through a cross section of the wire in unit time.* For example, if a charge ΔQ passes through the cross section A in figure 20.3 in a time Δt, then the current I in the wire is

$$I = \frac{\Delta Q}{\Delta t}$$

The unit of current is coulomb per second, which is called an *ampere*.

In order to assign a direction to the current, we stipulate that it is in the direction of positive charge motion. Or if $\hat{\mathbf{v}}$ is the unit vector in the direction of the average velocity of the directed charge motion, then

$$\mathbf{I} = \frac{\Delta Q}{\Delta t}\hat{\mathbf{v}} \qquad\qquad 20.1$$

(Of course, if $\Delta t \to 0$, $\Delta Q/\Delta t$ can be replaced by dQ/dt.) Notice that if the charge carriers are electrons, as they are in metals, ΔQ is negative, and so $\mathbf{I}$

$$\mathbf{I} = \frac{\Delta Q}{\Delta t}\,\hat{\mathbf{v}}$$

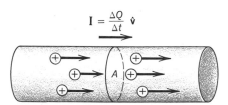

FIGURE 20.3 By definition, the current has the same direction as **v** for positive charges but is opposite to **v** for negative charges.

is in the opposite direction to the direction of charge motion. Since positive charges ordinarily move in the direction of the electric field, the current as we have defined it will be in the direction of the electric field. It will flow through a wire from points of high potential to points of lower potential, i.e., from + to −. *Current flows from high potential to lower potential.*

We can, for many purposes, think of the electrons (or other charge carriers) as being small particles of charge. These charges ordinarily undergo random thermal motion. As a result, in a metal, the electrons are moving, even though no directed current may exist in the metal. However, if an electric field exists in the metal, the motion of the electrons will be slightly biased by the force exerted on them by the field. Therefore the electrons, while undergoing rapid random motion (typically 10^5 m/s), will slowly drift in a particular direction under the action of the electric field (with a typical speed of 0.1 cm/s). This situation is shown (with the drift effect greatly exaggerated) in figure 20.4. When we speak of the velocity of the charges composing a current, it is this drift velocity to which we refer.

It is a simple matter to find a relation between the current in a wire and the properties of the charges within it. Suppose the charges have a drift velocity **v** and the charge on each is q. If we refer to the wire in figure 20.5, the time Δt taken for all the moving charges in the length ΔL of wire to pass through A (where **v** is perpendicular to the area) is just $\Delta L/v$. Since the length ΔL has a volume of $A\,\Delta L$ of wire associated with it, the number of charges passing through A in this time will be

$$n(A\,\Delta L)$$

where n is the number of charges per unit volume. Multiplying this number by the charge on each particle, q, we find the total charge passing through A in time Δt to be

$$\Delta Q = qn(A\,\Delta L)$$

We now make use of the definition of current, *20.1*, to find

$$\mathbf{I} = \frac{\Delta Q}{\Delta t}\hat{\mathbf{v}}$$

or

$$\mathbf{I} = \frac{qnA\,\Delta L}{\Delta L/v}\hat{\mathbf{v}}$$

which is

$$\mathbf{I} = \mathbf{v}qnA \qquad\qquad 20.2$$

Since I depends upon the cross section of the particular wire used, it is often more convenient to speak of the *current density* **J**, which is the current per unit area. We then have, assuming the charge flow to be uniform,

$$\mathbf{J} = \frac{I}{A}\hat{\mathbf{v}} = qn\mathbf{v} \qquad\qquad 20.3$$

As we should expect, the current density is proportional to q, n, and **v**. It should be pointed out once again that if the charges are negative, q will be negative, and **v** will be in a direction opposite to **J**.

20.3 RESISTANCE AND RESISTIVITY

Even in conductors, charges are not perfectly free to move. As indicated in figure 20.4, the charges follow a zigzag path. This path is the result of collisions or other interactions with the stationary portions of the atoms

The electron's drift velocity is much smaller than its instantaneous velocity

FIGURE 20.4 An impressed electric field causes the electron to drift with velocity **v** toward the right. What is the direction of **E**?

Current and current density are proportional to drift velocity

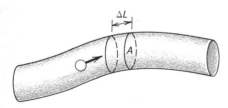

FIGURE 20.5 Although I is constant through each cross section of this wire, J will change from place to place.

constituting the conductor. During these collisions, or localized interactions, the moving charges lose much of the directed energy flow they acquired as a result of the electric field in the conductor. This lost energy almost always appears as heat in the conductor.

From an overall viewpoint, this conversion of electric energy to heat energy can be represented as being due to a friction force on the moving charges. It is often convenient to picture a viscous-type friction force to act on the moving charges even though the actual force is not this simple. Since viscous retarding forces are proportional to the speed of the object, one would expect, approximately, at least, that the drift velocity $\mathbf{v}$ of the charge q would be proportional to the electric force tending to make it move, namely, $\mathbf{E}q$. Of course, $\mathbf{E}$ is the directed electric field in the conductor which is causing the current. We should then write

$$\mathbf{v} \sim \mathbf{E}q$$

If the mobility μ is large, a small $\mathbf{E}$ will cause a large $\mathbf{v}$

If a charge can move very freely in the conductor, the force $\mathbf{E}q$ will give a relatively large $\mathbf{v}$. We say in this case that the *mobility* of the charge is high. The mobility μ is defined as the proportionality constant between $\mathbf{v}$ and $\mathbf{E}$ in this relation, and we have, then,

$$\mathbf{v} = \mu\mathbf{E} \qquad 20.4$$

Notice, as we have said, $\mathbf{v}$ will be large if the charge mobility μ is made large.

This expression for $\mathbf{v}$ may now be used in *20.3* to find the current density and current in a wire. We have upon substitution that

$$\mathbf{J} = \frac{I}{A}\hat{\mathbf{v}} = \mu n q \mathbf{E} \qquad 20.5$$

Definition of conductivity and resistivity

The quantity $\mu n q$ is often called the *conductivity* σ of the material. Its reciprocal, represented by ρ, is referred to as the *resistivity* of the substance. Both ρ and σ are properties of the material alone and do not depend on the size or shape of the conductor. In terms of the resistivity or conductivity, the relation for the current density becomes

$$\mathbf{J} = \frac{I}{A}\hat{\mathbf{v}} = \frac{\mathbf{E}}{\rho} = \sigma\mathbf{E} \qquad 20.6$$

From a practical standpoint we are usually more interested in the relation between the current in a wire and the voltage difference between its two ends. For example, if the electric field $\mathbf{E}$ in the wire segment of length L shown in figure 20.6 is uniform, the potential difference between the two ends of the wire is just

$$V = V_M - V_N = -\int_N^M \mathbf{E} \cdot d\mathbf{s} = EL$$

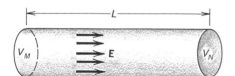

FIGURE 20.6 If $\mathbf{E}$ is uniform, show that $V_M - V_N = EL$.

Solving for E and substituting V/L for E in *20.6*, we find

$$V = I\frac{\rho L}{A}$$

This relation is more frequently written

$$V = IR \qquad 20.7$$

where, in this case of a straight uniform wire, $R = \rho L/A$. The quantity R is called the *resistance* of the wire.

Equation *20.7* can be used to define the resistance of *any* circuit element. If a voltage difference *V* exists between its two ends, and if a current *I* flows through it, the resistance of the element is defined to be

$$R = \frac{V}{I}$$

Clearly, the units of resistance are volts per coulomb per second or volts per ampere. This unit is called the *ohm* (Ω). If we now return to our expression for resistivity ρ in terms of the resistance R of a uniform wire,

$$R = \frac{\rho L}{A} \quad \text{or} \quad \rho = \frac{RA}{L} \qquad \qquad 20.8$$

We see that the units of resistivity are ohm $\cdot$ meters. Typical values for ρ are given in table 20.1.

Equation *20.7*, $V = IR$, is always true provided a steady current can be maintained through a resistance element by a fixed voltage V. The ratio V/I is *defined* to be the resistance of the element. This equation was first found experimentally by George Simon Ohm (1789–1854). He further implied that R is independent of V and I over reasonable ranges of V and I. We call *20.7* Ohm's law. It is correct only in general, however, if we discard Ohm's idea that R does not vary.

If one reviews the steps leading to *20.5* and *20.6*, it will be noticed that several dubious assumptions were made. For example, the assumption that

$$\mathbf{v} = \mu \mathbf{E}$$

is open to question if μ is considered to be a true constant. This assumption is known not to be true for certain materials, and so R in *20.7* will not be constant in those cases. Moreover, heating of resistance materials frequently changes their resistance. As a result, in the sense that Ohm thought R should be perfectly constant, Ohm's law often fails. However, *20.7*, which is one major portion of Ohm's law, is always true if V and I can be reproduced.

Illustration 20.1

Find the drift velocity for the electrons in a silver wire which has a radius of 1.0 mm and carries a current of 2 A.

Reasoning From *20.3* we have

$$I = qnvA$$

TABLE 20.1 Resistivities and Their Temperature Coefficients

Material	Resistivity ρ at 20°C ($\Omega \cdot$ m)	α at 20°C (per °C)
Silver	1.6×10^{-8}	3.8×10^{-3}
Copper	1.7×10^{-8}	3.9×10^{-3}
Aluminum	2.8×10^{-8}	3.9×10^{-3}
Tungsten	5.6×10^{-8}	4.5×10^{-3}
Iron	10×10^{-8}	5.0×10^{-3}
Graphite (carbon)	3500×10^{-8}	-0.5×10^{-3}
Glass	$\approx 10^{+11}$	
Quartz	$> 10^{+16}$	

Since silver is univalent, each atom can be assumed to contribute one electron. We know that one atomic weight of silver, 108 kg, has in it Avogadro's number of atoms, 6×10^{26} atoms/kmol.

The density of silver is found in the "Handbook of Chemistry and Physics" to be $10,500 \text{ kg/m}^3$. Therefore, 108 kg occupies $108/10,500 \approx 0.0103 \text{ m}^3$. Since n is the number of electrons per unit volume, we have, then,

$$n = \frac{6 \times 10^{26}}{0.0103 \text{ m}^3} = 6 \times 10^{28} \text{ per m}^3$$

The cross-sectional area of the wire is πr^2, and this is

$$A = \pi(10^{-3})^2 \approx 3.1 \times 10^{-6} \text{ m}^2$$

Substituting this and the other values in the equation for v, we find

$$v = \frac{2 \text{ C/s}}{(1.6 \times 10^{-19} \text{ C})(6 \times 10^{28} \text{ m}^{-3})(3 \times 10^{-6} \text{ m}^2)}$$

$$v = 7 \times 10^{-5} \text{ m/s}$$

Illustration 20.2

How large is the current density in the wire of the preceding illustration?

Reasoning The current density is defined to be

$$J = \frac{I}{A}$$

In our case $I = 2.0$ A, and $A = 3.1 \times 10^{-6} \text{ m}^2$. Therefore

$$J = 6.4 \times 10^5 \text{ A/m}^2$$

Illustration 20.3

What voltage difference is required to send a current of 2 A through 50 cm of the wire of illustration 20.1?

Reasoning We make use of Ohm's law,

$$V = IR$$

to find the voltage difference. To do this, R must first be found. From 20.8 we have

$$R = \frac{\rho L}{A} = \frac{(1.6 \times 10^{-8} \,\Omega \cdot \text{m})(0.50 \text{ m})}{3.1 \times 10^{-6} \text{ m}^2}$$

which gives $\quad R = 2.6 \times 10^{-3} \,\Omega$

Therefore $\qquad\qquad V = IR$

$$= (2.0 \text{ A})(2.7 \times 10^{-3} \,\Omega)$$

$$= 5.4 \times 10^{-3} \text{ V}$$

20.4 TEMPERATURE VARIATION OF RESISTANCE

It is found experimentally that the resistivity of materials changes with temperature. For metals, the resistance usually increases with increasing temperature. In the case of semiconductors and insulators, however, the resistivity frequently decreases with increasing temperature.

Over restricted temperature ranges the following relation is often found to apply:

Variation of resistivity with temperature

$$\rho = \rho_{\text{ref}}(1 + \alpha \, \Delta t) \qquad\qquad 20.9$$

In this relation ρ is the resistivity at temperature t, and ρ_{ref} is the resistivity at some reference temperature, t_{ref}. Also, $\Delta t = t - t_{\text{ref}}$ and α is an experimental constant called the *temperature coefficient of resistivity*. Typical values of α are given in table 20.1, where the reference temperature is taken to be 20°C.

20.5 CAPACITORS AND CAPACITANCE

When two pieces of metal are connected to the two terminals of a battery, charges move from the battery to the metal. Typical situations are shown in figure 20.7. Notice that in each case positive charge has been provided by the positive terminal, and negative charge is provided by the negative terminal. It is characteristic of batteries that these opposite charges on the two pieces of metal are equal in magnitude.

Devices such as those shown in figure 20.7 are called *capacitors*. We shall see later that they are important since they can be used to store charge. To describe the capabilities of any particular capacitor to hold charge, we

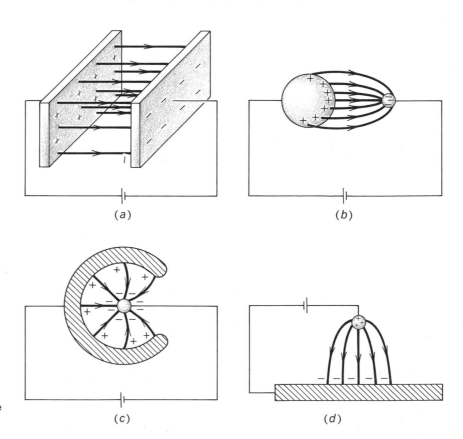

FIGURE 20.7 Four devices which have capacitance.

define the *capacitance* of a capacitor in the following way. If, when the capacitor is connected across a battery of voltage difference V, a charge of magnitude Q appears on either plate, the capacitance C is given by

Defining equation for capacitance

$$C = \frac{Q}{V} \qquad 20.10$$

Notice that it measures the charge on the capacitor for unit voltage across it and is therefore a measure of the capacitor's ability to hold charge. Because $V \sim Q$, the capacitance C depends only on the geometry and construction details of the device, as we shall see.

Capacitances are measured in farads

The unit of capacitance C is coulombs per volt. This unit is designated to be a *farad* (F). We shall soon see that the usual practical capacitances have values of the order of microfarads (μF), where $1\ \mu\text{F} = 10^{-6}$ F. To measure the capacitance of a device, one need merely measure the charge Q which runs onto one plate of the device when a voltage V is connected across it. Let us now compute the capacitances of some simple devices.

20.6 PARALLEL PLATE CAPACITOR

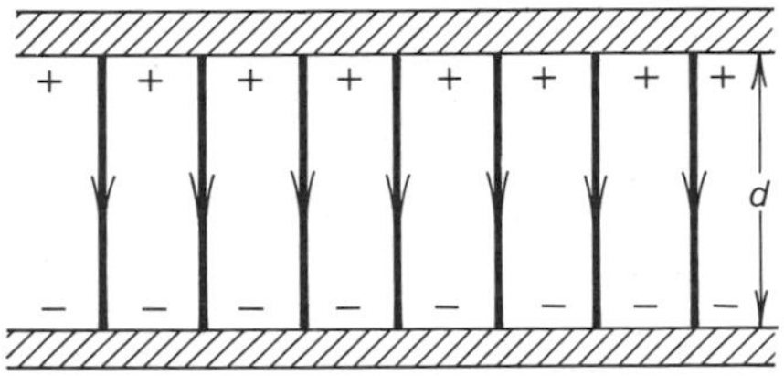

FIGURE 20.8 How does the capacitance of this device depend upon the plate separation *d*?

The computation of the capacitance of a device is made by finding the voltage difference between its two plates for a given charge on one of the plates. Knowing both Q and V, the value of C can then be found from *20.10*. In the case of the parallel plate capacitor shown in cross section in figure 20.8, we proceed as follows.

Suppose a positive charge σ per unit area exists on the upper plate as shown. (Of course, an equal and opposite charge must exist on the lower plate.) Using Gauss' law, as was done in Chap. 18, we find the electric field E between the plates to have a value given by

$$E = \frac{\sigma}{\epsilon_0}$$

Since E is constant between the plates, the definition of potential difference, namely,

$$V = V_{top} - V_{bot} = -\int_{bot}^{top} \mathbf{E} \cdot d\mathbf{s}$$

gives in this case

$$V = E \int ds = Ed$$

where d is the distance between the plates. Substituting for E, we find

$$V = \frac{\sigma d}{\epsilon_0}$$

This is the denominator of *20.10*.

The total charge Q on one plate of the capacitor is σ multiplied by the area A of the upper plate. Therefore

$$Q = \sigma A$$

and

$$V = \frac{\sigma d}{\epsilon_0}$$

from which *20.10* gives the capacitance C to be

Capacitance of a parallel plate capacitor

$$C = \frac{\epsilon_0 A}{d} \qquad 20.11$$

We see, as stated previously, C depends only on the geometry and construction details of the device. Since $\epsilon_0 = 8.85 \times 10^{-12}$ F/m (where do these units come from?), a parallel plate capacitor with 1-m^2 plates separated by 1 mm has a capacitance of only 8.85×10^{-9} F. It is instructive to take a radio capacitor apart and see how a capacitance of 10^{-7} F is achieved in a device smaller than a cigarette.

20.7 CYLINDRICAL CAPACITOR

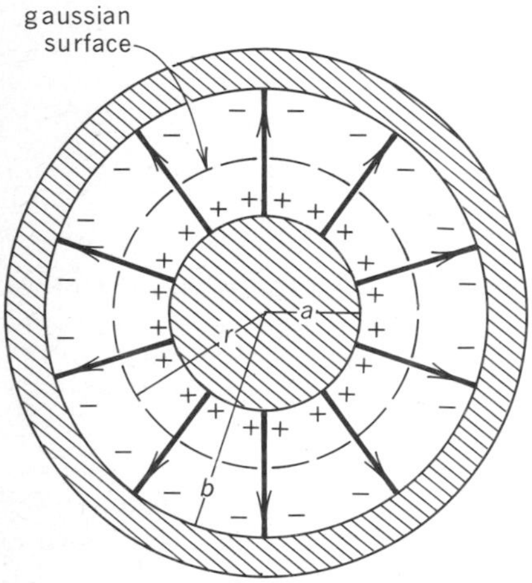

FIGURE 20.9 Can you prove that the charges on the two plates of this cylindrical capacitor are equal in magnitude?

Another configuration sometimes of importance as a capacitor is that shown in cross section in figure 20.9. It consists of two concentric metal cylinders. This device might in practice be anything from a coaxial cable to a configuration of elements in a vacuum tube. Or it might be used directly as a capacitor. In any event, we compute the capacitance by first assuming a definite charge on the inner cylinder. Let us call it λ per unit length.

To find the electric field between the cylinders we make use of Gauss' law. Taking a closed gaussian cylinder of radius r and length d, we have, from Gauss' law,

$$\int_S \mathbf{E} \cdot d\mathbf{A} = \frac{1}{\epsilon_0} \sum q$$

There is no flux through the ends of the cylinder, and **E** is constant on and perpendicular to the side surface. As a result,

$$E(2\pi r d) = \frac{1}{\epsilon_0} \lambda d$$

where the charge inside the gaussian surface is λd. Solving, we find

$$E = \frac{\lambda}{2\pi\epsilon_0 r}$$

We now use this value of E to find the potential difference between the cylinders. By definition,

$$V = V_{\text{outer}} - V_{\text{inner}} = -\int_a^b \mathbf{E} \cdot d\mathbf{r}$$

Since **E** and $d\mathbf{r}$ are parallel, we have

$$V = -\int_a^b E\, dr$$

Placing in the value found for E, this becomes

$$V = -\frac{\lambda}{2\pi\epsilon_0} \int_a^b \frac{dr}{r}$$

This then gives

$$V = -\frac{\lambda}{2\pi\epsilon_0} \ln \frac{b}{a} \qquad 20.12$$

It is seen that V is given here as negative. Since this is defined by our notation to be

$$V = V_{\text{outer}} - V_{\text{inner}}$$

and since the inner cylinder is considered positive, and therefore at the higher potential, this sign is correct. However, in the definition of capacitance we are not concerned with the polarity of the device. As a result we use the absolute value of V when computing capacitance. We then find

$$C = \frac{Q}{V} = \frac{\lambda L}{(\lambda/2\pi\epsilon_0) \ln (b/a)}$$

where it is assumed the capacitor has a length L.

The capacitance per unit length of the device is given by dividing C by L:

Capacitance of a cylindrical capacitor

$$\frac{C}{L} = \frac{2\pi\epsilon_0}{\ln (b/a)} \qquad 20.13$$

Here, as always, C depends only on geometrical and construction details.

20.8 SIMPLE SERIES CIRCUIT

Now that we understand the meaning of the three circuit elements, the battery, resistor, and capacitor, we are prepared to discuss what happens when they are connected together. We call such a group of connected elements a *circuit*. One of the simplest, but quite important, circuits consists of a battery and resistor connected together. As an example, consider the incandescent light bulb and battery connected as shown in figure 20.10. The curly wire inside the bulb is simply a wire of quite high resistance, much higher resistance than that of a lamp-cord wire, for example. As current flows through it, the high resistance causes a great deal of heat to be generated and the wire becomes white hot. This is the source of the light given off by the bulb.

It is convenient when dealing with electric circuits to have a schematic way for representing them. The circuit in figure 20.10(a) is commonly represented by a diagram such as that shown in (b). You will recognize the battery symbol there (—| |—), and you will recall that the long side of the symbol represents the positive terminal of the battery. Since the connecting wires from the battery to the filament have very small resistances in comparison to the resistance of the filament (perhaps 0.01 Ω in comparison to 200 Ω), we ignore their resistance. They are represented by straight lines on the diagram.

Resistance symbol is —\/\/\/—

Finally, the filament is represented by a symbol used for all resistances, —\/\/\/\—. You should study parts (a) and (b) of the figure so that you understand fully what the diagram in (b) represents. It should also be clear that, since positive charges move from the positive to negative terminal of a battery, current flows around the circuit in the direction shown.

Referring to either part (a) or (b) of figure 20.10, we see that point P is at the highest potential and that point N is at the lowest. They are the two terminals of the battery. To see how the electrical potential level varies from point to point in the circuit, refer to part (c) of the figure. There we have plotted the potential vs. position in the circuit. Since the zero of potential is

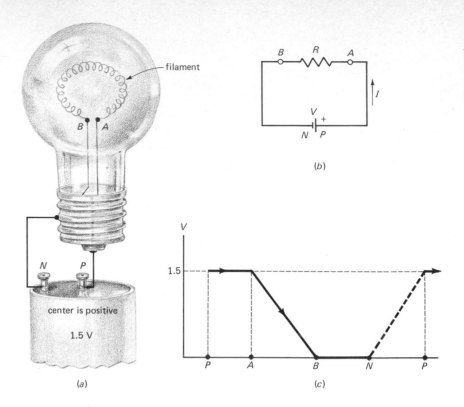

FIGURE 20.10 A simple series circuit.

We usually assume IR = 0 for connecting wires since R → 0 for them

Charges flowing through a resistor lose energy

ours to choose, we shall take the negative battery terminal (N) to be at zero level. If this is a 1.5-V battery, then point P will be at $+1.5$ V as indicated. Let us start at point P and go around the circuit in a counterclockwise direction.

Consider the wire from P to A. We assume its resistance to be so small that IR for it will be essentially zero. Then, $V = IR$ tells us that no appreciable potential drop occurs in it. It is for this reason that the potential (or electrical level) at P and A are the same in part (c) of figure 20.10. This approximation is usually made: there is no potential drop (IR drop) in the connecting wires of a circuit.

Coming now to section AB of the circuit, the light bulb filament, it is here that the major IR drop (potential drop) occurs. This is shown as the region AB in the graph of figure 20.10. Finally we go back to the negative terminal of the battery via section BN of the circuit. Since we are assuming the connecting wires to have negligible resistance, no IR drop occurs in this section, as indicated on the graph.

The very last portion of our circuit, from N to P, occurs inside the battery itself. Since the internal workings of batteries are quite complicated, we simply show a dashed-line potential rise up to the electrical level of the positive terminal at P. In any event, as we pass from N to P within the battery, we climb a potential hill to achieve the original starting level at P. Let us now circle the circuit, viewing the situation from the standpoint of a positive charge, one of the charges which make up the current.

Originally at P, the positive charge has considerable energy since it is at a high potential. It retains this energy as it moves from P to A. But in falling down the potential hill from A to B it loses energy. This lost energy is changed to heat in this case and gives rise to the white-hot filament. Once the charge reaches B, it is at the bottom level of the circuit and simply stays at this level until it reaches the negative battery terminal N. The battery then lifts the charge from the low electrical level at N to its high level at P. Now the charge has its original high potential energy and can once again circle the circuit. Notice that the energy supplied to the charges by the battery is lost as the charges fall down through the electrical potential hill in the resistor.

Since the battery supplies energy to the circuit, the energy needed to light the bulb, it continually loses energy. Eventually the chemicals within the battery which supply this energy become depleted. When that happens, we say that the battery has run down. No longer is it capable of lifting charge from a low to high level. It can no longer supply energy to the circuit.

20.9 ELECTRIC WORK AND POWER

As we saw in the previous section, the battery furnishes energy to the charges which flow in a circuit. The energy is then lost by the charges as they flow around the circuit—lost as heat in the resistor in the circuit we have considered. Let us now find the quantitative relations which apply to energy gain and loss in circuits of all types.

It is a simple matter to write down the energy gain or loss of a charge as it moves through a potential rise or drop of V volts. Because potential difference is the energy gained (or lost) by a unit positive charge as it moves through that potential difference, we know at once that qV is the energy gained (or lost) as the charge q rises (or falls) through a potential difference V. If we refer to figure 20.11, we see that a charge q will gain energy in the amount qV_0 as the battery lifts it from its negative to its positive terminal. The charge then flows from P to A to B back to N. In the process, it loses energy qV_0 as it falls through the voltage drop $IR_0 = V_0$ in the resistor. The energy furnished by the battery is dissipated in the resistor.

Usually we are most interested in power consumption in electric circuits. You will recall that power is defined to be work done (or energy gained or lost) per unit time. We already know that energy gain or loss in electricity is simply Vq if we are considering a charge q rising or falling between two points with potential difference V. For example, if a stream of charges totaling Δq flows through the resistor in figure 20.11 in time Δt, then the energy lost by these charges in this time will be $V_0 \Delta q$. The power loss will be energy $\div$ time or

$$\text{Power} = V_0 \frac{\Delta q}{\Delta t}$$

But $\Delta q/\Delta t$ is by definition the current I flowing through the resistor. Therefore the power loss in it is simply

$$\text{Power} = V_0 I$$

Similarly, the energy furnished to the charges totaling Δq as they flow through the battery and are lifted V_0 in potential is $V_0 \Delta q$. It takes a time Δt for these charges to flow through the battery, and so the power furnished by the battery is $V_0 \Delta q/\Delta t$. But since $\Delta q/\Delta t$ is simply the current I, we have here too that power $= V_0 I$.

This same reasoning can be applied to any situation where charges move through a potential difference V. If the current flowing through a

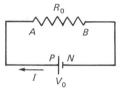

FIGURE 20.11 What happens to the power furnished to the circuit by the battery?

circuit element is I, and if the potential difference between the terminals of the element is V, then the power loss or gain in the element is given by

Electric power

$$\text{Power} = VI \qquad 20.14$$

If the charges move from + to − through the element (i.e., if they fall from a high potential to a lower potential), then there will be power lost in the element. Charges always give rise to power loss in resistors; they always go from high to low potential as they flow through a resistor. *Current always flows downhill electrically in resistors.*

In our study of mechanics, we gave the name watt (W) to the SI unit of power. We now see from *20.13* that a watt is a volt · ampere. The watt is a unit well known to you. We are accustomed to using 60-W bulbs and 1000-W heaters, for example. Usually a bulb or appliance will have marked upon it two pieces of data: (1) the potential difference which should be placed across its terminals if the proper current is to flow through it and (2) the power which the device will dissipate when so used.

How appliances are labeled

For example, a bulb might be labeled 120 V/60 W. This tells us the bulb is to be operated on 120 V. Under these conditions, it will dissipate 60 W. A 100-W bulb will be brighter than a 60-W bulb since it dissipates and gives off more energy each second. Notice also, since $I = V/R$, that if this same bulb is operated on 100 V rather than 120, less current will flow through it and it will not glow as brightly; it will dissipate less than its rated power. More will be said about this later. Let us now work a few examples so as to solidify our understanding of power.

Illustration 20.4

A current of 0.50 A flows through a 200-Ω resistor. How much power is lost in the resistor?

Reasoning We apply *20.14*, which is

Power loss in a resistor is I^2R

$$\text{Power} = VI = I^2R$$

where the latter form is found by use of Ohm's law, $V = IR$. Placing in the values, we find

$$\text{Power} = 50\ \text{W}$$

As indicated in our discussion of resistance earlier in this chapter, this lost power will appear as heat in the resistor. In this case 50 J of energy is lost each second, and so 50/4.185, or about 12 cal of heat, is generated each second.

Illustration 20.5

A bulb rated 120 V/90 W is operated from a 120-V power source. Find the current flowing through it and its resistance.

Reasoning Since power = VI we can find I immediately. It is

$$I = \frac{\text{power}}{V} = \frac{90\ \text{W}}{120\ \text{V}}$$

$$= \tfrac{3}{4}\ \text{A}$$

Since the potential drop across the bulb is 120 V and since the current through it is $\frac{3}{4}$ A, Ohm's law tells us

$$R = \frac{V}{I} = \frac{120 \text{ V}}{0.75 \text{ A}}$$

$$= 160 \text{ }\Omega$$

20.10 KIRCHHOFF'S LOOP RULE

We can gain the use of a very powerful tool for circuit analysis if we examine the circuit of figure 20.12. Before doing that, though, let us discuss two very important features of the circuit. First, as we have already pointed out, the current always flows through a resistor from the high potential end to the low potential end. In the present case, end A is V_0 volts higher than end B so the current I flows from A to B through the resistor. The magnitude of the current is, of course, related to the resistance R_0 through Ohm's law, $V_0 = IR_0$.

Second, a fundamental characteristic of a *series* circuit, such as this, is that the charges traveling around the circuit must all follow the same path. The charge moves through P, A, B, N, and back to P with no other path possible. Moreover, even though the valence electrons in a metal are free to move, they always distribute throughout the metal in such a way that there are no excess charges in the wire. The valence electrons exactly balance the positive ions. Therefore, when a quantity of charge enters one end of a wire, an equal charge exits from the other end. The situation is much like water in a filled pipe. When a little water enters one end, the exact same amount must exit immediately from the other. In that sense the current in a wire is analogous to the flow of fluid in a pipe.

With this in mind, if we refer again to figure 20.12, a current I flows out of the battery into the wire at P. This same current must flow out at A and into the resistor. There, too, the charges cannot accumulate and so an equal current must exit at B. Indeed, the current flowing past all points in this circuit, P, A, B, N, and again P, must have the value I. Charge cannot be compressed in a wire, resistor, or battery and so the current flowing into these elements must equal the current flowing out. As a result, *the current everywhere in a series circuit must be the same.*

In a series circuit, the current is everywhere the same

We can obtain another very important result if we monitor the potential drops and rises around a circuit. Suppose we start at point P in the circuit of figure 20.12 and follow the circuit from P to A and back through B to P. Since the electrostatic field is conservative, it follows that the potential (or

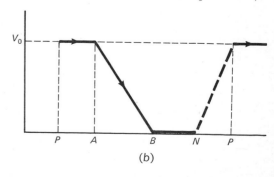

FIGURE 20.12 Why must the sum of the voltage changes around the circuit be zero?

electrical level) at P must have a unique value. Hence, when we add up the potential rises and drops we encounter in going along the path $PABN$ and again P, they must bring us back to the same electrical level at which we started. That is to say, the sum of the potential drops must exactly equal the sum of the potential rises through which we move as we trace a path around a complete circuit. This is, of course, exactly what we found when we drew the graph in part (b) of figure 20.12. Point P was at the same electrical level no matter how we arrived at it.

Let us now apply this result in detail to the circuit of figure 20.12. In so doing, let us take voltage rises to be positive and voltage drops to be negative. Starting at P and going clockwise around the circuit, there is no rise or drop in the wire PA. (The reason for this: $V = IR$ and, since R for this connecting wire is assumed zero, there is no voltage difference between its ends.) As we go through the resistance in the direction of the current, from A to B, we encounter a voltage drop of magnitude IR_0. (Recall that the current always goes from high to low potential through a resistor.) No electrical level change occurs as we pass along the resistanceless wire BN. Finally, we encounter a voltage rise of V_0 as we go from N to P through the battery. (Remember, the potential difference between the battery terminals is the voltage of the battery, V_0 in this case.) Now let us add all these voltage changes we encountered in our passage around the circuit. Calling rises positive and drops negative, we find $-IR_0 + V_0$. But, as explained earlier, these changes must add to zero since we started at P and returned to this same point. Therefore

$$-IR_0 + V_0 = 0$$

which gives

$$V_0 = IR_0$$

If we examine this result, we see it is simply a statement of Ohm's law. Clearly, the battery potential difference V_0 is applied directly to the two ends of the resistance R_0. Hence Ohm's law tells us at once that $V_0 = IR_0$, as we have just found using more involved reasoning. Although this more involved way of examining the circuit seems pointless in the present instance, we shall soon see that it is the most fruitful way of investigating more complicated situations. For that reason, we shall now summarize our procedure in what is *Kirchhoff's loop rule* called *Kirchhoff's loop* (or *circuit*) *rule*.

If voltage rises are taken positive and drops are taken negative, the algebraic sum of all the voltage rises and drops encountered as one traces a closed circuit must be zero.

This is a very powerful result which we shall use repeatedly in circuit analysis.

20.11 *RC* TIME CONSTANT

We are now prepared to discuss circuits which contain capacitance as well as resistance. Before applying Kirchhoff's loop rule to the circuit of figure 20.13, let us discuss its behavior qualitatively. When the switch is open as indicated, we assume the capacitor to be uncharged. (Note the symbol used for a *Symbol for a capacitor is* —||— capacitor: —||— or —|(—. It is meant to suggest the two plates of a parallel plate capacitor.) As soon as the switch is closed, charge will run from the battery to the capacitor so as to charge its plates. The direction of the

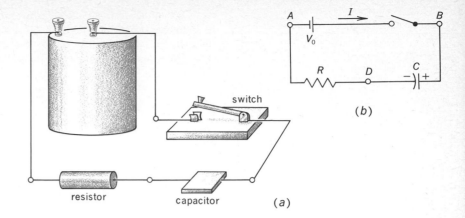

FIGURE 20.13 An *RC* circuit. When *I* approaches zero, what will be the voltages across the resistor and capacitor? Note the symbol for the capacitor.

resistor   capacitor   (a)

(b)

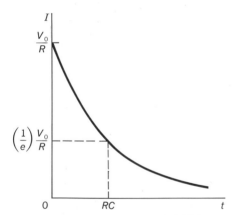

FIGURE 20.14 Notice that the initial current in the circuit is the same as though the capacitor were not there. Why?

current will, as usual, be out of the positive terminal of the battery, as indicated.

Soon, of course, the current must stop since the capacitor will become as fully charged as the battery V_0 can charge it. Note in this regard that charge does not flow through the capacitor since there is no way for the charge to jump from one plate to the other. However, as positive charges run onto the right-hand plate, an equal number leave the plate on the left, so that the current *seems* to flow through the capacitor. But in the end this flow must stop since only a limited charge can be placed on the capacitor plates. In any event, we see that the current will be fairly large when the switch is first closed; it will then decrease to zero as the capacitor becomes fully charged. If we were to take a guess as to how the current varies with time, we might draw a graph like the one shown in figure 20.14. Let us now examine the circuit quantitatively.

To analyze this circuit, we write down Kirchhoff's circuit rule for it, assuming the switch to be closed of course. We start at point *A* and move round the circuit through points *B*, *D*, and back to *A*. Going from *A* to *B*, we go from the negative to the positive side of the battery, and so this is a voltage rise of $+V_0$. In moving from *B* to *D* we go through the capacitor from its positive to its negative plate. Since the positive plate is always at the higher potential, we shall drop in potential by an amount $-V_c$ as we go through the capacitor. Finally, in going from *D* to *A*, we pass through the resistor. This is a voltage drop $-V_R$. It is a drop since we are going in the same direction as the current, and the current always goes from high to low potential through a resistor.

According to Kirchhoff's rule, the sum of these voltage changes must be zero. That is to say,

$$+V_0 - V_c - V_R = 0$$

We know from *20.10*, however, that

$$Q = CV_c \qquad \text{and so} \qquad V_c = \frac{Q}{C}$$

In addition, $V_R = IR$. Therefore this equation becomes

$$+V_0 - \frac{Q}{C} - IR = 0 \qquad\qquad 20.15$$

We can gain considerable information from this equation, and we shall do so before proceeding further. Notice that when the switch is first closed (that is, $t = 0$) there will as yet be no charge on the capacitor plates. Q will be zero. The equation then says $V_0 - IR = 0$, which tells us

$$I = \frac{V_0}{R} \qquad \text{at } t = 0$$

This is just Ohm's law for the simple series circuit containing a resistance. The initially uncharged capacitor does not impede the flow of current at all. The circuit behaves just as though the capacitor were a resistanceless connecting wire.

In the end, however, the capacitor will become fully charged and so the

Superconductors

Ordinary conductors of electricity become better conductors as the temperature is lowered. This was already known in 1835, from the measurements of Heinrich Lentz. There was some speculation in later years that perhaps the resistance of a metal would actually be zero at absolute zero, i.e., at $-273°C$, or 0 K. However, no one was able to carry resistance measurements to very low temperatures until much later than 1835, because no means existed for cooling objects to such low temperatures.

The attainment of very low temperatures received considerable impetus in 1883, when Wroblewski and Olzewski succeeded in liquefying air. With liquid air as a cooling agent, it was possible to carry out experiments at the boiling point of liquid oxygen ($-183°C$) and liquid nitrogen ($-196°C$). Soon thereafter, in 1898, James Dewar (1842–1923)

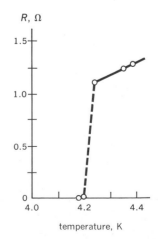

A lead ball floats in the intense magnetic field of two superconducting rings. (*Courtesy 500 Incorporated, Arthur D. Little, Inc.*)

current will stop. At that time, $t \to \infty$, we have $I = 0$ and so 20.15 becomes

$$V_0 - \frac{Q}{C} = 0 \qquad \text{or} \qquad Q = CV_0$$

But this is simply 20.10, the charge-voltage relation for a capacitor. It is easily seen why this result is obtained if we notice that, when $I = 0$, no IR drop occurs across the resistor in figure 20.13. The voltage drop across the capacitor must then be equal to the battery voltage V_0. As a result, 20.10 tells us $Q = CV_0$ as we have also found from 20.15.

To proceed further, we must use 20.15 to obtain a relation between current in the circuit and time. Unfortunately, time does not appear in the equation and, instead, the varying charge on the capacitor Q complicates the equation. We can eliminate Q in the following way.

---

succeeded in liquefying hydrogen, which has a boiling point of $-263°C$, or 10 K. There remained only one other gas which had not been liquefied, helium. The liquifaction of helium was finally achieved in 1908 by Heike Kamerlingh Onnes (1853–1926), and it was found that liquid helium boiled at 4.2 K.

Using his liquified helium to achieve lower temperatures than had been possible before, Kamerlingh Onnes set out to measure the resistance of metals at very low temperatures. The measurement technique was quite simple in principle, although quite complex in practice, because of the difficulty of maintaining the temperature constant. Since resistance R is defined by Ohm's law, Onnes needed only to measure the voltage drop V across a given cylinder of the metal when a definite current I flowed through it. Then R = V/I. For his measurements, Onnes used pure mercury as his metal. (It solidifies at $-39°C$, and so it is a solid.) When he carried out the requisite measurements in 1911, Onnes obtained the astonishing data shown in the graph.

Although the metal was steadily approaching a very low resistance as the temperature was being lowered, at 4.2+ degrees absolute the resistance suddenly decreased to zero. The mercury had become what we now refer to as a superconductor. Subsequent experiments indicated that a current would flow essentially forever in a ring of the superconductor, even though no source of emf was maintained. As far as experiment has been able to tell, the resistance of the superconductor is essentially zero.

Kamerlingh Onnes succeeded later in showing that lead, tin, and indium also become superconductors at 7.2, 3.7, and 3.4 K, respectively. Since that time, many other superconductors have been found. Most interesting has been the fact that certain alloys become superconductors at rather high temperatures. For example, the compound Nb_3Sn becomes superconducting at 17.9 K. Since no heat is generated when a current flows through a metal having no resistance, superconductors are now being used where this property is of great importance, for example, in large electromagnets. The theoretical reasons for superconductivity in terms of atomic structure have been clarified only recently, mostly since 1950.

Notice that Q, the charge on the capacitor, is furnished by the current flowing onto the capacitor plate. Hence the two must be related. Indeed, from the definition of current, the charge flowing out the end of the wire onto the positive capacitor plate is simply $\Delta Q/\Delta t$. Therefore we can write

$$I = \frac{\Delta Q}{\Delta t} \quad \text{or} \quad I = \frac{dQ}{dt}$$

Clearly, we could now eliminate dQ/dt from 20.15 but, unfortunately, the equation does not contain this quantity. Instead, it contains Q.

To remedy this situation we simply take the time derivative of the whole equation 20.15. In doing so we notice that the battery voltage V_0, the resistance R, and the capacitance C do not vary with time. They are constants. As a result

$$\frac{d}{dt}\left(V_0 - \frac{1}{C}Q - RI = 0\right)$$

becomes

$$0 - \frac{1}{C}\frac{dQ}{dt} - R\frac{dI}{dt} = 0$$

Or, upon replacing dQ/dt by I we find, after rerranging,

$$I = -RC\frac{dI}{dt}$$

Notice that this equation relates the two quantities of interest, I and t. It is a simple differential equation which you will find to be of nearly trivial difficulty when you encounter it in your mathematics classes. To solve it, we recall that dI/dt is simply the limiting form of $\Delta I/\Delta t$, so we may multiply the equation by dt to obtain

$$dt = -RC\frac{dI}{I}$$

We can integrate both sides of this equation from time $= 0$ when $I = I_0$ to some later time t when the current is I. Then the above equation gives

$$\int_0^t dt = -RC\int_{I_0}^I \frac{dI}{I}$$

which becomes

$$t = -RC(\ln I - \ln I_0)$$

or

$$\ln \frac{I}{I_0} = \frac{-t}{RC}$$

This can be put in nicer form by taking antilogs to give

$$\frac{I}{I_0} = e^{-t/RC}$$

Rearranging yields the final relation between current and time in a resistance-capacitance circuit while the capacitor is charging:

Charging current in an RC circuit

$$I = I_0 e^{-t/RC} = \frac{V_0}{R}e^{-t/RC} \qquad 20.16$$

where, in the latter form, I_0 has been replaced by V_0/R, the value we found previously for the current when $t = 0$. Notice that the equation gives $I = V_0/R$ when $t = 0$ and $I = 0$ when $t \to \infty$.

As we guessed, the current drops to zero asymptotically as the capacitor becomes fully charged. At the very first instant after closing the switch, there is yet no charge on the capacitor. The current is limited entirely by the resistor at that time, and so the initial current is merely V_0/R.

A rough measure of the time taken to charge the capacitor is given by a characteristic time, called the *time constant* of the circuit. It is the time taken for the current to drop to $1/e$ of its initial value; i.e.,

The time constant is RC

$$e^{-t/RC} \to e^{-1} \approx 0.37$$

This condition exists when $t = RC$. The quantity RC, a time, is designated the time constant. For a typical case $C = 10^{-2}\,\mu\text{F} = 10^{-8}\,\text{F}$ and $R = 10^6\,\Omega$; so the time constant would be $10^{-2}\,\text{s}$. This means that, in a circuit with these values for R and C, the current will drop to $1/e$ (about $\tfrac{1}{3}$) of its original value in a time of $10^{-2}\,\text{s}$. As we should expect, increasing R and C increases the time taken to charge the capacitor.

We can easily find how the charge on the capacitor changes with time. To do so, we use *20.15*. Substitution in it of the value we have found for I (*20.16*) yields, after rearranging,

During charging, a capacitor's charge increases as $1 - e^{-t/RC}$

$$Q = CV_0(1 - e^{-t/RC}) \qquad\qquad 20.17$$

This equation for Q is plotted in figure 20.15. Notice that when $t \to \infty$, $Q \to CV_0$, as it should. At a time equal to the time constant, the capacitor has acquired about 0.63 of its final charge.

20.12 DISCHARGE OF A CAPACITOR

If a fully charged capacitor with initial voltage V_0 is arranged in the circuit of figure 20.16, it can be discharged through the resistor by closing the switch. To find the current and charge as a function of time, we proceed exactly as we did in the preceding section. Starting at point A and proceeding to points B, D, and back to A, we write Kirchhoff's circuit rule. Since we are going through R in the direction of the current, there will be a voltage drop in it,

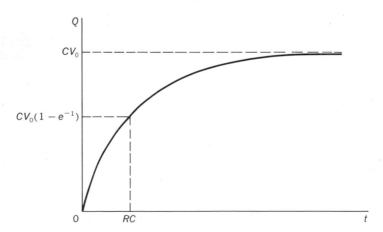

FIGURE 20.15 The charging curve for a capacitor.

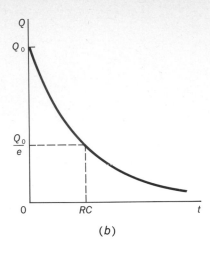

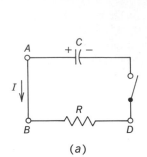

FIGURE 20.16   If R is 1 MΩ and C is 1 μF, what is the value of the time constant?

(a)

(b)

$-V_R$. Going from D to A, we pass from the negative to the positive plate of the capacitor, and this is a voltage rise, $+V_C$.

The sum of these voltage drops and rises must be zero around the closed loop. Therefore

$$V_C - V_R = 0$$

Placing in the values for these quantities, we find

$$\frac{Q}{C} - IR = 0 \qquad\qquad 20.18$$

which becomes, after differentiation,

$$\frac{1}{C}\frac{dQ}{dt} - R\frac{dI}{dt} = 0$$

In this case the current is flowing out of the capacitor, and so the rate of change of charge on the capacitor will be negative. As a result,

$$\frac{dQ}{dt} = -I$$

Substitution of this value in the preceding equation gives

$$-\frac{1}{C}I - R\frac{dI}{dt} = 0$$

After rearranging, this becomes

$$\frac{dI}{I} = -\frac{1}{RC}\,dt$$

As in the preceding section, this equation can be integrated to give

$$I = I_0 e^{-t/RC} \qquad\qquad 20.19$$

The charging and discharging currents obey the same relation

where $I_0 = Q_0/RC$ is the initial current which flows. The current therefore decreases exponentially much like the result found in figure 20.14. Again, the time constant is RC.

To find Q as a function of time, we notice from *20.18* that

During discharge, Q decreases exponentially

$$Q = RCI \qquad\qquad 20.20$$

so that

$$Q = I_0 RC e^{-t/RC}$$

Therefore, in this case, the charge decreases exponentially as a function of time. This variation is shown in part (*b*) of figure 20.16. Here, too, the time constant is *RC*.

20.13 COMBINATIONS OF CAPACITORS

In practical applications one sometimes finds it convenient to combine capacitors. Two different methods are indicated in figure 20.17. In (*a*) is shown what is called a *parallel* combination between *A* and *B* while in (*b*) the capacitors are said to be in series between *A* and *B*. In a series combination, as shown in (*b*), there is only one path by which one can go through the capacitors to get from *A* to *B*. But in a parallel combination, as in (*a*), it is possible to get from *A* to *B* by more than one path. For example, you can go from *A* to *B* via C_1 or C_2, or C_3.

We wish now to find what single capacitor has the same capacitance as the combination in each case. The single capacitor is then called the *equivalent capacitance* of the combination and is denoted C_{eq}. In part (*a*) each capacitor has the same voltage drop *V* across it. The charges on them are

$$Q_1 = C_1 V \qquad Q_2 = C_2 V \qquad Q_3 = C_3 V$$

As a result, the charge which is stored on their upper plates and is available at point *A* is

$$Q = Q_1 + Q_2 + Q_3$$
or
$$Q = V(C_1 + C_2 + C_3)$$

Looking now at the lower equivalent capacitor, we see that it should have the same charge *Q* for the voltage *V* if it is to act like the combination. As a result, we can write for it

$$Q = V C_{eq}$$

Comparing these two equations, we see that

Capacitances in parallel add directly

$$C_{eq} = C_1 + C_2 + C_3 \qquad\qquad 20.21$$

for capacitors in parallel. Extension to several capacitors is obvious.

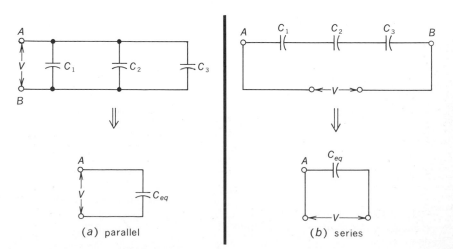

FIGURE 20.17 For the same voltage difference *V*, the equivalent capacitance furnishes the same charge at *A* as the combination does.

(*a*) parallel

(*b*) series

For the series combination shown in figure 20.17(b), the charges on C_1, C_2, and C_3 must all be the same. This follows from the fact that if the left plate of C_1 has a charge Q, so too must the right plate. But the charge on the right plate of C_1 must come from the left plate of C_2, and so C_2 must have the same charge. Similarly for C_3. If C_{eq} is to be equivalent, it too must have the same charge Q for the same voltage V.

The voltage V is the sum of the voltage drops across C_1, C_2, and C_3. Therefore

$$V = V_1 + V_2 + V_3$$

and so

$$V = \frac{Q}{C_1} + \frac{Q}{C_2} + \frac{Q}{C_3}$$

For the equivalent capacitor we have

$$V = \frac{Q}{C_{eq}}$$

Comparing these relations, we see that

Capacitances in series are combined by adding their reciprocals

$$\frac{1}{C_{eq}} = \frac{1}{C_1} + \frac{1}{C_2} + \frac{1}{C_3} \qquad 20.22$$

for capacitors in series. Additional capacitors would merely add additional terms to the equation.

Illustration 20.6
Find the equivalent capacitance of the device shown in figure 20.18(a).

Reasoning Successive stages in the reduction of this combination are shown in the figure. Notice that parallel and series combinations are reduced successively until the single equivalent capacitance is reached. As indicated, the combination is equivalent to a 2-μF capacitor.

20.14 ENERGY STORED IN A CAPACITOR

By the definition of potential difference, the work done in carrying a positive charge q up through a potential rise of V is merely qV. For example, if a

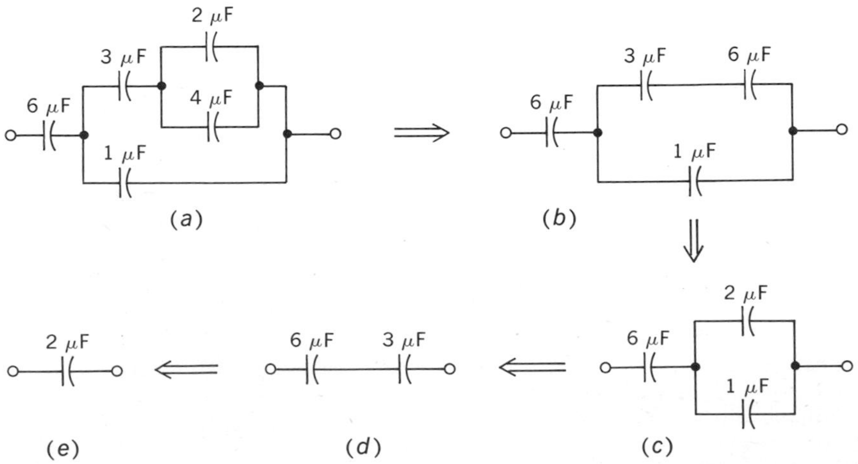

FIGURE 20.18 Successive steps in reducing a capacitance combination to its equivalent.

positive charge q is carried from the negative plate of a capacitor to the positive plate, work equal to qV must be done on it, where V is the difference in potential between the plates. In the process of carrying the charge from the negative to positive plate, it has been given a potential energy equal to the work done. It is a simple matter to extend this idea to find the potential energy of a charged capacitor since it is simply the work done in charging the capacitor.

Consider the capacitor to be initially uncharged. We can think of charging it by carrying small positive charge increments from one of the plates (which will be left negative by this process) to the other plate (which is thereby made positive). At a certain time during this process the voltage difference between the plates will be V. When the incremental charge dq is now lifted through this potential difference, a work dW is done, and this work is just

$$dW = V\,dq$$

Now we know that for a capacitor which has a charge q on it and a capacitance C,

$$q = CV$$

Using this relation to replace V in the preceding equation, we have

$$dW = \frac{1}{C}q\,dq$$

The total work done in charging the capacitor to a final charge Q is obtained by adding all the dW for all the charge increments dq. Therefore

$$\text{Work} = \int dW = \int_0^Q \frac{1}{C}q\,dq$$

This expression for the work is readily integrated since C is constant. We then find

$$\text{Work} = \frac{1}{2}\frac{Q^2}{C}$$

Therefore the potential energy stored in the capacitor is

Energy stored in a capacitor

$$\text{Energy} = \frac{1}{2}\frac{Q^2}{C} = \tfrac{1}{2}QV = \tfrac{1}{2}CV^2 \qquad\qquad 20.23$$

where the latter forms are obtained by use of $Q = CV$. Of course, the energy will be in joules.

In the mid-1800s, when the theoretical basis for electricity was being laid, men such as Faraday and Maxwell still believed in the "ether." This elastic material was thought to fill all of space which we now know as vacuum. They considered the energy stored in a capacitor to be elastic energy stored in the "ether" between the capacitor plates. Of course, we know now that their mental picture was wrong. In spite of this, it proves convenient (but not necessary) to consider the capacitor's energy to be stored in the region between its plates. Moreover, Maxwell proved that the potential energy stored in any electrostatic system could be considered stored in the region of the electric field. This, too, turns out to be a convenient representation.

In the interest of convenience it is customary (and certainly harmless) to consider an electric field to have energy stored in it. To find this energy in useful terms, let us examine the energy stored in a parallel plate capacitor as given in 20.23. We had that the energy is $Q^2/2C$. But $Q = \sigma A$ while $C = \epsilon_0 A/d$, where A is the plate area, d is the plate separation, and σ is the charge per unit area. Therefore

$$\text{Energy} = \frac{\sigma^2 A D}{2\epsilon_0}$$

Notice, however, that the electric field between the capacitor plates E is simply σ/ϵ_0, a fact we have used repeatedly. Replacing σ in terms of E yields

$$\frac{\text{Energy}}{Ad} = \tfrac{1}{2}\epsilon_0 E^2$$

But Ad is simply the volume in which the electric field exists, the volume between the capacitor plates. Hence we infer that the energy stored in unit volume of an electric field E is given by

Energy density in an electric field is $\tfrac{1}{2}\epsilon_0 E^2$

$$\frac{\text{Energy}}{\text{Volume}} = \tfrac{1}{2}\epsilon_0 E^2 \qquad\qquad 20.24$$

Although we have arrived at this result by consideration of a very simple case, Maxwell showed it to be consistent with experiment in all cases. We shall, then, whenever convenient, assume that energy in the amount $\tfrac{1}{2}\epsilon_0 E^2$ is stored in unit volume of an electric field E. This result will be particularly useful when we consider electromagnetic waves.

Illustration 20.7

According to the theory of relativity, a mass m_0 possesses an energy $m_0 c^2$. Suppose the mass energy of the electron is to be contained wholly in its electric field. If we picture the electron to be a charged spherical surface, what would be the radius of the electron sphere? (This radius is often referred to as the *classical* electron radius. One must consider this model for the electron to be only a crude approximation of reality.)

Reasoning The electric field outside the electron is radial and has the value

$$E = \frac{q}{4\pi\epsilon_0 r^2}$$

where q is the electron charge. Referring to figure 20.19, let us compute the energy stored in the field within the spherical shell shown, which has radius r and thickness dr. The volume of the shell is $4\pi r^2\, dr$. According to 20.24, the energy density is $\tfrac{1}{2}\epsilon_0 E^2$. Therefore the energy stored in this volume of the shell is

$$\Delta(\text{energy}) = (\tfrac{1}{2}\epsilon_0 E^2)(4\pi r^2\, dr)$$

If we place in the value for E, we find

$$\Delta(\text{energy}) = \frac{q^2\, dr}{8\pi\epsilon_0 r^2}$$

Integrating from the surface of the "electron" (radius a) to infinity gives the energy stored in the electron's field:

$$\text{Energy} = \frac{q^2}{8\pi\epsilon_0} \int_a^\infty \frac{dr}{r^2} = \frac{q^2}{8\pi\epsilon_0 a}$$

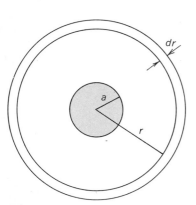

FIGURE 20.19 How would the calculation of the classical electron radius have to be modified if the electron were considered to have a uniform volume charge rather than a surface charge?

To find the classical radius of the electron, we equate this energy to the rest mass energy of the electron, m_0c^2. After solving for a, we find

$$a = \frac{q^2}{8\pi\epsilon_0 m_0 c^2}$$

Placing in the values $m_0 = 9.1 \times 10^{-31}$ kg and $q = 1.6 \times 10^{-19}$ C yields $a = 2.8 \times 10^{-15}$ m. In many experiments, the electron behaves as though it had a radius of this order of magnitude. However, as we shall see later, such detailed pictures of the electron are of questionable validity.

Chapter 20
Questions
and Guesstimates

1. What is the distinction between the terms *drift velocity* and *velocity?*
2. Current is never lost in a circuit involving steady currents. Arguing from the analogy between current and flow of water in a pipe, how would you try to convince a student that current is not "used up" as it flows through a resistor? What is lost as current flows through a resistor?
3. How do we know which end of a battery is at the higher potential in a schematic diagram of a circuit? Of a resistor? Is there any such rule for a capacitor?
4. It was shown in illustration 20.1 that a typical drift speed for electrons is about 10^{-4} m/s. How is it possible then for the lights in a room to light almost instantaneously when you flip on the switch?
5. In previous chapters we always assumed $E = 0$ in a metal. But in this chapter we talk about currents being caused by nonzero electric fields in the metal. Aren't we being inconsistent?
6. A wire of radius 0.20 mm is soldered to the end of a 0.80-mm-radius rod. A current of 0.5 A flows down the rod into the wire. How do the currents and current densities compare in the wire and in the rod? How do the drift velocities compare?
7. A lightening stroke consists of a spark jumping from one cloud to another or to the earth. What data would you need to find the current in the stroke? The current density? What data would you need in order to determine the direction of the current?
8. Why can't you light an ordinary house bulb by use of a 1.5-V flashlight battery? What happens if you try to light a flashlight bulb by use of the 120-V house voltage?
9. When you grasp the two wires leading from the two plates of a charged capacitor, sometimes you may feel a shock. The effect is much greater for a 20-μF capacitor than for a 0.02-μF capacitor even though both are charged to the same potential difference. What can you conclude from this regarding the cause of sensation of shock given to the body?
10. Suppose by some sort of magic we were given a wire in which the "electrons" were positive charges and the "nuclei" carried negative charges. Could we tell that the wire was different from an ordinary wire by carrying out a simple conduction experiment with it?
11. From qualitative considerations alone, why should the equivalent capacitance for two capacitors in parallel exceed each individual capacitance?
12. Two originally uncharged capacitors $C_1 > C_2$ are placed in series and charged. How do Q_1 and Q_2 compare? Explain your reasoning.
13. When you walk across a deep-pile carpet in very dry weather, your body sometimes becomes charged. In effect, you are one plate of a capacitor.

Where is the other plate? Where do the field lines that originate on your body terminate?

14. Is the energy stored in three dissimilar capacitors larger when the capacitors are charged in series or in parallel? Assume they are charged by the same battery. When they are charged in series, does the energy stored depend upon which capacitor is in the center?

15. Starting from the defining equation for current, show that when the positive charge moves in a direction opposite to the negative charge and with the same speed as the negative charge, the current is twice as large as it is when the positive charges stand still.

16. A capacitor consisting of two parallel plates is charged to a charge Q. It is then disconnected from the battery, after which the plates are slowly pulled apart until the plate separation is doubled. Does Q change? Does the energy stored in the capacitor change? Explain the reason and mechanism for any change.

17. Repeat question 16 if the capacitor remains attached to the battery as its plates are separated.

18. Make plausible the fact that the time constant of a capacitance-resistance system is large for R large. Why should it be large for C large?

19. Suppose two capacitors C_1 and C_2 are connected together in series but with a resistor R between them. Will they each still acquire a charge $Q_1 = Q_2 = VC_1C_2/(C_1 + C_2)$ when connected across a battery with voltage V?

20. A certain circuit consists of the following elements taken in order: battery, capacitance C_1, resistance R_1, capacitance C_2, resistance R_2, and then back to the battery, all in series. Is the time constant for this circuit $(R_1 + R_2)C_{eq}$ with $C_{eq} = C_1C_2/(C_1 + C_2)$?

21. Estimate the power consumption of a city (similar to one familiar to you) of 100,000 people.

Chapter 20
Problems

1. When a 1.5-V flashlight battery sends a current of 0.20 A through a flashlight bulb, how many electrons flow through the bulb each second?

2. A 60-W bulb connected to a 120-V power source has a current of 0.50 A flowing through it. How much charge passes through the bulb in an hour? How many electrons pass through the bulb in this time?

3. The charge on a silver ion is e while the mass of each ion is $108/6 \times 10^{26}$ kg. When a current passes from one electrode to another through a silver nitrate solution, silver ions plate out as atoms on the negative electrode. If the current is 0.20 A, how much silver will plate out in 10 min? Assume the current is carried through the solution by the silver ions.

4. Two copper bars are placed as electrodes in an acid solution. By means of an external battery, current is caused to flow through the solution from one electrode to the other. If a current of 0.20 A flows for 5 min, by how much does the mass of the positive electrode change in this time? The charge on the copper ion is 3.2×10^{-19} C and the mass of each ion is $63.5/6 \times 10^{26}$ kg. (*Hint:* The current is carried by copper ions traveling through the solution from one electrode to the other.)

5. A typical copper wire might have 2×10^{21} free electrons in 1 cm of its length. Suppose the drift speed of the electrons along the wire is 0.05 cm/s. How many electrons would pass through a given cross section

of the wire each second? How large a current would be flowing in the wire?

6. In a TV tube a beam of electrons is shot down the axis of the tube and hits the fluorescent screen at the tube's end. If the current carried by the beam is 1.6×10^{-3} A, how much charge strikes the end of the tube each second? If the speed of the electrons in the beam is 5×10^7 m/s, how long a length of beam strikes the screen in a second? How many electrons are there in a 1-cm length of the beam?

7. Number 10 copper wire can carry a maximum current of about 30 A before overheating. Its diameter is 0.26 cm. Find the resistance of a 1-m length of the wire. How large a voltage drop occurs along it per meter when it carries a current of 30 A (see table 20.1)?

8. A particular piece of wire is 10 m long and has a diameter of 0.20 cm. When its two ends are connected to the two terminals of a 1.5-V battery, a current of 0.70 A flows through the wire. Find the resistance of the wire and the resistivity of the material from which it is made.

9. A 50-cm-long metal rod consists of a copper sheath (inner diameter = 2 mm, outer diameter = 3 mm) with an iron core. What is the resistance of the rod? (*Hint:* Find the current that would flow through it when the potential difference is V.)

10. A current of 3.0 A flows down a straight metal rod that has an 0.20 cm diameter. The rod is 1.5 m long and the potential difference between its ends is 40 V. Find: (*a*) current density and (*b*) field in the rod; (*c*) resistivity of the material of the rod.

11. A 60-W bulb carries a current of 0.5 A when operating on 120 V. The temperature of its tungsten filament is then 1800°C. Find the resistance at its operating temperature. Find its resistance at 20°C.

12. If it were allowable to use the data of table 20.1 at low temperatures, at what temperature would the resistance of silver become zero? In practice, the resistance does not decrease as fast as one would predict from this at very low temperatures.

13. It is desired to make a 20.0-Ω coil of wire which has a zero thermal coefficient of resistance. To do this, a carbon resistor of resistance R_1 is placed in series with an iron resistor of resistance R_2. The proportions of iron and carbon are so chosen that $R_1 + R_2 = 20.00\ \Omega$ for all temperatures near 20°C. How large are R_1 and R_2?

14. A metal sphere of radius b is far distant from all other objects and carries a charge Q. (*a*) What is the absolute potential of the sphere? (*b*) Considering it to be one plate of a capacitor with the other plate being its surroundings (nearly infinitely far away), show that the capacitance of the sphere and its surroundings is $4\pi\epsilon_0 b$.

15. A spherical capacitor is to be constructed by using a metal sphere of radius b as one plate and a concentric spherical metal shell as the other plate. If the inner radius of the shell is $a > b$, show that the capacitance of the device is

$$C = \frac{4\pi\epsilon_0 ab}{a - b}$$

16. Using the result of problem 15, show that if the separation of the spheres is very small in comparison with their radii, the capacitance is given by the parallel-plate relation

$$C = \frac{\epsilon_0 A}{d}$$

Also, find the limiting form for C when the outer plate is infinitely far away. This is the capacitance of an isolated sphere.

17. An electric toaster is rated 900 W/120 V. How much current does the toaster draw? What is its resistance when in operation?

18. A stereo set rated at 20 W/120 V will draw how large a current when in operation? What is the apparent resistance of the set?

19. It is desired to heat a cup of coffee (200 cm³) by use of an immersion heater from 20 to 90°C in 0.5 min. How many calories are needed? How much power is needed? How much current would the heater draw from 120 V?

20. To determine the mechanical equivalent of heat (i.e., the relation between the joule and calorie) a student sends a current of 0.75 A through a heater immersed in a calorimeter. The potential difference across the heater is 12.00 V. During the 4.00 min she allows the current to flow, the contents of the calorimeter experiences a temperature rise of 2.73°C. She knows the calorimeter and its contents to be equivalent in heat capacity to 187 g of water. From these data, how many joules are equivalent to 1.00 cal?

21. A 60-W bulb is placed in a vat of oil. Assuming that all the heat and light from the bulb is captured in the oil, at what rate will the temperature of the oil rise? There is 12 kg of oil in the vat and $c = 0.70$ cal/g °C for the oil.

22. Lead storage batteries have an *energy density* of about 80,000 J/kg, that is, 1 kg of such batteries can furnish about 80,000 J of energy. For comparison purposes, gasoline as used in modern cars has an energy density of about $100 \times 80,000$ J/kg. Assuming a 12-V car battery has a mass of 15 kg and maintains a constant voltage of 12 V, for how many hours can the fully charged battery furnish a 5-A current before becoming discharged? Batteries are frequently rated in *ampere-hours,* the number of hours they can furnish a current of 1 A. What would be the ampere-hour rating of this battery?

23. How many times could the fully charged 12-V battery mentioned in the previous problem recharge a 5-μF capacitor before becoming discharged?

24. In the laboratory, a student charges a 2-μF capacitor by placing it across a 1.5-V battery. While disconnecting it, he holds its two lead wires in his two hands. Assuming the resistance of his body between hands is 60,000 Ω, what is the time constant of the series circuit composed of the capacitor and his body? How long does it take for the charge on the capacitor to drop to 1/2.718 of its original value? To 1/100? (*Hint:* Note that $0.01 = e^{-t/0.12}$ can be written as $100 = e^{t/0.12}$ and that $\ln 100 = 2.30 \log 100$.)

25. In a certain electronic device, a 10-μF capacitor is charged to 2000 V. When the device is shut off, the capacitor is discharged for safety reasons by a so-called "bleeder" resistor of 10^6 Ω placed across its terminals. How long does it take for the charge on the capacitor to decrease to 0.01 of its original value? (See hint given for previous problem.)

26. A very conscientious student is doing a physics experiment which involves charging a 2-μF capacitor by connecting it across a 1.5-V battery. To be sure it is fully charged, he waits 5 min. Assuming the resistance of the circuit is 0.2 Ω, how long must he really wait for the charging current to drop to 10^{-4} of its original value? (See hint for problem 24.)

27. A 12-V battery, a resistor with $R = 10^6$ Ω, and a 3.0-μF capacitor are connected in series with a switch. When the switch is closed, the capaci-

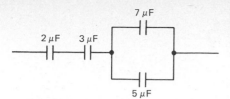

FIGURE P20.1

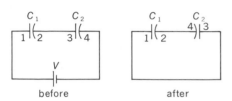

FIGURE P20.2

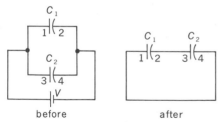

FIGURE P20.3

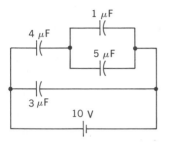

FIGURE P20.4

tor charges. Find the time taken for the capacitor to become (*a*) 50 percent charged; (*b*) 99 percent charged. Express both times in seconds and also in number of time constants elapsed.

28. Four capacitors with values 2, 3, 4, and 6 μF are connected in series. What is their equivalent capacitance?

29. Find the equivalent capacitance of the combination shown in figure P20.1.

30. Three capacitors (2, 3, and 4 μF) are connected in series with a 6-V battery. When fully charged, what is the charge on the 3-μF capacitor? What is the potential difference between the two ends of the 4-μF capacitor?

31. Two capacitors in parallel, 2 and 4 μF, are connected, as a unit, in series with a 3-μF capacitor. The combination is connected across a 12-V battery. Find the equivalent capacitance of the combination and the potential difference across the 2-μF capacitor.

32. Find the equivalent capacitance of the combination shown in figure P20.2. Also find the charge on the 4-μF capacitor.

33. Two capacitors, $C_1 = 3\ \mu$F and $C_2 = 6\ \mu$F, are connected in series and charged by connecting a battery of voltage $V = 10$ V in series with them. They are then disconnected from the battery, and the loose wires are connected together. What is the final charge on each?

34. Repeat problem 33 if, after being disconnected from the battery, the capacitors are disconnected from each other. They are now reconnected as shown in figure P20.3. What is the final charge on each?

35. If two capacitors $C_1 = 4\ \mu$F and $C_2 = 6\ \mu$F are originally connected to a battery $V = 12$ V, as shown in figure P20.4, and then disconnected and reconnected as shown, what is the final charge on each capacitor?

36. A 2-μF capacitor is charged to a voltage of 100 V. How much energy is stored in the capacitor? If it is now discharged through a 5-Ω resistor, how many calories of heat will be generated in the resistor?

37. An uncharged capacitor C is connected in series with a switch, a resistance R, and a battery of voltage V. The switch is now closed and the capacitor begins to charge. (*a*) What is the final energy stored in the capacitor? (*b*) How much energy was lost in the resistor during charging? (*c*) Compute the energy output of the battery during the charging process by integrating VI. (*d*) Compare the results of (*a*) + (*b*) with (*c*) and explain.

38. By use of the energy stored in the electric field between its plates, show that the energy stored in a spherical capacitor is given by $2\pi\epsilon_0 V^2 ab/(a - b)$ where a and b are the radii of the spherical plates and V is the potential difference between them.

39. A parallel plate capacitor C with initial charge Q_0 has oil between its plates. The oil is dirty and allows charge to leak from one plate to the other. It acts like a resistance R between the plates. Show that the charge Q on the capacitor varies in the following way with time: $Q = Q_0 \exp(-t/RC)$. (*Hint:* $-\Delta Q/\Delta t$ = current flowing through the oil.)

40. Consider a parallel plate capacitor with plate area A and charge Q. (*a*) Find the force on one plate because of the charge on the other. (Refer to problem 21 of Chap. 18.) (*b*) Compute the work done in separating the plates from essentially zero separation to a separation d. (*c*) Compare this work with the energy stored in the capacitor as given by 20.23 and explain. (Assume Q to remain unchanged.)

21 DC Circuits

The behavior of electric circuits is a subject of considerable importance. In this chapter we apply the methods of preceding chapters to steady (or direct) current circuits. It will be seen that these circuits can be described by means of Kirchhoff's two rules. The behavior of alternating-current (ac) circuits will be studied in a later chapter, after we have become familiar with the action of magnetic fields.

21.1 RESISTANCES IN SERIES AND PARALLEL

Many electric circuits can be simplified for analysis by noting that certain combinations of resistors can be replaced by a single equivalent resistor. The two principal examples of this technique are shown in figure 21.1(*a*) and (*b*). In both cases we wish to know the value of R_{eq} which, when connected across a battery, would give rise to the same current drawn from the battery as found for the combination.

The combination shown in (*a*) is called a *series* connection. In a series connection the same charge flows through each resistor in turn. Thus, in part (*a*), only one path through the resistors exists between A and B. To find the equivalent resistance of this series connection, we note that the voltage drops across R_1, R_2, and R_3 add to give V. We then have

$$V = IR_1 + IR_2 + IR_3$$

Applying Ohm's law to R_{eq} in the lower part of figure 21.1(*a*) gives

$$V = IR_{eq}$$

Equating these two expressions for V yields, after canceling I,

Resistors in series add directly

$$R_{eq} = R_1 + R_2 + R_3 \qquad\qquad 21.1$$

for a series combination.

We emphasize that this result applies only to the series situation shown in figure 21.1(*a*). Notice that for R_1, R_2, and R_3 to be in series from A to B, there must be only one path through the resistors from A to B. No batteries, capacitors, or branch-off wires can be present if *21.1* is to be applicable.

Resistors connected as shown in figure 21.1(*b*) are said to be in parallel. In such a circuit, several paths exist between A and B. If each path contains *Meaning of parallel resistors* a single resistor and nothing else, as shown, then the resistors are in parallel.

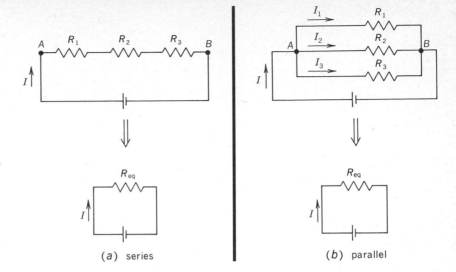

FIGURE 21.1 What is the general formula for R_{eq} for any number of resistors arranged in the ways shown?

(a) series

(b) parallel

Notice that no other element such as a battery or capacitor can exist in the same path as the resistance if the following discussion is to apply.

Referring to the figure, we see that I splits into three parts (I_1, I_2, and I_3) at point A. Since charge cannot accumulate at a point in a wire, all the charge entering A each second must leave immediately. This means that the current coming into point A (namely, I) must equal the sum of the currents leaving the point. Therefore we can write

$$I = I_1 + I_2 + I_3$$

But since all three resistors R_1, R_2, and R_3 are connected directly across the battery, Ohm's law tells us

$$V = I_1 R_1 \qquad V = I_2 R_2 \qquad V = I_3 R_3$$

from which

$$I_1 = \frac{V}{R_1} \qquad I_2 = \frac{V}{R_2} \qquad I_3 = \frac{V}{R_3}$$

Similarly, Ohm's law applied to the lower diagram in figure 21.1(*b*) tells us

$$V = IR_{eq} \qquad \text{or} \qquad I = \frac{V}{R_{eq}}$$

Replacing the currents in the equation for I by these values gives, after canceling V,

To add resistors in parallel, one adds their reciprocals

$$\frac{1}{R_{eq}} = \frac{1}{R_1} + \frac{1}{R_2} + \frac{1}{R_3} \qquad\qquad 21.2$$

for resistors in parallel.

With these two relations, one is frequently able to reduce a system of resistors down to a single resistor. As an example, consider the resistor network shown in figure 21.2. It may be reduced as indicated. The system of resistors may be replaced by a 9-Ω equivalent resistor.

It is instructive to find the current in the resistors of the system in figure 21.2(*a*) when a battery is connected between the points a and b. To be

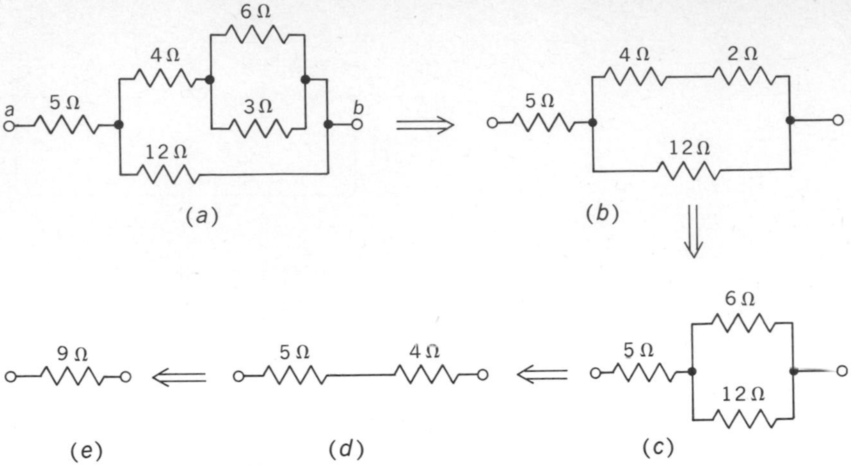

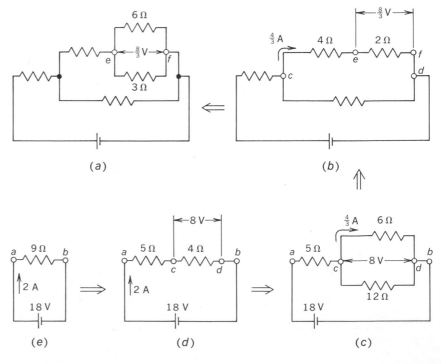

FIGURE 21.2 Successive steps in reducing a system of series and parallel resistors.

specific, suppose an 18-V battery is connected from a to b and it is desired to find the current which flows in the 3-Ω resistor. This question is answered by working backward through the equivalent circuits in figure 21.2.

Starting with (e) of figure 21.2, as shown in figure 21.3, the current in the 9-Ω resistor is 2 A. Moving to circuit (d), the current is still 2 A. In addition, from $V = IR$, we see that the voltage drop across the 4-Ω resistor is 8 V. This is also the voltage drop across the parallel combination in part (c).

In part (c) of figure 21.3 we see that the voltage drop across the 6-Ω resistor is 8 V. As a result, the current through it will be 8 V/6 Ω, or $\frac{4}{3}$ A. But the 6-Ω resistor is really the series combination of a 4- and a 2-Ω resistor, as shown in part (b) of figure 21.3. The voltage drop across the 2-Ω resistor is,

FIGURE 21.3 To find the current in the 3-Ω resistor, we move backward through the equivalent circuits of figure 21.2.

from $V = IR$, merely $\frac{8}{3}$ V. As a result, the voltage difference between points e and f in part (a) is $\frac{8}{3}$ V. This is also the voltage drop across the 3-Ω resistor. Therefore, according to $V = IR$, the current through the 3-Ω resistor is $\frac{8}{9}$ A. We could easily find the current in any of the other resistors of this circuit by these methods.

21.2 KIRCHHOFF'S RULES

Many circuits cannot be reduced by the series-parallel methods of the preceding section. Typical examples are given in figure 21.4. None of these circuits can be further reduced since the resistors involved are neither in simple series nor in simple parallel combination. For these circuits we can employ a more powerful method based upon *Kirchhoff's rules*.

Kirchhoff's loop rule

We have already encountered one of these two rules in the preceding chapter. It was *Kirchhoff's circuit,* or *loop, rule,* and stated: *The algebraic sum of all the voltage changes around a closed circuit is zero.* You will recall that this followed directly from the fact that the static electric field is a conservative field. If one follows any path through the field and returns to the original starting point, the total change in potential is zero. This rule applies in all situations where the electric field and currents are steady.

We have already made use of Kirchhoff's second rule without mentioning we were doing so. It simply takes cognizance of the fact that charge cannot accumulate in a wire. This means that the current flowing into a given point of a circuit must equal the current flowing out of the point. Or, if we call currents coming into a point positive and those leaving, negative, then *Kirchhoff's junction* (or *point*) *rule* can be stated as follows:

Kirchhoff's junction rule

The algebraic sum of all the currents coming into a point in a circuit is zero.

We shall now illustrate the power of these rules by working several examples. Follow through each step of the examples carefully since important features of circuit analysis are pointed out in them. Be particularly sure that you understand the reason for the algebraic sign given to each term in the equations.

Illustration 21.1

Find the current in the circuit of figure 21.5, as well as the power output of each battery and the power dissipated in the resistors.

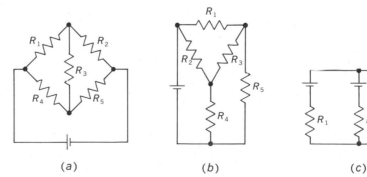

FIGURE 21.4 Circuits which cannot be further reduced by the method of series and parallel resistors.

(a)   (b)   (c)

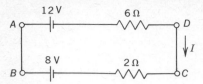

FIGURE 21.5 How will we know from our result that *I* has been chosen in the wrong direction?

Reasoning Our intuition would tell us in this case that since the 12- and 8-V batteries oppose each other, the two will be equivalent to a 4-V battery. In addition, we would expect the total resistance of the circuit to be 8 Ω. Since the 12-V battery would cause current to flow counterclockwise in the circuit, we should expect the current to flow in that direction. Its magnitude should be 4 V/8 Ω, or $\frac{1}{2}$ A. This result is easily confirmed using Kirchhoff's circuit rule.

In using Kirchhoff's rules we must assign currents to all the wires in the circuit. Although we think the current will flow counterclockwise, let us take it to be *I* and flowing clockwise, just to see what will happen. We shall now start at point *A* and move around the circuit by way of *B*, *C*, and *D* back to *A*. The circuit rule says the voltage changes encountered must add to be zero. In writing them, we shall call voltage drops negative and rises positive. In this connection, it will be recalled that the current always flows from the high- to the low-potential end of a resistor. For batteries, the positive end is always at the higher potential, no matter what the direction of the current may be.*

Starting at point *A* and going counterclockwise around the circuit, the first voltage change occurs at the 8-V battery. As we go from the + to the − side of the 8-V battery, we have a voltage change −8 V. Through the 2-Ω resistor we are going against the current, and therefore from low to high, a voltage rise of +2*I*. Similarly, against the direction of the current in the 6-Ω resistor we have +6*I*. And finally, back to *A* through the 12-V battery, we have +12 V, since we are going from the negative to the positive terminal. The total equation is therefore

$$-8 + 2I + 6I + 12 = 0$$

from which we find

$$I = -0.50 \text{ A}$$

The minus sign tells us that the current is in the opposite direction to that shown on the diagram. The direction we assume for *I* is immaterial since the sign of the answer will indicate its true direction.

As for the power balance in this circuit, the 12-V battery is furnishing power to the circuit in the amount $|VI|$, which is 6 W. The 8-V battery is being charged; i.e., charge is flowing back into it. It is therefore consuming power in the amount $|VI| = 4$ W. In addition, power is being dissipated in the resistors and is given by *VI* or I^2R for each. This dissipated power in the resistors is therefore equal to

$$(\tfrac{1}{4})(2) + (\tfrac{1}{4})(6), \text{ or } 2 \text{ W}$$

Notice that the total power consumed is $2 + 4 = 6$ W and that the 12-V battery furnishes exactly this amount of power to the circuit.

Illustration 21.2

For the circuit shown in figure 21.6, find the voltage of the unknown battery Ɛ. The ammeter, represented by the symbol —Ⓐ—, can be considered perfect (i.e., no resistance), and it reads the current in the wire to be $\frac{1}{2}$ A in the direction shown. (Recall that script E, namely Ɛ, is used for emf.)

FIGURE 21.6 How large is Ɛ, the emf of the battery, if the current read by the ammeter is $\frac{1}{2}$ A in the direction shown?

*There is one exception to this rule: When a battery having very high internal resistance is being discharged, the potential of the battery may reverse if external batteries exist in the circuit. This very unusual case is mentioned later.

Reasoning We first assign currents to all the wires. Since the current will be negative if we assign the wrong direction to it, we do not worry about its proper direction when making this assignment. They have been labeled I_1 and I_2 on the diagram. Notice that the current in all parts of the wire, P, A, B, C, is I_2, while in C, D, F, P, it is $\frac{1}{2}$ A. Writing the point-rule equation for point P, we have

$$\tfrac{1}{2} + I_1 - I_2 = 0$$

In essence, this equation simply says that the charge flowing into the point equals the charge flowing out. An identical, but not independent, equation could be written for point C.

We now write the circuit equation for loop $CDFPC$:

$$-\mathcal{E} - (\tfrac{1}{2})(2) + 4I_1 + 12 = 0$$

The ammeter is assumed perfect in writing this; i.e., it is assumed to have zero resistance. Similarly, for loop $CBAPC$ we have

$$-4 + 6I_2 + 4I_1 + 12 = 0$$

It is possible to write other loop equations, but they will not be independent. Since we have covered all the circuit elements in our equations, another loop could contain no new information.

In any case, we now have three equations involving the three unknowns I_1, I_2, and $\mathcal{E}$. Solving simultaneously, we find

$$I_1 = -1.1\,\text{A}$$
$$I_2 = -0.6\,\text{A}$$
$$\mathcal{E} = 6.6\,\text{V}$$

You should be sure that you understand these equations, especially the signs involved in them.

Illustration 21.3

Consider the circuit of figure 21.7. The voltmeter, represented by the symbol $-\!\overline{V}\!-$, reads 1.60 V, with the polarity indicated. Find the value of R. Assume the voltmeter to be perfect, i.e., to have infinite resistance. Also, find the charge on the capacitor.

Reasoning We can ignore the voltmeter since no current will flow through its infinite resistance. Moreover, when a steady state has been achieved (how large is the time constant of a 3-Ω 5-μF combination?), there will be no current flowing in the wire in which the capacitor exists. After labeling the currents as indicated, we can write Kirchhoff's rules.

For point d we have

$$I_1 + I_2 - I_3 = 0$$

For the loop $adcba$ we find

$$-4I_3 + 6 - 2I_1 = 0$$

and for loop $adba$ (remembering the voltmeter reads 1.6 V)

$$-8 - 1.6 + 6 - 2I_1 = 0$$

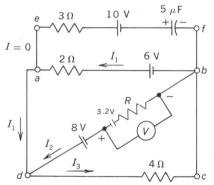

FIGURE 21.7 If the voltmeter reads 1.60 V with the polarity shown, how large is R? Assume the voltmeter to have infinite resistance.

Solving these equations simultaneously gives

$$I_3 = 2.4 \text{ A}$$
$$I_1 = -1.8 \text{ A}$$
$$I_2 = 4.2 \text{ A}$$

To find R we make use of $V = IR$. The voltmeter reads 1.6 V, and $I = I_2$ is 4.2 A. Therefore $1.6 = 3.2 - 4.2R$ and

$$R = 0.38 \ \Omega$$

In order to find the charge on the capacitor we write the loop equation for $aefba$, which is (recalling $I = 0$ for the capacitor wire)

$$-10 - V_C + 6 - (2)(-1.8) = 0$$

We designate the voltage difference across the capacitor by V_C and assume the polarity shown. (If this polarity is wrong, V_C will be negative.) Solving yields

$$V_C = -0.4 \text{ V}$$

and so the polarity is opposite to that shown. Since $Q = CV$, we find

$$Q = 2 \ \mu\text{C}$$

21.3 MEASUREMENT OF RESISTANCE

It seems relatively simple to measure the resistance of an element. One need only apply Ohm's law, $V = IR$, and the measurement of the current through and the voltage across a resistor will give its resistance. This is the basis for the so-called *ammeter-voltmeter** method for resistance measurement. A typical circuit is shown in figure 21.8. Actually, there are two possible arrangements, as shown. The two arrangements are not equivalent, as we shall now see.

In practice, one wishes the voltmeter to have an infinite resistance so that it will not disturb the circuit by passing current when it is connected between two points which are at different potential. In fact, however, all voltmeters pass some current. The resistance of a typical nonelectronic voltmeter is of the order of $10^4 \ \Omega$ times its maximum scale reading (this is usually designated as $10^4 \ \Omega/\text{V}$). Electronic voltmeters have much higher resistances, of the order of $10^7 \ \Omega$. If the voltmeter resistance is far larger than R_x in figure 21.8, the current which it passes will be negligible, and so either circuit in figure 21.8 would be allowable. However, if the voltmeter resistance is equal to R_x, as much current will flow through the voltmeter as through the resistance. In that case, only circuit (b) would be allowable since, in circuit (a), the ammeter would read much higher than the actual current through R_x.

A perfect ammeter would have zero resistance, so that it would not disturb the circuit when placed in series with it. This ideal cannot be realized. If the resistance of the ammeter is comparable with R_x, the arrangement of figure 21.8(b) is very poor, since the voltmeter will not read the voltage drop across R_x alone. On the other hand, if the ammeter resistance is small

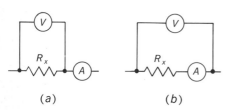

FIGURE 21.8 If R_x is large, which arrangement is preferable for computing its value from the ammeter and voltmeter readings?

*The construction of ammeters and voltmeters is discussed in detail in the next chapter. Now we simply consider a voltmeter to be a device which reads the voltage difference between its terminals, and an ammeter reads the current which flows through it.

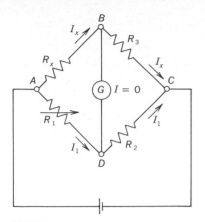

FIGURE 21.9 What factors influence the accuracy of the Wheatstone bridge in measuring R_x?

compared with R_x, both circuits in figure 21.8 will be allowable, provided the voltmeter resistance is high enough.

The ammeter-voltmeter method suffers from other disadvantages. It requires the use of two calibrated meters. Even with well-calibrated meters it is difficult to obtain more than three-digit accuracy in R_x since meters cannot easily be read to higher accuracy. As a result, other methods are required for very precise work. One of these is called the *Wheatstone bridge,* which we now describe.

The Wheatstone bridge requires no calibrated meters, but instead requires one calibrated variable resistor and two resistors whose ratio is accurately known, together with a sensitive current-detecting device. The Wheatstone bridge circuit is shown in figure 21.9. One adjusts the value of the variable resistance R_1 (the arrow through it is used to show it is variable) until no current flows through the galvanometer. The bridge is then said to be balanced. The galvanometer (a sensitive ammeter) is used only to indicate when no current flows, and need not be calibrated, but should be highly sensitive. When the bridge is balanced, the currents I_x and I_1 continue past the junction point as shown.

To obtain the working equation for the bridge we write the loop equations for loops *ABDA* and *BCDB*. If we notice that the voltage drop across the resistance of the galvanometer is zero since the current through it is zero, these equations are

$$-I_x R_x + I_1 R_1 = 0$$
and
$$-I_x R_3 + I_1 R_2 = 0$$

After eliminating I_1 and I_x from the equations by transposing the negative terms and dividing one equation by the other, we find

The Wheatstone bridge measures resistances in terms of known resistances

$$R_x = R_1 \frac{R_3}{R_2} \qquad\qquad 21.3$$

Clearly, R_x will be known to the same accuracy as R_1 and the ratio R_3/R_2. No calibrated meters are needed. However, the accuracy of the device depends seriously upon the sensitivity of the galvanometer. (Why?)

The method we have just described is only one example of a whole family of devices which use the same null principle (or balance). They are referred to as *bridges,* and vary widely. By replacing some of the resistors by capacitors or inductances, bridges can be constructed for the measurement of capacitance and inductance. Many of them use alternating rather than direct current. All these methods have the advantage that calibrated meters are not required for the measurement of the unknown quantity.

21.4 THE POTENTIOMETER

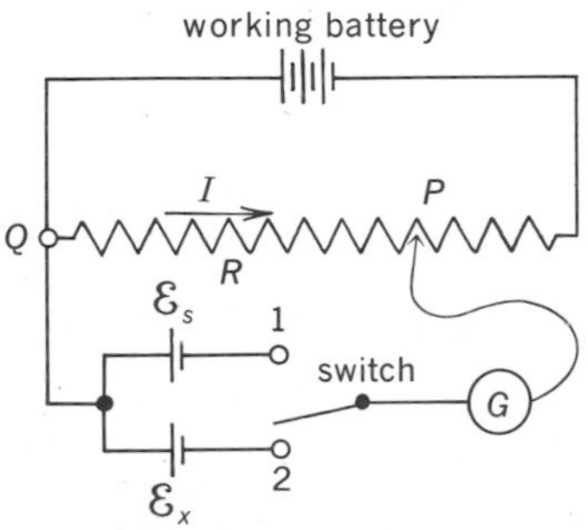

Another null-type instrument in common use is the potentiometer. This device is used to measure potential differences by comparison with a standard voltage source. A schematic diagram of its circuit is shown in figure 21.10. The working battery furnishes a steady current I to the resistor, and a variable potential difference is available between points P and Q since contact P can be moved back and forth.

One common method of operation is as follows. The switch is first thrown to position 1 so that the standard cell of known voltage $\mathcal{E}_s$ is con-

FIGURE 21.10 What are the advantages of the potentiometer?

nected. Point P is then moved until no current flows through the galvanometer. Under those conditions, the loop equation for the cell and R is merely

$$-IR_s + \mathcal{E}_s = 0 \qquad \text{or} \qquad \mathcal{E}_s = IR_s$$

where R_s is the resistance value at the null (or balance) condition.

If now the switch is thrown to position 2 and P moved to obtain a new balance (i.e., no current in the galvanometer), the new loop equation is

$$-IR_x + \mathcal{E}_x = 0 \qquad \text{or} \qquad \mathcal{E}_x = IR_x$$

where R_x is the new value of R. Dividing one equation by the other, one finds

$$\mathcal{E}_x = \mathcal{E}_s \frac{R_x}{R_s}$$

Millikan's Oil-Drop Experiment

The magnitude of the quantum of charge, the electronic charge e, was first measured accurately by R. A. Millikan and his coworkers (1909–1913). Millikan's experiment has a simplicity and directness which establish without doubt that charge is quantized. His apparatus is illustrated schematically in the figure.

Basically, the apparatus consists of two parallel metal plates (separation $\simeq 1$ mm and area $\simeq 100$ cm^2) with a known variable potential difference between them. Because of the potential difference V, an electric field $E = V/d$ exists between the plates, where d is the plate separation. If now a charged oil droplet (mass m and charge q) finds itself between the plates, a vertical electric force Eq will act on the droplet. By adjusting the potential difference between the plates, this force can be made equal and opposite to the gravitational force mg on the droplet. When this condition is achieved, the resultant force on the droplet is zero, and the droplet will remain motionless. Then

$$mg = Eq$$

from which

$$q = \frac{mg}{E}$$

Knowing the electric field E, the mass of the droplet m, and the acceleration due to gravity g, the charge q on the droplet can be found.

In practice, Millikan allowed very small oil droplets sprayed from an atomizer to fall through a small hole in the top plate. Many of these droplets contain excess charge generated by friction as the drops are produced in the atomizer. An intense light shining between the plates causes the droplets to sparkle, and so they can be observed by looking through the microscope, as indicated. Provision was also made for

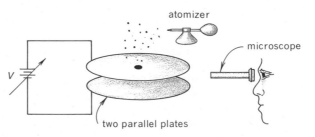

atomizer

microscope

V

two parallel plates

Therefore $\mathcal{E}_x$ is determined entirely in terms of the ratio of two resistances and the voltage of the standard cell. This method is easily capable of four-digit accuracy. Most commercial potentiometers have provision for adjusting the current I, so that the ratio $\mathcal{E}_s/R_s$ can be made unity or some other fixed constant. Then $\mathcal{E}_x$ is equal in magnitude to R_x, or some simple multiple of it, and the instrument is thereby made direct-reading.

A potentiometer measures emfs

In addition to its high accuracy, the potentiometer has the advantage that it draws no current from the voltage source being measured. As such, it is unaffected by internal resistance of the voltage source. It measures the zero current voltage of the source, the emf of the source.

Illustration 21.4

A very old dry cell may have a rather large internal resistance. Suppose a voltmeter reads its voltage to be 1.40 V while a potentiometer reads its

---

changing the charge on the droplet by subjecting the region between the plates to a weak x-ray beam when desired.

The electric field between the plates was calculated from the measured plate separation d and the potential difference between the plates, V. To find the mass of the oil particle, m, Millikan made measurements of the time taken for the droplet to fall a measured distance through the air in the absence of an electric field. Since the droplets were moving at constant speed—their terminal speed—through the distance chosen, the gravitational force mg was equal to the viscous friction force between the droplet and the air. Assuming that the viscous force was proportional to the speed of the droplet, v, he could write

$$mg \sim v$$

One of the major sources of error involved in the experiment was associated with the proportionality constant in this relation. To fair accuracy, however, one can use the so-called Stokes law form for this viscous force and write

$$mg = 6\pi\eta av$$

where η is the viscosity of air, and a is the radius of the droplet. Since the radius of the drop is related to its mass through the fact that the density of the oil, ρ, multiplied by the volume of the drop is equal to the drop's mass,

$$m = \tfrac{4}{3}\pi a^3 \rho$$

the mass of the drop can be found from a knowledge of ρ, η, and its terminal speed v.

As shown, knowing m and E, the charge on the droplet can be found. Millikan carried out thousands of measurements of this general type (slightly modified for higher accuracy). His result for q was always an integer multiple of 1.60×10^{-19} coulomb. He therefore concluded that charge was quantized, the magnitude of the charge quantum being 1.60×10^{-19} coulomb, which we designate by e. All charges thus far found in nature are integer multiples of this value. In particular, the charge on the electron is $-e$, while that on the proton is $+e$. The currently accepted value of e is 1.60219×10^{-19} coulomb.

---

voltage to be 1.55 V. What is (*a*) the emf of the cell, (*b*) its internal resistance, and (*c*) the current it would supply to a 5-Ω resistor? Assume the voltmeter resistance to be 280 Ω.

Reasoning The dry cell may be thought of as a pure voltage source ε and a resistor r in series. For a good battery, the value of the internal resistance r is usually negligibly small. However, this may not be true for an old battery.

We picture the battery as the circle in figure 21.11(*a*). When no current is being drawn from it, the potential difference between its terminals is just that of the pure voltage source, and this is called the emf of the battery. It is this quantity which the potentiometer measures since this instrument draws no current from the cell being measured. Therefore we find the emf of the battery to be 1.55 V.

When the 280-Ω voltmeter is connected across the battery as shown in figure 21.11(*b*), current flows from the battery. Assuming the voltmeter to read correctly, we know that the voltage difference between the battery terminals, $V_B - V_A$, is just the voltmeter reading, 1.40 V. It will be equal to the IR drop through the voltmeter. Therefore

$$I(280) = 1.40$$

from which

$$I = 0.0050 \text{ A}$$

If we write the loop equation for figure 21.11(*b*), we find (using *1.55* for ε)

$$+1.55 - I(r) - I(280) = 0$$

Knowing I to be 0.0050 A, we find the internal resistance r of the battery to be 30 Ω.

When this battery is connected across a 5-Ω resistor as shown in part (*c*) of figure 21.11, the current which will flow is readily found by use of the loop equation. The loop equation is

$$+1.55 - I(30) - I(5) = 0$$

Solving for I, we find

$$I = 0.044 \text{ A}$$

This is only about one-tenth as large as the current which would flow from a new dry cell under these same conditions.

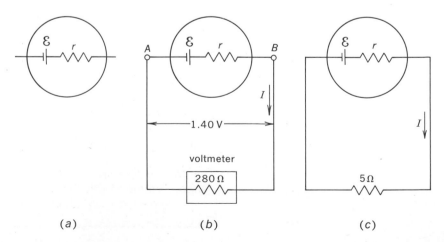

FIGURE 21.11 What is the maximum current which this battery can produce in a wire?

(*a*)   (*b*)   (*c*)

21.5 HOUSE CIRCUITS AND ELECTRICAL SAFETY

Let us turn our attention now to some very practical considerations affecting our lives daily. We live in a world of electrical appliances, and we should spend a short time discussing the electrical systems which power them. Even though most power systems in the world use alternating voltages, a topic to be considered in detail later, there are many features of the systems which do not depend upon this fact. Since we do not wish to postpone consideration of this topic further, we shall assume a steady (dc) voltage source in order to facilitate the discussion.

The power company which furnishes electrical energy to a house, store, or factory runs at least two wires to the building it supplies. Let us say it is your house. The potential difference between these two wires is about 120 V in the United States. (In many other countries, it is about 240 V.) These two lead-in wires to the house usually have a large diameter so that they can carry considerable current without heating up. (The larger the cross-sectional area of the wire, the less its resistance will be. Since heat generated is proportional to I^2R, the low resistance will ensure low heat dissipation.)

In most newer houses, the wires are capable of carrying about 30 A without undue heating. However, to protect against too large a current, a fuse or circuit breaker is placed in series with the wire. Its purpose is to disconnect the wire from the voltage source if greater than the allowed current is drawn from it. This procedure automatically disconnects any wire which is accidentally called upon to carry more than the safe current.

A typical house circuit consists of two parallel wires strung through the house from the 120-V source provided by the lead-in wires to the house. This is shown schematically in figure 21.12. Each light bulb, appliance, etc., is connected to the low-potential wire. When the switch to that appliance is closed, current runs through the device from the + to the − wire of the power system. The low-potential wire is usually grounded. That is to say, the wire is connected to a water pipe or some other metallic object connected to the moist earth. This keeps this wire at the same potential as the earth, usually assigned the value zero in practical electricity. The symbol for the ground connection is ⊪⊢.

Let us compute how much current is drawn by the 60-W bulb of figure 21.12 when it is turned on by closing the switch. Since power = VI, and since $P = 60$ W and $V = 120$ V in this case, we find the current through the bulb to be $I = 0.50$ A. Similarly, when turned on, the stove draws 10 A, the radio draws 0.167 A, and the 120-W bulb draws 1.0 A. If they are all turned on at once, a total of 11.667 A will pass through the fuse. Usually a house circuit would be fused for no less than 15 A, and so no danger exists in this case.

A house which has a large number of electrical appliances requires more

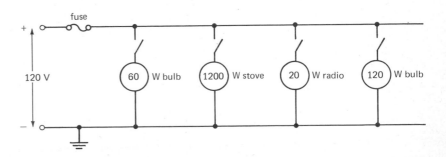

FIGURE 21.12 The household appliances shown act as resistors. Each is put into operation by closing the appropriate switch.

than one circuit. Most houses have several separate fused circuits such as the one shown in figure 21.12. Each of these starts at the source furnished by the lead-in wires to the house and runs to various portions of the house.

Most houses have a third wire furnished by the power company. Its electrical level is 120 V below ground potential. As a result, the potential difference between the two nongrounded wires is about 240 V. Appliances which consume large amounts of power (electric stoves and driers) are usually designed to operate from this higher voltage. The reason for this is as follows: Since power = VI, a 2400-W appliance would require a current of 20 A operating on 120 V but only a current of 10 A when operating on 240 V. This means that the current is less in a 240-V system and so the wires can be kept smaller without overheating. This is why most countries operate on 240 V rather than 120 V. Of course, the higher-voltage systems are more dangerous to use.

Since we use electrical apparatus daily, we should understand the elements of electrical safety. Electricity can kill a person in two ways: it can cause the muscles of the heart and lungs (or other vital organs) to malfunction or it can cause fatal burns.

Even a small electric current can seriously disrupt cell functions in that portion of the body through which it flows. When the electric current is 0.001 A or higher, a person can feel the sensation of shock. At currents 10 times larger, 0.01 A, a person is unable to release the electric wire held in his hand because the current causes his hand muscles to contract violently. Currents larger than 0.02 A through the torso paralyze the respiratory muscles and stop breathing. Unless artificial respiration is started at once, the victim will suffocate. Of course, the victim must be freed from the voltage source before he can be touched safely; otherwise the rescuer, too, will be in great danger. A current of about 0.1 A passing through the region of the heart will shock the heart muscles into rapid, erratic contractions (ventricular fibrillation) so the heart can no longer function. Finally, currents of 1 A and higher through body tissue cause serious burns.

The important quantity to control in preventing injury is electric current. Voltage is important only because it can cause current to flow. Even though your body can be charged to a potential thousands of volts higher than the metal of an automobile by simply sliding across the car seat, you feel only a harmless shock as you touch the door handle. Your body cannot hold much charge on itself, and so the current flowing through your hand to the door handle is short-lived and the effect on your body cells is negligible.

In some circumstances, the 120-V house circuit is almost certain to cause death. One of the two wires of the circuit is always attached to the ground, so it is always at the same potential as the water pipes in a house. Suppose a person is soaking in a bathtub; his body is effectively connected to the ground through the water and piping. If his hand accidentally touches the high-potential wire of the house circuit (by touching an exposed wire on a radio or heater, for example), current will flow through his body to the ground. Because of the large, efficient contact his body makes with the ground, the resistance of his body circuit is low. Consequently the current flowing through his body is so large that he will be electrocuted.

Similar situations exist elsewhere. For example, if you accidentally touch an exposed wire while standing on the ground with wet feet, you are in far greater danger than if you are on a dry, insulating surface. The electrical circuit through your body to the ground has a much higher resistance if your feet are dry. Similarly, if you sustain an electrical shock by touching a bare wire or a faulty appliance, the shock is greater if your other hand is touching the faucet on the sink or is in the dishwater.

As you can see from these examples, the danger from electrical shock can be eliminated by avoiding a current path through the body. When the voltage is greater than about 50 V, avoid touching any exposed metal portion of the circuit. If a high-voltage wire must be touched (for example, in case of a power-line accident when help is not immediately available), use a dry stick or some other substantial piece of insulating material to move it. When in doubt about safety, avoid all contacts or close approaches to metal or to the wet earth. Above all, do not let your body become the connecting link between two objects that have widely different electric potentials.

Chapter 21
Questions
and Guesstimates

1. A resistor is connected from point a to point b. How does one tell which it is from a to b, a potential drop or a potential rise? Repeat for a battery and for a capacitor.

2. Explain the following statement: for series resistors, the equivalent is always larger than the largest; for parallel resistors, the equivalent is always smaller than the smallest. What is the similar statement for capacitors?

3. Estimate the electric-power consumption of a nearby city. How large a current would have to be supplied to the city at 220 V to supply this energy? Explain why very high voltage ($\sim 10^5$ V) power transmission is of advantage.

4. Why would one connect two batteries in series? In parallel? Why should unlike batteries never be connected in parallel?

5. When a battery is being charged, its terminal potential is higher than its emf. The reverse is true when it is being discharged. Why? Is it possible to reverse the polarity of a battery, i.e., make the normally negative terminal positive with respect to the other terminal?

6. What would happen if a small child were to cut a lamp cord in two with a pair of wire-cutting pliers when the cord is plugged in? Is the child in any danger?

7. Parents frequently worry about their small children playing near electrical outlets. Discuss the various factors which determine how badly the child could be shocked.

8. Birds sit on high-tension wires all the time. Why aren't they electrocuted, since sometimes the wires have gaps where there is no insulation on them?

9. If a current of only a small fraction of an ampere flows in one hand and out the other, the person will be likely to be electrocuted. If the current flows into one hand and out the elbow above the hand, the person can survive even if the current is large enough to burn the flesh seriously. Explain.

10. Using an ohmmeter (basically a battery in series with a very sensitive ammeter), measure your resistance from one hand to the other. A current of about 0.03 A through the midsection of one's body is sufficient to paralyze the breathing mechanism. About how large a voltage difference between your hands is needed to electrocute you?

11. It is possible to measure a resistor precisely, provided one has available a single standard resistor and a potentiometer. How can this be done?

12. It is a difficult task to measure the resistance of an approximately $10^8 \, \Omega$ resistor. Why will the usual methods not prove accurate? Can you devise a way of measuring the resistance by use of a standard capacitor, a sensitive galvanometer, and a standard cell?

13. The terminal voltage of a battery is measured as a function of the current being drawn from it. Show that the slope of the V versus I graph will give the internal resistance of the battery.

14. For such purposes as electrocardiograms, brain-damage tests, lie detectors, etc., one wishes to measure the voltage differences between various portions of a person's body. Why cannot one use a simple voltmeter for this purpose? What precautions must one take?

15. Why is it very dangerous to set a 120-V-powered radio near the bathtub when you are bathing in it? Is a battery-powered radio equally dangerous?

16. Not long ago, car manufacturers changed from a 6-V electrical system to a 12-V system. Why?

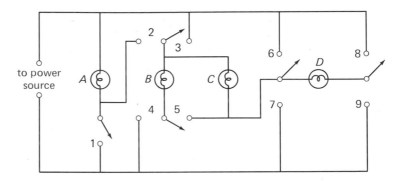

FIGURE P21.1

17. In the circuit shown in figure P21.1, A, B, C, and D are bulbs. Which connections must be closed to light the bulbs as indicated below?

1. A	11. A, B in parallel
2. B	12. A, C in parallel
3. C	13. A, B, C in parallel
4. D	14. A, B, C, D in parallel
5. A, B in series	15. B, C, D in parallel
6. A, C in series	16. A, B, D in parallel
7. B, C in series	17. A, C, D in parallel
8. C, D in series	18. B, C in parallel
9. B, C, D in series	19. B, D in parallel
10. A, C, D in series	20. C, D in parallel

21. A, D in parallel
22. A in parallel with C and D in series

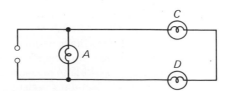

23. *A* in parallel with *B* and *C* in series
24. *A* in parallel with *B*, *C*, *D* in series
25. *B* in parallel with *C* and *D* in series

26. *AC* parallel combination in series with *B*

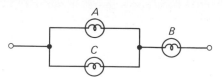

27. *BC* parallel combination in series with *A*
28. *BC* parallel combination in series with *D*
29. *AC* series combination in parallel with *D*
30. *BC* series combination in parallel with *D*

35.

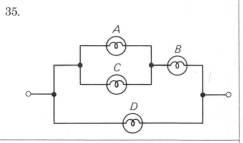

31.

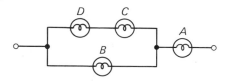

36.

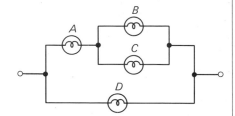

32.

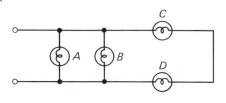

37.

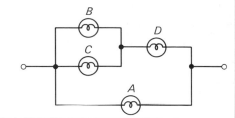

33.

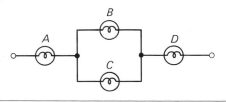

38.

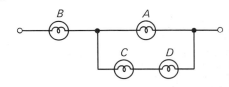

34.

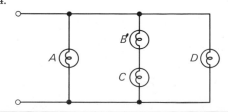

39.

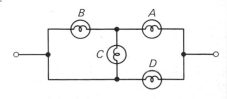

Chapter 21
Problems

1. Given three resistors with values 2, 3, and 4 Ω. By use of them, 11 additional different resistances can be made. What are they?
2. By using resistors R_1 and R_2 in various combinations, the following resistances can be obtained: 3.2, 4, 16, and 20 Ω. What are R_1 and R_2?
3. For the resistances shown in figure P21.2, what is the equivalent resistance?

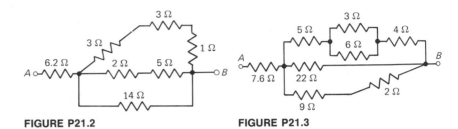

FIGURE P21.2   **FIGURE P21.3**

4. Find the equivalent resistance for the combination shown in figure P21.3.
5. If an 18-V battery is connected from A to B in figure P21.2, how large a current flows through the 6.2-Ω resistor? The 14-Ω resistor? The 5-Ω resistor?
6. If a 6-V battery is connected from A to B in figure P21.3, how large a current flows through the 7.6-Ω resistor? the 22-Ω resistor? the 9-Ω resistor?
7. For the circuit shown in figure P21.4 find R_{eq} together with I_1 and I_2. (*Hint:* Note that a, b, c, and d are all the same point from an electrical standpoint. Redraw the diagram so as to show this.)
8. For the circuit shown in figure P21.5, find R_{eq} together with I_1 and I_2. (*Hint:* At the outset, be sure you understand which resistors are in parallel. Do this again after each simplification you make in the circuit.)
9. Two batteries connected in parallel are connected across a 4.0-Ω resistor. Battery A has $\mathcal{E} = 6$ V and $r = 0.02\ \Omega$, while battery B has $\mathcal{E} = 6$ V and $r = 2.0\ \Omega$. Find the current through the resistor and the currents through each battery.
10. A battery is connected in series with a variable resistor and an ammeter that has negligible resistance. When the variable resistor is 10 Ω, the current is 2.000 A. When the resistance is 5 Ω, the current is 3.800 A. Find the emf and internal resistance of the battery.

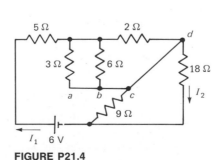

FIGURE P21.4

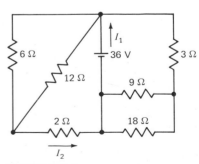

FIGURE P21.5

454   DC CIRCUITS

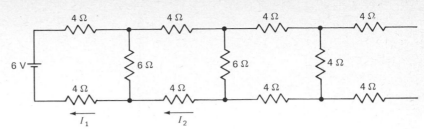

FIGURE P21.6

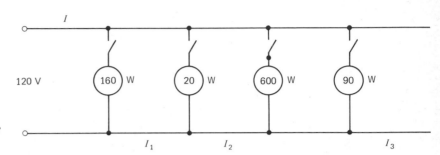

FIGURE P21.7

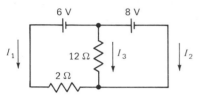

FIGURE P21.8

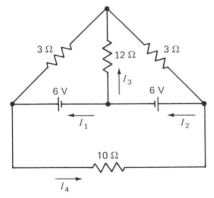

FIGURE P21.9

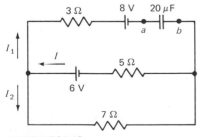

FIGURE P21.10

11. Two batteries of the same emf but with internal resistances r_1 and r_2 ($r_1 > r_2$) are connected in series aiding. The combination is connected across a variable resistance R. (a) For what value of R will the terminal voltage of battery 1 be zero? (b) What then will be the terminal voltage of the other battery?

12. A battery with emf $= \mathcal{E}$ and internal resistance r is connected across a resistor R. Show that the power dissipated in the resistor is $\mathcal{E}^2 R/(R + r)^2$. For what value of R will the power dissipated be a maximum? [Recall from calculus that a function $f(x)$ has an extremum when $df/dx = 0$.]

13. Find I_1 and I_2 for the circuit shown in figure P21.6.

14. The house circuit shown in figure P21.7 has appliances with the indicated wattages connected to it. When all the appliances are turned on, find I, I_1, I_2, and I_3.

15. Find I_1, I_2, and I_3 for the circuit shown in figure P21.8.

16. Find I_1, I_2, I_3, and I_4 for the circuit of figure P21.9.

17. If the 12-Ω resistor in figure P21.9 is replaced by a 3-V battery with positive terminal on the top, find I_1, I_2, I_3, and I_4.

18. For the circuit shown in figure P21.10, find I, I_1, I_2, and the charge on the capacitor. Which is at the higher potential, point a or point b?

19. A 4-Ω resistor is connected from a to b in figure P21.10, in parallel with the capacitor shown. Find I, I_1, and I_2 a few seconds after this has been done.

20. A 1400-Ω resistor is connected in series with a 200-Ω resistor and a 12-V battery. (a) By how much does the current in the circuit change when a 5000-Ω voltmeter is connected across (in parallel with) the 1400-Ω resistor? (b) Across the 200-Ω resistor? (c) What does the voltmeter read in (a)? (d) In (b)?

21. In order to measure a certain resistance using the ammeter-voltmeter method, use is made of an ammeter having a resistance of 0.0100 Ω and a voltmeter with resistance 10,000 Ω. If the resistance actually has a value of 500 Ω, what will be the two experimental results using the two possible

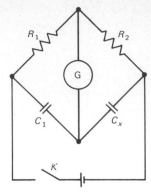

FIGURE P21.11

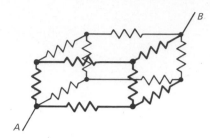

FIGURE P21.12

connections shown in figure 21.8? (*Hint:* Find the voltmeter and ammeter readings for a battery connected across the circuit.)

22. Figure P21.11 shows a possible bridge for measuring capacitance. If no current flows through the galvanometer when the key K is closed, the bridge is balanced. Show that the balance condition is

$$C_x = \frac{R_1}{R_2}C_1$$

This bridge method has several practical difficulties which make it undesirable.

23. The electric power station at Grand Coulee Dam in Washington has an estimated ultimate electric power production capability of 9.8×10^9 W. If this power were transmitted at 120 V, how large a current would be carried by the wires leading from the plant? At 10^6 V? How much heat would a current of 10,000 A release each second while flowing through a copper rod 1 m long and cross section 100 cm^2?

24. Find the equivalent resistance of the circuit shown in figure P21.12 if each resistor has a resistance R. (*Hint:* By symmetry, many of the resistors will have equal currents.)

22 Magnetic Fields

So far in our study of physics we have encountered two types of field, the electric and the gravitational field. In this chapter we are concerned with a third field, the magnetic field. Although we shall see in a later chapter that the electric and magnetic fields are not really distinctly different quantities, we develop the subject independently in this chapter. The forces on moving charges and on current-carrying wires as a result of magnetic fields are discussed. In addition, the expression for the torque on a current-carrying loop in a magnetic field is derived.

22.1 MAGNETIC FIELD PLOTTING

The basic features of simple magnetic fields are taught in grade school. Magnets were known by the ancients, and much of our qualitative terminology concerning magnetic fields is colored by this fact. In particular, we often speak of magnetic fields in terms of bar magnets, since this is the way the fields were first studied. For example, we know that the poles of a bar magnet experience forces when placed in a magnetic field. If a bar magnet is suspended by a delicate fiber, as shown in figure 22.1, a particular end of the magnet will always point approximately north on the earth provided no other magnetic objects are nearby. This end of the magnet is called the *north pole* of the magnet. The other end is the *south pole*. A device such as this is nothing more than a simple compass.

Further studies with bar magnets show that the north poles of two magnets repel each other. The south pole of a magnet is always attracted by the north pole of another magnet. If one tries to break the north pole off from a simple bar magnet, the effort proves unsuccessful. The broken magnet becomes two new bar magnets, each having a north and a south pole. These are qualitative features with which we are all familiar. More is told about their molecular causes in a later chapter.

Magnetic fields are easily plotted by means of compass needles, small bar magnets. The direction in which the compass needle points is taken to be the direction of the magnetic field. We can therefore determine the direction of the magnetic field at a point by observing the orientation of a small compass needle placed at the point. This fact is used in figure 22.2 to plot the magnetic field in the vicinity of a bar magnet. The magnetic field lines are drawn in such a way that a compass needle placed on the line will align itself tangentially to the line. Typical magnetic fields are shown in figure 22.3.

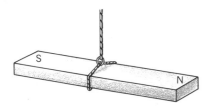

FIGURE 22.1 The north pole of a magnet is defined to be the pole which points north when the magnet is suspended as shown.

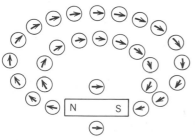

FIGURE 22.2 A compass needle points in the direction of the magnetic field.

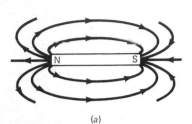

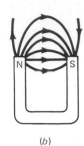

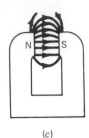

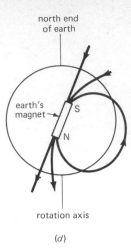

(a)　　　　(b)　　　　(c)　　　　(d)

FIGURE 22.3 Using the fact that a compass needle should line up along the field lines, you should be able to show that the lines drawn are reasonable.

Notice that the field lines emerge from north poles and enter south poles. Why is this always true? As we see from the figure, the earth acts like a huge magnet with the magnet's north pole being near the position of the earth's south pole.

22.2 CURRENTS AS SOURCES OF MAGNETIC FIELDS

Currents produce magnetic fields

An important step in the understanding of the nature of magnetic fields occurred in 1820 when Hans Christian Oersted discovered that currents in wires produce magnetic fields. This fact is easily demonstrated by the experiment illustrated schematically in figure 22.4(a). If no current flows in the wire, the compass needles line up parallel to each other, all pointing north. However, when a current flows in the wire, the needles line up as indicated. Experiments such as this show that the current-carrying wire has a magnetic field about it similar to that shown in figure 22.4(b). In this, as well as in later diagrams, the symbol ∘ indicates an arrow coming toward the reader, and ×

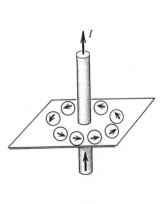

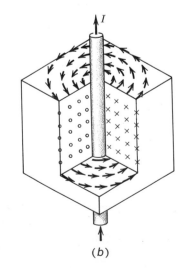

FIGURE 22.4 Notice that the magnetic field lines caused by a current have no ends; they circle back upon themselves.

(a)　　　　(b)

represents an arrow going away from the reader. The symbols are meant to suggest the tip and tail of the arrow.

It is well to learn a simple rule for remembering the direction of the magnetic field lines about a wire. Many people make use of the *right-hand rule* for this purpose. This rule states that if one grasps the wire with the right hand in such a way that the thumb points in the direction of the current, the fingers will circle the wire in the same sense as do the field lines. We illustrate this rule in figure 22.5. You will notice that it is very similar to the right-hand rule we learned for the direction assigned to torques and rotations.

The magnetic fields due to currents in curved wires, coils, solenoids, and other configurations are of great importance. They may be found from a knowledge of the magnetic field due to a portion of a straight wire. For example, when a compass is used to plot the magnetic field around a current-carrying loop of wire, the result shown in figure 22.6 is found. You should convince yourself that this is reasonable by applying the right-hand rule to a portion of the loop. It is interesting, and important, to notice in figure 22.6 that the loop's magnetic field is much like that of a bar magnet. In that sense, the current loop can be considered to have a north and south pole. We shall see later that this is one aspect of a very important and far-reaching similarity between bar magnets and current loops. Let us now begin the main topic of discussion for this chapter, the forces exerted on currents and moving charges by magnetic fields.

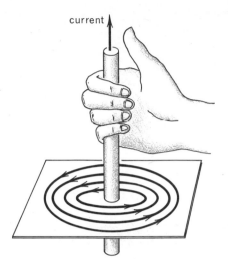

current

FIGURE 22.5 The right-hand rule for remembering the direction of a magnetic field caused by a current in a wire.

Currents experience a force due to a magnetic field

FIGURE 22.6 The current-carrying loop shown in (*a*) and (*b*) has a magnetic field much like that of the short bar magnet shown in (*c*).

22.3 FORCE ON CURRENTS AND MOVING CHARGES

It is a simple matter to show that a wire carrying a current through a magnetic field experiences a force. For example, a schematic diagram of such an experiment is shown in figure 22.7. There we see a wire carrying a current *I* through a magnetic field furnished by a magnet. The field is directed from right to left since the field lines come out of the north pole and enter the south pole. We indicate the field by the vector labeled **B**. When the experiment shown in the figure is carried out, it is found that the wire experiences a force which is proportional to both the current and the strength of the magnetic field. In figure 22.7(*a*), the force is directed upward, perpendicular to the surface of the table on which the wire lies.

The direction of the force on a wire carrying a current through a magnetic field may seem strange to you. It is not in the direction of the field

(*a*)

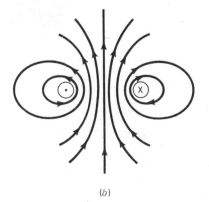

(*b*)

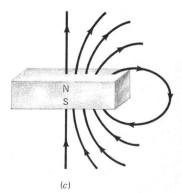

(*c*)

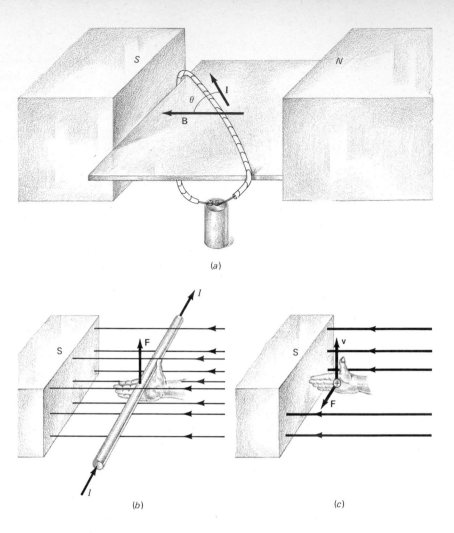

(a)

(b)   (c)

FIGURE 22.7 A simple right-hand rule
for remembering the direction of **F**.

lines and not in the direction of the wire. It is in fact perpendicular to both of
these. As we see in figure 22.7(a) and (b), the field lines (represented by **B**) and
the current **I** define a plane—the plane of the tabletop in figure 22.7(a). The
force on the wire is always perpendicular to the plane defined by **B** and **I**.

Variant of the right-hand rule   To find the direction of the force, many people use a variant of the
right-hand rule. It is shown in figure 22.7(b). *If one's right hand is held flat
with the fingers pointing in the direction of the field lines and the thumb
pointing in the direction of the current, then the palm of the hand will push in
the direction of the force.* In using the rule, note that the field lines and wire
determine a plane, parallel to your hand. The force is always perpendicular to
this plane.

Experiment shows that a similar situation exists when a positive charge
shoots through a magnetic field. As you might expect, the velocity vector for
the moving charge plays the same role as the current vector does for wires.
The force direction is reversed for a   You should examine part (c) of figure 22.7 to make sure that you can find the
negative charge   direction of the force shown there. Moving negative charges also experience a
force in a magnetic field; it is opposite in direction to the force a positive
charge would experience.

22.4 DEFINITION OF THE MAGNETIC INDUCTION *B*

We have already stated that the magnetic field is represented by the symbol **B** much like **E** was used to represent the electric field. To give quantitative meaning to the electric field, we defined **E** to be the force on unit positive charge. A similar definition is needed to specify the strength of the magnetic field. Our definition can be made in terms of the force on either a moving charge or on a current. The two definitions will be related, obviously, since current is a flow of positive charge. We shall state the definition in terms of the experimentally determined force on a current-carrying wire.

In figure 22.8 we see a wire carrying a current I through a magnetic field B. The angle between the field lines (represented by **B**) and the current vector **I** is called θ. Experiment shows that the force $\Delta\mathbf{F}$ on a length ΔL of the wire is related to the strength of the magnetic field B, the current I, and the angle θ in the following way:

$$\Delta F \sim (\Delta L)IB \sin \theta$$

We now *define* the units of B, the quantity which represents the strength of the magnetic field, by making this proportionality an equation. Then

Defining equation for B

$$\Delta F = (\Delta L)IB \sin \theta \qquad\qquad 22.1$$

B is measured in tesla

The units of B are clearly a N/A · m, which turns out to be a kg/C · s. We call this unit the tesla (T) although the older name for this unit, the weber/m² (Wb/m²) is still used frequently. Another unit in common usage, the *gauss* (G), is defined by 1 tesla = 10^4 gauss. Since the gauss is not a member of the SI, we should always convert it to tesla before using it in our equations. To give you a feeling for these units, we note that the earth's field

1 gauss = 10^{-4} T

is of the order of 1 gauss = 10^{-4} T while a strong bar magnet may produce a field of about 0.2 T. As you see, a magnetic field of 1 T is very strong indeed. The quantity B goes under several names. Its correct name is the *magnetic induction*. For reasons to become apparent later, it is also called the *magnetic flux density*. More colloquially, it is designated the magnetic field strength, although another quantity which we shall encounter later is also given this latter name.

Equation *22.1* can be put in more compact form if we recall that the vector cross product $\mathbf{A} \times \mathbf{B}$ gives rise to a vector of magnitude $AB \sin \theta$ which is perpendicular to the plane defined by **A** and **B**. In this notation

Force on a current segment of length ΔL

$$\Delta\mathbf{F} = (\Delta L)\mathbf{I} \times \mathbf{B} = I(\Delta\mathbf{L} \times \mathbf{B}) \qquad\qquad 22.2$$

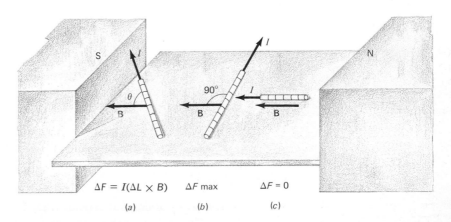

FIGURE 22.8 The force is perpendicular to the tabletop and is proportional to sin θ.

The latter form of this equation, in which the vector character of **I** is assigned to ΔL, is most often used. Recall that $\mathbf{A} \times \mathbf{B} \neq \mathbf{B} \times \mathbf{A}$, and so the force vector is in the correct direction only if the correct order is preserved in the cross product. It is always well to use the right-hand rule, discussed in the previous section, to check the direction of the force.

Notice that the force on the wire varies from maximum to zero as θ, the angle between **B** and **I**, changes from 90 to 0°. This is shown in parts (b) and (c) of figure 22.8. In particular, when the current is along one of the field lines, the force on it is zero. Maximum force is experienced when the field lines are perpendicular to the wire. Let us now apply equation 22.1 (or 22.2) to several situations so that we can better see its meaning.

Illustration 22.1

A certain power line is aligned in an east-west direction on the earth and carries a current of 20 A. Find the force per unit length on it due to the earth's magnetic field, which is about 1.0 G.

Reasoning Since the earth's magnetic field lines are in a direction essentially perpendicular to the wire, we have

$$\Delta F = IB \, \Delta L \sin \frac{\pi}{2}$$

therefore

$$\frac{\Delta F}{\Delta L} = IB$$

$$= (20 \text{ A})(10^{-4} \text{ T})$$

$$= 2 \times 10^{-3} \text{ N/m}$$

This is the force per unit length of the wire, i.e., the force per meter length. Notice that this force represents the weight of about 0.2 g of material.

Illustration 22.2

Find the force on the portion of the current-carrying wire shown in figure 22.9.

Reasoning The forces on the wires AC and DG are zero because these wires are parallel to the field lines. In that case $\sin \theta = 0$, so that $d\mathbf{L} \times \mathbf{B}$ is also zero. A different situation exists along the arc CD.

Consider the small element of arc $d\mathbf{L}$ shown in the figure. It has the direction of the current, as indicated. We see that the angle between $d\mathbf{L}$ and **B** is $180° - \phi$. Therefore the force on the little element of arc is

$$d\mathbf{F} = I \, d\mathbf{L} \times \mathbf{B}$$
$$= I \, dL \, B \sin (180° - \phi)\mathbf{k}$$

The unit vector **k** is used to designate the direction of $d\mathbf{F}$. Since $d\mathbf{L}$ and **B** define a plane, the plane of the page, the force vector must be perpendicular to the page. By use of the right-hand rule, we see that the force is into the page. Therefore **k** is a unit vector into the page.

Clearly, each little element of arc will contribute a force in the same direction, into the page. To find the total force we merely add (or integrate) these forces since they are in the same direction. Therefore

$$\mathbf{F} = \int d\mathbf{F} = \int \mathbf{k} IB \sin (180° - \phi) \, dL$$

where the integral extends from C to D.

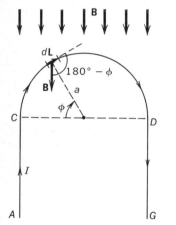

FIGURE 22.9 What is the force on sections *AC* and *DG*?

To simplify this we notice that $\mathbf{k}$, I, and B are all constant, and can be removed from the integral. Moreover,

$$\sin(180° - \phi) = \sin\phi$$

and so

$$\mathbf{F} = \mathbf{k}IB \int \sin\phi \, dL$$

To proceed further, we must relate ϕ and dL. This is easily done if we recall that an element of arc is $r \, d\theta$, where $d\theta$ is the angle subtended by the arc at the center of the circle to which it belongs. In this case

$$dL = a \, d\phi$$

and so we have

$$\mathbf{F} = \mathbf{k}IBa \int_0^\pi \sin\phi \, d\phi$$

The limits on ϕ are set by noticing that at one end of the arc, at C, $\phi = 0$. At the other end, at D, $\phi = \pi$. Upon carrying out the integral and placing in the limits, we find

$$\mathbf{F} = \mathbf{k}(2IBa)$$

In other words, the resultant force on the arc CD is into the page and has a magnitude $2IBa$. Notice that this is the same force that a straight wire between C and D would experience. Can you see why this is true?

Illustration 22.3

Find the total force on the circuit shown in figure 22.10 due to the magnetic field. The center of the circular arc is at P.

FIGURE 22.10 What is the resultant force on this closed circuit?

Reasoning The force on the straight-wire portion is simply

$$F_1 = ILB$$

and is directed toward the bottom of the page as shown. You should check this direction by use of the appropriate rules.

Next consider the force $d\mathbf{F}_2$ on the element of arc $d\mathbf{s}$. Since B and $d\mathbf{s}$ are perpendicular, we have

$$dF_2 = IB \, ds$$

The direction of $d\mathbf{F}_2$ is as shown. Since the various $d\mathbf{F}_2$ vectors for the various elements ds are in different directions, we must take components before adding. We can see that the horizontal component of $d\mathbf{F}_2$ will be canceled by an equal and opposite component due to a symmetric $d\mathbf{s}$ on the other side of the arc. (If we did not notice this, we could compute the horizontal force by use of the method we shall apply to the vertical force. The answer will be zero in either case.)

Each element of the arc gives a vertical force whose magnitude is

$$dF_{2y} = dF_2 \sin\theta$$

as shown. Since these component forces are all in the same direction, they can be added directly. Therefore

$$F_{2y} = \int \sin\theta \, dF_2 = \int IB \sin\theta \, ds$$

We recall that $ds = a\,d\theta$, and rewrite this as

$$F_{2y} = IBa \int_{\theta_0}^{180-\theta_0} \sin\theta\,d\theta$$

Notice the way we have chosen the limits. They are the smallest and largest values of θ as we traverse the arc. Performing the integration, we find

$$F_{2y} = IBa[\cos\theta_0 - \cos(180° - \theta_0)]$$

This, together with the facts that

$$F_{2x} = 0$$

and

$$F_1 = IBL$$

constitutes the solution to the problem. However, if we notice that

$$\cos\theta_0 - \cos(180° - \theta_0) = 2\cos\theta_0$$

and that

$$2(a\cos\theta_0) = L$$

In a uniform magnetic field, a current loop experiences no net force

we see that $F_{2y} = IBL$. It is equal and opposite to F_1. Therefore the total force on the circuit is zero. In general, the following is true: *The resultant force on a closed circuit due to a uniform magnetic field is zero.* Can you prove this?

22.5 FORCE ON A MOVING CHARGE

The force which a magnetic field exerts upon a moving positive charge q is obtained easily from equation *22.2*. We need only recall from our discussion of electric current that the drift velocity $\mathbf{v}$ of the charge q in a wire of cross section A is related to $\mathbf{I}$ by equation *20.2*, namely,

$$\mathbf{I} = qnA\mathbf{v} \qquad\qquad 20.2$$

where n is the number of free charges per unit volume. Substituting this expression for $\mathbf{I}$ into equation *22.2* gives

$$\Delta\mathbf{F} = (\Delta L)Anq\mathbf{v} \times \mathbf{B}$$

But $(\Delta L)A$ is simply the volume of the wire segment of length ΔL. So $n(\Delta L)A$ is the number of moving charges in the portion of the wire for which we are writing the force. Hence the force on a single charge is

$$\mathbf{F} = \frac{\Delta\mathbf{F}}{n(\Delta L)A} = q\mathbf{v} \times \mathbf{B} \qquad\qquad 22.3$$

Force on a moving charge

The direction of $\mathbf{F}$ is reversed if q is changed from positive to negative.

This relation has been tested repeatedly by experiment and is found to be correct. As we have pointed out in connection with our discussion of the direction of the force, the force is perpendicular to both $\mathbf{B}$ and $\mathbf{v}$. Since there

The magnetic force does no work

is no component of the force in the direction of motion, the force does no work upon the moving charge. It neither slows nor speeds it.

In spite of the fact that a steady magnetic field does no work upon a charge moving through it, the field does influence the motion of the charge. For example, suppose a positive charge q is shot into a uniform magnetic field as shown in figure 22.11. (The field shown there is out of the page—recall, the circles represent arrow tips coming toward you.) By use of the right-hand rule, you should be able to show that the force on the charge is directed

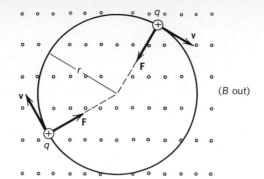

FIGURE 22.11 Since the force **F** is always perpendicular to **v**, it does no work on the charge. What does it do?

*A charged particle moving ⊥ to a uniform **B** follows a circular path*

perpendicular to both **v** and **B** as indicated. Notice that the force simply alters the direction of **v** without slowing or speeding the particle.

This situation is well known to us. It prevails when a ball at the end of a string is swung in a circle and when the sun pulls the earth into a near-circular path centered on the sun. Here, too, the force perpendicular to **v** causes the charged particle to move in a circular path. The magnetic force, $qvB \sin 90°$ in this case, causes the particle to move in a circle. It is this force which supplies the centripetal force, mv^2/r, needed for circular motion. As a result, we can equate these two forces

$$\frac{mv^2}{r} = qvB$$

to obtain

$$mv = qBr$$

As we see, this experimental arrangement can be used to find the momentum mv of a charged particle. One simply measures the radius of the circular path followed by the particle when it is shot perpendicularly into a known magnetic field.

Illustration 22.4

If the charge shown in figure 22.11 is actually a proton ($m = 1.67 \times 10^{-27}$ kg, $q = +e$) moving at a speed of 10^6 m/s, and if the radius of the circle is 4.0 cm, how large is the magnetic induction B?

Reasoning The force on the charge is $q\mathbf{v} \times \mathbf{B}$, and this force must furnish the centripetal force mv^2/r. Equating magnitudes,

$$\frac{mv^2}{r} = qvB \sin \theta$$

But θ, the angle between **v** and **B**, is $\pi/2$; so $\sin \theta = 1$. Solving

$$B = \frac{mv}{rq}$$

Placing in the values given, we find

$$B = 0.26 \text{ T} = 2600 \text{ G}$$

Illustration 22.5

A proton is shot with a speed of 10×10^5 m/s along a line that makes an angle of 37° to the x axis. If a uniform magnetic field of 0.20 Wb/m² exists parallel to the x axis, describe the motion of the proton.

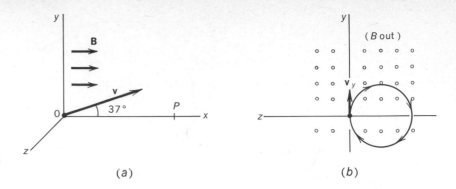

FIGURE 22.12 The proton will follow a helical path as it travels down the x axis with a constant v_x.

(a) (b)

Reasoning The situation is shown in figure 22.12(a). It is well in cases such as this to split the velocity vector into two components, one perpendicular and the other parallel to the field. Since the field is in the x direction, the x component of the velocity is unaffected by the field. This follows from the fact that the force due to v_x is $qv_x B \sin \theta$ where θ is the angle between v_x and B. This angle is zero so the force in question is zero. As a result, the particle moves in the x direction with constant speed, $v_x = v \cos 37° = 8 \times 10^5$ m/s.

The y component of $\mathbf{v}$, however, is perpendicular to the field. If we view the particle along the x axis as shown in figure 22.12(b) (looking toward the origin from $+x$ values), this is equivalent to the problem posed in figure 22.11. The particle will travel in a circle as a result of the interaction between v_y and B. Equating magnetic to centripetal force gives

$$\frac{mv_y^{\,2}}{r} = qv_y B$$

or

$$r = \frac{mv_y}{qB} = 0.31 \text{ m}$$

It will take a time

$$t = \frac{2\pi r}{v_y} = 3.2 \times 10^{-6} \text{ s}$$

for the proton to go around this circle. During that time, it will have traveled with speed $v_x = 8 \times 10^5$ m/s parallel to the x axis, and so, after completing the circle, it will be at point P on the axis [shown in figure 22.12(a)]. The distance OP will be

$$OP = v_x t = 2.6 \text{ m}$$

It should be clear from this that the proton will follow a helixlike path with radius 31 cm and pitch 2.6 m. What sort of path would an electron follow?

22.6 THE HALL EFFECT

In the absence of a gravitational field (or when we can neglect its very small effect) the force on a charge q will be

The Lorentz relation $$\mathbf{F} = q\mathbf{E} + q\mathbf{v} \times \mathbf{B} \qquad\qquad 22.4$$

This is often called the *Lorentz relation,* and shows the combined effect of electric and magnetic fields. If there is no electric field, as in the preceding sections, the first term is zero, and *22.3* results.

Let us now consider a situation in which both terms of *22.4* are of importance. Suppose we have a steady current I flowing in a uniform conducting strip, as illustrated in figure 22.13. One would expect the potential difference between symmetric points M and N to be zero since the potential drop from P to M should be the same as from P to N. This is in fact true. If now a magnetic field **B** is imposed perpendicular to the strip and into the page as shown in (b), the situation is quite different.

For discussion purposes, suppose the current is carried by positive charges q moving to the right through the strip. These charges, moving with velocity **v** perpendicular to a magnetic field **B**, will experience a force

$$\mathbf{F'} = q\mathbf{v} \times \mathbf{B}$$

You should check to see that the usual rules tell us the force is directed toward the upper side of the strip, as indicated in part (b) of figure 22.13. This force will cause the positive charges to be driven to the upper edge of the strip, and so point M will be positively charged and N will be negative. As a result, a downward electric field will be created, as shown.

Of course, this process cannot go on indefinitely. Eventually, **E** becomes large enough so that the total crosswise force on the charges becomes zero. At that time

$$\mathbf{F'} = -q\mathbf{E}$$

The charges will then move straight down the length of the strip, and no further buildup of charges on the edges of the strip will occur. However, there now exists an electric field **E** directed from M to N. There will therefore be a voltage difference between M and N given by

$$V = -\int \mathbf{E} \cdot d\mathbf{y} = Ed$$

where d is the width of the strip.

Definition of the Hall voltage   This voltage is called the *Hall voltage* V_H, after the man who first predicted and measured this so-called *Hall effect*.

To find an expression for V_H, we substitute in the equation

$$F' = qE$$

to give

$$qvB \sin \frac{\pi}{2} = qE$$

and find

$$E = vB$$

The Hall voltage is then merely

$$V_H = vBd \qquad\qquad 22.5$$

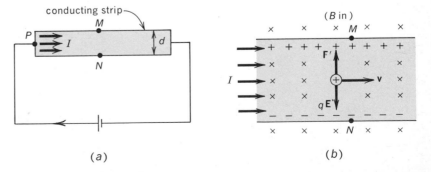

FIGURE 22.13 Can you show that the Hall voltage reverses sign if the charge carriers are negative instead of positive?

(a)   (b)

Equation *22.5* is of importance in several respects. Knowing V_H, B, and d, we can find v, the average drift speed of the charge carriers. Moreover, for positive charges moving in the direction of the current, point M is at a higher potential than N. Alternatively, if the current consists of negative charges moving in the reverse direction, then v will become negative in *23.5*, and the polarity of these two points should change. You should examine figure 22.13(b) and see whether it really is true that the point M will become negatively charged if the charge carriers are negative. It *is* true, of course, and this provides one of the few methods by which we can ascertain the sign of the charge carriers.

The sign of the Hall voltage gives the sign of the charge carriers

We can carry this one step further by noticing that the drift velocity of the charge carriers in a current was found in *20.3* to be

$$v = \frac{J}{nq}$$

where J is the current density and n is the number of current carriers per unit volume. Substituting this value for v in *22.5* gives

$$V_H = \frac{1}{nq} BJd \qquad 22.6$$

The Hall coefficient gives the number of charge carriers in unit volume

The factor $1/nq$ is often referred to as the *Hall coefficient.* Since all the quantities in this equation are measurable exept for n, it can be used to find the number of charge carriers per unit volume. For metals such as Na, Cu, Ag, and Au, the value for n given by this equation is close to the number of valence electrons per unit volume. In the case of semiconductors and the transition elements, the interpretation becomes more complex. We touch briefly on this in a later chapter. However, it should be noticed that the Hall voltage varies inversely as n, and so one would expect it to be larger for semiconductors than for metals.

Illustration 22.6

A silver wire is in the form of a ribbon 0.50 cm wide and 0.10 mm thick. When a 2-A current passes through the ribbon perpendicular to a 0.80-T magnetic field, how large a Hall voltage is produced? (The density of silver is 10,500 kg/m^3.)

Reasoning The atomic weight of silver is 108 kg/kmol, and so the number of atoms in 1 m^3 is

$$n = (6 \times 10^{26} \text{ atoms/kmol})\left(\frac{10{,}500 \text{ kg/m}^3}{108 \text{ kg/kmol}}\right) \cong 6 \times 10^{28} \text{ atoms/m}^3$$

We set the number of atoms equal to n, the number of charge carriers, since silver is monovalent. Substituting the numbers in *22.6*, we find, after setting $J = I/(d)(\text{thickness})$,

$$V_H \approx 1.7 \times 10^{-6} \text{ V}$$

As we see, the Hall voltage is very small in good conductors.

22.7 TORQUE ON A CURRENT LOOP

Many electromechanical devices make use of the fact that a current-carrying coil of wire is caused to rotate by a magnetic field. Electric motors and

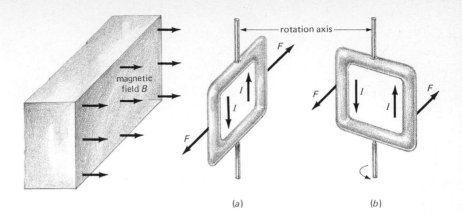

FIGURE 22.14 The magnetic field causes the coil to rotate about the axis. What effects do the forces on the top and bottom of the loop cause?

(a) (b)

moving-coil meters are typical examples. It was pointed out in illustration 22.3 that no net force acts on a current loop in a uniform magnetic field. In spite of this, a torque does exist on such a coil. This is easily seen from figure 22.14. Although the turning effect is zero when the coil is in the orientation shown in (a), a torque is seen to exist for orientation (b). Even though the forces **F** are unchanged in the two cases, the lever arm from the axis is zero in (a) while nonzero in (b). Let us now find the quantitative relation for the torque in terms of I, B, and the coil geometry.

Referring to figure 22.15 we note that each of the two forces F gives a torque $F \times$ lever arm which is $Fa \sin \theta$, where θ is the angle shown in (b). Therefore the torque on the coil is

$$\text{Torque} = 2Fa \sin \theta$$

But F is simply the force on the vertical side of the coil. If the vertical side has a length b, and if the current is I, each vertical wire will contribute a force BIb to F. But there are N loops on the coil, so $F = NBIb$ and the torque becomes

Torque on a coil

$$\tau = (2ab)(NI)(B \sin \theta)$$

Notice that $2ab$ is simply the area of the coil. We can therefore write

$$\tau = (\text{area})(NI)(B \sin \theta) \qquad 22.7a$$

Although we have derived *22.7a* for a very special shaped coil, it turns out that the equation is true for all flat coils. Since NI is the current flowing around the coil, we see that the important features of the coil (aside from its orientation) are the area of the coil and the current in it. In view of this, it is customary to define a quantity called *magnetic moment* of a current loop:

Magnetic moment

$$\mu = \text{magnetic moment} = (\text{area})(I)$$

There is a definite advantage to thinking of a current loop as a bar magnet characterized by its magnetic moment, as we now shall see.

We have already pointed out that a current loop has a magnetic field similar to that of a bar magnet. This is shown once again in figure 22.15(c). Notice that the coil acts like a short, fat bar magnet with north and south poles as indicated. This similarity is further pointed out in figure 22.15(d). Moreover, if we refer back to parts a and b, we see that the coil itself experiences a torque similar to that upon its equivalent bar magnet. For example, in part a the coil will act like a bar magnet with its north pole on

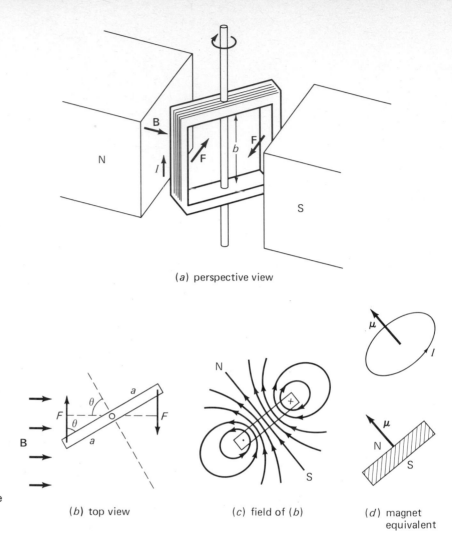

(a) perspective view

(b) top view (c) field of (b) (d) magnet equivalent

FIGURE 22.15 The coil in (a) will experience a torque and for this purpose is equivalent to the short bar magnet shown in (d).

the far side of the coil. Since this side is close to the north pole of the magnet causing the external field, the coil will be rotated in the direction indicated there.

To make full use of his analogy, one gives direction to the magnetic moment μ and makes it a vector. This is shown in figure 22.15(d). We see that $\boldsymbol{\mu}$ is taken perpendicular to the area of the loop and in the direction given by the right-hand rule; i.e., if the fingers circle in the direction of I, the thumb points along $\boldsymbol{\mu}$. As a result, the magnetic moment vector $\boldsymbol{\mu}$ points out of the north pole of the equivalent magnet. This has the following important consequence:

> *When a current loop is placed in a magnetic field, it rotates so as to align its magnetic moment vector with the magnetic field vector.*

One can appreciate why this is true by recalling that a compass needle is simply a bar magnet and that the field direction is defined to be that direction along which the needle aligns. *We shall find it convenient from time to time to think of a current loop as a magnet with magnetic moment μ.*

The equation for the torque on a current loop can be expressed in a very compact form if we note the similarity between

$$\tau = IAB \sin \theta = \mu B \sin \theta$$

and the expression for the cross product, $|\mathbf{A} \times \mathbf{B}| = AB \sin \theta$. We then have for the torque on a current loop

$$\tau = \mu \times \mathbf{B} \qquad\qquad 22.7b$$

Recalling that the direction of $\mu \times \mathbf{B}$ is given by the cross-product notation, we see that its direction is into the page in figure 22.15(d). Obviously we have chosen our cross-product notation to agree with the known fact that the coil experiences a torque that tries to align μ with $\mathbf{B}$. As a result, equation 22.7b not only gives the magnitude of the torque, but it gives its direction as well. Equation 22.7b is also valuable in that it points out at once that θ is the angle between μ and $\mathbf{B}$.

22.8 METER MOVEMENTS AND MOTORS

The fact that a coil of wire which is carrying a current in a magnetic field experiences a torque that tries to align μ with $\mathbf{B}$. As a result, equation 22.7b instruments a coil of wire (containing N loops) is suspended between the poles of a magnet somewhat as shown in figure 22.16. Very often the pole pieces of the magnet are so shaped that, as the coil turns on its axis, the magnetic field from the magnet remains essentially tangential to the area of the loop. That is to say, from 22.7, the torque on the coil is for $\theta = 90°$,

$$\tau = INAB$$

Notice that the torque is proportional to the current in the coil if B is constant, and so the torque can be used as a measure of the current.

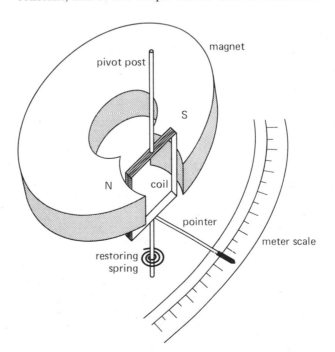

FIGURE 22.16 The coil rotates until the magnetic-field-caused torque on the coil is balanced by the restoring torque of the spring.

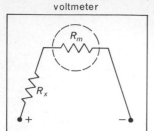

voltmeter

FIGURE 22.17 To make a voltmeter from a sensitive movement, we place a large resistor in series with it.

A voltmeter should have very high resistance

An ammeter should have very low resistance

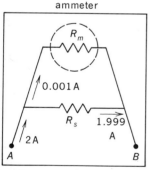

ammeter

FIGURE 22.18 Only a small portion of the current goes through the movement of an ammeter. Most of it goes through the shunt resistor R_s.

As the coil turns under the action of this torque, it distorts the spring and comes to rest at the position where the torque due to the spring balances the magnetic torque on the coil. Since the distortion of the spring is proportional to the rotation of the coil, and also proportional to the magnetic torque, the rotation of the coil can be used as a measure of the current through the coil. The rotation of the coil is read by means of a needle or some other device fastened to the coil. (In ac meters, the magnet may be an electromagnet, in which case B is also approximately proportional to I. The scale on this type of meter would not be linear.)

Meter movements such as the one shown in figure 22.16 can be used to construct a voltmeter or ammeter. Usually, the movement is very sensitive, deflecting full scale for perhaps a small fraction of a milliampere of current through the coil. The appropriate circuits for transforming such a movement into an ammeter or voltmeter are shown in figures 22.17 and 22.18. These circuits are enclosed in the meter box, of course.

Since a voltmeter must have a very high resistance so that it will not short-circuit the voltage it is to measure, a very large resistor R_x is placed in series with the meter movement. The meter movement's resistance is designated R_m. Suppose one had a movement whose resistance $R_m = 60\,\Omega$ and which deflects full scale for a current of 0.0010 A. In order to make a voltmeter from it which would deflect full scale for a voltage of 100 V, one would need the series resistor R_x to satisfy the following equation:

$$100 = (R_x + 60)(0.0010)$$

This is merely Ohm's law written for the meter in the case where there is 100 V causing the full-scale deflection current of 0.0010 A. Solving, we find $R_x = 10^5\,\Omega$. Clearly, the meter will have a high resistance, which is desirable.

In the case of an ammeter, we place the meter directly in the current circuit. We want it to have very low resistance so that it will not disturb the circuit. To achieve this we place a *shunt* resistor R_s in parallel with the movement. If the meter movement discussed above is to be used, and if we want the ammeter to deflect full scale for a current of 2 A, the currents must be as shown in figure 22.18. Making use of the loop rule, we find

$$-1.999R_s + 0.001R_m = 0$$

Since $R_m = 60\,\Omega$, we find that the shunt resistor R_s should be 0.03 Ω. The resistance of the meter between points A and B will be slightly less than this (why?), and so the meter has a very low resistance, as we wish it to have.

Another familiar device which makes use of the torque exerted on a coil by a magnetic field is the motor. There are many kinds of motors, but we shall only concern ourselves with the simplest direct-current type. As shown in figure 22.19, current is sent through the coil of the motor and the coil is mounted in a magnetic field. As the torque exerted by the magnetic field on the coil causes it to turn, the shaft AA' of the motor turns. Connection to the rotating coil is made by brushes sliding along the metal slip rings designated R in the figure.

The current-carrying coil acts like a short bar magnet as discussed previously in connection with figure 22.15. It is convenient to discuss the torque on the coil in terms of the equivalent bar magnet. You should use the right-hand rule to assure yourself that the end of the coil labeled N is really a north pole. In a practical motor, the coil is wound on an iron core. As we shall learn in later chapters, this causes the torque on the coil to be increased by more than a hundredfold.

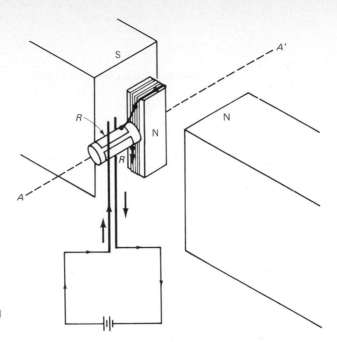

FIGURE 22.19 A simple dc motor. With the slip ring as shown, which way should the motor rotate?

In the figure, the north pole of the coil is repelled by the outer magnet, and if the coil is given a slight, counterclockwise rotation (as viewed along AA' from A to A'), the coil will be made to rotate by the mutual repulsion of the poles. After rotating 180°, the north pole of the coil will be adjacent to the south pole of the outer magnet and would be retarded from rotating away from that position. However, when it reaches that position, the sliding contacts on the split slip ring slide over the gap and the current flowing through the coil reverses. This in turn reverses the poles of the coil, and so once again the situation shown in figure 22.19 is achieved. Repulsion is maintained, and as a result rotation continues.

There are various modifications of such a motor. Most motors consist of several coils wound with their planes through AA' but at various angles to each other. Each coil has current flowing through it for only a small portion of a cycle during the time when its orientation to the field is right for obtaining maximum torque. Such a motor gives a much more uniform torque than one could obtain from a single coil.

22.9 MAGNETIC INTERACTION AT HIGH PARTICLE SPEEDS

When we wrote and used *22.3* for the force on a moving charge, namely,

$$\mathbf{F} = q\mathbf{v} \times \mathbf{B} \qquad 22.3$$

we tacitly assumed that $v \ll c$, so that relativistic effects are unimportant. Let us now examine the proper form of this equation in the case of high particle speeds. Since *22.3* is the defining equation for B, it is, by definition, true at all particle speeds. The only possibility of difficulty occurs in the meaning of $\mathbf{F}$ at relativistic speeds. In that case we must be very careful to use the relativistic form for Newton's second law, namely,

$$\mathbf{F} = \frac{d\mathbf{p}}{dt}$$

where $\mathbf{p}$ is the momentum of the particle. The preferred defining equation for $\mathbf{B}$, equivalent to 22.3, is

$$\frac{d\mathbf{p}}{dt} = q\mathbf{v} \times \mathbf{B} \qquad\qquad 22.8$$

Let us now apply this equation to the motion of a particle in a circular path under the action of a magnetic field.

We follow a procedure very similar to that used when we derived an expression for centripetal force in Chap 9. Here, however, we concentrate on $\mathbf{p}$ rather than $\mathbf{v}$. To find $d\mathbf{p}/dt$, let us first write

$$\mathbf{p} = p_x\mathbf{i} + p_y\mathbf{j}$$

where the z direction is taken perpendicular to the plane of motion.

Upon differentiating, we find

$$\frac{d\mathbf{p}}{dt} = \mathbf{i}\frac{dp_x}{dt} + \mathbf{j}\frac{dp_y}{dt}$$

from which

$$\left|\frac{d\mathbf{p}}{dt}\right| = \sqrt{\left(\frac{dp_x}{dt}\right)^2 + \left(\frac{dp_y}{dt}\right)^2}$$

In our particular situation, the particle is moving around a circular path, so that

$$p_x = p \sin \omega t \qquad \text{and} \qquad p_y = p \cos \omega t$$

Differentiating these values with respect to time and substituting in the preceding equation, we find

$$\left|\frac{d\mathbf{p}}{dt}\right| = p\omega = \frac{pv}{r}$$

Notice that pv/r is simply mv^2/r if we take m to be the relativistic mass of the particle.

Let us now apply our result to the case where $\mathbf{v}$ and $\mathbf{B}$ are mutually perpendicular. Then equation 22.8 becomes

$$\frac{mv^2}{r} = qvB$$

Even at relativistic speeds for circular motion, $mv^2/r = qvB$ if m is the relativistic mass

This is our usual equation for circular motion in a magnetic field. We see that it applies even at relativistic speeds providing we use the relativistic mass $m = m_0/\sqrt{1 - (v/c)^2}$.

Chapter 22
**Questions
and Guesstimates**

1. A silver coin has a multitude of "free" valence electrons moving within it. When the coin is placed in a magnetic field, each moving charge experiences a force. But careful measurements show that the coin experiences no net force due to the constant field. Explain why.
2. Glib teachers sometimes say "The earth's north pole is a south pole and vice versa." What does the teacher mean?
3. A wire carries a current out along the $+x$ axis in a region where the

magnetic field is in the $+y$ direction. (*a*) What is the direction of the force on the wire? (*b*) If the wire is replaced by a beam of particles shooting in the $+x$ direction, what can be said about the direction of the force? (*c*) Why can't we determine from an experiment of this type what the sign of the charge carriers in a wire is?

4. When a beam of electrons is shot into a certain region of space, the electrons travel in a straight line through the region. Can we conclude there is no electric field in the region? Magnetic field?

5. In a certain experiment a beam of electrons is shot out along the positive x axis. It is found that the beam deflects toward positive y values in the xy plane. If this deflection is the result of a magnetic field, in what direction is the field? Repeat for an electric field.

6. It is found that a beam of charged particles is deflected as it passes through a certain region of space. By taking measurements on the beam, how could you determine which caused the deflection, a magnetic or an electric field?

7. A proton is shot from the coordinate origin out along the $+x$ axis. There is a uniform magnetic field in the $+y$ direction. (*a*) Describe the motion of the proton paying particular attention to the quadrants in which it travels. (*b*) Repeat for an electron. (*c*) Repeat if the proton velocity is such that $v_x = v_y$ and $v_z = 0$. (*d*) Repeat if $v_x = v_y = v_z \neq 0$.

8. Suppose you had no magnet or iron available when you landed on the moon. You could, however, do any other electrical experiment you wished. How would you determine the characteristics of the moon's magnetic field? It is assumed that, at the start, you do not know if an electric field exists on the moon.

9. We know that in a TV tube electrons are shot from one end of the tube to the other, where they strike the fluorescent screen. Suppose your little brother insists that his general science teacher says that protons are used, not electrons. How could you prove to him that he was wrong without dismantling the set?

10. Suppose you were given a material which was poorly conducting but still conducted enough to obtain a measurable current through it. How could you decide whether the current was being carried by positive or by negative charges or by both? Give as many ways as you can.

11. Design a method for reversing the direction of motion of an electron beam without changing the speed of the electrons.

12. It is proposed to furnish the propulsion force to a spaceship in the following way. By use of a nuclear reactor or some other means, electricity will be furnished. Large currents will be sent through copper bars in the ship, and the forces exerted on these bars by the earth's magnetic field will propel the ship. What objections do you see to such a plan?

13. A current-carrying loop is suspended by a thin thread in a region where there is a northward-directed magnetic field. (*a*) How will the loop orient in the field? (*b*) What kind of motion will it undergo if it is turned slightly from this orientation and released?

14. Cosmic rays (i.e., charged particles coming to the earth from outer space) are unable to reach the surface of the earth unless they have very high energy. One reason for this is that they have to penetrate the earth's atmosphere. However, for particles coming toward the equator along a radius of the earth, magnetic effects are also important. Explain why, being careful to point out why particles can reach the poles of the earth without encountering this difficulty.

15. When observations are taken of cosmic rays at the equator and above the earth's atmosphere, an *east-west* effect is found. One finds that the cosmic rays approach the earth preferentially from the west. What does this indicate about the sign of the charge on these cosmic particles? (In doing this, recall that the magnetic pole at the north end of the earth is really a south pole since it attracts the north poles of compass needles.)

16. Consider the relation $\mathbf{F} = q\mathbf{v} \times \mathbf{B}$. Is $\mathbf{F}$ always perpendicular to $\mathbf{v}$? To $\mathbf{B}$? Is $\mathbf{v}$ always perpendicular to $\mathbf{B}$?

17. Give an order-of-magnitude estimate of the displacement of the electron beam on a TV screen because of the earth's magnetic field.

Chapter 22
Problems

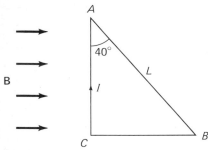

FIGURE P22.1

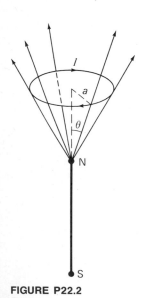

FIGURE P22.2

1. At the equator, the earth's magnetic field is nearly horizontal directed from the southern to northern hemisphere. Its magnitude is about 0.50 G. Find the force (direction and magnitude) on a 20-m wire carrying a current of 30 A parallel to the earth from east to west. Repeat for north to south.

2. In Nebraska, the horizontal component of the earth's field is 0.20 G and the field is inclined at an angle of 70° to the horizontal. (We term this angle the *dip angle*.) If a vertical wire carries a current of 30 A upward there, what is the magnitude and direction of the force on 1 m of the wire?

3. For the circuit shown in figure P22.1, find the magnitude and direction of the force on wire (*a*) *AC*, (*b*) *BC*, (*c*) *AB*, and (*d*) show that the total force on the loop is zero.

4. A circular loop (radius = *b*) lies in the *xz* plane and carries a current *I*. (*a*) A uniform magnetic field $B\mathbf{j}$ is impressed on it. Use integration to find the magnitude of the force on one quarter of the loop. (*b*) Repeat for a magnetic field $B\mathbf{i}$.

5. As shown in figure P22.2, a long bar magnet has poles at its two ends. Near the poles, the magnetic field is radial. At the position of the circular current loop shown the radial field is *B*. What is the magnitude and direction of the resultant force on the current loop?

6. As shown in figure 22.3, a metal bar of mass *m* can slide on two bare wires along a horizontal table. A current *I* is maintained in the wire and bar. The whole system is immersed in a vertical downward magnetic field *B*. If the friction force acting on the bar is *f*, and small enough so the bar will move, find the acceleration of the bar. Will it move toward the right or toward the left?

7. Repeat problem 6 if the magnetic field lines are parallel to the tabletop and make an angle of θ to the line of the bar.

8. In Vermont the dip angle for the earth's magnetic field (angle the field lines make with the horizontal) is 74° and the component of the field parallel to the earth's surface is 0.16 G. If an electron is shot with a speed 10^6 m/s vertically upward there, how large a force acts on the electron and what is the direction of the force? How large an acceleration will this force cause?

9. Repeat problem 8 if the electron is shot horizontally northward.

10. The parallel-plate device shown in figure P22.4 gives rise to an electric field $\mathbf{E}$ perpendicular to a magnetic field $\mathbf{B}$. If the charge q has just the proper speed, it will pass straight through between the plates undeviated. The device is capable of selecting particles of this unique velocity, and it

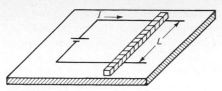

FIGURE P22.3

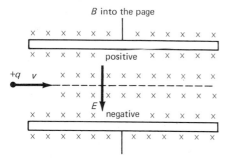

B into the page

+q v

positive

E

negative

FIGURE P22.4

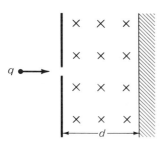

q

d

FIGURE P22.5

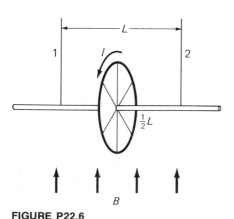

L

1   I   2

½L

B

FIGURE P22.6

is called a *velocity selector*. Show that the particle must have a speed $v = E/B$ if it is to move on a straight-line path. (*Hint:* The magnetic and electric forces on the charge must balance.)

11. A particle of charge 1.6×10^{-19} C and speed 3×10^4 m/s is shot perpendicularly into a magnetic field with $B = 200$ G. If the particle is a sodium ion ($m = 23 \times 1.66 \times 10^{-27}$ kg), find the radius of the circular path it will follow. This general procedure is used in a *mass spectrograph* to separate isotopes of elements. It is discussed in Chap. 37.

12. Assuming a circular path around the earth can be found upon which the earth's field is horizontal and constant at 0.50 G, how fast must a proton be shot in order to circle the earth? In what direction? Neglect relativistic effects. Is this allowable? ($R_{earth} = 6.4 \times 10^6$ m.)

13. A proton is accelerated through a potential difference of 10^5 V. It then enters perpendicular to a magnetic field and follows a circle of 30 cm radius. Find the value of B in the field.

14. A proton that has a kinetic energy of 200 keV follows a circular path in a field $B = 500$ G. What is (*a*) the radius of the circle and (*b*) the frequency of rotation of the proton in the orbit?

15. A particle with charge q and mass m is shot with kinetic energy K into the region between two plates as shown in figure P22.5. If the magnetic field between the plates is B and as shown, how large must B be if the particle is to miss collision with the opposite plate?

16. The magnetic field in a certain region of space is given by $\mathbf{B} = 0.080\mathbf{i}$ T. A proton is shot into the field with velocity $2 \times 10^5\mathbf{i} + 3 \times 10^5\mathbf{j}$ m/s. Give the radius and pitch for the helical path the proton follows.

17. A coil of wire is wound on an evacuated, hollow glass tube with inner diameter 2.0 cm. The coil produces a uniform magnetic field of 3×10^{-2} T parallel to the axis of the tube. Protons with speed 5×10^5 m/s are shot into the tube at a point on its axis. What is the maximum angle to the axis that a particle can be moving as it enters the tube if it is not to hit the wall as it spirals down the tube?

18. In a certain region the magnetic field is uniform and is given by $\mathbf{B} = B\mathbf{i}$. At what angle to the x axis should a charge q be shot if it is to follow a helical path for which the diameter and pitch of the helix are to be equal?

19. A device used to measure magnetic fields makes use of the Hall effect. When in a magnetic field of 200 G it gives a Hall voltage of 16 μV. If, with the same current and orientation, it gives a Hall voltage of 23 μV in an unknown field, what is the magnitude of the unknown field?

20. In a certain Hall-type experiment a current of 0.25 A is sent through a metal strip having thickness = 0.20 mm and width = 0.50 cm. The Hall voltage is measured to be 0.15 mV when a magnetic field of 2000 G is used. (*a*) What is the number of charge carriers (assumed $q = e$) per unit volume and (*b*) what is the drift speed of these carriers?

21. For the current loop shown in figure P22.1 find (*a*) the direction and magnitude of its magnetic moment and (*b*) the torque on the loop.

22. The circular current loop of radius b shown in figure P22.6 is mounted rigidly on the axle as indicated. In the absence of an external magnetic field the tensions in the two supporting cords are equal and are T_0. (*a*) What will be the tensions in the two cords when the vertical magnetic field B is present? (*b*) Repeat if the field is parallel to the axis and to the right.

23. In the Bohr model of the hydrogen atom the electron follows a circular path centered on the nucleus. Its speed is 2.2×10^6 m/s and the radius of

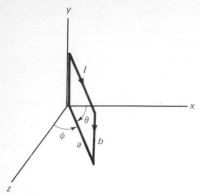

FIGURE P22.7

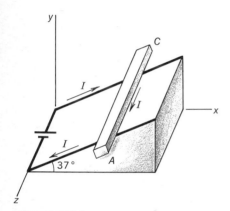

FIGURE P22.8

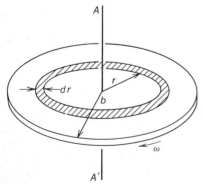

FIGURE P22.9

the orbit is 5.3×10^{-11} m. (*a*) Show that the effective current in the orbit is $ev/2\pi r$. (*b*) What is the magnetic moment due to this orbital motion? (*c*) Show that $\mu = (e/2m)\mathbf{L}$ where $L = mrv$ is the angular momentum of the electron in its orbit.

24. A rigid circular loop of radius r and mass m lies in the xy plane on a flat table and has a current I flowing in it. At this particular place, the earth's magnetic field is $\mathbf{B} = B_x\mathbf{i} + B_z\mathbf{j}$. How large must I be before one edge of the loop will lift from the table?

25. A circular loop of wire of radius r lies in the xy plane and carries a current I. Impinging on it is a magnetic field given by $\mathbf{B} = B_x\mathbf{i} + B_y\mathbf{j} + B_z\mathbf{k}$. Find the vector torque which acts on the coil due to the magnetic field. Two answers are possible; give both.

26. Find the torque which acts on the current loop shown in figure P22.7 if a magnetic field $\mathbf{B} = B\mathbf{i}$ is impressed upon it. If allowed to move, will the loop turn so as to increase or decrease the angle ϕ?

27. In figure P22.8 the bar AC is 40 cm long and has a mass of 50 g. It slides freely on the metal strips at the edges of the incline. A current I flows through these strips and the bar, as indicated. There is a magnetic field $B_y = -0.20$ T directed in the $-y$ direction. How large must I be if the rod is to remain motionless? Neglect the slight overhang of the rod.

28. If the current flowing in the wire of problem 27 is actually 2.50 A, find the acceleration of the rod along the incline.

29. A certain galvanometer movement has a resistance of 40 Ω and deflects full scale for a voltage of 100 mV across its terminals. How can it be made into a 3-A ammeter?

30. If a meter movement deflects full scale to a current of 0.010 A and has a resistance of 50 Ω, how can it be made into a 4-A ammeter?

31. How can the meter movement of problem 30 be made into a 20-V voltmeter?

32. How can the meter movement of problem 30 be made into an ammeter having two ranges, 10 and 1.0 A?

33. How can the meter movement of problem 30 be made into a voltmeter having two ranges, 12 and 120 V?

34. The flat disk shown in figure P22.9 carries an excess charge on its surface of σ C/m². Consider the disk to rotate around the axis AA' with angular speed ω rad/s. Show that the rotating charge is equivalent to currents $\sigma\omega r\,dr$ flowing in all the circular rings, such as the one shown, which serve to make up the disk. If a magnetic field $\mathbf{B}$ is directed perpendicular to the rotation axis AA', show that the torque which acts on the disk is

$$\tau = \frac{\sigma\omega\pi B b^4}{4}$$

35. In a large nuclear accelerator protons are accelerated to kinetic energies of 1×10^9 eV. They are made to travel in circular orbits by an impressed magnetic field. (*a*) What is the relativistic mass ratio m/m_0 for these protons? (*b*) To two significant figures, what is their speed? (*c*) If the magnetic field is 0.9 T, how large is the radius of their orbit?

36. A beam of electrons falls through a very high voltage such that their speeds are approximately c, the speed of light. They enter a region perpendicular to a 0.80-T magnetic field and describe circular paths with a radius of 40 cm. (*a*) What is the relativistic mass of such an electron? (*b*) What is its kinetic energy? (*c*) Through how large a potential difference did it fall?

23 Sources of Magnetic Fields

In this chapter we present the experimental law which allows us to compute the magnetic field resulting from a steady current. The law can be stated in two different but equivalent ways. Each statement of the law has its own particular advantages, and we investigate them. The magnetic fields resulting from several common current distributions are determined.

23.1 AMPÈRE'S CIRCUITAL LAW

We saw in the preceding chapter that a current-carrying wire produces a magnetic field which circles the wire, as shown in figure 23.1. When this field is experimentally investigated, it is found to obey a particularly simple relationship. Of course, from the cylindrical symmetry of the wire, the magnitude of B is constant on a circle of radius r centered on the axis of the wire. B is, as we might expect, proportional to I, the current in the wire. Moreover, the magnitude of the field decreases in a simple way with increasing r. It is found that

The magnetic field circling a long straight wire decreases as 1/r

$$B = \frac{\mu_0 I}{2\pi r} \qquad\qquad 23.1$$

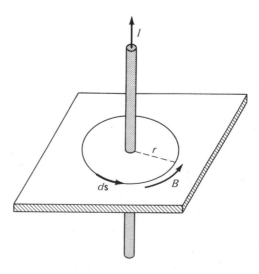

FIGURE 23.1 The field circles the wire as shown. How does B depend on I and r?

where the constant $\mu_0 = 4\pi \times 10^{-7}$ if the units used for B, I, and r are tesla, amperes, and meters, respectively. It is called the *permeability of free space*.

The relation given in *23.1* can be used to lead us to an important method for computing magnetic fields. To see what is, consider the circular path shown in figure 23.1. For the *ds* shown, the scalar product $\mathbf{B} \cdot d\mathbf{s}$ is the component of $\mathbf{B}$ parallel to *ds* multiplied by *ds*. If we add these quantities for each *ds* around the circle, we obtain $\int_{\text{circle}} \mathbf{B} \cdot d\mathbf{s}$ and the relation we wish to state is as follows:

$$\int_{\text{circle}} \mathbf{B} \cdot d\mathbf{s} = \mu_0 I$$

We can see that this reduces to *23.1* by noting that $\mathbf{B}$ is parallel to *ds* so $\mathbf{B} \cdot d\mathbf{s}$ is simply $B\, ds$. Further, B is constant on the circle and so the above equation becomes

$$B \int_{\text{circle}} ds = \mu_0 I$$

But the integral of *ds* around the circle is $2\pi r$ and so we find, as in *23.1*, that $B = \mu_0 I / 2\pi r$.

We can go a step further and show that the relation

$$\int \mathbf{B} \cdot d\mathbf{s} = \mu_0 I$$

applies to any path that encircles the wire. To show this, consider the solid-line path illustrated in figure 23.2. We need first to evaluate

$$\mathbf{B} \cdot d\mathbf{s} = B\, ds \cos \theta$$

To do this we notice that, since $\mathbf{B}$ is tangent to the circle shown in figure 23.2, $ds \cos \theta$ is also tangent to the circle. In the limit of *ds* very small, we can write

$$ds \cos \theta = r\, d\phi$$

where $d\phi$ is the angle subtended by *ds* at the wire and by $ds \cos \theta$ on the circle. As a result, the integral around the solid-line path becomes

$$\oint \mathbf{B} \cdot d\mathbf{s} = \oint Br\, d\phi$$

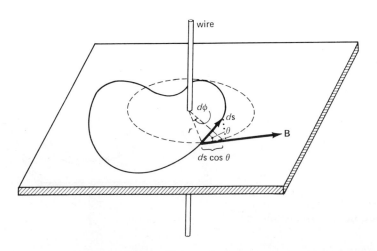

FIGURE 23.2 Can you show that $\int \mathbf{B} \cdot d\mathbf{s} = 0$ if the path does not enclose the wire?

where the circle on the integral means the integral is to be taken along a *closed* path. If we now substitute the experimental value for B and remove the constants from the integral, this becomes

$$\oint \mathbf{B} \cdot d\mathbf{s} = \frac{\mu_0 I}{2\pi} \oint d\phi$$

But from figure 23.2, we see that the sum of all the $d\phi$'s subtended by the various ds's composing the closed path will turn out to be 2π. Putting in this value, we find

$$\oint \mathbf{B} \cdot d\mathbf{s} = \mu_0 I$$

which is the same as found for the circular path. It will be left as an exercise for the student to show that a similar procedure can be applied to a path which does not enclose the wire. In that case it is found that

$$\oint \mathbf{B} \cdot d\mathbf{s} = 0$$

These are all examples of a quite general law first deduced by Ampère. It states: Considering a line integral along a closed path,

Ampère's circuital law
$$\oint \mathbf{B} \cdot d\mathbf{s} = \mu_0 \times \text{current enclosed} \qquad\qquad 23.2$$

In words: The sum of all the $\mathbf{B} \cdot d\mathbf{s}$'s around a closed loop is equal to the total current encircled by the loop multiplied by μ_0. We must be careful to note, however, that the current must be encircled by the loop; currents that do not pass through the area of the loop are not to be included. Furthermore, circled currents for which $\mathbf{B}$ circles in the same direction that we traverse the loop are to be considered positive, while those currents in the opposite direction must be considered negative.

The total current enclosed can be expressed in a more compact way:

*The enclosed current is ∫**J**·d**A** over the area bounded by the enclosing path*
$$\text{Current enclosed} = \int_S \mathbf{J} \cdot d\mathbf{A}$$

where the surface integral of the current density $\mathbf{J}$ is to extend over the area enclosed by the closed path. If a single wire carries a current I through this enclosed area, the integral of $\mathbf{J} \cdot d\mathbf{A}$ over the cross section of the wire gives I in the wire. Notice that *23.2* equals zero, as stated above, if the path of integration encloses no current. This general relation is known as *Ampère's circuital law*. It is of fundamental importance in the computation of magnetic fields.

23.2 B OF A LONG STRAIGHT WIRE

To illustrate the application of Ampère's circuital law, let us use it to find the magnetic field both inside and outside a long straight wire. We shall go into considerable detail in this application in order that the procedure may be seen clearly. In order to find the field at point P shown in figure 23.3(a), we *If possible, choose an integration path* must choose a closed path through P over which the integral of *23.2* is to be *upon which B is constant* evaluated. Notice that we have

$$\oint \mathbf{B} \cdot d\mathbf{s} = \mu_0 \int \mathbf{J} \cdot d\mathbf{A} \qquad\qquad 23.2a$$

and so a path upon which B is constant is desirable. On such a path, B can be removed from the integral.

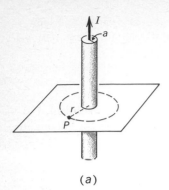

(a)

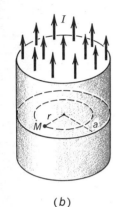

(b)

FIGURE 23.3 Why choose the paths shown for the integration in Ampère's law?

Referring to figure 23.3(a), we see that a circle through P and concentric to the wire provides a desirable path for the integration. From the symmetry of such a path, B should have the same magnitude at all points on it. Using that path for the integration, we note that $\mathbf{B}$ is tangential to the path and is constant so that

$$\int \mathbf{B} \cdot d\mathbf{s} \qquad \text{becomes} \qquad B \int ds$$

Further, since $\int \mathbf{J} \cdot d\mathbf{A}$ is just the current passing through the area of the loop, we replace it by I. We then have that $B \oint ds = \mu_0 I$. The integral over ds is the length of the integration path, $2\pi r$, and so we obtain

$$B = \frac{\mu_0 I}{2\pi r} \qquad\qquad 23.1$$

which is merely 23.1.

To find the field at a point M inside the wire as shown in figure 23.3(b), we also take a circular integration path as indicated. Since B is constant on this path, we find, from 23.2a, that

$$B \oint ds = \mu_0 \int \mathbf{J} \cdot d\mathbf{A}$$

Or, after integrating around the path,

$$2\pi r B = \mu_0 \int \mathbf{J} \cdot d\mathbf{A}$$

The quantity $\int \mathbf{J} \cdot d\mathbf{A}$ is simply the current which flows through our circular integration path of radius r.

If we are to proceed further, some assumption must be made concerning the flow of charge through the wire. Experimentally, one finds the charge flow to be evenly distributed across the wire's cross section unless I is varying at very high frequency. Therefore we assume the current density $\mathbf{J}$ to be uniform and perpendicular to the cross section of the wire. Then

$$J = \frac{I}{\pi a^2}$$

where I is the current in the wire, and a is the wire's radius. We therefore find

$$2\pi r B = \mu_0 \int J \, dA$$

$$= \mu_0 \frac{I}{\pi a^2} \int dA$$

The integral remaining is to be taken over the area *enclosed by the path* of integration. As a result we find

$$2\pi r B = \mu_0 \frac{I}{\pi a^2} \pi r^2$$

The field inside a circular wire is due entirely to the current inside the radius considered

which gives

$$B = \frac{\mu_0 I}{2\pi} \frac{r}{a^2} \qquad r \leq a \qquad\qquad 23.3$$

Notice that the field drops to zero at the center of the wire. As a check, we notice that the values predicted by 23.1 and 23.3 are the same when $r = a$.

Illustration 23.1

Two long straight wires carry currents I_1 and I_2 into the page in figure 23.4. Find the magnetic field at point P.

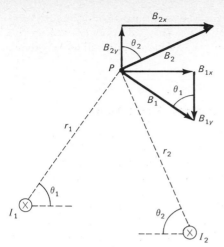

FIGURE 23.4 Where will B due to these wires be zero?

Reasoning The fields B_1 and B_2 due to the two wires are shown in the figure. From *23.1* we have

$$B_1 = \frac{\mu_0 I_1}{2\pi r_1} \quad \text{and} \quad B_2 = \frac{\mu_0 I_2}{2\pi r_2}$$

Taking x and y components, we find

$$B_x = B_{1x} + B_{2x} = \frac{\mu_0}{2\pi}\left(\frac{I_1}{r_1}\sin\theta_1 + \frac{I_2}{r_2}\sin\theta_2\right)$$

and

$$B_y = \frac{\mu_0}{2\pi}\left(\frac{I_2}{r_2}\cos\theta_2 - \frac{I_1}{r_1}\cos\theta_1\right)$$

The resultant field can be found in the usual way to be

$$B = \sqrt{B_x{}^2 + B_y{}^2}$$

Illustration 23.2

Find the force on a long straight wire carrying current I_1 due to a second parallel wire with current I_2 at a distance a away.

Reasoning Wire 1 experiences a force because it is in the magnetic field created by the current in wire 2. This field and the force involved are shown in figure 23.5. You should check to see that the indicated directions are correct. The value of B at the position of I_1 is

$$B = \frac{\mu_0 I_2}{2\pi a}$$

Since **B** is perpendicular to wire 1, the force on a length L of wire 1 is simply

$$F = I_1 L B$$

Or the force per unit length is

$$\frac{F}{L} = \frac{\mu_0 I_1 I_2}{2\pi a} \qquad\qquad 23.4$$

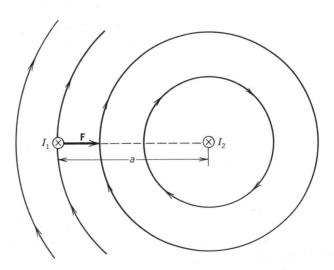

FIGURE 23.5 The current I_2 in wire 2 produces a magnetic field at the position of wire 1. As a result I_1 experiences a force **F** as shown. What happens if I_1 and I_2 are in opposite directions?

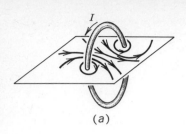

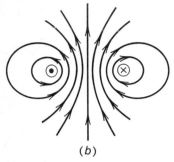

(a)

(b)

FIGURE 23.6 Two views of the magnetic field about a current-carrying loop.

Notice that this expression contains μ_0 as well as the measurable quantities a, F, and L. Defining $\mu_0 = 4\pi \times 10^{-7}$ SI units, this equation is used as the defining equation for the unit of current, the ampere. More will be said about this later.

23.3 FIELD OF A LONG STRAIGHT SOLENOID

The magnetic field resulting from a current in a solenoid, a cylindrical coil, can be pictured most easily after one has examined the field in a loop or compact coil. When a current flows in a circular loop, the magnetic field is found to be directed as in figure 23.6. You should convince yourself, using the right-hand rule, that the lines of flux will circle the wire. [In part (b) only the cross section of the wire is shown; the current goes into the page at the position of the ⊗.] The field due to a coil composed of N such loops would look exactly the same but would be N times as large. We shall return to a quantitative discussion of the field for a loop, later in this chapter.

A solenoid can be thought of as being a cylindrical stack of current-carrying loops. The solenoid of four turns shown in figure 23.7(a) is more loosely wound than many solenoids. However, it shows clearly how the fields of the various loops combine to produce a more or less uniform field inside the solenoid. The situation inside a more tightly wound, much longer solenoid is shown in part (b) of figure 23.7. Notice that the field of a solenoid is rather uniform on its cross section, unlike the field of a single loop. [The ends of the solenoid in (b) are assumed far beyond the small section shown.]

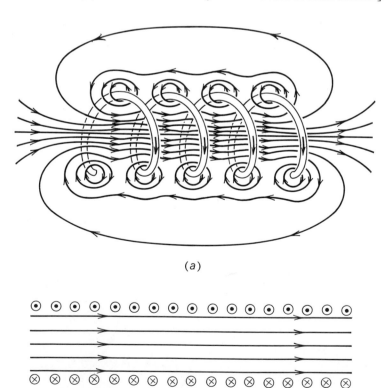

(a)

(b)

FIGURE 23.7 In (b) only a small central section of a long solenoid is shown.

To find the value of B inside a solenoid by use of Ampère's law, we must notice two things. First, as one might guess from figure 23.7, the magnetic field is directed lengthwise along the axis of a tightly wound, long solenoid. Second, if the solenoid is made long, the field lines emerging from the end of the solenoid will fan out widely as they come back around to enter the other end. This indicates that the magnetic field outside the solenoid is many times weaker than inside. We shall soon approximate the situation by considering the field outside the solenoid to be negligibly small.

To make use of Ampere's circuital law in finding the field at point P within the solenoid shown in figure 23.8, we must choose a suitable integration path. Because of the cylindrical symmetry of the situation, one might at first be tempted to choose a circle through P with center on the cylinder axis as an integration path. However, we notice that such a path has no current flowing through its enclosed area. We leave it for the student to prove that the use of such a path merely shows that B has no component tangential to such a circle.

It proves convenient to take the rectangular path $ACDFA$ in figure 23.8 as the path of integration. This path is particularly simple for the following reasons:

1. Outside the solenoid, along DF, we can approximate the field by setting it equal to zero. Therefore

$$\int_D^F \mathbf{B} \cdot d\mathbf{s} = 0$$

2. Alongside FA and CD, $\mathbf{B}$ is either zero (outside the solenoid) or perpendicular to $d\mathbf{s}$ (inside the solenoid), and so

$$\int_C^D \mathbf{B} \cdot d\mathbf{s} = \int_F^A \mathbf{B} \cdot d\mathbf{s} = 0$$

3. Inside the solenoid, $\mathbf{B}$ will be constant and in the same direction as $d\mathbf{s}$ on a line such as AC. Therefore

$$\int_A^C \mathbf{B} \cdot d\mathbf{s} = B \int_A^C ds = Bd$$

where d is the length of side AC.

If we use the rectangle as our path of integration and recall these facts, we can evaluate

$$\oint \mathbf{B} \cdot d\mathbf{s} = \mu_0 \int \mathbf{J} \cdot d\mathbf{A}$$

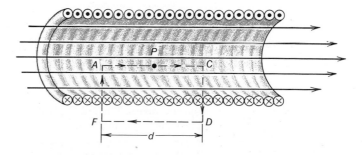

FIGURE 23.8 If we used a circle concentric to the solenoid axis as our path for integration, what result should we find?

to give

$$\int_A^C \mathbf{B} \cdot d\mathbf{s} + \int_C^D \mathbf{B} \cdot d\mathbf{s} + \int_D^F \mathbf{B} \cdot d\mathbf{s} + \int_F^A \mathbf{B} \cdot d\mathbf{s} = \mu_0 \int \mathbf{J} \cdot d\mathbf{A}$$

or

$$Bd + 0 + 0 + 0 = \mu_0(nd)I$$

The term on the right represents the current coming through the area of the rectangle. If there are n wires per unit length of the solenoid, the number coming through the rectangle will be nd. Since each carries a current I, the current through the area is simply ndI.

Solving for B, the field inside the long solenoid, we find

B is axial and uniform in the central region of a long solenoid

$$B = \mu_0 nI \qquad\qquad 23.5$$

Notice that B does not depend upon the position of the point within the solenoid provided we are far from the ends of the solenoid. We therefore conclude that B is uniform in the central region of a very long solenoid. This proves to be a convenient method for obtaining a uniform field. In a typical case

$$I = 2\,\text{A}$$
$$n = 100 \text{ loops/cm} = 10^4 \text{ loops/m}$$

so that
$$B \approx 0.025\,\text{T} = 250\,\text{G}$$

To obtain fields much stronger than this, ferromagnetic materials must be used, in a way to be discussed in Chap. 25, unless one uses superconductors in place of ordinary wires.

23.4 THE TOROID

If a solenoid is bent into the form of a circle so as to join its two ends, one obtains a *toroid*. This device is shown in figure 23.9. (There should be many more loops than shown on the toroid.) To evaluate the field at point P inside the coil, we note that the field forms concentric circles inside the coil. (You should use the right-hand rule to show that the field due to any of the loops of the coil causes the field to be in the direction shown.) If we take our path of integration along the circle of radius r shown, $\mathbf{B}$ will be constant along the path and tangential to it. Therefore

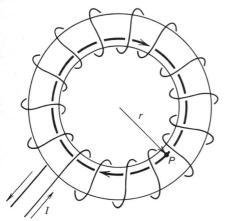

FIGURE 23.9 The magnetic field circles inside the toroid as indicated. What is its functional dependence on r?

$$\oint \mathbf{B} \cdot d\mathbf{s} = \mu_0 \int \mathbf{J} \cdot d\mathbf{A}$$

gives
$$B(2\pi r) = \mu_0 NI$$

In this expression N is the total number of loops on the toroid, and I is the current in each. Solving for B yields

B inside a toroid is parallel to the toroid axis

$$B = \frac{\mu_0 NI}{2\pi r} \qquad\qquad 23.6$$

Notice that in this case B varies from point to point within the toroid. We shall have occasion to refer again to toroids when we discuss magnetic flux measurements. The chief advantage of a toroid is that it has no ends, and so end-effect considerations do not complicate the situation.

23.5 BIOT-SAVART LAW

Although Ampère's circuital law is completely general for steady currents, it suffers some serious practical disadvantages. In general, to use it effectively, one must have a sufficiently simple magnetic field so that B can be removed from inside the integral sign. This becomes impractical even in the case of a simple coil. For this reason we seek an alternative method for computing B due to a current. Such a method is presented by the *Biot-Savart law*.

The Biot-Savart law can be related directly to Ampère's circuital law, and so it is really not a new principle. However, the derivation of one law from the other is beyond the scope of this text, and we merely state the law as an experimental fact. If a small length of wire $d\mathbf{L}$ carries a current I in the direction of $d\mathbf{L}$, then the magnetic field $d\mathbf{B}$ at a distance r due to this incremental length is given by

The Biot-Savart law

$$d\mathbf{B} = \frac{\mu_0 I}{4\pi} \frac{d\mathbf{L} \times \hat{\mathbf{r}}}{r^2} \qquad 23.7$$

where the symbols are as shown in figure 23.10. By $\hat{\mathbf{r}}$ we mean a unit vector in the direction of r. As usual, the vector (cross) product $d\mathbf{L} \times \hat{\mathbf{r}}$ means to take the product of the magnitudes of $d\mathbf{L}$ and $\hat{\mathbf{r}}$ and multiply by $\sin\theta$, where θ is the angle between them. The vector result has the direction given by the right-hand-screw rule.

Notice that the Biot-Savart law, *23.7*, is an inverse square law. It is more complicated than the central-force Coulomb and gravitation laws, however, because of the vector cross product. You should examine the vectors $d\mathbf{L}$, $\mathbf{r}$, and $d\mathbf{B}$ in figure 23.10 so that you are familiar with the directions involved. Unlike Ampère's law, the Biot-Savart law represents the field in terms of contributions from the infinitesimal elements composing the current circuit. It is this property of the law which makes it more useful than Ampère's law in certain circumstances. In other circumstances, for example, the toroid, this property makes the Biot-Savart law very difficult to use.

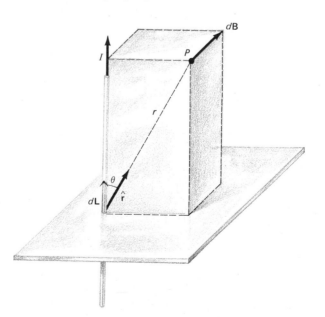

FIGURE 23.10 The vectors **r** and $d\mathbf{L}$ define a plane. We see that **B** is perpendicular to this plane, in accordance with the right-hand rule.

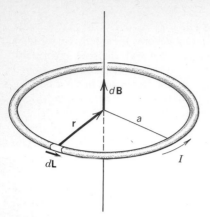

FIGURE 23.11 Identify each of the geometrical quantities in *23.7*, the Biot-Savart law, with the corresponding quantity in this figure.

One feature of the Biot-Savart law is of particular interest. Because $d\mathbf{B} \sim d\mathbf{L} \times \hat{\mathbf{r}}$, $dB = 0$ if $d\mathbf{L}$ and $\hat{\mathbf{r}}$ are in line. Hence a current element contributes no field to points that lie along the line of the current element.

23.6 CIRCULAR LOOP: CENTER POINT

As an illustration of the use of the Biot-Savart law, let us find the magnetic field at the center of a circular current loop. The situation is shown in figure 23.11. To find the field $d\mathbf{B}$ due to the current element $d\mathbf{L}$, we can write *23.7* as

$$dB = \frac{\mu_0 I}{4\pi} \frac{dL}{a^2} \sin \frac{\pi}{2}$$

since $d\mathbf{L}$ is perpendicular to $\mathbf{r}$.

Notice that each element of arc of the current loop gives a $d\mathbf{B}$ which is along the axis of the loop. Since all the $d\mathbf{B}$'s are in the same direction, we can add the fields due to all the $d\mathbf{L}$'s to find

$$B = \frac{\mu_0 I}{4\pi a^2} \int_0^{2\pi a} dL$$

or

$$B = \frac{\mu_0 I}{2a} \qquad\qquad 23.8$$

This, then, is the field at the center of a single circular current loop.

Before leaving this example, it should be pointed out that Ampère's law is impractical in the present case. The difficulty becomes apparent if one attempts to find a path of integration upon which B is constant. No such path, enclosing an area through which the current passes, is possible.

23.7 CIRCULAR LOOP: AXIAL POINT

We wish to find B at the point P on the axis of a circular current loop as shown in figure 23.12. Before solving the present problem by use of the Biot-Savart law, you should convince yourself that the use of Ampère's law is not practical in this instance. The symbols which appear in *23.7* are illustrated in the figure. Be sure you understand the direction assigned to dB. When you do, you will notice that the components of $d\mathbf{B}$ perpendicular to the axis will cancel, as one may easily show by considering the field due to an element opposite to $d\mathbf{L}$.

From *23.7* we have

$$d\mathbf{B} = \frac{\mu_0 I}{4\pi} \frac{d\mathbf{L} \times \hat{\mathbf{r}}}{r^2}$$

In the present case $\mathbf{r}$ and $d\mathbf{L}$ are perpendicular, and so this becomes

$$dB = \frac{\mu_0 I}{4\pi} \frac{dL}{r^2}$$

But from figure 23.12 it is clear that the $d\mathbf{B}$ due to the various $d\mathbf{L}$ will have different directions. We must therefore take components and add the components to find the effect of the whole current loop.

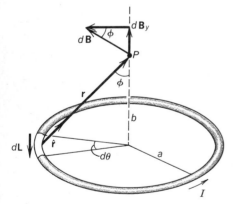

FIGURE 23.12 What is the direction of the resultant field at *P*?

As mentioned above, the components of $d\mathbf{B}$ perpendicular to the axis cancel. We need consider only the axial components dB_y. From the figure

$$dB_y = dB \sin \phi = \frac{\mu_0 I}{4\pi} \frac{dL}{r^2} \sin \phi$$

$$B_y = \int dB_y = \frac{\mu_0 I}{4\pi} \int \frac{dL \sin \phi}{r^2}$$

where the integral is to include all the dL composing the current loop. In moving around the loop, both r and $\sin \phi$ are constant, so that

$$B_y = \frac{\mu_0 I \sin \phi}{4\pi r^2} \int dL$$

$$= \frac{\mu_0 I \sin \phi}{4\pi r^2} 2\pi a$$

This expression can be simplified by replacing $\sin \phi$ and r from the dimensions given in the diagram. We then obtain

$$B_y = \frac{\mu_0 I a^2}{2(a^2 + b^2)^{3/2}} \qquad\qquad 23.9$$

Can you show that, when $b = 0$, this reduces to the result of the preceding section? The general expression for B at *any* point outside a current loop is, in principle, found in the same way as we have done for an axial point. In practice, however, the geometry becomes quite complex.

Illustration 23.3

Find the magnetic field at the point P shown in figure 23.13.

Reasoning Since the two side wires, MN and its companion, are in direct line with P, the angle between $\Delta\mathbf{L}$ and $\hat{\mathbf{r}}$ is zero for these wires. Therefore $\Delta\mathbf{L} \times \hat{\mathbf{r}}$ is zero and so they contribute nothing to B at point P. We need consider only the wire NS. Taking $\Delta\mathbf{L}$ and $\hat{\mathbf{r}}$ as shown in the figure, the Biot-Savart law tells us

$$d\mathbf{B} = \frac{\mu_0 I}{4\pi} \frac{d\mathbf{L} \times \hat{\mathbf{r}}}{r^2}$$

The direction of all the $d\mathbf{B}$'s due to the $d\mathbf{L}$'s of the length NS is into the page, and so we can add them without taking components. Therefore

$$B = \left| \frac{\mu_0 I}{4\pi} \int \frac{d\mathbf{L} \times \hat{\mathbf{r}}}{r^2} \right| = \frac{\mu_0 I}{4\pi} \int \frac{\sin \theta \, dx}{r^2}$$

where dL has been replaced by dx.

We have three variables under the integral, x, r, and θ. If we replace r^2

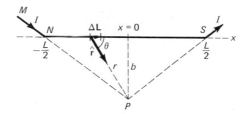

FIGURE 23.13 What is B at point P due to the wire segment MN?

by $x^2 + b^2$ and $\sin \theta$ by $b/\sqrt{x^2 + b^2}$, we find

$$B = \frac{\mu_0 I b}{4\pi} \int_{-L/2}^{L/2} \frac{dx}{(x^2 + b^2)^{3/2}}$$

Using the tables to evaluate this integral we obtain

$$B = \frac{\mu_0 I L}{2\pi b \sqrt{L^2 + 4b^2}}$$

As a check on our result, we note that if $L \gg b$, this reduces to the expression for b outside a very long straight wire.

23.8 SOLENOID: END EFFECTS

In Sec. 23.3 we found B inside a long solenoid by use of Ampère's law. We found the field to be uniform throughout the solenoid cross section as long as we could ignore end effects. With the aid of the result obtained in the last section, we can carry the computation into the region near the ends of the solenoid. However, we shall be restricted to axial points of the solenoid.

Consider the solenoid shown in figure 23.14. It is essentially a stack of circular current loops. The field at any point on the axis of such a loop has been found in *23.9*. If we concern ourselves with the loops within the length dy, their number will be $n\,dy$, where n is the number of current loops per unit length of the solenoid. These $n\,dy$ loops give a field dB_y at point P, which is $n\,dy$ times larger than the field of a single loop given in *23.9*. Therefore

$$dB_y = \frac{\mu_0 I a^2 n\,dy}{2(a^2 + y^2)^{3/2}}$$

where b in *23.9* has been replaced by its equivalent shown in figure 23.14, namely, y.

To find the total field at P we need to integrate this expression over the whole length of the solenoid, i.e., from $-(L - g) \leq y \leq g$. We then have

$$B_y = \frac{\mu_0 n I a^2}{2} \int_{-(L-g)}^{g} \frac{dy}{(a^2 + y^2)^{3/2}}$$

This integral is easily found in the usual integral tables, or can be evaluated by making the substitution (see figure 23.15)

$$y = a \tan \theta$$

so that

$$dy = a\,\frac{d\theta}{\cos^2 \theta}$$

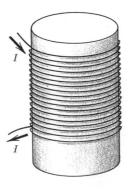

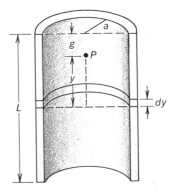

FIGURE 23.14 To find B at point P, we add the fields due to the current loops which compose the solenoid.

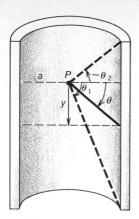

FIGURE 23.15 A diagram to show the physical significance of the angular transformation used to evaluate the integral leading to *23.10*.

We shall use the substitution to illustrate a point. The integral transforms as follows. Since

$$\cos \theta = \frac{a}{\sqrt{a^2 + y^2}}$$

the expression for B_y becomes

$$B_y = \frac{\mu_0 nI}{2} \int \cos \theta \, d\theta$$

In order to find the limits on this integral, let us refer to figure 23.15, where the significance of our substitution is shown. We set $\tan \theta = y/a$; so θ is the angle shown. The limiting values on θ are θ_1 and θ_2, as in the figure. After carrying out the integral, we find

$$B_y = \frac{\mu_0 nI}{2}(\sin \theta_1 - \sin \theta_2) \qquad 23.10$$

Notice that, as drawn in figure 23.15, θ_2 is actually a negative angle. Therefore, if the solenoid is very long in each direction from point P,

$$\theta_1 \approx 90° \qquad \text{and} \qquad \theta_2 \approx -90°$$

so that

$$B_y \approx \mu_0 nI$$

which is the value found for a long solenoid in Sec. 23.3.

If P is exactly at the end of the solenoid and if the other end is far away, we have

$$\theta_1 \approx 90° \qquad \text{and} \qquad \theta_2 \approx 0$$

The axial B at the end of a long solenoid is one-half the interior value

so

$$B_y = \tfrac{1}{2}\mu_0 nI \qquad \text{(end)}$$

Therefore we see that the field is half as large at the center of the end face of the solenoid as in the middle. The variation of B along the axis of the solenoid is shown in detail in figure 23.16. Notice the horizontal scale to be in terms of distance from the center of the solenoid.

It should be seen from these results that the important feature of a solenoid when considering the uniformity of its field is the ratio of its length to its diameter. If this ratio is large, the field is essentially uniform throughout most of the solenoid. Although these considerations apply strictly only to

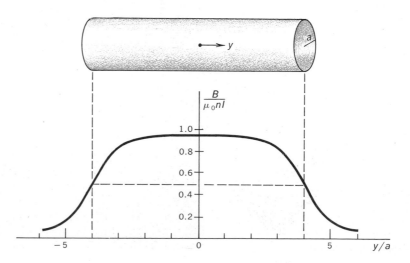

FIGURE 23.16 Variation of B along the axis of a solenoid of which the radius is one-eighth the length.

axial points since they are based on 23.9, they are also qualitatively correct over the whole cross section of the solenoid.

Illustration 23.4

Find the magnetic field outside an infinite current sheet, carrying current into the page as shown in figure 23.17.

The Electron and Thomson's Atom

We have previously discussed the investigations of Crookes, Perrin, and others into the nature of cathode rays (page 390). They found that cathode rays consist of negatively charged particles. However, essentially nothing was known to them about the mass m or charge q of the particles. Data concerning these quantities were first obtained by J. J. Thomson (1856–1940) in about 1897. Thomson made use of the fact that charged particles moving through a magnetic field are deflected.

As we saw in Chap. 22, if a charged particle moving with speed v enters a uniform magnetic field of strength B, its velocity being perpendicular to **B**, the particle will describe a circle of radius $r = mv/Bq$, where m and q are the mass and charge of the particle. Moreover, the direction in which the circle is traversed indicates whether the particle

Circular path of electrons in a uniform magnetic field. (*From Oldenburg and Holladay, "Introduction to Atomic and Nuclear Physics," 4th ed., McGraw-Hill Book Company, New York, 1967. Courtesy of K. T. Bainbridge.*)

Reasoning Suppose the current passing through the unit width of the sheet is j. Since the magnetic field around a wire carrying current into the page would circle the wire clockwise, we infer that the field will be directed toward the right above the sheet and toward the left below it. Moreover, the field should be symmetrical above and below the sheet, and, because the sheet is infinite, the field should be uniform.

---

is negatively or positively charged. The speed of the particles, v, can be measured by passing the particles through a velocity selector consisting of crossed electric and magnetic fields, as was also pointed out in Chap. 22. Making use of both these techniques, Thomson was able to measure r and v and thereby compute the ratio of charge to mass, q/m, for the particles in the cathode rays. He also confirmed the fact that the particles were negatively charged. The value of q/m was found to be independent of the type of gas in the tube.

As we have pointed out in our discussion of Faraday's electrolysis experiments (page 352), an estimate of the charge carried by ions in solution had already been made by Stoney. In addition, Stoney had named the smallest charge found on an ion, the electron. Thomson associated this unit of charge with the charge q of the cathode rays, and used it to find the mass of the particles in the rays from the value of q/m found in his own measurements. Thomson found m to be about 2000 times smaller than the mass of the lightest atom, hydrogen. We should note that an alternative was possible to Thomson. Upon assuming the mass of the cathode-ray particles to be the same as that of a hydrogen atom, the charge on the particles would be about 2000 times larger than the unit of charge found in electrolysis experiments. Thomson considered this to be an unlikely alternative.

Although the ratio q/m for the cathode rays was independent of the gas used in the cathode-ray tube, later, more difficult experiments showed that positive particles travel through the tube in a direction opposite to the negative cathode rays. Their speed was much less than that of the cathode rays, and their value of q/m was several thousand times smaller than for the cathode rays. In addition, the value of q/m for the positive particles depended upon the type of gas in the tube.

By using the same value of q as for Stoney's electron, the mass of the positive particles was computed, and was found to be comparable with the mass of the gas atoms being used.

With these facts in mind, Thomson pictured the cathode rays to consist of negative particles, namely, Stoney's electrons, torn loose from the originally neutral gas atoms. This led him to conceive of the atom as a more or less spherical, positively charged mass in which were embedded enough very small negative electrons so that the atom as a whole carried no net charge. His concept of the atom became quite widely accepted in the years following 1897. We shall see in a later chapter, however, that in 1911 Thomson's atom model was shown to be wrong by Ernest Rutherford (1871–1937). Thomson's concept of the nature of the cathode rays, though, has been fully justified by many other pieces of evidence adduced since 1897.

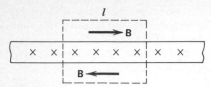

FIGURE 23.17 The infinite sheet carries current into the page. A current jl flows through the area of the rectangle if j is the current per unit width of the sheet.

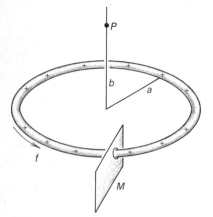

FIGURE 23.18 The charged loop rotating with frequency f constitutes a loop current.

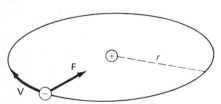

FIGURE 23.19 Bohr pictured the electron as circling the nucleus.

Using Ampère's law and the path of integration shown, we have

$$Bl + Bl = \mu_0 jl$$

from which

$$B = \frac{\mu_0 j}{2}$$

Illustration 23.5

A thin rod is bent into the form of a circle as shown in figure 23.18. The rod carries a charge λ per unit length. If the circle is rotated around its axis at a speed of f rotations per second, find the value of B at point P on the axis due to it.

Reasoning The moving charge caused by the rotation of the circle constitutes a current. To find the current we note that by definition the current is the charge passing through a given cross section per second. At a point such as M, the charge flowing through the area shown, per second, is the total charge on the circle multiplied by f, the number of times the circle rotates through the area in a second. Therefore

$$I = (2\pi a\lambda)f$$

We therefore have reduced the problem to that of a circular current loop. From *23.9* we find

$$B = \frac{\mu_0 a^2 (2\pi a\lambda f)}{2(a^2 + b^2)^{3/2}}$$

or

$$B = \frac{\pi \mu_0 a^3 \lambda f}{(a^2 + b^2)^{3/2}}$$

Illustration 23.6

In 1913, Niels Bohr postulated the following picture of the hydrogen atom, which formed one of the basic steps in the development of quantum theory. The electron revolves in a circular orbit about the nucleus as center, as shown in figure 23.19. The allowed orbit was postulated to be that for which the angular momentum of the electron, mvr, was $h/2\pi$, where h is a universal constant called Planck's constant (6.63×10^{-34} J·s). Find the magnetic field caused at the nucleus by the rotation of the electron in its circular orbit.

Reasoning This problem is much like the preceding one. We must find the frequency with which the electron rotates in the orbit to find the equivalent current due to it. According to Bohr,

$$mvr = \frac{h}{2\pi}$$

where

$$m = \text{electron mass}$$
$$v = \text{its velocity}$$
$$r = \text{path radius}$$

Two unknowns, v and r, exist in this relation. We need yet another equation.

The centripetal force holding the electron in orbit is furnished by the electrostatic attraction between the electron and nucleus. Therefore we can write

$$\frac{mv^2}{r} = \frac{1}{4\pi\epsilon_0}\frac{q^2}{r^2}$$

where q is the charge on the electron or proton.

Solving these two equations for v and r, we find

$$v = \frac{q^2}{2\epsilon_0 h}$$

and

$$r = \frac{\epsilon_0 h^2}{\pi m q^2}$$

But the frequency of rotation will be

$$f = \frac{v}{2\pi r} = \frac{mq^4}{4\epsilon_0^2 h^3}$$

The equivalent current will be $I = qf$, and so, from 23.8,

$$B = \frac{\mu_0 r^2 I}{2r^3} = \frac{\mu_0 I}{2r}$$

which, after substitution, yields

$$B = \frac{\pi\mu_0 q^7 m^2}{8\epsilon_0^3 h^5}$$

Placing in the numerical values, one finds

$$B \approx 14\ \text{T}$$

As we see, the electron gives rise to a very sizable field. We must interject a word of caution, however. Physicists now believe Bohr's theory to be in serious error. This is discussed in more detail when we consider the structure of atoms in later chapters.

Chapter 23
Questions
and Guesstimates

1. A loop of wire carrying a current of 5 A clockwise is concentric and coplanar, with a large loop carrying a current of 10 A counterclockwise. Describe the forces on each loop. Repeat the question if the two loops are coplanar but their areas do not overlap.
2. A serious problem encountered when one tries to achieve very high magnetic fields with a solenoid has to do with the fact that the magnetic field of loops in the solenoid exerts forces on the neighboring loops. These forces require the solenoid to be strongly constructed. Describe these forces; in particular, are they explosive or compressive?
3. What determines which method one should use in computing the magnetic field in a certain situation, Ampère's law or the Biot-Savart law?
4. In order to minimize magnetic field effects on surrounding components in electronic devices, the wire carrying current to an element is often twisted together with the wire carrying current away. What is the basis for this procedure?
5. A long, straight piece of copper tube carries a current lengthwise. What can you say about the magnetic field inside the tube? Outside the tube?
6. Suppose a long, straight, uniformly wound solenoid has an oval cross

section rather than one that is circular. What can you say about the magnetic field inside it? What if its cross section is rectangular with one side being much smaller than the other?

7. Give an example to show that the magnetic field is not, in general, a conservative field.

8. Suppose in a certain conductor both positive and negative charges moved in such a way that the same amount of negative charge moved to the left through a cross section as the amount of positive charge which moved to the right. Show that the magnetic field is not zero but actually twice as large as it would be if only the negative charges were moving.

9. A spiral spring supporting a mass m at its end is being used by a student to measure the mass from the amount the mass stretches the spring. Suppose a wire is connected to the top of the spring and another to the bottom so that current can be sent through the spring. If the current is suddenly turned on, will the spring length increase or decrease?

10. It is believed that the earth's magnetic field is the result of currents flowing in the metallic central core of the earth. How must these currents flow to give rise to the observed field? Could currents in the upper atmosphere be responsible for the field?

11. Design an experiment to determine the direction of current flow in a lightning flash.

12. A beam of electrons shoots straight down a TV tube so as to strike a central spot on the TV screen. (*a*) Describe the magnetic field generated by the beam and (*b*) indicate whether the magnetic field generated by the moving charges tends to focus or defocus the beam.

13. A lighted fluorescent light bulb carries a current from one end to the other. How could you compute the magnetic field outside a straight fluorescent tube? What approximations might you make in various circumstances?

14. Estimate the magnetic field at a distance of 1 m from a lamp cord which lights a 60-W lamp. Give an answer which is in error by less than 0.01 G.

Chapter 23
Problems

1. A long straight wire carries a current of 5 A out along the x axis. Find the magnitude and direction of $\mathbf{B}$ at the point $(3, 2, 0)$ m.

2. Two long straight wires carry currents of 5 A out along the x axis and y axis, respectively. Find the direction and magnitude of $\mathbf{B}$ at the point $(40, 20, 0)$ cm.

3. Two long parallel straight wires are 40 cm apart and carry currents of 5 A each in the same direction. Find the value of B midway between the two wires. Repeat if the current directions are opposite.

4. Two long parallel straight wires lie in the xy plane parallel to the y axis. One wire is coincident with the y axis and the other passes through the point $x = 20$ cm, $y = z = 0$. Both wires carry currents of 5 A in the $+y$ direction. Find $\mathbf{B}$ at the point $(30, 0, 0)$ cm. Repeat for $(5, 0, 0)$ cm.

5. For the two wires described in problem 4, find the components of $\mathbf{B}$ at the point $(10, 0, 5)$ cm.

6. For the two wires described in problem 4, find the components of $\mathbf{B}$ at the point $(0, 0, 40)$ cm.

7. As shown in figure P23.1, a straight wire of radius a carries a current I_1 along the axis of a metal tube with inner radius b and outer radius c. The tube carries a current I_1 in a direction opposite to that in the wire. Find B

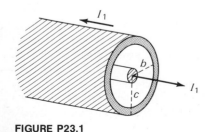

FIGURE P23.1

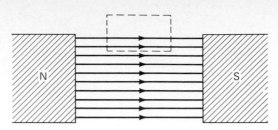

FIGURE P23.2

for $a < r < b$. Repeat for $r > c$. Coaxial cables used in television and other electronic devices arrange the lead wires in this fashion so as to minimize magnetic effects.

8. Repeat problem 7 if the current in the outer tube is I_2. Repeat if I_1 and I_2 are in the same direction.

9. A long straight wire carries a current I out along the x axis while a second wire carries a current $\frac{1}{2}I$ out along the y axis. Where is their combined magnetic field zero?

10. A certain hollow solenoid is 50-cm long, has a radius of 5 mm, and has 400 loops of wire on it. A current of 3 A is sent through it. (*a*) How fast should an electron be moving if it is to follow a circular path of 2-mm radius inside the solenoid? (*b*) How fast must the electron be moving if it is shot at an angle of 20° to the axis and is to follow a helical path with 2-mm radius? What is the pitch of the helix in that case?

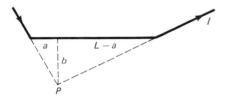

FIGURE P23.3

11. A long straight wire carries a current I out along the x axis. It is desired to have an electron follow a helical path with the wire as axis. Find the ratio of v_x to $v_\perp^2$, the velocity component perpendicular to the wire, for this to occur.

12. Use the Biot-Savart law to show that the field outside a long straight wire is $\mu_0 I / 2\pi r$.

13. A long straight metal tube has inner radius a and outer radius b. It carries a current I lengthwise spread uniformly over its cross section for $a \leqslant r \leqslant b$. Find the magnetic field in each of the following regions: (*a*) $r < a$; (*b*) $a \leqslant r \leqslant b$; (*c*) $r > b$.

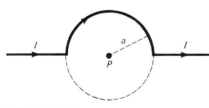

FIGURE P23.4

14. A very long straight rod of outer radius b carries a surface charge σ C/m². The rod is spun with frequency f on its axis. (*a*) Find the magnetic field for $r < b$. (*b*) About what is the field for $r > b$?

15. It is impossible to have a uniform magnetic field that diminishes to zero abruptly in the way shown in figure P23.2. In practice the field must fringe out in curved lines at its edge. Prove that the field shown in the figure is incorrect by applying Ampère's law to the dashed path indicated.

16. For the situation shown in figure P23.3, show that **B** at point P is into the page and given by

$$\frac{\mu_0 I}{4\pi b}\left(\frac{L-a}{\sqrt{b^2 + (L-a)^2}} + \frac{a}{\sqrt{b^2 + a^2}}\right)$$

17. Making use of the result given in problem 16, find the magnitude of B at the center of a square loop of wire with side length c and current I.

18. Starting from the Biot-Savart law, show that B at point P in figure P23.4 is $\mu_0 I / 4a$.

19. Refer to figure P23.5. Find B at the center point if the angle θ is (*a*) 0°, (*b*) 40°, (*c*) 90°.

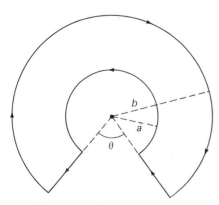

FIGURE P23.5

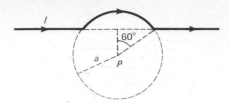

FIGURE P23.6

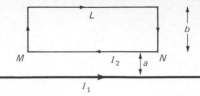

FIGURE P23.7

20. In figure P23.6, show that the magnitude of **B** at point P due only to the current flowing in the circular arc shown is $\mu_0 I/6a$. The field due to the two end wires is the subject of the next problem. Do not assume the result for a complete circle is known. Start directly from the Biot-Savart law so that you obtain practice using it.

21. Find the magnitude of B at point P in figure P23.6. Assume the end wires are very long. You can make use of the result of problem 20. If you are ingenious, you can use the result of illustration 23.3 to simplify the problem.

22. A small coil with n loops and of area A is suspended inside a solenoid of length L and a total of N loops on it. The axis of the coil makes an angle θ to the axis of the solenoid. If a current I flows through each, find the torque on the inner coil.

23. For the situation shown in figure P23.7, find the force experienced by side MN of the rectangular loop. Also, find the torque on the loop.

24. Following the method of Sec. 23.1, prove that $\oint \mathbf{B} \cdot d\mathbf{s} = 0$ if the integration path encloses no current.

25. A bundle of 1000 parallel long straight wires form a cylinder of radius b. Each wire carries a current I so that the total current passing down the cylinder is $1000I$. Assuming the wires to be uniformly packed, find the force per unit length on one such wire at a radius $r < b$. What is the direction of the force? A similar situation occurs in a beam of moving charges and leads to what is called the *pinch effect*.

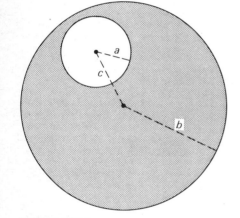

FIGURE P23.8

26. As shown in figure P23.8, a long straight metal rod has a very long hole of radius a drilled parallel to the rod axis. If the rod carries a current I, show that B has the value (a) $\mu_0 I a^2/2\pi c(b^2 - a^2)$ on the axis of the rod and (b) $\mu_0 I c/2\pi(b^2 - a^2)$ on the axis of the hole. (*Hint:* Superpose the fields from two cylindrical conductors to obtain the required field.)

27. For the situation of problem 26, find the field for a point P on the line of centers at a radius (a) $R > b$ from the center of the rod and (b) $R < b$ but not within the hole.

28. So-called *Helmholtz coils* are sometimes used to obtain a uniform magnetic field in situations where a solenoid would be impractical. Two coils, placed as shown in figure P23.9, are used. Show that b at point P

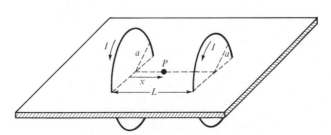

FIGURE P23.9

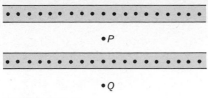

•P

•Q

FIGURE P23.10

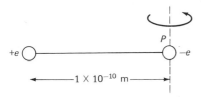

+e ⊙———————————⊙ −e

P

|←——1 × 10⁻¹⁰ m——→|

FIGURE P23.11

has value

$$B = \frac{\mu_0 I a^2}{2}\left[\frac{1}{(a^2 + x^2)^{3/2}} + \frac{1}{(2a^2 + x^2 - 2ax)^{3/2}}\right]$$

if $L = a$. To show that the field is uniform in this region, show that $\partial B/\partial x$ and $\partial^2 B/\partial x^2$ are both zero when P is at the center between the coils.

29. The two infinite plates shown in cross section in figure P23.10 carry jA of current out of the page per unit width of plate. Find the magnetic field at points P and Q.

30. Referring to the situation described in problem 29 and shown in figure P23.10, find the force per unit area on the lower plate because of the current in the upper plate.

31. As a very rough approximation to the true situation, we can approximate the HCl molecule as shown in figure P23.11. Assuming its rotation frequency to be 10^{12} Hz, find the magnetic field its rotation causes at point P.

32. A circular disk of radius a carries a uniform surface charge σ C/m². The disk rotates about an axis through its center and perpendicular to its face. Its angular speed is ω rad/s. Find the magnetic field it generates at the center point of the disk.

24 Magnetic Induction Effects

When a change occurs in a magnetic field which passes through the area enclosed by a wire loop, a current is caused to flow momentarily in the loop. This is one example of a potential difference induced by use of a magnetic field. In this chapter we consider this phenomenon in some detail and state the laws which apply to it. These laws are then applied to several cases of interest.

24.1 INDUCED EMFS

If the magnetic field through a closed loop of wire is changed, a current is found to flow in the loop. This is only a transient current and exists only as long as the magnetic field continues to change. Since currents are caused to flow through ordinary wires by batteries and other sources of electric energy, which we have named "sources of emf," we conclude that the changing magnetic field causes a source of emf to exist in the coil. We call this an *induced emf*. Typical experiments obtained with a coil of wire connected to a galvanometer are shown in figure 24.1.

A changing magnetic field can induce an emf

You will notice in each of the experiments shown in figure 24.1 that a current flows through the coil only when the magnet is moving. Seemingly a battery effect, an induced emf, occurs in the coil each time the strength of the magnetic field in the region of the coil is changed. The emf exists, and the current flows, only when the change is occurring. Moreover, the induced battery effect, or emf, reverses polarity depending upon whether the magnetic field is increasing or decreasing, as we see by comparing parts (*b*) and (*d*). Equally intriguing is the fact, seen by comparing (*b*) and (*e*), that the direction of the magnetic field is important; even though the field is increasing in strength in each of these cases, the emf is reversed.

There are two ways that we can approach the analysis of this effect. One approach views the situation from the standpoint of what an individual charge "sees" as the magnetic field changes. Using that approach, together with the fact that only relative motion has meaning, the effect can be interpreted in the following way. In (*b*) of figure 24.1, the coil of wire can be considered moving toward the north pole of the stationary magnet. The charges in the wire, since they are moving in the magnetic field of the magnet, experience a force $q\mathbf{v} \times \mathbf{B}$ as discussed in Chap. 22. The free charges in the wire flow under the action of this force and give rise to the observed current.

Cause of the induced emf

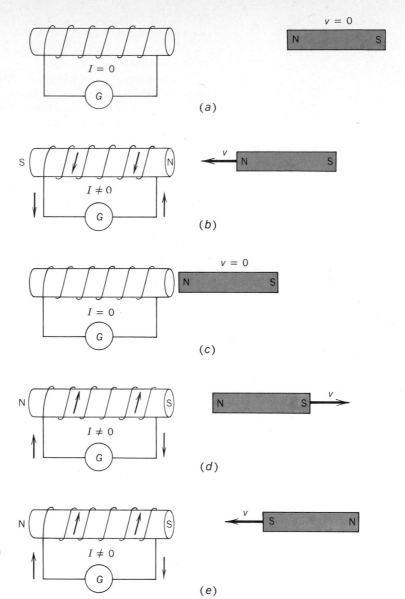

FIGURE 24.1 Current is induced in the coil only when the flux through it is changing. How can we predict which way the current will flow? Notice that the induced current causes the solenoid to act like a bar magnet with the polarity indicated.

This approach has the advantage that it shows us at once how induced emfs are related to phenomena we have already studied. We shall use this way of looking at the situation from time to time. However, it turns out that in most practical and engineering applications of the phenomenom, another approach is more advantageous. For this reason we shall use it. Since it involves the concept of magnetic flux lines, we shall digress for a moment to review what we mean by flux.

You will recall that, in Chap. 18 while discussing Gauss' law, we made use of electric field lines and electric flux. We agreed to draw a number of lines equal to E through unit area erected perpendicular to the electric field lines. With that definition, the number of lines $\Delta\phi$, or flux, through an area ΔA was taken to be $\mathbf{E} \cdot \Delta\mathbf{A}$ where the area vector $\Delta\mathbf{A}$ is taken perpendicular to the surface under consideration. The total electric flux through an area is

$$\phi_{\text{electric}} = \int \mathbf{E} \cdot d\mathbf{A}$$

These ideas are carried over into magnetism. We have been drawing magnetic field lines and they represent B. In analogy with the electric field, we agree now to draw B lines through unit area erected perpendicular to the field lines as shown in figure 24.2(a). The flux (i.e., the number of field lines or

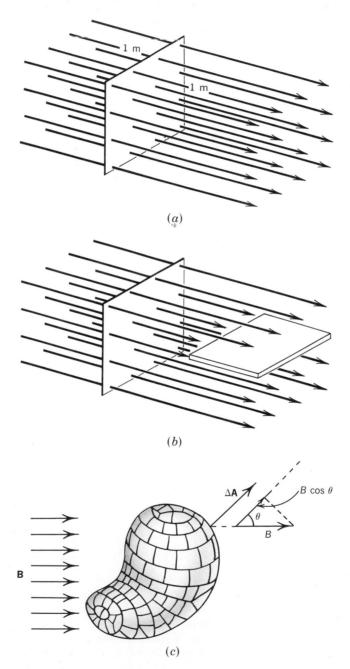

(a)

(b)

FIGURE 24.2   (a) We agree to draw B flux lines through a unit area perpendicular to the magnetic field lines. (b) The flux through an area depends upon the orientation of the area. (c) $\Delta\phi = \mathbf{B} \cdot \Delta\mathbf{A}$. (c: after Edward M. Purcell, "Electricity and Magnetism," p. 22, vol. 2, Berkeley Physics Course, McGraw-Hill Book Company, New York, 1965. Courtesy of Education Development Center, Inc., Newton, Mass.)

(c)

flux lines) going through an area depends upon the orientation of the area as shown in parts (*b*) and (*c*) of the figure. As indicated in part (*c*), only the component of **B** perpendicular to the surface (i.e., parallel to the area vector) is used in determining the flux. As a result, we can write

$$\Delta\phi_{\text{magnetic}} = B \cos \theta \, \Delta A = \mathbf{B} \cdot \Delta\mathbf{A}$$

The total flux through an area (i.e., the number of lines passing through the area) is then

Magnetic flux through an area

$$\phi = \int_{\text{area}} \mathbf{B} \cdot dA \qquad\qquad 24.1$$

where the integral is to be taken over the area in question.

Now that we know what is meant by magnetic flux, let us return to the experiments illustrated in figure 24.1 and in figure 24.3.

Michael Faraday (1791–1867) was the first to carry out detailed experiments such as these to show that the induced emf $\mathcal{E}$ which appears in a coil of wire containing N loops is

Faraday's law

$$\mathcal{E} = -N\frac{d\phi}{dt} \qquad\qquad 24.2$$

where ϕ is the magnetic flux through the coil. We call this *Faraday's law*. The significance of the negative sign will be brought out a little later in our discussion. We notice that the induced emf is zero unless the number of lines

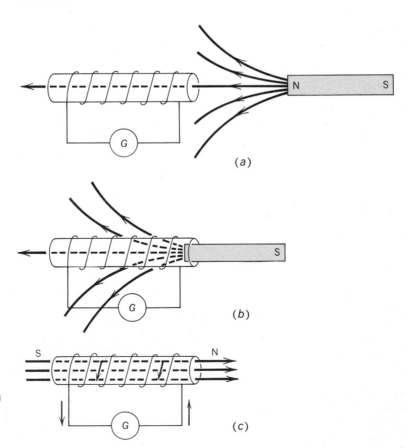

FIGURE 24.3 When the magnet is being moved from part (*a*) to part (*b*), the current in the coil flows in the direction shown in part (*c*).

of flux through the coil is changing. This agrees with the experimental results shown in figures 24.1 and 24.3.

It is instructive to examine the direction in which the induced current flows in the experiments of figures 24.1 and 24.3. By reference to figure 24.3 we see that the magnetic field produced by the induced current in the coil is in such a direction as to minimize or oppose the external change of flux through the coil. When the flux through the coil is increasing and toward the left, the induced current causes a flux toward the right in an effort to cancel the increasing flux. Referring back to figure 24.1, we see that a similar situation exists in each of these cases. We may therefore state the rule, called *Lenz's law: A change in flux through a loop will induce an emf in the loop. The direction of the current produced by the induced emf will be such that the flux generated by the current will tend to counterbalance or oppose the original change in the flux through the loop.* The negative sign in *24.2* indicates that the direction of the induced emf is such as to stop the change in flux.*

There is another way in which we can describe Lenz's law that provides convenience in certain applications. Notice in figure 24.3(c) that the induced emf causes the solenoid to generate a field much like that from a bar magnet. Notice further that the north pole of this induced "magnet" is positioned so that it opposes the motion of the north pole of the bar magnet towards the solenoid. The induced north pole of the solenoid repels the approaching north pole of the bar magnet. A similar situation is seen to exist in figure 24.1. In this sense, too, the induced emf opposes the change that is occurring. Therefore Lenz's law is frequently stated as follows: *The induced emf is in such a direction as to oppose the change that causes it.* This way of looking at the law shows at once that the energy resident in the induced emf is provided as the work done by the agent causing the change, the person moving the magnet in this case.

Illustration 24.1

A coil with N loops is wound outside a solenoid as shown in figure 24.4. The current in the solenoid is varying sinusoidally with time according to the relation

$$i = i_0 \sin 2\pi ft$$

where f is its frequency. This is the type of current supplied by the power company. Assuming the solenoid to have n loops per meter, find the induced potential difference between the terminals AB of the coil.

*This is a consequence of the conservation of energy. If the flux generated by the induced current were in such a direction as to augment the flux, the induced current could continue to induce more current, without end.

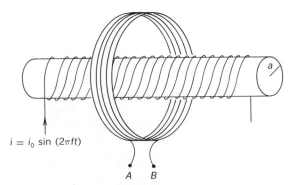

$i = i_0 \sin (2\pi ft)$

A B

FIGURE 24.4 How does the induced emf between points A and B depend upon the radius of the large coil? The radius of the solenoid?

Reasoning The induced potential difference will be equal to the induced emf. We shall assume the solenoid to be long enough so that B inside it is given by 23.5, namely,

$$B = \mu_0 n i$$

Since B outside the solenoid is negligibly small, lines of flux exist only in a part of the area of the coil. If the radius of the solenoid is a, then the flux through it is $\pi a^2 B$. This is the total flux through the coil as well. Therefore

$$\phi = \pi a^2 \mu_0 n i$$

From 24.2 we find the induced emf in the coil to be

$$\mathcal{E} = -\pi \mu_0 N n a^2 \frac{di}{dt} \qquad\qquad 24.3$$

Substituting the value given for i,

$$\mathcal{E} = -2\pi^2 \mu_0 f N n a^2 i_0 \cos 2\pi f t$$

Notice that the induced emf oscillates sinusoidally 90° out of phase with the current; moreover, $\mathcal{E}$ is proportional to f and to N.

24.2 MUTUAL INDUCTANCE

As illustrated in the preceding example, a changing current through one coil can produce an induced emf in a neighboring coil. In any general situation, the induced emf in coil 2, the so-called secondary coil, will be

$$\mathcal{E}_2 = -N_2 \frac{d\phi_2}{dt}$$

where N_2 is the number of loops on the secondary coil.

The flux ϕ_2 which appears in this expression is the result of the magnetic field B arising from the current i_1 in the other coil, the primary coil. Since B is proportional to i_1, we have

$$\phi_2 \sim i_1$$

A moment's consideration will convince you that the only other factors which influence ϕ_2 are geometric and coil-construction factors. They include the closeness together of the coils, the number of loops on the primary, their relative orientations and sizes, etc.

It is customary to define a constant M which includes all these geometric factors. We define it in such a way that

$$\mathcal{E}_2 = -\frac{d(M i_1)}{dt}$$

Unless the coils possess iron cores,* the geometric quantity M may be considered constant. In the case of M constant,

$$\mathcal{E}_2 = -M \frac{di_1}{dt} \qquad\qquad 24.4$$

---

*The effect of iron or other ferromagnetic materials is to invalidate the relation $B \sim i$. This is discussed subsequently. Even in that case, for many purposes an average value for M can be used and considered constant.

The quantity M in *24.4* is designated the *mutual inductance* of the two coils. By comparing *24.4* and *24.3* we see that M for the coils discussed in that example is given by

$$M = \pi\mu_0 Nna^2$$

As we should expect, M contains the construction details of the coils. Although the units of M are properly volt · seconds per ampere, this unit is given the special name of the *henry* (H). In the case of illustration 24.1, if $n = 10^4$ loops/m, $N = 100$ loops, and $a = 0.02$ m, we find

$$M = 1.6 \times 10^{-3} \, \text{H}$$

We shall see later that much larger inductances are possible if the coils are wound on iron cores.

24.3 SELF-INDUCTANCE

A coil of wire can induce an emf in itself. This follows directly from the fact that when a current flows in a coil, the current causes a flux through the same coil. If the current changes, the flux also changes, and an emf is induced in the coil.

As in the two-coil case, the induced emf depends upon the geometry of the coil. To the same approximation as in the two-coil case, we can write that the emf induced by a coil in itself is given by

$$\mathcal{E} = -L\frac{di}{dt} \qquad\qquad 24.5$$

The constant L is called the *self-inductance* of the coil and is measured in henrys.

We can easily find the self-inductance of a solenoid. Suppose the solenoid contains n loops/m length, is b m long, and has a radius of a. The flux through the solenoid at a current i is (if we ignore end effects)

$$\phi = \pi a^2 B = \pi a^2 \mu_0 ni$$

The induced emf in the solenoid is given by *24.2*, and is

$$\mathcal{E} = -(nb)\frac{d\phi}{dt}$$

or

$$\mathcal{E} = -\pi\mu_0 a^2 n^2 b\frac{di}{dt}$$

Comparison of this relation with *24.5* indicates that the self-inductance of the solenoid is

$$L = \pi\mu_0 a^2 n^2 b$$

Illustration 24.2

Find the self-inductance of the toroid, half of which is shown in figure 24.5. The total number of loops on the toroid is N.

Reasoning We found in the preceding chapter that B inside a toroid is given by

$$B = \frac{\mu_0 Ni}{2\pi r}$$

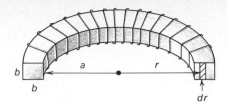

FIGURE 24.5 Half a toroid. What is its self-inductance?

Since B is not constant over the cross section, we must integrate to obtain the flux.

The flux through the shaded rectangular area in figure 24.5 is

$$d\phi = B(b\ dr)$$

or

$$d\phi = \frac{\mu_0 Nib}{2\pi}\frac{dr}{r}$$

To find the total flux we must integrate this from $a \le r \le (a+b)$. That is,

$$\phi = \frac{\mu_0 Nib}{2\pi}\int_a^{a+b}\frac{dr}{r}$$

This yields the result

$$\phi = \frac{\mu_0 Nib}{2\pi}\ln\left(1 + \frac{b}{a}\right)$$

Substituting this in *24.2* gives

$$\mathcal{E} = -\frac{\mu_0 N^2 b}{2\pi}\ln\left(1 + \frac{b}{a}\right)\frac{di}{dt}$$

After comparison with *24.5* we find

$$L = \frac{\mu_0 N^2 b}{2\pi}\ln\left(1 + \frac{b}{a}\right)$$

It is left as an exercise to show that if $b/a \ll 1$, this reduces to the result for a solenoid having similar dimensions.

24.4 THE *LR* CIRCUIT

Most electric circuits have loops of wire, and therefore self-inductance, in them. Let us now examine what happens when a switch is first closed in a circuit which contains a battery, resistor, and inductance coil. The circuit is shown in figure 24.6. Notice the symbol used for the inductance coil L. Do not confuse it with the somewhat similar symbol used to designate R. In practice, R might merely be the resistance of the coil itself.

Before writing the Kirchhoff loop equation for this circuit, let us discuss the effect of the inductance. There will be induced in it an emf

$$\mathcal{E} = -L\frac{di}{dt}$$

If i is increasing, this induced emf will try to stop the increase in i and the consequent increase in flux through itself. We therefore conclude that the inductance coil acts like a battery, with an emf $L\ di/dt$ opposing V_0.

With this polarity consideration in mind, the loop equation for the circuit of figure 24.6 becomes

$$+V_0 - iR - L\frac{di}{dt} = 0$$

We can simplify this differential equation by changing variables so that

$$y = \frac{V_0}{R} - i$$

and

$$-dy = di$$

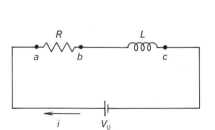

FIGURE 24.6 Since i is increasing, the inductance coil will have induced in it an emf in such a direction as to decrease i. It acts like a battery opposing V_0.

The loop equation then becomes

$$y + \frac{L}{R}\frac{dy}{dt} = 0$$

After rearranging, this equation becomes

$$\frac{dy}{y} = -\frac{R}{L}dt$$

We now integrate from $t = 0$ (where $y = y_0$) to t (where $y = y$):

$$\int_{y_0}^{y}\frac{dy}{y} = -\frac{R}{L}\int_{0}^{t}dt$$

which gives

$$\ln y - \ln y_0 = -\frac{R}{L}t$$

or

$$\ln \frac{y}{y_0} = -\frac{R}{L}t$$

After taking antilogs, this yields

$$y = y_0 e^{-Rt/L}$$

Replacing y by its value in terms of i,

$$i = \frac{V_0}{R} - y_0 e^{-Rt/L}$$

The value of y_0 can be found by noticing that $i = 0$ when the switch was first closed at $t = 0$. Therefore, when $t = 0$,

$$0 = \frac{V_0}{R} - y_0$$

and so

$$y_0 = \frac{V_0}{R}$$

The expression for the current is therefore given by

$$i = \frac{V_0}{R}(1 - e^{-Rt/L}). \qquad 24.6$$

This function is plotted in figure 24.7. As one would expect, the inductance coil does not allow the current (and flux) to rise infinitely fast when the switch is first closed. Instead, the current rises gradually to its equilibrium value V_0/R, as given by Ohm's law. Notice that, as in the case of the

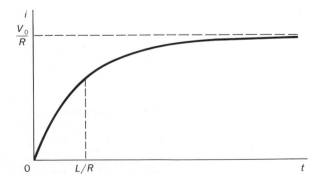

FIGURE 24.7 Variation of current in an LR circuit. Notice the values given for the maximum current and the time constant.

capacitance-resistance circuit, one can define a time constant for this circuit as well. It is L/R, the time taken for the current to reach $(1 - e^{-1})$ of its final value.

24.5 ENERGY IN A MAGNETIC FIELD

As we saw in the previous section, an inductance coil prevents a battery from instantaneously establishing a current through it. The battery has to do work against the coil as the current is caused to flow. We shall now compute how much work is done and, in so doing, arrive at an expression for the energy stored in a magnetic field.

When an inductance coil is connected to a battery, causing current to flow in the inductance coil, work is done on the coil by the battery. From the definition of self-inductance L, we have

$$\mathcal{E} = -L\frac{di}{dt}$$

The induced emf $\mathcal{E}$ between the terminals of the coil opposes the flow of current through it. We saw in Chap. 20 that the power expended in sending a current I through a potential difference V is VI. Since the induced emf results in a potential difference $V = \mathcal{E}$ between the terminals of the coil, the instantaneous power furnished to the coil by the battery is iV or

$$\text{Power} = iL\frac{di}{dt}$$

To find the total work W done by the battery on the coil we must recall that

$$\text{Power} = \frac{dW}{dt}$$

and so

$$\frac{dW}{dt} = iL\frac{di}{dt}$$

In order to find the work done as the current in the coil is changed from zero to a value I we multiply through by dt and integrate.

$$\int dW = L\int_{i=0}^{I} i\,di = \tfrac{1}{2}LI^2$$

We have removed the self-inductance L from beneath the integral since it is constant.

Upon carrying out the integration we arrive at the indicated result,

$$W = \tfrac{1}{2}LI^2 \qquad\qquad 24.7$$

This work done on the inductor must be stored as energy in the inductor. We shall see in Chap. 26 that this energy can be retrieved. We therefore conclude that the energy stored in an inductor is given by *24.7*.

It will be recalled that a similar expression was obtained in Sec. 20.13 for the energy stored in a capacitor, namely, $\tfrac{1}{2}CV^2$. We showed that this energy could be considered stored in the electric field. The energy stored per unit volume was found to be $\tfrac{1}{2}\epsilon_0 E^2$. A similar expression can be found for the energy stored in a magnetic field.

To do this, let us consider the case of a long solenoid. We found in Sec.

24.3 that the inductance for a solenoid was

$$L = \pi\mu_0 a^2 n^2 b$$

where b is the length of the solenoid. The field within the solenoid was found in the preceding chapter to be

$$B = \mu_0 nI$$

from which
$$I = \frac{B}{\mu_0 n}$$

If we substitute these values in 24.7, we find

$$W = \tfrac{1}{2}\pi a^2 b \frac{B^2}{\mu_0}$$

Since $\pi a^2 b$ is merely the volume of the solenoid, the energy stored per unit volume will be

$$\text{Energy density} = \frac{B^2}{2\mu_0} \qquad\qquad 24.8$$

Although this relation has been derived for a special case, more advanced texts show that it has general validity.

24.6 ROTATING COILS; GENERATORS

Electric generators and motors make use of coils which rotate in a magnetic field. As the coil rotates, the flux through it changes, and therefore an emf is induced in it. To obtain an expression for this induced emf, consider the situation shown in figure 24.8.

If the angle between the field **B** and the area vector **A** for the loop is θ, then the flux through the loop will be

$$\phi = \mathbf{B} \cdot \mathbf{A}$$

This may be rewritten as

$$\phi = AB \cos \theta$$

We assume, however, that the loop rotates about the vertical axis shown with angular speed ω, so that

$$\theta = \omega t$$

After making this substitution, we find

$$\phi = AB \cos \omega t$$

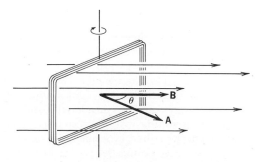

FIGURE 24.8 The flux through the loop is $AB \cos \theta$.

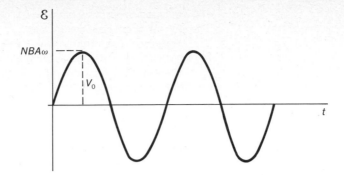

FIGURE 24.9 An alternating emf is induced in a coil rotating in a uniform magnetic field.

The induced emf in the coil is given by *24.2* and is

$$\mathcal{E} = -N\frac{d\phi}{dt}$$

or

$$\mathcal{E} = ABN\omega \sin \omega t$$

where N is the number of loops on the coil. The induced emf is sinusoidal in time and is of the form shown in figure 24.9. This is the usual type of alternating potential difference, and we shall have more to say about such ac voltages in Chap. 26. We shall usually represent the emf as

$$V = V_0 \sin 2\pi f t$$

where we have replaced the angular frequency in radians per second, ω, by the frequency in revolutions or cycles per second, f, through the relation $\omega = 2\pi f$. The voltage supplied by the power company has this form. It is generated by rotating coils in a magnetic field. In most cases, f is maintained accurately at 60 Hz by the power company.

24.7 INDUCED EMFS AND ELECTRIC FIELDS

In the preceding sections of this chapter the concept of induced emfs has been presented from a purely experimental standpoint. It is easily found experimentally that a changing flux through a coil of wire causes a potential difference equal to the induced emf to appear between the two ends of the coil. We shall now begin to build a more general mathematical structure by means of which we can more easily deal with the results of less simple experiments.

First let us note that the existence of an induced emf implies the existence of an induced electric field. To show this, consider the single loop shown in figure 24.10. We assume that the loop is in a uniform but changing magnetic field **B** directed into the page. Equation *24.2* tells us that an emf $\mathcal{E}$ will be induced in the loop of such a magnitude that

$$\mathcal{E} = -\frac{d\phi}{dt}$$

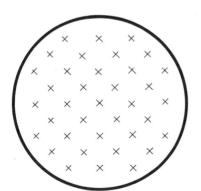

FIGURE 24.10 The changing flux though the loop induces an electric field in the vicinity of the loop. It is this electric field which causes an induced current to flow in the loop.

A changing flux causes an electric field

In the particular situation under discussion, all points on the loop are equivalent because of the circular symmetry of the problem. It is therefore not correct to picture the induced emf $\mathcal{E}$ as being localized at any particular point on the loop. For many purposes, it is as though a single battery with emf $\mathcal{E}$ were placed in the loop. However, a more realistic picture would be that of a whole series of infinitesimal batteries in series extending around the loop. The sum of their emfs would be $\mathcal{E}$.

What we are actually saying by this crude type of model is as follows: Between the ends of a little segment of the loop separated by a length Δs, there is an emf $\Delta\mathcal{E}$. The sum of these emfs around the entire loop is equal to the induced emf in the loop, namely, $-d\phi/dt$. Expressed as an equation,

$$\sum \Delta\mathcal{E} = -\frac{d\phi}{dt}$$

where the sum is to extend around the loop.

But $\Delta\mathcal{E}$ is simply the potential difference ΔV between the two ends of the little length element Δs. We can therefore replace $\Delta\mathcal{E}$ by ΔV to give

$$\sum \Delta V = -\frac{d\phi}{dt}$$

Further, since ΔV is by definition the work done in carrying a unit positive charge from one end of Δs to the other, you will recall that it is related to the electric field $\mathbf{E}$ by the relation $\Delta V = -E_s \Delta s$, where E_s is the component of $\mathbf{E}$ along Δs. Substituting this value for ΔV we obtain, after passing to the limit of small Δs,

$$\oint E_s \, ds = \frac{d\phi}{dt} \qquad 24.9$$

Now $E_s \, ds = \mathbf{E} \cdot d\mathbf{s}$ and $\phi = |\int \mathbf{B} \cdot d\mathbf{A}|$. But we must specify the orientation of vector $\mathbf{A}$. By convention, it is taken in the direction a right-hand screw goes if rotated in the same sense as the integral on its circumference is taken. We then find the above equation becomes (verify the sign using the right-hand rule and area vector convention)

An important relation between induced electric field and changing magnetic field

$$-\oint \mathbf{E} \cdot d\mathbf{s} = \frac{d}{dt} \int \mathbf{B} \cdot d\mathbf{A} \qquad 24.10$$

where the integral on the right is to extend over the area enclosed by the loop.

Although the discussion leading to *24.10* was based upon the idea of a circular loop, this is not a necessary condition. In fact, if you review the preceding steps, you will see that none of the critical arguments made use of the fact that the loop had any particular shape. We therefore conclude *24.10* to be true for any loop. It is one form of Faraday's law.

It might in fact be asked whether the loop as a physical entity need be present at all for *24.10* to be true. By placing charges in changing magnetic fields, experiment shows that an electric field given by *24.10* exists even in the absence of a physical loop. In applying *24.10* to such cases, the loop used as a path for integration is chosen wherever convenient.

Illustration 24.3

Suppose a long solenoid has n loops per unit length and carries a varying current

$$i = i_0 \sin 2\pi ft$$

Find the induced electric field E both inside and outside the solenoid.

Reasoning The solenoid is shown in cross section in figure 24.11. Let us assume the current to be in the direction shown and increasing so that $\mathbf{B}$ is

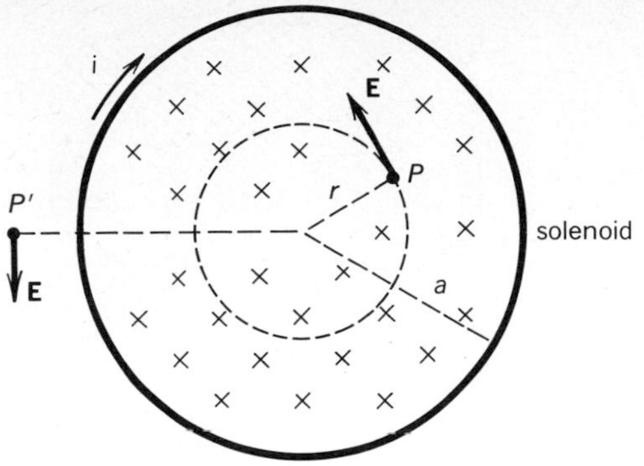

FIGURE 24.11 If the current in the solenoid is in the direction shown and increasing, the electric fields at P and P' will be as indicated. Can you use the right-hand rule and Lenz' law to verify these directions?

increasing into the page. We know from the preceding chapter that

$$B = \mu_0 ni$$

inside the solenoid and nearly zero outside.

To find the electric field at point P we need to apply *24.10* to a path which passes through P. It is convenient to take as our path the dashed circle shown in figure 24.11 since, from the circular symmetry, E will be constant on it. Equation *24.10* then becomes

$$-\oint \mathbf{E} \cdot d\mathbf{s} = \frac{d}{dt} \int \mathbf{B} \cdot d\mathbf{A}$$

$$-E \oint ds = \frac{d}{dt}\left(B \int dA\right)$$

where B is assumed constant on the enclosed area. Of course, we are only evaluating the component of $\mathbf{E}$ tangential to the path. (Why?)

The above relation becomes

$$-E(2\pi r) = \frac{d}{dt}\pi r^2 \mathbf{B}$$

After substituting the following value for B,

$$B = \mu_0 ni_0 \sin 2\pi ft$$

this becomes

$$-2\pi rE = \pi\mu_0 ni_0 r^2(2\pi f)\cos 2\pi ft$$

From this

$$-E = \pi\mu_0 ni_0 fr \cos 2\pi ft$$

So we see that E increases in proportion to r and varies sinusoidally with time. Physically, why should E increase with f as indicated by the equation we have obtained?

We leave it as an exercise for the student to show that at point P' the electric field is

$$-E = \frac{1}{r}\pi a^2\mu_0 ni_0 f \cos 2\pi ft$$

Notice that $E \sim 1/r$ outside the solenoid.

24.8 BETATRON ACCELERATION

The fact that a changing magnetic flux produces an electric field is basic to the operation of many high-energy particle accelerators. Since the principle

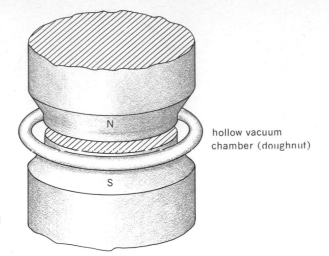

FIGURE 24.12 The changing flux through the central area induces a tangential electric field within the doughnut-shaped vacuum chamber. Particles accelerated by this electric field are held in a circular orbit by the perpendicular magnetic field.

hollow vacuum chamber (doughnut)

was first successfully applied to the acceleration of electrons (or β particles) in a device called the *betatron*, this method of acceleration is often given that name. The general idea involved is shown in figure 24.12.

A large electromagnet is used to produce a changing flux through a circular loop defined by the doughnut-shaped vacuum chamber. From *24.9* we see that there will be an electric field E along the circular length of the doughnut, i.e., circling the magnet poles, given by

$$2\pi a E = \frac{d\phi}{dt}$$

where a is the radius of the doughnut. Any charged particle inside the vacuum chamber will experience a force qE and will accelerate. Ordinarily, the charged particle would shoot out of the vacuum chamber and become lost.

However, if the magnetic field at the position of the doughnut is just proper to satisfy the relation

Centripetal force = magnetic force

or*
$$\frac{mv^2}{a} = qvB$$

the charge will travel in a circle within the doughnut. By proper shaping of the magnet pole pieces, this relation can be satisfied (see the last problem at the end of this chapter). As a result, the charge will move at high speed along the loop within the doughnut. Each time it goes around the loop, it has, in effect, fallen through a potential difference equal to the induced emf, namely,

$$\mathcal{E} = \frac{d\phi}{dt}$$

Its energy after n trips around the loop will be

$$q(n\mathcal{E})$$

---

* This relation is applicable even at relativistic speeds if m is taken to be the relativistic mass, as was pointed out in Sec. 22.9.

A typical betatron is capable of accelerating electrons to energies of 100 MeV, or 1.6×10^{-11} J. (Recall in this connection that an electron volt is the energy acquired by a charge of 1.6×10^{-19} C when it falls through a potential of 1 V.) In other words, the charge has effectively fallen through a potential difference of $\approx 10^8$ V.

Illustration 24.4

In a certain betatron the electron completes 2×10^5 trips around the loop before being ejected against a metal plate to produce x-rays. If, during this time, $d\phi/dt = 400$ V, what are the energy and speed of the ejected electron?

Reasoning Each time around the loop the electron effectively falls through a potential difference of

$$\mathcal{E} = \frac{d\phi}{dt} = 400 \text{ V}$$

After 2×10^5 trips it has effectively fallen through

$$(2 \times 10^5)(400) = 8 \times 10^7 \text{ V}$$

Its energy will therefore be 80 MeV.

The electron will be moving at relativistic speeds, and so we must compute its speed by use of the general formula for kinetic energy, namely,[*]

$$K = mc^2 - m_0 c^2$$

In this expression $K = 80 \times 10^6 \times 1.6 \times 10^{-19}$ J and

$$m = \frac{m_0}{\sqrt{1 - (v/c)^2}}$$

Placing in the values, one finds $v = 0.99998c$. Can you show that $m \approx 160 m_0$ at this speed?

24.9 POTENTIAL AND CURRENTS

Induced voltages are somewhat different from the voltage differences produced by static or stationary charges. In particular, if induction effects resulting from changing flux are present, it is difficult to define a unique potential difference between two points in space. This is easily seen by considering the simple case treated earlier, the case of a solenoid with changing current and flux. Consider the situation shown in figure 24.13.

In electrostatics we showed that the work done in carrying a positive test charge from one point to another was independent of the path followed. This meant that zero work was done in taking a test charge from some point, around a path, and back to the original point. We called such a field, the electrostatic field in that case, a conservative field. Let us now consider moving a test charge from point P in figure 24.13 along a path and back to P.

The path through R is quite simple. No flux change is occurring through this loop, and so there is no induced emf in it. Therefore the work done in carrying the test charge from P to R and back is zero, as we should expect it to be.[†]

---

[*] See Appendix 8 if you have not yet studied Chap. 8.
[†] To show this mathematically, use 24.9.

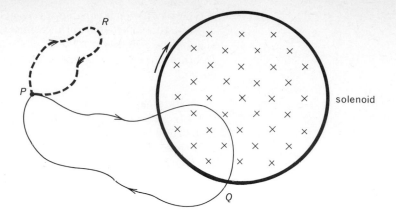

FIGURE 24.13 Starting from point P, zero work is done in carrying a positive test charge back to P via R. This is not true for the path $P \rightarrow Q \rightarrow P$, since the induced emf is not zero around this latter path.

In the presence of changing magnetic flux the electric field is not a conservative field

solenoid

An entirely different situation exists if one carries the test charge from P to Q and back to P by the path indicated. This path encloses a *changing* flux since we are assuming the current in the solenoid to change. There is, therefore, an induced emf $= \mathcal{E}$ in this loop, and so the work done in carrying the charge q from P to Q and back to P on this loop is $q\mathcal{E}$. As a result, the test charge is not at the same potential when it gets back to P as it was when it left P. The potential at point P is therefore not unique since it depends upon the path by which the test charge is brought to P from infinity. For this reason the electrostatic concept of potential must be used with care when one is dealing with *changing* magnetic fluxes.

24.10 MOTIONAL EMFS

Under certain special circumstances it is quite easy to show that the induced emf in a circuit arises because of the motion of charges in a steady magnetic field. A case in point might be the induced emf in a loop as it is moved from a region where $B = 0$ to a region where $B > 0$; or a loop could be rotating in a constant field B. In such instances an emf exists even though B is constant. To understand such situations, recall that a charge q moving with velocity $\mathbf{v}$ in a magnetic field $\mathbf{B}$ experiences a force

$$\mathbf{F} = q\mathbf{v} \times \mathbf{B}$$

This force will exist even if the charge is in a wire. Let us consider the case shown in figure 24.14.

We consider the wire AC to be pulled to the right with constant velocity $\mathbf{v}$. It makes contact with the wire $DGHK$ at points A and C so that a wire loop exists, $ACGH$. We shall measure the position of the moving wire by the distance x shown in the figure. A steady magnetic field $\mathbf{B}$ exists, directed into the page. The induced emf in the loop is given by

$$\mathcal{E} = -\frac{d\phi}{dt} = -\frac{d}{dt}Bbx$$

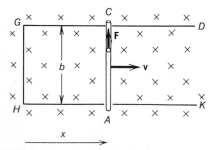

FIGURE 24.14 In this case the induced emf in the loop can be interpreted in terms of the forces on the charges in the moving wire.

where bx is the area of the loop. Since x changes with time, this becomes

$$\mathcal{E} = -Bb\frac{dx}{dt} = -Bbv \qquad 24.11$$

516 MAGNETIC INDUCTION EFFECTS

But we can compute this emf in another way. Each charge in the wire AC experiences a force in the direction shown* because it is moving through the magnetic field. The force is

$$\mathbf{F} = q\mathbf{v} \times \mathbf{B}$$

After dividing through by q this becomes

$$\frac{\mathbf{F}}{q} = \mathbf{v} \times \mathbf{B}$$

Now F/q is the electric force on unit charge and is therefore the electric field strength E. You may not feel that this is our usual electric field, but it is consistent with our definition of E. We therefore write

$$\mathbf{E} = \mathbf{v} \times \mathbf{B} \qquad\qquad 24.12$$

and treat it as any other electric field. If we do so, we can compute the potential difference† between points A and C due to this field, i.e., the emf in this wire. It is, from our definition of $V_A - V_C$,

$$V_A - V_C = \int_C^A \mathbf{E} \cdot d\mathbf{s}$$

where $d\mathbf{s}$ is a little length increment directed downward along AC. Since E is constant and since $|\mathbf{v} \times \mathbf{B}| = vB$ in the present case, this becomes

$$V_A - V_C = vBb$$

Notice that this is exactly the value obtained for the induced emf in *24.11*.

If we check the directions involved, we find from Lenz' law that the induced emf is counterclockwise around the loop. This agrees with the direction the charges are caused to move by the force on those in the moving wire. It therefore appears that, in this case at least, the induced emf can be *Motional emfs are the result of charge* thought of as the result of charge motion in a magnetic field. We designate *motion in a magnetic field* emfs of this type to be motional emfs.

Illustration 24.5

If the resistance of the circuit in figure 24.14 at any instant is R, find the force needed to pull the wire with speed v. Discuss the destination of the power furnished.

Reasoning   We saw in the preceding section that the induced emf is vBb. Therefore the current in the circuit is given by

$$\mathcal{E} = IR$$

to be

$$I = \frac{vBb}{R}$$

The wire AC will carry this current upward. Any wire carrying a current in a magnetic field experiences a force, ILB in this case, so

$$F = \frac{vB^2b^2}{R}$$

---

*This is strictly true only if the drift velocity of the charges in the wire is zero. For a more detailed treatment see E. P. Mosca, *Am. J. Phys.* **42**, 295 (1974).

†We are allowed to speak about potential difference here since B is constant, and therefore the complication discussed in the preceding section does not arise.

Using the general rules, we find this force to be toward the left. It is necessary for the pulling agent to exert an equal force toward the right. This pulling agent expends power of magnitude

$$\text{Power} = Fv = \frac{v^2B^2b^2}{R}$$

This power must appear in the wire circuit as heat. We can easily check this since the heat produced per second is merely I^2R. Placing in the value found previously for I, we have

$$\text{Power lost to heat} = \frac{v^2B^2b^2}{R}$$

which agrees with the power input, as it should.

Illustration 24.6

A fan blade of length $2a$ rotates with frequency f Hz perpendicular to a magnetic field B. Find the potential difference between the center and end of the blade.

Reasoning Notice that the emf is a motional emf and is generated in the moving blade. We can find this emf in either of two ways. Let us first do it directly from

$$\mathcal{E} = -\frac{d\phi}{dt}$$

To apply this equation we must have a closed path, and so we *imagine* the dashed circuit shown in figure 24.15 to slide along the blade. As the contact at point D slides along, the area A of the loop ODC increases. We have

$$\mathcal{E} = -\frac{d(BA)}{dt} = -B\frac{dA}{dt}$$

The rate of change of the area is easily found by noting that in the time taken for 1 rev, namely, $1/f$, the upper part of the blade sweeps out an area of πa^2. Therefore

$$\mathcal{E} = -B\frac{\pi a^2}{1/f} = -\pi Ba^2 f$$

This is the induced emf in the loop, and it is also the potential difference from O to D.

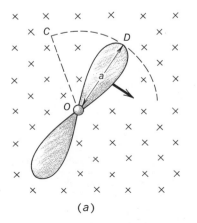

FIGURE 24.15 Can you show that the voltage difference between the two ends of the rotating blades is zero?

(a)

(b)

A second approach is indicated in part (b) of figure 24.15. The speed v of a point on the blade a distance r from the center is, from our motion relations in Chap. 9,

$$v = r\omega = 2\pi rf$$

It was shown in the preceding section that an electric field will exist at this point in the blade, and it is, because v and B are perpendicular,

$$E = vB = 2\pi rfB$$

To find the potential difference between points O and D, we use the definition of potential difference, namely,

$$V_D - V_O = -\int_0^D E_r\, dr$$

$$= -\int_0^a 2\pi fBr\, dr$$

This gives
$$V_D - V_O = -\pi fBa^2$$

which is the same result as found by the other method.

Can you show that no potential difference exists between the two ends of the blade? As a matter of interest, in the earth's magnetic field ($\approx 10^{-4}$ T), a blade 20 cm in radius moving at 10 rev/s would exhibit a potential difference of about 10^{-4} V. Notice that here, too, we were dealing with a constant magnetic field, so that it was meaningful to speak of potential difference.

Chapter 24
Questions
and Guesstimates

1. Two circular current loops lie on a table. One loop has a battery and switch in it, while the other is just a closed wire loop. Describe what will happen in the closed loop when the switch is suddenly closed; opened. Consider both cases, when the loops do not and do overlap. Draw a current vs. time graph in each case.
2. An inductance coil is connected in series with a switch and a 6-V battery. When will the bigger spark jump across the switch, when it is pulled open or when it is pushed closed? With large inductances, the induced voltage becomes large enough to become dangerous.
3. It is possible to wind a coil of wire on a straight wooden rod in such a way that the coil has essentially zero self-inductance. How?
4. A particular series circuit consists of a switch, a 6-V battery, and an inductance coil. A voltmeter can be connected across various portions of the circuit. Is it possible for the voltmeter to read more than 6 V even momentarily?
5. When the switch is closed in a series circuit that contains a battery and inductance coil, energy is stored in the coil. What happens to this energy when the switch is suddenly opened?
6. What happens in the secondary coil in figure P24.1 when the switch in the primary coil circuit is (a) pushed closed? (b) when pulled open?

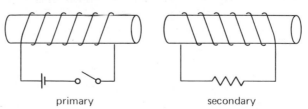

FIGURE P24.1   primary   secondary

FIGURE P24.2

FIGURE P24.3

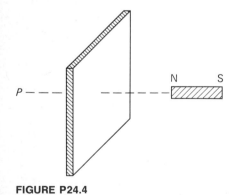

FIGURE P24.4

7. Suppose you are given two identical flat coils. How should the two coils be placed so that their mutual inductance is (*a*) largest? (*b*) smallest? When connected in series by a flexible wire, how should they be positioned to make the self-inductance (*c*) largest? (*d*) smallest?

8. A small coil is placed inside a long solenoid. How does the mutual inductance of the two change with the orientation of the coil?

9. A copper ring lies on a table. There is a hole through the table at the center of the ring. If a bar magnet is held vertically by its south pole high above the table, and is then released so that it drops through the hole, describe (*a*) the forces which act on the magnet and (*b*) the current in the ring, both as a function of time.

10. A very long copper pipe is oriented vertically. Describe the motion of a bar magnet which is dropped lengthwise down the pipe. Why does the magnet reach a terminal speed?

11. Discuss the possibility of using induced emfs in earth satellites to power the various pieces of electronic equipment. Satellites travel with very high speed through the earth's magnetic field.

12. *Eddy currents:* When a series of loops of wire swings past the north pole of a magnet, as shown in figure P24.2, currents are induced in the loops. Show the direction of the currents at two different times, and show the force on the loops. Similar *eddy* currents occur in solid pieces of metal. Explain.

13. Induced currents are used to damp the motion of coils on very sensitive galvanometers. These coils are very close to friction-free, and swing for a long time if allowed to do so. However, by connecting the two ends of the coil together, the coil will stop swinging at once. Why?

14. As shown in figure P24.3, a metal ring sits on the end of a solenoid and is held in place there. An alternating current (produced by a alternating emf) is sent through the solenoid. The ring becomes hot. Why? A metal plate also becomes hot when held above the solenoid. Explain how eddy currents are induced in it and cause it to heat.

15. Two identical size rings, one made of copper and the other of plastic, are placed in the same magnetic field. When the magnetic field changes, how do the induced emf's in the rings compare? The induced currents? The induced electric fields?

16. A metal plate separates point *P* from the bar magnet in figure P24.4. When the system is at rest, *P* experiences the full magnetic field from the magnet. But if the magnet is suddenly moved closer to *P*, the field at *P* increases to its full value only after some time has passed. Why the delay? In fact, if the metal sheet is large and is a superconductor, the magnetic field at *P* does not change at all. Why? This general phenomena is used to shield sensitive equipment from swiftly varying magnetic fields. Why will it not provide shielding from steady fields?

17. A copper ring is set as shown in figure 24.3 above a solenoid (which has an iron core to increase its field). When the current is turned on in the solenoid, the copper ring flies upward. Explain why. Be particularly careful about directions.

18. Motors do work on external objects. Explain clearly how this energy is transferred from the electric current to the rotating portion of the motor.

19. Electric generators transform mechanical work into electric energy. Explain how the energy is transferred.

20. Estimate the emf induced in a 1.0-H coil placed in a series circuit with a

2.0-Ω resistor and a 12-V battery when the switch is first closed. First opened.

21. Estimate the self-inductance of a lamp cord 3 m long.

---

1. A magnetic field given by $B = B_x = 0.2 \sin(\omega t)$ T is impressed upon a flat coil of area A and resistance R that has N loops of wire on it. If the angle between the area vector and the x axis is θ, find both the induced emf and the current in the coil.

2. A tiny flat coil (N_c loops, area A_c, resistance R) is suspended inside a long solenoid (area A_s and n loops/m) with its axis (or area vector) at an angle θ to the solenoid axis. The field within the solenoid varies in the following way: $B = B_0 \sin \omega t$. Find both the induced emf and current in the coil.

3. A circular coil of wire (N loops, radius a) lies on a table. On top of it and bisecting it lies a long straight wire carrying a current $i = i_0 \sin \omega t$. Find the induced emf in the coil. Repeat if the straight wire is perpendicular to the tabletop. (*Hint:* If you cannot obtain the answer by inspection, you are on the wrong track.)

4. A long straight wire carrying a current $i = i_0 \sin \omega t$ lies on the axis of a solenoid (n turns/m and radius a). Find the emf induced in the solenoid by the current in the straight wire.

5. A circular coil with N loops and area A rotates with rotation frequency f about a diameter as axis. Impressed upon it is a constant magnetic field B directed perpendicular to the rotation axis. As the coil rotates, the angle θ between its area vector and $\mathbf{B}$ varies as $\theta = 2\pi ft$. Find the flux through the coil at time t and obtain from it the emf generated in the coil.

6. A flat loop of wire (area A) has its two ends connected together so as to form a metal ring. It is placed in a constant, uniform field $\mathbf{B}$ in such a way that its area vector makes an angle θ to the lines of $\mathbf{B}$. The loop now rotates about a diameter so that $\theta = 2\pi ft$, where f is the frequency of rotation. If the resistance of the coil is R, find the current through the coil as a function of time. The current-carrying coil experiences a torque in the magnetic field. How large is the torque? This large a torque must be applied to the coil in order to cause it to rotate.

7. A coil of N loops is wound tightly on the center section of a long solenoid with n loops/m. The cross-sectional areas of both are essentially the same, A. Find the induced emf in the coil due to a current $i = i_0 \cos \omega t$ in the solenoid. What is the mutual inductance of the combination?

8. A small flat coil (N loops and area A_c) is placed inside a long solenoid (n loops/m and area A_s) with its area vector making an angle θ to the solenoid axis. Find the induced emf in the coil due to a current $i = i_0 \cos \omega t$ in the solenoid. What is the mutual inductance of the combination? By making θ variable, it is possible to construct a variable mutual inductance this way.

9. Two uniform coils are wound tightly on a glass tube, one on top of the other, so as to form two solenoids each having n loops/m and with nearly identical cross sections. However, viewed from the end, one solenoid is wound clockwise away from the viewer while the other is wound counterclockwise. They are connected in electrical series by joining the two coils at one end of the tube. The other two end wires run to the terminals

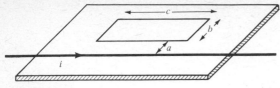

FIGURE P24.5

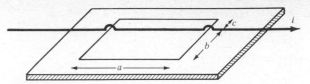

FIGURE P24.6

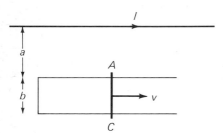

FIGURE P24.7

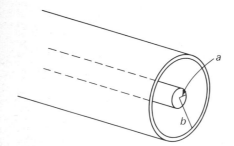

FIGURE P24.8

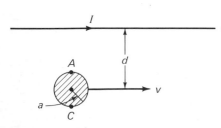

FIGURE P24.9

of a power source which sends a current $i_0 \sin \omega t$ through them. Find the flux through either solenoid. What is the self-inductance of the unit consisting of the two coils in series? This is a common way of constructing a noninductive coil.

10. It is desired to send an alternating current $5 \cos (400t)$ A through a solenoid that has a self-inductance of 3 mH. Assuming the resistance of the solenoid to be negligible, what voltage must be impressed across the solenoid terminals to provide this current?

11. A long straight wire lies on a table close to a rectangular loop of wire, as shown in figure P24.5. If a current i flows in the wire, how large is the flux which goes through the loop? What is the mutual inductance of the combination?

12. For the situation shown in figure P24.6, if the current in the long straight wire is i, what is the net flux through the rectangular loop? What is the mutual inductance of the combination? What would the answers to these two questions be if $b = c$?

13. The coaxial cable shown in figure P24.7 has an inner wire of radius a and an outer metal sheath with inner radius b. (a) When a current I flows down the inner wire and back through the sheath, how much flux threads the gap for a unit length of the cable? (b) What is the self-inductance of a unit length of cable?

14. The bar AC in figure P24.8 moves with constant speed v as shown. What is the induced emf in the loop due to the constant current I in the long straight wire?

15. As shown in figure P24.9, a metal sphere moves with speed v parallel to a long wire that carries a constant current I. Show that the induced emf between points A and C on the sphere is $(\mu_0 Iv/2\pi) \ln [(d + a)/(d - a)]$.

16. Two long parallel wires have radii equal to b and are a distance d between centers. Their far ends are connected together so the wires form two sides of the same loop. For a section far from the ends, show that the self-inductance per meter length is $(\mu_0/\pi) \ln [(d - b)/b]$ if the flux within the wires is neglected.

17. A 20-mH coil is connected in series with a 2000-Ω resistor, a switch, and a 12-V battery. What is the time constant of this circuit? How long after the switch is closed will it take for the current to reach 99 percent of its final value?

18. An inductor L, a resistance R, and an emf $= V_0$ are connected in series with a switch. The switch is pushed closed at $t = 0$. Find the following: (a) I at $t = 0$; (b) induced emf at $t = 0$ and (c) at $t = \frac{1}{2}(L/R)$; (d) I at $\frac{1}{2}(L/R)$.

19. In a series circuit consisting of an inductance coil L, a resistor R, a switch, and a battery V, find the following at a time equal to the time constant after the switch is closed: induced emf in the inductor, power output of battery, power loss in the resistor, power being stored in the inductor.

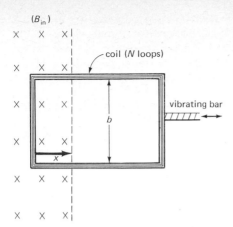

(B_in)

coil (N loops)

b

x

vibrating bar

FIGURE P24.10

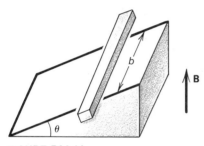

b

θ

B

FIGURE P24.11

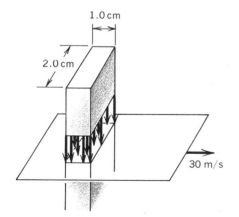

1.0 cm

2.0 cm

30 m/s

FIGURE P24.12

20. A ballistic galvanometer can be used to measure a pulse of charge sent through it. Suppose a coil with N loops is connected across the terminals of such a galvanometer. The total resistance of coil and galvanometer is R. If the coil is now suddenly thrust into the region between the poles of a U-type magnet, the flux through the coil will suddenly change from zero to ϕ. In the process, a momentary current $i = N(d\phi/dt)/R$ will flow through the galvanometer. Show that the total charge pulse Q which flows through the galvanometer is given by $Q = N\phi/R$. By measuring Q, one can determine ϕ and, therefore, B for the magnet.

21. An airplane with a 20-m wingspread is flying at 250 m/s straight south parallel to the earth's surface. If the earth's magnetic field in that region has a horizontal component of 0.20 G and a dip angle of 70°, what is the induced potential difference between the plane's wing tips?

22. Magnetic transducers are often used to monitor small vibrations. For example, the end of a vibrating bar is attached to a coil which in turn vibrates in and out of a uniform magnetic field B, as shown in figure P24.10. Show that the speed of the end of the bar, dx/dt, is related to the emf induced in the coil by emf $= NBb(dx/dt)$.

23. Refer back to figure 24.14. Suppose a voltage source is inserted in the side GH so as to send a clockwise current through the loop. The bar AC has mass m and starts from rest. (a) Find the bar's speed as a function of t if the current in the circuit has a constant value I. (b) If the source voltage is constant and equal to V_0, show that the limiting speed of the rod is V_0/Bb. (c) What is the current in the circuit when the terminal speed has been reached? Assume only the bar has resistance and it is R.

24. Refer back to problem 13 and the coaxial cable shown in figure P24.7. (a) How much magnetic field energy is stored in a unit length of the cable? (b) Show that this is also the energy given by $\frac{1}{2}LI^2$.

25. The bar of approximate length b shown in figure P24.11 slides from rest down the incline along the contacts as shown. If **B** is directed in the vertical direction, find an equation for the induced emf between the ends of the sliding rod as function of time. Is the emf clockwise or counterclockwise when viewed as shown? Assume negligible friction and negligible current in the circuit.

26. The magnetic field between two magnets is to be determined by measuring the induced voltage in the loop of figure P24.12 as it is pulled through the gap at uniform speed $v = 30$ m/s. (Instead of voltage, one would usually measure the charge as outlined in problem 20.) Plot a graph of voltage vs. time, assuming no fringing at the magnets. Compute the maximum voltage in terms of B.

27. A coil of wire with N loops swings with speed v into a uniform magnetic field B as shown in figure P24.13. If the ends of the coil are connected together and the resistance of the coil is R, find the force exerted on the coil by the field when in the position shown. What does the force do to the coil? This is the principle behind magnetically damped balances and similar devices.

28. A long solenoid that has 800 loops per meter length carries a current $i = 3 \sin (400t)$ A. Find the electric field inside the solenoid at a distance of 2 mm from the solenoid axis. Consider only the field tangential to a circle having its center on the axis of the solenoid.

29. A long solenoid with n loops per meter is filled with a metal core having resistivity ρ. Find the eddy current density that flows in the core at a distance r from the solenoid axis when the current in the solenoid is

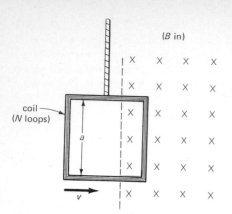

(B in)

coil
(N loops)

a

v

FIGURE P24.13

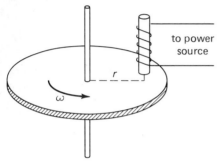

to power
source

r

ω

FIGURE P24.14

$i_0 \sin (2\pi ft)$. Assume the magnetic field due to the core itself is negligible.

30. The device shown in figure P24.14 is an eddy current brake. When current passes through the solenoid, the metal disk that rotates below it experiences a retarding torque. What is the functional dependence of the torque on the current I through the solenoid, the speed of rotation ω of the disk, and the resistivity ρ of the metal plate?

31. Refer back to figure 24.14. Suppose the rod of mass m and resistance R is moving with speed v_0 when the force pulling it is removed. (a) Show that its speed will decrease as $v = v_0 \exp (-t/T)$ where $T = mR/B^2b^2$. (b) Show that the total heat energy developed in the rod by the current as the rod comes to rest is $\frac{1}{2}mv_0^2$. Assume the resistance of the stationary wires to be zero.

32. Suppose in problem 25 that the resistance of the bar is R, its mass is m, and that the resistance of the rest of the circuit is negligible. (a) Show that the bar reaches a terminal velocity $gT \sin \theta$ where $T = mR/B^2b^2 \cos^2 \theta$. (b) Show that the general equation for v is $gT \cdot \sin \theta (1 - e^{-t/T})$.

33. A cylindrical vacuum tube is placed inside a solenoid so that the solenoid fits snugly around it. Show that the force on an electron within the tube has two distinct parts resulting from the increasing flux in the solenoid. What are they?

34. In the betatron, the situation is somewhat like that of problem 33. However, the magnetic field is weaker at the orbit B_0 than it is in the area enclosed by the orbit B_i. Show that the average value of B_i must be twice as large as B_0. *Hint:* For the electron to move in orbit,

$$p = rqB_0$$

and if the electron starts from rest as B starts from zero,

$$qE = \frac{dp}{dt} \quad \text{with} \quad 2\pi rE = \frac{d\phi_i}{dt}$$

25 Dielectric and Magnetic Materials

In the previous chapters on electricity we have grouped substances into conductors and nonconductors. It was sufficient to state the resistivity or resistance of the materials in order to describe the electrical behavior of the systems of which they were a part. There are other situations and materials, however, which cannot be treated this way. We shall see in this chapter how the results of previous chapters must be altered when the dielectric and magnetic properties of the substances present become important.

25.1 ELECTRIC DIPOLES

Both the electric and magnetic properties of substances are profoundly influenced by the presence of atomic and molecular dipoles. An electric dipole is simply two equal but opposite charges $\pm q$ separated by a small distance a. Such a dipole together with its electric field is shown in figure 25.1(a). We can of course compute the electric field and potential in the region near the dipole by use of the methods of previous chapters. Referring to 25.1(b), you should be able to write down at once that the absolute potential at point P is given by

$$V = \frac{q}{4\pi\epsilon_0}\left(\frac{1}{r_1} - \frac{1}{r_2}\right) = \frac{q(r_2 - r_1)}{4\pi\epsilon_0 r_1 r_2}$$

Potential far from a dipole

Usually one is interested in atomic- or molecular-size dipoles, in which case any point of observation such as P will be far enough away so that

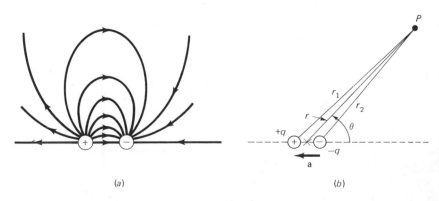

FIGURE 25.1 The dipole moment is defined to be $\mathbf{p} = q\mathbf{a}$.

(a)

(b)

$r \gg a$. In that case we can set $r_1 r_2 \approx r^2$ and $r_2 - r_1 \approx a \cos \theta$, yielding

$$V \approx \frac{qa \cos \theta}{4\pi\epsilon_0 r^2} \qquad 25.1$$

The quantity qa is called the *dipole moment* of the dipole and is represented by the symbol p. Frequently, direction is assigned to the dipole moment, and p becomes a vector given by

$$\mathbf{p} = q\mathbf{a} \qquad 25.2$$

Notice $\mathbf{a}$ is a vector directed from the negative to the positive end of the dipole.

To find the electric field due to a dipole, we can proceed directly from 25.1. You will recall that we have shown $E_x = -\partial V/\partial x$ and so on for E_y and E_z. In the present situation it is convenient to use polar rather than rectangular coordinates. Using a similar approach in polar coordinates, one can show that the electric field in the r and θ directions is given by

$$E_r = -\frac{\partial V}{\partial r} = \frac{qa \cos \theta}{2\pi\epsilon_0 r^3}$$

and

Dipole field decreases as $1/r^3$

$$E_\theta = -\frac{1}{r}\frac{\partial V}{\partial \theta} = \frac{qa \sin \theta}{4\pi\epsilon_0 r^3} \qquad 25.3$$

It is important to notice that the electric field of a dipole decreases as $1/r^3$ rather than as $1/r^2$ as found for a point charge. This causes dipole effects often to be negligible in comparison to the effects of isolated charges.

When an electric dipole is placed in an external electric field $\mathbf{E}$, the dipole experiences a torque which tends to align the dipole with $\mathbf{E}$. This is easily seen from figure 25.2, where we see a dipole $\mathbf{p}$ between the plates of a capacitor. Recalling that torque is (force)(radius)(sin θ), one sees from figure 25.2(b) that the torque about the center of the dipole is simply

$$\text{Torque} = (qE)(\tfrac{1}{2}a)(\sin \theta) + (qE)(\tfrac{1}{2}a)(\sin \theta)$$

or

$$\tau = pE \sin \theta$$

where p is the dipole moment. This may be written in more compact form by using $\mathbf{p}$ and $\mathbf{E}$ as vectors:

Torque on a dipole

$$\boldsymbol{\tau} = \mathbf{p} \times \mathbf{E} \qquad 25.4$$

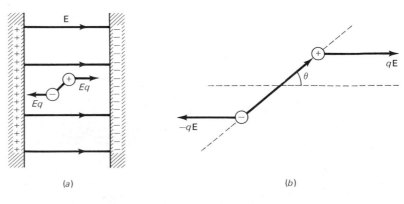

FIGURE 25.2 A dipole tends to line up with an impressed electrical field.

(a)

(b)

The important point for our purposes, however, is that *the dipole is caused to line up along any impressed electric field.* In the next section, we shall apply these concepts to the behavior of molecules and atoms.

25.2 ATOMIC AND MOLECULAR ELECTRIC DIPOLES

Many molecules act like tiny electric dipoles. Several of these are shown schematically in figure 25.3. The dipole moment for each is given in a unit called the debye (D). It is not a member of the SI, but it is a unit in frequent use: $1\ \mathrm{D} = \frac{1}{3} \times 10^{-29}\ \mathrm{C} \cdot \mathrm{m}$.

Chemists have been quite active in the measurement of molecular dipole moments since the values they obtain give them clues as to the structure of the molecule. Molecules which have nonzero dipole moments are called *dipolar molecules*.

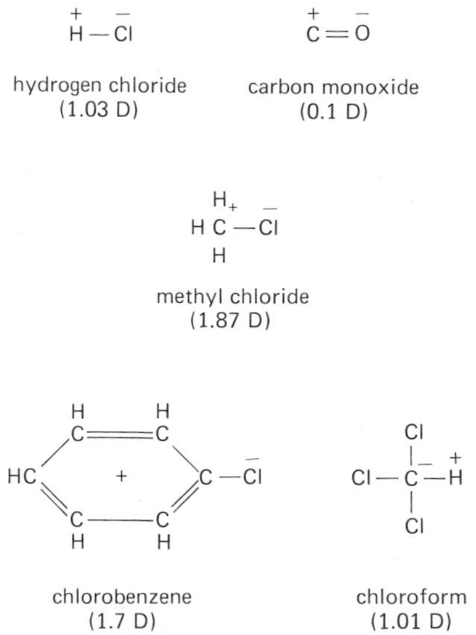

In a typical molecule, the dipole moment is of the order of 1 D. For example, if the separated charge in the HCl molecule of figure 25.3 is taken to be the electronic charge, 1.6×10^{-19} C, then, since the dipole moment of this molecule is 1.03 D,

$$qa = (1.6 \times 10^{-19}\ \mathrm{C})(a) = (1.03)(\tfrac{1}{3} \times 10^{-29})\ \mathrm{C} \cdot \mathrm{m}$$

where a is the charge separation. One then finds $a \approx 0.5 \times 10^{-10}$ m. In many cases, a is smaller than the known separation of the atoms, and so the situation corresponds more to an unequal sharing of charge between the atoms. For HCl the value we have found for a is only about half the known distance between the atom centers.

When a gas, liquid, or solid which contains dipolar molecules is placed in an electric field, the dipoles try to line up with the field. However, the random thermal motion of the molecules tends to disorient them. As a result, at ordinary temperatures and fields, the molecules do not even come close to total alignment. However, on the average, the positive charge of each molecule is displaced somewhat in the direction of the field, while the average position of the negative end of the dipole is shifted in the opposite direction. Keep this in mind since we shall soon have occasion to make use of it.

Not all molecules act like dipoles. In fact, molecules such as methane, carbon disulfide, carbon tetrachloride, and many more have zero dipole moment. Evidently, in these molecules the positive and negative charges are effectively centered on the same point. The same situation exists for isolated atoms. Even though the atom's negative electrons do not coincide with the positive nucleus, the electron cloud is centered on the nucleus. As a result, the dipole moment qa of the atom is zero since the effective charge separation a is zero.

Even though an atom or molecule may have no permanent dipole moment, it still is affected by an impressed electric field. This can be seen by reference to figure 25.4. We see at the left an atom (or molecule) for which the negative and positive charge centers coincide. When an electric field is impressed, as shown at the right, the positive charge is shifted in the direction of **E**, while an opposite effect occurs for the negative charge. The molecule (or atom) is then said to possess an *induced dipole*.

The important point to notice here, though, is that the same overall effect occurs for both polar and nonpolar molecules. When an atom or molecule of any type is placed in an electric field, positive charge shifts in the direction of the field while negative charge shifts in the opposite direction.

1 debye $= \frac{1}{3} \times 10^{-29}\ C \cdot m$

hydrogen chloride
(1.03 D)

carbon monoxide
(0.1 D)

methyl chloride
(1.87 D)

chlorobenzene
(1.7 D)

chloroform
(1.01 D)

FIGURE 25.3 Typical dipolar molecules ($1\ \mathrm{D} = \frac{1}{3} \times 10^{-29}\ \mathrm{C} \cdot \mathrm{m}$).

Induced dipoles

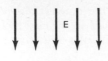

This is true for all substances and is fundamental to their dielectric properties. Let us now see how this atomic and molecular behavior evidences itself in the bulk electrical properties of materials.

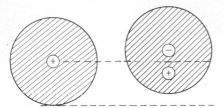

FIGURE 25.4 An impressed electric field causes a charge shift in all atoms and molecules. Electric fields induce dipoles in atoms.

25.3 DIELECTRICS AND DIELECTRIC CONSTANT

There are two different approaches we could take to the explanation of the electrical properties of solids, liquids, and gases. We could try to describe in detail the electric fields and charges within the material. The field would vary widely from place to place in the material as we moved from the vicinity of an electron to that of a nucleus, for example. To describe the electric field variation within even a single atom is a problem of great complexity. When we consider the fields of many atoms all in thermal motion, we see that simplifications are highly desirable.

For most electrical applications, such a detailed examination of a material is neither necessary nor desirable. Certainly the erratic motion and field of a single atom or molecule will have no noticeable effect on the overall properties of the material. Only those motions and behaviors which are typical of large numbers of atoms and molecules acting together can be of concern. Hence we are led to consider only the average effects and average fields throughout the material. The highly variable and erratic fields found between points only an atom or two apart will not concern us. We shall restrict our discussion to the average field in regions large enough so that many atoms or molecules are contained therein. With that in mind, let us refer to figure 25.5.

Definition of dielectric

A *dielectric* is any material which contains no appreciable number of free charges; it is therefore any nonconductor. In figure 25.5(*a*) we show a slab of dielectric. We assume it to have no net charge. Even though the electric field varies widely from point to point within it, we concern ourselves only with an average field in regions large enough to contain many atoms. This average field is zero since no average net charge exists in the dielectric.

Suppose now an external electric field is applied to the slab, as shown in figure 25.5(*b*). As we discussed in the previous section, all the atoms and molecules of the slab will have their positive charge shifted slightly to the right while the negative charge will shift to the left. As shown in the figure, this causes a layer of positive charge to appear on the right side of the slab and a layer of negative charge to appear on the left. This type of charge is

Dielectrics are polarized by an electric field

called *polarization* or *bound* charge, and the dielectric is said to be *polarized*. The terminology "bound" occurs since the charges in the layers are still tightly bound to atoms and molecules. Unlike free charge on a conductor, it cannot leave the dielectric. We shall represent this bound surface charge density by the symbol σ_b. Let us now see what effect this polarization charge has on the electric field.

Consider the parallel plate capacitor shown in figure 25.6. When a charge σ_f is placed on a unit area of its plates,* the field E_0 between the plates can be found by use of Gauss' law as we have done many times. Using the box of end area A shown in cross section in part (*a*) of the figure, we have

$$\epsilon_0 \int \mathbf{E}_0 \cdot d\mathbf{A} = \Sigma \, q_i$$

*The subscript f stands for "free" because the charges can be conducted off the metal.

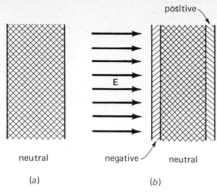

(a)   (b)

FIGURE 25.5 When a dielectric is placed in an electric field, positive charge shifts in the direction of the field, while negative charge shifts in the opposite direction. This causes charge layers to appear on the surfaces.

For the same charge, dielectrics cause the field to be decreased

Since the flux of E_0 through the sides of the box is zero and also through the end in the metal, we find

$$\epsilon_0[0 + 0 + 0 + 0 + 0 + E_0 A] = \sigma_f A$$

where E_0 is the field between the plates. Solving for E_0 gives the familiar result $E_0 = \sigma_f/\epsilon_0$. The subscript zero on E_0 is to alert us that this is the result for an empty capacitor.

Let us now consider the effect of placing a dielectric slab between the plates as in figure 25.6(b). (We show a slight gap between the slab and plate for ease of discussion.) The original electric field E_0 between the plates due to σ_f causes polarization charges to appear on the surfaces of the dielectric slab. Notice that positive σ_f induces negative σ_b close to it. Since the gap shown in the figure is highly exaggerated, it is apparent that the included charge $-\sigma_b$ and free charge σ_f on the metal plate combine to give an effective charge on the left plate $\sigma = \sigma_f - \sigma_b$. In effect, then, the dielectric slab partly cancels the charge originally on the capacitor plates. We could guess that the field within the dielectric E_d should no longer be given by σ_f/ϵ_0 but should be

$$E_d = \frac{\sigma_f - \sigma_b}{\epsilon_0}$$

This guess can also be arrived at by examining the field lines in figure 25.6(b). As we see, some of the field lines leaving the metal plate terminate on the face of the dielectric slab. This, too, indicates E_d is smallest for dielectrics which are the most polarizable. For those materials σ_b may actually become nearly as large as σ_f, in which case E_d would become nearly zero.

The best way to find the field within the dielectric is to apply Gauss' law to the gaussian surface shown in cross section in the figure. Proceeding as before we find

$$\epsilon_0[0 + 0 + 0 + 0 + 0 + E_d A] = (\sigma_f - \sigma_b)A$$

from which

$$E_d = \frac{\sigma_f - \sigma_b}{\epsilon_0} \qquad\qquad 25.5$$

showing that our guess is justified.

Unfortunately 25.5 is seldom useful since σ_b, being a charge bound to the dielectric, cannot be measured easily. To put the relation in a more useful

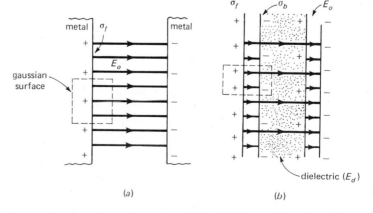

(a)   (b)

FIGURE 25.6 The bound charge on the surface of the dielectric causes the field inside the dielectric E_d to be less than E_0.

form, we make use of the fact that, for many dielectrics (called homogeneous, isotropic, linear dielectrics), one finds that $\sigma_b \sim E_d \sim E_0$. This is not unexpected since E_0 causes the polarization, and so both the bound charge and E_d might well be expected to be proportional to E_0. For such dielectrics it is customary to write

Electric susceptibility

$$\sigma_b = \chi \epsilon_0 E_d \qquad\qquad 25.6$$

where χ (Greek chi) is called the *electric susceptibility*. We shall not use χ as such but simply note it in passing.

We can now replace σ_b by $\chi \epsilon_0 E_d$ and σ_f by $\epsilon_0 E_0$ in 25.5 to give

$$E_d = E_0 - \chi E_d$$

from which

$$E_d = \frac{E_0}{1 + \chi} \equiv \frac{E_0}{k}$$

or

$$\frac{E_0}{E_d} = k \qquad\qquad 25.7$$

Dielectric constant k

The quantity k, the ratio of the field in vacuum to the field in the dielectric, is called the *dielectric constant* of the material. Typical values are given in table 25.1. Because k is a simple ratio, it has no units.

Referring to the dielectric constants given in table 25.1, we note that vacuum has $k = 1.00$. As we would expect, $E_d = E_0$ since this is simply the limiting case of no dielectric at all. The few molecules present in air cause k to be 1.006, indicating $E_{\text{air}} = E_0/1.006$. Most materials have a dielectric constant in the range 2 to 8. However, hydrogen-bonding molecules such as water and alcohol have exceptionally high dielectric constants. For example, $E_{\text{water}} = E_0/81$. The molecules of these substances are bound together in large dipolar clumps by their hydrogen bonds, and these exceptionally large dipole groups give rise to very large polarization effects. Metal is actually not a dielectric in the usual sense since it possesses free charges. However, a sheet of metal could be used in the experiment shown in figure 25.6. Since E in the metal is zero (the induced charge equals σ_f), we would have $k = E_0/E_d \to \infty$ in this case.

Dielectrics often decrease E_0 to E_0/k

Although our discussion has been restricted to the case of a plane slab of dielectric in a uniform field, we shall see that our result has wider application. In many simple situations the effect of a dielectric is to decrease the electric field by a factor $1/k$. Before examining this point further, we shall see how one measures dielectric constants.

TABLE 25.1   Dielectric Constants (20°C)

Material	k
Vacuum	1.00000
Air	1.006
Paraffin	2.1
Petroleum oil	2.2
Benzene	2.29
Polystyrene	2.6
Ice (−5°C)	2.9
Mica	6
Acetone	27
Methyl alcohol	38
Water	81
Metal	∞

25.4 MEASUREMENT OF DIELECTRIC CONSTANT

Dielectric constants are most frequently measured by means of capacitance measurements. You will recall that if a charge Q is placed on one plate of the capacitor and if V is the potential difference between the plates, then $C = Q/V$. Let us now consider a parallel plate capacitor which can be measured under two conditions: first with vacuum (or, in practice, air) between the plates and then with the material under investigation between the plates.

Suppose in both cases a charge Q is placed on the plates. The charge density on the plates will be Q/A, where A is the plate area. In the case of vacuum between the plates E will be σ/ϵ_0, which is $Q/A\epsilon_0$. With the dielectric between the plates, the field will be $1/k$ as large. If the plate separation is d, the potential difference between the plates will be (from $V = Ed$)

$$V_v = \frac{Q}{A\epsilon_0}d \qquad V_d = \frac{Q}{kA\epsilon_0}d$$

for vacuum   for dielectric

Then, since $C = Q/V$, we find

$$C_v = \frac{A\epsilon_0}{d} \qquad C_d = k\frac{A\epsilon_0}{d}$$

for vacuum   for dielectric

Relation for C in vacuum- and dielectric-filled capacitors We see that the effect of the dielectric has been to increase the capacitance of the capacitor by a factor equal to the dielectric constant. Therefore, if we measure the vacuum-filled capacitor and find its capacitance to be C_v and if we repeat the measurement with dielectric between the plates obtaining C_d, then

$$k = \frac{C_d}{C_v} \qquad\qquad 25.8$$

25.5 GAUSS' LAW FOR DIELECTRICS

The two fundamental relations involving electric fields were Coulomb's law and Gauss' law. We now wish to see how the presence of a dielectric can be accounted for by these laws. To do this, we shall first find the necessary modification of Gauss' law. Then we shall use it to derive Coulomb's law. You will recall that the two are not independent.

If you refer back to our discussion of Gauss' law in Chap. 18, you will see that nothing we said restricted it to metals and vacuum. The result was

$$\epsilon_0 \int \mathbf{E} \cdot d\mathbf{A} = \sum q$$

The integral is to extend over a closed surface, while the right-hand side was to be the sum of *all* the charges within the gaussian surface. Our only complication in the case of dielectrics is that the charges could be of two kinds, free charges (q_f) and bound charges (q_b). Let us write Gauss' law in terms of them:

$$\epsilon_0 \int \mathbf{E} \cdot d\mathbf{A} = \sum (q_f + q_b)$$

The difficulty with this relation is that q_b, the bound charge, is not easily obtained from experiment. We would like, therefore, to replace it in terms of E and the dielectric constant much as we replaced σ_b in 25.6. Although it is not difficult to show how this replacement can be justified, the necessary steps are lengthy. For that reason we simply resort to analogy here and state that substitutions similar to 25.6 and 25.7 can be made. In so doing, we can eliminate the bound charge from the equation and obtain

Gauss' law for dielectrics
$$\epsilon_0 \int\limits_{\substack{\text{closed}\\ \text{surface}}} k\mathbf{E} \cdot d\mathbf{A} = \sum_{\text{inside}} q_f \qquad\qquad 25.9$$

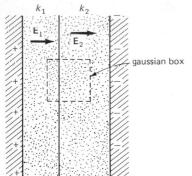

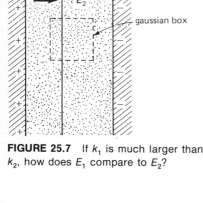

FIGURE 25.7 If k_1 is much larger than k_2, how does E_1 compare to E_2?

This is the form Gauss' law takes in the presence of a linear, isotropic dielectric.*

Equation 25.9 involves the integral not of E but of kE over the closed gaussian surface. If the dielectric constant is different at different points on the surface, then the value of k appropriate to each dA must be used. By E we mean the average electric field at the area dA. The right-hand side of the equation is simply the total free charge within the gaussian surface. Notice that the polarization charge need not be considered. Its effects are contained in k.

Illustration 25.1

Two dielectrics with constants k_1 and k_2 are placed between two infinite charged plates as shown in figure 25.7. Use Gauss' law to find the ratio of the electric fields in the two dielectrics.

Reasoning Apply Gauss' law to the box of end area A shown in cross section. No field lines emerge from its sides, and so the integral of $kE \cdot dA$ is zero there. The flux is into the left end, and so the integral there is $-k_1E_1A$. On the opposite end the integral is k_2E_2A. Notice that we use the dielectric constant at the position where E is being used. There is no free charge within the gaussian box. Therefore Gauss' law

$$\epsilon_0 \int k\mathbf{E} \cdot d\mathbf{A} = \sum q_f$$

becomes in this case

$$0 + 0 + 0 + 0 - \epsilon_0 k_1 E_1 A + \epsilon_0 k_2 E_2 A = 0$$

from which

$$\frac{E_1}{E_2} = \frac{k_2}{k_1}$$

25.6 COULOMB'S LAW FOR DIELECTRICS

The force which a point charge Q_1 exerts in vacuum on a second point charge Q_2 when separated by a distance r is given by Coulomb's law

$$F = \frac{1}{4\pi\epsilon_0} \frac{Q_1 Q_2}{r^2} \qquad \text{(vacuum)}$$

If Q_2 is a unit positive test charge, then the force on it is defined to be the electric field strength resulting from Q_1. We might expect that if the charges are immersed in a dielectric with constant k, the electric field and force would be decreased by a factor $1/k$. This is, indeed, true, as we shall now show.

Consider a charge Q immersed in an infinite dielectric, as shown in figure 25.8. To find the electric field E at point P, we take a spherical gaussian surface as shown. Then, Gauss' law

$$\epsilon_0 \int k\mathbf{E} \cdot d\mathbf{A} = \sum q_f$$

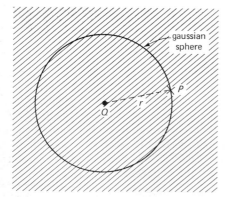

FIGURE 25.8 Find E at point P if Q is immersed in an infinite dielectric.

---

*The quantity $\epsilon_0 k$ is sometimes represented by the symbol ϵ and is called the *electrical permittivity* of the material. Since k is unity for vacuum, ϵ_0 is called the *permittivity of free space*.

becomes, since from symmetry E is constant on the surface,

$$\epsilon_0 k E \int dA = Q$$

Point charge field in a dielectric

But the integral is simply the area of the sphere, $4\pi r^2$, and so

$$E = \frac{1}{4\pi\epsilon_0}\frac{Q}{kr^2} \qquad\qquad 25.10$$

As we expected, the dielectric has decreased E by a factor $1/k$.

Since E is the force on a unit charge placed at P, a point charge Q' will experience a force

$$F = \frac{1}{4\pi\epsilon_0}\frac{QQ'}{kr^2} \qquad\qquad 25.11$$

The Coulomb force is decreased by a factor 1/k

This is the form Coulomb's law takes for charges immersed in an effectively infinite dielectric. Notice that the force is decreased by a factor of $1/k$. Since water has $k = 81$, forces between ions in water solution are only about one-eightieth as large as they would be in vacuum. This is why water is so effective in dissolving ionic substances. The ions can readily escape each other by use of thermal energy.

25.7 MAGNETIC PROPERTIES

Now let us leave the topic of dielectric properties and turn to the subject of how materials behave in magnetic fields. Unlike the situation we found for dielectric properties, most materials alter magnetic fields very, very little. Only the ferromagnetic substances, primarily iron, nickel, cobalt, gadolinium, and dysprosium, cause large changes in magnetic fields. For this reason we can usually consider all but these materials to behave the same as vacuum in a static magnetic field.

To understand the magnetic properties of materials, we must first understand the origin of magnetic fields. As we saw in the previous chapters, magnetic fields are generated by electric currents. To the best of our knowledge, this is the *only* source of magnetic fields. Therefore, if a substance alters a magnetic field, we should be able to explain this alteration in terms of currents within the substance. In this chapter we shall not be concerned with the long-range currents which ordinarily flow through wires since we have already discussed them in previous chapters. We shall, instead, be interested in currents within atoms and nuclei which give rise to the magnetic properties characteristic of the substance itself.

Currents are the source of magnetic fields

Many atoms and fundamental particles act like tiny bar magnets. For example, the individual electron, proton, neutron, and many atoms containing them have magnetic fields characteristic of a bar magnet. Each acts as though it has a north and a south pole. But, as we have seen previously, and as we show again in figure 25.9, a current loop acts like a bar magnet. Part (c) of the figure shows how we might picture the generation of the magnetic field of a proton. If the proton is a charged sphere, then, if it were rotating, it would act like a stack of current loops. Each loop would generate a magnetic field, and the particle would act like a tiny bar magnet. We say that the particle has *spin* and that it possesses a *spin magnetic dipole moment*.

Spin magnetic dipole moment

We do not know if such an interpretation is correct in detail. The fact that even the neutron (which has no net charge) has a magnetic field leads us

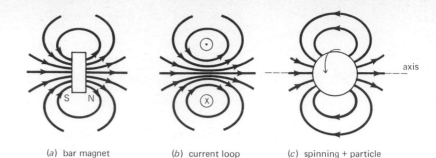

FIGURE 25.9 All magnetic effects can be traced to charge motion.

(a) bar magnet   (b) current loop   (c) spinning + particle

to suspect that the fundamental particles have internal structure and motion. As of yet, we know very little about this structure, however. But we firmly believe that the magnetic fields of all particles must be caused by charge motion and, in effect, current loops.

Atoms, since they are composed of electrons, protons, and neutrons, also have magnetic fields associated with them. We shall see in later chapters that in many atoms the basic particles within them arrange themselves in such a way that their magnetic fields cancel each other. But in addition to that effect, the electrons may, for many purposes, be thought of as circling around the atomic nucleus. These moving charges constitute atomic-sized electric currents which also give rise to magnetic effects.

The point we wish to emphasize strongly is the following: Atoms can be thought of as containing tiny current loops; the magnetic effects of all substances are dependent upon the behavior of these atomic currents. In the following sections we shall discuss the two fundamental types of magnetic effects in atoms, diamagnetism and paramagnetism. We shall see that, in certain very special atoms, paramagnetic effects become exceptionally strong and give rise to the strong magnetic properties we usually associate with iron.

25.8 DIAMAGNETISM

One of the most fundamental properties of all current loops is that an emf can be induced in them by a changing magnetic field. According to Lenz' law, the induced current caused by the induced emf is such as to oppose the change in magnetic field. For example, if the magnetic flux through a loop of wire is originally zero, then the loop opposes any change in this condition. If one tries to impress a field B upon the loop, the induced current in the loop will generate an opposite field. As a result, B through the loop is always *decreased* by the loop itself.

B is decreased by diamagnetic effects

In the case of wires, this effect is only transient. It lasts only while the flux through the loop is changing. Such is not the case for atomic current loops. The reason for this can be traced back to the law of conservation of energy. For simplicity, consider the electrons in atoms to orbit in circles about the nucleus. (We shall see later that this is a useful but not very realistic picture of the atom.) If the induced emf in one of these electron loops opposes the motion of the electron around the loop, the electron will lose energy and thereby be slowed down. Since the electron is unable to acquire energy easily from outside, its loss in energy will persist even after the induced emf has become zero. As a result, the effect of the induced emf is lasting, and so it is as though the induced current persisted even after the flux has stopped changing.

From this we see that all atoms will, because of this effect, tend to reduce the magnetic field impressed upon them. In other words, the magnetic field is reduced in magnitude when substances showing this effect are introduced into the region. We call this effect *diamagnetism*. Substances in which this effect is dominant are called *diamagnetic substances*. However, this is a very small effect and is often overshadowed by another effect to be discussed in the next section.

25.9 PARAMAGNETISM

The magnetic properties of each individual atom are caused primarily by the nucleus and the effects of the electrons which orbit the nucleus. Depending upon the type of nucleus involved, its constituent protons and neutrons may or may not cause it to act like a current loop. In terms of the terminology introduced in Sec. 22.7, the nucleus may or may not have a magnetic moment μ different from zero. (You will recall that the magnetic moment $\mu = IA$, where I is the current in the loop and A is the loop's area.) The nuclear magnetic moment is quite small: it is of the order of 10^{-3} as large as the magnetic moment caused by a single electron in the atom as it orbits the nucleus.

As we would expect, the electron's orbital motion about the nucleus causes the atom to act like a current loop. But, in many atoms, the various electrons circulating about the nucleus do so in such a way as to cancel each other's orbital magnetic field. In addition, even though each electron acts like a tiny current loop because of its spin, the electrons are oriented within many atoms in such a way as to cancel each other's spin magnetic moment as well. As a result, some atoms (carbon, copper, and lead, for example) have no permanent magnetic moment. These atoms show only the diamagnetic effect discussed in the preceding section. In other atoms, all but a very few of the electrons have their magnetic moments canceled. These uncompensated electrons give the atom a permanent magnetic moment. Such atoms exhibit the property of *paramagnetism*.

Paramagnetic effects increase B

When a current loop or atom with a magnetic moment is subjected to a magnetic field, the field tends to align the magnetic moment with the field, as we saw in Sec. 22.7. As a result, when a magnetic field is impressed upon a substance containing atoms with permanent magnetic moments, the atomic bar magnets tend to align with the field. Although the thermal motion of the atoms prevents this alignment from being complete, on the average each atom remains aligned more than disaligned. By reference to figure 25.10 we

FIGURE 25.10 The aligned magnetic dipole moment (current loop) augments the field which aligns it.

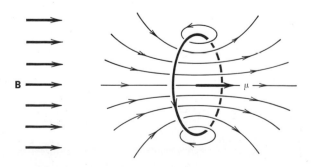

see that the aligned current loop (or atom) tends to increase the magnetic field which is causing the alignment. Paramagnetic materials are therefore seen to *increase* the magnetic field into which they are placed. This is exactly opposite to the situation we found for diamagnetic materials.

Of course even those atoms which have permanent magnetic moments are subject to the diamagnetic effect. However, being paramgnetic as well, the combination of the two effects complicates the situation. Frequently, the paramagnetic tendencies predominate. To separate the two effects we note that the paramagnetic effect depends upon the alignment of permanent magnetic moments. Increased temperature with its more violent thermal motion makes alignment more difficult. As a result, the paramagnetic effect decreases with increasing temperature. Diamagnetism occurs at both low and high temperatures equally well. We see, therefore, that the two effects show different temperature dependencies, and this fact can be used to study the two effects separately.

Paramagnetic effects are temperature-dependent

TABLE 25.2   Susceptibilities*

Material	Temperature (°C)	$\chi_m \times 10^5$
Vacuum		zero
Air	STP	0.04
Oxygen (gas)	STP	0.18
Aluminum	20	2.1
Cerium	18	130
Ferric chloride	20	306
Oxygen (liquid)	−219	490
Copper	18	−0.96
Lead	18	−1.6
Mercury	18	−2.8
Carbon (diamond)	20	−2.2
Carbon (graphite)	20	−9.9

*Values χ' given in the older data tables are related to our definition of χ_m through the relation $\chi_m = (4\pi\rho/M)\chi'$, where ρ is the density and M the molecular weight of the substance. Other tables define χ' in such a way that our $\chi_m = 4\pi\rho\chi'$.

25.10 MEASUREMENT OF MAGNETIC EFFECTS

To measure the magnetic properties of a material one can make use of several methods. The one to be described here is based upon the results given in the previous chapter. Suppose we possess a hollow toroidal coil, such as the one shown schematically in figure 25.11. Suppose further that a sinusoidal current is sent through the toroidal coil by a sinusoidal voltage source, represented by —$\bigcirc$—. If now a secondary coil is wound upon the toroid as indicated, the changing flux (due to the changing current in the toroid) will induce a voltage in the secondary. In particular, a current $i = i_0 \sin 2\pi ft$ will give rise to a field $B = B_0 \sin 2\pi ft$, and this changing field will induce a voltage $V = V_0 \cos 2\pi ft$ in the secondary.* Since V_0 will be proportional to B_0, the induced voltage can be used as a measure of the field in the toroid.

Suppose the voltage-measuring device measures a voltage V_0 when the toroidal coil is empty. If the coil is now filled with a certain material, then B_0 will change to B and the induced voltage will change from V_0 to V. However, V_0 and V are measures of B_0 and B within the toroid. As a result, we can write the proportion

$$\frac{B}{B_0} = \frac{V}{V_0}$$

*The cosine function occurs in V since $V \sim dB/dt$.

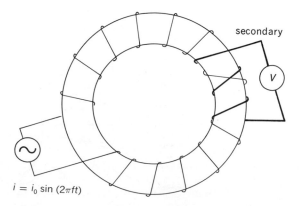

secondary

$i = i_0 \sin (2\pi ft)$

FIGURE 25.11   The voltage induced in the secondary coil can be used to determine the magnetic properties of the material inside the toroid.

If V is twice as large as V_0, for example, we know that the material has caused the magnetic field to increase by a factor of 2.

We characterize substances by this ratio,

Relative permeability k_m

$$k_m = \frac{B \text{ in presence of material}}{B \text{ in vacuum}}$$

It is called the *relative permeability* of the material. For vacuum, its value is unity. For diamagnetic substances, k_m is less than unity since they tend to decrease B. Since paramagnetic substances increase B, they have k_m values greater than unity.

In the case of most materials, k_m is of the order of 1.00000 ± 0.00001, and so the change from B_0 to B is of the order of only 1 part in 10^5. (Since this change is so small, the toroid system described here is usually of too low sensitivity to give accurate results. Other, more sensitive techniques are used.) It is therefore more convenient to define a new constant, the magnetic susceptibility χ_m, by the relation

$\chi_m = k_m - 1$ is of the order of 10^{-5} for most materials, except iron, cobalt, and nickel

$$\chi_m \equiv k_m - 1 = \frac{B}{B_0} - 1 \qquad 25.12$$

Typical values of χ_m are given in table 25.2. Since χ_m and k_m are defined in terms of B/B_0, they have no units.

Notice from the data in table 25.2 that some substances increase the flux within the toroid (positive χ_m) while others decrease the flux (negative χ_m). Those substances having negative χ_m values are *diamagnetic*, while those with positive values are *paramagnetic*.

Diamagnetic materials decrease B, while paramagnetic materials increase B

Another quantity frequently used in discussions of magnetic materials is the *permeability* of the material, μ. It is defined by

Definition of permeability μ

$$\mu = k_m \mu_0 \qquad 25.13$$

You will recall that $\mu_0 \equiv 4\pi \times 10^{-7}$ in the SI and is called the permeability of free space (i.e., vacuum). Notice that k_m is unity for vacuum, and so the permeability of vacuum is μ_0, as its name implies. Since the ratio μ/μ_0 is the ratio of the permeability of the material to that of vacuum, it should be clear why k_m is called the relative permeability.

It is also customary to define a quantity H called the *magnetic field strength*. For the simple cases we are considering, this definition can be written

In a toroid, the field strength H is given by B/μ

$$B = \mu H \qquad \text{or} \qquad B_0 = \mu_0 H \qquad 25.14$$

In more complicated situations than the toroid containing an isotropic material, the relation between B and H is more involved.

25.11 FERROMAGNETIC MATERIALS

The elements iron, nickel, and cobalt and their alloys, as well as a few lesser-known materials, react especially strongly in a magnetic field. They are said to be *ferromagnetic*. Whereas the magnetic susceptibility χ_m for most materials is only a very small fraction, the ferromagnetic materials have χ_m values of several hundred or larger. In terms of our toroid experiment, most materials change the magnetic field B by only a small fraction of a percent. The ferromagnetic materials, however, increase B by a factor of several hundred. We shall now investigate the reason for this.

Ferromagnetic materials have χ_m values of order 10^2

Iron, nickel, and cobalt are among the so-called transition elements. They are peculiar atoms in the sense that their electron structures are not completely orthodox. In particular, the next outer electron shell contains electrons, even though the inner shell is still not filled. As a result, the usual cancellation of electron spin magnetic moments does not take place in these elements. Consequently, these atoms have unusually large magnetic moments.

The forces between these rather complex atoms are not simple. They are strong enough, however, to cause the atoms in small regions to align with their magnetic moments all in the same direction. Of course, if the temperature is raised high enough, the thermal energy available will cause the atoms to break loose from each other, and the alignment is lost. The temperature at which this happens is called the *Curie temperature*. Its value is 770°C for pure iron and 358°C for pure nickel. Above this temperature these elements are much like any other paramagnetic element.

At temperatures below the Curie temperature, ferromagnetic materials are characterized by the presence of small regions called *domains,* in which the atoms all have their magnetic moments aligned. A schematic representation of this is shown in figure 25.12. The material shown might be a bar of demagnetized iron. It has no overall magnetic field associated with it because the domains are randomly oriented. Sizes of the domains vary, but they may become as large as a fraction of a millimeter in linear dimension.

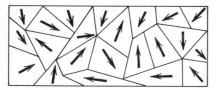

FIGURE 25.12 A schematic diagram showing the domains in an unmagnetized bar of iron. The arrows give the alignment direction of the magnetic moment in each domain. Domain size is much exaggerated.

If now an external magnetic field is imposed on the iron (suppose, for example, it is inside a toroid, and the current in the toroid is turned on), the domains aligned with the field direction begin to grow at the expense of the nonfavorably aligned domains. As a result, more atom magnetic moments become aligned in the direction of the field, and they cause an augmentation of the field. The domains grow by means of the motion of the domain boundaries, or walls. Typically, a domain wall will cease to move when it strikes an impurity or another nearly favorably aligned domain. Since the domain will not continue to grow further until a much larger applied magnetic field causes it to break loose, the atoms in the sample are only partly aligned with the field at low values of the field.

As the external field is increased, however, the alignment of the sample increases. Finally, at very high fields, essentially all the atoms are aligned with the field. The effect of this process on the total magnetic field within the toroid is shown in figure 25.13. We have plotted there B_0, the field which would exist in the toroid if it were vacuum-filled, against B, the actual field observed. This type of graph is called a *magnetization curve*. Notice that the ferromagnetic material has increased the field by a factor of about 10^3. At very high applied fields, the magnetic moments are essentially all aligned with the field. The maximum B contributed by the ferromagnetic material occurs at that point, and is called the *saturation field*. It appears in this case to be about 1.8 T.

25.12 THE HYSTERESIS CURVE

As one might expect, the curve of figure 25.13 is not completely reversible. This is a result of the fact that the motion of the domain walls shows an effect much like friction. When the impressed magnetic field is reduced to zero, the domains do not become completely disoriented. Instead they retain a portion of their original alignment. A typical case is shown in figure 25.14.

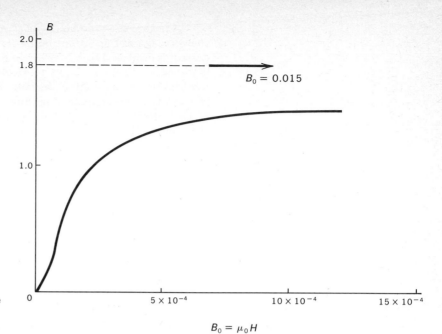

FIGURE 25.13 The magnetization curve for a common transformer iron. Numerical values are in teslas.

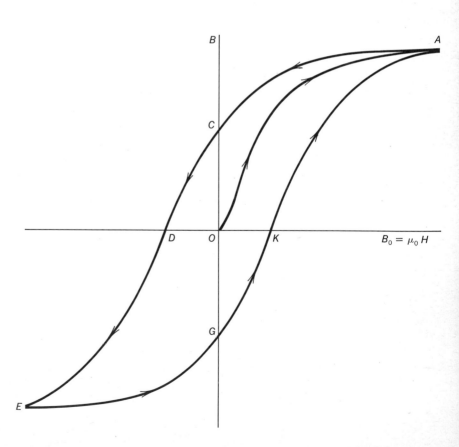

FIGURE 25.14 Why is not the curve OA reversible for ferromagnetic materials? Is it reversible for paramagnetic and diamagnetic materials?

In figure 25.14 the sample has been magnetized in a toroid by following the curve OA as the current in the toroid is increased. After reaching A, the current is decreased to zero. This decreases B_0, the field expected in the case of a vacuum-filled toroid, to zero. However, because of the remaining aligned domains in the material of the core, B has the value indicated by point C when B_0 is zero. The remaining value of B at point C is called the *retentivity* of the material. To reduce B in the toroid to zero one must forcibly disalign the domains by using a reverse external field. Not until B_0 reaches the negative value indicated by point D on the curves does B within the solenoid become zero. The external field needed to reach point D is called the *coercive force*. For a permanent magnet one would wish this value, as well as that of the retentivity, to be high. Why?

One can magnetize the material in the reverse direction by going to still larger negative values of B_0. B in the toroid follows curve DE. If the current is then decreased to zero, B follows portion EG of the curve. Finally, point A can be reached again by applying a sufficiently large positive value of B_0 so that the material follows curve GKA. Repetition of the previous current cycle will carry the iron around this curve repeatedly.

The area of the hysteresis loop measures energy loss

The total curve shown in figure 25.14 is called the *hysteresis curve* for the material. Although we do not show it here, the area enclosed by this curve is directly porportional to the energy lost to heat as the sample is carried around the cycle. For ac devices such as transformers, where energy loss and heat production are undesirable, one would like a material which has a hysteresis curve of small area. It is possible, by proper alloy composition and heat treatment, to produce ferromagnetic materials of widely different hysteresis characteristics. A great deal of metallurgical research is directed to this end.

25.13 GAUSS' LAW FOR MAGNETISM

Magnetic lines of flux are continuous

No one has ever detected a single isolated magnetic pole. Magnetic fields appear always to be the result of current loops or of atomic effects which can be so described. This means that, no matter how tiny or how large a volume one may choose, a flux line will not terminate within the volume since the lines of flux from current loops have no beginning and no end. Therefore we conclude that if a line of flux enters a volume, it must also emerge from that volume. As a result, the algebraic sum of the number of flux lines coming out of the surface of a volume must be zero. Expressed as an equation,

Gauss' law for magnetism

$$\int_{\substack{closed \\ surface}} \mathbf{B} \cdot d\mathbf{A} = 0 \qquad\qquad 25.15$$

This relation is similar to the equation we called Gauss' law in electrostatics, namely,

$$\epsilon_0 \int_S \mathbf{E} \cdot d\mathbf{A} = \sum q$$

The right-hand side of *25.15* is zero because there are no free poles in magnetism which would correspond to the charges q in electrostatics. Equation *25.15* is often referred to as Gauss' law for magnetism.

Equation *25.15* can be used to tell us how the magnetic field changes as it goes from one material to another. For example, if the material at the bottom (1) in figure 25.15 is iron and that at the top (2) is air, we can compare

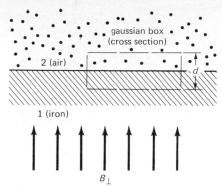

FIGURE 25.15 The flux coming out through the sides of the box can be made negligible by shrinking the dimension D nearly to zero.

the components of B perpendicular to the interface (the so-called normal components $B_\perp$) in the following way. If we choose a small box at the interface, as indicated in figure 25.15, the flux through its four small sides can be made negligible by shrinking the box to nearly zero thickness, just thick enough to enclose the interface. Since

$$\int_{\text{box}} \mathbf{B} \cdot d\mathbf{A} = 0$$

we have

$$- \int_{\text{bottom}} B_{1\perp}\, dA + \int_{\text{top}} B_{2\perp}\, dA + \int_{\text{sides}} B_\perp\, dA = 0$$

where the minus sign is appended since the flux is *into*, not out of, the bottom surface. Further, since

$$\int_{\text{sides}} B_\perp\, dA = 0$$

we have

$$\int_{\text{top}} B_{2\perp}\, dA = \int_{\text{bottom}} B_{1\perp}\, dA$$

Now the top and bottom area of the box may be taken as small as we choose; so the last equation holds only if

$$B_{1\perp} = B_{2\perp}$$

*The normal component of **B** is continuous at a surface*

at all points on the surface. This tells us that the normal component of B is continuous. That is to say, the component of B perpendicular to the surface is the same on both sides of the surface.

25.14 BOUND CURRENTS IN MAGNETIC MATERIALS

A great deal of insight into the behavior of devices which contain ferromagnetic materials can be obtained if we make use of a model used first by Ampère. He recognized that the magnetic effects of materials must be due to current loops. If one pictures the loops to be rectangular, then the end view of a magnetized bar of iron should be somewhat as shown in figure 25.16(a). (There should, of course, be many more loops than shown.) The loops represent the atomic currents aligned in the magnetic field.

Ampère noticed that the current loops cancel each other everywhere within the iron. For example, an upward current in one side of a loop is canceled by a downward current in the loop next to it. Only at the surface does this cancellation not occur. It is as though a giant current loop existed circling the bar as in (b), of figure 25.16. In this representation, the magnetized bar can be replaced by a series of current loops wound upon it. The bar, in effect, is a solenoid with a current of atomic origin flowing around it. This conclusion is reinforced by the similarities between the fields of a solenoid and bar magnet, as shown in figure 25.16(c).

It should be recognized that the bound current at the surface of a magnetized material is of atomic origin. Since the moving charges are bound to the atoms, they cannot be made to flow away in wires. Even so, they act like any other current in generating magnetic fields. Their usefulness can be seen by an example.

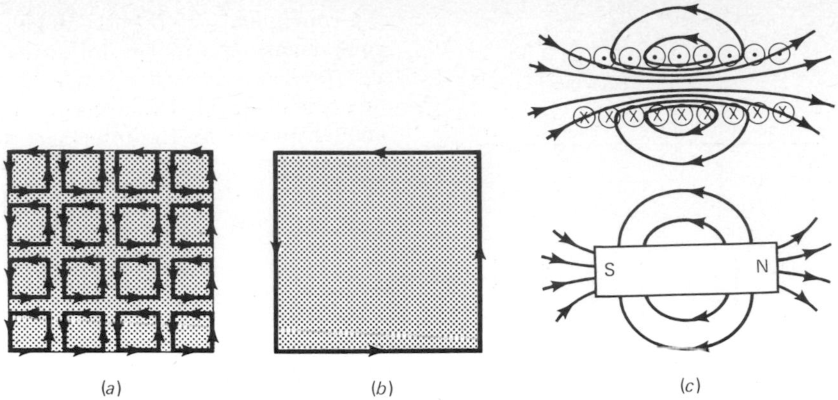

FIGURE 25.16 The magnetic field of a magnetized bar can be interpreted in terms of bound magnetization currents.

(a)   (b)   (c)

Suppose one has a yoke of iron such as the one shown in figure 25.17. A current I is sent through a coil wound on one side of it. The magnetic field of the coil magnetizes the iron in the direction indicated by the line labeled B as shown. But the magnetized iron gives rise to a magnetic field which can be represented by a bound current. The field of the bound current will be that of a toroid wound on the yoke as core. Since this latter field is many times larger than that generated by the current I, we see that the field B will be very strong and will circulate around the yoke as shown. This is but one example of how the concept of bound currents can be used to tell us the effects of magnetization.

There is one other point we should mention in regard to this example: the field lines follow through a highly magnetic material. This is typical of all situations of this type. Since B is strong within highly magnetized materials, the lines are very densely packed within them. In effect, this causes the magnetic field lines to concentrate in and to follow through ferromagnetic materials.

Field lines follow and concentrate in ferromagnetic materials

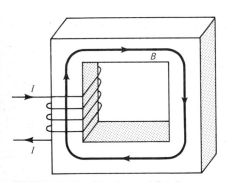

FIGURE 25.17 The lines of B follow around the contours of the iron yoke.

25.15 AMPÈRE'S CIRCUITAL LAW FOR MAGNETIC MATERIALS

It will be recalled that Ampère's circuital law was given by *23.2a* as

$$\oint \mathbf{B} \cdot d\mathbf{s} = \mu_0 \int \mathbf{J}_f \cdot d\mathbf{A} \qquad 23.2a$$

where $\mathbf{J}_f$ is the ordinary (or "free") current density; it does not include the bound currents. The right-hand side of this equation merely represents μ_0 times the current encircled by the integration path on the left. We obtained this relation, tacitly assuming the path of integration to lie in vacuum. Of course, for most practical purposes, we could also apply it to regions containing nonferromagnetic materials since these materials alter B only very slightly. Trouble occurs, however, if the integration path passes through ferromagnetic materials.

This fact is easily seen if we consider our usual toroid experiment. If we take the dashed-line circle shown in figure 25.18 as our path of integration, we have

$$\oint \mathbf{B}_0 \cdot d\mathbf{s} = \mu_0 N i \qquad 25.16$$

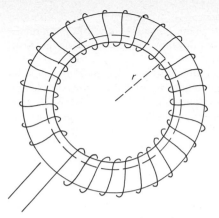

FIGURE 25.18 What does Ampère's circuital law say for this situation when the toroid has an iron core?

H is defined such that this equation is always true

where N is the total number of loops on the toroid, and i is the current in its windings. We use the subscript zero on B_0 in order to remind ourselves that the result is applicable only to a vacuum (or, to a good approximation, nonferromagnetic-material-) -filled toroid.

Clearly, B_0 cannot be replaced by B in this expression since, when an iron core is placed in the toroid, B becomes very large, while $\mu_0 N i$ does not change. Therefore, in applying Ampère's circuital law to ferromagnetic materials, we cannot use it in its form as given by 23.2. In the case of a toroid, at least, we see that the proper relation to use when a ferromagnetic core is present is obtained by replacing B by B_0.

It will be recalled we have defined a quantity

$$H = \frac{B_0}{\mu_0}$$

Since even in the presence of a ferromagnetic core the relation

$$\oint \mathbf{B}_0 \cdot d\mathbf{s} = \mu_0 \int \mathbf{J}_f \cdot d\mathbf{A} \qquad 25.17$$

applies to a toroid, the relation may also be written

$$\oint \mathbf{H} \cdot d\mathbf{s} = \int \mathbf{J}_f \cdot d\mathbf{A} \qquad 25.18$$

This is the form in which Ampère's circuital law is usually written when magnetic materials are present. *The quantity H is defined in such a way that 25.18 is always true.* In all these relations J_f is the real, or free, current density; so H does not contain the effects of the currents one might picture to be associated with the magnetic moments of the atoms. B, on the other hand, is the result of both real and bound currents. As we have indicated previously, the relation between B and H can become quite complicated in more complex situations than the toroid. However, when ferromagnetic materials are absent, 23.2, 25.17, and 25.18 are essentially equivalent since $B \approx B_0$.

Chapter 25
Questions
and Guesstimates

1. Distinguish clearly between the terms *free charge* and *bound charge*. Do these types of charges differ in the electric fields which they produce?
2. What would you expect the dielectric constant of mercury gas to be? Should it be close to infinity as in the case of the liquid metal, or should it be close to unity as in the case of a gas?
3. The dielectric constant of a substance composed of polar molecules such as methyl alcohol in the glassy state is in the neighborhood of 2. When changed to liquid, its dielectric constant increases to a value over 10 times larger. Such behavior is not found for nonpolar liquids. Explain the cause of this behavior.
4. Suppose one made a synthetic dielectric by dispersing metal "hairs," or "whiskers," in a plastic. Assuming the metal whiskers were well enough dispersed so that the material was still a nonconductor, what would you predict for its dielectric constant?
5. Reasoning entirely from the fact that induced bound charge appears at the metal-dielectric interface, explain why the electric field due to a charged metal sphere is much smaller when the sphere is immersed in an infinite dielectric.

6. A free atom, a dipolar molecule, and a small particle of dielectric all move into the region of stronger field when placed in an inhomogeneous electric field. Their centers of mass do not move when placed in a uniform electric field. Explain these results.

7. It is possible to tell whether or not a molecule has a permanent dipole by examining the temperature dependence of its dielectric constant when the molecules are in the gas phase. The dielectric constant of the dipolar gas decreases with increasing temperature while k is nearly temperature-independent for a nonpolar gas. Explain why this should be true.

8. A parallel plate air capacitor is charged by connecting it across a battery of voltage V_0. The battery is then disconnected and the region between the plates is filled with an insulating oil. (a) Does the charge on the plates change? (b) The potential difference between the plates? (c) The capacitance? (d) The field between the plates? Repeat if the battery remains connected to the plates as oil is placed between them.

9. An unmagnetized needle is suspended at its center by a very delicate thread. When a magnetic field is turned on at an acute angle θ to the line of the needle, the behavior depends upon the material of the needle. A paramagnetic needle will tend to rotate so as to decrease θ, while the opposite will occur for a diamagnetic material. Explain why.

10. A magnet will attract an unmagnetized nail made from iron, and that nail will attract another, and so on. Explain why this is true.

11. An initially unmagnetized iron rod is held in a north-south direction but with the north end and the line of the rod at an angle of about 60° below the horizontal at a point somewhere in the middle of the United States. If the higher (i.e., south) end of the rod is hit sharply several times with a hammer, the rod becomes a magnet. Explain why, and point out in particular which poles are on which ends. How can the rod be demagnetized?

12. A very strong bar magnet is slid across a table toward a bar of iron. The two bars are in line, the north pole of the bar magnet approaching the other bar. At first the bar is repelled by the magnet, but when they become quite close, it is attracted. Explain what must be going on.

13. When an iron nail is laid down on a tabletop close to a bar magnet and released, the nail accelerates toward the magnet and strikes it. (a) What is the source of the nail's kinetic energy as it accelerates? (b) Relate this to the fact that you must do work to pull the nail loose and take it back to its original position.

14. What is the net force that acts on a bar magnet when it is placed in a uniform magnetic field? On a bar of unmagnetized iron? Why, then, is an iron bar attracted by a magnet?

15. Attempt to draw the magnetization curve for each of the following on the same graph: iron, diamond, and aluminum. What difficulty do you encounter? Draw the graph for the latter two only.

16. Current theories lead us to suspect that the neutral particle called the neutron contains still undiscovered charged particles called *quarks*. If this proves to be true, how will it help us to understand better the fact that neutrons have a magnetic moment?

17. Magnetic flux lines are said to seek out and follow through iron rather than going through air. Draw the magnetic field around a horseshoe magnet. Now place a bar of iron between the poles. Show the induced poles on the bar and draw the magnetic field. Repeat for a string of

several iron tacks between the poles of the magnet. Why do the lines tend to go through the iron rather than the air?

18. Why is diamagnetism, unlike paramagnetism, relatively temperature-insensitive?

Chapter 25
Problems

1. A certain capacitor has a capacitance of 50 pF with air between its plates and 370 pF with a plastic between its plates. What is the dielectric constant of the plastic?

2. A chemist wishes to measure the dielectric constant of an oil by use of a concentric cylinder capacitor. The capacitances of the air-filled and oil-filled capacitor are 75 and 172 pF, respectively. What is the dielectric constant of the oil?

3. After a parallel-plate air-filled capacitor has been charged to 70 V potential difference, the capacitor is disconnected from the voltage source. A sheet of plastic, $k = 3.50$, is then slipped between its plates so as to fill the air gap. What is now the potential difference between the plates assuming that no charge left the capacitor's metal plates?

4. The plate separation between the two plates of a parallel plate capacitor is 0.20 cm. When the potential difference between the capacitor plates is 100 V, how large is E between the plates if the capacitor is (a) air-filled? (b) Has a dielectric with $k = 3.0$ between its plates?

5. The dielectric strength of polyethylene (i.e., the electric field value above which a spark will jump through the material) is about 1.8×10^5 V/cm. The dielectric constant for polyethylene is 2.3. If polyethylene fills the gap between two metal plates 0.20 cm apart, how large a voltage difference can be applied to the plates before breakdown will occur?

6. Using the fact that the energy stored in a capacitor is $\frac{1}{2}CV^2$, together with the expression for the capacitance of a dielectric-filled parallel plate capacitor found in this chapter, show that the energy stored in a unit volume of dielectric due to the presence of an electric field is $\frac{1}{2}k\epsilon_0 E^2$.

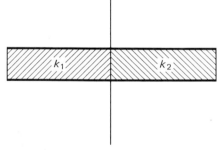

FIGURE P25.1

7. Show that the capacitance of the capacitor of figure P25.1 is

$$C = \frac{\epsilon_0 A}{2d}(k_1 + k_2)$$

where A is the area of one plate and d is the plate separation. Assume d is much smaller than shown. Reasoning from your result, can this capacitor be thought of as two capacitors in parallel?

8. The parallel plate capacitor shown in figure P25.2 is constructed of two equal thickness dielectrics (each being d) with dielectric constants k_1 and k_2. Assume the plate area A to be large and d to be small. Show that the capacitance is $(\epsilon_0 A/d)(k_1 k_2)/(k_1 + k_2)$. Could this result be obtained by thinking of the capacitor to be two capacitors in series?

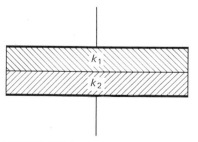

FIGURE P25.2

9. Two large parallel metal plates with area A and gap spacing d have between them a plastic sheet (dielectric constant k) of thickness $b < d$. Show that the capacitance of the device is

$$\frac{\epsilon_0 A k}{kd - (k-1)b}$$

10. Starting from the expression for the electric field due to a point charge in an infinite dielectric, show that the potential difference between two

points at r_1 and r_2 from the point charge is

$$-\frac{Q}{4\pi\epsilon_0 k}\left(\frac{1}{r_1} - \frac{1}{r_2}\right)$$

11. Using the result of the previous problem, find the work needed to pull a sodium ion $(+e)$ from a position 2 Å from the center of a chlorine ion $(-e)$ to infinity if the two are immersed in water. Repeat for benzene. For comparison purposes, the average thermal energy for each ion is $\frac{3}{2}kT$ or about 0.04 eV. Express your answer in electron volts.

12. Show that the bound charge density on the surface of a dielectric is $\sigma_b = \epsilon_0(k - 1)E_d$ where E_d is the field just inside the dielectric.

13. Show that the bound and free-surface charge densities for a capacitor plate and the dielectric next to it are related by $\sigma_b/\sigma_f = (k - 1)/k$.

14. A parallel plate capacitor has its gap filled with a plastic sheet that has thickness d and dielectric constant k. When a potential difference V is impressed across it, what is (a) the electric field in the dielectric, (b) the value of σ_f for its plates, and (c) the surface charge density σ_b for the dielectric?

15. One sometimes draws the electric field for a parallel plate capacitor in the way shown in figure P25.3; it appears that the electric field ends abruptly at the edge. By applying Faraday's law, $-\oint \mathbf{E} \cdot d\mathbf{s} = \dfrac{d}{dt}\int \mathbf{B} \cdot d\mathbf{A}$, to the dashed path indicated, show that this is not a realistic diagram. In practice, fringing occurs at the edges.

16. Before the dielectric is placed between its plates the capacitor shown in figure P25.4 is charged by a battery of voltage V. The battery is then removed and a dielectric is placed as shown so as to fill half the space between the plates. (a) Compare E_a and E_d, the final fields in the air gap and dielectric, respectively. (b) Compare σ_{fa} and σ_{fd}, the free charge densities above the air gap and dielectric, respectively.

17. A parallel plate capacitor with area $A = 50\text{ cm}^2$ and gap width $d = 2.0\text{ mm}$ is charged to a potential difference of 40 V when its gap is filled with a plastic sheet with dielectric constant $k = 3$. After the battery is removed, the plastic sheet is slid from between the plates. Neglecting friction forces on the sheet, how much work was required to remove the sheet? (*Hint:* How much did the energy of the capacitor change?)

18. Two identical electric dipoles lie in line on the same line with their centers a distance R apart with R being very large. Show that the force on one dipole due to the other is $3p^2/2\pi\epsilon_0 R^4$.

19. An electric dipole $\mathbf{p}$ is suspended freely in a uniform electric field $\mathbf{E}$ so that $\mathbf{p}$ can align with $\mathbf{E}$. Suppose the dipole is now disaligned by an angle θ_0 and released. Show that for θ_0 small (a) the restoring torque on the dipole is proportional to θ and that (b) the dipole undergoes simple harmonic rotational motion with period $2\pi\sqrt{I/pE}$, where I is the moment of inertia of the dipole about an axis through its center of mass. (c) Assuming $t = 0$ when the dipole is released, what is the equation for θ as a function of time?

20. A long solenoid is wound on a wooden core. If the current in the solenoid's 1000 loops/m is 2.0 A, how large is B within the solenoid? Repeat if the core is carbon. If it is iron for which $\mu = 200\,\mu_0$.

21. If the magnetic field within a hollow toroid is about 10^{-3} T, about what

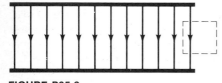

FIGURE P25.3

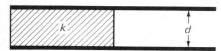

FIGURE P25.4

would B within the toroid be if it were filled with the iron whose characteristics are shown in figure 25.13?

22. As indicated in the text, the field of a bar magnet can be considered caused by a surface current flowing on the surface of the magnet. If a bar magnet is to act like a solenoid whose interior field is 0.3 T, how large a surface current must flow on each centimeter length of the bar?

23. A hollow solenoid has a self-inductance of 3.5 mH when air-filled. But when an iron rod is slipped into it so as to fill its interior, the self-inductance rises to 1.3 H. What is k_m for the iron being used? What is χ_m for it?

24. Referring to the iron of figure 25.13, sketch a graph showing k_m for the iron as a function of B_0. What is the maximum value of k_m? At what value of B_0 does the maximum occur?

25. Using Ampère's circuital law, show that, if no real current is flowing, the component of $\mathbf{H}$ parallel to a boundary between two magnetic materials is constant as one passes from the one material to the other. (*Hint:* Apply the integral to the path shown in figure P25.5.)

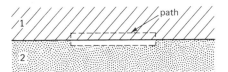

FIGURE P25.5

26. Using the fact that, in rotational motion, Δ work $= \tau(\Delta\theta)$, show that the potential energy of an electric dipole $\mathbf{p}$ in a uniform electric field $\mathbf{E}$ is given by $-\mathbf{p} \cdot \mathbf{E} +$ const. How is the dipole oriented when its potential energy is zero according to this relation if the constant is to be zero?

27. Using the fact that, in rotational motion, Δ work $= \tau(\Delta\theta)$, show that the potential energy of a magnetic dipole with moment $\boldsymbol{\mu}$ in a uniform magnetic field $\mathbf{B}$ is given by $-\boldsymbol{\mu} \cdot \mathbf{B} +$ const. How is the dipole oriented when its potential energy is zero according to this relation if the constant is to be zero? (Do not confuse the magnetic moment μ with the same symbol used for magnetic permeability.)

28. By use of the Maxwell-Boltzmann distribution law and the result of problem 27, show that the ratio of the number of magnetic dipoles completely in line with the field N_+ to the number in the opposite direction N_- is, for a paramagnetic material,

$$\frac{N_+}{N_-} = e^{2(\mu B / kT)}$$

A typical atomic magnetic moment is 10^{-23} A $\cdot$ m^2. Evaluate this ratio at $B = 1$ T and $T = 300$ K. (*b*) Evaluate the ratio N_+/N_- for an electric dipole with $p = 1.5$ D in an electric field 10,000 V/m at a temperature of 300 K. (Refer to problem 26.)

29. Using the fact, proven in problem 25, that the tangential component of $\mathbf{H}$ is constant across an interface between two materials, find the ratio of B inside a bar of iron composing the core of a long solenoid to B just outside it at a point halfway between the ends of the bar. Ignore the small field due directly to the wires on the solenoid and call the relative permeability of the iron k_m.

30. Compute the self-inductance per meter length of a long iron-core solenoid. Using the expression found in Chap. 24 for the energy stored in an inductance, show that the energy stored in unit volume of magnetic material where the field is B can be written as $B^2/2\mu$.

AC Circuits

Practical applications of inductances and capacitances occur widely. They form the basis for most electronic circuits. In this chapter we show how these circuit elements behave in typical ac situations. It will be found that their behavior can be described by writing down Kirchhoff's loop equation for each circuit. A relation similar in form to Ohm's law applies to each of these circuit elements. These results are used to illustrate various types of behavior of circuits containing resistance, inductance, and capacitance. In particular, the phenomenon of electric resonance is investigated.

26.1 RMS QUANTITIES

The voltage furnished by the power company is sinusoidal in form. A typical voltage vs. time curve is shown in figure 26.1. We can write the equation of this voltage*

$$v = v_0 \sin 2\pi ft$$

where f is the frequency (in cycles per second or hertz) of repetition of the voltage. We call v_0 the *amplitude* of the voltage, and its significance is shown in the figure. You will recall that $2\pi f$ is ω, the angular frequency in radians per second. We see at once from figure 26.1 that the voltage is positive as much as

*We shall use lowercase letters to represent varying voltages and currents.

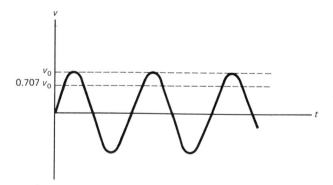

FIGURE 26.1 For an alternating current, the effective, or rms, current is $v_0/\sqrt{2}$.

it is negative, so that the average of v over one cycle is zero. Clearly, the average value of any sinusoidal quantity is zero, whether it is a voltage or current.

In spite of this fact, we know that alternating voltages and currents are useful. This is the result of the fact that most applications involve the square of the current and voltage. For example, the joule heating effect of a current is given by i^2R, and so one is interested in the average value of i^2, not i, and this average is not zero since i^2 is always positive. It turns out that most usages of alternating currents and voltages depend upon the square of v and i rather than the first power. For that reason, let us now compute the average value of v^2.

Sinusoidal i and v average to zero, while i^2 and v^2 do not

We have, from the definition of the average of a quantity (equation *4.9*),

$$\langle v^2 \rangle = \frac{\int_0^\tau v_0^2 \sin^2 (2\pi ft)\, dt}{\int_0^\tau dt}$$

where $\tau = 1/f$, and is the time for one cycle. Carrying out the integration, one finds for the average of v^2

$$\langle v^2 \rangle = \tfrac{1}{2}v_0^2$$

$$\sqrt{\langle v^2 \rangle} \equiv V \equiv v_{\mathrm{rms}} = \frac{v_0}{\sqrt{2}} \qquad\qquad 26.1$$

rms values are amplitude divided by $\sqrt{2}$

This type of average of a sinusoidal function is called the *root mean square* (rms), or *effective*, value. Most alternating-current (ac) ammeters and voltmeters are calibrated to read this type of average. When speaking of meter readings we assume this to be the quantity measured. We represent rms values by capital letters, V and I.

The advantage in using the rms voltage and current is the result of the following fact. Equations involving average power are similar for direct and alternating current if rms values are used for the alternating quantities. For example, the average power dissipated in a resistor is

$$P_{av} = \langle i^2R \rangle = \langle i^2 \rangle R \equiv i^2_{\mathrm{rms}}R \equiv I^2R$$

Average power is I^2R if rms values are used

and so the direct-current (dc) formula $P = I^2R$ applies to alternating currents as well, provided rms values are used.

Illustration 26.1

How large a dc current is needed to produce the same heating effect in a resistor as a sinusoidal current of amplitude 2.0 A?

Reasoning An ordinary ac meter will read the rms current, namely,

$$I = \frac{2.0}{\sqrt{2}} = 1.41\ \text{A}$$

Since

$$\text{Power} = I^2R$$

for both ac and dc conditions if rms current is used, the equivalent dc current must be the same as the rms current, or 1.41 A.

26.2 SINGLE-ELEMENT CIRCUITS

To begin our study of alternating-current (ac) circuits, let us examine the behavior of the individual circuit elements (resistor, inductor, and capacitor) when an ac current flows through them. The simplest element is the resistor, and so we consider it first.

Suppose one arranges to have a current given by

$$i = i_0 \sin 2\pi ft$$

flow through the resistor R shown in figure 26.2(a). The potential drop from A to B across this resistor will be

$$v_R = iR$$
$$= i_0 R \sin 2\pi ft \qquad\qquad 26.2$$

Therefore the potential difference across the resistor varies sinusoidally and is *in phase* with the current. That is to say, both the current and v_R reach their maximum values at the same time, as shown in part (b) of figure 26.2. To find the rms value of v_R we note that

$$(v_R)_{\text{rms}} = R(i_0 \sin 2\pi ft)_{\text{rms}}$$

or
$$V_R = IR \qquad\qquad 26.3$$

Ohm's law applies between rms V and I

where V_R and I are written to represent the rms values of v_R and i. Notice that *26.3* is simply Ohm's law as found for dc circuits, and so we conclude Ohm's law relates the rms current and potential for a resistor.

Next let us consider the situation shown in figure 26.3. A current

$$i = i_0 \sin 2\pi ft$$

flows through an inductor L. To find the potential drop from point A to point B, v_L, we recall that*

$$v_L = L\frac{di}{dt}$$

After making use of the equation for i, this becomes

$$v_L = (2\pi fL)i_0 \cos 2\pi ft \qquad\qquad 26.4$$

Notice that v_L is *not* in phase with the current, as we see from figure 26.3(b). It reaches its maximum value when the current through it is zero. Although this might seem strange at first, it is easily justified in the following way. Since $v_L \sim di/dt$, the value of v_L will be largest when the current is changing most rapidly. (Recall that inductors oppose the change in current.) Since di/dt is simply the slope of the sinusoidal function representing i, the induced voltage v_L will be maximum when the sine curve has its largest slope. Inspection of the curve for i shows at once that this occurs when the curve goes through zero. Therefore v_L has its maximum value when i is zero. We see from this that v_L and i are $\frac{1}{4}$ cycle (or 90°) out of phase.

To find the rms value of v_L we note that the rms value of $i_0 \cos 2\pi ft$ is equal to the rms value of $i_0 \sin 2\pi ft$, which is simply I. (Why?) Therefore, from *26.4*, we find

$$V_L = (2\pi fL)I$$

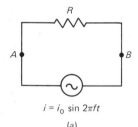

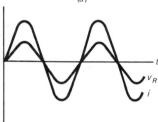

$$i = i_0 \sin 2\pi ft$$

(a)

$$v_R = i_0 R \sin 2\pi ft$$

(b)

FIGURE 26.2 The current in a resistor is in phase with the voltage across the terminals. The symbol $\ominus$ represents an alternating voltage source.

---

* You should be able to show that the usual negative sign in this relation is not needed if v_L is defined to be the potential *drop* from A to B.

This may be placed in an Ohm's law form by defining the *inductive reactance* of the element

$$X_L = 2\pi fL$$

in which case we have

$$V_L = IX_L \qquad 26.5$$

The inductive reactance X_L replaces R in this analog to Ohm's law. Its units are volts per ampere, which are equivalent to ohms. As one can infer from 26.5, X_L is a measure of the inductance's ability to impede the flow of alternating current. Can you justify physically why this impeding effect should be proportional to f and to L?

A third type of circuit element often found in ac circuits is capacitance. Suppose a current

$$i = i_0 \sin 2\pi ft$$

flows into a capacitor, as shown in figure 26.4. Since the two plates of the capacitor must have equal and opposite charges,* the current flowing into one plate must equal the current flowing out of the other. As a result, the current appears to flow through the capacitor.

We know that, for a capacitor,

$$q = Cv_C$$

and

$$\frac{dq}{dt} = i$$

$$= i_0 \sin 2\pi ft$$

where q is the charge on the capacitor, and v_C is the potential drop across it. Integration of the latter equation yields

$$q = \frac{-1}{2\pi f} i_0 \cos 2\pi ft$$

where the constant of integration, representing a constant charge on the capacitor, has been set equal to zero. Since $q = Cv_C$, we find

$$v_C = -\frac{1}{2\pi fC} i_0 \cos 2\pi ft \qquad 26.6$$

This is graphed in figure 26.4(*b*)

*An exception occurs at frequencies higher than one can achieve in any common circuit. When the wavelength of the electromagnetic wave accompanying the oscillation becomes comparable with the plate separation of the capacitor, the electric field within the capacitor can no longer be treated as an approximately electrostatic field.

For rms quantities, if $X_L = 2\pi fL$, then $V_L = IX_L$

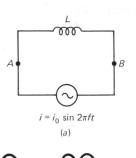

$i = i_0 \sin 2\pi ft$

(a)

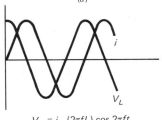

$V_L = i_0 (2\pi fL) \cos 2\pi ft$

(b)

FIGURE 26.3 The voltage across the inductance leads the current through it by 90°, or one-quarter cycle. Notice the symbol used for inductance.

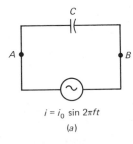

$i = i_0 \sin 2\pi ft$

(a)

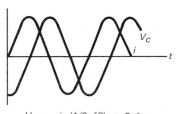

$V_C = - i_0 (1/2\pi fC) \cos 2\pi ft$

(b)

FIGURE 26.4 The voltage across a capacitor reaches its maximum one-quarter cycle later than the current does. Both the voltage and the charge on the capacitor are in phase.

As in the case of the inductor, the potential drop here is also not in phase with the current. In addition, because of the minus sign in *26.6*, v_C and v_L are also not in phase. The variations of all four quantities, v_L, v_C, v_R, and i, are shown in figure 26.5. It is seen that v_C is $\frac{1}{4}$ cycle (or 90°) out of phase from i and that v_L and v_C are $\frac{1}{2}$ cycle (or 180°) out of phase with each other.

To find the rms value of v_C we note that the rms value of $-i_0 \cos 2\pi ft$ is the same as I, the rms value of i. (Why?) Therefore we find

$$V_C = \frac{1}{2\pi fC} I$$

which can be placed in an Ohm's law form by writing $X_C \equiv 1/2\pi fC$.

An Ohm's law form also applies to capacitors

$$V_C = IX_C \qquad 26.7$$

The quantity X_C is called the *capacitive reactance* and is a measure of the impeding effect of the capacitor on the current. Why, physically, should it depend upon $1/fC$? The units of X_C are the same as those for R and X_L, namely ohms.

Illustration 26.2

Find the reactance of a 10-mH inductor at a frequency of $200/\pi$ Hz. Repeat for $2 \times 10^6/\pi$ Hz.

Reasoning The reactance of an inductor is $2\pi fL$. We have

$$\text{At } f = \frac{200}{\pi}\text{s}^{-1}: \qquad X_L = 2\pi\frac{200}{\pi}(10 \times 10^{-3}) = 4\ \Omega$$

$$\text{At } f = \frac{2 \times 10^6}{\pi}\text{s}^{-1}: \qquad X_L = 40{,}000\ \Omega$$

X_L increases with frequency

Notice that the reactance (or current-stopping effect) of an inductor increases linearly with frequency. Inductors have high reactance at high frequencies. This is the result of the fact that the current must change very rapidly at high frequencies, and the induced emf in an inductor increases with increasing rate of flux change through it.

Illustration 26.3

Find the reactance of a 0.10-μF capacitor at a frequency of $200/\pi$ Hz and at $2 \times 10^6/\pi$ Hz.

FIGURE 26.5 The voltages across R, L, and C are not all in phase with the current. Although V_R is in phase with i, both V_L and V_C are one-quarter cycle (or 90°) out of phase with i.

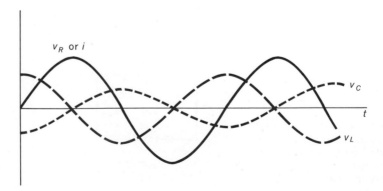

Reasoning The reactance of a capacitor is $1/2\pi fC$. In the present case this becomes

At $f = \dfrac{200}{\pi}$ s^{-1}: $\qquad X_C = \dfrac{\pi}{2\pi(200)(0.1 \times 10^{-6})} = 25{,}000\,\Omega$

At $f = \dfrac{2 \times 10^6}{\pi}$ s^{-1}: $\qquad X_C = 2.5\,\Omega$

X_C decreases with frequency

We see that a capacitor has a very high reactance at low frequencies and passes the current more readily at high frequencies. This is reasonable since we know that in the limit of $f = 0$ the capacitor will allow no current to flow when a steady state has been reached. Of course, the behavior at high frequencies is due to the fact that the capacitor is not able to build up a near-equilibrium charge during the time available to it.

26.3 THE SERIES *LRC* CIRCUIT

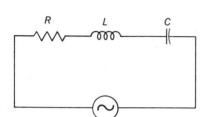

FIGURE 26.6 A simple series circuit. Why do not the potential differences read by a meter across R, L, and C add to give the reading across the oscillator?

We are now in a position to discuss the behavior of a series circuit which contains inductance, capacitance, and resistance, as well as an oscillating voltage source. Such a circuit is shown in figure 26.6. Let us suppose the current in this circuit to be

$$i = i_0 \sin 2\pi ft$$

since this is the current considered in the preceding section. The voltage drops across R, L, and C were found in the preceding section for such a current, and so the results found there can be applied directly to the present case. We shall first find the voltage of the source which gives rise to the assumed current.

If we write the loop equation for the circuit of figure 26.6, we find at once that

$$v = v_R + v_L + v_C$$

where v is the potential difference across the terminals of the voltage source. Replacing v_R, v_L, and v_C by their values given in *26.2*, *26.4*, and *26.6*,

$$v = i_0[R \sin (2\pi ft) + (X_L - X_C) \cos (2\pi ft)]$$

We see at once from this that, since v_R, v_L, and v_C are not in phase, these potential differences do not always add as positive numbers. As a matter of fact, inspection of figure 26.5 shows at once that v_L and v_C always add so as to cancel each other. This is also shown by the factor $(X_L - X_C)$ in the above equation for v. For this reason, we shall not be surprised to learn later that a voltmeter placed across the voltage source does not read the sum of its readings when placed across the individual elements R, L, and C.

The equation for v can be put in a much nicer form by use of the trigonometric relation

$$A \sin \theta + B \cos \theta = \sqrt{A^2 + B^2} \sin (\theta + \phi)$$

with $\qquad \tan \phi = \dfrac{B}{A}$

By use of this we find

$$v = i_0[R^2 + (X_L - X_C)^2]^{1/2} \sin (2\pi ft + \phi) \qquad\qquad 26.8$$

with $\qquad \tan \phi = \dfrac{X_L - X_C}{R}$

We see from this that a sinusoidal voltage source gives rise to a sinusoidal current in an LRC series circuit. However, the current and voltage are out of phase by an angle ϕ, the phase angle.

It is of interest to find the expression for the rms source voltage from 26.8. Since the rms value of $i_0 \sin(2\pi ft + \phi)$ is simply I, this equation yields

$$V = IZ \tag{26.9}$$

An Ohm's law form applies to a series RLC circuit, with R replaced by Z

where

$$Z \equiv [R^2 + (X_L - X_C)^2]^{1/2}$$

The quantity Z is called the *impedance* of the circuit, and its units are ohms. We see that the series RLC circuit, like the individual circuits, can be described in terms of an analog to Ohm's law, *26.9*.

Illustration 26.4

Find the voltages across the various elements shown in the circuit in figure 26.6 if $v = (100\sqrt{2})\sin 1000t$ V. Assume $R = 1000\,\Omega$, $L = 2.0$ H, and $C = 1\,\mu$F.

Reasoning The rms voltage across the oscillator is 100 V. In addition since $2\pi f = 1000\ \text{s}^{-1}$, we have

$$X_L = 2000\ \Omega \qquad \text{and} \qquad X_C = 1000\ \Omega$$

from which

$$Z = 1410\ \Omega$$

Since

$$I = \frac{V}{Z}$$

we have

$$I = \frac{100}{1410} = 0.071\ \text{A}$$

We can now use this current to find the voltage drops across the separate elements since I is the same everywhere in a series circuit.

Using *26.3*, *26.5*, and *26.7* gives

$$V_R = (0.071)(1000) = 71\ \text{V}$$
$$V_L = (0.071)(2000) = 142\ \text{V}$$
$$V_C = (0.071)(1000) = 71\ \text{V}$$

In addition, we had that the source voltage was 100 V. Clearly, rms voltages do not add directly since

$$V_R + V_L + V_C = 284\ \text{V}$$

Meter voltages do not add directly in an ac circuit

and not 100 V. The reason for this, of course, is that these voltages are not in phase, as was pointed out in connection with figure 26.5.

26.4 VECTOR REPRESENTATIONS

The results of the preceding section can be summarized usefully by means of a vector diagram. To see how such a diagram is possible, let us consider the impedance of the series RLC circuit given by *26.9* to be

$$Z^2 = R^2 + (X_L - X_C)^2 \tag{26.10}$$

This equation can be thought of as the theorem of Pythagoras written for a right triangle, with Z as its hypotenuse and R and $(X_L - X_C)$ as its legs. This triangle is shown in figure 26.7.

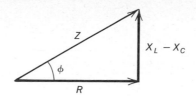

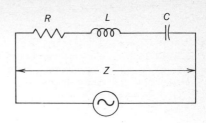

FIGURE 26.7 The impedence Z of a series *RLC* circuit is given by the vector triangle shown.

*A vector **Z** may be found by considering its x and y components to be R and $X_L - X_C$*

The triangle of figure 26.7 also shows the phase angle ϕ between the driving voltage and the current for the circuit. We found in *26.8* that

$$\tan \phi = \frac{X_L - X_C}{R}$$

One notices that the angle labeled ϕ in figure 26.7 has this tangent, and so the phase angle is as shown.

Another interesting relation is obtained if one multiplies through *26.10* by I^2. One then has

$$I^2 Z^2 = I^2 R^2 + I^2 (X_L - X_C)^2$$

which is simply

$$V^2 = V_R^2 + (V_L - V_C)^2 \qquad 26.11$$

Voltages can also be represented by vectors

As with the impedances given in *26.10*, the voltages given by *26.11* can also be represented by a vector triangle. This is shown in figure 26.8. One can use these vector diagrams along with the fact that

$$i = i_0 \sin 2\pi ft \quad \text{and} \quad v = v_0 \sin (2\pi ft + \phi)$$

to summarize many of the facts we have found concerning *RLC* circuits. A few examples of this will now be given.

Illustration 26.5

A circuit operating at $360/2\pi$ Hz contains a 1-μF capacitor and a 20-Ω resistor. How large an inductor must be added in series to make the phase angle for the circuit zero? What will then be the current in the circuit if the applied voltage is 120 V?

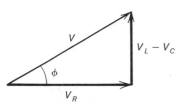

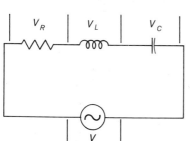

FIGURE 26.8 The voltage V across a series combination of *R*, *L*, and *C* is represented by this diagram. You should be able to show from it how ϕ and the relative voltages change with frequency. What is the significance of the frequency for which $V_L = V_C$?

Reasoning From figure 26.7, the condition $\phi = 0$ means $X_L = X_C$. The circuit is then said to be at resonance for reasons we shall discuss presently. Setting $X_L = X_C$ gives

$$2\pi \frac{360}{2\pi} L = \frac{1}{360\,C}$$

from which

$$L = \frac{1}{(360)^2 \times (1 \times 10^{-6})} = 7.7 \text{ H}$$

Notice that a very large inductance is needed for resonance at low frequencies.

When $X_L = X_C$ we see from figure 26.7 that $R = Z$. Therefore

$$I = \frac{V}{Z} = \frac{V}{R} = \frac{120 \text{ V}}{20\,\Omega} = 6 \text{ A}$$

FIGURE 26.9 The vector diagram for an inductance coil which has resistance. What should the vector diagram for a good transformer look like?

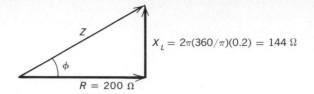

$$X_L = 2\pi(360/\pi)(0.2) = 144\ \Omega$$

Illustration 26.6

Find the phase angle at $360/\pi$ Hz for an inductance of 0.20 H which has a resistance of 200 Ω.

Reasoning The vector diagram for this inductor is shown in figure 26.9. We see at once that

$$\tan \phi = \frac{144}{200} = 0.72$$

from which

$$\phi = 36°$$

Illustration 26.7

In a certain RLC series circuit the voltage across the source is 80 V, while across the resistor it is 20 V. What is the phase angle between the current and the source voltage for this circuit?

Reasoning Referring to the vector diagram of figure 26.10, we see

$$\cos \phi = \frac{20}{80} = \frac{1}{4}$$

From this we have $\phi = 76°$. Since we were given no data concerning which is larger, X_L or X_C, it is impossible to say whether ϕ is plus or minus.

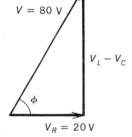

$V = 80$ V

$V_L - V_C$

ϕ

$V_R = 20$ V

FIGURE 26.10 If the resistance is 5 ohms, what is the impedance of the circuit?

26.5 RESONANCE

We have seen in *26.9* that $V = IZ$ for a simple RLC series circuit such as the one shown in figure 26.6. If we solve this Ohm's law equation for I and replace Z, the circuit impedance, by its value from *26.9*, we find

$$I = \frac{V}{\sqrt{R^2 + (X_L - X_C)^2}} \qquad 26.12$$

Since $X_L = 2\pi fL$ and $X_C = 1/2\pi fC$, both X_L and X_C vary with the frequency of the driving voltage. As a result, the current I in the circuit will be a function of the oscillator frequency.

Resonance in a series circuit occurs when $X_L = X_C$

A situation of particular interest occurs if the frequency is chosen so as to make $X_L = X_C$. In that case the denominator of *26.12* takes on its minimum value, and the current in the circuit becomes maximum. At that frequency, the so-called *resonance frequency* of the circuit, the current would become infinite if the circuit resistance R could be made zero. This is not possible, of course. The variation of the current I as a function of frequency predicted by *26.12* is shown in figure 26.11 for various values of R, the circuit resistance. As expected, the resonance phenomenon becomes more apparent when R is small.

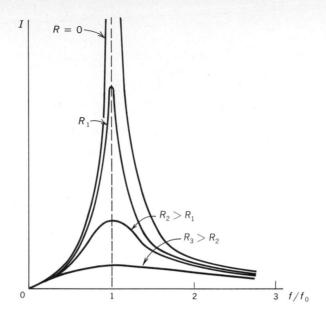

FIGURE 26.11 Behavior of the current in an RLC circuit as the frequency is varied. At the resonance frequency f_0, $X_L = X_C$. What value does this give for the resonance frequency?

The resonance frequency f_0 is found from the condition that, at resonance,

$$X_L = X_C$$

or

$$2\pi f_0 L = \frac{1}{2\pi f_0 C}$$

from which the resonance frequency is given to be

$$f_0 = \frac{1}{2\pi}\sqrt{\frac{1}{LC}} \qquad\qquad 26.13$$

A very close analogy can be drawn between the RLC series circuit and the mechanical system consisting of a mass at the end of a spring. When a driving force is applied to the mass, the response of the system varies as a function of the frequency of the driving force. As we saw in Chap. 12, such a system resonates when the driving force has a frequency

$$f_0 = \frac{1}{2\pi}\sqrt{\frac{k}{m}}$$

where k is the spring constant, and m is the mass. Comparing these two expressions for the resonance frequency, it is apparent that L and $1/C$ in the electric circuit take the place of m and k in the mechanical system. (Why should an inductance behave like an element having inertia, and why should a capacitance be somewhat like an "electric spring" with spring constant $1/C$?)

26.6 POWER CONSUMPTION

It is of interest to compute the average power furnished to a series RLC circuit by the oscillator. By definition, the power at any instant is

$$P = vi$$

or

$$P = v_0 i_0 \sin\left(2\pi ft + \phi\right) \sin 2\pi ft$$

To find the energy furnished by the source in one cycle, we must integrate P through the time interval $0 \le t \le 1/f$, since $1/f$ is equal to the period. The average power will therefore be this energy divided by the time for one cycle, $1/f$. We have

$$P_{av} = v_0 i_0 f \int_0^{1/f} \sin(2\pi ft + \phi) \sin 2\pi ft \, dt$$

This integral is most easily evaluated if we expand the first factor, using

$$\sin(x + y) = \sin x \cos y + \cos x \sin y$$

to give

$$P_{av} = v_0 i_0 f \int_0^{1/f} [\cos \phi \sin^2 2\pi ft + \sin \phi \sin 2\pi ft \cos 2\pi ft] \, dt$$

But

$$\sin 2\pi ft \cos 2\pi ft = \tfrac{1}{2} \sin 4\pi ft$$

and so the integral of the last term over one cycle will be zero. We therefore find

$$P_{av} = v_0 i_0 f \cos \phi \int_0^{1/f} \sin^2 2\pi ft \, dt$$

which gives

$$P_{av} = \tfrac{1}{2} v_0 i_0 \cos \phi \qquad\qquad 26.14a$$

This may be written in terms of the rms values, and is then

$$P = VI \cos \phi \qquad\qquad 26.14b$$

Average power in an ac element

We see that the power expended in an RLC circuit is simply the usual expression, VI, multiplied by the factor $\cos \phi$, where ϕ is the phase angle between V and I. We call the factor $\cos \phi$ the *power factor*. Since for a circuit in which R is zero, the value of $\phi = 90°$, the power factor is zero unless resistance is present. We can easily show that the power loss given by *26.14b* is the power loss in the resistor. From figure 26.10 one sees that $V \cos \phi$ is simply V_R, which in turn is IR. Therefore,

$$P = VI \cos \phi = I^2 R$$

Clearly, all the power loss occurs in the resistor, none occurring in a pure capacitor or inductor.

26.7 POWER TRANSMISSION AND THE TRANSFORMER

A transformer is a mutual inductance designed to change a given ac voltage into a larger or smaller ac voltage. It consists of two coils of wire wound on an iron yoke, somewhat like the system shown in figure 26.12. The primary coil has impressed across it an ac voltage V_P. This causes a current to flow through the primary coil, and this current gives rise to flux in the iron, as shown.

Since the flux follows the iron yoke, it passes through the second coil as well. Of course, the flux is changing all the time, because V_P varies sinusoidally. This varying flux through the secondary induces an emf of magnitude V_S in the secondary. Clearly, V_S also varies sinuosidally.

Most transformers have very little resistance in their wires. Hence, the current through the primary is impeded only by the inductance of the primary itself. The behavior of a circuit consisting of an inductor connected

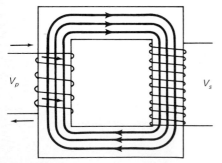

FIGURE 26.12 An iron-core step-up transformer.

V_p

V_s

directly to an alternating-voltage source was discussed in detail in Sec. 26.2. For that circuit, the driving voltage equals the induced emf in the inductor. In the present case we have from Faraday's law

$$V_P = N_P \frac{\Delta\phi}{\Delta t}$$

where N_P is the number of turns on the primary coil and $\Delta\phi/\Delta t$ is the rate of change of flux through it. However, since the flux follows the iron yoke, all the flux will go through the secondary coil as well. The induced emf in the secondary is clearly

$$V_S = N_S \frac{\Delta\phi}{\Delta t}$$

Division of this expression by the equation for the voltage across the primary coil yields

$$\frac{V_S}{V_P} = \frac{N_S}{N_P} \qquad\qquad 26.15$$

This is the transformer equation, and it tells us how the secondary voltage is related to the primary voltage. The two voltages are in the same ratio as the number of turns on the coils. When N_S is larger than N_P, V_S will be larger than V_P. This is called a *step-up* transformer since the voltage is increased by it. The reverse case is called a *step-down* transformer.

If the secondary circuit is not closed, current cannot flow in it. Hence, there is no power loss in the secondary coil when it is not in use. Moreover, we showed in Sec. 26.6 that there was also no power loss in an inductor which has no resistance. This fact makes it possible for the power company to keep their transformers running throughout a city even when no one is using the electricity they are providing. The transformers themselves consume very little energy.

However, if current is drawn from the secondary, to run a heater, for example, energy is being consumed by the heater. This energy must be fed into the primary of the transformer so that it can be delivered to the secondary. Under these conditions, the loss in power at the secondary causes the primary to act as though it had resistance.

There are many uses for transformers. Every plug-in type radio and every TV set contains one or more of them. It is necessary to transform the 120-V house-line voltage to lower voltages to operate the tubes and transistors. In addition, the TV picture tube requires a voltage about 100 times as large as the line voltage. A transformer is needed to provide this.

Another use of transformers has to do with power transmission. Many power companies provide power to cities which are perhaps 100 km from the generators. This proves to be quite a problem. Suppose that in a city of 100,000 people each person is using 120 W of power. This would be the equivalent of one or two lighted light bulbs for each person. The power consumed is (120)(100,000) W, and at a voltage of 120 V we have

$$\text{Total power} = VI$$
$$(120)(100,000) = 120I$$
$$I = 100,000 \text{ A}$$

where we have assumed the power factor to be unity.

Since an ordinary house wire can safely carry only about 30 A without overheating, the power company would need the equivalent of about 3000 of these wires to carry power to the city. Although this is not impossible, the cost of the copper alone would be tremendous. The power companies get around this difficulty quite nicely by noticing that the important quantity is VI and not I alone. They therefore choose to transmit power over long distances at very high voltages. In the above example, if V had been 100,000 V, we would have

$$(120)(100,000) = 100,000I$$

or
$$I = 120 \text{ A}$$

It is for this reason that the power companies use high-tension or high-voltage difference lines to transmit power over large distances.

Of course, they would not dare to have such high voltages wired directly to a house. The danger from electrocution and fire would be tremendous. Instead, they use step-down transformers to convert these high voltages to the normal voltage used in houses in the United States, about 120 V.

26.8 DECAY OF OSCILLATIONS

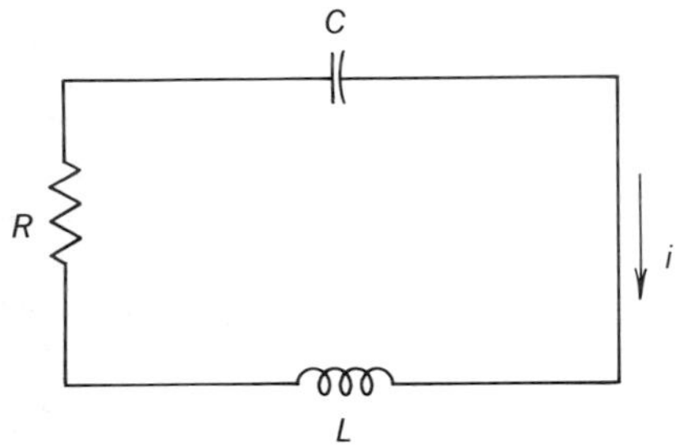

Let us now consider what happens in a series RLC circuit that has no voltage source in it, and let us say that R is negligibly small. As a typical example, suppose a charge Q is placed on the capacitor in the circuit of figure 26.13 while a switch in the circuit is open so that no current can flow. If now the switch is closed, the charge will begin to flow off the capacitor and a current will start to flow. Let us say its direction is clockwise. The charge will not flow off the capacitor instantaneously because the rising current induces an opposing emf in the inductor. However, in spite of this, the current does rise. As a result, an energy $\frac{1}{2}Li^2$ becomes stored in the inductor.

Eventually, the capacitor loses all its charge, and so its original stored energy $Q^2/2C$ has been lost. At that instant, an equivalent energy $Li^2/2$ must be stored in the inductor. Clearly, even though the capacitor is now uncharged, the current in the circuit is quite large. Since the inductor will not let the current drop to zero instantaneously, current will continue to flow clockwise and will charge the capacitor opposite to its original charge. The current in the circuit will not stop until the capacitor has acquired a charge Q, equal and opposite to its original charge. At that time, all the energy of the circuit will again be stored in the capacitor and none will be stored in the inductor. The entire process will now repeat itself with a counterclockwise current flowing in the circuit. Evidently, the current in the circuit will oscillate indefinitely, clockwise and counterclockwise.

From the law of energy conservation, the original energy of the circuit $Q^2/2C$ oscillates back and forth between capacitor and inductor, i.e., from $Q^2/2C$ to $LI^2/2$ back to $Q^2/2C$. At intermediate times the energy stored in the capacitor will be $q^2/2C$ and it will be $Li^2/2$ in the inductor. The sum of these must be $Q^2/2C$, the original energy of the circuit. As a result we can write

$$\tfrac{1}{2}Li^2 + \tfrac{1}{2}\frac{q^2}{C} = \text{const} \qquad\qquad 26.16$$

at all times.

If resistance is present in the circuit, energy will be lost from the circuit as a result of joule heating. This means that the current in the inductance and charge on the capacitor must eventually decay to zero. In such a case,

FIGURE 26.13 What determines whether or not the current in this circuit will be oscillatory?

26.16 would not be true. Let us now solve the general case, and then apply the solution to these special cases.

The Kirchhoff loop equation for the circuit of figure 26.13 is

$$-Ri + \frac{q}{C} - L\frac{di}{dt} = 0$$

Upon taking the derivative with respect to time and using the fact that $i = -dq/dt$, we have

$$L\frac{d^2i}{dt^2} + R\frac{di}{dt} + \frac{i}{C} = 0$$

We do not solve this differential equation directly, but instead simply state its solution and show that the solution is correct. Since we expect the current eventually to decay to zero, let us try a solution of form

$$i = i_0 e^{\lambda t} \qquad\qquad\qquad 26.17$$

where i_0 and λ are constants whose values we do not yet know. This form has the additional advantage that $e^{\sqrt{-1}\gamma t} = \cos \gamma t + \sqrt{-1} \sin \gamma t$, and so an exponential solution is capable of representing both an oscillating and decaying current. Substituting *26.17* in the differential equation gives (after canceling i_0 and $e^{\lambda t}$)

$$L\lambda^2 + R\lambda + \frac{1}{C} = 0$$

This equation determines what value we must assign to λ if the expression given for i in *26.17* is to be valid.

Solving for λ by use of the quadratic formula, we obtain

$$\lambda = -\frac{R}{2L} \pm \frac{R}{2L}\sqrt{1 - \frac{4L}{R^2C}} \qquad\qquad 26.18$$

Or after defining

$$\alpha \equiv \frac{R}{2L} \qquad \text{and} \qquad \beta \equiv \sqrt{\left(\frac{R}{2L}\right)^2 - \frac{1}{LC}}$$

this becomes

$$\lambda = -\alpha \pm \beta$$

We therefore have two possible solutions,

$$i = i_0 e^{-\alpha t} e^{\beta t}$$

and

$$i = i_0 e^{-\alpha t} e^{-\beta t}$$

Depending upon what was happening in the circuit at time $t = 0$, either or both of these may be correct solutions. To cover all the possibilities, we take as our solution*

$$i = i_{01} e^{-\alpha t} e^{\beta t} + i_{02} e^{-\alpha t} e^{-\beta t} \qquad\qquad 26.19$$

Let us consider the case where $i = i_0$ at time $t = 0$. This condition can be satisfied if we set $i_{01} = i_{02} = i_0/2$. Then

$$i = \tfrac{1}{2} i_0 e^{-\alpha t}(e^{\beta t} + e^{-\beta t})$$

The behavior of the current predicted by this equation will depend upon the values of α and β. We now investigate the most important possibilities.

---

*Since each of these terms is a solution of the original equation, their sum is also a solution, as one may verify by substitution.

Case 1   R = 0

In this case $\alpha = 0$ and

$$\beta = \sqrt{-\frac{1}{LC}}$$

But this means β is an imaginary number, and since $\sqrt{1/LC}$ is the resonance frequency ω_0 of the circuit, we have

$$\beta = j\omega_0$$

where $j = \sqrt{-1}$. Our equation becomes

$$i = \tfrac{1}{2}i_0(e^{j\omega_0 t} + e^{-j\omega_0 t})$$

However, as one may easily show by expanding the exponentials and comparing with the expansion for $\cos\theta$, this is equivalent to

$$i = i_0 \cos\omega_0 t \qquad\qquad 26.20$$

In other words, the current in the resistanceless circuit follows a simple cosine curve. This is shown in part (a) of figure 26.14.

The charge on the capacitor can be found from the fact that

$$i = \frac{dq}{dt}$$

That is, q is the inverse derivative of $i_0 \cos\omega_0 t$, or

$$q = \frac{i_0}{\omega_0}\sin\omega_0 t \qquad\qquad 26.21$$

We see that q and i are 90° out of phase, one being a maximum when the other is a minimum. This is a necessary condition if *26.16* for the total energy of the circuit is to be true.

Our result tells us that when the capacitor has its maximum charge, it has all the energy of the circuit stored in it, since i, and the energy stored in the inductor, are zero. A quarter of a period later the situation is just *In a resonant circuit the energy* reversed. The capacitor is uncharged, and the current is maximum. All the *oscillates between capacitor and* energy is stored in the inductor. This process repeats over and over; the *inductor* energy of the circuit is stored alternately in the inductor and capacitor.

Case 2 $0 < \dfrac{R}{2L} \ll \sqrt{\dfrac{1}{LC}}$

In this case β is still very close to the value given in the previous case, $j\omega_0$. However, α is no longer zero. We have, following the same procedure as used previously to obtain *26.20*,

$$i = i_0 e^{-\alpha t} \cos\omega_0 t \qquad\qquad 26.22$$

The oscillation damps down as $e^{-\alpha t}$, with The current in the circuit is oscillatory in character because of the *$\alpha = R/2L$* factor $\cos\omega_0 t$. However, it is multiplied by a *damping factor,* $e^{-\alpha t}$, which causes the amplitude of the current to decrease exponentially. This is not unexpected since this circuit has $R \neq 0$. Energy is lost as heat in the resistor, and so the circuit energy, and current, must also decrease. The behavior predicted by *26.22* is shown in figure 26.14(b).

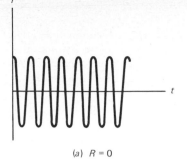

(a) $R = 0$

(b) underdamped
$$0 < R/2L \ll \sqrt{1/LC}$$

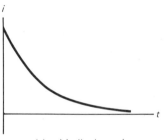

(c) critically damped
$$R/2L = \sqrt{1/LC}$$

FIGURE 26.14 Behavior of an *RLC* circuit without a power source. The scales in (b) and (c) are not the same.

Case 3 $\dfrac{R}{2L} = \dfrac{1}{LC}$

In this case, called *critical damping*, $\beta = 0$ and no oscillation occurs. From 26.19 we have

$$i = i_0 e^{-\alpha t} \qquad \text{with } \alpha = \frac{R}{2L} \qquad\qquad 26.23$$

The current simply falls exponentially to zero, as shown in figure 26.14(*c*). Under this condition, the current no longer oscillates. If R is made still larger, we say the circuit is overdamped. In such cases, the current slowly decreases to zero without oscillating.

Illustration 26.8

Two series circuits both have $L = 1\,\mathrm{H}$, $C = 1\,\mu\mathrm{F}$, but have different resistances. One (circuit A) has a resistance of $20\,\Omega$, while the other (circuit B) has $R = 20{,}000\,\Omega$. Compare their decay characteristics.

Reasoning The decay constants are as follows:

$$\alpha_A = 10\,\mathrm{s}^{-1} \qquad \alpha_B = 10{,}000\,\mathrm{s}^{-1}$$

In addition, for both circuits the natural resonance frequency (i.e., the frequency with which it oscillates freely) is

$$\omega_0 = 2\pi f = \frac{1}{\sqrt{LC}} = 10^3\,\mathrm{s}^{-1}$$

If the circuit oscillates, it will go through about 160 oscillations per second; so its frequency is $f \sim 160\,\mathrm{Hz}$.

Circuit A has a decay factor e^{-10t}, and so its current will drop to $1/e$ of its original value in a time of $0.10\,\mathrm{s}$. During this time the circuit will oscillate about 16 times. It will not decay quite as rapidly as the circuit illustrated in figure 26.14(*b*).

Circuit B has a decay factor of $e^{-10,000t}$, and so it will decay to $1/e$ of its original value in $10^{-4}\,\mathrm{s}$. Since its oscillation period is of the order of $6 \times 10^{-3}\,\mathrm{s}$, the circuit will not oscillate at all. It is overdamped.

Chapter 26
Questions
and Guesstimates

1. In some cities in the world low-frequency ac voltage (considerably less than 60 Hz) is used. The electric lights in these cities can be seen to flicker rapidly. Explain the cause of this flickering.
2. For which of the following uses would dc and ac voltage be equally acceptable: incandescent light bulbs, electric stove, electrolysis, TV set, fluorescent light, neon-sign transformer, battery charging, toaster, electric clock?
3. Give the physical reasons behind the facts that the impedance of a capacitor decreases with frequency while the impedance of an inductor increases with frequency.
4. What is the power factor for a circuit that contains only (*a*) resistance, (*b*) inductance, (*c*) capacitance, (*d*) inductance and capacitance?
5. Two code words sometimes used by people who work with ac circuits are ICE and ELI. The former is interpreted to mean: In a capacitive circuit

the current leads the voltage. What does this mean? What does the other code word mean?

6. A variable-frequency constant-voltage power source is connected across a resistance. (*a*) How does the current in the circuit vary as a function of the frequency? (*b*) Repeat for an inductor. (*c*) For a capacitor. (*d*) An inductor and capacitor in series. (*e*) An inductor and resistor. (*f*) A capacitor and resistor.

7. A typical electrocardiogram (ECG) graph is shown in figure P26.1. This is a graph of the voltage difference between the left leg and left arm. From the graph, estimate the average voltage, rms voltage, and the relation between peak voltage and rms voltage for this waveform. Why doesn't a simple galvanometer deflect in this way when attached to these two points on your body?

8. Consider the energy relations for the vibration of a mass *m* at the end of a spring (constant *k*)

$$\text{Total energy} = \tfrac{1}{2}mv^2 + \tfrac{1}{2}kx^2$$

and the vibration of an *LC* resonant circuit,

$$\text{Total energy} = \tfrac{1}{2}Li^2 + \tfrac{1}{2}\frac{q^2}{C}$$

Using the similarity between these relations, one can say that the inductance *L* acts like an inertial mass *m* and the capacitance acts like a spring of constant $1/C$. Basing your thoughts on this analogy, point out as many similarities as you can between the two systems.

9. A dc voltmeter is connected across the terminals of a variable-frequency oscillator. How would the meter behave as the frequency is slowly increased from 0.01 to 100 Hz?

10. Why would it be unwise to use 1000-V ac lines in a home, even though this would be more economical from a wiring standpoint?

11. If by magic you had containers of positive and negative charge which you could ladle out in small portions, how could you use these charges to build up a large oscillation in an *LC* circuit? If you could ladle only a limited amount at a time, what effect would a slow increase in resistance of the circuit have?

12. Comment upon the following statement published in a daily newspaper.

A warning that home electric appliances can cause fatal injuries has been sounded by City Health Director, J. R. Smith. His comments followed the death of an 18-year-old boy who accidentally electrocuted himself by inserting a fork in a toaster. Dr. Smith pointed out that even adults can be killed by such electrical

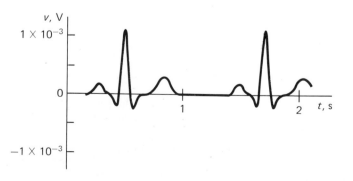

FIGURE P26.1

shocks. Ordinary house current is 100 volts but the voltage is increased if the current is grounded, he said.

13. The loop equation for a series RLC circuit was found to be $-Ri + q/C - L(di/dt) = 0$. If we multiply through this equation by i, the first term is $-Ri^2$, which represents the power loss in the resistor. What is the meaning of each of the other terms? What does the equation tell us about power?

14. Show that the following equations apply to an ideal transformer (i.e., one with no internal losses):

$$V_P I_P \cos \phi_P = V_S I_S \cos \phi_S \quad \text{and} \quad N_P I_P \cos \phi_P = N_S I_S \cos \phi_S$$

If no current is being drawn by the secondary, is $I_P = 0$?

15. Suppose you are given a box with four terminals on it, two on each side, and told that it contains a 120-V to 6-V step-down transformer. How could you tell which are the primary terminals? What would probably happen if you connected the secondary terminals to 120 V?

16. The devices shown in figure P26.2 are called filters. When an ac voltage is put into the device, the output ac voltage depends upon the frequency of the oscillating voltage. One of these devices lets the input voltage pass through undisturbed if the oscillation frequency is high. The other passes only low-frequency voltages. Explain which is which.

17. In a radio, a resonant circuit is tuned to vibrate with the same frequency as the station to which one is listening. Using a 10^{-9}-F capacitor, about how many loops of wire would one need to wind on one's finger in order to obtain a circuit tuned to a frequency of 10^6 Hz?

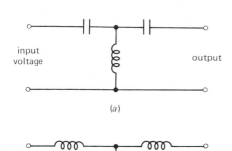

input voltage

output

(a)

input

output

(b)

FIGURE P26.2

Chapter 26 Problems

Note: Voltages and currents are rms unless stated otherwise.

1. A certain power source furnishes a voltage $v = v_0 \sin 2\pi ft$. It is connected across a pure resistance R. In terms of v_0 and R, (a) what is the rms source voltage? (b) The rms current in the circuit? (c) The average power loss in the resistor?

2. A pure resistance R is connected in series with an ac voltage source. If the current in the circuit is $i = i_0 \sin 2\pi ft$, in terms of R and i_0, (a) What is the rms current in the circuit? (b) The rms voltage of the source? (c) The equation of the source voltage? (d) The average power loss in the resistor?

3. How large a current is caused to flow through a 0.5-H essentially resistanceless inductor by a 40-V source at a frequency of $200/\pi$ Hz? At $200,000/\pi$ Hz?

4. When a certain resistanceless inductor is connected directly across a $(200/\pi)$-Hz voltage source, the current in the circuit is 0.30 A. An ordinary ac meter reads the source voltage to be 80 V. What is the inductance of the inductor? How much current will flow through it if the source frequency is decreased to $2/\pi$ Hz?

5. If an 80-V, $(200/\pi)$-Hz voltage source is connected directly across a certain capacitor, the current in the circuit is 1×10^{-3} A. What is the capacitance of the capacitor? How much current will flow in the circuit if the source frequency is $200,000/\pi$ Hz?

6. A 0.5-μF capacitor is connected directly across a 20-V, $(200/\pi)$-Hz source. Find the current it will draw from the source. Repeat for a source frequency of $200,000/\pi$ Hz.

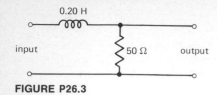

0.20 H

input

50 Ω

output

FIGURE P26.3

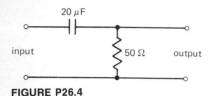

20 μF

input

50 Ω

output

FIGURE P26.4

7. Although one of the circuits shown in figure P26.2 is preferable, the circuit shown in figure P26.3 is capable of separating a low-frequency voltage from a high-frequency voltage when the two voltages are superimposed on the same line. It is a "low-pass" filter. Suppose a 20-V source is used as input; find the output voltage if the input frequency is $20/\pi$ Hz. Repeat for a frequency of $3000/\pi$ Hz. Assume infinite resistance at the output end.

8. Repeat problem 7 for the circuit of figure P26.4. This is a "high-pass" filter.

9. An inductance coil (0.20 H, 6.0 Ω) is connected in series with a 0.50-μF capacitor and a 100-V, $(200/\pi)$-Hz source. Find the current in the circuit, the phase angle between current and driving voltage, the average power dissipated in the coil, and the power factor for the circuit.

10. Repeat problem 9 for a frequency of $2000/\pi$ Hz.

11. A pure 20-Ω resistor is connected in series with an inductance coil having appreciable resistance. When the combination is connected across a $(200/\pi)$-Hz, 40-V power source, the voltage drop across the resistor is 5.0 V. When the combination is connected across a 12-V battery, the drop across the resistor is 8.0 V. Find the resistance and inductance of the coil.

12. A capacitor is connected in series with a pure 20-Ω resistor, and the combination is connected across a 40-V, $(200/\pi)$-Hz voltage source. The voltage drop across the resistor is found to be 30 V. Find the value of the capacitor and the voltage drop across it.

13. Find the resonant frequency for a series circuit containing a 2.0-μF capacitor, a 0.10-H inductor, and a 60-Ω resistor. If the voltage of the power source is 120 V, how much current will flow when resonance is achieved?

14. In a series RLC circuit, resonance occurs at 1×10^5 Hz. At that time, the potential difference across the 100-Ω resistance is 40 V while the potential difference across the pure inductor is 30 V. Find the value of both L and C.

15. A 90-Ω resistor, a 2-μF capacitor, and a pure inductor are connected in series across a 100-V power source. The individual potential differences across the resistor and capacitor are both equal to 70 V. Find the frequency of the power source and the value of the inductance.

16. A certain series circuit consists of a 100-V, $(200/\pi)$-Hz power source, a 5-μF capacitor, a 50-Ω resistor, and an essentially resistanceless coil. The potential difference across the resistor is 20 V. Find the impedance of the circuit together with the inductive reactance of the coil. Two answers for X_L are possible; find both. What are the phase angles in the two possible cases?

17. A certain RLC series circuit has a power factor of one-third at a certain frequency. If $R = 200$ Ω, what reactance must be added in series to change the power factor to unity? If the circuit is originally inductive, should the added reactance be an inductor or a capacitor?

18. A series RLC circuit with R_1, L_1, and C_1 resonates at the same frequency as a second series RLC circuit with R_2, L_2, and C_2. Suppose all of these elements $(R_1, R_2, L_1, L_2, C_1, C_2)$ are now connected in series. Show that this new circuit will have the same resonance frequency as the individual circuits had.

19. A series RLC circuit is connected across a voltage $v_0 \sin (2\pi ft)$. (a) Show that the charge on the capacitor obeys the relation $q = -(v_0/2\pi fZ) \cdot$

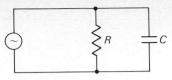

FIGURE P26.5

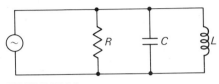

FIGURE P26.6

$\cos (2\pi ft - \phi)$. (*b*) Show that the amplitude of q will be maximum when $2\pi f = \sqrt{(1/LC) - (R^2/2L^2)}$.

20. In the parallel circuit shown in figure P26.5, the voltage source is $v = v_0 \sin (2\pi ft)$. (*a*) Show that the current drawn from the source is $(v_0/R) \sin (2\pi ft) + (v_0/X_c) \cos (2\pi ft)$. (*b*) Show that the current can be written as $(v_0/Z) \sin (2\pi ft + \phi)$ where $1/Z = \sqrt{(1/R)^2 + (1/X_c)^2}$ and $\tan \phi = R/X_c$.

21. In the parallel circuit shown in figure P26.6, the voltage is $v = v_0 \sin (2\pi ft)$. Show that the current drawn from the source is $(v_0/Z) \sin (2\pi ft + \phi)$ with $1/Z = \sqrt{(1/R)^2 - [(1/X_c) - (1/X_L)]^2}$ and $\tan \phi = R[(1/X_c) - (1/X_L)]$.

22. An oscillation is started in an *RLC* circuit. When the source of power is removed, the amplitude of oscillation drops down to $1/e$ of its original value in a time of 0.0010 s. If $L = 0.030$ H and $C = 0.010$ μF, how large is R for the circuit? About how many oscillations will the circuit undergo in 0.0010 s?

23. It is desired to construct a "ringing circuit," i.e., one which oscillates by itself after the power source is removed, which has a frequency of 10^5 Hz and undergoes 10^3 oscillations before the amplitude of the oscillation has dropped to $1/e$ of its original value. A 10^{-3}-H coil is available together with many different capacitors. What value of capacitor should be used, and what resistance should the circuit have?

24. A coil having both inductance and resistance is connected in series with a 2.0-μF capacitor and a 90-V power source whose frequency is $200/\pi$ Hz. The potential difference across the coil is 60 V, while the potential difference across the capacitor is 80 V. Find the resistance and inductance of the coil.

25. The *quality factor Q* of an *RLC* circuit measures the sharpness of its resonance and is defined by

$$Q = \frac{2\pi f_0 L}{R}$$

where f_0 is the resonance frequency of the circuit with resistance R and inductance L. Show that, if f is the frequency near f_0 at which the current in the circuit has dropped to $1/\sqrt{2}$ of its value at resonance (assuming V remains constant), then

$$Q \cong \frac{f_0}{2(f - f_0)}$$

so it is very large for a very sharp resonance. This latter relation is only true if $(f - f_0)/f_0 \ll 1$. [*Hint:* f is the frequency at which $Z/Z_0 = \sqrt{2}$. Note that $2\pi fL = 2\pi(f_0 + \Delta f)L$ and $1/2\pi fC \cong (1/2\pi f_0 C)(1 - \Delta f/f_0)$ where $\Delta f = f - f_0$.]

27 Waves

One aspect of electricity and magnetism that is of great importance to us is electromagnetic waves. We encounter these waves in many forms, ranging from radio waves, through infrared waves and light, to the very short waves we know as x-rays. In this chapter we shall begin a study of waves, not only electromagnetic waves, but waves of all types. We shall see that many facets of wave behavior are independent of the exact type of wave involved. Let us now see what features all waves have in common.

27.1 WAVE TERMINOLOGY

Many of the features of waves are easily understood by reference to waves on a string. To begin, suppose a disturbance is sent down a stretched string as shown in figure 27.1. The pulse shown there is initiated by a sudden up and down motion of the hand that holds the string. This disturbance travels with speed v along the string. We should note two very important features of such a pulse. First, it carries energy down the string. When the pulse strikes a given point on the string, it causes that portion of the string to momentarily acquire both kinetic and potential energy. This energy was given to the pulse by the source that initiated it. The energy moves with speed v down the string.

Second, the pulse on the string is a record of the history of what the pulse source has done. We can see at once by reference to figure 27.1 that the hand moved so as to initiate the pulse at a definite time in the past. Indeed, what the pulse source was doing at a time t ago is shown by the string at a distance $x = vt$ from the pulse source. In other words, the string at a distance x from the source is performing the same motion that the source initiated at a time $t = x/v$ previously.

Let us now see what happens when the source at the end of the string vibrates with simple harmonic motion. The situation is shown in figure 27.2. As we expect, the string shows the past history of the way its end was vibrated. The up and down motion of the string's end is transmitted down the string with speed v. Because of this the string has a sinusoidal form at a given instant and this sinusoidal pattern on the string travels to the right with speed v. As it moves, it carries energy down the string, energy furnished to the string by the vibration source.

There are certain words we use to describe such a wave. The points A and C, the tops of the wave, are called the *wave crests*. Points such as B are

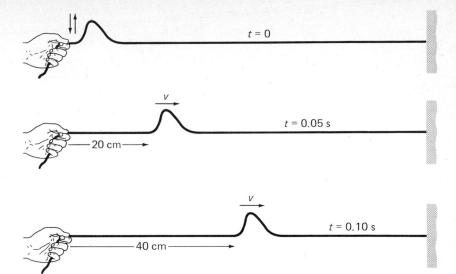

FIGURE 27.1 A pulse carries energy down the string. What is the speed of this wave?

called *wave troughs*. We call the maximum displacement of the string from its equilibrium position the *amplitude* of the wave. The amplitude of the wave in figure 27.2 is shown as y_0. Notice that it is only half the total vertical displacement of the string.

The distance between two adjacent crests on the wave, between A and C for example, is called the *wavelength* of the wave. It is indicated as the distance λ (Greek lambda) in figure 27.2. The wavelength λ of a wave is the distance between any two equivalent adjacent points along the wave. As we can see from figure 27.2, one wavelength of a wave is sent out by the wave source as it executes one complete vibration.

Consider the motion of point P on the string in figure 27.2. During the time that the source sends out a complete wave, one complete wave must pass through point P to make way for it. As a result, point P of the string will undergo one complete cycle of motion in the same time that the source takes to undergo one complete vibration. We see from this that the period of the vibrating source is the same as the period for the vibration of a point in the path of the wave. This time taken for a complete vibration of a point in the path of the wave is called the period of the wave and is represented by τ (Greek tau). We see, then, *the period of a wave is equal to the period of the vibrating source that produces the wave.**

*This assumes the conditions shown in figure 27.2, namely that the source and point in the wave path are at rest relative to each other.

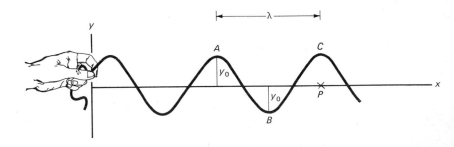

FIGURE 27.2 A source that vibrates with simple harmonic motion sends a sinusoidal wave down the string.

But you will recall that there is a simple relation between the period of vibration τ of an object and the frequency f of the vibration. It was that $f = 1/\tau$. For example, a source that takes 0.01 s to undergo one complete vibration has a frequency of $f = 1/0.01$ s $= 100$ s$^{-1} = 100$ Hz. Because the period τ is the same for both the source and the wave, we can state at once that *the frequency of a wave is equal to the frequency of its source*. Often the symbol ν (Greek nu) is used in place of f for the frequency of a wave. We therefore have that

$$\nu = f = \frac{1}{\tau} \qquad\qquad 27.1$$

A very important relation exists between the wavelength λ of the wave and the frequency. Referring again to figure 27.2, we see that a length λ of the wave is sent out during the time τ that the wave source takes to undergo one complete vibration. Therefore, the wave moves a distance λ in a time τ and so we find, from $v = x/t$, that $v = \lambda/\tau$ where v is the speed of the wave. Thus

$$\lambda = v\tau \qquad \text{and} \qquad \lambda = \frac{v}{\nu} \qquad\qquad 27.2$$

This relation is true for all waves, not just for waves on a string.

27.2 EQUATION OF A TRAVELING WAVE

If you look at the wave on a string shown in figure 27.3, you can see that its y coordinate is a sinusoidal function of x. You might guess that the wave could be represented by an equation such as the following:

$$y = y_0 \sin \frac{2\pi x}{\lambda} \qquad \text{(incomplete)} \qquad\qquad 27.3$$

This seems reasonable because it gives $y = 0$ at $x = 0$, and at $x = \lambda$, and at $x = 2\lambda$, and so on. In other words, it is a sine function with wavelength λ and amplitude y_0. However, it is not a complete equation for the wave.

The trouble lies in the fact that this equation is true only for the instant shown in figure 27.3. As time goes on, the wave moves to the right and so, at a given value of x, the y displacement of the string must change. At

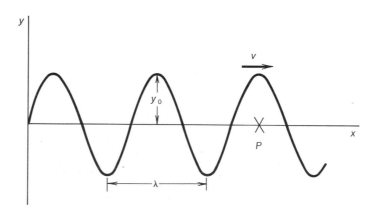

FIGURE 27.3 What will this wave on a string look like one-quarter period later?

point P, for example, the string oscillates up and down as the wave passes through it. This reminds us that each point along the wave path undergoes motion that reflects the motion the source underwent some time in the past. We shall see that this fact allows us to arrive at the correct equation for the wave.

Suppose the source vibrates the end of the string up and down according to the familiar simple harmonic motion equation

$$y = y_0 \sin 2\pi f t \qquad \text{(source)} \qquad 27.4$$

As a result, the string at a distance x away from the source (at point P in figure 27.3, for example) will undergo this same motion but after a time delay. The time delay is equal to the time it takes the wave to travel the distance x from the source to P. This time is given by $x = vt$ to be x/v.

We see, then, that a point P a distance x away from the source will undergo the same vibration as the source but at a time x/v later than when it occurred at the source. Therefore the vibration at P will obey equation 27.4 but with t replaced by $t - (x/v)$. As a check, we notice that the vibration that occurred at the source when $t = 0$ will arrive at P when $t = x/v$. It follows, then, that the wave motion of the string is given by *27.4* with t replaced by $t - (x/v)$:

Equation of a traveling wave

$$y = y_0 \sin \left[2\pi \nu \left(t - \frac{x}{v} \right) \right] \qquad 27.5a$$

where $\nu = f$, the frequency of the wave.

To see that this equation contains within it our preliminary guess stated in *27.3*, we note that $v = \lambda \nu$, which gives

$$y = y_0 \sin \left[2\pi \left(\nu t - \frac{x}{\lambda} \right) \right] \qquad 27.5b$$

At $t = 0$ (and also when νt equals any integer) this complete equation reduces to the incomplete equation we had guessed. But, in addition, *27.5* represents the translational motion of the wave along the string. We call a wave that travels out through space from a source a *traveling wave,* and *27.5* is the equation for a traveling wave on a string. It also applies to any other wave traveling with constant speed and amplitude along a line. In the next section we shall see how the motion of the wave can be related to the tension in and mass of the string.

27.3 WAVE EQUATION FOR A STRING

In order to obtain more insight into the motion of a wave on a string, let us write $F = ma$ for a small segment of the string. We shall say that the mass per unit length of the string is ρ and that the tension in the string is T. Moreover, we shall consider only very small displacements of the string, and so the situation shown in figure 27.4 is greatly exaggerated. Let us apply $F_y = ma_y$ to the segment of the string between points A and B. The original length of the segment was Δx, and so its mass is $\rho \, \Delta x$. Furthermore,[*]

We apply $F = ma$ to a small segment of the string

$$a_y = \frac{\partial}{\partial t} v_y = \frac{\partial}{\partial t} \frac{\partial y}{\partial t} = \frac{\partial^2 y}{\partial t^2}$$

[*] We use the partial differentiation notation because v_y and y are functions of both t and x. The notation $\partial v_y / \partial t$ means to take the derivative of v with respect to t while all other variables (such as x) are considered to be constant. If this notation frightens you, replace all ∂'s by d's in your mind.

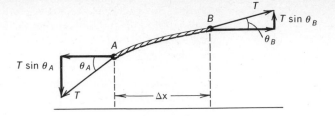

FIGURE 27.4 Notice that $\tan \theta_A$ and $\tan \theta_B$ are merely the slopes at A and B of the y vs. x curve defined by the string.

Therefore we have $F_y = ma_y$ to be

$$F_y = \rho \, \Delta x \frac{\partial^2 y}{\partial t^2}$$

It now remains for us to find an expression for F_y in terms of T, the tension in the string.

As one can see from figure 27.4, the resultant force in the y direction on the string segment shown is

$$F_y = T \sin \theta_B - T \sin \theta_A$$

But since we are considering only small displacements, θ_A and θ_B will be very small. You will recall that for small angles the tangent and the sine are the same; so we can write

$$F_y = T(\tan \theta_B - \tan \theta_A)$$

We now make use of the fact that the tangents at B and A are just the slopes of the curve defined by the string at these two points, and since the slope is $\partial y / \partial x$, we have

$$F_y = T\left[\left(\frac{\partial y}{\partial x}\right)_B - \left(\frac{\partial y}{\partial x}\right)_A \right]$$

It is now appropriate to recall that the derivative of a function is defined as

$$\frac{\partial f}{\partial x} = \lim_{\Delta x \to 0} \frac{f(x + \Delta x) - f(x)}{\Delta x}$$

We can obtain this form by dividing and multiplying the right-hand side of the equation for F_y by Δx. Then, after dividing through by T,

$$\frac{F_y}{T} = \left[\frac{(\partial y / \partial x)_B - (\partial y / \partial x)_A}{\Delta x} \right] \Delta x$$

Associating $(\partial y / \partial x)_B$ with $f(x + \Delta x)$ and $(\partial y / \partial x)_A$ with $f(x)$, the quantity in brackets becomes the derivative of $\partial y / \partial x$ when we let $\Delta x \to 0$. Upon doing so we find

$$F_y = T \, \Delta x \frac{\partial^2 y}{\partial x^2}$$

We can now equate this expression for F_y to that we obtained previously from $F_y = ma_y$. The result is, after some simplification,

The wave equation

$$\frac{\rho}{T} \frac{\partial^2 y}{\partial t^2} = \frac{\partial^2 y}{\partial x^2} \qquad\qquad 27.6$$

This is called the *wave equation* for a wave on a string. Any wave moving down the string must obey (or satisfy) this relation. In particular, the wave given by *27.5* must satisfy this equation.

We can gain useful information by substituting y from *27.5* into *27.6*. Upon doing so we find that

$$-\frac{\rho}{T}4\pi^2\nu^2 y_0 \sin\left[2\pi\nu\left(t-\frac{x}{v}\right)\right] = -\frac{4\pi^2\nu^2 y_0}{v^2}\sin\left[2\pi\nu\left(t-\frac{x}{v}\right)\right]$$

from which we discover that (after canceling like factors and simplifying)

Speed of a wave on a string

$$v = \sqrt{\frac{T}{\rho}} \qquad\qquad 27.7$$

This expression for the speed of the wave on a string tells us that the speed increases with increasing tension and is smallest for strings that have a large mass per unity length. This fact is easily verified by experiment.

It is interesting to recast *27.6*, the wave equation, in terms of our discovery concerning the speed of the wave. After substituting for ρ/T, it becomes

The wave equation

$$\frac{1}{v^2}\frac{\partial^2 y}{\partial t^2} = \frac{\partial^2 y}{\partial x^2} \qquad\qquad 27.8$$

We shall have reason to refer back to this equation when we discover other types of waves that obey an equation of similar form.

Illustration 27.1

Λ 200-cm length of wire weighs 0.60 g. If the tension in the wire is caused by hanging a 500-g mass from it, what will be the wavelength of the wave sent down it by a 400-Hz vibration?

Reasoning To find the speed of the wave on the string, we make use of the fact that

$$\rho = \frac{0.60\times 10^{-3}\,\text{kg}}{2.0\,\text{m}} = 3\times 10^{-4}\,\text{kg/m}$$

and $\qquad\qquad T = (0.50)(9.8)\,\text{N} = 4.9\,\text{N}$

Therefore $\qquad v = \sqrt{4.9/3\times 10^{-4}} = 128\,\text{m/s}$

Since $v = \lambda\nu$, we find at once that $\lambda = 32$ cm.

27.4 ENERGY TRANSMISSION ALONG STRINGS

When a wave is sent down a string in the positive x direction, the wave is propagated along because each string segment pulls upward or downward on the segment adjacent to it at a slightly larger value of x and, as a result, does work upon the string segment to which the wave is traveling. For example, in figure 27.5(a), the portion of the string at A is going upward, and will pull the portion at B upward as well. In fact, at any point along the string, each segment of the string is pulling on the segment just adjacent to its right, causing the wave to propagate. It is by this process that energy is sent down the string.

To calculate how much energy is propagated down the string each

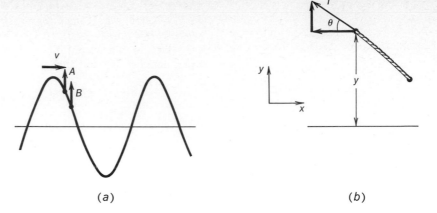

FIGURE 27.5 As the wave moves toward the right, the string at point A does work on the string at B.

(a)　　　　　　　　　　(b)

second, we need only compute the work done by a segment of the string on its neighbor segment each second. First we shall compute the work done during one cycle. The work per second can be found by multiplying this result by v. Referring to figure 27.5(b), the component of T in the direction of the displacement is $T \sin \theta \equiv T_y$. For small displacements, $\sin \theta = \tan \theta = \partial y/\partial x$; so this can be written

$$T_y = -T\frac{\partial y}{\partial x}$$

(The negative sign appears because, as shown in the figure, the slope is negative when T_y is positive.) In a time dt, this force will act through a distance

$$dy = v_y\, dt = \frac{\partial y}{\partial t}\, dt$$

Therefore the work done by the force in time dt is

$$dW = T_y\, dy = -T\frac{\partial y}{\partial x}\frac{\partial y}{\partial t}\, dt$$

But we know that in the case we have been considering, a sinusoidal wave,

$$y = y_0 \sin\left[\omega\left(t - \frac{x}{v}\right)\right]$$

where we have replaced $2\pi v = 2\pi f$ by ω, and so

$$\frac{\partial y}{\partial x} = -\frac{\omega y_0}{v}\cos\left[\omega\left(t - \frac{x}{v}\right)\right]$$

and

$$\frac{\partial y}{\partial t} = \omega y_0 \cos\left[\omega\left(t - \frac{x}{v}\right)\right]$$

Substituting these values gives

$$dW = \frac{\omega^2 y_0^2 T}{v}\cos^2\left[\omega\left(t - \frac{x}{v}\right)\right] dt$$

To find the work done in one cycle, this must be integrated over $0 \le t \le \tau$, where $\tau = 1/v = 2\pi/\omega$. Since the work done will be the same at any place along the string, we pick the point $x = 0$ for convenience. Then

$$\frac{\text{Work}}{\text{Cycle}} = \frac{\omega^2 y_0^2 T}{v} \int_0^{2\pi/\omega} \cos^2 \omega t \, dt$$

$$= \frac{\pi \omega y_0^2 T}{v}$$

The power transmitted down the string (i.e., the energy sent down it per second) is found by multiplying this result by ν, the frequency. We then find

The intensity of the wave is proportional to the square of the amplitude

$$\text{Power} = \frac{2\pi^2 \nu^2 y_0^2 T}{v} = 2\pi^2 \nu^2 y_0^2 \rho v \qquad 27.9$$

Notice that the power, which we call the intensity of the wave, is proportional to the square of the amplitude of the wave.

Illustration 27.2

How much power is transmitted down a string having $\rho = 3 \times 10^{-4}$ kg/m and $T = 5.0$ N by a 200-Hz vibration of 0.20-cm amplitude?

Reasoning Since $v = \sqrt{T/\rho}$,

$$\text{Power} = 2\pi^2 \nu^2 y_0^2 \rho v$$

becomes

$$\text{Power} = 2\pi^2 \nu^2 y_0^2 \sqrt{\rho T}$$

Substituting the values given, we find

$$\text{Power} \approx 0.125 \text{ W}$$

27.5 COMPRESSIONAL WAVES

Waves on a string are important examples of what we call *transverse waves*. The line defined by the string is necessarily the line of propagation of the waves, i.e., the line out along which the waves travel. But the individual particles of the string vibrate perpendicular (or transverse) to the line of propagation and so we term this type of wave a transverse wave. Later we shall learn that electromagnetic waves are also transverse. Now, however, we shall investigate the behavior of compressional waves, a type of wave that is not transverse.

Consider the spring shown in figure 27.6. It is suspended by threads so it can move horizontally. Suppose an oscillating force is applied to its end as indicated. The force will alternately compress and stretch the spring, thereby sending a series of compressions and stretched regions (called *rarefactions*) down the spring. As we see, the force causes a *compressional wave* to move down the spring. This type of wave is also called a *longitudinal wave*. Clearly, *in a compressional wave, the particles in the path of the wave move along the line of propagation of the wave.*

To describe the behavior of the spring quantitatively, we notice in figure 27.6 that the supporting threads would be exactly vertical if the spring was undistrubed. The disturbance passing down the spring causes displacements of the elements of the spring from their equilibrium positions. These displacements are indicated in the figure by the displacements of the threads from the vertical. It is therefore a simple matter to graph the displacements

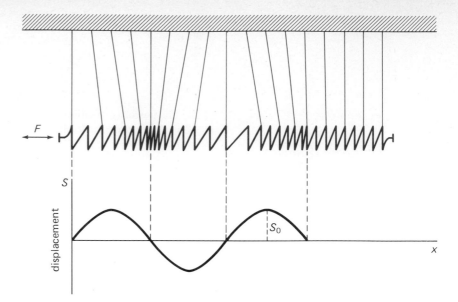

FIGURE 27.6 A compressional wave is sent down a spring by a sinusoidal displacing force acting at the spring's end.

of the spring elements from their equilibrium positions and this is done in the lower part of the figure.

Let us denote the displacement of a spring element from its equilibrium position by s. Notice that s takes the place of y in our description of a wave on a string. (We do not use y now because we do not want you to confuse the transverse motion that occurs on the string with the longitudinal motion that occurs in the compressional wave on the spring.) If the end of the spring is vibrated sinusoidally, then the wave on the spring will also be sinusoidal. This is the type wave shown in the lower part of the figure. Its equation is

$$s = s_0 \sin\left[\omega\left(t - \frac{x}{v}\right)\right] \qquad 27.10$$

where v is the speed of the wave along the spring and s_0 is the amplitude of the wave. We can rewrite this equation in various other ways by substituting from the relations $\omega = 2\pi\nu$ and $\lambda = v/\nu$. Of course ν, the frequency of the wave, equals the frequency of the force that vibrates the spring.

We found that waves on a string obeyed a wave equation of form $(\rho/T)(\partial^2 y/\partial t^2) = \dfrac{\partial^2 y}{\partial x^2}$. It is shown in Appendix 9 that a similar equation applies to the compressional wave on a spring; it is

$$\left(\frac{\rho}{k}\right)\frac{\partial^2 s}{\partial t^2} = \frac{\partial^2 s}{\partial x^2} \qquad 27.11$$

where k is the spring constant for a unit length of the spring and ρ is its mass per unit length. As was true for the wave on a string, we can find the speed of the wave by substituting the equation for the wave, *27.10*, in the wave equation *27.11*. Upon doing so, we find that

$$v = \sqrt{\frac{k}{\rho}} \qquad 27.12$$

As we might have suspected, the speed of the wave is largest for a stiff spring that has a small mass per unit length.

It is shown in Appendix 9 that the power transmitted down the spring is

$$P = 2\pi^2\nu^2 s_0^2 \rho v \qquad 27.13$$

This equation is identical to *27.9* for the power transmitted by a string. In both cases, the power varies as the square of both the frequency and the wave amplitude.

27.6 COMPRESSIONAL WAVES IN MATTER

Suppose the spring of the previous section is replaced by a long metal rod. In a sense, the rod is simply a very stiff spring. It is reasonable to think, then, that we can understand compressional wave motion in the rod by comparing it to a spring. We can carry out a computation almost identical to that displayed in Appendix 9 and find the following results: the wave equation for compressional waves on a rod is

$$\frac{\rho}{Y}\frac{\partial^2 s}{\partial t^2} = \frac{\partial^2 s}{\partial x^2} \qquad 27.14$$

and the speed of the wave is

$$v = \sqrt{\frac{Y}{\rho}} \qquad 27.15$$

where Y is Young's modulus and ρ is the mass density for the material of the rod.

Similarly, for a compressional wave traveling in the x direction through an extended fluid such as water, oil, or air, it is found in Sec. 27.12 that the wave equation is

$$\frac{\rho}{B}\frac{\partial^2 s}{\partial t^2} = \frac{\partial^2 s}{\partial x^2}$$

where B is the bulk modulus* of the material. The speed of the wave is given by

$$v = \sqrt{\frac{B}{\rho}} \qquad 27.16$$

Typical speeds are given in table 27.1. Furthermore, in all of these cases the power transmitted by the wave is given by

$$\frac{\text{Power}}{\text{Area through which it passes}} = 2\pi^2\nu^2 s_0^2 \rho v \qquad 27.17$$

Once again the transmitted power is proportional to the squares of the frequency and amplitude of the wave. Let us now turn our attention to one of the most important types of compressional waves, sound waves.

---

*The bulk modulus was defined in Sec. 13.3. When a pressure increment ΔP is applied to a fluid with original volume V, the volume changes by $-\Delta V$. The bulk modulus B is then defined by $\Delta P = -B(\Delta V/V)$.

TABLE 27.1 Speed of Compressional Waves

Material	Speed (m/s)	Material	Speed (m/s)
Air (STP)	331	Copper	4,760
Hydrogen (STP)	1,284	Glass (Pyrex)	5,640
Mercury (25°C)	1,450	Steel	$\approx$5,900
Water (25°C)	1,478	Molybdenum	6,250
Lucite	2,680	Beryllium	12,890

27.7 SOUND WAVES

Sound waves are simply compressional disturbances sent out through the air or, in a more general sense, through any material. The normal human ear responds well to such disturbances only in the approximate frequency range of 20 to 18,000 Hz. Compressional waves of still higher frequency are called *ultrasonic waves*. Those with frequencies less than about 20 Hz are called *infrasonic waves*. We shall deal here primarily with compressional waves in air and other gases. Our discussions will apply to a wide range of frequencies and will not be restricted to the audible range.

To understand the nature of sound waves, let us refer to figure 27.7 where we see a vibrating piston sending sound waves through a tube filled with air. As the piston moves to the right, it compresses the air and sends a compression down the tube. Later, when the piston moves back towards the left, the air to the right of it is expanded; at that time a rarefaction is sent down the tube. The end result is that a longitudinal wave consisting of alternate compressions and rarefactions is sent down the tube.

This situation is much like that which occurred with the spring in figure 27.6. If you look back at that figure, you can see clearly that the wave on the spring consists of alternate compressions and rarefactions. Just as the elements of the spring oscillate back and forth around their equilibrium positions, so too do the air particles in the tube. Only the wave and the energy that it carries moves down the tube. The air particles undergo no permanent displacement; they simply vibrate back and forth over a very small distance.

We describe the disturbance, the sound wave, by specifying the displacement of the air particles from their equilibrium positions. As with the spring, we denote by s the displacement of an air particle from its equilibrium position. The value of s for a given air particle varies sinusoidally as the sinusoidal wave sent out by the piston passes the particle in question. Taking

FIGURE 27.7 The oscillating piston sends a sound wave through the gas.

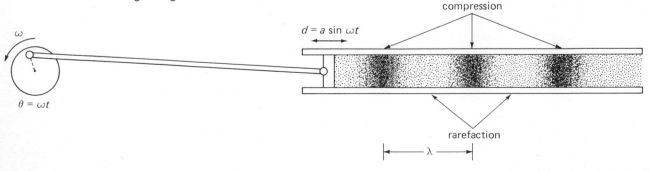

the x coordinate along the axis of the tube, the equation of the traveling sound wave in the tube is

$$s = s_0 \sin\left[\omega\left(t - \frac{x}{v}\right)\right]$$

In practical situations, the piston shown in figure 27.7 would be replaced by a loudspeaker, a tuning fork, a guitar string, a person's mouth, or some other means for causing motion of the air molecules. A loudspeaker consists of a flexible diaphragm driven by an alternating current in an electromagnet. As the diaphragm oscillates back and forth, it sends a sound wave into the air. The vibrating tuning fork prong and the vibrating string on a guitar accomplish the same task. Our vocal cords vibrate when we speak or sing and these, too, generate sound waves. All sound sources have one feature in common: they generate alternate compressions and rarefactions in the air.

The sound waves sent out by a loudspeaker are not usually confined to a tube. Instead, they spread out in all directions from the source. To better understand this feature of wave motion, refer to figure 27.8(a) page 580. It shows a water wave traveling out from a central source. A diagrammatic way for representing the situation in (a) is shown in (b). As we see, the wave crests take the form of larger and larger circles as they move away from the source. At great distances from the source, the circles are so large that they have very little curvature. Such a wave crest from a very distant source appears as a nearly straight line as it sweeps past a point on the water. Waves at large distances from their source have negligible curvature and are called *plane waves*.

The waves in figure 27.8 carry energy away from the source. Because the energy moves in the direction of propagation of the waves, the energy travels out along radial lines such as those labeled *rays* in the figure. Notice that the rays are of necessity perpendicular to the wave crests, or wavefronts. Moreover, at large distances from the source, the crests form essentially straight lines. Because the rays are perpendicular to the crests, the rays are parallel at a great distance from the wave source.

We have an analogous situation in the case of sound waves in air. But in this three-dimensional situation, the wave crests (or wavefronts) are spherical surfaces centered on the sound source. These spherical waves have decreasing curvature as they move farther away. At a distant point from the source, the wave surface appears essentially flat; the waves are then essentially flat planes and we refer to them as *plane waves*. As before, the rays are perpendicular to the wavefronts and so the rays are parallel in the case of a plane wave.

There is one other feature we should notice about the waves shown in figure 27.8; the amplitude of the wave decreases with increasing distance from the source. This is simply a reflection of the fact that the energy carried by the wave is spread out over an increasingly larger wavefront as the wave proceeds. Hence a unit length of a wave crest contains less energy as the crest continues to move away from the source. This feature did not concern us in the case of a wave on a string, spring, or bar because all the energy propagated is along the same line. But in the case of spherical waves, the energy becomes spread ever more thinly as the surface area of the wave continues to increase. Only in the case of plane waves can this decrease in amplitude be ignored. In that case the rays are parallel and so the energy is carried forward along a single direction; it does not spread over a larger wavefront as it moves along.

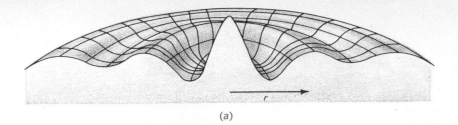

(a)

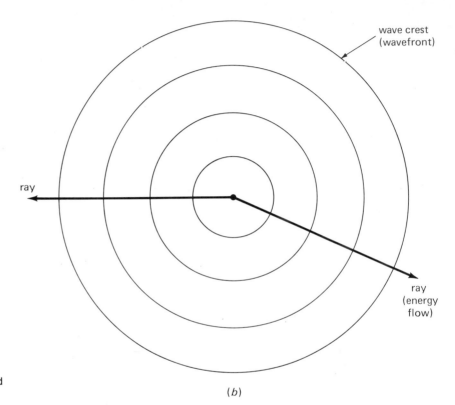

wave crest
(wavefront)

ray

ray
(energy
flow)

(b)

FIGURE 27.8 (a) A wave source at the center sends circular waves across the surface of the water. (b) A diagram used to represent a situation such as (a).

27.8 SPEED OF SOUND

Let us consider a sound source that sends a sound wave out along a line. The source might be a loudspeaker so constructed that it sends out a beam of sound in the x direction. The situation is shown in figure 27.9. We have plotted there a graph of the air-particle displacement from equilibrium as a function of distance along the path of the beam. Remember that the particles in the path of the wave oscillate back and forth along a line *parallel* to the x axis. The graph gives the values of s that exist at a certain instant.

The wave shown in figure 27.9 is a traveling wave much like the compressional waves discussed previously. Its equation is of the form

$$s = s_0 \sin\left[\omega\left(t - \frac{x}{v}\right)\right] \qquad 27.18$$

where s_0 is the amplitude of the wave and v is its speed, the speed of sound.

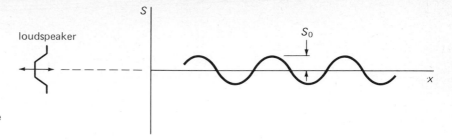

FIGURE 27.9 The sound wave beam travels with constant amplitude along the x axis.

We shall show in Sec. 27.13 that unidirectional sound waves obey the following wave equation:

$$\frac{\rho}{B}\frac{\partial^2 s}{\partial t^2} = \frac{\partial^2 s}{\partial x^2} \qquad 27.19$$

where B is the bulk modulus of the air or other material through which the sound is moving. You will recognize this as identical to the wave equation for a spring except that now ρ is the density of the air and the spring constant k is replaced by the bulk modulus B. We can easily see that B is a Hooke's law spring constant by noting that it is given by $\Delta P = -B\frac{\Delta V}{V}$. In other words, B measures how much pressure ΔP is required to compress a volume V by an amount ΔV.

We can find the speed of the sound waves by substituting *27.18* into the wave equation in the usual way. Upon doing so, you will find that

$$v = \sqrt{\frac{B}{\rho}} \qquad 27.20$$

which is the same as *27.16* given previously for the speed of compressional waves. Unfortunately, *27.20* for the speed of sound is not very convenient because B depends upon the pressure and nature of the gas. We therefore seek to find B from a consideration of the gas law.

By its definition, if a pressure increase ΔP causes a volume V of gas to change to $V + \Delta V$, then

$$B = -\frac{\Delta P}{\dfrac{\Delta V}{V}}$$

Because P and V are related through the gas law, we should be able to simplify this expression. However, it is not convenient to use the gas law directly because a new variable T would be encountered. Another approach would be to note that for an ideal gas $PV =$ constant for an isothermal change and that $PV^\gamma =$ constant for an adiabatic change.* In most instances of compressional waves in gases, the compressions are rapid enough so that the change is essentially adiabatic. We shall therefore try to relate ΔP and V through $PV^\gamma =$ constant.

If we take the derivative of this equation with respect to P we have that

$$V^\gamma + \gamma PV^{\gamma-1}\frac{\partial V}{\partial P} = 0$$

---

*Recall that $\gamma = c_p/c_v$ for an ideal gas.

which simplifies to

$$\frac{\partial V}{\partial P} = -\frac{V}{\gamma P} \qquad\qquad 27.21a$$

But $\partial V/\partial P$ is just the change in V with P and this is simply $\Delta V/\Delta P$ in our previous notation. We therefore have that

$$\frac{\Delta V}{\Delta P} = -\frac{V}{\gamma P} \qquad \text{or} \qquad -\frac{\Delta P}{\Delta V/V} = \gamma P \qquad 27.21b$$

But this is the defining expression for the bulk modulus and so we have arrived at the result that $B = \gamma P$ for an ideal gas.

We can now substitute this value for B in the equation for the speed of sound, *27.20*. The result is

Speed of sound
$$v = \sqrt{\frac{\gamma P}{\rho}} \qquad\qquad 27.22$$

Illustration 27.3
Find the approximate speed of sound in air under standard conditions.

Reasoning Air is mostly nitrogen and oxygen, both diatomic molecules. For such molecules, $\gamma = 1.40$ (see table 15.3) and $\rho = 1.29 \text{ kg/m}^3$ under standard conditions. Therefore

$$v = \sqrt{\frac{(1.40)(1.01 \times 10^5 \text{ N/m}^2)}{1.29 \text{ kg/m}^3}} = 331 \text{ m/s}$$

This is the measured value for the speed of sound in air under standard conditions.

Illustration 27.4
Find the dependence of the speed of sound in an ideal gas on temperature.

Reasoning Assuming the gas as a whole to be at equilibrium, we can apply the ideal gas law to it (see *14.5*),

$$PV = (m/M)RT \equiv nRT$$

Substituting this in 27.21 gives

$$v = \sqrt{\frac{\gamma nRT}{\rho V}}$$

Since $\rho = \text{mass/volume} = m/V$, we have

$$v = \sqrt{\gamma \frac{n}{m} RT}$$

We recall that n is the number of kilomoles of the gas, and so, if M is the molecular weight of the gas, $n = m/M$. Using this value, v becomes

$$v = \sqrt{\frac{\gamma}{M} RT} \qquad\qquad 27.23$$

This can be put in a very convenient form if we write t for the Celsius temperature and restrict the result to $t \ll 273$. Then

$$T = 273 + t = 273\left(1 + \frac{t}{273}\right)$$

582 WAVES

In our case $t/273 \ll 1$, and so

$$\sqrt{T} = \sqrt{273\left(1 + \frac{t}{273}\right)} \approx \left(1 + \frac{t}{2 \times 273}\right)\sqrt{273}$$

where the expansion $(1 + x)^{1/2} = 1 + \frac{1}{2}x + \cdots$ has been used. Therefore

$$v = \sqrt{\frac{\gamma}{M}R(273)}\left(1 + \frac{t}{546}\right)$$

But the quantity in the square root is merely the velocity at 0°C, and so

$$v = v_0 + \frac{v_0}{546}t$$

Numerical value for the speed of sound in air

In the case of air, $v_0 = 331$ m/s, and so

$$v = 331 + 0.61t \qquad \text{m/s}$$

One should notice that 27.23 predicts the speed of sound to be independent of pressure provided the temperature remains constant.

27.9 INTENSITY OF SOUND

As we have seen, waves carry energy along their line of propagation. The intensity of a wave such as sound is defined in terms of the energy carried through unit area each second. For example, in figure 27.10 a sound wave is traveling to the right through the area. We define the intensity of the wave to be the power passing through the unit area erected perpendicular to the direction of propagation.

We have noticed previously that the power carried by a wave is proportional to the squares of the wave amplitude and frequency. Sound waves are no exception. As we shall see in Sec. 27.12, the intensity I of a sound wave is given by

$$I = 2\pi^2 \rho v s_0^2 \nu^2 \qquad\qquad 27.24$$

Its units are W/m².

Typical sound intensities are given in table 27.2. You will notice that the human ear responds to a wide range of sound intensities. A sound that is just audible has an intensity of about 10^{-12} W/m² while an intensity of greater than about 1 W/m² produces the sensation of pain. As you might expect from this, the ear does not judge the loudness of sounds in direct

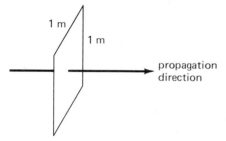

FIGURE 27.10 The intensity of the wave is equal to the energy carried through the unit area in unit time. Notice that the area must be perpendicular to the direction of propagation.

TABLE 27.2 Approximate Sound Intensities

Type of sound	Intensity (W/m²)	Intensity level (dB)
Pain-producing	1	120
Jackhammer or riveter*	10^{-2}	100
Busy street traffic*	10^{-5}	70
Ordinary conversation*	10^{-6}	60
Average whisper*	10^{-10}	20
Rustle of leaves*	10^{-11}	10
Barely audible sound	10^{-12}	0

*For a person near the source of the sound.

proportion to their intensities. After all, we usually don't think that ordinary conversation is 10,000 times louder than a whisper. Yet this is the factor by which their intensities differ.

To correspond more closely with the way the ear judges loudness of sound, an intensity-level scale, called the *decibel scale*, is defined. The intensity level measured in *decibels* (dB) is defined by

$$\text{Intensity level} = 10 \log \frac{I}{I_0} \qquad 27.25$$

where I is the intensity of the sound under consideration and I_0 is a reference intensity usually taken to be $10^{-12}\ \mathrm{W/m^2}$. Table 27.2 also contains the intensity levels of common sounds.

Illustration 27.5

Find the intensity of a 45-dB sound.

Reasoning From *27.25*,

$$4.5 = \log \frac{I}{I_0}$$

from which

$$\frac{I}{I_0} = 10^{4.5} = 10^{0.5} \times 10^4 = 3.16 \times 10^4$$

Therefore

$$I = I_0 \times 3.16 \times 10^4 = 3.16 \times 10^{-8}\ \mathrm{W/m^2}$$

27.10 THE DOPPLER EFFECT

An interesting and important effect occurs when the wave source and observer are in relative motion. One finds that if the observer and wave source are in motion relative to each other, the apparent frequency which the observer assigns to the waves depends upon the speed of this relative motion.* To understand this, consider the situation shown in figure 27.11. The wave source, a loudspeaker in this case, is moving along a line toward the observer with speed V_s relative to the earth. Suppose the speed of the waves relative to the earth is v. The speed of the observer relative to the earth is denoted V_0. During one oscillation of the source, one complete wave is sent out. The time taken for one oscillation is 1/frequency, and we shall designate the frequency of the source by ν_0. Since the speed of the wave is v, the crest of a wave will travel a distance v/ν_0 during one cycle. As a result, the crests would be

---

*Recall that our earlier assertion that the source and wave frequencies are the same was restricted to situations where they were at rest relative to each other.

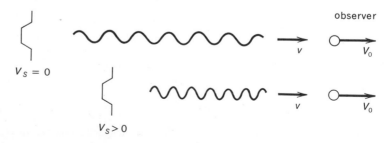

FIGURE 27.11 The forward motion of the loudspeaker compresses the wave. As a result, more wave crests pass the observer in unit time, and the frequency observed is higher. This is an instance of the Doppler effect.

separated by this distance if the source were at rest. In that case, the wavelength of the waves would be

$$\lambda_0 = \frac{v}{\nu_0}$$

However, since the source is actually moving with velocity V_s as shown in the figure, the distance moved by the source will be V_s/ν_0 during the time of one oscillation. Therefore the second wave crest will be emitted that much closer to the first. The actual wavelength then becomes

$$\lambda = \lambda_0 - \frac{V_s}{\nu_0}$$

from which

$$\lambda = \frac{v - V_s}{\nu_0}$$

Since the observer is moving with velocity V_0 relative to the earth along the line between source and observer, the speed of the wave crests relative to the observer will be $v - V_0$. In unit time, a portion of the wave $v - V_0$ long will pass the observer. Since the wavelength of the wave (the distance between its crests) is $(v - V_s)/\nu_0$, the length $v - V_0$ of the wave will contain

$$\frac{v - V_0}{\lambda} \qquad \text{or} \qquad \frac{v - V_0}{v - V_s}\nu_0$$

wave crests. This many wave crests will pass the observer each second. He will therefore observe a wave having a frequency equal to this number, namely,

General expression for the Doppler frequency shift

$$\nu = \frac{v - V_0}{v - V_s}\nu_0 \qquad\qquad 27.26$$

In making use of 27.26, one must be careful about the signs. We have assumed velocities in the $+x$ direction to be positive in its derivation, and the observer to be at a larger value of x than the source. If either the source or observer is moving in the $-x$ direction, its velocity must be considered negative. In particular, if the source is approaching the observer, who in turn is at rest on the earth, then

$$\nu = \frac{v}{v - V_s}\nu_0$$

and the observer will notice the wave to have a frequency higher than ν_0. If the source is receding from a stationary observer, then

$$\nu = \frac{v}{v + V_s}\nu_0 \qquad\qquad 27.27$$

and the apparent frequency of the wave is lowered.

This general effect is called the *Doppler effect,* after Christian Johann Doppler, who showed in 1842 that this effect should be observed for light waves as well as for sound waves. However, we should not expect 27.26 to apply to light waves, since, in its derivation, we assumed the waves could be given a definite speed relative to the earth. We know, however, that the speed of light is the same for all observers, even if they are in relative motion. In spite of this fact, 27.27 is found to be valid for light waves also if V_s is the relative speed of separation of source and observer and if $V_s \ll c$.

An interesting situation arises if the speed of the sound source approaches or equals the speed of sound. Then, from 27.26, we see that the

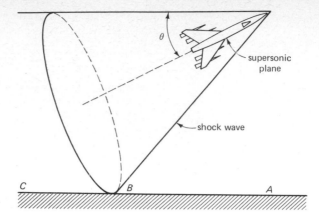

FIGURE 27.12 The sonic boom has already hit point C and is moving through point B toward A.

sound frequency ν approaches infinity. This simply means that a nearly infinite number of wave crests reaches the listener in a very short time. We can easily understand this by referring once again to figure 27.11.

Suppose the moving source has a speed equal to the speed of sound. Then all the wave crests will lie upon one another. They, together with the sound source itself, will all pass a given point at the same time. All the energy of the sound waves would be compressed into a very small region in front of the sound source. Consequently, this very concentrated region of sound energy, a shock wave, would cause an extremely large sound as it passes the point. Basically this is the origin of the sonic boom which accompanies supersonic aircraft.

When an airplane moves through the air with a speed near that of sound, the noise and air disturbances originating from the plane are built up into a shock wave, which, as we have just seen, is simply a region of very dense sound energy. The exact shape of the shock wave depends upon the speed of the airplane. In general, the shock zone covers the surface of a cone, as illustrated in figure 27.12. The angle θ of the cone depends upon the ratio of the speed of the plane v_p to the speed of sound v in the following way (provided $v < v_p$):

$$\sin \theta = \frac{v}{v_p}$$

As the plane's speed becomes larger with respect to v, the angle of the cone decreases. The ratio v_p/v is often called the *Mach number*. In this terminology, a plane traveling at Mach 2 is moving at twice the speed of sound.

Since the shock wave is a region of very concentrated sound energy, it can cause severe damage when it strikes something. The familiar sonic boom is the result of the conical shock-wave surface passing over the earth. For example, in figure 27.12, the sonic boom will soon strike point A; it is currently striking point B; it has already hit point C. Depending upon the intensity of the wave, its effects will be more or less damaging.

27.11 RELATION BETWEEN PRESSURE AND s*

There are two different ways of describing a sound wave. The one we have been using focuses on the displacements of the gas particles in the path of the

*Optional section.

wave. A particle at a given point x will oscillate about its equilibrium point as the wave passes by. We have already seen that this oscillatory displacement is given by $s = s_0 \sin\left[\omega\left(t - \dfrac{x}{v}\right)\right]$.

But there is another way we can describe the wave. We can describe it in terms of the pressure of the gas in the path of the wave. The pressure within the compressions and rarefactions must be different from P, the average gas pressure. To find a relation between the pressure and s, the displacement, let us consider a sound wave traveling down a tube with cross-sectional area A as shown in figure 27.13. Let us examine the behavior of the volume of gas that is between points x_1 and x_2 when the gas is not disturbed.

When the wave enters the region shown in figure 27.13, at a certain instant the displacement of the particle whose equilibrium position is at x_1 will be s_1, as shown. A similar displacement s_2 occurs for the gas at x_2. Notice in this case that the volume of gas originally between x_1 and x_2 is $A\,\Delta x$ where A is the tube's cross section. But at the instant shown, the volume has been displaced and expanded to a new volume. The new volume is, from the figure, $[(s_2 + \Delta x) - s_1]A$. We therefore find that the wave has changed the volume by an amount

$$\Delta V = (s_2 - s_1)A$$

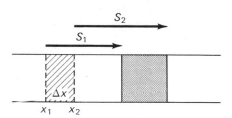

FIGURE 27.13 When the wave passes through the region, the volume between x_1 and x_2 changes to the heavily shaded volume.

The fractional change in volume is $\Delta V/V$, which is, in this case,

$$\frac{\Delta V}{V} = \frac{(s_2 - s_1)A}{(\Delta x)A} = \frac{s_2 - s_1}{\Delta x}$$

But $(s_2 - s_1)/\Delta x$ is simply the rate of change of s with x, namely $\partial s/\partial x$. We therefore find that

$$\frac{\Delta V}{V} = \frac{\partial s}{\partial x} \qquad\qquad 27.28$$

Because of this change in volume, there must be a change in the pressure of the gas. We have already found in *27.21* that

$$\frac{\Delta V}{V} = -\frac{\Delta P}{\gamma P}$$

Equating this to the value found in *27.28* gives

$$\frac{\partial s}{\partial x} = -\frac{\Delta P}{\gamma P}$$

This can be simplified by noting that the velocity of sound is given by $v = \sqrt{\gamma P/\rho}$. Upon substituting for γP by use of this relation we find that

$$\Delta P = -\rho v^2 \frac{\partial s}{\partial x} \qquad\qquad 27.29$$

We remind you that ΔP is the increase in pressure at point x due to the passage of the wave through point x.

We can put this relation in a different form by noting that $s = s_0 \sin[\omega(t - x/v)]$. Differentiating to find $\partial s/\partial x$ and substituting in *27.29* leads to the following result:

$$\Delta P = \rho\omega v s_0 \cos\left[\omega\left(t - \frac{x}{v}\right)\right] \qquad\qquad 27.30$$

This is the traveling pressure wave that travels down the tube of figure 27.13. Notice that its amplitude $(\Delta P)_0$ is $\rho\omega v s_0$. Further, the pressure wave is a cosine function while the displacement wave is a sine function. The two waves are one-quarter cycle out of phase. If you look back to figure 27.6 for a wave on a spring, you can see that there, too, the displacement is out of phase with the tension in the spring. Where the displacement is a maximum, the tension in the spring is zero. Similarly in a sound wave; the maxima of the displacement wave coincide with the zero points on the pressure wave.

Illustration 27.6

In order to be so loud as to cause pain to the ear, a sound wave must have a pressure amplitude of about 30 N/m². Find the amplitude for the motion of the air particles in such a wave.

Reasoning   From *27.30*, the amplitudes are related through

$$(\Delta P)_0 = \rho(2\pi\nu)vs_0$$

In our case $\rho = 1.29\,\text{kg/m}^3$, $v = 331\,\text{m/s}$, and $(\Delta P)_0 = 30\,\text{N/m}^2$. Substitution gives $s_0 = 0.011/\nu$ m. For a sound frequency of 1000 Hz, this gives $s_0 = 0.01$ mm. Notice how very small the amplitude is for even such a very loud sound.

27.12 TWO DERIVATIONS*

Intensity of Sound

Now that we know how the displacement and pressure in a wave are related, we can find the energy that is transported by the wave. To do so, consider the situation shown in figure 27.14. Let us take the cross-sectional area of the sound beam to be A. Then the force exerted on the area at x due to the pressure of the gas to its left will be $(P + \Delta P)A$ where P is the average pressure and ΔP is the pressure increase due to the wave.

The air at x is being displaced with a speed $\partial s/\partial t$. As a result, the force $A(P + \Delta P)$ is doing work on it and the rate of doing work (the power transported) is

$$\frac{dW}{dt} = A(P + \Delta P)\frac{\partial s}{\partial t}$$

But we found in the previous section that $\Delta P = (\Delta P)_0 \cos\left[\omega(t - x/v)\right]$ while $s = s_0 \sin\left[\omega(t - x/v)\right]$. We differentiate s to give

$$\frac{\partial s}{\partial t} = \omega s_0 \cos\left[\omega\left(t - \frac{x}{v}\right)\right]$$

Therefore we have that

$$\frac{dW}{dt} = P\omega s_0 A \cos\left[\omega\left(t - \frac{x}{v}\right)\right] + \omega s_0(\Delta P)_0 A \cos^2\left[\omega\left(t - \frac{x}{v}\right)\right]$$

*Optional section.

FIGURE 27.14   The sound wave carries energy through the cross section A.

propagation direction

The power carried by the wave is the average over one cycle of this instantaneous power. Because the average of the cosine is zero, the first term is zero. However, the second term involves the average of the cosine squared, and this is $\frac{1}{2}$, as we saw in Sec. 26.1. We therefore find that

$$\text{Average power} = \tfrac{1}{2}\omega s_0 (\Delta P)_0 A = \pi \nu s_0 (\Delta P)_0 A$$

From the last section we know that $(\Delta P)_0 = 2\pi \nu s_0 \rho \nu$ and so we have that

$$\text{Average power} = 2\pi^2 \rho \nu \nu^2 s_0^2 A$$

But the intensity of the wave, I, is just the average power transported through unit area. To obtain it, we divide by A and find for the intensity of the sound wave

$$I = 2\pi^2 \rho \nu s_0^2 \nu^2 \qquad\qquad 27.31$$

Wave Equation

To find the wave equation for sound waves we shall write $F = ma$ for the mass of air between x_1 and x_2 in figure 27.15. Suppose a wave is passing through the region that was originally between x_1 and x_2. Then the force on the left end of this region is $A[P + (\Delta P)_1]$ while the retarding force at the other end is $A[P + (\Delta P)_2]$. The net force on the air in this region is therefore

$$F = A[(\Delta P)_1 - (\Delta P)_2]$$

This force will be equal to the mass of this region multiplied by the acceleration of its center of mass.

The equilibrium volume of this region is $A\,\Delta x$ and so its mass is $\rho A\,\Delta x$. Denoting the displacement of its mass center from its equilibrium position by s, the acceleration is $\partial^2 s/\partial t^2$. We can therefore write $F = ma$ for the region in the following way:

$$A[(\Delta P)_1 - (\Delta P)_2] = \rho A\,\Delta x \frac{\partial^2 s}{\partial t^2}$$

which gives

$$-\frac{\partial(\Delta P)}{\partial x} = \rho \frac{\partial^2 s}{\partial t^2}$$

where we have replaced $[(\Delta P)_2 - (\Delta P)_1]/\Delta x$ by $\partial(\Delta P)/\partial x$.

To eliminate ΔP from this relation we make use of 27.28 and the definition of the bulk modulus:

$$\frac{\Delta V}{V} = \frac{\partial s}{\partial x} \qquad \text{and} \qquad \Delta P = -B\frac{\Delta V}{V}$$

They yield

$$\Delta P = B\frac{\partial s}{\partial x} \qquad\qquad 27.32$$

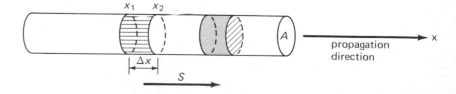

FIGURE 27.15 To obtain the wave equation, we apply $F = ma$ to the gas originally between x_1 and x_2.

Making this substitution we find that

$$\frac{\partial^2 s}{\partial x^2} = \frac{\rho}{B}\frac{\partial^2 s}{\partial t^2}$$

27.33

which is the wave equation for unidirectional sound waves.

Chapter 27
Questions
and Guesstimates

1. Distinguish clearly between transverse and longitudinal waves. Give examples of each. In what ways are both types of waves similar?
2. If you examine the strings on a violin, guitar, or piano, you will see that the strings differ in their construction depending on what tone they are to give off. Why aren't all of the strings made in the same way?
3. A long string consists of a heavy cord tied to the end of a strong thread. The two parts have the same tension. Which of the following will be the same in the two portions: speed, wavelength, frequency, period, amplitude, power of the wave?
4. How could you measure the speed of a wave along a string? A sound wave through air?
5. Our treatment for the wave on a string also applies to a wave on a chain, a string of beads, etc. The only requirement is that the discontinuities in the string, chain, beads, etc., be much smaller than the wavelength of the wave. Show where our derivation woud break down for a string of beads if λ were of the order of the bead diameter.
6. List as many cases as you can where the energy of a system is proportional to the square of the amplitude of motion. Can you think of any cases where this is not true? What characteristic feature is common to all the amplitude-squared systems?
7. Air molecules are in motion at all times. Why then do we not hear a constant background of sound in our ears from this cause?
8. As stated in the text, *27.23* predicts the speed of sound in an ideal gas to be essentially independent of pressure if the temperature remains constant. Is this not contrary to *27.22*, which states

$$v = \sqrt{\frac{\gamma P}{\rho}}$$

9. The sound from a nearby explosion can break a person's eardrum. Explain. How is this related to the fact that it is dangerous to slap someone on the ear?
10. Give a qualitative argument to show why the speed of sound should be larger in a solid metal than in air.
11. Describe clearly why a bell ringing inside a vacuum chamber cannot be heard on the outside.
12. Would you expect a sound heard under water to have the same frequency as when heard in air if the sources vibrate identically? Explain.
13. When a firecracker explodes in a large room such as a gymnasium, the sound persists for some time and then dies out. Explain what happens to the sound energy given off by the explosion.
14. The reverberation time of a room is the time taken for a sound, for example, the tone from an organ pipe, to die out after the sound source is shut off. Explain why the reverberation time is shorter when the windows are open. What other means can be used to control the reverberation time?

15. There is on the market a device which uses intense ultrasonic waves in water to wash dirt loose from cloth and other objects. Explain how this works, and list its advantages and disadvantages.
16. When making underwater investigations, it is common to use *sonar* to "see" underwater. Pulses of sound are sent out through the water, and the pulses reflected back to the sending point are observed. Explain how one can make inferences about the surroundings by such a method. In what ways is sonar similar to radar? Bats use a form of sonar for their flights through pitchblack caves. How does this work?
17. How does police radar equipment detect speeders?
18. A long line of fourth-grade children form a human chain by joining hands. They stretch out the chain as well as they can. If the children at one end of the chain suddenly start to move perpendicular to the line of the chain, about how fast will the disturbance move down the line?

Chapter 27
Problems

1. A wave along a string has the following equation in SI units:
$$y = 0.02 \sin (30t - 4.0x)$$
Find its amplitude, frequency, speed, and wavelength.

2. The equation of a compressional wave is, in SI units,
$$\Delta P = 0.001P \cos (30{,}000t - 50x)$$
where ΔP is the excess pressure in the wave and P is the pressure of the gas. Find its amplitude, frequency, speed, and wavelength.

3. For the wave shown in figure P27.1, find its amplitude, frequency, and wavelength if its speed is 300 m/s.

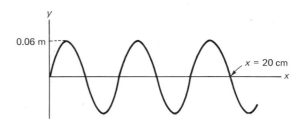

FIGURE P27.1

4. If the frequency of the wave on a string shown in figure P27.1 is 150 Hz, what is its speed? Assuming the string to have a mass of 0.20 g per meter length, how much energy is sent down the string each second?

5. A cord 15 m long hangs from a window ledge and supports a 2-kg carton of milk. The cord weighs 40 g. How long does it take a transverse pulse to travel from the window ledge to the milk carton?

6. The speed of a wave on a string is given by $v = \sqrt{T/\rho}$. Show that the right-hand side of this equation has the units of speed. Repeat for $v = \sqrt{B/\rho}$.

7. A traveling wave on a string has a frequency of 30 Hz and a wavelength of 60 cm. Its amplitude is 2 mm. Write the equation for the wave in SI units.

8. The equation of a sound wave traveling through water, expressed in SI units, is
$$s = 3 \times 10^{-4} \sin (2000t - 1.38x)$$

Find the equations for the speed and acceleration of the water molecules due to the wave. Show that a Hooke's law–type force is needed to cause this motion.

9. Using the fact that hydrogen gas consists of diatomic molecules with $M = 2\,\text{kg/kmol}$, find the speed of sound in hydrogen at $27°\text{C}$.

10. Using the result given in *27.23*, compute the speed of sound in nitrogen under STP and compare it to the value given for air in table 27.1. How can you explain this difference?

11. Two equal volumes of H_2 and N_2 gas both at STP are connected together so that the gases mix. Find the speed of sound in the gas mixture at STP. The densities of H_2 and N_2 at STP are 8.9×10^{-5} and $1.25 \times 10^{-3}\,\text{g/cm}^3$, respectively.

12. From the fact that the speed of sound in steel is about 5900 m/s, compute the bulk modulus of steel. The density of steel is $7.9\,\text{g/cm}^3$.

13. The bulk modulus of water is $2.1 \times 10^9\,\text{N/m}^2$. Use it to find the speed of sound in water.

14. What is the intensity of a 48-dB sound? What is the intensity of a sound that has twice this intensity level?

15. What is the intensity level of a sound that has an intensity of $4 \times 10^{-6}\,\text{W/m}^2$? What is the intensity level for a sound that has twice this intensity, namely $8 \times 10^{-6}\,\text{W/m}^2$?

16. What is the amplitude of motion for the air in the path of a 60-dB 800-Hz sound wave? Assume $\rho = 1.29\,\text{kg/m}^3$ and $v = 330\,\text{m/s}$.

17. What is the pressure amplitude for the situation described in the previous problem? What fraction of standard pressure is this?

18. A 500-Hz sound wave with $1 \times 10^{-4}\,\text{cm}$ amplitude passes through air under standard conditions. What is the maximum fractional increase in the pressure due to the wave?

19. Suppose the compressional vibration in a gas were isothermal in character rather than adiabatic. Find the expression equivalent to *27.22* for the speed of sound in that case.

20. Assuming CO_2 gas at STP to behave as an ideal gas, find the speed of sound in it at STP.

21. A 1000-Hz sound source is to be used to illustrate the Doppler effect. How fast must the source be moving toward the stationary observer if the frequency of the sound is to be raised by 10 percent? Take the speed of sound to be 330 m/s.

22. A loudspeaker is mounted on an oscillating table such that the speaker vibrates back and forth along the direction of the sound propagation. If the position of the source is given by $x = 0.20 \sin 2\pi ft$ in SI units, how large must the oscillation frequency be if the sound frequency is to vary by ± 5 percent?

23. A flexible steel cable of total length L and mass per unit length ρ hangs vertically from a support at one end. Show that the speed of a transverse wave down the cable is $v = \sqrt{g(L - x)}$, where x is measured from the support. How long will it take for a wave to travel down the cable? (*Hint for second part:* Since $v = dx/dt$, one knows x as a function of t.)

24. Derive the wave equation for longitudinal waves on a rod.

28 Electromagnetic Waves

There are many seemingly different types of electromagnetic waves; they include radio waves, radar, infrared rays, light, ultraviolet rays, x rays, and gamma rays among others. The connection between these various waves was first suspected by James Clerk Maxwell in 1865. At the time, radio waves were unknown, as were several of the others listed. Maxwell was investigating the equations that describe electricity, primarily Ampère's circuital law and Faraday's law. As we shall see in Sec. 28.8, these laws can be used to obtain two wave equations, one that involves the electric field and another that involves the magnetic field. You will recall from the last chapter that the coefficient in the wave equation gives the speed of the waves. Maxwell thereby was able to state that these still undiscovered waves should have a speed in vacuum equal to $1/\sqrt{\epsilon_0\mu_0}$. Astonishingly, this speed turns out to be 3×10^8 m/s, the speed of light in vacuum. Maxwell therefore inferred that light waves are electric and magnetic field waves. Let us now see how such waves can be generated.

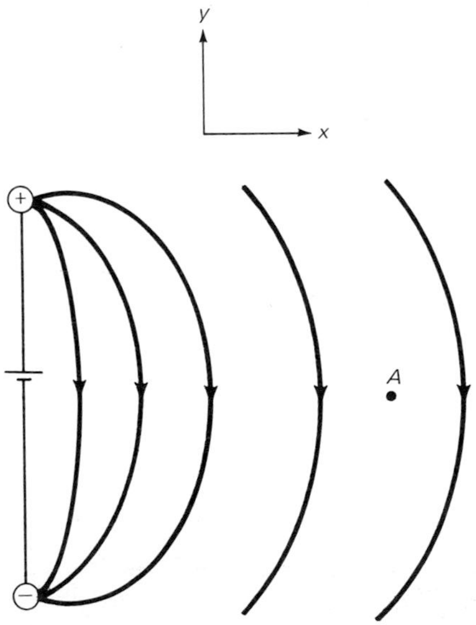

28.1 THE ELECTRIC FIELD WAVE

Radio waves, one type of electromagnetic radiation, are generated by oscillating charges on a transmitting antenna. To see how this comes about, refer to figure 28.1. We see there an electric dipole, two equal magnitude charges of opposite sign. Notice how the electric field lines along the x axis tell us of the presence of the charges. In this case the charges are supplied by a battery as indicated.

But suppose the battery is replaced by an alternating voltage source. Then the charges on the dipole will change with time. For a sinusoidal voltage, the upper charge q would vary as

$$q = q_0 \sin 2\pi ft$$

where f is the frequency of the voltage source. The direction of both the dipole and the electric field about it will reverse with a frequency f.

We can think of the electric field as being the disturbance sent out by the dipole source much like a displacement wave on a string is the disturbance sent down the string by an oscillating source. At a certain instant the electric field sent out along the x axis will be as shown in figure 28.2. The field shows the history of the charge on the dipole. The downward-directed electric fields were sent out at times when the top of the antenna was positive; the

FIGURE 28.1 A portion of the instantaneous electric field close to two charged balls. If the charges oscillate back and forth between the balls, the electric field at point A will alternately point up and down.

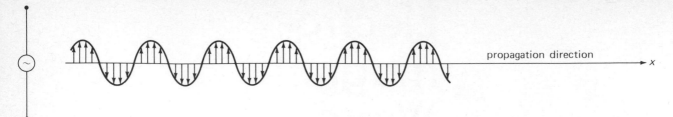

propagation direction

x

FIGURE 28.2 The alternating charges on the dipole antenna send an electric field disturbance out into space.

upward-directed fields were sent out a half-cycle later, when the top of the dipole was negative.

In the case of a radio station, the dipole (or antenna) is often simply a long wire. If you visit the transmitting site for the station, you will see the antenna as a long wire stretched between two towers or as a vertical wire held by a single tower. Charges are placed on the antenna by an ac voltage from a transformer system. The electric field wave sent out by the antenna blankets the earth around it, as shown in figure 28.3. At a point such as A in the path of the wave, the electric field reverses periodically as the wave passes through the point. The frequency of the oscillating electric field at A is the same as the frequency of the source.

Now that we see the qualitative aspects of the electric field wave, our prior knowledge of waves allows us to write the equation for this traveling wave. Denoting the electric field by E_y because it is perpendicular to the x direction, we have the following equation for the electric field wave traveling in the x direction:

$$E_y = E_{0y} \sin\left[2\pi\nu\left(t - \frac{x}{v}\right)\right] \qquad 28.1$$

In this equation v is the speed of the wave and ν is its frequency equal to the frequency of the oscillating charges that caused it. As before, we shall often write $2\pi\nu = \omega$. By writing E_y with a constant amplitude E_{0y}, we restrict ourselves to regions far from the station where the wave is nearly a plane wave so that its intensity does not change much with distance.

We should note two very important features of the electric field wave. First, it is a transverse wave because the vibrating quantity, the electric field, is directed perpendicular to the line of propagation. Second, the wave requires no material for its transmission. Unlike waves on a string (that require a string) and sound waves (that require air), the electric field and its wave can exist in vacuum. The electric field wave can move through vacuum, empty space, and it is capable of transporting energy even in the absence of all materials.

28.2 THE MAGNETIC FIELD WAVE

It is easy to see that the radio station antenna necessarily generates a magnetic field wave as it generates the electric field wave. To see this, refer to figure 28.4. At the radio station, charges are sent up and down the antenna in

FIGURE 28.3 The electric field wave from the antenna blankets the area even quite distant from the station.

E

radio station

A

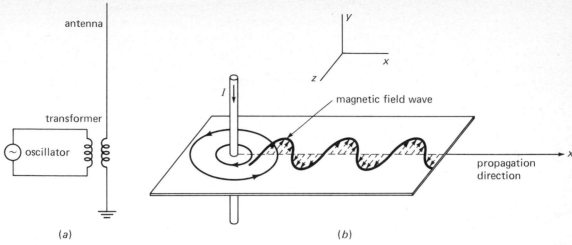

antenna

transformer

oscillator $\sim$

magnetic field wave

I

propagation direction

(a) (b)

FIGURE 28.4 As charge rushes up and down the antenna in (a), a magnetic wave is sent out as shown in (b).

(a) to produce the alternating charges we have been discussing. This charge movement constitutes an alternating current in the antenna and, because a magnetic field circles a current, an oscillating magnetic field is produced as shown in (b). As with the oscillating electric field, the magnetic field also travels out along the x axis as a transverse wave.

Notice, however, that the magnetic field is in the z direction while the electric field was in the y direction. The magnetic field (the **B** field) is perpendicular to the **E** field. We depict this situation in figure 28.5. The two waves are drawn in phase (i.e., they reach their maxima together), and we shall see later that this is true for waves far removed from the antenna. We call the combined electric and magnetic field waves an *electromagnetic wave* (an em wave). As stated previously (and we shall prove it later), em waves travel through vacuum with the speed of light, $c = 3 \times 10^8$ m/s. Their speed through air is only slightly less than this.

Illustration 28.1

Radio station WJR in Detroit operates at a frequency of 760 kHz. What is the wavelength of the radio wave it sends out?

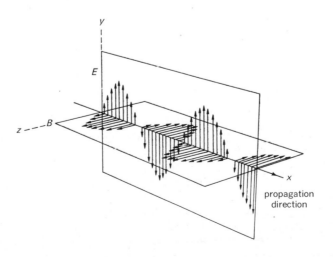

FIGURE 28.5 The oscillating antenna sends out a magnetic field wave perpendicular to the electric field wave.

Reasoning We make use of the general relation $\lambda = v/\nu$, equation 27.2. In the present case $v = 3 \times 10^8$ m/s and $\nu = 7.6 \times 10^5$ Hz. Therefore

$$\lambda = \frac{v}{\nu} = \frac{3 \times 10^8 \text{ m/s}}{7.6 \times 10^5 \text{ s}^{-1}} = 395 \text{ m}$$

Television frequencies are considerably higher and give rise to wavelengths of only a few meters.

28.3 RECEPTION OF RADIO WAVES

Television sets and radios are very sensitive electronic devices designed to detect electromagnetic waves which have wavelengths in the radio range. We cannot go into detail here about the construction of such devices; instead, we shall simply point out how they detect and tune to radio waves. There are two basic fields present in an electromagnetic wave, the E and B fields, and the wave can be detected by means of either of them.

To detect the electric field part of the wave, one need only place a long wire (called the receiving antenna) in the path of the wave. Referring to figure 28.6(a), we see that the E field will induce charges to oscillate in the antenna wire. When E_y is positive, the top of the antenna will be positive. An instant later, the antenna's polarity will reverse as the E vector due to the wave reverses. This repeated action causes current to flow up and down the antenna wire. As it does so, it induces a voltage in the LC circuit coupled to it by means of a mutual inductance. (Although the figure shows the same coil for both the mutual and self-inductance, this need not be the actual situation.)

The induced voltage in the LC circuit will cause resonance in the circuit provided the circuit is properly tuned. Let us clarify that point. Each radio and TV station is assigned its own individual frequency. The station sends out waves of that frequency only. Since the receiving antenna is buffeted by waves from many stations, some means must be found to select only the wave from the station one wishes to receive. In an ordinary radio, the selection is done by adjusting C within the resonant circuit until its resonant frequency is the same as the station's frequency. When this selection has been made, the circuit responds strongly to the station in question while effectively ignoring all others. The current and voltage variations in the resonant circuit are then amplified and modified within the radio or TV to produce finally the sound and picture presented to us by these devices.

We can also detect the electromagnetic wave by means of its oscillating B field. Since B is varying rapidly, the wave induces an emf in a loop such as the one shown in figure 28.6(b). Notice that the loop must be properly oriented so that the flux lines of B go through the loop. (It is for this reason that small radios which use this type of antenna show markedly different reception in different orientations.) The induced voltage in the loop antenna is impressed on an LC circuit. Tuning is accomplished as with the other means of detection.

One might well ask why all electromagnetic waves, light and x-rays included, cannot be detected by radio-type devices. The reason for this is quite simple. Very-high-frequency waves require LC resonant circuits which are thus far impossible to build. You will recall that the resonance frequency of a circuit is given by $1/(2\pi \sqrt{LC})$. To make the resonance frequency very high, both L and C must be very small. In the cases of infrared waves, light waves,

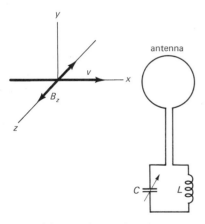

(a) electric wave detection

(b) magnetic wave detection

FIGURE 28.6 Two methods for detecting radio waves: (a) electric wave detection; (b) magnetic wave detection.

and x-rays, two tiny wires lying side by side already have L and C values which are too large. We shall see in later chapters that a circuit of atomic size would be needed to detect these waves. Indeed, we shall learn that individual atoms and molecules, in effect, become the resonant circuit for detecting these very-high-frequency electromagnetic waves.

28.4 TYPES OF ELECTROMAGNETIC WAVES

As stated previously, one can predict the speed of electromagnetic waves by use of Faraday's and Ampère's laws. This speed turns out to be the same as the speed of light, c. When this was pointed out by Maxwell in 1865, radio and similar waves had not yet been produced experimentally. As a result, even the basic existence of electromagnetic waves was unknown. But as time went on and it became possible to produce electromagnetic waves in the radio range, a large body of experimental evidence was amassed to show that heat and x radiation, as well as light, are electromagnetic waves.

The basic differences between these various types of electromagnetic waves are the result of their differing wavelengths. Since all electromagnetic radiation travels through vacuum with the speed of light, c, 27.2 tells us that

$$\lambda = \frac{c}{\nu}$$

for electromagnetic radiation. Hence a difference in λ implies a difference in the frequency ν of the radiation. We give an overall survey of the various types of electromagnetic radiations in figure 28.7. You should examine this chart carefully and be familiar with the wavelength ranges involved. The methods used to measure these wavelengths are discussed in Chap. 32. Let us now discuss briefly the nature of each of these types.

Radio Waves

We have already discussed these waves in some detail. Their wavelengths range from a meter or so to extremely large values. If one wished to obtain a wave with $\lambda = 10^8$ km, the distance from the earth to the sun, how frequently should one reverse the charges on the antenna?

Radar Waves (Microwaves)

These are merely short-wavelength radio waves. The shortest wavelength given in figure 28.7 for microwaves represents the lower limit of wavelengths one can generate electronically at present. Notice that, at a frequency of

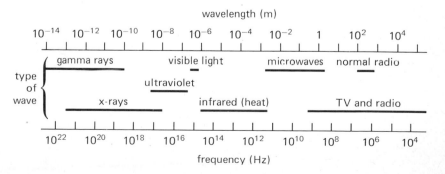

FIGURE 28.7 Types of electromagnetic radiation.

10^{12} Hz, light can travel only a distance of 3×10^{-4} m during one oscillation. Since we know that material particles and energy cannot travel faster than the speed of light, this means that we must use an antenna shorter than this length, 0.03 cm, if we are to be able to charge it during this short time. This fact should indicate why very-short-wavelength waves are difficult to produce electronically.

Infrared Radiation

Radiation in the range of wavelengths between that of visible light, 7×10^{-7} m, and radar waves is denoted by this name. This radiation is readily absorbed by most materials. The energy contained in the waves is also absorbed, of course, and appears as heat energy. In addition to the light we receive from the sun, the earth receives from the sun a large amount of this radiation as well.

Visible Light

The wavelengths of the visible portion of electromagnetic radiation extend only from about 4×10^{-7} to 7×10^{-7} m. We classify various wavelength regions in this range by the names of colors. The sensitivity of the normal human eye to wavelengths in this region is shown in figure 28.8. You should learn the approximate wavelength regions for the various colors. In addition, the following units are frequently used for small distances and wavelengths:

$$1 \text{ micrometer } (\mu\text{m}) = 10^{-6} \text{ m}$$
$$1 \text{ nanometer } (\text{nm}) = 10^{-9} \text{ m}$$
$$1 \text{ angstrom } (\text{Å}) = 10^{-10} \text{ m}$$

Ultraviolet

This is radiation with λ shorter than visible violet light but still longer than about 10 nm. At the shorter wavelengths, it is not distinct from x-rays.

x-Rays

These are electromagnetic radiation with $\lambda \lesssim 10$ nm. Usually, one reserves this classification for the radiation given off by electrons in atoms which have been bombarded. This process is discussed in more detail in Chaps. 35 and 36.

γ-Rays

These are electromagnetic radiation given off by nuclei and in nuclear reactions. They differ from x-rays only in their manner of production. We discuss γ-rays more fully in our study of nuclear physics.

The foregoing discussion completes the spectrum of electromagnetic radiation. Notice that it encompasses waves with wavelengths extending

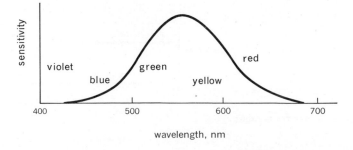

FIGURE 28.8 Sensitivity curve for the eye. Human beings are most sensitive to greenish-yellow light.

from longer than 10^6 m to those shorter than 10^{-15} m. Even though they are all electromagnetic waves, these waves differ considerably in their mode of interaction with matter. Much of the remainder of this book is concerned with various aspects of this problem.

28.5 RELATION BETWEEN E AND B IN ELECTROMAGNETIC WAVES

Now that we understand the qualitative features of electromagnetic waves, let us proceed to their more quantitative aspects. In this section we shall prove that the **E** and **B** waves are in phase. Moreover, we shall find a numerical relation between the values of **E** and **B**. To do this we shall apply Faraday's law in its integral form (see equation *24.10*)

$$-\oint \mathbf{E} \cdot d\mathbf{s} = \frac{\partial}{\partial t} \int \mathbf{B} \cdot d\mathbf{A}$$

to a particular path in an electromagnetic wave.

Consider the **E** wave shown in figure 28.9. Let us look at the small section of the wave in the *xy* plane at point *P*. This section is shown magnified in the lower part of the figure. From the fact that the magnetic field wave is in a direction perpendicular to the page, its changing magnetic field should induce an emf in the hypothetical loop *ADCBA*. We now apply Faraday's law to this loop. Then

$$-\oint \mathbf{E} \cdot d\mathbf{s} = \frac{\partial}{\partial t} \int \mathbf{B} \cdot d\mathbf{A}$$

FIGURE 28.9 We apply Faraday's law to the path *CBADC*.

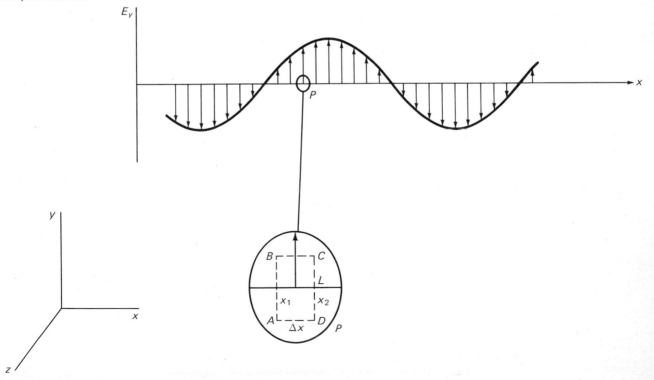

becomes

$$\int_A^D \mathbf{E} \cdot d\mathbf{s} + \int_D^C \mathbf{E} \cdot d\mathbf{s} + \int_C^B \mathbf{E} \cdot d\mathbf{s} + \int_B^A \mathbf{E} \cdot d\mathbf{s} = -\frac{\partial}{\partial t} \int \mathbf{B} \cdot d\mathbf{A}$$

We notice from the figure that $\mathbf{E}$ is in the y direction and so the integrals become

$$0 + (E_y)_2 L + 0 - (E_y)_1 L = -\frac{\partial}{\partial t} \int \mathbf{B} \cdot d\mathbf{A}$$

or

$$L[(E_y)_2 - (E_y)_1] = -\frac{\partial}{\partial t} \int \mathbf{B} \cdot d\mathbf{A}$$

where $(E)_1$ and $(E)_2$ are the values of E at x_1 and x_2 respectively.

But $\int \mathbf{B} \cdot d\mathbf{A}$ is just the flux through the area of the loop. Because we are assuming the loop to be of infinitesimal size, $\mathbf{B}$ will be nearly constant on the area and so we have that

$$\int \mathbf{B} \cdot d\mathbf{A} = B_z(L \, \Delta x)$$

We therefore find that

$$L[(E_y)_2 - (E_y)_1] = -\frac{\partial}{\partial t}[B_z(L \, \Delta x)]$$

However L and Δx do not depend on time and so they may be removed from the integral. After dividing through by $L \, \Delta x$ we have

$$\frac{(E_y)_2 - (E_y)_1}{\Delta x} = -\frac{\partial B_z}{\partial t}$$

The left-hand side of this equation is simply $\partial E_y / \partial x$ because we are letting $\Delta x \to 0$. And so we arrive at the following relation between the electric field and the magnetic field:

$$\frac{\partial E_y}{\partial x} = -\frac{\partial B_z}{\partial t} \qquad\qquad 28.2$$

In the case we are considering, where E_y is sinusoidal and is given by 28.1,

$$E_y = E_{0y} \sin\left[\omega\left(t - \frac{x}{v}\right)\right] \qquad\qquad 28.1$$

we can substitute for E_y and find that

$$-\frac{\partial B_z}{\partial t} = -(\omega/v)E_{0y} \cos\left[\omega\left(t - \frac{x}{v}\right)\right]$$

To find B_z itself, we can now integrate this with respect to t and find that[*]

$$B_z = \frac{1}{v}E_{0y} \sin\left[\omega\left(t - \frac{x}{v}\right)\right]$$

or, after using 28.1,

$$E_y = vB_z \qquad\qquad 28.3$$

[*] The constant of integration, representing a steady field, has been set equal to zero.

This is a very important relation because it tells us the following:

1. The **B** field and **E** field are in phase (they are both sine functions in this case).
2. The **B** and **E** fields are mutually perpendicular.
3. Their magnitudes are related through $E_y = vB_z$, where v is the speed of electromagnetic waves.

Illustration 28.2

Suppose an electromagnetic wave has an electric field amplitude of 10^{-5} V/m. (This would be a rather weak radio signal.) What is the amplitude of the **B** wave?

Reasoning Using equation *28.3* with $v = c = 3 \times 10^8$ m/s we find that

$$\mathbf{B}_{max} = \mathbf{E}_{max}/v = 3.3 \times 10^{-14} \text{ T}$$

Notice how very weak both the **E** and **B** fields are. It is obvious that the ordinary radio is a very sensitive detection device.

28.6 ENERGY IN ELECTROMAGNETIC WAVES

Radio waves from a radio station are able to cause a disturbance in a distant radio. To do so, they must carry energy from the station to the radio. Similarly, we receive heat energy from the sun. It is carried to us by electromagnetic waves emitted by the sun. As another example, light and x-rays cause a chemical reaction in photographic film to produce an image. In order to do this, the radiation has to carry energy into the film. Let us now find an expression for the energy transported by a plane electromagnetic wave such as the wave shown in figure 28.10.

By definition, a plane electromagnetic wave is a wave in which E and B are constant (at a given instant) on any plane perpendicular to the direction of propagation of the wave. This is illustrated in figure 28.10. We can find the energy stored in the volume occupied by this wave by use of the following facts, proved earlier.

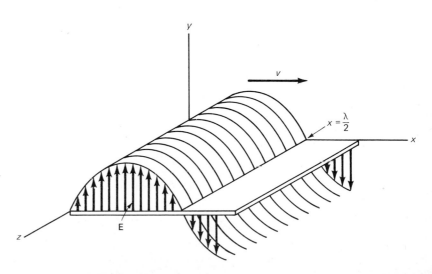

FIGURE 28.10 For a plane electromagnetic wave moving along the positive x axis, E will have the same value at all points on a plane parallel to the yz plane. For example, at the instant shown, $E = 0$ everywhere on the plane $x = \lambda/2$.

1. The energy stored in unit volume of space when the electric field is E was given by *20.24* to be $\epsilon_0 E^2/2$.
2. The energy stored in unit volume of space where the magnetic field is B was given by *24.8* to be $B^2/2\mu_0$.

Let us apply these two facts to the volume of the box illustrated in figure 28.11 in order to find the average energy per unit volume of a plane electromagnetic wave in free space.

It is assumed in figure 28.11 that a plane electromagnetic wave propagates in the x direction with speed v. At a certain instant the wave is situated as shown in the figure. The box illustrated is one wavelength long, and has cross-sectional area A. If we find the energy resident in this box, it will give us the *average* energy of the wave, since the box contains one wavelength. We can write for the wave within the box at this instant (which for convenience we take as $t = 0$)

$$E_y = E_{0y} \sin \frac{2\pi v x}{v} = E_{0y} \sin \frac{2\pi x}{\lambda}$$

$$B_z = B_{0z} \sin \frac{2\pi x}{\lambda}$$

The energy stored in a thin slab of width Δx, such as the one shown in figure 28.11(*b*), will be

$$\tfrac{1}{2}\left(\epsilon_0 E_y{}^2 + \frac{B_z{}^2}{\mu_0}\right) A\,\Delta x$$

where $A\,\Delta x$ is the volume of the slab. To find the energy stored in the box we integrate this expression from $x = 0$ to $x = \lambda$ as follows (after substituting for E_y and B_z):

$$\text{Energy in box} = \tfrac{1}{2}A \int_0^\lambda \left(\epsilon_0 E_{0y}{}^2 + \frac{B_{0z}{}^2}{\mu_0}\right) \sin^2\left(\frac{2\pi x}{\lambda}\right) dx$$

This gives

$$\text{Energy in box} = \frac{A\lambda}{4}\left(\epsilon_0 E_{0y}{}^2 + \frac{B_{0z}{}^2}{\mu_0}\right)$$

If we divide by the volume of the box, λA, we obtain the energy density in the wave. In addition, we found in 28.3 that $B_{0z} = E_{0y}/v$ for an electromagnetic wave. Moreover, we will show in Sec. 28.8 that the speed v of an electromagnetic wave is $1/\sqrt{\epsilon_0 \mu_0}$ in free space. Therefore

$$\text{Energy density} = \frac{(\lambda/4)(\epsilon_0 E_{0y}{}^2 + \epsilon_0 E_{0y}{}^2)}{\lambda}$$

from which

$$\text{Energy density} = \tfrac{1}{2}\epsilon_0 E_{0y}{}^2 \qquad\qquad 28.4$$

The average energy per unit volume of a plane electromagnetic wave is $\tfrac{1}{2}\epsilon_0 E_0{}^2$

This is the energy stored in a unit volume of a plane electromagnetic wave. Notice that this energy is stored, half and half, in the electric and magnetic fields.

It is instructive to find the intensity of the wave, the energy incident per unit time on unit area of surface perpendicular to the direction of propagation of such a wave. Since a length c (c being the velocity) of the wave will strike the surface in 1 s, the volume of the wave which strikes a unit area of the surface in unit time will be $c \times 1$. We therefore find

The intensity of electromagnetic radiation is proportional to $E_0{}^2$

$$\text{Intensity} = \frac{\text{energy incident}}{(\text{area})(\text{time})} = \tfrac{1}{2}c\epsilon_0 E_{0y}{}^2 \qquad\qquad 28.5$$

Notice that the intensity is proportional to the square of the amplitude of the wave. This proportionality between intensity and amplitude squared was

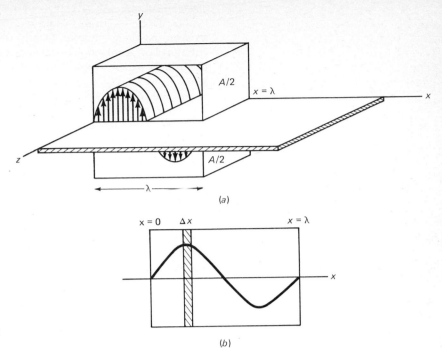

FIGURE 28.11 How much energy is stored in the electromagnetic field within the box shown in (a)?

found to be true for other waves as well as for electromagnetic waves. Equation 28.5 can be put in a somewhat more symmetrical form by placing $E_{0y} = B_{0z}/\sqrt{\epsilon_0\mu_0}$ and $c = 1/\sqrt{\epsilon_0\mu_0}$ in this relation to obtain

$$\text{Intensity} = \frac{1}{2}\frac{E_{0y}B_{0z}}{\mu_0} = \tfrac{1}{2}E_{0y}H_{0z} \qquad 28.6$$

In the more general case of any type of electric and magnetic energy flow, this quantity, which represents the flow of energy through unit area, is called the *Poynting vector* **S**. It is taken to have the same direction as the propagation direction of the energy.

Illustration 28.3

The average energy reaching the earth from the sun is about 330 cal/m² · s. Assuming this to be the result of a single plane wave (not true, of course; this is some sort of average wave), what will be the amplitude of the electric field in the wave?

Reasoning Since 330 cal = 1400 J, the energy incident per unit area per second is 1400 J. Using this in 28.5, we find

$$E_{0y} = 1000 \text{ V/m}$$

Since $B_{0z} = E_{0y}/c$, we find

$$B_{0z} = 0.33 \times 10^{-5} \text{ T}$$

28.7 MAXWELL'S EQUATIONS

We wish now to find the wave equation that applies to electromagnetic waves. Only then can we prove the correctness of our previous assertion that electromagnetic waves travel with the speed of light. Unfortunately we cannot obtain a suitable wave equation from the laws of electricity as we have stated them thus far. One of the laws we have been using, Ampère's circuital law, is incomplete and electromagnetic waves are not predicted by the form in which we know it. This was the state of knowledge in 1865 when Maxwell was carrying out his monumental work in electricity theory. He discovered the inadequacy of Ampère's law as then known. Let us see what the complete laws of electricity are and how they can be used to obtain the wave equation for electromagnetic waves.

We begin by summarizing the basic equations of electric and magnetic fields as we have learned them. They are

Gauss' Law

$$\int_{cs} \epsilon_0 \mathbf{E} \cdot d\mathbf{A} = \sum q_i$$

Gauss' Law (magnetism)

$$\int_{cs} \mathbf{B} \cdot d\mathbf{A} = 0$$

Faraday's Law

$$\oint \mathbf{E} \cdot d\mathbf{s} = -\frac{d}{dt}\int \mathbf{B} \cdot d\mathbf{A}$$

Ampère's Law

$$\oint \frac{1}{\mu_0}\mathbf{B} \cdot d\mathbf{s} = \int \mathbf{J} \cdot d\mathbf{A}$$

These four equations, as written, were found applicable to *free space*. In the presence of simple dielectrics, ϵ_0 should be replaced by ϵ in the first relation. If simple ferromagnetic materials are present, the last relation is still true if μ_0 is replaced by μ and if there are no poles of magnets in the vicinity. Because of this latter complication, we restrict the discussion which follows to nonferromagnetic materials.

You will recall that the first two of these relations, Gauss' law for electric charges and Gauss' law for magnetism, are based upon the concept of flux lines. The left side of each equation gives the net number of field lines coming out of a closed surface. Since electric field lines originate on charges, we can also count the number of lines by finding the total charge within the surface. This latter counting procedure gives rise to the right-hand side of the first equation. However, we know that magnetic field lines are due only to current loops; no isolated magnetic poles have ever been found. As a result, the same number of magnetic field lines must enter and leave any closed volume. That is why the second equation tells us that the net number of magnetic field lines leaving the closed volume is zero.

If we now look for a similar correspondence between the latter two equations, Faraday's and Ampère's laws, we note at once that at least a superficial similarity exists. The left side of each tells us to take the integral of **E** or **B** around a closed loop. Apparently the quantity on the right-hand side of the equation in each case is generating a circular E field or B field, as shown in figure 28.12. We are therefore led to seek a similarity in the right-hand sides of these two equations.

Ampère's law tells us that the B field is generated by electric charges which flow through the area encircled by the integration path since the right-hand side of that equation is simply $\int \mathbf{J} \cdot d\mathbf{A}$. It is obvious that no such term can exist in the corresponding relation, Faraday's law, since there is no such thing as an isolated magnetic pole (or "charge"). A current of magnetic poles does not exist in nature. Therefore, if we were to guess Faraday's law

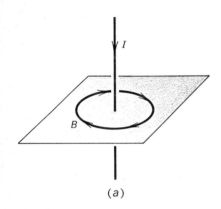

(a)

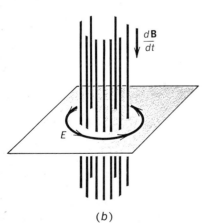

(b)

FIGURE 28.12 The change in magnetic flux corresponds to the electrical current I, and it induces a field E much in the same way that I causes the field B.

from a knowledge of Ampère's law, we would guess

$$\oint \mathbf{E} \cdot d\mathbf{s} = 0$$

But experiment shows this guess to be wrong. Our guess says electric field lines should never circle upon themselves. That is indeed the way it would be if E could only be generated by point charges since such fields are always radial. But experiment shows that E can also be generated by changing magnetic fields and that is why

$$\oint \mathbf{E} \cdot d\mathbf{s} = -\frac{d}{dt} \int \mathbf{B} \cdot d\mathbf{A}$$

An inquisitive person, having reached this point in the discussion, might ask why Ampère's law does not take the form

$$\oint \mathbf{B} \cdot d\mathbf{s} = \mu_0 \int \mathbf{J} \cdot d\mathbf{A} + (\text{const})\frac{d}{dt} \int \mathbf{E} \cdot d\mathbf{A}$$

After all, we can give a good reason why the $\oint \mathbf{E} \cdot d\mathbf{s}$ relation has no $\mathbf{J}$ term in it (there are no magnetic poles to flow). Why does not the $\int \mathbf{B} \cdot d\mathbf{s}$ equation have a d/dt term in it?

At this point we should remember that all four of these basic equations of electricity are experimental laws. We sometimes have a tendency to become so involved with the mathematics that we lose sight of a very important fact: All these mathematical relations are simply our attempts to summarize known experimental facts. Perhaps the four laws we have given are not precisely true; perhaps there are other experiments which could be done which would show that Ampère's law should contain still another term, something like the $\int \mathbf{E} \cdot d\mathbf{A}$ term we now guess should be there. It is this point which we now investigate.

The missing term we are searching for in Ampère's law is to be added to the term $\mu_0 \int \mathbf{J} \cdot d\mathbf{A}$, which is simply $\mu_0 I$. The term in question must be dimensionally correct, and so it must represent μ_0 times a current of some sort. However, by comparison with Faraday's law, we guess that the term must involve

$$\frac{d}{dt} \int \mathbf{E} \cdot d\mathbf{A}$$

The units of this term as it stands can be seen easily by noting from Coulomb's law that $\mathbf{E}$ has the form

$$\frac{1}{\epsilon_0} \times \frac{(\text{charge})}{(\text{distance})^2}$$

from which the term itself is of the form

$$\frac{d}{dt} \times \frac{1}{\epsilon_0} \times \frac{(\text{charge})}{(\text{distance})^2} \times (\text{distance})^2 = \frac{1}{\epsilon_0}\frac{d}{dt}(\text{charge})$$

Because d/dt (charge) is current, this term will take on the form of a current, as desired, if we multiply it by ϵ_0.

We are therefore led to suspect that the missing term is simply

$$\epsilon_0 \frac{d}{dt} \int \mathbf{E} \cdot d\mathbf{A}$$

Maxwell arrived at the same conclusion by a more sophisticated line of reasoning. Because this term has the units of current, it is often called the *displacement current*. (The origin of this name is a story in itself. It involves Maxwell's picture of what an electric field really was. Since his picture was wrong, even though the results he obtained were correct, we shall not try to justify the name.) With this added term, Ampère's law becomes

$$\oint \mathbf{B} \cdot d\mathbf{s} = \mu_0 \int \mathbf{J} \cdot d\mathbf{A} + \mu_0 \epsilon \frac{d}{dt} \int \mathbf{E} \cdot d\mathbf{A}$$

where ϵ_0 is replaced by ϵ so as to make the equation applicable to dielectrics as well as to vacuum. Whether or not Ampère's law as expressed above is correct can only be decided by experiment. We shall soon see that it is fully justified by experiment.

It is now appropriate to state the four equations known as Maxwell's equations. They consist of the two Gauss' law equations, Faraday's law, and the modified Ampère's law.

Maxwell's equations

$$\epsilon_0 \int_{cs} \mathbf{E} \cdot d\mathbf{A} = \sum q \qquad\qquad 28.7a$$

$$\int_{cs} \mathbf{B} \cdot d\mathbf{A} = 0 \qquad\qquad 28.7b$$

$$\oint \mathbf{E} \cdot d\mathbf{s} = -\frac{d}{dt} \int \mathbf{B} \cdot d\mathbf{A} \qquad\qquad 28.7c$$

$$\oint \mathbf{B} \cdot d\mathbf{s} = \mu_0 \int \mathbf{J} \cdot d\mathbf{A} + \mu_0 \epsilon \frac{d}{dt} \int \mathbf{E} \cdot d\mathbf{A} \qquad\qquad 28.7d$$

These four equations were first stated and used in a coherent way by Maxwell. They form the basis for all computations involving electromagnetic waves. As stated previously, these waves include radio, radar, light, and ultraviolet waves, as well as heat radiation and x-rays.

Illustration 28.4

To see why the extra term in Ampère's law had not been discovered previously, let us apply the law to the situation in figure 28.13. We see there a wire carrying a current $i = i_0 \sin 2\pi ft$ through a long straight wire that has a resistance of 400 Ω per meter of length. Find the relative magnitude of the new term.

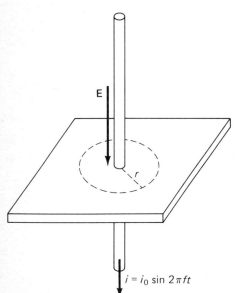

E

$i = i_0 \sin 2\pi ft$

FIGURE 28.13 Because of the *iR* drop along the wire, there is an electric field parallel to the wire.

Reasoning There will be an *iR* drop of 400*i* V/m along the wire. As a result, an electric field will exist parallel to the wire as shown and $E = 400i$ V/m. Let us apply Ampère's law to the circle of radius r shown. Then it becomes

$$2\pi rB = \mu_0 \left[i + \epsilon \frac{d}{dt} (400i \times \pi r^2) \right]$$

from which

$$2\pi rB = \mu_0 \left(i + 400\pi r^2 \epsilon \frac{di}{dt} \right)$$

Substituting the given value for the current yields

$$2\pi rB = \mu_0 \left(i + 800\pi^2 \epsilon r^2 f \cos 2\pi ft \right)$$

In a case where $r = 1$ cm and $f = 100$ Hz, this becomes

$$2\pi r B = \mu_0 \left[i_0 \sin 2\pi f t + 7 \times 10^{-10} i_0 \cos (2\pi f t) \right]$$

The second term is negligibly small in comparison to the first. Hence it is not surprising that it escaped experimental detection. However, at much higher frequencies, the term becomes sizable and is important.

28.8 WAVE EQUATION FOR ELECTROMAGNETIC WAVES

You will recall that in Sec. 28.5 we applied Faraday's law to the **E** portion of an electromagnetic wave in an effort to find a relation between **E** and **B**. The result was

$$\frac{\partial E_y}{\partial x} = - \frac{\partial B_z}{\partial t} \qquad\qquad 28.2$$

We shall now apply Ampère's law to the **B** wave to find a similar relation. By comparing the two results, we shall be able to write the wave equation.

Consider the **B** wave shown in figure 28.14. As shown in the enlargement, we take a path L long and Δx wide as our infinitesimal loop for

FIGURE 28.14 To find a relation between E_y and B_z we apply Ampère's law to the infinitesimal path shown in the insert.

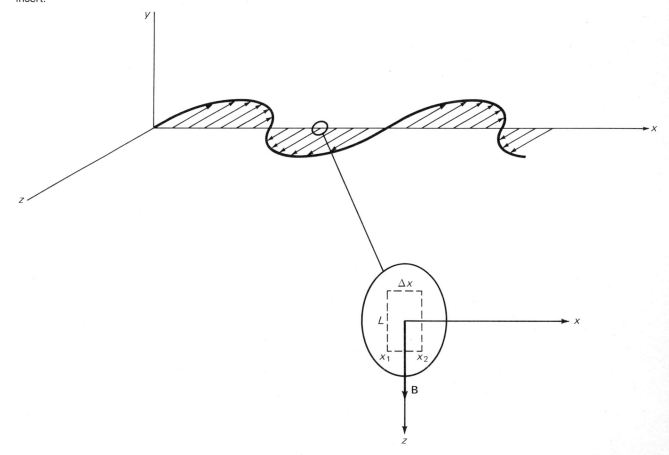

integration. Ampère's law

$$\oint \mathbf{B} \cdot d\mathbf{s} = \mu_0 \int \mathbf{J} \cdot d\mathbf{A} + \mu_0 \epsilon \frac{\partial}{\partial t} \int \mathbf{E} \cdot d\mathbf{A}$$

becomes, because $\mathbf{J} = 0$,

$$\int_{\substack{\text{bottom}}} B_x \, dx - \int_{\substack{\text{right} \\ \text{side}}} B_z \, dz - \int_{\substack{\text{top}}} B_x \, dx + \int_{\substack{\text{left} \\ \text{side}}} B_z \, dz = 0 + \mu_0 \epsilon \frac{\partial}{\partial t} \int \mathbf{E} \cdot d\mathbf{A}$$

Now $B_x = 0$ and $E = E_y$ is essentially constant on this small area $L \, \Delta x$. Therefore the equation becomes

$$-(B_z)_2 L + (B_z)_1 L = \mu_0 \epsilon \frac{\partial}{\partial t}(EL \, \Delta x)$$

Dividing through by $L \, \Delta x$ and noting that $[(B_z)_2 - (B_z)_1]/\Delta x$ is $\partial B_z/\partial x$ in our limiting case of $\Delta x \to 0$, we find that

$$\frac{\partial B_z}{\partial x} = -\mu_0 \epsilon \frac{\partial E_y}{\partial t} \qquad\qquad 28.8$$

To find an equation that involves only E_y, we try next to eliminate B_z between equations 28.2 and 28.8. This may be done by differentiating 28.2 with respect to x, and 28.8 with respect to t. The two equations then become

$$\frac{\partial^2 E_y}{\partial x^2} = -\frac{\partial^2 B_z}{\partial t \, \partial x} \qquad \text{and} \qquad \mu_0 \epsilon \frac{\partial^2 E_y}{\partial t^2} = -\frac{\partial^2 B_z}{\partial t \, \partial x}$$

After subtracting one equation from the other, we find one form of the wave equation:

$$\frac{\partial^2 E_y}{\partial x^2} = \mu_0 \epsilon \frac{\partial^2 E_y}{\partial t^2} \qquad\qquad 28.9$$

A similar equation can be found for B_z.

As we have seen previously, in 27.8 for example, the coefficient of the time derivative term is always $1/v^2$, where v is the speed of the wave. Because both the B_z and E_y wave equations contain the factor $\mu_0\epsilon$, we conclude that the speed of electromagnetic waves through a material of permittivity ϵ is given by

$$v = 1/\sqrt{\mu_0 \epsilon} \qquad\qquad 28.10$$

In the case of vacuum we have that $v = 1/\sqrt{\mu_0 \epsilon_0}$. Using the values $\mu_0 = 4\pi \times 10^{-7}$ and $\epsilon_0 = 8.85 \times 10^{-12}$, all in SI units, we find the speed of electromagnetic waves in vacuum to be 2.998×10^8 m/s, the speed of light. It was in this way that Maxwell found that his (then hypothetical) electromagnetic waves have the speed of light, c. This established the connection between electricity and the many forms of electromagnetic waves known to us.

28.9 ELECTROMAGNETIC WAVES

In the preceding section we showed that Maxwell's equations predict that the electric field wave in vacuum will travel with speed $1/\sqrt{\epsilon_0 \mu_0} = 2.998 \times 10^8$ m/s, the speed of light. If the material is a dielectric, the speed was shown to be $1/\sqrt{\epsilon\mu_0}$. Since, generally, $\epsilon > \epsilon_0$, the speed v in a dielectric is smaller than c, the speed in vacuum. We return to this latter point in our study of light waves.

Now let us consider the magnetic field associated with the wave. From 28.2 we have

$$\frac{\partial B_Z}{\partial t} = -\frac{\partial E_y}{\partial x}$$

If we substitute for E_y its value from 28.1, this becomes

$$\frac{\partial B_Z}{\partial t} = \frac{2\pi f}{v} E_{0y} \cos\left[2\pi f\left(t - \frac{x}{v}\right)\right]$$

This can be integrated with respect to t to give

$$B_Z = \frac{1}{v} E_{0y} \sin\left[2\pi f\left(t - \frac{x}{v}\right)\right]$$

where the constant of integration, representing a constant external magnetic field, has been taken to be zero. Since $1/v$ was found to be $\sqrt{\epsilon\mu_0}$, we find

$$B_Z = \sqrt{\epsilon\mu_0} E_{0y} \sin\left[2\pi f\left(t - \frac{x}{v}\right)\right] \qquad 28.11$$

Notice that the amplitude of the B wave is related to the amplitude of the E wave by the factor $\sqrt{\epsilon\mu_0}$, which is $1/v$.

The magnetic field is in phase with the electric field, and $B = E/c$

Equation 28.11 tells us that the magnetic field, perpendicular to the electric field, is in phase with it and has an amplitude $\sqrt{\epsilon\mu_0}E_{0y} = E_{0y}/v$. This is the situation depicted in figure 28.5. Now we see that the pictures are confirmed by our results. We therefore can picture a radio wave, or any other electromagnetic wave, as a combination of an electric and a magnetic field wave. These two waves are oriented as shown in figure 28.5. They travel through space and dielectrics with the speed of light, namely, $1/\sqrt{\mu_0\epsilon}$.

We should notice an additional feature of these waves. Nowhere in our use of Maxwell's equations did we postulate an antenna to send waves. It was postulated that an oscillatory electric field existed in space. This varying field and the associated displacement current, according to Ampère's law, generate an oscillating magnetic field. The oscillating magnetic field, according to Faraday's law, regenerates the oscillating electric field. It therefore appears that the electromagnetic wave regenerates itself as it moves through space. Even so, the energy carried by the wave must originate at the source of the wave.

28.10 DEFINITION OF ELECTRICAL UNITS

We are now prepared to state *precisely* the definitions of the quantities we have been using in our study of electricity. The delay has been necessitated by the fact that the velocity of light as expressed in terms of ϵ_0 and μ_0 forms the basis for these definitions. Now that we have shown that the velocity of electromagnetic radiation in vacuum, $1/\sqrt{\epsilon_0\mu_0}$, is merely c, we can make use of this fact to define values for ϵ_0 and μ_0.

The permeability of free space, μ_0, is arbitrarily defined to be $4\pi \times 10^{-7}$ N $\cdot$ s^2/C^2. Using this value, we can define ϵ_0 in terms of the measured speed of light, c. Thus, since

$$\mu_0 = 4\pi \times 10^{-7}\,\text{N} \cdot \text{s}^2/\text{C}^2$$

and

$$c^2 = 1/\epsilon_0\mu_0 = (2.998 \times 10^8\,\text{m/s})^2$$

we have

$$\epsilon_0 = 8.85 \times 10^{-12}\,\text{C}^2/\text{N} \cdot \text{m}^2$$

In order to define the unit of current, the ampere, we make use of the fact that a wire carrying a current experiences a force when placed in a magnetic field. If we consider two long parallel straight wires through which the same current I flows, the wires will experience forces because each is in the magnetic field of the other. When the separation of the wires is d, the magnetic field at one, because of the current of the other, is simply

$$B = \frac{\mu_0 I}{2\pi d}$$

as we found in 23.1. The field will be perpendicular to the second wire, and so the force on a length L of the wire will be

$$F = BIL$$

or, after substituting for B and dividing through by L,

$$\frac{F}{L} = \frac{\mu_0 I^2}{2\pi d}$$

Since the force on a unit length of the wire, F/L, can be measured, together with the separation of the wires, d, the current I can be evaluated in terms of known quantities. When F is measured in newtons and μ_0 is given the value $4\pi \times 10^{-7}\,\text{N} \cdot \text{s}^2/\text{C}^2$, the current is in the unit we define to be the ampere. Hence the ampere is defined directly in terms of force and length, both fundamental units.

The coulomb of charge is defined to be the charge carried through a cross section of a wire in 1 second when the current in the wire is 1 ampere. As a result, the definition of the coulomb is based directly upon the same measurement used to define the ampere.

To define the unit of flux density B, we make use of the relation

$$d\mathbf{F} = I\,d\mathbf{L} \times \mathbf{B}$$

If a 1-m length of wire carries a current of 1 A perpendicular to a magnetic field, and the force on that wire is 1 N, then the value of the flux density B is, by definition, 1 T. This is the same as $1\,\text{Wb/m}^2$ or $1\,\text{N} \cdot \text{s/C} \cdot \text{m}$.

Definition of the other quantities used in electricity has already been made in terms of the quantities defined above, together with force, length, and time units. We do not repeat them all here. However, it should be pointed out that we have succeeded in defining all the electrical units in terms of definite experiments involving the measurement of forces, lengths, and times. As a result, anyone who is able to duplicate our units for these three basic quantities will be able to duplicate our electrical units as well.

Chapter 28
Questions
and Guesstimates

1. What vibrates in the following types of waves: electromagnetic waves, waves on a string, sound waves, light waves, γ-ray waves, water waves?
2. From your experience at this stage, what can you say about the relative ability of the following types of electromagnetic radiation to penetrate matter, giving examples from your everyday experience and general knowledge: x-rays, radio waves, light waves, ultraviolet waves?

3. An electromagnetic wave travels along the x axis with its E vector in the y direction. It passes over a loop of wire lying in the xy plane. What effect does the wave have on this loop? Repeat if the loop lies in the xz plane. In the yz plane.

4. A piece of metal can be melted by placing it in the center of a solenoid through which a very-high-frequency current is flowing. Explain how this *induction heating* occurs. Extend your reasoning to explain why so-called *microwave ovens* are capable of cooking various materials.

5. Electromagnetic waves from most of the radio stations in the world are passing through the region around you. How does a radio or TV set select the particular station you want to listen to? When you turn the dial on a radio, what is happening inside to select the various stations?

6. There are two types of radio and TV receiving antennas in use; one picks up the electric part of the electromagnetic wave, and the other picks up the magnetic. Examine a pocket transistor radio or a table radio and see which method is used. Is it possible to use both?

7. Using a transistor radio which you know to have a coil-type antenna, determine whether the transmitting antenna of a local station is vertical or horizontal. Check you answer by looking at the station's transmitter tower(s).

8. From time to time in the movies or on TV one sees the good guys trying to locate a clandestine radio transmitter by driving through the neighborhood with a device which has a slowly rotating coil on top. Explain how the device works.

9. It is claimed that in the vicinity of a very powerful radio-transmitting antenna, one can sometimes see sparks jumping along a wire farm fence. What do you think of this claim?

10. It is sometimes stated that in an electromagnetic wave, the magnetic field generates the electric wave while the electric field generates the magnetic wave. What justification can you give for this statement?

11. What are the physical meanings of the Maxwell equations in the case of static fields without currents?

12. A cylindrical wire has a uniform resistance. Initially, no current flows in it. As the current through the wire is slowly increased, describe the displacement current which flows, if any.

13. A highly charged thundercloud induces a nearly equal and opposite charge on the earth below it. Describe the variation of the displacement current at a point just above the earth as the cloud passes swiftly overhead.

14. Does the energy density of an electromagnetic wave depend upon its frequency? What must be true if the energy density in a light wave is to be the same as that in a radio wave?

15. Reasoning from the fact that energy and momentum of a particle with speed v are related by $U = mc^2$ and $p = mv$, show that one might suspect electromagnetic radiation to have momentum given by

$$p = \frac{U}{c}$$

As we shall see in later chapters, this supposition is correct.

16. According to Maxwell's result, the speed of an electromagnetic wave depends upon the properties of the material through which it travels. We shall see in our study of light that v usually decreases as the frequency of the wave decreases. What physical property of the material appears to be

varying with frequency? In fact, the low-frequency, or dc, value of ϵ for *polar materials* is completely inappropriate for waves of very high frequency. Why should ϵ of a polar material vary with frequency?

17. A straight wire (area = 0.10 cm^2) carries an ac current (60 Hz) such that the rms voltage drop along the wire is 0.010 V/m. Estimate, to an order of magnitude, the displacement current flowing along the wire.

Chapter 28
Problems

1. A radio station sends out an electromagnetic wave at the rate of one crest every 10^{-6} s. (Its frequency is 10^6 Hz.) How far does a wave crest move out from the antenna in a time of 10^{-6} s? How far apart are the crests in the station's wave? (Assume the speed of a crest to be the speed of light in free space.)

2. When you are tuned to a radio station which is 161 km (100 mi) away, how long does it take for an electromagnetic signal to travel from the station to you? If the station operates at 10^6 Hz, how many wave crests does it send out each second? How far does a wave crest travel in a second through free space. How far apart in space are the wave crests?

3. At a certain place on the earth the electric field portion of a plane electromagnetic wave is found to vary in the following way, expressed in SI units:

$$E = 2 \times 10^{-4} \cos (5 \times 10^6\, t)$$

If the place in question is at $x = 0$ and the wave is moving in the $-x$ direction through air, write the equation of the wave in terms of x and t.

4. The electric field portion of a particular electromagnetic wave is given by the equation (all in SI)

$$E = 10^{-4} \sin (6 \times 10^5\, t - 0.01x)$$

Find the frequency of the wave and its speed. What is the maximum value of the electric field in the wave?

5. At a particular place, the electric field due to a distant radio station is given by (in SI)

$$E = 10^{-4} \cos (1 \times 10^7\, t)$$

What is the frequency of the wave? How large a potential difference would be induced by it between the ends of a 3-m-long wire oriented in the direction of the field?

6. If the electric field at a certain place due to an electromagnetic wave is given by (in SI)

$$E = 10^{-4} \cos (5 \times 10^6\, t)$$

what will be the equation for the magnetic field at that place due to the wave?

7. The amplitude of the electric field wave from a radio station should be greater than about 10^{-5} V/m if a radio is to respond well to it. What is the intensity of such a wave?

8. A certain radio is capable of detecting a radio signal of 10^{-13} W/m^2. What is the amplitude of the B field in such a wave?

9. The magnetic field of an electromagnetic wave obeys the following relation in a certain region:

$$B = 10^{-12} \sin (5 \times 10^6\, t)$$

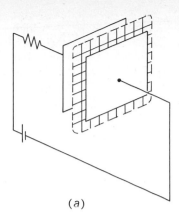

(a)

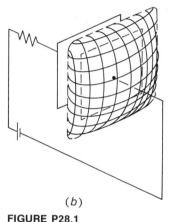

(b)

FIGURE P28.1

all in SI units. How large an emf would it induce in a 300-turn coil of 20-cm^2 area oriented perpendicular to the field?

10. Television frequencies are of the order of 10^8 Hz, while radio frequencies are of order 10^6 Hz. Using these as typical frequencies, find the ratio of the emf generated in a loop antenna by the TV wave to that generated by a radio wave if both have equal electric field intensities.

11. At the limit of detectability, the B field of a radio wave is given in SI units by the following equation:

$$B = 3 \times 10^{-14} \sin (6 \times 10^6\, t)$$

Find the emf it induces in a coil antenna with 200 loops and 0.3-cm^2 cross-sectional area wound on a core with permeability of 100.

12. The HCl molecule acts somewhat like two oppositely charged balls connected by a spring. Its natural frequency of vibration when the spring stretches back and forth is about 8×10^{13} Hz. Find the wavelength of the wave the vibrating molecule sends out. What type of wave is this?

13. Show that the two terms on the right-hand side of Ampère's law as written in equation *28.7d* have the same units.

14. Two circular plates, each of area A, placed one above the other and separated by a distance b, act as a parallel plate capacitor. Show that B between the plates is given by $\mu_0 Ir/2A$ when a current I is flowing into the capacitor. The distance r is the radial distance measured from the line of centers of the plates to the point being considered. (*Hint:* Apply Ampère's law to a loop of radius r between the plates.)

15. Maxwell showed that Ampère's law without the displacement current gave dissimilar results for the integral of **B** along the perimeter of the two shaded areas shown in figure P28.1(*a*) and (*b*), even though both perimeters were identical. (It is assumed that the capacitor is still charging.) Show that both cases give identical results for

$$\oint \mathbf{B} \cdot d\mathbf{s}$$

provided the displacement current is included.

16. Starting from Maxwell's equations, show that in a region of space where $q = 0$ the displacement current is also zero, provided no electric fields from outside enter this region of space. Why is this latter condition necessary?

17. Using a method analogous to that used in obtaining *28.9*, find the wave equation for B_z, namely,

$$\frac{\partial^2 B_z}{\partial x^2} = \mu_0 \epsilon \frac{\partial^2 B_z}{\partial t^2}$$

18. Prove that, for any point in a plane electromagnetic wave, the electric field energy stored per unit volume is equal to the magnetic field energy stored in the same volume.

19. Show that the electric and magnetic field vectors given by *28.1* and *28.11*, together with the fact that $E_x = B_x = E_z = B_y = 0$ in a plane wave, satisfy all four of Maxwell's equations. Do this by direct substitution.

29 Reflection and Refraction

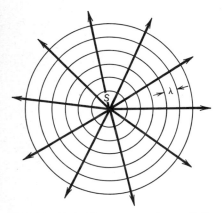

FIGURE 29.1 The rays, coming radially from the source, show the direction of energy flow. What do the circles represent?

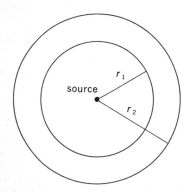

FIGURE 29.2 In a steady-state situation, the energy flowing through the sphere of radius r_1 must equal the energy flow through the sphere of radius r_2.

The intensity of a spherical wave decreases as $1/r^2$

Lenses and mirrors form an important part of optical systems. In this chapter, we see how the concept of waves can be used to describe the action of these devices. Although we discuss, primarily, the reflection and refraction of electromagnetic waves, similar reasoning may be used to describe reflection of other waves as well. Once we have mastered the subjects of reflection and refraction, we shall be able to extend our previous discussion of waves to the phenomena of diffraction and interference.

29.1 SPHERICAL AND PLANE WAVES

Visualize an ideal point source of waves S which is capable of sending out waves of wavelength λ uniformly in all directions. A cross section of the wavefront diagram for such a source is shown in figure 29.1. You will recall that the wavefronts (indicated by the circles in the figure) show the positions of the wave crests. These wave crests form concentric spherical shells about the source S, and each shell is a distance λ from its neighbor shell.

We represent the motion of the wavefronts by the rays which radiate out from the source. The rays are necessarily perpendicular to the wavefronts. They may be thought of as showing the flow of energy out from the wave source. This energy flow (through unit area perpendicular to the ray) per second is called the *intensity* of the wave. (We have already computed this quantity for a plane electromagnetic wave, in *28.5.*) It is a simple matter to see how the intensity I changes with radius r in the case of a spherical wave such as shown in figure 29.1. For this purpose we examine two spherical shells at radii r_1 and r_2, as shown in figure 29.2.

If we assume the situation to be such that a steady flow of energy has been achieved, the energy flowing through the spherical shell of radius r_1 must equal the energy flowing through the spherical shell of radius r_2. Calling I_1 and I_2 the intensities at these two places, we have

$$\text{Energy through } 1 = \text{energy through } 2$$
$$I_1(4\pi r_1^2) = I_2(4\pi r_2^2)$$

This gives at once the relation

$$\frac{I_1}{I_2} = \frac{r_2^2}{r_1^2} \qquad 29.1$$

The intensity of a spherical wave decreases as the inverse square of the radius.

A limiting case of considerable importance is obtained if the radius is very large. In that case, even though points 1 and 2 may be quite far apart, their separation will still be small in comparison to r_1 or r_2. Then, since $r_1 \approx r_2$, $I_1 \approx I_2$ and so I is nearly independent of position. Moreover, when r is very large, the rays are nearly parallel since the curvature of the spherical surface is very small. In the limit of $r \to \infty$, the wavefronts and rays will be as shown in figure 29.3. Notice that the wave surfaces are essentially flat planes. We call this type of wave a *plane wave*. Since $r \to \infty$, the intensity of such a wave changes hardly at all over any reasonably small distance.

One other point should be mentioned about the figures. They assume the source to be sending out a perfect sinusoidal-type wave. This is nearly correct for some sound waves. However, light waves and most sound waves are not necessarily of this type. We are justified, however, in our present approach by the following fact: Any wave, no matter how complex, can be represented to within any accuracy desired by a sum of pure sinusoidal waves. (That is to say, we can represent the wave by a Fourier series. We shall learn about this representation in Sec. 30.1.) As long as the wave does not change its shape as it travels through space, we can keep track of its motion by following any one of the sinusoidal waves of which it is composed.

A particular wave of importance is the *monochromatic wave*. As its name implies, it is a single-(mono) color (chromatic) wave. Or more generally, it is a wave which consists of a number of waves, all of the same wavelength but not necessarily in phase. The yellow light given off by a sodium vapor lamp is very nearly monochromatic. By sending multicolor light through light filters which absorb all except one small band of wavelengths, one can also obtain nearly monochromatic waves. In addition to these monochromatic sources, lasers are now used as extremely pure single-wavelength sources.

I is constant for a plane wave

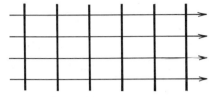

FIGURE 29.3 When the radii of spherical waves becomes very large, the wavefronts appear as flat planes in a limited region. What can one say about the rays and the intensity in such a case?

29.2 LAW OF REFLECTION

Consider a plane wave incident upon a plane reflecting surface as shown in figure 29.4(a). We call the angle $i = \theta$, the angle between the ray and the normal to the reflecting surface, the *angle of incidence*. If we assign a velocity v to the wave, it will be directed along the ray as indicated in (b). The velocity vector can be resolved into two components, one parallel to the surface and one perpendicular to the surface, as in the figure. Experiment shows (and theory can be shown to predict) that the velocity component perpendicular

FIGURE 29.4 Upon reflection, the component of v perpendicular to the surface is reversed. What relation is implied between the angle of incidence i and the angle of reflection r?

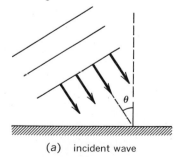

(a) incident wave

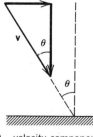

(b) velocity components

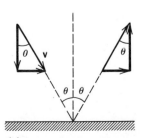

(c) the reflection process

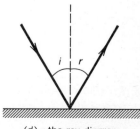

(d) the ray diagram

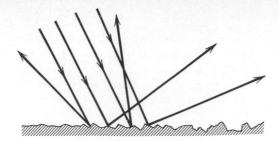

FIGURE 29.5 Even in diffuse reflection, $i = r$ for the individual rays.

to the surface is reversed upon reflection. The other component remains unchanged. This results in the situation in figure 29.4(c); the reflected wave is also a plane wave.

One can easily see that the angles labeled θ in figure 29.4(c) are all equal. As a result, we conclude that the reflected ray will have an angle of reflection r which equals the angle of incidence i, where these angles are defined in figure 29.4(d). We shall base much of our study of reflection on this simple fact: *The angle of incidence equals the angle of reflection.*

The angle of incidence equals the angle of reflection

This is always true for an individual ray even at a rough surface since, if we look at the surface with high enough magnification, it will look somewhat like the one shown in figure 29.5. As we see, a bundle of rays is reflected off in many directions from such a surface. However, as indicated, for each of the rays $i = r$ even though it does not apply to the bundle as a whole. We call such type reflection *diffuse reflection*. Although we have considered only the case of plane waves, the angles of incidence and reflection are equal for the rays in nonplanar waves as well. Can you justify this?

29.3 PLANE MIRRORS

Let us now consider the formation of an image by the reflection of light waves from a plane mirror. (Even though we shall use light waves for this particular discussion, the development will be seen to apply to other wave types as well.)

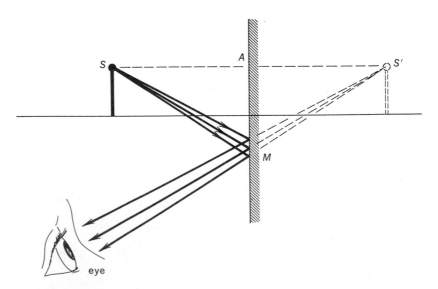

FIGURE 29.6 Using the fact that $i = r$, show that triangle SMS' is isosceles and that the image S' is as far behind the mirror as S is in the front.

When one sees an image of an object in a mirror, one is really seeing the light emitted (or reflected) from the object. A typical case is shown in figure 29.6. The rays of light coming from the source S obey the usual law of reflection at the mirror, and enter the eye of the observer. Assuming the rays to have traveled in a straight line, the observer concludes they came from S'. He therefore sees an image of S at point S'.

Since the angle of incidence equals the angle of reflection, the triangle $SMS'S$ is isosceles. Therefore $SA = S'A$, and so the image of S, namely, S', is as far behind the mirror as S is in front of it. By similar reasoning we find that the image of the entire object is as shown. We conclude from this that a plane mirror gives rise to an image the same size as the object. This type of image, one through which the observed rays do not actually pass, is called a *virtual* or *imaginary* image. In other words, the rays reaching the eye do not really come from the point where we see the image. There is no possibility whatsoever that a sheet of paper placed at S' behind the mirror would have a lighted object appear on it. The mind merely imagines that the light comes from S'. It is always true, of course, that the image of an actual object seen by reflection in a plane mirror is a virtual image. The image is always exactly as far behind the mirror as the object is in front, as demonstrated in the figure.

A virtual image is an image from which the light does not actually come

Illustration 29.1

Find the positions of the images formed by the two plane mirrors shown in figure 29.7(a). Where must one look to see each?

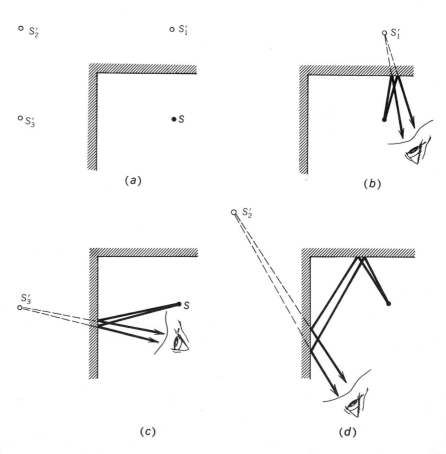

(a)

(b)

(c)

(d)

FIGURE 29.7 Must the eye be placed as shown to see the various images?

Reasoning. There are three possible images, shown as S'_1, S'_2, and S'_3. Notice that S'_1 and S'_3 are the ordinary images of S. However, in a sense at least, image S'_2 is an image of an object placed at the position of S'_1 or S'_3. We see from part (*d*) of figure 29.7 the process by which the image is formed. Are the positions shown for the eye in each case the only possible ones for seeing the image?

29.4 SPHERICAL MIRRORS: CONCAVE MIRROR

A plane mirror is a limiting form of a spherical mirror, i.e., a spherical mirror for which R, the radius of curvature, approaches infinity. We wish now to treat the general case of a spherical mirror. Two varieties are possible, concave and convex mirrors. A concave mirror is shown in figure 29.8. We shall discuss the convex mirror in due course.

In figure 29.8 we see two rays which come from the tip of the object S. Ray 1 strikes the center of the mirror, and is reflected as shown, with angles i and r equal. Ray 2, when extended, goes through C, the center of the sphere of which the mirror is a part. Since all radii are perpendicular to the surface of the sphere, ray 2 is reflected straight back on itself. The reflected rays, A and B, when viewed by eye, appear to come from point S'. Since these two, as well as all other rays, appear to come from S', one therefore sees an image of S at S'.

We wish to find an algebraic expression relating the object distance p, the image distance p', and the radius of curvature of the mirror R. These distances are all shown in figure 29.8. Calling the length of the object S and the length of the image S', we see from the shaded triangles in the figure that

$$\tan r = \frac{S'}{p'} \qquad \text{and} \qquad \tan i = \frac{S}{p}$$

But the law of reflection tells us these two angles are equal; so we can set

$$\frac{S'}{p'} = \frac{S}{p}$$

from which

$$\frac{S'}{S} = \frac{p'}{p} \qquad\qquad 29.2$$

A second relation can be obtained for S'/S by noting from the figure that

$$\tan \theta = \frac{S}{R - p} \qquad \text{and} \qquad \tan \theta = \frac{S'}{p' - R}$$

from which

$$\frac{S'}{S} = \frac{p' - R}{R - p}$$

Equating this expression for S'/S to the one given in *29.2*, and rearranging, we find

The mirror equation

$$\frac{1}{p} + \frac{1}{p'} = \frac{2}{R} \qquad\qquad 29.3$$

This is one form of the *mirror equation*. It relates the object distance p to the image distance p' and the radius of curvature of the mirror. We shall see later that a clear image is obtained only if the rays are restricted to the central portion of the mirror. When light is reflected from more than a small fraction

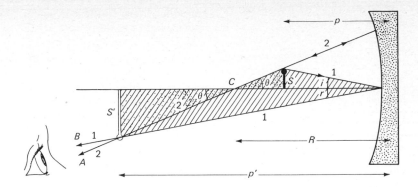

FIGURE 29.8 What is the algebraic relation between p, p', and R for a concave mirror such as this?

of a hemisphere, the image is not well defined, and *29.3* is of qualitative value only.

An interesting case arises if the object S is very far from the mirror. In that case $1/p \to 0$, and so *29.1* gives $p' = R/2$. The image of a distant object is found at a point halfway between the mirror and the center of its sphere. This point is called the *focal point,* and its distance from the mirror is designated f, the *focal length.* As we saw earlier in this chapter, if the source of light is far away, the rays of light caused by it are essentially parallel. Light rays from the sun, for example, would appear parallel, as shown in figure 29.9. They would be focused so as to form an image of the sun at the focal point of the mirror, as shown. In terms of the focal length, $f = R/2$, *29.3* becomes

A distant object is imaged at the focal point

The focal length of a spherical mirror is R/2

$$\frac{1}{p} + \frac{1}{p'} = \frac{1}{f}$$

29.4

Because of the symmetry between p and p' in *29.4*, we see that the image and object can be reversed in position. One important consequence of this is that the direction of light rays can be reversed and still represent a possible physical situation. For example, in figure 29.9, light coming from a source placed at the focal point would be reflected as parallel light (i.e., as a plane wave). This is a convenient method for obtaining plane waves (or parallel light) without resorting to a distant light source.

It is often of importance to find the relative size of the image formed by a mirror. This can be found by use of *29.2*. The ratio S'/S, the ratio of the

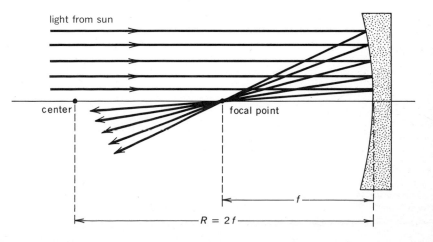

FIGURE 29.9 How can one make use of a distant light source to find the focal length and radius of curvature of a spherical mirror?

image height to the object height, is often called the *magnification* of the system and is given by 29.2 to be

$$M = \text{magnification} = \frac{S'}{S} = \frac{p'}{p} \qquad 29.5$$

Clearly, the images of distant objects will be greatly reduced in size.

A serious problem occurs if a spherical mirror is too highly curved. Rays which strike the edges of such a mirror are not properly focused. This is easily seen by reference to figure 29.10; the reflection path of the parallel rays shown there can be drawn by setting the incidence and reflection angles equal. Instead of converging to the focal point as they should, the rays reflected from the edge of the mirror do not converge to a single point, as you can see from the figure. Because of this defect, only the central portion of a spherical mirror should be used if sharp images are to be obtained with it. One can correct this defect of spherical mirrors, called *spherical aberration,* by shaping the mirror to be a paraboloid of revolution. However, even in the case of paraboloids, the image will not be sharp if the object is too far away from the axis of the mirror.

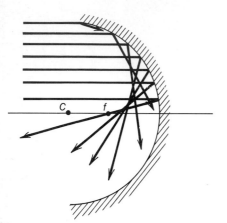

FIGURE 29.10 Rays striking a highly curved spherical mirror are not properly focused by the outer portions of the mirror. What shape should the mirror have for proper focusing?

Illustration 29.2

An image of a tree 20 m from a concave mirror ($R = 50$ cm) is formed by the mirror. Find the position and relative size of the image.

Reasoning Using the mirror equation, we have (taking all dimensions in meters)

$$\frac{1}{20} + \frac{1}{p'} = \frac{2}{0.5}$$

from which

$$p' = \frac{1}{3.95} \approx 0.253 \text{ m}$$

Notice that the image will be only 0.3 m from the focal point of the mirror.

From the magnification equation,

$$\frac{S'}{S} = \frac{0.253}{20} = 0.013$$

The image is slightly more than 1 percent as high as the object.

29.5 SPHERICAL MIRRORS: CONVEX MIRROR

A convex mirror is shown in figure 29.11. The two rays drawn from the tip of the source S are seen to come from S' when viewed from the extreme left of

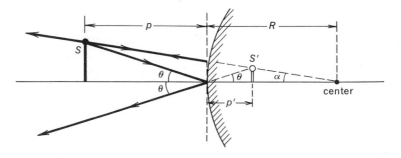

FIGURE 29.11 In order for the mirror equation 29.3 to apply to this case, both R and p' must be taken as negative.

the diagram. You should check their construction. Construction of other rays also shows that S' is an image of S. Let us now find a relation between p, p', and R for this situation.

You should be able to show that the angles labeled θ in figure 29.11 are equal. Call S and S' the object and image heights; we have from the figure that

$$\tan \theta = \frac{S}{p} \quad \text{and} \quad \tan \theta = \frac{S'}{p'}$$

from which

$$\frac{S'}{S} = \frac{p'}{p} \tag{29.6}$$

(In passing we note that this is the magnification equation, *29.5*, and so the magnification equation is the same for concave and convex mirrors.) Similarly, for the angle labeled α, one finds

$$\tan \alpha = \frac{S}{p + R} \quad \text{and} \quad \tan \alpha = \frac{S'}{R - p'}$$

so that

$$\frac{S'}{S} = \frac{R - p'}{R + p}$$

Equating this expression for S'/S to that found in *29.6*, and rearranging, we have

$$\frac{1}{p} - \frac{1}{p'} = -\frac{2}{R}$$

This equation differs from *29.3* for a concave mirror because of the two minus signs. We can write it in a form identical with *29.3* if we make the following stipulations:

Rules for the signs to be used in the mirror equation

1. Radii of convex mirrors will be taken to be negative; radii of concave mirrors will be taken to be positive.
2. When the image distance p' is behind the mirror (i.e., in the region where the light does not penetrate), p' will be taken negative for all mirrors. It will be taken positive on the lighted side of the mirror.
3. Object distances will be called positive if the object is real. (We shall later encounter a case of a virtual object, in which case the object distance will be taken as negative.)

With these stipulations, *29.3* applies to all mirrors:

$$\frac{1}{p} + \frac{1}{p'} = \frac{2}{R} = \frac{1}{f} \tag{29.3}$$

Since the focal length f is taken as $R/2$, convex mirrors will have negative focal lengths. (Notice that in this case, too, a distant object, $p \to \infty$, is imaged at the focal point.)

Illustration 29.3

Where must an object be placed with respect to a convex mirror having a 20-cm focal length if the image is to be half as far behind the mirror as the object is in front?

Reasoning Since the image is to be behind the mirror, p' must be negative. Therefore

$$p' = \frac{-p}{2}$$

In addition, since this is a convex mirror, f is negative, $-20\,\mathrm{cm}$. Using the mirror equation

$$\frac{1}{p} + \frac{1}{p'} = \frac{1}{f}$$

we find

$$\frac{1}{p} - \frac{2}{p} = -\frac{1}{20}$$

from which

$$p = 20\,\mathrm{cm}$$

29.6 RAY DIAGRAMS FOR MIRRORS

It is often convenient to sketch diagrams showing the positions and details of the images formed by mirrors. Although there are rules which tell one whether an image is real or imaginary and erect or inverted, these rules are easily forgotten. In addition, a sketch can often point out an error of sign or arithmetic in the use of the mirror equation. For these, as well as other reasons, we make use of so-called *ray diagrams* to locate images.

Once the focal point and center of curvature of a mirror are located, three rays from the object can be drawn by use of a straightedge alone. These three rays are shown in the several examples of figure 29.12.

Three simple rays for locating images

1. Ray 1 in each case leaves the object parallel to the axis of the mirror. As we know, parallel rays are reflected through the focal point (in the case of concave mirrors) or from the focal point (from convex mirrors).
2. Ray 2 leaves the object headed toward the center of curvature. It hits perpendicular to the surface of the mirror and is reflected straight back on itself.
3. Ray 3 leaves the object headed for the focal point [or from it, in case (c)]. It is essentially the reverse of ray 1, and is reflected back parallel to the axis.

You should study each of the five examples shown in figure 29.12 so that you are able to draw ray diagrams for all possible situations, in addition to those shown.

It is of interest to notice from figure 29.12 that a convex mirror always gives a virtual (imaginary) image of a real object. (Recall that an image lying behind the mirror is virtual.) In addition, the image is always upright and diminished in size. For a concave mirror, the image is always inverted and real unless the object is less than a distance f from the mirror. If $p < f$, the image is virtual and erect for a concave mirror.

Illustration 29.4

If a concave mirror has $f = 20\,\mathrm{cm}$, where must the object be placed if the image is to be virtual and twice the object size?

Reasoning From the magnification equation,

$$\frac{S'}{S} = 2 = \frac{p'}{p}$$

However, p' must be negative since the image is to be virtual, behind the mirror. The mirror equation gives

$$\frac{1}{p} + \frac{1}{-2p} = \frac{1}{20\,\mathrm{cm}}$$

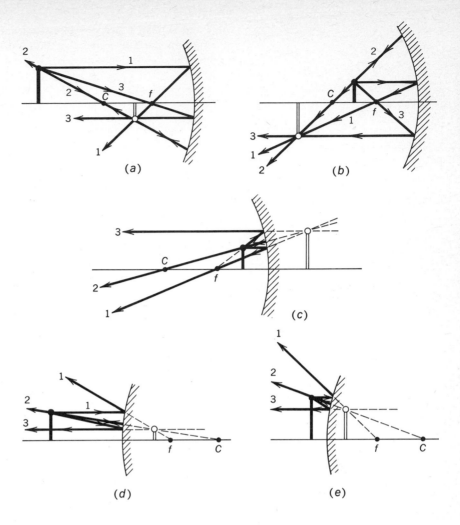

FIGURE 29.12 You should be able to draw sketches such as this for any object distance.

Solving, we find

$$p = 10 \text{ cm}$$

This answer can be checked easily by means of a ray diagram.

29.7 REFRACTION OF WAVES

Bending of a ray as it passes from one material to another is called refraction

If a beam of light passes from one material to another and if the speed of the light is different in the two materials, the light beam will be *refracted,* with a refraction angle θ_2, as shown in figure 29.13. To see why this is so, first consider a series of plane waves passing through the interface between two materials, as in figure 29.14(a). Assuming v_2, the speed in the lower material, to be less than v_1, the speed in the upper material, the wavefronts will be closer together in material 2. This must be true since (except in the formation of shock waves) the number of wave crests passing points A and B per second must be the same, and equal to the frequency f of the wave. (If this were not true, it would mean that wave crests were piling up or disappearing between

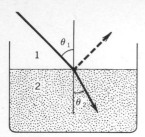

FIGURE 29.13 A light ray passing from one material to another is refracted as shown if $v_1 > v_2$. Also, part of the incident beam is reflected.

A and *B*.) Therefore the time taken for a wave crest to traverse one wavelength in material 1, namely, λ_1/v_1, must equal the similar time in the lower material, λ_2/v_2. Consequently, the ratio of the distance between wavefronts will be

$$\frac{\lambda_1}{\lambda_2} = \frac{v_1}{v_2}$$

The more general case where the angle of incidence is θ_1 is shown in part (b) of figure 29.14. As in part (a), the wavelengths in the two media are in proportion to the speeds. Further, from a consideration of the two small triangles having the common side d in figure 29.14(b), we have

$$\sin \theta_1 = \frac{\lambda_1}{d} \quad \text{and} \quad \sin \theta_2 = \frac{\lambda_2}{d}$$

After eliminating d from these two equations by simultaneous solution, one finds

$$\frac{\sin \theta_1}{\sin \theta_2} = \frac{\lambda_1}{\lambda_2}$$

Or, in terms of v_1 and v_2, this becomes

Snell's law (general form)

$$\frac{\sin \theta_1}{\sin \theta_2} = \frac{v_1}{v_2} \qquad\qquad 29.7$$

This relation is one form of *Snell's law*. In its more common form it is usually written for the case where material 1 is vacuum. In that case, $v_1 = c$, and

Snell's law (common form)

$$\frac{\sin \theta_1}{\sin \theta_2} = \frac{c}{v_2} \equiv \mu \qquad \text{(vacuum)} \qquad 29.8$$

The quantity μ is called the *absolute index of refraction* of material 2 (or more commonly, just the index of refraction of material 2). It is seen that the absolute index of refraction of a material is the ratio of the speed of light in vacuum to the speed of light in the material. Typical values for μ are listed in table 29.1. Notice that, for light, μ is always larger than or equal to unity. Of course, the index of refraction of vacuum is c/c, or unity.

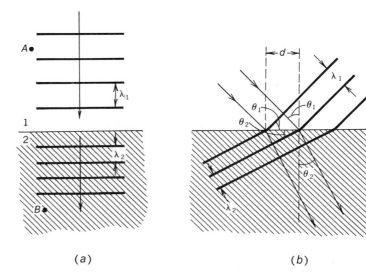

FIGURE 29.14 If the speed of the wave is less in material 2 than in material 1, the waves are refracted as shown in part (b).

(a)

(b)

A more convenient form of Snell's law is obtained by replacing $v_1 = c/\mu_1$ and $v_2 = c/\mu_2$ in 29.7. One then finds

Snell's law for two media

$$\mu_1 \sin \theta_1 = \mu_2 \sin \theta_2 \qquad 29.9$$

where μ_1 and μ_2 are the absolute refractive indices of materials 1 and 2. Sometimes the ratio μ_2/μ_1 is called the *relative* index of refraction.

Illustration 29.5

What is the effect of a thick uniform layer of substance upon a light beam transmitted through it?

Reasoning The general situation is shown in figure 29.15, where the layer in question is material 2 with refractive index μ_2. Applying Snell's law to the top interface gives

$$\sin \theta_2 = \frac{\mu_1}{\mu_2} \sin \theta_1$$

At the second interface we have

$$\sin \theta_3 = \frac{\mu_2}{\mu_3} \sin \theta_2$$

Replacing $\sin \theta_2$ by its value in terms of θ_1, we find

$$\sin \theta_3 = \frac{\mu_1}{\mu_3} \sin \theta_1$$

or
$$\mu_1 \sin \theta_1 = \mu_3 \sin \theta_3$$

This is the result we would obtain if the beam went directly from material 1 to material 3, and so the intermediate layer does not alter the angle of the emerging beam. It does displace the beam, however, as you will find in one of the problems at the end of this chapter.

TABLE 29.1 Refractive Indices

Material	$c/v = \mu$
Air (STP)	1.0003
Water	1.33
Ethanol	1.36
Fused quartz	1.46
Benzene	1.50
Crown glass	1.52
Sodium chloride	1.53
Polystyrene	1.59
Carbon disulfide	1.63
Flint glass	1.66
Methylene iodide	1.74
Diamond	2.42

29.8 TOTAL INTERNAL REFLECTION

When a beam of light is transmitted from a material with high index of refraction to one of lower index, the beam can, under certain conditions, be totally reflected. This becomes obvious upon referring to figure 29.16. At some

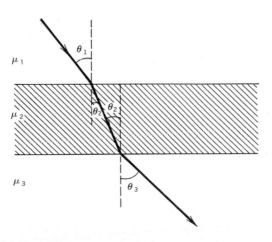

FIGURE 29.15 What is the relation between $\sin \theta_1$ and $\sin \theta_3$?

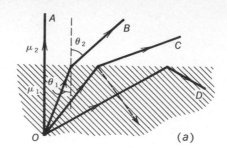

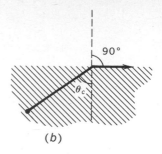

FIGURE 29.16 When $\theta_1 > \theta_c$, the beam is totally internally reflected.

(a)

(b)

critical angle of incidence θ_c, shown in part (b) of the figure, the angle of refraction is 90°. For incidence angles greater than θ_c, the beam is entirely reflected.

The critical angle is easily found by means of Snell's law. When $\theta_1 = \theta_c$, the angle of refraction is 90°. Therefore, because $\sin \theta_2 = 1$,

Total internal reflection can occur only if
$$\mu_1 > \mu_2$$

$$\sin \theta_c = \frac{\mu_2}{\mu_1} \qquad \mu_1 > \mu_2$$

Fermat's Principle

The law of reflection of light was well known in ancient times. It was first noticed by Hero of Alexandria in about A.D. 100 that the law of reflection could be derived from a general principle. He showed that when light from a point source was reflected from a mirror to a second point, the path taken by the light was the shortest possible such path between the two points. This principle of shortest path, or least path, might be stated as follows: When a beam of light from a source S is reflected by a mirror to a point P, the path taken by the light ray is the shortest possible path from S to the mirror to point P.

It was not until nearly 1600 years later that Pierre de Fermat (1601–1665) extended Hero's result to include the case of refraction. Fermat's principle of least time may be stated in the following way: A beam of light which starts from point A and goes to point B will follow the path between the two points which requires the shortest traveling time. We shall now illustrate Fermat's principle by using it to derive the laws of reflection and refraction.

Consider the beam of light shown in part (a) of the figure. We wish to find the angles α and β at which the beam is incident on and reflected from the mirror as it travels from A to B. To do this, we find the time t taken for the beam to travel the general path indicated. Calling the speed of light c, the time taken for it to go from A to B is clearly

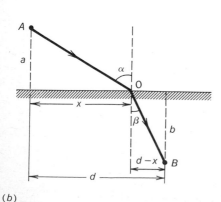

(a)

(b)

$$t = \frac{\text{distance}}{\text{speed}} = \frac{\sqrt{a^2 + x^2} + \sqrt{b^2 + (d - x)^2}}{c}$$

where the distances a, b, d, and x are as shown in the figure.

The path which requires minimum (or maximum) time can be found by the usual calculus method for locating an extremum, namely, by equating dt/dx to zero. Doing this, we find

$$\frac{x}{\sqrt{a^2 + x^2}} - \frac{d - x}{\sqrt{b^2 + (d - x)^2}} = 0$$

We see from this that if μ_1 is much larger than μ_2, θ_c becomes small. Typical values for θ_c are (assuming $\mu_2 = 1$) 49° for water, 42° for crown glass, and 24° for diamond. Because of this small value of θ_c in diamond, a beam of light can become temporarily trapped within a properly shaped diamond crystal. It is this fact which causes diamonds to sparkle, i.e., emit light in all directions. Can you explain why?

Total internal reflection makes it possible to "pipe" light around corners. By using a gently curved rod of transparent material such as glass or plastic, the light can be confined to travel along the rod. It is also found possible to use a bundle of fibers to "pipe" an image from place the place. This is particularly advantageous when one wishes to view an image formed at a rather inaccessible position.

29.9 IMAGE FORMATION BY REFRACTION

Suppose an object emits light which passes through a spherical interface into a second material, as shown in figure 29.17. We shall take the material to the

---

But from the figure we see that these two terms are simply $\sin \alpha$ and β, and so we arrive at the result

$$\sin \alpha = \sin \beta$$

In other words, the beam follows a path which makes the angle of incidence equal to the angle of reflection. This is, of course, the well-known law of reflection.

Part (b) of the figure shows the situation for a ray refracted as it travels from A to B. The speed of the ray is c/μ_1 in the upper material and c/μ_2 in the lower. From the figure we find that the time taken for the beam to travel from A to B is

$$t = \frac{\overline{AO}}{v_1} + \frac{\overline{OB}}{v_2} = \frac{\sqrt{a^2 + x^2}}{c/\mu_1} + \frac{\sqrt{b^2 + (d-x)^2}}{c/\mu_2}$$

Finding the extremum by setting $dt/dx = 0$, we obtain

$$\frac{\mu_1 x}{\sqrt{a^2 + x^2}} - \frac{\mu_2(d-x)}{\sqrt{b^2 + (d-x)^2}} = 0$$

After recognizing $\sin \alpha$ and $\sin \beta$ in this expression, we find

$$\mu_1 \sin \alpha = \mu_2 \sin \beta$$

which is Snell's law.

Fermat's principle provides a basic means for finding the path taken by light beams. In later years a similar principle was sought which would apply to the motion of particles, i.e., balls, etc. The most widely used statement of the suitable principle was found by Hamilton (1805–1865) and is called the principle of least action. He found that the proper quantity to deal with in particle motion is the product of time and a function of the energy. Both Fermat's and Hamilton's principles are possible ways of stating the basic laws of optics and particle dynamics.

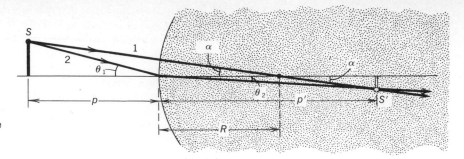

FIGURE 29.17 What assumptions are made about θ and α in obtaining the image position from 29.12?

left of the interface to have index of refraction μ_1, and the material on the right to have a refractive index $\mu_2 > \mu_1$. Two rays from the tip of the object are shown. Ray 1 is a special ray which is incident normal to the surface. Ray 2, however, is a more general ray, since the length of the object and its orientation relative to the surface need not be as shown in the figure. In any event, an image of the top of the object is formed at S' where these, as well as other, rays from the top of the object converge.

To find the position of the image we designate the image and object lengths by S' and S, respectively. As one can see from the figure

$$\tan \theta_1 = \frac{S}{p} \quad \text{and} \quad \tan \theta_2 = \frac{S'}{p'}$$

If the angles θ_1 and θ_2 are small (which will be the case if the dimensions of the portion of the interface are small compared with R), the tangents can be replaced by the sines of the angles. Then, after dividing one equation by the other, we find

$$\frac{\sin \theta_1}{\sin \theta_2} = \frac{S}{S'}\frac{p'}{p}$$

However, from Snell's law, the left side of this equation can be replaced by μ_2/μ_1 to give

$$\frac{S}{S'} = \frac{p\mu_2}{p'\mu_1} \tag{29.10}$$

We obtain another relation for S/S' by noting in figure 29.17 that

$$\sin \alpha \approx \frac{S}{p + R} \quad \text{and} \quad \sin \alpha \approx \frac{S'}{p' - R}$$

In writing this we have assumed that α is small enough so that the tangent and sine are nearly equal. After equating these two expressions for $\sin \alpha$ and rearranging, it is found that

$$\frac{S}{S'} = \frac{p + R}{p' - R} \tag{29.11}$$

Upon equating this with our preceding expression for S/S', we find, after some algebra,

Image formation by a spherical surface

$$\frac{\mu_1}{p} + \frac{\mu_2}{p'} = \frac{\mu_2 - \mu_1}{R} \tag{29.12}$$

which is the general refraction equation for a spherical interface. It assumes that the rays from the object are paraxial, i.e., not very divergent from a line through the center of curvature.

Although the present development was carried out for $\mu_2 > \mu_1$, one can carry through the development in the opposite case in the same way. Equation *29.12* is still found to apply. If the radius of curvature is such that the center of curvature is on the side from which the light comes, *29.13* is seen to apply provided R is taken to be negative. In all cases, the image distance p' is positive if the image is on the side to which the light is going; otherwise p' must be taken as negative. We therefore have the following rules for the use of *29.12*:

Sign conventions for the refraction equation

1. R is positive if the center of curvature is on the side to which the light is going; it is negative otherwise.
2. p' is positive if the image is on the side to which the light is going; it is negative otherwise.
3. p is positive if the object is on the side from which the light is coming; it is negative otherwise.

Spherical lenses and surfaces are subject to what is called *spherical aberration*. This arises if our assumption of paraxial rays is not valid. To avoid distorted and fuzzy images from this cause, only the central portion of a spherical lens should be used. For this reason one usually "stops down" (or uses only a small central portion of) a lens if well-defined images are to be obtained. In the process, less light is allowed to enter the lens. As a result, a compromise must usually be made between intensity and definition.

Illustration 29.6

As seen from straight above in the air, how far beneath the surface of a lake does a fish appear to be when it is actually at a depth d?

Reasoning The situation is shown in figure 29.18. As long as the object is viewed from above, the rays will be paraxial, and *29.12* will apply. Recognizing that $R \to \infty$ in this case and taking μ for water to be 1.33, *29.12* becomes

$$\frac{1.33}{p} + \frac{1}{p'} = 0$$

Solving for p' and recalling that $p = d$, we find $p' = -d/1.33 = -0.75d$. The image distance is negative because the image is on the side of the interface from which the light is coming.

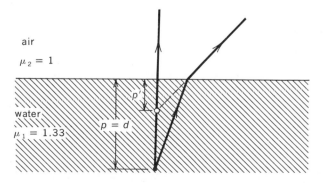

FIGURE 29.18 In order for this figure to correspond to *29.17* it should be turned 90° clockwise. Notice that the image is virtual. What will be the sign of p'?

Illustration 29.7

It is desired to produce an image of the sun at the center of a glass sphere. Can this be done if the sphere is in air?

Reasoning We wish $p \to \infty$ and $p' = R$. Placing these values in *29.12*, we find

$$\frac{\mu_2}{R} \overset{?}{=} \frac{\mu_2 - \mu_1}{R}$$

Since $\mu_1 \approx 1$, this relation cannot be satisfied. The higher the index of refraction of the material of the sphere, μ_2, the closer one could come to fulfilling this objective.

29.10 THIN LENSES

A thin lens is composed of portions of two spheres joined together to form a thin region, as shown in figure 29.19. We shall confine our attention to lenses which are thin enough so that the thickness of the lens is negligible in comparison with the radius of curvature of its surfaces. The curvatures have been made excessively large in the figure so that geometric detail may be seen more easily.

Suppose, first, that the material of the thin lens is not bounded by surface 2, but extends indefinitely to the right. From *29.12* the image formed by the first surface would be at a distance p'_a from the surface:

$$\frac{\mu_1}{p} + \frac{\mu_2}{p'_a} = \frac{\mu_2 - \mu_1}{R_1}$$

We therefore have

$$\frac{\mu_2}{p'_a} = \frac{\mu_2 - \mu_1}{R_1} - \frac{\mu_1}{p}$$

Now the second surface, of radius R_2, will use this image as its object. Notice that the object is virtual; i.e., it is on the side of the surface to which the light is going. We must therefore take the object distance to be negative in this case. Calling the image distance for this surface p' and assuming the distance of separation of the surfaces to be negligible, we have, from *29.12*,

$$\frac{\mu_2}{-p'_a} + \frac{\mu_1}{p'} = \frac{\mu_1 - \mu_2}{R_2}$$

Upon substituting the value found previously for μ_2/p'_a, we find, after some rearrangement,

$$\frac{\mu_1}{p} + \frac{\mu_1}{p'} = (\mu_1 - \mu_2)\left(\frac{1}{R_2} - \frac{1}{R_1}\right)$$

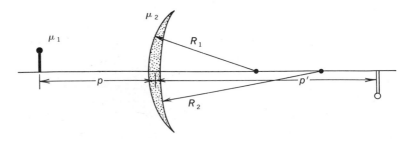

FIGURE 29.19 By definition, R_1, R_2, p, and p' are positive when the situation is as shown in this diagram.

This simplifies to give

$$\frac{1}{p} + \frac{1}{p'} = \left(\frac{\mu_2}{\mu_1} - 1\right)\left(\frac{1}{R_2} - \frac{1}{R_2}\right) \qquad 29.13$$

Let us state the sign convention to be used with this equation:

1. If the center of curvature lies on the side of the lens to which light is going, R is positive; otherwise it is negative.
2. To have positive distances, the object should lie on the side from which light is coming, and the image should lie on the side to which light is going; otherwise the distance is negative.

Notice that the positive positions are the usual positions for object and image.

An interesting case occurs if the object is infinitely far from the lens. In that case the rays are parallel. According to *29.13*, the image will appear at $1/p'$ given by

$$\left(\frac{\mu_2}{\mu_1} - 1\right)\left(\frac{1}{R_1} - \frac{1}{R_2}\right) = \frac{1}{f} \qquad 29.14$$

This equation defines what we mean by the *focal length f* of a lens. It is the image length when the object is at infinity. Equation *29.14*, which allows one to compute the focal length of a lens from its geometry, is called the *lens-maker's equation*. We see that, if either one of the factors in parentheses becomes negative, f will be negative. In such a case the lens diverges the light. By substituting f in *29.13*, we obtain the *thin-lens formula*

In the thin-lens formulas, f must be taken negative for a diverging lens

$$\frac{1}{p} + \frac{1}{p'} = \frac{1}{f} \qquad 29.15$$

It must be recalled that f is to be taken as positive for a converging lens and negative for a diverging lens.

Whether or not a lens is converging can be seen by simple inspection of the system. Consider the plane wave striking the lens system shown in figure 29.20. We note at once that if the lens has a higher index of refraction than

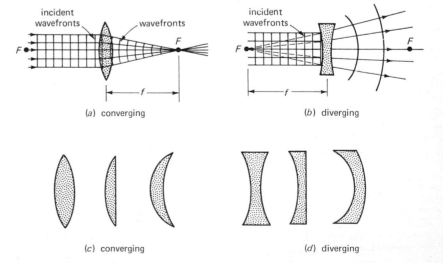

(a) converging   (b) diverging

(c) converging   (d) diverging

FIGURE 29.20 The situations shown assume μ_2, the index of refraction of the lens, to be larger than μ_1, the index of refraction of the surrounding material. In the opposite case, the lenses in (a) and (c) should be interchanged with those in (b) and (d).

the surrounding material (the usual case), the thickest portion of the lens holds the wave back most. This follows since a high index of refraction implies a low speed of light. As a result, a lens which is thickest in the middle causes the wave to converge, and a lens which is thinnest in the middle causes the wave to diverge. In the rare case, where $\mu_2 < \mu_1$, the reverse will be true. One should also notice in figure 29.20 the positions of the focal points F. Since simple thin lenses are completely reversible (i.e., it does not matter which side faces the light source), two focal points are shown for each lens.

29.11 RAY DIAGRAMS FOR LENSES

We saw, previously, that simple ray diagrams provide a quick method for checking the position and character of images formed by mirrors. This is also true in the case of lenses. In this case, too, there are three rays which are easily drawn using only a straightedge. Several typical ray diagrams are shown in figure 29.21.

Three rays are easily drawn to locate the image formed by a thin lens

1. Ray 1 in each case approaches the lens parallel to the axis. For a converging lens it is *converged to* the focal point F. For a diverging lens it is *diverged from* the focal point F.

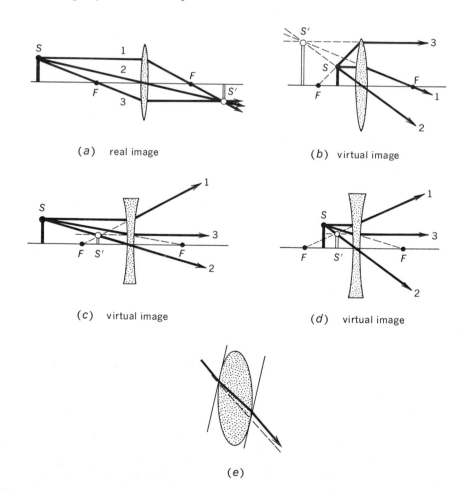

(*a*) real image

(*b*) virtual image

(*c*) virtual image

(*d*) virtual image

(*e*)

FIGURE 29.21 You should be able to trace the three rays shown for all positions of the object. Part (*e*) shows why ray 2 goes through the lens undeviated.

2. As shown in part (e), a ray through the center of the lens enters and leaves through parallel faces. As we saw in illustration 29.5, such a ray is undeviated. This is ray 2.
3. Ray 3 is the reverse of ray 1. For a converging lens it passes through F and then the lens. Upon leaving the lens it is parallel to the axis. In the case of a diverging lens, it is heading for F on the far side of the lens, and is then bent parallel to the axis.

We can easily see one other point from figure 29.21. In each case ray 2 and the axis form similar triangles based on S and S'. From these similar triangles we have at once

The magnification relation for lenses is the same as for mirrors

$$\frac{S'}{S} = \frac{p'}{p} \qquad\qquad 29.16$$

This equation for the magnification of the image is identical with the similar equation found for mirrors. We disregard the signs of the various quantities when using this relation. Only the absolute value of the ratio is important for us.

Illustration 29.8
A converging lens of 20-cm focal length is to be used to form a real image of an object. The image is to be 3 times as large as the object. Where should the object be placed?

Reasoning By reference to figure 29.21(a) and (b), we see that the object must be on the left side of F to give a real image. Therefore both p and p' are positive. From *29.16*

$$p' = 3p$$

Substituting in the thin-lens equation gives

$$\frac{1}{p} + \frac{1}{3p} = \frac{1}{20}$$

Solving for p gives

$$p = 27 \text{ cm}$$

Illustration 29.9
A diverging lens has a 20-cm focal length. Where will it image an object which is 40 cm from the lens?

Reasoning From figure 29.21(c) and (d) we see that the image will certainly be virtual and upright. Recalling that the focal length of a diverging lens is negative, we have, upon using the lens equation,

$$\frac{1}{40} + \frac{1}{p'} = \frac{1}{-20}$$

which gives

$$p' \approx -13 \text{ cm}$$

The image is 13 cm to the left of the lens.

Illustration 29.10
If a lens having $\mu = 1.50$ has a focal length of 20 cm in air, what will be its focal length when it is immersed in water?

Reasoning The focal length of a lens is given by *29.14*:

$$\frac{1}{f} = \left(\frac{\mu_2}{\mu_1} - 1\right)\left(\frac{1}{R_1} - \frac{1}{R_2}\right)$$

We can write this twice, once with $\mu_1 = 1.00$ and $f = 20$ cm, and once with $\mu_1 = 1.33$ and $f = f_w$. After dividing one equation by the other, we find

$$\frac{f_w}{20 \text{ cm}} = \frac{(\mu_2/1.00) - 1}{(\mu_2/1.33) - 1}$$

Putting in the value $\mu_2 = 1.50$ and solving for f_w gives

$$f_w - 80 \text{ cm}$$

Chapter 29
Questions
and Guesstimates

1. Consider a concave mirror and an object at infinity. Where is the image formed? Is it upright or inverted? Is it real or imaginary? Is it larger or smaller than the object? Answer these questions as the object is slowly moved in toward the mirror. In particular, note the positions where any of the answers change.

2. Repeat question 1 for the case of a convex mirror.

3. Repeat question 1 for a converging lens.

4. Repeat question 1 for a diverging lens.

5. Neither the mirror nor the lens equation applies to any but rays near the axis of the mirror or lens. Where was this approximation made in the derivation of each of these relations?

6. As we saw in this chapter, the index of refraction $\mu = c/v$. Maxwell's theory for electromagnetic waves was shown in Chap. 28 to give $k = \epsilon/\epsilon_0 = (c/v)^2$, and so the dielectric constant k should be equal to the refractive index squared for a material. This is nearly true for nonpolar liquids and glasses but is completely wrong for water, alcohol, etc. Where does the trouble lie? (*Hint:* If the dielectric constant of water, for example, is measured at very high frequencies, it becomes much smaller.)

7. Explain, using a wavefront diagram, why a lens can be either converging or diverging, depending upon the material in which it is embedded.

8. Can an empty water glass focus a beam of light? A full water glass? Is it possible to start a fire by accident if a bowl of water is set in a sunlit window?

9. A "solar furnace" can be constructed by using a concave mirror to focus the sun's rays on a small region, the furnace region. How would you expect the temperature of the furnace to vary with area of the mirror and focal length of the mirror?

10. A spherical air bubble in a piece of glass acts like a small lens. Explain. Is it converging or diverging?

11. How can one determine the focal length of a converging lens? Of a diverging lens?

12. Repeat question 11 for mirrors.

13. Two plane mirrors are placed together so that they form a right angle. An object is then placed between them. How many images are formed? Repeat for an angle between the mirrors of 30°.

14. About how much longer does it take for a pulse of light from the moon to reach the earth because of the presence of air rather than a vacuum above the earth?

15. Estimate how much the index of refraction of water changes as its temperature is increased from 0 to 100°C.

16. Newton believed light consisted of a stream of particles and that the "light corpuscles" were strongly attracted by the water surface as light went from air to water. How would this lead to the observed refraction effect? Why did the observed speed of light negate this idea?

17. In various science museums (as well as in some unexpected places), a room is so designed that a person can whisper at one particular point in the room and be heard clearly at a certain distant point. How must the room be constructed so as to achieve this effect?

Chapter 29
Problems

Note: When possible, check your answers with a ray diagram.

1. When you stand with your nose 20 cm in front of a plane mirror, for what distance must you focus your eyes in order to see your nose in the mirror? If your right eye is blue and your left is green, your image behind the plane mirror will have a green right eye and blue left eye. Explain.

2. If an object is placed between two parallel mirrors, an infinite number of images results. Suppose the mirrors are a distance $2b$ apart and the object is put at the midpoint between the mirrors. Find the distances of the images from the object.

3. A certain concave mirror is found to form an image of the sun at a distance of 8.0 cm from the mirror. What is the focal length of this mirror? Its radius of curvature? If an object is placed 24 cm from the mirror, where will its image be formed? Is the image real? Repeat for 4.0 cm.

4. An object is placed 21 cm in front of a concave mirror which has a 14-cm radius of curvature. Find the position of the image. Is it real? Repeat for object distances of 7 cm and 1.0 cm.

5. An object 2 cm high is placed at the center of a concave mirror which has a 100-cm radius of curvature ($p = 100$ cm). Find the position and size of the image. Is it real? Is it upright? Repeat for an object distance of 40 cm.

6. When a 2-cm-high object is placed at a distance of 50 cm from a concave mirror, a real image is formed at the position of the object. Find the focal length of the mirror. Is the image erect or inverted? What is the size of the image?

7. Where must the object for a concave mirror of focal length f be placed if a real image is to be formed which is 3 times larger than the object? Will the image be inverted?

8. Using a convex mirror with 100-cm radius of curvature, where will the image be formed if the object distance is 200 cm? Is the image real? Is it upright? Repeat for object distances of 50 cm and 10 cm.

9. Where must the object for a convex mirror of focal length f be placed if the image is to be half as tall as the object? Is the image real? Is it upright? Is it possible to obtain an image having the same height as the real object?

10. A converging lens images the sun at a distance of 20 cm from the lens. What is the focal length of the lens? If an object is placed 100 cm from the lens, where will its image be formed? Is it real? Upright? Repeat for object distances of 25 cm and 10 cm.

11. An object is placed 30 cm from a converging lens with 10-cm focal length. Find the position of the image. Is it real or virtual? Erect or inverted? Repeat for a 5-cm object distance.

12. Where must an object be placed in the case of a converging lens of focal length f if the image is to be real and the same size as the object? Where must it be placed so the image is virtual and 3 times as large as the object?

13. Repeat problem 11 if the lens is diverging.

14. A narrow beam of light strikes a glass plate ($\mu = 1.55$) at an angle of 37° to the normal. Find the angle the beam makes to the normal inside the glass.

15. At what angle to the vertical must a submerged fish look if it is to see a fisherman seated on the distant shore of a still pond?

16. A narrow beam of light strikes a glass plate ($\mu = 1.60$) at an angle of 53° to the normal. If the plate is 2 cm thick, what will be the lateral displacement of the beam after it emerges from the plate?

17. A fish is near the center of a spherical, water-filled fish bowl. Where would a child's nose, which is at a distance of one bowl radius from the surface of the bowl, appear to be to the fish? Call the radius of the bowl R. Where would the child see the fish to be?

18. It is desired to make a planoconvex lens (converging lens with one flat side) of focal length 20 cm from plastic for which $\mu = 1.50$. What must be the radius of curvature of the second surface?

19. In certain cases when an object and screen are separated by a distance D, two positions x of the converging lens relative to the object will give an image on the screen. Show that these two values are

$$x = \frac{D}{2}\left(1 \pm \sqrt{1 - \frac{4f}{D}}\right)$$

Under what conditions will no image be found?

20. A plastic rod with index of refraction 1.50 has one end polished flat while the other is rounded so as to have a radius of curvature of 20 cm. If a light source is placed 50 cm outside the rounded end, about how long must the rod be if the image is to appear at the same distance beyond the flat end?

21. Repeat problem 20 if both ends of the rod are rounded so as to have a 20-cm radii of curvature.

22. An object is at a depth d in water. When one views the object from the air above at an incidence angle θ, show that the apparent depth of the object is $3d \cos\theta / \sqrt{7 + 9\cos^2\theta}$.

23. A glass sphere with 10-cm radius has a 5-cm radius spherical hole at its center. A narrow beam of parallel light is directed radially into the sphere. Where, if anywhere, will the sphere produce an image? The index of refraction of the glass is 1.50.

24. A 20-cm-diameter glass sphere ($\mu = 1.50$) has a cubical hole at its center. The cube is 10 cm on each edge. A narrow parallel beam of light enters the sphere radially and perpendicular to a face of the cube. Where, if anywhere, will the sphere produce an image?

30 Interference of Waves

Two sets of waves can combine with each other to produce a resultant wave. The process by which the combined wave is produced is called *interference*. We see in this chapter that interference can either enhance or destroy the effect of the individual waves. Exactly what happens depends upon the phase relations between the two sets of waves.

30.1 SUPERPOSITION OF WAVES

We shall need in this chapter to answer the question of what happens when two or more separate waves arrive at the same point at the same time. Determining the answer to this question could be extremely difficult if it were not for the fact that most wave disturbances add linearly. It is observed experimentally that, at small and moderate amplitudes at least, the following *superposition principle* holds for all common wave phenomena: *When several wave disturbances arrive at a point simultaneously, the resultant disturbance is the vector sum of the separate disturbances.*

Many waves obey the superposition principle

As examples, we could cite the fact that when two radio stations send waves into the same region, the resultant electric field is the vector sum of the two fields. Or when two stones are dropped in a pool of water, the surface of the water displaces an amount equal to the vector sum of the individual displacements of the waves when they cross. Many similar examples could be given, and they seem almost self-evident. However, at extremely high electric fields in material bodies, superposition does not hold. Nor does it hold at very high intensities in sound waves. These nonconforming cases are difficult problems to treat and, fortunately, are seldom encountered.

One great simplification made possible by the superposition principle can be stated this way: No matter how complicated a wave disturbance may be, it can be represented as the sum of a group of sinusoidal-type waves to within experimental error. To those who are knowledgeable in mathematics, this statement is equivalent to saying that one can represent any periodic function by a Fourier series. For example, the rather queer-shaped wave shown in figure 30.1 can be thought of as being the sum of an infinite number of sine waves, as follows:

$$y = 2y_0 \left[\sin\left(\frac{2\pi x}{\lambda}\right) - \frac{1}{2}\sin\left(\frac{2 \cdot 2\pi x}{\lambda}\right) + \frac{1}{3}\sin\left(\frac{3 \cdot 2\pi x}{\lambda}\right) - \cdots \right]$$

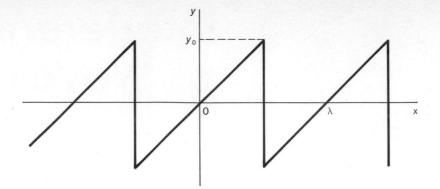

FIGURE 30.1 This wave may be
represented in the range
$-\lambda/2 < x < \lambda/2$ by the relation

$$y = y_0 \sum_{n=1}^{\infty} (-1)^{n-1} \frac{2}{n} \sin\left(2\pi n \frac{x}{\lambda}\right)$$

In other words, an electromagnetic wave whose electric field varies as shown in figure 30.1 can be treated as though it were the result of an infinite number of sinusoidal waves. The nth wave would have the form

$$2y_0 \frac{1}{n} \sin \frac{n2\pi x}{\lambda}$$

Complicated waves can be considered to be the sum of sine and cosine waves

This means that we can describe the behavior of even the most complicated wave by use of sine and cosine waves. All our discussion is concerned with this simple wave, with the understanding that a more complicated wave can be described in terms of the simple results we obtain. We usually require one thing of our complicated waves, however. They must be *coherent*. *Coherent waves have the same form, the same frequency, and a fixed-phase difference.*

Definition of coherency

In order for two waves to be *coherent,* the waves must have a definite nonrandom phase relation to each other. Two identical loudspeakers vibrating with the same frequency give off coherent waves. The waves may not be in phase if the loudspeaker diaphragms do not vibrate outward at the same time, but they are still coherent since the vibrations have a very definite, constant-phase relationship to each other. However, light waves from two sodium lamps will not be coherent. Even though they give out the same frequency waves, these waves result from the actions of a large multitude of atoms. The individual waves sent out by the atoms add to give the highly complex wave observered to come from the source. In order for these two resultant waves from the two sources to be coherent, the atoms in both sources must be acting in identical ways. This is impossible (except under very special conditions), and so the two sources cannot be coherent. In our discussion of interference, we shall require the waves to be coherent for reasons which we shall see presently.

30.2 INTERFERENCE OF SOUND WAVES

We begin our discussion of interference with a description of a sound-wave interferometer. Although this device, as we describe it, has very little practical importance, it is analogous to other more complex devices used in practical applications of sound and light. Suppose that one has a pipe system such as that shown in Figure 30.2. A pure sine wave is sent in the pipe at the left by a loudspeaker. The sound splits, half the sound intensity going up through section A, while the remaining half goes through the lower section.

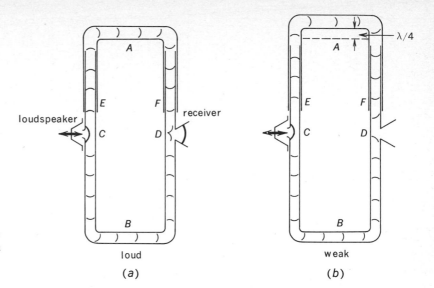

FIGURE 30.2 A sound wave from the loudspeaker is split into two parts. When they are reunited at D, either a loud or a weak sound results, depending upon the path lengths traveled by the two parts.

Each pipe carries half the sound, and this sound is a wave motion in the air, a series of compressions and rarefactions.

Eventually the two waves are reunited at the outlet on the right, at D, where one places a sound detector such as the ear or a microphone. One will observe that the sound emitted at D can be made loud or very faint, depending upon the position of the sliding pipe EAF. Moreover, as the pipe at A is slowly pulled upward, the sound intensity at D becomes alternately large and small. We shall now investigate the reasons for this interference phenomenon.

When a compression of the air is caused by a rightward movement of the loudspeaker diaphragm, a region of high pressure starts into the pipe at C. The region of high pressure, the compression, causes compressions to move in both pipes, toward A and toward B. We shall say that the original compression in the entrance pipe at C has split into two equal parts and that one part went up toward A, the other down toward B. Since the compression propagates through the pipe with the speed of sound, the compressions will reach point D simultaneously *provided that the pipe length through A, L_A, is the same as through B, L_B*. They will then reunite at D, giving the original compression, and this will then exit from the pipe system at D.

Assume the loudspeaker to be sending out a pure sound, a sinusoidal wave, consisting of alternate compressions and rarefactions. If $L_A = L_B$, the two portions of the original compression will always meet at D. The same is true for the two halves of the rarefactions. Hence, compressions and rarefactions identical to the originals will exist at point D, and the sound will be loud. This fact is represented in figure 30.3(a) where the pressure in the A wave at point D is plotted as a function of time and where this is added to the pressure in the B wave at point D to obtain the total pressure at D as a function of time.

On the other hand, suppose that the length along section A of the pipe is somewhat longer than along section B. In that case the half of the compression traveling through A will arrive at point D somewhat later than the half of the compression which traveled through B, since it had farther to travel. If L_A is half a wavelength longer than L_B, the A wave will be one-half wavelength behind the B wave when they meet at D. Hence a compression in

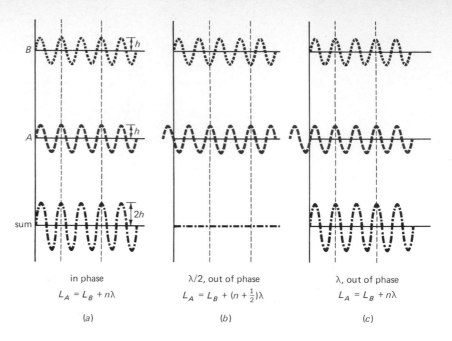

FIGURE 30.3 Identical (i.e., coherent) waves can reinforce or cancel each other, depending upon their relative phase.

in phase	$\lambda/2$, out of phase	λ, out of phase
$L_A = L_B + n\lambda$	$L_A = L_B + (n + \frac{1}{2})\lambda$	$L_A = L_B + n\lambda$
(a)	(b)	(c)

the A wave will meet the rarefaction of the B wave at point D. Since the compression will tend to increase the pressure at D and the rarefaction will tend to decrease the pressure at D, the waves will cancel and the pressure will actually remain unchanged. As a result, no sound will be emitted at D. This is illustrated in figure 30.3(b).

When one wave is half a wavelength behind another wave, we say that the waves are a half wavelength, or 180°, out of phase with each other. This terminology, 180°, is based upon the fact that a vibration is in many ways akin to a particle moving in a circle, as was pointed out in Chap. 12. One full vibration is taken as equivalent to one full rotation around the circle, i.e., to 360°. Hence, a retardation of $\frac{1}{2}\lambda$ is equivalent to 180°, $\frac{1}{4}\lambda$ is equivalent to 90°, and so on.

Two coherent waves interfere constructively if they are in phase

If L_A is now increased still further by pulling the pipe at A upward, the A wave will be held back still more. When L_A is one whole wavelength longer than L_B, a crest of the A wave will again reach D at the same time as a crest of the B wave. Although these crests did not start together at point C, the A crest occurring one compression prior to the B crest, this is of no concern when they reach point D. The wave will appear as in figure 30.3(c), and so the original intensity sound will be produced at D.

Similarly, if L_A is increased until its length is $1\frac{1}{2}$ wavelengths longer than L_B, the situation in figure 30.3(b) will arise once again. No sound will be heard at D. Moreover, it is clear that no sound will be heard at D whenever $L_A = L_B + (n + \frac{1}{2})\lambda$, where n can be any integer, including zero. When two

Destructive interference occurs when the waves are 180° out of phase

waves exactly cancel each other in this way, we say that there is complete *destructive interference*. Clearly, reinforcement, i.e., constructive interference, of the waves occurs whenever $L_A = L_B + n\lambda$, where n is any integer, including zero. Of course, if L_B is greater than L_A, these interferences will also occur.

It is not necessary to have a pipe system such as this to obtain interference. All that is needed is two coherent waves which can be brought together. For example, the two prongs of a tuning fork constitute coherent sources. The two sets of waves generated by a 1000-Hz tuning fork can easily be observed to interfere with each other; one need only slowly rotate the fork about an axis along its handle to hear alternate regions of loudness and near silence. We shall find many similar examples involving light waves. In each of these examples we will see that interference is achieved by splitting a beam into two nearly equal beams and then bringing them back together. If one of these beams has been held back through a half a wavelength, it will cancel the other beam when they are recombined. A similiar situation exists if one of the beams has been retarded by $3\lambda/2$, $5\lambda/2$, etc. But if one of the beams is retarded by λ, 2λ, 3λ, etc., with respect to the other, they will reinforce when recombined.

30.3 ADDITION OF TWO WAVES

Now that we have seen the qualitative features of interference between two waves, let us put the situation in mathematical terms. We have two waves of equal amplitude and frequency whose equations can be written

$$y_B = y_0 \sin \omega t$$
$$y_A = y_0 \sin (\omega t - \phi)$$

30.1

where ϕ is the phase difference between the two waves. In terms of path lengths, since 2π rad is equivalent to a path difference of λ, we have

To relate phase differences, one uses

$$\frac{\phi}{2\pi} = \frac{\delta}{\lambda}$$

$$\frac{\text{Phase difference}}{2\pi} = \frac{\text{path difference}}{\lambda}$$

or

$$\phi = 2\pi \frac{\delta}{\lambda}$$

30.2

where δ is the path difference between the two waves (precisely, the excess of path A over path B).

The sum of these two waves is

$$y = y_0[\sin \omega t + \sin (\omega t - \phi)]$$

This may be simplified by use of the fact that

$$\sin \theta_1 + \sin \theta_2 = 2 \sin \left(\frac{\theta_1 + \theta_2}{2}\right) \cos \left(\frac{\theta_1 - \theta_2}{2}\right)$$

and we find

$$y = 2y_0 \cos (\tfrac{1}{2}\phi) \sin (\omega t - \tfrac{1}{2}\phi)$$

30.3

When two coherent waves are added, the resultant wave has the same frequency as the original waves

This tells us at once that the resultant wave is sinusoidal with the same frequency ω. However, its amplitude is changed by a factor $2 \cos \tfrac{1}{2}\phi$.

To check this result we notice that, when the phase angle

$$\phi = 0, 2\pi, 4\pi, \ldots$$

the amplitude of the oscillation is $2y_0$. This corresponds to the result found in the preceding section for $\delta = 0, \lambda, 2\lambda$, etc., and of course *30.2* shows these two conditions to be identical. In addition, for

$$\phi = \pi, 3\pi, 5\pi, \ldots$$

$\cos \tfrac{1}{2}\phi$ is zero. This corresponds to the condition found for destructive interference in the preceding section.

We recall from our previous results that the intensity of a wave is proportional to the square of its amplitude. Making use of this fact, we can write the intensity of the resultant wave as the square of *30.3*; i.e.,

$$I = 4y_0^2 \cos^2 (\tfrac{1}{2}\phi) \sin^2 (\omega t - \tfrac{1}{2}\phi)$$

As we should expect, the intensity of the wave disturbance shows the same general behavior as the amplitude. In practice, the oscillations in time are so rapid that most recording instruments measure the average intensity. Recalling that the average of $\sin^2 \theta$ is $\tfrac{1}{2}$, we have

$$I_{av} = I_0 \cos^2 \tfrac{1}{2}\phi \qquad\qquad 30.4$$

where we have written $2y_0^2$ as I_0. As we see, the intensity is zero for $\phi = \pi$, 3π, 5π, etc., and this corresponds to $\delta = \tfrac{1}{2}\lambda$, $\tfrac{3}{2}\lambda$, $\tfrac{5}{2}\lambda$, etc.

30.4 YOUNG'S DOUBLE-SLIT EXPERIMENT

One of the first definitive experiments to show that light is a wave phenomenon was reported in 1803 by Thomas Young. In his earliest experiments he obtained two coherent beams of light by allowing the light from a single source (a hole in a window shutter upon which sunlight was shining) to fall upon two very narrow, close-together slits in a playing card. The situation is much like that in figure 30.4. By analogy with the behavior of water waves striking a hole in a barrier, it was believed that the slits acted as new light sources and sent out light at all angles. (This general idea was first proposed by Huygens, and is called *Huygens' principle,* which is used more extensively later.) In any event, Young pictured the light to radiate from the two sources as shown.

From our foregoing discussion we should expect interference to occur between the waves from the two slits in the region to the right of the slits in figure 30.4. To observe this effect, one lets the light from the slits fall upon a screen, as shown diagrammatically in figure 30.5. (Actual dimensions might be $D = 100$ cm, $d = 0.02$ cm, and distance between fringes = 0.2 cm. The figure is definitely not to scale.) When this is done, a series of dark and bright fringes is seen on the screen. A typical pattern is shown in figure 30.6. Let us now interpret this result in terms of our knowledge of the interference between waves.

The essential geometry of the double-slit system is shown in figure 30.7. At the point P indicated on the screen, a distance x from the center, the light from slit 1 has traveled a distance δ farther than that from slit 2. From the discussion of the preceding section, point P will be a point of high light intensity if $\delta = 0$, 1λ, 2λ, 3λ, ... In those cases the light waves from the two slits will reach P in phase so as to give constructive interference. However, if $\delta = \lambda/2$, $3\lambda/2$, $5\lambda/2$, ..., point P will be a dark spot since complete destruc-

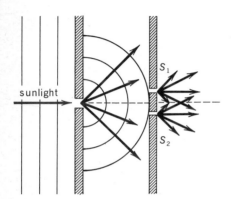

FIGURE 30.4 According to Huygens' principle, the two slits S_1 and S_2 act as coherent sources of light, and we expect interference to occur in the region to the right of the slits.

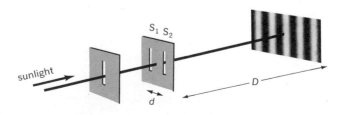

FIGURE 30.5 The two slits S_1 and S_2 act as coherent sources of light which give rise to the interference fringes on the screen. (See text for typical dimensions.)

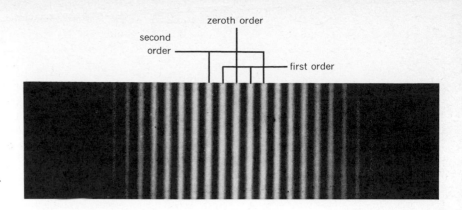

zeroth order

second order

first order

FIGURE 30.6 Interference fringes produced by a double-slit system. (After Jenkins and White.)

tive interference will occur. This is the reason for the alternate bright and dark fringes observed on the screen.

To obtain a mathematical description of the fringe locations, we note that, since in practice $D \gg x$, D is essentially equal to the hypotenuse of the large triangle in figure 30.7. Then, from the two similar (shaded) triangles, we can write

Maxima occur when $\delta = n\lambda$

$$\frac{\delta}{d} = \frac{x}{D} \qquad\qquad 30.5$$

We can use this equation to locate the bright and dark fringes if we recall

$$\delta = 0, \lambda, 2\lambda, 3\lambda, \ldots \qquad \text{bright fringes}$$

$$\delta = \tfrac{1}{2}\lambda, 3(\tfrac{1}{2}\lambda), 5(\tfrac{1}{2}\lambda), \ldots \qquad \text{dark fringes}$$

Minima occur when $\delta = (n + \tfrac{1}{2})\lambda$

When $\delta = 0$, 30.5 gives the location of the central (or *zeroth-order*) bright fringe. For $\delta = \lambda$, we find the position of the *first-order* bright fringe, and so on. This terminology is shown in figure 30.6.

The intensity of light in the fringe system should be given by *30.4*, where we need to recall from *30.2* that $\phi = 2\pi\delta/\lambda$. Replacing δ by its value from *30.5*, we find

$$I_{\text{av}} = I_0 \cos^2 \frac{\pi x d}{\lambda D} \qquad\qquad 30.6$$

As seen from figure 30.6, the intensity of the successive fringes is not constant, and so this expression is only qualitatively correct. That is to say, experiment shows that I_0 is a slowly varying function of x. However, it does give the maxima and minima positions quite accurately. The maxima occur when the argument of the cosine is $n\pi$ and so

$$\frac{\pi x d}{\lambda D} = n\pi \qquad \text{or} \qquad x = n\frac{\lambda D}{d}$$

FIGURE 30.7 Since in practice $D \gg x$, we can write *30.5* for the shaded similar triangles.

at the maxima. Similarly, at the darkest positions

$$\frac{\pi x d}{\lambda D} = (n + \tfrac{1}{2})\pi \qquad \text{or} \qquad x = (n + \tfrac{1}{2})\frac{\lambda D}{d}$$

at the minima.

Illustration 30.1

In a particular double-slit experiment, yellow light from a sodium arc is used in place of sunlight. If $D = 100$ cm, $x = 0.50$ cm for the second-order bright fringe, and $d = 0.023$ cm, find the wavelength of the sodium light.

Reasoning At the position of the second-order bright fringe, $\delta = 2\lambda$. Making use of *30.5*, we find

$$\frac{2\lambda}{d} = \frac{x}{D}$$

from which $\qquad \lambda = 5.8 \times 10^{-5}\,\text{cm} = 580\,\text{nm} = 5800\,\text{Å}$

This is close to the accepted value, 589 nm. Young was the first to obtain an accurate measure of the wavelengths of light. Of course, since he used sunlight, his fringes were colored and very broad. Should the portion of a fringe closest to the center be blue or red?

Illustration 30.2

If the mercury blue line ($\lambda = 436\,\text{nm}$) and the mercury green line ($\lambda = 546\,\text{nm}$) are used in the arrangement of illustration 30.1, what will be the separation of the blue and green first-order fringes?

Reasoning In both cases $\delta = \lambda$. Using the two values of δ in *30.5* to compute two values for x, we find

$$x_{\text{blue}} = 0.189\,\text{cm} \qquad x_{\text{green}} = 0.237\,\text{cm}$$

which gives a separation of 0.048 cm.

30.5 PHASOR ADDITION OF WAVES

We saw in Sec. 30.2 how the disturbance resulting from two waves can be added by a straightforward mathematical approach. Although this method is quite satisfactory for the addition of two waves, it becomes rather cumbersome for the addition of several waves. We shall have occasion to add the waves from a large number of sources in this and the next chapter. For that reason we now investigate a semigraphical method for the addition of waves.

Consider a wave disturbance represented by an equation of form

$$y_1 = y_0 \sin \omega t$$

We can easily represent y by a rotating vector, which we call a *phasor*. The representation of y is shown in figure 30.8. Notice that the vector of magnitude y_0 (the phasor) makes an angle ωt with the axis. As time goes on, t will increase, and the tip of the phasor will rotate counterclockwise around the circle, with angular speed ω. Its vertical projection, labeled y_1 in the figure, is equal to the wave disturbance we wish to represent. As the phasor rotates, y_1 represents the magnitude of the wave disturbance at any time. We therefore have a graphical means for representing the wave disturbance.

Suppose we have a second wave disturbance given by

$$y_2 = y_0 \sin (\omega t + \phi)$$

We can represent it by a similar diagram, with ωt replaced by $\omega t + \phi$. These two phasors are represented in figure 30.9(a). The resultant disturbance due to y_1 and y_2 is

$$y = y_1 + y_2$$

By examination of figure 30.9(b) we see that $y_1 + y_2$ is merely the y component of a vector of length R drawn as shown. It is the vector sum of the two phasors used to represent y_1 and y_2. Furthermore, as y_1 and y_2 rotate, so too will vector R. Since the angle ϕ must be preserved and ωt increases uniformly

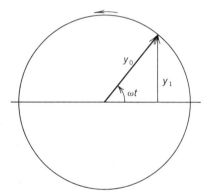

FIGURE 30.8 As *t* increases, the phasor of length y_0 rotates counterclockwise on the circle. Hence, y_1 varies as $y_0 \sin \omega t$.

The y component of the rotating phasor represents the sinusoidal disturbance

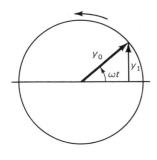

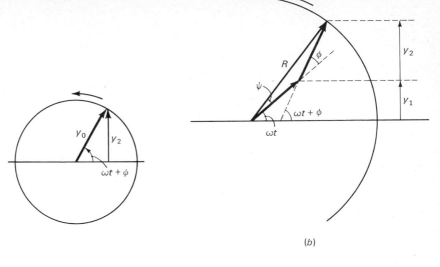

(a)

(b)

FIGURE 30.9 R is the resultant of the two phasors shown in part (a). It also rotates with angular speed ω, and its y component is the sum of the two disturbances Y_1 and Y_2.

with time, the three vectors will rotate as a rigid unit, with the tip of R traversing the circle counterclockwise with speed ω. The rotating vector R has a vertical projection equal to $y_1 + y_2$, and so it can be used to represent this sum.

We therefore conclude that

$$y = y_1 + y_2 = R \sin(\omega t + \psi)$$

where ψ is the angle which the R phasor makes with the y_1 phasor. It is the phase angle of the resultant disturbrance. Since R is the amplitude of the disturbance, the resultant intensity of the combined two disturbances is proportional to R^2. It is seen, then, that the resultant of two phasors is the amplitude of the sum of the disturbances represented by the phasors. The phase angle of the resultant disturbance is the angle which the resultant phasor makes with the y_1 phasor. Moreover, the resultant disturbance is sinusoidal, and has the same frequency as the component disturbances.

The reasoning used for the addition of two waves can be extended to several waves provided they have identical frequencies. In figure 30.10, we show the sum of the following waves:

$$y = y_0[\sin(\omega t) + \sin(\omega t + \phi) + \sin(\omega t + 2\phi)$$
$$+ \sin(\omega t + 3\phi) + \sin(\omega t + 4\phi)]$$

where ϕ is the phase-angle difference between each wave. We can write the result

$$y = R \sin(\omega t + \psi)$$

where ψ is the phase angle of the resultant wave. This type of representation can easily be extended to a large number of waves, and we shall have occasion to do so in the next chapter. For now, however, we will only make qualitative use of the phasor method.

In order to obtain some practice with the phasor method, we shall apply it to the double-slit situation. As we have seen, the phase angle is given by *30.2*:

$$\phi = 2\pi\frac{\delta}{\lambda} \qquad\qquad 30.2$$

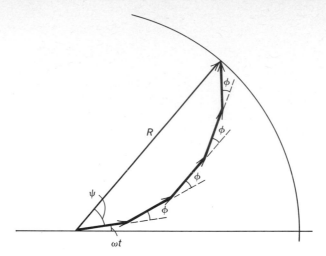

FIGURE 30.10 The amplitude of the resultant of the five waves shown is R. Its phase angle is ψ.

In terms of the position on the screen, x, this becomes, after use of *30.5*,

$$\phi = 2\pi \frac{d}{D} \frac{x}{\lambda}$$

Let us now work directly from a consideration of the two phasors which represent the disturbances from the two slits. The phase angle ϕ between them is simply $2\pi\delta/\lambda$, which is the same as $2\pi xd/\lambda D$. We can therefore think of ϕ as being proportional to displacement along the screen, x. Typical phasor diagrams for various values of ϕ are shown in figure 30.11. The results for values of ϕ larger than 2π are obtained by returning to part (a) of the figure and adding λ to each value of δ given there. For example, part (b) would then represent $\phi = 360 + 45°$ and $\delta = (1 + \frac{1}{8})\lambda$. Notice that the intensity (and R) will be zero at $\delta = \lambda/2, \lambda + \lambda/2$, etc., while R will be maximum at $\delta = 0, \lambda, 2\lambda$, etc., as we found previously. The values found for R by use of the phasor diagrams can then be used to find the light-amplitude variation as shown in the graph.

Illustration 30.3

Suppose the disturbances shown in figure 30.9(a) are given by

$$y_1 = y_0 \sin \omega t$$

and

$$y_2 = y_0 \sin \left(\omega t + \frac{\pi}{4} \right)$$

Find the resultant disturbance by use of the phasor method.

Reasoning We wish to find R in figure 30.9(b). Let us imagine the figure redrawn for the instant $\omega t = 0$. Then the x and y components of R will be, at $t = 0$,

$$R_x = y_0 + y_0 \cos \frac{\pi}{4} = y_0(1.707)$$

$$R_y = 0 + y_0 \sin \frac{\pi}{4} = y_0(0.707)$$

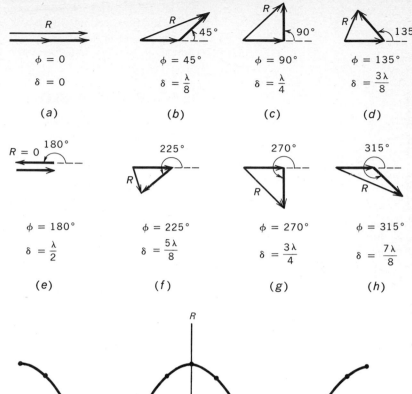

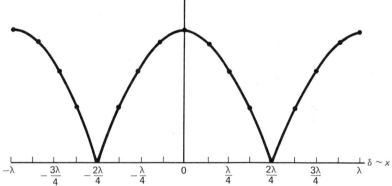

FIGURE 30.11 Phasor diagrams for a double-slit pattern. Notice that R becomes maximum at $\delta = 0, \lambda, 2\lambda, \ldots$, while $R = 0$ at $= \lambda/2, (\lambda + \lambda/2), \ldots$. The values for R found by use of the phasor diagrams are plotted in the graph.

Therefore

$$|R| = \sqrt{R_x{}^2 + R_y{}^2} = 1.85y_0$$

The phase angle ψ will be given by

$$\tan \psi = \frac{R_y}{R_x}$$

and

$$\psi = 22.5°$$

Therefore

$$R_y = 1.85y_0 \sin(\omega t + 22.5°)$$

This result can be checked by use of the algebraic method summarized by *30.3*. Recognizing that $\phi = -45°$ in this case gives

$$R_y = 2y_0 \cos(22.5°) \sin(\omega t + 22.5°)$$

But cos (22.5°) is 0.924, and so the answer checks. One advantage of the phasor method used here is that it can be applied readily to the addition of disturbances of different amplitudes. How would the example given here be changed if $y_2 = 0.5 \sin (\omega t + \pi/4)$?

30.6 THE TRIPLE SLIT

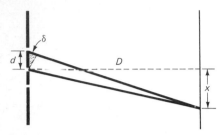

FIGURE 30.12 The triple slit. What will δ be at the position of the first maximum?

The interference pattern caused by three small parallel slits separated by equal distances d is also easily found from consideration of a phasor diagram. We see from figure 30.12 that δ, the path-length difference for the rays from two adjacent slits, is given by

$$\frac{\delta}{d} = \frac{x}{D}$$

Here too we assume $x \ll D$ and $d \ll D$. The phase-angle difference ϕ between the two rays is

$$\phi = 2\pi \frac{\delta}{\lambda} = 2\pi \frac{d}{\lambda D} x$$

and so $\phi \sim x$, as with the double slit. In figure 30.13 are shown the phasor diagrams for increasing values of ϕ for this case. The R values found from the phasor diagrams are plotted in the graph.

The first minimum occurs when
$\phi = 120°$, *or* $\delta = \lambda/3$

Notice that zero amplitude is *not* achieved when $\delta = \lambda/2$. In that case the disturbance from the first slit cancels that from the second and leaves the third slit's disturbance uncompensated. It is for this reason that the secondary maximum at $\lambda/2$ occurs as shown in the lower part of figure 30.13. Although the maxima at $\pm\lambda$ are shown equal to the central maximum, this is not found to be true in practice. The intensity diminishes as one proceeds away from the central maximum. The intensity pattern is, of course, proportional to the square of the y values shown.

30.7 MICHELSON INTERFEROMETER

A very important case of interference occurs when a wave is reflected by two parallel surfaces. We describe this situation in terms of light waves, but it will apply to other types of waves as well. Before describing the Michelson interferometer, a device of great historical and practical importance, we shall consider an idealized system which can only be approximated in practice. It is shown in figure 30.14.

A beam of parallel light is partly reflected from a very thin semitransparent plane. The remainder of the beam (we shall assume it to be exactly half) is reflected by the lower plane. It is apparent that rays such as A and B in part (b) of the figure will fall on top of each other. Since they have traveled different distances, much like the sound waves in the tubes we discussed earlier, they will exhibit interference effects. In particular, if the path-length difference for rays A and B is

$$0, \lambda, 2\lambda, 3\lambda, \ldots$$

they will interfere constructively, and one will observe brightness. But if the path-length difference is

$$\tfrac{1}{2}\lambda, 3(\tfrac{1}{2}\lambda), 5(\tfrac{1}{2}\lambda), \ldots$$

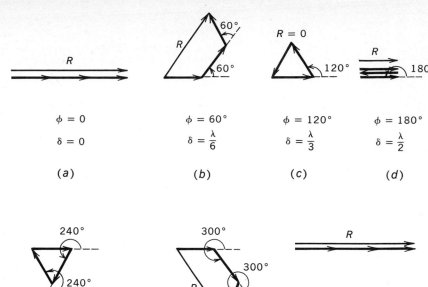

$\phi = 0$

$\delta = 0$

(a)

$\phi = 60°$

$\delta = \dfrac{\lambda}{6}$

(b)

$\phi = 120°$

$\delta = \dfrac{\lambda}{3}$

(c)

$\phi = 180°$

$\delta = \dfrac{\lambda}{2}$

(d)

$R = 0$

$\phi = 240°$

$\delta = \dfrac{2\lambda}{3}$

(e)

$\phi = 300°$

$\delta = \dfrac{5\lambda}{6}$

(f)

$\phi = 360°$

$\delta = \lambda$

(g)

FIGURE 30.13 The phasor diagrams for the triple slit predict the amplitude to vary with δ as shown. Since δ is proportional to x, the lower portion of the figure shows the resultant amplitude as a function of $(d/D)x$.

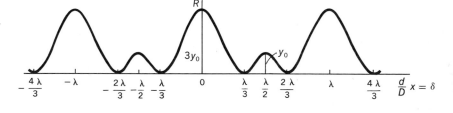

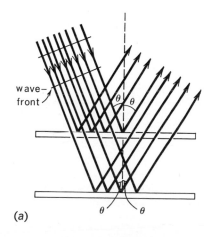

FIGURE 30.14 Rays of light reflected from the bottom plane can interfere with those reflected from the upper plane. If $\theta = 0$, what values of d will give constructive interference? Destructive interference?

(a)

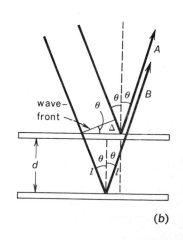

(b)

they will be 180° out of phase and will cancel each other. Darkness will then be observed.

In order to find the path difference, we see from figure 30.14(b) that ray B travels the following distance more than ray A:

$$\delta = \text{path difference} = 2l - \Delta$$

We are assuming rays A and B to fall exactly upon each other. But from the geometry of the figure we have

$$l = \frac{d}{\cos \theta}$$

and

$$\Delta = 2(d \tan \theta) \sin \theta = \frac{2d \sin^2 \theta}{\cos \theta}$$

After substituting this into the expression for δ, we find

$$\delta = \frac{2d}{\cos \theta}(1 - \sin^2 \theta) = 2d \cos \theta \qquad 30.7$$

Applying the conditions given above, we have

For Brightness $\qquad 2d \cos \theta = n\lambda$

For Darkness $\qquad 2d \cos \theta = (2n + 1)(\tfrac{1}{2}\lambda)$ $\Big\}$ $n = 0, 1, 2, \ldots$ $\qquad 30.8$

This is easily checked by noting that, for normal incidence, $\theta = 0$ and $\cos \theta = 1$. The path difference will just be $2d$, the factor 2 occurring since beam B must go both down and up through the region between the plates. Equation 30.8 says that if this distance is a whole number of wavelengths long, the two beams will be in phase, and constructive interference will occur.

The actual situation in a Michelson interferometer is shown in figure 30.15. A beam of light from the source strikes a semitransparent mirror at P. Part A of the beam goes to mirror A, and is reflected back to P. We are interested in that part of the beam which is then reflected at P in the manner

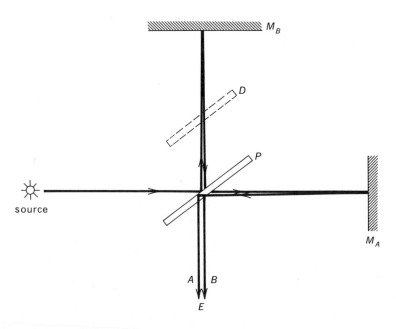

FIGURE 30.15 The Michelson interferometer. The extra glass plate at D is used to equalize the path in glass for the two beams. We have not shown some of the reflected beams which do not equalize with A and B.

shown for beam A. The other part of the incident beam is reflected by P to mirror B. Upon returning to P, part of the beam will be transmitted as shown for beam B. Since beams A and B have traveled different distances, we expect interference between them when they are brought together again.

If leg A of the interferometer (the path to mirror A) is the same length as leg B, the path difference δ for the two beams will be zero, and they will interfere constructively. Beams A and B will therefore join to give brightness. Suppose now mirror B is slowly moved away from P. The path length for ray B will lengthen by *twice* the amount the mirror is moved (twice because the ray travels down and back). After B has been moved back a distance $\lambda/4$, the path difference between beams A and B will be $2(\lambda/4)$ and $\delta = \lambda/2$. The two beams will interfere destructively, and darkness will be observed at point E.

If now mirror B is moved back still farther, the path difference will continue to increase. Eventually, δ will increase to λ, and brightness will again be seen at E. This process can be continued. When brightness occurs for the nth time at E, one knows that mirror B has been moved back a distance of $n(\lambda/2)$. Obviously, by observing the emitted light at E, one can measure displacements of mirror B to accuracy of better than $\lambda/2$. It is by this general method that the former meter standard was used to measure wavelengths of light. The reverse process is also possible, of course; one can measure distances in terms of the wavelength of light.

In practice, one does not observe total darkness or brightness if one places his eye at E and observes rays A and B. We have been assuming the rays to be perfectly parallel and incident normal to mirrors A and B. This corresponds to the case $\theta = 0$ in the two-plane-reflection situation discussed earlier in this section. Actually, this is true only for the center point of mirror P (if the system is well adjusted). If one observes light coming from other than the center point, θ is no longer zero, and so the intensity of light observed will depend upon the angle from which the light departs from the central ray. When properly adjusted, one observes a series of circular bright fringes centered on plate P. Fringes of other shapes are also possible, however, as is shown in detail in most intermediate optics texts.

30.8 SPEED OF LIGHT IN GASES

An interferometer may be used to measure the speed of light in gases. To do this, one places an evacuated tube of length L in one leg of the interferometer, as shown in figure 30.16. This tube will contain n_V wavelengths of light of wavelength λ_V in vacuum, where

$$n_V = \frac{L}{\lambda_V}$$

If the tube is now filled with some other material in which the speed of light is v_g, the wavelength of the light will change as indicated by our relation

$$v_g = f\lambda_g$$

where f is the frequency of the wave. The tube will now be n_g wavelengths long,

$$n_g = \frac{L}{v_g/f}$$

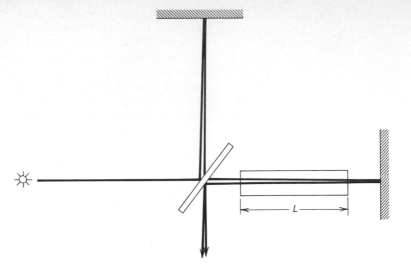

FIGURE 30.16 By placing an evacuated tube in one leg of a Michelson interferometer, one can determine the index of refraction of a gas. What is the optical path length equivalent to the length L of the tube?

It is customary to place this in a slightly different form by substituting for v_g in terms of the index of refraction μ. We recall that $\mu = c/v_g$, and so

$$n_g = \frac{\mu L}{c/f} = \frac{\mu L}{\lambda_V}$$

By comparing this expression with our result for n_V, we see that the path length for the light has effectively been increased by a factor μ when the vacuum was filled with gas. One often refers to the quantity μL as the *optical path length* equivalent to the path L in vacuum. In the present situation it means that the light appears to travel a distance μL as it traverses the tube of length L.

The optical path length is μL

Effectively, then, when gas is slowly admitted to the tube in figure 30.16, the length of that leg of the interferometer is increased to a value μL. This is equivalent to moving the mirror back a distance

$$\Delta x = \mu L - L = (\mu - 1)L$$

As shown in the preceding section, as the mirror is moved back a distance $(\mu - 1)L$, the field of view will change from bright to dark to bright Δn times (we say, Δn fringes are observed):

$$\Delta n = \text{no. of fringes} = \frac{2(\mu - 1)L}{\lambda} \qquad 30.9$$

The factor 2 appears since the light goes through the tube twice.

If we count the number of fringes which pass as the gas is slowly let into the tube, we can solve for the index of refraction of the gas, μ. We find

$$\mu - 1 = \frac{\lambda \, \Delta n}{2L} \qquad 30.10$$

Since $\mu = c/v_g$, knowing μ, we can find v_g. A modification of this general method can be used to measure the index of refraction of a plate of glass. How could this be done?

30.9 CHANGE OF PHASE ON REFLECTION

An interesting effect can be observed if a planoconvex lens (such as the one shown in figure 30.17) is placed on top of an optically flat surface (a good glass plate will do). If the lens has a very large radius of curvature (10 m, for example) and if monochromatic light is used to view the lens from above as indicated, then a series of circular fringes is observed. These fringes, called *Newton's rings,* are shown in part (*b*) of the figure.

The rings are an interference effect caused by the two beams shown, one reflected from the bottom plate, the other reflected from the bottom surface of the lens. One would think that the center point should be bright since the separation there is very nearly zero and both rays travel the same distance. Early workers tried in vain to push the plate closer to the lens and, by careful polishing, to achieve a bright center spot. However, the better the contact between the two pieces, the more clear it became that the center point was dark, not bright. We see that the center is dark in part (*b*) of figure 30.17. This is now known to be the result of the fact that the ray reflected from the lower surface suffers a 180° phase change upon reflection. It is, in effect, held back through $\lambda/2$ by the reflection process. It therefore interferes destructively with the other ray.

Although we do not investigate this change in phase effect in detail until Chap. 33, it is possible to state the result simply at this time. When light is reflected *at near-normal incidence* from a surface, *a 180° phase change is observed if the reflecting material has a higher refractive index than the material into which the light is reflected.* In the case of Newton's rings, this condition is clearly satisfied since the two materials are glass and air. Notice that the upper ray, being reflected by the air into the glass, suffers no phase change. This is typical of reflections by a less optically dense (low-μ) material into a more optically dense (high-μ) material.

We do not examine Newton's rings quantitatively since the geometry is somewhat involved. However, it is clear that the thickness of the air space increases by $\lambda/2$ as one proceeds from one circular bright fringe to the next-larger bright fringe. (Notice that the distance is $\lambda/2$ since the lower ray travels down and back, thereby traveling a total distance λ farther than the ray reflected from the upper surface.) As shown in figure 30.18, interference fringes similar to Newton's rings can be used to test the contours of a nearly flat surface.

Condition for 180° phase change upon reflection

FIGURE 30.17 The ray reflected from the lower side of the lens interferes with the ray reflected from the glass plate and gives rise to Newton's rings. Why is the center point dark? [Part (*b*) courtesy of Bausch & Lomb Optical Co.]

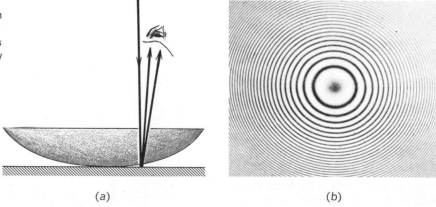

(a) (b)

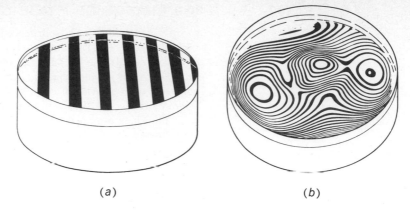

FIGURE 30.18 In each of the two cases shown, glass plates are sitting upon perfectly flat surfaces. Interference fringes formed by reflections at the contacting surfaces allow us to learn how flat the upper surface is. In (*a*) the upper surface is nearly flat. But in (*b*) the upper surface is uneven. (From Konrad B. Krauskopf and Arthur Beiser, "The Physical Universe," 3d ed., fig. 9.11, p. 218, McGraw-Hill Book Company, New York, 1973. By permission of the publishers.)

(*a*)   (*b*)

30.10 INTERFERENCE IN THIN FILMS

You have probably noticed the bands of color which sometimes occur on the surface of a slightly oily pool of water. These bands are really interference fringes caused by rays reflected from the upper and lower surfaces of a thin film of oil on the water's surface. The colors one observes on soap bubbles also originate from a similar process. Let us now see how these interference bands come about.

Suppose we have a very thin wedge-shaped structure such as the one shown in figure 30.19. This might represent a small portion of an oil film, or a sliver of glass. In any case we shall assume it to have index of refraction μ and to have air above it. The material below it will be considered to have an index of refraction less than μ. As a result, ray U suffers a 180° phase change upon reflection, while ray L does not. Since the actual path difference between the two rays is zero at the left end of the wedge, the rays U and L are 180° out of phase there because of the phase change upon reflection, and give darkness D at that point.

At the first bright fringe B from the left edge of the wedge, the rays must be in phase. This will be the case if ray L has been held back an additional half wavelength while traversing the distance t_1 twice, down and up. We saw in Sec. 30.8 that the optical path length is μ multiplied by the actual distance traveled, and so t_1 causes the L ray to be held back a distance $\mu(2t_1)$. As we have just shown, this must be equal to $\lambda/2$, so that t_1 must be given by

$$\mu(2t_1) = \frac{\lambda}{2} \qquad \text{or} \qquad t_1 = \frac{\lambda}{4\mu}$$

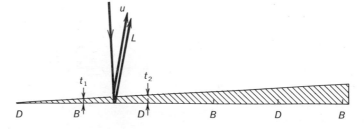

FIGURE 30.19 What is the thickness of the wedge at the bright spot on the right end?

Similarly, we must have

$$\mu 2t_2 = \lambda \qquad \text{or} \qquad t_2 = 2\frac{\lambda}{4\mu} \qquad \cdots$$

We notice that the position of the fringes depends upon the wavelength of the light. Therefore the bright fringes for different colors will occur at different points. This results in the colored fringes one observes in thin-film interference.

One application of thin-film interference is in the elimination of unwanted reflections. A thin film of optical thickness just sufficient to produce total destructive interference between the reflected rays is put on a surface. Therefore the surface will be a nonreflecting surface. This technique is used in coating lenses and other optical devices to eliminate undesirable reflections. Of course, since the required thickness of the coating depends upon wavelength, a surface cannot be made a perfect nonreflector for all wavelengths by this technique.

Chapter 30
Questions
and Guesstimates

1. What is meant by coherent waves? Must they both be sinusoidal? Need they have the same frequency? In optical experiments, how are two coherent beams produced?

2. The sum of two sine waves with the same frequency is a sine wave. In view of this, prove that the sum of a large number of arbitrary sine waves, all of the same frequency, is a sine wave.

3. When light composed of many colors is used in a Michelson interferometer, one obtains visible fringes only when the two legs of the interferometer are of nearly equal length. For large path differences the fringes become blurred together. Explain why this occurs.

4. Why is it impossible to obtain an interference maximum in a double-slit experiment if the slit separation is less than the wavelength of the light being used?

5. Very thin films are sometimes deposited on glass plates. The thickness of the film can be controlled by observing the change in color of white light reflected from the surface as the film's thickness is increased. Explain.

6. Mercury light consists of several distinct wavelengths. Suppose in a double-slit experiment, filters are placed over the slits so that $\lambda = 436$ nm (blue) light goes through one slit and $\lambda = 546$ (green) light goes through the other. Will one be able to see an interference pattern on the screen?

7. What change will occur in a Young's double-slit experiment if the whole apparatus is immersed in water rather than air? What change would one observe in a Newton's rings arrangement if the space between plate and lens is filled with water?

8. When properly carried out, interference in thin films of material is easily observed. However, as the film is made thicker, the fringes become more closely spaced. As a result, it is very difficult to obtain interference effects with a thick plate of glass. Explain.

9. Reasoning from the phasor diagram, show why it is impossible to obtain destructive interference for more than an instant from two equal-amplitude waves whose frequencies are not the same.

10. Using two pieces of flat glass (microscope slides are ideal), press them together in various ways and estimate how close together the surfaces are from observation of the interfering reflected light. (You can see the

interference pattern easily in any lighted room *provided* you get the plates close enough together.)

11. Two cars sit side by side in a large, vacant parking lot with their horns blowing. Would you expect to be able to notice interference effects from the two sound sources? What if the horns were replaced by two clarinets playing the same note?

12. Musicians often tune their instruments by listening for "beats" between their tone and the one they are tuning to. For example, if the frequency of a violin string is 330.5 Hz and it is being tuned to a piano note of 330.0 Hz, the combined tone of the two instruments will fluctuate in intensity: it will be large (i.e., beat) every 2.0 s. When the instruments are nearly exactly in tune, the time between beats becomes very long. Explain how this phenomenon arises.

13. Radio and TV reception from a distant station is interrupted by an airplane flying over the receiver. Explain why the plane affects reception. Why is the effect not noticed for all planes in the area between the station and the receiver?

14. Instructions for connecting the two bass speakers of a stereo system usually include the following: "Place the speakers side by side and listen to them. Now reverse the lead wires to one of them and listen. Choose the connection which gives the loudest sound." Why should there be a difference in sound intensity if one set of leads is reversed?

15. Explain the following statement: The difference in thickness of a thin film between the positions of two bright fringes in an interference pattern is zero or $\lambda/2\mu$, where λ is the wavelength of light being used and μ is the refractive index of the film material.

16. Suppose in a Young's double-slit experiment the two slits are replaced by four, as shown in figure P30.1. For certain values of D, the central point P is dark rather than bright. Explain.

17. Referring back to figure 30.18, what can you say about the thickness of the air gap along a single fringe countour? What can you say about the gaps at two adjacent fringes? At two fringes separated by a single fringe of the same type?

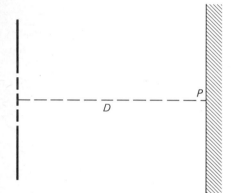

FIGURE P30.1

Chapter 30
Problems

1. The two identical loudspeakers shown in figure P30.2 are driven from the same oscillator to give off a tone with frequency 200 Hz. As speaker A is moved back along the line indicated, an observer at P hears a strong tone when the speaker is at the positions marked S but a weak tone when the speaker is at positions W. How far apart are the W positions?

2. In order to measure the speed of sound, the two loudspeakers shown in figure P30.2 are driven at 400 Hz by the same oscillator. As speaker B is moved away from the observer at P, the sound heard at P is strongest when the speaker is at the positions marked S. These positions are found to be 82 cm apart. What is the speed of sound one would calculate from these data?

FIGURE P30.2

3. Add the following two waves: $y_1 = 20 \sin \omega t$ and $y_2 = 20 \sin (\omega t + 60°)$.

4. Add the following two waves: $y_1 = 30 \sin \omega t$ and $y_2 = 30 \cos \omega t$.

5. In a Young's double-slit experiment using blue light with wavelength 4400 Å (440 nm), the bright interference fringes on a screen 2.0 m away are separated by a distance of 0.15 cm. How far apart are the slits?

6. In a certain Young's double-slit experiment for which $D = 1.00$ m and $d = 0.10$ cm, the bright fringes are 0.050 cm apart. What wavelength of light is being used?

7. Two wavelengths of light λ_1 and λ_2 are sent through a Young's double-slit apparatus simultaneously. What must be true concerning λ_1 and λ_2 if the third-order λ_1 bright fringe is to coincide with the fourth-order λ_2 fringe?

8. Two waves are being sent down a string simultaneously: $y_1 = 0.50 \sin (\omega t - kx)$ and $y_2 = 0.50 \sin (\omega t - kx + 60°)$. Find the resultant wave on the string.

9. Two waves are being sent along a string in opposite directions. Their equations are $y_1 = y_0 \sin (\omega t - kx)$ and $y_2 = y_0 \sin (\omega t + kx)$. Find the resultant wave on the string. What will be the equation for the resultant y at the point $x = \pi/2k$?

10. Two parallel glass plates are originally in contact and viewed from directly above with 500-nm light (green) reflected nearly perpendicularly by the surfaces. As the plates are slowly separated, darkness is observed at certain separations. What are the first four of these values?

11. Referring to figure 30.18(a), the right-hand edge is in contact. If the light used is blue light with $\lambda = 4400$ Å, about how far apart are the plates at the last dark band on the left? Assume the space between the plates is air-filled.

12. Repeat problem 11 if the gap between the plates is filled with a liquid having an index of refraction of 1.50.

13. Suppose in figure 30.17(b) that the light used to produce the Newton's rings is the mercury green line, 546 nm, and that the radius of the seventh dark ring is 1.5 cm. (a) How large is the air gap at this position? (b) If the gap is now filled with water, how big is the gap at the new position of this ring?

14. A very thin film of transparent plastic ($t = 3.17 \times 10^{-6}$ m, $\mu = 1.50$) covers the two slits in a Young's double-slit experiment that uses a helium-neon laser as light source ($\lambda = 633$ nm). (a) Is the central maximum of the interference pattern still at the same position as without the film? (b) What happens at the center position as the film is removed from one slit? (c) From both slits?

15. To determine the pitch of a high-precision screw, the screw is used to move one mirror in a Michelson interferometer. It is found that one revolution of the screw results in 2023 fringes ($\lambda = 5460$ Å, green) passing the field of view. How far does one turn of the screw move the mirror?

16. One leg of a Michelson interferometer has in it an evacuated glass tube 2.0 cm long. A gas is slowly let into the tube, and the number of times the field of view changes from bright to dark back to bright is counted to be 210 (that is, 210 fringes pass). If yellow light with $\lambda = 579$ nm is being used, what is the refractive index of the gas?

17. A very thin wedge of plastic shows interference fringes when illuminated perpendicularly with white light. Two adjacent blue fringes ($\lambda = 450$ nm) are separated by 0.40 cm. If the index of refraction of the plastic is 1.48, what is the difference in thickness of the wedge between the positions of

these two fringes? Two answers are possible. How could you determine which is correct?

18. An oil slick on a puddle shows interference fringes in the light of the sun overhead. The fringes between two points A and B are as follows: yellow, green, blue, red, yellow. Assuming λ for yellow light to be 580 nm and the index of refraction of the oil to be 1.50, find the difference in thickness between A and B. Repeat if the color sequence is yellow, red, yellow, green, blue, green, yellow.

19. Using the phasor method, find the sum of the following two waves: $y_1 = y_0 \sin \omega t$ and $y_2 = 2y_0 \sin (\omega t + 40°)$.

20. A certain Young's double-slit experiment has slits of different size so one slit lets through a beam with twice the *intensity* of the other. (a) Assuming the amplitude of the stronger beam to be y_0, what is the amplitude of the weaker one? (b) What is the combined amplitude at a position where the phase difference is δ?

21. Using the phasor method, find the sum of the following three waves: $y_1 = y_0 \sin \omega t$, $y_2 = 2y_0 \sin (\omega t + 30°)$, and $y_3 = 3y_0 \sin (\omega t + 60°)$.

22. Light of wavelength λ strikes four identical slits which are d apart. Using the phasor method, sketch the interference pattern one would observe on a screen for this system. Assume $D \gg d$ and $D \gg x$.

23. As shown in figure P30.3, the signal from a radio station ($\lambda = 300$ m) reaches a receiver by two different paths, one direct and the other by reflection from the ionosphere (a layer of ions near the top of the atmosphere formed by the sun's radiation hitting air molecules). At least how high must the ionospheric layer be if destructive interference between the two beams is to occur? Assume no phase change upon reflection (not strictly true). This type of interference contributes to the "fading" one sometimes notices in radio reception.

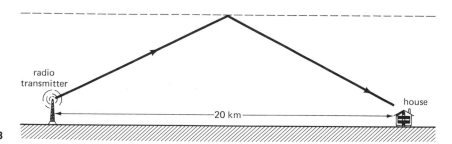

FIGURE P30.3

24. Expensive cameras have nonreflecting coated lenses. Typically, the lens surface is covered with a thin layer of magnesium fluoride ($\mu = 1.25$). How thick must this layer be for destructive interference to occur between the light reflected from the two surfaces of the layer when 5000-Å light is used? Why would a layer of Lucite ($\mu = 1.48$) be unsuitable for this purpose?

25. A beam of light with wavelength λ is sent through two parallel closely spaced glass plates, as shown in figure P30.4. As the separation of the plates is slowly increased, the transmitted beam is alternately bright and dim. For what values of the plate separation will the transmitted light be bright? (In practice, the inner sides of the plates are lightly silvered to emphasize the reflected rays. The device is called a Fabry-Perot interferometer.)

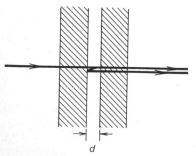

FIGURE P30.4

26. In the Lloyd's mirror experiment illustrated in figure P30.5, light striking the screen directly from the source S interferes with that coming from the image of the source, S'. Since a change of phase occurs upon reflection, the point at $x = 0$ is dark. Find the values of x which give brightness. Assume $D \gg d$ and $D \gg x$.

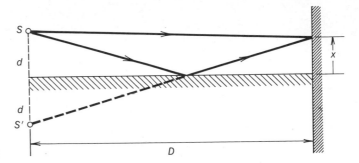

FIGURE P30.5

27. Two small sources of coherent sound waves are positioned as shown in figure P30.6. Where will one fine maximum sound intensity at the position of the screen? Assume $D \gg d$.

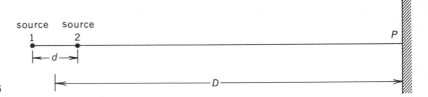

FIGURE P30.6

28. Two sound sources send out waves of slightly different frequency, ν and $\nu + \Delta\nu$. The equations for the sound waves reaching a listener's ear are

$$y_1 = y_0 \sin 2\pi\nu t \quad \text{and} \quad y_2 = y_0 \sin [2\pi(\nu + \Delta\nu)t]$$

Show that they combine to give a wave

$$y = 2y_0 \cos (\pi \Delta\nu \, t) \sin [2\pi(\nu + \tfrac{1}{2}\Delta\nu)t]$$

Notice that $2y_0 \cos (\pi \Delta\nu \, t)$ is a slowly varying amplitude of the combined wave. If $\Delta\nu = 4$ Hz, sketch the amplitude as a function of t. How many times each second is loudness (or a "beat") heard? What general conclusion can you draw from this? The phenomenon of beats is used for accurate comparison of nearly equal frequencies.

31 Diffraction

In this chapter we continue the discussion of interference of waves. However, the major concern here is with interference effects which give rise to the bending of waves into the zone which should be shadow if light traveled in straight lines. We shall see that waves passing through an aperture give rise to an interference pattern rather than casting a sharp shadow. Several examples of this are studied, and the applications of this phenomenon are pointed out.

31.1 INTRODUCTORY CONSIDERATIONS

We saw in the preceding chapter that waves do not always cast sharp shadows. For example, in Young's double-slit experiment, one does not observe two distinct beams of light from the two slits. The light rays bend out of their straight-line paths, and light enters the regions which would otherwise be shadow. This bending of light and other waves into the region which would be shadow if rectilinear propagation prevailed is called *diffraction*.

In the case of the double slit, diffraction is the result of interference between the waves from the two slits. It will be seen that diffraction is simply the result of interference between coherent waves.

Although the discussion in this chapter is carried out in terms of light waves, the student should be careful to notice that the effect is common for all types of waves. For example, sound waves are heard even behind obstacles which hide the source because the waves are diffracted around the obstacle. Water waves on a smooth pond do not cast clear "shadows" of objects in the pond because the waves are diffracted. These examples of diffraction are so commonplace that we often do not question their origin. They differ from diffraction effects found with light waves chiefly because of the influence of wavelength upon diffraction. As you continue to read this chapter, it would be well to relate each diffraction effect discussed in terms of light to a similar effect one could observe with sound or water waves. Let us now consider a principle which is basic to our discussion.

In order to discuss Young's double-slit experiment, we assumed that each slit acted as a new source for light. Such an assumption is fully justified for water waves since a wave hitting a hole in a wall will certainly cause the hole to act as a new wave source. This is easily demonstrated by the experiment shown in figure 31.1. Similarly, one would expect the air particles at a hole being struck by a sound wave to act as a new sound source. Huygens

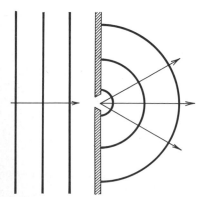

FIGURE 31.1 When a water or sound wave strikes a hole in a barrier, the hole acts as a new wave source.

Huygens' principle

generalized this idea to state what is now known as *Huygens' principle: Each point of a wavefront acts as a new source of waves.*

Although Huygens' principle seems quite reasonable for waves in material media such as water, air, etc., the application of it to electromagnetic waves is not so obvious. However, the results obtained by use of this principle in the case of light are in agreement with experiment. In fact, the early workers in the field of optics who used Huygens' principle to interpret their results believed that light traveled through a material medium they called the *ether*. It was only later that the ether was discarded as a necessary material for light transmission. Even prior to that time, however, it had been shown that Huygens' principle is a necessary consequence of the wave equation. We therefore can use it with confidence.

In our discussion of diffraction, we distinguish two separate types of experiments. The first of these, called *Fraunhofer diffraction*, is characterized by the fact that parallel rays (i.e., plane waves) are used. We can obtain waves of this sort by using lenses (in the case of light) or a distant source and a distant observation point. These two possibilities are illustrated in figure 31.2(a) and (b), where the hole is assumed to have the form of a long narrow slit. Remember, in connection with (a), that parallel light can be produced by placing the source at the focal point; recall also that parallel rays are focused to the focal point. In part (a), the light intensity observed at various points on an observation screen at a distance f from the lens is also shown. Notice that brightness is observed at $\theta = 0$ with alternate bright and dark fringes observed on each side of the central bright fringe.

The second class of diffraction effects occurs when parallel rays are not

Fraunhofer diffraction uses parallel light, while Fresnel diffraction does not

FIGURE 31.2 Typical examples of Fraunhofer and Fresnel diffraction.

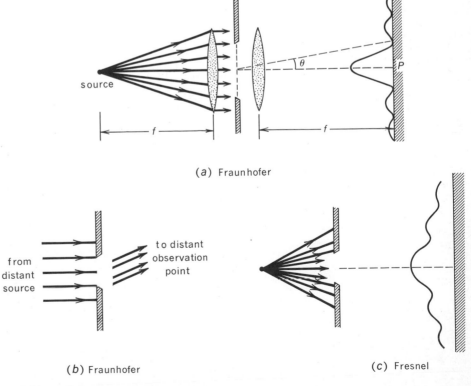

(a) Fraunhofer

(b) Fraunhofer

(c) Fresnel

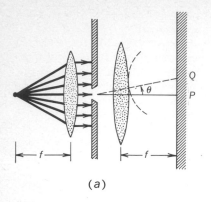

(a)

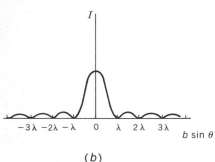

$$-3\lambda \quad -2\lambda \quad -\lambda \quad 0 \quad \lambda \quad 2\lambda \quad 3\lambda$$
$$b \sin \theta$$

(b)

FIGURE 31.3 Experimental arrangement for observing Fraunhofer diffraction from a single slit of width b. The observation screen or film is at the focal point of the second lens (not to scale). A sketch is also given to show the light intensity pattern; the peaks actually are in the ratios of $1:0.045:0.016:0.008$.

FIGURE 31.4 In analyzing the single-slit pattern qualitatively, we section the slit into portions whose rays differ by $\lambda/2$ in path length. Why?

used. For example, either or both the source or observation point might be relatively close to the slit and the rays not made parallel by use of lenses. This is called *Fresnel diffraction*. A typical distribution of observed light intensity in this case is shown in figure 31.2(c). Of course, the pattern in part (c) must change to that of part (a) as the source and observation point are moved farther from the slit. Under some conditions the center point of the Fresnel diffraction pattern can actually become a point of minimum intensity. Fresnel diffraction is somewhat more complicated than Fraunhofer diffraction, and so we restrict our discussion to the latter.

31.2 FRAUNHOFER DIFFRACTION AT A SINGLE SLIT

Consider the typical Fraunhofer arrangement shown in figure 31.3. Plane waves (i.e., parallel rays) strike the slit as shown. That portion of the wave striking the slit acts as a large number of sources which send out waves in all directions. The second lens, in effect, selects out all parallel rays and focuses them on the screen placed one focal length away. For example, those parallel rays coming from the slit at an angle θ to the axis are all focused on the screen at the point labeled Q. Notice also, as shown in figure 31.3(a), that the effect of the lens is to bend the wavefront into the circular arc shown. This wave converges onto a single point on the screen. The optical path length for the direct ray is equalized with the other rays by means of the increased path length in the lens. For this reason, we can ignore the lens in our dicussion and consider only the parallel rays. The lens will bring them together on the screen without changing their relative phases.

Let us refer to figure 31.4, where we see various sets of parallel rays which emanate from the plane waves striking the slit.

These rays, according to Huygens' principle, are drawn as though each point on the wavefront striking the slit acted as a new ray source. In part (a) of figure 31.4, where the diffraction angle θ is zero, we see that all these rays are of equal length. When brought to a focus at the observation point, these waves from the various parts of the slit will be in phase. We therefore expect to observe brightness at the center point, where $\theta = 0$. This is the situation shown at point P in figure 31.2(a) and at $\sin \theta = 0$ in figure 31.3(b).

A quite different situation exists in part (b) of figure 31.4. Here the observation angle θ is so chosen that the ray from the top of the slit (ray C) travels a distance $\lambda/2$ farther than the ray from the middle (ray B). Notice

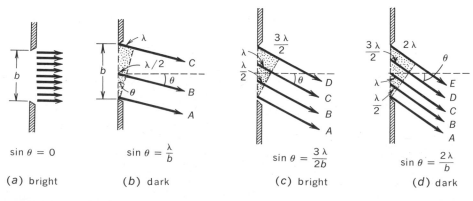

$$\sin \theta = 0$$

(a) bright

$$\sin \theta = \frac{\lambda}{b}$$

(b) dark

$$\sin \theta = \frac{3\lambda}{2b}$$

(c) bright

$$\sin \theta = \frac{2\lambda}{b}$$

(d) dark

that the wave traveling along ray C will just exactly cancel the wave coming along ray B since they are $\lambda/2$, or $180°$, out of phase. Similarly, a ray just slightly below ray C will be canceled by the corresponding ray from the lower half of the slit. A similar situation exists for all points in the two halves of the slit. We therefore conclude that the intensity observed at the angle θ shown in figure 31.4(b) will be zero. To find this angle we note that the shaded triangle yields the relation $\sin \theta = \lambda/b$. In a typical case for light, $b \approx 5 \times 10^{-4}$ m and $\lambda = 5 \times 10^{-7}$ m, from which $\theta \approx 4°$. Clearly, the diffraction pattern will be confined to the region close to the central maximum.

When the angle θ becomes as large as shown in figure 31.4(d), darkness again will prevail. This follows from the fact that the rays from the region of the slit between A and B are, here too, a half wavelength ahead of the corresponding rays from the region between B and C. These two portions of the slit therefore cancel each other's effect. Similarly, the region between C and D cancels the region between D and E. As a result, each ray from the slit is canceled by another ray, and darkness is observed at this angle.* To find the angle, we make use of the shaded triangle, from which we have $\sin \theta = 4(\lambda/2b)$.

The intermediate case shown in figure 31.4(c) is also of interest. Here too, the rays from the region between A and B cancel those from the region BC. However, the rays from the next region, between C and D, are not canceled. We therefore find brightness at this observation angle. It will be seen in the following sections that the case illustrated in part (c) gives the position of the intensity maximum to a good approximation.

To summarize the results of our qualitative considerations, we can refer to figure 31.3(b). Although the heights of the intensity maxima shown there are not to scale, their relative positions are correctly shown. (The relative intensities are given in the figure legend.) We see that the central maximum is twice as wide as the fringes on each side of it. The side fringes are spaced in such a way that $b \sin \theta$ is an integer multiple of λ at the minima. A photograph of this intensity pattern is shown in figure 31.5.

*The question is sometimes asked: How does the slit know it should be divided up this way? Would not some other method of division give a different answer? What is the answer?

(a)

(b)

(c)

FIGURE 31.5 A photograph of a Fraunhofer single-slit diffraction pattern. The slit width is shown in (a), while (b) and (c) are for short- and long-exposure times. (After Jenkins and White.)

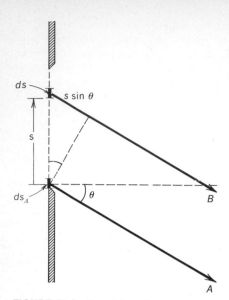

FIGURE 31.6 By adding the contributions of all elements such as ds, the amplitude of the resultant wave from the slit can be found.

31.3 SINGLE-SLIT DIFFRACTION: QUANTITATIVE STUDY

In the preceding section we succeeded in obtaining the general features of the diffraction pattern for a single slit. Our reasoning was mainly qualitative, and so we were unable to obtain the light intensities within the pattern. Let us now analyze the situation quantitatively by a straightforward summation of the waves coming from the various portions of the slit. To do this, we recall that the magnitude of a plane wave at any time t and phase ϕ can be written

$$y = y_0 \sin (2\pi ft - \phi)$$

Consider now the situation shown in figure 31.6. Let us split the slit into a large number of very narrow incremental slits of width ds. The instantaneous magnitude of the wave disturbance from the bottommost portion of the slit (coming along ray A) will be represented as

$$dy_A = (y_0 \, ds_A) \sin 2\pi ft$$

We take the amplitude of this wave as $y_0 \, ds_A$ since the amplitude will certainly be proportional to the width of the narrow incremental slit.

A wave traveling along ray B will be a distance $x = s \sin \theta$ behind the wave which comes along A, since it must travel that much farther. We know, however, that the path length difference δ is related to a phase angle ϕ through $\phi = 2\pi \, \delta/\lambda$. Therefore $\phi = 2\pi s \, (\sin \theta)/\lambda$, and so the wave disturbance along ray B will have the form

$$dy = (y_0 \, ds) \sin \left[2\pi \left(ft - \frac{s \sin \theta}{\lambda} \right) \right]$$

Since ray B is a purely arbitrary ray, this equation will apply to the disturbance from any portion of the slit.

In order to find the resultant disturbance from the slit at an angle θ, we must sum contributions such as dy for all portions of the slit. Since we are dealing with infinitesimal values of ds, this sum is just the integral over s as s goes from zero to the width of the slit b. The resultant disturbance is therefore

$$y = \int_0^b y_0 \sin \left[2\pi \left(ft - \frac{s \sin \theta}{\lambda} \right) \right] ds$$

After carrying out the integration over s, keeping t and θ constant, we find

$$y = \frac{y_0 \lambda}{2\pi \sin \theta} \left\{ \cos (2\pi ft) - \cos \left[2\pi \left(ft - \frac{b \sin \theta}{\lambda} \right) \right] \right\}$$

This equation can be simplified considerably by use of the trigonometric relation

$$\cos \theta - \cos \phi = -2 \sin \frac{\theta + \phi}{2} \sin \frac{\theta - \phi}{2}$$

It then becomes

$$-y = \frac{y_0 \lambda}{\pi \sin \theta} \sin \left(\frac{\pi b \sin \theta}{\lambda} \right) \sin \left[2\pi \left(ft - \frac{b \sin \theta}{2\lambda} \right) \right] \qquad 31.1$$

The quantity of interest to us is the intensity of the wave, which, in turn, is proportional to the amplitude squared. From *31.1* we see that the wave amplitude is

$$\frac{y_0 \lambda}{\pi \sin \theta} \sin \left(\frac{\pi b \sin \theta}{\lambda} \right)$$

This then yields

$$I \sim \left(\frac{\lambda}{\sin\theta}\right)^2 \sin^2\left(\frac{\pi b \sin\theta}{\lambda}\right)$$

or

$$I = I_0\left(\frac{\sin u}{u}\right)^2$$

where

$$u \equiv \frac{\pi b \sin\theta}{\lambda} \qquad\qquad 31.2$$

We can see in the following way that *31.2* agrees with our previous results for the location of the dark fringes. Since $\sin u$ is zero for $u = \pi, 2\pi, 3\pi$, etc., it is apparent that I will be zero for

$$\frac{\pi b \sin\theta}{\lambda} = \pi,\ 2\pi,\ 3\pi,\ \ldots$$

From this, the dark fringes will occur at

Minima occur when $\sin\theta = n\lambda/b$

$$b \sin\theta = \lambda,\ 2\lambda,\ 3\lambda,\ \ldots$$

which is the result shown previously, in figures 31.3 and 31.4.

In addition, *31.2* can be used to estimate the positions of the maxima in intensity. Although the variation of the denominator must be considered for an exact solution, I will be close to a maximum when $\sin u = 1$, its largest value. This will occur when u is an odd-integer multiple of $\pi/2$. Therefore the positions of the maxima will be given approximately by

$$u = 3\left(\frac{\pi}{2}\right),\ 5\left(\frac{\pi}{2}\right),\ \ldots$$

The maxima are given approximately by
$b \sin\theta = n(\lambda/2)$, with $n = 3, 5, 7, \ldots$

which means

$$b \sin\theta = 3\left(\frac{\lambda}{2}\right),\ 5\left(\frac{\lambda}{2}\right),\ \ldots$$

Why have we omitted the value $u = \pi/2$ in listing the positions of the maxima?

Illustration 31.1

How narrow must a slit be if the first diffraction minimum is to occur at an angle of 30°? Take $\lambda = 5000$ Å.

Reasoning The first intensity minimum will occur when u in *31.2* is π. We then have

$$u = \frac{\pi b \sin\theta}{\lambda} = \pi$$

Solving for b,

$$b = \frac{\lambda}{\sin\theta}$$

In our particular case $\sin\theta = \frac{1}{2}$ and $\lambda = 5 \times 10^{-5}$ cm, so that

$$b = 10 \times 10^{-5}\text{ cm}$$

How difficult would it be to produce a slit of this width? Can you think of any troubles one might have in photographing the diffraction pattern for such a slit? Can you estimate the exposure time within an order of magnitude?

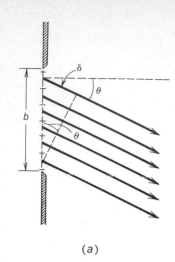

(a)

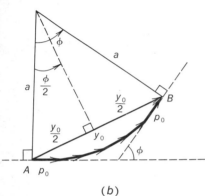

(b)

FIGURE 31.7   The phasor resultant y_0 in part (b) gives the amplitude of the diffracted wave at angle θ.

31.4 SINGLE-SLIT DIFFRACTION: GRAPHICAL METHOD

The phasor method outlined in the preceding chapter can also be used to find the diffraction pattern for a single slit. Although we solved this problem quantitatively in the preceding section, we shall use it as an illustration of the phasor technique, which will later be used for more complicated problems. Our approach is as follows. We section the slit into a large number N of narrow pseudoslits of equal width, b/N. The contribution of each pseudoslit will be represented by a phasor. To find the resultant wave amplitude, we add the N phasors.

For convenience in drawing, the slit in figure 31.7(a) has been split into six pseudoslits, although many more should be shown. If we disignate the phase difference between the first and last phasors as ϕ, it will be the angle labeled as such in figure 31.7(b). It is related to the path difference δ shown in part (a) by the usual relation

$$\phi = 2\pi\frac{\delta}{\lambda} \qquad\qquad 31.3$$

(As a check, notice that when $\delta = \lambda$, the phase difference is 2π, as it should be.) Moreover, simple geometrical considerations show that the other angles are as labeled in the figure.

The resultant of these phasors gives the amplitude of the resultant wave. It has a magnitude y_0, as indicated in the figure. From the right triangle with apex angle $\phi/2$, we have at once that

$$\tfrac{1}{2}y_0 = a \sin \tfrac{1}{2}\phi \qquad\qquad 31.4$$

We must now relate a, the radius of the arc formed by the phasors, to the lengths p_0 of the individual phasors.

Because the number of phasors from A to B in figure 31.7 is really very large, the phasors constitute an arc of a circle of length Np_0. Since a is the radius of this arc, we have

$$\frac{\text{Arc length}}{2\pi a} = \frac{Np_0}{2\pi a} = \frac{\phi}{2\pi}$$

which yields

$$a = \frac{Np_0}{\phi}$$

Substituting this value in *31.4* gives

$$y_0 = Np_0\frac{\sin u}{u}$$

where

$$u = \frac{\phi}{2} = \frac{\pi\delta}{\lambda}$$

The latter equality is obtained by using *31.3* to replace ϕ.

This value for u is readily seen to be the same as the similar quantity used in the preceding section. From figure 31.7(a) we have

$$\delta = b \sin \theta$$

which gives

$$u = \frac{\pi b \sin \theta}{\lambda}$$

To find the intensity of the wave diffracted at angle θ, we use the fact that it is proportional to $y_0{}^2$. We then have

$$I = I_0 \left(\frac{\sin u}{u} \right)^2$$

with

$$u = \frac{\pi b \sin \theta}{\lambda} \qquad \qquad 31.2$$

as was found in the preceding section.

In practice, of course, a slit is actually a rectangle, and so one would expect diffraction effects at its ends, as well as from its sides. This results in essentially two single-slit diffraction patterns at right angles to each other. A photograph of the pattern caused by a slit of width b and length l is shown in figure 31.8. Notice that the fringes caused by the effective slit of largest width are closest together.

31.5 RESOLVING POWER OF A SINGLE SLIT

We have seen in preceding sections that diffraction effects cause waves to bend around obstacles such as the edges of slits. This has profound consequences in nature. It means, as we shall see in Chap. 36, that a basic uncertainty must always exist in our knowledge of a physical situation. Because of diffraction effects, we are unable to measure precisely both the position and momentum of a particle at the same time. As a result, we shall never be able to learn, even in principle, the exact behavior of a given particle. More is said about this after we become acquainted with the wave nature of material particles, in later chapters. For now, we restrict our discussion to the limits which diffraction effects place on our ability to discern detail in objects.

Referring to figure 31.9, consider two distant objects S and S' which are to be viewed by focusing their images on a screen. The focusing instrument might be a telescope or a camera lens, for example. In any event, we shall consider the viewing instrument to be such that, in effect, a slit exists and that the beam of light is restricted to a certain region by it. If no diffraction

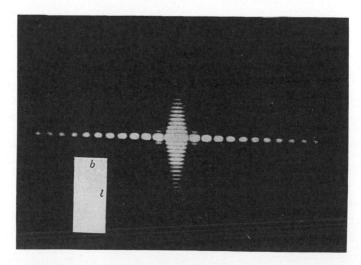

FIGURE 31.8 Diffraction pattern from a rectangular slit. The relative dimensions and orientation of the slit are shown by the rectangular insert. (After Jenkins and White.)

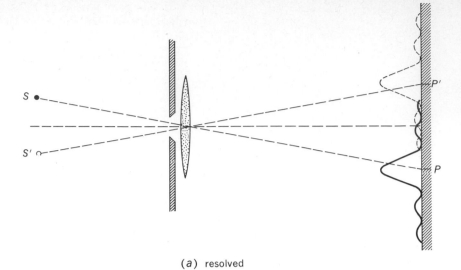

(a) resolved

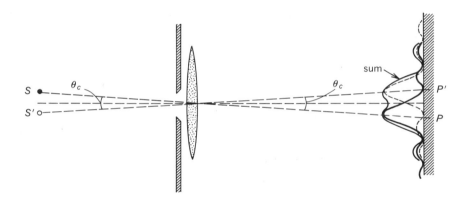

FIGURE 31.9 What is the limiting condition under which two images are said to be resolved?

(b) limit of resolution

Images are always blurred by diffraction effects

effects occurred, one would note at P and P' two distinct images of the two objects. However, as we have seen, the slit gives rise to a diffraction pattern, as indicated. Each source is imaged as a central bright image with maxima and minima at each side. The sum of these two intensity patterns for the two slits gives the observed illumination at the screen. In the case shown in part (a) of figure 31.9, the two images are well enough separated so that they can be distinguished, or *resolved*.

Condition for limiting resolution

This will not be the case if the sources are closer together, as in figure 31.9(b). The case illustrated there is that in which the maximum for one image occurs at the first minimum of the other. Under this condition, the angle subtended by the sources at the slit, θ_c, is equal to the angle from the central maximum to the first minimum. We designate this condition to be the case of *limiting resolution*. From it we define the limiting *angle of resolution* of the system to be θ_c. (This angle is sometimes called the *resolving power*. However, we avoid this name since it implies better resolution for larger values of θ_c, which is not actually the case.)

We saw in the preceding sections that the first minimum occurs at an angle θ_c, given by (for a slit of width b)

$$b \sin \theta_c = \lambda$$

However, in these cases θ_c will be small enough so that

$$\sin \theta_c \approx \theta_c$$

and so we have

$$\theta_c = \text{limiting angle of resolution} = \frac{\lambda}{b} \qquad \text{(slit)} \qquad \qquad 31.5$$

The angular separation of two objects must be greater than λ/b if the objects are to be resolved

Notice that, for good resolution, one should have a large slit, and the wavelength used should be small.

The case of a circular opening is more complicated than the long slit. That is, however, the more important case since most optical devices such as cameras, telescopes, and microscopes use circular openings. We merely state the result for the limiting angle of resolution of a circular opening:

$$\theta_c = 1.220 \frac{\lambda}{D} \qquad \qquad 31.6$$

where D is the diameter of the circular aperture. We shall have occasion to use it in our discussion of telescopes and microscopes in the next chapter.

31.6 THE DOUBLE-SLIT DIFFRACTION PATTERN

In the previous chapter we discussed the interference of light produced by two slits. The Young's double-slit experiment was described in terms of only two rays, one from each slit. This will be allowable if the slits are so narrow that all the rays from one of the slits have path-length differences which are negligible in comparison to λ. In terms of the single-slit diffraction pattern, this means that each slit is narrow enough so that the central maximum of the single-slit diffraction pattern covers the whole field of view. Since the first minimum of the single-slit pattern occurs where $\sin \theta = \lambda/b$ (see illustration 31.1), we tacitly assumed the slit width b to be of the order of λ or smaller. Now, however, we wish to describe what happens when the single-slit diffraction pattern is comparable in size to the double-slit pattern. In such a case, b, the slit width, and a, the distance between slit centers, are comparable.

Suppose we have a single slit of width b which has the diffraction pattern given in figure 31.10(a). Suppose further that the interference pattern for two extremely narrow slits (width $\ll b$) separated by a distance a is as shown in figure 31.10(b). If now these two slits are replaced by two slits of width b, the pattern shown in part (c) is obtained. We can understand it in the following way.

Each slit of width b gives rise to wave amplitudes at the screen which are similar to the pattern shown in figure 31.10(a). However, when the two sets of waves join together, they reinforce at certain places while canceling at other places. As a result, the combined pattern is not simply two patterns like figure 31.10(a) added together. To find the actual pattern, we must refer to part (b) of the figure. This shows us where the waves from the two slits reinforce and where they cancel. The combined pattern must be zero wherever either of the patterns (a) or (b) is zero.

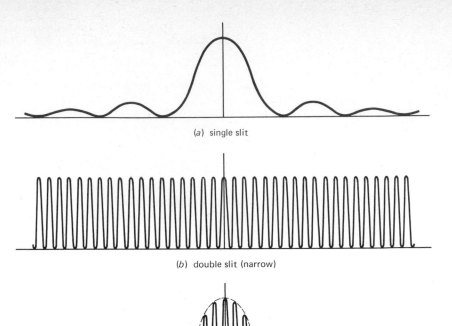

(a) single slit

(b) double slit (narrow)

(c) combined

FIGURE 31.10 The double-slit pattern is the product of the single-slit pattern and the simple double-slit pattern.

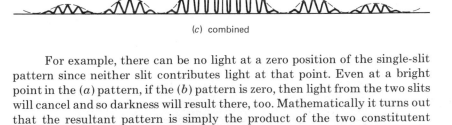

For example, there can be no light at a zero position of the single-slit pattern since neither slit contributes light at that point. Even at a bright point in the (a) pattern, if the (b) pattern is zero, then light from the two slits will cancel and so darkness will result there, too. Mathematically it turns out that the resultant pattern is simply the product of the two constitutent patterns. This is shown in part (c) of figure 31.10. Photographs showing this effect for several ratios of b/a are shown in figure 31.11.

Illustration 31.2

Two slits are a distance a apart and have individual widths b. What must be the ratio of b to a if the single slit's first minimum occurs at the position of the double slit's third-order maximum? (In that case, the third maximum would be missing.)

Reasoning Since we are dealing with parallel rays, the situation is as shown in figure 31.12. For the third maximum in the double-slit pattern, $\delta = a \sin \theta = 3\lambda$, and so $\sin \theta = 3\lambda/a$. But the first minimum for the single-slit pattern occurs where (see illustration 31.1) $\sin \theta = \lambda/b$. If we equate these two expressions for $\sin \theta$ we find

$$\frac{3\lambda}{a} = \frac{\lambda}{b}$$

from which

$$\frac{b}{a} = \frac{1}{3}$$

This is the situation shown in the second part of figure 31.11.

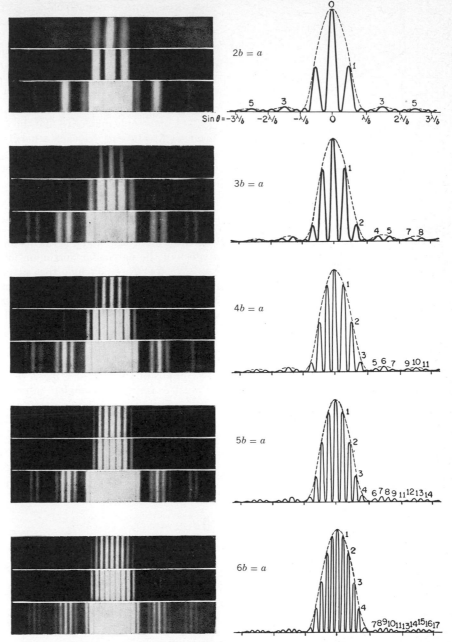

FIGURE 31.11 Photographs (at three different exposure times) and intensity curves for double-slit diffraction patterns. In interpreting them, consider the slit width b to be constant. The slit separation a increases from top to bottom. (After Jenkins and White.)

In the figure, the following labels appear beside each intensity curve: $2b = a$, $3b = a$, $4b = a$, $5b = a$, $6b = a$. For the top curve the horizontal axis is labeled $\sin\theta = -3\lambda/b \quad -2\lambda/b \quad -\lambda/b \quad 0 \quad \lambda/b \quad 2\lambda/b \quad 3\lambda/b$.

31.7 THE DIFFRACTION GRATING*

None of the interference and diffraction fringes discussed prior to this point can be used for high-precision wavelength determination. The fringes have not been sharp and narrow enough to allow one to locate the fringe position

---

*I am indebted to Professors W. H. Colbert and G. Moneti for valuable suggestions concerning this development.

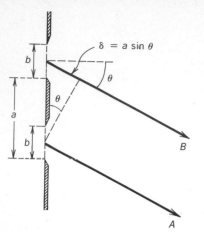

$\delta = a \sin \theta$

FIGURE 31.12 The double-slit pattern is obtained by summing the contributions from the two slits.

with certainty. A device which is not subject to this criticism is the *diffraction grating,* and we shall see that it can be used for accurate wavelength measurements.

Let us start by considering the diffraction pattern of N parallel slits a distance d apart and of individual width b. The situation is shown in figure 31.13. We can simplify the discussion by assuming the slits are thin enough so that the central maximum of the single-slit pattern fills the whole field of view. We can then ignore the complicating effect of the single-slit pattern.

It is a simple matter to find the angle θ_n at which the beams from the various slits all reinforce each other. If, in figure 31.13(a), the length designated by Δ is a whole number of wavelengths long, namely $n\lambda$, then the beams will all reinforce. From the geometry of the small triangle near the bottom slit, we have that, for reinforcement of all the beams,

$$\sin \theta_n = \frac{\Delta}{d} = \frac{n\lambda}{d}$$

where n is an integer and d is the distance between slits. We call n the *order number* of the particular maximum in question.

We therefore conclude that at certain angles, θ_n, the beams will all reinforce. As a result, very strong maxima, the diffraction grating fringes, are observed at these angles, θ_n. The relation for θ_n was given above. It is usually written in the following form, called the *grating formula,*

Grating formula

$$n\lambda = d \sin \theta_n \qquad\qquad 31.7$$

These fringes can be made extremely narrow and sharp if the number of slits composing the grating is made very large. To see how this comes about, we must examine the complete interference pattern for N slits.

To find the Fraunhofer interference pattern for N slits we make use of phasors. Assuming each slit to contribute a phasor of amplitude p_0, the phasor diagram is as shown in figure 31.13(b). We wish to find y_0, the resultant of the N phasors.

First we notice from part (a) of the figure that

$$\delta = (N - 1) d \sin \theta$$

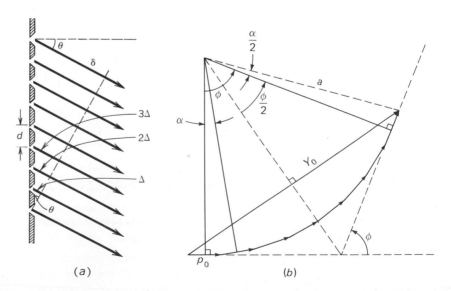

FIGURE 31.13 Although the diffracted light will be at maximum at many other places, the major maxima are obtained for $\Delta = n\lambda$, where n is an integer. Why?

(a)

(b)

and, because

$$\text{Phase angle} = 2\pi \frac{\text{path difference}}{\lambda}$$

we have that

$$\phi = \frac{2\pi(N-1)\,d\sin\theta}{\lambda} \qquad\qquad 31.8$$

Now let us look at part (b) to find y_0. We see that

$$\tfrac{1}{2}y_0 = a\sin\left(\tfrac{1}{2}\phi + \tfrac{1}{2}\alpha\right) \qquad\qquad I$$

Also, from the uppermost triangle, we have

$$\tfrac{1}{2}p_0 = a\sin\tfrac{1}{2}\alpha \qquad\qquad II$$

We also notice from the same figure that

$$\phi = (N-1)\alpha \qquad\qquad III$$

Solving for a in II and for α in III, we can substitute in I to obtain

$$y_0 = p_0\frac{\sin\dfrac{N\phi}{2(N-1)}}{\sin\dfrac{\phi}{2(N-1)}}$$

Upon using equation 31.8 for ϕ, this becomes

$$y_0 = p_0\frac{\sin(\pi N\,d\sin\theta/\lambda)}{\sin(\pi\,d\sin\theta/\lambda)} = p_0\frac{\sin Nk}{\sin k}$$

where $k \equiv \pi\,d\sin\theta/\lambda$. Because the intensity is equal to the square of the amplitude, we find

$$I = p_0{}^2\left(\frac{\sin Nk}{\sin k}\right)^2 \qquad\qquad 31.9$$

Typical interference patterns (plots of I vs. $\sin\theta$) for various numbers of slits are shown in figure 31.14. Actual photographs of the patterns formed by various numbers of slits are shown* in figure 31.15. When the number of slits is very large, one refers to the combination of the slits as a *diffraction grating*. A typical grating would consist of 10,000 slits, with $d = 10^{-4}$ cm. As indicated in figure 31.14, the maxima from a diffraction grating will be extremely sharp and well defined.

The principal maxima given in figure 31.14 are obtained from 31.9 as the values when $k = n\pi$, where $n = 0, 1, 2$, etc. Although both numerator and denominator of 31.9 go to zero at these values of k, the limit of the ratio is finite and equal to N^2. As a result, the intensities at the major maxima are given by $I_{\max} = N^2 p_0{}^2$. Of course, the relation $k = n\pi$, that applies at the principal maxima, reduces to 31.7, the grating formula. Can you show this to be true?

It is of value to consider the qualitative reason for the sharpening of the diffraction pattern as the number of lines in a grating is increased. This may be done by noticing what happens to the first minimum next to the zeroth-order fringe as N is increased. From figure 31.16 we can see that, when $\delta = \lambda$, darkness will prevail. This follows from the fact that the ray from the center

---

*The slits used for the photos shown in figure 31.15 were wider than we have assumed. As you can see, the first minimum of the single-slit pattern occurs at about the position of the seventh-order fringe in the double-slit pattern.

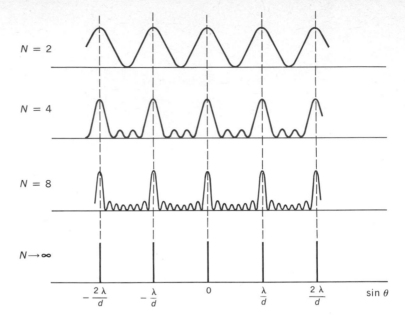

$N = 2$

$N = 4$

$N = 8$

$N \rightarrow \infty$

$-\dfrac{2\lambda}{d}$ $-\dfrac{\lambda}{d}$ 0 $\dfrac{\lambda}{d}$ $\dfrac{2\lambda}{d}$ $\sin\theta$

FIGURE 31.14 Notice that the principal maxima for a series of N slits with spacing d occur when the path difference between adjacent slits is $n\lambda$, where n is an integer.

slit will be $\lambda/2$ behind the ray from the bottommost slit. They will cancel. Similarly, the rays from the two slits just above these will cancel. And so on for all the pairs of rays in the two halves of the grating. We therefore conclude that the first minimum will occur when

$$\delta = \lambda$$

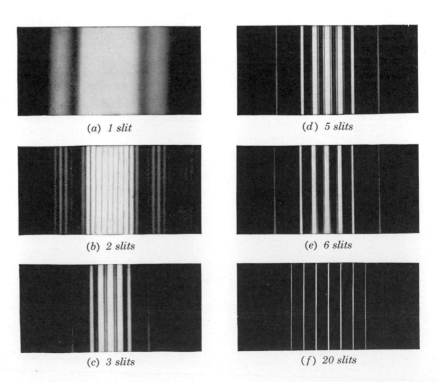

(a) 1 slit

(b) 2 slits

(c) 3 slits

(d) 5 slits

(e) 6 slits

(f) 20 slits

FIGURE 31.15 Fraunhofer diffraction patterns from the indicated number of slits. Are the slits equally spaced in all the cases shown? (After Jenkins and White.)

or since $\delta = (N - 1) d \sin \theta$, when

$$\sin \theta = \frac{\lambda}{(N - 1) d} \cong \frac{\lambda}{N d} \qquad 31.10$$

In other words, since $\theta \approx \sin \theta$ at small angles, the width of the maximum will decrease as $1/N$.

Our discussion of the diffraction grating has been in terms of slits in an opaque screen. However, everything we have said could equally well apply to thin rectangular regions of high transmission (or mirror reflection), alternating with regions which are opaque (or which reflect diffusely). In practice, diffraction gratings are most often made by ruling grooves on a perfectly flat piece of metal or glass. The metal ruling can be used as a reflection grating. So-called *replica* gratings can be made by laying down a film of plastic over the grooves, which then replicates the groove pattern. When the film is stripped loose, it will constitute a transmission grating. Original rulings are very difficult to make with precision and are too expensive for ordinary student use. For that reason, the gratings you encounter in the laboratory will probably be replicas.

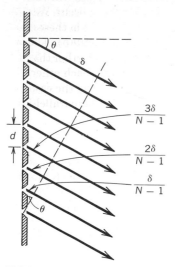

Illustration 31.3

Under certain conditions, a diffraction order may be missing. What must be the relation between b and d for the second-order diffraction maximum to be missing from the grating diffraction pattern?

Reasoning The second-order maximum occurs when $n = 2$ in *31.7*. Therefore

$$\sin \theta_2 = \frac{2\lambda}{d}$$

If a minimum from the single-slit diffraction pattern occurs at this same angle, this fringe will be missing. According to *31.2*, these minima will occur when $\sin u = 0$, or where

$$\frac{\pi b \sin \theta}{\lambda} = \pi, \, 2\pi, \, \ldots$$

Therefore, for the two to coincide, $\theta = \theta_2$, and we have

$$\frac{\pi b \sin \theta_2}{\lambda} = n\pi \qquad n = 1, \, 2, \, \ldots$$

Or, after substituting $2\lambda/d$ for $\sin \theta_2$,

$$\frac{2b}{d} = n$$

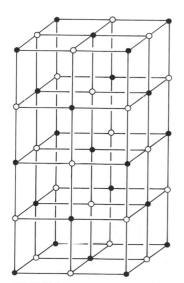

FIGURE 31.16 At what angle does the first minimum in the interference pattern occur?

from which $\qquad 2b = nd \qquad n = 1, \, 2, \, \ldots$

if the second-order fringe is to be missing. Only $n = 1$ is meaningful. Why?

31.8 DIFFRACTION BY CRYSTALS

A crystal is an ordered array of atoms or groups of atoms. For example, the sodium chloride crystal can be represented as shown in figure 31.17. (Actually, of course, the atoms are larger and more complicated than the diagram

FIGURE 31.17 In this diagram of the sodium chloride lattice, the solid circles represent Na^+, while the open circles represent Cl^-.

indicates.) Notice that the atoms appear to lie in planes. This becomes more obvious if we examine the crystal in cross section, as shown in figure 31.18.

When a system of waves is incident on a crystal, the atoms in these layers are capable of absorbing and immediately reradiating these waves. In effect, the waves reflect from these layers. Since the path length will vary for rays reflected from the various layers, interference effects can occur. We shall examine a simple system and see what quantitative law applies in these cases.

Consider the two layers of atoms shown in figure 31.19(b). The ray reflected from the lower layer must travel a distance $2d \sin \theta$ farther than the upper ray. In order for the reflected rays a and b to reinforce each other, this extra path length must be an integer multiple of λ. That being the case, not only rays a and b, but also rays reflected at this angle from all parallel-layer planes in the crystal, will reinforce. As a result, when

The Bragg relation

$$n\lambda = 2d \sin \theta \qquad n = 1, 2, \ldots \qquad 31.11$$

the waves will be strongly reflected by the crystal. This is called the *Bragg relation* (after W. H. and W. L. Bragg, who first used it extensively) for the maxima of waves diffracted by a crystal.

In connection with *31.11*, we notice that it has a *superficial* resemblance to the grating equation *31.9*. *They are not the same.* The angle θ in *31.11* is defined as 90° minus the θ used in *31.9*. In addition, the factor 2 does not appear in *31.9*. Be careful not to confuse these two relations.

Although it is fairly common to diffract other types of waves, by far the largest application is to x-ray diffraction. We shall see in a later chapter how x-rays are generated. For now, we only state that a beam of short-wavelength electromagnetic radiation can be directed at a crystal, as shown in figure 31.20. The method illustrated is called the Laue method. If the incident beam contains many wavelengths, *31.11* is satisfied by those wavelengths for which a set of planes at angle θ exist in the crystal. Usually, a fairly large number of planes and wavelengths satisfy this relation, and so a series of bright spots are found on the photographic film at the angles determined by *31.11*. A typical Laue photograph is shown in figure 31.21(a).

If a crystalline powder is used in place of a single crystal in figure 31.20, the orientation of the crystal planes will be random. In that case a series of diffraction rings is found, as shown in figure 31.21(b). Both these photographs can be used to determine the structure of the crystals. Since the diffraction angle θ is known in each case, it is possible from a knowledge of λ to find the lattice spacing d by means of *31.11*. Conversely, the wavelength of the x-rays can be measured if the crystal spacing is known.

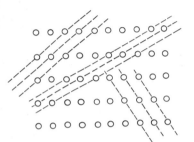

FIGURE 31.18 Many parallel-layer systems of atoms are possible in a crystal. Three of them are shown in the figure.

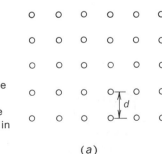

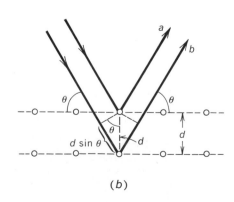

FIGURE 31.19 The atoms in crystals lie in evenly spaced planes. A simple example is shown in part (*a*). When the atomic planes reflect x-rays, as shown in part (*b*), the reflected rays give rise to interference effects.

(*a*)

(*b*)

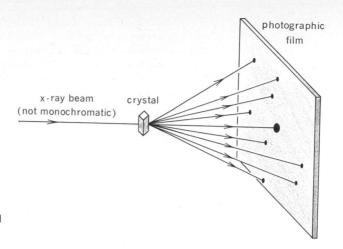

FIGURE 31.20 Production of a Laue-type x-ray diffraction pattern. What determines at which angles the diffracted beams will be found?

Illustration 31.4

In a certain x-ray diffraction experiment, a first-order image is noted at a diffraction angle of 5° for a crystal-plane spacing of 2.8×10^{-10} m. What is the x-ray wavelength?

Reasoning Making use of the Bragg relation *31.11*, with $n = 1$, $\theta = 5°$, and $d = 2.8 \times 10^{-10}$, we find

$$\lambda = (5.6 \times 10^{-10})(0.087) = 0.49 \times 10^{-10} \text{ m}$$
$$= 0.49 \text{ Å}$$

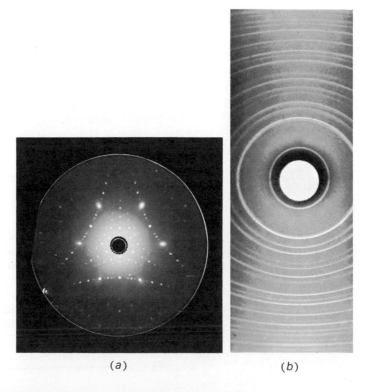

FIGURE 31.21 When a beam of x-rays is shone on a single crystal, a Laue diffraction picture is obtained, as shown in part (*a*). If a crystalline powder is used instead of a single crystal, a Debye-Scherrer diffraction pattern is obtained, as shown in part (*b*).

(*a*)   (*b*)

Until the diffraction of x-rays was first measured by von Laue in 1912, it was not known for sure that they were wavelike in character. Attempts to obtain diffraction effects using ordinary slits had failed. You should be able to explain why such attempts had been unsuccessful.

Chapter 31
Questions
and Guesstimates

1. Distinguish clearly between the terms *diffraction* and *interference*. Can you think of an example where diffraction occurs without interference? Where interference occurs without diffraction?

2. If you were given a diffraction grating of unknown spacing, how could you determine the spacing of the lines on it?

3. A student sends the light from a mercury arc through a diffraction grating and observes the first-order 4358-Å blue line at an angle of 69°. He knows this light also contains green and yellow wavelengths but is unable to see these colors. What could cause this effect?

4. Two slits are at right angles to each other so as to form a plus sign, thus, +. Describe as much as you can about the interference and diffraction pattern one would observe for such a situation.

5. Describe qualitatively the interference pattern one would observe for a single-square slit.

6. Would it be just as reasonable if diffraction gratings were called "interference gratings"?

7. How would the Fraunhofer diffraction pattern for a single slit change as (*a*) the slit width is slowly changed from extremely small to large? (*b*) As the wavelength is slowly changed from very large to small?

8. If you look through a sheer curtain at a distant light at night you will see an interference pattern. From the pattern, estimate the relative spacings of the vertical and horizontal threads in the sheer. Check by direct examination of the curtain.

9. A telephone pole will cast a sharp shadow in the light from a distant source. But the pole does not cast a sharp shadow for the sound from a distant car horn. Why is there this difference?

10. The center point in a Fraunhofer diffraction pattern from any number of parallel slits is always a bright point. Why is this true? Why is it not true for Fresnel diffraction?

11. From figure 31.14 it appears that for a set of N slits, where N is an even number, there will be $N - 1$ minima between the principal maxima. Give a qualitative argument showing why this should be true.

12. Describe an experiment by which you could determine the interference pattern of a double slit for 200-Hz sound if you were given a small loudspeaker capable of sending out sound of this frequency.

13. A light bulb hangs in a hallway 10 m from a door at the end of the hall. The door has a crack in it about 1 mm wide. Even though light from the bulb going through the crack can be seen to illuminate faintly the wall opposite the door in the dark room behind it, the diffraction-pattern characteristic of a single slit is not seen. Why not?

14. Two tiny loudspeakers sit side by side about 1 m apart and send sound of wavelength 50 cm out across a large, flat field. What sort of interference effects would you expect for them? Suppose 10 speakers at this same separation were used. How would the interference effects be changed?

15. To a first approximation the opening in the eye acts like a slit. If diffraction was the controlling factor, about how far away from you

could a car be for you to still be capable of resolving its headlights into two distinct sources?

16. What basis is there for the following statement: Interference from a thin film is much like double-slit interference, while x-ray interference from a crystal is similar to the interference effect of a diffraction grating.

Chapter 31
Problems

Note: Whenever possible, work out the problems from consideration of ray diagrams and not by substitution in a formula.

1. If yellow light from a sodium arc ($\lambda = 589$ nm) is used in a Fraunhofer single-slit diffraction experiment, how wide must the slit be if the first minimum is to occur at an angle of 5°? Would it be difficult to carry out such an experiment?

2. When yellow light from a sodium arc ($\lambda = 5890$ Å) is sent through a single slit of width 0.10 mm in a Fraunhofer-type diffraction experiment, at what angle will the first-order minimum occur?

3. Using the mercury arc green light ($\lambda = 5460$ Å) in a Fraunhofer single-slit experiment, the third-order minimum occurs at an angle of 0.5°. How wide is the slit?

4. Upon sending the mercury arc blue light ($\lambda = 436$ nm) through a single slit in a Fraunhofer diffraction experiment, the fourth-order minimum is found to occur at an angle of 0.3°. How wide is the slit?

5. In a single-slit Fraunhofer diffraction experiment, the slit width is 0.050 cm and the red light from a helium-neon laser ($\lambda = 6328$ Å) is used. How far from the center maximum is the third-order maximum found on a screen 10 m away from the slit?

6. To determine the width of a very narrow slit, a girl sends red light from a helium-neon laser ($\lambda = 633$ nm) through the slit in a single-slit diffraction experiment. Since she (rightly) does not wish to view the laser light directly, the diffracted beam is allowed to strike a screen 5.0 m from the slit. She measures the distance from the central maximum to the second-order maximum to be 3.0 cm. How wide is the slit?

7. The width of the central maximum in a Fraunhofer single-slit experiment is 4.0 cm when using sodium yellow light (589 nm). What will be the width if the light is changed to mercury blue light (436 nm)?

8. Mercury arc light contains three very strong wavelengths, 436 nm, 546 nm, and 579 nm, which correspond to blue, green, and yellow, respectively. Find the angular positions of the first- and second-order minima in the blue single-slit diffraction pattern and compare them with the position of the first-order maximum in the yellow pattern. Assume a slit width of 0.01 cm and Fraunhofer diffraction.

9. The headlights of a distant truck are viewed by a man's eye. If the eye opening has a diameter of 0.30 cm, how far away will the truck be if the two headlights are just resolved? Assume that the limiting factor is diffraction caused by the eye opening, that the effective $\lambda = 500$ nm, and that the truck light separation is 150 cm. What can you conclude from your result?

10. An image of a photographic slide is projected onto a screen 3 m away by a lens with 2-cm diameter. The lens is 12 cm from the slide. Assuming the lens to be perfect so that diffraction limits its imaging ability, how close together can two tiny spots on the film be if they are to be resolved on the screen? Assume $\lambda = 500$ nm and Fraunhofer conditions.

11. Microwaves of $\lambda = 3$ cm strike a metal plate in which a slit of width b exists. Assuming a Fraunhofer arrangement, find how large b should be if the first-order diffraction maximum is to occur at an angle of about 60°.

12. Using microwaves in an approximately Fraunhofer situation, the interference pattern formed from two slits in a metal plate shows a first-order maximum at an angle of 20°. If the slits are 10 cm between centers, what is the wavelength of the microwaves?

13. It is desired to obtain an essentially "pure" double-slit interference pattern (i.e., without diffraction effects). The slit separation is 1.5 mm. How narrow must the slits be if the first diffraction minimum is to be at the tenth bright fringe of the interference pattern? Assume Fraunhofer conditions.

14. What must be the ratio of the slit width b to the slit separation a in a Fraunhofer double-slit experiment if the third-order double-slit maximum is to be missing? What other maxima would also be missing?

15. To calibrate a diffraction grating, a student sends red light from a helium-neon laser (6328 Å) through the grating in a Fraunhofer-type experiment. The first-order maximum occurs at an angle of 38°. What is the grating spacing? At what angle does the second-order maximum occur?

16. The sodium arc yellow light is actually a doublet composed of two wavelengths, 5889.95 and 5895.92 Å. Compute the angular separation between these two lines in first order for a grating with 5000 slits/cm. Repeat for the second order.

17. Suppose a diffraction grating gives a first-order maximum at 30° for a particular wavelength of light. If now the whole experiment is carried out under water, at what angle should the same maximum occur?

18. Modern grating spectrometers often use reflection gratings. These are mirrors on the surface of which are a series of reflecting lines (equivalent to the slits of the transmission grating). This situation is shown in figure P31.1, where the distance between the centers of the reflecting lines is d. Find the grating equation for this device, i.e., the angles at which interference maxima occur.

19. Steel sheds often have a corrugated metal surface with corrugation repeating every 10 cm or so. Under appropriate conditions this type of wall can act as a reflecting diffraction grating for sound waves (see problem 18). What wavelength waves at normal incidence will give rise to a first-order maximum at an angle of 30° to the normal?

20. A beam of monochromatic x-rays ($\lambda = 0.48$ Å) is used to obtain Bragg reflection from a crystal. Strong reflection is observed at an incidence angle of 20°. What are the possible layer plane spacings which could give rise to this maximum?

21. The first-order Bragg reflection occurs at a reflection angle of 67° for a certain set of crystal layer planes and a monochromatic beam of x-rays. At what angle will the second-order reflection from these planes be found?

22. A "crystal" for a laboratory experiment using 3 cm microwaves consists of a stack of plastic sheets with a rectangular array of metal spheres placed on each sheet. The vertical distance between these "atom" planes is 5 cm. A microwave beam is incident on this "crystal" from above. At what angles of incidence will the microwave beam be reflected strongly?

23. Show that the grating equation, equation *31.9*, is consistent with the

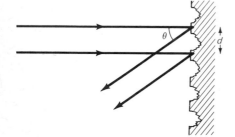

FIGURE P31.1

Young's double-slit equation, *30.5*, in the Fraunhofer limit, i.e., for $d \ll D$ and $x \ll D$.

24. Show that the multislit pattern predicted by *31.8* reduces in the case of two slits to the double-slit pattern given by *30.6* in the limit of $x \ll D$.

25. Use the limiting methods of calculus to show that *31.8* gives $I = (Np_0)^2$ at the major maxima.

26. A certain diffraction grating is made up of alternating equal-width transparent and opaque strips. Show that the maxima for $n = 2, 4, 6, \ldots$ will be missing for this grating.

27. Use calculus methods to show that *31.2* predicts the maxima in the single-slit diffraction pattern to occur when $\tan u = u$. Show that the first two roots of this equation are $u = 0$ and $u = 4.5$. (You can do this graphically by plotting both $\tan u$ and u both vs. u on the same graph and finding the intersections of the two plots.)

28. Show that the angular separation between two spectral lines in the nth order for a diffraction grating is given by

$$\frac{n\,\Delta\lambda}{d\cos\theta}$$

provided the wavelength separation of the lines, $\Delta\lambda$, is small.

32 Optical Instruments

The principles of wave motion, interference, and diffraction which we have been studying find important application in optical instruments. We outline the construction of some of these instruments in this chapter. It is found that diffraction seriously affects the ability of optical devices to discern detail. Since many optical measuring and viewing devices make use of polarized light, a discussion of polarization is also included in this chapter.

32.1 SIMPLE MAGNIFIER

One of the simplest optical instruments is the magnifying glass. Since it is one of the basic constituents in many optical devices, we describe its characteristics. It is a converging lens, and its focal length is designated f. In practice, one uses the magnifier in order that an enlarged image of the object being viewed may be formed on the retina of the eye. (The retina serves as the recording surface for the eye. It is equivalent in function to the film in a camera.) We can understand this most easily by reference to the two diagrams shown in figure 32.1.

A magnifying glass allows one to observe an enlarged virtual image of an object

As seen in the upper part of figure 32.1, the size of the image formed on the retina grows as the object is brought very close to the eye. But the human eye is unable to focus well on objects closer than a distance γ (≈ 25 cm). If we use a converging lens in front of the eye, as shown in figure 32.1, we can view the virtual image of the object formed by the lens. Although the image is at a distance γ from the eye, the object is much closer. As a result, the image on the retina of the eye is much larger, and more detail within the image can be seen.

Two methods are used to measure the magnifying effect in this case. The ordinary, or *linear, magnification M* was defined in Chap. 29 to be the ratio of the image length to the object length. This was shown to be equivalent to the ratio of the image distance p' to the object distance p. In the present case

$$M = \frac{\gamma}{p} \approx \frac{\gamma}{f} \qquad\qquad 32.1$$

Definitions of linear and angular magnification

The latter form is used since, in practice, the object is placed just slightly inside the focal point, and so $p \approx f$.

The second method for describing the magnification is to use a quantity

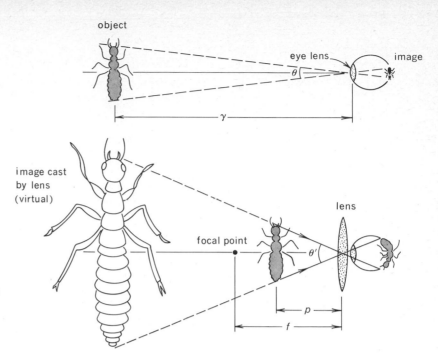

object
eye lens
image
θ
γ

image cast
by lens
(virtual)

lens

focal point
θ'

p

f

FIGURE 32.1 The virtue of the simple magnifying glass is that it enables one to see an object clearly when it is closer than $\gamma \approx 25$ cm to the eye. For clarity, the image on the retina is drawn much too large.

called the *angular magnification*. It is defined to be the ratio of the angle θ', the angle subtended by the viewing device with the lens, to the angle θ, the angle subtended without the lens. In the present case we have, from a consideration of figure 32.1 and upon assuming small angles so that $\sin \theta' \approx \theta'$, and calling the length of the object S,

$$\text{Angular magnification} = \frac{\theta'}{\theta} \approx \frac{S/p}{S/\gamma} = \frac{\gamma}{p} \approx \frac{\gamma}{f}$$

As we see, the two definitions give the same results under the present conditions.

A typical simple magnifying glass might have a focal length of about 5 or 10 cm. Since $\gamma \approx 25$ cm, it would provide a magnification of between 2.5 to 5. In other words, if all other factors remained constant, such a lens would allow one to observe detail with dimensions as small as one-fifth as large as would be possible with the naked eye. Usually, however, other factors must also be considered. Among these are blurring of the image due to spherical and chromatic aberrations of the lens. In order to remove these aberrations, good eyepieces use a complex combination of lenses in place of a simple lens. Even in the case of perfect lenses, there is a limit to the detail resolvable. We discuss the causes of this limit in the next section.

32.2 LIMITS OF RESOLUTION

When light passes through a lens, the lens acts like a circular aperature. In the preceding chapter we saw that the aperture gives rise to a diffraction pattern. Because of this, the image of a point becomes a diffraction disk. We saw (*31.6*) that the images of two objects could be resolved as distinct entities

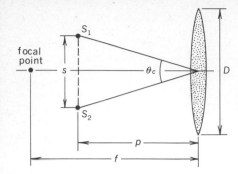

FIGURE 32.2 The two sources can just be resolved in the image formed by the lens if $\theta_c = 1.22\lambda/D$.

if they subtended an angle θ_c or larger at the aperture, where

$$\theta_c = 1.22\frac{\lambda}{D} \qquad\qquad 32.2$$

In this expression λ is the wavelength of the light being used, and D is the diameter of the circular aperture. We actually stated this equation for the case of plane incident waves. However, it turns out to be nearly correct in all cases.

To illustrate the significance of this limiting angle in the case of a simple lens, let us refer to figure 32.2. The lens shown there is being used as a simple magnifier, and will form a virtual image of the two sources s_1 and s_2 at some point to the left of the focal point. Equation *32.2* states that these sources will be resolved in the image only if

$$\theta_c \geq 1.22\frac{\lambda}{D}$$

Since it would be of more value in the present case to state this limit of resolution in terms of the allowable distance between the sources s, we seek a relation between θ_c and s. This is easily found if we assume θ_c is small enough so that the angle is nearly equal to its sine or tangent. In that case

$$\theta_c = \frac{s}{p} \cong \frac{s}{f}$$

Using this value for θ_c in *32.2*, we find that the sources will be resolved if

$$s \geq 1.22\frac{f\lambda}{D} \qquad\qquad 32.3$$

Of course, the factor 1.22 is no longer exact because of the approximations we have made. However, in most cases the focal length of the lens f will be larger than the lens diameter D. In fact, a lens with f much smaller than D is not commonly found. Therefore we can state with confidence that, since $f/D \geq 1$, two objects can be resolved by even the most advantageous lens only if

$$s > \lambda$$

An optical system cannot resolve detail smaller than λ

As a rule of thumb, then, we can state that *an optical system cannot resolve detail with dimension less than the wavelength of light used.* In most cases, this limit of resolution is not approached because of other unfavorable factors.

The eye itself has a limit to the detail which it can resolve. If we consider it to be a simple lens with aperture 0.3 cm and minimum object distance 10 cm, then *32.3* tells us it can only resolve details with dimensions s given by

$$s \geq (1.22)(33\lambda) \approx 40\lambda$$

In practice, the human eye is not quite capable of resolving objects as close together as *32.3* would allow. We must therefore conclude that defects in the eye-lens system, or the "graininess" of the retina, are responsible for this additional loss in resolution.

Although our present discussion has been carried out in terms of light waves, the same results apply to other wave disturbances. One need only replace the eye by a photocell, for example, and ultraviolet radiation could be used. Or, using a magnetic lens to focus electrons, the same considerations apply to the electron microscope. In that case one would use the wavelength of the probability wave (discussed in Chap. 35) assigned to the electrons in place of λ. We shall see in a later chapter that the electron microscope has the ability to resolve detail in the 10^{-10}-m range because the electron wavelength is of that order of magnitude.

Illustration 32.1

The large radio-telescope reflecting mirror shown in figure 32.3 has an aperture diameter of 64 m and focuses radio waves from space with wavelengths of 21 cm. What is its limiting angle of resolution? How small a detail can it "see" on the moon?

Reasoning The diffraction formula for a circular aperture will apply here too. From *32.2* we have

$$\theta_c = 1.22 \frac{21 \times 10^{-2}}{64} = 4.0 \times 10^{-3} \, \text{rad}$$

Since the distance to the moon is 3.8×10^8 m, two objects on the moon separated by a distance s will subtend an angle

$$\theta = \frac{s}{3.8 \times 10^8 \, \text{m}}$$

on the earth. Replacing θ by θ_c, we see that the finest detail observable is

$$s = 1.5 \times 10^6 \, \text{m}$$

From the fact that the radius of the moon is about 1.8×10^6 m, it is clear that even this huge radio telescope is of little value in exploration of the moon.

32.3 THE SIMPLE TELESCOPE

The simplest telescope consists of two lenses, an objective lens and magnifying (or eyepiece) lens. As shown in figure 32.4, the objective lens with focal length f_0 forms an image of the distant object at which the telescope is aimed. Since the object distance is large, the image will be formed very close to the focal point of the objective lens. This image is then examined in the usual way by the eyepiece acting as a magnifying glass.

In finding the magnification of this system one generally uses the angular magnification since the object distance and object size are often nearly infinite. As stated in Sec. 32.1, the angular magnification is the ratio of the angle θ' subtended at the eye when viewing through the device to the angle θ subtended by the object at the eye in the absence of the lens system.

FIGURE 32.3 The reflecting telescope at Parkes, Australia, has an aperture diameter of 64 m. It can be rotated so as to "see" most of the visible sky. Although it is designed to function best at the microwave wavelength of the 21-cm hydrogen line, it is capable of detecting wavelengths as small as a few centimeters.

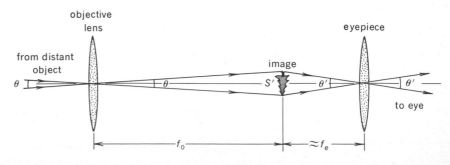

FIGURE 32.4 How would one choose f_0 and f_e to make a more powerful telescope than the one shown here?

If the rays shown coming to the objective lens are from the extreme edges of the object, then θ is equivalent to the angle so labeled in figure 32.4. As shown in the figure, θ' is determined by the position of the image relative to the eyepiece lens. In practice, the image will be viewed at a position close to the focal point of the eyepiece lens. As a result, the distance from the eyepiece lens to the image is essentially equal to f_e, the focal length of the eyepiece lens.

Since the angles involved will be small, we can equate the angles to their sines and tangents. Doing so, we see from figure 32.4 that

$$\theta = \frac{S'}{f_0} \quad \text{and} \quad \theta' = \frac{S'}{f_e}$$

From this, the angular magnification becomes

For good magnification the focal length of the telescope objective should be large

$$M = \frac{\theta'}{\theta} = \frac{f_0}{f_e} \qquad\qquad 32.4$$

We see that a powerful telescope should have a short-focal-length eyepiece lens and a long-focal-length objective lens.

Another factor which influences the construction of a telescope is the limit of resolution of the device. From *32.2* we have that the objective lens will be able to resolve detail which subtends an angle θ_c or larger, where

$$\theta_c = 1.22 \frac{\lambda}{D}$$

In order to make θ_c as small as possible, the diameter D of the objective lens should be very large. One of the largest existing refracting telescopes, the one at the Yerkes Observatory, has a lens diameter of about 1.0 m and a focal length of about 19 m. It is capable of resolving objects which subtend an angle*

$$\theta_c = 1.22 \frac{\lambda}{1.0} \cong 5 \times 10^{-7} \text{ rad}$$

In an effort to obtain better resolution than this, reflecting telescopes have been constructed. These replace the lens by a converging parabolic mirror. The image formed by such a mirror does not suffer from spherical and chromatic aberrations. Moreover, large mirrors are more feasible to make than lenses. The Mt. Palomar reflecting telescope has an aperture diameter of 5 m and a magnification of about 3500. Of course, large objective lenses and mirrors have the additional advantage that they gather more light. As a result, fainter objects can be seen by their use.

Illustration 32.2

How far apart must objects be on the moon if they are to be resolved by the Yerkes telescope?

Reasoning Using the fact that $\theta_c \approx 5 \times 10^{-7}$ rad and that the earth-to-moon distance R is 3.8×10^8 m, the distance s between just resolvable objects will be given by

$$\theta_c = \frac{s}{R}$$

from which

$$s \approx 200 \text{ m}$$

*We do not consider the resolving power of an eyepiece lens since, for the most exacting work, the image is recorded photographically.

In practice, the image is also blurred by the nonuniformity in the atmosphere above the earth.

32.4 THE COMPOUND MICROSCOPE

In the microscope, use is made of the fact that an objective lens can give an enlarged real image of an object. This image is then examined by an eyepiece magnifying glass. The simplest microscope of this type is shown in figure 32.5(a) and (b). In (b) we have drawn the appropriate ray diagram so that the image locations and character are clearly seen. Notice that the eyepiece magnifying lens forms a virtual image at a distance of about γ from the eye. (Recall that γ is the closest distance at which an eye can comfortably view an object.) One obtains the image in this position by changing the microscope-object distance until a clear image is seen.

To find the magnification obtained with this microscope, we notice that the objective lens causes a magnification

$$M_0 = \frac{p_0'}{p_0} \approx \frac{p_0'}{f_0}$$

FIGURE 32.5 For high magnification in a microscope, how should the focal lengths of the lenses be chosen?

We have replaced p_0 by the focal length of the objective lens, f_0, since one usually has the object quite close to the focal point. The magnification

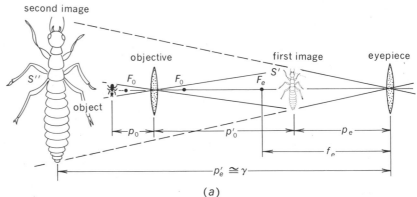

(a)

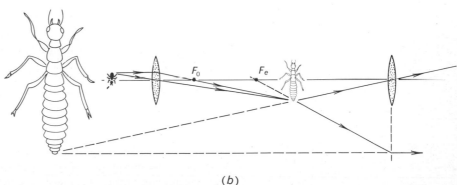

(b)

resulting from the eyepiece lens is

$$M_e = \frac{p'_e}{p_e} \approx \frac{\gamma}{f_e}$$

In writing this we are merely duplicating the derivation of the magnification of a magnifying glass as given in Sec. 32.1.

The total magnification of the system will be the product of M_0 and M_e. (Why not their sum?) It is

Both the objective and eyepiece lenses in a microscope should have short f's

$$M = \frac{\gamma p'_0}{f_0 f_e} \qquad\qquad 32.5$$

Since this expression indicates that the focal lengths of the two lenses should be small for maximum magnification, both f_0 and f_e are usually small compared with the length of the microscope itself. As a result, p'_0 will not differ much from the length of the microscope, about 18 cm. A typical value for γ would be about 25 cm.

If one has perfect lenses, the limit of resolution of the microscope will be determined by the diffraction patterns associated with the images. As we have shown for the simple magnifier, the detail which can be resolved by a microscope will be of the same order as the wavelength of the light. To obtain very high resolution, one would use blue (or even ultraviolet) light since λ is smaller in these cases. However, except in the very best microscopes, the limit on resolution is set by the aberrations of the lenses. The best microscopes use very complex combinations of lenses for the objective and eyepiece in order to eliminate the effects of lens defects.

32.5 THE GRATING SPECTROMETER AND SPECTRA

Precise measurement of wavelength of light is a widely used tool in science and engineering. As we shall see in our study of atoms and molecules, the wavelengths of light emitted and absorbed by a substance tell us a great deal about the structure of the material. One of the most widely used methods for measuring wavelength is the grating spectrometer, one version of which we now describe.

As we saw in the preceding chapter, a diffraction grating gives extremely sharp fringes at the diffraction angles, satisfying the grating equation

$$n\lambda = d \sin \theta$$

where d = distance between slits
θ = diffraction angle (shown in figure 32.6)
n = order number of fringe

Knowing n, d, and θ, the wavelength of the light giving rise to the fringe can be computed. A typical experimental setup for student use is shown in figure 32.6. It functions as follows.

The source of light illuminates the slit, and this slit is at the focal point of a lens called the collimating lens. As a result, parallel light comes from the collimator, as shown, and is incident perpendicular to the grating. If the light from the source is monochromatic, strong beams of light exit to the right of the grating at angles which satisfy the grating equation. By swinging the telescope to the appropriate angle, the beam of light enters the telescope, and the objective lens of the telescope forms an image of the slit of the collimator.

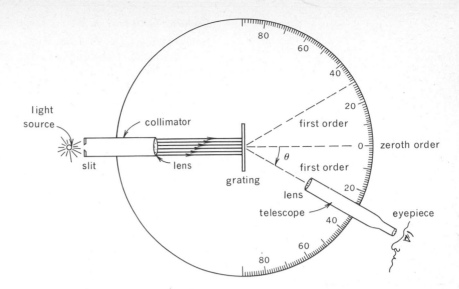

FIGURE 32.6 In the grating spectrometer, one can replace the eye and eyepiece by a camera and record the spectral lines (the slit images) on film. At what angle would the second-order image occur if the first-order is exactly at 30°?

This slit image occurs at a precise angle, the angle satisfying the grating equation. From its position, the wavelength of the light can be computed.

As we have indicated, each wavelength of light coming from the collimator results in an image of the slit at the eyepiece of the telescope. These slit images appear as *lines* in the field of view, and are commonly called *spectral lines*. They are actually images of the slit, one image for each wavelength in a given diffraction order. The positions of the slit images, or spectral lines, are determined by the grating equation. Very frequently the eyepiece system is replaced by photographic film, and the spectral lines (slit images) are recorded on it. A diagram of a typical set of spectral lines is given in figure 32.7. We shall see later that the pattern and wavelengths of the spectral lines emitted by an atom are characteristic of that atom, and afford us information about the atom in question. The particular example shown in figure 32.7 is for the light from a mercury arc.

Illustration 32.3

The visible sodium arc spectrum consists of two yellow lines of wavelength 5890 and 5896 Å. Using a grating of 4000 lines/cm, find the angles for the first two diffraction orders.

Reasoning We refer to the grating equation, making use of the fact that $d = 10^{-3}/4$ cm.

First Order, n = 1 $\sin\theta = 4 \times \lambda \times 10^3$

from which $\theta_{5890} = 13°37.6'$ and $\theta_{5896} = 13°38.5'$.

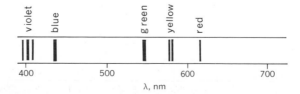

FIGURE 32.7 When a spectrometer is used to photograph a slit illuminated by a mercury arc, several images of the slit (or spectral lines) appear on the photograph, as shown here.

Second Order, n = 2 $\sin \theta = 8 \times \lambda \times 10^3$

from which $\theta_{5890} = 28°6.7'$ and $\theta_{5896} = 28°8.6'$.

Notice that the angular separation is very small but increases in the higher orders.

32.6 RESOLVING POWER OF A GRATING

As shown in the preceding example, the angular separation between two spectral lines may be very small indeed. It becomes important, therefore, to see if the grating is really capable of resolving the images of the slit. Basically, what we ask, then, is, What is the angular difference between the position of the bright fringe and the minimum on each side of the fringe? The situation is shown schematically in figure 32.8.

In figure 32.8(a) the condition for the intense nth-order diffraction image is shown. The path difference between the waves from adjacent slits is $n\lambda$, and since there are N slits, the extreme path difference is $(n\lambda)N$. If the extreme path difference is increased by a length λ (i.e., if θ is increased by $\Delta\theta$), the path length for each slit in the lower half of the grating will be $\lambda/2$ shorter than that for a corresponding slit in the upper half. As a result, the waves from all the slits will cancel, and the minimum adjacent to the spectral line is achieved. Two spectral lines must be separated by at least this angle $\Delta\theta$ in order to be resolved. We now find an expression for $\Delta\theta$ in terms of $\Delta\lambda$, the difference in wavelenghs for resolution.

From the grating equation we have

$$n\lambda = b \sin \theta$$

If we take the derivative with respect to θ, we find

$$n\frac{d\lambda}{d\theta} = b \cos \theta \qquad\qquad 32.6$$

The dispersion of a grating is proportional to $n/b \cos \theta$ The quantity $1/(d\lambda/d\theta)$, which is $\Delta\theta/\Delta\lambda$, measures the change in angle due to a change in wavelength. Because it is a measure of how much the grating

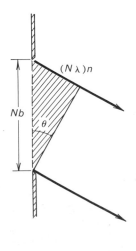

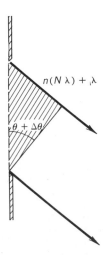

(a) brightness   (b) darkness

FIGURE 32.8 The angular distance $\Delta\theta$ from the diffraction-line center to its side minimum is the change in angle which increases the extreme path difference by λ. In part (b), each slit in the lower half of the grating cancels the wave from a slit in the upper half. Why?

separates the spectral lines, it is called the *dispersion* of the grating. As one would expect, it is proportional to n/b. It is possible to obtain another expression for $d\lambda/d\theta$ and then eliminate $\cos\theta\,d\theta$ from these relations. To do this we notice from figure 32.8 that

$$nN\lambda = Nb\sin\theta$$

and

$$nN\lambda + \lambda = Nb\sin(\theta + \Delta\theta)$$

After subtracting the first relation from the second and using the trigonometric identity

$$\sin(\theta + \Delta\theta) = \sin\theta\cos\Delta\theta + \cos\theta\sin\Delta\theta$$

we find

$$\lambda = Nb(\Delta\theta)\cos\theta \qquad 32.7$$

In writing this, use has been made of the fact that $\Delta\theta$ is very small; so $\sin\Delta\theta = \Delta\theta$ and $\cos\Delta\theta = 1$.

Let us now rewrite *32.6*, using the symbol $\Delta\theta$ for $d\theta$ and $\Delta\lambda$ for $d\lambda$. It then becomes

$$n\,\Delta\lambda = b(\Delta\theta)\cos\theta$$

which can now be used in simultaneous solution with *32.7*. Dividing it by *32.7*, we find that the difference $\Delta\lambda$ between the wavelengths which can just be resolved is given by

Theoretical limit for resolution by a grating

$$\frac{\Delta\lambda}{\lambda} = \frac{1}{Nn} \qquad 32.8$$

We see, therefore, that the lower limit on the difference in λ which can be resolved becomes smaller if the number of lines in the grating N is increased and as one goes to higher orders. This is to be expected since we know that an increase in N sharpens the pattern and that the lines are further spaced at higher orders.

One often defines a quantity called the *resolving power* of a grating:

$$\text{Resolving power} = \frac{\lambda}{\Delta\lambda} = Nn \qquad 32.9$$

Illustration 32.4

Can the grating of illustration 32.3 resolve in the first order the two lines of the sodium doublet? The grating has a width of 2.0 cm.

Reasoning Since the grating in question has 4000 lines/cm, we see that $N = 8000$. Its resolving power in the first order is 8000. Therefore, for $\lambda = 5893$ Å, we have

$$\Delta\lambda = \frac{5893\text{ Å}}{8000}$$

$$= 0.74\text{ Å}$$

Since the two lines in question are 5893 and 5896 Å, they can be resolved easily by this grating.

32.7 PRISM SPECTROMETER

The index of refraction of all materials changes with wavelength. This fact allows one to make use of a prism to separate wavelengths. One typically finds that the index of refraction μ of a transparent material in the visible portion of the spectrum obeys a *dispersion relation* (the *Cauchy formula*)

The Cauchy formula

$$\mu = A + \frac{B}{\lambda^2}$$

32.10

where A and B are experimental constants. If the constant B is large, the index of refraction changes most strongly with wavelength. As we shall see,

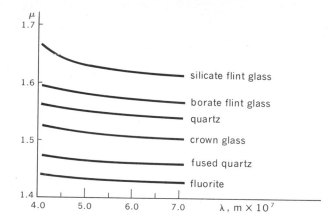

FIGURE 32.9 Dependence of μ on λ. If all other factors were equal, which material would best disperse the spectrum?

Electrostatic and Magnetic Lenses

Such electronic devices as the TV tube, the oscilloscope, and the electron microscope require a beam of electrons to be focused. Optical lenses cannot be used for this purpose. Instead, one uses an electrostatic or a magnetic lens. These devices operate as follows.

A typical electrostatic lens is shown in part (a) of the figure. The lens consist of two metal cylinders A and B at different potentials. Notice that the electric field is directed from B to A, and recall that the force on the electron is opposite in direction to the field. When the electrons from the source are in A, their speed and momentum are low, and so they are strongly deflected by the electric field at the edge of A. However, in going from A to B, they are greatly accelerated, and so when they pass into B, their momentum is much increased. The field

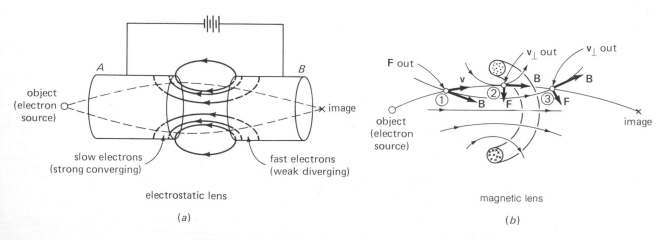

the larger the change in μ with λ (i.e., the larger the *dispersion* of the material $d\mu/d\lambda$), the better this material will be able to separate wavelengths when used as a prism. The variation of μ with λ for several materials is shown in figure 32.9.

When light is sent through a prism, as illustrated in figure 32.10, the refraction at the two surfaces will depend upon μ, according to Snell's law. As we see from figure 32.9, the index of refraction is larger for small λ (blue light) than it is for large λ (red light). As a result, the *angle of deviation D* will be larger for blue than for red light. The colors will therefore be separated by the prism, as shown.

In order to make use of a prism for dispersing wavelengths, one adopts an optical system much like that used for the grating spectrometer. A typical setup might be like the one in figure 32.11. We show a photographic

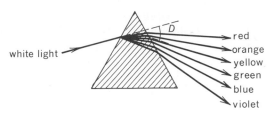

FIGURE 32.10 The angle of deviation D by a prism is not the same for all wavelengths of light. Hence the prism disperses white light into its constituent colors.

near B is therefore much less effective in causing deflection than is the field in A. As a result, the deflection caused at the A edge predominates. You should examine the figure to assure yourself that the deflections are as indicated. How can the focal length of this lens be changed?

The principle of operation of a magnetic lens is shown in part (b) of the figure. As shown there, the lens consists of a coil of wire which gives rise to the magnetic field indicated. (The coil is often combined with iron shielding so as to shape the field more effectively.) To see how this lens operates, examine the force on the electron at the points ①, ②, and ③ shown. At point ① the rule $\mathbf{F} = q\mathbf{v} \times \mathbf{B}$ shows the force on the electron to be out of the page. As a result, when the electron reaches points such as ② and ③, it has a component of $\mathbf{v}$, $v_\perp$, out of the page. The force on this component of $\mathbf{v}$ is toward the axis and gives rise to the focusing action of the lens. In addition to the focusing action, the lens also causes the electrons to rotate around the axis of the lens, and therefore the image formed by the lens is rotated about the axis.

It is possible to use electrostatic and magnetic lenses in "electron optical" devices similar to optical lens systems. For example, the electron microscope uses magnetic lenses and electrons in an optical system similar to that of a light microscope. As we shall see in Chap. 35, electrons behave like waves in such a device, and so the calculations for the resolution of optical devices apply to the electron microscope as well. It will be shown in Chap. 35 that the electron wavelength can be made very small ($<10^{-10}$ m), and so the resolution possible with the electron microscope is much higher than for a light microscope.

FIGURE 32.11 An image of the slit is
obtained on the photographic plate of
the prism spectrometer. If the light had
contained more than one wavelength,
multiple images would have been found
in the photograph.

plate as the receiver in this case. However, a telescope and eye or a photocell
might equally well be used. In any case, a series of images of the slit would
result, one for each wavelength of the spectrum. Unlike the diffraction
grating, all the light of one wavelength is imaged at the same place in the case
of a prism. It is this fact which makes the prism the more desirable dispersing
element when high intensity is of primary concern.

32.8 POLARIZED WAVES

Many optical and electromagnetic devices are designed to operate with
polarized waves. We have already used these waves in our discussion of the
electromagnetic radiation from an antenna. For an antenna with charge
oscillating along the y axis, the wave radiated along the x axis appears as
shown in figure 32.12. The electric field vector is always parallel to the xy
plane. A wave such as this is said to be *plane-polarized*. Its electric field

*A wave is plane-polarized if its electric
vector is always parallel to a plane*

vector is always directed in the $\pm y$ direction in this case. For this wave, E can
be specified completely by giving E_y, since E_x and E_z are zero.

Most electromagnetic radiation is not plane-polarized. The light emitted from a source, for example, may not be polarized. Its electric vector can
and does vibrate in all directions, with one exception: the oscillating electric
field has no component in the direction of propagation of the wave. This is
illustrated in figure 32.13(a), where the beam of light is assumed to be

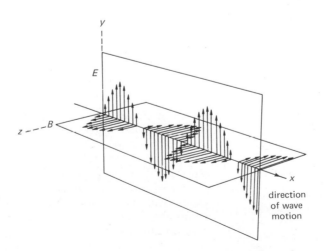

FIGURE 32.12 The wave shown here is
plane-polarized. We take its plane of
polarization to be the plane of the E
wave.

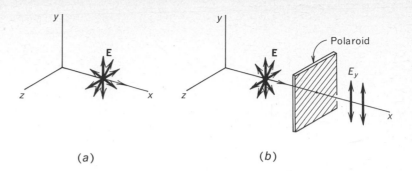

FIGURE 32.13 The Polaroid sheet allows only the *y* component of the electric field to pass through. This sheet consists of oriented dichroic crystals.

(a)

(b)

traveling along the *x* axis. As mentioned before, the form of this wave will be extremely complex since it is the result of many unrelated events in the atoms of the light source.

There are several important ways for changing this unpolarized beam of light into a plane-polarized beam. One very common method is by use of oriented dichroic crystals in a plastic film. (This composite substance has the name *Polaroid*.) *Dichroic substances* have the property of transmitting light with *E* vibrating in only one direction. As a result, plane-polarized light is obtained when unpolarized light is passed through a sheet composed of oriented dichroic crystals. (In Polaroid, the crystals used are iodoquinine sulfate.) This situation is shown in figure 32.13(*b*). If the Polaroid sheet is thick enough, E_z of the wave striking the sheet is nearly completely absorbed, while a large fraction of the *y* component of *E* is transmitted.

Polaroid consists of sheets containing oriented dichroic crystals

Another method for obtaining plane-polarized light is by reflection. When a beam of light is reflected from a surface, it is found that the component of *E* parallel to the surface is more strongly reflected than the other component. In fact, at one particular angle of incidence on a dielectric (the *Brewster angle*), the reflected beam consists entirely of light whose electric vector is parallel to the surface. This situation is illustrated in figure 32.14. It turns out that the polarizing angle (or Brewster angle) is such that the reflected and refracted rays are at an angle of 90° to each other.

To find the relation between the polarizing angle (the angle of incidence for plane polarization) and the index of refraction of the reflecting substance, we use Snell's law. We have

$$\mu = \frac{\sin i}{\sin r}$$

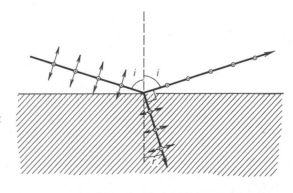

FIGURE 32.14 When light is reflected at the Brewster (or polarizing) angle, the reflected ray is completely polarized. Encircled dots represent the component of *E* perpendicular to the page.

where i is the incidence angle, and r is the refraction angle. At the polarizing angle, $i = \theta_p$, and from figure 32.14,

$$\theta_p + 90° + r = 180°$$

Therefore

$$\sin r = \cos \theta_p$$

and we find

$$\mu = \tan \theta_p \qquad\qquad 32.11$$

Light reflected at the Brewster angle is plane-polarized

This relation is called *Brewster's law.*

Although polarized beams of light used to be obtained by reflection, this method is cumbersome, and is seldom used today. However, we encounter this phenomenon in many everyday circumstances. Light reflected from water, oily concrete, etc., is partly polarized by the process. This fact is not usually noticed since our eyes are unable to distinguish the state of polarization of light. We shall see in a later section how the state of polarization of a beam can be ascertained.

32.9 DOUBLE REFRACTION AND THE NICOL PRISM

In certain crystals one finds the index of refraction of the material to depend upon the state of polarization of the light. These substances are said to be *doubly refracting* since the two different indices give rise to two refracted beams rather than one. A well-known case is that of a calcite crystal. An experimental situation which illustrates this effect is shown in figure 32.15.

Doubly refracting crystals produce two plane-polarized beams

As shown, a ray of unpolarized light separates into two plane-polarized rays as it passes through the doubly refracting crystal. One ray has the same index of refraction for all directions of travel through the crystal, and is called the ordinary, or O, ray. The other ray, the *extraordinary,* or E ray, travels with different speeds (i.e., different μ's) in different directions. There is only one direction in which the O and E rays coincide, and this direction is called the *optic axis* of the crystal.

Prior to the development of Polaroid, most optical instruments using polarized light made use of calcite crystals to obtain the plane-polarized beam. To do this, the calcite crystal was cut in half, as indicated in figure 32.16. The two halves were then cemented together with a material called Canada balsam. This material has an index of refraction such that the

FIGURE 32.15 Double refraction in a calcite crystal separates the unpolarized light into two plane-polarized components.

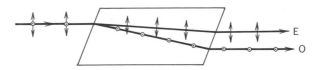

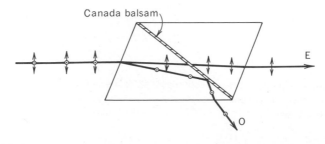

FIGURE 32.16 The Nicol prism.

ordinary ray is totally internally reflected, as shown. The split crystal then transmits only the E ray, which is plane-polarized. This device is called a *Nicol prism*. Because of the convenience and low cost of Polaroid, the Nicol prism is much less widely used than formerly.

32.10 USES OF POLARIZED LIGHT

To determine whether or not a particular beam of light is plane-polarized is a simple matter. For example, suppose a beam of light has been plane-polarized by a Polaroid sheet (called the *polarizer*), as shown in figure 32.17. If now a second Polaroid (called the *analyzer*) is placed in the beam, the light transmitted will depend upon the orientation of the Polaroid. If the two Polaroids have their axes parallel, as in part (*a*), they both transmit light in the same plane, and so the beam will be transmitted. However, if, as shown in (*b*), the axes of the Polaroids are perpendicular (the Polaroids are said to be *crossed*), no light will be transmitted through the analyzer. Hence, if rotation of the analyzer causes extinction of the beam, the light is plane-polarized.

An interesting case occurs if one interposes a sugar solution in the beam of light between two crossed Polaroids. It is found that the presence of the solution causes light to be transmitted through the analyzer. However, if the analyzer is rotated, a new position is found where the beam is extinguished. Evidently, the sugar solution has rotated the plane of polarization of the plane-polarized light. Substances which do this are said to be *optically active*. Since the angle through which the plane of polarization is rotated is proportional to the concentration of the optically active substance, this effect is often used to measure concentrations of these substances.

Most transparent substances become doubly refracting under stress. This fact can be used to measure the stresses within materials. When an object under nonhomogeneous stress is placed between crossed Polaroids (crossed so that, in the absence of the sample, no light will be transmitted), patterns such as the one shown in figure 32.18 are observed. The alternate dark and bright lines are closest together at the places where the stress is most uneven. As one can see, this provides an experimental method for

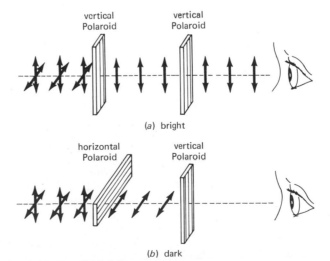

FIGURE 32.17 The unpolarized light is polarized by the first Polaroid, the polarizer. In part (*a*) the second Polaroid, the analyzer, transmits the light. In part (*b*), however, the analyzer and polarizer are crossed, and the beam is completely stopped by the analyzer.

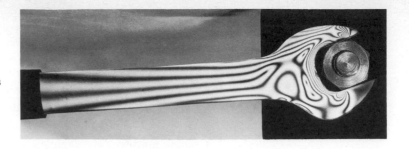

FIGURE 32.18 A strained transparent object viewed through crossed Polaroids shows alternate dark and bright bands. The bands are contour lines of constant stress, and the stress is greatest where they are closest together.

analyzing stress distributions in complex applications. The analysis of patterns such as the one shown is a well-developed branch of engineering, and is treated in texts on optical stress analysis.

There are many other uses of polarized light. For example, many materials examined by biologists under the microscope are doubly refracting. Much detail, unobservable in ordinary light, is clearly brought out when the object is viewed through crossed Polaroids. Although these and other uses of polarized light are of great interest, we cannot explore them further here. The reader is referred to the book entitled "Polarized Light" * for an interesting treatment of the subject.

Chapter 32
Questions
and Guesstimates

1. Clearer images are obtained in optical instruments when only a small portion of the lens is used. In the case of the pinhole camera, no lens is needed. To see how this is possible, draw a small bright object about 1 mm high at a distance of 10 cm from a 1-cm opening in a large opaque screen. Show how the bright spot cast by the object on a screen 5 cm behind the opening decreases in size as the opening is made smaller. Show that in the limit of a pinhole opening, two objects 1 cm apart and both 10 cm from the opening will give rise to well-defined images on the screen.
2. The wavelengths of light emitted by hot mercury vapor, for example, are called spectral lines. Explain clearly why they are called lines.
3. What limits the fineness of detail which one can see with a cheap magnifying glass?
4. One can buy a cheap microscope for use by children. Invariably, the images seen in such a microscope have colored edges. Why is this so?
5. Suppose a reflecting-type (e.g., metal-mirror) telescope could be fitted with a detector so that it could be used equally well with all radiation from radio waves to x-rays. What type of radiation would give the best resolution of detail in the object being examined? What difficulties arise if radiation other than visible light is used?
6. It is possible, using visible light, to measure detail smaller than the wavelength of the light by making use of interference techniques. Explain an example of this.
7. Does the size of the opening in front of the objective lens of a telescope or microscope influence the magnification of the instrument?
8. When a beam of light is sent through a spectrometer, compare the intensity of the lines observed in the spectrum formed by a prism and grating spectrometer. Give reasons for the differences.

*W. A. Shurcliff and S. S. Ballard, "Polarized Light," D. Van Nostrand Company, Inc., Princeton, N.J., 1965.

9. Grating spectrometers can be used for infrared, visible, ultraviolet, and x-ray spectra, while glass-prism spectrometers are restricted mostly to the visible and near infrared. Explain why.

10. What happens to the light energy which is not transmitted by a polarizer if the polarizer is a sheet of Polaroid? If it is a Nicol prism? Can you see any drawback to the use of Polaroid when intense beams of light are used?

11. How can one determine whether a beam of light is polarized? Whether it is composed of two beams, one polarized and the other not?

12. Using a commerical camera with a 5-mm-diameter lens opening, i.e., aperture diameter, the proper exposure time for a scene is $\frac{1}{60}$ s. About what would be the exposure time for a pinhole camera with a 0.50-mm-diameter pinhole and the same type of film?

13. You have available a long cylindrical cardboard mailing tube and two lenses with focal lengths 60 to 10 cm that can be fitted into the tube. Use these to design a toy telescope.

14. How do Polaroid sunglasses decrease eye strain? Discuss the principle of their operation.

Chapter 32
Problems

1. A lens forms an image of the sun at a distance of 8 cm from the lens. About how large an angular magnification could one obtain with this lens?

2. By about what factor is the length of an image on the retina of the eye increased if viewed through a 6-cm-focal-length converging lens rather than directly? Assume the object to be held as close to the eye as possible for the direct observation.

3. Show that the size of the image on the retina of your eye varies inversely with the distance of the object from the eye.

4. The eye of an insect might be about 0.5 mm in diameter. Assuming the eye to be a scaled-down version of the human eye, what would the equivalent focal length of the eye lens have to be to form an image of a distant object on the retina at a distance of 0.4 mm behind the effective lens center? What difficulties are presented by a human-type eye on such a small scale?

5. A certain diffraction grating has 5000 lines/cm. Considering only the limitation due to diffraction, could the lines be distinguished in a photograph taken by a camera which has a lens with a 1.5-cm-diameter opening? Assume the object can be photographed at a distance of 8.0 cm from the lens. How large is θ_c for this camera? Assume $\lambda = 5 \times 10^{-5}$ cm.

6. Referring to figure 32.11, the objective lens has a diameter of 2.0 cm. How large must the angle between two spectral lines be if they are to be resolved on the photographic plate? Assume the limiting effect to be diffraction due to the objective. The lines have wavelengths close to 5000 Å.

7. In a certain radar system, signals of 3-cm wavelength are reflected off airplanes at a distance of 16 km from the receiver. The receiver consists of a portion of a sphere and the diameter of its open end is 1.0 m. If the system is diffraction-limited, how close can two airplanes be at 16 km from the receiver if they are to be seen as two separate entities?

8. A camera which has a lens opening (aperture) of 1.2 cm diameter photographs a scene adequately when the exposure time is 0.01 s. If the aperture is decreased to 0.4 cm, what exposure time should be used?

9. By what factor is the light intensity increased in a telescope if the diameter of the objective lens is changed from 0.50 to 4.0 cm? (Assume that the other dimensions remain constant.)

10. In a microscope, the objective has a 3-cm focal length, while the eyepiece has a 5-cm focal length. What is the magnifying power of the microscope?

11. A boy makes a microscope by cementing a 5.0-cm-focal-length lens to one end of a 10-cm-long tube and a 3.0-cm-focal-length lens to the other. (a) If he uses the 3.0-cm lens as the eyepiece, about how far in front of the objective must he place the specimen he is looking at? (b) What will be the approximate magnifying power of his microscope?

12. A telescope at the Yerkes Observatory has an objective lens with a focal length of 19 m. When observing the moon, how many meters on the moon correspond to a 1.0-mm length on the image cast by the objective lens? (Distance to moon $= 3.8 \times 10^8$ m.)

13. You will notice in figure 32.4 that the telescope inverts the object. This is an objectionable point if one wishes to view the opera from a distant seat in an opera house. Instead, one can use an opera glass (i.e., galilean telescope) such as the one pictured in figure P32.1. For the one shown there, locate the position of the final image of a distant object. Is it real or virtual? Erect or inverted?

14. For the opera glass shown in figure P32.1, what is the angular magnification one would expect when used to view a distant object?

15. An object is placed 50 cm in front of a converging lens, $f = 25$ cm, which in turn is 80 cm in front of a plane mirror. Find all the images formed of the object.

16. Repeat problem 15 if the plane mirror is 40 cm from the lens. State whether each image is real or virtual.

17. Show that the effective focal length (f) of two thin lenses (f_1 and f_2) separated by a negligible distance is given by

$$\frac{1}{f} = \frac{1}{f_1} + \frac{1}{f_2}$$

18. The value of $1/f$ with f in meters is called the *power* of a lens in *diopters*. For thin lenses in close combination, the lens powers add as shown in the previous problem. What focal-length lens combinations can be achieved by placing the following lenses in close combination: $+2.0$, -5.0, and $+3.0$ diopters?

19. A *nearsighted* person is unable to see distant objects clearly. The eye-lens system is unable to form an image of a distant object on the retina. A certain eye is able to see clearly only those objects closer than 60 cm. What focal-length thin lens should be used just in front of the eye to correct the difficulty? (*Hint:* The lens must image the distant object at 60 cm from the eye.)

20. *Farsighted* people are unable to see clearly objects that are close to the eye. A certain eye is able to see clearly only those objects that are 100 cm away and farther. What focal-length corrective lens will allow the person to see objects at 25 cm from the eye? (See the previous problem for a devious hint.)

21. Show that if the apex angle of a prism A is very small (i.e., the prism is very thin) and a beam of light strikes the prism perpendicular to one of the faces, the deviation D of the beam is given by $D = (\mu - 1)A$.

22. A converging lens ($f = 20$ cm) stands 30 cm in front of a concave mirror

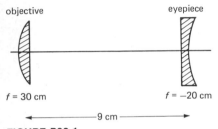

objective   eyepiece

$f = 30$ cm $f = -20$ cm

←————— 9 cm —————→

FIGURE P32.1

of 40-cm radius of curvature. Where does the system form images of an object placed 40 cm in front of the lens?

23. Repeat problem 22 if the lens-to-mirror separation is 50 cm.

24. The constants A and B in the Cauchy equation for a certain glass are 1.5020 and 5.2×10^5 Å^2. What is the index of refraction of the glass at 5000 Å? What is the dispersion at this wavelength?

25. Suppose the index of refraction of glass has been measured as a function of λ and found to be

λ (nm)	656.3	589.0	486.2
μ	1.514	1.517	1.524

If these data obey the Cauchy relation, how should they be plotted to obtain a straight line? Make such a plot and determine A and B. What should μ be at 500 nm?

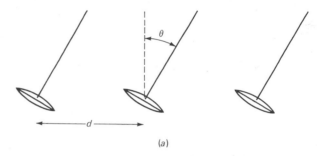

(a)

FIGURE P32.2 The NRAO three-element interferometer. Three fully steerable 85-ft radio telescopes are used in this interferometer. Two are movable along a 5000-ft roadway, and the third, the most distant and smallest white disk, is fixed in place. The longest spacing between two telescopes is 9000 ft. (Courtesy of National Radio Astronomy Observatory, Green Bank, West Virginia.)

(b)

26. A lens is made from the glass for which data are given in problem 24. If the lens images the sun at a distance of 12.000 cm when red light (656 nm) from the sun is used, where will an image of the sun appear when 486 nm light is used? This defect of a lens is called *chromatic aberration*.

27. So-called *circularly polarized light* is obtained by combining two coherent, equal-amplitude, plane-polarized beams whose planes of polarization are perpendicular to each other and are 90° out of phase. Show that when the two beams $E_y = E_0 \sin\left[\omega(t - x/v)\right]$ and $E_z = E_0 \cos\left[\omega(t - x/v)\right]$ are combined, their resultant at a particular point in space is an electric field vector whose magnitude is constant but whose direction rotates on a circle perpendicular to the x axis. *Elliptically polarized light* can be obtained by adding beams of unequal amplitude.

28. Two coherent x-directed plane-polarized beams of equal amplitude are combined. One has its **E** vector in the y direction while that for the other is in the z direction. If the two beams are in phase, what is the state of polarization of the resultant beam? What is the amplitude of the resultant beam if the original beams had amplitude E_0?

29. A beam of polarized light has an intensity I. It then passes through an analyzer with the direction of vibration transmission oriented at an angle θ to the direction of polarization of the incident light. Show that the transmitted intensity is $I \cos^2 \theta$.

30. If a number of radio telescopes such as the ones shown in figure P32.2(b) are placed in a straight line as shown in the figure, they can be used as a unit to obtain high resolution. In effect, they act as the reverse of a diffraction grating and respond strongly to only a very limited angular range. Referring to figure P32.2(a), show that they select for observation an angle and wavelength related in the following way: $n\lambda = d \sin \theta$. The individual telescopes are mounted on a track so that d may be altered.

31. For the radio-telescope array described in the previous problem, show that the resolution angle (measured from the maximum to the first minimum) is given by

$$\Delta\theta = \frac{\lambda}{Nd \cos \theta}$$

where N is the number of telescopes in the array.

33 Resonance of Waves

Although sound, light, and electromagnetic waves in general are of great importance to us, it has become clear since 1900 that wave phenomena are even connected with particle motion. In all these cases, it is important to consider the resonance of waves. In this chapter we consider the resonance of mechanical and electromagnetic waves. These forms of resonance will be found to have much in common with each other. The correspondence between waves will prove of value to us when we extend the present concepts to the wavelike behavior of particles, in later chapters.

33.1 WAVES ON A STRING: EXPERIMENTAL

Before beginning a quantitative study of wave resonance phenomena, it is of value to review a few experimental facts concerning waves on a string. It is these facts which we describe quantitatively in later sections of this chapter. They will be seen to be typical of wave resonance phenomena in many diverse systems.

Suppose a string is passed over a massive pulley as shown in figure 33.1, with a weight attached to its end. The tension T in the string will remain essentially constant in such an arrangement provided we do not try to lift or lower the weight with too large an acceleration. As a result, the speed of a wave disturbance sent down the string will be constant. It is given by 27.7 as

$$ v = \sqrt{\frac{T}{\rho}} $$

where ρ is the mass per unit length of the string.

If, now, the hand holding the left end of the string is suddenly moved up and then back down, a pulse is sent along the string, as shown in parts (b) to (d) of figure 33.1. The behavior of the pulse when it strikes the pulley depends upon the exact situation which obtains there. If the pulley is rather massive and if the string does not slip easily on the pulley, the end of the string at the pulley will remain motionless when the pulse reaches it. Since this portion of the string does not move, it must be true that the forces exerted on it by the incoming pulse and the pulley surface are equal and opposite. If there were no end to the string, the incoming pulse would exert a force on the portion of the string just ahead of it, and thereby would cause the pulse to propagate along the string. At the pulley, however, the equal and opposite force exerted by the

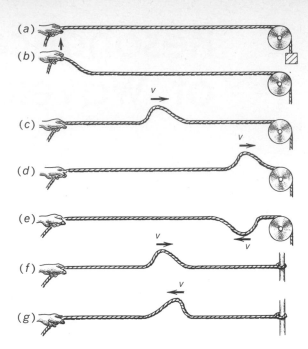

FIGURE 33.1   Reflection of wave pulses on a string.

At a fixed end, the wave is inverted by the reflection process

Inversion does not occur at a free end

pulley on the string causes an inverted pulse to travel along the string in the reverse direction. This situation is shown in part (*e*) of figure 33.1.

As we have just indicated, a pulse is always inverted (i.e., turned upside down) when reflected from a fixed end of a string. This is not true, however, if the end of the string is free to move up and down. Such a situation can be approximated by placing a freely sliding slip ring at the string's end, as shown in part (*f*) of figure 33.1. When a string is terminated in this way, the reflected wave is not inverted, and appears as shown in figure 33.1(*g*).

Suppose now that a series of sinusoidal pulses is sent down the string by means of a vibrating tuning fork or some other fixed frequency device, as shown in figure 33.2. When one does this, one almost always finds that the very small pulses sent down the string by such a vibrator merely cause the string to vibrate slightly, scarcely noticeably, as shown in figure 33.2(*a*). However, if one slowly increases the length of the string by pulling the string over the pulley, the string's behavior changes markedly. For a certain length of the string, the string begins to vibrate back and forth very strongly, its vibration amplitude being much larger than the vibration amplitude of the source vibrator. The vibration pattern of the string might appear as shown in figure 33.2(*b*). Under these conditions, the string is said to be *in resonance.*

At resonance, nodes and antinodes appear along the string

Notice that, at resonance, certain points along the string remain motionless. These points are called *nodes.* Other points, midway between the nodes, vibrate with maximum amplitude, and these are called *antinodes.* As we shall see in the next section, the distance between two adjacent nodes is $\lambda/2$, where λ is the wavelength of the sinusoidal wave being sent down the string.

If now the string is lengthened still more, the wide-amplitude vibration ceases, and the string remains nearly motionless, as shown in figure 33.2(*c*). However, when the string is lengthened still further, resonance is again achieved at a certain critical length, as shown in figure 33.2(*d*). As one might

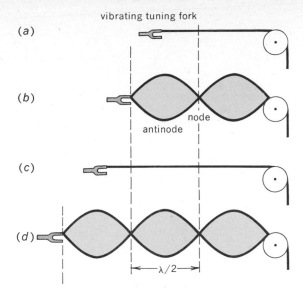

vibrating tuning fork

(a)

(b)

antinode   node

(c)

(d)

$\leftarrow \lambda/2 \rightarrow$

FIGURE 33.2 At what other lengths will the string resonate?

Resonance occurs when L = n(λ/2)

surmise from these figures, the string will always resonate when its full length is an integer multiple of $\lambda/2$. Let us now begin a quantitative study of waves on a string so as to see more clearly why this behavior occurs.

33.2 REFLECTION OF WAVES ON A STRING

In Chap. 27 we discussed the motion of waves of various types. They all obeyed a partial differential equation, the wave equation. We found a solution of that equation to be as follows. If a sinusoidal disturbance of amplitude y_0 propagates along the positive x axis with speed v and angular frequency ω, its magnitude y_i as a function of x and t is

$$y_i = y_0 \sin\left(\omega t - \frac{\omega x}{v}\right) \qquad 33.1$$

Since the angular frequency ω is related to the frequency ν in hertz and wavelength λ through the relations

$$\omega = 2\pi\nu = \frac{2\pi v}{\lambda}$$

33.1 can be written in several alternative ways.

Let us first consider a sinusoidal wave traveling along a string as shown in figure 33.3(a). In Chap. 27 we assumed that the wave could continue on for an infinite distance. This will not be the case here, however. The string must have an end, and we must consider what happens as the wave reaches it. There are two limiting cases: the end could be perfectly free or held rigidly in place. For a stretched string, the latter case is most often encountered, and it is the one we consider.

When the wave strikes the fixed end at $x = 0$ in figure 33.3, the energy of the wave must be reflected if we assume the support to be rigid and therefore nonabsorbing. This means there will be a reflected wave with frequency and amplitude equal to that of the incident wave. It will be going

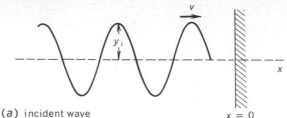

(a) incident wave $x = 0$

(b) reflected wave $x = 0$

(c) resultant wave

FIGURE 33.3 The wave on a string suffers a 180° phase change upon reflection at a fixed end. This amounts to inverting the wave. What happens to diagrams (b) and (c) as time progresses?

in the $-x$ direction, however, and so its equation of motion will be

$$y_r = y_0 \sin\left(\omega t + \frac{\omega x}{v} + \phi\right)$$ 33.2

In writing 33.2, we have changed the sign of v because the direction of propagation has been reversed. Moreover, since a phase change may occur upon reflection, we add a phase angle ϕ to the reflected wave. It now becomes necessary to evaluate this phase angle.

The phase angle may be evaluated by noticing that the resultant motion of the string will be the sum of y_i and y_r. We have

$$y = y_0\left[\sin\left(\omega t + \frac{\omega x}{v} + \phi\right) + \sin\left(\omega t - \frac{\omega x}{v}\right)\right]$$ 33.3

We now make use of the boundary condition that at $x = 0$, the fixed end point of the string, y must be zero at all times. This can be true only if, at $x = 0$, the two terms in the brackets in 33.3 are equal and opposite. In other words,

$$\sin(\omega t + \phi) = -\sin \omega t$$

Since $\sin \theta = -\sin(\theta + \pi) = -\sin(\theta + 3\pi) = \cdots$

we conclude that ϕ equals π, 3π, etc. All values being physically indistinguishable, we choose the simplest one and set $\phi = \pi$.

In other words, at the fixed end of a string, the wave suffers a 180° phase change upon reflection. ($\phi = 0$ for a *free* end.) This reverses the sign of the reflected wave, as shown in figure 33.3(b), and ensures that the reflected

A 180° phase change occurs upon reflection from a stationary boundary

and incident waves will add to give zero motion at the fixed end. The resultant motion of the string will then be

$$y = y_0 \left[\sin \left(\omega t - \frac{\omega x}{v} \right) - \sin \left(\omega t + \frac{\omega x}{v} \right) \right]$$

and is shown in figure 33.3(c) for the instant at which figure 33.3(b) is valid.

Before proceeding further with our analysis, let us see qualitatively what will happen to the waves in figure 33.3(b) as time goes on. The solid-line wave moves to the left and the dashed-line wave moves to the right. As this happens, the two waves are always equal and opposite at the positions of the dashed lines labeled N. They will exactly cancel each other there, and so the string will not move at all at these points, the nodes. However, at the points labeled A, the waves always reinforce each other, and so the string will vibrate widely there. These points are the antinodes. Let us now see how our mathematical analysis points out this behavior.

To put the equation for the resultant motion in a more convenient form, we make use of the trigonometric identity

$$\sin x - \sin y = 2 \sin \frac{x - y}{2} \cos \frac{x + y}{2}$$

which gives

$$y = \left(2y_0 \sin \frac{\omega x}{v} \right) \cos \omega t \qquad\qquad 33.4$$

We see from this that the string vibrates sinusoidally with frequency ω. But now the amplitude of motion (the factor in parentheses) is a function of position. This amplitude factor is plotted as a function of x in figure 33.4. In plotting this we note that $\omega/v = 2\pi/\lambda$, so that the amplitude factor is

$$2y_0 \sin \frac{2\pi x}{\lambda}$$

Notice that at x values of $\lambda/2$, λ, $3\lambda/2$, etc., the amplitude factor is zero. At these points, the *nodes* mentioned previously, the string must remain stationary. To find the motion at other values of x we must multiply the amplitude factor by $\cos \omega t$. As a result, the string will vibrate back and forth between the limits imposed by the amplitude factor. This results in a vibration of the string as shown in figure 33.4(b). The $\pm$ amplitude factor curve forms the envelope for the vibration of the string. The nodes N are always $\lambda/2$

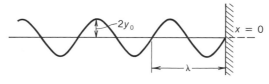

(a) the amplitude factor

FIGURE 33.4 The vibration of the string as a result of the incident and reflected waves is found by multiplying the amplitude factor by $\cos \omega t$. Consequently, the string vibrates sinusoidally within the limits shown in part (b).

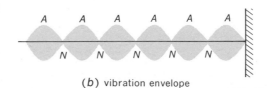

(b) vibration envelope

apart, as one sees from figure 33.4. Points A, the antinodes, are also $\lambda/2$ apart. This type of vibration, in which a wave oscillates back and forth between an envelope as shown, is called a *standing wave*. It will be seen to be a very important form of wave motion.

Definition of a standing wave

33.3 RESONANCE OF A STRING

Since a string has two ends, not just one, our story is not yet complete. The incident wave in the preceding section must have come from somewhere, and the reflected wave cannot keep going toward the left in figure 33.3 forever. One way out of this dilemma would be to place the other end of the string at a node. We could then imagine that a wave exists in the string traveling to the right. It reflects from the right-hand end and forms the node-antinode pattern shown in figure 33.4(*b*) as it combines with the incident wave. When this reflected wave (call it wave 2) strikes the left-hand end, it will again be reflected. Let us call this reflected wave, wave 3. Our previous considerations insist that this end be a node in the standing-wave pattern. This will be possible since the original wave and wave 2 cancel at this point. Moreover, wave 3 will be joined by wave 4, the reflection of wave 3, and these waves will cancel in the same way the first two did.

We see, then, that a wave can form a stable standing-wave pattern provided the fixed ends of the string are at the positions of nodes. This means that the ends must be an integer number of $\lambda/2$ apart as shown in figure 33.5. Standing-wave patterns such as this are called *resonant modes of motion* of the system. (Do not confuse the two words *mode* and *node!*)

A string resonates in such a way that
$$L = n(\lambda/2)$$

One small detail still remains. How is energy supplied to the wave? Not only was energy needed to start the wave in the first place, but since energy losses occur in the string and by friction with the air, energy must be added from outside to compensate for the loss. In order to add energy to the system, we must drive it with a force which is of the same frequency as the string oscillation frequency. This force must be applied in such a way that it does not disturb the vibration pattern of the string. If a vibrator is used, it should be attached to the string at a point which has the same amplitude of vibration as the vibrator has. In most cases of standing waves, the energy source vibrates with rather small amplitude, and so it will be placed close to a node.

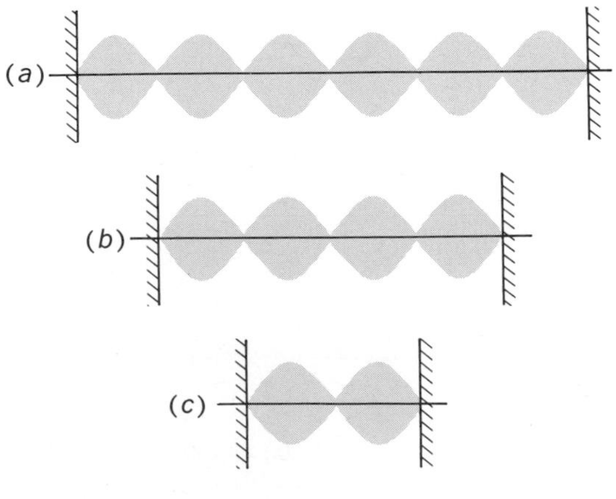

(a)

(b)

(c)

FIGURE 33.5 A standing-wave pattern is possible in a string if the fixed supports are placed at wave nodes. In other words, they must be $n(\lambda/2)$ apart, where n is an integer.

We see from what has been said that a string will vibrate in a standing-wave pattern (i.e., it will resonate) if the supports are $n(\lambda/2)$ apart, where n is an integer. This means the condition for resonance is the following:

$$L = n\frac{\lambda}{2} \qquad n = 1, 2, 3, \ldots \qquad\qquad 33.5$$

where L is the distance between the ends of the string. Usually, L will be fixed, and one is interested in the frequencies to which the string will vibrate. We can find these resonance frequencies ν_n of the string by recalling that $\lambda = v/\nu$, where v is the speed of the wave. Putting this value in 33.5 yields

Eigenfrequencies for a stretched string

$$\nu_n = n\frac{v}{2L} \qquad n = 1, 2, 3, \ldots \qquad\qquad 33.6$$

The resonance frequencies of the string are often called the *eigenfrequencies* of the system.

Equation *33.6* tells us that the stable vibration patterns and frequencies of a string are *quantized*. Not all frequencies of vibration give stable patterns or standing waves. Only those frequencies which satisfy *33.6*, the eigenfrequencies, give standing waves. It is of interest to notice in this case that the eigenfrequencies are evenly spaced. Each one differs from its predecessor by an amount $v/2L$. The quantized eigenfrequencies are multiples of this lowest eigenfrequency.

The results we have just obtained for the resonance frequencies of a string are well known to musicians. They express these results in terms of a *fundamental* vibration of a string, together with its overtone vibrations. For example, when a violin or guitar string vibrates in one segment as shown in figure 33.6(a), it gives off a tone which is called the fundamental. When vibrating in two, three, or more segments, as shown in the other parts of figure 33.6, the string gives off tones of higher frequencies, called the overtones of the string. A violin string, when used in the customary way, vibrates in several resonance patterns at the same time, all superposed one upon the other. As a result, the string gives off a sound which consists of a mixture of the fundamental and the overtones.

Illustration 33.1

A string 3 m long is found to resonate in four segments [as in figure 33.6(d)] for a frequency of 50 Hz. To what other frequencies will it resonate?

Reasoning The eigenfrequencies of the string are

$$\nu_n = n\frac{v}{2L}$$

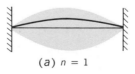

(a) $n = 1$

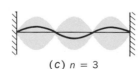

(c) $n = 3$

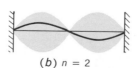

(b) $n = 2$

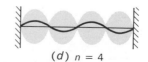

(d) $n = 4$

FIGURE 33.6 Resonant modes of motion in a string must have nodes at the two fixed ends. Shown are the four simplest standing waves possible. They are sometimes called the *fundamental*, the *first overtone*, the *second overtone*, etc.

Resonant modes of motion for various values of n are shown in figure 33.6. In our case, $n = 4$. From this fact we have

$$50\,\text{s}^{-1} = 4\,\frac{v}{2L}$$

so that

$$\frac{v}{2L} = 12.5\,\text{s}^{-1}$$

In passing, we note that $v = 75\,\text{m/s}$. Using this value in the equation for ν_n gives

$$\nu_n = 12.5n$$

as the resonance frequencies of the string. The resultant standing waves of the vibrating string are shown in figure 33.6. If this were the string of a muscial instrument, the vibrations for $n = 1, 2, 3, \ldots$ would be known as the fundamental frequency, the first overtone, the second overtone, etc.

33.4 VIBRATION OF A DRUMHEAD

A similar, but much more complicated, vibration takes place when we consider the vibration of a sheet rather than a string. One device which is familiar to everyone is the drumhead. It is frequently a circular sheet, with its outside perimeter held stationary. The perimeter of the drumhead is analogous to the ends of the string. Consequently, the one-dimensional string is replaced by a two-dimensional sheet. Moreover, for a circular drumhead, the problem has circular symmetry, and so polar coordinates would be used. Unfortunately, the motion in this case has a rather complicated mathematical form, which is most simply expressed in terms of Bessel functions. Since most of you are not yet acquainted with these functions, we consider a rectangular drumhead whose motion can be described in terms of sine and cosine functions. The motion in question can be found by solving the equation for waves in two dimensions.

The wave equation in two dimensions can be derived by a method much like that used in Chap. 27 for the one-dimensional equation for waves on a string. We shall only state its solution here. Referring to figure 33.7 for the coordinates involved, it is found that the solutions for the y displacement of a sheet in the xz plane are of the form

The m, n eigenfunction for a rectangular drumhead

$$y_{mn} = A_{mn} \sin \frac{n\pi x}{L_x} \sin \frac{m\pi z}{L_z} \cos \omega_{mn} t \qquad 33.7$$

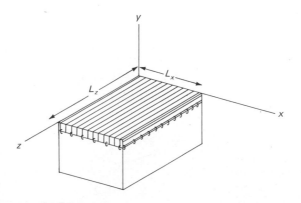

FIGURE 33.7 The drumhead is held fixed at its rectangular perimeter. We can specify its motion by giving y as a function of x, z, and t.

where m and n are integers. You will recall that one integer was needed to specify the motion of a string. For a string, if $n = 1$, the string vibrated in one segment. For $n = 2$, it vibrated in two segments, and so on. In the present case, two integers, m and n, are needed to specify the modes of motion. The various modes of motion are dependent upon the values of m and n. We call the mathematical description of these modes of motion, given in *33.7*, the *eigenfunctions* for the motion.

The eigenfunction solutions given in *33.7* are simply the product of two amplitude functions such as those obtained for a string. We can therefore infer (rightly, it turns out) that the solution represents the resonance of two types of stringlike waves. One standing-wave pattern is generated by waves traveling parallel to the x axis and being reflected from the two edges at $x = 0$ and $x = L_x$. This pattern is multiplied by a second similar standing-wave pattern originating from waves traveling along the z axis. As one would expect, to fit the boundary conditions of nodes on the perimeter of the drumhead, only certain eigenfrequencies are found allowable. There is one for each of the eigenfunctions given in *33.7*, and they are

Eigenfrequencies for the vibration modes

$$\nu_{mn} = \frac{v}{2}\sqrt{\left(\frac{m}{L_z}\right)^2 + \left(\frac{n}{L_x}\right)^2} \qquad 33.8$$

where $m = 1, 2, \ldots$, and $n = 1, 2, \ldots$.

To see what the standing waves look like, it is interesting to plot the eigenfunctions of *33.7* for various combinations of m and n. This is done in figure 33.8. The eigenfrequencies of each mode are given in terms of the fundamental frequency ($m = n = 1$), assuming $L_x = L_z$. Notice that, as in the case of a string, the value of m (or n) is equal to the number of segments into which the z (or x) coordinate is divided by the nodal lines and the perimeter. As one sees, the frequencies are no longer integer multiples of the fundamental frequency, in this case.

Although two-dimensional resonance patterns of this type are not

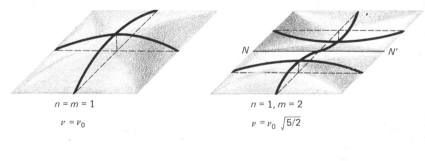

$$n = m = 1 \qquad\qquad\qquad n = 1, m = 2$$
$$\nu = \nu_0 \qquad\qquad\qquad \nu = \nu_0 \sqrt{5/2}$$

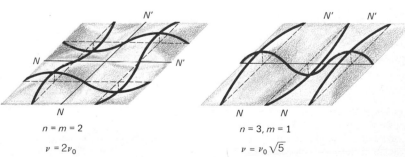

$$n = m = 2 \qquad\qquad\qquad n = 3, m = 1$$
$$\nu = 2\nu_0 \qquad\qquad\qquad \nu = \nu_0 \sqrt{5}$$

FIGURE 33.8 The first few resonant modes of vibration of a square membrane or drumhead. The nodal lines are represented at NN'.

usually found separately when a rectangular surface vibrates, the actual vibration can be synthesized by adding solutions such as this. This is the two-dimensional analog of the proposition stated earlier that, no matter how complicated, any physical wave can be represented as a sum of sinusoidal waves. In the present terminology, this can be stated in the following way: *The vibration pattern of a vibrating membrane or sheet of any type can be expressed as an appropriate sum of the eigenfunction solutions for the sheet in question.*

33.5 LONGITUDINAL RESONANCE OF A ROD

The preceding sections of this chapter have been devoted to the resonance of transverse waves. A similar situation exists in the case of longitudinal waves. We can infer this at once from the fact that both types of waves satisfy wave equations of the same type, as was shown in Chap. 27. Of course, we could arrive at this same conclusion by going through the qualitative considerations we used early in this chapter. The only differences which would appear would be in the boundary conditions to the problem. Usually, the end of a rod is bounded by material less rigid than itself, and so we should expect an antinode, rather than a node, at its end.

Suppose a rod is subjected to a sinusoidal driving force at one end, as shown in figure 33.9. If the rod is free to vibrate at its end, a standing longitudinal wave can exist in the rod only if the ends of the rod are antinodes. This means that the vibration pattern for the rod must appear as one of the modes of vibration shown in figure 33.9 and similar modes. (*Recall that the graph represents the horizontal displacement of the particles in the rod. These displacements are plotted vertically for convenience.*) If we notice that the distance between adjacent antinodes is $\lambda/2$, we see that the resonance condition is

The ends of the rod are antinodes, and so $L = n(\lambda/2)$

$$ L = n\frac{\lambda}{2} \qquad n = 1, 2, \ldots $$

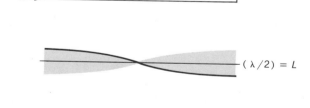

FIGURE 33.9 The resonant modes of motion for longitudinal vibration of a bar have antinodes at the ends. Where should the bar be clamped in each case to produce the desired vibration? (Notice that we plot the vibration vertically even though the actual vibration of the particles in the rod is horizontal.)

To find the frequencies to which the bar will resonate, we replace λ by v/ν:

$$\nu_n = n\frac{v}{2L} \qquad 33.9$$

The subscript n has been added to ν in order to distinguish the various resonance frequencies. In this situation, too, the stable frequencies of vibration are quantized.

Illustration 33.2

A metal rod 2.00 m long is clamped at a point 50 cm from one end. It is now set into longitudinal vibration by pulling a rosin-covered cloth lengthwise along it. The rod gives off a sound of approximately 2500 Hz. What is the approximate speed of longitudinal waves in the rod?

Reasoning Since the rod is clamped at the quarter point, that point must be a node. This means its vibration must correspond to that shown in the middle example of figure 33.9. Therefore

$$2\frac{\lambda}{2} = L \qquad \text{or} \qquad \lambda = L$$

with $L = 2.00$ m. But we have the general relation

$$\lambda = \frac{v}{\nu}$$

so that, using $\nu = 2500$ Hz, we find

$$v = 5000 \,\text{m/s}$$

33.6 RESONANCE OF SOUND IN PIPES

The most important type of longitudinal (or compressional) waves is sound waves. These waves can achieve resonance when confined to pipes, tubes, and other hollow vessels. Let us first consider the resonance of sound waves in a hollow pipe which is open at both ends. The situation is illustrated in figure 33.10, where the sound source could be a tuning fork or a loudspeaker driven by a sinusoidal voltage.

Since both ends are open, the air particles at the ends are relatively free to move. We therefore expect motion antinodes to exist near the ends. With this restriction, we can draw the resonant modes at once, as is done in figure 33.10. Notice that the resonance frequencies are evenly spaced, with quantum jumps of $v/2L$ between adjacent frequencies.

Antinodes exist near open ends of pipes, while nodes appear at closed ends

When the tube is closed at one end, a motion node must exist there, since the end does not allow the air molecules to move. An antinode will still exist at the open end. The resonances in this case are also shown in figure 33.10. Notice that here too the eigenfrequencies are evenly spaced. The quantum jump between adjacent frequencies is still $v/2L$, as one can see from the figure. However, the fundamental frequency is only half as large in this case as it was in the case of the pipe open at both ends.

Many other types of cavity can be made to resonate to sound waves. For example, when one blows across the lip of a bottle, a definite resonance frequency is often observed. Resonators such as this are rather difficult to treat mathematically since the reflections and interferences of the waves

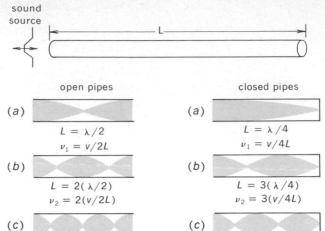

FIGURE 33.10 Resonant modes for sound waves in two types of tubes. What is the general relation for ν_n in each case?

open pipes

(a) $L = \lambda/2$
$\nu_1 = v/2L$

(b) $L = 2(\lambda/2)$
$\nu_2 = 2(v/2L)$

(c) $L = 3(\lambda/2)$
$\nu_3 = 3(v/2L)$

(d) $L = 4(\lambda/2)$
$\nu_4 = 4(v/2L)$

closed pipes

(a) $L = \lambda/4$
$\nu_1 = v/4L$

(b) $L = 3(\lambda/4)$
$\nu_2 = 3(v/4L)$

(c) $L = 5(\lambda/4)$
$\nu_3 = 5(v/4L)$

(d) $L = 7(\lambda/4)$
$\nu_4 = 7(v/4L)$

within the volume are complicated. Even the simple organ pipe, although behaving approximately like the tubes discussed here, is complicated by the presence of the whistle at one end which excites the vibration in the tube. In spite of that complication, however, the resonance frequencies of organ pipes are spaced in the way we found for simple tubes. Such is not the case for resonators of more complicated shape.

Illustration 33.3

Can any of the resonance frequencies of a pipe open at both ends coincide with a resonance frequency of a closed pipe if the pipes are of the same length?

Reasoning In the case of an open pipe the resonance frequencies are given by

$$\nu_{on} = n \frac{v}{2L}$$

while for a closed pipe they are (figure 33.10)

$$\nu_{cm} = \frac{v}{4L} + m \frac{v}{2L}$$

For these to be equal one would have

$$n \frac{v}{2L} \overset{?}{=} \frac{v}{4L} + m \frac{v}{2L}$$

which gives $n \overset{?}{=} \frac{1}{2} + m$.

Since both n and m are integers, we clearly cannot obtain n (an integer) by adding $\frac{1}{2}$ to m (an integer). Therefore the two pipes would have no frequencies in common.

33.7 RESONANCE OF ELECTROMAGNETIC WAVES: PARALLEL PLATES

Since electromagnetic waves satisfy the wave equation, we should expect resonance for these waves under appropriate conditions. A one-dimensional situation which approximates the case of a wave on a string is achieved when one considers an infinite plane wave traveling along the x axis. We assume it to be plane-polarized as well, and so at any given instant it can be represented as shown in figure 33.11(a).

Suppose now we confine this wave by placing an infinite metal plate (perpendicular to the x axis) in its path. The wave will strike the plate and be reflected back upon itself. Since the electric field parallel to the surface must be zero at the surface of the metal (why?), the reflected and incident waves must exactly cancel at the reflecting surface. This is the same situation which occurred with the string, and, as in that case, it means the wave must undergo a 180° phase change upon reflection. We therefore see the reason for the change in phase of light upon reflection, which was first mentioned in Chap. 30. (The situation is not quite so simple in the case of reflection from dielectrics.) In any event, the reflected and incident waves combine to give a node (in E) at the reflecting metal surface.

As in the case of the string, the combination of the incident and reflected waves will cause a series of nodes and antinodes to exist. If a second infinite metal plate is so placed as once again to reflect the reflected wave, a standing wave may or may not result. Since the position of this second plate must also be at a node, it will give rise to a standing wave only if it is placed at one of the nodes formed by the incident and reflected waves. When this is done, the standing-wave pattern shown in figure 33.11(b) is formed.

The condition for the resonances shown in the figure is that the distance between plates be $n(\lambda/2)$, where n is an integer. As a result, for a fixed-plate separation L_x, the eigenfrequencies for resonance will be

$$\nu_n = \frac{c}{\lambda} = \frac{c}{2L_x/n} = \frac{nc}{2L_x}$$

where c is the velocity of electromagnetic waves. We see that the resonance frequencies are evenly spaced, being integer multiples of $c/2L_x$.

Although one might think that evidence for this type of wave resonance would be found primarily in the microwave region where wavelengths are of

For standing electromagnetic waves reflected perpendicular to metal plates, the plates are at nodes

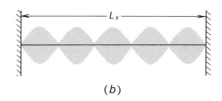

(a)

(b)

FIGURE 33.11 A standing electromagnetic wave can be obtained by reflecting a plane wave at normal incidence between two parallel metal plates. One then observes a set of parallel nodal planes, $\lambda/2$ apart, parallel to the metal plates.

the order of a few centimeters, this type of resonance has also been found in the optical region. By reflecting a beam of parallel plane-polarized light from a flat metal mirror, one should be able to produce nodal planes parallel to the mirror at distances of $n(\lambda/2)$ from the mirror. A photographic film placed at the nodal planes should not be affected by the light, and it should darken at the antinodes. This was first shown to be experimentally correct in 1890, by Wiener, and represents another striking proof of our wave concept of light.

33.8 RESONANCE IN TUBES AND CAVITIES

As we saw in the preceding section, the resonance of electromagnetic waves between infinite parallel metal plates was analogous to the resonance of waves on a string. We might now seek the electromagnetic analog to a vibrating rectangular membrane or drumhead. As in the membrane case, we need two sets of waves, one reflecting back and forth along the x axis, the other along the z axis. This can be accomplished by appropriate plane electromagnetic waves moving in the x and z directions within an infinite rectangular metal tube, such as in the one shown, in part, in figure 33.12.

The boundary conditions for this metal tube are (for waves polarized in the y direction)

$$E_y = 0 \begin{cases} x = 0,\, x = L_x \\ z = 0,\, z = L_z \end{cases}$$

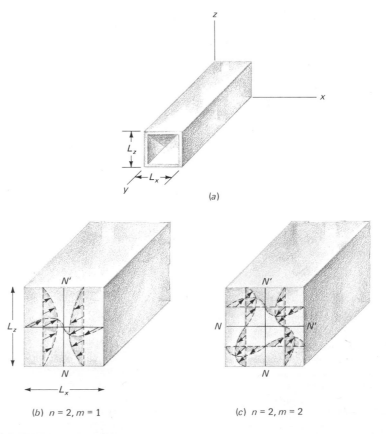

FIGURE 33.12 Resonance pattern for an infinite metal tube, when E_y is the only component of the field.

(a)

(b) $n = 2, m = 1$

(c) $n = 2, m = 2$

These are exactly the same as the boundary conditions we had on the y displacement of the drumhead in Sec. 33.4. Since both of these problems must satisfy the same wave equation and similar boundary conditions, their solutions must be of the same form. As a result, we can write immediately, from 33.7 and 33.8, that

$$E_{mn} = A_{mn} \sin \frac{n\pi x}{L_x} \sin \frac{m\pi z}{L_z} \cos \omega_{mn} t$$

and

$$\nu_{mn} = \frac{c}{2} \sqrt{\left(\frac{m}{L_z}\right)^2 + \left(\frac{n}{L_x}\right)^2} \qquad 33.10$$

The resonance pattern consists of nodal planes inside the pipe and parallel to the sides of the pipe. They are, as usual, $\lambda/2$ apart. Cross sections of these resonance patterns are typified by those shown in figure 33.12. Of course, the fields reverse with frequency ν_{mn}, and so the arrows continually reverse direction as time goes on.

If we cut off the two ends of the tube and place metal plates on its ends so that it becomes a rectangular box of length L_y, then we can have standing waves in three dimensions. To obtain the form of these standing waves, a three-dimensional wave equation must be solved. We shall only state the result for the eigenfrequencies, by analogy with the two-dimensional case:

$$\nu_{mnl} = \frac{c}{2} \sqrt{\left(\frac{m}{L_z}\right)^2 + \left(\frac{n}{L_x}\right)^2 + \left(\frac{l}{L_y}\right)^2} \qquad 33.11$$

where m, n, and l are integers.

As one would expect, three sets of nodal planes for E exist parallel to the three perpendicular sides of the box or cavity. Their spacing along the x axis is given by

$$n\frac{\lambda}{2} = L_x$$

and there are similar relations for the y and z directions. In any event, a cavity such as this will resonate to the frequencies given in 33.11. Typical frequencies are of the order of 10^{10} Hz.

Resonant cavities for electromagnetic waves are most widely used in microwave electronics. They can take the place of LC resonance circuits and have the same type of use. These cavities can also be used to select particular frequencies from a wide frequency spectrum. They are also used to determine the dielectric properties of materials at high frequencies. When the cavity is filled with dielectric, the velocity c must be replaced by the value appropriate to the dielectric, $1/\sqrt{\mu_0\epsilon}$. Furthermore, the energy loss in the dielectric appears as losses in the cavity, and can also be measured.

Chapter 33
Questions
and Guesstimates

1. A string can resonate only in segments of definite length. How long is a segment, and what is the relation between this length and the frequency and speed of the wave on the string?
2. A string is vibrated by a tuning fork of frequency ν. Under what conditions will the string resonate? What variables could be changed to cause resonance, assuming the frequency to be fixed? How could one determine the speed of the waves on the string?
3. A variable frequency oscillator sends waves down a string (of length L) whose ends may be considered nodal positions. At a distance of $L/5$ from

one end, a tiny clamp holds the string nearly motionless although it still allows wave energy to pass by. Describe the standing waves one will notice on the string as the frequency of the oscillator is slowly increased from a very low value.

4. Two identical vibrators are attached at opposite ends of a stretched string. They are adjusted so that when used one at a time, they will cause the string to resonate under transverse motion. When both vibrators are vibrating, will the string resonate? If so, under what conditions?

5. Is it possible for two identical waves traveling in the same direction down a string to give rise to a standing wave?

6. Is it possible to obtain a definite set of nodes and antinodes on a string if the incident wave loses half its energy on reflection, so that the amplitude of the reflected wave is less than that of the incident wave?

7. The little pieces of a string near the antinodes have a great deal of kinetic energy. Since the parts of the string near the nodes have very little kinetic energy, how did the energy get from the source to the antinodes?

8. At resonance in a string, the reflected wave cancels out the vibration of the incident wave at the nodes. Was energy destroyed? What happened to it?

9. The speed of sound in air is about 330 m/s. Estimate the frequency at which a 15-cm-long test tube will resonate when one blows across its lip.

10. Estimate the resonance frequency of a small Coca-Cola bottle. You can check your answer experimentally by comparing the sound obtained when you blow across it with the sound from a loudspeaker connected to a variable-frequency oscillator.

11. All common metals expand when heated. Try to devise a method for monitoring the temperature of a wire by a vibration-resonance technique. Steel wire lengthens about 0.001 percent for each degree change in temperature. Do you think the vibration method is feasible?

12. The pioneering physicist Helmholtz analyzed sounds to see if they contained various frequencies, by use of hollow ball-like devices such as the one shown in figure P33.1. As indicated, one end of the ball was placed in his ear and the other end was placed near the source of sound. Explain the theory behind these so-called Helmholtz resonators.

13. Examine the construction of a child's whistle and explain why it gives off a loud sound of definite frequency. In what ways is it like a trumpet or a tuba?

14. The notes sounded by a flute, clarinet, and "sweet potato" are selected by covering and opening holes along the instrument. Explain what is going on to cause the sound to change.

15. Suppose on some distant planet there exist humanoids whose hearing mechanisms are designed as follows. From the outside, their heads look like our own. However, a 1-cm-diameter hard-surfaced cylindrical hole passes through the head from ear to ear. At the midpoint of the channel a thin circular membrane acts like a drumhead separating the two halves of the channel. These beings experience the sensation of sound when this drumhead vibrates. What can you infer about their hearing abilities and the ways they will communicate orally with each other?

16. From a small height drop a metal bar, glass rod, or a ruler end first onto a solid floor. From the sound given off estimate its fundamental resonance frequency and from this estimate the speed of a compressional (sound) wave in the material from which it is made.

incident sound

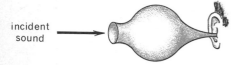

FIGURE P33.1

17. A steel guitar string is tuned to 330 Hz. Make an order-of-magnitude estimate of how much the frequency of the string changes when its temperature is lowered 20°C.

Chapter 33
Problems

1. A certain guitar string has a fundamental frequency of 198 Hz. To what other frequencies will it resonate? If the string is 70 cm long, what is the speed of the wave on the string?

2. Draw the four simplest resonance modes for a string fastened at its two ends. If the string has its fundamental frequency at 60 Hz, what will be the three other resonance frequencies for which you have drawn diagrams?

3. A string resonates in its fundamental to a frequency of 100 Hz. If a node is caused to exist one-third of the way from one end by holding that point with a finger, to what frequencies will the string then resonate?

4. You probably know that a violinist changes the tone given off from a string by moving a finger along the string. If the fundamental of a string is 264 Hz, what will be the fundamental and first two overtones when the violinist's finger is placed one-fourth of the way from the upper end?

5. A 160-cm-long string has two adjacent resonances at frequencies of 85 and 102 Hz. (*a*) What is the fundamental frequency of the string? (*b*) What is the length of a segment at the 85-Hz resonance? (*c*) What is the speed of the waves on the string?

6. A prankster in the chemistry lab wishes to make a 1000-Hz whistle from a glass tube open at both ends. He intends simply to blow across its end. About how long should he make the tube? What will be its resonance frequency if he closes one end with his finger? (The speed of sound can be taken as 330 m/s.)

7. The Lincoln Tunnel under the Hudson River in New York is about 2600 m long. To what sound frequencies will it resonate? What, if any, practical importance do you think this has?

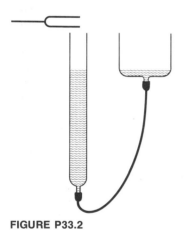

FIGURE P33.2

8. Figure P33.2 shows a simple way to find the speed of sound. At certain heights of water in the tube, the tube resonates to the tuning fork above it. If two adjacent resonances occur when the water height is 13.0 cm and 47.0 cm when a 500-Hz tuning fork is being used, what is the speed of sound in air?

9. A 60-cm-long organ pipe is open at one end and closed at the other. (*a*) What are the fundamental and first two overtone resonance frequencies when the speed of sound is 340 m/s? (*b*) By what percent do the resonance frequencies change if the temperature of the air in the pipe is increased by 10°C? Is it a rise or a decrease in frequency?

10. A cylindrical tube resonates to the following consecutive frequencies: 375 Hz, 525 Hz, and 675 Hz. (*a*) What is the fundamental resonance frequency of the tube? (*b*) Is the tube open at just one end or is it open at both ends?

11. A 40-cm-long brass rod is dropped one end first onto a hard floor but is caught before it topples over. According to a bystander who claims the gift of perfect pitch, the rod emits a tone of frequency 3000 Hz. If he is right, what is the speed of sound in brass?

12. A metal bar clamped at its center resonates in its fundamental to longitudinal waves of frequency 4000 Hz. (*a*) What will be its fundamental resonance frequency when the clamp is moved to its end? (b) What will then be its first two overtone frequencies?

13. The equation for a particular standing wave on a string is $y = 0.15 \sin 5x \cos 300t$ meters. Find the following: (a) amplitude of vibration at the antinode, (b) distance between nodes, (c) wavelength, (d) frequency in Hz, (e) speed of the wave.

14. (a) Write the equation for the fundamental standing sound wave in a tube that is closed at both ends if the tube is 80 cm long and the speed of the wave is 330 m/s. Represent the amplitude of the wave at an antinode by s_0. (b) Repeat for the next two higher resonance frequencies.

15. Twenty-centimeter-wavelength microwaves are sent from a rather distant source straight toward a metal plate and are reflected straight back upon themselves. Where, in front of the plate should a microwave receiver be placed for (a) maximum response and (b) minimum response?

16. Using qualitative reasoning, show that two superimposed waves of the same frequency and amplitude traveling in the same direction cannot give rise to a standing wave. Prove your result mathematically.

17. A rather stiff wire is bent into a circular loop of diameter D. It is clamped by knife edges at two points opposite each other. A transverse wave is sent around the loop by means of a small vibrator which acts close to one clamp. Find the resonance frequencies of the loop in terms of the wave speed v and D.

18. Two identical loudspeakers, placed at $x = 0$ and at $x = 30$ m, send sound toward each other. A man notices that the sound is large when he is at $x = 15$ m but the sound nearly disappears when he moves to $x = 16.5$ m. What is the wavelength of the sound coming from the speakers? With what frequency are they vibrating?

19. A wave given by $y_1 = A \sin(\omega t - \omega x/v)$ is sent down a string. Upon reflection it becomes $y_2 = -\frac{1}{2}A \sin(\omega t + \omega x/v)$. Show that the resultant of these two waves on the string can be written as a combination of a standing wave and a traveling wave.

20. A narrow pipe is bent into the form of the circumference of a circle of radius R. In the center of the pipe a tiny sound source is positioned so as to send sound clockwise around the circle. If the speed of the sound in the pipe is v, for what frequencies of sound will the sound intensity in the pipe be very large?

21. A thin film of transparent dielectric (index of refraction = 1.50) is bounded on its two sides by evaporated metal plates. One plate is semitransparent so that electromagnetic radiation can be sent into the dielectric. If a beam of wavelength λ is sent in perpendicular to the surface, how thick must the dielectric be if the wave is to resonate between the two metal surfaces? More than one answer is possible. Give them all.

22. In a certain device a hollow cubical metal box (or cavity) is used as a resonator for microwaves. When air-filled, the cavity resonates primarily to microwaves of wavelength 2.5 cm. If the cavity is filled with oil of index of refraction 1.50, to what wavelength microwaves will it resonate?

23. A square cardtable top made of a thin plastic sheet is fastened to a rigid frame on its edges. When tapped lightly at its center, it gives off a tone of about 200 Hz. Assuming it to be resonating in the mode shown in figure 33.8 with $n = m = 1$, what is the speed of the wave along the cardtable top? The tabletop is 85 cm on a side. If one pushes down steadily on the center of the table and taps the table in the center of one of its quad-

rants, which mode of motion should result predominantly? What frequency sound should it give off?

24. A large bass drum with drumheads on its two sides has a separation between drumheads of 40 cm. To what frequency should the drumhead resonate if the fundamental resonances of the drumhead and air are to coincide?

25. Find the four lowest resonance frequencies of an electromagnetic wave in a metal tube under the conditions applicable to *33.10*. Repeat for a metal box and *33.11*. Assume all sides are equal to 20 cm.

26. One way of describing thermal motion in solids is to consider the motion to consist of a superposition of all the resonating sound-wave modes of motion within the solid. If a solid cube of side b has its edges along the x, y, and z directions, show that the resonance frequencies of an x-directed plane sound wave are given by $n(v/2b)$.

27. Referring to the previous problem, the thermal energy which gives rise to the specific heat of a solid is considered to consist of the energy of the resonant sound waves. Each resonance is associated with one degree of freedom and carries an energy kT. Although one must work in three dimensions to be accurate, even a consideration of one dimension shows why the shape and size of the solid is not usually important. Assuming the same type wave as in problem 26 and $b = 1.0$ cm, find the total number of resonant frequencies possible in one dimension if all wavelengths down to 2×10^{-10} m (the approximate distance between atoms) are possible. Compare this to the number of waves with $\lambda > 2 \times 10^{-4}$ cm. Assume $v = 2 \times 10^3$ m/s.

Quantum Phenomena

Starting with a discovery by Planck in 1900, a great change began in the field of physics. His discovery gave the first hint that a large portion of physics was still unknown, and marked the beginning of modern physics. As we shall see in this chapter and the next, the years following Planck's discovery were filled with further discoveries of previously unsuspected fundamental laws of nature. We now discuss these very important developments.

34.1 THE ULTRAVIOLET CATASTROPHE

In the year 1900 Max Planck solved a problem which had been vexing physicists for years, and in so doing opened a Pandora's box filled with surprises for mankind. The problem in question had to do with the electromagnetic radiation given off from a very small hole in a hot furnace, i.e., from a so-called *blackbody*.* Careful measurements of the energy radiated by such a blackbody as a function of wavelength had been made. The results are shown in figure 34.1(a). In part (b) of the figure, the same results are plotted as a function of frequency ν rather than wavelength λ. (Recall that $\nu = c/\lambda$, where c is the speed of light.) In (b) the quantity $U(\nu)\,d\nu$ is the energy radiated per square meter per second from the oven in the frequency region of width $d\nu$ centered at ν. $U(\nu)$ is called the spectral *energy density* of the radiation. Several empirical relations had been determined from these experimental results.

As one can see, the wavelength at which maximum radiation occurs is shifted toward the blue as the temperature of the oven is raised. In particular, the wavelength at which the maximum occurs, $\lambda_{\max}$, is related to the absolute temperature T of the blackbody by *Wien's law,* i.e.,

$$\lambda_{\max}T = \text{const} \qquad\qquad 34.1$$

This agrees with the common observation that a white-hot furnace is hotter than one which is red hot.

A second relation, *Stefan's law,* was also formulated in an effort to describe these curves. It states that the total energy radiated per second per

---

*This name arises because, by definition, a blackbody absorbs all radiation incident upon it. In this case, any radiation striking the small hole in the oven has a nearly zero chance of being reflected back in the reverse direction. In addition, the radiation within the oven is at equilibrium, and any radiation coming out of the hole is characteristic of the blackbody at this temperature.

unit surface area is proportional to T^4. In mathematical terms,

$$U \equiv \int_{\nu=0}^{\infty} U(\nu)\,d\nu = (\text{const})T^4 \qquad 34.2$$

Other relations were proposed to describe the shape of the curves of figure 34.1, but only one had a theoretical basis—and that equation failed to predict even the maximum in the curve. Let us see how classical reasoning was used to try to explain this radiation spectrum.

It was recognized that the radiation coming from the oven was emitted by the hot atoms at the walls of the oven. Since it was known that oscillating charges in an antenna or dipole could emit electromagnetic radiation, it was natural to assume that something of the same sort was happening in the atoms of the oven. In particular, it was assumed that the atoms could be treated as oscillating dipoles with various oscillation frequencies. By use of Maxwell's equations it can be shown that a dipole oscillating with frequency ν will emit an energy ΔW_e each second given by

$$\Delta W_e = (\text{const})\nu^2 \langle \epsilon \rangle$$

The quantity $\langle \epsilon \rangle$ is the average energy of the oscillating dipole.

Since the atoms within the oven wall must, as a whole, be at equilibrium, they should emit as much energy as they absorb from the radiation hitting the wall. One can show that if the energy density of the radiation striking an oscillating dipole is $U(\nu)$, then the dipole will absorb an energy each second equal to

$$\Delta W_a = (\text{const})U(\nu)$$

At equilibrium, the two quantities ΔW_e and ΔW_a must be equal. Therefore

Maxwell's equations predict $U(\nu) \sim \nu^2 \langle \epsilon \rangle$

$$U(\nu) = (\text{const})\nu^2 \langle \epsilon \rangle \qquad 34.3$$

If we knew the average energy $\langle \epsilon \rangle$ of an oscillator, it could be placed in this equation, and $U(\nu)$ would be known.

We saw in our study of heat that, for each vibrational degree of freedom

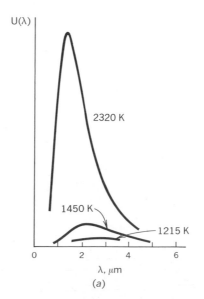

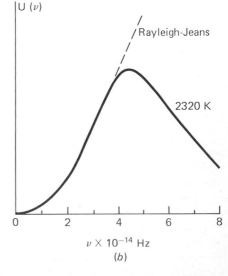

FIGURE 34.1 Measured spectrum of electromagnetic energy from a blackbody (a perfect radiator).

of a molecule, the molecule has an average energy kT, where k is Boltzmann's constant. Carrying over this result to the present case, it means that $\langle \epsilon \rangle \sim T$, and so *34.3* becomes

Setting $\langle \epsilon \rangle \approx kT$ gives the Rayleigh-Jeans law

$$U(\nu) = (\text{const})\nu^2 T \qquad \qquad 34.4$$

This is the *Rayleigh-Jeans law,* named after the men who first derived it in 1900. Obviously, it is wrong. As shown in figure 34.1(*b*), the relation says that $U(\nu)$ should increase without limit as ν becomes larger. But we saw in figure 34.1 that the radiation reaches a maximum at some value of ν and then decreases toward zero as ν increases further. Although the Rayleigh-Jeans law fits the experimental data quite well at small values of ν, it does not predict a maximum value for $U(\nu)$ and is completely wrong at large frequencies, i.e., at short wavelengths.

The Rayleigh-Jeans law predicts $U(\nu) \to \infty$ as $\nu \to \infty$, and therefore fails at large ν

This failure of the Rayleigh-Jeans law at short wavelengths is called the *ultraviolet catastrophe* since the trouble occurs in the ultraviolet and shorter wavelengths. Although many attempts have been made to formulate a better theory for the radiation from a blackbody, reasoning along this line has been singularly unsuccessful. In the next section we shall see how Planck stumbled onto a way out of the dilemma.

34.2 PLANCK'S HYPOTHESIS

One of those who were trying to formulate a proper theory of the radiation from a blackbody was Max Planck. He had already developed an empirical relation which successfully described the experimental data illustrated in figure 34.1. Seeking to justify his relation, he noticed that he could do so by evaluating $\langle \epsilon \rangle$, the average vibrational energy of an oscillator, in a rather arbitrary way.

Prior to that time, the average energy was found in the following way. As we saw in Sec. 14.8, the probability of a molecule having an energy ϵ is given by the Maxwell-Boltzmann distribution formula to be proportional to

$$e^{-\epsilon/kT}$$

where k is Boltzmann's constant, and T is the absolute temperature.* Following the procedure we outlined in Sec. 4.5 for finding averages, we have

$$\langle \epsilon \rangle = \frac{\displaystyle\int_0^\infty \epsilon e^{-\epsilon/kT}\, d\epsilon}{\displaystyle\int_0^\infty e^{-\epsilon/kT}\, d\epsilon} \qquad \qquad 34.5$$

By use of the integral tables we find from this that

$$\langle \epsilon \rangle = kT$$

as stated in the preceding section.

Planck assumed ϵ could take on only values $n\epsilon_0$, where n is an integer

Planck noticed that a much different result is obtained if one arbitrarily states that the oscillator can take on only energies 0, ϵ_0, $2\epsilon_0$, $3\epsilon_0$, and so on.

---

*This follows from *14.13* by writing it in the following form:

$$n = (n_2 e^{+E_2/kT})e^{-\epsilon/kT}$$

and dividing by the total number of molecules, N. Then n/N is the probability of any given molecule having an energy ϵ, and since $(n_2/N)e^{E_2/kT}$ is a constant, the above result is found.

In place of the integrals of *34.5*, we should now have sums, i.e.,

$$\langle \epsilon \rangle = \frac{\sum\limits_{n=0}^{\infty} n\epsilon_0 e^{-n\epsilon_0/kT}}{\sum\limits_{0}^{\infty} e^{-n\epsilon_0/kT}} \qquad 34.6a$$

These sums can be evaluated (see problem 28), and the result turns out to be

Planck's values for $\langle \epsilon \rangle$ and $U(\nu)$

$$\langle \epsilon \rangle = \frac{\epsilon_0}{e^{+\epsilon_0/kT} - 1} \qquad 34.6b$$

If we now place this value for $\langle \epsilon \rangle$ in *34.3*, we find

$$U(\nu) = (\text{const})\frac{\nu^2 \epsilon_0}{e^{\epsilon_0/kT} - 1} \qquad 34.7$$

Now this equation should, if correct, duplicate the curves of figure 34.1. It will in fact do so if one sets

$$\epsilon_0 = h\nu$$

where h is an experimental constant needed to fit the data. It is now known as *Planck's constant,* and has the value 6.63×10^{-34} J $\cdot$ s. Equation *34.7* can then be written

Planck's radiation law

$$U(\nu) = (\text{const})\frac{h\nu^3}{e^{h\nu/kT} - 1} \qquad 34.8$$

To agree with experiment, ϵ_0 must be $h\nu$, where h is Planck's constant

and is known as *Planck's radiation law.* It has proved to be in very good agreement with experiment. Let us review exactly what Planck had done to arrive at it.

First we must recongize that Planck's assumption that $E = n\epsilon_0 = nh\nu$ is at the heart of the matter. Planck assumed that the atomic oscillators which act as radiation sources could have only certain discrete energies. These energies had to be integer multiples of ϵ_0, which, in turn, was equal to h (a previously undiscovered constant of nature) multiplied by the frequency of the oscillator. In other words, an oscillator vibrating with a frequency ν cannot have all possible amplitudes and energies; its allowed energies are quantized. The energy difference between each allowed energy is $h\nu$, and the allowed energies are

$$\epsilon_n = nh\nu \qquad n = 0, 1, 2, \ldots \qquad 34.9$$

We call n a *quantum number.* (Later work shows that the correct relation replaces n by $n + \frac{1}{2}$.)

Radiation is quantized

Second, since the oscillator can take on only certain discrete energies, it must lose or gain energy in pieces (or quanta) of size $h\nu$. It is customary under these circumstances to say that the energy is *quantized,* and the smallest energy change possible is called the energy *quantum.*

Planck's work on this subject was published in 1900. It was far too radical for most physicists of the time, and even Planck himself considered the quantum concept to be a mathematical artifice. The actual existence of quantized allowed energy levels for oscillators was considered to be unrealistic and unacceptable to most people. They (and Planck included) continued to search for a more realistic theory for radiation from a blackbody. However, in 1905, Einstein showed that Planck's quantum concept could be applied to an entirely different phenomenon. We discuss this in the next section.

Illustration 34.1

For the sake of argument, let us say that Planck's quantum condition applies (as it does) to all oscillating systems, including masses on the ends of springs. Suppose a 20-g mass hangs at the end of a spring, with constant 2.0 N/m. What are the amplitudes of vibration for the allowed vibrations?

Reasoning The energy of the oscillating spring system is equal to its potential energy when its displacement is maximum. Calling the amplitude x_0, the energy of the oscillator ϵ_n is

$$\epsilon_n = \tfrac{1}{2}kx_0{}^2$$

where $k = 2.0$ N/m, the spring constant. But according to Planck,

$$\epsilon_n = n\epsilon_0 = n(h\nu) \qquad n = 0, 1, 2, \ldots$$

where ν is the frequency of oscillation. We know from our study of springs that

$$\nu = \frac{1}{2\pi}\sqrt{\frac{k}{m}} = \frac{5}{\pi} \qquad \text{Hz}$$

Therefore, using $h = 6.63 \times 10^{-34}$ J $\cdot$ s, we find

$$\epsilon_n = n(6.63 \times 10^{-34})\frac{5}{\pi} \approx n(10.6 \times 10^{-34}) \qquad \text{J}$$

Equating this to $\tfrac{1}{2}kx_0{}^2$ gives

$$x_0 \approx \sqrt{10.6n} \times 10^{-17} \qquad \text{m}$$

Therefore, as n changes from n to $n + 1$, the amplitude of vibration changes by

$$\Delta x_0 \approx (3.3 \times 10^{-17})(\sqrt{n+1} - \sqrt{n}) = (3.3 \times 10^{-17})\sqrt{n}\left(\sqrt{1 + \frac{1}{n}} - 1\right)$$

$$\approx \frac{1.6 \times 10^{-17}}{\sqrt{n}} \qquad \text{m}$$

where use has been made of the fact that $\sqrt{1 + x} \approx 1 + x/2$. Notice how terribly small this displacement is; far too small to measure. The amplitude x_0 changes by jumps of the order of 10^{-17} m. We see, therefore, that the eigenstates (or allowed amplitudes of vibration) are much too closely spaced for us to recognize that they are discrete states. This turns out to be true for all nonatomic oscillators, and so it is impossible to check Planck's supposition using ordinary laboratory oscillators.

34.3 THE PHOTOELECTRIC EFFECT

Another unexplained set of experimental results at the beginning of this century had to do with the *photoelectric effect*. This effect is observed in the following way. One seals two metal electrodes in a glass bulb, as shown in figure 34.2. All air is evacuated from the bulb, or tube, and the tube is connected in series with a battery and sensitive galvanometer as shown. Of course, no current will flow in the circuit since the two pieces of metal in the tube do not touch, and so the circuit is open. The resistance of that portion of the circuit is infinite.

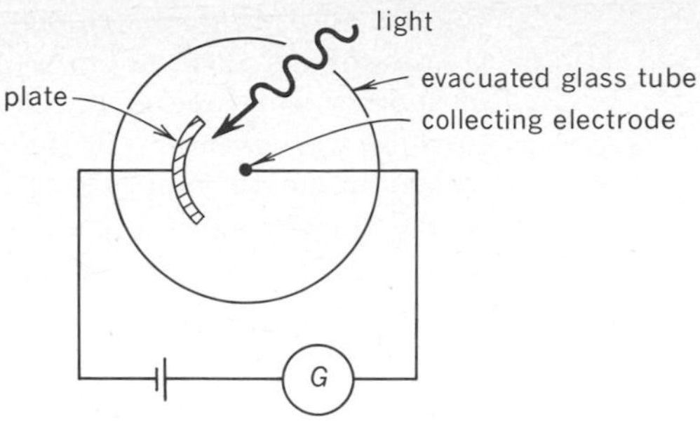

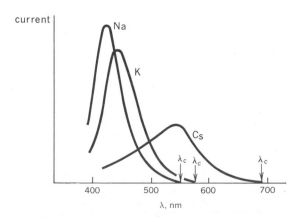

FIGURE 34.2 When light with short enough wavelength is incident on the plate, a current is found to flow in the circuit. Why must the battery polarity be as shown?

However, if one shines light upon one of the surfaces (which we call the plate), a steady current flows in the circuit. We must qualify this statement somewhat by stating that it depends upon the wavelength of the light used as to whether or not a current flows. The variation of current with the wavelength of the light used is shown in figure 34.3. Notice that only wavelengths *shorter* than a certain critical wavelength (λ_c) can cause this effect. For most metals, λ_c is in the blue or ultraviolet portion of the spectrum.

The photoelectric effect: light of high frequency causes electrons to be emitted from solids

Various measurements show conclusively that the light causes *electrons* to be emitted from the plate. These electrons then move across the tube to the positive collector and carry current through the tube. One might at first think of several simple explanations for this effect, but all the explanations proposed before 1905 were proved to be wrong. For example, since light carries energy with it, one might suppose that the light gives enough thermal energy to the electrons of the plate so that they are able to escape. This cannot be correct, however, for the following reasons:

1. No matter how weak the beam of light is, if $\lambda < \lambda_c$, electrons are found to be emitted the instant the light reaches the plate. Since time is needed to raise the energy of the electrons high enough for emission, one would expect a dependence between light intensity and time to leave the plate. No such dependence is found.

Emission is instantaneous

2. A very intense beam of light, with $\lambda > \lambda_c$, will cause electrons to leave the plate only after it has heated the plate red hot. Even the weakest beam with $\lambda < \lambda_c$ causes emission of electrons at once.

A critical wavelength exists

3. Although the number of electrons emitted from the plate is proportional to the intensity of the light beam, the kinetic energies of the electrons emitted do not depend upon the light intensity.

FIGURE 34.3 The current in the circuit of figure 34.2 varies with wavelength as shown. Data for three different metals—sodium, potassium, and cesium—are shown. What is the meaning of the λ_c value indicated in each case?

4. One can measure the maximum kinetic energy of the emitted electrons in the tube of figure 34.2 by observing how large a negative voltage difference must be used in place of the battery in order to stop all the electrons from reaching the collector. (In that case $Vq = \frac{1}{2}mv_{\max}^2$.) It is found that the maximum kinetic energy K of the electrons varies in the following way with frequency ν of the light waves (recall $\nu = c/\lambda$):

$$K = A\nu - B \qquad\qquad 34.10$$

where A and B are constants.

No satisfactory explanation of these facts has ever been found based upon pre-1900 physical ideas.

It was in 1905 that Albert Einstein discovered the correct interpretation for these experimental results. He generalized Planck's results in the following way, so as to obtain a *model for what light is*. Planck assumed that an oscillator could change energy by losing or gaining energy quanta of magnitude $h\nu$, where ν is the natural frequency *of the oscillator*. Einstein postulated that when an oscillator loses a quantum of energy $h\nu$, the oscillator emits a pulse of light having energy $h\nu$, and that the frequency of the light emitted *is equal to this same* ν. In Einstein's view, light consists of energy pulses called quanta, or *photons,* and the quantity of energy carried by each photon is $h\nu = hc/\lambda$, where λ is the wavelength of the light. Let us now apply this model, or concept of light, to the photoelectric effect.

Einstein's postulate: light consists of quanta (photons), each with energy $h\nu$

A beam of light of wavelength $\lambda = c/\nu$ consists of a stream of particlelike photons. Each photon carries an amount of energy $h\nu$. When the light strikes a surface, the photons collide with the electrons of the material. If an energy ϕ is needed to tear an electron loose from the material, a photon must have at least this amount of energy if it is to knock an electron loose. (The quantity ϕ is called the *work function* of the material.) This explains λ_c, the critical wavelength for electron emission, quite naturally. If the photons have energy such that

$$h\nu > \phi \qquad \left(\text{that is, } \frac{c}{\lambda} > \frac{\phi}{h}\right)$$

then the photons will be capable of knocking electrons loose. Otherwise, current will not flow. We therefore have

Unless $h\nu > \phi$, the work function, electrons will not be emitted

$$\lambda_c = \frac{hc}{\phi}$$

In fact, a measurement of λ_c should allow one to determine the work function of the material.

At any wavelength $\lambda < \lambda_c$, each photon will have more than enough energy to free an electron. This additional energy will appear as kinetic energy of the emitted electron. Considering only the fastest emitted electrons (some will have more energy lost than others before getting free), we have

Photon energy = work to break free + kinetic energy of electron

That is,
$$h\nu = \phi + K$$

The photoelectric equation   or
$$K = h\nu - \phi \qquad\qquad 34.11$$

Comparison of this equation, the *photoelectric equation,* with *34.10* shows them to be identical in form. In fact, the experimental constant A in *34.10* is measured to be exactly equal to h, as *34.11* predicts. We therefore see that Einstein's theory is a complete success.

Neither Planck's radiation theory nor Einstein's photoelectric theory would be very acceptable if they stood alone. Each makes a supposition concerning energy quanta which is contrary to the way laboratory objects and light were at that time believed to behave. The fact that both theories make use of the same concepts and the same physical constant to achieve explanations for two entirely different phenomena leads one to suspect that these theories and new concepts have validity. Since that time, other evidence has been accumulated, and physicists recognize that electromagnetic radiation, when interacting with matter, behaves like quanta, with energy $h\nu$. This does not, of course, invalidate our previous ideas concerning the way in which light travels through space, since light's quantum nature is not evident except when it strikes matter. Its motion through space is clearly wavelike, and can best be described in terms of waves.

Light travels through space like a wave, but acts like a particle when interacting with matter

Illustration 34.2

When yellow light from a sodium lamp is incident on a certain photocell, a negative potential of 0.30 V is needed to stop all the electrons from reaching the collector. What potential will be needed to stop the electrons if light with $\lambda = 4000$ Å (i.e., 400 nm) is used?

Reasoning The wavelength of Na light is 5893 Å. Equating the energy of the photons to ϕ plus the kinetic energy of the emitted electrons, we have

$$h\nu = \phi + K$$

But to stop the electrons, a repelling field must be used such that

$$Vq = K$$

where V is the applied negative voltage (the *stopping potential*), and q is the electronic charge. For the two wavelengths concerned we then have

$$h\frac{c}{\lambda_1} = \phi + V_1 q$$

$$h\frac{c}{\lambda_2} = \phi + V_2 q$$

Subtracting one equation from the other, we find

$$hc\left(\frac{1}{\lambda_1} - \frac{1}{\lambda_2}\right) = q(V_1 - V_2)$$

In our case

$$\lambda_1 = 5.893 \times 10^{-7}\,\text{m} \qquad \text{and} \qquad \lambda_2 = 4.0 \times 10^{-7}\,\text{m}$$

while $V_1 = 0.30$ V

Placing these values in the equation and solving for V_2, we find $V_2 = 1.30$ V.

Illustration 34.3

Find the work function for the material of the plate in the preceding example.

Reasoning Using the photoelectric equation, we have

$$h\frac{c}{\lambda} = \phi + Vq$$

or

$$(6.63 \times 10^{-34} \, \text{J} \cdot \text{s}) \left(\frac{3 \times 10^8 \, \text{m/s}}{5.9 \times 10^{-7} \, \text{m}} \right) = \phi + (0.3 \, \text{V})(1.6 \times 10^{-19} \, \text{C})$$

Solving for ϕ gives

$$\phi = 2.9 \times 10^{-19} \, \text{J} = 1.8 \, \text{eV}$$

You will recall that the conversion factor between joules and electron volts is the quantum of charge, 1.6×10^{-19} C. Most metals have work functions several times as large as this value. However, various oxides and more complex compounds have work functions in this range.

34.4 COMPTON SCATTERING

More evidence for the quantum nature of light

We must now consider the nature of photons somewhat further. It is impossible to describe light and electromagnetic radiation in the way we should describe a chair or house or some other visible object. In the case of electromagnetic radiation we must be content to describe it in terms of what it does. For example, we know very well that light undergoes diffraction and interference and therefore appears wavelike under the circumstances where these effects are noticed. That is to say, electromagnetic radiation can be ade-

---

Radiation Pressure

When a beam of light is shined onto a surface, the light beam exerts a pressure upon the surface. This phenomenon was first observed experimentally in 1900, some years after its existence had been predicted. It is possible to show by two different theoretical approaches that light, as well as other electromagnetic radiation, should exert a pressure on a surface upon which the radiation is incident. The earliest theory of this effect was based upon the electromagnetic theory of light, a subject which we discussed in Chap. 28. A second approach became feasible after 1906, the year in which Albert Einstein showed that a beam of light sometimes acts as though it is composed of a stream of particles. This latter approach is by far the simplest of the two, and it is the one we discuss.

Einstein showed that when a beam of light strikes matter, the light behaves as though it consists of a stream of particles. These light particles are called light quanta, or photons. Each photon moves with the speed of light in the direction of propagation of the light beam. Since the photon's speed is identical with that of light, the photon must have zero rest mass. The photon energy is found to be $h\nu$, where ν is the frequency of the light (i.e., the ratio of the speed to the wavelength of the light), and h is a universal constant called Planck's constant ($h = 6.63 \times 10^{-34}$ J $\cdot$ s).

One may think of the light beam as being a stream of photons, each of energy $h\nu$, striking a surface. We can proceed to compute the light-radiation pressure on the surface in a way similar to the method used to compute the pressure of a gas. To use that method, however, we must know the momentum of a photon. It is easily obtained if one recalls that momentum is mv and energy is mc^2 for a particle having $m_0 = 0$. In the present case, since the photon energy is $h\nu$ and its

quately described, as far as its motion through space is concerned, by a wave description. However, as we have seen, when the radiation interacts with matter so as to cause an energy transfer from the matter to the radiation, or vice versa, it is best described in terms of energy quanta, or photons. We conclude from all this that electromagnetic radiation is best looked upon as a distinct means of energy transport which possesses its own peculiar properties. These properties are partly those of particlelike quanta and partly those of waves as we know these quantities in the laboratory.

It is natural to ask about the mass one should attribute to photons. One immediately sees an obstacle to our concept of the photon when it is recognized that light, and therefore the light photon, travels with the speed of light. According to relativity theory, the mass of a particle is given by

$$m = \frac{m_0}{\sqrt{1 - (v/c)^2}}$$

as was shown in Chap. 8. In this expression m_0 is the rest mass of the particle. We note that if the particle speed v is the same as the speed of light, as it will be for a photon, then

$$m = \frac{m_0}{0} \rightarrow \infty$$

Comets' tails extend away from the sun, an effect believed due in part to the pressure of solar radiation. (*Photograph from the Mount Wilson and Palomar Observatories.*)

speed is c, we have

$$\text{Photo momentum} = mc = \frac{mc^2}{c} = \frac{h\nu}{c}$$

Suppose now that a beam of photons (light) strikes a plane perpendicular to the direction of propagation of the light beam. Suppose, further, that the beam is completely reflected without change, so that the photons can be considered to undergo elastic collisions with the surface. Each photon will undergo a momentum change of $2h\nu/c$ as it strikes the wall. (Why is the 2 necessary?) If n photons strike the unit area per second, then the change in momentum per second is $2nh\nu/c$. This, by definition, is equal to the force on the unit area, i.e., the pressure. Therefore we find the radiation pressure to be

$$P = \frac{2nh\nu}{c}$$

This may be put in a more convenient form by noting that $nh\nu$ is simply the radiation energy striking the unit area each second. Calling this quantity Q, we have

$$P = \frac{2Q}{c}$$

As a typical example, a square meter of the earth receives an energy of 1.4×10^3 J/m² · s in the direct rays of the sun. This gives rise to a very small pressure, 4.7×10^{-6} N/m². In spite of its small magnitude, radiation pressure is rather easily observed, and quantitative agreement between theory and experiment is found.

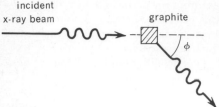

incident
x-ray beam   graphite

ϕ

FIGURE 34.4 Scheme of Compton's experiment.

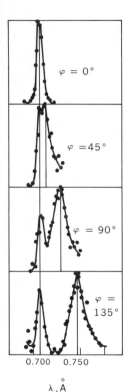

$\varphi = 0°$

$\varphi = 45°$

$\varphi = 90°$

$\varphi = 135°$

0.700   0.750

λ, Å

FIGURE 34.5 These are typical results found by Compton for scattering at various angles φ. How can we explain the fact that some of the scattered radiation has the same wavelength as the incident beam (at about 0.71 Å)?

Compton equation

There is only one way in which we can avoid the result of infinite mass m for the photon. If the photon's rest mass m_0 is zero, then m becomes indeterminate. We therefore conclude that the photon has zero rest mass, and all its energy and momentum result from its motion with speed c. Recalling that Einstein's mass-energy relation is

$$\text{Energy} = mc^2$$

and since the energy of a photon is $h\nu$, we can write

$$h\nu = mc^2$$

From this we find the relativistic momentum of the photon to be

$$p = mc = \frac{h\nu}{c} \tag{34.12}$$

The momentum of a photon is a meaningful quantity since it can be measured directly by allowing electromagnetic radiation to collide with matter. We shall now discuss an experiment which illustrates this point.

In an experiment reported in 1923, A. H. Compton directed a beam of x-rays at a block of graphite and measured the wavelength of the x-rays scattered at various angles ϕ to the incident beam, as shown in figure 34.4. He found that two types of scattered radiation can exist. One type had the same wavelength as the incident beam. The other one had slightly longer wavelength, as illustrated in figure 34.5. He (and, simultaneously, P. Debye) explained these unexpected results in the following way.

Consider a collision of a photon and an electron, as shown in figure 34.6. The components of the net momentum before collision must equal the components after collision. We can therefore write (using *34.12*, after replacing ν/c by $1/\lambda$)

$$\frac{h}{\lambda} = \frac{h}{\lambda'} \cos\phi + mv_x \tag{34.13}$$

and

$$0 = \frac{h}{\lambda'} \sin\phi - mv_y \tag{34.14}$$

We have called the photon wavelength after collision λ', and the mass of the electron is m.

Since the electron and photon appear to be unable to absorb energy internally, the kinetic energy must also be conserved during the collision. Therefore (assuming nonrelativistic speeds for the electron[*])

$$h\frac{c}{\lambda} = h\frac{c}{\lambda'} + \tfrac{1}{2}m(v_x^2 + v_y^2) \tag{34.15}$$

We have only three unknowns in these equations, v_x, v_y, and λ'. They can therefore be solved simultaneously to give λ' as a function of ϕ. Doing this, one finds, after neglecting v^2/c^2 in comparison with unity,

$$\lambda' = \lambda + \frac{h}{mc}(1 - \cos\phi) \tag{34.16}$$

where m is the rest mass of the scattering particle, an electron in the present case.

Notice that *34.16* predicts $\lambda' = \lambda$ when $\phi = 0$. As ϕ increases, λ' also

---

[*] The same result (*34.16*) is found by using the relativistic form for the kinetic energy of the electron, as you will find in problem 29 at the end of this chapter.

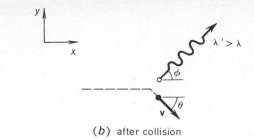

FIGURE 34.6 When a photon collides with an electron, both energy and momentum are conserved. Here the axes are chosen so that no motion occurs in the z direction.

photon (λ)   electron

(*a*) before collision

$\lambda' > \lambda$

(*b*) after collision

increases. This is exactly the behavior shown by the data of figure 34.5. Moreover, substitution of the appropriate numerical values in *34.16* gives the correct experimental value for λ'. We therefore conclude that, insofar as the shifted wavelength portion of the scattering is concerned, the photon concept is valid.

The interpretation of the other portion of the scattered radiation, that with unchanged wavelength, is as follows. In the foregoing discussion we assumed the electron to be free. This will be approximately true for the outer electrons of the atoms in the graphite. However, the electrons close to the nucleus of the atoms will be tightly held by the atom. As we shall see later, these electrons are able to absorb only certain special energies. For the x-rays under consideration, the inner electrons cannot be dislodged from the atom and cannot be given additional energy. Therefore the photon acts as though it collided with a massive object, the atom as a whole. Although the result of *34.16* should still apply, *m* in this case is the mass of the atom, some 20,000 or more times larger than the electron mass. As a result, λ and λ' are essentially identical for these scattered photons.

Illustration 34.4

Find the velocity of the recoil electrons in the experiment of figure 34.5 if $\phi = 180°$.

Reasoning From figure 34.5 we see that $\lambda \approx 0.71 \times 10^{-10}$ m. Using *34.16*, we have

$$\lambda' = \lambda + \frac{6.63 \times 10^{-34}\,\text{J} \cdot \text{s}}{(9.1 \times 10^{-31}\,\text{kg})(3 \times 10^8\,\text{m/s})} \times 2$$

so that $\qquad \Delta\lambda = \lambda' - \lambda = 0.049 \times 10^{-10}$ m

We know that the energy lost by the photon will equal the kinetic energy of the electron. Therefore

$$\frac{hc}{\lambda} - \frac{hc}{\lambda'} = \tfrac{1}{2}mv^2$$

where we are assuming $v \ll c$.

This can be simplified by noting that

$$\frac{hc}{\lambda} - \frac{hc}{\lambda'} = \frac{hc}{\lambda}\left(1 - \frac{1}{1 + \Delta\lambda/\lambda}\right)$$

$$\approx \frac{hc}{\lambda}\left[1 - \left(1 - \frac{\Delta\lambda}{\lambda}\right)\right]$$

$$= \frac{hc\,\Delta\lambda}{\lambda^2}$$

As a result

$$v^2 = \frac{2hc\,\Delta\lambda}{\lambda^2 m} = 4.2 \times 10^{14}\,\mathrm{m^2/s^2}$$

and

$$v \approx 2.1 \times 10^7\,\mathrm{m/s}$$

Notice that $v \ll c$, and so use of nonrelativistic formulas for kinetic energy and momentum was justified. For high-energy γ-rays, however, v will approach the speed of light. Even in the case of $v \approx c$, *34.16* is correct if m is considered to be the rest mass. However, in the present example, the kinetic energy could not be written $\frac{1}{2}mv^2$. How should we carry out the computation in this case? (See problem 29.)

34.5 THE NUCLEAR ATOM

One of the fundamental questions still unanswered in 1900 was concerned with the structure of atoms. As we have seen (page 492), the charge and mass of the electron were known approximately. The masses of the various atoms were also known, together with the chemist's estimates of the number of electrons in each. Since atomic masses are of the order of 10^3 times larger than the mass of the electrons in them, and since the atom is electrically neutral, it was believed in 1900 that the major portion of the atomic mass carries a positive charge.

The most widely accepted model for atomic structure in 1900 was the Thomson model. In this model, the atom was considered to be a roughly spherical solid of diameter about 10^{-10} m. The density of the sphere was pictured to be reasonably uniform, as was the positive-charge density throughout it. Here and there, throughout the volume of the atom, were located the atomic electrons. Because of this intimate mixing of positive and negative charge within the atom, all portions of the atom were nearly electrically neutral.

Thomson's model for the atom is disproved by scattering experiments

This concept of the atom was seriously tested in 1903 by Lenard. He shot a beam of electrons through a thin metal foil and measured the properties of the electrons which were able to pass through the foil. According to the Thomson model, the situation could be depicted as shown in figure 34.7. The films used by Lenard were actually several thousand atom layers thick, and so his foils were actually much thicker than the one shown. He therefore expected that the electron beam would lose a great deal of its energy as it went through the foil. His experiment showed quite different results.

Lenard found that most of the electrons which he shot at the foil went straight through, undeviated. Although his experiments were rather crude, the effect was clearly shown. He concluded from this that the Thomson model for the atom was wrong. Instead, he proposed that the atom consists of very small positive and negative aggregates of charge. The major portion of the atom, according to his proposal, was empty space, except for a few electrons floating in it. Most of the mass, and all its positive charge, were pictured to be in the small aggregates of positive charge.

These rather crude measurements inspired E. Rutherford and his associates to design and carry out more precise investigations. The most complete measurements of this type were carried out by Geiger and Marsden in Rutherford's laboratory in 1911. Their experiment differed in several respects from Lenard's. They used α particles emitted from various radioactive materials rather than electrons as their bombarding particles. Ingenious

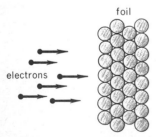

FIGURE 34.7 The foils used by Lenard and later workers were much thicker than the one shown here. According to the Thomson model, the atoms would be pictured as spheres with nearly uniform density.

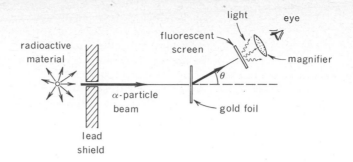

FIGURE 34.8 A schematic diagram of the Geiger-Marsden apparatus to study the passage of α particles through a gold foil.

The α-particle scattering experiment of Geiger and Marsden

The nuclear model of the atom is confirmed by experiment

experiments by Rutherford and his associates had previously shown α particles to be helium nuclei. These particles, being about 2000 times more massive than electrons, should not be deflected much by collisions with electrons. As a target foil, they used a thin sheet of gold. Gold has two advantages over other metals; it is easily hammered into very thin foils, and its atoms are very heavy, so that one should expect them to remain nearly immobile after collision with the much smaller α particles.

A schematic diagram of the Geiger-Marsden experiment is shown in figure 34.8. The radioactive α-particle emitter was shielded by lead, and only a portion of the α particles was allowed to emerge as a narrow beam. This beam was incident on a foil, as shown. To detect the α particles, use was made of a small fluorescent screen. When an α particle strikes the screen, a pulse of light is given off. (This is similar to the fluorescent action of the TV screen on the end of the TV tube which gives off light when electrons strike it.) By placing the screen at various angles to the direct-through beam, one could count the number of α particles scattered at various angles.

The results of Geiger and Marsden confirmed the results of Lenard's measurements. Their very detailed investigations showed clearly that the gold foil did not affect the greater share of the α particles. A small portion of the α particles was strongly scattered, however. Some of them were reflected nearly straight back, indicating that they had struck a very massive object essentially head on. These results indicate clearly that the major portion of the mass of the atom is highly concentrated in a very small fraction of the volume of the atom.

From the results of these experiments, as well as additional evidence, part of which is presented in the next section, we have come to accept the *nuclear model* for the atom. We conclude that the atom consists of a tiny nucleus ($\approx 10^{-14}$ m in diameter) which contains all the mass of the atom except for the mass of its electrons. The nucleus also contains a positive charge Ze, where Z is the atomic number of the element and e is the charge quantum, 1.60×10^{-19} C. Scattered throughout the region outside the nucleus (a sphere $\approx 10^{-10}$ m in diameter) are the Z electrons of the atom. Clearly, since the ratio of the nucleus radius to the atom radius is about 10^{-4}, the major portion of the atom is empty space.

34.6 RUTHERFORD SCATTERING

Although the qualitative arguments presented in the preceding section indicate that we should accept a nuclear model for the atom, scientists are reluctant to accept conclusions based on qualitative reasoning alone. They much prefer to defer acceptance of a model until it can be shown that a mathematical treatment of the model leads to quantitative prediction of experimental results. For this reason, we should like to have a quantitative theory for scattering of α particles by the nuclear atom so that we could compare its predictions with the known experimental results. The quantitative theory for this situation was first provided by E. Rutherford. We now present a somewhat abbreviated version of his theory.

Rutherford's theory to explain the results of Geiger and Marsden

Gravitational Mass of the Photon

We have seen in this chapter that light of frequency v can be considered to consist of photons, each photon having an energy hv, where h is Planck's constant. As shown by Einstein, energy and mass are related in such a way that, in this case,

$$hv = mc^2$$

where m is the inertial mass of the photon. (The photon, of course, has zero rest mass, and so m is entirely the result of the motion of the photon, with speed c.)

One might well ask, since inertial and gravitational mass appear to be identical for objects with nonzero rest mass, whether the photon acts in such a way as to have a gravitational mass $M = hv/c^2$. The following experiment, first performed in 1960 by Pound and Rebka, indicates that the photon does indeed appear to have gravitational mass and that it is equal to its inertial mass, hv/c^2. Their experiment is based upon the idea that a photon of gravitational mass M should acquire an energy MgL as it falls through a vertical distance L.

Suppose a photon is falling toward the earth. Its kinetic energy at a height L above the earth will be hv. After the photon has fallen through the distance L, its kinetic energy will be increased by an amount MgL. In writing this we assume L to be small enough so that both M and g are substantially constants. Since the original kinetic energy of the photon was hv, its kinetic energy at the surface of the earth will be

$$hv' = hv + MgL$$

But since $hv = mc^2$, if the inertial and gravitational masses are the same, we can replace M and obtain

$$hv' = hv + hv\frac{gL}{c^2}$$

or

$$\frac{v' - v}{v} = \frac{gL}{c^2}$$

Since L will be of the order of 10 meters in a practical experiment, the fractional change in frequency $(v' - v)/v$ will be of order 10^{-15}. A test of our assumption concerning the equality of inertial and gravitational masses can be made by checking this relation for $(v' - v)/v$. Although this fractional change in frequency is extremely small, Pound and Rebka were able to measure it by use of an effect discovered about three years prior to 1960, the Mossbauer effect. (This effect is discussed in Sec. 37.13.) To the precision of the measurements carried out to date, the expected change in frequency of the photons is found. We

Rutherford assumed that the only force acting between the α particle and the nucleus was the coulomb electric repulsion between the two particles. Since they carry charges $2e$ and Ze, respectively (the atomic number Z being 79 for gold), the coulomb force is

Rutherford assumed a nuclear atom with coulomb forces only

$$F = \frac{(2e)(Ze)}{4\pi\epsilon_0 r^2}$$

therefore conclude that the photon can be assigned a gravitational mass equal to its inertial mass, $h\nu/c^2$.

G. A. Rebka, Jr., is shown adjusting photomultipliers at the lower end of the Pound falling-photon experiment. (*From Kittel, Knight, and Ruderman, "Mechanics," Berkeley Physics Course, vol. 1, McGraw-Hill Book Company, New York, 1965. Courtesy of R. V. Pound.*)

where r is the distance between the α particle and the nucleus. Notice that Rutherford does not consider the particle as actually touching the nucleus. The nuclear diameter is assumed so small that the electric repulsion will deflect the particle before it gets close enough to come into contact with the nucleus.

In figure 34.9 we show a typical collision process. It is assumed that the incident α particle, when still far from the nucleus, is traveling along a straight line which passes at a distance b from the nucleus. The distance b is *Definition of impact parameter* called the *impact parameter* for the collision. As the particle approaches the nucleus, it is deflected along the curved path and, when far past the nucleus, follows the final straight line shown. The collision process has deflected the α particle through the angle θ. Let us now find how θ depends upon the impact parameter b. In doing this, we assume the nucleus to be so massive that it remains essentially motionless during the collision.

We can write two basic equations for this collision. The first of these expresses the fact that the change in the y-directed momentum of the particle is the result of the impulse exerted on it by the y component of the coulomb force. That is to say,

$$(\text{Impulse})_y = (\Delta mv)_y$$

But since impulse is $F \, \Delta t$, we have

$$(\text{Impulse})_y = \int F_y \, dt$$

$$= \frac{(2e)(Ze)}{4\pi\epsilon_0} \int \frac{\sin \phi}{r^2} \, dt$$

Since the initial y momentum was zero, $(\Delta mv)_y$ is simply the final y momentum, namely, $mv_0 \sin \theta$. We therefore have that

$$(\text{Impulse})_y = (\Delta mv)_y$$

Impulse equation becomes $\qquad \dfrac{(2e)(Ze)}{4\pi\epsilon_0} \displaystyle\int \frac{\sin \phi}{r^2} \, dt = mv_0 \sin \theta$ $\qquad\qquad$ *34.17*

A second relation can be obtained from the conservation of angular momentum. Since the coulomb force is a central force, the angular momentum of the α particle about the nucleus must be constant. The original angular momentum (when r is large and $\phi \rightarrow 0$) was $mv_0 b$. At a more general position, such as the one shown in figure 34.9, its angular momentum is

$$mr^2\omega = mr^2 \frac{d\phi}{dt}$$

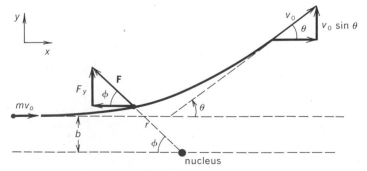

FIGURE 34.9 What two equations hold for an α particle undergoing Rutherford scattering by a nucleus?

Since the angular momentum does not change, this gives

Conservation of angular momentum

$$mv_0b = mr^2\frac{d\phi}{dt} \qquad\qquad 34.18$$

If we solve *34.18* for $1/r^2$ and substitute in *34.17*, we find

$$\frac{Ze^2}{2\pi\epsilon_0v_0b}\int \sin\phi\,d\phi = mv_0\sin\theta \qquad\qquad 34.19$$

But the integral must be evaluated for the whole collision process. Before the collision, $\phi = 0$, and after the collision, $\phi = \pi - \theta$; so these are the limits on the integral. Upon evaluation of the integral, *34.19* yields

The scattering angle equation

$$\tan\frac{\theta}{2} = \frac{e^2Z}{2\pi\epsilon_0v_0^2mb} \qquad\qquad 34.20$$

In obtaining this, use has been made of the fact that

$$\tan\frac{\theta}{2} = \frac{\sin\theta}{1+\cos\theta}$$

Equation *34.20* is one form of the Rutherford scattering formula. It tells us at what angle an α particle will be scattered if its impact parameter is b. Let us now see how close an α particle must come to the nucleus (i.e., how large b must be) if it is to be scattered through an angle of 6° or larger. In order to answer this question we must know the speed of the particle, v_0. The α particles used by Geiger and Marsden had an energy of 7.7 MeV; so their speed can be obtained from

$$\tfrac{1}{2}mv^2 = (7.7\times10^6)(1.6\times10^{-19})\,\text{J}$$

One finds from this that

$$v_0 \approx 1.9\times10^7\,\text{m/s}$$

which is small enough so that relativistic effects are negligible. Substitution of this in *34.20*, together with $\theta = 6°$, gives

$$b = 3\times10^{-13}\,\text{m}$$

Compare this with the radius of the atom, about 10^{-10} m. Obviously, most of the particles incident on the film will have impact parameters larger than this and will not be scattered appreciably.

Upon comparing *34.20* with the experimental results, one finds good agreement, except for very high energy particles scattered at very large scattering angles. It can be concluded from these measurements that the coulomb force satisfactorily describes the force field near a nucleus to radii of less than 10^{-14} m. In recent years it has been possible to carry out scattering experiments with particles of extremely high energy, and thereby approach and even enter the nucleus. From these measurements we have learned that the nucleus is of the order of 10^{-15} m in radius and is only roughly representable by a uniform sphere.

The nuclear atom model explains the experimental results

The radius of the nucleus $\approx 10^{-15}$ m

Chapter 34
Questions
and Guesstimates

1. What familiar examples can you give of the fact that the wavelength of the radiation emitted by a hot object decreases as the temperature of the object increases?
2. Devise an optical method for determining the temperature of a glowing object.
3. Certain regions of the sky appear to contain so-called "radio stars" which

have their maximum electromagnetic emission in the region of microwave wavelengths. What can one conclude about the temperature of such a star?

4. How does the photon picture of light explain the following features of the photoelectric effect: (a) critical wavelength; (b) current proportional to light intensity; (c) current flows the instant light is turned on; (d) stopping potential is inversely proportional to wavelength?

5. How can one measure the work function of a metal? Planck's constant?

6. Make a list of experiments in which light behaves as a wave and a list of experiments in which its quantum character is important. Is there any experiment in your list which can be explained from both standpoints?

7. Could an intense beam of radio waves focused on a metal plate cause photoelectric emission? Explain.

8. It is much easier to observe the Compton effect when x-rays rather than light waves are used. By reference to the billard-ball analogy, explain why.

9. When a metal ball is suspended from an insulated thread and held in a beam of x-rays, the ball acquires a positive charge. Why?

10. When light of a given wavelength strikes a surface and causes the emission of photoelectrons, not all the emitted electrons have the same velocity. All velocities between zero and a definite maximum value are found. Why is there a definite maximum? Why are not all the electrons emitted with this velocity?

11. A beam of light is shone onto a flat surface. In which case will the pressure exerted by the beam be greatest, when the surface is highly reflecting or when it is highly absorbing?

12. Suppose Planck's constant was 0.01 J · s instead of 6.63×10^{-34} J · s. How would a pendulum then behave assuming the pendulum to have a period of 2 s and its bob to have a mass of 50 g? Repeat for values of 0.1 J · s and 1 J · s.

13. Ultraviolet light causes sunburn while visible light does not. Explain why. Some people insist they sunburn easiest when their skin is wet. Do you see any reason for this?

14. An invisible fixed object hangs in an open window. The window has an area A. Using a peashooter, 10,000 peas are shot at random at the window opening. Of these, only 9500 go through; the rest hit the invisible object. What is the cross-sectional area of the object? What relation does this experiment have to the Rutherford experiment?

15. The average temperature of the universe has been measured by use of radio telescopes. By measuring the intensity of radiation from outer space as a function of wavelength, a tentative temperature of 3 K has been obtained. Explain the principle of the method.

16. Estimate the power change for a local radio station antenna system as it changes from one quantized oscillation energy state to an adjacent state. What energy photons does the station radiate? What wavelength photons? What frequency?

Chapter 34
Problems

1. According to the data of figure 34.1, an oven at 1500 K should have its maximum radiation intensity at a wavelength of about 2 μm. The average temperature of outer space is about 3 K. At what wavelength should its maximum radiation intensity appear?

2. Assuming from figure 34.1 that a blackbody at 1500 K has its maximum

radiation intensity at about 2 μm, how hot must an object be to have its maximum at 550 nm in the center of the visible region? For comparison purposes, the melting point of tungsten is 3643 K.

3. The planet Venus is about the same size as the earth. However, its average temperature is about 500°C, while the earth's average temperature is about 22°C. Find the ratio of the energy radiated into space by the earth to that radiated by Venus.

4. Show that the Rayleigh-Jeans radiation law does, indeed, predict that the total energy radiated by an object is infinite. [*Hint:* Integrate $U(\nu)$ from $0 \leqslant \nu \to \infty$.]

5. Show that in the limit of low frequencies (that is, $h\nu \ll kT$) Planck's radiation law reduces to the Rayleigh-Jeans law.

6. Show that in the region of high frequencies where $h\nu \gg kT$ that Planck's radiation law reduces to

$$U(\nu) = (\text{const})\nu^3 e^{-h\nu/kT}$$

This is often called Wien's formula. Notice that it predicts a rapid decrease in radiation intensity at high frequencies.

7. The two nitrogen atoms in a nitrogen molecule act much like two equal masses connected by a spring. The molecule's natural vibration frequency is 7×10^{13} Hz. According to Planck, what is the difference in energy between its allowed energies of vibration? Express your answer in electron volts. To have this much energy, through how large a potential difference would an electron have to fall? Notice that the energy difference is measurable for atomic systems.

8. The natural frequency of vibration for the CO molecule as the springlike bond between its two atoms stretches and compresses is 6.5×10^{13} Hz. With what energies (in eV) can this molecule vibrate? Use the correct $n + \frac{1}{2}$ rather than Planck's n. Notice that these energies are measurably different.

9. The critical wavelength for photoelectric emission from a certain substance is 400 nm. What is the work function (in eV) of this material?

10. It is often convenient to remember that photons of wavelength 1240 nm have energies of 1.0 eV. Show that this is true. What is the energy of a 3100-Å photon? Other energies can be found by proportion.

11. An energy of 13.6 eV is needed to tear loose the electron from a hydrogen atom, i.e., to ionize the atom. If this is to be done by striking the atom with a photon, what is the longest wavelength photon which can accomplish it? (Assume all the photon energy to be effective.)

12. A typical laboratory helium-neon laser beam might have a power of 1 mW. Its wavelength is 633 nm. (*a*) How many photons strike a surface in its path each second? (*b*) How many photons does a 50,000-W radio station antenna emit each second if the station operates at a frequency of 750 kHz?

13. Referring back to figure 34.3, what is the work function of cesium? Of sodium?

14. The photoelectric threshold for potassium is 570 nm. (*a*) What is the work function for potassium? (*b*) What would be the stopping potential when violet light with 400-nm wavelength is used?

15. A helium-neon laser for laboratory use might have a beam power of 0.5 mW. Assume the beam is completely absorbed in a 2-g silver coin. (*a*) How many silver atoms are there in the coin? (*b*) Assuming one free electron per atom, and that the energy is all taken up by these electrons,

how long would the beam have to strike the coin to give each free electron an average energy of 4.6 eV, the work function for silver?

16. When light with $\lambda = 500$ nm is incident on a surface, the photoelectric stopping potential is found to be 0.15 V. What is the work function of the surface in electron volts?

17. Ultraviolet radiation with $\lambda = 300$ nm strikes a surface which has a work function of 3.7 eV. What will be the photoelectric stopping potential in this case?

18. The energy needed to break apart the atoms in the CN molecule (its dissociation energy) is 7.61 eV. What maximum wavelength radiation would be capable of causing this molecule to dissociate? In what region of the spectrum is this?

19. Many of the molecules in our bodies have large numbers of carbon-carbon bonds in them. The energy needed to disrupt such a bond is about 2.8 eV. How short must the wavelength of light be which is just capable of tearing apart such bonds? Which type of radiation causes sunburn: infrared or ultraviolet?

20. A photon with $\lambda = 2.000$ Å strikes a free electron and rebounds straight back. What is its new wavelength? Repeat for a photon with $\lambda = 5000$ Å.

21. The Compton equation can be written as $\lambda - \lambda' = (h/mc)(1 - \cos \phi)$. The factor h/mc is the wavelength shift that occurs when $\phi = 90°$. It is called the *Compton wavelength*. Evaluate it for scattering from (a) electrons and (b) protons. In the case of electrons, what percentage change is this if the incident radiation has a wavelength (c) 500 nm (visible light) and (d) 0.050 nm (0.5-Å x-rays)?

22. A photon with $\lambda = 0.5$ nm strikes a free electron head on and is scattered straight backward. Assuming the electron to be at rest initially, what is its speed after the collision?

23. A photon with $\lambda = 0.5$ nm is moving along the x axis when it strikes a free electron (initially at rest) and is scattered so as to move along the y axis. What are the x and y components of the electron's velocity after collision?

24. A photon strikes a free electron at rest and is scattered straight backward. If the electron's speed after collision is αc where $\alpha \ll 1$, show that the electron's kinetic energy is a fraction α of the photon's initial energy.

25. A 2.00-MeV gamma ray strikes a free proton head on and rebounds straight back in a Compton-type collision. Find the energy of the proton after collision. Assume initially it was at rest.

26. In a Rutherford scattering experiment, we might take as a criterion for a "collision" that a collision occurred if the alpha particle is scattered by more than 1°. How large an impact parameter corresponds to this angle? What is the ratio of the impact parameter to the atomic radius in this case? Assume 7.7-MeV alpha particles incident upon gold atoms with a radius of 10^{-10} m.

27. In the experiments of Geiger and Marsden, the impact parameter corresponding to a deflection of 1° is about 2×10^{-12} m. Assuming 3×10^{17} atoms in each square centimeter of gold foil, about what fraction of the alpha particles were deflected through angles larger than 1°?

28. Show that the sums given in *34.6a* for $\langle \epsilon \rangle$ do indeed give *34.6b*. In doing this, make the substitution that $x = e^{-\epsilon_0/kT}$ and note that the resultant series are common binomial series.

29. Derive the Compton equation *34.16,* using the relativistic expressions for the energy and momentum of the electron.

35 Introduction to Atoms and Spectra

The 15 years from 1911 to 1926 were exciting ones for those who realized the discoveries which were taking place in physics. Shortly after the acceptance of Rutherford's model of the atom, great progress was made in our understanding of the nature of atoms and the radiation they emit. This progress was tremendously accelerated by the discovery that particle and wave phenomena are intimately linked. In this chapter we review these discoveries and show how they led to the new perspective of physics as embodied in the concepts of quantum mechanics.

35.1 THE HYDROGEN SPECTRAL SERIES

As we have stated previously in this text, measurement is fundamental in science. Although there are some notable exceptions, progress in science usually proceeds in the following way. Those scientists who are most capable in carrying out experiments (the so-called experimentalists) design and conduct experiments to measure the various phenomena of nature. Since, in general, they do not know what results their experiments will produce, they can only hope that the results will eventually prove useful. Often the results they discover are not easily understood, and progress is restricted until enough results are obtained to allow some order or unifying principle to be discerned in them.

A typical example of this is apparent in the development of our understanding of the solar system. Astronomers carried out measurements for many years before Kepler and others were able to unify their results in terms of Kepler's three laws. These were simply summaries of the experimental data in compact mathematical form. No explanation for the data was known. Finally, Newton was able to show how the experimental laws could be explained in terms of a previously unknown law of nature, the law of gravitation.

By 1900 an analogous situation was evolving. For many years physicists had been measuring the radiation given off by atoms. At the turn of the century, the simplest of these experimental results, to be discussed subsequently, had been codified in mathematical terms. In other words, this field of physics had reached the stage astronomy was in at the time of Kepler's laws. However, no basis for the observed experimental relations had yet been found. Let us now see how the physical laws governing radiation from atoms were discovered.

Since hydrogen is the simplest of all atoms, it seems wise to begin the study of atoms by first examining the light which it gives off. When hydrogen gas is placed in an electric discharge tube, and a discharge is caused in it by means of a high potential across the tube, the gas becomes luminous and gives off a bluish-red light. This light can be analyzed by sending it through a spectrograph, as discussed in connection with the grating-and-prism spectrograph in Chap. 32. One then finds on the photographic plate a series of lines (actually, images of the slit), each line representing a wavelength of light given off by the light source. In figure 35.1 we show the spectral lines one finds in the visible and near-ultraviolet portion of the spectrum for hydrogen. This series of spectral lines is called the *Balmer series*, after the man who first found a mathematical relation to characterize it.

Luminous hydrogen gas gives off an ordered array of spectral lines

Notice that the spectral lines form an ordered array. The lines become closer and closer together as the wavelength decreases. In the region below about 370.0 nm, they are so closely spaced that the spectrometer resolution is not good enough to show the individual lines. Finally, at 364.6 nm, the series of lines ends. An equation to fit this ordering of spectral lines was first found by J. J. Balmer in 1885. His formula for this, the Balmer series, is

Balmer series (visible)

$$\frac{1}{\lambda} = R\left(\frac{1}{2^2} - \frac{1}{p^2}\right) \qquad p = 3, 4, \dots \qquad 35.1$$

The quantity R is a measured constant needed to fit the data. It is determined directly from the measurement of wavelength, and since wavelengths can be measured with great accuracy, this constant is known quite exactly. The constant R is called the *Rydberg constant,* and its value is $1.0967800 \times 10^7 \, \text{m}^{-1}$. To illustrate the utility of *35.1*, let us evaluate a few wavelengths with it.

p = 3   Placing the value $p = 3$ in *35.1*, one solves for λ and finds $\lambda = 656.3$ nm. Referring to figure 35.1, this is seen to be the first line of the Balmer series.

p = 4   Using this value for p, one finds $\lambda = 486.2$ nm, the second line of the Balmer series.

p = 10 This value for p gives $\lambda = 380.0$ nm, and is the eighth line of the series.

p = ∞   The wavelength corresponding to this value of p is 364.6 nm, the so-called *series limit* wavelength.

The series limit occurs for p = ∞

We see from this that each line in the Balmer series is given by a particular integer value of p in the Balmer formula, *35.1*. The shortest wavelength of this infinite series of lines, the series limit, is obtained when p is set equal to infinity. It is found that the Balmer formula fits the observed wavelength data to very high precision. There seems to be something very fundamental about this extremely simple empirical relation.

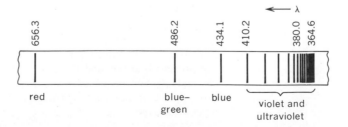

FIGURE 35.1 The Balmer series of spectral lines for hydrogen (λ in nm).

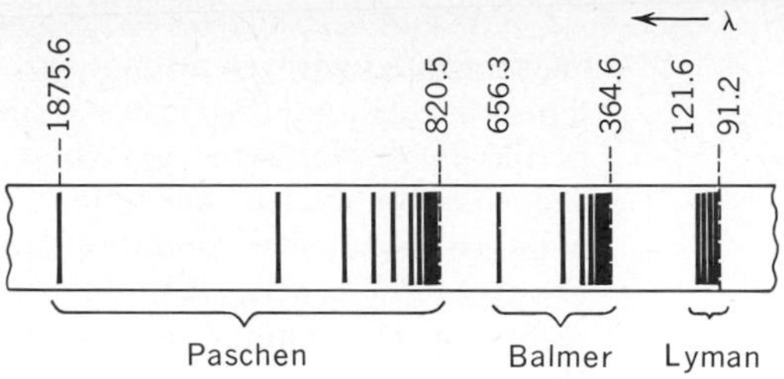

FIGURE 35.2 The three spectral series of shortest wavelength given off by hydrogen atoms (λ in nm).

Later work showed that the Balmer series of lines was not the only series which exists in the radiation from excited hydrogen atoms. Another series is found in the ultraviolet extending between the limits 121.6 and 91.2 nm. This is called the *Lyman series,* named after its discoverer. Other series are found in the infrared region of the spectrum, named after their discoverers, Paschen and Brackett. The relative locations of three of these series are shown in figure 35.2; the Brackett series lies at still longer wavelengths.

It was found that these series, too, could be represented by simple formulas. In particular, one finds

Hydrogen spectral series

Lyman

$$\frac{1}{\lambda} = R\left(\frac{1}{1^2} - \frac{1}{p^2}\right) \qquad p = 2, 3, \ldots$$

Balmer

$$\frac{1}{\lambda} = R\left(\frac{1}{2^2} - \frac{1}{p^2}\right) \qquad p = 3, 4, \ldots$$

Paschen

$$\frac{1}{\lambda} = R\left(\frac{1}{3^2} - \frac{1}{p^2}\right) \qquad p = 4, 5, \ldots$$

Brackett

$$\frac{1}{\lambda} = R\left(\frac{1}{4^2} - \frac{1}{p^2}\right) \qquad p = 5, 6, \ldots$$

35.2

Other longer-wavelength series are also known, and these are represented by similar formulas.

As we see, there is a striking regularity in the emission spectrum of hydrogen. This fact was well known in 1911, the time of Rutherford's work on the nuclear atom. At that time no physical reason for these experimental results was known. The time was ripe for a theory to relate the observed spectrum to Rutherford's model of the atom. We shall see, in the remainder of this chapter, that the complete explanation of the relations 35.2 was not achieved until a whole new perspective of nature's laws had been achieved. Attempts to explain the emission spectra of atoms led us to discover that newtonian mechanics is only an approximation to the exact laws of mechanics needed to describe nature. We shall now see how the more exact concepts of quantum mechanics were developed.

35.2 BOHR'S THEORY OF THE HYDROGEN ATOM

The first successful attempt to explain the emission spectrum of hydrogen was made by Niels Bohr in 1913. He was at that time a twenty-eight-year-old postgraduate student working as a visiting scientist in Rutherford's laboratory. Bohr was of course familiar with the success of Rutherford's model for the atom and, in addition, was acquainted with the quantum concepts of Planck and Einstein. He determined to unite these concepts into a practical model for light emission from the hydrogen atom.

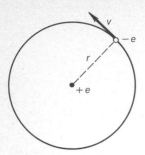

FIGURE 35.3 Bohr pictured the electron to move in a circular orbit around the hydrogen nucleus.

As had been suggested previously by others, Bohr considered the single electron in a hydrogen atom as circling the nucleus with speed v, as shown in figure 35.3. From Rutherford's measurements, he knew that the nucleus carries a positive charge $+e$, where $-e$ is the charge on the electron. Further, the hydrogen nucleus has a mass about 1840 times larger than the electron mass; so the nucleus should remain essentially at rest as the electron orbits about it. The centripetal force needed to hold the electron in orbit is furnished by the coulomb attraction between nucleus and electron. We can therefore write

$$\frac{mv^2}{r} = \frac{e^2}{4\pi\epsilon_0 r^2}$$

where m is the mass of the electron, and ϵ_0 is the permittivity of free space. For later use, we write this

Kinetic energy of the electron

$$\tfrac{1}{2}mv^2 = \frac{e^2}{8\pi\epsilon_0 r} \qquad\qquad 35.3$$

The Discovery of Radioactivity

One of the topics which interested physicists greatly about 1890 was the fact that some materials fluoresce. Upon exposure to bright sunlight, certain substances were found to continue to glow for a short time when moved quickly to darkness. Henri Becquerel (1852–1908) was investigating this phenomenon in 1896 when he encountered, quite by accident, an unsuspected property of commercial salts of uranium. It happened in this way.

On February 26 and 27, 1896, Becquerel was planning to expose a uranium salt to the sun's rays and then see whether or not radiation from the salt would penetrate through black paper and darken a photographic plate. He suspected that this might happen because of the highly penetrating nature of x-rays generated in materials held near the glow of a discharge tube. However, the sun did not shine brightly these two days, and so Becquerel put the plate and salt, separated from each other by black paper, away in a drawer. When, on March 1, he decided to develop the plate, expecting only very slight darkening if any, he found the plate to be completely darkened where it had been close to the uranium salt. This surprising result caused him to pursue the matter further. He found that all salts of uranium behaved in the same way. Even when confined to darkness for several months, they still caused any nearby photographic plate to be darkened. Hence the effect appeared intrinsic to uranium, and exposure of the salt to light was not necessary to produce the effect. He further found that heating, grinding, and dissolving the salts did not change the emitted radiation. The radiations emitted by the uranium salts responsible for the darkening of the photographic plate were found to be exceedingly penetrating. They could penetrate not only paper, but also glass, and even aluminum, although they were considerably attenuated by the latter. Becquerel concluded that the more uranium the sample contained, the more intense the emitted radiations. These radiations, he believed, were similar to, if not identical with, the x-rays discovered by Roentgen (see Chap. 5). Later investigations by others confirmed Becquerel's results, and

The difficulty with this model is as follows. As the electron rotates around the nucleus, it, together with the positive nucleus, acts like an oscillating dipole, or antenna. To see this, refer to figure 35.3. When the electron is at the top of the circle, the system resembles an antenna charged negatively at its upper end. After half a cycle, the charge orientation has reversed. For this reason the atom should radiate energy in the form of electromagnetic radiation with frequency equal to the rotation frequency of

showed that uranium salts were not the only radioactive substances. In fact, only a little over a year later, Marie and Pierre Curie succeeded in isolating two much more highly radioactive substances, radium and polonium. Their discovery is discussed in Chap. 37. It therefore became apparent that the atoms of certain substances spontaneously emit a radiation similar to the x-rays which could be generated in a discharge tube. These substances are said to possess the property of natural radioactivity.

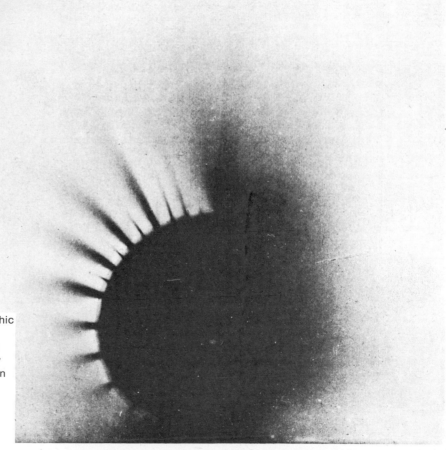

One of Becquerel's original photographic plates, dated 27 April, 1901. Direct radiation from a radium source (in the square block in the inset) was partially intercepted by two screens (also shown in the inset), producing a sunburst pattern of exposure. The general blackening on the right was due to secondary radiation. (*Courtesy of The New York Public Library.*)

the electron around the nucleus. As the atom loses energy, the electron should fall closer to, and eventually into, the nucleus. The atom would then be destroyed. Moreover, as r decreases, according to *35.3*, v should increase. Consequently, the frequency of the radiation given off by the atom should increase steadily until the atom is destroyed. Both these conclusions are contradicted by experiment.

In the first place, hydrogen atoms are quite stable and do not collapse. Second, the spectrum of radiation given off by hot hydrogen atoms consists of definite sets of spectral lines, as we have seen. This does not agree with the conclusion that a continuous range of frequencies should be emitted. These spectral lines had been accurately measured, and as we showed in the preceding section, empirical formulas had been discovered to represent them. Bohr knew all this, and set out to reconcile the idea of an orbiting electron with the experimental facts.

It appeared to Bohr that the idea of an orbiting electron had to be retained. He postulated that, since everyone knew that the atom did not radiate energy and collapse, *there are certain orbits in which the electron is stable and does not radiate.* This, of course, contradicts what one would predict from an application of Maxwell's equations to this electrical system. Bohr's only excuse for the assumption was that no one had actually tested the equations for antenna radiation by atomic systems, and so there was no proof that they applied. Furthermore, as we shall see, Bohr's best excuse was that he got the right answer.

Bohr's first postulate

Let us call the radii of Bohr's stable orbits r_1, r_2, etc. An electron moving in such an orbit has both kinetic and electric potential energy. Its potential energy is Vq, where $q = -e$, and V is the absolute potential $e/4\pi\epsilon_0 r$. Therefore

Potential energy of the electron

$$\text{Potential energy} = \frac{-1}{4\pi\epsilon_0}\frac{e^2}{r}$$

The potential energy is negative since the electron is attracted by the nucleus. The total energy of the electron, when in the pth orbit, is therefore

Total energy in pth orbit

$$E_p = \tfrac{1}{2}mv_p{}^2 - \frac{e^2}{4\pi\epsilon_0 r_p}$$

If we now make use of *35.3* for the kinetic energy, we find

$$E_p = -\frac{e^2}{8\pi\epsilon_0 r_p} \qquad\qquad 35.4$$

As the electron moves closer to the nucleus, r will decrease and the electron's energy will decrease (i.e., become more negative). Bohr now postulated that *when the electron falls from orbit p to orbit j, the energy which it loses, $E_p - E_j$, is radiated as a quantum of light.* He therefore wrote

Bohr's second postulate

$$h\nu_{pj} = E_p - E_j$$

or

$$h\nu_{pj} = \frac{e^2}{8\pi\epsilon_0}\left(\frac{1}{r_j} - \frac{1}{r_p}\right) \qquad\qquad 35.5$$

This relation can be put in a more suitable form by using the relation $\nu = c/\lambda$ to obtain, after dividing by hc,

$$\frac{1}{\lambda_{pj}} = \frac{e^2}{8\pi\epsilon_0 hc}\left(\frac{1}{r_j} - \frac{1}{r_p}\right) \qquad\qquad 35.6$$

We now have a relation for the wavelengths of light which should, if Bohr is correct, be emitted by hydrogen atoms. As we have shown earlier, these wavelengths had been measured with high accuracy, and the following empirical relation had been found to describe them:

$$\frac{1}{\lambda_{pj}} = R\left(\frac{1}{j^2} - \frac{1}{p^2}\right) \qquad 35.7$$

where $j < p$, and p and j are integers. The Rydberg constant R is 10,967,800 m^{-1}.

A striking similarity exists between 35.6, Bohr's theoretical equation, and 35.7, the experimental result. In fact, Bohr found he could make their forms coincide exactly if he set

Bohr's third postulate

$$mvr_n = \frac{nh}{2\pi} \qquad 35.8$$

where n is an integer and h is Planck's constant. Using 35.3 to eliminate v from this relation, one can substitute in 35.6 to obtain

Bohr succeeded in deriving the hydrogen spectral series

$$\frac{1}{\lambda_{pj}} = \frac{e^4 m}{\epsilon_0^2 h^3 c}\left(\frac{1}{j^2} - \frac{1}{p^2}\right) \qquad 35.9$$

The constant in 35.9 agrees within experimental error with the Rydberg constant. This is a truly astonishing result. Bohr had been able to arrive at the emission spectrum of hydrogen by use of the Rutherford model, together with the following assumptions: (1) Light is emitted when the electron jumps from one stable orbit to a lower stable orbit (35.5). (2) Certain stable orbits exist, and their radii are given by 35.8.

Bohr's theory of the hydrogen atom is successful in obtaining a physical model which leads to the observed experimental results for the light emitted from hydrogen atoms. It is also noteworthy in that it gives a basis for the quantum, or photon, character of light. The photon is emitted from the atom as the electron falls from one orbit to another. It is a pulse of radiated energy.

Bohr was unable to give a reason for the stable orbits

However, the theory has a very serious drawback. Bohr was unable to give any reason for the existence of stable orbits and for the condition 35.8, which selects them. The situation is much like that of a student who knows the answer to a problem and then obtains it by a method he cannot justify. No one feels very happy about such a situation.

In spite of this drawback, 35.8 appears to be fundamental. Notice that Planck's constant occurs here once again. Moreover, mvr is the angular momentum of the electron in its circular orbit. Therefore 35.8 is equivalent to

An alternative postulate

assuming that *angular momentum is quantized in units of $h/2\pi$*. This is reminiscent of Planck's discovery that energy (in the case of an oscillator) is quantized in units of $h\nu$. Bohr's theory is of value to us, not so much because it described the hydrogen spectrum, but because it provided the clues which eventually led to a satisfactory theory of atomic structure. We shall see how this was done in the following sections.

Illustration 35.1

How large are the orbits in Bohr's model of the hydrogen atom?

Reasoning Combining 35.8 and 35.3 to eliminate v, we obtain

$$r_n = n^2 \frac{h^2 \epsilon_0}{\pi m e^2} = 0.53 \times 10^{-10} n^2 \text{ m} \qquad n = 1, 2, 3, \dots \qquad 35.10$$

Therefore the first orbit has a radius of 0.53 Å; the second has a radius 4 times as large; the third, 9 times as large; etc. Since the radius of a hydrogen atom is known to be of the order of an angstrom, this result is very reasonable.

Illustration 35.2

How fast is the electron moving in its various orbits?

Reasoning   Making use of *35.8*, we have

$$v_n = \frac{nh}{2\pi r_n m}$$

Using the value of r_n found in the preceding example, this becomes

$$v_n = (2.2 \times 10^6 \,\text{m/s})/n$$

Notice that this speed is small compared with the speed of light.

35.3 ENERGY LEVELS IN HYDROGEN

The Bohr model of the hydrogen atom as developed in the preceding sections is not strictly correct, as we show later in this chapter and in the next. In particular, Bohr's concept of well-defined orbits is found to be a gross over-simplification. However, the model does succeed in unifying a number of aspects of atomic behavior and, if one recognizes its limitations, proves to be a convenient starting point for discussion. It is for that reason that we have begun our study of atomic structure using Bohr's model for hydrogen. Let us now summarize the results Bohr found for the stable energy levels of the hydrogen atom.

It will be recalled that Bohr's model assumed that the single electron in a hydrogen atom could exist stably in orbits with radii (*35.10*)

$$r_n = n^2 \frac{h^2 \epsilon_0}{\pi m e^2}$$

and total energy (*35.4*)

The hydrogen atom can have only certain energies

$$E_n = -\frac{e^2}{8\pi\epsilon_0 r_n} = -\frac{1}{n^2}\frac{me^4}{8\epsilon_0^2 h^2} \qquad\qquad 35.11$$

Placing in the appropriate values, this becomes

$$E_n = -\frac{2.18 \times 10^{-18}\,\text{J}}{n^2} = -\frac{13.6}{n^2}\,\text{eV} \qquad n = 1,2, 3, \ldots \qquad 35.12$$

In other words, when the electron is in the smallest Bohr orbit, $n = 1$ and $E_1 = -13.6\,\text{eV}$.

When $n \to \infty$, we see that $r \to \infty$, and so the electron is free from the nucleus. The atom is ionized. From *35.12* we find that $E \to 0$ as $n \to \infty$. This is to be expected since we take the potential energy of a charge to be zero when the charge is at infinity. If the *freed electron* has kinetic energy, its total energy will be larger than zero. For a free electron, Bohr's computation no longer applies, and the free electron can take on any amount of kinetic energy larger than zero.

For the ionized atom n = ∞

These facts may be summarized in an *energy-level diagram,* as shown in figure 35.4. In it, the energy of the electron is measured vertically, and so the diagram is primarily an energy scale. We represent the values of E_n from

35.12 by the horizontal lines along this scale. For example, the $n = 1$ line appears at -13.6 eV because the hydrogen electron (and therefore the atom) has that energy when the electron is in the first Bohr orbit. Similarly, the $n = 2$ energy level occurs at $-13.6/4$ eV, and so on.

The structure of the diagram in figure 35.4 shows that only the energy levels below $E = 0$ are quantized. Depending upon which energy state the atom is in, the energy of the atom must coincide with one of the horizontal lines (or levels) shown in the diagram. Even though the energy levels with n larger than about 10 are very closely spaced, the levels are still discrete. The atom cannot have energies intermediate to these levels. However, if the electron is torn free from the atom and is at rest, both its potential and kinetic energies are zero. This is the meaning of the $n = \infty$ level. It is possible for the free electron to have kinetic energy greater than zero, in which case it would be in the continuum of levels shown above the zero energy. In this region the energy is not quantized, and the electron can take on any energy.

The energy-level diagram in figure 35.4 can be used to illustrate the origin of the various spectral series observed in the emission spectrum of hydrogen. According to Bohr, the atom emits light when it falls from one of the higher energy levels to a lower level. In his model, when the atom is in the $n = 4$ energy level, the electron is in the fourth circular orbit (shown in figure 35.5). If the electron falls to the second orbit, for example, the atom loses energy $E_4 - E_2$, and this energy is radiated as a quantum of light, with the light frequency $\nu = c/\lambda$ given by

Light emission occurs when the atom undergoes transition to a lower energy level

$$h\nu = E_4 - E_2$$

This energy difference can be visualized directly from figure 35.4 as the arrow extending from the $n = 4$ level to the $n = 2$ energy level of the atom.

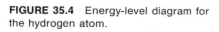

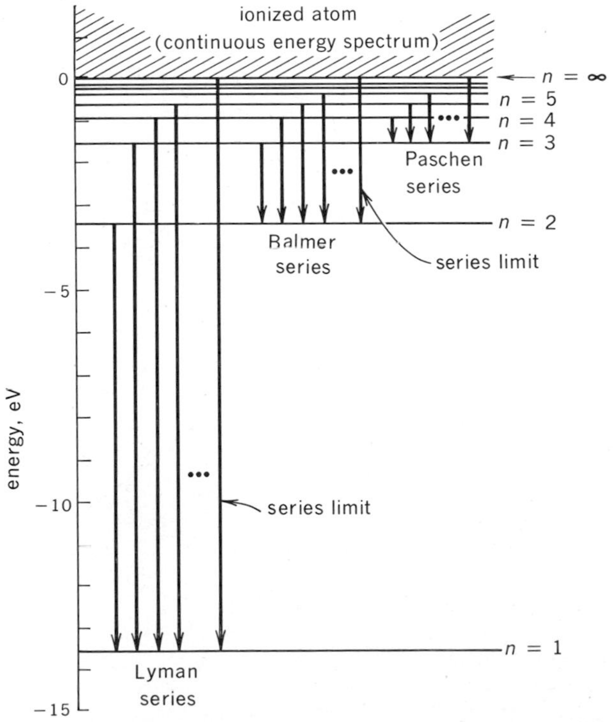

FIGURE 35.4 Energy-level diagram for the hydrogen atom.

It will be recalled that the Balmer formula was

$$\frac{1}{\lambda} = R\left(\frac{1}{2^2} - \frac{1}{p^2}\right) \qquad p = 3, 4, \ldots$$

According to Bohr, the quantity p is the energy level from which the atom falls, while the quantity 2 tells us that the atom ends up in the $n = 2$ energy state. As a result, we see that the Balmer series results when the atom falls from higher energy states to the $n = 2$ state (or the electron falls to the $n = 2$ orbit). This series of transitions is labeled Balmer series in both figures 35.4 and 35.5. Notice that the merging together of the spectral lines near the series limit is the result of the fact that the higher energy levels of the atom are increasingly more closely spaced. The series limit results when an electron falls from the $n \to \infty$ level, the outside of the atom, to the $n = 2$ level.

Similar considerations apply to the Lyman and Paschen series. As shown in figures 35.4 and 35.5, the Lyman series of lines are emitted when the atom falls to the *ground state,* i.e., the lowest ($n = 1$) energy level. The Paschen series results from transitions to the $n = 3$ level. To what set of transitions does the Brackett series correspond?

The lowest possible energy state is called the ground state

It is of interest to notice that the Lyman series wavelengths are approaching x-ray wavelengths. Let us digress for a moment to discuss the origin of x-rays in atomic spectra. In an atom with a large nuclear charge, the innermost electrons will be much more strongly bound than is the electron in hydrogen. If one assumes that a calculation similar to Bohr's can be carried out for the inner electrons of these more massive atoms, an energy-level diagram similar to figure 35.4 should be applicable to the transitions of these electrons as well. However, because of the much greater nuclear charge, the energies involved will be much larger than for hydrogen. These suppositions turn out to be qualitatively correct, as we shall see in the next chapter. Since the energy difference between these lower levels is much larger in atoms heavier than hydrogen, these heavy atoms will have their Lyman series in the x-ray region. We therefore conclude that x-radiation can result from transitions of the inner electrons within atoms of high atomic number.

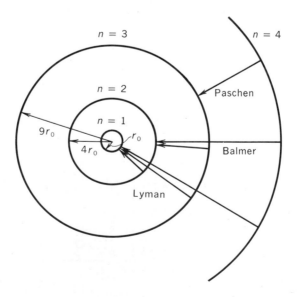

FIGURE 35.5 The electron orbits in hydrogen according to Bohr's model.

35.4 ABSORPTION SPECTRUM OF HYDROGEN

Normally, hydrogen atoms are in their ground state; i.e., the electron is in the $n = 1$ level. Since this is the lowest possible energy state for the atom, it is impossible for the atom to emit light. In order to have light emission, the electron must in some way be raised to a higher energy level. Light is then given off when the electron falls back down to the lower levels.

There are several mechanisms by which an atom can be excited to a higher energy level. If two atoms collide with enough energy (perhaps in a flame or other hot gas), the collision can throw the atom to a higher energy state. Or electrons, ions, or other particles can be accelerated to high speeds in a discharge tube, in which case the atoms are thrown to excited energy states upon collision.

Only excited atoms can give off light

In the present instance we are concerned with the collision of photons with the atoms. Except in the case of the Compton effect (where the change in λ is quite small), a photon either loses all its energy (and vanishes) or it loses none of its energy in such a collision. The primary factor determining which of these two possibilities will prevail is the following. If the energy of the colliding photon is exactly equal to the energy difference between the $n = 1$ level and some other level, the photon can be absorbed. Otherwise, it must continue with its original energy.

The reason for this is quite simple. Since the electron in an atom can exist in only one of the discrete energy levels, it can take on only increments of energy which will transfer it from one level to another. For example, if the atom is in the $n = 1$ level, as it normally is, it can absorb only energies equal to the energy differences between the $n = 1$ level and the higher levels. As we see in figure 35.4, these transitions correspond to energies which (in emission) give the Lyman series of lines. Therefore photons with wavelength equal to that of the first line of the Lyman series (121.6 nm) will have enough energy to excite the atom from the $n = 1$ to the $n = 2$ state. They can therefore be absorbed by the atom.

Atoms absorb only those wavelengths which they emit

Similarly, photons with wavelength equivalent to any of the lines in the Lyman series can be absorbed by hydrogen atoms in the ground state. No other intermediate wavelength photons can be absorbed since their energies will not correspond to an allowed transition for the electron. However, photons with wavelengths less than the Lyman series limit, 91.2 nm, can be absorbed. These photons have enough energy to excite the electron into the continuum, the region of continuous energy levels. Photons with this much energy tear the electron completely loose from the atom (i.e., ionize it) and give additional kinetic energy to the electron. This type of photon absorption process is similar to photoelectric emission of electrons from a solid, and is referred to as the *atomic photoelectric effect*.

Hydrogen in the ground state absorbs only the Lyman series

From what has been said, we can state what will happen when a continuous band of wavelengths (a continuous spectrum) of radiation is passed through a gas of atomic hydrogen. Most of the wavelengths will not be absorbed since their photons do not have the proper energies to excite the atom to an allowed energy state. However, wavelengths corresponding to lines in the Lyman series will be absorbed, since the corresponding photons have the proper energy to excite the atom to an allowed energy state. We call such an absorption spectrum a *line absorption spectrum*. Wavelengths shorter than the Lyman series limit will be absorbed since these photons will ionize the atom and carry the electron into the continuum. The absorption in this wavelength region is called a *continuous absorption spectrum* since a whole range, or band, of wavelengths is absorbed.

Finally, we should note that absorption lines corresponding to the Balmer series lines do not exist, except perhaps extremely weakly. The reason for this is as follows. We know the Balmer series corresponds to transitions between the $n = 2$ and higher levels. Since very few electrons are normally in the $n = 2$ state, only a very few atoms are capable of having an electron knocked from the $n = 2$ state to higher states. Therefore photons corresponding to these energies will not be strongly absorbed. Of course, in highly excited hydrogen gas, the situation becomes more favorable for detecting absorption at the Balmer line wavelengths. Why?

Illustration 35.3

What fraction of the atoms of hydrogen gas at 527°C have their electrons in the $n = 2$ state?

Reasoning The Maxwell-Boltzmann distribution law (14.13) tells us that the ratio of the number N_2 of particles with energy E_2 to the number N_1 with energy E_1 is*

$$\frac{N_2}{N_1} = e^{-(E_2 - E_1)/kT}$$

where k is Boltzmann's constant. In our case, $E_1 = -13.6$ eV and $E_2 = -3.4$ eV. Changing these to joules, we find

$$\frac{N_2}{N_1} = e^{-(E_2 - E_1)/kT} = e^{-148} = 4.5 \times 10^{-65}$$

We conclude from this that, for all practical purposes, the number of atoms in the $n = 2$ state is zero.

Illustration 35.4

What must be the resolving power of a diffraction grating if it is to be able to resolve the seventh from the eighth line of the Balmer series?

Reasoning We recall from Chap. 32 that the resolving power is $\lambda/\Delta\lambda$. It is necessary to find $\Delta\lambda$, the separation of the two spectral lines. Since

$$\frac{hc}{\lambda_7} = E_9 - E_2 \quad \text{and} \quad \frac{hc}{\lambda_8} = E_{10} - E_2$$

we have $\lambda_7 = 3.837 \times 10^{-7}$ m and $\lambda_8 = 3.799 \times 10^{-7}$ m

The resolving power is therefore

$$\frac{\lambda}{\Delta\lambda} = \frac{3.82}{0.038} = 100$$

How many lines must the grating have?

35.5 THE CORRESPONDENCE PRINCIPLE

We must now make an effort to join the classical Maxwell picture of a radiating electric dipole with Bohr's picture of the hydrogen atom. This is

---

*Actually, this ratio should be multiplied by the degeneracy ratio 8:2 since the atom has 8 states with the energy E_2 and only 2 states with energy E_1, as we shall see in the next chapter. However, this factor is of no concern for the present order-of-magnitude computation.

important since, theoretically at least, the hydrogen orbits could be large enough so the orbiting electron would cause the atom to approximate a dipole of macroscopic size. In this latter case, Maxwell's equations are known to give correct results. The question then arises as to whether Bohr's model gives the correct result for such large dipoles.

Maxwell's equations predict that a charge which travels in a circular orbit about a central charge should emit electromagnetic radiation continuously. The radiation should have the same frequency as the orbiting frequency of the charge. In the case of the hydrogen atom, the orbiting frequency is

$$\nu_{orb} = \frac{v_n}{2\pi r_n} = \frac{me^4}{4n^3h^3\epsilon_0{}^2} \qquad 35.13$$

where use has been made of the values found in illustrations 35.1 and 35.2 for v_n and r_n. Therefore classical theory insists that the hydrogen atom should emit light of this frequency.

Bohr's theory, on the other hand, does not, in general, predict radiation of this frequency. We have already shown that Bohr's result is correct for an atomic-size system. It is now important to know if it is capable of describing macroscopic systems. Bohr noticed that if one considers transitions between energy levels in his model, for which n is large, the emitted frequency will be equal to that given by classical theory. To see this, we make use of 35.12. The energy emitted when the electron jumps from the $n + 1$ level to the nth level is

$$h\nu = \frac{me^4}{8\epsilon_0{}^2h^2}\left[\frac{1}{n^2} - \frac{1}{(n+1)^2}\right]$$

Rearranging, this becomes

$$\nu = \frac{me^4}{8h^3\epsilon_0{}^2n^2}\left[1 - \frac{1}{(1+1/n)^2}\right]$$

Now *if n is large,* the quantity $1/n \ll 1$, and so this becomes, after use of the expansion $1/(1+x)^2 = 1 - 2x + \cdots$,

$$\nu \approx \frac{me^4}{8h^3\epsilon_0{}^2n^2}\left(1 - 1 + \frac{2}{n}\right) = \frac{me^4}{4h^3\epsilon_0{}^2n^3}$$

which is the same as the classical result.

Definition of the correspondence principle

Bohr therefore concluded (and this is called the *correspondence principle*) that classical and quantum theory agree at large quantum numbers, i.e., for n large. In that case the system becomes of macroscopic size,* and so our classical macroscopic theories should apply. The fact that Bohr's quantum theory agrees with the classical result for large quantum numbers (where classical theory can be tested and is found to be correct) is a point which strongly favors quantum theory. In the atom domain (where classical theory had been unable to explain experiment), the quantum theory also meets with success. It therefore appears that the quantum approach has wider validity than classical theory. In view of this fact, it is not surprising that further development of Bohr's ideas caused profound changes in our understanding of the laws of physics.

---

* Notice that "macroscopic" results are obtained in the present instance when the following approximation is justified: $1 - 1/n = 1$. How large would be the radius of the orbit if the error incurred by this approximation were only about 1 percent?

35.6 DE BROGLIE WAVES

Bohr's model for the hydrogen atom was open to criticism because he was unable to justify satisfactorily his choice of orbits. He found by trial and error that the following choice (35.8)

$$mvr_n = \frac{nh}{2\pi}$$

35.14

De Broglie attempted to justify Bohr's orbit selection

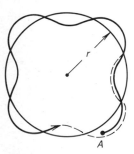

FIGURE 35.6 If the orbit length $2\pi r$ is an integral number of wavelengths, the wave will reinforce itself when it returns to the starting point A. In the case shown, $2\pi r = 4\lambda$.

gave the correct result. Further progress in understanding must necessarily include a justification for this relation. The first successful attempt to justify the relation was made by Louis de Broglie, in 1923. (He was at that time thirty-one years old, and this work constituted his doctoral thesis.)

De Broglie noticed that Bohr's quantum condition, *35.14*, could be interpreted in terms of wave resonance. We recall from our study of resonance of waves that quantum restrictions on resonance frequencies and wavelengths arise quite naturally from the boundary conditions. In the present case of Bohr orbits, we could explain them quite simply in terms of wave resonances if the electron were a wave. To see this, suppose a wave is constrained to travel around the circular path (a Bohr orbit), as in figure 35.6. For the case shown, the wavelength of the wave is just long enough so that the orbit is four wavelengths long. As a result, when the returning wave reaches the starting point, A, it will be in phase with itself. Therefore the wave will be in resonance with itself.

It is thus seen that a wave will resonate on a circular path provided the path length is one whole wavelength long, λ, or any integer multiple of λ. The resonance condition is

$$n\lambda = 2\pi r_n \qquad n = 1, 2, \ldots$$

Let us now see what happens if we replace r_n by Bohr's formula for it, *35.14*. We then find

$$n\lambda = \frac{nh}{mv}$$

or after canceling n from each side,

$$\lambda = \frac{h}{mv}$$

35.15

Notice that the wavelength of the resonating wave (which we have simply speculated about) is given as Planck's constant divided by the momentum of the Bohr electron. One wonders if there is any meaning to this relation.

For further insight into *35.15* let us return to the case of Einstein's photons, used to describe some particle aspects of light waves. We saw in *34.12* that the momentum p of a photon was given by

$$p = \frac{h\nu}{c}$$

34.12

However, since ν/c is simply $1/\lambda$, this can be written

$$p = \frac{h}{\lambda}$$

or

$$\lambda = \frac{h}{p} \qquad \text{(photon)}$$

35.16a

This is identical with *35.15* for the Bohr electron if one recalls that $mv = p$. De Broglie postulated, therefore, that material particles have associated with them a wavelength given by

De Broglie postulated the electron to have wave properties

$$\lambda = \frac{h}{p} \qquad \text{(particle)} \qquad\qquad 35.16b$$

We refer to this wavelength as the *de Broglie wavelength.*

At the time of de Broglie's hypothesis, the only evidence available for the wave properties of electrons (and other particles) was the case we have discussed, the Bohr atom. To review what we have said, if the electron behaves like a wave, one could conceive of the electron wave resonating in certain orbits about the nucleus. If we associated these resonances with the orbits postulated by Bohr, the electron wave would need to have a wavelength given by *35.16*. This equation has a striking similarlity to the relation expressing the momentum of light photons in terms of wavelength. However, the similarity in the two cases, electromagnetic radiation (light) and electrons, is very slight indeed. For example, the photon possesses no rest mass at all, while the rest mass of the electron is finite and well known. In addition, the photon and electron are far different in respect to charge, spin, and other properties. As a result, the de Broglie wavelength concept for material particles was looked upon as a curious hypothesis until further evidence was found to support it.

Illustration 35.5

An electron in a TV tube might have a speed of 5×10^7 m/s. Neglecting relativistic effects, what is the de Broglie wavelength associated with this electron?

Reasoning This is easily answered by substituting in *35.16*. One finds

$$\lambda = 0.145 \times 10^{-10} \text{ m} = 0.145 \text{ Å}$$

Apparently, the wavelength associated with an electron is in the x-ray range of lengths. (We do not mean to imply that de Broglie waves are related to electromagnetic waves. They most certainly are *not* electromagnetic in nature.)

35.7 THE DAVISSON-GERMER EXPERIMENT

The existence of de Broglie waves was first postulated in 1923 to explain Bohr's selection of orbits. This concept was hardly more attractive than Bohr's mere statement of the existence of allowed orbits. Until such time as further confirmatory evidence was found, de Broglie's ideas could not be taken too seriously. A second piece of experimental evidence favoring de Broglie waves was found in 1927, and this evidence was of such a nature as to be conclusive.

Further evidence for the wave properties of particles

Davisson and Germer were investigating the scattering of a beam of electrons by a metal crystal (nickel). Their apparatus, enclosed in a vacuum chamber, is sketched schematically in figure 35.7. A beam of electrons is given a known energy by accelerating the electrons through the potential difference V. As shown in the figure, measurements were made of the number of electrons scattered by a nickel crystal upon which the beam was incident. The unexpected result was that the electrons reflected very strongly at certain special angles and not at others. These results were reported as unexplained by Davisson and Germer.

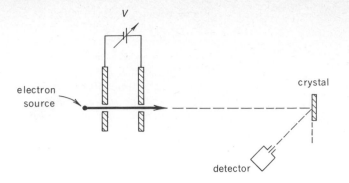

FIGURE 35.7 Davisson and Germer measured the numbers of electrons reflected from the crystal at various angles.

It was then suggested to the two investigators (by Elsasser) that perhaps this was an evidence of de Broglie's radical ideas. They therefore undertook further measurements with properly oriented crystals to see if one could apply Bragg's law (for the reflection of x-rays) to explain the data. You will recall from Chap. 31 that interference of waves reflected from various planes within a crystal could be described by the following relation, Bragg's law: *If the spacing between crystal planes is d and if the waves have wavelength λ, then strong reflection should occur at angles θ given by*

$$n\lambda = 2d \sin \theta \qquad n = 1, 2, \ldots$$

where θ is the angle between the reflected beam and the reflecting crystal plane.

Since Davisson and Germer knew the value of d and the positions of strong reflection θ of the electrons, they could compute a λ value. In addition, the momentum of the electrons could be found since

$$\tfrac{1}{2}mv^2 = Ve$$

or
$$p = mv = \sqrt{2Vme}$$

where V is the potential difference through which the beam is accelerated. The de Broglie wavelength is given by

$$\lambda = \frac{h}{p}$$

Davisson and Germer's experiments agree with de Broglie's postulate

and so it too could be obtained. Davisson and Germer found that these two wavelengths were identical. In other words, the electrons are reflected in the same way as their de Broglie waves should be reflected. This is very direct evidence for de Broglie's idea that electrons have wave properties.

As the years have gone by, it has been found that neutrons, protons, atoms, and molecules, as well as other particles, show the same wave effects that one is able to obtain with electrons. We are therefore compelled to

Particles have associated with them a wavelength h/mv, the de Broglie wavelength

believe that particles, when moving through space, behave like waves of wavelength h/mv. In this expression h is Planck's constant, while m and v apply to the particle in question. Why this behavior had not previously been noted for macroscopic particles is discussed in the following illustration.

Illustration 35.6

Describe the diffraction pattern one would obtain by shooting gunshot ($m = 0.10 \, \text{g}$, $v = 200 \, \text{m/s}$) through a slit which is 0.20 cm wide.

Reasoning The wavelength of the de Broglie wave associated with the gunshot is

$$\lambda = \frac{h}{p} = \frac{6.6 \times 10^{-34}}{(10^{-4})(2 \times 10^2)} = 3.3 \times 10^{-32} \text{ m}$$

From this fact alone, and the knowledge that interference and diffraction effects become large only if λ is comparable with the slit width or separation, one would conclude that interference effects would be negligible. However, to show this clearly, let us find the angle θ between the straight-through beam and the first diffraction maximum. This maximum occurs when

$$\sin \theta = \frac{3(\lambda/2)}{\text{slit width}} = 2.5 \times 10^{-29}$$

In other words, the diffraction angles will be so small that all the particles will travel essentially straight through the slit. Straight-line motion results, and the wave effects are unobservable. This situation always occurs for macroscopic experiments, and it is for this reason that the de Broglie wave effects are unobservable in the motion of macroscopic particles.

35.8 SCHRÖDINGER'S EQUATION

The developments discussed previously indicate that one should properly discuss the motion of atomic-size particles (at least) in terms of some sort of waves. These waves apparently tell us something about the position at which we may expect to find the particle. We have something analogous (but different in substance) to the relation between light waves and light photons. The light waves tell us where the photons go.

Light waves tell us where the photons go

For example, suppose light waves are sent through a slit. The slit causes a diffraction pattern to appear on a screen beyond the slit. Where the light intensity is high, the number of photons hitting the screen is large. Where no light strikes the screen, no photons are found to come. We can calculate the light intensity from a consideration of wave interference. The interference pattern we so calculate tells us the number of photons which will strike various portions of the screen. The important quantity is the *intensity* of the light. This quantity is proportional to the *square* of the amplitude of the wave.

De Broglie waves tell us where particles go

We might therefore expect that the de Broglie wavelength of a particle plays the same role for material particles as the wavelength of light waves plays for photons. If this is true, we should be able to describe the motion of particles in terms of their de Broglie waves by means of a wave equation. The intensity of the waves, i.e., their amplitude squared, should tell us where the particles would be found. Let us now follow this line of thought to see what result it gives. Notice that now we think of a beam of material particles somewhat analogous to a beam of photons in a light beam.

The one-dimensional wave equation for wave disturbances has the form (as we saw in Chap. 27)

$$\frac{\partial^2 \Psi}{\partial x^2} = \frac{1}{v^2} \frac{\partial^2 \Psi}{\partial t^2} \qquad\qquad 35.17$$

where v is the speed of the waves along the x axis, and Ψ is the amplitude of the wave at any instant. (We later use Ψ to represent de Broglie waves.) In all the previous cases involving standing waves, we have eventually separated out the time-oscillation feature of the wave. This we did since we were always interested only in the standing-wave pattern, and this, of course, does not vary with time. Although it is not necessary, we concern ourselves only with standing de Broglie waves, and so, at the outset, we separate out the time dependence of Ψ.

We suspect that Ψ will depend upon time in the usual way, namely,

$$\Psi = \psi \sin \omega t \qquad\qquad 35.17a$$

where $\omega = 2\pi v/\lambda$, with λ the wavelength of the waves. Putting this in *35.17* and recalling that ψ still depends on x but not on t, we find

$$\sin \omega t \frac{\partial^2 \psi}{\partial x^2} = -\left(\frac{\omega^2}{v^2}\right)\psi \sin \omega t \qquad\qquad 35.17b$$

This yields

$$\frac{\partial^2 \psi}{\partial x^2} = -\left(\frac{\omega^2}{v^2}\right)\psi$$

Discovery of the Neutron

All particles of atomic size exhibit a wave nature. They need not be charged. One of the basic uncharged particles, the neutron, is sometimes used in experiments, similar to x-ray diffraction experiments, to determine crystal structures. We now relate how this very important particle was discovered.

By the year 1920 it had been clearly shown that the masses of the nuclei of atoms were very nearly integer multiples of the mass of the hydrogen nucleus. The hydrogen nucleus, the proton, carries a positive charge, equal in magnitude to the charge on the electron. However, a nucleus such as that of helium has a mass 4 times larger than that of the proton, while carrying a charge of only twice that of a proton. In fact, all nuclei show a similar discrepancy, there being more units of mass than charge when the proton charge and mass are used as the basic units.

For several years, the prevailing attitude was that the nucleus must contain electrons as well as protons. The discrepancy between the number of units of positive charge and mass was considered to be the result of the electrons in the nucleus balancing part of the charges on the protons. Since the mass of the electron is less than $\frac{1}{1800}$ that of a proton, the mass of the electron could be neglected. Although this explanation is attractive in many respects, it soon encountered difficulties. First, as we see in this chapter, the electron has associated with it a definite wavelength, which is a function of its energy. This fact indicated that the wavelength of the electron was far too large for it to fit into the nucleus. Second, the known angular-momentum properties of nuclei are not easily reconciled with a model consisting only of protons and electrons.

It became apparent, therefore, in the mid-1920s, that a new particle which has the mass of a proton but no charge probably exists within nuclei. Unsuccessful attempts were made to knock this particle out of nuclei by striking various nuclei with fast-moving protons and α

However, since $\omega = 2\pi v/\lambda$, and since in the case of de Broglie waves, $\lambda = h/p$, with p the momentum of the particle, this becomes

$$-\frac{\partial^2 \psi}{\partial x^2} = \frac{4\pi^2 p^2}{h^2}\psi$$

Usually, it is more convenient to replace the momentum of the particle by the potential and total energy of the particle. We have

$$\text{Total energy} = \text{potential energy} + \text{kinetic energy}$$

or

$$E = V + \tfrac{1}{2}mv^2 = V + \frac{p^2}{2m} \qquad\qquad 35.18$$

where the symbols are as defined by the equation. Solving for p^2 and substituting, we find the following equation for de Broglie waves:

$$\frac{\partial^2 \psi}{\partial x^2} = -\frac{8\pi^2 m}{h^2}(E - V)\psi \qquad\qquad 35.19$$

Schrödinger's wave equation describes the motion of particles

Neutron radiograph of a grasshopper. (*Argonne National Laboratory.*)

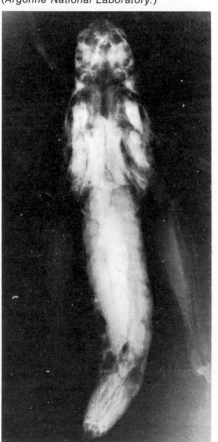

particles. It was not until 1930 that Bothe and Becker noticed a very penetrating radiation given off by beryllium when it was bombarded by α particles. They assumed that the penetrating radiation was γ-rays. Even when it was discovered that this radiation caused protons to be ejected from paraffin, the γ-ray explanation of the radiation was still retained, for these rays were known to be capable of knocking electrons out of materials.

The γ-ray hypothesis for this radiation soon became suspect, however. Measurement of the energy of the protons ejected from the paraffin indicated that the γ-ray would have a much higher energy than any previously known. J. Chadwick first proposed that this radiation was the neutral particle which was suspected of being inside nuclei and, in 1932, he bombarded both hydrogen and nitrogen with this unknown radiation. He was able to measure the recoil energy of the nitrogen and hydrogen nuclei after collision with the radiation. Assuming the incident radiation to be particles, he wrote down the collision equations, namely, the conservation-of-momentum and the conservation-of-energy equations for the particles, if such they were, in the unknown radiation. From the conservation-of-energy equation he obtained the energy of the particles. From the momentum equation he obtained their momentum. He then had two simultaneous equations involving *m* and *v* which he could solve for the mass and the speed of the particles in the unknown radiation.

The particle mass which Chadwick computed from his measurements was found to be only slightly larger than the mass of the proton. Later measurements showed the ratio of this mass to the proton mass to be about 1.001. As a result, Chadwick concluded that these new particles, which he called neutrons, were the long-sought neutral particles which exist within nuclei. We now have available many more pieces of evidence to confirm his supposition.

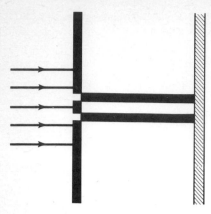

(a) λ << slit width

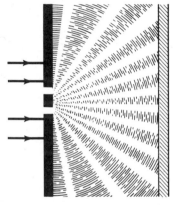

(b) λ comparable to slit dimension

FIGURE 35.8 When the wavelength associated with a particle is much smaller than the slit width, clear images of the slits are formed, as shown in part (a). However, when λ is comparable to the slit width, typical wave-interference phenomena are observed, as shown in part (b).

Classical ideas fail when the particle λ is comparable to dimensions being measured

This equation is called *Schrödinger's* (time-independent) *equation,* after the man who first proposed it, in 1926. The three-dimensional equation is similar to this. Notice that V in this equation is the potential energy of the particle, not the electric potential.

It should be pointed out before proceeding further that *we have not derived* Schrödinger's equation. We have simply guessed that such an equation might be of value for particle waves by using analogies with other wave phenomena about which we know. Whether or not 35.19, Schrödinger's equation, is correct can be decided only by experiment. Investigation of this equation shows that, for macroscopic particles, it gives rise to the same results one obtains from classical mechanics and Newton's laws. This is a result of the fact pointed out in illustration 35.6, that the wavelengths of particles of macroscopic size are so small as to make observations of wave effects impossible. However, during the years since 1926 a tremendous amount of evidence has been accumulated to show that Schrödinger's equation is, in addition, capable of representing the behavior of atomic and nuclear systems, for which classical mechanics and Newton's laws no longer hold.

Let us examine a simple experiment to see exactly where the classical methods of mechanics do or do not apply. For this purpose, consider a modified double-slit experiment, as shown in figure 35.8. A uniform beam of light is incident on the two slits, as indicated, and the light passing through the slits strikes the screen behind them. As we learned in Chap. 31, if the slit widths and slit separation d are much larger than the wavelength of the light, clear shadows are cast. Two bright spots would be noticed on the screen, and these would be rather clear-cut images of the slits. This is shown in figure 35.8(a).

Similarly, if a parallel *beam of electrons* is incident on two slits, they will behave as shown in figure 35.8(a) provided that the wavelength associated with the electrons is much smaller than the slit separation. Hence, the situation shown in figure 35.8(a) is exactly what one would predict from classical particle mechanics. Either a beam of baseballs or a beam of electrons would equally well pass through the holes and hit the screen within a well-defined region. Thus, classical newtonian mechanics is valid when the particle wavelength is much smaller than the geometrical dimensions involved in the experiment.

If we consider, however, the behavior of a light beam when the slit separation is comparable with the wavelength of the light, we observe a wide interference pattern on the screen. As shown in figure 35.8(b), images of the slits are no longer observed. Similarly with the electron beam: If the wavelength associated with the particles is comparable to the slit separation, the electron beam spreads and hits the screen in an interference pattern, as illustrated in part (b) of figure 35.8. The intensity of the interference pattern in the case of the light waves is analogous to the number of electrons hitting the screen in the case of the particles. No particles strike where the intensity of light with identical λ is zero, and a maximum number of particles strike the screen where the light intensity would be maximum. This behavior is completely different from what newtonian mechanics would predict. Hence, classical mechanics is not applicable in this situation. *Classical mechanics becomes invalid when the particle wavelength becomes comparable with the geometrical dimensions involved in the experiment.*

It would appear, however, that the behavior of the light beam is always adequately described in terms of wave phenomena, at least in this experiment. Since the particle behavior can be described in terms of the associated wave in both these cases, while newtonian mechanics can describe only the case shown in part *a*, it would appear that the wave viewpoint is more generally applicable. This in fact proves to be the case.

We are therefore forced to admit that all matter has certain wavelike properties. Each particle is characterized by a wavelength, its de Broglie wavelength. In cases where the wavelength is large enough so that interference and diffraction effects are observable, the methods of classical physics must be replaced by the more exact treatment embodied in the use of wave (or quantum) mechanics. Typically, one must use quantum-mechanical methods when dealing with atomic-size particles. Let us now review the general approach one follows when using the Schrödinger representation of quantum mechanics.

Our prescription for describing the locations of the particles in a beam of particles (in one dimension) is as follows. We consider the total energy of a single particle E to be constant since its energy is conserved. We replace V by the potential energy of the particle as a function of x. Then *35.19* is solved for ψ. The intensity of the wave is found by squaring ψ. (In many cases, ψ is a complex number. We then multiply it by its complex conjugate rather than squaring it.) We interpret the intensity to be proportional to the number of particles which will be found at the point under consideration. If we have only one particle instead of a large group, or beam, of particles, the intensity will be proportional to the number of times we should find the particle at the given point if we determine its position on a large number of independent observations.

The wave properties of particles become measurable when their de Broglie wavelength h/p becomes large enough

ψ^2 is proportional to the intensity or number of particles

35.9 PARTICLE IN A TUBE (OR BOX)

Before going further, let us illustrate the use of Schrödinger's equation by solving a simple example. Suppose one has a hollow tube of length L placed on the x axis, as shown in figure 35.9. We make the tube thin enough so that only the x coordinate will be needed to locate a particle within it. Since the potential energy V of the particle will be assumed constant inside the box, we arbitrarily set it equal to zero. In addition, the walls are assumed to be impervious to a particle, and so V will rise to infinity at the walls. Let us now solve Schrödinger's equation for a particle inside this one-dimensional tube.

There are two ways we could approach this problem, and we shall use each in turn. Before solving Schrödinger's equation directly, let us use our knowledge of wave resonance to obtain the basic features of the solution. (This situation is particularly advantageous for such a method of solution.) We will be looking for the resonances of the particle wave in the box (or tube)

FIGURE 35.9 Can a particle confined to a narrow tube have any and all values of kinetic energy?

much like we looked for the resonance modes of motion for a string or electromagnetic wave.

It will be recalled that the resonance forms are determined by the conditions which must apply at the ends of the string or the walls of a cavity. (These are called the *boundary conditions*.) Once we know where nodes or antinodes must exist, we can state which wavelengths will give rise to resonance. We therefore seek the boundary conditions which must apply in the present situation.

Since the ends of the tube are impervious to the particle, the particle has no chance of being found outside the tube. The chance of finding the particle therefore drops to zero at the points $x = 0$ and $x = L$. Since we have agreed that ψ^2, the amplitude of the particle's wave, is a measure of the likelihood of the particle's appearance at a point, ψ must be zero at $x = 0$ and $x = L$; the chance of finding the particle there drops to zero. Hence, the two ends of the tube must be the positions of nodes. This, then, supplies the boundary conditions which tell us how the particle's wave will resonate in the tube.

Since the ψ wave must resonate with nodes at the two ends of the tube, the particle's wave has the resonant forms shown in figure 35.10. From it we see that the particle's wavelength λ must be related to the length L of the tube through the equation

$$L = n\frac{\lambda}{2} \qquad n = 1, 2, 3, \ldots \qquad 35.20$$

These will be the allowed states (or ways) in which the particle can exist stably in the tube. They are analogous to the allowed stable orbits in Bohr's theory of the hydrogen atom.

Equation *35.20* has some serious and perhaps surprising consequences. Since the wavelength of the particle is related to the particle's momentum through de Broglie's relation,

$$\lambda = \frac{h}{p} = \frac{h}{mv}$$

we see that *35.20* places a condition on the velocity of the particle. In particular, if we replace λ by h/mv and solve for v we find

$$v_n = n\frac{h}{2mL}$$

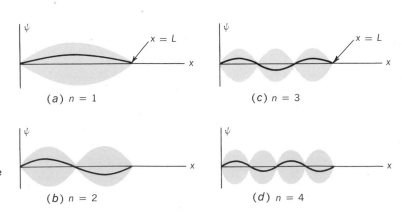

(a) n = 1

(c) n = 3

(b) n = 2

(d) n = 4

FIGURE 35.10 The first few resonance forms for the wave of the particle in a box are as shown.

where the subscript n is placed upon v to indicate that its value depends upon n. Notice the meaning of this result: The particle can exist in the box in a stable way only if its speed is an integer multiple of $h/2mL$. Not all speeds of the particle are allowed. Let us see what this implies for the energy of the particle in the box.

The potential energy of the particle is constant, and we have set it equal to zero. Therefore the particle's energy is all kinetic and will be

$$E_n = \tfrac{1}{2}mv_n{}^2$$

or, after substituting for v_n,

$$E_n = n^2 \frac{h^2}{8mL^2} \qquad\qquad 35.21$$

Notice that the energy of the particle is quantized. For a particle to exist in a stable way within the box, it can possess only the energies given by *35.21*. This, too, is analogous to the allowed energy levels of the electron in the hydrogen atom.

Now that we see physically what the situation is for the particle in a linear box, let us see how this may be described mathematically. Starting with Schrödinger's equation, *35.19*, and after setting $V = 0$, we have

$$\frac{\partial^2\psi}{\partial x^2} = -\left(\frac{8\pi^2 m}{h^2}E\right)\psi$$

Let us try a solution of form

$$\psi = \psi_0 \sin \frac{2\pi x}{k}$$

where k is a constant. Substitution of this in the equation gives

$$\frac{-4\pi^2}{k^2}\psi_0 \sin \frac{2\pi x}{k} = \frac{-8\pi m}{h^2}E\psi_0 \sin \frac{2\pi x}{k}$$

which allows us to solve for k:

$$k = \frac{h}{\sqrt{2mE}} \qquad\qquad 35.22$$

A solution for ψ is therefore

$$\psi = \psi_0 \sin\left(\frac{2\pi x \sqrt{2mE}}{h}\right) \qquad\qquad 35.23$$

This is the equation for the standing wave within the box.

The important quantity is $I = \psi^2$ (the intensity of the wave) since this quantity is proportional to the chance of finding the particle at various positions in the box. We have

$$I = \psi_0{}^2 \sin^2\left(\frac{2\pi x \sqrt{2mE}}{h}\right) \qquad\qquad 35.24$$

The wave function ψ must satisfy the boundary conditions

Our problem is not yet solved, however. As with all solutions of the wave equation, we must be certain it fits the boundary conditions appropriate to our problem. In this particular case we know that the particle cannot leave the ends of the box. Therefore I must drop to zero at $x = 0$ and $x = L$. The standing wave must have nodes at the two ends of the box.

We see at once that the boundary condition at $x = 0$ is automatically satisfied by *35.24* since sin 0 = 0. However, in order for I to be zero at $x = L$, the argument of the sine must be π, 2π, 3π, etc., when x is set equal to that value, L. Therefore

$$\frac{2\pi L \sqrt{2mE}}{h} = n\pi \qquad n = 1, 2, \ldots \qquad \qquad 35.25$$

To satisfy the boundary conditions, E must be quantized

Since L, h, and m have well-defined values, we see that *35.25* puts a condition on the total energy of the particle. Solving for E,

$$E_n = n^2 \frac{h^2}{8mL^2} \qquad n = 1, 2, \ldots \qquad \qquad 35.26$$

where a subscript has been placed on E to show that its values depend on n. This is identical to the result we found in *35.21*.

Equation *35.26* indicates that the allowed energies for a particle in a tube are quantized. The particle can take on only the energy $h^2/8mL^2$ and n^2 times this value, where n is a *quantum number* taking on the values 1, 2, ... These energies are shown in the energy-level diagram of figure 35.11. Since $V = 0$, E is entirely kinetic energy. Notice that the particle cannot have zero energy. This follows from the fact that, since $n = 0$ would correspond to ψ^2 being identically zero, the smallest acceptable value for n is unity, since for $n = 0$ there would be no chance at all of finding the particle in the box. This may seem strange since we are not accustomed to thinking that a gas molecule's energy, when confined to a tube, must be quantized, and cannot be zero. Let us see how large the *zero-point energy* (that is, E_1) of a particle would be.

We have, after placing in the appropriate value for h^2, that

$$E_n = n^2 \frac{5.5 \times 10^{-68} \, \text{J}^2 \cdot \text{s}^2}{mL^2}$$

To find the largest imaginable value for E_n, let us set $L = 10^{-3}$ m and use the mass of an electron for m. This gives

$$E_n = 6 \times 10^{-44} n^2 \, \text{J}$$

This is an extremely small energy for small values of n. For comparison purposes, the average thermal energy of a particle, $\frac{3}{2}kT$, is about $2 \times 10^{-23}T$ J, where T is the absolute temperature. Clearly, the quantization of energy for a particle in a box has no practical importance for laboratory-size boxes since the energy levels are so closely spaced. The energy difference between adjacent levels is too small to measure, and so, for practical purposes, the particle energies appear continuous. In order to reach measurable energies, n must be very large. The correspondence principle tells us that, in that limit, our classical ideas should agree with the results of this, the quantum theory. On the other hand, as we shall see in illustration 35.7 below, the zero-point energy and level spacing have a value of several electron volts for atomic-size boxes and are of considerable importance.

The quantization of E can be observed easily only for atomic-size systems

Let us now see what quantum theory predicts about the location of the particle in the box. Our result was, from *35.24* and *35.25*,

$$I = \psi_0^2 \sin^2 \frac{n\pi x}{L} \qquad \qquad 35.27$$

This tells us the chance, or probability, of finding the particle at a position x within the box. Plots of I/ψ_0^2 for various values of n are shown in figure 35.12.

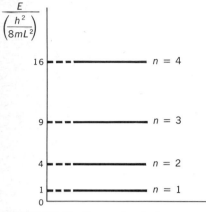

FIGURE 35.11 Energy levels for a particle in a tube.

If the particle is in the $n = 1$ state, it is most likely to be found near the center of the tube. In the $n = 2$ state, it is most probable to find the particle at the one-quarter or three-quarter point in the tube. And so on, for the other values.

In all practical laboratory-size tubes, the energy of the particle will be such that n is extremely large. The peaks in the ψ^2 function will be very closely spaced. In fact, from *35.27*, we see that the peaks are separated by a distance Δx large enough for

At large n values, classical mechanics agrees with quantum mechanics

$$\frac{n\pi \Delta x}{L} = \pi$$

Therefore

$$\Delta x = \frac{L}{n}$$

With n very large, the distance Δx will be immeasurably small, and so we shall not be able to tell that certain points are forbidden to the particle. Moreover, since the pattern for ψ^2 changes as n changes, the variation in energy of a particle will cause the pattern to become diffuse. It should be clear from this that, as $n \to \infty$ and the classical region is considered, the probability of finding the particle will be the same for all points within the tube.

Illustration 35.7

Find the energy levels of an electron confined to a box 1 Å long, which is a box of atomic size.

Reasoning   From *35.26* we have

$$E_n = \frac{n^2 h^2}{8mL^2}$$

In the present case this becomes

$$E_n = 6.0 \times 10^{-18} n^2 \text{ J}$$
$$= 38n^2 \text{ eV}$$

Notice that in this case the energy levels are separated by a sizable energy when compared with the energies of atoms. We conclude from this that the quantization of energy will be important for electrons and other small particles confined to regions of atomic size.

Illustration 35.8

Suppose an electron is truly free to move within a thin wire 50 cm long. What would be the wavelength of the light emitted by the electron as it falls from the $n = 1000$ level to $n = 999$?

Reasoning   The energy levels are given by *35.26*. Putting in the appropriate values yields

$$E_n = n^2(2.2 \times 10^{-37}) \text{ J}$$

When the electron falls from the higher to the lower level, it will emit radiation of wavelength λ, given by

$$h\frac{c}{\lambda} = E_{1000} - E_{999}$$

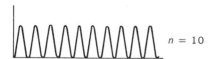

$n = 10$

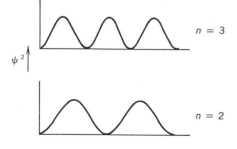

$n = 3$

ψ^2

$n = 2$

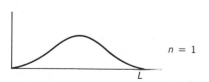

$n = 1$

L

FIGURE 35.12   The position function for a particle in a tube of length L. What does the correspondence principle tell us about these diagrams?

This gives

$$\lambda = \frac{hc}{(2.2 \times 10^{-37}\,\text{J})[1000^2 - (1000 - 1)^2]}$$

$$= \frac{hc}{(2.2 \times 10^{-31}\,\text{J})[1 - (1 - 10^{-3})^2]}$$

$$\approx \frac{hc}{(2.2 \times 10^{-31}\,\text{J})[1 - 1 + (2 \times 10^{-3})]}$$

or finally,

$$\lambda = \frac{hc}{4.4 \times 10^{-34}\,\text{J}} = 4.5 \times 10^8\,\text{m}$$

Apparently the energy levels in this laboratory-size system are so close together that the emitted wavelength is very long, equivalent to an antenna vibrating at about 1 Hz. Can you think of any difficulties one might encounter in trying to observe the discrete nature of this emission spectrum?

Chapter 35
Questions
and Guesstimates

1. Suppose one bombards hydrogen gas atoms at room temperature with electrons of various energies. As the energy of the bombarding electrons is increased, which wavelength of radiation will first be emitted by the gas?
2. What wavelengths of radiation will hydrogen gas atoms emit when bombarded with electrons of energy 12.2 eV?
3. It is possible to obtain a beam of electromagnetic radiation weak enough so that individual photons can be detected separately by means of a photocell. If such a beam is sent through a Young's double slit, what will the photocell show as it is moved through the region where the interference pattern is ordinarily found?
4. Suppose x-rays of wavelength 2.0 Å are used in a Bragg reflection experiment from a crystal. How would the reflection pattern compare with that for reflection of electrons with a de Broglie wavelength of 2.0 Å?
5. Why does not the hydrogen gas prepared by students in the laboratory glow and give off light?
6. Given a glass tube containing two electrodes sealed through its two ends. The gas inside is either hydrogen or helium. How can you tell which it is without breaking the tube? If the gas is at high pressure, what difficulty might you have?
7. When white light is passed through a vessel containing hydrogen gas, it is found that wavelengths of the Balmer series as well as the Lyman series are absorbed. We conclude from this that the gas is very hot. Why can we draw this conclusion? (This is actually the basis for one method of measuring the temperature of a hot gas.)
8. When a photon is emitted by an excited hydrogen atom, the atom recoils. Why? Shouldn't the emitted photon therefore have less energy than the difference in energy between the two states? Why don't we usually concern ourselves about this?
9. The potential energy of the electron in the hydrogen atom is negative. Why? As we have seen, the kinetic energy of the electron in orbit has a *magnitude* only half that of the potential energy. How would the emission spectrum be different if these two energies were equal in magnitude?
10. Refer back to figure 35.7. If strong reflection is observed when the detector is in the position indicated, where might the crystal reflection plane be for this reflection maximum?
11. In some dye molecules which contain a large number of double bonds,

one of the electrons in the molecule acts like a free electron restricted to a certain closed-circuit path within the molecule. Suppose the path in a particular molecule to be circular, with a perimeter of length L. What qualitative statements could one make about the behavior of the electron?

12. Suppose an electron is shot toward a negative metal plate in a vacuum tube. What could one say about the variation of the de Broglie wavelength of the electron as it crosses the tube? Assume the electron stops before striking the negative plate.

13. The diameter of a nucleus is about 10^{-15} m. Estimate the least energy a proton must have if it is to be a part of the nucleus. (*Hint:* The nucleus acts like a box for the proton.)

Chapter 35
Problems

1. Using the Balmer formula with $R = 1.097 \times 10^7\, \text{m}^{-1}$, compute the wavelength of the seventh line of the Balmer series.

2. Using the Lyman formula with $R = 1.097 \times 10^7\, \text{m}^{-1}$, compute the wavelength of the third line of the Lyman series.

3. The energy levels of the hydrogen atom are given by $E_n = -13.6/n^2$ eV. What wavelength is emitted as the atom falls from the tenth to the ninth level? What type of electromagnetic radiation is this?

4. Starting from the energy levels of the hydrogen atom, $E_n = -13.6/n^2$ eV, find the wavelength of the third line in the Balmer series.

5. Starting from the energy levels of the hydrogen atom, $E_n = -13.6/n^2$ eV, find the wavelength of the fourth line in the Lyman series.

6. Lithium atoms have a nuclear charge $3e$ and three electrons. Doubly ionized lithium has lost two of its three electrons. Show that the Bohr computation finds the energy levels of doubly ionized lithium to be given by $E_n = -122/n^2$ eV.

7. According to the previous problem, the energy levels of doubly ionized lithium are given by $E_n = -122/n^2$ eV. Find the wavelength of the equivalent of the first line of the Balmer series for this ion.

8. Electrons with energy 11.6 eV are shot into a gas of hydrogen atoms. Referring to the energy-level diagram of figure 35.4 and to figures 35.1 and 35.2, what wavelengths of radiation will be emitted by the gas?

9. Electrons with energy 12.6 eV are shot into a gas of hydrogen atoms. What wavelengths of radiation will be emitted by the gas?

10. Monochromatic light with $\lambda = 9.72 \times 10^{-8}$ m (the third line of the Lyman series) is passed through a container of hydrogen-gas atoms. What wavelengths will the gas radiate?

11. Light from the sun which reaches the earth's surface has lost much of its energy with wavelength 656 nm, the first line of the Balmer series. What other wavelengths would you expect to be partly missing from sunlight? Would you expect this same effect to occur for light from a light bulb sent through a bottle of hydrogen gas in the laboratory? Why?

12. (*a*) Calculate the recoil speed of a hydrogen atom due to its emission of a photon with wavelength 656 nm, the first line of the Balmer series. (*b*) Find the ratio of this recoil energy of the atom to the difference in energy between the two states that gave rise to the emission line.

13. Eight years before Bohr's theory of the hydrogen atom was proposed, the *Ritz combination principle* was discovered. One example of it is as follows: The frequency of the second line in the Lyman series equals the

sum of the frequency of the first Lyman line and the frequency of the first Balmer series line. (a) Using Bohr's theory, show that this result should be true. (b) What would the principle predict for the third line of the Lyman series?

14. Assume the angular momentum of the earth's rotation about the sun obeys Bohr's condition that the angular momentum equals $nh/2\pi$. What would be the value of the quantum number n in this case? What does the correspondence principle imply if such is the case?

15. Classical electromagnetism predicts the following if the Bohr orbit is large enough so its diameter is comparable to the distance light will travel in the time taken for the electron to make one complete orbit: the "antenna" will radiate not only at its fundamental frequency ν_0 but also at frequencies $n\nu_0$ where n is any integer and ν_0 is the frequency of the electron in orbit. Show that these frequencies of emitted radiation are predicted by Bohr's theory in the region where the correspondence principle applies. (In the Bohr atom the orbital velocity is so small that these higher frequency radiations are very weak.)

16. A proton is accelerated through a potential difference of 1000 V. What is its de Broglie wavelength?

17. A helium nucleus ($m = 4 \times 1.67 \times 10^{-27}$ kg, $q = 2e$) is accelerated through 1000 V. What is its de Broglie wavelength?

18. The average kinetic energy of a free electron in a metal is $3kT/2$ at high temperatures. (a) At what temperature would the electron's average de Broglie wavelength be 0.5 nm? (b) Repeat for a helium atom that has an energy $3kT/2$ and a mass $4 \times 1.67 \times 10^{-27}$ kg.

19. The nuclei of atoms have radii of order 10^{-15} m. Consider a hypothetical situation of a proton confined to a narrow tube with length 2×10^{-15} m. What will be the de Broglie wavelengths which will resonate in the tube? To what momentum does the longest wavelength correspond? Assuming relativistic effects to be negligible, to what energy (in eV) does this correspond?

20. As a very crude model of an atomic nucleus, suppose it consists of noninteracting protons and neutrons traveling in circular paths within the nucleus. Since the radius of a typical large nucleus might be 5×10^{-15} m, assume particles in the ground state to have orbit radii of 5×10^{-15} m. What must be the de Broglie wavelength for a neutron which resonates in such an orbit in its ground state? The kinetic energy (in eV) of the neutron? Neglect relativistic effects.

21. The perimeter of the benzene molecule is a hexagon, each side of the hexagon having a length of 1.40 Å. Since the molecule has three double bonds, it is not totally unreasonable to assume that one electron in the molecule can circulate freely around this perimeter much as though it were a free electron restricted to a hexagonal path. Using wave-resonance reasoning and de Broglie's wavelength, show that the energy levels for such an electron should be (to this approximation)

$$E_n = 7.1 \times 10^{17} \frac{n^2 h^2}{m}$$

with all quantities in the SI system. (Hint: This is not a box with ends.)

22. If the result of the computation in problem 21 was correct, at what wavelengths would you expect benzene to absorb light? Does this contradict the fact that benzene is a crystal-clear liquid?

23. Consider a beam of electrons shot toward a crystal, as shown in figure

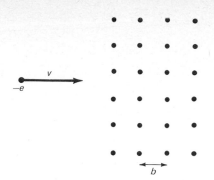

FIGURE P35.1

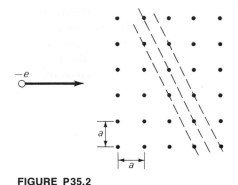

FIGURE P35.2

P35.1. The crystal spacing is b, as indicated. For what de Broglie wavelengths will the electron beam be strongly reflected straight back upon itself? For what electron kinetic energies? It is found by experiment that electrons having these energies are unable to move through such a crystal in the direction shown. Evaluate the energies in electron volts for $b = 2 \times 10^{-10}$ m.

24. A beam of electrons is shot at a crystal, as shown in figure P35.1. What must be the de Broglie wavelengths of the electron beam if the beam is not to reflect straight back upon itself? The electron energies? Evaluate the energies in electron volts for $b = 2 \times 10^{-10}$ m.

25. Rock salt forms a cubic lattice with the sides of each cube having a length $a = 5.63$ Å. An electron beam is incident on a rock salt crystal as shown in figure P35.2. (*a*) What is the spacing between the lattice planes indicated by the broken lines? (*b*) Through what smallest potential should the electrons be accelerated if they are to be reflected strongly off the planes indicated by the dashed lines?

26. The electrons in figure P35.2 have been accelerated through a potential of 2000 V and are incident as shown on a cubic crystal with lattice spacing a as shown. (*a*) What is the de Broglie wavelength for these electrons? (*b*) What values of a would result in strong reflection of the beam straight back on itself? (*c*) Strong reflection off the planes indicated by the dashed lines?

27. A particle is confined within a cubical box that has edge length b. One corner of the box is at the coordinate origin and the adjacent sides lie along the positive x, y, and z axes. Where is the particle most likely to be found when it is (*a*) in its lowest energy state? (*b*) In the state for which $n_x = n_y = n_z = 2$ where the n's are the quantum numbers for the x, y, and z wave functions?

28. An electron is confined to a tube of length L. The electron's potential energy in one-half of the tube is zero, while the potential energy in the other half is 10 eV. If the electron has a total energy $E = 15$ eV, what will be the ratio of the de Broglie wavelength of the electron in the 10-eV region of the tube to that in the other half of the tube? Sketch the wave function ψ for the electron along the tube.

36 Atoms and Molecules

Bohr's model for atoms, though successful in some respects, failed in others. With the advent of quantum mechanics and its embodiment in Schrödinger's equation, an accurate theory of atomic structure became possible. The picture of the atom so obtained differs markedly from Bohr's model. In this chapter, we discuss atomic and molecular structure in terms of the quantum theory. Such new concepts as the uncertainty principle and electron spin are also presented.

36.1 SCHRÖDINGER'S EQUATION AND THE HYDROGEN ATOM

In the preceding chapters we found out a great deal about the hydrogen atom. However, our discussion has been incomplete in several respects. Using the concept of de Broglie waves, we have located the positions of the Bohr orbits as being the resonances of the waves. This is nearly equivalent to locating the positions of the maxima in an interference or diffraction pattern for light. Since light can also be found at other positions than these, we suspect that the electron in the hydrogen atom will not always be found at the Bohr orbit positions. To find the complete behavior of the electron in the hydrogen atom, we must solve Schrödinger's equation.

The time-independent Schrödinger equation *35.19* can be extended to three dimensions:

Schrödinger's equation in three dimensions

$$\frac{\partial^2 \psi}{\partial x^2} + \frac{\partial^2 \psi}{\partial y^2} + \frac{\partial^2 \psi}{\partial z^2} = \frac{-8\pi^2 \mu}{h^2}(E - V)\psi \qquad 36.1$$

where μ is the mass of the electron.*

If the atom is isolated, E, the total energy, will remain constant. However, V is the potential energy of the electron in the field of the nucleus, and is

$$V = -\frac{e^2}{4\pi\epsilon_0 r}$$

where

$$r = \sqrt{x^2 + y^2 + z^2}$$

---

* More precisely, if motion of the nucleus is also considered, $\mu = m_e m_N/(m_e + m_N)$, where m_e and m_N are the mass of electron and nucleus.

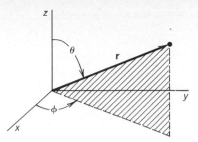

FIGURE 36.1 To be single-valued, a function f of θ, ϕ, and r must be such that $f(\phi) = f(\phi + 2\pi m)$, where m is an integer.

The solution of *36.1*, using this potential, is far from simple. We cannot give it in detail here; our aim is to outline the most important features of the solution. As we shall see, significant differences exist between it and the Bohr theory.

The difficulty involved in solving the Schrödinger equation in this case arises because the form of the potential, V, does not result in simple sine- and cosine-term solutions of *36.1*. In particular, the potential is a function of the magnitude of the radius vector from the nucleus to the electron, and so the problem possesses spherical rather than rectangular symmetry. It is therefore more convenient to carry out the solution in terms of spherical coordinates ϕ, θ, and r. These coordinates are shown in figure 36.1.

By solving the Schrödinger equation in the ϕ, θ, and r coordinate system, one obtains ψ, the wave function, in the following form:

$$\psi = f_1(\phi)f_2(\theta)f_3(r)$$

In other words, the effects of the three coordinates on the wave function can be separated into three different functions. The ϕ function is of the form $e^{\pm im\phi}$, where $i = \sqrt{-1}$ and m is a constant determined by the boundary conditions. You are probably familiar with this function. However, the θ and r functions are more complex. It is for this reason that we cannot pursue further the quantitative solution of Schrödinger's equation in this text. Instead, we shall simply discuss the solution in qualitative terms.

As we saw in the previous chapter, the standing waves for the ψ function must satisfy the boundary conditions of the problem. For the problem of a particle in a linear tube, the wave function had to have nodes at the two ends of the tube. Only then would the wave resonate in the tube. For the case of a wave traveling around a circular path, resonance will occur only if the path length is a whole number of wavelengths long. When we considered wave resonance in two dimensions, such as waves on a rectangular drumhead, two sets of boundary conditions had to be satisfied. For resonance of waves in a cubical cavity, a three-dimensional problem, three sets of boundary conditions need be satisfied.

ψ has three resonance coordinates in the hydrogen atom

Since the hydrogen atom presents us with a three-dimensional problem, we expect three sets of boundary conditions must be satisfied for resonance of the ψ wave. The boundary condition on the ϕ coordinate is very much like the boundary condition for a wave traveling on a circular path. Figure 36.1 shows that ϕ measures the dependence of ψ on the angle around the z axis, and therefore the function $f_1(\phi)$ must repeat itself every 2π rad. Hence

$$f_1(\phi) = f_1(\phi + 2\pi) = f_1(\phi + 4\pi) = \cdots = f_1(\phi + 2m\pi) = \cdots$$

ψ must be periodic in ϕ

where m is an integer. (Do not confuse this use of m with the symbol for mass.) We see from this that the boundary condition on the ϕ coordinate gives rise to a set of quantum numbers $m = 0, 1, 2, \ldots$. More will be said about m in a minute.

The boundary condition on r is not quite as obvious. There is no fixed edge to the atom, and so a node as such does not exist. However, we do know that ψ must decrease to zero as $r \to \infty$. If it did not, then the atom's electron would be most likely to exist infinitely far from the nucleus (since the volume available there approaches infinity). Of course, such is not the case. This condition on ψ at $r \to \infty$ describes the resonance of the ψ wave in the r coordinate direction and gives rise to another set of quantum numbers which is represented by n. It is called the *principal* quantum number and can take on the values $1, 2, 3, \ldots$.

ψ must go to zero at $r \to \infty$

Finally, the θ coordinate supplies a third set of quantum numbers designated by the symbol *l* and called the *orbital* quantum number. This quantum number is the result of the boundary condition on the θ portion of the ψ wave. In order for the wave function to be finite and consistent with the quantum numbers *m* and *n*, it turns out that *l* is an integer which can take on the values 0, 1, 2, . . . , $n - 1$. Let us now examine the physical significance of these quantum numbers.

The principal quantum number *n* describes the energy of the atom. It is found from the solution that the free atom can take on only the following energies:

$$E_n = -\frac{\mu e^4}{8\epsilon_0^2 h^2}\frac{1}{n^2} = -\frac{13.6}{n^2}\text{ eV} \qquad 36.2$$

where $n = 1, 2, 3, . . . , \infty$. Notice that the Schrödinger solution gives exactly the same atomic energy levels as Bohr arrived at. We are therefore assured that the wave solution will be capable of representing the spectral data summarized in the Balmer and similar series.

The orbital quantum number *l* is a measure of the angular momentum of the atom. It will be recalled that the angular momentum of a mass μ traveling with velocity **v** in a circle of radius *r* is simply $\mu\mathbf{v} \times \mathbf{r}$. We represent the angular momentum by a vector **L** chosen in the way shown in figure 36.2. The wave solution shows that the atom can have several values of angular momentum for each value of *n* and the energy E_n. Bohr's theory, on the other hand, attributed a unique value to *v* and *r* for each value of *n* (that is, *n* specified the orbit in which the electron was to exist). We are forced to abandon Bohr's simple picture of circular orbits if we are to accept the wave solution since this latter result tells us that $\mu\mathbf{v} \times \mathbf{r}$ can take on several values for each value of *n*.*

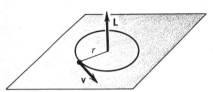

FIGURE 36.2 The angular momentum vector **L** is perpendicular to the plane of the orbit and directed according to the right-hand-screw rule. Its magnitude can taken on only the values ħ $\sqrt{l(l+1)}$.

The wave solution shows that when the atom is in the *n*th energy level, it can take on any one of the following angular momenta:

$$L = \frac{h}{2\pi}\sqrt{l(l+1)} \qquad l = 0, 1, 2, . . . , n - 1 \qquad 36.3$$

(The quantity $h/2\pi$ is very often written as ħ, called h bar.) One can easily see why *L* cannot become infinitely large for an atom in an energy level *n*. For circular motion, $L = \mu v r$. But the energy of the atom also depends on *v* and *r*. At a given energy, the product *vr* is specified by the energy. As a result, the maximum value of *L* is also specified by the energy and therefore by the quantum number *n* which determines the energy of the atom.

The third quantum number *m* is called the *magnetic* quantum number. It specifies the orientation direction of the atom's angular momentum vector **L**. (We shall see that the orientation of **L** is measured by use of magnetic fields and this is why *m* is called the magnetic quantum number.) In particular, if we refer to figure 36.3, the *z* component of **L** is found to be special in the wave solution. It can take on only certain discrete values, and these are given by

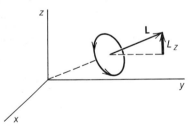

FIGURE 36.3 Both **L** and its *z* component are quantized.

$$L_z = m\hbar \qquad \text{with } m = 0, \pm1, \pm2, . . . , \pm l$$

---

*Later workers showed that, if Bohr's approach was extended so as to include ellipitcal orbits, this difficulty could be partly overcome. For example, if the electron moves back and forth along a very narrow ellipse, *r* would be very small most of the time. As a result, **L** could be nearly zero even though the energy of the atom was very large. We shall see later why we do not accept such detailed pictures of the atom today.

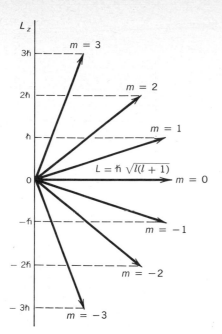

FIGURE 36.4 The angular momentum vector **L** is quantized in space in such a way that L_z is an integer multiple of $\hbar$. In the case illustrated, $l = 3$.

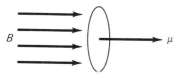

FIGURE 36.5 The current loop aligns with the magnetic field.

Since $L = \hbar \sqrt{l(l + 1)}$ and since m can never exceed l, it is clear that our equations specify $L_z < L$ as it must be. Figure 36.4 shows the allowed orientation of **L** (and of the electron's orbit because **L** is perpendicular to the orbital plane) for the case when $l = 3$.

We see then that the Schrödinger (or wave) solution of the hydrogen atom differs in three ways from the Bohr theory:

1. The Bohr theory involved only one quantum number, n, the principal quantum number. It specified not only the energy of the atom but also the exact circular orbit in which the electron moved. The wave solution involves three quantum numbers, n, l, and m.
2. In an energy state n, the Bohr theory specifies the angular momentum **L** to be $n\hbar$. The wave solution gives the values $\hbar \sqrt{l(l + 1)}$ for it with $l = 0, 1, 2, \ldots, n - 1$. As a result, Bohr's idea of circular orbits must be abandoned because such orbits allow only one value for **L** for a given energy.
3. Bohr's theory allows all orientations for the electron orbit while the wave solution shows that only certain orientations are allowed.

There are other important differences that we shall point out soon. But now let us delve further into the matter of the orientation of the electron's orbital plane.

36.2 EFFECTS OF MAGNETIC FIELD

It may seem strange to you that we have given preference to a particular direction in space, the z direction. The importance and meaning of this preference becomes clear when we consider the effect of a magnetic field on the atom. The z direction is taken to coincide with the direction of the magnetic field. In the absence of a field, it is impossible to determine the orientation of the atom and so it is meaningless to discuss what the z direction should be in that case. But when the atom is placed in a magnetic field, the predicted orientation properties of **L** are rather easily measured. To see how, let us consider the effects of a magnetic field on the hydrogen atom.

As we saw in Chap. 25, a current loop acts like a magnetic dipole. When the current loop is placed in a magnetic field, it aligns with the magnetic field as shown in figure 36.5. In the case of an atom, the electron orbiting the nucleus acts like a current loop. The orbit, too, should orient in a magnetic field with its magnetic dipole moment in the direction of the field. In other words, the plane of the orbit should be perpendicular to the z direction, the field direction.

Of course this would be true only if the dipole had its lowest energy. If it has thermal energy, it will have enough energy to disalign from the field. (Recall that work must be done to overcome the torque that forces the magnetic dipole to line up with the field.) In the case of a very weak field, scarcely any alignment will exist, at least according to classical ideas.

But the wave solution tells us that the atom does not behave in this way. The angular momentum vector **L** shows us the direction of the magnetic dipole moment due to the orbital motion of the electron. According to classical theory, **L** should tend to line up with **B**, the z direction. Even though wave mechanics agrees with this in a rough way, it tells us that not all alignments are possible. The atom can only take up those orientations given

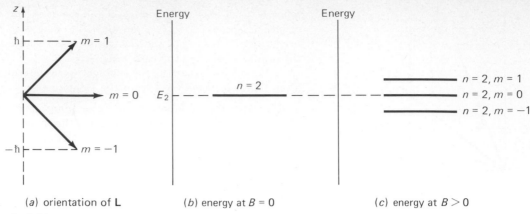

FIGURE 36.6 A magnetic field causes the $n = 2$ energy level to subdivide.

by the fact that $L_z = m\hbar$. Indeed, as we saw in figure 36.4, the magnetic dipole *cannot* line up exactly with the field.

There is good experimental evidence to support this quantization of the atom's orientation. One of the most compelling confirmations is based on the emission spectrum observed for atoms in a magnetic field. For example, let us consider a hydrogen atom that is in the $n = 2$ energy state. From the rules we learned previously that relate n, l, and m, we have for $n = 2$ that l can take on the values 0 and 1. Because **L** $= 0$ when $l = 0$, let us consider the more interesting case where $l = 1$ and **L** $= \hbar \sqrt{2}$. In that case m can take on the values of 0 and ± 1. The allowed orientations of the atom in the $l = 1$ state are shown in figure 36.6(a).

Nature of the Radiation from Radioactive Substances

The discovery of radioactivity in 1896 raised the question of the nature of the radiation being emitted by these substances. Since radium and polonium were very difficult to obtain, much of the early work was done with the much less radioactive substances uranium and thorium. As early as 1899 Ernest (Lord) Rutherford (1871–1937) showed that the radiation from uranium contained at least two components. One of the radiations was capable of passing through only about $\frac{1}{500}$ cm of aluminum, while the other component could penetrate a millimeter of the same substance. He called the easily stopped component of the radiation α- (alpha) rays and the other component β- (beta) rays. A third, still more penetrating radiation was discovered in 1900, and was named γ (gamma) radiation. All naturally radioactive minerals are found to emit all three of these radiations.

Within a very short time after its discovery, the β radiation was found to behave in the same way as cathode rays, and was therefore assumed to be composed of swiftly moving electrons. The γ radiation could not be deflected by either electric or magnetic fields, and was therefore presumed to be uncharged. Although more penetrating than the x-rays available at that time, the γ-rays showed all the properties of x-rays. It was therefore concluded that γ-rays were very high energy x-rays. Over the years, additional evidence has completely confirmed the suppositions concerning these two radiations.

The nature of α-rays was not so easily discerned. Because of their

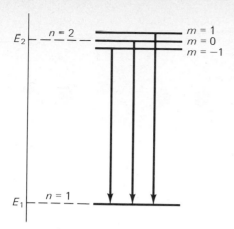

E_2 —— $n = 2$ ———— $m = 1$
$m = 0$
$m = -1$

E_1 —— $n = 1$ ————

FIGURE 36.7 The first line of the Lyman series becomes three lines when a magnetic field is imposed.

Because work must be done to rotate a magnetic dipole away from the field direction, the atom will possess different energies in its various orientations. In part (b) of the figure we show the $n = 2$ energy level in the absence of a magnetic field. But when a field is present, as in (c), the three orientations shown in (a) no longer have identical energies; energy must be added to the atom to disorient it. As a result, the $n = 2$ level in (b) splits into the three levels shown in (c); each of the three orientations in the field has its own unique energy. We shall see in the next section how this splitting of the energy levels leads to changes in the emission spectrum of hydrogen.

36.3 THE ZEEMAN EFFECT

Let us consider what happens to the first line of the Lyman series when a hydrogen atom is subjected to a magnetic field. You will recall that this line is emitted when the atom falls from the $n = 2$ level to the $n = 1$ level. We have just seen that the magnetic field splits the $n = 2$ level into three distinct levels. The $n = 1$ level does not split because for it $l = 0$ and so the angular momentum of the atom is zero; the atom in the $n = 1$ level in effect has zero orbital motion and so its orbital magnetic dipole moment is zero. As a result, the $n = 1$ and $n = 2$ levels for an atom in a strong magnetic field are as shown in figure 36.7.*

Ordinarily the first line of the Lyman series has a wavelength given by

* You might first think that the $m = 1$ orientation, being the case when **L** is aligned with the field, should have the lowest energy. However, the orbiting particle is negative and so the magnetic moment itself is opposite in direction to **L**. Therefore when **L** is aligned, the magnetic moment is opposite to the field.

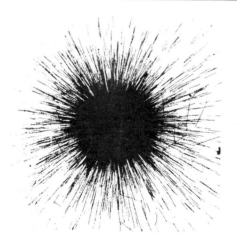

Tracks of a particles from radium; the central spot is about 10^{-4} m across. (Courtesy C.F. Powell and G.P.S. Occhialini, "Nuclear Physics in Photographs," Oxford University Press, New York, 1947.)

very short range of travel (only a few centimeters in air), it was difficult to experiment with them. Moreover, early experiments indicated that they were not deflected by electric and magnetic fields. However, in 1903 Rutherford succeeded in deflecting the rays in electric and magnetic fields and showed that the ratio of the charge to the mass of the particles (if particles they were) was half what one would expect for hydrogen ions. He concluded that the particles were positively charged ions having a mass at least twice that of hydrogen ions.

In 1903 two of Rutherford's associates, Ramsey and Soddy, presented preliminary evidence to the effect that, when α particles were absorbed in water, the water acquired helium gas. Since helium has 4 times the mass of hydrogen and a valence of 2 rather than 1, its ions could show the charge-to-mass ratio found by Rutherford. Finally, in 1908, Rutherford and Geiger presented evidence to show that the particles were doubly charged. Moreover, in 1909, Rutherford and Royds showed that lead, after bombardment with α particles, gave off helium gas when heated, whereas a control did not. Other, more direct collection methods have been used since then. They all confirm the idea that α particles are nothing more than helium atoms from which two electrons have been stripped. We therefore conclude that naturally radioactive materials give off three types of radiation, electrons (β particles), x-rays (γ-rays), and doubly ionized helium atoms (α particles).

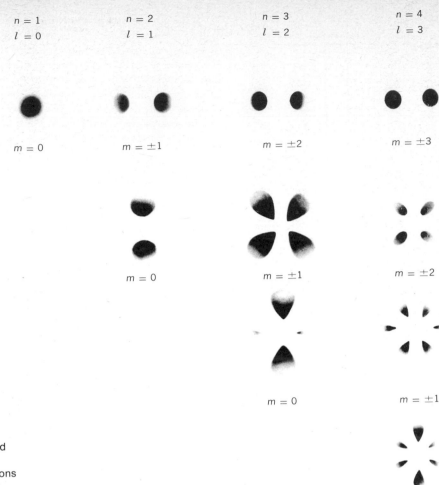

| $n = 1$ | $n = 2$ | $n = 3$ | $n = 4$ |
| $l = 0$ | $l = 1$ | $l = 2$ | $l = 3$ |

$m = 0$ $m = \pm 1$ $m = \pm 2$ $m = \pm 3$

$m = 0$ $m = \pm 1$ $m = \pm 2$

$m = 0$ $m = \pm 1$

$m = 0$

FIGURE 36.8 The z axis is vertical and in the plane of the page through the center of the atom in these cross sections showing ψ^2 for the various states of the atoms. To obtain the three-dimensional ψ^2 patterns, rotate these diagrams about the z axis in each case. (Not to scale.)

$E_2 - E_1 = hc/\lambda$. But in a magnetic field E_2 is split into three separate levels with only the center level being equal to the original E_2. Now three separate transitions can occur as the atom falls from the $n = 2$ to the $n = 1$ state, one transition corresponding to each of the vertical arrows shown in figure 36.7. As a result, the $n = 2$ to $n = 1$ transition gives rise to the emission of three spectral lines instead of one.

This splitting of spectral lines that occurs for atoms in a magnetic field is called the *Zeeman effect*. As you would expect, the energy gap between the component lines is proportional to the magnetic field. In a field of 0.2 T, the gap is only about 1×10^{-5} eV, an energy much smaller than the value $E_2 - E_1 = 10.2$ eV. As a result, the splitting of the spectral lines is very difficult to detect unless a spectrometer with good resolution is used. Even so, precise measurements show without a doubt that the wave theory accurately predicts the observed spectral line splitting.

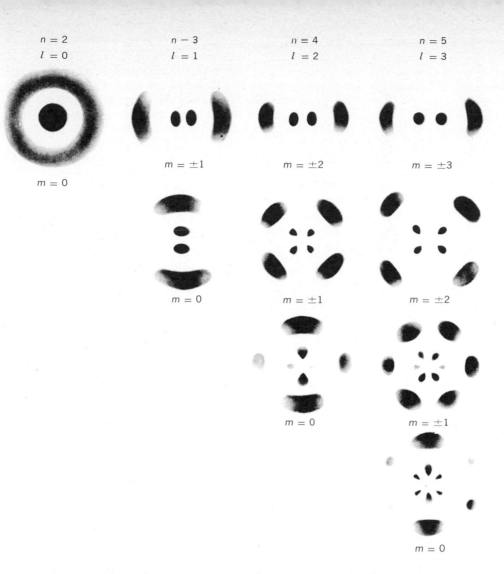

$$n = 2 \quad\quad n = 3 \quad\quad n = 4 \quad\quad n = 5$$
$$l = 0 \quad\quad l = 1 \quad\quad l = 2 \quad\quad l = 3$$

$m = 0$

$m = \pm 1 \quad\quad m = \pm 2 \quad\quad m = \pm 3$

$m = 0 \quad\quad m = \pm 1 \quad\quad m = \pm 2$

$m = 0 \quad\quad m = \pm 1$

$m = 0$

36.4 ELECTRON DISTRIBUTION IN HYDROGEN

ψ^2 gives the probable positions of the electron in the atom

We wish now to describe the position of the electron in a hydrogen atom. To do this, we recall that the intensity ψ^2 of the de Broglie wave tells us about the location of the particle. In particular, if we were to measure the position of the electron in many different hydrogen atoms (or in the same atom on many different trials), the chance, or probability, of finding the electron at a given position is proportional to ψ^2 at that position. It is therefore important to see how ψ^2 varies from point to point around the nucleus.*

The value obtained for ψ using Schrödinger's equation depends upon n, l, and m, the three quantum numbers. This dependence is best shown by means of figure 36.8. In that figure, the z axis of the atom is taken to be

*More precisely, ψ^2 should be replaced by ψ times its complex conjugate, as mentioned previously.

vertical and in the plane of the page. The figure then shows the ψ^2 function on the plane of the page cutting through the center of the atom. It turns out that ψ^2 is independent of the angle which measures rotation of the atom about the z axis through its center. Therefore the three-dimensional representation of ψ^2 can be obtained by rotating the diagram about the z axis.

Notice that the electron is likely to be found in many positions besides those given by Bohr's orbits. Although the states given by n, l, m values of $(1, 0, 0)$, $(2, 1, 1)$, $(3, 2, 2)$, etc., approximate the orbits of the Bohr model, the other states with different l and m values are quite different from Bohr's predictions. As one can easily see from figure 36.8, these distributions become quite complex. We cannot go further than this in our treatment of these distributions. However, it should be mentioned that the lobes of the more complicated patterns have significance in determining how various atoms combine chemically. Another factor that influences the chemistry of atoms is the so-called spin of the electron, a topic discussed in the next section.

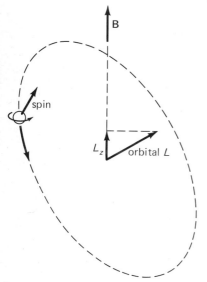

Bohr's concept of fixed orbits is not correct

FIGURE 36.9 Both the orbital and spin-angular momenta are quantized.

36.5 ELECTRON SPIN

When spectroscopists began to investigate the Zeeman effect, they found that a magnetic field caused some spectral lines to break into more components than we have predicted. This matter was finally explained with the discovery that the electron itself has a magnetic moment, i.e., acts like a bar magnet. If you will refer back to figure 25.9(c) you will see that the spinning of a charged particle will cause the particle to act like a current loop. With this in mind, the magnetic moment of the electron is referred to as its *spin magnetic moment*.

Associated with this magnetic moment is angular momentum and it is found that this spin angular momentum of the electron is also quantized in units of $\hbar$. Let us represent the spin angular momentum by $\mathbf{S}$. As with the orbital angular momentum, $\mathbf{L}$, the z component of $\mathbf{S}$ is quantized with its allowed values being $S_z = \frac{1}{2}\hbar$ and $-\frac{1}{2}\hbar$. We summarize this by use of a new quantum number m_s stating that

$$S_z = m_s\hbar \qquad \text{with } m_s = \pm\tfrac{1}{2}$$

We call m_s the *spin quantum number*.

If you refer to figure 36.9 you will see how this affects the atom in a magnetic field. The hydrogen atom has two magnetic dipoles associated with it, that due to the orbital angular momentum and that due to the spin.* We have already discussed how the orbital magnetic dipole causes the atomic energy levels to be split by a magnetic field. But now, because the electron can take up two orientations in the field (we say it can be aligned or anti-aligned), the magnetic field divides each of these already split levels into two more. As a result, each of the levels shown in figure 36.7 is actually split into two levels.

Because of this interaction of the electron spin with the field and the resultant splitting of energy levels, additional spectral line splitting is seen in a magnetic field. We will not delve further into this complexity. For our purposes it is sufficient to know that the wave-mechanical picture of the hydrogen atom necessitates the use of four quantum numbers: the principal

---

*We shall learn in the next chapter that the nucleus of the atom also possesses a magnetic dipole moment.

quantum number n, the orbital quantum number l, the magnetic quantum number m, and m_s, the spin quantum number.

36.6 THE PERIODIC TABLE AND THE EXCLUSION PRINCIPLE

The concept of electron spin led to the first reasonable explanation for the structure of the periodic table of elements. This explanation was given by Pauli in 1925. Basic to his explanation was the following assumption, called the *Pauli exclusion principle: In a system containing several electrons, no two electrons can be in the same state, characterized by the same four quantum numbers, n, l, m, and m_s*. He combined this assumption with the energy-level scheme for the hydrogen atom in the following way.

Pauli's exclusion principle

Assume that the atoms possess an ordering of energy states qualitatively similar to that in hydrogen. Four quantum numbers are needed to describe an electron in any state. They are, together with their restrictions:

No two electrons in an atom can have the same value for n, l, m, and m_s

Principal Quantum Number	$n = 1, 2, 3, \ldots$
Orbital Quantum Number	$l = 0, 1, 2, \ldots, (n - 1)$
Magnetic Quantum Number	$m = 0, \pm 1, \pm 2, \ldots, \pm l$
Spin Quantum Number	$m_s = \pm \frac{1}{2}$

The question we must now answer is, How do the electrons arrange themselves in the various atomic states when more than one electron exists in an atom? For example, there are six electrons in each carbon atom. In which energy levels and electronic states are they to be found? This question can be answered by making use of the following three rules, which we have already discussed:

A neutral atom has a number of electrons equal to its atomic number Z.

In an unexcited atom, the electrons are in the lowest possible energy states.

No two electrons in an atom can have the same four quantum numbers (the exclusion principle).

Let us now use these rules to determine the electronic structure of the unexcited atoms in the periodic table.

Hydrogen (Z = 1)
Its single electron will be in the $n = 1$ level. This is the lowest possible energy level, and no violation of the exclusion principle occurs.

Helium (Z = 2)
Its two electrons can both exist in the $n = 1$ level since they can have the following, nonidentical quantum numbers:

	n	l	m	m_s
Electron 1:	1	0	0	$\frac{1}{2}$
Electron 2:	1	0	0	$-\frac{1}{2}$

However, since these are the only combinations of quantum numbers possible for $n = 1$, a third electron cannot enter this level. The level is filled.

Lithium (Z = 3)

This atom has three electrons, and so the third must go into the $n = 2$ level. We have

	n	l	m	m_s
Electron 1:	1	0	0	$\frac{1}{2}$
Electron 2:	1	0	0	$-\frac{1}{2}$
Electron 3:	2	0	0	$\frac{1}{2}$

Since this third electron is in the second energy level, it is much more easily removed from the atom than the first two are. Hence lithium loses one electron in chemical reactions and is univalent.

Obviously there are quite a few possible combinations for the quantum numbers when $n = 2$. If you count them, you will find there are eight, as follows:

n	l	m	m_s
2	0	0	$\pm\frac{1}{2}$
2	1	0	$\pm\frac{1}{2}$
2	1	+1	$\pm\frac{1}{2}$
2	1	−1	$\pm\frac{1}{2}$

Therefore the $n = 2$ level can accommodate eight electrons, and so the atoms from $Z = 3$ to $Z = 10$ occur in this group. Notice that the last of these atoms, Ne, will be inert since this group of energy states is now filled.

We can proceed in a manner similar to this to build up the whole atomic table. For example, sodium with $Z = 11$ will have a valence of 1 since the eleventh electron must now go into the $n = 3$ energy level, from which it can easily be removed. Although difficulties occur for the most complex atoms, these difficulties arise because the higher energy levels are split by electron-electron interactions and tend to overlap. Unfortunately, the solution of Schrödinger's equation for atoms more complex than hydrogen is extremely difficult, and only approximate solutions are available.

36.7 THE UNCERTAINTY PRINCIPLE

The quantum theory has met with profound success in describing physical phenomena. Not only does it yield classical newtonian mechanics in the region of macroscopic particles; it also describes adequately the behavior of atomic particles, a feat which classical physics is unable to duplicate. We must therefore conclude that the de Broglie wave concept is more realistic than our classical ideas concerning particle motion. Once we accept this concept, however, we are forced to accept a rather startling principle along with it, the *Heisenberg uncertainty principle*. We now investigate this principle.

Suppose we wish to represent a single free particle in terms of a de Broglie wave. If we know the particle's momentum, p, we can write λ for the de Broglie wave to be h/p. The wave representing the particle's position would then appear as in figure 36.10(a). Since the wave runs from $-\infty$ to $+\infty$, the de Broglie wave tells us nothing about where the particle is. How, then, can we represent a particle at a particular position in space by means of de Broglie waves?

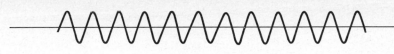

(a)   Single sinusoidal de Broglie wave

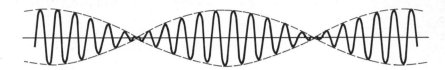

(b)   Wave in (*a*) combined with another having slightly
different λ to produce beats. The envelope of the
resultant wave is sinusoidal with λ much longer
than the λ's of the original waves.

FIGURE 36.10 How do the diagrams provide evidence for the uncertainty relations?

(c)   A particle-like pulse obtained
by adding many waves of different λ's

A clue to the answer to this question is given in figure 36.10(*b*). Here we plot the wave resulting from the combination of two waves which differ slightly in wavelength and frequency. For example, if one wave has a frequency $\omega_0 - \Delta\omega$, it is given by

$$y_1 = y_0 \sin\left[(\omega_0 - \Delta\omega)\left(t - \frac{x}{v}\right)\right]$$

and the second with frequency $\omega_0 + \Delta\omega$ is given by

$$y_2 = y_0 \sin\left[(\omega_0 + \Delta\omega)\left(t - \frac{x}{v}\right)\right]$$

The resultant wave y is the sum of these two. Making use of the relation that

$$\sin A + \sin B = 2 \sin\frac{A + B}{2} \cos\frac{A - B}{2} \qquad\qquad 36.4$$

this gives

$$y = 2y_0 \sin\left[\omega_0\left(t - \frac{x}{v}\right)\right] \cos\left[\Delta\omega\left(t - \frac{x}{v}\right)\right] \qquad\qquad 36.5$$

The cosine factor in equation *36.5* has a frequency $\Delta\omega$ which is much smaller than ω_0, and so it acts as an oscillation of much longer wavelength imposed upon the original wave as shown in figure 36.10(*b*). This phenomenon is often called the phenomenon of *beats,* since the original oscillation alternately grows and diminishes in amplitude.

Although the sum of two de Broglie waves produces regions where the particle represented by them is more likely than others, these positions still extend to infinity. In order to obtain a wave function which is highly localized, such as the one shown in figure 36.10(*c*), one must add a very large number of de Broglie waves together. These individual waves have different wavelengths. Therefore, to obtain a localized wave in order to represent the position of a particle, the wavelength to be associated with the particle becomes very uncertain, because many λ's are used in representing the particle. Since λ is uncertain, and since $\lambda = h/p$, p also will be uncertain.

We are therefore faced with the following fact of quantum-mechanical life: *If the momentum of a particle is accurately known, then its position is unknown. If the position of a particle is accurately known, then its momentum is unknown.* In fact, a mathematical treatment of our qualitative ideas shows that the uncertainty in the x component of a particle's momentum Δp_x and its position uncertainty Δx must conform to the following relation:

Heisenberg's (p, x) uncertainty relation

$$(\Delta p_x)(\Delta x) \gtrsim \hbar \qquad 36.6$$

This relation is called the *Heisenberg uncertainty relation.* From it we see that if $\Delta p_x = 0$, then Δx, the position uncertainty, is infinite; if $\Delta x = 0$, then Δp_x must be infinite. Position and momentum of a single particle cannot be exactly known simultaneously.

Similar reasoning leads to a second uncertainty relation, involving the energy of a particle and the time at which it had the energy. If ΔE is the uncertainty in our knowledge of the energy of a particle, and if the time interval during which the particle had the energy $E \pm \Delta E/2$ is

$$t_0 - \frac{\Delta t}{2} \leq t \leq t_0 + \frac{\Delta t}{2}$$

Heisenberg's (E, t) uncertainty relation

then
$$(\Delta E)(\Delta t) \gtrsim \hbar \qquad 36.7$$

This is the second of Heisenberg's uncertainty relations. In the next section we offer physical justification for them.

36.8 EXPERIMENTAL FACTORS INFLUENCING UNCERTAINTY

If the de Broglie wave concept of particles is true, then, apparently, the Heisenberg uncertainty relations are true. These uncertainties are, in the last analysis, not theoretical, but experimental, considerations. If the uncertainty relations are really true, we should not be able even to *imagine* an experiment which would contradict them. As of this date, no experiment has been suggested which contradicts these relations. Let us now see where the experimental difficulty lies.

One might think that it would be possible to build a superaccurate microscope which should locate the position of a particle on a very finely divided scale. (Recall, we are just imagining an experiment. If it shows promise, we can think about doing it.) However, we know that diffraction effects limit the detail one can see with a microscope. This is true for both light microscopes and particle (electron) microscopes (described in Chap. 32). As we saw in Chap. 32, detail smaller than approximately λ cannot be resolved. Therefore, no matter how perfect the microscope may be, it cannot locate a particle closer than about a distance λ, and so $\Delta x \approx \lambda$.

We can reduce Δx to an extremely small value by using very short wavelength x-rays or particles in our microscope. However, a new difficulty arises. This can be seen in figure 36.11, where, for illustration purposes, it is assumed that a microscope employing photons is used. In order for the particle to be observed, it must be struck by this high-energy photon. It is

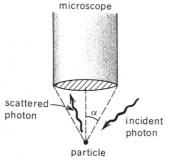

FIGURE 36.11 When the photon strikes the particle, it imparts momentum to it.

Diffraction effects limit the resolving power in both optical and particle microscopes

therefore necessary to consider the momentum of the particle to be changed during the process of observation. This change in momentum could be computed from the law of momentum conservation if the direction of the original photon beam and the scattering direction were known accurately. However, the scattering direction is uncertain for the following reason.

As we see from figure 36.11, the photon could enter the microscope at any angle up to α. Therefore the horizontal momentum of the photon is uncertain by an amount equal to the largest value the horizontal momentum could have, namely, $p \sin \alpha$. But $\sin \alpha$ is approximately equal to the ratio of the lens radius to the object distance, a number of order unity. Therefore

$$\Delta p_x = p \sin \alpha \approx p$$

where p is the momentum of the photon, h/λ.

If we now combine our estimates of the smallest possible values for Δp_x and Δx (namely, h/λ and λ), we shall have

$$(\Delta p_x)(\Delta x) \gtrsim h > \hbar$$

It therefore appears that even the best-designed experiment, using the most accurate microscope imaginable, is unable to violate Heisenberg's (x, p) uncertainty relation. A similar fate is found for all other experiments designed to show that a violation is possible.

To test the (E, t) uncertainty relation, we might also try to use our supermicroscope. It has already been shown that $\Delta p_x \approx p$, where $p = h/\lambda$, the momentum of the photons (or particles) used in the microscope. This implies that ΔE, the uncertainty in the energy of the particle being investigated, is of the order of E or larger, where $E = hc/\lambda$ is the photon energy.

When the particle under investigation was being struck by the photon with wavelength λ, the location of the particle was uncertain by at least a distance λ. As a result, the time at which the photon struck the particle will be uncertain by at least the time taken for the photon to move a distance λ. Therefore the time uncertainty will be given by

$$\Delta t \gtrsim \frac{\lambda}{c}$$

Combining this with the value for ΔE, we find

$$(\Delta E)(\Delta t) \gtrsim h > \hbar$$

We conclude from these and many similar thought experiments that the Heisenberg uncertainty relations are correct. Our ability to know the exact state of a particle is restricted by them. As we saw, this uncertainty arises, basically, because of the wave nature of photons and matter. Since quantum mechanics includes this previously unsuspected uncertainty, it appears that this approach to physics is much more general than Newton's classical physics.

Quantum mechanics includes the uncertainty principle; classical mechanics does not

Illustration 36.1

As shown in the preceding chapter, the angular momentum of the hydrogen electron when in the ground state ($n = 1$) is (according to Bohr's model)

$$mvr = \hbar \qquad \text{or} \qquad mv = \frac{\hbar}{r}$$

How accurately can the position of the electron be found when the atom is in this energy state?

Reasoning We make use of the relation

$$(\Delta p_x)(\Delta x) \gtrsim \hbar$$

To first find the smallest value of Δx, let us assume the linear momentum mv_x to be uncertain by an amount $mv = \hbar/r$. This corresponds to saying we have no idea at all in which direction the electron is going, so that the momentum uncertainty is equal to its magnitude in this energy state. We then have

$$(\Delta x)\left(\frac{\hbar}{r}\right) \gtrsim \hbar$$

so that

$$\Delta x \gtrsim r$$

In other words, the position of the electron cannot be measured more accurately than the radius of its Bohr orbit. From this we see that Bohr's model is very unrealistic, since we shall never be able to measure whether or not the electron is in the orbit he postulates. The smeared position found on the basis of the wave theory (see figure 36.8) shows this uncertainty in position quite clearly.

Illustration 36.2

As a general rule, an electron thrown to a high-energy state in an atom will fall back to its ground state in about 10^{-8} s. Find the uncertainty in the frequency of the light which it will emit in the process.

Reasoning The uncertainty results from the uncertainties in the two energy levels involved. Since the electron will be in the ground state for a long time after dropping to it, the uncertainty in the ground-state energy E_g is very small. This follows from

$$(\Delta E_g)(\Delta t_g) \geq \hbar$$

and the fact that Δt_g can be very large. (Recall that t is the time at which a system has an energy E. The value of t must lie within the time interval Δt. For the ground state, this time interval is very large.) However, the electron is in the upper state E_n only for a time of 10^{-8} s, and so $\Delta t_n \approx 10^{-8}$ s. Using this in the uncertainty relation, we find

$$(\Delta E_n)(10^{-8}\text{ s}) \geq \hbar$$

so that

$$\Delta E_n \gtrsim 10^8 \hbar \text{ J}$$

For the transition from state n to state g, we have

$$E_n - E_g = \frac{hc}{\lambda}$$

Taking derivatives, we find (assuming E_g to be constant, i.e., exactly known)

$$dE_n = -\frac{hc}{\lambda^2} d\lambda$$

If we now set $dE_n \cong \Delta E_n = 10^8 h/2\pi$, we have

$$d\lambda \cong \Delta\lambda = \frac{-10^8 \lambda^2}{2\pi c}$$

For $\lambda = 5 \times 10^{-7}$ m, this becomes

$$\Delta\lambda = -1.3 \times 10^{-14}\text{ m}$$

which represents the smallest possible width of the spectral line. Of course, this does not guarantee that λ can actually be measured so accurately. This is simply the best conceivable accuracy, and it may or may not be possible to approach it with present-day equipment.

36.9 SPECTRA OF ATOMS

When a substance is vaporized so that a gas of individual atoms is obtained, the atoms can be caused to emit light. Light is emitted when any electron within the atom falls from a high to a lower energy state. Of course, in most cases the atoms are in their ground states; the electrons within each atom have fallen to their lowest possible energy levels. Such an atom cannot give off light. However, if a spark is sent through the gas or particles are shot into it, collisions of the atoms will throw some of the electrons to a higher energy state. The atoms then emit electromagnetic radiation as the electrons fall back down to lower states.

Since most atoms contain many electrons, the interactions between these electrons seriously affects the energy of the atoms. The Bohr energy levels we have found for hydrogen, a single electron atom, are only qualitatively applicable to these atoms. Instead, the atoms have many more discrete energy levels than does the hydrogen atom. As a result, very complex spectra are emitted by the atoms as electrons make transitions among this multitude of energy levels. Refer to figure 36.12 for an example of a typical spectrum. Notice that in this, a very small portion of the iron spectrum, the number of lines is very large, and no simple order seems to prevail among them.

The atoms of each element of the periodic table emit spectral lines which are unique to that element. No other element emits the spectral lines shown for iron in figure 36.12. For this reason we can identify the atoms present in a substance by vaporizing its atoms and noting the spectra emitted by the excited atoms. This forms the basis for the widely used technique of spectroscopic analysis in which the composition of materials is determined from their atomic emission spectra.

Emission spectra are used to characterize atoms

A rough idea of the wavelengths of atomic emission spectra can be obtained in the following way. We know that the higher energy levels in the Bohr atom are closely spaced. We should also expect to find very closely spaced levels in more complex atoms since the Bohr picture should give a rough idea of the higher energy levels in these atoms as well. For this reason we would expect low-energy photons (i.e., long wavelengths) to be emitted by all atoms as electrons make transitions between these outer, closely spaced levels. Let us now see what we should expect when the innermost electrons make transitions within complex atoms.

FIGURE 36.12 A small portion of the iron spectrum. (*O. Olderburg and N. Rasmussen, "Modern Physics for Engineers," fig. 7.3, p. 97, McGraw-Hill Book Company, New York, 1966. By permission of the publishers.*)

3000 Å   3100 Å

For hydrogen and one-electron ions, the spectra are quite simple, since their energy levels are given by the relation

E_n for one-electron atoms

$$E_n = -\frac{13.6Z^2}{n^2} \text{ eV}$$

In other atoms, the innermost electrons are largely influenced by the nearby nucleus, and so this relation applies rather well for them too. One should notice in particular that the charge of the nucleus is very influential in determining these inner levels. For example, in the case of $Z = 100$, the energies involved are 10,000 times larger than for hydrogen. Consequently, photons emitted when an electron falls from the $n = 2$ level to the $n = 1$ level *in a high-Z element* will have energies in the range of 100,000 eV. This corresponds to a wavelength* of about 0.12 Å, a wavelength in the x-ray region.

Although the atoms of heavy elements are capable of emitting x-rays, they do not do so unless properly excited. In order for an x-ray photon to be emitted, an electron must fall from an outer level of the atom to a vacancy in an inner level. Since unexcited atoms have no vacancies in these levels, x-rays are not emitted by them. The atom must first be excited in such a way that an electron in the $n = 1$ or $n = 2$ level is thrown out of the level, thereby providing a vacancy into which an outer electron can fall. This is usually accomplished with an x-ray tube like the one shown in figure 36.13. (We note in passing that the $n = 1$ and $n = 2$ levels are not much influenced by nearby atoms if the atom has a large number of electrons. For this reason, our present discussion applies to solids and liquids as well as gases.)

Characteristic x-rays are emitted by high-Z atoms

As shown in figure 36.13, electrons emitted from the hot filament are accelerated through potential differences of the order of 10^5 V. When these high-energy electrons strike the high-Z atoms in the target, electrons are knocked out of the inner levels of the atoms. As other electrons fall into the vacancies, x-ray photons are emitted. The x-rays so generated have wavelengths characteristic of the energy differences between the various levels within the atom. That is, the emitted photons carry an energy equal to the difference in energies between the two levels which act as starting point and end point for the electron that falls into the vacancy. X-rays emitted by this process are referred to as *characteristic x-rays*.

Another type of x-ray emitted from a target when it is bombarded by electrons is referred to as *bremsstrahlung*, from the German "braking radiation." As the name implies, these x-rays are emitted by the bombarding electrons as they are suddenly slowed upon impact with the target. We know that any accelerating charge emits electromagnetic radiation (a charge oscillating on an antenna, for example). Hence these impacting electrons also emit radiation as they are strongly decelerated by the target. Since the rate of deceleration is so large, the emitted radiation is correspondingly of short wavelength and so the bremsstrahlung is in the x-ray region. However, unlike the characteristic x-rays, the bremsstrahlung has a continuous range of wavelengths. This reflects the fact that the deceleration process can occur in a nearly infinite number of different ways, so that the energy released varies widely from one impact to another.

In figure 36.14 is shown a graph of the radiation emitted from a molybdenum target when bombarded by 35,000-eV electrons. The two sharp peaks are the characteristic x-rays emitted as electrons fall to the $n = 1$ level

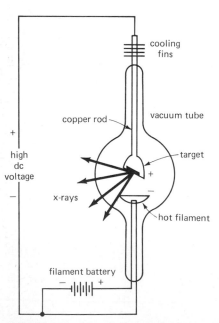

FIGURE 36.13 Electrons emitted by the hot filament bombard the target. The target then emits x-rays.

*To make the conversion rapidly recall that energy $\sim 1/\lambda$ and that 1240 nm corresponds to 1 eV.

of this atom from its $n = 2$ and $n = 3$ levels. The shorter wavelength, of course, corresponds to the higher energy transition, i.e., the $n = 3$ to $n = 1$ transition. Bremsstrahlung is the cause of the lower-intensity radiation spread over all wavelengths longer than λ_m. Since the energy of the electrons in the impacting beam was 35,000 eV, the emitted photons cannot have energies larger than this value. Using our conversion based upon 1240 nm being equivalent to 1 eV, we find that 35,000 eV corresponds to about 0.35 Å. As we see from figure 36.14, the highest-energy bremsstrahlung does indeed have this wavelength.

Illustration 36.3

From the data in figure 36.14, find the energy difference between the $n = 1$ and $n = 2$ levels in molybdenum.

Reasoning As we saw in our discussion of figure 36.14. the long-wavelength peak in the figure, 0.70 Å, results from the $n = 2$ to $n = 1$ transition. Therefore the photon of wavelength at 0.70 Å carries away the energy lost by an electron as it falls from the $n = 2$ to the $n = 1$ level. Since 12,400 Å corresponds to 1 eV, 0.70 Å corresponds to an energy of 12,400/0.70, or about 18,000 eV. Therefore the energy difference between these two levels in molybdenum atoms must be about 18,000 eV.

36.10 COHERENCY

Let us now investigate the nature of a beam of light. As we have seen, light and other electromagnetic radiations are emitted by atoms and molecules as they fall from an excited state to a lower-energy state. In each transition, a photon is emitted. The frequency of the wave in the energy pulse is given by energy $= h\nu$, but the wave in the pulse extends only a short distance in space. We have tried to represent the situation schematically in figure 36.15(a).

As we show there, a monochromatic light beam consists of a stream of light quanta. Each one is of a finite length and contains a limited number of wavelengths in its length (more than shown, however). These pulses all have nearly the same wavelengths since they result from a single atomic transi-

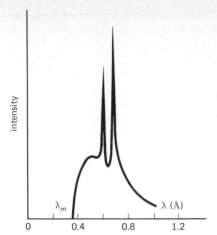

FIGURE 36.14 X-rays emitted from a molybdenum target when bombarded by 35,000-eV electrons.

FIGURE 36.15 (a) Schematic representation of an ordinary light beam (incoherent photons). (b) By use of a double slit or reflection from a thin film, it can be split into two coherent beams.

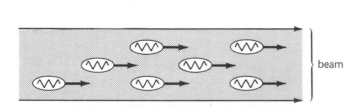

(a) incoherent photons

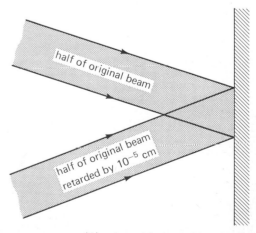

(b) coherent beams

tion. However, each is emitted by a different atom. As a result, there is no definite relation either in time or position between the pulses. They were emitted at random by the atoms and so their waves have random phase relations with respect to each other. *We call a beam such as this, where the photons have random phases (and/or) wavelength, an incoherent beam or pulse.*

Consider what happens as such a beam strikes a screen or the retina of one's eye. Two or more photons may strike the same spot at once. If their waves are in phase, they will add their amplitudes and reinforce. But, if they are 180° out of phase, they will cancel each other. Since the pulses are at random phase relative to each other, both reinforcement and cancellation (and everything in between) will occur as time goes on and the photons strike the screen. If we were able to record the light intensity and show the details of its variation to a time as short as 10^{-10} s, we would record a highly variable light intensity on the screen. This would be a record of the cancellations and reinforcements of the photons as they strike the screen.

But most light-measuring devices, including the eye, record only an average intensity over a period of time much longer than 10^{-10} s. As a result an incoherent beam of photons appears bright but not nearly as bright as it could be if all the photons reinforced each other. And even though cancellation within the beam is taking place, it occurs for such a short time that darkness is not perceived. Even so, it is possible to produce destructive interference with a monochromatic beam, as we shall now see.

Suppose the beam of figure 36.15(a) is split into two parts by means of a double slit or reflection from the two surfaces of a thin film. Let us further suppose that each photon consists of a wave pulse which is 30 cm long,* i.e., the pulses in the figure are 30 cm long. If the beams are now rejoined [as shown in figure 36.15(b)] after one has traveled only a short distance farther than the other, the two halves of each photon will be rejoined, slightly out of phase. In effect, each photon has been sliced in half lengthwise. The amplitude is smaller in each but otherwise the photons are unchanged. One-half has been held back relative to the other; then the two halves are rejoined. As long as the path-length difference is small in comparison with the length of a wave pulse, the half beams will show obvious interference effects when they are rejoined. If the path difference is $n\lambda$ (where n is an integer), constructive interference and brightness will exist at the screen. If the path difference is $n\lambda + \frac{1}{2}\lambda$, destructive interference will occur and darkness will be observed on the screen. This is, of course, exactly what happens in the interference experiments described in Chap. 30.

The two beams shown in Fig. 36.15(b) give rise to visible interference effects at the screen. This is a result of the fact that the wave pulses hitting the screen from the two beams are identical and have a fixed phase relation to each other. We say that the two beams are coherent. *Coherent beams consist of wave pulses corresponding identically in the two beams and which maintain a fixed phase relation to each other.*

It is interesting to note that the beams lose their coherency if one beam is retarded too far. Each wave pulse is limited in length. If the path-length difference for the two beams is too large, the pulses will no longer be joined on top of each other when the beams are rejoined. Instead, unrelated photons will be brought together, and only random phase differences will exist.

*For example, suppose it takes an atom 10^{-9} s to fall from one state to another. During this time, a pulse is being emitted. The leading end of the pulse will travel a distance $= ct = 0.30$ m in this time, and so the pulse will be 30 cm long.

Long-term cancellation and reinforcement will no longer be observed. *The individual wave pulses in an ordinary beam of light are not coherent; the wave in one pulse has phase relations that differ widely from those in the waves in the other pulses.* In the next section we shall learn about a light source which gives out coherent wave pulses.

36.11 THE LASER

Let us now turn our attention to a remarkable type of device which makes use of the fact that, under very special circumstances, atoms can be made to emit light waves which are all in phase with each other. In almost all light sources the atoms act independently; the emission of a photon by one atom is not coordinated with the emission by other atoms. As a result, the light beam consists of a complex mixture of electromagnetic waves from the various atoms. Of course these waves are not all in phase with each other, and so they sometimes cancel and sometimes add. This causes the light beam to be much less intense than it would be if all the atoms emitted their waves in phase. A very intense and coherent beam would result if all the atoms could be persuaded to emit their waves together in phase since the waves would all reinforce each other. One light source comes close to achieving this; it is called a *laser.*

There are many types of lasers available, but they all operate on the principle from which they get their name, *l*ight *a*mplification by *s*timulated *e*mission of *r*adiation. We shall describe a helium-neon gas laser. This gives a continuous light beam. Much more intense beams can be achieved from other types of lasers. The basic outline of the helium-neon gas laser is shown in figure 36.16 together with an illustration of the narrow pencil-like quality of the beam it gives off.

The heart of the laser is a glass tube containing helium and neon gas at relatively low pressure. At the ends of the tube are extremely flat glass plates accurately parallel to each other. Each plate is coated to act like a mirror; but one end plate is coated lightly enough so that about 1 percent of the light striking it from inside leaks through the mirror and leaves the tube.

Not just any two gases will work in a laser. The two chosen in this case have a very special relation to each other. Each has its own function in the tube. One type of atom, helium, acts as an energy source for the other, neon in this case. In turn, the helium atoms must be given energy. This is done by means of a very-high-frequency electrical discharge in the tube. Basically, the helium in the tube is acting like the gas in any gas-discharge tube, such as a mercury-vapor lamp. The high voltage on the tube produces a gas discharge, and many energetic ions and electrons are made to shoot around inside the tube. The excited helium atoms then give off the usual helium spectrum as the electrons fall back to the $n = 1$ level. However, the light from this effect is only a very small fraction of the total light emitted from the tube.

Of course the electrons in the helium atoms are thrown up to all the possible energy levels and states of the atom by means of collisions inside the tube. It turns out that one state which has an energy 20.6 eV higher than the $n = 1$ state is what is called a *metastable state*. In such a state, the electron resists falling to the lower states, and so the atom exists in this state for an abnormally long time. However, the energy-level structure of a neon atom is just right so that, upon collision between the neon and helium atoms, the excited helium atom gives up its energy to the neon atom. Actually, the

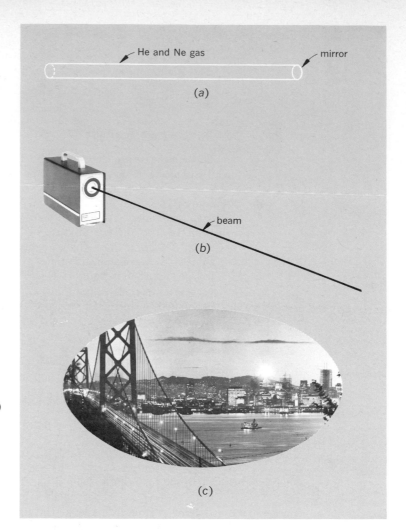

FIGURE 36.16 Even though the Electro Optics Associates laser shown in part (b) has an output of only 0.0005 W, its narrow, pencil-like beam shines brightly a distance of more than 2 mi away. For comparison, notice the less bright light coming from the high-intensity lamps on the San Francisco–Oakland Bay Bridge. (*Electro Optics Associates, Palo Alto, Calif.*)

excited neon atom has an energy slightly greater than that of the helium atom, and this excess energy is furnished by the kinetic energy it acquires upon collision.

What has happened is this. The helium atoms are excited in the gas discharge. Some find themselves in a metastable state and wander around until they collide with a neon atom. During the collision, the neon atom takes the energy from the helium atom and is then in an excited state. This state turns out to be a rather stable state in the neon atom, and so the electron in it does not fall to a lower state immediately. As time goes on, the helium atoms excite a large number of neon atoms and they are then simply waiting to radiate their energies. When they do, they fall to a level 1.96 eV lower and, in the process, emit electromagnetic radiation with wavelength 633 nm, in the red. This in itself is no advantage since we would then simply recover the energy the helium atoms alone would have radiated. The advantage occurs because we can now control the neon atoms and make them emit their waves in phase.

In spite of the fact that the neon atoms are reluctant to fall to the lower energy state, some eventully do so. As a result, a pulse of electromagnetic radiation with $\lambda = 633$ nm is released in the tube. On its way through the tube, it passes other excited neon atoms and subjects them to an oscillating electromagnetic field of frequency identical to the frequency of the light which the excited atom should emit. Such an oscillating field is very effective in causing the excited neon atom to fall to its lower state. It does this in such a way that the photon it sends out has its wave exactly in phase with the radiation which *stimulated* it to emit the photon.

Lasers emit coherent waves

As these waves are reflected back and forth between the ends of the tube, they are vastly augmented by identical waves which they stimulate other neon atoms to emit. Consequently there exists in the tube a very strong *coherent* wave with wavelength 633 nm. A small portion of this radiation leaves the tube through the slightly leaky mirror at one end, and this is the laser beam. Unlike in an ordinary source, the neon atoms have all added their radiation together so that the waves are all cooperating. The resultant beam is composed of waves which are in phase and coherent and hence is very intense.

Lasers emit a fine pencil beam

In addition, the fact that the beam reflects back and forth many times between the parallel mirrors causes the light rays to come straight out from the end of the tube. Any which diverge from the axis of the tube are lost out the sides during the many trips back and forth through the tube. The fact that the beam is a fine pencil of rays is also of great importance. Unlike light from a bulb, the energy does not spread out in space. Instead, it flows out into space through a thin cylinder and maintains its strength over very long distances. For example, the laser shown in figure 36.16 has a power output of only 0.0005 W, but the light energy is confined to such a narrow pencil that the intercepted beam far outshines the high-power bulbs in the foreground. Laser beams sent to the moon and reflected back to the earth are currently used to make measurements on the moon.

In recent years very powerful lasers with outputs of many watts have found widespread use in many situations. You no doubt are familiar with their use in eye surgery, tumor destruction, and similar medical applications. Surgeons now make use of them routinely. The more powerful lasers are currently used for welding metals by means of the intense heat generated as the beam is absorbed in matter. Many other uses could be cited, and the list grows almost daily. For example, the communication of information over large distances by means of laser beams is nearly reality at this time. In research, the laser has already become a useful tool. It is interesting to realize that it would have been impossible to conceive this very useful light source without the vast amount of basic knowledge of atoms, their energy levels, and behavior amassed by many scientists over nearly a half century. This is a typical example of how increasing knowledge of nature leads us to better ways of utilizing its laws.

36.12 SPECTRA OF SOLIDS

The previous discussions have been concerned with individual atoms and their well-defined energy levels. When an electron within an atom falls from one level to another, the photon emitted has a unique energy and wavelength. Each of these separate wavelengths gives rise to a line in the spectrum of the element concerned. The spectra which we have been describing consist of well-defined spectral lines and are called *line spectra*. This is not the type of spectrum one observes from an incandescent lamp or from red-hot solids and liquids. We shall now examine the reasons for this difference.

Suppose a large number N of isolated atoms are brought closer and closer together. Eventually they become close enough so that the outer electrons of the atoms begin to interact with those of neighbor atoms. The energies of these outer electrons are thereby altered because of the presence of the neighbor atoms. When completely isolated, the N atoms each had identical energy levels. Their $n = 3$ levels, for example, were all at the same energy. However, when they are close together, the interaction between the atoms causes these levels to shift in such a way that they are no longer identical. The combined $n = 3$ levels of the N atoms forms a band of energy levels.

In solids, the highest atomic energy levels broaden into bands

This is illustrated in figure 36.17, where the energy levels of sodium atoms are plotted as a function of r, the internuclear distance of the atoms. Notice how the levels widen as the atoms are brought closer together. The equilibrium separation of the atoms in solid sodium is at 3.67 Å, as indicated in the figure. Notice that the $n = 1$ level is not shown. It lies at about -80 eV and is not broadened appreciably in the solid.

The conventional energy-level diagram for solid sodium, similar to the energy-level diagrams we have used in the past, is obtained from the energy levels as they appear at $r = 3.67$ Å. This diagram is shown in figure 36.18.

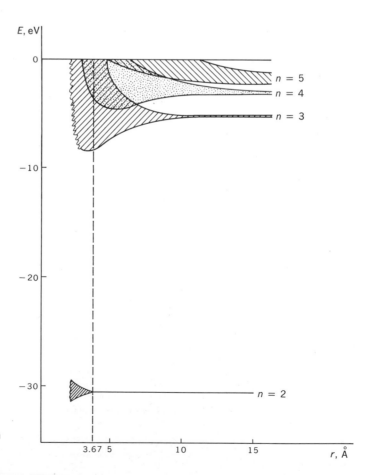

FIGURE 36.17 The energy levels of solid sodium are those which exist for an internuclear distance of 3.67 Å.

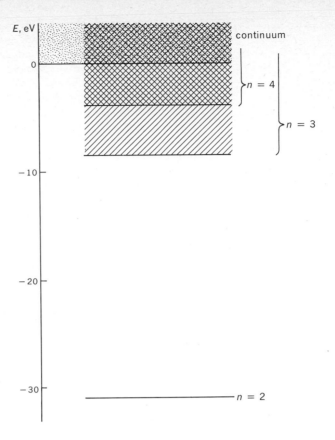

FIGURE 36.18 Energy-band diagram for solid sodium.

A continuous spectrum is observed for light emitted by solids

Notice that the $n = 3$ level has broadened into a band so wide that it extends up into the continuum. The same is true of the $n = 4$ and higher levels. This has a profound influence on the spectrum of the solid.

Suppose an $n = 2$ or $n = 1$ electron is excited to the $n = 3$ level. It can have any energy in the band. As a result, when it falls back down to the $n = 2$ or $n = 1$ level, the energy it emits depends upon the exact position in the $n = 3$ band from which it fell. As another example, in figure 36.18 for sodium metal, we see that when the valence electron (which is the single electron in the $n = 3$ level for sodium) is excited to the $n = 4$ level, it can fall back to the $n = 3$ level and emit light. However, because of the width of the $n = 3$ level, it can emit any energy between 0 and about 8 eV in the process. Therefore the emission line given off can have any wavelength longer than about 150 nm. A similar situation exists for all other transitions from the higher levels of sodium.

Because of this, excited atoms in solids do not give off a discrete line spectrum of light. Instead, they give off a large range of wavelength of radiation, and this type of radiation is called (for visible wavelengths) *white light*. The optical emission spectrum of a solid is a *continuous spectrum*.

A discrete line emission spectrum also exists for solid sodium. It results from transitions from the higher levels to the $n = 2$ and $n = 1$ levels.* The energies here are quite large, and so these emitted wavelengths correspond to

---

*Transitions to these levels can only occur, of course, if an $n = 1$ or $n = 2$ electron has been excited to a higher level so as to leave a vacancy in the $n = 1$ or $n = 2$ levels. In an x-ray tube this excitation is accomplished by bombarding the metal with high-energy electrons.

x-rays. When a transition occurs between the $n = 2$ and $n = 1$ level, the emitted x-ray photon has an energy of about 75 eV. Since both the levels involved are not appreciably broadened, this x-ray emission line will be sharp. However, transitions from the $n = 3$ to $n = 2$ level can give rise to photon energies lying between 22 and 30 eV. This transition will therefore result in a wide line, or band, in the long x-ray region. Similar situations exist in the cases of all solids.

One further point should be mentioned in connection with the energy-level diagrams of figures 36.17 and 36.18. If an electron in an atom happens to be in a level which overlaps a higher, empty level, that electron proves to be essentially free from its original atom. It is then capable of moving freely through the solid, and the material will be a conductor, i.e., a metal.

However, if the electron in the highest energy state of the atom exists at the top of a level which does not overlap higher energy levels, this electron will be held firmly to its original atom. Such a material will be a nonconductor of electricity. An intermediate case exists if the energy levels do not overlap but are close enough so that the gap between them is of the order of thermal energies. These materials are called semiconductors. In that case, electrons can be freed from their parent atoms by thermal energy of the solid. Such a material will be a poor conductor, but its conductivity will increase with increasing temperatures. These and similar topics are discussed in courses dealing with the physics of the solid state.

Illustration 36.4

The K-α x-ray emission line of tungsten occurs at $\lambda \approx 0.21$ Å. What is the energy difference between the $n = 1$ and $n = 2$ levels in this atom?

Reasoning The nomenclature used is as follows. Transitions ending in the $n = 1$ level are called the K series; those ending in the $n = 2$ level are called the L series; those ending in the $n = 3$ and $n = 4$ levels are the M and N series, respectively. We designate the lines in a given series as a function of increasing energy as the α, β, γ, etc., lines. In our case, the K-α line means it is the result of a transition from the $n = 2$ to $n = 1$ level. We have

$$E_2 - E_1 = h\nu$$

Since $\nu = c/\lambda$, this becomes

$$E_2 - E_1 = \frac{(6.63 \times 10^{-34})(3 \times 10^8)}{0.21 \times 10^{-10}} \, J$$

or
$$E_2 - E_1 = 9.5 \times 10^{-15} \, J = 59,000 \text{ eV}$$

In order to obtain this emission line, an electron must be knocked out of the $n = 1$ level. This will require somewhat more than 59 keV since the $n = 2$ level is already filled in this atom.

36.13 SPECTRA AND ENERGY LEVELS OF MOLECULES

Although a semiquantitative solution of Schrödinger's equation for the case of even the simplest molecules is beyond the scope of this book, we state below a few facts which the solution yields. When two hydrogen atoms with

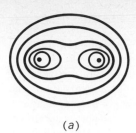

(a)

(b)

FIGURE 36.19 In part *(a)* are shown contour lines of electron probability in the hydrogen molecule. The probability of finding an electron is constant along a contour line. In part *(b)*, the probability is proportional to the darkness of the cloud.

electrons *having spins in opposite directions* are brought together, they tend to attract each other. This attraction causes them to bind together in the H_2 molecule. As we might suspect, since the lowest energy state of the two atoms can hold two electrons, *provided the electrons have spins in opposite directions,* the two electrons of the molecule are shared by the two atoms. It is this sharing which gives rise to the binding energy, the energy needed to tear the atoms loose from each other. The electron distribution for the ground state of the molecule, seen in figure 36.19, shows this mutual sharing of the electrons quite clearly.

It should be reasonably obvious that two helium atoms will not bind together in this way to form a molecule. Both atoms have filled ground-state levels. If they were to combine so as to form a molecule, they would have to share their electrons. But because of the Pauli exclusion principle, if the electron from one atom joins another atom, it must go into a higher energy level. As a result, the combination requires energy to put it together, and so, given time, it will spontaneously fall to its lower energy state, consisting of two separate atoms.

All atoms which bind together to form molecules have similar potential energy curves. Restricting ourselves to diatomic molecules, the energy of the system as a function of the distance between the two atoms is as shown in figure 36.20. At large separation distances the atoms attract slightly. As they are brought closer together, the attraction increases, and their potential energy decreases. Finally, after reaching the separation a_0, they begin to repel each other because of the repulsive force between the positive nuclei, now quite close. Moreover, as the electron clouds of the atoms penetrate deeply into each other, the Pauli exclusion principle effect forces some of them into higher energy states. This requires additional energy to push the atoms closer. In interpreting this graph we should recall our analogy between the bead on a wire and the system to which this graph applies. (See Secs. 7.12 and 7.13.) Of course, a_0 is the equilibrium separation of the atoms.

The spectra of molecules can be separated into two parts: (1) the spectrum resulting from the transitions of the electrons of the individual atoms, and (2) the spectrum due to the vibrational and rotational motion of the nuclei of the atoms composing the molecule. The latter effect is absent for free atoms since it involves the relative motion of two or more atoms.

Since the outer electrons of an atom give rise to the binding effects with other atoms, their energies will be different from what they were in the free atom. The inner electrons, however, are mainly influenced by the atom of which they are a part. Therefore the lower electronic energy levels in which these inner electrons reside will be essentially unchanged from what they were in the free atom. As a result, much of the atomic spectra of the

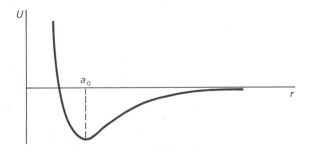

FIGURE 36.20 The potential energy of a diatomic molecule as a function of the separation of the nuclei. What would be the separation at $T = 0$ K?

constituent atoms of the molecule will be the same as it was for the free atoms.

In addition to the electronic energy levels of the constituent atoms, the molecule as a whole will have rotational and vibrational energies. Consider, first, the vibration of the molecule. The vibration of the system occurs in the potential energy curve of figure 36.20. It will vibrate (like a bead) about the equilibrium atomic separation a_0. Since the potential curve near a_0 is nearly parabolic, the system behaves like two masses separated by a spring, as shown in figure 36.21. The spring constant of the spring is k. (Can you show that $k = d^2U/dr^2$ evaluated at $r = a_0$?) If one of the masses, M, is taken so large that it nearly stands still, while the smaller one, m, vibrates back and forth, we saw in our study of mechanics that its classical frequency of vibration ν_c is*

$$\nu_c = \frac{1}{2\pi}\sqrt{\frac{k}{m}}$$

All amplitudes of vibration are possible classically.

The vibrational energy of a molecule is quantized

If we solve Schrödinger's equation for the spring system, the result is found to be quite different. It is found that the energy of the vibrator is quantized; the allowed energy levels are

$$E_n = (n + \tfrac{1}{2})h\nu_c \qquad n = 0, 1, 2, \ldots \qquad 36.8$$

Notice that the vibrator has a zero-point energy, since, when $n = 0$, E_0 is not zero. The zero-point energy is required by the uncertainty principle. If $E = 0$, then Δp would be zero. As a result, $\Delta p\,\Delta x = 0$, which violates the uncertainty

A zero-point energy exists

principle. Therefore E cannot be zero. This means that, even at absolute zero of temperature, motion exists within the molecule. For this reason we avoided stating that all motion ceases at $T = 0$ K in our discussion of thermodynamics. The vibrational-energy-level diagram is shown in figure 36.21(b). As we shall see, these energies are considerably smaller than the energies involved in the electronic-energy-level diagrams.

In addition to its vibrational energy, a molecule can possess rotational

*If M and m are comparable, the relation for ν_c is still correct provided m is replaced by the so-called reduced mass, $\mu = mM/(m + M)$.

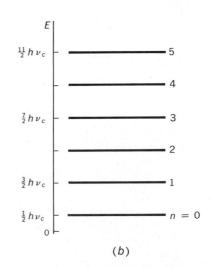

FIGURE 36.21 Because the potential energy function is parabolic near a_0, a diatomic molecule behaves like a spring system in that neighborhood. The allowed energy levels are shown in (b).

(a)

(b)

energy. As we saw in our study of classical mechanics, the kinetic energy of rotation is $\frac{1}{2}I\omega^2$, or since $\omega = 2\pi\nu$, the rotational energy is $2\pi^2I\nu^2$. If a diatomic molecule is rotating about an axis AA' through the center of mass of the molecule, as shown in figure 36.22, its moment of inertia about that axis must be used for I. Notice that, since most of the mass of the molecule is contained in the two nuclei, the molecule shown is nearly equivalent to two point masses. Classically, the rotational energy of the molecule can assume any value.

The quantum-mechanical solution, however, shows that the magnitude of the angular momentum of the rotating molecule, namely, $L = I\omega$, is quantized. It turns out that the quantization rule is the same as the rule for the angular momentum of the orbiting electron. In other words,

$$L = \sqrt{j(j+1)}\hbar \qquad j = 0, 1, 2, \ldots \qquad\qquad 36.9$$

In terms of the rotational energy, this means (since $E = L_{\text{rot}}^2/2I$)

$$E_j = \frac{\hbar^2}{2I}j(j+1) \qquad j = 0, 1, 2, \ldots \qquad\qquad 36.10$$

It is seen from this that the rotational energy of the molecule is quantized, and the levels are spaced as shown in figure 36.22(b). These energy levels are even more closely spaced than the vibrational levels. Moreover, the molecule can make transitions only from one level to its next nearest level.* As a result, the change in energy in a given transition will be $E_j - E_{j-1}$.

Three basic sets of emission spectral lines are observed for a molecule of this type. The emission spectrum in the visible and ultraviolet is caused by electronic transitions within the individual atoms. Those spectral lines which

* Basically, this *selection rule* is a result of the law of conservation of rotational momentum. When a photon is emitted, it must carry off the angular momentum lost by the molecule in the transition. Since the photon has a definite angular momentum (that is, $\hbar$), this fact leads to the selection rule given.

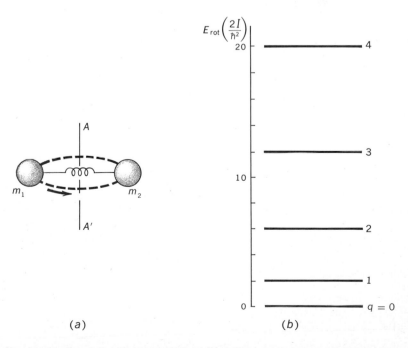

FIGURE 36.22 The rotational energy of the diatomic molecule is quantized; its energy levels are shown in part (b).

(a)

(b)

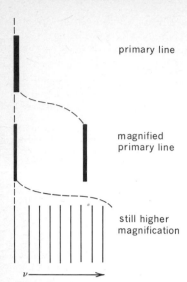

primary line

magnified primary line

still higher magnification

ν ——————→

FIGURE 36.23 The structure of an electronic emission line of a molecule. The vibrational energy splits the line, and the component lines are further split by the rotational transitions.

occur in the infrared are the result of vibrational-energy transitions. The rotational-energy-level transitions are of very low energy and have wavelengths as long as a centimeter. However, as we shall soon see, the vibration and rotation transitions can also cause splitting of the electronic emission lines.

An interesting feature of the rotational emission lines is that they are evenly spaced on a frequency scale. To see this we note that, as stated previously, transitions occur only between adjacent levels. Therefore

$$h\nu = E_j - E_{j-1} = \frac{\hbar^2}{2I}[j(j+1) - (j-1)j]$$

or

$$h\nu = \frac{\hbar^2}{I}j$$

and therefore ν is proportional to j.

Often, all three types of transitions occur at the same time. As a result, the electronic emission line is split by the vibration emission spectrum, and these spectral lines are split by the rotational spectral lines. This is illustrated schematically in figure 36.23. An actual photograph of a molecular emission line is shown in figure 36.24.

Illustration 36.5

From the data of figure 36.24, find the natural vibration frequency of the CN molecule, assuming the two main lines are for a vibrational transition from state n to state $n - 1$.

Reasoning The vibrational energy levels are given by *36.8*. For the transition between adjacent vibration levels we have

$$E_n - E_{n-1} = h\nu_c$$

Referring to figure 36.24, the energies associated with the two main spectral lines are

$$E_1 = \frac{hc}{\lambda_1} = h\frac{3 \times 10^8 \, \text{m/s}}{3.8834 \times 10^{-7} \, \text{m}}$$

$$E_2 = h\frac{3 \times 10^8}{3.8714 \times 10^{-7}}$$

The energy difference is

$$E_2 - E_1 = 3 \times 10^8 h \left(\frac{10^7}{3.8714} - \frac{10^7}{3.8834}\right)$$

$$= \frac{3 \times 10^{15} h}{3.8714}\left(1 - \frac{1}{1 + \frac{0.0120}{3.8714}}\right)$$

FIGURE 36.24 A portion of the emission spectrum of molecular CN. Notice that wavelength increases toward the right, frequency increases toward the left. The dark lines are positions of high light intensity. (*After Oldenburg and Rasmussen.*)

3871.4 Å   3883.4 Å

Or, after expanding and keeping only the first term,

$$E_2 - E_1 = \frac{3 \times 10^{15} h}{3.8714} \left(\frac{0.0120}{3.8714} \right) = 3.24 h \times 10^{12}$$

all in SI units. Equating this to $E_n - E_{n-1}$ yields

$$\nu_c = 3.24 \times 10^{12} \, s^{-1}$$

This is a typical vibration frequency for a molecule. It gives rise to a wavelength of about 10^{-4} m, which is in the far infrared. Infrared spectroscopy is of great importance in determining the force constants and atomic arrangement of molecules.

Chapter 36
Questions
and Guesstimates

1. Compare the Bohr picture of the hydrogen atom with the picture which results from the solution of Schrödinger's equation. What are the similarities and differences?
2. Bohr's basic quantum condition was that the angular momentum could take on only the values $n\hbar$. How does this compare with the result obtained from the Schrödinger equation?
3. Does the correspondence principle apply equally well to the Bohr and Schrödinger solutions for the isolated hydrogen atom?
4. Discuss how our world would be affected if nature suddenly changed in such a way that Planck's constant became 10^{32} times larger than it is. Consider the situation from two different aspects: (a) quantization of energy of oscillators; (b) the uncertainty principle.
5. Explain clearly why x-ray emission lines in the range of 1 Å are not observed from an x-ray tube using a low-atomic-number metal as the target in the tube.
6. Why do hot solids give off a continuous spectrum, while hot gases give off a line spectrum?
7. A steel company suspects that one of its competitors is adding a fraction of a percent of a rare-earth element to its (the competitor's) product. How can the element quickly be identified and its concentration determined?
8. Explain the observed valences (as indicated) of Li^{+1}, Be^{+2}, B^{+3}, C^{+4}, N^{-3}, O^{-2}, F^{-1}, Ne^{0}.
9. Can we prove experimentally that Bohr was wrong when he postulated that the electron in a hydrogen atom follows a circular path?
10. Construct a figure such as the one in figure 36.4 for the case where $l = 4$.
11. In which of the diagrams of figure 36.8 is the electron distribution cloud a solid sphere? A spherical shell? A torus?
12. To obtain very sharp spectral lines from a light source (a mercury arc, for example), the gas must be very dilute, so that few collisions occur. Explain why the spectral lines should broaden as the frequency of collision is increased. (Two lines of reasoning are possible: one is based upon the collision process itself, and the other reasons from the uncertainty relations.)
13. Explain clearly what determines whether a material will give off sharp spectral lines or a continuous spectrum. Can the same material give off both simultaneously?
14. A flask contains hydrogen gas at room temperature. We saw, in illustra-

tion 35.1, that the fraction of atoms in any but the $n = 1$ state is negligible. Therefore they must all be in the $n = 1$ state, and so $\Delta E = 0$. Since they are in that state all the time, the value of Δt is also zero. Therefore

$$(\Delta t)(\Delta E) = 0$$

and the uncertainty principle is violated. What is wrong with our reasoning?

15. In the helium atom, the two electrons are in the same level but avoid each other well enough so that their interaction is of only secondary importance. Estimate the ionization energy (in electron volts) for helium, i.e., the energy required to tear one electron loose. Also, estimate the energy needed to tear the second electron loose. Which of these two values is most reliable?

16. The ionization energies for lithium, sodium, and potassium are 5.4, 5.1, and 4.3 eV, respectively, while those for helium, neon, and argon are 24.6, 21.6, and 15.8 eV, respectively. Explain in terms of atomic structure why these values are to be expected.

17. Estimate how much energy a photon must have if it is to be capable of expelling an electron from the innermost level of a gold atom.

18. When the hydrogen atom is in the $l = 0$ state its angular momentum and magnetic moment are zero. It then shows no Zeeman splitting. Early workers tried to explain this in terms of an elliptic orbit for the electron where the ellipse had degenerated to a straight line through the center of the atom. Would this type of "orbit" give zero orbital angular momentum? Would it be consistent with the distributions shown in figure 36.8?

Chapter 36
Problems

1. In the hydrogen atom, if the quantum number $n = 4$, what values can l take on? For $l = 3$, what values can m take on?

2. How many different states are available to a hydrogen atom when its quantum number $n = 4$?

3. What is the orbital angular momentum of a hydrogen atom (in terms of $h/2\pi$) when it is in the $l = 1$ configuration? What are the possible L_z values it can take on in this case? What angles can $\mathbf{L}$ make with the z axis when $l = 1$?

4. What is the orbital angular momentum of a hydrogen atom (in terms of $h/2\pi$) when it is in the $l = 2$ configuration? In this circumstance, what possible values can L_z take on? What possible angles can $\mathbf{L}$ make with the z axis when $l = 2$?

5. Find the ratio of the orbital portion of the angular momentum to the z-component spin portion of the angular momentum for a hydrogen atom in the $l = 1$ state.

6. Referring to figure 36.8, for which of the configurations shown does the atom have complete spherical symmetry? What can you generalize from this? These are the so-called "s states" of an atom.

7. Referring to figure 36.8, for which of the configurations does the electron have Bohr-like orbits?

8. Consider a hydrogen atom in the $n = 3, l = 2$ state. Taking into account both the orbital and spin angular momenta, what possible values (in terms of $h/2\pi$) can the z component of the atom's angular momentum have?

9. The average translational kinetic energy of a particle of mass m due to thermal motion is $3kT/2$. This corresponds to an average momentum of about $\sqrt{3kTm}$. At any instant, a ball sitting on a table might be moving with about this momentum due to the thermal impacts of the molecules touching it. Assuming the uncertainty in momentum of a ball to be this large, about what is the minimum uncertainty in position of the ball? (Take $m = 10$ g and $T = 300$ K.) Repeat for a molecule of mass 10^{-20} g.

10. The radius of a typical atomic nucleus is about 5×10^{-15} m. Assuming the position uncertainty of a proton in the nucleus to be 5×10^{-15} m, what will be the smallest uncertainty in the proton's momentum? In its energy in electron volts?

11. If Planck's constant had been 660 J $\cdot$ s instead of 6.6×10^{-34} J $\cdot$ s, our world would be much more complicated. In that case, the de Broglie wavelength of a 100-kg football player running at 5 m/s would have been how large? About what would have been the least uncertainty of his location according to an opposing player?

12. Starting from Bohr's relation $E_n = -13.6Z^2/n^2$, what minimum energy would be required to knock an electron out of the $n = 1$ level in tungsten? What wavelength x-ray would be emitted as an electron falls from the $n = 2$ to the $n = 1$ level in this atom? Experimentally, one finds this latter value to be 0.21 Å.

13. The following values are found for the gold atom's x-ray emission spectrum:

$$\text{K-}\alpha \text{ line} = 0.18 \text{ Å}$$
$$\text{L-}\alpha \text{ line} = 1.3 \text{ Å}$$

From these values, obtain the energy gap (in eV) between the $n = 1$ and $n = 2$ levels as well as the gap between the $n = 2$ and $n = 3$ levels. (See illustration 36.4 for terminology.)

14. Moseley's law states that the square root of the K-α x-ray frequency should be proportional to the atomic number Z of the element from which the x-rays come. Justify this result in terms of the Bohr energy levels.

15. The spring constant for the stretching vibration of the H_2 molecule is 570 N/m. What are the allowed vibrational energies for this molecule? Give your answer in both joules and electron volts. Thermal energy at room temperature is about $\frac{1}{40}$ eV. Find the ratio of the thermal energy to the gap energy between vibrational levels.

16. The distance between nuclei in the N_2 molecule is 1.12 Å. Determine the moment of inertia of this molecule about its center. What are the rotational energy levels for this molecule? At about what level (that is, j value) will the rotational energy equal thermal energy, about $\frac{1}{40}$ eV?

17. The rotational energy levels for the Cl_2 molecule are given by

$$E_j = 4.7 \times 10^{-24}j(j + 1) \text{ J}$$

Find the moment of inertia of the molecule about its center and the length of the Cl-Cl bond.

18. As we saw in Chap. 14, the mean free path of an air molecule is about 10^{-6} cm and its speed is about 500 m/s. The lifetime of an excited vibration state of the molecule is about equal to the time between molecular collisions. Assuming these assertions to be correct, what is the inherent uncertainty in the vibration energy of the excited state?

19. Suppose an excited atom emits a photon wave packet that extends for 1 m in space along its line of motion and has a wavelength of 500 nm. (a) How many wavelengths long is the wave packet? (b) What is the inherent uncertainty in the momentum of the photon? (c) To about what uncertainty in λ does this correspond? (d) About what is the ratio of the width of this spectral line to its wavelength?

20. When two incoherent beams of light of equal intensity I_0 are added together, the resultant beam has an intensity $2I_0$. (a) Show that if, instead, the beams are coherent and in phase, the resultant intensity is $4I_0$. (b) Show that the resultant intensity of N incoherent beams when added together is NI_0. (c) Show that the result for N coherent beams in phase is N^2I_0. (*Hint:* Recall that the intensity is proportional to the amplitude squared.)

21. Show that for the Bohr atom the magnetic dipole moment μ is related to the momentum $\mathbf{L}$ by $\mu = (q/2m)\mathbf{L}$ where m is the mass of the electron and q is its charge.

22. Suppose the angular momentum of a sphere rotating about its axis is quantized in units of $\hbar$. If the sphere is a uniform solid of radius 0.5 cm and mass 4 g, (a) what is its rate of rotation (in rev/s) when it is in its lowest state, that is, $L = \hbar$? (b) What is its angular momentum (in units of $\hbar$) when the sphere is rotating at 1 rev/s? (c) What are the rotational energy levels of the sphere?

23. Referring to figure 36.24, several of the consecutive rotational lines have the following wavelengths: 3872.7, 3873.4, 3874.1, 3874.8, 3875.5 Å. Find the energy gap in electron volts between the rotational levels which give rise to these lines.

24. Schrödinger's equation can be written as three interdependent equations in the three spherical coordinates ϕ, θ, r. The ϕ equation is

$$\frac{d^2\psi_\phi}{d\phi^2} = -m^2\psi_\phi$$

where $\psi = \psi_\phi\psi_\theta\psi_r$. Show that a solution of this equation is

$$\psi_\phi = \sin m\phi + \cos m\phi$$

Why does m have to be an integer in order for this solution to be reasonable?

37 Nuclei of Atoms

At the center of each atom is a positively charged nucleus which contains all but a very small fraction of the mass of the atom. In this chapter we begin our study of the nucleus and its properties. Natural radioactivity is discussed, and the behavior of the radiation emanating from radioactive nuclei found in nature is investigated. Methods for measuring the mass of nuclei are described, and the concept of the mass defect is introduced. Current theories of nuclear structure are also examined.

37.1 THE ELECTRON, PROTON, AND NEUTRON

We begin our study of the atomic nucleus by reviewing the properties of the electron, proton, and neutron. In so doing, we acquire a familiarity with the basic properties which are later used to describe the nucleus. As we study the nucleus it will be possible for us to understand its structure better by comparing its properties with those of these three basic particles.

From the work of Faraday, Millikan, and many others, we know that the magnitude of the electron and proton charge is $e = 1.602 \times 10^{-19}$ C. The proton's charge is positive $(+e)$, while the charge of the electron is negative $(-e)$. As pointed out on page 760, where the discovery of the neutron was discussed, the neutron has no charge.

The rest masses of these particles are known accurately. Knowing the charge on the electron, its mass can be determined by observing its deflection in a magnetic field. The pertinent relation is given in Sec. 22.5. A similar method may be used to find the proton's mass. Since the neutron has no charge, its mass must be found in other ways. One method for finding the neutron mass by use of the law of conservation of momentum is described on page 760. There are also other methods. The results found for the electron, proton, and neutron are

$$m_e = 9.11 \times 10^{-31} \, \text{kg}$$
$$m_p = 1.673 \times 10^{-27} \, \text{kg}$$
$$m_n = 1.675 \times 10^{-27} \, \text{kg}$$

Notice that the neutron and proton masses are nearly, *but not exactly,* the same.

Another important property of a particle is its angular momentum. If we consider rotation of a finite-size *nonrelativistic* particle about an axis through its mass center, then (as we shall see in illustration 37.1) a consideration of Schrödinger's equation tells us that the angular momentum about that axis is quantized. In particular, calling the axis in question the z axis, the z component of the nonrelativistic particle's angular momentum is

$$(\text{Angular momentum})_z = n\hbar \qquad n = 1, 2, \ldots$$

This relation does not give us the total angular momentum of the particle, however, since the particle can rotate about more than one axis through its mass center. Solution of Schrödinger's equation in three dimensions gives the result that the total angular momentum of the particle can be written

$$\text{Total angular momentum} = \sqrt{n(n + 1)}\,\hbar$$

These results for the angular momentum of a rotating (or spinning) particle are *not* correct for particles such as the electron and proton. In the case of these particles, relativistic effects become important, and Schrödinger's equation is no longer adequate. A modified equation to include relativistic effects was proposed by Dirac in 1928. Solution of this equation shows that the total *spin angular momentum* of an electron can have only the single value $\hbar(\sqrt{3}/2)$. Designating this quantity as S, we have

$$S = \frac{\sqrt{3}}{2}\hbar \qquad\qquad 37.1$$

Further, not only must the spin angular momentum have this very definite value, the *direction* of the angular momentum vector is quantized.

The angular momentum vector of the electron can take up only those orientations whose z component S_z is*

$$S_z = \pm\tfrac{1}{2}\hbar \qquad\qquad 37.2$$

This restricts the angular momentum vector to the surface of a cone, as shown in figure 37.1. A simple reason may be given for the fact that the vector cannot orient in line with the field. If it did, there would be no uncertainty at

---

*Here again it is important to realize that physics is an experimental science. In order to *measure* the orientation of the angular momentum vector, one places the particle in a magnetic field. The direction of the field defines the z direction, and so a unique direction for a reference system exists. In the absence of the field, measurements are not possible, and questions concerning orientation in that situation cannot be answered.

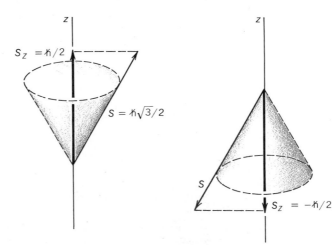

FIGURE 37.1 The angular momentum vector of the electron can have only $S_z = \pm\,\hbar/2$, This restricts the vector to the surfaces of the two cones shown.

all in the angular momentum of the particle and a violation of the uncertainty principle for the angular coordinate ϕ, namely,

$$\Delta p_\phi \, \Delta \phi \geq \hbar$$

would result. In spite of the fact that the spin vector does not align perfectly with an impressed field, it is common to say that the spin vector can take up two orientations only, parallel or antiparallel to the magnetic field. One should always interpret this statement to mean that the z component of the spin angular momentum can have only two values, namely, those given in 37.2, $\pm \hbar/2$.

The spin quantum number m_s

The spin angular momentum of a particle is characterized by its *spin quantum number,* or simply, its *spin m_s*. This number is defined in such a way that it satisfies the relation

$$S = \sqrt{m_s(m_s + 1)}\,\hbar \qquad\qquad 37.3$$

where S is the total angular momentum of the particle. You will recognize this to be a generalization of the equation given earlier in this section (and derived in illustration 37.1) for the allowed spin angular momentum values for nonrelativistic particles. In the case of the electron, $S = \sqrt{3}\,\hbar/2$, and so $m_s = \frac{1}{2}$. We say that the electron has spin $\frac{1}{2}$. The spins of the proton and neutron are also $\frac{1}{2}$.

The spins of the electron, proton, and neutron are $m_s = \frac{1}{2}$

Illustration 37.1

A particle (perhaps a sphere) rotates with angular speed $\omega = d\phi/dt$ about a fixed axis through its mass center. What does Schrödinger's equation predict for its motion?

Reasoning We assume the particle to be constrained to rotate freely about its axis. Then the motion of the particle can be completely specified in terms of ϕ, the rotation angle about this axis. Schrödinger's equation is most easily used in this case if it is expressed in spherical coordinates (r,θ,ϕ) rather than x, y, z coordinates. The equation in this case (with r and θ constant) turns out to be

$$\frac{\hbar^2}{2I}\frac{\partial^2 \psi}{\partial \phi^2} = -\tfrac{1}{2}I\omega^2\psi$$

where I is the moment of inertia of the particle, and ω is its (as yet) arbitrary angular speed.

A solution of this equation is

$$\psi = A \sin k\phi + B \cos k\phi$$

where

$$k = \frac{I\omega}{\hbar}$$

Since ψ must be a single-valued function of ϕ, it must repeat itself when ϕ changes by 2π. Therefore k is restricted to the values 0, 1, 2, ... As a result,

$$\frac{I\omega_n}{\hbar} = n \qquad n = 0, 1, 2, \ldots$$

The subscript n has been added to ω in order to indicate that ω is related to n. Solving for the angular momentum $I\omega_n$, we find

$$\text{Angular momentum} = n\hbar \qquad n = 0, 1, 2, \ldots$$

We therefore find the angular momentum about this axis to be quantized in units of $\hbar$. As pointed out in the text, this result is correct only for nonrelativistic particles. It does not, therefore, apply to the electron, proton, and other similar particles.

37.2 MAGNETIC MOMENTS OF PARTICLES

If one pictures the electron to be a charged sphere, its rotation about an axis through the center of the sphere should give rise to magnetic effects. In particular, the rotating sphere can be thought of as being composed of current loops about the axis. Each loop acts as a tiny magnet, a magnetic dipole, directed along the axis of rotation. As a result, this model of a spinning electron predicts the electron to have a magnetic dipole moment μ.

Although we cannot accept such a model in detail, it is found that the electron does possess a magnetic moment. Its moment can be measured to high accuracy by a method called *electron spin resonance*. This method makes use of the fact that the electron can exist in two different energy states when placed in a magnetic field B. These two energy states correspond to the electron in its two possible alignments in the magnetic field.

When the z component of the magnetic moment μ_z is opposite to the magnetic field direction, it has a potential energy $2B\mu_z$ larger than when it is aligned with the field. This can be seen from figure 37.2. As shown in Sec. 22.7, the torque on the magnetic dipole μ_z is given by

$$\tau = \mu_z B \sin \theta$$

where θ is the angle in the figure. Since the work done by a torque τ in an angular displacement $d\theta$ is $\tau\, d\theta$, the work done in turning the dipole from the aligned position ($\theta = 0$) to the antialigned position ($\theta = \pi$) is

$$\text{Work} = \mu_z B \int_0^\pi \sin \theta \, d\theta$$
$$= 2\mu_z B$$

This shows that the energy difference between the two electron dipole moment orientations is $2B\mu_z$.

Suppose now that a sample containing essentially free electrons is placed in a magnetic field. Two energy levels exist for the electrons, the difference in energy between the two levels being $\Delta E = 2\mu_z B$. If electromagnetic radiation of such a frequency that

$$h\nu = \Delta E = 2\mu_z B \qquad\qquad 37.4$$

is sent through the sample, this radiation will excite the electrons in the lower energy state to the higher state. As a consequence, radiation of this frequency will be strongly absorbed. In practice, these frequencies are in the radio-, or microwave, region. To determine μ_z, one simply varies the radiowave frequency until the highly absorbed frequency is found. For this frequency, *37.4* applies. Knowing B and ν, the value of μ_z can be computed. A similar technique (called *nuclear magnetic resonance*) can be used to find μ_z for nuclei and other particles.

The measurements of magnetic moments in the way outlined gives μ_z. One finds

For the Electron $\qquad \mu_z = 9.27 \times 10^{-24} \, \text{J/T}$

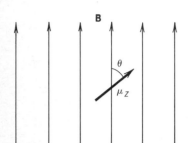

FIGURE 37.2 The work done in reversing the direction of μ_z may be found by integrating $\tau\, d\theta$ over $0 \leq \theta \leq \pi$, where τ is the torque on the dipole.

The value predicted by Dirac's theory is

μ_z for the electron is 1 Bohr magneton

$$\mu_z \equiv \frac{e\hbar}{2m} \equiv 1 \text{ Bohr magneton} \qquad 37.5$$

which agrees with the measured value. In this expression, m is the electron mass. The quantity $e\hbar/2m$ is often called the *Bohr magneton*, since it is also the unit of magnetic moment predicted for the orbital motion of the electron, using Bohr's theory for hydrogen.

Of course, μ_z is only the z component of the magnetic dipole moment. The total dipole moment μ is larger than this:

$$\mu = \sqrt{3}\mu_B$$

where μ_B is the Bohr magneton.

The magnetic dipole moment of the proton and of the neutron is much smaller than that of the electron. In fact, one would expect from classical ideas that the proton's magnetic moment should be smaller by the ratio of the electron mass to the proton mass. This follows from the fact that the two particles have similar angular momenta, and so the proton should, because of its larger mass, be spinning more slowly. The magnetic moment should, of course, be proportional to the rate at which the particle spins. Although this leads to a correct order-of-magnitude estimate of μ for the proton, it does not give the exact value. In addition, since the neutron has no charge, the classical model predicts zero magnetic moment for it.

Determinations of the magnetic moment of particles such as these are actually done by measuring the z component of the moment. For this, and other reasons, one usually finds the value of μ_z given for these particles. This fact is frequently obscured by the common tendency to refer to μ_z as the magnetic moment of the particle, and the subscript z is often omitted. In any case, experiment shows

For the Proton $\qquad\qquad \mu_z = +2.79\mu_N$

For the Neutron $\qquad\qquad \mu_z = -1.91\mu_N$

where $\qquad\qquad \mu_N = \dfrac{e\hbar}{2M_p} = 5.05 \times 10^{-27} \text{ J/T} \qquad 37.6$

and M_p is the mass of the proton. The quantity μ_N is called the *nuclear magneton*, in analogy with the Bohr magneton. Notice that, even though the neutron has no charge, it still has a magnetic moment. One suspects from this that the neutron may have a rather complex internal structure.

Illustration 37.2

Compute the frequency at which nuclear magnetic resonance occurs for a proton in a field of 1000 G.

Reasoning From *37.4* we have at resonance

$$h\nu = 2\mu_z B$$

In the present case

$$\mu_z = (2.79)(5.05 \times 10^{-27}) \text{ J/T}$$
$$B = 0.10 \text{ T}$$

which gives $\qquad\qquad \nu = 4.26 \times 10^6 \text{ Hz}$

a frequency in the radio range. An electron in this same magnetic field would resonate at a much higher frequency. Why?

37.3 ATOMIC NUMBER (Z) AND MASS NUMBER (A)

As we saw in the last chapter, the chemical behavior of an atom is determined by the electrons that exist in the relatively vast reaches of space outside the nucleus. All atoms of a given chemical type have the same number of electrons and electronic structure. Each carbon atom has 6 electrons and each gold atom has 79 electrons, for example. We designate the number of electrons in a neutral atom by Z and it is called the *atomic number* for that chemical species. For carbon $Z = 6$ and $Z = 79$ for gold. The atomic numbers of the elements are given in Appendix 6A and B.

Because the atom as a whole is neutral, the nucleus of an element of atomic number Z must have a charge $+Ze$ to counterbalance the charge of the atom's Z electrons. This charge is ascribed to Z protons, each with charge $+e$, within the nucleus. But the mass of the nucleus (except in the case of hydrogen) is larger than the mass of the Z protons. This additional mass is due to neutrons in the nucleus. We therefore conclude that, for an element of atomic number Z, the nucleus contains Z protons together with additional neutrons.

It is convenient when discussing the masses of nuclei to use a mass unit called the atomic mass unit, u, whose exact definition will be given shortly. The following relationship exists:

$$1\,u = 1.66057 \times 10^{-27}\,kg$$

In terms of this unit, the masses of various particles are

$$\text{Electron mass} = 0.0005486\,u$$
$$\text{Proton mass} \;\; = 1.007276\,u$$
$$\text{Neutron mass} = 1.008665\,u$$

Notice that, on this scale, the proton and neutron masses are about unity.

Because the nucleus is composed of protons and neutrons, and because each of these particles has a mass close to 1 u, we would expect the nuclear mass to be nearly an integer when measured in atomic mass units. For example, the common helium nucleus contains two neutrons and two protons and so its mass should be (and is) about 4 u. Similarly, the mass of the common carbon atom nucleus (which contains six neutrons and six protons) is 12 u. With this in mind, we assign a mass number A to each nucleus; the *mass number A* of an atom and its nucleus is equal to the number of nucleons (protons + neutrons) in the nucleus and this is approximately equal to the mass (in u) of the nucleus.

37.4 NUCLEAR MASSES; ISOTOPES

The masses of nuclei have been measured to high precision. These measurements are carried out by use of *mass spectrometers;* a schematic diagram of one type is given in figure 37.3. In this apparatus, ions of the element under consideration are allowed to escape from the ion source. It is necessary for the atoms to be ionized so that the atom can be accelerated in an electric field. (The whole apparatus is in vacuum; so collisions of the ions with air molecules can be ignored.) The accelerating electric field is provided by the potential difference V. After being accelerated through the potential difference V, the ion beam is collimated by means of slits such as S_2. Leaving S_2, the ions are moving with speed v, and are deflected into a circular path by

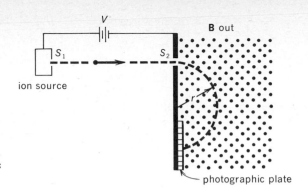

FIGURE 37.3 In the mass spectrograph, positive ions are deflected by a magnetic field.

the magnetic field shown. The radius r of this path is measured by noting the positions at which the ions strike a photographic plate or some other detector.

To find the working equation for the mass spectrometer, we note that the energy furnished by the accelerating potential is

$$Vq = \tfrac{1}{2}mv^2$$

In this equation, q and m are the charge and mass of the ion. When the ion enters the magnetic field, the centripetal force is furnished by the magnetic field, and so we can write

$$\frac{mv^2}{r} = qvB$$

Upon solving these two equations simultaneously for m, we find

$$m = \frac{r^2B^2q}{2V} \qquad\qquad 37.7$$

We can therefore compute the mass m of the ion if r, B, q, and V are known. To find the mass of the nucleus, one subtracts the mass of the electrons associated with the ion from m.

When the mass spectrometer is used to measure the nuclear mass, an interesting effect is observed. Very frequently, one finds that a certain element will give rise to two or more different beams in the spectrometer. That is to say, particles will appear at the detector at two or more very well defined radii. From 37.7 one concludes that nuclei of the *same element* may have different masses.

As an illustration, when chemically pure chlorine is sent through the mass spectrometer, it appears to consist of two different types of nuclei. They are, together with the relative percent of each:

Species 1   Mass = 34.97 u   Percent = 75.4
Species 2   Mass = 36.97 u   Percent = 24.6

Both these species have exactly the same chemical behavior, and so their atomic electron structures are the same. Therefore their nuclear charges must be the same, equal to the atomic number Z multiplied by the charge quantum e. We call nuclei such as this, having the same charge but different masses, *isotopes* of the element in question.

Definition of isotope

In the atomic chart of elements, one usually finds listed the *atomic masses* as determined by chemical means. For this reason the atomic mass given there is an average value of the isotope masses found in nature. For example, the average mass of the two isotopes in chlorine is

$$m_{av} = (35)(0.754) + (37)(0.246) = 35.5\,u$$

and this is the value given in the atomic chart (Appendix 6A).

Definition of atomic mass unit, u

The atomic masses (or "weights") of the elements are given in Appendix 6B. Notice that these are the masses of the nuclei *plus the atomic electrons*. These masses are given in *atomic mass units,* and this unit is defined as follows. One isotope of carbon (carbon 12, the most abundant isotope) is arbitrarily assigned the value of *exactly* 12 u. All other masses are then measured in this unit by comparison. On this scale, the proton and neutron masses are approximately 1 u. As stated previously, $1\,u = 1.6606 \times 10^{-27}\,kg$.

Illustration 37.3

The atomic masses given in the periodic table are for the nucleus plus the atom's electrons. What fraction of the atomic mass of ^{235}U (uranium with mass number = 235) is due to its electrons?

Reasoning The atomic mass of ^{235}U is 235 u. Since the atomic number of uranium is 92, it has 92 electrons. Using the fact that the mass of the electron is $9.1 \times 10^{-31}\,kg$ or 0.00055 u, we have that the fraction of the mass due to electrons is

$$\frac{(92)(0.00055)\,u}{235\,u} = 2.15 \times 10^{-4}$$

Therefore, for many purposes, the mass of the electrons can be ignored.

37.5 SIZE OF NUCLEI

There are several methods by which we can estimate the size of the nucleus of an atom. One obvious method is to shoot particles of various types at the nucleus and see how they are reflected. The method would be similar to that treated by Rutherford for α-particle scattering, discussed in Chap. 34. However, one must use very high energy particles in order to overcome the coulomb repulsion of the nucleus if the bombardment is to be done with protons or α particles. Scattering measurements with these particles as well as with neutrons have been made. The results of these measurements show that the nucleus cannot be pictured as a simple hard sphere of uniform constitution.

In spite of the fact that the nucleus has no sharp cutoff radius for its charge or its mass, as would be the case for a hard sphere, the edges of the nucleus are well enough defined so that a meaningful approximate radius may be given. As one would expect, bombardment with charged particles will measure primarily the charge distribution in the nucleus, while bombardment with neutrons will measure another average-size parameter. Other methods also can be used for measuring the nuclear radius. They all agree approxi-

Nuclear radius is of the order of $5 \times 10^{-15}\,m$

mately with each other, and from them one infers the nuclear radius to be

$$R_0 \approx 1.2 \times 10^{-15}\,A^{1/3}\,m$$

where A is the mass number of the atom concerned.

It is reasonable that the nuclear radius should vary as $A^{1/3}$. This follows from the fact that the mass of the nucleus is proportional to its mass number. If one assumes the nuclei of the various elements to have the same mass density ρ, then

$$A \sim \text{mass} = (\tfrac{4}{3}\pi R_0{}^3)\rho$$

and so one would have

$$R_0 \sim A^{1/3}$$

The fact that R_0 does vary in this way, at least for the larger nuclei, indicates that the nuclei of the elements have about the same density.

Illustration 37.4

Find the mass density ρ of the gold nucleus.

Reasoning The mass of a gold nucleus is nearly equal to its atomic mass, given in Appendix 6B to be 197 u. The volume of the nucleus is

$$\text{Volume} = \tfrac{4}{3}\pi R_0{}^3 = 7.2 \times 10^{-45}\,A\ \text{m}^3$$

Because $A = 197$, we have that

$$\rho = \frac{\text{mass}}{\text{volume}} = \frac{(197\,\text{u})(1.66 \times 10^{-27}\,\text{kg/u})}{(7.2 \times 10^{-45})(197)\,\text{m}^3} \cong 2.3 \times 10^{17}\,\text{kg/m}^3.$$

The atomic nuclei have about equal densities, $\approx 10^{17}$ kg/m³

Notice that because the atomic number ($A = 197$) is nearly equal to the atomic mass (197 u), the 197's cancel, and so this is the approximate density within all nuclei. Such extremely high densities are never encountered on a large scale on earth. Only in the interior of certain stars (the *white dwarfs*) are such high densities found. In these stars the electron shells of the atoms have been collapsed by the huge gravitational forces at the star's center.

37.6 STABLE NUCLEI

Examination of the masses of the nuclides given in Appendix 6B shows an interesting fact. The masses of the nuclides are very close to being integers. They range from $n - 0.10$ to $n + 0.09$, where n is an integer. Most of the nuclide masses do not depart nearly this much from an integer value. Since the mass of a neutron is 1.009 u and the mass of a proton is 1.007 u, it appears reasonable to consider the nucleus to be made up of neutrons and protons.

The nuclear charge is $+Ze$, where Z is the atomic number of the element. This indicates that the nucleus contains Z protons, each proton contributing a charge of $+e$ and a mass of 1 u to the nucleus. Since the mass number A of the nucleus is often larger than Z, we conclude that the nucleus also contains $A - Z$ neutrons, each neutron having a mass of 1 u.

It is of interest to examine the number of protons and neutrons in the nuclei of elements found in nature. This is shown graphically in figure 37.4. Notice that for small nuclei, the proton number and neutron number are nearly the same. As one goes to higher-atomic-number elements, the number of neutrons exceeds the number of protons.

In order to classify the nuclei in terms of their mass and charge, it is customary to designate an element whose symbol is E as

$$^A_Z E$$

For example, the standard for the atomic mass unit scale of masses is the isotope of carbon, C, which has a nuclear charge of $Ze = +6e$ and a mass

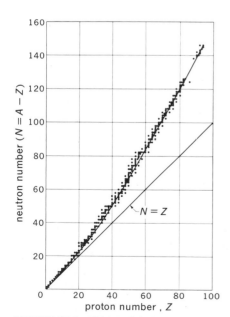

FIGURE 37.4 The nuclei shown as dots on this chart are stable or have half-lives in excess of 1000 years. (*After Oldenburg and Holladay.*)

number $A = 12$. We therefore designate it $^{12}_{6}C$. Frequently, we refer to this isotope as carbon 12. Similarly, oxygen 16 is $^{16}_{8}O$ and uranium 235 is $^{235}_{92}U$.

If we accept the idea that the nuclei consist of protons and neutrons, one can go a long way in describing their properties. As we have seen, the masses of the nuclei are nearly integers, and so, roughly, the concept of proton and neutron building blocks each having unit mass is attractive. However, the masses of both these particles are slightly in excess of 1 u, and so one would think that all the nuclei should have masses in excess of integer values. This is not true, as we have seen. Any theory of the nucleus must explain this fact. As it turns out, the explanation is reasonably straightforward, and we give it below.

Another property of the nuclei which must be explained by any reasonable theory is the nuclear angular momentum, or spin. The spins of nuclei are measured in the same way as we have described for the proton in Sec. 37.1. It so turns out that all measured nuclear spins can be explained in terms of the spins of the constituent protons and neutrons. For example, the spin of *heavy hydrogen*, or *deuterium* $^{2}_{1}H$, is 1. This is explained if one assumes that the spin $\frac{1}{2}$ neutron and spin $\frac{1}{2}$ proton are aligned with each other to give a total spin of unity. (The only other possible value is zero, since only two spin orientations are allowed, as we saw previously.) Notice that a nucleus composed of two protons and an electron, both spin $\frac{1}{2}$ particles, could not give the proper nuclear spin, although they would give the proper mass and charge. For this, as well as other reasons, we do not believe electrons exist within nuclei.

A very serious problem is posed by the neutron-proton nuclear model. Since protons repel each other by the coulomb force, it is impossible to construct a stable nucleus from these particles if the only forces available are electric and gravitational. The coulomb force would cause it to explode since the gravitational forces are many orders of magnitude smaller than the electric forces. In order to explain the stability of nuclei, we must postulate a third force, an attractive force between nucleons, i.e., neutrons and protons.

A nucleon-nucleon attractive force must exist to explain the stability of nuclei

37.7 NUCLEAR BINDING ENERGY AND MASS DEFECT

It is reasonably clear that to explain the existence of nuclei there must be a strong attractive force which exists between the nucleons. It is a short-range force. Two nucleons do not begin to attract each other appreciably until they are about 4×10^{-15} m apart. The attractive force rises rapidly as the particles come still closer. Moreover, it is found that the nuclear force is approximately independent of whether the particles involved are two neutrons, two protons, or a neutron and a proton.

The nuclear force has a range of about 4×10^{-15} m

As a rough approximation to the true situation (and we must admit at the outset that the true state of affairs is not completely known), we can picture the potential energy curve for a nucleus as shown in figure 37.5. The upper curve shows the potential energy of a proton as it is brought up to, and made a portion of, a stable nucleus. The parameter r is the distance between centers of the nucleus and proton. At large distances from the nucleus, the proton experiences the coulomb repulsion, and work must be done to bring it to decreasing r. When the nuclear radius r_0 is approached, the nuclear attractive force increases sharply and overcomes the coulomb repulsion. The proton is strongly attracted into the nucleus in the region $r = r_0$ and so the proton's potential energy drops sharply as it is brought closer to the nucleus. (Recall the bead on a wire analogy to the potential energy curve.) Once inside

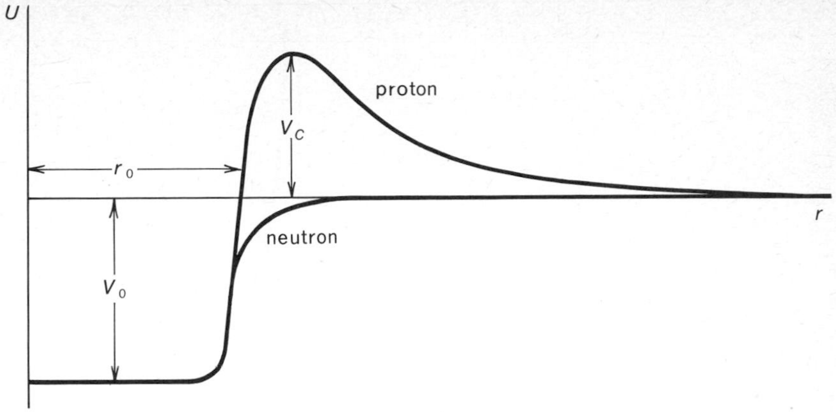

FIGURE 37.5 Form of the potential energy of a neutron and proton in a stable nucleus. Typical values might be $V_0 \simeq 8$ MeV, $V_q \simeq 10$ MeV, and $r_0 \approx 5 \times 10^{-15}$.

the nucleus, it is believed that the particle is at a reasonably constant potential.

A similar diagram can be drawn for the neutron, and it too is shown as the lower curve in figure 37.5. Since the neutron is not charged, the coulomb-repulsion portion of the potential is missing. There appears to be no region in which the neutron is repelled by the nucleus. For this reason, neutrons can be more easily shot into a nucleus than can protons.

Let us now consider what happens to the energy of a group of widely separated neutrons and protons as they are assembled into a stable nucleus. When the protons and neutrons are far separated, their combined energy state may be taken to be zero. However, as they are assembled into a nucleus, each particle will move in along r in figure 37.5, and when in the nucleus, will have lost an energy comparable with V_0. We therefore conclude that the energy of the particles, when assembled into a stable nucleus, is less than the original energy of the particles. This energy must be *given back* to the nucleus if it is to be torn apart into its original protons and neutrons. We call this energy difference between the free and bound nucleons the *binding energy* of the nucleus.

Nucleons lose energy when assembled into nuclei; this is the binding energy

According to Einstein, this loss in energy of the nucleons, ΔE, is equivalent to a loss in mass ΔM through the relation

$$\Delta E = (\Delta M)c^2$$

where c is the speed of light.* We see here the explanation for the fact pointed out earlier that the nucleus has a mass less than that of its constituent neutrons and protons. Mass was lost as the nucleus was assembled since energy was lost in the process. This mass loss is called the *mass defect* of the nucleus. The magnitude of ΔM per nucleon, $\Delta M/A$, for various nuclei is shown in figure 37.6. We have occasion to refer to this figure again when we discuss nuclear fission and fusion.

Energy loss is related to mass loss, the mass defect

37.8 LIQUID-DROP MODEL OF THE NUCLEUS

Now that we understand the qualitative features of nuclear binding, we are prepared to discuss various models for nuclear structure. Unfortunately, our

---

*It is convenient to remember the following conversion factor: 1 u of mass is equivalent to 931.5 MeV of energy.

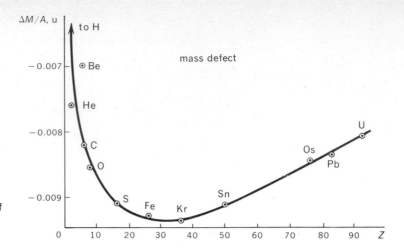

FIGURE 37.6 The mass discrepancy per nucleon for nuclei near the center of the periodic table is larger than that for elements at either end of the table.

lack of knowledge concerning the exact behavior of the nuclear force prevents us from devising an accurate theory for the nucleus. At the present time, the theory of the nucleus is still incomplete. For this reason, the theory we describe in this section, as well as in the next, must be considered to be only first approximations.

Two basic models for the nucleus are widely used. The *liquid-drop* nuclear model recognizes that, to a certain extent, the nucleons (i.e., the particles composing the nucleus) are held together much like water molecules in a drop of water. This model has been found most valuable for nuclei having a large number of nucleons, as one might expect. As we have seen from scattering experiments, the volume of the nucleus is given by

$$V = \tfrac{4}{3}\pi R_0{}^3 A$$

where A is the mass number of the nucleus, and R_0 is a fundamental radius of value about 1.2×10^{-15} m. We infer from this that the A nucleons in a nucleus pack tightly, much like molecules pack in a droplet.

The chief justification for the liquid-drop model, however, is the fact that it can predict approximately the nuclear binding energy. To do this, one considers three types of energy in a spherical nucleus:

1. The *volume energy,* due to the attractive forces between nucleons, is calculated on the basis of nearest-neighbor contacts. This is allowable since the nuclear force is of such short range. If each nucleon is adjacent to p other nucleons as neighbors, and if the binding energy between neighbors is ϵ_N, then the volume energy will be

$$(p\epsilon_N)A$$

2. In 1 above we assumed all the nucleons to be in the interior of the drop. However, those on the surface will have fewer than p neighbors. The number of surface nucleons will be proportional to the surface area of the drop, that is, $\sim V^{2/3} \sim A^{2/3}$. Since these surface nucleons give less binding energy than they should, we must subtract a binding energy (the *surface energy*) proportional to their number. We write it

$$-c_s A^{2/3}$$

where c_s is a constant.

3. There is *coulomb repulsion energy* between the nucleons. It will be a negative binding energy since it does not bind, but disrupts, the nucleus. This is simply the work done in bringing Z protons (the number of protons in the nucleus) from infinity to a sphere of volume V. It is found in illustration 37.5 that this energy is

$$-c_c \frac{Z(Z-1)}{A^{1/3}}$$

where c_c is a constant.

The total binding energy of the nucleus is the sum of these three terms:

$$\text{Binding energy} = p\epsilon_N A - c_s A^{2/3} - c_c \frac{Z(Z-1)}{A^{1/3}} \qquad 37.8$$

We can relate the binding energy to the mass defect per nucleon $\Delta M/A$ through the relation

$$\text{Binding energy} = |\Delta M| c^2 \qquad 37.9$$

It is possible to compare the experimental binding energy (obtained from the data of figure 37.6) with the values predicted by *37.8*. Three arbitrary constants ($p\epsilon_N$, c_s, and c_c) can be used in fitting the theoretical relation to experiment. A fairly reasonable fit is obtained, although wiggles in the experimental curve are not duplicated by the theoretical curve. In view of the number of arbitrary constants involved, the agreement found does not represent a firm confirmation of the liquid-drop model.

Another feature of the liquid-drop model is attractive, however. We know that if a drop is vibrating strongly enough, it may split into pieces, as shown in figure 37.7. This idea has been applied to the breaking apart of nuclei with some success, as we shall see in a later section. It pictures the very high atomic mass nucleus to have considerable internal kinetic energy. As the droplike nucleus vibrates, it distorts. Under certain conditions, it splits apart and undergoes fission.

Illustration 37.5

Assuming the nucleus to be a uniformly charged spherical shell of radius R_0, find the work done against electrostatic forces in giving it a charge Ze.

Reasoning The potential of a charged sphere of radius R_0 is

$$V = \frac{q}{4\pi\epsilon_0 R_0}$$

Suppose the sphere to have zero charge initially and then be charged by bringing to it charge increments each of magnitude e. No work is needed to bring up the first charge e since there is as yet no charge on the sphere. To

FIGURE 37.7 The liquid-drop model of the nucleus shows how internal vibrations can lead to nuclear fission.

bring up the second charge e, the work done is

$$e\frac{e}{4\pi\epsilon_0 R_0}$$

In bringing up the third charge, a work

$$e\frac{2e}{4\pi\epsilon_0 R_0}$$

is done. Similarly, the nth charge requires a work

$$e\frac{(n-1)e}{4\pi\epsilon_0 R_0}$$

Summing all such values from 1 to $n = Z$, we have

$$\text{Work} = \frac{e^2}{4\pi\epsilon_0 R_0}[0 + 1 + 2 + 3 + \cdots + (Z-1)]$$

Since the sum in brackets is simply $\frac{1}{2}Z(Z-1)$, this becomes

$$\text{Work} = \frac{e^2 Z(Z-1)}{8\pi\epsilon_0 R_0}$$

37.9 THE SHELL MODEL OF THE NUCLEUS

One feature of nuclear structure which is not explained by the liquid-drop model is the following. When one examines the isotopes which appear in the table of elements, it is found that certain combinations of neutrons and protons appear to give more stable nuclei than others. In particular, if the neutron number or proton number, or both, are certain "magic" numbers, the nucleus is exceptionally stable. These numbers are 2, 8, 20, 28, 50, 82, and 126. The existence of these more stable configurations is reminiscent of the situation in atoms, where elements of atomic numbers 2, 10, 18, 36, 54, and 86 are stable and unreactive.

In an effort to formulate a theory for the nucleus which is analogous to the Schrödinger theory for the atom, one must first know the potential of a particle near and in the nucleus. This potential is not known exactly. Fortunately, its exact form is not critical for a reasonable approximation. In the *shell model* of the nucleus, each nucleon is assumed to move in a static averaged force field caused by the other nucleons. Since the proton and neutron each have the same spin as the electron, the Pauli exclusion principle applies to them, and so only two neutrons or two protons can occupy the same quantum state. Because the neutron and proton are different particles, they have their own separate sets of energy levels. The proton's levels are at higher potential energy than the neutron's because the coulomb repulsion between protons increases their potential energies somewhat.

When this problem is solved by means of Schrödinger's equation, one finds that the proper stable nuclear states are not found. In order to obtain the proper result, one must assume a large contribution to the energy due to magnetic effects. One must assume a strong interaction between the magnetic moment of each nucleon and its magnetic field because of its orbital motion in the nucleus. If this is done, the proper stable states for the nucleus (i.e., the filled-shell states) occur quite naturally. This is a strong point in favor of the shell model.

Other factors also favor the shell model. It is noticed that nuclei which contain even numbers of neutrons and/or even numbers of protons are more stable than average. According to the shell model, a neutron sublevel is filled when it contains two neutrons with opposite spins. The same applies to the proton sublevels. Since the addition of an extra proton or neutron to a filled sublevel system will considerably increase the energy of a nucleus, the resulting nucleus will be less stable than its predecessor. This reasoning is confirmed by the fact that there are 160 stable isotopes which have even numbers of both neutrons and protons (*even-even* nuclei), while there are only four with odd numbers of both protons and neutrons (*odd-odd* nuclei). In addition to this, the shell model is capable of predicting the proper angular momentum for nuclei.

Although most properties of nuclei are fairly well described by either the liquid-drop or shell model of the nucleus, a combined treatment seems called for. In particular, the static nuclear potential used in the shell model should be replaced by a time-dependent potential. This is done, in part at least, by the *collective model* of the nucleus. As one would expect, this model is capable of describing the nucleus more adequately. However, it still suffers from the fact that our knowledge of the force field between nucleons is not complete.

Even-even nuclei are most stable

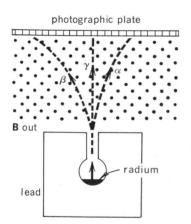

FIGURE 37.8 The radiation from radium is separated into three components by a magnetic field. The β rays actually follow a more curved path than shown.

37.10 RADIOACTIVITY

The fact that there is a competition within a nucleus between the nuclear attractive force and the coulomb repulsion leads one to think that not all combinations of neutrons and protons in nuclei will lead to stable configurations. This is, in fact, true. Even certain naturally occurring elements are not stable. They slowly decompose by emission of β particles (electrons not originally in the nucleus, but created in the process of decomposition) and α particles (helium nuclei), and are said to be *radioactive*. A typical case is that of radium. If a small amount of radium is placed in the center of a block of lead, a beam of radiation is found to emerge from a thin hole drilled in the block. This situation is shown in figure 37.8. (In practice, lead is found to be very effective in stopping radiation. We shall learn the reasons for this later. Since radioactive materials are dangerous to one's health, an experiment such as this should be carried out only with appropriate protection.)

If the beam of radiation from the radium is allowed to pass into a magnetic field, the beam splits into three components, as indicated in figure 37.8. (The β particles are bent much more than shown.) From the directions in which the rays are bent, we conclude that one component has no charge (the γ-rays) and the two others have opposite charges. Since these radiations were originally unidentified, they were given the designation α, β, and γ-rays. We now know that the γ-rays are x-rays; the α particles are helium nuclei, $^4_2$He; the β particles are negative electrons. All elements with atomic number greater than 83 are radioactive, and spontaneously emit these radiations.

It is found experimentally that the number of nuclei which decay (i.e., decompose) per unit time is proportional to the number of nondecayed nuclei present. In terms of a mathematical relation, if n nuclei still have not decayed at time t, then the rate at which nuclei will decay is given by

$$\frac{dn}{dt} = -\lambda n \qquad 37.10$$

where λ is a proportionality constant called the *decay constant*. The negative sign is necessary since n decreases with time, and dn/dt is therefore negative.

Rewriting *37.10*, we have

$$\frac{dn}{n} = -\lambda \, dt$$

which can be integrated to yield

$$\ln n = -\lambda t + \text{const}$$

To evaluate the constant, we notice that when $t = 0$, the number of nuclei was n_0. Therefore the constant is $\ln n_0$, and so the relation becomes, after taking antilogs,

Decay relation

$$n = n_0 e^{-\lambda t} \qquad\qquad 37.11$$

A plot of this relation is shown in figure 37.9.

The Discovery of Radium and Polonium

In preceding chapters we described how many of the phenomena of physics were discovered. Often these discoveries have been the result of accident, an unexpected experimental result being recognized as significant by an observant investigator. Other discoveries have been the result of tedious but well-designed experiments instigated specifically to produce the discovery which was eventually made. The discovery we now discuss is of this type.

Becquerel's discovery in 1896 that uranium atoms spontaneously emit radiation (Sec. 35.2) served as the impetus for a number of investigations having as their purpose the discovery of other radioactive atoms. Foremost among these investigators was Marie Sklodowska Curie (1867–1934) and, later, her husband, Pierre Curie (1859–1906). Madame Curie carried out a very systematic study of various compounds in search of other radioactive elements besides uranium. She, and simultaneously G. C. Schmidt, discovered that thorium was also radioactive. Further, she found that certain minerals (pitchblende, chalcolite, and uranite) emit radiation much more strongly than one would expect from their known concentrations of uranium and thorium. She enlisted the aid of her husband, who had been doing research in other aspects of physics at the time, and Marie and Pierre Curie set out to find the nature of the atom responsible for this radioactity.

Preliminary work in 1897 and 1898 indicated to the Curies that the radioactive material in pitchblende was present only in very small quantities. Starting with approximately a ton of pitchblende, which was initially about $2\frac{1}{2}$ times more radioactive than uranium, they carried out a systematic chemical separation of the elements of which it was composed. After each chemical separation, that portion of the products which was radioactive was further separated, the other portion being discarded. Two types of elements appeared to carry the radioactive material, a substance which separated chemically along with bismuth and an element which behaved chemically like barium. Since neither bismuth nor barium is radioactive, the Curies concluded that two new elements, one chemically similar to bismuth and the other similar to barium, were responsible for the radioactivity. They named the one similar

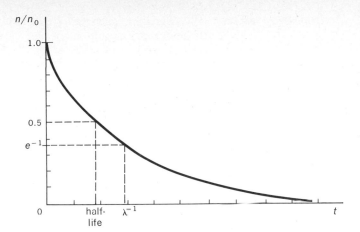

FIGURE 37.9 Decay of a radioactive element. If n_0 nuclei were present at $t = 0$, the number of the original nuclei present at time t is n.

This huge open-pit mine in New Mexico yields uranium minerals from which radium and polonium are ultimately extracted. (*The Anaconda Company.*)

to bismuth polonium, after Poland, the native country of Marie Curie. The other element, similar chemically to barium, they named radium. After long, tedious chemical separations the Curies were able to concentrate the radium portion of the ore until it had a radioactivity 900 times that of uranium. The polonium fraction was concentrated until its radioactivity was 400 times that of uranium. Later other investigators, using spectroscopic and mass spectrometer techniques, fully confirmed the Curies' supposition that they had discovered two new elements in the periodic table. We now recognize polonium and radium as elements 84 and 88, respectively, in the table of elements. These were the first of several high-atomic-number elements to be discovered by using radioactivity as the tool to detect their existence.

The time taken for half the nuclei to decay is called the *half-life* $T_{1/2}$ of the material. For radium, the half-life is 1620 years. To see how $T_{1/2}$ and λ are related, we rewrite *37.11* at time $t = T_{1/2}$, at which time $n/n_0 = \frac{1}{2}$. Then

$$\tfrac{1}{2} = e^{-\lambda T_{1/2}}$$

Upon taking logs of both sides we find

$$\lambda T_{1/2} = 0.693 \qquad\qquad 37.12$$

which is a convenient relation between the two quantities.

An interesting feature of the decay law given in *37.11* is that it predicts nuclear decay to be independent of the past history of the nucleus. For example, suppose at time $t = 0$ there are n_0 nuclei. After one half-life there will be

$$n_1 = n_0 e^{-\lambda T_{1/2}} = \tfrac{1}{2}n_0$$

nuclei remaining undecayed. When two half-lives have passed, the number remaining will be

$$n_2 = n_0 e^{-2\lambda T_{1/2}} = (n_0 e^{-\lambda T_{1/2}})e^{-\lambda T_{1/2}} = \tfrac{1}{2}n_1$$

In other words, no matter when in the history of a radioactive nucleus one begins to time events, after one half-life further has passed, the number of nondecayed nuclei will have decreased by a factor of $\frac{1}{2}$.

Because of this behavior, it is impossible to say how old a particular nucleus is. Given a milligram of radium, half of it will decay in 1620 years, no matter how old the radium is. This situation is commonplace in instances where independent probabilities are concerned. For example, it matters not at all how many times a coin has been flipped previously; its chance of coming up heads on the next trial is still $\frac{1}{2}$. We can picture a radium nucleus as having within it an α particle which tries to escape many times per second but whose chance of escape is very small. The chance that it will escape on the next trial, and lead to decay of the nucleus, is completely independent of the number of previous trials.

Illustration 37.6

A *curie* (Ci) of radioactive material is the amount of that material in which 3.70×10^{10} disintegrations occur per second. How many kilograms of radium corresponds to 1 Ci? The half-life of radium is 1620 years, or 5.1×10^{10} s.

Reasoning To find the decay constant λ we make use of the fact (*37.12*) that

$$\lambda = \frac{0.693}{T_{1/2}} = 1.36 \times 10^{-11}\,\text{s}^{-1}$$

From *37.10* we find

$$\frac{dn}{dt} = -n\lambda$$

or since we are interested in 1 Ci,

$$3.70 \times 10^{10}\,\text{s}^{-1} = (1.36 \times 10^{-11}\,\text{s}^{-1})n$$

from which

$$n = 2.72 \times 10^{21}\,\text{atoms}$$

This is the number of radium atoms needed. From Appendix 6A or 6B, the atomic mass of radium is 226 u, which is 3.75×10^{-25} kg. Therefore the mass of this number of atoms is

$$\text{Mass} = (2.72 \times 10^{21})(3.75 \times 10^{-25}\,\text{kg}) = 1.02 \times 10^{-3}\,\text{kg}$$

Originally, the curie was intended to be the radioactivity of one gram of radium. However, its definition was based on an erroneous value for radium's half-life; hence the slight discrepancy. In practice, microcuries of radioactive material are usually encountered, not curies. The SI unit which replaces the curie is the becquerel (Bq) where 1 Ci = 37 GBq. It will be discussed further in the next chapter.

37.11 RADIOACTIVE SERIES

When a nucleus emits a γ-ray, it falls to a lower energy state

When a nucleus emits a particle or γ-ray, the nucleus is changed in some way. The most simple situation is the emission of a γ-ray. Since a γ-ray is nothing more than a photon, it is an indication that the nucleus has changed its energy state. The situation is analogous to the emission of an x-ray by an atom. In both systems, no fundamental change in the constitution of the system occurs; the atom and nucleus are still essentially the same after the emission. However, a rearrangement of some kind must have occurred within the system so that its energy level became lower. We shall return later to the meaning of these emitted γ-rays.

A quite different situation exists when an α particle is emitted. In this case the nucleus changes its charge and its mass. It loses two units of positive charge and 4 u of mass. This can be illustrated by the reaction

$$^{A}_{Z}X \rightarrow ^{4}_{2}\alpha + ^{A-4}_{Z-2}Y$$

Notice that, in a symbolic equation such as this, the atomic numbers on one side of the relation must add to equal the sum on the other side. This is the result of the conservation of charge. The conservation of mass requires the sums of the mass numbers to be the same on the two sides of the equation. In the present instance,

$$Z = 2 + Z - 2$$
and
$$A = 4 + A - 4$$

Notice that an α particle is simply a helium nucleus.

The emission of a β particle (an electron) *in effect* changes a neutron within the nucleus to a proton.[*] As a result, the atomic number of the nucleus (its charge) is increased by one unit of charge. Since the mass of an electron is only 0.00055 u, it is ignored in writing the typical decay scheme, namely,

$$^{A}_{Z}X \rightarrow ^{0}_{-1}\beta + ^{A}_{z+1}Y$$

Notice that here too the two sides of the relation must balance.

Three distinct sets of radioactive elements exist free on the earth. Each set, or decay *series,* starts from the decay of a separate nucleus existing on the earth. For example, one of these series, the uranium decay series, starts with $^{238}_{92}$U. It proceeds as follows:

$$^{238}_{92}\text{U} \rightarrow ^{4}_{2}\alpha + ^{234}_{90}\text{Th}$$

The half-life of this decay is 4.5×10^9 years. Its very large value explains why there is still some ^{238}U left on earth.

---

[*] Since the nucleus does not contain electrons, this is, of course, a gross oversimplification of the actual process which goes on.

The thorium, ^{234}Th, formed when the ^{238}U decays, is also radioactive. It decays by β emission. Therefore

$$^{234}_{90}\text{Th} \rightarrow {}^{0}_{-1}\beta + {}^{234}_{91}\text{Pa}$$

The protactinium in turn decays to ^{234}U:

$$^{234}_{91}\text{Pa} \rightarrow {}^{0}_{-1}\beta + {}^{234}_{92}\text{U}$$

Several other steps occur in this series before the final stable element of the series is reached. In this case it is an isotope of lead, ^{206}Pb. This series is shown in detail in figure 37.10. Notice that in the latter stages of the decay scheme, alternative possibilities for decay exist.

The two other decay series found on earth are similar to the one illustrated. They are summarized in table 37.1. Notice that they all start with a very long half-life element and eventually decay to a stable isotope of lead. Presumably, other decay series existed on earth at earlier times, but have decayed too rapidly to be detected at this late date.

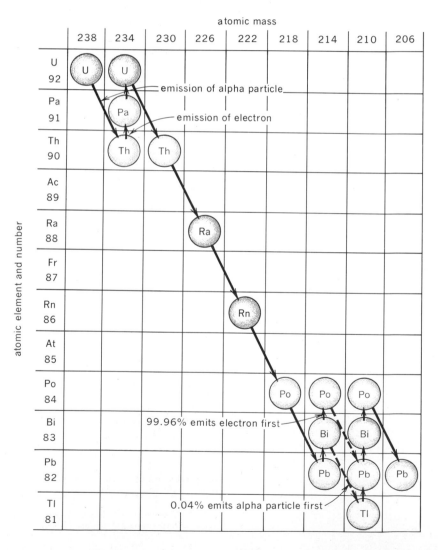

FIGURE 37.10 A typical radioactive series. It is called the uranium series since the parent nucleus is uranium.

TABLE 37.1 The Natural Radioactive Series

Series	Starting Element	Half-life (years)	Stable End Product
Uranium	$^{238}_{92}U$	4.51×10^9	$^{206}_{82}Pb$
Thorium	$^{232}_{90}Th$	1.39×10^{10}	$^{208}_{82}Pb$
Actinium	$^{235}_{92}U$	7.07×10^8	$^{207}_{82}Pb$

One other point should be mentioned in regard to these decay schemes. Along with the emission of α and β particles, γ-rays are emitted. These do not change either Z or A for the nucleus, and so they are not shown in the decay scheme. As stated earlier, they are emitted as the nucleus settles to a lower energy state.

Illustration 37.7

Assuming the age of the earth to be 10^{10} years, what fraction of the original amount of uranium 238 is still in existence on the earth?

Reasoning The half-life of ^{238}U is 4.51×10^9 years. We know that

$$\frac{n}{n_0} = e^{-\lambda t}$$

But $\lambda T_{1/2} = 0.693$, and so $\lambda = 1.54 \times 10^{-10}$ per year. Therefore

$$\text{Fraction} = e^{-1.54} = 0.214$$

37.12 ORIGIN OF γ-RAYS

When a nucleus undergoes radioactive decay, it is often left in an excited energy state. To reach the ground state, it could emit other particles or, more frequently, one or more γ-rays. If the nucleus makes a transition from state E_n to E_q, then it will emit a γ-ray photon of frequency

$$h\nu = E_n - E_q$$

This is completely analogous to the emission of a photon by an atom as its electronic structure adjusts to a lower energy state. Of course, γ-ray photons are basically the same as light and x-ray photons. However, the term γ-ray is usually given to photons emitted from the nucleus, while an identical photon emitted during an atomic electron transition is called an x-ray in the γ-ray region of wavelengths.

The wavelengths of γ-rays emitted by nuclei can be measured in several ways. They may be reflected off crystals of known structure, as discussed previously for x-ray diffraction, and their wavelength determined by use of the Bragg relation. Another method makes use of *internal conversion electrons*. These are electrons thrown out of the atom as photoelectrons when the nucleus interacts with the atomic electrons in such a way as to transfer to the electron the energy normally carried away by the γ-ray. (Roughly, one can think of the nucleus emitting the γ-ray, and this at once striking an electron and losing all its energy to the electron. Such a detailed picture of the process is not supported by experimental evidence.) The emitted electron will have an energy less than that of the γ-ray, less by an amount equal to the binding energy of the electron in the atom. As we shall see in the next section, the energy of high-energy electrons can be measured accurately, and so, knowing the binding energy of the electron, the γ-ray energy can be computed. Other methods for energy measurement based in part upon the interaction of the γ-rays with matter through the Compton and photoelectric effect also exist.

As in the case of atoms, emitted photons give us a tool for determining the energy-level structure of nuclei. Even the stable nuclei in the list of elements can be investigated by this method. To do this, unstable nuclei of other elements are produced by bombardment in a nuclear reactor (discussed in a subsequent section) or by some other means. These nuclei eventually decay to a stable nucleus. But since this isotope is usually in an excited state just after its formation, it emits γ-rays to return to its ground state. The energy-level schemes of most stable nuclei have now been measured in this as well as in other ways. Unfortunately, an exact theory to explain the existence of the known energy levels of the nuclei has not yet been found.

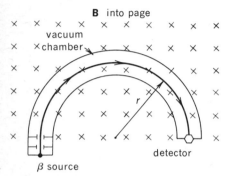

B into page

vacuum
chamber

r

detector

β source

FIGURE 37.11 In the β-ray spectrometer, the β particle's momentum is determined by deflection in a magnetic field.

37.13 EMISSION OF β PARTICLES

Many radioactive nuclei emit β particles, fast-moving electrons. The momenta of these usually relativistic particles are commonly measured by means of the β-ray spectrometer. This device is illustrated schematically in figure 37.11. The β particle is caused to move in a circular path by an impressed magnetic field B. Equating magnetic force to centripetal force, one has

$$\frac{mv^2}{r} = Bqv$$

from which the particle momentum $p = mv$ is

$$p = Bqr \qquad 37.13$$

A measurement of r and B yields the momentum of the particle.

To relate p to the kinetic energy K of the particle, one makes use of the relativistic equation for K, namely,

$$K = mc^2 - m_0c^2$$

where m_0 is the rest mass of the β particle. To relate p to mc^2 one notes that

$$m^2 = \frac{m_0{}^2}{1 - (v^2/c^2)}$$

where v is the particle's speed. This relation can be rearranged to give

$$m^2c^4 = m_0{}^2c^4 + m^2c^2v^2$$

Recognizing that $p = mv$, one has

$$mc^2 = \sqrt{m_0{}^2c^4 + p^2c^2}$$

which then gives the kinetic energy in the form

$$K = \sqrt{m_0{}^2c^4 + p^2c^2} - m_0c^2 \qquad 37.14$$

Therefore, having measured p by use of the β-ray spectrometer, the β-particle energy can be found.

Unlike the case of γ-ray emission, where only γ-rays of definite energies corresponding to differences in energy states of the nucleus are found, β

particles of widely varying energies are emitted. A typical β-particle energy spectrum is shown in figure 37.12. This is not what one would expect since, if a β particle is emitted, one would think it should carry away a reproducible energy corresponding to the difference in energy between the initial and final states of the nucleus.

Another puzzling fact about β-particle emission is the following. Observation of the recoil of the nucleus after β-particle ejection further indicates that the linear momentum of the two objects is not reasonable. In particular, the momentum of the ejected electron is not equal and opposite to the recoil momentum of the nucleus. To explain this, it was postulated that a second, undetected particle is emitted with the β particle. This particle should have zero rest mass and zero charge; it was given the name *neutrino*. As we shall see in the next chapter, this particle was eventually found in the mid-1950s. Although the neutrino has zero mass and zero charge, it must possess angular momentum (or spin) in order to preserve the law of conservation of angular momentum during β decay.

A neutrino must also be emitted with each β particle

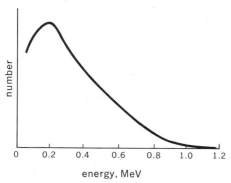

FIGURE 37.12 The energy distribution for β particles emitted from $^{210}_{83}$Bi.

37.14 EMISSION OF α PARTICLES

Radioactive nuclei often emit α particles. These emitted particles have energies characteristic of the nucleus from which they come. Among the natural radioactive substances, these energies range from about 4 to 9 MeV. It was recognized quite early that this range is too low. We can see that by reference to the potential energy diagram for a nucleus and α particle, as shown in figure 37.13.

In figure 37.13 is shown the energy diagram for the α particle-nucleus system as a function of distance of the α particle from the nucleus. The external portion of this potential function is reasonably well known from the results of experiments in which particles are shot at the nucleus. As indicated, the top of the curve at point A has an energy of about 25 MeV.

From classical physics one would reason as follows about an α particle which was capable of reaching point A and escaping from the nucleus. After reaching point A, it would fall down the curve toward B and (being repelled by the positive nucleus it left behind) would fly out to infinity. At that point, the particle would have zero potential energy, and so all the energy it had at A, 25 MeV, would be kinetic energy when the particle got far away. Instead, experiment shows that the emitted α particle has only about one-fifth of this

Classical ideas cannot explain α-particle emission

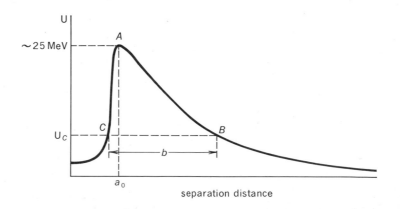

FIGURE 37.13 According to quantum theory, an α particle can tunnel through the barrier from C to B.

energy. The explanation for this was provided in 1928, and represented one of the first major triumphs of quantum theory.

Quantum theory predicts that an α particle can move from point C to B in figure 37.13 even though it does not have enough energy to surmount the barrier. The α particle tunnels through the barrier, and this is called the tunnel effect. To show why this is possible, one can make reference to the uncertainty principle in the form

According to quantum theory the α particle can tunnel through the barrier

$$(\Delta E)(\Delta t) \geq \hbar$$

For present purposes, this relation means that measurement is *incapable of disproving* any process which requires a particle to acquire an energy ΔE for a time Δt provided $(\Delta E)(\Delta t) < \hbar$. That is to say, the uncertainty principle places a limit on our ability to measure precisely, and if $(\Delta E)(\Delta t) < \hbar$ for a process, we shall be unable to measure that such is the case.

Suppose point A in figure 37.13 is at an energy ΔU higher than point C. Then, if a particle is to move from C to B, its energy will have to exceed the value U_c by at least ΔU. The time taken for the particle to move from C to B through the barrier would be b/v, where v is the speed of the particle.

The Mössbauer Effect

When a nucleus emits a γ-ray photon of frequency ν, it makes a transition from an excited energy state to a lower state. If the lower state is the ground state of the nucleus, an unexcited nucleus should absorb a photon of this same energy when it is thrown to its excited state. However, in almost all cases, the emitted photon has a lower frequency than the photon which will excite the nucleus. The reason for this is as follows.

We saw in Chap. 34 that a photon has a momentum $h\nu/c$. When a free nucleus emits a photon, the nucleus will recoil with a momentum equal and opposite to the photon's momentum. As a result, part of the energy lost by the nucleus as it falls to its ground state is given to the nucleus as recoil kinetic energy. The photon will have too little energy to raise an unexcited nucleus to its excited state. In general, photons emitted by free nuclei will not have enough energy to be absorbed by unexcited nuclei of the same type.

As an example, when radioactive ^{57}Co decays to ^{57}Fe, the ^{57}Fe nucleus is in an excited state. It then emits a γ-photon of energy 14.4 keV. In the process, the nucleus recoils with energy 2×10^{-3} eV. Although this energy is small, it is larger than the uncertainties in the energy of the photon and in the energy difference between the two energy states of the ^{57}Fe nucleus. As a result, the emitted photon has about 2×10^{-3} eV too small an energy to excite a nucleus in the ground state to its excited state. Therefore the γ-rays emitted by the excited nucleus are not absorbed by the unexcited nuclei.

It was found by Mössbauer, in the late 1950s, that under certain conditions the nucleus could be prevented from recoiling. In certain special crystals, the nuclei are bound so rigidly in the crystal lattice that the crystal as a whole recoils when the γ-ray is emitted. You should be able to show from the conservation of momentum that the energy carried off by the recoiling particle is inversely proportional to the mass of the recoiling particle. As a result, the massive crystal receives very little energy as it recoils after the γ-ray is emitted. Therefore, when γ-ray

Therefore the time for which the particle would need an excess energy ΔU would be $\Delta t = b/v$. According to our considerations of the uncertainty relation, our limit of measurement is such that we cannot disprove any process for which

$$(\Delta E)(\Delta t) < \hbar$$

In this case, this condition is

$$(\Delta U)\left(\frac{b}{v}\right) < \hbar$$

Clearly, when the barrier is too high (ΔU too large) and too wide (b too large), we shall be able to disprove experimentally any process which allows the α particle to go from C to B if its expected energy is U_c.

However, in the case in question, where the barrier width b is of the order of perhaps 10^{-15} m and ΔU is about $20 \times 10^6 \times 1.6 \times 10^{-19}$ J, we have

$$(\Delta U)\left(\frac{b}{v}\right) \approx \frac{3 \times 10^{-27}}{v} \text{J} \cdot \text{m}$$

---

emission occurs in a crystal of this type (where the crystal as a whole recoils rather than the decaying nucleus), the emitted γ-ray has an energy almost identical with the energy difference between the two nuclear-energy states. As a result, these γ-rays have the proper energy to be absorbed by the unexcited nuclei under consideration.

This discovery by Mössbauer that γ-rays emitted from certain crystals have energies essentially equal to the energy difference between the two nuclear-energy states is of considerable importance. For example, if Mössbauer-type γ-rays from nuclei in one crystal are shot into another identical crystal, they will be absorbed. However, if the energy states of the nuclei in the second crystal are even slightly altered (by impressed fields, for example), the γ-rays will not be absorbed. Moreover, the change in the energy levels of the second crystal's nuclei can be measured by giving either the emitting or absorbing crystal a slight relative motion. To see this, consider the following argument.

Suppose the energy levels in the absorbing nuclei have been changed in such a way that the energy difference between levels is no longer $h\nu_0$ but $h\nu'$, instead, with $\nu_0 < \nu'$. The emitted γ-rays of frequency ν_0 are therefore no longer absorbed. However, if the emitter crystal is moving toward the absorber crystal with speed v, then the frequency of the emitted γ-rays which strike the absorber crystal will no longer be ν_0 because of the Doppler effect. As we saw in 27.11, the frequency of the signal reaching the absorber will be $\nu_0 c/(c - v)$.

If the emitter's speed v is adjusted so that this received frequency is equal to ν', then the photons will have the proper energy to be absorbed. As a result, a knowledge of the proper v allows one to compute how much the nuclear energy levels have been shifted.

The Doppler effect may also be used to measure changes in frequency of the photon itself. Use of this fact is made in the Pound-Rebka experiment described on page 736.

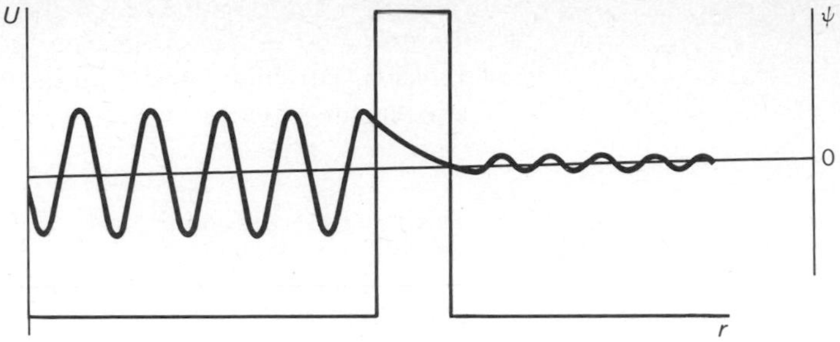

FIGURE 37.14 The wave function does not drop to zero at the edge of a finite potential barrier. This gives rise to the phenomenon of tunneling.

If we take $v \approx 3 \times 10^7$ m/s, we find

$$(\Delta U)\left(\frac{b}{v}\right) \approx 10^{-34} \, \text{J} \cdot \text{s}$$

This value is nearly equal to $\hbar$, and so we see that we are at the measurement limit for deciding experimentally whether or not an α particle can move from C to B. We therefore conclude that, because of the inherent uncertainty in energy of a particle, we cannot rule out the possiblity that an α particle will be able to tunnel through the barrier from C to B.

This result is shown more clearly by solving Schrödinger's equation for the potential well of figure 37.13. A simplified potential, that shown in figure 37.14, is quite easily solved using Schrödinger's equation. The resultant wave functions are shown. Since the potential does not go to infinity at the barrier, the wave function does not drop to zero there. Instead, the wave function decays exponentially through the barrier and is nonzero even on the outside of the barrier. Since the square of the wave function represents the chance of finding the particle, one sees that tunneling through the barrier is predicted by quantum theory.

Classically, an α particle with energy U_c would merely vibrate around inside the nucleus of figure 37.13. It has no chance at all of reaching point A in figure 37.13 since its energy is only U_c. However, as we have seen, the quantum-mechanical solution indicates that the α particle can escape from the nucleus by tunneling through the barrier from C to B.

Upon reaching point B, the α particle acquires kinetic energy as it falls down the rest of the potential energy curve and flies out to infinity. On this picture, as observed, the emitted α particle will have only the energy characteristic of point B, not of point A.

37.15 NUCLEAR FISSION

Readioactive decay and γ-ray emission by nuclei are examples of unstable nuclei adjusting to a more stable configuration. As we have seen many times in this text, physical systems tend to fall to their lowest possible energy state. Nuclei are no exception. A striking example of this is found in the processes called nuclear *fission* and *fusion*.

We recall that energy is liberated as isolated nucleons are brought together to form a nucleus. Since this energy must be furnished to the nucleus once again if its nucleons are to be torn apart from each other, we call this energy the binding energy of the nucleus. It is convenient to discuss the average energy lost by each individual nucleon in the nucleus. This quantity is plotted vs. the mass number of the elements in figure 37.15. We see from the graph, for example, that each neutron and proton composing the ^{4}He nucleus lost 7 MeV of energy as they were brought together to form the nucleus. That energy was presumably lost early in the history of the universe when the elements were first formed.

Figure 37.15 indicates two ways that we can obtain additional energy today. If we could fuse the small nuclei such as H, He, Li, etc., together to form more massive nuclei such as Ca and Fe, the constituent protons and neutrons will lose additional energy in the process. This follows since the graph shows us that each nucleon in Ca, for example, has lost more energy than in He and Li. Hence, if the nucleons in He and Li are brought together to form a Ca nucleus, they will lose energy (i.e., give off energy) in the process. This type of process is used to produce energy in the fusion reaction, the energy source for the hydrogen bomb. We shall discuss this process in the next chapter.

An alternate energy source can be devised using the nuclei at the large A end of the graph. Notice that each of the nucleons in U, for example, has more energy than it would have if it were in one of the smaller nuclei such as Fe or Cd. If a nucleus such as U could be persuaded to split into half, the two halves would then be nuclei in the range close to Cd. The nucleons would then find they had too much energy for this size nucleus, and so the two new nuclei would give off energy by emitting γ-rays (or in other ways) until the nucleon energy had decreased to its appropriate value. This process, by which a very large nucleus splits into two or more moderate-size nuclei, is called a *fission reaction*. Since the process gives off energy, the fission reaction can be used to produce energy. It is the process utilized in the uranium bomb and in present-day nuclear reactors. Let us now discuss this very important process.

Fission as an energy source

Although it is possible to split any nucleus of high atomic number into two nearly equal fragments, and thereby release energy, this is not practical in most cases. The difficulty lies in the fact that nearly all the high-atomic-number nuclei require considerable outside energy to cause them to split. As we found in our discussion of the liquid-drop model of the nucleus, if the drop vibrates violently enough, it will split into parts. However, for most nuclei, one must furnish several MeV of energy to the nucleus to cause it to undergo violent enough vibration to split.

Spontaneous fission rarely occurs in nuclei found in nature

We have a similar situation in chemistry. Heat is given off when wood is

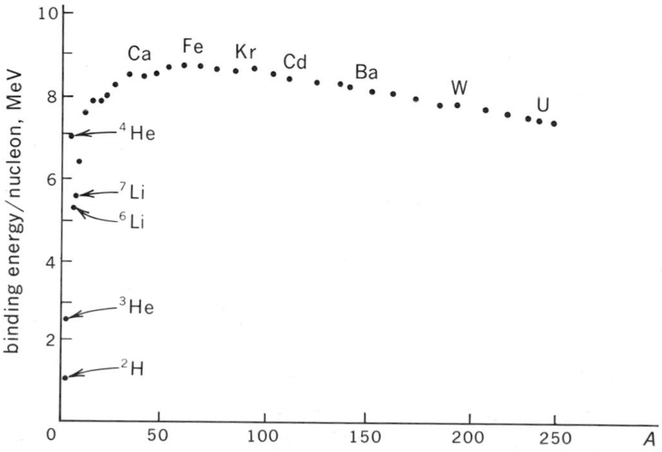

FIGURE 37.15 Binding energy per nucleon for representative elements. Note the very high stability of ^{4}He.

burned in oxygen. Obviously, then, the reactants have more energy than the products, and hence the reaction should be possible. However, wood does not combine spontaneously with oxygen at room temperature—at least not to any great extent. The wood-oxygen chemical reaction must first be started by some external means, a hot flame from a match, for example. Once it has started, the reaction produces enough heat to keep itself going, provided that the geometric arrangement of the piece of wood is such that the heat generated does not escape too easily.

As it turns out, the nuclei of most of the heavy elements cannot easily be split. By striking them with very-high-energy particles, splitting does sometimes occur. Even though more energy is given off when the fission reaction occurs than was needed to start it, this energy is not easily utilized to keep the reaction going. Hence, even though all heavy nuclei are potential sources of energy, it is impractical to obtain the energy given off by the fission reaction in nearly all cases.

There are, however, a few instances in which the fission reaction proves to be of practical importance. The first of these was discovered by Hahn and Strassman in 1939. With the aid of others, they found that an isotope of uranium, ^{235}U, would capture (or attach itself to) a slowly moving neutron and would then undergo spontaneous fission.* Here was a fission reaction which did not require high-energy bombardment. Moreover, when each ^{235}U nucleus splits, the reaction products contain about three neutrons, and of course a large amount of energy is given off. Each ^{235}U nucleus gives off about 200 MeV of energy as it splits.

A typical reaction for the fission process might be as follows:

$$n + {}^{235}_{92}\text{U} \rightarrow {}^{140}_{56}\text{Ba} + {}^{92}_{36}\text{Kr} + 4n + \text{energy}$$

where n represents a neutron. Many other possible reaction products exist, and in general many different *fission fragments* (i.e., reaction products) are produced. In addition, β particles and γ-rays, as well as fast-moving fragments and neutrons, result from the fission. These high-energy products are given off by the fission fragments as they decay to stable isotopes.

Since the fission of one ^{235}U nucleus gives rise on the average to about 3 neutrons, and since neutrons induce ^{235}U nuclei to undergo fission, a self-sustaining reaction is possible. Suppose one has a mass of ^{235}U so large that the number of neutrons which escape from its surface is negligible compared with the total number of neutrons. Then, when a single neutron enters a ^{235}U nucleus, it will give rise to, let us say, three neutrons as the nucleus undergoes fission. These three neutrons in turn cause three more nuclei to split, thereby liberating a total of $3^2 = 9$ neutrons. These 3^2 neutrons cause the fission of other nuclei to produce 3^3 neutrons, and so on. This process is illustrated in figure 37.16 and is called a *chain reaction*. After q steps in the chain reaction have occurred, 3^q neutrons will be available. If each step of the reaction takes 0.01 s, at the end of 1 s the total number of neutrons will be $3^{100} \approx 10^{48}$. Since 235 kg of ^{235}U contain only 6×10^{26} atoms, it is clear that a reaction such as this could occur with explosive violence.

The fission chain reaction forms the basis for operation of nuclear reactors and the fission (old-style) bomb. In these practical applications, several complications arise. To maintain a steady, nonexplosive reaction in a reactor, each fission process should cause one additional fission process (not

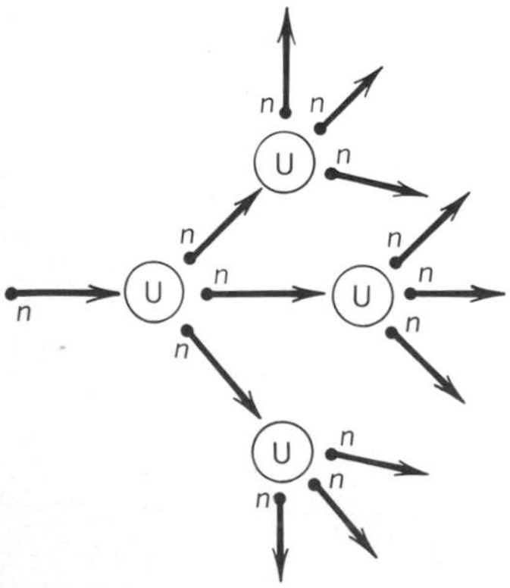

FIGURE 37.16 A chain reaction can be initiated by a single neutron.

*This isotope of uranium constitutes only 0.7 percent of the mixture of uranium isotopes occurring in nature.

two, or the reaction will explode, nor less than one, or the reaction will die out). In order to retain enough neutrons in the reaction chamber, the size of the fissionable material must be large enough so that not too many neutrons stray through its surface and become lost to the reaction. There is a *critical size* for the fissionable material. If too little material is available, not enough neutrons can be retained in it to produce a self-sustaining chain reaction. Similarly, a nuclear bomb must have a larger-than-critical size in order to produce the necessary violent chain reaction.

A second important consideration has to do with the fact that the ability of neutrons to be captured by a ^{235}U nucleus depends upon the speed of the neutrons. Slow neutrons are much more likely to cause fission than are fast neutrons. For this reason, a large part of the total volume of a nuclear reactor consists of a moderator, a nonreactive material used to slow down the neutrons emitted in the fission process. Since neutrons have a mass of 1 u, and since, upon collision, a particle is slowed best by particles of nearly the same mass, the moderating material in reactors usually consists of low-atomic-weight substances. Common examples are carbon, water, and hydro-carbon plastics.

37.16 NUCLEAR REACTORS

The reactor in a nuclear power station serves the same purpose as the furnace in a steam generator. It acts as an intense source of heat, and that heat is used to generate steam. The steam in turn is used to drive the turbines of the electric-generator system. A schematic diagram of a typical reactor is shown in figure 37.17.

The heart of the reactor consists of the fissionable material, the fuel, sealed in cylindrical tubes. Originally, uranium 235 was the principal reactor fuel. Now, however, other fissionable materials are also in use as fuel rods. These rods are immersed in a material such as water, carbon, a hydrocarbon, or some similar low-atomic-mass material. This material, the moderator, slows the fission-produced neutrons and reflects them back into the fissionable material. (Water made from the isotope ^{2}H rather than ^{1}H is often used since it is less apt to remove neutrons from the reaction.) In the design shown, the moderator also acts as the heat-exchange fluid to carry heat away from the fuel rods.

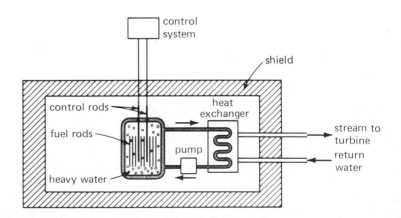

FIGURE 37.17 A schematic diagram of a reactor.

If a much larger than critical mass is involved, the reaction will build up at a fast rate and an explosion will occur. This is desirable, of course, if one is making a nuclear weapon. However, in the nuclear reactor, one wishes the reaction to proceed smoothly so that a steady but nonexplosive source of energy results. In practice, the number of reacting neutrons in a reactor is controlled by the use of neutron-absorbing rods. For example, cadmium rods readily absorb neutrons, thereby removing them from the reaction. Hence, if such rods are put into the reactor, the nuclear reaction will slow down. The reaction rate is readily adjusted by positioning control rods of this sort in the reactor.

When a nucleus undergoes fission within the fuel rod, highly unstable intermediate Z nuclei are formed. These undergo extensive radioactive decay and eject high-energy particles in the process. As these particles are slowed, their energy is changed to heat, thereby heating the reactor system. This heat is then carried away to a heat exchanger by the heavy water.

In the heat exchanger, the heat is transferred to ordinary water in a steam-boiler system. Steam is generated and this steam is then used to power electric turbines. As we see, the steam does not come in direct contact with the reactor core. For this reason, its level of radioactivity is low. But the fluid which circulates through the core is bombarded by radiation from the fission products. Like all other portions of the core material, it is often highly radioactive.

When the material in the fuel rods has been used for many months, its original fissionable material is much depleted. The fuel rods are then removed and replaced by new ones. Unfortunately we still have no really satisfactory disposal method for the waste material in the old rods. This material consists of highly radioactive, fairly long-lived fission products. It takes centuries for the radioactivity to decay to harmless levels. Disposal of this waste is one of the major drawbacks of nuclear reactors.

However, reactors also can provide us with radioactive materials for medical, industrial, and other uses. Many of the radiation sources presently used by hospitals, industry, and research laboratories are made by placing suitable materials within the core of the reactor. In addition, research reactors exist in many parts of the world. The intense radiation in their cores can be "piped" outside the reactor to act as powerful beams of radiation. As we see, the fission process has vast potential as well as hazards for humankind.

Chapter 37
Questions and Guesstimates

1. Consider the artificially produced element neptunium, $^{237}_{93}\text{Np}$. How many electrons, protons, and neutrons does one atom of this element contain?
2. How do chemists decide whether or not an atom is chlorine, for example? Why cannot their techniques distinguish between isotopes of the same element? Would the optical electronic spectra of two isotopes of the same element be the same?
3. The rotational and vibrational molecular spectra of Cl_2 are not quite the same if different isotopes of chlorine are used in making the molecule. Explain why this is true.
4. When a nucleus at rest emits a β particle, the nucleus recoils in the opposite direction. Why does not the attraction of the negative β particle for the positive nucleus cause the nucleus to be attracted so as to move in the same direction as the β particle?

5. The isotope neptunium 237, $^{237}_{93}$Np, is made in nuclear reactors. It decays by emitting in succession the following particles: α, β, α, α. What is the mass number of the resulting nucleus? What is its value for Z and what element is this?

6. Helium gas is obtained on earth by separating it from natural gas. Why is it unlikely that this gas was formed at the time that the earth was formed? Where does the gas come from?

7. Researchers often plot the intensity of radiation I from a radioactive source in the form $\ln I$ vs. t. What should the graph look like for (*a*) a single radioactive substance and (*b*) a substance that contains originally equal amounts of two isotopes with half-lives of 2 min and 2 days? Assume that they both decay to nonradioactive substances.

8. Half-lives for radioactive decay are essentially temperature-independent. Shouldn't the decay be faster at high temperatures because the particles within the nucleus have more energy to escape and are vibrating faster so they make more escape attempts in unit time?

9. The mass of an electron is 0.00055 u, while that of a proton is 1.0073, and that of a neutron is 1.0087 u. Is it feasible (from energy considerations alone) for a proton and electron, both at rest, to combine and form a neutron? Would it matter if they were at rest close together or far apart?

10. The radioactive-decay product of radium, radon gas, has a half-life of only 3.8 days. Why has not all the radon long since vanished from the surface of the earth?

11. A certain amount of radium is sealed in a glass tube. By accident, a small hole is left in the tube. The tube is let sit on a shelf in a cupboard. A year later the interior of the whole cupboard is found to be radioactive. Explain how this occurred.

12. A nuclear reactor produces energy chiefly in the form of heat energy. Explain how this heat is generated as a result of nuclear fission.

13. The artificially produced isotope ^{102}Ag has a half-life of 73 min. It decays in two alternative ways. Part of the nuclei emit a positive electron, a positron. The rest capture one of the electrons in the first Bohr orbit and take it into the nucleus. In what way are the two equivalent? What would one notice in the laboratory for each process?

14. A small amount of radium is sealed in an evacuated tube. When the tube is later broken open, it is found to contain some gas. The mass spectrometer shows the gas to consist of a very low molecular weight species and a very high molecular weight species. What are they? Which would be more abundant?

15. Explain clearly how the potential energy curve for a particle within a stable nucleus differs from that of an unstable nucleus.

Chapter 37
Problems

1. Consider the isotope barium 138. (*a*) What is its nuclear charge? (*b*) How many protons are in its nucleus? (*c*) How many neutrons?

2. A certain element has a nuclear charge of $+47e$ and its mass number is 107. (*a*) What element is this? (*b*) How many neutrons are in its nucleus? (*c*) What fraction of the atomic mass is due to its electrons?

3. What is the approximate radius of the ^{226}Ra nucleus? The ^{22}Na nucleus?

4. The radius of a carbon atom is 0.75×10^{-10} m. Find the ratio of the volume of the ^{12}C nucleus to the volume of the C atom.

5. The earth's radius is 6.3×10^6 m and its density is about 5.5 g/cm^3. If the earth were to shrink enough so its density was about equal to the density within a nucleus (2×10^{17} kg/m^3), how large would the earth's radius be?

6. The total mass of the universe is thought to be of the order of 10^{54} g. How large would the radius of the universe be if it were compressed until its density was 2×10^{17} kg/m^3, the density within nuclei? Find the ratio of this radius to the radius of the sun, 7×10^{10} cm.

7. In a certain mass spectrograph, a velocity selector is used to obtain a beam of ions with $v = 3.0 \times 10^5$ m/s. Find the radius of the path which a ^{12}C ion (univalent) will follow if B within the spectrograph is 0.070 T.

8. For the spectrograph described in problem 7, what will be the difference in the radii of the paths followed by a ^{12}C and a ^{14}C ion?

9. The following isotopes of neon are found in nature:

^{20}Ne   abundance = 90.9 percent

^{21}Ne   0.3

^{22}Ne   8.8

Find the atomic mass of Ne as given in the periodic table.

10. Uranium found on earth has two principal isotopes, ^{238}U and ^{235}U. Find the approximate percentages of each from the fact that the masses of the two isotopes are 238.051 and 235.044 u, while the chemical mass is 238.030 u.

11. The atomic masses listed in Appendix 6B include the mass of the electrons. The mass of $^{12}_{6}$C is listed as 12 u exactly. What is the nuclear mass of this isotope in atomic mass units? In kilograms?

12. From the data in Appendix 6B and the neutron mass, find the mass defect for the isotope $^{12}_{6}$C. What is the binding energy per nucleon (in eV) for this nucleus?

13. The mass defect curve given in figure 37.6 tells us that the mass loss per nucleon for tin, Sn, is about 0.00915 u per nucleon. From this fact and the known masses of the neutron and $^1_1$H, estimate the atomic mass of $^{120}_{50}$Sn (in u) and compare with the value given in Appendix 6B.

14. The half-life of ^{60}Co, a radioactive element produced in reactors for use as a medical and commercial radiation source (1.33- and 1.17-MeV γ-rays), is 5.3 years. How many atoms of ^{60}Co are there in 1 g of the material? What is the decay constant for the material? How many disintegrations occur each second in 1 g of the material? Express the radioactivity of 1 g in curies.

15. Strontium 90 is a radioactive fission product from nuclear fission reactors and bombs. Since its half-life is quite long (about 28 years or 8.8×10^8 s), it is a persistent contaminant and presents serious disposal problems. Answer the questions posed in problem 14 in the case of this material, ^{90}Sr.

16. The half-life of ^{90}Sr, a radioactive contaminant left in the wake of fission-type bombs, has a half-life of about 28 years. How long does it take for this type of contamination to decay to $\frac{1}{100}$ of its original value?

17. The end decay product of ^{238}U is ^{206}Pb. The half-life of ^{238}U is about 4.5×10^9 years. On earth we find ancient rocks which have ^{238}U and ^{206}Pb intermixed in such a way that it appears that about 40 percent of the original ^{238}U has decayed. How old are the rocks? (This is one way in which we can estimate the time since the earth solidified.)

18. A radiation counter placed near a radioactive source yields the following data:

Radiation (counts/s)	250	226	204	165	110	51	11
Time (h)	0	1	2	4	8	16	32

(*a*) Use a graph of ln *I* vs. *t* to find the half-life of the substance. (*b*) What is its decay constant? (*c*) At what time would the count rate have dropped to 0.1 percent of its original value?

19. The radioactive element $^{90}_{39}$Y (yttrium 90) emits a beta particle. What is the resultant isotope?

20. The radioactive element $^{66}_{29}$Cu emits a beta particle. What is the resultant isotope?

21. Plutonium 246, $^{246}_{94}$Pu, emits in succession two beta particles and two alpha particles and several γ-rays. What is the resultant isotope?

22. Actinium 231, $^{231}_{89}$Ac, emits in succession two beta particles, four alphas, one beta, and one alpha plus several γ-rays. What is the resultant isotope?

23. The carbon 14 isotope formed in the upper atmosphere by action of extraterrestial high-energy radiation on the air has a half-life of 1.81×10^{12} s. It emits a β particle with energy 0.156 MeV. Find the speed of the particle with this energy. Do not neglect relativistic effects.

24. Radon 220 emits a 6.29-MeV alpha particle. (*a*) Find the speed of the alpha particle. (*b*) What is the recoil speed of the remaining nucleus?

25. Cobalt 60 emits a 1.33-MeV gamma ray. By what fraction does its nuclear mass decrease in the process?

26. A radioactive isotope of mercury, $^{203}_{80}$Hg, emits a 0.279-MeV gamma ray. By what fraction does the nuclear mass decrease in the process?

27. A beam of beta particles follows a circle with radius = 20 cm when shot into a beta ray spectrometer that has a field of 0.10 T. What is the kinetic energy (in MeV) of the beta particles?

28. A nuclear power station reactor using ^{235}U as fuel has an output of 10^7 W. How much uranium is consumed per hour if the energy released in the fission of one ^{235}U nucleus is 200 MeV and if the reactor has an overall efficiency of 10 percent?

29. Thorium 232 emits alpha particles with energy 4.0 MeV. Assuming only the coulomb energy to be of importance, how close to the center of the nucleus was the α particle when its acceleration process began? (Thorium has $Z = 90$.)

30. Assume a neutron to be a classical, nonrelativistic solid spherical particle (untrue) of radius 1×10^{-15} m. If its angular momentum due to spin around an axis through its center is $\hbar$, find the angular speed ω of the neutron. What is the linear speed of a point on its equator?

31. Compute the frequency at which an electron would undergo magnetic resonance (similar to nuclear magnetic resonance but called, in this case, electron paramagnetic resonance) in the earth's field of about 0.50 G.

32. One microcurie of a radioactive substance which emits particles is placed at the center of a sphere of radius 50 cm. How many particles go through 1 cm² of the sphere's surface area each second? (Neglect absorption effects of the air even though this might be a poor assumption in many cases. One particle is emitted for each distintegration.)

33. One milligram of radium is sealed in a 2.0-cm^3 bottle. Assuming that no helium gas can escape from the bottle, how much helium will there be in the bottle after 1 year? (The half-life of radium is 1620 years, while the half-lives of the remaining members of the series do not exceed a few days, until ^{210}Pb is reached. Since ^{210}Pb has a half-life of 21 years, one can approximate the situation by assuming the series to terminate with this isotope.)

34. One milligram of radium is sealed in a bottle. It decays to radon, and the radon in turn decays with a half-life of 3.8 days. After about a month, how much radon (in grams) will there be in the bottle? (*Hint:* The total amount of radium will remain essentially constant. At equilibrium, the same number of nuclei of radium decay to radon as the number of radon nuclei which decay in unit time.)

38 High-Energy Physics

Much of present-day research in physics is concerned with the behavior of nuclei and other particles as they collide at high energies. In this chapter we discuss some aspects of the problems one encounters at such high energies. It will be seen that nuclear reactions carried out in this range give rise to new particles which are still not fully understood. A short review of the instrumentation used for the production and detection of these new particles is also presented.

38.1 COLLISIONS

In this chapter we are interested in the interactions between particles and nuclei. Since these interactions take place in collisions, we must first discuss the nature of collisions and the methods used to describe them. This topic is not completely new to us since we investigated the collision of particles which interact with coulomb forces, Rutherford scattering, in Chap. 34. Very often, however, the exact nature of the collision is not known or is not easily described in detail. For that reason, collisions are often described in terms of a rather general parameter, called the *collision cross section* σ.

Although we use a pictorial model to present the concept of collision cross section, we shall later see that the model is not a necessary part of the definition. Suppose we send a beam of particles (or radiation) of initial number per unit area per second (or intensity), I_0, through a very thin film of material, as shown in figure 38.1. The film is assumed so thin that no particle in the film is in the shadow of another particle. Suppose further that the intensity of the beam emerging from the film is I. Because the film is so thin, $I \cong I_0$. The fraction of the beam that is transmitted is I/I_0 and the fraction stopped is $1 - I/I_0$. We see from figure 38.1 that each absorbing particle casts a shadow within the beam. If the cross-sectional area of each shadow is σ, then the N particles in unit area will have a total shadow area $N\sigma$. This is a fraction $N\sigma/1$ of the total unit area, and this fraction must be equal to the fraction of the beam stopped, namely, $1 - I/I_0$. Therefore

$$N\sigma = 1 - \frac{I}{I_0}$$

from which the particle's collision cross section σ is

$$\sigma = \frac{1 - I/I_0}{N} \qquad\qquad 38.1$$

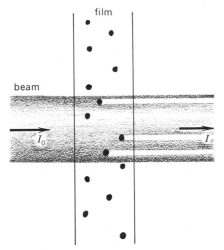

FIGURE 38.1 The absorber decreases the beam intensity from I_0 to I.

In stating this experimental definition of σ, we assume that the film is thin enough so that one particle has a negligible chance of being in the shadow of any other particle. The units of σ are those of area. Commonly, the unit used in nuclear studies is the *barn*, where 1 barn = $10^{-28}\,\text{m}^2$.

1 barn = $10^{-28}\,m^2$

It is easy to write down the way the intensity of such a beam decreases as the beam traverses an absorbing material. The fraction of the beam lost in a thickness dx of absorber is

$$-\frac{dI}{I} = \text{fraction lost} = \frac{\text{area obscured}}{\text{total area}}$$

$$= \frac{\sigma \times [(\text{no. of absorbers})/(\text{unit volume})](\text{volume of film})}{\text{area of film}}$$

$$= \frac{\sigma \times n \times (A\ dx)}{A}$$

where n is the number of absorbers per unit volume of the material, and A is the area of the film. (Why does the negative sign appear in this equation?) One then has

Definition of σ

$$-\frac{dI}{I} = \sigma n\ dx \qquad\qquad 38.2$$

This expression is the preferred definition of σ.

We integrate this expression from the left side of the film (where $I = I_0$ and $x = 0$) to a position x, where the beam intensity is I. This gives

$$\ln\frac{I}{I_0} = -\sigma n x$$

from which

The beam intensity decays exponentially

$$I = I_0 e^{-\sigma n x} \qquad\qquad 38.3$$

Clearly, the beam intensity decreases exponentially through the film, as shown in figure 38.2.

Equation *38.3* is applicable only to the type of absorption we have envisaged in our starting assumptions. The major assumption is as follows. When an incident particle strikes an absorbing particle, it is completely removed from the incident beam. In other words, the particle is not simply slowed down a little by a collision; it is either stopped or scattered appreciably from its original direction of motion. As we shall see later, a beam of α particles is not stopped in this way in general. Instead, each α particle is deflected hardly at all, and loses only a small fraction of its energy in each collision as it passes through air, for example. Such a particle stops when it

α-particle and electron beams do not follow the exponential law

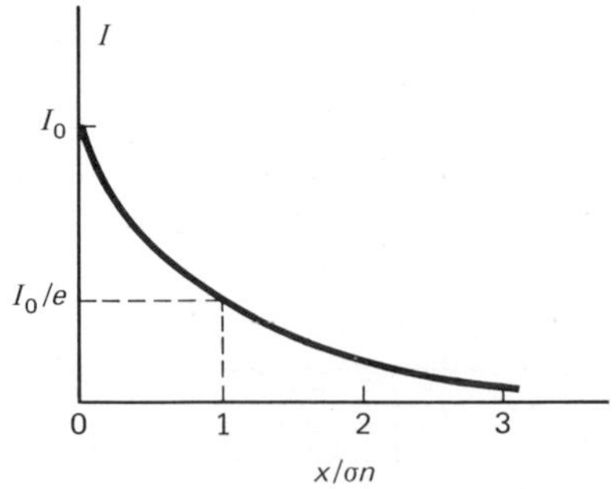

FIGURE 38.2 Intensity vs. depth in absorber for a beam which obeys the relation $I = I_0 e^{-\sigma n x}$. Typically the beam would be monoenergetic photons (light, x-rays, and γ-rays).

loses all its energy. Most of the particles require thousands of collisions before this happens, and so they all stop at about the same thickness of absorber.

Another word of caution should be given concerning too literal an interpretation of σ as a cross section. For example, in Rutherford-type scattering, the incident particles are scattered entirely by the coulomb forces and hardly ever come close to the actual cross section of the scatterer. As a result, σ is very large for gold nuclei if the bombarding particles are α particles. However, if neutrons are used, the coulomb force does not act, and so the scattering and cross section are correspondingly much smaller. Similarly, σ will become large if the bombarding particle has a favorable energy and structure for being absorbed by a nucleus. We see, then, that σ is much more than a mere geometrical cross section. Its actual significance may be quite complex in certain instances. However, its definition, 38.2, is still applicable.

We often extend the idea of cross section to include reactions as well as absorption and scattering. The reaction cross section is defined in terms of the number of particles which undergo reaction as the beam passes through an absorber. Calling I the incident number of particles per unit area per second, and $-dI$ the number which have caused reactions after passing through a thickness dx of absorber, 38.2 becomes

$$-\frac{dI}{I} = \sigma n \, dx$$

from which

$$-\sigma = \frac{1}{In}\frac{dI}{dx}$$

38.4

The basic difference between σ defined in this way and by 38.2 is in the meaning of dI. In 38.4 it represents that portion of the incident beam which causes a reaction in the length dx.

38.2 ABSORPTION OF γ-RAYS

We saw in the previous section that the intensity of a beam decreases exponentially as the beam penetrates an absorber provided the following is true: The particles in the beam must be lost from the beam when they undergo collision. This condition is satisfied for photons in a beam of electromagnetic radiation, and so 38.3 and figure 38.2 apply to such a beam. In the present section we will be concerned primarily with γ-rays and x-rays although much of the discussion applies to all electromagnetic radiation.

Gamma rays are electromagnetic photons. They have no rest mass and no charge. In spite of this they possess angular momentum, and their spin number is unity. Because of their lack of charge, the coulomb force interaction is absent. As a result, they are not stopped as easily as α, β, and similarly charged particles. There are three primary processes by which γ-rays, photons, are lost from an incident beam.

1. *Photoelectric absorption.* In this process a photon strikes an electron and gives all its energy to the electron. The net result is that the photon disappears and a high-energy electron is ejected from an atom. This process is very evident for γ-rays with energies less than about 2 MeV, as shown in figure 38.3.
2. Some of the photons are scattered out of the incident beam by Comp-

absorption
coefficient
α, cm^2/g

0.15

0.10

0.05

total

pair

photoelectric

Compton

0

10^{-1} 1   10 10^2

photon energy, MeV

FIGURE 38.3 As a γ-ray beam penetrates a material its intensity decreases according to $I = I_0 e^{-\alpha\rho x}$. Given here are the values of α in cm^2/g for the case of γ-rays penetrating the lead.

ton-type collisions. As we saw in Chap. 34, this type of scattering is similar to the collision of two balls, one being the photon and the other an electron. As shown in figure 38.3, this process also is largest at low γ-ray energies.

3. At photon energies in excess of 1.02 MeV, another photon absorption process occurs. Under certain conditions, which we cannot present here because the theory involved is extremely complex, a photon in the vicinity of a nucleus will disappear. (See problem 20 at the end of this chapter.) In its place will appear two charged particles, an electron and a positron. As we shall see later, a positron is identical with an electron in mass and spin. However, its charge is $+e$ rather than $-e$. This phenomenon, called *pair production,* is shown schematically in figure 38.4. Notice that, to conserve charge, such a process requires the two particles to have equal and opposite charges. What must be true in order that angular and linear momenta be conserved? Clearly, the photon from which the pair was created must have an energy at least $2mc^2$, where m is the electron mass. It is for this reason that pair production is not important below 1.02 MeV, the energy equivalent of $2m$.

Pair production predominates at high energies

The relative effects of these three absorption mechanisms are shown in figure 38.3. For the curves given there, the absorber is lead. Since a photon is absorbed (or scattered from the beam) in a single encounter, the intensity of a γ-ray beam should decrease in the manner given by *38.3.* This is conveniently written

γ-rays obey the exponential absorption law

$$I = I_0 e^{-\alpha\rho x} \qquad\qquad 38.5$$

where ρ is the density of the absorbing material, and α is called the *mass absorption coefficient.* To a very rough first approximation, α is independent of atomic number for γ-rays between 1 and 10 MeV. For lower-energy γ-rays, however, the dependence of α on atomic number becomes much more pronounced. Since $\alpha\rho$ is the important quantity, high-density elements are most effective in stopping γ-rays.

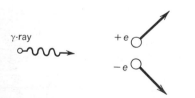

γ-ray

$+e$

$-e$

FIGURE 38.4 In pair production, what can one say about the vector velocities of the two particles created?

Illustration 38.1

From figure 38.3 estimate the thickness of water needed to decrease a 5-MeV γ-ray beam to $1/e$ of its original intensity.

Reasoning In this energy range, α is fairly insensitive to atomic number. Therefore, from figure 38.3, we find $\alpha \simeq 0.04 \text{ cm}^2/\text{g} = 0.004 \text{ m}^2/\text{kg}$. Also, from 38.5, to have

$$\frac{I}{I_0} = e^{-1}$$

we must set

$$\alpha \rho x = 1$$

Since $\rho = 1000 \text{ kg/m}^3$ for water, we find, upon solving for x, that the required thickness is 0.25 m.

38.3 CENTER-OF-MASS SYSTEM

Unlike photon beams, charged-particle beams of high energy do not usually decrease exponentially in intensity as the beam traverses an absorber. Instead, each particle loses only a small fraction of its energy in each collision. As a result, the particles all stop at about the same depth in the absorber. A typical example is shown in figure 38.5 where the tracks left by α particles in a cloud chamber are shown. Notice that the particles all stop after traveling about the same distance.

In the next section we shall examine this type of behavior in detail. Before doing that, though, let us discuss collisions between particles.

When we discussed Rutherford scattering of α particles by gold nuclei,

FIGURE 38.5 Cloud-chamber photograph of α rays from thorium C and C'. (Courtesy of C. T. R. Wilson.)

we assumed the nuclei were massive enough so that we could describe them as remaining motionless. In many of the collisions we consider in this chapter, this simplification is not possible. For that reason, we seek a method for treating most expeditiously the collision of particles of comparable size; call their masses m_1 and m_2.

Suppose a moving particle (m_1, v_1) collides with a particle at rest (m_2, $v_2 = 0$) as shown in figure 38.6. We know, from the conservation of momentum, that the momentum of the system after collision is the same as it was before. This is equivalent to stating that the motion of the center of mass of the system is constant along a straight line before, during, and after the collision. Use may be made of this fact to simplify computations.

The speed of the center of mass is easily found from the initial conditions, $v_2 = 0$ and $v_1 = v_1$. From the definition of the x coordinate of the mass center we have

$$x_{cm} = \frac{m_1 x_1 + m_2 x_2}{m_1 + m_2}$$

Upon taking the time derivative of this expression, assuming the m's to be constant, we have

$$\frac{dx_{cm}}{dt}(m_1 + m_2) = m_1 \frac{dx_1}{dt} + m_2 \frac{dx_2}{dt}$$

Calling the speed of the mass center V, and since $v_2 = 0$, this becomes

$$V = \frac{m_1}{m_1 + m_2} v_1 \qquad\qquad 38.6$$

The larger the ratio m_2/m_1, the smaller is V. Recall that V is unchanged by the collision.

Let us now compare the kinetic energies of the system as seen by two observers, one in the laboratory frame of reference (who sees the situation depicted in figure 38.6) and the other in the frame of one who rides along with the center of mass. The kinetic energy as seen in the laboratory frame *before* collision is

Laboratory Frame $\qquad\qquad K_L = \tfrac{1}{2}m_1 v_1^2$

while in the center-of-mass frame, it is

Center-of-Mass Frame $\qquad\qquad K_{cm} = \tfrac{1}{2}m_1(v_1 - V)^2 + \tfrac{1}{2}m_2(V)^2$

The latter expression is obtained by noting that the velocity of m_2 relative to the moving mass center is $-V$, and that of m_1 is $v_1 - V$.

By rearranging the expression for K_{cm}, one finds

$$K_{cm} = \tfrac{1}{2}m_1 v_1^2 + \tfrac{1}{2}m_1 V^2 + \tfrac{1}{2}m_2 V^2 - m_1 v_1 V$$

But from *38.6* we can replace $m_1 v_1$ to give

$$K_{cm} = \tfrac{1}{2}m_1 v_1^2 - \tfrac{1}{2}(m_1 + m_2)V^2 \qquad\qquad 38.7$$

We notice that the last term in this equation is that portion of the kinetic energy associated with motion of the mass center. This portion of the kinetic energy must be preserved in any collision since, as pointed out earlier, the law of conservation of momentum requires the motion of the center of mass to be the same before and after collision. Therefore the energy given in *38.7*, being

In a collision, the mass center moves with constant velocity

v_1 $\qquad$ $v_2 = 0$

before collision

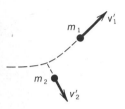

after collision

FIGURE 38.6 No matter how complicated the collision, the momentum of the system does not change.

Only the kinetic energy in the center-of-mass system is available for reaction

the initial laboratory-frame kinetic energy minus the mass-center-motion kinetic energy, is the energy available for reaction during a collision.

The result given in *38.7* may be put in a more convenient form. Since K_{cm} is the energy available for reaction and since K_L is the total energy before collision, the fractional energy available for reaction as viewed by the laboratory observer is (after a little arithmetic)

$$\text{Lab energy fraction available} = \frac{K_{cm}}{K_L} = \frac{m_2}{m_1 + m_2} \qquad 38.8$$

As one would expect, the more massive the second particle in comparison with the first, the greater the fraction of energy available for reaction.

38.4 INTERACTION OF CHARGED PARTICLES WITH MATTER

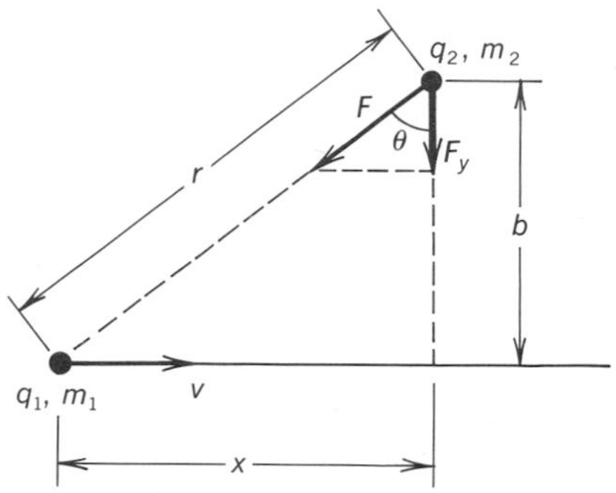

FIGURE 38.7 As it flies by, the swiftly moving particle imparts a small fraction of its energy to the particle m_2. In the diagram, q_1 and q_2 are assumed to have opposite signs.

In the preceding section we discussed the case of two particles colliding at high energy. When a particle passes through matter, it undergoes many "collisions," rather than just a single one. We place quotation marks around the word *collision* because, for the most part, the interaction between a charged particle and matter is electrical, rather than direct mechanical, collision. Although occasionally a proton, electron, α particle, or other charged particle will interact with a nucleus, the vast majority of the interactions occur because of coulomb forces between the charged particle and the electrons and nuclei of the material through which the particle is passing. Let us therefore find an approximate relation for the energy lost through coulomb forces by a charged particle as it passes close to an electron or a nucleus.

Referring to figure 38.7, we see a swiftly moving particle of mass m_1 and charge q_1 moving with speed v past a second particle. The second particle will be assumed to move slowly in comparison with the first and the deflection of q_1 will be assumed negligible. As q_1 goes by q_2, it will exert an impulse on q_2 in the direction of F_y. (The impulse due to F_x will be zero since F_x will reverse direction as q_2 passes by.) This impulse will be

$$\text{Impulse} = \int F_y\, dt = \frac{q_1 q_2}{4\pi\epsilon_0} \int \frac{\cos\theta\, dt}{r^2}$$

where the integral is to be taken over the total time of the collision.

To express all variables in terms of θ, we notice that

$$\frac{1}{r} = \frac{\cos\theta}{b}$$

Since $x = b\tan\theta$ we have that

$$dx = \frac{b\, d\theta}{\cos^2\theta}$$

But $v = dx/dt$, and so dx can be replaced by $v\, dt$ to give

$$dt = \frac{dx}{v} = \frac{b\, d\theta}{v\cos^2\theta}$$

Upon substitution we find

$$\text{Impulse} = \frac{q_1 q_2}{4\pi\epsilon_0 bv} \int_{-\pi/2}^{\pi/2} \cos\theta\, d\theta = \frac{q_1 q_2}{2\pi\epsilon_0 bv}$$

But the impulse is equal to the change in momentum of q_2, p. Assuming the particle to have negligible energy to begin with, the energy gained by q_2 is

Loss of energy per collision

$$\Delta E = \frac{p^2}{2m_2} = \frac{q_1{}^2 q_2{}^2}{8\pi^2 \epsilon_0{}^2 b^2 m_2 v^2}$$

38.9

This is, of course, equal also to the energy lost by the incident particle. Notice that m_2 is the mass of the stationary particle, not of the incident particle.

The Neutrino

One of the most difficult particles to detect is an elusive particle called the neutrino. Thirty years before the neutrino (or "little neutron") was actually detected, physicists firmly believed in its existence. If it did not exist, the laws of conservation of momentum and energy would not be valid in one type of radioactivity. In particular, it is found that, when a radioactive nucleus ejects a β particle (a fast electron), the energy and momentum of the emitted β particle are not reproducible. The emitted β particles are observed to have energies anywhere from nearly zero up to a definite maximum energy. It turns out that energy and momentum cannot be conserved under these conditions unless another particle is emitted from the nucleus simultaneously with the β particle. Rather than abandon the laws of conservation of energy and momentum, it seemed more attractive to believe that an unobserved particle was emitted along with the β particle. This view was first suggested by Wolfgang Pauli in 1927. Support for Pauli's suggestion was forthcoming when, in 1934, Enrico Fermi showed that one could base an accurate theory for β-particle emission, β decay, upon Pauli's concepts. It was Fermi who gave the name neutrino to the particle emitted with the β particle. Since the particle must be uncharged if charge is to be conserved in β decay, it was expected that the neutrino would not interact strongly with matter. Moreover, in the 1940s, several investigators examined the recoil of the nucleus upon emission of a β particle and were able therefrom to determine the energy and momentum of the neutrino which was supposedly emitted at the same time. They concluded that the neutrino's speed was essentially the same as the speed of light and that its rest mass was essentially zero. Even though the neutrino had zero charge and zero rest mass, it would have to possess angular momentum (or spin) for angular momentum to be conserved in the process of β decay.

All these properties of the neutrino were known only by inference, on the assumption that the conservation laws must be true. Unfortunately, verification of these inferences is not easy, since the neutrino very seldom interacts with matter. In fact, a neutrino is now known to travel an average equivalent distance of about 1000 light-years through solid matter before it undergoes a collision. Direct detection of the neutrino must search out and find one of these very rare collisions.

Although the laws of conservation of energy, momentum, and nuclear angular momentum insist that the neutrino must exist, we still need direct evidence. This was finally obtained in the mid-1950s by F. Reines and C. C. Cowan and associates. They based their work on the theoretical prediction that a neutrino would react with a proton to yield a neutron and a positive electron. The positive electron would unite almost at once with a negative electron to give off two γ-rays in a

For a given collision parameter b, we see that the incident particle will lose energy in proportion to the square of its charge and inversely as the square of its speed. From this we conclude that protons will lose energy much more rapidly than electrons, since, for the same energy, the speed of the latter will be much higher. However, the foregoing discussion has neglected relativistic and radiation effects, which are important at very high energy. Since both these effects are more pronounced for electrons than for more massive particles, we expect *38.9* to fail at lower energies for electrons than for

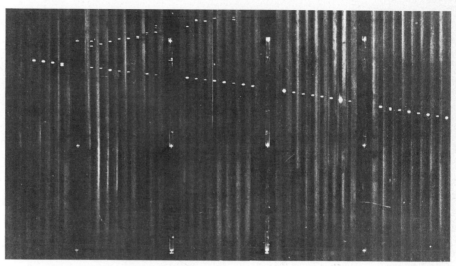

Indirect record of a neutrino. Lacking charge and rest mass itself, a neutrino creates a mu meson within the Columbia University spark chamber; and the meson leaves a long straight track. (*Brookhaven National Laboratory.*)

way which we discuss in this chapter. Moreover, the neutron would be allowed to react with a cadmium nucleus to produce more γ-rays. These latter γ-rays would be emitted about 15μs after the first pair of γ-rays. To detect the neutrino, the detection of the two emitted sets of γ-rays in the proper relative directions at the proper relative times would constitute an observation of the neutrino reaction.

To carry out this experiment, Reines and Cowan used a nuclear reactor to produce an intense beam of neutrinos from radioactive β decay. As the beam passed through 1 cm^2 of area, it carried about 10^{13} neutrinos/s through the area. The intricate detection system for observing the γ-rays was shielded from other radiation coming from the reactor by walls of concrete, water, and lead. When the experiment was tried, the detector counted several reactions per minute with the nuclear reactor in full operation. Since the sequence of γ-rays was exactly that predicted for the neutrino reaction, and since the reaction stopped as soon as the reactor was shut off, there is little doubt that the neutrino was actually being detected. We find, therefore, that our faith in the laws of conservation of energy and momentum is justified. The neutrino was finally detected thirty years after its invention in our efforts to preserve these laws. In later years, other methods for detecting neutrinos were devised. An indirect observation of a neutrino is shown in the accompanying photograph.

protons. In fact, *38.9* is not even qualitatively correct for electrons with energies in excess of about 1 MeV.

When a particle travels through most materials, nearly all the charged particles it encounters are electrons. For this reason one can usually replace m_2 and q_2 by the mass and charge of the electron. Moreover, the mass of the incident particle does not enter into *38.9*, and so protons and electrons of equal speed should, accordingly, lose the same energy per interaction. In addition, one notices that the energy loss of a particle as it traverses a material will be proportional to the number of particles it encounters in each centimeter of its path. Since most of these stopping particles are electrons, the stopping ability of a substance will be proportional to the number of electrons per unit volume of the absorber. For this reason, high-Z elements are usually the best absorbers.

These facts are compactly summarized in terms of the *stopping power* S of a material for a particle. It is defined to be the energy loss by the incident particle per unit path length in the stopping material. The units of S are conveniently taken to be MeV/cm. In tabulating these values it is customary to list $S/\rho Z^2$, where ρ is the density of the material in g/cm³ and Ze is the charge on the incident particle. Typical values for $S/\rho Z^2$ are shown in figure 38.8. Notice that they have been plotted against the kinetic energy of the incident particle (in MeV) divided by the mass of the incident particle (in u). This is done because ΔE depends on v^2 and not $\frac{1}{2}m_1 v^2$, as shown by *38.9*.

As shown in figure 38.8, even for electrons, the stopping power as given on such a plot does not differ too much from that of a proton below K/m_1 values of about 10^3 MeV/u. At first thought this might seem to contradict our previous assertion that the crude theory for stopping power given above does not apply to electrons in excess of 1 MeV. However, since the mass of the electron is about $\frac{1}{1850}$ u, we see that a value of $K/m_1 = 10^3$ MeV/u is equivalent to $K \approx 0.5$ MeV for an electron.

The quantity S, the stopping power, gives us valuable information concerning the energy loss of a particle as it traverses matter. To find how far a particle goes through matter before stopping, the *range R* of the particle, one makes use of S in the following way. From the definition of S, a particle

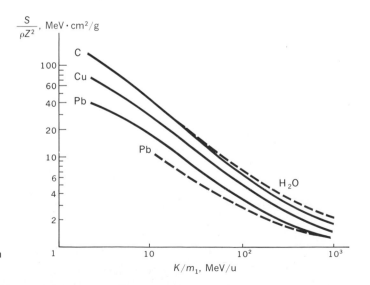

FIGURE 38.8 Stopping power of several substances for protons, α particles, and similar particles. The dashed lines are for electrons. m_1 and Ze are the mass (in u) and charge of the incident particle.

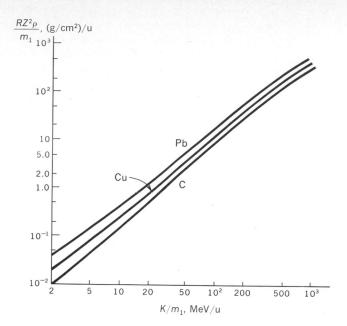

$\dfrac{RZ^2\rho}{m_1}$, (g/cm²)/u

Pb

Cu

C

K/m_1, MeV/u

FIGURE 38.9 Approximate range of charged particles (protons and α particles) in three materials.

loses an energy $S\,dx$ in traversing the distance dx. If we integrate $S\,dx$ along the complete path of the particle (from the point where energy loss commences to the point where the particle stops), this integral should be the total initial energy of the particle. Expressed as an equation,

$$K = \int_0^R S\,dx \qquad\qquad 38.10$$

In order to evaluate *38.10*, and thereby obtain a relation between the energy and range, one must know how S varies with energy from zero energy to K. No simple relation for S covers this whole range. As a result, the integral of *38.10* is not easily carried out. Therefore, even though a theoretical relation between S and R does exist, R is most easily found directly from experiment. Typical values of R can be found from figure 38.9. As in the case of S, it is convenient to plot the variation of R against energy divided by the mass of the particle in u. Further, the ranges for various particles can be shown by the same curves if $Z^2\rho R/m_1$ is plotted rather than R. (Recall that ρ is the density of the absorbing material, Ze is the charge of the *incident particle*, and m_1 is its mass in u.) Such a plot is shown in figure 38.9.

R and 1/S vary roughly as 1/ρ

Illustration 38.2

Estimate the range of an 8-MeV α particle in (*a*) carbon, (*b*) aluminum, and (*c*) air.

Reasoning Referring to figure 38.9, we note that, since in our case $K/m_1 = 2$ MeV/u, the range in carbon is given by

$$\frac{RZ^2\rho}{m_1} \approx 0.010 \ (\text{g/cm}^2)/\text{u}$$

Since $m_1 = 4$ u, $Z = 2$, and $\rho \approx 2.0$ g/cm³, we find

$$R \approx 0.0050 \text{ cm}$$

Similarly, since aluminum has an atomic number of 13, while the atomic numbers of carbon and copper are 6 and 29, respectively, we interpolate between the curves to find for aluminum that

$$\frac{RZ^2\rho}{m_1} \approx 0.015 \ (\text{g/cm}^2)/\text{u}$$

Since $\rho \approx 2.70 \text{ g/cm}^3$ for aluminum, we have

$$R \approx 0.0056 \text{ cm}$$

Notice how very small the range of even these high-energy α particles is.

In the case of air, the major constituent is nitrogen, which has an atomic number close to that of carbon. Since the density of air is 0.00129 g/cm^3, we have

$$R \approx 7.9 \text{ cm}$$

In view of this fact, can you think of any difficulty which necessarily complicated the scattering experiments of Geiger and Marsden?

Typical α-particle ranges are a few centimeters in air

38.5 RADIATION DETECTORS

Now that we understand how radiation interacts with matter, we can profitably discuss the operation of radiation detectors. Nearly all detectors of high-energy radiation operate on the principle that the radiation will furnish energy to the electrons of the material through which it is passing. We now discuss the operation of several general classes of detectors.

Cloud, Bubble, and Spark Chambers

When a high-energy particle passes through a gas or liquid, it forms ions. These ions then give a trace of the path taken by the particle. In the *Wilson cloud chamber,* the particle moves through a supersaturated vapor, leaving ions in its path. Since droplets form most easily on ions, the supersaturated vapor condenses in droplets on the ions along the particle's path. These ion tracks show not only the path taken by the particle but also something about the nature of the particle. Typical tracks are shown in figure 38.10. As pointed out in the figure legend, each particle has its own distinctive type of track.

Since high-energy particles are not readily stopped by vapor, the Wilson cloud chamber has been replaced in part by the *bubble chamber.* In

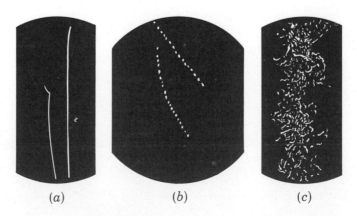

FIGURE 38.10 In part (*a*) are shown the tracks of two α particles. Two β-particle tracks are shown in part (*b*). A γ-ray beam leaves no track directly, but its path is revealed in part (*c*) by the β particles which it has ejected.

(a) $\qquad$ (b) $\qquad$ (c)

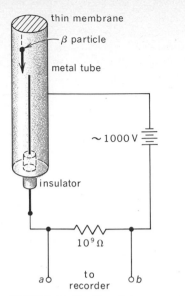

FIGURE 38.11 The Geiger counter.

this device one uses a superheated liquid. After the passage of an ionizing particle through it, the liquid begins to boil along the particle's path. The result is much like the examples shown in figure 38.10, but with much shorter path lengths.

A further modification of this idea is the *spark chamber*. Parallel metal plates with air gaps between them replace the bubble or vapor chamber. When a high potential is applied between adjacent plates, a spark jumps between the plates, preferentially along the ion track left by the high-energy particle as it passes through the plates. The particle's path is shown on a photograph by the light given off from the spark. (See p. 847 for an example of such a photograph.)

Ionization and Geiger Counters

It is possible to measure the charges liberated by an ionizing particle as it moves through a gas. The *Geiger counter*, illustrated in figure 38.11, operates in the following way. Under ordinary conditions, no charge exists in the gas within the metal tube. No current is able to pass from the center wire to the metal tube, and therefore no current flows in the circuit. When an ionizing particle enters the tube, the ions and electrons move across the tube under the influence of the electric field between the cylinder and the central wire. The field is made large enough so that the ions and electrons ionize other gas atoms as they move across the tube, causing an avalanche of charge. As a result, the current across the tube is much larger than would result from the original ions alone. Soon after the particle has passed through, all the ions have been collected, and the current stops. Therefore each ionizing particle gives rise to a current pulse in the resistor. The resulting voltage pulses are applied to a recording electronic system which then gives a record of the number of ionizing particles which entered the counter.

Scintillation Counters

When an electron hits the fluorescent screen of a TV tube, light is given off. This effect is used in the scintillation counter. Special transparent crystals and plastics (the *scintillators*) are capable of giving off light when ionizing radiation passes through them. As shown in figure 38.12, the light given off by the scintillator is incident upon a photocell (actually, a photomultiplier tube). The pulse of current which then passes through the photomultiplier circuit because of the light pulse is used to count the particle.

Other types of radiation detectors exist also. Solid-state devices which

FIGURE 38.12 In the scintillation counter, light from the scintillator causes photoemission of electrons in the phototube. In the first stage, electrons accelerate through the potential V_0, and each liberates several secondary electrons. This multiplicative process continues, and results in a comparatively large output-current pulse, much larger than indicated in the figure.

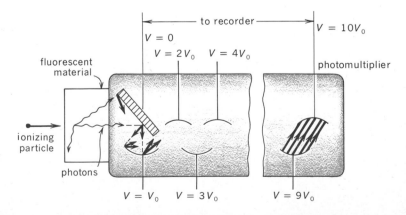

pass a current pulse when radiation is absorbed are now widely used. In addition, the paths of particles are often recorded directly by the ions formed as they pass through a thick photographic emulsion. These and other detection devices allow the experimentalist a fairly wide choice of detectors to fit his or her particular experimental conditions.

38.6 RADIATION UNITS

We now wish to discuss the units in which radiation is measured. There are four different classes of units designed to measure four different facets of radiation. They measure (a) source activity, (b) exposure, (c) absorbed dose, and (d) biological effectiveness. Let us now take up each of these in turn.

Source Activity

The activity of a radioactive source is equal to the number of disintegrations that take place in the source in unit time. As an equation

$$\text{Activity} = \text{disintegration rate} = \left| \frac{dn}{dt} \right| \qquad 38.11$$

where n is the number of undecayed nuclei at time t.

The SI unit for activity is the *becquerel* (Bq) and is the number of disintegrations per second. For example, we learned in illustration 37.6 that in 1 g of radium there are 37×10^9 disintegrations/s. Hence the activity of a 1-g radium source is 37×10^9 Bq. A microgram of radium would have an activity of 37,000 Bq.

You will recall from illustration 37.6 that a unit called the curie (Ci) is also used for activity. One curie is the activity of about 1 g of radium; the exact relation is

$$1 \, \text{Ci} = 37 \times 10^9 \, \text{Bq}$$

There is a simple relation between the activity and the decay constant for a source. The decay law given in *37.10* states that

$$\frac{dn}{dt} = -\lambda n$$

from which we find that

$$\text{Activity} = \lambda n = 0.693 \frac{n}{T_{1/2}} \qquad 38.12$$

where the latter form is found by use of the fact that $\lambda T_{1/2} = 0.693$.

Illustration 38.3

Strontium 90 has a 28-year half-life. What is the activity of a 1-g strontium 90 source?

Reasoning In applying equation *38.12*, we know that $T_{1/2} = 28$ years $= 8.8 \times 10^8$ s and n is the number of atoms in 1 g of strontium 90. Because one kmol (90 kg) of strontium contains 6.02×10^{26} atoms, we have that

$$n = (0.001/90) \times 6.02 \times 10^{26} = 6.7 \times 10^{21}$$

Therefore,

$$\text{Activity} = \frac{0.693n}{T_{1/2}} = \frac{0.693 \times 6.7 \times 10^{21}}{8.8 \times 10^8 \,\text{s}} = 5.3 \times 10^{12}\,\text{Bq}$$

Exposure

The exposure is a measure of the ionization caused by x-ray and γ-ray beams. It is defined only for these beams and only if their photon energies are less than 3 MeV. Its unit is the roentgen (R), which is defined in the following way.

The beam is sent through air at STP for a certain time. Positive ions are formed in the air by the beam. If Q coulombs of positive ions are formed in m kilograms of the air through which the beam passes, then the exposure during that length of time is

$$\text{Exposure in R} = \frac{Q}{2.58 \times 10^{-4}\,m} \qquad \text{C/kg}$$

Therefore *an exposure of 1 R will produce 2.58×10^{-4} coulomb of positive ions per kg of standard air.*

Absorbed dose

The absorbed dose is a measure of the energy absorbed from the radiation beam in unit mass of biological material through which the beam is passing. Its most common unit is the *rad* (rd). Suppose a beam passes through a mass m kilograms of biological tissue and loses an energy U joules in that mass during the duration of the beam. Then the absorbed dose in rads given to the mass m is defined by

$$\text{Absorbed dose in rd} = \frac{U}{0.01m} \qquad \text{J/kg} \qquad\qquad 38.13$$

In other words, *1 rad is equivalent to an absorbed energy of 0.01 J/kg.* The SI unit for absorbed dose, called the *gray* (Gy), is increasingly used: $1\,\text{Gy} = 100\,\text{rd}$.

The rad and gray units are applicable to absorbed dose of all types of ionizing radiation. However, the definition was originally chosen in such a way that an exposure of flesh to 1 R of x-rays gives rise to an absorbed dose of 1 rd. As a result, the rad and roentgen are often used interchangeably when dealing with the very special case of x-rays absorbed in flesh.

Biological Effectiveness

As we have seen, the rad (and gray) measures the beam energy absorbed in a unit mass of biological material. This type of measure is satisfactory from a physical standpoint but it is inappropriate for many biological applications. The difficulty arises because the biological changes caused by 1 rd of absorbed radiation varies depending on the type of radiation involved and the material being irradiated. To circumvent this difficulty, a quantity called the *relative biological effectiveness* (RBE) of a radiation is defined.*

A beam of 200-keV x-rays is taken as a standard for defining RBE. We define the RBE of a beam interacting with a biological system in the following way:

$$\text{RBE} = \frac{\text{biological effect of 1 rd of radiation}}{\text{biological effect of 1 rd of 200-keV x-rays}}$$

*The RBE is sometimes called the *quality factor* (QF).

TABLE 38.1 Typical RBE Values

Radiation	Typical RBE
200-keV x-rays	1.00000
γ-rays	
1-MeV	0.7
4-MeV	0.6
Electrons	1.0
Protons (1 to 10 MeV)	2
Neutrons	2–10
α-particles	10–20

By its very nature, the RBE is a rather imprecise quantity. Despite that fact, typical values for various types of radiation can be stated and a few such values are given in table 38.1. The appropriate RBE values for a given type of radiation also depend on the object being irradiated. For example, neutrons are particularly effective in causing eye cataracts and an RBE of about 30 is found in that instance.

To quantify the biological effect of a given absorbed dose, a unit called the *rem* is introduced. It is defined as follows:

$$\text{Biologically equivalent dose in rem} = (\text{RBE}) \times (\text{dose in rd}) \quad \textit{38.14}$$

As an example, we see from the table that an absorbed dose of 1 rd of 4-MeV γ-rays is equivalent to 0.6 rem. Similarly, 3 rd of 5-MeV protons is equivalent to 6 rem. Moreover, 3 rd of 5-MeV protons is equivalent to 6 rd of 200-keV x-rays.

Illustration 38.4

Cobalt 60 γ-rays have an RBE = 0.7. A tumor that is ordinarily given a dose of 1000 rd from a cobalt source is to be treated with neutrons having an RBE = 3. How many rads are needed from the neutron beam?

Reasoning The cobalt 60 dose is equivalent to (from *38.14*)

$$(0.7 \text{ rem/rd})(1000 \text{ rd}) = 700 \text{ rem}$$

Applying the same equation to the neutron beam gives

$$700 \text{ rem} = (3 \text{ rem/rd})(\text{neutron dose})$$

and so the biologically equivalent neutron dose would be $700/3 = 233$ rd.

38.7 RADIATION DAMAGE

Since radiation can tear apart molecules, it is capable of damaging materials. One of the most common types of radiation damage is due to the ultraviolet rays in sunlight. These lead to sunburn and tanning of the skin. The high-energy photons disrupt skin molecules upon impact and cause these easily observed effects. In this case, the damage is usually of little importance. Most of the sun's ultraviolet rays are absorbed by the ozone in the upper atmosphere so normal exposure to the sun's rays need not be avoided. However, in recent years we have become aware that a serious hazard could arise if we deplete the ozone layer with manufactured chemicals. There is danger then that the increased ultraviolet radiation reaching us could increase the incidence of skin cancer.

We are continuously exposed to other radiation in addition to sunlight. Nearly all materials contain a slight amount of radioactive substances. As a result, your body is unavoidably exposed to a low level of background radiation. Typically, each person experiences a background radiation dose of about 0.1 rem each year. Let us now examine the effects of different levels of radiation dose upon the body.

High levels of radiation covering the whole body disrupt the blood cells so seriously that life cannot be maintained. For whole-body doses in excess of 500 rem, death is likely to occur. Even a whole-body dose of 100 rem can cause radiation sickness of a very serious, although nonfatal, nature. Blood abnormalities occur for doses in the range of 30 rem and above. At still lower whole-body doses, the overall effects on the body are less apparent but nevertheless can cause serious consequences.

Even very low radiation doses reaching the reproductive regions of the body are potentially dangerous. The giant molecules in our bodies which carry reproductive information can be disrupted by a single radiation impact. If enough of these molecules are damaged, defective reproduction information will be furnished to a fetus as it develops. As a result, birth abnormalities will occur. Even though there is some evidence that a low level of reproduction abnormalities may be beneficial to mankind, most birth defects are not desirable. For this reason, *no one of child-bearing age should be exposed to unnecessary radiation of the reproductive organs*. Of course a properly given arm x-ray, for example, presents no such danger.

In addition to causing birth abnormalities, low levels of radiation present two other hazards. First, there appears to be a delayed cancer effect. Although cancer may not appear at once, low levels of radiation may cause cancer to develop many years later. Second, a child is particularly vulnerable to radiation. Because the child is growing rapidly, any cell mutations caused by radiation could have serious consequences. For this reason, most doctors are reluctant to prescribe x-ray scans for children unless absolutely necessary.

There is no "safe" limit of body exposure to radiation. It can only be said that radiation should be kept to the least value possible within reason. For example, since we are all subjected to a background radiation of about 0.1 rem/year, there is no reason to disrupt our lives to avoid radiation doses less than this. Even though a person who lives in the mountains may experience an annual background dose 0.05 rem higher than at sea level, the difference is not large enough to warrant moving. In the last analysis, one must often make a compromise between radiation safety and other considerations. Despite that fact, maximum occupational doses are of value and have been specified. As a rough rule, the maximum yearly dose, except for the eyes and reproductive organs, is about 5 rems.

38.8 PARTICLE ACCELERATORS

Now that we have seen how fast-moving particles behave and are detected, it is appropriate to discuss the machines used to accelerate the particles. A charged particle can be accelerated by simply allowing it to fall through a large potential difference. This must be done in vacuum, of course, so that negligible energy is lost to collisions. Unfortunately, it is very difficult to maintain potential differences larger than a few million volts since sparking occurs between electrodes. A typical accelerator for this rather low energy range is the Van de Graaff generator.

Van de Graaff Generator

Basically the machine, shown in figure 38.13, is quite simple. A high-voltage generator G (about 50,000 V) places positive charge on a belt made of an insulating material. As the belt moves, it carries the charge up to the metal sphere, or dome, at the top of the machine. At this point the charge is conducted off the belt, and it flows to the outside of the metal sphere. The upper limit for operation of this device is reached when sparking from the sphere to tank and leakage of the charge along the belt drain off as much charge as the belt carries up. In practice, the potential difference between the sphere and the tank cannot be made much higher than 10×10^6 V.

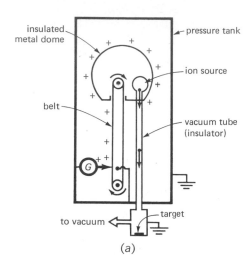

FIGURE 38.13 In a Van de Graaff generator the moving belt carries the charge up to the insulated metal dome. Charged particles are released at the ion source and accelerate down the vacuum tube to hit the target. In operation, the huge metal tank in part (*b*) is lowered. (High Voltage Engineering Corporation.)

(a)

(b)

Therefore a potential difference of a few million volts is available between the sphere and a target material which is grounded to the tank, as shown in the figure. An evacuated tube made of glass or other insulator extends from the sphere to the target. The bombarding particles, hydrogen ions (protons), for example, are released from a source at the sphere. They fall down the tube through the potential difference V of a few million volts. If the particles are protons, their energy at the target will be V eV. If the particles are doubly charged, say helium nuclei, they will have an energy $2V$ eV when they strike the target. (Why?)

The Van de Graaff machine is a very useful device for accelerating particles to an energy of a few million electron volts. It provides a steady, rather intense beam of particles which have a constant known energy. For the energy range up to about 3 MeV it is a widely used and valuable instrument. Since sparking becomes a serious problem at very high voltages, the machine is not widely used for energies higher than about 5 MeV although modern tandem arrangements give the equivalent of 30 MeV.

Linear Accelerator

To reach still higher particle energies, other techniques must be used. In the linear accelerator, a long series of cylindrical tubes placed end to end is used to accelerate charges down the tube axes. Perfectly synchronizd oscillators provide differences in potential between each adjacent tube so that as the particle falls from tube to tube, it is given additional energy. The device shown in figure 38.14 is capable of accelerating electrons to energies of 20 billion eV, that is, 20 GeV. Its cost was about $100 million.

Circular Accelerators

Another way for achieving high energies is to have a particle move in a circular path, falling across the same potential difference each time it goes around the circle. The particle is held in a circle by a magnetic field, and you will recall that this field neither adds energy to nor subtracts energy from the particle. If the particle falls through a potential difference of 5×10^5 V on each traversal of the circular path, its energy after 1000 traversals will be 500 MeV. A typical device of this sort is shown in figure 38.15. It accelerates protons to energies of about 500 GeV. Since the tools used in nuclear and

FIGURE 38.14 The Stanford University linear accelerator. By the time an electron has traveled two miles through the small white tube shown above, its energy is 20 GeV (20,000 MeV). An air view shows more clearly the physical size of the installation. (Courtesy of Stanford Linear Accelerator Center, Stanford University.)

(a)

(b)

FIGURE 38.15 (a) Aerial view of the National Accelerator Laboratory near Batavia, Illinois. The main accelerator, which is contained in an underground tunnel 4 miles in circumference, can be seen in the background; the three experimental areas are in the foreground. (b) The main accelerator tunnel is shown during construction in late 1970. (Photographs courtesy of National Accelerator Laboratory.)

particle physics are often large and expensive, the future of research with such tools is closely tied to the national economy and society in general.

38.9 NUCLEAR REACTIONS AND THE CONSERVATION LAWS

There are two primary reasons for building particle accelerators. As we shall see in a later section, when high-energy particles collide, new particles are sometimes created. The study of these new particles gives us new insight into the laws of nature. A second use of accelerators is to initiate nuclear reactions. It is this aspect which we shall now discuss.

Any process which changes the character or energy of a nucleus can be written as a *nuclear reaction*. For example, radium 226 decays by emitting an α particle. This decay can be written

$$^{226}_{88}\text{Ra} \rightarrow {}^{4}_{2}\alpha + {}^{222}_{86}\text{Rn}$$

Several factors influence whether or not a particular nuclear reaction will occur. The most obvious of these is the fact that the total energy of the reactants (mass energy included) must equal the total energy of the reaction products. For the reaction stated above, radium decaying to radon, this means that the mass energy of ^{226}Ra must equal the total mass energy of the α particle and radon plus the kinetic energies of these products.

By examining the energy balance in a proposed nuclear reaction, one can often rule out the possibility of the reaction. Let us consider one of the very first induced nuclear reactions. Rutherford, in 1918, carried out the following reaction by shooting α particles from radium at nitrogen nuclei:

$$^{17}_{7}\text{N} + {}^{4}_{2}\alpha \rightarrow {}^{17}_{8}\text{O} + {}^{1}_{1}\text{H}$$

where ${}^{1}_{1}\text{H}$ is simply a proton. When we examine the mass balance in this equation, we find

$$\text{Mass before} = 14.00307 + 4.00260 = 18.00567 \text{ u}$$
$$\text{Mass after} = 16.99914 + 1.00797 = 18.00711 \text{ u}$$

As we see, an amount of mass equal to 0.00144 u was *created* during the reaction.

In order to create mass we must have energy available. This energy must have been carried into the reaction by the α particle. Since 1 u of mass is equivalent to 931 MeV of energy (Sec. 38.10), the α particle must have carried an energy of

$$(931 \text{ MeV})\left(\frac{0.00144}{1}\right) = 1.34 \text{ MeV}$$

into the reaction in the center-of-mass system. Since the α particle in question has an energy of about 4.6 MeV, it is capable of providing both the center-of-mass energy and the reaction energy. (How large an energy must the α particle have in the laboratory reference frame?) In any event, the 4.6-MeV α particle is able to cause this reaction. Reasoning similar to this can decide the minimum energy needed to cause a reaction.

Other factors must also be considered when the feasibility of a reaction is being considered. The fact that charge is always conserved tells us that the

Charge is conserved in all reactions

algebraic sum of the charges of the reactants must equal the algebraic sum of the charges of the reaction products. In addition to energy and charge, angular momentum must also be conserved. Therefore the vector sum of the angular momenta of the reactants must equal the vector sum of the momenta after reaction. Since the angular momenta of the various elementary particles and nuclei are quantized and are integer or half-integer multiples of $\hbar$, this is equivalent to saying that the spins of the products must be obtainable from the spins of the reactants. (The proton, neutron, and electron all

Spin is conserved

have angular momenta of $\frac{1}{2}\hbar$, and so their spins are $\frac{1}{2}$. Spins of other particles will be listed later.)

As we now see, solid guidelines to the feasibility of a nuclear reaction are given by the conservation laws. These reactions must never violate the principles of conservation of mass energy, linear momentum, angular momentum, and charge. In spite of these guidelines, various other factors enter into the possibility of a reaction. For example, what prevents a mercury nucleus from being radioactive? Answers to questions such as this can be given only by reference to the internal structure of the nucleus itself. In the next section, we shall discuss a very important class of nuclear reactions, fusion reactions.

38.10 FUSION REACTIONS

We saw in the last chapter that nuclear reactors make use of the fact that energy is released when heavy nuclei undergo fission. Since this reaction uses scarce nuclear fuel and also produces undesirable radioactive waste, we desperately need a more bountiful, cleaner energy source. Fusion reactions hold great promise for the future.

It was pointed out in the preceding chapter that fusing together of light nuclei to produce nuclei near $A = 50$ is accompanied by the release of energy. To illustrate the magnitudes involved, consider the formation of a helium nucleus by the fusing together of two neutrons and two protons. Since the individual masses are*

$$2 \text{ protons} = 1.007825 \times 2 = 2.015650 \text{ u}$$
$$2 \text{ neutrons} = 1.008665 \times 2 = 2.017330 \text{ u}$$

From this the expected helium mass is

$$\text{Expected mass} = 4.032980 \text{ u}$$

The measured helium mass is

$$\text{Measured mass} = 4.002604 \text{ u}$$

which gives a mass loss of

$$\text{Loss in mass} = 0.030376 \text{ u}$$

Since

$$\text{Energy} = (\Delta m)c^2$$

we have, for masses in u,

$$\text{Energy (eV)} = \frac{(\Delta m)(1.67 \times 10^{-27})(9 \times 10^{16})}{1.6 \times 10^{-19}}$$
$$= (931 \times 10^6)(\Delta m)$$

1 u is equivalent to 931 MeV From this we see that *1 u of mass is equivalent to 931 MeV of energy.* This is a useful fact to remember. Applying it, we find that the energy released in the fusion of two neutrons and two protons to form a single helium nucleus is

$$\text{Energy} = 931 \times 0.0304 = 28 \text{ MeV}$$

In 1 kg of He there are $6 \times 10^{26}/4$ nuclei, since the atomic mass of He is 4 kg/kmol. Therefore the energy liberated when 1 kg of helium is formed is

$$\text{Energy} = (28 \text{ MeV})(1.5 \times 10^{26}) = 4.2 \times 10^{33} \text{ eV}$$

or

$$\text{Energy} = 6.7 \times 10^{14} \text{ J}$$

*Since the final result for the mass difference will be the same whether or not the electrons associated with the protons are considered, the masses given for the proton and helium nuclei are actually those for the nuclei plus the atomic electrons. These are the values listed in Appendix 6B.

Nuclear fusion reactions provide about 10^7 times as much energy per unit mass as chemical reactions

For comparison purposes, the energy liberated when 1 kg of carbon is burned in oxygen is about 8×10^6 cal, or about 3.3×10^7 J. From this one sees that the energy released in chemical reactions is about a factor 10^{-7} as large as that released in nuclear fusion reactions.

Although the source of energy in the sun and stars is a fusion process, the fusion reaction has not yet been made a practical, steady-energy source on earth. It is, however, the source of energy for the hydrogen and other types of fusion bombs. The difficulty in obtaining a steady fusion reaction is as follows.

As pointed out in the illustration which follows, the energy required to shoot two protons close enough together so that they will fuse is of the order of 1 MeV. This energy is easily attainable using the particle accelerator machines described earlier in this chapter. However, the efficiency of these machines is far too low to make their use in this way practical. To achieve a practical reaction method one must make use of the thermal energies generated by the reaction itself. Let us see what sort of temperatures are needed to furnish energies of 1 MeV, the energy needed to carry out these fusion processes.

We recall from Chap. 14 that the average kinetic energy of a particle at temperature T is $\frac{3}{2}kT$. Setting this equal to 10^6 eV, or 1.6×10^{-13} J, one has

$$(\tfrac{3}{2})(1.38 \times 10^{-23})T = 1.6 \times 10^{-13}$$

from which

$$T = 7.7 \times 10^9 \text{ K}$$

Extremely high temperatures are needed to maintain a fusion reaction

This temperature is far in excess of those produced in any existing stable process on earth. Work is progressing, however, to achieve very high temperatures in highly ionized gases (*plasmas*) confined to a region in space by magnetic fields. As yet no method has been devised to produce and contain gases at a temperature high enough to maintain a steady fusion reaction. But present research indicates that commercial utilization of the fusion reaction is probably feasible. If this proves correct, we will be able to utilize the hydrogen in the oceans as a nearly endless energy source.

Illustration 38.5

When two protons are about 5×10^{-15} m apart, or less, the nuclear attractive force overpowers the electrostatic repulsion of the charges. How large an electrostatic potential energy do two protons have when they are 4.5×10^{-15} m apart?

Reasoning The potential energy of two charges is

$$U = \frac{1}{4\pi\epsilon_0}\frac{q_1q_2}{r}$$

In our case this becomes, using SI units,

$$U = (9 \times 10^9)\frac{e^2}{4.5 \times 10^{-15}} = 2 \times 10^{24}e^2 \text{ J}$$

or

$$U = 3.2 \times 10^5 \text{ eV}$$

This energy must be furnished to the protons if they are to be joined together. Actually, an energy at least twice this large will be required since, as shown by *38.8*, only half the energy of an incident particle would be available for reaction.

38.11 PARTICLE PHYSICS

By the start of the Second World War, near 1940, scientists had discovered what were then known as the basic particles. They consisted of the electron, proton, neutron, and the massless photon and neutrino. One additional particle, the *meson,* had been postulated by the Japanese physicist Yukawa in 1935. (He needed such a particle in his theory for the origin of the nuclear force.) The meson, unlike the massless photon, was predicted to have a mass of about 250 electron masses, a mass equivalent to about 150 MeV.

It was expected that if a nucleus could be bombarded with particles having energy of about 300 MeV or larger, this particle could be created. Prior to 1940, however, accelerators capable of energies this high were impossible to construct. During the Second World War, scientists diverted their attentions to scientific pursuits that led to the development of radar, nuclear energy, computers, sophisticated electronics, and previously unknown materials such as semiconductors and synthetic fibers and plastics, all items of immediate practical interest. It was not until the late 1940s that research time and money became available for research with less promise for short-term practical application.

Using the technology and government financial resources developed during the war, it was possible after peace was achieved to begin construction of the huge particle accelerators we have mentioned previously. Even before that, in the late 1940s, high-energy particles called *cosmic rays* that strike the earth from outer space had been shown to create a new particle in collisions. This particle was originally believed (wrongly) to be Yukawa's meson.

But, as large accelerators and their intense beams of high-energy particles became available, it was found that a number of new particles are created in high-energy collisions. Among these are not only Yukawa's meson but a multitude of other particles as well. To this day new particles are being discovered as more intense and energetic beams become available. Current theory even indicates that the proton, neutron, and other particles are composed of still smaller particles called *quarks.* They are postulated to carry fractional charges. Although much research has been carried out to detect isolated quarks, the search has thus far been fruitless.

At the present time there is no satisfactory proved theory for the multitude of new particles that have been found. Theories for their interrelation exist, but discoveries are occurring so rapidly that no single theory has remained acceptable, unchanged or unextended, for long. You can obtain a good idea of the present situation by referring to a recent readable article on the subject.* By the time you read it, it, too, will probably be out of date.

In spite of this chaos, there are some results which we know to be important. They involve fundamental facts applicable to all particles. A final—or at least moderately acceptable—theory of these particles will make use of them. Let us see what some of these basic facts are.

1. All particles obey the conservation laws of energy, linear momentum, and charge. When particles decay into other particles, or when new particles are formed in bombarding reactions, these laws still apply.
2. All particles obey the law of conservation of angular momentum; i.e., spin is conserved. Moreover, spin is quantized. A particle can have an angular momentum (pictured as being due to its spin on its axis) of $m_s(h/2\pi)$, where $m_s = 0, \frac{1}{2}, 1, \frac{3}{2}, \ldots$ and h is Planck's constant. We call m_s the spin of the

---

*S. L. Glashow, Quarks with Color and Flavor, *Sci. Am.,* October 1975, p. 38.

particle, and each particle always has the same value for m_s. The electron, proton, and neutron all have $m_s = \frac{1}{2}$. They are called spin-$\frac{1}{2}$ particles. The photon has $m_s = 1$. Mesons have zero spin, and so on. In any process, the angular momentum of the reactants must equal that of the products.

3. Baryon number is conserved. Spin-$\frac{1}{2}$ particles with masses equal to or larger than the proton are called *baryons*. Each baryon is assigned a number, either $+1$ or -1, in a way described in the next paragraph. In any process, the baryon number at the start must be the same as at the end.

4. Each particle has an antiparticle. For every particle there is another particle which is "identical but opposite" to it. Not only does an electron exist, but so does an antielectron, which we call the positron. The positron has the same mass as the electron, but it has opposite charge $+e$. When an electron meets a positron, they annihilate each other. The proton also has an antiparticle, the antiproton. For the neutron there is an antineutron, and so on. In assigning baryon numbers, particles are $+1$ and antiparticles are -1.

5. All spin-$\frac{1}{2}$ particles obey Fermi-Dirac statistics. By this we mean that no two of these particles can have the exact same set of quantum numbers in any one system. For electrons in the atom, this reduces to the Pauli exclusion principle discussed previously. However, the exclusion principle also applies to spin-$\frac{1}{2}$ particles in the nucleus and in other particle aggregates.

In addition to these facts, there are others less well established. Other conserved quantities have been proposed; strangeness and iso spin are two of them. Despite their difficulties in solving the problem of the fundamental particles, physicists still remain in good humor. Perhaps as an unconscious mechanism to keep spirits high, they describe properties of particles in such whimsical terms as "strangeness," "charm," "color," and so on. Nevertheless, theirs is research of the highest caliber and most intense effort. The study of the basic particles probes the deepest depths of our universe. We expect that a way will soon be found to bring order out of the present state of chaos.

Chapter 38
Questions
and Guesstimates

1. Why is a given thickness of lead a better absorber of 1-MeV α, β, and γ-rays than is the same thickness of water? Why is water a better shielding barrier against neutrons than is lead?

2. Tritium is the following isotope of hydrogen, ^{3}H. It has an atomic mass of 3.016, while the atomic mass of ^{1}H is 1.0078. Using the fact that the neutron mass is 1.00867, what do you predict about the stability of tritium? Repeat for ^{2}H, deuterium, which has a mass of 2.0141.

3. One nucleus has a reaction cross section for neutron absorption which is twice that for another nucleus. Does this mean one nucleus is twice as big as the other? What *does* it mean?

4. A positron and an electron moving with negligible speed collide and annihilate each other. What can be said about the directions taken by the two emitted photons? When a photon materializes into two particles, an electron and positron, in a process called pair production, what can one say about the motion of the emitted particles?

5. What reasons do we have to think that the following reaction is impossible?

$$^1_0n \rightarrow\ _{-1}^{0}e + ^1_1\text{H}$$

6. Why is a Geiger counter not a suitable detector for fast neutrons? How can fast neutrons be detected with the aid of a counter which measures ionization?

7. Van de Graaff generators have been built to accelerate ions of atoms as large as nitrogen. What advantage would these generators have over the usual ones designed to accelerate deuterons?

8. After a proton, for example, leaves an accelerating machine, and during the process in which it strikes a target nucleus, the center of mass of the proton-nucleus system must move in such a way that the momentum of the total mass, considered located at the mass center, must remain constant. Compare the observations of an observer moving on the mass center with an observer at rest in the laboratory as two protons collide head on. Do they agree on the kinetic energy involved in the collision? What effect would it have on their interpretation of the reaction if the two particles stuck together after collision?

9. Before an x-ray survey is made of the gastrointestinal tract, the patient must drink a solution of a barium compound. Why?

10. When an x-ray photograph is taken of a person's arm, the bones are clearly shown on the photograph. Why do they show up in this way? After all, the arm is no thicker where the bone is than elsewhere.

11. Van de Graaff generators are often contained in sealed tanks filled with gas under high pressure in order to prevent sparking. Why is a higher voltage needed to cause a spark in air at 10 atm than in air at normal pressures? (*Hint:* Discuss how a single ion gives rise to a multitude of other ions when a spark is initiated.)

12. Why is the fusion reaction so much more difficult to initiate than the fission reaction?

13. Is there any possibility of causing the fusion of two small nuclei without making them collide with extremely high energy? Defend your answer.

14. It is possible for a man working with x-rays (or γ-rays) to burn his hand so seriously that he must have it amputated, and yet the man may suffer no other consequences. However, an x-ray overexposure so slight as to cause no observable damage to his body could cause one of his subsequent offspring to be seriously deformed. Explain why.

15. Most radiologists feel that women beyond childbearing age can safely be exposed to much more x-radiation than young women. How can they justify such an opinion?

16. Low-energy (soft) x-rays are used to treat skin cancer, while high-energy (hard) x-rays are used to treat cancer deep within the body. Why are soft x-rays not used for this latter purpose, even though they can penetrate deeply enough to kill the cancerous region?

Chapter 38
Problems

1. All perfect gases under standard conditions contain 2.687×10^{19} molecules/cm³. (This is called *Loschmidt's number*.) A certain beam of x-rays decreases in intensity by a factor of $\frac{1}{2}$ as the beam traverses 23 cm of a gas at STP. What is the cross section of the gas molecules for these x-rays? Give your answer in barns. Do not use *38.1*. Why not?

2. If the cross section for γ-ray absorption in a gas at STP is 490 barns, what fraction of the γ-ray beam can penetrate 2 cm into the gas? See problem 1 for a hint.

3. The cross section for the absorption of 100-pm x-rays in aluminum is 616 barns. Find (a) the number of aluminum atoms/cm^3 in aluminum ($\rho_{Al} = 2.70 \, \text{g/cm}^3$) and (b) the distance such an x-ray beam will travel through aluminum before being reduced to 10 percent of its original value ($A_{Al} = 27$).

4. The density of gold is 19.3 g/cm^3, and its cross section for absorption of 1.0-Å x-rays is 2×10^4 barns. Find (a) the number of gold atoms/cm^3 in gold and (b) the thickness of gold for such an x-ray beam to be reduced to 10 percent of its original value.

5. The mass absorption coefficient for 10-pm x-rays in aluminum is 0.16 cm^2/g. Find the thickness of aluminum needed to reduce a beam of such x-rays to 10 percent of its original value. ($\rho_{Al} = 2.70 \, \text{g/cm}^3$.)

6. The mass absorption coefficient for 0.10-Å x-rays in lead is 3.7 cm^2/g. Find the thickness of lead needed to reduce a beam of such x-rays to 10 percent of its original value. ($\rho_{Pb} = 11{,}300 \, \text{kg/m}^3$.)

7. As viewed from the laboratory reference frame, a 4.00-MeV alpha particle strikes a deuterium nucleus, ^2_1H. How much energy is available for reaction in the process? Repeat if the deuteron is replaced by a ^{238}U nucleus.

8. The ionization energy of hydrogen is 13.6 eV. What minimum energy must a bombarding deuteron (^2_1H) have if it is to be able to ionize a stationary free hydrogen atom? Repeat if the bombarding particle is an electron. All quantities are to be measured in the laboratory reference frame.

9. As shown in illustration 38.2, the range of an 8-MeV alpha particle in air at STP is about 7.9 cm. About what would the pressure have to be (in atmospheres) if the range was to be 1000 mi (1.61×10^6 m)? Pressures must be less than this in large circular particle accelerators since the particles circle the machine many times during the acceleration process.

10. Using figure 38.9, estimate the range of a 20-MeV deuteron in polystyrene plastic. (Polystyrene has a density close to 1.0 g/cm^3 and is mostly composed of carbon.) Repeat for a 20-MeV alpha particle.

11. Estimate the range of a 200-MeV alpha particle in sandy soil whose density is about 2 g/cm^3 and whose composition is basically SiO$_2$. Make use of figure 38.9.

12. Referring to figure 38.5, suppose the tracks shown there are in air and that the actual tracks were as long as the dimension in the photograph. What would have been the energy of the high-energy alpha particles shown in the figure?

13. Neutrons are best slowed by collision with particles of equal mass. Suppose a neutron with speed 2×10^7 m/s strikes a free, stationary proton head on. What will be the final speed of the neutron? Repeat if it collides elastically with a free, stationary gold atom.

14. What is the ratio of the mass of a 500-GeV proton to its rest mass? What is the value of v/c for the particle?

15. Find the ratios v/c and m/m_0 for a 50-GeV electron.

16. Iodine 131 is used to treat thyroid disorders because, when ingested, it localizes in the thyroid gland. Its half-life is 8.1 days. What is the activity of 1 μg of iodine 131?

17. Phosphorous 32 has a half-life of 14.3 days and is used in medicine because it tends to localize in bone. What is the activity of 1 g of phosphorous 32?

18. How many grams of iron 59 are there in a 1-mCi sample of it? Its half-life is 46.3 days.

19. The isotope tritium, $^3_1$H, has a half-life of 4600 days. How many grams of tritium are there in a sample that has an activity of 1×10^9 Bq?

20. How much is the temperature of water raised when given a radiation dose of 3 rads?

21. How large a radiation dose must be deposited in lead to raise its temperature 3°C? For lead, $c = 0.031$ cal/g · °C.

22. The yearly average exposure to medical x-rays in the U.S. is about 70 millirems per person. (*a*) To how many rads of 200-keV x-rays is this equivalent? (*b*) To how many rads of 5-MeV protons?

23. To give a lethal dose of 500 rem, how long would one need to be exposed to 4-MeV gamma rays that give a dose of 20 rad/s?

24. One step in the radioactive series of figure 37.10 was

$$^{222}_{86}\text{Rn} \quad \rightarrow \quad ^{218}_{84}\text{Po} \quad + \quad ^4_2\text{He} \quad + \text{energy}$$
$$222.01753 \qquad 218.00893 \qquad 4.00260$$

The isotope masses (including electrons) are given below the reaction. Assuming all the energy is given to the alpha particle, what should be its energy in MeV? The observed energy is 5.49 MeV for the fastest alpha particle. Where might the rest of the energy have gone?

25. Uranium 238 has a half-life of about 4.5×10^9 years and decays according to the reaction

$$^{238}_{92}U \quad \rightarrow \quad ^{234}_{90}\text{Th} \quad + \quad ^4_2\text{He} \quad + \text{energy}$$
$$238.05077 \qquad 234.04358 \qquad 4.00260$$

The isotope masses (including electrons) are also given. Assuming all the energy becomes kinetic energy of the alpha particle, what will its energy be in MeV? Its actual energy is 4.19 MeV. How can you account for the discrepancy?

26. When polonium 209 decays by alpha-particle emission, it also emits an 0.80-MeV gamma ray together with the 5.30-MeV alpha particle. The reaction is

$$^{210}_{84}\text{Po} \rightarrow \quad ^4_2\text{He} \quad + \quad ^{206}_{82}\text{Pb} \quad + \gamma$$
$$? \qquad 4.00260 \qquad 205.97447$$

with the atomic masses shown. (*a*) Knowing the kinetic energy of the α particle to be 5.30 MeV, find the approximate recoil energy of the lead atom. (*b*) Calculate the expected atomic mass for ^{210}Po. The measured mass is 209.9829 u.

27. It is desired to split a deuteron into a proton and neutron by means of a γ-ray. What is the minimum energy required, and what wavelength must the γ-ray have? (Masses of the nuclei of the deuteron, proton, and neutron are 2.0136, 1.0073, and 1.0087 u.)

28. Show that pair production, the formation of an electron and positron from a γ-ray, cannot take place in free space since in that case both energy and momentum cannot be conserved. One exception exists to this rule. What is it?

29. An approximate form for the force between nucleons is e^{-kr}, with $k = 10^{15}$ per meter. At what value of r will the force have fallen to 10^{-10} of its original value?

30. An electron and a positron having negligible kinetic energy combine with each other and disintegrate into two photons. What energy and wavelength are associated with each photon? What will be their relative directions of motion?

31. A positron and electron approach each other with equal energies, 0.50 MeV, and collide head on. Find the energies and wavelengths of each of the two γ-rays which result when they annihilate each other.

32. Suppose 1 kg of deuterium (heavy hydrogen, ^{2}H) is combined to form 1 kg of helium according to the reaction

$$\underset{2.0141}{{}^{2}_{1}\text{H}} \quad + \quad \underset{2.0141}{{}^{2}_{1}\text{H}} \quad \rightarrow \quad \underset{4.0026}{{}^{4}_{2}\text{He}}$$

where the atomic masses are given. (a) How much energy is liberated? (b) If the confined helium has a specific heat capacity of 0.75 cal/g · °C, by how much does its temperature increase as this energy is added to it?

33. Neutrons are frequently detected by allowing them to be captured by a boron nucleus. This reaction is

$$\underset{1.0087}{{}^{1}_{0}n} \quad + \quad \underset{10.0129}{{}^{10}_{5}\text{B}} \quad \rightarrow \quad \underset{7.0160}{{}^{7}_{3}\text{Li}} \quad + \quad \underset{4.0026}{{}^{4}_{2}\text{He}}$$

where the mass of the atoms is given without subtracting the electron masses. Energy given off in the reaction appears as kinetic energy of the products. The resulting swiftly moving helium ions (α particles) are then counted, using ordinary techniques. Compute the speed of the α particle, assuming the original reactants to be essentially at rest. (*Hint:* Compute the energy liberated in the reaction, and then write down the laws of conservation of energy and momentum.)

Appendix 1
Conversion Factors

Conversions which are exact are signified by an asterisk *. For brevity, the conversion factors are written in a dimensionless form. To use them, notice that the symbolism 2.54 cm/in means there are 2.54 centimeters in 1 inch. A quantity multiplied by any one of these factors changes only the units of the quantity. For example,

$$30 \text{ in} = (30 \text{ in})\left(2.54\frac{\text{cm}}{\text{in}}\right) = 76.2 \text{ cm}$$

and

$$5 \text{ cm} = (5 \text{ cm})\left(\frac{1}{2.54 \text{ cm/in}}\right) = 1.97 \text{ in}$$

Length
* 2.54 cm/in
* 0.3048 m/ft
* 1.609344 km/mi
  9.461 $\times$ 10^{15} m/light-year

Time
* 86,400 s/day
  3.16 $\times$ 10^7 s/year

Mass
1.6606 $\times$ 10^{-27} kg/u
6.022 $\times$ 10^{26} u/kg

Speed
* 0.3048 (m/s)/(ft/s)
  3.6 (km/h)/(m/s)
  60 (mi/h)/88 (ft/s)
  1.609 (km/h)/(mi/h)

Force
* 10^5 dyn/N
  4.45 N/lb
  0.225 lb/N
  1 kg weighs 2.21 lb at $g = 9.80$ m/s^2

Pressure
* 1 Pa/(N/m^2)
* 1.01325 $\times$ 10^5 (Pa)/atm
* 1.01325 bars/atm
* 0.10 (Pa)/(dyn/cm^2)
  6.895 $\times$ 10^3 (Pa)/(lb/in^2)
  133.32 (Pa)/mmHg at 0°C
  76 cmHg/atm
  14.7 (lb/in^2)/atm
  1 torr/mmHg

Work and Energy
* 10^7 ergs/J
* 4.184 J/cal
* 3.60 $\times$ 10^6 J/kWh
* 1054 J/Btu
  1.6022 $\times$ 10^{-19} J/eV
  6.242 $\times$ 10^{18} eV/J
  0.239 cal/J
  0.738 ft $\cdot$ lb/J
  23.06 (kcal/mol)/(eV/molecule)
  1 u $\rightarrow$ 931.5 MeV

Power
746 W/hp
550 (ft $\cdot$ lb/s)/hp

Electrical
1.6022 $\times$ 10^{-19} C/electron charge
96,485 C/faraday
* 10^4 G/T

Appendix 2
Physical Constants and Data

Speed of light	$c = 2.997925 \times 10^8 \, \text{m/s}$
Gravitational constant	$G = 6.67 \times 10^{-11} \, \text{N} \cdot \text{m}^2/\text{kg}^2$
Avogadro's number	$N_A = 6.022 \times 10^{26} \, \text{particles/kmol}$
Boltzmann's constant	$k = 1.38066 \times 10^{-23} \, \text{J/K}$
Gas constant	$R = 8314 \, \text{J/kmol} \cdot \text{K}$
	$= 1.9872 \, \text{kcal/kmol} \cdot \text{K}$
Planck's constant	$h = 6.6262 \times 10^{-34} \, \text{J} \cdot \text{s}$
Electron charge	$e = 1.60219 \times 10^{-19} \, \text{C}$
Electron rest mass	$m_e = 9.1095 \times 10^{-31} \, \text{kg}$
	$= 5.486 \times 10^{-4} \, \text{u}$
Proton rest mass	$m_p = 1.6726 \times 10^{-27} \, \text{kg}$
	$= 1.007276 \, \text{u}$
Neutron rest mass	$m_n = 1.6749 \times 10^{-27} \, \text{kg}$
	$= 1.008665 \, \text{u}$
Permittivity constant	$\epsilon_0 = 8.85419 \times 10^{-12} \, \text{C}^2/\text{N} \cdot \text{m}^2$
Permeability constant	$\mu_0 = 4\pi \times 10^{-7} \, \text{N/A}^2$
Standard gravitational acceleration	$g = 9.80665 \, \text{m/s}^2 = 32.17 \, \text{ft/s}^2$
Mass of earth	$5.98 \times 10^{24} \, \text{kg}$
Average radius of earth	$6.37 \times 10^6 \, \text{m}$
Average density of earth	$5.57 \, \text{g/cm}^3$
Average earth-moon distance	$3.84 \times 10^8 \, \text{m}$
Average earth-sun distance	$1.496 \times 10^{11} \, \text{m}$
Mass of sun	$1.99 \times 10^{30} \, \text{kg}$
Radius of sun	$7 \times 10^8 \, \text{m}$
Sun's radiation intensity at the earth	$0.032 \, \text{cal/cm}^2 \cdot \text{s} = 0.134 \, \text{J/cm}^2 \cdot \text{s}$

A3.1 POWERS OF TEN

It is often inconvenient to write numbers such as 1,420,000,000 and 0.00031. These numbers can be written $(1.42)(10^9)$ and $(3.1)(10^{-4})$, as we now show.
  Let us consider the following identities:

$$10 = 10^1$$
$$100 = (10)(10) = 10^2$$
$$1000 = (10)(10)(10) = 10^3$$
$$1,000,000,000 = 10^9$$

If we wish to write the number 4,561,000,000, we have at once that this number is equivalent to multiplying 4.561 nine times by 10. (Each time we multiply by 10 we move the decimal point one place to the right.) Hence

$$4,561,000,000 = (4.561)(10^9)$$

In general, if we move the decimal place on a number q places to the left, we must multiply the number by 10^q if it is to remain unchanged. For the example just given, $q = 9$.
  The procedure for writing numbers less than unity is somewhat similar. We make use of the fact that

$$0.1 = \frac{1}{10} = 10^{-1} = (1)(10^{-1})$$

$$0.01 = \frac{1}{100} = \frac{1}{10^2} = 10^{-2} = (1)(10^{-2})$$

$$0.001 = \frac{1}{1000} = \frac{1}{10^3} = 10^{-3} = (1)(10^{-3})$$

$$0.000,000,01 = \frac{1}{10^8} = 10^{-8} = (1)(10^{-8})$$

Clearly, when we move the decimal point q places to the right, we must multiply by 10^{-q}. In the case originally treated, $0.00031 = (3.1)(10^{-4})$, q was 4, since the decimal point was moved four places.

A3.2 CONVERSION OF UNITS

Suppose we want to know how many inches are equivalent to 2 mi. Problems such as this are easily done provided we know the proper conversion factors. In this case we know there are 5280 ft/mi and 12 in/ft. We then proceed as follows:

$$s = 2\,\text{mi}$$
$$= (2\,\text{mi})\left(5280\,\frac{\text{ft}}{\text{mi}}\right) = 10,560\,\text{ft}$$
$$= (10,560\,\text{ft})\left(12 \cdot \frac{\text{in}}{\text{ft}}\right) = 126,720\,\text{in}$$

Notice that the units of the conversion factors are treated like algebraic symbols.
  As another example, let us find the number of hours in 60,000 s.

$$t = 60{,}000\,\text{s}$$

$$= (60{,}000\,\cancel{s})\left(\frac{1}{60}\,\frac{\min}{\cancel{s}}\right) = 1000\,\min$$

$$= (1000\,\cancel{\min})\left(\frac{1}{60}\,\frac{\text{hr}}{\cancel{\min}}\right) = 16.7\,\text{h}$$

Here we have made use of the factor 60 s/min, which has been inverted to give $\frac{1}{60}$ min/s.

A3.3 TRIGONOMETRY

In this text we need only the basic definitions of trigonometry, together with a very few relations between functions. As shown in figure A3.1, the sine, cosine, and tangent functions are defined in terms of a right triangle.

$$\sin\theta = \frac{a}{c} \qquad \cos\theta = \frac{b}{c} \qquad \tan\theta = \frac{a}{b}$$

Although we seldom use them, the cosecant, secant, and cotangent are

$$\csc\theta = \frac{c}{a} \qquad \sec\theta = \frac{c}{b} \qquad \cot\theta = \frac{b}{a}$$

From the fact that in any right triangle such as the one shown, $a^2 + b^2 = c^2$, we have upon substitution

$$\sin^2\theta + \cos^2\theta = 1$$

A few other formulas are also used in the text, and we list them here:

$$\sin(x \pm y) = \sin x \cos y \pm \cos x \sin y$$
$$\cos(x \pm y) = \cos x \cos y \mp \sin x \sin y$$
$$\sin 2\theta = 2\sin\theta\cos\theta$$
$$\cos 2\theta = \cos^2\theta - \sin^2\theta = 1 - 2\sin^2\theta$$

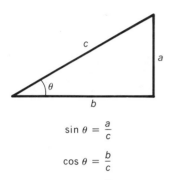

$$\sin\theta = \frac{a}{c}$$

$$\cos\theta = \frac{b}{c}$$

$$\tan\theta = \frac{a}{b}$$

FIGURE A3.1

A3.4 SERIES EXPANSIONS
($|x| < 1$ unless otherwise indicated)

$$(1 \pm x)^n = 1 \pm nx + \frac{(n)(n-1)}{2!}x^2 \pm \frac{n(n-1)(n-2)}{3!}x^3 + \cdots$$

$$(1 \pm x)^{-n} = 1 \mp nx + \frac{n(n+1)}{2!}x^2 \mp \frac{n(n+1)(n+2)}{3!}x^3 + \cdots$$

$$(1 \pm x)^{1/2} = 1 \pm \tfrac{1}{2}x - \tfrac{1}{8}x^2 \pm \tfrac{1}{16}x^3 - \cdots$$

$$(1 \pm x)^{-1/2} = 1 \mp \tfrac{1}{2}x + \tfrac{3}{8}x^2 \mp \tfrac{5}{16}x^3 + \cdots$$

$$\left.\begin{aligned}
\sin x &= x - \frac{x^3}{3!} + \frac{x^5}{5!} - \cdots \\[4pt]
\cos x &= 1 - \frac{x^2}{2!} + \frac{x^4}{4!} - \cdots \\[4pt]
e^x &= 1 + x + \frac{x^2}{2!} + \frac{x^3}{3!} + \cdots
\end{aligned}\right\} \text{ for all real } x$$

$$\ln(1 \pm x) = \pm x - \tfrac{1}{2}x^2 \pm \tfrac{1}{3}x^3 - \tfrac{1}{4}x^4 \pm \cdots$$

Appendix 4
Trigonometric Functions

angle (deg)	sine	cosine	tangent	angle (deg)	sine	cosine	tangent
0°	0.000	1.000	0.000				
1°	.018	1.000	.018	46°	0.719	0.695	1.036
2°	.035	0.999	.035	47°	.731	.682	1.072
3°	.052	.999	.052	48°	.743	.669	1.111
4°	.070	.998	.070	49°	.755	.656	1.150
5°	.087	.996	.088	50°	.766	.643	1.192
6°	.105	.995	.105	51°	.777	.629	1.235
7°	.122	.993	.123	52°	.788	.616	1.280
8°	.139	.990	.141	53°	.799	.602	1.327
9°	.156	.988	.158	54°	.809	.588	1.376
10°	.174	.985	.176	55°	.819	.574	1.428
11°	.191	.982	.194	56°	.829	.559	1.483
12°	.208	.978	.213	57°	.839	.545	1.540
13°	.225	.974	.231	58°	.848	.530	1.600
14°	.242	.970	.249	59°	.857	.515	1.664
15°	.259	.966	.268	60°	.866	.500	1.732
16°	.276	.961	.287	61°	.875	.485	1.804
17°	.292	.956	.306	62°	.883	.470	1.881
18°	.309	.951	.325	63°	.891	.454	1.963
19°	.326	.946	.344	64°	.899	.438	2.050
20°	.342	.940	.364	65°	.906	.423	2.145
21°	.358	.934	.384	66°	.914	.407	2.246
22°	.375	.927	.404	67°	.921	.391	2.356
23°	.391	.921	.425	68°	.927	.375	2.475
24°	.407	.914	.445	69°	.934	.358	2.605
25°	.423	.906	.466	70°	.940	.342	2.747
26°	.438	.899	.488	71°	.946	.326	2.904
27°	.454	.891	.510	72°	.951	.309	3.078
28°	.470	.883	.532	73°	.956	.292	3.271
29°	.485	.875	.554	74°	.961	.276	3.487
30°	.500	.866	.577	75°	.966	.259	3.732
31°	.515	.857	.601	76°	.970	.242	4.011
32°	.530	.848	.625	77°	.974	.225	4.331
33°	.545	.839	.649	78°	.978	.208	4.705
34°	.559	.829	.675	79°	.982	.191	5.145
35°	.574	.819	.700	80°	.985	.174	5.671
36°	.588	.809	.727	81°	.988	.156	6.314
37°	.602	.799	.754	82°	.990	.139	7.115
38°	.616	.788	.781	83°	.993	.122	8.144
39°	.629	.777	.810	84°	.995	.105	9.514
40°	.643	.766	.839	85°	.996	.087	11.43
41°	.658	.755	.869	86°	.998	.070	14.30
42°	.669	.743	.900	87°	.999	.052	19.08
43°	.682	.731	.933	88°	.999	.035	28.64
44°	.695	.719	.966	89°	1.000	.018	57.29
45°	.707	.707	1.000	90°	1.000	.000	∞

Appendix 5
Table of Exponential Functions

x	e^x	e^{-x}	x	e^x	e^{-x}
0.00	1.0000	1.00000	2.10	8.1662	0.12246
0.01	1.0101	0.99005	2.20	9.0250	0.11080
0.02	1.0202	0.98020	2.30	9.9742	0.10026
0.03	1.0305	0.97045	2.40	11.023	0.09072
0.04	1.0408	0.96079	2.50	12.182	0.08208
0.05	1.0513	0.95123	2.60	13.464	0.07427
0.06	1.0618	0.94176	2.70	14.880	0.06721
0.07	1.0725	0.93239	2.80	16.445	0.06081
0.08	1.0833	0.92312	2.90	18.174	0.05502
0.09	1.0942	0.91393	3.00	20.086	0.04979
0.10	1.1052	0.90484	3.10	22.198	0.04505
0.20	1.2214	0.81873	3.20	24.533	0.04076
0.30	1.3499	0.74082	3.30	27.113	0.03688
0.40	1.4918	0.67032	3.40	29.964	0.03337
0.50	1.6487	0.60653	3.50	33.115	0.03020
0.60	1.8221	0.54881	3.60	36.598	0.02732
0.70	2.0138	0.49659	3.70	40.447	0.02472
0.80	2.2255	0.44933	3.80	44.701	0.02237
0.90	2.4596	0.40657	3.90	49.402	0.02024
1.00	2.7183	0.36788	4.00	54.598	0.01832
1.10	3.0042	0.33287	4.10	60.340	0.01657
1.20	3.3201	0.30119	4.20	66.686	0.01500
1.30	3.6693	0.27253	4.30	73.700	0.01357
1.40	4.0552	0.24660	4.40	81.451	0.01228
1.50	4.4817	0.22313	4.50	90.017	0.01111
1.60	4.9530	0.20190	4.60	99.484	0.01005
1.70	5.4739	0.18268	4.70	109.95	0.00910
1.80	6.0496	0.16530	4.80	121.51	0.00823
1.90	6.6859	0.14957	4.90	134.29	0.00745
2.00	7.3891	0.13534	5.00	148.41	0.00674

Appendix 6A
Periodic Table of the Elements

The values listed are based on $^{12}_{6}C = 12$ u exactly. For artificially produced elements, the approximate atomic weight of the most stable isotope is given in brackets.

period	series	I	II	III	IV	V	VI	VII	VIII			0
1	1	1 H 1.00797										2 He 4.003
2	2	3 Li 6.939	4 Be 9.012	5 B 10.81	6 C 12.011	7 N 14.007	8 O 15.9994	9 F 19.00				10 Ne 20.183
3	3	11 Na 22.990	12 Mg 24.31	13 Al 26.98	14 Si 28.09	15 P 30.974	16 S 32.064	17 Cl 35.453				18 Ar 39.948
4	4	19 K 39.102	20 Ca 40.08	21 Sc 44.96	22 Ti 47.90	23 V 50.94	24 Cr 52.00	25 Mn 54.94	26 Fe 55.85	27 Co 58.93	28 Ni 58.71	
	5	29 Cu 63.54	30 Zn 65.37	31 Ga 69.72	32 Ge 72.59	33 As 74.92	34 Se 78.96	35 Br 79.909				36 Kr 83.80
5	6	37 Rb 85.47	38 Sr 87.62	39 Y 88.905	40 Zr 91.22	41 Nb 92.91	42 Mo 95.94	43 Tc [98]	44 Ru 101.1	45 Rh 102.905	46 Pd 106.4	
	7	47 Ag 107.870	48 Cd 112.40	49 In 114.82	50 Sn 118.69	51 Sb 121.75	52 Te 127.60	53 I 126.90				54 Xe 131.30
6	8	55 Cs 132.905	56 Ba 137.34	57-71 Lanthanide series*	72 Hf 178.49	73 Ta 180.95	74 W 183.85	75 Re 186.2	76 Os 190.2	77 Ir 192.2	78 Pt 195.09	
	9	79 Au 196.97	80 Hg 200.59	81 Tl 204.37	82 Pb 207.19	83 Bi 208.98	84 Po [210]	85 At [210]				86 Rn [222]
7	10	87 Fr [223]	88 Ra [226]	89-103 Actinide series†								

*Lanthanide series

57 La 138.91	58 Ce 140.12	59 Pr 140.91	60 Nd 144.24	61 Pm [147]	62 Sm 150.35	63 Eu 152.0	64 Gd 157.25	65 Tb 158.92	66 Dy 162.50	67 Ho 164.93	68 Er 167.26	69 Tm 168.93	70 Yb 173.04	71 Lu 174.97

†Actinide series

89 Ac [227]	90 Th 232.04	91 Pa [231]	92 U 238.03	93 Np [237]	94 Pu [242]	95 Am [243]	96 Cm [247]	97 Bk [247]	98 Cf [251]	99 E [254]	100 Fm [253]	101 Md [256]	102 No [254]	103 Lw [257]

Appendix 6B
An Abbreviated Table of Isotopes

The values listed are based on $^{12}_{6}C = 12$ u exactly. Electron masses are included.

atomic number Z	symbol	average atomic mass	element	mass number A	relative abundance (%)	mass of isotope
1	H	1.00797	Hydrogen	1	99.985	1.007825
				2	0.015	2.014102
2	He	4.0026	Helium	3	0.00015	3.016030
				4	100—	4.002604
3	Li	6.939	Lithium	6	7.52	6.015126
				7	92.48	7.016005
4	Be	9.0122	Beryllium	9	100—	9.012186
5	B	10.811	Boron	10	19.78	10.012939
				11	80.22	11.009305
6	C	12.01115	Carbon	12	98.892	12.0000000
				13	1.108	13.003354
7	N	14.0067	Nitrogen	14	99.635	14.003074
				15	0.365	15.000108
8	O	15.9994	Oxygen	16	99.759	15.994915
				17	0.037	16.999133
				18	0.204	17.999160
9	F	18.9984	Fluorine	19	100	18.998405
10	Ne	20.183	Neon	20	90.92	19.992440
				22	8.82	21.991384
11	Na	22.9898	Sodium	23	100—	22.989773
12	Mg	24.312	Magnesium	24	78.60	23.985045
13	Al	26.9815	Aluminum	27	100	26.981535
14	Si	28.086	Silicon	28	92.27	27.976927
				30	3.05	29.973761
15	P	30.9738	Phosphorus	31	100	30.973763
16	S	32.064	Sulfur	32	95.018	31.972074
17	Cl	35.453	Chlorine	35	75.4	34.968854
				37	24.6	36.965896
18	Ar	39.948	Argon	40	996	39.962384
19	K	39.102	Potassium	39	93.08	38.963714
20	Ca	40.08	Calcium	40	96.97	39.962589
21	Sc	44.956	Scandium	45	100	44.955919
22	Ti	47.90	Titanium	48	73.45	47.947948
23	V	50.942	Vanadium	51	99.76	50.943978
24	Cr	51.996	Chromium	52	83.76	51.940514
25	Mn	54.9380	Manganese	55	100	54.938054
26	Fe	55.847	Iron	56	91.68	55.934932

atomic number Z	symbol	average atomic mass	element	mass number A	relative abundance (%)	mass of isotope
27	Co	58.9332	Cobalt	59	100	58.93319
28	Ni	58.71	Nickel	58	67.7	57.93534
				60	26.23	59.93032
29	Cu	63.54	Copper	63	69.1	62.92959
30	Zn	65.37	Zinc	64	48.89	63.92914
31	Ga	69.72	Gallium	69	60.2	68.92568
32	Ge	72.59	Germanium	74	36.74	73.92115
33	As	74.9216	Arsenic	75	100	74.92158
34	Se	78.96	Selenium	80	49.82	79.91651
35	Br	79.909	Bromine	79	50.52	78.91835
36	Kr	83.30	Krypton	84	56.90	83.91150
37	Rb	85.47	Rubidium	85	72.15	84.91171
38	Sr	87.62	Strontium	88	82.56	87.90561
39	Y	88.905	Yttrium	89	100	88.90543
40	Zr	91.22	Zirconium	90	51.46	89.90432
41	Nb	92.906	Niobium	93	100	92.90602
42	Mo	95.94	Molybdenum	98	23.75	97.90551
43	Tc	*	Technetium	98		97.90730
44	Ru	101.07	Ruthenium	102	31.3	101.90372
45	Rh	102.905	Rhodium	103	100	102.90480
46	Pd	106.4	Palladium	106	27.2	105.90320
47	Ag	107.870	Silver	107	51.35	106.90497
48	Cd	112.40	Cadmium	114	28.8	113.90357
49	In	114.82	Indium	115	95.7	114.90407
50	Sn	118.69	Tin	120	32.97	119.90213
51	Sb	121.75	Antimony	121	57.25	120.90375
52	Te	127.60	Tellurium	130	34.49	129.90670
53	I	126.9044	Iodine	127	100	126.90435
54	Xe	131.30	Xenon	132	26.89	131.90416
55	Cs	132.905	Cesium	133	100	132.90509
56	Ba	137.34	Barium	138	71.66	137.90501
57	La	138.91	Lanthanum	139	99.911	138.90606
58	Ce	140.12	Cerium	140	88.48	139.90528
59	Pr	140.907	Praseodymium	141	100	140.90739
60	Nd	144.24	Neodymium	144	23.85	143.90998
61	Pm	*	Promethium	145		144.91231
62	Sm	150.35	Samarium	152	26.63	151.91949
63	Eu	151.96	Europium	153	52.23	152.92086
64	Gd	157.25	Gadolinium	158	24.87	157.92410
65	Tb	158.924	Terbium	159	100	158.92495
66	Dy	162.50	Dysprosium	164	28.18	163.92883

atomic number Z	symbol	average atomic mass	element	mass number A	relative abundance (%)	mass of isotope
67	Ho	164.930	Holmium	165	100	164.93030
68	Er	167.26	Erbium	166	33.41	165.93040
69	Tm	168.934	Thulium	169	100	168.93435
70	Yb	173.04	Ytterbium	174	31.84	173.93902
71	Lu	174.97	Lutetium	175	97.40	174.94089
72	Hf	178.49	Hafnium	180	35.44	179.94681
73	Ta	180.948	Tantalum	181	100	180.94798
74	W	183.85	Tungsten	184	30.6	183.95099
75	Re	186.2	Rhenium	187	62.93	186.95596
76	Os	190.2	Osmium	192	41.0	191.96141
77	Ir	192.2	Iridium	193	61.5	192.96328
78	Pt	195.09	Platinum	195	33.7	194.96482
79	Au	196.967	Gold	197	100	196.96655
80	Hg	200.59	Mercury	202	29.80	201.97063
81	Tl	204.37	Thallium	205	70.50	204.97446
82	Pb	207.19	Lead	208	52.3	207.97664
83	Bi	208.980	Bismuth	209	100	208.98042
84	Po	[210]	Polonium	210		209.98287
85	At	*	Astatine	211		210.98750
86	Rn	*	Radon	211		210.99060
87	Fr	*	Francium	221		221.01418
88	Ra	[226]	Radium	226		226.02536
89	Ac	*	Actinium	225		225.02314
90	Th	[232.038]	Thorium	232	100	232.03821
91	Pa	[231]	Protactinium	231		231.03594
92	U	[238.03]	Uranium	233		233.03950
				235	0.715	235.04393
				238	99.28	238.05076
93	Np	*	Neptunium	239		239.05294
94	Pu	*	Plutonium	239		239.05216
95	Am	*	Americium	243		243.06138
96	Cm	*	Curium	245		245.06534
97	Bk	*	Berkelium	248		248.070305
98	Cf	*	Californium	249		249.07470
99	Es	*	Einsteinium	254		254.08811
100	Fm	*	Fermium	252		252.08265
101	Md	*	Mendelevium	255		255.09057
102	No	*	Nobelium	254		254
103	Lw	*	Lawrencium	257		257

* The atomic masses of unstable elements are not listed unless the isotope given constitutes the major isotope.

Appendix 7
Integration

The concepts of integration and differentiation, the basic tools of calculus, were conceived by Sir Isaac Newton (1642–1727). He needed these tools in order to describe the motion of the planets and show how their motion was a consequence of gravitational forces between material objects. Although you will study both of these processes in detail in your calculus courses, a brief introduction to integration at this time will facilitate our study of physics.

Integration is basically addition. It is convenient to think of it as being the addition of small areas. For that reason, we discuss it in terms of the area under a curve. Taking a very special case first, let us find the shaded area under the curve of figure A7.1 between the points $x = a$ and $x = b$. [We of course know this area is simply $G(b - a)$; so we can easily check our result.] Since we shall find it useful in more complex situations, let us compute the area in the following way.

Referring to figure A7.2, we have split the desired area into N rectangular subareas, each of width Δx. Calling the average value of $f(x)$ in the nth rectangle $f(x_n)$, we can write the area of the nth rectangle as $f(x_n)\,\Delta x$. [It may seem unreasonable to you to write $f(x_n)$ for the value of $f(x)$ in this region since it is obviously just G. However, we are looking forward to more complex cases we shall encounter later, and so we are formulating the solution in a somewhat more general fashion than is necessary in this particular case.] To find the total area I we simply add the areas of all these rectangles to obtain

$$\text{Total area} \equiv I = f(x_1)\,\Delta x + f(x_2)\,\Delta x + \cdots + f(x_n)\,\Delta x + \cdots + f(x_N)\,\Delta x$$

$$A7.1$$

This sum can be written in more compact form in terms of the summation sign Σ to give

$$\text{Total area} = I = \sum_{x_n=x_1}^{x_n=x_N} f(x_n)\,\Delta x \qquad A7.2$$

It is important to remember that this shorthand symbolism is simply the sum indicated.

Now if we make the width of the rectangles exceedingly small, i.e., if we let $\Delta x \to 0$, then we indicate this fact in the following way:

$$\text{Total area} = I = \lim_{\Delta x \to 0} \sum_{x_1}^{x_N} f(x_n)\,\Delta x \qquad A7.3$$

It is allowable to take Δx as small as we wish since the area in question will be equal to the sum of the areas of the rectangles, no matter how thin the

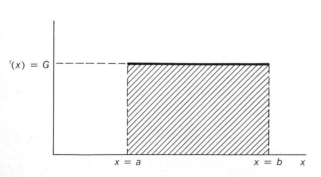

FIGURE A7.1

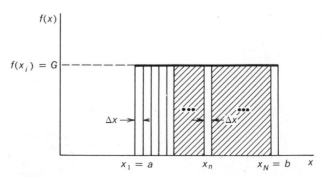

FIGURE A7.2

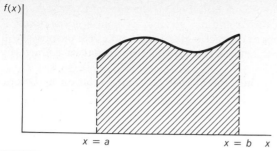

FIGURE A7.3

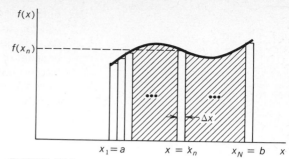

FIGURE A7.4

rectangles may be. The utility of letting $\Delta x \to 0$ is seen in our next example. For now we simply do it and state that, by definition, the quantity

$$I = \lim_{\Delta x \to 0} \sum_{x_1}^{x_N} f(x_n) \, \Delta x$$

is what we call the *integral of* $f(x_n)$ *from* x_1 *to* x_N. It is represented as follows:

$$\text{Total area} = I = \lim_{\Delta x \to 0} \sum_{x_1}^{x_N} f(x_n) \, \Delta x \equiv \int_{x_1}^{x_N} f(x) \, dx \qquad A7.4$$

The symbol $\int$ is called an *integral sign,* and the end values of x, namely, x_1 and x_N, are called the limits of integration. In our case $x_1 = a$ and $x_N = b$, and so the result is

$$\text{Total area} = I = \int_a^b f(x) \, dx \qquad A7.5$$

Do not be confused by the symbolism. The integral $\int_a^b f(x) \, dx$ is simply a sum, namely,

$$I = \int_a^b f(x) \, dx \equiv \lim_{\Delta x \to 0} [f(x_1) \, \Delta x + f(x_2) \, \Delta x + \cdots + f(x_N) \, \Delta x] \qquad A7.6$$

In our particular example, this sum is easily evaluated since $f(x_1) = f(x_2) = \cdots = f(x_N) = G$. Therefore

$$I = \lim_{\Delta x \to 0} (\Delta x + \Delta x + \cdots + \Delta x)G$$

But these Δx's are simply the bases of the rectangles from $x = a$ to $x = b$, and so their sum is nothing but $b - a$. As a result, the integral becomes

$$I = (b - a)G$$

which is the total area, as it should be.

Let us now apply this procedure to finding the area under a more complex curve, such as the one shown in figure A7.3. In the same way as before, we split the area into a series of rectangles as shown in figure A7.4. One important difference is noticed, however. Because of the curvature of the function $f(x)$'s curve, the tops of the rectangles do not quite coincide with the curve itself. As a result, the sum of areas of the rectangles will differ from the true area under the curve. However—and this is the important point—if the width of the rectangles Δx is taken very small, so that $\Delta x \to 0$, then the number of rectangles will approach infinity, and the tops of these rectangles will touch the curve at a nearly infinite number of points. As a result, if $\Delta x \to 0$, the sum of the areas of the rectangles, for all practical purposes, will be equal to the area under the curve. It is for this reason that we defined the integral only in this limit.

We therefore have in the general case the same relation as before, namely.

$$\text{Total area} = I = \lim_{\Delta x \to 0} \sum_{x_1}^{x_N} f(x_n) \, \Delta x$$

$$\equiv \int_a^b f(x) \, dx \qquad\qquad A7.7$$

As we see, our previous definition of the integral applies to any continuous function $f(x)$. The integral is, by definition, the area under the curve of $f(x)$ from $x = a$ to $x = b$. Or alternatively, it may be thought of as the value of the following sum:

$$I = \int_a^b f(x) \, dx \equiv \lim_{\Delta x \to 0} \sum_{x_n=a}^{x_n=b} f(x_n) \, \Delta x \qquad\qquad A7.8$$

The area under the curve of figure A7.4 can actually be found in several ways. It can be measured directly, for example. Or it could be computed directly from the sum if the value of $f(x)$ is known at a large number of values of x. In many cases, the sum can be arrived at by other mathematical means. Moreover, the value of the sum, the integral, may often be found in *integral tables,* provided the function $f(x)$ is not too complicated. A short table of integrals is given below. Most engineers and scientists own a more complete compilation of integrals. An excellent book of integrals is that of Dwight.*

When one opens a book of integrals one finds equations such as the following:

$$\int x^2 \, dx = \tfrac{1}{3}x^3$$

This listing is meant to tell us: If $f(x) = x^2$, then

$$\int_a^b f(x) \, dx = \tfrac{1}{3}b^3 - \tfrac{1}{3}a^3$$

As another case, one finds listed

$$\int \frac{dx}{1 + x} = \ln (1 + x)$$

which means
$$\int_a^b \frac{dx}{1 + x} = \ln (1 + b) - \ln (1 + a)$$

Or in terms of a sum, it means

$$\lim_{\Delta x \to 0} \sum_{x_n=a}^{x_n=b} \frac{\Delta x}{1 + x_n} = \ln (1 + b) - \ln (1 + a)$$

*H. B. Dwight, "Tables of Integrals," The Macmillan Company, New York, 1949.

Similarly,

$$\int \sin \theta \, d\theta = -\cos \theta$$

means

$$\int_a^b \sin \theta \, d\theta = -\cos b - (-\cos a)$$

As we see, tables of integrals are simply statements of the results for areas under curves for given functions $f(x)$. Or they may be interpreted as the values of sums such as those above. We refer to such tables to evaluate areas and sums we may encounter.

Some Useful Integrals

In the following equations c is a constant and n is an integer, with both + and − values allowed.

$$\int c \, dx = cx \qquad\qquad\qquad A7.9$$

$$\int cx \, dx = \frac{c}{2}x^2 \qquad\qquad\qquad A7.10$$

$$\int cx^n \, dx = \frac{c}{n+1}x^{n+1} \qquad\qquad\qquad A7.11$$

$$\int \frac{c \, dx}{x} = c \ln x \qquad\qquad\qquad A7.12$$

$$\int \frac{c \, dx}{x^2} = -\frac{c}{x} \qquad\qquad\qquad A7.13$$

$$\int \frac{c \, dx}{x^{n+1}} = -\frac{c}{nx^n} \qquad n \neq 0 \qquad A7.14$$

$$\int (x + c)^{n-1} \, dx = \frac{1}{n}(x + c)^n \qquad n \neq 0 \qquad A7.15$$

$$\int \sqrt{x} \, dx = \tfrac{2}{3}x^{3/2} \qquad\qquad\qquad A7.16$$

$$\int (x + c)^{n/2} \, dx = \frac{2(x + c)^{1+n/2}}{n + 2} \qquad n \neq 0 \qquad A7.17$$

$$\int e^{cx} \, dx = \frac{1}{c}e^{cx} \qquad\qquad\qquad A7.18$$

$$\int \cos cx \, dx = \frac{1}{c}\sin cx \qquad\qquad\qquad A7.19$$

$$\int \sin cx \, dx = -\frac{1}{c}\cos cx \qquad\qquad\qquad A7.20$$

Appendix 8A
Summary of Relativity Results

When you measure the position of an object, you do so relative to some fixed point or coordinate system. The coordinate system relative to which you take measurements is called the *reference frame* for your measurements. Most often, our reference frames are at rest on the surface of the earth, and, except for such small effects as the rotation of the earth about its axis and about the sun, the earth's surface can be considered a nonaccelerating reference frame. Reference frames that are not accelerating are called *inertial reference frames*. In them, Newton's law of inertia (an object at rest remains at rest, etc.) as well as all the other fundamental laws of physics apply.

Einstein based his theory of relativity on two basic postulates: (*a*) identical experiments in all inertial reference frames will give identical results, and (*b*) the speed of light in vacuum is c ($\cong 3 \times 10^8$ m/s) no matter what the motion of the light source may be. Reasoning from these two postulates he was able to prove the following:

1. No material object can be accelerated to speeds in excess of c. (The speed of light is a limiting speed.)
2. A clock of any type moving with speed v past an observer will appear to that observer to tick out a time $t_0 \sqrt{1 - (v/c)^2}$ during a time interval t_0 ticked out by a clock at rest relative to the observer. (Moving clocks tick too slowly.)
3. An object moving with speed v past an observer appears to that observer to be shortened along the line of motion by a factor $\sqrt{1 - (v/c)^2}$. (Length contraction.)
4. An object moving with speed v past an observer appears to that observer to have a mass $m = m_0 / \sqrt{1 - (v/c)^2}$, where m_0 is the *rest mass* of the object, its mass measured by an observer relative to whom the mass is at rest. As an object's speed approaches c, the mass of the object appears to approach infinity.
5. Mass is one form of energy. When an energy ΔE of any kind is given to an object, the object's mass increases by Δm where $\Delta E = (\Delta m)c^2$. A mass m is equivalent to an energy mc^2.
6. The kinetic energy of an object is $(m - m_0)c^2$ where $m = m_0 / \sqrt{1 - (v/c)^2}$. For $v \lll c$, this reduces to $\frac{1}{2}m_0 v^2$.
7. The momentum of an object is mv where $m = m_0 / \sqrt{1 - (v/c)^2}$.

It should be noted that all of these effects depend on the relativistic factor $\sqrt{1 - (v/c)^2}$. Only when $v \to c$ does this factor depart much from unity, and only then do the effects of relativity become large. At speeds less than about $0.1c$, the nonrelativistic mechanics we have been learning is usually a satisfactory approximation to reality.

The light sphere discussed in the text obeys the following equations in the stationary and moving frames, respectively:

$$x^2 + y^2 + z^2 = c^2t^2 \qquad\qquad 8.2$$

and

$$x'^2 + y'^2 + z'^2 = c^2t'^2 \qquad\qquad 8.3$$

We seek relations between the unprimed and primed coordinates.

As the simplest possibilities, let us assume (to be tested later) that

$$\begin{aligned}
x' &= P(x - vt) \\
t' &= Qt + Rx \\
y' &= y \\
z' &= z
\end{aligned} \qquad\qquad A8.1$$

where P, Q, and R are constants. If these transformations are correct, then 8.2 should result when they are substituted in 8.3.

If we substitute these values for x', y', and z' in 8.3, we obtain

$$P^2(x - vt)^2 + y^2 + z^2 = c^2(Qt + Rx)^2$$

Upon expanding and collecting like terms, we find

$$(P^2 - R^2c^2)x^2 + y^2 + z^2 + (P^2v^2 - Q^2c^2)t^2 - 2(QRc^2 + P^2v)xt = 0 \quad A8.2$$

This is to be compared with 8.2,

$$x^2 + y^2 + z^2 - c^2t^2 = 0 \qquad\qquad 8.2$$

Since these two equations must be true for all values of x^2, y^2, z^2, and t^2, the coefficients of each term in 8.2 must equal the corresponding coefficient in A8.2. Therefore

$$\begin{aligned}
P^2 - R^2c^2 &= 1 \\
P^2v^2 - Q^2c^2 &= -c^2 \\
QRc^2 + P^2v &= 0
\end{aligned}$$

Solving these three simultaneous equations for P, Q, and R, we find, writing $v/c \equiv \beta$,

$$P = Q = \frac{1}{\sqrt{1 - \beta^2}} \equiv \gamma$$

$$-R = P\left(\frac{v}{c^2}\right)$$

If we place these quantities in A8.1, we find

$$\begin{aligned}
x' &= \gamma(x - vt) \\
y' &= y \\
z' &= z
\end{aligned}$$

$$t' = \gamma\left(t - \frac{\beta x}{c}\right)$$

where $\beta = v/c$, and $\gamma = 1/\sqrt{1 - \beta^2}$. These equations tell the stationary observer how the moving coordinates, the primed coordinates, are related to the values of x, y, z, and t that he measures. They are called the Lorentz-Einstein relations. Although we have derived them in a somewhat arbitrary way (since we knew the answer before we started), the transformations can be proved to be unique, with no others possible.

Appendix 9
Waves on a Spring

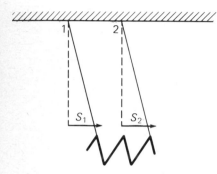

FIGURE A9.1

Power

A small section of the spring shown in figure 27.6 is redrawn in figure A9.1. As we see, this portion of the spring has been stretched an amount $s_2 - s_1$. Calling its original length Δx, its fractional increase in length is $(s_2 - s_1)/\Delta x$. We can define a stretching modulus for the spring in the following way:

$$k = \text{modulus} = \frac{F/A}{\Delta L/L_0} = \frac{T}{(s_2 - s_1)/\Delta x}$$

where T is the tension in the portion of the spring under consideration. The modulus so defined is simply the Hooke's law constant for unit length of the spring.

Solving this equation for T, we find that

$$T = k\frac{s_2 - s_1}{\Delta x}$$

But $(s_2 - s_1)/\Delta x$ is simply the rate of change of s with x, namely $\partial s/\partial x$. It is apparent, then, that the tension in the spring is given by

$$T = k\frac{\partial s}{\partial x}$$

To find the power transmitted down the spring, we find the work done by the tension at a given point. Consider the small element of the spring shown in figure A9.1. The tension at its left end does work on the element as s_1 is increased by an amount Δs. The work it does is

$$\Delta W = -T\,\Delta s = -k\frac{\partial s}{\partial x}\,\Delta s$$

The minus sign appears because the tension does negative work when the displacement increment is positive.

The displacement increment Δs is equal to $v_s\,dt$ where v_s is $\partial s/\partial t$. Therefore

$$\Delta W = -k\frac{\partial s}{\partial x}\frac{\partial s}{\partial t}\,dt$$

Knowing that $s = s_0 \sin\left[\omega(t - x/v)\right]$, we can take the appropriate derivatives of s and substitute in the equation for ΔW. The result is

$$\Delta W = \frac{ks_0{}^2\omega^2}{v}\cos^2\left[\omega\left(t - \frac{x}{v}\right)\right]dt$$

This expression is almost identical to that obtained in Sec. 27.4 for the similar problem involving energy transfer down a string. The only difference is the replacement of T by k. Proceeding exactly as was done there, we find that the power transmitted down the spring is

$$\text{Power} = 2\pi^2 v^2 s_0{}^2 \rho v$$

Wave Equation

To find the wave equation for the spring, refer to the portion of the spring shown in figure A9.1. We wish to write $F = ma$ for it. To do so, we recall that we have found the tension in the spring to be

$$T = k\frac{\partial s}{\partial x}$$

The net force on the element shown is

$$F = T_2 - T_1 = k\left[\left(\frac{\partial s}{\partial x}\right)_2 - \left(\frac{\partial s}{\partial x}\right)_1\right]$$

where T_1 and T_2 are the tensions at the left and right ends respectively of the element. This can be rewritten as

$$F = k\left[\frac{(\partial s/\partial x)_2 - (\partial s/\partial x)_1}{\Delta x}\right]\Delta x = k\frac{\partial^2 s}{\partial x^2}\Delta x$$

in the limit of Δx very small.

We now substitute this value for F in $F = ma$ with $m = \rho\,\Delta x$ and $a = \partial^2 s/\partial t^2$ and find

$$\frac{\rho}{k}\frac{\partial^2 s}{\partial t^2} = \frac{\partial^2 s}{\partial x^2}$$

This is the wave equation for compressional waves in a spring with modulus (or force constant per unit length) k and mass per unit length ρ.

Answers to Odd-Numbered Problems

Chapter 1

1. 3.407×10^3; 2.91×10^{-2}; 6.37×10^2; 1×10^4;
   1×10^{-4}; 1.00×10^{-4}; 1.37×10^{-4}
3. $4.2 \times 10^{-5}\,\text{m}$
5. $1 \times 10^{-4}\,°\text{C}$
7. $1.83\,\text{m}$
9. 3220 cubits
11. $[C_1] = LT^{-1}$, $[C_2] = LT^{-2}$, $[C_3] = L$
13. length/(time)^2; acceleration due to gravity
15. 5×10^7
17. 1.3×10^{23}
19. $8.5 \times 10^{22}\,\text{atoms/cm}^3$

Chapter 2

1. 31.8 m directed from start to end
3. (6.8 m, 18.8 m); (-10.0 m, 17.3 m);
   (-6.8 m, -18.8 m)
5. 5.39 blocks; 21.8°N of E
7. 72 N; 236.3°
9. 31.6 N; 145°
11. (-2 m, 1 m)
13. (-14, 11)
15. $-2\mathbf{i} + 3\mathbf{j} + 5\mathbf{k}$; 6.2
17. $-4\mathbf{i} + 6\mathbf{j} + 3\mathbf{k}$; 7.8
19. 166 N at 51°; 166 N at 231°
21. 35.5°
23. $-\mathbf{i} + 2.5\mathbf{j} - 3.5\mathbf{k}$
25. $2\mathbf{i} + 6\mathbf{j} - \mathbf{k}$

Chapter 3

1. 192 N
3. 158 N, 18.4°
5. 144 N, 81 N
7. $0.75W$
9. 500 N, 400 N, 670 N, 530 N
11. $T_1 = T_2 = T_3 = W_2 = W_1/(2 \cos \theta_1)$
13. ... (no answer required)
15. 22 m · N; zero
17. 700 N, 600 N
19. $1.80W$, $1.69W$, $1.62W$
21. 410 N; $H = 140$ N, $V = 300$ N
23. $2.77F$; $-0.45F$, $3.47F$
25. 36.4 N, 50 N, 30 N
27. $\theta = \arctan(d/2h)$

Chapter 4

1. 0.0222 mi/s; zero
3. 62 mi/h; zero
5. 6.7 m/min east; zero; 13.3 m/min west
7. 0.10 s; 300 m/s^2
9. 3.3×10^{-11} s; 2.4×10^{17} m/s^2
11. 17 s; 24 m/s; 204 m
13. 0.2 m/s^2; 828 m
15. -0.45 m/s^2, -0.22 m/s^2, -1.5 m/s^2; 130 m
17. 20.4 m; 2.04 s
19. 2.15 s; -16.1 m/s
21. 3.55 m/s; 0.64 m
23. 3.03 m; 2.02 s
25. 0.113 m
27. 3.1×10^{-16} m; 0.196 m
29. 22.3 m; 68.3 m
31. 120 m/s
33. ...
35. 2 m/s; -4 m/s^2; slowing
37. $16t$ m/s^2; 317 m/s
39. 13.1 m; $\frac{2}{3}$ maximum

Chapter 5

1. 20 N
3. 4800 N
5. 18.8 m/s
7. 1.43×10^{-13} N; 1.6×10^{16}
9. $\tan \theta = F_y/F_x$
11. 15.6 lb
13. 25.5 N
15. 77.3 N; 0.83
17. $0.45\,g = 4.41$ m/s^2
19. 0.53
21. 10 N; 8.5 N
23. $1/2 \tan \theta$
25. 2 m/s^2; 0.20
27. 3.9 N
29. 11.8 N; 0.5
31. 12 ft/s^2
33. $(F/2m) - (f/m)$
35. 1.3 kg; 12.2 N
37. 3.5 m/s^2 down the incline

Chapter 6

1. 1960 J
3. 58.8 J
5. ...
7. 7.35 m; 0.44 N; 9.5 m/s
9. 56 hp
11. 0.44 hp
13. 16.9 km/gal = 10.5 mi/gal
15. 3.64×10^{-14} N
17. 5000 N; 8 s; 0.255
19. 16.4 N
21. 12.5 m/s; 8.85 m/s
23. 1450 J
25. 73.5 J; zero
27. $v\sqrt{m/A}$
29. 3.7 m/s; 3.1 m/s
31. 6.2 N; 0.98 m/s
33. $7.3\mathbf{j}$ m/s; $11.8\mathbf{i} + 7.3\mathbf{j}$ m/s
35. 19.6 N/m
37. 0.294 J
39. $x_0\sqrt{k/M}$; $(x_0/3)\sqrt{5k/M}$
41. 0.124 m; 0.095 m
43. $\frac{1}{2}mv_f^2 - \frac{1}{2}mv_0^2$

Chapter 7

1. $m(v_f + v_0)$
3. $-mv_0\mathbf{i}$; $mv_f\mathbf{j}$; $m(v_f\mathbf{j} + v_0\mathbf{i})$
5. Mv; Mv/t_s; 25,000 N
7. 25 N · s; 125 J
9. $0.020t$ N · s; 0.020 N
11. 0.15 m/s
13. 15 m/s
15. $1.5v_0\mathbf{i} - v_0\mathbf{j}$
17. $-v_0(\mathbf{i} + 2\mathbf{j})$
19. $-Bv\mathbf{i}/(M + 2m)$; $Bv\mathbf{i}/(M + 2m)$; zero; $\pm Bv\mathbf{i}/(M + 2m)$
21. $Mv_0/(M + kt)$
23. $At_0^2/2$; $At_0^2/2M$
25. 1.26×10^{-10} m
27. $x = 1.62L$
29. ...
31. $L(2AL + 3B)/(3AL + 6B)$
33. 0.10 m
35. 2 u
37. velocity v_0 in opposite direction
39. $-\frac{1}{2}v_0\mathbf{i} + v_0\mathbf{j}$; no
41. 21 m/s
43. Zero; 1.85 m/s
45. 6.2×10^7 J; 1.11×10^4 m/s
47. 0.80 m
49. 8.75×10^6 m/s; 7.3×10^6 m/s
51. $-(20y/z)\mathbf{i} - (20x/z)\mathbf{j} + (20xy/z^2)\mathbf{k}$

Chapter 8

1. 5×10^{-13}
3. $1000 - 5 \times 10^{-10}$ s
5. 2048
7. $0.87c$
9. $0.95c$
11. Zero; $0.54c$
13. c; 37°
15. 0.946; 4 s
17. 1×10^{10} years; 6.1×10^{15} m; $0.988c$
19. $(1 - 5 \times 10^{-7})c$
21. 1.80×10^{-13} J; 1.13 MeV
23. 3.7×10^{-15} kg
25. 4.2×10^6
27. 22.3; $0.9990c$
29. ...

Chapter 9

1. 0.35 rad, 0.56 rev; 23°, 0.064 rev; 120°, 2.09 rad
3. 0.0167 rev/min; 1 rev/min; 1.0167 rev/min; 0.0236 rev/min northeast
5. 0.040 rev/s²; 8 rev
7. 2.25 rad/s²; 72 rev
9. 17.3 s; 1.09 rad/s²; 17.0 m/s; 320 m/s²; $\sqrt{(320)^2 + (0.98)^2}$ m/s²
11. 3.6×10^7 m/s²; 3.6×10^6
13. 10 m/s
15. 14 m/s
17. 0.55 deg
19. $T_1 = (\frac{1}{2}m/\sin\theta)(r\omega^2 + g\tan\theta)$; $T_2 = (\frac{1}{2}m/\sin\theta)(r\omega^2 - g\tan\theta)$
21. $-1.4M^2 \times 10^{-9}$ N
23. $GM[AL/a(a + L) + BL]$
25. $GM_e m/R_e$
27. 2740 N
29. 3.58×10^7 m
31. $RBt + RCt^2$; $RB + 2RCt$
33. $0.17R_e$
35. ...

Chapter 10

1. −16, 12; −16, 12; zero, 20
3. $-24\mathbf{i} - 18\mathbf{j} + 6\mathbf{k}$; $24\mathbf{i} + 18\mathbf{j} - 6\mathbf{k}$
5. $2.17b$; $1.13b$
7. $7ML^2/48$
9. $(7/5)Mr^2$
11. For a single axis, I for a group of objects equals the sum of the individual I's
13. 0.98 rad/s²
15. mgb/I; zero; $0.80mgb/I$
17. 0.178 kg · m²
19. 3.3 rad/s²

21. $b^2[(g/a)(m_1 - m_2) - (m_1 + m_2)]$
23. 3.23 m
25. 3.71 s
27. $(3g/2L) \cos \theta$; $(3g/2) \cos \theta$; $(3g/2) \cos^2 \theta$

Chapter 11
1. $1.98 \times 10^{-46} \text{ kg} \cdot \text{m}^2$; $1 \times 10^{12} \text{ rev/s}$; $2.6 \times 10^{33} \text{ J}$
3. 5300 J; $340 \text{ kg} \cdot \text{m}^2/\text{s}$; left
5. $14.1 \text{ N} \cdot \text{m}$
7. $1.4 \text{ kg} \cdot \text{m}^2$
9. 0.966
11. $0.78 \text{ kg} \cdot \text{m}^2$
13. ...
15. 0.86 rev/s
17. $1.02 \text{ N} \cdot \text{m}$
19. $a = F/M$; $\alpha = 2F/MR$
21. $v_p/v_a = r_a/r_p$
23. 0.75 rev/s; yes, decreased
25. $(R_0/R_s)^2 \omega_0$
27. 0.10 rev/s
29. 1.5 rad/s
31. $-\frac{1}{2}b$
33. $11R/4$
35. 1.18 m
37. 0.588 m; 0.96 m/s
39. $6\sqrt{gb/199}$; counterclockwise

Chapter 12
1. 1.25 Hz
3. 0.40 m; 0.111 Hz; 8.97 s; 17.2°; $0.49 \text{ N/kg} \cdot \text{m}$
5. $2 \cos (0.10t) \text{ rad/s}$; $-0.2 \sin (0.10t) \text{ rad/s}^2$
7. 0.58 m/s, -0.45 m/s^2; 0.60 m/s; zero
9. $y = 0.05 \cos (2.89t) \text{ m}$
11. $0.15 \cos (6\pi t) \text{ m}$; 53.3 m/s^2; 0; $53.3 \cos \theta \text{ m/s}^2$
13. ...
15. 0.40 s
17. $2\pi f = (a) \sqrt{k_1 k_2/m(k_1 + k_2)}$; $(b) \sqrt{2k_1/m}$;
    $(c) \sqrt{(k_1 + k_2)/m}$
19. 0.40 Hz
21. 0.102 kg; 0.102 kg; 0.788 Hz; 0.788 Hz
23. 1.42 Hz
25. $111 \text{ kg} \cdot \text{m}^2$; $36 \text{ kg} \cdot \text{m}^2$
27. $0.40 \sqrt{m_1/(m_1 + m_2)}$
29. $(1/2\pi) \sqrt{6k/m}$; $L\theta_0 \sqrt{3k/2m}$
31. Zero; $3x_0^4/8$

Chapter 13
1. $6.76 \times 10^{21} \text{ cm}^{-3}$
3. $7.1 \times 10^{20} \text{ kg/m}^3$
5. 6100 N
7. 6×10^{-7}
9. $1.87 \times 10^4 \text{ N/m}^2$

11. 9.1 cm
13. $\frac{1}{2}\rho_w gH^2L$; $\frac{1}{6}\rho_w gH^3L$
15. $4.6 \times 10^6 b^2 \text{ m/s}$
17. 880 kg/m^3; $2.58 \times 10^{-6} \text{ kg}$
19. 21.6 g (0.21 N)
21. 0.89
23. ...
25. ...
27. ...
29. $P_0 + 500[1 - (A_2/A_1)^2]v_2^2$; $v_2(A_2/A_1)$
31. 7.7 m/s; 3 m
33. $P_a + P - 0.65v^2$; $\sqrt{(P + \rho gh)/0.65}$
35. $(1/2\pi) \sqrt{A\rho_f g/M}$

Chapter 14
1. $2.82 \times 10^{-26} \text{ kg}$
3. $9.0 \times 10^{-26} \text{ kg}$; 2.7×10^{25}; 353 m/s
5. $1.40v_0$; v_0; $0.58v_0$
7. $2.5 \times 10^{-5} \text{ kg}$; $3.1 \times 10^{-8} \text{ m}^3$
9. $1.22 \times 10^4 \text{ m/s}$
11. $2.63V_0$
13. $v_0[1 + (\rho_w gh/P_a)]$
15. 176°C
17. $1.20P_0$
19. $7.3 \times 10^{-8} \text{ m}$
21. 540 m/s
23. $1.4 \times 10^5 \text{ K}$; $1 \times 10^4 \text{ K}$; $6 \times 10^3 \text{ K}$
25. $2.1 \times 10^{-22} \text{ kg}$; 4500

Chapter 15
1. 2400 cal; $1.00 \times 10^4 \text{ J}$; 264 cal; 1100 J
3. 1500 g
5. 75.7°C
7. 179 cal; 94°C
9. 0.0094°C
11. 87°C; 83°C
13. 1120 kg; $9.9 \times 10^5 \text{ cal}$
15. $0.172 \text{ cal/g} \cdot °\text{C}$
17. $0.062 \text{ cal/g} \cdot °\text{C}$; $0.087 \text{ cal/g} \cdot °\text{C}$
19. 43 g
21. 14.5 g
23. 0.083°C
25. 1.986 cm
27. ...
29. $0.99772\omega_0$; $0.99772(\frac{1}{2}I_0\omega_0^2)$
31. $3.4 \times 10^{-5} \text{ cal/g} \cdot °\text{C}$
33. 800 cal/s
35. 40,000Am

Chapter 16
1. 1670 J; 1670 J
3. $15 \times 10^5 \text{ J}$; 27°C

5. 169 J; 2086 J

7. 91 K

9. $-150°C$

11. 661

13. 1920 cal

15. 3650 cal

17. 41 hp

19. $-12,000$ J; zero; 3000 J; zero; 9000 J

21. $6P_0V_0$; $3P_0V_0$; about 0.18

23. $(P_0V_0/334)\ln(V_1/V_0)$ grams gained

25. $9 \times 10^{22}kT_1$; $9 \times 10^{22}kT_2$; $9 \times 10^{22}k(T_1 - T_2)$; zero

27. 9.9 cal/K

29. $mc\ln(T_2/T_1)$

31. $(mR/M)\ln(V_2/V_1)$; zero

33. . . .

35. 8; 32; 1.1×10^{15}

37. Zero; 2.22×10^{-23} J/K

Chapter 17

1. 9.2×10^{-8} N

3. 1.53×10^{-13}

5. $-1 \times 10^{12}Q^2\mathbf{i}$

7. 8.1×10^3 N/C at 124°; 1.3×10^{-15} N

9. $0.42b$

11. $1.58 \times 10^9 Q$; $-3.84 \times 10^8 Q$

13. $q = \frac{1}{2}Q$

15. $-9.6 \times 10^{20}\mathbf{i}$ N/C; 4.8×10^{-19} J; 3.0 eV

17. $(cL/4\pi\epsilon_0)\mathbf{i}$

19. $E_x = -\lambda/4\pi\epsilon_0 b$; $E_y = \lambda/4\pi\epsilon_0 b$

21. $(A/4\pi\epsilon_0)[(L/b) - \ln(1 + L/b)]$

23. Zero

25. $-A/8\epsilon_0 b$; zero

27. 3.5×10^5 m/s

29. $0.36mv_0{}^2/qE$; $0.96mv_0{}^2/qE$

31. $(Q/2\pi\epsilon_0 a^2)[1 - b/\sqrt{a^2 + b^2}]$

33. $(\lambda q/8\pi^2\epsilon_0 a^2 m)^{1/2}$

Chapter 18

1. Zero; zero; 200 A

3. Zero; zero; $2\pi BL$; $2\pi BL$

5. $Q/4\pi\epsilon_0 r^2$; zero; $Q/4\pi\epsilon_0 r^2$

7. $\lambda_1/2\pi\epsilon_0 r$; zero; $(\lambda_1 + \lambda_2)/2\pi\epsilon_0 r$; $-\lambda_1$; $\lambda_1 + \lambda_2$

9. $\rho b^2/2\epsilon_0 r$; $\rho r/2\epsilon_0$; at surface; on axis

11. σ/ϵ_0

13. . . .

15. $-\frac{1}{3}\mu C$; $\frac{1}{9}\mu C$

17. 0.01

19. $kb/\epsilon_0 r$; k/ϵ_0

21. $\sigma/2\epsilon_0$; $\sigma^2/2\epsilon_0$; attractive

23. . . .

Chapter 19

1. 25 V; S

3. Zero; $-20,000$ V

5. Negative; 3470 V; B

7. 20,000 eV; 8.4×10^7 m/s

9. -60 V; lower

11. 2×10^5 V/m; 2.4×10^5 V

13. $x = 18.75$ cm, -75.0 cm; ± 0.375 m; $x = -1.72$ m, $y = z = 0$

15. 1650 V; 5.6×10^5 m/s

17. 18,000 V; 18,000 V

19. 2.39 MV; 5.2×10^{-14} m

21. . . .

23. 1.44×10^{-15} m

25. $Aa/\pi\epsilon_0(a^2 + b^2)^{1/2}$

27. $-2Ax/(x^2 + y^2)$; $-2Ay/(x^2 + y^2)$; zero

29. . . .

31. $(\sigma/2\epsilon_0)[1 - b(a^2 + b^2)^{-1/2}]$

33. $(Q + Q')/4\pi\epsilon_0 r$; $(Q + Q')/4\pi\epsilon_0 b$;
$(1/4\pi\epsilon_0)[(Q/b) + (Q'/b) + (Q'/r) - (Q'/a)]$

35. $a(E_2 - E_1)$

Chapter 20

1. 1.25×10^{18}

3. 0.135 g

5. 1×10^{20}; 16 A

7. $3.2 \times 10^{-3} \Omega$; 0.096 V

9. 0.532Ω

11. 240Ω; 26.6Ω

13. 18.18 and 1.82Ω

15. . . .

17. 7.5 A; 16Ω

19. 14,000 cal; 1950 W; 16.3 A

21. $1.71 \times 10^{-3}°C/s$

23. 3.3×10^9 times

25. 46 s

27. 2.1 s; 0.69; 13.8 s; 4.6

29. $1.09 \mu F$

31. $2.0 \mu F$; 4.0 V

33. Zero

35. 9.6 and $14.4 \mu C$

37. $\frac{1}{2}CV^2$; $\frac{1}{2}CV^2$; CV^2

39. . . .

Chapter 21

1. 5, 6, 7, 1.2, 1.33, 1.71, 9, 3.71, 4.33, 5.2, 0.923Ω

3. 9.0Ω

5. 2.0 A; 0.40 A; 0.80 A

7. 12Ω; 0.50 A; 0.167 A

9. 1.49 A; 1.48 A; 0.0148 A

11. $r_1 - r_2$; $\mathcal{E}(1 - r_2/r_1)$

13. 12.0Ω; 0.50 A; 0.167 A

15. 7.0, -7.67, 0.67 A

ANSWERS TO ODD-NUMBERED PROBLEMS

17. 1.0, -1.0, -2.0, 0 A
19. 0.235, -0.454, 0.689 A
21. 476.2 Ω; 500.01 Ω
23. 8.17×10^7 A; 9.8×10^3 A; 41 cal/s

Chapter 22

1. 3×10^{-2} N, down; zero
3. $IBL \cos 40°$, in; zero; $IBL \cos 40°$, out
5. $2\pi \, aIB \sin \theta$ straight up
7. $ILB \sin \theta$ per m, upward
9. 8.9×10^{-18} N, east; 9.8×10^{12} m/s²
11. 0.36 m
13. 0.152 T
15. $(1/dq)\sqrt{2mK}$
17. 1.65°
19. 287.5 G
21. $0.246L^2I$ into page; $0.246L^2IB$ from A to C
23. $\frac{1}{2}ver = 9.3 \times 10^{-24}$ A · m²
25. $\pm \pi r^2 I(B_x \mathbf{j} - B_y \mathbf{i})$
27. 4.6 A
29. 0.0334 Ω in parallel
31. 1950 Ω in series
33.

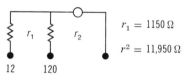

$r_1 = 1150 \, \Omega$

$r^2 = 11{,}950 \, \Omega$

12   120

35. 2.065; 2.6×10^8 m/s; 6.2 m

Chapter 23

1. 5×10^{-7} **k** T
3. Zero; 1×10^{-5} T
5. 8.0×10^{-6} **i** T
7. $2 \times 10^{-7} I_1/r$; zero
9. Along line through origin and in x, y plane at 63°
   to x axis
11. $2\pi m/\mu_0 qI$
13. Zero; $\mu_0 I(r^2 - a^2)/2\pi r(b^2 - a^2)$; $\mu_0 I/2\pi r$
15. . . .
17. $4\mu_0 I/\pi \sqrt{2c}$
19. $\frac{1}{2}\mu_0 I\left(\frac{1}{a} - \frac{1}{b}\right)$ into page; $0.44I \, \mu_0\left(\frac{1}{a} - \frac{1}{b}\right)$;

   $0.375I \, \mu_0\left(\frac{1}{a} - \frac{1}{b}\right)$

21. $0.21\mu_0 I/a$
23. $\mu_0 I_1 I_2 L/2\pi a$; zero
25. $500\mu_0 I^2 r^2/\pi b$; radially inward

27. $\dfrac{\mu_0 I}{2\pi(b^2 - a^2)}\left[\left(\dfrac{b^2}{R}\right) - \dfrac{a^2}{R \pm c}\right]$;

   $\dfrac{\mu_0 I}{2\pi(b^2 - a^2)}\left[R \pm \dfrac{a^2}{R \pm c}\right]$

29. Zero; $\mu_0 j$
31. 1.01×10^{-3} T

Chapter 24

1. $-0.2\omega NA \cos \theta \cos (\omega t)$; $\mathcal{E}/R$
3. Zero; zero
5. $BA \cos (2\pi ft)$; $(2\pi fBNA) \sin (2\pi ft)$
7. $\mu_0 i_0 nNA\omega \sin (\omega t)$; $\mu_0 nAN$
9. Zero; zero
11. $(\mu_0 ic/2\pi) \ln (1 + b/a)$; $(\mu_0 c/2\pi) \ln (1 + b/a)$
13. $(\mu_0 I/2\pi) \ln (b/a)$; $(\mu_0/2\pi) \ln (b/a)$
15. . . .
17. 10^{-5} s; 4.6×10^{-5} s
19. $0.63V/R$; $-0.37V$; $0.63V^2/R$; $0.40V^2/R$; $0.23V^2/R$
21. 0.274 V
23. (a) $IBbt/m$; (c) zero
25. $Bbgt \sin \theta \cos \theta$; ccw viewed from above
27. $N^2a^2B^2v/R$; to stop the motion
29. $(\pi\mu_0 nfi_0 r/\rho) \cos (2\pi ft)$
31. . . .
33. $q\mathbf{v} \times \mathbf{B}$; $\frac{1}{2}qr(dB/dt)$

Chapter 25

1. 7.4
3. 20 V
5. 36,000 V
7. Yes
9. . . .
11. 0.089 eV; 3.14 eV
13. . . .
15. . . .
17. 1.1×10^{-7} J
19. $\theta = \theta_0 \cos (\sqrt{pE/It})$
21. 1.4 T
23. 371; 370
25. . . .
27. $\boldsymbol{\mu}$ is perpendicular to **B**
29. k_m

Chapter 26

1. $v_0/\sqrt{2}$; $v_0/R\sqrt{2}$; $v_0^2/2R$
3. 0.20 A; 2.0×10^{-4} A
5. 3.1×10^{-8} F; 1.0 A
7. 19.7 V; 0.83 V
9. 0.0203 A; 89.93°; 2.5×10^{-3} W
11. 10 Ω; 0.393 H
13. 356 Hz; 2.0 A
15. 885 Hz; 0.033 H
17. 566 Ω; capacitor
19. . . .
21. . . .
23. 2.54×10^{-9} F; 0.20 Ω
25. . . .

Chapter 27

1. 0.02 m; 4.78 Hz; 7.5 m/s; 1.57 m
3. 0.06 m; 3750 Hz; 0.08 m
5. 0.175 s
7. $y = 2 \times 10^{-3} \sin (188t - 10.5x)$ m
9. 1326 m/s
11. 457 m/s
13. 1450 m/s
15. 66 dB; 69 dB
17. 0.029 Pa; 2.9×10^{-7}
19. $\sqrt{B/\rho} = \sqrt{P/\rho}$
21. 30 m/s
23. $2\sqrt{L/g}$

Chapter 28

1. 300 m; 300 m
3. $2 \times 10^{-4} \cos (5 \times 10^6 t + 0.017x)$
5. 1.6×10^6 Hz; 3×10^{-4} V
7. 1.33×10^{-13} W/m^2
9. $-3 \times 10^{-6} \cos (5 \times 10^6 t)$ V
11. $-1.08 \times 10^{-7} \cos (6 \times 10^6 t)$ V
13. ...
15. ...
17. ...
19. ...

Chapter 29

1. 40 cm
3. 8.0 cm; 16.0 cm; 12 cm; yes; -8 cm, no
5. 100 cm, 2 cm, yes, no; -200 cm, 10 cm, no, yes
7. $1.33f$; yes
9. $p = f$; no; yes; no
11. 15 cm, real, inverted; -10 cm; virtual, erect
13. -7.5 cm, virtual, erect; -3.3 cm, virtual, erect
15. 49°
17. $2R$ from surface in air; at the center
19. $4f > D$
21. 267 cm or 600 cm
23. Virtual; 15 cm from center

Chapter 30

1. 1.65 cm
3. $34.7 \sin (\omega t + 30°)$
5. 5.87×10^{-4} m
7. $\lambda_1/\lambda_2 = \frac{4}{3}$
9. $2y_0 \cos (kx) \sin (\omega t)$; zero
11. 1.32×10^{-6} m
13. 1.91 μm; 1.44 μm
15. 5.52×10^{-4} m
17. Zero or 1.52×10^{-7} m
19. $2.84 y_0 \sin (\omega t + 27°)$
21. $5.55 y_0 \sin (\omega t + 40°)$

23. 1.23 km
25. $\frac{1}{2} n\lambda$
27. Radius $= D \sqrt{(d/n\lambda)^2 - 1}$

Chapter 31

1. 6.76×10^{-6} m
3. 1.88×10^{-4} m
5. 0.044 m
7. 0.0296 m
9. 7.4 km
11. 5.2 cm
13. 0.15 mm
15. 1.03×10^{-6} m; impossible
17. 22.1°
19. 0.05 m
21. Refraction angle $= 38.6°$
23. ...
25. ...
27. ...

Chapter 32

1. About 3
3. ...
5. No; 4.1×10^{-5} rad
7. 586 m
9. 64
11. 17 cm; 3.3
13. 420 cm left of eyepiece; erect, virtual
15. 50 cm, 110 cm, and -32 cm from lens
17. ...
19. -60 cm
21. ...
23. From lens, 40, 70, and -28 cm
25. 1.5020, 5.20×10^5 Å^2; 1.5228
27. ...
29. ...
31. ...

Chapter 33

1. $198n$ Hz; 277 m/s
3. $300n$ and $150n$ Hz
5. 17 Hz; 32 cm; 54.4 m/s
7. $0.063n$ Hz
9. 142 Hz, 426 Hz, 710 Hz; 1.8% increase
11. 2400 m/s
13. 0.15 m; 0.628 m; 1.26 m; 47.7 Hz; 60 m/s
15. $0.05 + 0.10n$ m; $0.10n$ m
17. $nv/\pi D$
19. $\frac{1}{2}A \sin (\omega t - \omega x/v) - A \sin (\omega x/v) \cos \omega t$
21. $n\lambda/3$
23. 240 m/s; 400 Hz

25. 1.06, 1.68, 2.12, 2.70 $\times$ 10^9 Hz; 1.30, 1.84, 2.25, 2.49 $\times$ 10^9 Hz

27. 10^8; ratio = 10^{-4}

Chapter 34

1. 0.10 cm
3. 0.021
5. ...
7. 0.29 eV; 0.29 V
9. 3.11 eV
11. 91 nm
13. 1.80 cV; 2.25 eV
15. 1.1 $\times$ 10^{22}; 190 days
17. 0.44 eV
19. 440 nm; ultraviolet
21. 2.4 pm; 1.32 $\times$ 10^{-15} m; 4.8 $\times$ 10^{-4}; 4.8
23. $v_x = 1.46 \times 10^6$ m/s; $v_y = -1.46 \times 10^6$ m/s
25. 8.7 $\times$ 10^3 eV
27. 3.8 $\times$ 10^{-2}
29. ...

Chapter 35

1. 3.84 $\times$ 10^{-7} m
3. 3.89 $\times$ 10^{-5} m; infrared
5. 9.5 $\times$ 10^{-8} m
7. 7.3 $\times$ 10^{-8} m
9. 102.8, 121.6, and 656.3 nm
11. All lines of the Lyman and Balmer series; no
13. $L(3) = L(1) + B(2)$; $L(3) = L(2) + P(1)$; $L(3) = L(1) + B(1) + P(1)$
15. ...
17. 3.2 $\times$ 10^{-13} m
19. 4 $\times$ 10^{-15} m; 1.66 $\times$ 10^{-19} kg · m/s; 51 MeV
21. ...
23. $2b/n$; $n^2h^2/8b^2m$; $9.4n^2$ eV
25. 0.2518 nm; 7.4 V
27. ($\frac{1}{2}b$, $\frac{1}{2}b$, $\frac{1}{2}b$); at ($\frac{1}{4}b$, $\frac{1}{4}b$, $\frac{1}{4}b$), ($\frac{1}{4}b$, $\frac{3}{4}b$, $\frac{1}{4}b$), etc.; at the centers of the eight cubes that make up the original cube

Chapter 36

1. 0, 1, 2, 3; 0, ±1, ±2, ±3
3. $\hbar\sqrt{2}$; 0, $\pm\hbar$; 45, 90, 135°
5. $2\sqrt{2}$
7. Those with $|m| = l \neq 0$
9. 9.5 $\times$ 10^{-24} m; 3.0 $\times$ 10^{-13} m

11. 1.32 m; about 1.3 m
13. 69,000 eV; 9500 eV
15. $(n + \frac{1}{2})(8.7 \times 10^{-20})$ J = $(n + \frac{1}{2})(0.55)$ eV; 0.046
17. 2.0 $\times$ 10^{-10} m
19. 2 $\times$ 10^6; 1 $\times$ 10^{-34} kg · m/s; 4 $\times$ 10^{-14} m; 8 $\times$ 10^{-8}
21. ...
23. 5.8 $\times$ 10^{-4} eV

Chapter 37

1. 56e; 56; 82
3. 7.3 $\times$ 10^{-15} m; 3.4 $\times$ 10^{-15} m
5. 192
7. 0.54 m
9. 20.18
11. 11.9967 u; 1.9921 $\times$ 10^{-26} kg
13. 119.90 u
15. 6.7 $\times$ 10^{21}; 7.9 $\times$ 10^{-10}/s; 5.3 $\times$ 10^{12}/s; 140 Ci
17. 3.32 $\times$ 10^9 years
19. $^{90}_{40}$Zr
21. $^{238}_{92}$U
23. 0.64c
25. 2.4 $\times$ 10^{-5}
27. 5.5 MeV
29. 6.3 $\times$ 10^{-14} m
31. 1.4 $\times$ 10^6 Hz
33. 3.1 $\times$ 10^{-11} kg

Chapter 38

1. 1120 barns
3. 6 $\times$ 10^{22}; 0.062 cm
5. 5.3 cm
7. 1.33 MeV; 3.93 MeV
9. 4.9 $\times$ 10^{-8} atm
11. 1.5 cm
13. Zero; 1.98 $\times$ 10^7 m/s
15. 1–5 $\times$ 10^{-11}; 97,680
17. 1.05 $\times$ 10^{16} Bq = 0.28 MCi
19. 2.9 $\times$ 10^{-9} kg
21. 38,900 rad
23. 42 s
25. 4.28 MeV
27. 2.33 MeV; 5.6 $\times$ 10^{-13} m
29. 2.3 $\times$ 10^{-14} m
31. 1.01; 1.24 $\times$ 10^{-12} m
33. 9.2 $\times$ 10^6 m/s

Index

Index

Cylindrical capacitor, 416

Dalton's law of partial pressure, 285
Damping:
    electrical, 560
    magnetic, 523
    mechanical, 239
Davisson-Germer experiment, 757
Dc (direct current) circuits,
      438−456
Dc series circuits, 417−421
de Broglie wavelength, 757
de Broglie waves, 756, 762
Debye (unit) 527
Debye-Scherrer diffraction, 677
Decay constant, radioactivity, 820
Decay law:
    absorption, 840
    radioactivity, 820
Decay time:
    ac circuit, 560
    damped oscillator, 239
Decibel scale, 583
Degrees of freedom, 296
Density:
    definition of, 250
    nuclear, 813
    table, 250
Detail, smallest observable, 683
Detectors of radiation, 850
Diagrams, potential energy, 123
Diamagnetism, 534
Dichroic substances, 695
Dielectric constant:
    definition of, 528
    measurement of, 530
    table, 530
Dielectric polarization, 528
Dielectrics, 525−533
    and Coulomb's law, 532
    energy in, 545
    Gauss' law for, 531
Diffraction, 660−680
    by crystals, 675
    Debye-Scherrer, 677
    definition of, 660
    double-slit, 669
    electron, 757
    Fraunhofer vs. Fresnel, 661
    of particles, 758, 762
    single-slit, 662−669
    x-ray, 675

Diffraction grating, 671
    resolving power of, 690
Diffuse reflection, 616
Dimensional analysis, 7
Diopter (unit), 700
Dip angle, 476
Dipole:
    electric, 525
    energy in $\mathbf{B}$ field, 547
    energy in $\mathbf{E}$ field, 547
    magnetic, 533, 535, 538
    torque on, 526
Dipole moment, 525, 527
Direct current (dc) circuits,
      438−456
Discovery of x-rays, 71
Disorder in thermodynamics,
      332
Dispersion of grating, 690
Dispersion of materials, 692
Displacement current, 606
Displacement vector, 12
Distribution functions, 278−281
Domains, magnetic, 538
Doppler effect, 584
Double refraction, 696
Double-slit diffraction, 669
Double-slit interference, 642, 647
Drift velocity, 410
Driven oscillations, 240
Drumhead, resonance of, 710
Dynamics of rotation, 187−200
Dyne (unit), 65

Earth:
    age of, 825
    magnetic field of, 458
Eddy current brake, 524
Eddy currents, 520
Effective current and voltage, 549
Efficiency, thermodynamic, 326,
      328
Eigenfrequency, 709, 711
Eigenfunction, 711
Elastic collisions, 111
Elastic energy, 95
Electric charge, 342
Electric conductor, 356
Electric current:
    definition of, 409
    flow direction, 409
Electric dipole, 525

Electric field:
    of charge sheet, 381
    in conductor, 372, 386
    conservative field, 386
    of cylinder, 375
    definition of, 347
    energy in, 432
    and flux, 365
    and induced emf, 511
    of loop, 355
    at metal surface, 378
    motion of charges in, 359
    of parallel plates, 374
    due to point charge, 348
    from potential, 396
    of rod, 350−351
    sketches of, 356
    of sphere, 370
Electric field diagrams, 356
Electric field strength, 348
Electric filters, 565, 566
Electric flux, 365, 368
Electric generator, 510
Electric potential (*see* Potential,
      electric)
Electric power, 419, 557
Electric strength of air, 404
Electric susceptibility, 530
Electrical conductivity, 411
Electrical damping, 560
Electrical insulator, 356
Electrical safety, 449, 450
Electrical units, 609
Electrolysis, 352
Electromagnetic (em) waves,
      593−613
    from blackbody, 723
    energy in, 601
    intensity of, 602
    relation of $\mathbf{B}$ to $\mathbf{E}$, 599
    resonance, 715
    spectrum, 597
    speed, 608
    types of, 597
    wave equation, 607
Electromotive force (emf), 408
    induced, 500−524
    motional, 516
Electron:
    and cathode rays, 390
    charge on, 342, 446
    classical radius of, 432
    and Faraday, 352

Electron (*Cont.*):
  magnetic moment of, 808
  mass of, 805, 810
  spin of, 780, 806, 807
  and Thomson's atom, 492
  wave nature of, 756
Electron diffraction, 757
Electron distribution in H, 779
Electron microscope, 692
Electron spin resonance, 808
Electron volt (unit), 82, 155, 398
Electrostatic precipitator, 404
Electrostatics, 342 – 364
Elements, table of, 874
Em wave (*see* Electromagnetic
    waves)
Emf (electromotive force), 408
  induced, 500 – 524
  motional, 516
Emission spectra, complex, 787
Energy:
  available for reaction, 845
  in capacitor, 430
  of charged sphere, 406
  in dielectrics, 545
  elastic, 95
  in electric field, 432
  in electromagnetic waves, 601
  gravitational potential, 91
  in inductor, 509
  interconversion with mass, 155
  internal (thermodynamics), 316
  kinetic translational, 87
  in magnetic field, 509
  relativistic, 153 – 155
  rotational, 204 – 210
  of spring, 96
  in wave on string, 573
  work and, 82 – 98
  zero-point, 766, 798
Energy diagrams, 122
  related to force, 123
Energy level diagram, 750
Energy levels:
  of H atom, 750, 774
  of molecules, 796
  particle in a tube, 763
  Planck oscillator, 724
  rotational, 799
  in solids, 794
  vibrational, 798
Entropy, 332
  and irreversible process, 334

Entropy (*Cont.*):
  and order, 332
  and second law of
      thermodynamics, 333
Equation of continuity, 259
Equation of SHM (simple harmonic
    motion), 232
Equilibrium:
  first condition for, 24
  of rigid body, 197
  second condition for, 29
  stable vs. unstable, 125
  static, 24, 29, 197
  thermodynamic, 319
Equipartition theorem, 296 – 297
Equipotentials, 389
Equivalence principle, 180
Equivalent capacitance, 429
Equivalent optical path, 652
Equivalent resistance, 438
Erg (unit), 82
Errors of measurement, 14
Escape velocity, 130, 185, 286, 288
Exclusion principle, 781
Expectation value, 331
Exponential functions, 873
Exposure, unit for, 853
Eye:
  resolution of, 684
  sensitivity of, 598

F = Ma, 63, 121
Fabry-Perot interferometer, 658
Farad (unit), 415
Faraday, M., and electrolysis, 352
Faraday (unit), 352
Faraday's law:
  integral form, 512
  magnetism, 503
  Maxwell equation, 604
Farsighted condition, 700
Fermat's principle, 626
Ferromagnetic materials, 537
Field:
  of charge sheet, 381
  of charged cylinder, 375
  of charged sphere, 370
  in a conductor, 372, 386
  conservative, 90
  electric (*see* Electric field)
  gravitational, 89, 346
  magnetic (*see* Magnetic fields)

Field (*Cont.*):
  at metal surface, 378
  between parallel plates, 374
Field lines, electric, 356
Filters, electric, 565, 566
First law of thermodynamics, 316
Fission, nuclear, 830
Five motion equations, 49
Flow tube, 259
Fluid, definition of, 249
Fluid flow, 258 – 263
Fluid pressure, 254 – 257
Flux:
  electric, 365, 368
  magnetic, 503
  from point charge, 368
Flux density, 461
Focal length:
  lens, 631
  mirror, 619
Focal point:
  lens, 631
  mirror, 619
Force:
  buoyant, 257
  centrifugal, 179
  centripetal, 170
  conservative, 89
  electromotive (*see* Electromotive
      force)
  gravitational, 172
  magnetic (*see* Magnetic force)
  nuclear, 814
  relativistic, 152
  as a vector, 16
Force units, 64, 65
Forced oscillations, 240
Fourier synthesis of waves, 637
Frames of reference, 132
Fraunhofer diffraction, 661
Free-body diagram, 25, 67
Free fall acceleration, 51
Freedom, degrees of, 296
Freely falling bodies, 51
Frequency, related to period, 227
Fresnel diffraction, 661
Friction coefficient, 77
  table of, 77
Friction forces, 34, 68, 76
Fundamental vibration, 709
Fusion, latent heat of, 304
  table of values, 302
Fusion reaction, nuclear, 156, 860